Jörg Freyhof, Baran Yoğurtçuoğlu, Arash Jouladeh-Roudbar, Cüneyt Kaya

Handbook of Freshwater Fishes of West Asia

Jörg Freyhof, Baran Yoğurtçuoğlu,
Arash Jouladeh-Roudbar, Cüneyt Kaya

Handbook of Freshwater Fishes of West Asia

DE GRUYTER

Authors

Jörg Freyhof
Museum für Naturkunde, Leibniz Institute
for Evolution and Biodiversity Science,
Invalidenstr. 43,10115 Berlin, Germany
joerg.freyhof@mfn.berlin

Arash Jouladeh-Roudbar
Lilla Jularpsvägen 53
SE-243 93 Höör, Sweden
jouladeh@gmail.com

Baran Yoğurtçuoğlu
Hacettepe University, Biology
Department, Beytepe Campus, 06800,
Çankaya/Ankara, Türkiye
yokbaran@gmail.com

Cüneyt Kaya
Recep Tayyip Erdogan University
Faculty of Fisheries, 53020,
Rize, Türkiye
cuneyt.kaya@erdogan.edu.tr

ISBN 978-3-11-167764-4
e-ISBN (PDF) 978-3-11-167781-1
e-ISBN (EPUB) 978-3-11-167824-5
DOI https://doi.org/10.1515/9783111677811

Library of Congress Control Number: 2024951485

Bibliographic information published by the Deutsche Nationalbibliothek
The Deutsche Nationalbibliothek lists this publication in the Deutsche Nationalbibliografie; detailed bibliographic data are available on the internet at http://dnb.dnb.de.

Suggested citation format
Freyhof, J., Yoğurtçuoğlu, B., Jouladeh-Roudbar, A. & Kaya, C.
Handbook of Freshwater Fishes of West Asia, Berlin, Boston: De Gruyter, 2025. https://doi.org/10.1515/9783111677811

Front cover image: *Cyprinion kais* by Jörg Freyhof
Back cover image: *Turyancay* by Jörg Freyhof
Typesetting: Compuscript Ltd. Shannon, Ireland
Printing and binding: CPI books GmbH, Leck

www.degruyterbrill.com
Questions about General Product Safety Regulation:
productsafety@degruyterbrill.com

MIX
Papier | Fördert gute Waldnutzung
FSC® C083411

Contents

The authors

Jörg Freyhof is a research scientist at the Museum für Naturkunde, Leibniz Institute for Evolution and Biodiversity in Berlin, Germany, and Regional Chair of the IUCN SSC Freshwater Fish Specialist Group. He is a world authority on freshwater fishes, focusing on conservation, biodiversity, biogeography, evolutionary ecology, and taxonomy. He has described more than 130 new species of freshwater fishes and is the author of some 350 scientific publications cited more than 20,000 times. These include the *Handbook of European Freshwater Fishes* and *Freshwater Fishes of the Arabian Peninsula.*

Arash Jouladeh-Roudbar is an independent researcher on aquatic biodiversity and fish taxonomy. His studies focus on the diversity, taxonomy, evolutionary history, and biogeography of the freshwater fauna of West Asia, and he has worked extensively on the freshwater fauna of Iran over the past decade. His academic background, with over 50 peer-reviewed scientific publications and the description of 15 new species, makes him a young member of the scientific community in West Asia. Arash is a science enthusiast interested in the global conservation of freshwater ecosystems and is active in international environments for the protection of aquatic fauna and flora.

Baran Yoğurtçuoğlu is a conservation scientist at Hacettepe University, Department of Biology in Ankara, Türkiye, and Regional Chair of the IUCN SSC Freshwater Fish Specialist Group. He studies the diversity, distribution, and interactions of freshwater fishes in West Asia, focusing on how evolutionary history and ecological processes, including species interactions and climate change, influence their distribution patterns and diversity profiles. He supports the documentation and conservation of freshwater fish biodiversity, particularly in the Mediterranean biodiversity hotspot, by leading and participating in several projects.

Cüneyt Kaya is a fish taxonomist at the Faculty of Fisheries at Recep Tayyip Erdogan University in Rize, Türkiye. Freshwater fishes have always been his main academic and personal interest, and he has always been motivated to pursue a career working with them. His main interests are the conservation, diversity, and distribution of freshwater fishes. His research also focuses on the effects of ecological change, including invasive species and habitat change, on aquatic fauna. During his career, he has published more than 80 peer-reviewed scientific papers and described 50 new species of freshwater fishes.

The authors at Arsuz, Türkiye, from left: Jörg Freyhof, Arash Jouladeh-Roudbar, Baran Yoğurtçuoğlu, Cüneyt Kaya. © R. Sanda

https://doi.org/10.1515/9783111677811-202

The Bouleuterion (council house) of the metropolis Aphrodisias in the Roman province of Caria. © M. Oral

Summary

The *Handbook of Freshwater Fishes of West Asia* is a thorough, open-access guide written by four international experts. The book offers a readily accessible and detailed overview of the highly diverse freshwater fishes of the region, which extends from the Turkish Bosporus and the southern Caucasus through the Levant, Iran, and the Arabian Peninsula. The book documents all 632 species of freshwater fish in the region, including 597 native species, 467 of which are endemic and 35 non-native species. What is fascinating is that 41 % of the species have been described since the year 2000.

This handbook goes far beyond a simple species catalogue. Built on about 20 years of dedicated research and exploration, it is packed with data, complete with illustrations of each species and detailed maps showing their distribution by site records and hydrobasin scale. It also provides key information on each species' conservation status. The IUCN Red List assesses 583 species. Of these, 12 are classified as extinct, while 238 are considered threatened (CR, EN, or VU).

In addition to species portraits, the book provides practical advice for anyone interested in studying freshwater fishes. This includes identification methods based on visible morphological characters and discussions of potential challenges related to taxonomic uncertainties and evolving species concepts. The book also draws attention to the critical role of extensive fieldwork, such as photographing fish, collecting DNA samples, and handling specimens ethically.

The authors address the many complex aspects of taxonomy and species classification. They formulate numerous research questions and recognise ever-evolving context of species concepts. They also underline the significant conservation challenges posed by human-induced threats such as water abstraction, dam construction, pollution, climate change, and the introduction of non-native species.

Despite these threats, the handbook highlights the resilience of West Asia's freshwater fish biodiversity. While many ecosystems are under immense pressure, the survival of most native species offers hope for future conservation efforts. The book emphasises that these ecosystems will be further burdened by the increasing demand for water in this globally water-scarce region, driven by human population growth and economic development. Ultimately, the *Handbook of Freshwater Fishes of West Asia* aims to inspire regional and international audiences by portraying the beauty and diversity of the freshwater fishes of the region. It serves as an essential tool for raising awareness of the critical importance of freshwater ecosystems, which are often overlooked in conservation programmes, and provides valuable documentation for conservationists, researchers, and enthusiasts alike.

https://doi.org/10.1515/9783111677811-203

Foreword

With freshwater fishes in severe decline worldwide and species being lost at an alarming rate, the arrival of this formative text on fishes from the West Asian region is timely and extremely valuable. Arid regions have lost a proportionally higher percentage of their freshwater biodiversity due to human activities, and fishes cannot compete with humans for scarce water resources. Many species in the region have been lost without being studied, and having a comprehensive guide to not only catalogue the taxa from this region but also serve as an in-depth resource and educational tool for both professional and amateur ichthyologists is extremely welcome. The authors of this book are leading experts in the region, and their combined years of experience and study are demonstrated by the level of detail to the species from the region and habitats found here, as well as the myriad of threats that the ecosystem and species face. Readers from outside the region will appreciate the surprising variety of habitats found across West Asia and learn a great deal about the ecology of freshwaters in arid habitats when using this handbook. Students and fish enthusiasts, as well as anyone interested in the natural environment of this region, now have a valuable tool to aid study and a comprehensive reference to refer to.

Kathy Hughes & Brian Zimmermann, Global Chairs of the IUCN SSC Freshwater Fish Specialist Group

Door in Saada bath palace in Tehran, Iran.

https://doi.org/10.1515/9783111677811-204

Foreword

The *Handbook of Freshwater Fishes of West Asia* is the first attempt to bring information on all the freshwater fishes of that area under a single cover. Because of the generally arid climate, the aquatic fauna of West Asia is usually considered depaurate, with little diversity. In fact, this only reflects the lack of knowledge. Jörg Freyhof, Baran Yoğurtçuoğlu, Arash Jouladeh-Roudbar, and Cüneyt Kaya show that this is an erroneous perception. They recognise 597 native species present today in these waters. This number will certainly increase as several areas are still poorly surveyed or even ichthyological *terrae incognitae* (or maybe better *aquae incognitae*!). Not less than 259 species have been discovered since 2000. Several other species that were described by earlier authors and later considered indistinguishable have been re-examined and revalidated.

The authors have travelled in most countries of West Asia and have collected and observed fishes in most river drainages. They have reviewed the published data, they have examined museum specimens, and they have consulted with numerous colleagues. During these 18 years, they also published 180 papers and one book on the fish fauna of West Asia. They are probably the most qualified ichthyologists to compile such a book. Almost all species from West Asia are figured in colour, each is accompanied by up-to-date nomenclature, diagnoses, information on distribution, habitat, biology, and conservation status.

Aquatic biodiversity has long been overlooked in conservation strategies and it is only in recent years that their importance and distinctness has been paid due attention. In the mostly arid landscapes of West Asia, global warming has led to even smaller precipitations, to reduced flow in rivers, to lower level of lakes and springs and increased salinity, even to desiccation. Water abstraction for agriculture, homes, and industrial usages contribute their share in habitat disturbance. Introduced fishes have predated and eaten to extinction native species, modified habitats, and imported their pathogens. These and other factors result in the rarefaction, extirpation, and even extinction of several fish species. In fact, 238 (41 %) of the freshwater fish species of West Asia are threatened and 12 of the named species are already extinct. Not to mention those that became extinct before we even knew of their existence.

In our quickly changing world, hard, hands-on scientific data are vital to manage and conserve aquatic biodiversity. Communication between all involved is needed, from scientists to conservationists, to decision-makers, and to the public. This information must be compiled and made widely available. The *Handbook of the Freshwater Fishes of West Asia* is an important tool to disseminate knowledge, especially as it is available as an open-access version allowing everybody to download it for free. It will help promote more research and contribute to the conservation of West Asian aquatic biodiversity.

Maurice Kottelat, National University of Singapore

The Old City of Jerusalem with the "Dome of the Rock" on top of the Al-Aqsa Mosque.

https://doi.org/10.1515/9783111677811-205

Preface

Why an open-access book? The cost and logistics of buying a heavy book like *Freshwater Fishes of West Asia* is a challenge. Only those who are enthusiastic, work in the field, or love printed books would invest in ordering a copy, which they can do. This book is published as open-access to reach far beyond this circle. We hope it will be read by a much broader audience in all parts of the world. This includes those who may have had only a marginal interest in fish and their environment before downloading this book.

Why is this book being written at a time when few people read anymore and when most, if not all, information is available on the internet? Because both of these statements are simply untrue. Very little of the information presented in the *Handbook of Freshwater Fishes of West Asia* is readily available online—and if it is available at all, it is embedded in hundreds of journal publications and local language books. In fact, this book is the only one that makes the vast amount of information available to everyone. Above all, this book is urgently needed. More and more people today are interested in nature and biodiversity, and freshwater biodiversity, including fish, is declining rapidly. This process is accelerating enormously, especially in West Asia. This book is the last opportunity to inform the wider public, including those in the conservation, management, and scientific communities, about what we still have before much of it is gone.

Only this book provides detailed information on the freshwater fishes of West Asia that is available to experts and non-experts alike. As it is published as an open-access book, it is easily accessible to all interested parties free of charge. It brings together the relevant and authoritative information needed to understand and work on freshwater biodiversity in the region and far beyond. In addition, for the first time, all species are illustrated in colour, and detailed distribution maps are presented. There are also many critical notes for future research, as there is still much to be done. Hopefully, this book will soon be outdated as new research becomes available.

The authors of this book have devoted much of their professional careers to studying freshwater fishes and their environments, and it was only by working together that they could gather information dense enough to make the *Handbook of Freshwater Fishes of West Asia* possible. It has been a significant effort to compile all the distribution records, images, and additional information to publish this book in a convenient and accessible format. The information presented is based on the combined fieldwork of the authors over many years and on the contributions (published and unpublished) of many other colleagues whose names are listed in the acknowledgments. Work on this book began in 2006, and over the next 19 years, we went to the field, examined material, prepared keys and diagnoses, and compiled images and distribution details.

This book aims to provide critical insights and knowledge about freshwater fishes and their habitats to a broad audience and raise awareness of the unique and wonderful world of freshwater fishes and their range of eclectic habitats, sometimes in remote mountainous areas but just as often in springs, rivers, and streams close to villages and towns.

This book fills a critical gap. Freshwater fish are an often forgotten but highly endemic and endangered element of biodiversity. Fish and their ecosystems are usually ignored when setting conservation priorities, planning infrastructure projects, and managing water resources. The authors hope to change this situation.

This book is an essential guide for anyone interested in freshwater fishes and the broader issues of biodiversity and ecology. It is also an essential resource for those studying and working in wetland and aquatic species conservation; those planning, developing, and managing protected natural areas and resources; those responsible for environmental impact assessments; those planning land use management; and those approving investment projects in freshwater ecosystems. It will also be of similar interest to naturalists and citizen scientists who want to know which fishes are found in the rivers, streams, and lakes in their area; those who study fishes and their diversity; and international readers interested in the biodiversity of West Asia and freshwater fishes in general.

https://doi.org/10.1515/9783111677811-206

Acknowledgements

This book would not have been possible without the help and contributions of many people. Special thanks again to Maurice Kottelat and Brian Zimmermann for writing the forewords. Maurice Kottelat also allowed the reusing and adaptation of major text elements from the *Handbook of European Freshwater Fishes.* Text elements of Fishes of Arabia were adapted from the *Handbook of the Freshwater Fishes of the Arabian Peninsula*, and many thanks to Johannes Els for allowing us to use some of the elements from this book. We, of course, take responsibility for any oversight.

We are pleased to acknowledge the colleagues and friends who facilitated our work, supported us with specimens, and information, and assisted in the field (in alphabetical order): Keyvan Abbasi (Bandar Anzali, Iran), Ashgar Abdoli (Tehran, Iran), Younis S. Abdullah (Sulaimani, Iraq), Omar Fadhil Al-Sheikhly (Bagdad, Iraq), Atheer Ali (Basra, Iraq), Nisreen Alwan (Abu Dhabi, UAE), Michel Bariche (Beirut, Lebanon), Esra Bayçelebi (Rize, Türkiye), Nina Bogutskaya (Dolsko, Slovenia), Jörg Bohlen (Libechov, Czech Republic), Kai Borkenhagen (Büsum, Germany), Soheil Eagderi (Tehran, Iran), Fitnat Güler Ekmekçi (Ankara, Türkiye), Johannes Els (Sharjah, UAE), Giorgi Epitashvili (Tbilisi, Georgia), Füsun Erk'akan (Ankara, Türkiye), Hamid Reza Esmaeili (Shiraz, Iran), Özcan Gaygusuz (İstanbul, Türkiye), Hamid Reza Ghanavi (Lund, Sweden), Daniel Golani (Jerusalem, Israel), Menachem Goren (Tel Aviv, Israel), Salim Serkan Güçlü (Isparta, Türkiye), Fatih Gündüz (Elazığ, Türkiye), Nashat A. Hamidan (Amman, Jordan), Iraj Hashemzadeh Segherloo (Shahr-e Kord, Iran), Shaig Ibrahimov (Baku, Azerbaijan), Ali İlhan (İzmir, Türkiye), Bella Japoshvili (Tbilisi, Georgia), Şerife Gülsün Kırankaya (Düzce, Türkiye), Fareed Krupp (Frankfurt, Germany), Fahrettin Küçük (Isparta, Türkiye), Tatia Kuljanishvili (Prague, Czech Republic), Irmak Kurtul (İzmir, Türkiye), Boris Levin (Cherepovets, Russia), Hamed Mousavi-Sabet (Rasht, Iran), Nabil Musa (Sulaimani, Iraq), Namig Mustafayev (Baku, Azerbaijan), Alexander Naseka (Dolsko, Slovenia), Müfit Özuluğ (İstanbul, Türkiye), Rahman Patimar (Gonbad, Iran), Samvel Pipoyan (Yerevan, Armenia), Gülşah Saç (İstanbul, Türkiye), Hasan Musa Sarı (İzmir, Türkiye), Adwan Shehab (Damascus, Syria), Vendula Slechtova (Libechov, Czech Republic), Davut Turan (Rize, Türkiye), Erhan Ünlü (Diyarbakır, Türkiye), Saber Vatandoust (Babol, Iran), Vedat Yeğen (Isparta, Türkiye), Shama Zaki (Muscat, Oman), and numerous fishers, anglers, and fisheries officials for providing help and licencing our work.

Many thanks to Ambily Nair and Rajeev Raghavan (Cochin, India) for reading the manuscript. Our work also required the examination of specimens accumulated by the generations of ichthyologists who worked before us. Many of their specimens still exist in natural history museums in various parts of the world. We thank the following colleagues and institutions for providing access to the material under their care (in alphabetical order): Michel Bariche (AUB, Beirut), Füsun Erk'akan (HUIC, Ankara), Menachem Goren (TAU, Tel Aviv), Fabian Herder (LIB Bonn), Marie-Hélène Hubert & Kamal Khidas (CMNFI, Ottawa), James McLane (BMNH, London), Hamed Mousavi-Sabet (VMFC, Rasht), Mikhail Nazarkin (ZIN, St. Petersburg), Müfit Özuluğ (IUSHM, İstanbul), Anja Palandacic (NMW, Vienna), Moritz Sonnewald (SMF, Frankfurt), Ralf Thiel (ZMH, Hamburg), Davut Turan & Esra Bayçelebi (FFR, Rize), Saber Vatandoust (VMFC, Babol).

This book has benefited greatly from the numerous photographs and drawings generously provided by friends and colleagues. We are happy to thank the following for providing illustrations used in this book (in alphabetical order): Keyvan Abbasi (Bandar Anzali, Iran), Asghar Abdoli (Tehran, Iran), Younis S. Abdullah (Sulaimani, Iraq), Sadi Aksu (Eskişehir, Türkiye), Omar Fadhil Al-Sheikhly (Baghdad, Iraq), Atheer Ali (Basra, Iraq), Archiv Aqualog (Rodgau, Germany), Michel Bariche (Beirut, Lebanon), Kai Borkenhagen (Büsum, Germany), Saygun Dura (Istanbul, Türkiye), Soheil Eagderi (Tehran, Iran), Ulrich Eichelmann (Vienna, Austria), Johannes Els (Shajah, UAE), Hamid Reza Esmaeili (Shiraz, Iran), Giorgi Epitashvili (Tbilisi, Georgia), Hans Esterbauer (Vienna, Austria), Matt Ford (Berlin, Germany), Matthias Geiger (LIB, Bonn Germany), Nashat A. Hamidan (Amman, Jordan), Zhou Hang (Guangzhou, China), Andreas Hartl (Dorfen, Germany), Jonny Jensen/JJPhoto.dk (Kastrup, Denmark), Sonke Johnsen (NOOA, USA), Münevver Kaba Oral (Rize, Türkiye), Seishi Kimura (Mie, Japan), Fareed Krupp (Frankfurt, Germany), Katariina Kurina (Tartu, Estonia), Anton Lamboj (Pottendorf, Austria), Rene Lennigk (Weil am Rhein, Germany), Boris Levin (Cherepovets, Russia), Gary Lyon (Muscat, Oman), Hamed Mousavi-Sabet (Rasht, Iran), Eugenio Miotti (Buritto, Italy); Theodor T. Nalbant (Bucharest, Romania), Alexander Naseka (Dolsko, Slovenia), Barbara Nicca (Winterthur, Switzerland), Yakov Oksman (Tel Aviv, Israel), Müfit Özuluğ (İstanbul, Türkiye), Johannes Pfleiderer (Duisburg, Germany), Gianluca Polgar (Rome, Italy), Michael Pope (Kuwait), Rahman Patimar (Gonbad, Iran), Samvel Pipoyan

https://doi.org/10.1515/9783111677811-207

(Yerevan, Armenia), John Ernest Randall (Honolulu, USA), Martin Reichard (Brno, Czechia), Akbarjon Rozimov (Tashkent, Uzbekistan), Akbarjon Rozimov (Tashkent, Uzbekistan), Zoltan Sallai (Turkeve, Hungary), Radek Sanda (Prague, Czechia), Johannes Schöffmann (St. Veit/Glan, Austria), Erwin Schraml (Augsburg, Germany), Andreas Spreinat (Göttingen, Germany), Sven Tränkner (SGN, Frankfurt, Germany), Sako Tuniev (Sochi, Russia), Alycia Koharu Uyeoka (San Francisco, USA), Ekaterina Vasil'va (Moscow, Russia), Saber Vatandoust (Babol, Iran), Hubert Wischmann (Mühlberg, Germany), Tse Chung Yi (Hong Kong), Gerald Zauner (Engelhartszell, Austria). Five pictures were taken from Shutterstock: Fernando M. Elkspera/shutterstock_2366550807, Ion Mes/shutterstock_2141170165, dwi putra stock/1595226994, Carlos Aguilera/1337013701, and feathercollector/ 305645615. Thanks to Jon-Paul Rodriguez (IUCN SSC) and Kathy Huges (IUCN/WI FFSG), who helped secure the necessary permissions to reproduce their organizations' logos in this book. It is impossible to estimate how many pleasant and helpful people we met during our research and how much unpublished information from numerous anonymous local experts is included. To all of them, we owe an outstanding debt of gratitude.

The Jameh Mosque of Isfahan in Iran is just one of the many cultural heritage sites in West Asia.

A few personal words at the outset

From the outside, West Asia is not a hostile place for freshwater biodiversity. It is largely an arid region facing increasing water stress caused by unsustainable water use and climate change. Water stress is evident to anyone who looks at the landscapes of West Asia with open eyes. Not every visit to a water body is a pleasant one. Dams, pollution, and water extraction are major challenges in the region. Water means prosperity, and there is no water to lose for biodiversity. Economies in West Asia are developing rapidly. This development comes at a cost borne mainly by nature and has a massive impact on freshwater ecosystems. Not surprisingly, freshwater biodiversity is declining at twice the rate of marine and terrestrial biodiversity. Travelling through West Asian countries and observing the many changes in rivers and streams is often an emotional challenge. While working on this book, marshes and lakes were drying up, springs were filled with non-native fish, streams were dying from heavy pollution, and in many cases, new dams were being built, desiccating the rivers and streams below.

Demand for fresh water in the region will increase sharply due to growth in human population and economic development. At the same time, climate change will lead to more frequent massive floods, severe droughts, and extreme heat, reducing the region's water resources. More and more landscapes will lose their ability to support agriculture because there will not be enough water or it will be simply too hot. West Asia's leaders have often built their economies on fossil fuels. While they are not significant producers of climate gases, they provide fossil fuels to others to drive climate change. At the same time, the region has witnessed and continues to witness significant wars and political crises such as the Syrian and Yemeni civil wars, the Russian–Georgian war, the Israeli–Palestinian wars, the Iranian crisis and their wars against the West, the wars against the Kurds and other minorities, the Iran–Iraq war, various wars in Iraq, the Nagorno–Karabakh conflict, the war against the Islamic State (ISIS), and several other smaller and larger never-ending conflicts that punish the people of the region and threaten their culture. All this makes West Asia a region of "endless sociopolitical crisis". In addition, the water and climate crises will fuel instability, emigration, nationalism, and terror in the region. All this takes money and energy away from poverty reduction, sustainability, climate change mitigation, and biodiversity conservation.

West Asia is a beautiful part of the world. The friendly people, diverse cultures, and fantastic landscapes we encounter during fieldwork make every visit an unforgettable experience. West Asia is the first cradle of civilization, and more than 10,000 years of human history can be seen in many places. Standing at one of the crystal clear springs with its endemic fishes, full of aquatic plants, or having tea in a small rest house on the banks of a free mountain river will open any nature lover's heart. To descend from the steppe into a green river valley is to find that the water's edge becomes a paradise. Despite all the doom and gloom, freshwater fish are very resilient, and despite many threats, aquatic biodiversity is still surviving surprisingly well. Many, many beautiful and intact places remain. Not to forget, only some West Asian rivers have been as destroyed as those in much of Central Europe.

Majestic Zagros Mountains in Iran. We can show only a few of the many impressive landscapes of West Asia in this book.

https://doi.org/10.1515/9783111677811-208

In West Asia, only 12 (2.0 %) freshwater fish species are thought to have become extinct worldwide, while most of the 597 native species have survived. This is not what we would refer to as an "extinction crisis." Still, we know the situation will change dramatically as climate change significantly increases drought, temperature, and water demand. Indeed, there is no room for fatalism. Economic development is freeing most people from their daily struggle to survive, young people are much better educated than ever, and awareness of nature is increasing, as is generosity and the desire to live in harmony with nature. In addition, the widespread availability of knowledge through the development of the internet has democratised access to information and enabled individuals to learn and grow as never before. The citizens of West Asia are discovering their biodiversity treasures and are ready to protect, restore, and provide spaces for biodiversity. This open-access book is aimed to be a part of this development. An increasing number of protected areas are being designated, and the global biodiversity crisis has raised awareness among the people of West Asia. We are optimistic that our book will not be seen as a book of memories, the last documentation of species before they become extinct. Are we at a tipping point where biodiversity continues to survive human impact today but is likely to erode at an increasing rate in the near future? Or are we at a tipping point where we will be conserving and restoring biodiversity at a growing rate in the near future? Fish have received more attention than any other element of freshwater biodiversity. This book documents the biodiversity heritage of the wonderful West Asian region, a document of our time for those who create the future.

We acknowledge the many local fishers, such as those on this Hirmand Canal in Jarikeh, Iran. They fish that we could not find and provided valuable information.

Introduction: About this book

Coverage

Geographical coverage. There is no agreed definition of the geographical area called West Asia, which includes or excludes the countries of the southern Caucasus, Iran, Israel, Türkiye, Egypt, and a few others. For this book, we follow a partly biogeographical and partly practical approach. West Asia, as covered in this book, includes all of Anatolia (the Asian part of Türkiye), Cyprus, the countries of the Caucasus south of its main rim (Armenia, Azerbaijan, Georgia, parts of Russia), Iran, Iraq, Syria, and the nations of the Levant (Lebanon, Israel, Jordan, Palestine), the Arabian Peninsula (Bahrain, Kuwait, Oman, Qatar, Saudi Arabia, United Arab Emirates, Yemen), including the coastal islands and the Socotra archipelago. In the Caucasus, the Greater Caucasus is considered to be the boundary that begins (approximately) at the Black Sea in the west, near the Russian city of Novorossiysk. Then, it runs along the entire Caucasus divide to Baku (Azerbaijan) on the Caspian Sea. The upper reaches of the Terek in Georgia are excluded as they are part of Europe, as are Türkiye north of the Bosporus and Azerbaijan north of Baku. In the Aegean, the boundary follows the political border between Greece and Türkiye. However, islands such as Lesbos and Rhodes were part of the Anatolian mainland during the last Ice Age. The entire territory of Iran is covered in this book, including the Sistan and Mashkid basins and the lower parts of the Helmand and Hari rivers. The middle and upper parts of these rivers in Pakistan, Turkmenistan, and Afghanistan are excluded. Strictly speaking, the Helmand (and others in the Sistan basin) and Hari are Central Asian rivers.

Toponymy (names of places, regions, rivers, etc.) follows the spelling in local languages or transcriptions as used in local maps. For features that extend across areas where different languages are spoken, the English name of the feature (if any) is used. Many of these names have several different spellings when transcribed from Arabic, Farsi, or other languages. We have tried to find a corrected and widely accepted transcription, but this has yet to be largely successful, as the names are strictly different in different languages and would always be translated differently. We use English instead of local names for

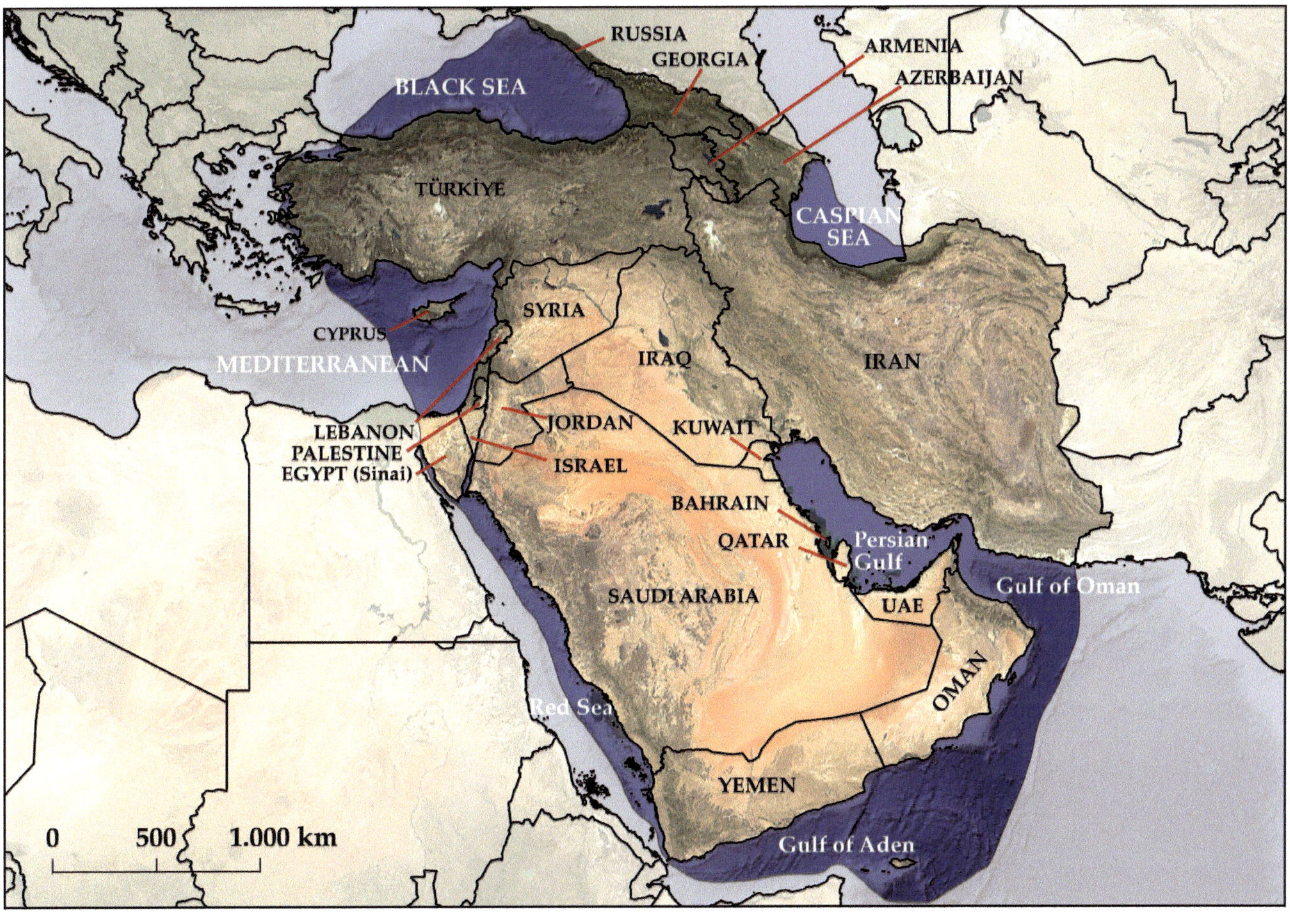

Figure 1. This book covers all freshwater fishes of West Asia known by early 2025 within the boundaries shown here.

https://doi.org/10.1515/9783111677811-001

countries, regions, large islands, major cities, and other well-known places.

Species coverage. All native freshwater fish species recorded in West Asia are included in the following accounts, including primary, secondary, diadromous, and sporadic species (see below for definitions). Accidental and vagrant species are excluded because many coastal species are occasionally found above the brackish water line. For the above purposes, records of occurrence have only been accepted if they are associated with reliable locality data and identification. No clear boundary exists between freshwater and marine/brackish water species. Particularly in the Persian Gulf basin, many marine fishes occur surprisingly high up in brackish rivers. Some species included in this book (such as some *Alosa*, *Chelon*, *Platichthys*, *Ponticola*, *Mesogobius*, and others) may only occur locally or rarely in freshwater. The reader of this book should be aware that most fish caught in brackish or coastal waters may not be covered. We often mention (some) species known from brackish waters in the family chapters and may include them in identification keys. Non-native species are only included if they have established self-sustaining populations or are stocked so regularly that there is a realistic chance of finding them. Species such as *Atractosteus spatula*, *Pangasianodon hypophthalmus*, and several others have been released from aquaria and aquaculture. These non-native species have occasionally been recorded and included in regional or national lists, although their establishment has yet to be proven. Such species are excluded from the scope of this book. If the reader discovers a non-native fish not covered in this book, it is always worth consulting the aquarium literature. The order of the families follows Near & Thacker (2024).

Osteomugil speigleri; Oman, about 100 mm SL, an accidental species common in coastal waters in the Indo-West Pacific.

Freshwater fish

Ecologists classify freshwater fishes according to their tolerance to salt water.

Primary divisions are families whose members are strictly intolerant of salt water, both in the present and in their evolutionary past. Examples include all Cypriniform fishes, of which only some species may have become tolerant to elevated salinities but not salinities as high as in the sea.

Secondary division families are considered to be of marine origin but whose members are now found mainly, wholly or partly in freshwater environments. Some of the freshwater species are able, or their ancestors are thought to have once been able, to tolerate seawater, at least for short periods; others are euryhaline, e.g., able to live in fresh or salt water. Examples of secondary division families are the Cichlidae and the Aphaniidae. In addition to the above categories, which apply to whole families, many individual fish species occur in freshwater, although they belong predominantly to marine families. They are referred to as follows:

Diadromous species are those that migrate between freshwater and seawater at different stages of their lives, either to spawn in the sea (catadromous, such as the eel *Anguilla anguilla*) or in freshwater (anadromous, such as the shad *Alosa agone*). Individuals of both groups can be relatively flexible in their migratory patterns, with catadromous species often remaining in the sea. In contrast, anadromous species might complete their life cycle in freshwater, frequently becoming land-locked in lakes or reservoirs. Anadromous fishes all belong to freshwater families.

Sporadic species are those that usually occur in estuaries and appear to be indifferent to salinity (e.g., they are euryhaline). Examples include several species of the Mugilidae.

Accidental or vagrant species are marine fish occasionally caught in inland waters. Their occurrence is unpredictable, and there are usually only a few records of the species in freshwater, often just above the tidal zone or in remnant pools close to the coast.

How to use this book

Keys. Keys are provided for the identification of freshwater fish genera found in West Asia. Species-level keys are provided for genera with more than two species, based on characters that non-specialists can easily observe without dissection or sophisticated techniques; however, a hand lens or dissecting microscope may be required. No molecular methods are needed to use the keys. An identification key consists of a series of questions that enable the user to establish the identity of an individual fish and is made up of a series of couplets (numbered 1, 2, 3, etc.). Each pair contains two alternative descriptions (numbered 1a and 1b, 2a and 2b, etc.). Each description includes one or more statements so that the two descriptions of the couplet always present alternative states of the same sign(s). To use a key, start with couplet 1. First, read sentence 1a and then 1b. If a sentence describes more than one character, read them all. Now, decide which phrase describes your fish. Each sentence concludes with a clue such as "go to 2" or "genus/species X." If the sentence describing the fish you want to identify ends with "go to 2," you should now read couplet 2 (both sentences again) and repeat the process until you have a correct name for your fish. In genus-level keys, if there is an exceptional species with respect to the character defined in a couplet, this species is indicated in parentheses in the sentence leading to the genus.

Characters, character states, and traits

A **character** is any morphological, colour (including pattern), ethological, molecular, or other characteristic that can diagnose a taxon (species, genus, family).

A **character state** is one of several alternative states that a given character may have. Character states are fixed for each species and, therefore, present in all individuals. For example, "interorbital pores present" is a character state of the character "presence or absence of interorbital pores" or "five" is a character state of the character "number of bars on flank."

A **trait** is a genetically determined manifestation of a character (or other physical feature) that is not fixed in a species or population, e.g., one that cannot be described in terms of discrete alternative states (character "states" and therefore cannot be used directly to diagnose a taxon). Traits vary along a continuum so that different individuals of a given species or population may exhibit different traits of the same character. However, different traits of the same character may occur in varying combinations in different species or populations. For example, the presence, absence, width, shape, etc., of a vertical bar at the base of the tail may all be characteristics of the character "mark at the base of the tail." In such a case, various combinations of the attributes of a given character may constitute different states of that character. For example, in species A, the bar may be absent, broad, or vertical in different individuals, whereas in species B, it may always be present, but either as a wide bar or as an oval spot.

Example. Let us take the following imaginary key:

1a - Two dorsal fins.
.... 2
1b - One dorsal fin.
.... 4

2a - Second dorsal fin with 7–8 rays.
.... Species Ab
2b - Second dorsal fin with 18–25 rays.
..........................3

First, read couplet 1 (both sentences 1a and 1b). Look at your fish. How many dorsal fins does it have, 1 or 2? If it has one dorsal fin, go to couplet 4. If it has two dorsal fins, go to couplet 2. Now read the two sentences in couplet 2. The second dorsal fin has a number of rays. Count them. If your fish has, for example, 20 rays in the second dorsal fin, go to couplet 3. If, for example, your fish has eight rays in the second dorsal fin, the key indicates that it belongs to species Ab. When running a key, a given species may fulfil the conditions of both sets of a couplet, or it may appear twice in the key. Suppose you come up with the following couplet:

6a - Caudal peduncle depth 20–26 % HL (head length).
..........................7
6b - Caudal peduncle depth 25–44 % HL (head length).
..........................10

Imagine that the material in front of you has a caudal peduncle depth of 21–28 % HL. On closer analysis, it may turn out that very few individuals have a caudal peduncle depth less than 25 % HL; if so, go to pair 10. However, if individuals with a caudal peduncle depth less than 25 % HL are common or represent most of the population, then both couplets 7 and 10 should be tested.

The key does not work! Why not? There are times when a key does not work. For example, it may lead you to a species completely different from the one you are trying to identify (of course, you started by looking at the picture), or it may ask questions that do not apply to your fish (e.g., the number of scales, when your fish has no scales). There are several possible explanations for this:

1. You may have an aberrant (malformed or deformed) individual; some characters may have been lost due to damage or injury; or your fish may be a senescent individual with an abnormal shape or a juvenile that has not yet developed the diagnostic characters. Try to check these characters on another individual.
2. You may be trying to identify a species not described in this book. The fish fauna of West Asia is still being catalogued; indeed, there are species unknown to the authors at the time of writing. Alternatively, your species may be a marine species found accidentally in freshwater, an introduced species, or a species that has escaped from an aquaculture facility or aquarium. In such situations, try to consult a specialist.
3. The key may be incorrect. If this is the case, please accept our apologies. For some species, we have only been able to examine a small number of individuals, or we have taken characters from published species descriptions. The key and diagnosis may not reflect the full variability of character states in a species for various reasons. In this case, we encourage the reader to publish the information and/or suggest a better key to improve our knowledge for future generations of ichthyologists.

A word of caution. Today, fish are usually sequenced for their mitochondrial DNA first and then compared with the sequence of other species. If both are different, morphological characters are examined to see if the molecular groups can also be distinguished externally. This morphological description is sometimes based on a very limited number of individuals and populations, usually because there were simply no more fish available to study. While almost all authors do their best to study enough individuals, there are some "black sheep" in fish taxonomy. Because it is easier to find morphological differences when only a few individuals, ideally from a single population, are examined, some authors intentionally limit their material, often ignoring published data. As species based solely on mtDNA distances are poorly defined and often not accepted by the scientific community, careless handling of morphological data discredits the science of taxonomy and should be avoided.

For example, a new species was described based on seven individuals from one locality. The original description distinguished it from just five individuals of another species that is widespread in the same area. A large previous study published morphological data from 65 individuals from this species. The new species (7 individuals) differs from the known species (5 individuals) in having a shorter head (20–24 % vs. 24–30 % SL; in the 65 individuals: 21–27 % SL), a shorter pectoral fin (10–15 % vs. 16–19 % SL; in the 65 individuals: 15–21 % SL), and 25–29 circumpeduncular scales (vs. 23–24; in the 65 individuals: 23–28, usually 24 or 26). The length of the pectoral fin seems to distinguish the two species, but the other characters do not when a more extensive series is examined.

This example shows why a key may not work. We often need to learn more about the variability of morphological characters in many species, and some character states published may no longer distinguish species when they are better studied.

Species description. The following headings explain the nature of the information provided in each species account.

Species names. According to the International Code of Zoological Nomenclature, all names used for the listed species are valid scientific names. A few species, such as some *Atherina*, *Oxynoemacheilus*, *Paraphanius*, *Ponticola*, and others, still need to have scientific names. This is because they have only recently been discovered by science, and the formal process of describing the species and assigning a valid name is time-consuming and requires the resolution of very complex nomenclatural and systematic issues. Common names also present a variety of problems. Many species do not have a species-specific common name (by which we mean a real one, in local lay use, not one coined by scientists, especially for a technical glossary). They are only referred to by a collective name, such as the different species of *Garra*. Of course, names mentioned in the scientific literature or official documents may be completely unknown to the general public. We have tried to give a single common name for each species. Where an English language name is commonly used, we have listed this as the vernacular name. As specific names based on local languages are rarely available, introducing English names as preferred vernacular names is beneficial. We have, therefore, tried to assign such names to all species based on their character states or geographic range. We have usually used names suggested by previous authors, but we have not hesitated to create new names where we felt that the earlier names could have been more satisfactory. We have never used personal names as part of a common name (e.g., *Garra gallagheri* is the Black garra, not Gallagher's garra).

Diagnosis. A diagnosis is a summary of characters applied to identify and distinguish a species from other similar species. There are usually many more characters that distinguish species than those listed in our diagnoses, but we focus on those that are the easiest or most reliable to use. In some cases, a single character may be sufficient to identify a species, but a combination of characters is required in many cases. If a fish exhibits some of the character states listed in a diagnosis but not others, it should be concluded that the individual does not belong to the diagnosed species. Very often, it is possible to identify species at a glance based on general appearance, but general appearance is often very difficult to describe in words. Increasingly, the artificial intelligence of citizen science platforms, largely based on general appearance, is making it easier to identify fish from images. Each diagnosis starts with an explicit statement that the species in question is distinguished from another species or a group of other species (the **reference group**) by the characters to be described. The size of the reference group varies according to the context, as does the usefulness of the characters and character states.

For example, suppose the reference group is "all species of the genus known from the Aegean basin." In that case, this means that the diagnosis can be used to reliably identify a fish from within the Aegean basin (e.g., the Anatolian Aegean basin) but not from outside the area covered by this book. Some additional species may occur outside Türkiye and/or outside the Aegean basin, and these may share the same character states but may be distinguished by additional characters not listed here. However, whenever possible, without including too many technical details and without writing long and cumbersome diagnoses, the reference group has been extended as far as possible. In the diagnoses, the different characters and character states are separated by slashes (/) and each character is preceded by a symbol indicating the "efficiency" of the character in identifying a species within the reference group. **The symbol "●" means that this character state is not observed in any other member of the reference group and that its presence alone allows identification of the species; whereas "○" means that this character state is shared with some (not all) species of the reference group**, but its presence alone does not allow identification of an individual fish. Therefore, it is the combination of all the characters given that distinguishes the species.

A word of caution. Where there is doubt as to whether a particular character state is unique to the diagnosed species, we have listed the character as "○" rather than "●." For example, we have generally not included fin ray counts, scale counts, and morphometric characters as "●" unless they are clearly distinctive because they are often close to or overlap with those of other species. Ambiguities are possible due to slightly different approaches to counting and measuring. Sizes are usually given as standard length (SL) (see Morphological features) and correspond to the largest size recorded based on literature or personal observation. However, this information must be treated with caution, as the true maximum size of some large or rare species may be considerably larger than that reported in the literature. Some sizes are given as total length (TL). Bilaterally symmetrical structures are written in the singular (e.g., "the eye is large," not "the eyes are large") unless syntax or common sense requires the plural. In keys and diagnosis, male and female are used in the singular.

Distribution. First, an overview of the native and non-native distribution in West Asia is given, followed by a summary of the global distribution. Including an area or country in the distribution is for general information only and does not imply that a species is evenly distributed within that area. Of course, the distribution of many species still needs to be better understood.

Habitat. The habitat types given are those in which the species has been found. This does not mean a given species could not colonise other habitats if they become available. Also, habitats can change significantly over a year, as floods and droughts can change the water table. For this reason, habitat descriptions are usually brief and general. West Asia is made up of a wide variety of freshwater ecosystems. These include waters in warm, wet, and humid forests, such as along the northern Black Sea and southern Caspian coasts; wadis in full deserts, typical of the Arabian Peninsula; high mountain lakes, such as in the Caucasus; and places remarkably below sea level, such as in the Dead Sea basin. Most areas are strongly influenced by seasonal rainfall patterns, with little or no rain in summer and more rain in winter. As a result, many areas in Anatolia and the Levant are classified as having a Mediterranean flow regime, with streams flowing from autumn to early summer, often with high flash floods in winter. In summer, surface water frequently stops flowing, leaving only pools of stagnant water. However, this type of flow regime is extreme in the arid areas of the Arabian Peninsula, where streams receive only infrequent rainfall, usually in winter, often associated with very high floods. During the dry season, groundwater continues to flow below the surface in the alluvial gravel that fills the channels and locally emerges as springs. In the arid and semi-arid landscapes of Arabia, Mesopotamia, Central Anatolia, and the Levant, streams are often spring-fed, and permanent water may be seasonally limited to the springs.

Biology. For most freshwater fish species in West Asia, even the most basic biological data still need to be included. Adequate biological information is generally available for species outside West Asia, particularly in Europe. Much of the available biological data are derived from observations outside West Asia, so it may only partially represent the region's fishes. Our descriptions of the biology of individual species usually begin with reproductive biology, including data on age at first spawn, frequency of females spawning in each season, and longevity, in addition to basic data on the spawning season, timing, and behaviour. In freshwater species, water temperature and day length often trigger spawning. The biology section usually ends with the type of food consumed. Even where data are available, it is impossible to give quantitative or detailed qualitative data on the food consumed. More data are needed for most species to correlate the type of food consumed with the prey available, and most species appear to prey indifferently on invertebrates of appropriate size. However, there is a strong correlation between (i) observed and available prey (common prey is usually consumed), (ii) prey size and the size of the fish (fish usually prefer large prey), and (iii) the prey chosen and the differences in predation risk that foraging fish select when searching for different types of prey (fish prefer to feed in safe places).

Conservation status. The global IUCN Red List status is given, followed by a brief description of the main threats to the species, if any, and related comments. The status of species endemic to West Asia is based on the results of the IUCN 2024 Assessment of the Status and Distribution of Freshwater Biodiversity in the region. Details of the categories and criteria and the assessment methodology are available in the IUCN Red List of Threatened Species database (http://www.iucnredlist.org). It is also important to note that the IUCN Red List is a tool for assessing extinction risk rather than determining conservation needs or prioritising conservation actions. During an assessment, the Red List category assigned to a species is not unchallengeable and may change over the years. Status is influenced by the recognition of new or previously unknown threats or opportunities that have been identified and are causing a visible decline or increase in the number of individuals of a species or are likely to cause a future decline or increase. Therefore, the conservation statuses given in this book may need to be updated in the coming years.

Remarks. Other information that is of potential interest, including open questions and research needs.

Further reading. References are given to provide published sources where more detailed accounts of a species can be found, and notes indicating what information is emphasised in each source where appropriate. The reader will find the full reference list in the bibliography section at the end of the book. Readers should be aware, however, that for many species, much of the information presented is based on field experience and laboratory research by the authors, some of which are presented here for the first time. The species bibliographies still need to be completed. The sources listed are the most relevant, contain important information, or provide references to additional publications or critical compilations, etc. In some cases, there may be only a single reference, which may only sometimes be fully authoritative.

Figures. Each species is illustrated with a figure in the left lateral view, as is standard in fish taxonomy. Only sometimes, the left side was unsuitable, then the image was rotated, and the right side of the body was shown. Species known to be highly variable are often illustrated with several figures. The origin and size of the individuals depicted are given where known. **Identification should be made only based on the characters presented in the diagnoses and not by comparing an individual with the pictures.** The general appearance of fish may vary seasonally (spawning, feeding), or by size and age, or environmental parameters (e.g., pattern and colour are more contrasting in individuals living in clear water than in turbid water). There is also variation between individuals and between sexes. The light conditions during photography and the condition of the individual fish (alive or dead, fresh or preserved, in or out of water, etc.) significantly influence appearance. It is impossible to illustrate every species under all these conditions, so we have chosen figures that will best facilitate the diagnosis of the species. While freshly preserved specimens are the best for comparison purposes, the colours of some fish fade or change as soon as they die, sometimes as soon as they are out of the water or under stress. We have tried not to make this a "book of dead fish" but to show living individuals whenever possible. However, there are a few species for which no live photographs are available, usually because the species is extinct.

Maps. Each species chapter consists of a distribution map with dots and hydrobasin layers as well as the species' latest IUCN categories. The dots indicate site-scale records for a species, e.g., where the species has been found at least once. This does not mean the species is still found there, especially as many records are decades old. In other cases, the species may have been found elsewhere, but the record(s) have yet to be published or sourced by the authors. Accurate mapping of freshwater fish species is challenging but essential for future conservation planning and research. The maps presented in this book are based on publicly available data downloaded from GBIF (www.GBIF.org), additional published records, museum records, authors' field records and the individuals whom they have examined, as well as information provided by colleagues. The background of the points is layers of hydrobasins. These show the distribution of the species according to the published knowledge and experience of the authors and experts consulted. For some species, there are few or many hydrobasins without

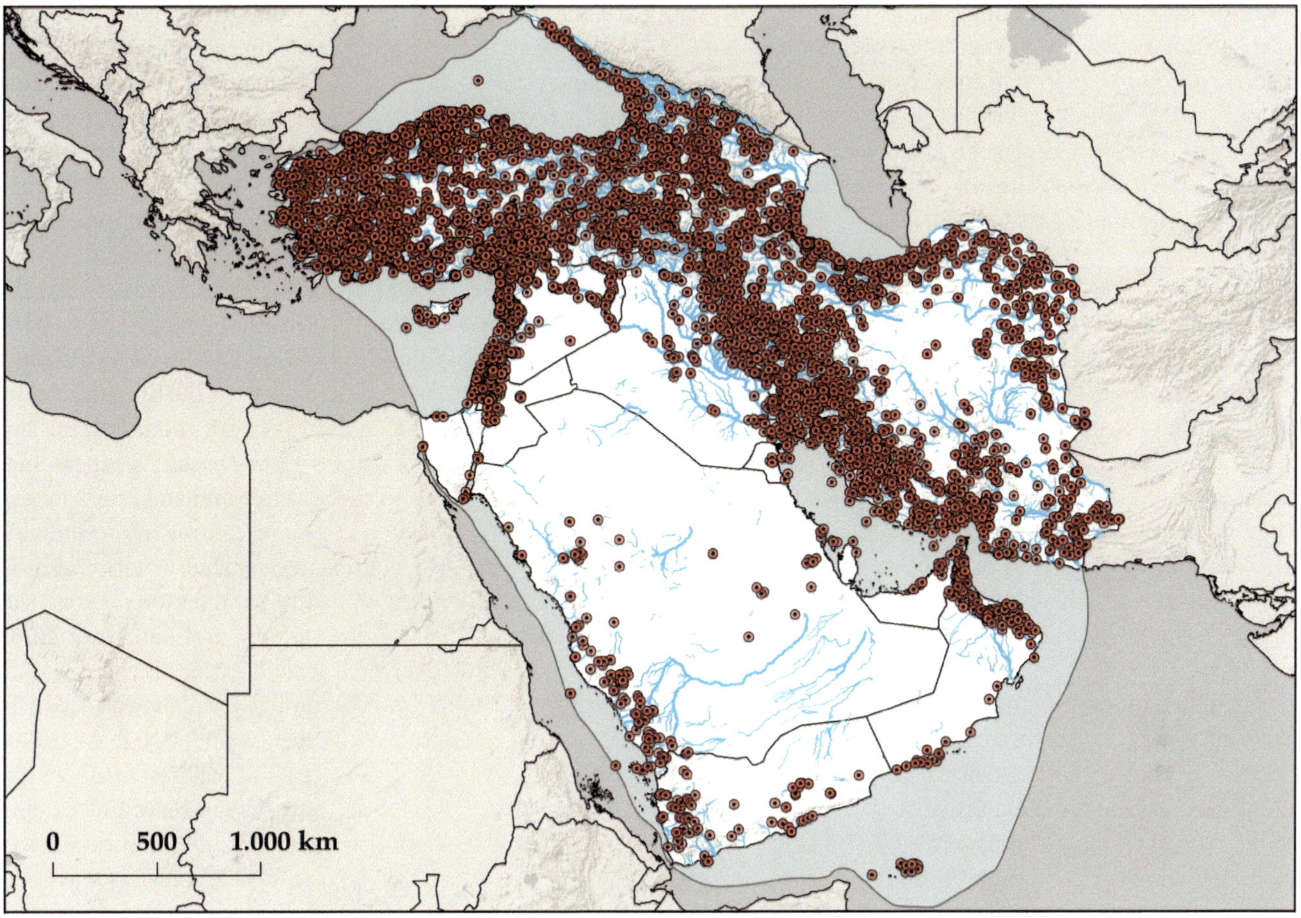

Figure 2. Precisely 26,003 site-scale records of freshwater fishes were used to map the species for this book. This map also shows areas without permanent waters and areas with very little research efforts (e.g., lower and middle Euphrates).

site-scale records, which simply shows how poorly the distribution of these species is documented.

On the other hand, hydrobasins without site-scale records also indicate that the species may have a much smaller range than previously thought. For example, several species known to be widespread in the northern Black Sea basin rivers were also considered widespread in the southern Black Sea basin (e.g., *Abramis brama*, *Leuciscus aspius*, *Rutilus lacustris*). However, many empty hydrobasins have been investigated, and these species have a very limited distribution in the southern Black Sea basin. All maps have been checked by the authors and regional experts.

The maps are produced using QGIS 3.28.15-Firenze version software. The distribution of each species was mapped to river and lake sub-basins as defined by the HydroBASINS framework, which consists of a series of vector-based polygon layers delimiting sub-basin boundaries at a global scale. HydroBASINS provides these delineations at 12 different levels of resolution and incorporates key details of hydrological connectivity. While IUCN Red List assessments typically map species distributions at HydroBASINS level 08, our methodology used a more detailed approach by using HydroBASINS up to level 12, which represents finer sub-basins, to achieve greater spatial precision in species distribution mapping. This increased resolution was applied in two scenarios: (1) when highly detailed spatial data were available and (2) for narrowly distributed endemic species, allowing for a more accurate representation of their geographic ranges. The maps show the distribution of all species native to West Asia, distinguishing the current native range (green), areas where they have been non-native (purple), and areas where they are extirpated (red). For species that also occur outside West Asia, only the West Asian range is shown. A simplified approach we have taken to mapping such species is illustrated in Figure 3.

Literature cited. The bibliography section lists only published references mentioned in the text.

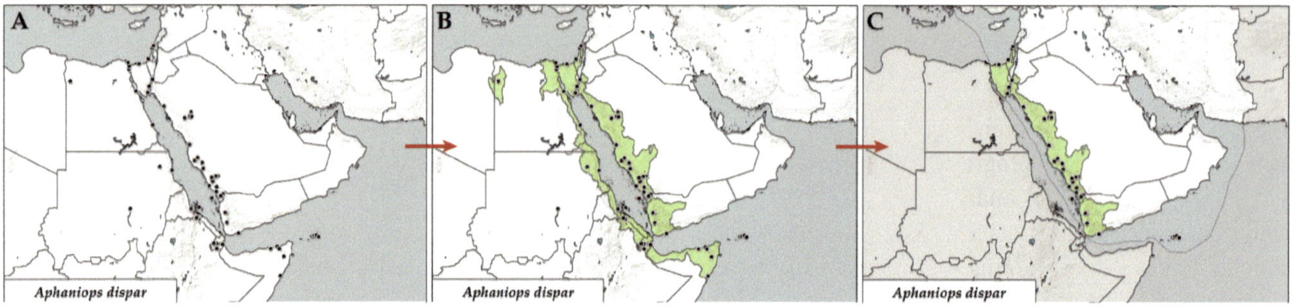

Figure 3. Illustration of the range mapping approach for species with a global range extending beyond West Asia. **a**, Global occurrence records are displayed as black dots; **b**, the Global range of the species, highlighted in green; **c**, the range of the species in West Asia, focusing only on the book coverage for consistency.

Visiting fish markets like this one in Hor Sosangerd (near the Hor Alazim wetland), Iran, enables contact with friendly, local experts. It is a long tradition in ichthyology and remains a good source of information and difficult-to-catch fish.

Fieldwork and data collection

Catching fish is challenging, but organising fieldwork to produce useful results is even more so. Catching fish is only part of the job; taking photographs of fish and habitats, collecting samples for DNA analysis, and collecting voucher material are also important. Finally, critical fieldwork components include documenting ecological and habitat conditions, collecting GPS data, and making all data available to the public. Catching fish depends largely on the choice of fishing gear designed for the specific habitat conditions. Different habitats require different gear, and different fish species and size groups may require different techniques and equipment. Planning fieldwork also involves applying for permits, assembling comprehensive sampling and recording equipment for the activities to be undertaken, and assembling adequate staff to carry out the proposed work within the time allotted. It is beyond the scope of this book to provide detailed instructions on observing and collecting freshwater fish in the field. Several comprehensive manuals guide how to catch fish in specific environments and how to fix and preserve specimens for various purposes. These topics are only briefly discussed here.

Electrofishing is the least invasive method as the fish are not harmed, and those not required for further study can be released. However, electrofishing is relatively labor-intensive, special permits may be necessary, and the equipment is expensive and physically heavy. In many situations, nets or traps may sufficiently effectively catch the target fish.

Electrofishing is a method for catching fish in small- to medium-sized waters. It is animal-friendly and hurts fish much less than any other method.

Small aquarium to photograph fishes in the field. Note the black background used in the field.

Photographing fishes. Photography is an essential part of any ichthyological survey. The aim should be to photograph several representative individuals of each species from each collection site. The optimal way to photograph fish is in the field using a field aquarium as soon as possible after collection. In this way, the fish are in good physical condition, and their natural colours can still be seen. Live and preserved fish can be photographed similarly, with the hand positioned underwater in the field aquarium and photographed by another person. Photographing fish on the bottom, in the grass, or on a person's hand will always result in poor-quality images. Highly specialised professional equipment is available for photographing fishes in the field, including special aquariums in which the fish can be fixed in any position under standard lighting conditions. Such equipment is ideal for large expedition teams with ample time and storage space. Underwater photographs, while aesthetically pleasing, often need to show fish in full detail, especially in locations where the water is murky or where fish are shy and not easily approached. **Further reading.** Garcia-Melo et al. 2019 (photographing fish).

Salmo brunoi photographed alive in the field. Note that the hand and fingers positioning the fish should not be seen behind the fins; this allows the fish to be cut out from the background using appropriate software. Below is the same image cleaned up by an image editing software.

Photographing live fish may be more time-consuming than photographing preserved fish. Still, images of live fish are more attractive, and you will find a wider range of uses for the photos than simply associating them with the voucher specimens. For photographic purposes, it is preferable to choose fully grown individuals with complete scale coverage and complete, undamaged fins. It is also interesting to photograph juveniles or individuals of different sexes. The usual procedure is as follows:

Add the reference number to the picture, e.g., on the bottom of the glass, to identify it from the image later.

- Place the live fish in the prepared field aquarium. An anaesthetised fish may be easier to handle. Fish can be anaesthetised with MS222, clove oil, or chlorobutanol. Avoid photographing dead fish. Even if the fish is freshly dead, the eye will fade, which can be spotted by experienced observers. Photograph the left side of the fish if it is undamaged.
- An operator immobilises the fish by gently pressing it against the front glass of the aquarium. Do not press too hard; the fish usually gives up trying to escape after a few seconds. Ensure the fins are in a natural position and you

cannot see the operator's hand behind a transparent fin. Position the fish's body parallel to the outer glass.

- Benthic fish such as loaches may look more natural if photographed on a rock rather than floating freely in the water column. Even nervous fish will remain calm if you gently redirect them a few times and sing a lullaby to them.
- Make sure the glass and water are clean and free of bubbles. The second operator photographs the fish through the glass. Ensure the fish remains parallel to get a good image of its side body.
- Take several shots to ensure that the fish's fins are in a neutral position and that its mouth is not open to breathe.
- Use a high f-number (small aperture) to ensure that different parts of the fish, from the eyes to the caudal fin, remain in sharp focus throughout the image. Start with f/8 or f/11 and adjust as needed. Higher f-numbers may require slower shutter speeds or increased ISO to maintain proper exposure, so a tripod can help prevent blur.
- Check the in-camera image to ensure that the entire fish is visible and in a natural position, that the image is not over- or underexposed, that there are no obvious reflections (at least not on the fish), and that the image, especially the eye, is in focus.
- If light reflection from the front glass is unavoidable, use a polarizing lens filter. Such a filter will slow down the shutter speed, so using a tripod will again improve the quality.

Figure 4. Safety comes first, both during fieldwork and when handling preserved fish in the laboratory.

Euthanizing and preserving fish. With a little experience, and especially some knowledge of the fauna of a particular area, most species can be identified alive, and euthanasia is not necessary. Sometimes, fish will be required to be sacrificed for examination. Preserving the fish immediately in the field is preferable to ensure the best possible condition of the specimen and, thus, optimal identification. Remember that fish are living creatures that can feel pain and should not be treated carelessly.

Euthanasia. There are many national regulations regarding the euthanasia of fish. For best results, fish should be euthanised with an overdose of anesthetic (MS222, clove oil, or chlorobutanol) and, when dead, carefully washed and immediately placed in formalin. Fish should never be left to die naturally. A fish forgotten in a plastic bottle, with damaged fins, which has died from suffocation or careless handling and/or has been fixed several hours after death, is often difficult to identify and useless for scientific research. Animal welfare laws must be followed as they do not allow unnecessary stress to be inflicted on vertebrates such as fish. Fish should not be transported in a bottle or other container for any length of time, as stress can cause them to lose colour, making identification more difficult, not to mention the possibility of death or the larger fish eating the smaller ones.

Fixation. Fixation is the preparation of a specimen with chemicals (preservatives) to prevent decay and allow for later examination and long-term preservation and storage. Proper fixation of freshly caught individuals is very important for correct identification. Formalin is the only fixative that should be used. Ethanol does not provide adequate fixation for identification and long-term storage. Formalin is a 35–40 % aqueous formaldehyde solution (a gas). Formalin should first be diluted 1:10 with water (to give an approximately 4 % formaldehyde solution); clean water from the stream from which the fish were collected should be used. For fixation, use a bucket or large container where the specimens lie flat and their fins extended.

Larger numbers of fish (even several species) from the same locality may be fixed in the same container. The volume of liquid should be more than twice the volume of the specimens. Ensure they do not move in the container during transport to prevent fins from breaking. If a fish has died in an unnatural position or with its mouth open, remove it from the formalin, rinse it thoroughly, turn it upright, or close its mouth and return it to the formalin bath. Be careful not to remove scales or break fins accidentally. Always work near a water source (a river, running water, or a large bucket of clean water) and immediately and thoroughly rinse any tools that come into contact with formalin. Never bring formalin into a room or vehicle. When working with preserved fish, preventing formalin from coming into contact with a person's skin or eyes is important. If contact does occur, the affected area should be rinsed thoroughly with plenty of water. Contact with the eyes will cause immediate and severe burning. Containers used to fix fish, especially larger species, should always be watertight and kept out of the reach of children and animals. It is best to avoid smoking when working with formalin. Formaldehyde is a carcinogen (can cause cancer in living tissue) and should not be inhaled. Nitrile or Neoprene gloves should be used, as even rubber or latex gloves may not protect the hands.

Larger fish (more than 150–250 mm long, depending on body shape and width) should be injected with pure formalin into the abdominal cavity. Alternatively, the body should be cut open slightly on the right side to allow immediate fixation of the viscera. The formalin used should be disposed of following local or national regulations. Do not dispose of it where it may be exposed to humans or animals. Used formalin that is no longer completely clear should be discarded. Fish from different locations should never be stored in the same container. Each container should be immediately labelled with the following information: province, river, locality, geographical coordinates, name of collector(s), and date. The label must always be placed inside the container and never glued or attached to the outside (it will eventually fall off, and specimens without locality data or uncertain locality data are useless). Use strong, waterproof, archival paper and write clearly with a waterproof pigment or archival ink or a soft pencil. Cheap paper without fibers (e.g., photocopy paper) is unsuitable as it disintegrates within minutes. Ballpoint pens should never be used. Do not try to save money or time at this crucial stage.

Preservation and shipping. Preserved fish should be kept in formalin for at least a week. They may then be rinsed in water and transferred to a 25 % ethanol solution for one week. They are then transferred to a 50 % ethanol solution for a further week and finally stored in a 70 % ethanol solution. Isopropanol or methylated spirits are unsuitable as they alter the structure, bleach, dehydrate, or harden the fish. When shipping fish samples, it is preferable to keep fish samples in formalin in their original **leak-proof** containers. This is not possible if they are sent by post, courier, or air, in which case the formalin should be drained and replaced with water after a few days in formalin. The specimens can remain in this water (which retains some formalin) for 1–2 weeks without damage. Poor-quality containers should be avoided at all costs. Each container should be sealed in a plastic bag and placed in a strong box. If this method is inconvenient, wrap the specimens in cheesecloth or strong white (unstained) tissue soaked in formalin (or in ethanol if the specimens have already been transferred to ethanol), place them in a sealed plastic bag with the labels inside, and then put in a strong box; if the fish have strong spines, remember that the spines may puncture plastic bags and formalin may leak. Use several layers of plastic bags and place newspaper, cardboard, or polystyrene between them.

Figure 5. Tissue collections are essential for studies on fish, providing the genetic data necessary for analysis. Fin clips are commonly used for this purpose in freshwater fish.

Preserving tissue for DNA extraction. Preservation of tissue samples for DNA analysis is standard practise in field research. The same general rules for preserving whole fish should be applied, except that tissues sampled for DNA analysis should never be exposed to formalin, only ethanol. Voucher specimens are the individuals from which the tissue for DNA extraction was taken, and the preservation of vouchers is an essential part of scientific documentation.

- Euthanise the fish with an overdose of anesthetic (MS222, clove oil, or chlorobutanol).
- Cut the pectoral or pelvic fin or part of it from the right side of the fish and place it in pure ethanol.
- Fins preserved in pure ethanol for a few days or 1 week can be kept and transported dry for some time. However, this requires rapid and complete desiccation, which usually takes only a few minutes as the ethanol evaporates quickly. Long-term storage of dried fins is not recommended as we still need to learn more about the decomposition processes in the dried tissues.
- Label the fish with the same number as the tube number. We use paper strips with the number printed four times; cut and fold the strip and place it behind the gill cover. You can cut one of the small numbers and put it in the tube with the fin, so you do not have to write a new label. Always put the number inside the tube!
- Preserve the fish in approximately 4 % formalin, as mentioned above.

Tissue preserved in ethanol can theoretically be stored for several decades without decaying. However, ethanol can evaporate very slowly from the vials. This reduces the concentration of ethanol, and the DNA samples may start to degrade. Deep freezing is one solution to reduce the speed of this process; an alternative is to replace the ethanol regularly. However, there needs to be more experience to advise on how to stop the process of long-term DNA degradation in tissue samples. There may be occasions when the voucher specimen cannot be preserved because it is too large, belongs to the fishers, or for other reasons. In these cases, it is preferable to take photographs of each voucher specimen. Small fish up to a few centimeters in length can be preserved in either 15- or 50-ml tubes of pure ethanol, whereas larger fish must be sampled individually. They must be euthanised by an overdose of anesthetic, as described above. A piece of tissue, usually the right pectoral or ventral fin, is removed and preserved in pure ethanol for larger specimens. It is the collector's responsibility to ensure that the tissue is labelled so that the tissue and voucher can be unambiguously linked later. Cutting fins from live fish is illegal in many countries. Different numbers of individually labelled specimens are ideal for other purposes, depending on the species and location. Typically, 5–10 individuals are appropriate for most study objectives.

Collecting DNA samples in areas where ethanol is not available

Ethanol is not always available, and the lack of ethanol is not a major obstacle. The main reason for using ethanol is to dry the fin tissue as quickly as possible. Fins can also be air-dried. Dehydration of the fins must be achieved as soon as possible, and the fins must remain dry during storage or transport. Even minor decomposition of dried tissues leads to DNA degradation and may result in loss of scientific material. Dehydrating chemicals without direct contact with the tissue sample(s) is often the solution. Using any type of cologne containing at least 80 % alcohol by weight is also effective in preserving fin tissues for some time. Placing the fins in ethanol and freezing them for long-term storage is highly recommended.

Publishing data. Data and conclusions should be made available to the scientific community or interested researchers, preferably by publication in a peer-reviewed journal or through an online platform such as ResearchGate (www.researchgate.net), which offers DOIs for citation. By making your data available through these channels, you increase the impact and visibility of your research, facilitate collaboration, and support the principles of open science. Journals specialising in publishing biodiversity data, such as the Biodiversity Data Journal (https://bdj.pensoft.net) or more specifically Freshwater Metadata Journal (http://www.freshwaterjournal.eu/), as well as dedicated ichthyology journals that accept manuscripts describing the distribution of freshwater fishes, are particularly suitable for this purpose. Individual species records must be submitted with GPS coordinates, ideally in decimal format, to be added to existing maps or used in other analyses such as species distribution modelling. For species that are more difficult to identify, lateral colour photographs should accompany the data to help validate identifications. Uploading records and images to citizen science platforms such as www.inaturalist.org or https://observation.org is recommended, as these make the records publicly available through GBIF (www.GBIF.org).

Further reading. Neumann 2010 (field techniques).

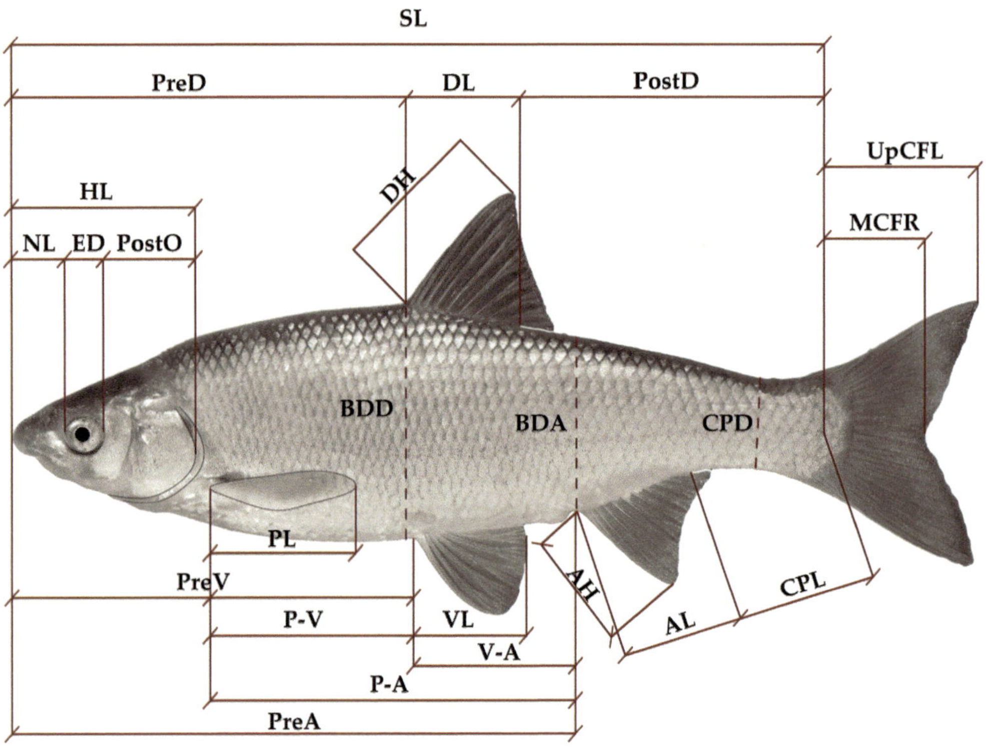

Figure 6. Morphometric characters: AH, anal fin height; AL, anal-fin base length; BDA, body depth at anal-fin origin; BDD, body depth at dorsal-fin origin; CPD, depth of caudal peduncle; CPL, length of caudal peduncle; DH, dorsal-fin height; DL, dorsal-fin base length; ED, eye diameter; HL, head length; MCFR, length of middle caudal-fin ray; NL, snout length; P-A, distance between pectoral and anal-fin origins; PL, pectoral-fin length; PostD, postdorsal length; PostO, postorbital distance; PreA, preanal length; PreD, predorsal length; PreV, prepelvic length; P-V, distance between pectoral and pelvic-fin origins; SL, standard length; UpCFL, length of upper caudal fin lobe; V-A, distance between pelvic and anal-fin origins; VL, pelvic-fin length.

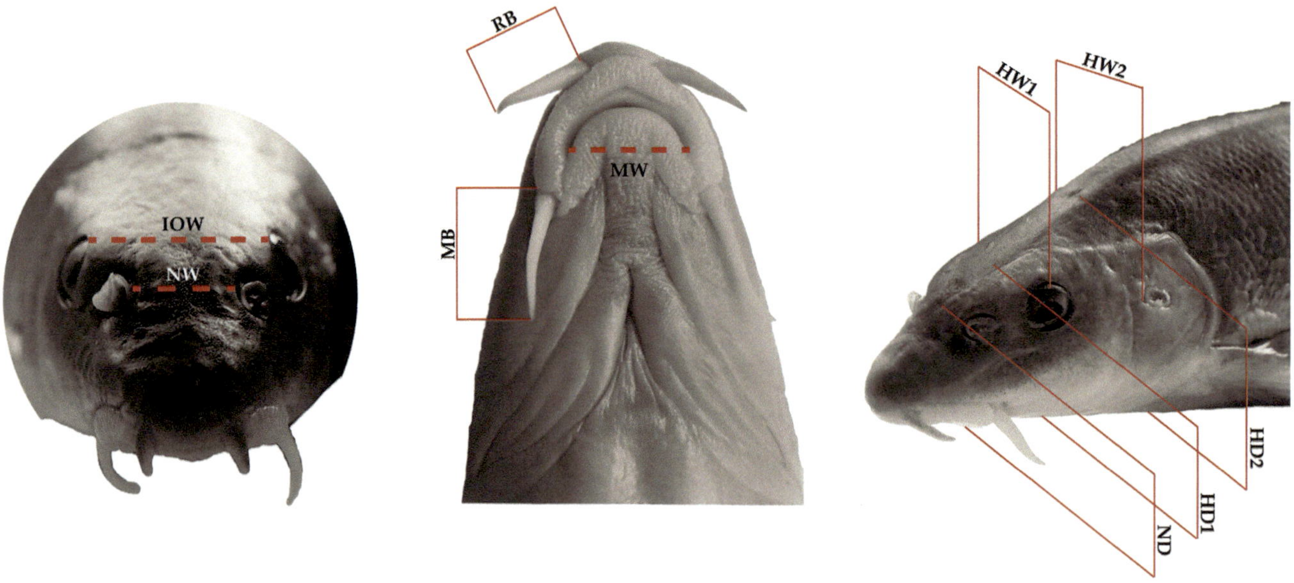

Figure 7. Morphometric characters: HD1, head depth at eye; HD2, head depth at nape; HW1, head width at eye; HW2, head width at nape; IOW, interorbital width; MB, length of maxillary barbel; MW, mouth width; ND, snout depth; NW, snout width or internasal; RB, length of rostral barbel.

Morphological characters

In the species chapters, terms are abbreviated as caudal-fin base, dorsal-fin origin, and similar terms, e.g., caudal base and dorsal origin.

Measurements. The standard measurements used are shown in Figures 6 and 7. All are taken as a straight line from point to point, not across the body curves or as projections along the longitudinal axis.

Standard length (**SL**) is measured from the foremost point of the body (usually the tip of the snout or upper lip) to the base of the median caudal-fin rays at the end of the hypural complex. The position of the end of the hypural complex is only sometimes obvious. Note that the base of the caudal fin rarely coincides with the last scales, as scales often extend onto the caudal fin. The end of the hypural complex is easily seen as a fold when the fin is bent from side to side (Figure 8). It is often located in front of the last 2–3 scales on the caudal peduncle, a point where the caudal fin can easily be bent against the body of the fish.

Total length (**TL**) is measured from the body's foremost point to the caudal fin's rearmost point. For most fish, SL is used as the reference length; for a few groups (e.g., eels), TL is used as the reference length. Although commonly used in fisheries, the fork length (from the anteriormost point of the head to the tip of the median caudal ray) is not used in ichthyology.

Head lenght (**HL**) is the distance from the body's foremost point to the opercular membrane's posteriormost point. Dorsal HL is measured from the foremost point of the body to the posteriormost point of the skull along the dorsal midline (occiput). Except for SL and the TL, which are given in millimeters, all other measurements are given either as a percentage (e.g., % SL or % HL) or as a ratio (e.g., HL four times in SL).

Dorso- and pelvic-hypural distance are measured, as shown in Figure 9.

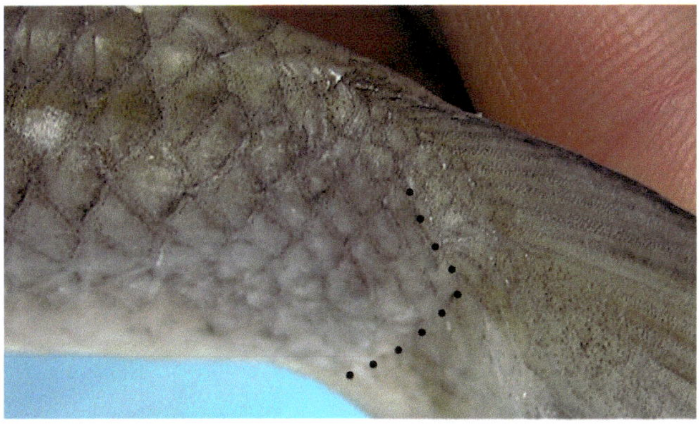

Figure 8. Locating the end of the hypural complex is achieved by bending the caudal against the corpus of the fish.

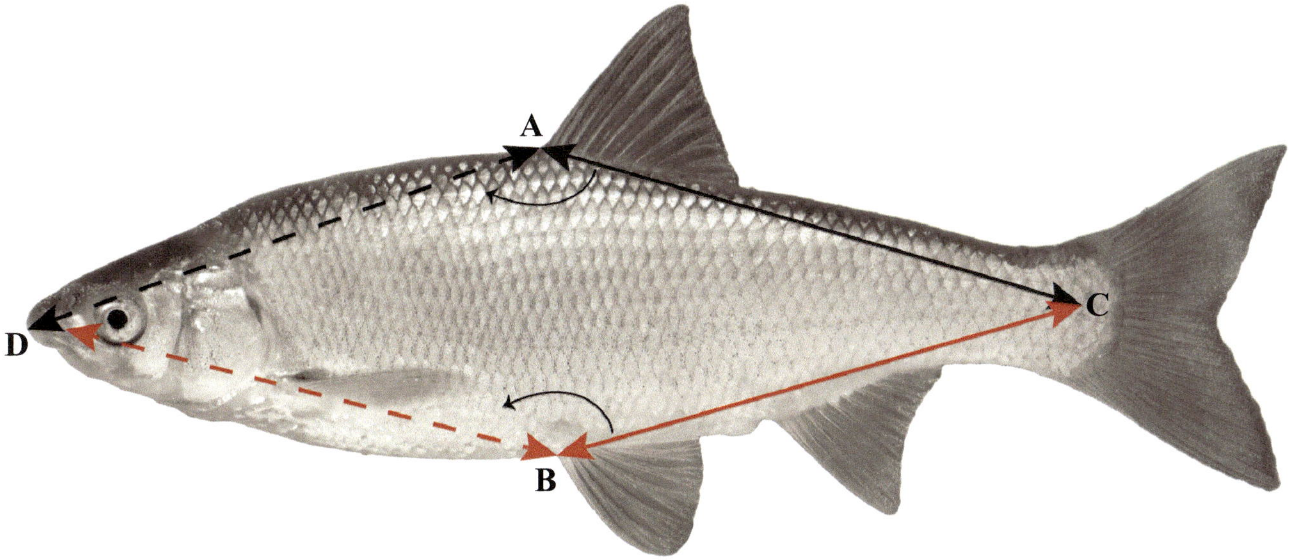

Figure 9. Dorso-hypural distance is measured from the origin of the dorsal fin to the base of the median caudal ray (A–C) and reported forward (A–D). The pelvic-hypural distance is measured from the origin or pelvic fin to the base of the median caudal ray (B–C) and reported forward (B–E).

Sources of data, data transformation, causes of errors, limitations of data

The data used to compile the species accounts come from various sources. Some are from our field observations and live, freshly preserved, or museum specimens. In such cases, we have ensured that the data are presented consistently. For some species, data have been obtained from the literature. Wherever possible, we have tried to verify such data with specimens; however, this was only sometimes achievable. Some authors may use their own methods of measuring or counting scales and rays or use different terminology. To make accurate comparisons, it is necessary to standardise these data. There are several systems for reporting morphometric (measured) and meristic (counted) characters. For some characters (e.g., dorsal or anal fin ray counts), the translation from one system to another is reasonably clear. For others, the translation requires interpretation of the data and a certain degree of speculation. For example, this may be the case with lateral line scale counts. We would have preferred to present them all in a single format (distinguishing scales on the body from those on the caudal fin base, e.g., 28–31+2–3), but converting total counts (including scales on the caudal fin base) to standard counts would have introduced a degree of subjectivity and potential error. We, therefore, decided to keep the data in the original format, which explains an apparent lack of consistency between some of the reports. We believe that the lack of consistency is less damaging than the risk of introducing errors.

It is impossible to describe the full range of variability in a character. There are always individuals with anomalous character traits due to injuries, poor health, etc. Moreover, very small or very large individuals may exhibit different appearances. Documenting all these subtleties here would be tedious and space-consuming if they are known at all. When measuring fish, it is important to remember that, besides rigid structures, they also contain soft tissues that can be deformed or damaged by tools. Furthermore, measurements of soft tissues are only partially reproducible. Therefore, specimens must be properly preserved, although this is only sometimes the case.

Whenever possible, conclusions drawn from morphometric characters should be based on a series of specimens rather than a single individual. If a diagnosis indicates 18–23 % and the specimen examined shows 17 %, this does not automatically exclude it from a particular species, as the sample you have studied may not include the full range of variation in the species. Some scale or fin ray counts exhibit very little variability within a population or species, whereas others may show considerable variability. Again, conclusions should be based on several specimens. Unique or rare counts have been ignored in diagnoses (e.g., if an author reported having counted dorsal fin rays in 589 specimens and observed 7½ rays in 578 and 8½ rays in 11, we have ignored the low-frequency 8½ value).

Figure 10. Common types of caudal fin: from left: rounded; truncated; emarginate; forked.

Figure 11. Dorsal fins of salmoniform fishes; also seen in most catfishes: **a**, rayed dorsal fin; **b**, adipose dorsal fin without rays.

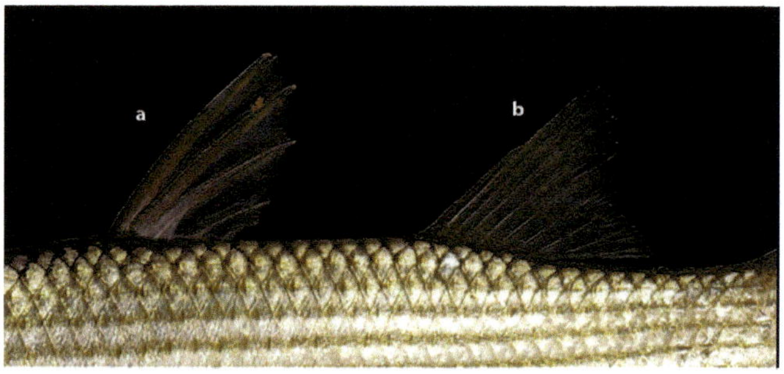

Figure 12. Dorsal fins of mullets, silversides, and others: **a**, first, spinous dorsal fin; **b**, second dorsal fin with unbranched and branched rays.

Figure 13. Dorsal fins of cichlids and many other percomorph fishes: **a**, first, spinous part of fin; **b**, second, soft part of fin.

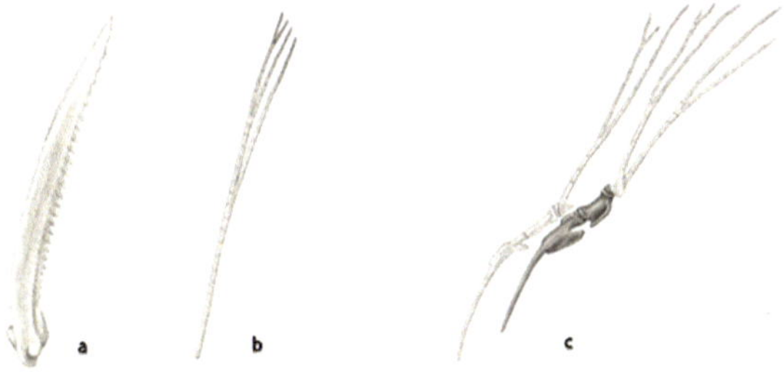

Figure 14. Main types of fin rays: **a**, spinous ray, here serrated posteriorly; **b**, schematised soft, segmented branched ray; **c**, posterior-most dorsal-fin rays and anal-fin rays showing the last two rays on a single pterygiophore (grey), which is counted as 1½ rays (after Kottelat & Freyhof 2007).

Fins and fin rays. The dorsal, caudal, and anal fins are called unpaired fins, and the pectoral and pelvic fins are called paired fins. Different shapes of the caudal fin are shown in Figure 10. In fishes with two dorsal fins (Figures 11 and 12), the anterior one often consists only of spines, and the second one often consists of a single anterior spine followed by soft (or segmented) rays. Anatomically, true spines are median (unpaired) structures, never branched, never segmented, and generally hard and pointed. Soft rays comprise a right and a left part, usually segmented and branched (Figure 14b). In some species (e.g., many cyprinids), some anterior rays may be fused into unsegmented, hard, and inflexible rays, which may also be posteriorly serrate (e.g., have a series of indentations or teeth along their posterior margins). These are called "spinous" ("similar to or shaped like a spine").

The base of a fin is the region where it attaches to the body. In some fishes, part of the rays and membranes are covered by scales, making it difficult to see the base of the ray. The origin of a fin is the insertion point of its most anterior ray (e.g., the most anterior point at the base of the fin). The heights of the dorsal and anal fins are measured from the origin of the fin to the uppermost (or lowermost) point on the fin. The respective positions of the fins are often given in the form "dorsal origin in front of anal origin" or "dorsal origin above pelvic base," meaning that the origin of the dorsal fin on the back is in front of a vertical line through the origin of the anal fin or above the base of the pelvic fin.

Fin ray counts may include only soft or branched rays, unbranched and branched rays counted separately, or the total number of rays. Each method is explicitly stated. The last two branched dorsal rays and anal rays are carried by a single pterygiophore (the bones on which the rays articulate; Figure 14c) in many fish families. Anatomically speaking, the last pterygiophore consists of two fused pterygiophore bones. These two rays are counted as "1½." Therefore, 13½ branched dorsal fin rays indicate that the dorsal fin contains 14 branched rays, the last two of which share the same pterygiophore (or sometimes appear as a single ray split to the base). Authors have been inconsistent in reporting fin ray counts or have often failed to explain their method. As a result, the same 13½ count may appear in the literature as 13 if the last two rays are counted as one or 14 if they are counted as two. We prefer the ½ notation because it immediately indicates that the author knows the problem with the last two rays. For caudal rays, "9+8 branched caudal rays" means nine branched rays in the upper lobe of the caudal fin and eight in the lower lobe. Usually, there is one principal unbranched ray above and one below the branched rays, along with several rudimentary rays in front that are not counted.

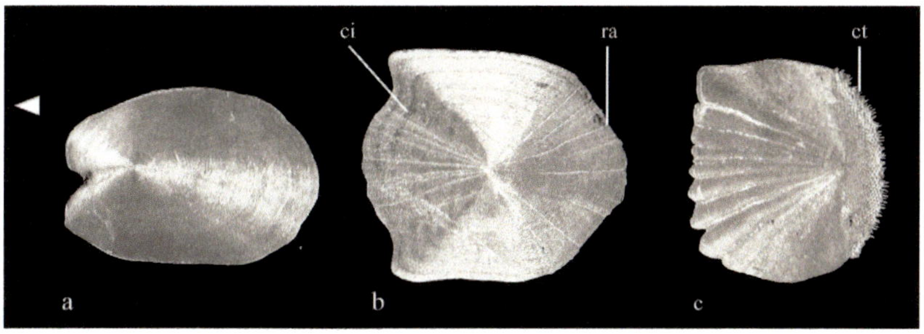

Figure 15. Main types of scales: **a,** *Esox lucius*, cycloid; **b,** *Squalius cephalus*, cycloid; **c,** *Sander lucioperca*, ctenoid. *ci*, circuli; *ct*, ctenii; *ra*, radii. The arrow points toward the head (from Kottelat & Freyhof 2007).

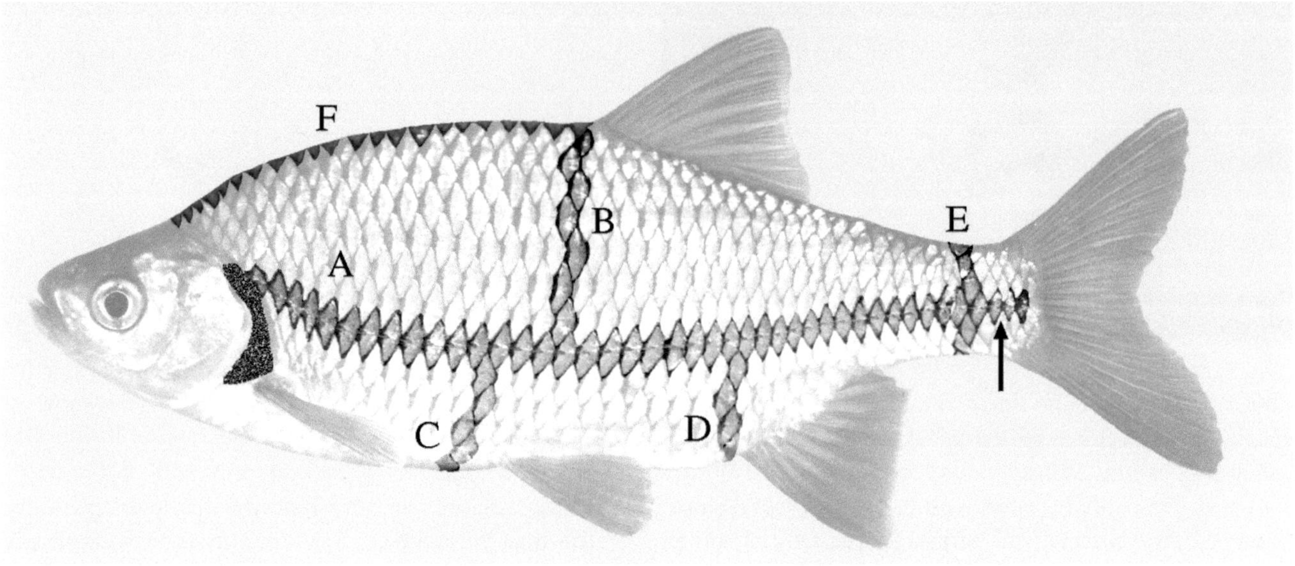

Figure 16. Illustration of principal scale counts. (A) Lateral-line scales (midlateral row); (B) transverse scales between lateral line and origin of dorsal fin; (C) transverse scales between lateral line and midline of belly; (D) transverse scales between lateral line and origin of anal fin; (E) circumpeduncular scales; (F), predorsal scales; (arrow) scales on caudal-fin base (from Kottelat & Freyhof 2007).

Scales. The lateral line scale count (Figure 16) is the number of scales on the lateral line. In the absence of lateral line pores in all scales, the midlateral series is the number of transverse scale rows counted at the mid-height of the flank. The most anterior scale counted is the one in contact with the shoulder girdle. **Counting ends at the base of the caudal fin.** Scales on the caudal fin are not counted, even if they are well-developed and porous. In such cases, they are best indicated with a "+." For example, the lateral line scale count of the fish in Figure 16 is given as 38+3, meaning that there are 38 scales along the lateral line of the body and 3 on the caudal fin. The total lateral line scale counts given in the text are those where the authors do not distinguish between scales on the body and those on the caudal fin. In this instance, the same fish has a total of (38+3) 41 scales. Fish can lose scales due to various factors, including predatory attacks and during courtship or spawning. This also happens when fish are handled frequently, for example, at sites used for annual censuses or research. Lost scales are replaced, but replacement scales often show growth or positional anomalies. Scales may also be deformed, fused, split, displaced, etc. In some species, counts will show minimal variation, whereas in others, the variability is greater. Where there is a great deal of variability, it is usually due to small, deciduous scales that are partially embedded in the skin or irregularly arranged scale rows. Counts should be checked on several specimens; extreme or anomalous counts should be disregarded, as should counts based on damaged individuals.

Transverse scale counts indicate the number of scale rows between the lateral line and the origin of the dorsal fin (or the first dorsal fin if more than one) and between the lateral line and the midline of the abdomen (anterior to either the pelvic fins or the origin of the anal fin, depending on the context). For these purposes, the scale on the dorsal or ventral midline (e.g., immediately anterior to the dorsal and anal fins) is recorded as ½. For example, the transverse scale count for the fish shown in Figure 16 is ½7/1/6½, meaning one scale immediately anterior to the dorsal (predorsal row), seven scales between the lateral line and the predorsal scale, one pore scale in the lateral line, six scales between the lateral line and the preanal scale, in addition to the preanal scale itself. The transverse scale count in front of the pelvic fins is ½7/1/5½. The predorsal scale count includes all scales on the dorsal midline in front of the dorsal fin origin (usually between the occiput and the dorsal fin origin). Predorsal scales are sometimes difficult to count accurately as the rows are not always regular. Circumpeduncular scale counts represent the number of scale rows crossing a line around the caudal peduncle at its narrowest point (16 in Figure 16, with a transverse count of ½3/1/3½).

Figure 17. *Coptodon zillii* has two dark-grey stripes, four red stripes, six dark-grey-bars, three dark-grey bands on the forehead, and one below the eye. A large ocellus, a black blotch with a white margin, is in the dorsal fin.

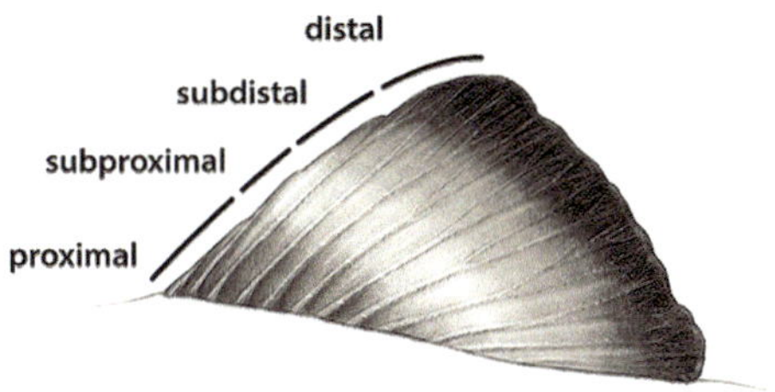

Figure 18. Positions of colour marks on fins (from Kottelat & Freyhof 2007).

Colour marks. Bars refer to vertical marks, and stripes denote longitudinal marks. Other elongated marks are called bands (Figure 17). Usually, round marks that are more or less the size of a pupil or smaller are called spots. Even smaller round marks are called dots, and larger ones are blotches. However, some inconsistencies exist, especially when round marks of different sizes are present. In such cases, spots mean smaller marks, and blotches mean larger marks. Irregularly shaped marks are called blotches. Ultimately, there are gradual differences between dots, spots, blotches, and bars, and no clear-cut definition exists to distinguish them. For example, an "irregularly shaped, short bar" and a "vertically elongated, large blotch" might be different descriptions for the same colour mark. On the fins, a proximal band is located along the base of the fin, a distal band along the outside margin of the fin, a median band in the middle of the fin, and subdistal and subproximal bands are located in intermediate positions.

Figure 19. Principal types of mouth position: from left: superior, terminal, subterminal, and inferior.

Other characters used for identification. The nomenclature for mouth positions and gill arches is shown in Figures 19. There are, of course, subtle differences between different mouth positions. It is often not trivial to decide whether a mouth is terminal (when the tip of the lower lip or jaw reaches the vertical of the upper lip, upper jaw, or rostral cap) or subterminal (when the upper lip, jaw, or rostral cap protrudes from the lower jaw or lip). The same difficulty arises when distinguishing between subterminal and inferior or superior and terminal mouth positions.

Gill rakers are the anterior bony projections usually present on all gill arches (Figure 20). There is a different number of rakers on each arch of the same individual, and all counts given here refer to the outer gill rakers on the first (anteriormost) gill arch on the right side of the head. Unless otherwise stated, these counts encompass all gill rakers, including formed rudiments. It is sometimes given as A+B+C, where A is the number of gill rakers on the upper limb (epibranchial), C is the number on the lower limb (ceratobranchial), and B is the angle between the upper and lower limbs (if any). It is sometimes necessary to cut and remove the gill arch to count the gill rakers. In ichthyology, dissection is always done on the right side, leaving the left side intact for identification, photography, etc.

The rostral barbel is located above the upper lip or upper jaw (see Figure 21); the maxillary barbel is situated on the lower jaw or at the corner of the mouth. A nasal barbel is a barbel-like extension of the anterior nostril (most fish have a nare on each side of the head, each with an anterior and posterior nostril) and a barbel-like extension on the lower lip of some cobitid loaches.

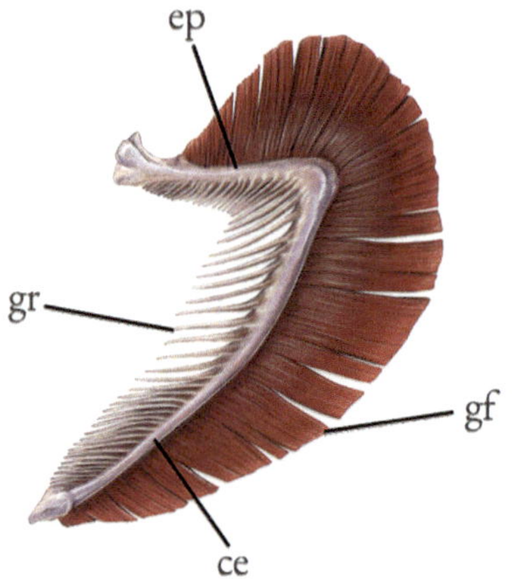

Figure 20. First gill arch, inner view. ce, ceratobranchial or lower limb; ep, epibranchial or upper limb; gf, gill filament; gr, gill rakers (after Freyhof et al. 2020).

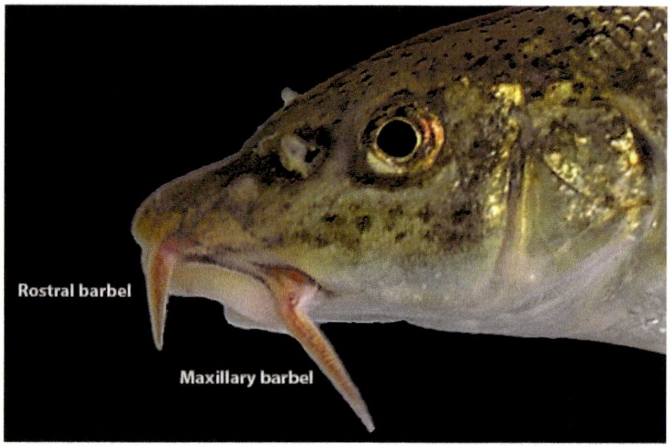

Figure 21. Head of a *Barbus* showing the two barbels typical for most cyprinids.

Taxonomic work relies heavily on fish collections, especially those containing type specimens, which are essential as references.

Taxonomy and nomenclature

Taxonomy is the theory and practise of describing the diversity of organisms and organising that diversity into a system that reflects their evolutionary relationships. Giving names to objects and living things seems natural and important to humans because it allows us to communicate with each other. This explains why most things that play a role in human life, positive or negative, have been given names, and animals are no exception. Every language, country, and region has its own set of names. Biologists have developed a nomenclature (naming) system that gives every animal species one (and only one and unique!) name to achieve high accuracy. Carl Linnaeus (1707–1778) first developed the current naming system, and the 10th edition of his **Systema Naturae**, published in 1758, is considered the starting point of modern zoological nomenclature. This system is now governed by a set of rules known as the **International Code of Zoological Nomenclature** (or "the Code"). The Code prescribes a system of naming, including rules for accepting valid names and deciding between duplicate names. This is vital to all biologists because unambiguous communication depends on names, and the name is the unique identifier that links to information on all aspects of the species. It is important to note that strict rules for describing and naming new species are set out in the International Code of Zoological Nomenclature. In modern times, for example, every newly described species must be associated with a "name-bearing type specimen," the **holotype**, a reference for the corresponding scientific name. Often, holotypes are associated with **paratypes**; specimens are usually collected with the holotype, which allows a better understanding of the variability of characters in a particular species. Holotypes and paratypes are of inestimable value and are kept in natural history museums where experts can freely examine them. **Further reading.** ICZN 1999 (International Code of Zoological Nomenclature).

Species and populations. Species is the "currency unit" in most fields of biology, and it is not surprising that its definitions and concepts have been (and continue to be) the subject of much debate. Here is just a summary of some of the basic concepts. The species concept we follow in this book is the **Evolutionary Species Concept** (ESC), which is defined as follows: a species is an entity composed of organisms that maintain their identity distinct from similar entities through time and space, in addition to having its independent evolutionary fate and historical tendencies. Fate, in this context, refers to each species' unique evolutionary pathways, including diversification, adaptation, and extinction. Species are discrete entities in nature. They participate in natural processes, evolve, have an origin, can give rise to other species (speciation), and will have an end (extinction). Of course, the future of a living species cannot be predicted, but we know that it will have a fate; the concept also applies to fossil species whose fates are already known. Because species are made up of individuals, they can never be fully defined, nor can all members of a species be expected to fit exactly into a definition. However, like individuals, species can be described and diagnosed, evolutionary lineages can be identified, and changes over time can be observed (given enough time and appropriate tools). Experience shows that most of the characters that define a species tend to vary, often in ways that are empirically predictable to some extent. It is, therefore, not surprising that most diagnoses include exceptions. Species are known from samples of individuals and samples of characters. In some cases, it can

Species have evolved from reproductively isolated populations, and there is no strict boundary between the two categories. The *Alburnus chalcoides* complex is an example of several previously recognised species consolidated into one. This may be revised in the future.

be difficult to distinguish whether differences between two populations result from phenotypic adaptations to particular environmental conditions or evolutionary adaptations that are genetically fixed.

In taxonomic discussions, a population was originally defined as a segment of a species whose members breed mostly or exclusively among themselves, usually due to physical isolation. This is a common challenge, especially in species with populations that are isolated from each other, such as many freshwater fishes. It can be difficult to decide whether two discrete groups of closely related individuals are two distantly related populations of a single species or two closely related species. The widely accepted criterion is that gene flow occurs between populations of a species but not between species. However, there are many exceptions and complexities, so broad generalisations should be avoided, and each case should be evaluated individually. The definitions of the terms "population" and "species" have not evolved in a coordinated way, and there is sometimes a grey area between them, especially when fine-scale molecular methods are used. Therefore, the rule of thumb among taxonomists is that fine-resolution molecular characters (such as those provided by increasingly available genomic methods) should be used cautiously and only when they are congruent with another set of characters (morphological, ecological, ethological, etc.). Conceptually, there is a continuum in the variability of different species populations, whereas in practise, there is a gap between the respective variabilities of two species. Of course, reconciling theory with the facts observed in nature is difficult. In West Asia, we have several groups of species (*Salmo*, Iranian *Glyptothorax*, *Esmaeilius*, some *Oxynoemacheilus*, *Alburnus*, *Barbus*, and *Garra*) where very different opinions have been published as to whether different populations are conspecific or not. We have tried our best to resolve some of these cases. Still, as this involves treating some species described by colleagues as synonyms, these colleagues frequently reject options (often without giving reasons). This is one of the reasons why the species list in this book differs from other publications, such as simple country checklists. **Further reading.** Cracraft 1989 (Phylogenetic Species Concept [PSC]); Mayden & Wood 1995 (ESC, ESU); Mayden 1997 (hierarchy of species concepts); Kottelat 1997 (PSC); Kullander 1999 (species); Mayden 2002 (species as individuals).

Nomenculature. Each species has a name comprising two words, a **generic name** (in the first position, starting with a capital letter) and a **specific name** (in the second position, beginning with a lower-case letter), for example, *Leuciscus aspius*. Each combination of two words is unique and applies to a single species. The generic name indicates affinities (or relationships) with other species. For example, several species share the generic name *Leuciscus*, suggesting they are quite similar and share a common ancestor in the group's phylogenetic (evolutionary) history. A third name (the subspecies name) was formerly used to indicate that an organism belonged to a more or less geographically distinct form of its species. Subspecies is a category that is no longer used by ichthyologists (the subspecies category is not defensible under the evolutionary species concept, and the boundaries are arbitrary under other species concepts). The name of a species is often followed by a combination of the name of one or more persons and a year, e.g., *Leuciscus aspius* (Linnaeus, 1758). This indicates the author of the species' first description (formally called the original description) and the year that description was published. If the author's name is given without parentheses, the species was placed in the same genus in the original description as it is today. If the author's name is given in parentheses, the species was originally placed in a different genus. In our

Franz Steindachner (Vienna) described *Nemacheilus angorae* in 1897. It was later placed in the genera *Orthrias* and *Barbatula* and is now recognised as a species in *Oxynoemacheilus*.

example, Linnaeus originally described *Leuciscus aspius* (Linnaeus, 1758) as *Cyprinus aspius*. Indication of author and year is not mandatory and is only justified in taxonomic publications if needed. Authors' names are not used in the species accounts below but are listed in the Appendix of this book. **We wish to emphasise the importance mistakenly attached to the unnecessary inclusion of authors' names in non-taxonomic literature.**

Many readers will notice that the nomenclature used in this book sometimes differs from that used in earlier sources. These changes are explained in several scientific publications written in recent years. This does not mean that the older publications were wrong; rather, as time passes, new scientific discoveries are made, new concepts are advanced (and old ones are sometimes discarded), and our understanding of evolution and the relationships between species evolves. This evolution of our knowledge is reflected in changes in the names of some species. Discoveries are certainly still to come, so the nomenclature in this book is likely to change in the future. Ideally, the system will one day be perfected, but that day is unlikely to come in our lifetime! In this book, we have used the nomenclature available in July 2025. Based on current scientific knowledge, these are the valid and correct names; other names are no longer valid and should not be used. By publishing this book in an open-access electronic format, we can more easily revise and update the nomenclature in future editions, ensuring that the content remains current with ongoing scientific developments. **Further reading.** Kottelat 1997 (species concepts); ICZN 1999 (Code).

The biological species concept. Textbooks still often refer to the biological species concept—the idea that different species do not interbreed in nature—although practicing taxonomists have long abandoned it because it is not testable. It can only be used to determine the distinctness of species if the species occur in sympatry, and only for a small number of species on Earth, as the most common form of speciation is allopatric. Most species for which we need to determine distinctness are found in allopatry, a situation that is particularly pronounced among freshwater taxa. We should also remember that the biological species concept only works in natural systems. Species A and B brought together in captivity, or species A introduced in the habitat of species B, represent artificial situations. Furthermore, many congeneric species found in sympatry hybridise, as demonstrated by frequent cases of introgression. This means that hybrids also occur in nature and that genetic exchange between species is still possible after speciation. **Further reading.** Mayden 1997 (hierarchy of species concepts); Kottelat 1997; Kunz 2012 (species concepts); Kullander 1999 (species).

Hybrid fertility. Hybrid fertility is often used as the main criterion for deciding whether two populations are conspecific, but this is an outdated approach. Hybrid infertility occurs at different ages in different lineages and even within a single lineage. Within some groups, species that diverged 2–100 million years ago can still produce fertile offspring; in other groups, species that diverged much more recently cannot. The latter is observed, for example, when there are chromosomal incompatibilities. In fishes, hybrid infertility is rare at the species level but more common at the genus, subfamily, or family level. **Further reading.** Cracraft 1989 (Phylogenetic Species Concept); Mayden 1997 (hierarchy of species concepts); Kottelat 1997 (hybrid fertility and species

A hybrid between *Squalius orientalis* and *Chondrostoma colchicum*. Such hybrids are often fertile. To our knowledge, only two species of possible hybrid origin occur in West Asia: *Coregonus sevanicus* and *Chondrostoma esmaeilii*.

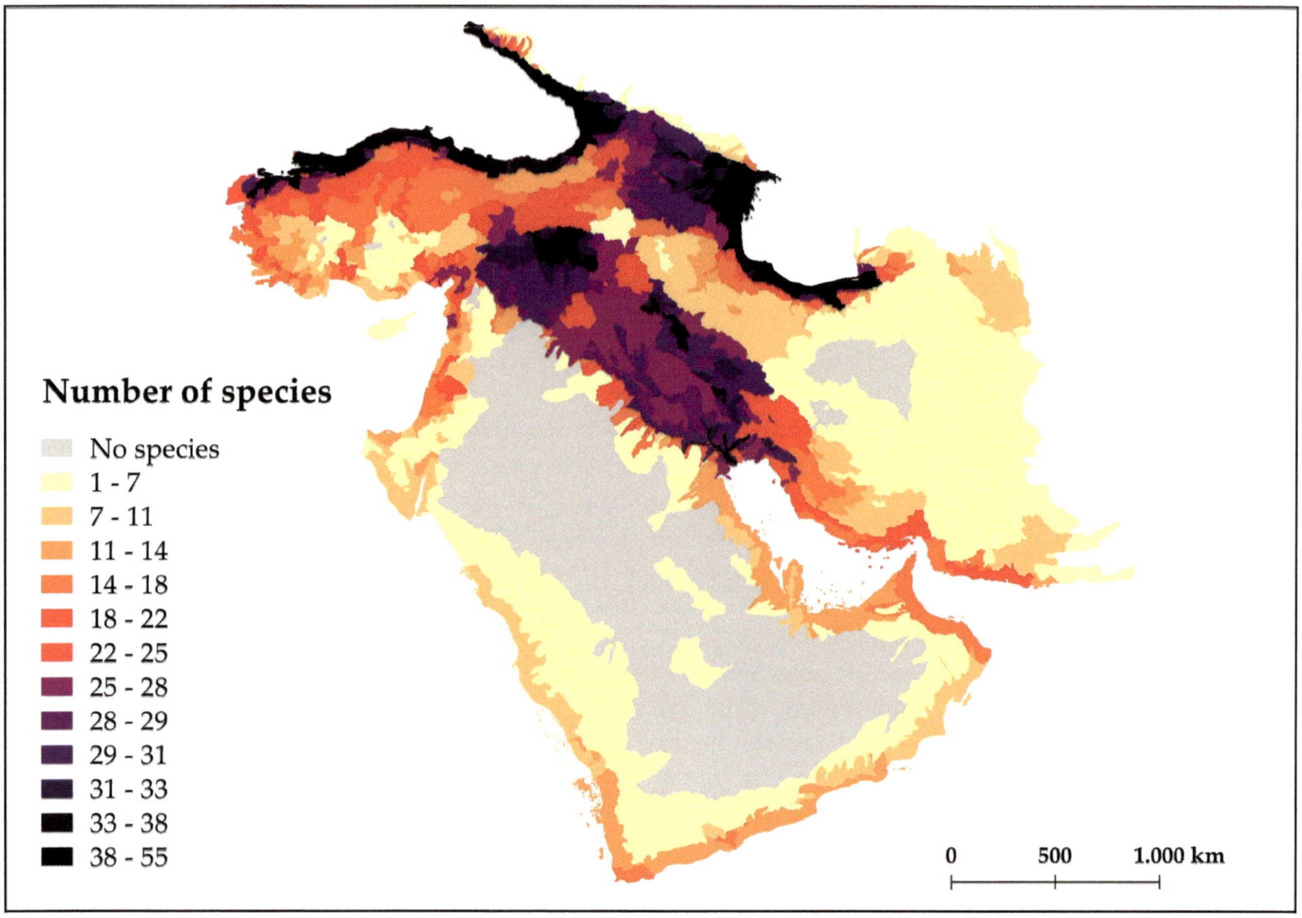

Figure 22. Species richness of native freshwater fishes in West Asia.

concepts); Geiger et al. 2014 (consistency of species and COI sequences).

Diversity of freshwater fishes. In the region covered by this book, we recognise 632 different freshwater fish species (July 2025), including non-native, diadromous, and marine species that regularly enter freshwater habitats. Figure 22 covers 597 native species, including five known but not yet described species and excludes 35 non-native species. There are numerous unresolved taxonomic issues, many of which are mentioned in the various species remarks, and this number is very unlikely to be stable. The effort to identify freshwater fish species in the area has been massive in recent years, and 259 native species (41 %) have been described since 2000 alone. This trend will certainly continue in the coming years. Several scientists in the region build their careers solely on the description of new species, so additional descriptions are expected following the publication of this book. However, species diversity in several genera may already be overestimated, and revisions will reveal some as synonyms in the future. Many marine species occasionally enter freshwater bodies, particularly in the lower Shatt al-Arab/Arvand and adjacent marshes in the northernmost Persian Gulf region. In this region alone, 130 marine species have been recorded, but freshwater habitats are not an essential part of their life cycle, so these species are excluded from this guide. Of course, we know that the line drawn between marine and freshwater fish is artificial, and other authors might include more marine species in the freshwater lists. Furthermore, the fauna will undoubtedly experience the invasion and introduction of several additional non-native species, such as the *Ameiurus* catfish, the Asian *Misgurnus*, and the Amur sleeper *Perccottus glenii*. Therefore, as with any book project, the species count will likely be outdated when the book is in your hands.

Total number of species covered by this book	632	100 %
Native species	597	94 %
Endemic species	467	74 %
Non-native species	35	6 %
Established non-native species	29	5 %
Cypriniformes	437	69 %
Species described since the year 2000	259	41 %

Figure 23. Information board on fish diversity at Eflatunpınarı in Türkiye to raise awareness for the spring.

Conservation status. The IUCN Red List status of each species is given, followed by a short description of the main threats to the species, if any, and related comments. The details of the evaluations are available from the IUCN Red List of Threatened Species webpage (http://www.iucnredlist.org). Most freshwater fishes in West Asia were twice assessed for extinction risk and conservation status using the International Union for Conservation of Nature (IUCN) criteria (Table 1). All assessments consider the global extinction risk. It is important to note that although several species are classified as Least Concern throughout the region, many of the isolated populations may be under significant threat. Of the 583 species assessed in the most recent assessments, 12 species (2 %) are considered extinct (including one extinct in the wild) and 238 species (41 % of assessed species) are in a threat category (assessed as Critically Endangered, Endangered, or Vulnerable) (Figure 24). The distribution of threatened species is given in Figure 25. It is important to note that the IUCN Red List is a tool designed to assess extinction risks, and not to determine conservation measures needed or even less a tool to fix priorities.

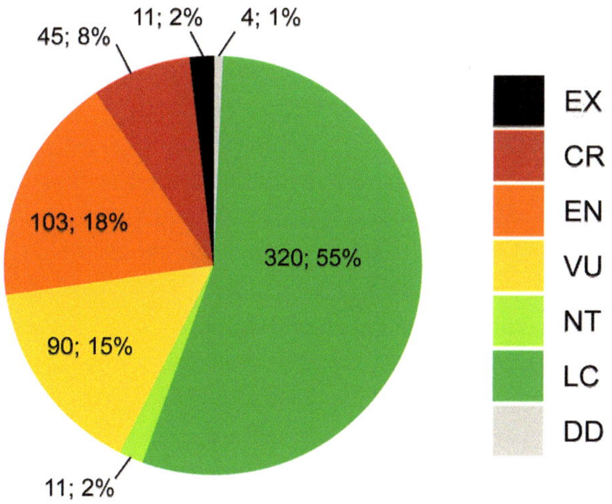

Figure 24. Relative proportions of the conservation status of 583 species of native freshwater fishes in West Asia. EX includes species that are extinct in the wild ($n = 1$), and CR includes potentially extinct species ($n = 2$).

Some important definitions: The **extent of occurrence** (EOO) is the area within the shortest continuous boundary within which the species is known to occur. It is defined as the convex polygon that includes all the known localities of a species. Of course, for freshwater organisms, this is of little value as they are unable to survive more than a few seconds on dry land; it results in a species restricted to a few coastal marshes around the Mediterranean (e.g., *Aphanius fasciatus*) having an extent of occurrence larger than the whole area of the Mediterranean.

The **area of occupancy** (AOO) is the area within the extent of the species' occurrence. Naturally, a species will usually only occur in some places within its extent of occurrence. For example, it will be absent in unsuitable (or, in the case of fish, terrestrial) habitats. The AOO includes only the habitats where the species normally forages and reproduces and the areas essential for its survival. For example, the area of occupancy of a homing species with an obligatory single spawning ground is the area of that spawning ground. It may be only a few square meters in extent (see also "location" below). In lacustrine habitats, a deep-water species may occasionally be observed in shallow coastal waters; these coastal waters are part of its extent of occurrence but not of its area of occupancy; the reverse would be true for a benthic littoral species occasionally observed to be pelagic in the middle of the lake. Indeed, knowledge about the exact distribution of most species is missing, and the river length is usually taken as a proxy for the AOO

A **location** (in the Red List context) is defined as a geographically or ecologically distinct area in which a single threatening factor can rapidly affect all individuals of the taxon present. A location may include part of one or many subpopulations. The location should be distinct from the locality.

The categories are as follows:

EXTINCT (EX): A taxon is Extinct when there is no reasonable doubt that the last individual has died. A taxon is presumed Extinct when exhaustive surveys in known and/or expected habitat, at appropriate times (diurnal, seasonal, annual), throughout its historic range have failed to record an individual. Surveys should be over a time frame relevant to the taxon's life cycle and life form. Available information leading us to believe that a species is extinct is given in the species account. [Extinct is used when the species has totally disappeared. If it has disappeared only in part of its range, it is referred to as extirpated].

EXTINCT IN THE WILD (EW): A taxon is Extinct in the Wild when it is known only to survive in cultivation, in captivity, or as a naturalised population (or populations) well outside the past range.

CRITICALLY ENDANGERED (CR): A taxon is Critically Endangered when the best available evidence indicates that it meets any of the criteria A to E for Critically Endangered (see below and Table 1), and it is therefore considered to be facing an extremely high risk of extinction in the wild.

ENDANGERED (EN): A taxon is Endangered when the best available evidence indicates that it meets any of the criteria A to E for Endangered (see below and Table 1). It is, therefore, considered to be facing a very high risk of extinction in the wild.

VULNERABLE (VU): A taxon is Vulnerable when the best available evidence indicates that it meets any of the criteria A to E for Vulnerable (see below and Table 1). It is, therefore, considered to be facing a high risk of extinction in the wild.

NEAR THREATENED (NT): A taxon is Near Threatened when it has been evaluated against the criteria but does not qualify for Critically Endangered, Endangered, or Vulnerable now but is close to qualifying for or is likely to qualify for a threatened category in the near future.

LEAST CONCERN (LC): A taxon is Least Concern when it has been evaluated against the criteria and does not qualify for Critically Endangered, Endangered, Vulnerable, or Near Threatened. Widespread and abundant taxa are usually included in this category.

DATA DEFICIENT (DD): A taxon is Data Deficient when there is inadequate information to make a direct or indirect assessment of its risk of extinction based on its distribution and/or population status. A taxon in this category may be well studied, and its biology is well known, but appropriate data on abundance and/or distribution are lacking. Data Deficient is therefore not a category of threat. Listing of taxa in this category indicates that more information is required and acknowledges the possibility that future research will show that a threatened classification is appropriate. It is important to make positive use of whatever data are available. In many cases, great care should be exercised in choosing between DD and a threatened status. If the range of a taxon is suspected to be relatively circumscribed, and a considerable period of time has elapsed since the last record of the taxon, threatened status may well be justified.

NOT EVALUATED (NE): A taxon is Not Evaluated when it has not yet been evaluated against the criteria.

The categories **CR**, **EN**, and **VU** are defined by quantified criteria:

A Reduction of the size of the population (reduction of the number of individuals, the area of occupancy, the extent of occurrence, quality of habitat, an increase of exploitation, pollution, parasites, competitors, or other stresses; introductions).

B Small extent of occurrence or area of occupancy.

C Small population size related with some degree of continuing decline.

D Very small population size.
E A quantitative analysis shows the probability of extinction in the wild within a short number of generations or years.

For all criteria, the threshold values are different according to the different category levels. The criteria and the values are listed in Table 1.

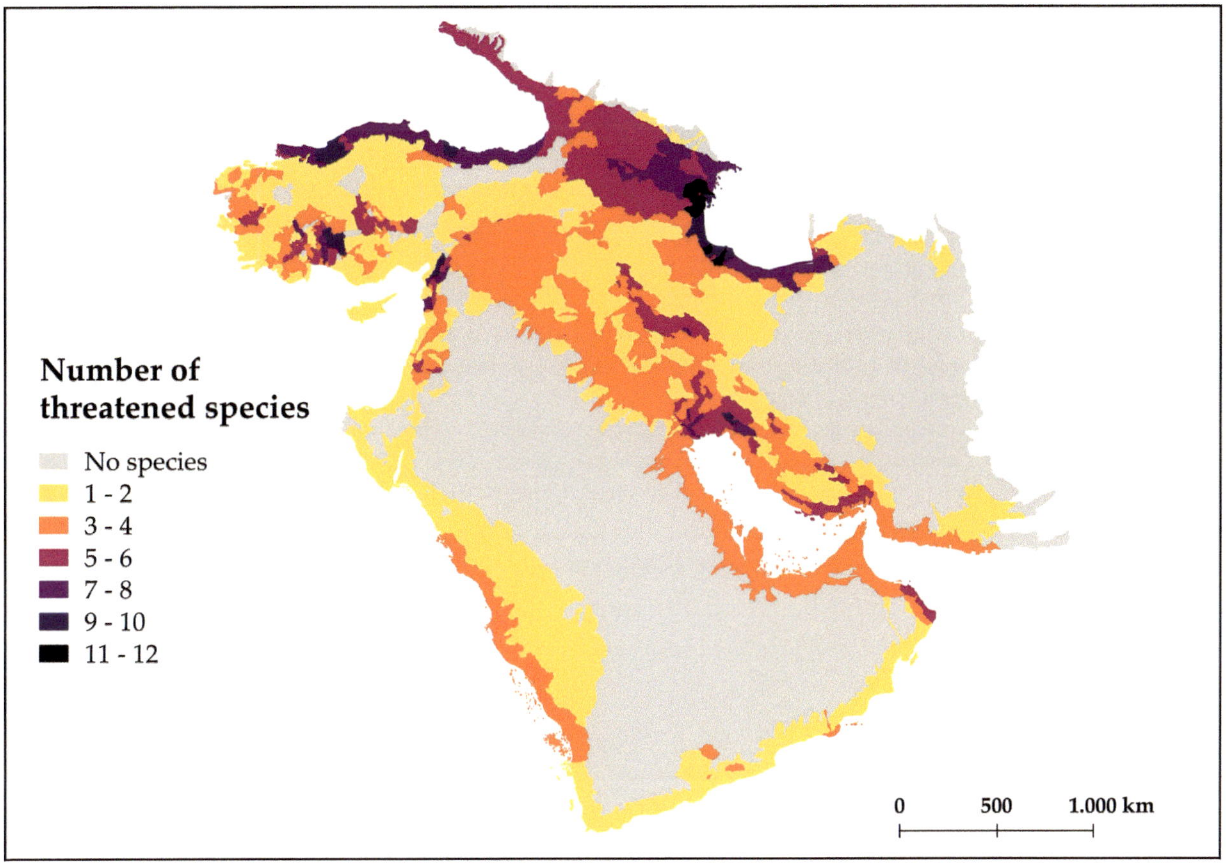

Figure 25. Species richness of threatened (VU, EN, CR) freshwater fishes in West Asia.

The entrance to Azraq Wetland Reserve in Jordan is dedicated to the conservation of birds and the endemic killifish *Aphaniops sirhani*.

Table 1. Summary of the five criteria (A–E) used to evaluate if a species belongs in a category of threat (Critically Endangered, Endangered, or Vulnerable) (from IUCN 2001).

Use any of the criteria A–E	Critically Endangered	Endangered	Vulnerable
A. Population reduction Declines measured over the longer of 10 years or 3 generations			
A1	>90 %	>70 %	>50 %
A2, A3, A4	>80 %	>50 %	>30 %
A1. Population reduction observed, estimated, inferred, or suspected in the past where the causes of the reduction are clearly reversible AND understood AND ceased based on and specifying any of the following:			
	(a) direct observation		
	(b) an index of abundance appropriate to the taxon		
	(c) a decline in AOO, EOO, and/or habitat quality		
	(d) actual or potential levels of exploitation		
	(e) effects of introduced taxa, hybridisation, pathogens, pollutants, competitors, or parasites.		
A2. Population reduction observed, estimated, inferred, or suspected in the past where the causes of reduction may not have ceased OR may not be understood OR may not be reversible, based on (a) to (e) under A1.			
A3. Population reduction projected or suspected to be met in the future (up to a maximum of 100 years) based on (b) to (e) under A1.			
A4. An observed, estimated, inferred, projected, or suspected population reduction (up to a maximum of 100 years) where the time period must include both the past and the future and where the causes of reduction may not have ceased OR may not be understood OR may not be reversible, based on (a) to (e) under A1.			
B. Geographic range in the form of either B1 (extent of occurrence) OR B2 (area of occupancy)			
B1. Extent of occurrence	<100 km^2	<5000 km^2	<20,000 km^2
B2. Area of occupancy and 2 of the following 3:	<10 km^2	<500 km^2	<2000 km^2
(a) Severely fragmented or # locations	=1	<5	<10
(b) Continuing decline in any of: (i) extent of occurrence; (ii) area of occupancy; (iii) area, extent, and/or quality of habitat; (iv) number of locations or subpopulations; (v) number of mature individuals			
(c) Extreme fluctuations in any of: (i) extent of occurrence; (ii) area of occupancy; (iii) number of locations or subpopulations; (iv) number of mature individuals			
C. Small population size and decline			
Number of mature individuals	<250	<2500	<10,000
and either **C1** or **C2**:			
C1. An estimated continuing decline of at least:	25 % in 3 years	20 % in 5 years	10 % in 10 years
up to a maximum of 100 years	or 1 generation	or 2 generations	or 3 generations
C2. A continuing decline and (a) and/or (b):			
(a i) # mature individuals in all sub-populations:	<50	<250	<1000
(a ii) or % individuals in one sub-population at least	90 %	95 %	100 %
(b) extreme fluctuations in the number of mature individuals			
D. Very small or restricted population			
Either:			
(1) number of mature individuals	<50	<250	<1000
OR			
(2) restricted area of occupancy	na	na	AOO < 20 km^2 or # of locations <5
E. Quantitative analysis			
Indicating the probability of extinction in the wild to be at least:	50 % in 10 years or 3 generations (100 years max)	20 % in 20 years or 5 generations (100 years max)	10 % in 100 years

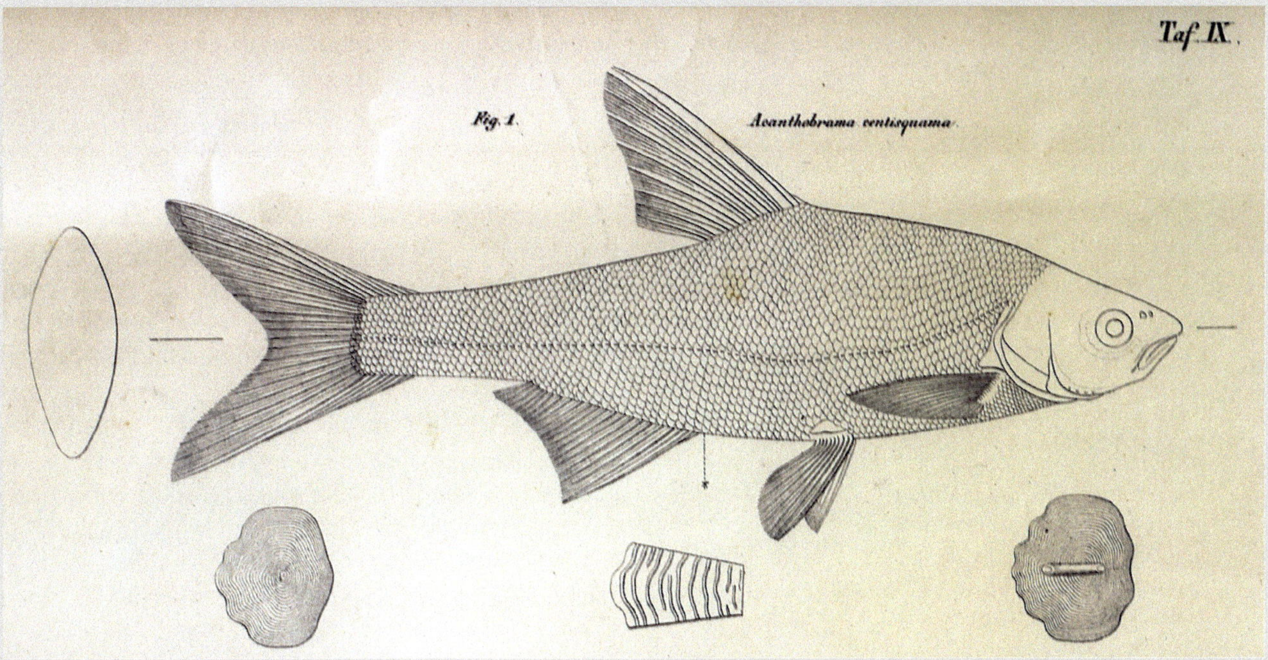

Extinct *Acanthobrama centisquama*; Orontes, Türkiye (from Heckel 1843).

Extinct freshwater fishes in West Asia

A dozen freshwater fish species are thought to be extinct, meaning they are no longer found alive in the wild. Eleven extinct species, and one more extinct in the wild sound a lot, but only 2 % of the native species are known from West Asia. The other 98 % have survived despite massive habitat change, severe water stress, and the presence of many non-native species. Despite considerable stress, West Asia's freshwater fishes have shown amazing resilience and adaptability to anthropogenic stressors. This may be due to the evolutionary and biogeographical past of the species, which have survived several difficult climatic and geological periods. Furthermore, *Esmaeilius persicus* has survived in captivity, and we cannot completely exclude that this species, *Cobitis amphilekta*, and *C. kellei*, may be rediscovered in the future, as not all potential habitats have been revisited yet. This is also true for the species that were found to be possibly extinct. Others, such as *Tristramella sacra*, may be conspecific with surviving species and not extinct. More research is needed to search for lost fishes and to resolve remaining taxonomic questions. This does not mean that freshwater fishes are invulnerable to stressors. Fish will disappear when all the water is gone. There are still springs, small streams, and enough rain to create perennial water bodies, but efforts to withdraw all water for human use have failed. However, the future is bleak for much of West Asia, a major disaster area due to climate change.

Extinct and Extinct in the wild

Acanthobrama centisquama, Alburnus adanensis, Alburnus akili, Anatolichthys splendens, Cobitis amphilekta, Cobitis kellei, Esmaeilius persicus, Mirogrex hulensis, Pseudophoxinus handlirschi, Rutilus sojuchbulagi, Salmo ischchan, Tristramella sacra

Possibly extinct

Acanthobrama tricolor, Caecocypris basimi, Oxynoemacheilus galilaeus, Salmo aestivalis, S. gegarkuni (surviving in non-native range)

Proper waste disposal is poorly managed in many regions of West Asia.

Threats to freshwater fish in West Asia. Freshwater fish face several threats, most related to increasing human development. These include uncontrolled water abstraction, dam construction, habitat loss, domestic and industrial wastewater pollution, agricultural run-off, and invasion by non-native species. Although these issues do not pose significant threats in all parts of the region, they are particularly relevant to fish populations, which are often small and located in geographically isolated habitats. In such cases, even relatively low levels of environmental stress to these small populations can significantly impact their overall health and likelihood of survival.

Water abstraction. West Asia is the first region in the world to run out of water effectively. In the arid parts of the area, surface water and groundwater are abstracted in large quantities and rarely sustainably, making it the main threat to most fishes and humans in arid and semi-arid landscapes. Large parts of Central and Western Anatolia, Iran, and the Levant, where pumps abstract surface water from streams and rivers, are most affected by water abstraction. In smaller streams, digging large holes in the streambed is common so water can be abstracted even when the stream is dry in late summer. Water is also often abstracted by pump trucks and transported to more distant locations. In West Asia, it is very common and natural for sections of streams and rivers to run dry in summer. However, dams and weirs impeding runoff that would otherwise be “lost” to human use leave little or no water flowing downstream. This reduces habitat availability for freshwater fish, even in ecosystems adapted to seasonal drought, where fish survival often depends on small refugia. Continued over-abstraction of water, coupled with the increasing frequency and severity of droughts, leads to the desiccation of these refugia and the extirpation (and extinction) of fish.

Pumping water from every river, stream, and spring is common in arid parts of West Asia, and surprisingly, there have been so few freshwater fish extinctions in the region.

Much of West Asia is losing groundwater reserves at an alarming rate, and the region has one of the highest water deficits in the world, second only to India. Water resources must meet the needs of intensive agriculture and a growing population. For example, the Iraqi marshes, the Turkish Sultan marshes, the Eşmekaya marshes, the Lakes Hotamış and Acıgöl, the Jordanian Azraq marshes, the Lebanese Ammiq marshes, and many others have all almost or completely dried up. The lowering of groundwater levels is affecting streams throughout the region, many of which have dried up; the Anatolian Küçük Menderes is one example. Another example is the loss of the Qweiq River, which once flowed through the Syrian city of Aleppo but has virtually disappeared. Today, only two very small headwaters remain of what was once a great river. Other examples include the once extensive spring areas of Ras Al Ain in northern Syria, which have almost completely dried up, and the famous Barada spring near Damascus, along with virtually the entire Damascus hydrological basin, as most of the water is extracted for the expanding city of Damascus.

All the countries considered here have water policies, but these are only sometimes enforced in a way that preserves or protects biodiversity. Apart from Israel, which introduced a "Water for Biodiversity" policy in the early 2000s, we are unaware of any country where a water policy is being implemented to ensure that enough water remains in lakes, marshes, streams, and rivers to meet the needs of biodiversity. In Israel and the Arabian Peninsula, seawater is increasingly desalinated in large quantities, reducing the pressure on freshwater supplies. Studies show that streams and springs can quickly recover when the amount of water withdrawn is reduced. However, desalination is expensive, requires access to seawater, is powered by fossil fuels, and should not be considered the sole answer to the region's water needs. The rapidly growing water needs of many West Asian countries cannot be met by further exploitation of water resources, except by developing desalination facilities or reallocating water resources from agriculture. Increased innovative efforts and financial support are needed to create desalination systems powered by solar or wind energy, not only to conserve freshwater biodiversity but also to benefit the overall water needs of West Asia. **Further reading.** Shacham 2003 (Israel water program); Allan 2001, Voss et al. 2013 (water stress); GegenStrömung 2011, International Rivers 2014 (dams in Türkiye).

This huge spring at Göksu, in the Mardin province of Türkiye, has fallen victim to water abstraction and drought. Much of West Asia will receive less rainfall in the future.

Thousands of dams are massively impacting the rivers of West Asia, and countries continue to invest in more dams, ignoring the negative environmental consequences.

Fish passes are often thought to compensate for the negative effects of dams. However, like the one in the picture, most are impossible for fish to use and do not support upstream migration. Nor do they compensate for the complete transformation of the river into a novel lake ecosystem dominated by non-native species.

Dams and reservoirs. Determining the exact number of dams and weirs in the region is impossible. Different sources give different figures, even for Türkiye, where information is available. Türkiye has more than 2000 dams and weirs and plans to build another 1700 within its borders, making it one of the most active dam-building countries in the world, with hardly a river in the country unaffected. Other countries, such as Iran and the countries of the Caucasus, have also followed Türkiye's lead in exploiting rivers for hydroelectric power. Iran, Iraq, and countries in the Levant have built dams on almost all suitable rivers. The impacts of dams on biodiversity and society remain largely unknown.

Clean water is a limiting factor for many human activities. Surprisingly, domestic, industrial, and agricultural pollution is a widespread threat to West Asia's freshwaters.

West Asia's reservoirs evaporate large amounts of water needed elsewhere and significant amounts of methane, but their contribution to climate change is poorly understood. The region is home to few long-distance migratory fish species, the most important of which are sturgeons. However, their need for free-flowing rivers has rarely been considered. The main threat associated with dams is not the dam itself, which prevents fish from migrating, but the reservoir, which transforms a flowing river into a stagnant lake, a habitat unsuitable for many fish species. While most rivers in the region are now dammed, it is usually a single dam, or in many cases, a few dams, rather than a "cascade of dams" that essentially turns the river into a series of lakes. However, this scenario is slowly changing as many new dams and weirs are being built across the region. The construction of new dams, especially for hydropower and water storage, is a major concern for freshwater fish conservation throughout the region. Hydropower is widely presented as a "green technology," leading many countries to aim to exploit their full hydropower potential.

Pollution. Throughout West Asia, particularly in the vicinity of urban areas, pollution is one of the major threats to freshwater fish. Most rivers and streams are heavily polluted downstream of urban areas, mainly by sewage, such as the Kura downstream of Tbilisi in Georgia, the Tigris downstream of Diyarbakir, and the Shatt Al-Arab in Iraq. However, it is in the rivers of Western Anatolia, such as the Bakırcay, Gediz, Küçük, and Büyük Menderes, that water pollution is most widespread and severe. The Küçük Menderes has virtually disappeared, and the lower part is filled with sewage from towns and industries. The Gediz and Bakırcay are so polluted that only *Gambusia holbrooki* seems to exist in the middle and lower parts of the main rivers. There are also high pollution levels in many other

Fieldwork over many years has shown a steady decline in habitat size and quality in many places. This is why so many fish species are listed as CR, EN, and VU in an IUCN threat category.

areas; for example, the upper Köprüçay, south of Isparta, is largely polluted with sewage, as is the Orontes and many coastal rivers in Syria. However, data on these rivers' chemical and biological water quality are sparse, and there are few continuous monitoring programs for open surface waters. As a result, little is known about the region's extent and intensity of urban and agricultural pollution. Pollution can be cleaned up relatively quickly, and we hope readers will prove us wrong in the years to come when all these rivers are clean again.

Climate change. West Asia is predicted to become much drier and warmer. Published scenarios suggest a more challenging future for freshwater fish in much of the region, with many areas already drying up and many once widespread fish species now being restricted to small refuges. Dramatic reductions in river flows (due to water abstraction and increased frequency of droughts) will cause significant environmental, economic, and political problems that only increase in the future unless there is a radically different approach to water management across the region. The long-term effects of climate change on the unique and endemic freshwater fishes of West Asia are not difficult to imagine, as climate change will only accelerate the ongoing desiccation of the region's springs, lakes, and rivers.

The desertification of the Fertile Crescent in West Asia has long been underway, and the process has been accelerated by water extraction and climate change

The Qweiq in Syria was still a major river in the 19th century. Today, even its reservoirs are dry, and finding traces of the river is difficult. Water comes now from the Euphrates to the city of Aleppo.

In some regions of West Asia, such as Central Anatolia, marble quarrying is a major source of fine sediment in streams, threatening freshwater biodiversity.

Climate endgame

We humans are driving the Earth into a Pliocene, possibly Miocene climate scenario. Temperatures more than 2°C above pre-industrial levels have not been sustained at the Earth's surface since the Pleistocene epoch (or more than 2.6 million years ago). Despite 30 years of effort and some progress under the United Nations Framework Convention on Climate Change, anthropogenic greenhouse gas emissions (GHG) continue to rise. Even if anthropogenic GHG emissions were to decline soon (which seems unrealistic), there is feedback in the carbon cycle and potential tipping points that could lead to high GHG concentrations often missing from models. Examples include the thawing of Arctic permafrost releasing methane and CO_2, the loss of carbon from intense droughts and fires in the Amazon, and the apparent slowing of mitigating feedback such as the capacity of natural carbon sinks. These are unlikely to be proportional to warming, as is sometimes assumed. Instead, abrupt and/or irreversible changes may be triggered at a temperature tipping point. This is particularly worrying as human societies are locally adapted to specific climatic niches and the cumulative effects of warming are very likely to overwhelm societal adaptive capacity. Climate change will also directly trigger other catastrophic risks, such as international conflicts, or intensify the spread of infectious diseases and spillover risks. These could be powerful multipliers of extreme hazards. Climate change could aggravate vulnerabilities and cause multiple indirect stresses (such as economic damage, land loss, and water and food insecurity) that combine into system-wide synchronous failures. Extreme temperatures can affect the yields of major cereal crops, and deadly heat could also significantly affect populated areas in West Asia. There is a striking overlap between current vulnerable states and future areas of extreme warming, highlighting the political fragility of our region, but the 'four horsemen' of the climate change endgame are likely to be famine and malnutrition, extreme weather events, armed conflict, and vector-borne diseases. **Further reading.** Kemp et al. 2022 (climate change scenarios).

Introduction of non-native species. Many freshwater fish species are threatened by invasive non-native species, with many fish assemblages being replaced by communities dominated by non-native fish species. We know that 35 non-native freshwater fish species have been introduced, and 29 species have become established in West Asia. This does not include the unknown number of cichlids introduced from Lake Malawi into the warm spring of Nahal Amal in Israel or the many tropical species introduced by aquarium hobbyists into the hot springs of the upper Sakarya drainage in Türkiye. Four species of Xenocyprididae are or have been regularly stocked but have yet to become established in the region. Similarly, rainbow trout (*Oncorhynchus mykiss*), despite being the most commonly farmed and escaped freshwater fish species, has yet established very few populations recorded in the region. Species such as *Carassius auratus*, *Clarias gariepinus*, *Coptodon zillii*, *Cyprinus carpio*, *Gambusia holbrooki*, *Hemiculter leucisculus*, *Heteropneustes fossilis*, *Lepomis gibbosus*, *Oreochromis aureus*, *Poecilia "latipinna"*, *Pseudorasbora parva*, and *Rhinogobius* sp. have expanded their ranges within the region. They are believed to be negatively impacting native fish communities where they occur. Although there are very few specific studies of the impact of these non-native species, it appears that they are out-competing native species, and the situation is particularly alarming in many regions as native habitats shrink due to habitat degradation, water scarcity, and environmental changes. Research on the impact and distribution of non-native species is strongly recommended to understand their effects, behaviour better, and, in particular, ways and means to limit their spread.

The rapid spread of invasive species is a widespread threat throughout West Asia. Here, *Gambusia holbrooki* is in the habitat of *Anatolichthys transgrediens*.

Family Petromyzontidae

Lampreys

Lampreys are not fishes. They are a distinct lineage of vertebrates traditionally associated with fishes. Lampreys evolved in the marine environment and originated in the Ordovician period, at least 450 million years ago. This makes them survivors of five of the Earth's great extinction events. Lampreys are often thought to be related to hagfish (Myxiniformes), which are a sister group to lampreys. Ammocoetes, the filter-feeding worm-like larval stage of modern lampreys, have long influenced hypotheses of vertebrate ancestry. The life history of modern lampreys, from a superficially amphioxus-like ammocoete to a specialised predatory adult, was thought to exemplify the origin of vertebrates. However, recent analyses of immature specimens of Palaeozoic stem lampreys show that they lack an ammocoete larval stage. Ammocoetes are specialisations of the life history of modern lampreys rather than relics of vertebrate ancestry. Since modern hagfish also lack a larval stage, the last common ancestor of hagfish and lampreys was probably a macrophagous predator without a filter-feeding larval stage. The earliest known record of the ontogenic transition from ammocoete to adult is *Mesomyzon* from Early Cretaceous lake beds in China. All modern lamprey lineages have evolved in freshwater within the last 200 million years. The derived state of an ammocoete and the prolonged filter-feeding larval phase undoubtedly delayed the need for the endocrine function of a thyroid gland in iodine-poor freshwater environments. This is one of the mechanisms that allowed lampreys to evolve in freshwater.

Lampreys are jawless and eel-like animals with seven roundish gill openings behind the eye but lacking paired fins. They have a very distinctive oral disc armed with rows of horny teeth, which they use to attach themselves to surfaces and prey. The dentition of the oral disc is an important character in identifying genera (Figure 26–27). All lampreys are found in cooler waters above 20° northern and southern latitudes. Petromyzontidae are found in the north, while a few species of other families are found in the southern hemisphere.

Lampreys spawn in spring and die shortly after spawning. They spend several years in a larval stage (ammocoete). The duration of the ammocoete stage is still controversial, but we know that lamprey species in West Asia undergo metamorphosis between late summer and early winter. Juveniles are always shorter than adult ammocoetes (larvae shrink during metamorphosis). Migrating *Petromyzon marinus* are attracted by steroidal pheromones released by stream-dwelling larvae, which indicate habitat suitability. These compounds are detected at extremely low concentrations and are released in large quantities. Laboratory experiments show that each *Petromyzon* larva activates at least 400 l of water per hour. Spawning lampreys cease their normal daylight avoidance and spawn on sunny days. The male arrives first at the spawning site and releases a bile acid that acts as a sex pheromone, inducing the ovulated female to search for him. Spawning generally occurs in fast-flowing water, often at the head of riffles in streams. During spawning, the male attaches to the female's neck and wraps his body around her. The female then deposits the eggs in the gravel with her belly.

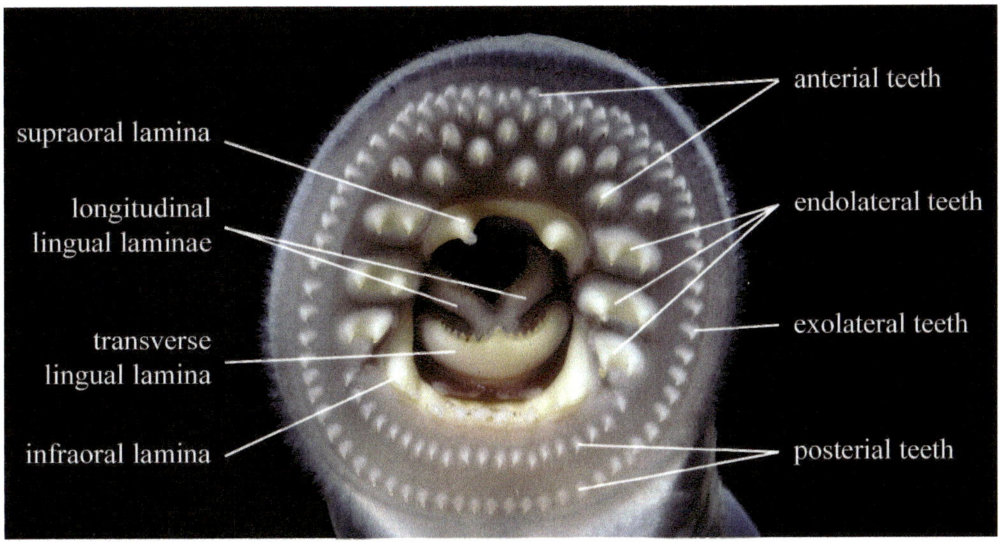

Figure 26. Oral disc of *Lethenteron camtschaticum* and dentition terminology (from Kottelat & Freyhof 2007, © N. Akiyama).

https://doi.org/10.1515/9783111677811-002

Figure 27. Mouth of Petromyzontidae: from left: *Caspiomyzon wagneri*; *Lampetra ninae*; *Petromyzon marinus* (© A. Hartl).

Of the four species from West Asia, two are anadromous: *Caspiomyzon wagneri* and *Petromyzon marinus*. *Caspiomyzon* is a predatory species, and *Petromyzon* is parasitic. The two species of *Lampetra* in West Asia are sedentary; they do not feed after metamorphosis. Instead, they spawn in the spring after metamorphosis. Other species of *Lampetra* are found in Europe.

Further reading. Vladykov & Kott 1979 (diversity, distribution); Hardisty & Potter 1971–1982, Hardisty 1979, Hardisty 1986a (biology); Youson & Sower 2001 (thyroid hormones); Gills et al. 2003 (phylogeny); Sorensen et al. 2005 (pheromone); Gess et al. 2006 (fossils); Lang et al. 2009 (phylogeny); Docker et al. 2015 (review); Miyashita et al. 2021 (evolution of ammocoetes); Hughes et. al. 2025 (generic concept).

Key to lampreys in West Asia

1a - Supraoral lamina narrow, with a single tooth or two adjacent teeth; anterial, posterial, and exolateral teeth numerous, in regular backward-curved radiating rows.
..................2

1b - Supraoral lamina transversely elongate, with a single tooth at each end; if present, anterial, posterial, and exolateral teeth not in regular backward-curved radiating rows.
..................3

2a - Infraoral lamina with 4–6 large, blunt teeth; oral disc width less than body width.
..................*Caspiomyzon wagneri*

2b - Infraoral lamina with 7–8 large, sharp teeth; oral disc width greater than body width.
..................*Petromyzon marinus*

3a - Posterial teeth absent or only two single unicuspid teeth.
..................*Lampetra lanceolata*

3b - Posterial teeth usually present in an interrupted or complete row, 4–23 unicuspid and bicuspid teeth.
..................*Lampetra ninae*

Caspiomyzon wagneri; Tajan, Iran; ~400 mm TL.

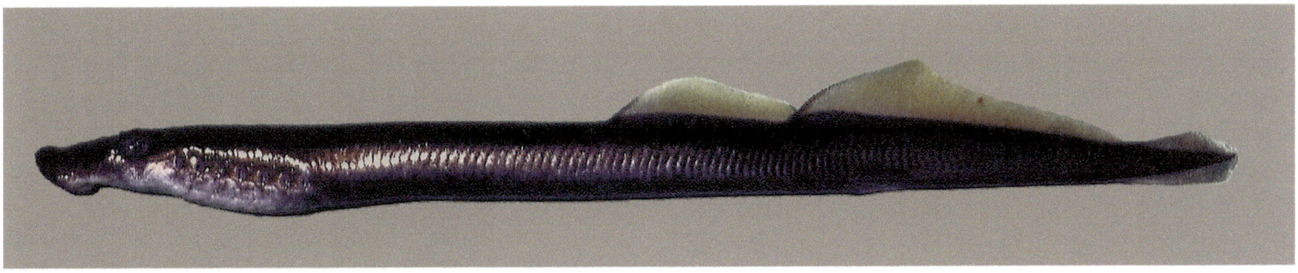

Caspiomyzon wagneri; Samur, Russia; 247 mm TL. © A. Naseka.

Caspiomyzon

The genus *Caspiomyzon* comprises three species: one migratory and feeding as adults, and the other two are resident brook lampreys, not feeding as adults. While *C. wagneri* is widespread, especially in the southern Caspian basin, both *Caspiomyzon* brook lampreys (*C. hellenicus*, *C. graecus*) are locally endemic to the Mediterranean basin (Greece and Albania). Both have been described in the European genus *Eudontomyzon*, and their placement in *Caspiomyzon* raises some interesting biogeographic questions, as *Caspiomyzon* does not occur in the Black Sea basin. A migratory lamprey historically known from the Black Sea basin was initially placed in the genus *Eudontomyzon*, but when rediscovered, it was identified as a species of *Lampetra*.

Caspiomyzon wagneri

Common name. Caspian lamprey.

Diagnosis. Distinguished from other species of lampreys in West Asia by: • many small teeth in regular backwards-curved radiating rows / ○ all teeth low, blunt, button-shaped / ○ supraoral lamina short, with one, rarely two closely connected teeth / • supraoral lamina tooth blunt / • infraoral lamina with 4–6 large teeth / ○ oral disc width less than body width / ○ ammocoetes with 53–68 trunk myomeres. Size up to 575 mm TL.

Distribution. Caspian Sea and most (even relatively small) tributaries. In West Asia, Kura drainage, up to Astara, lower Aras (Azerbaijan), and coastal Caspian rivers east to Gorgan. In Iran, now restricted to a few rivers, between Sefid to Gorgan. Also found in European Caspian basin. Migrated to uppermost tributaries of Volga (1500 km), but spawn along all rivers where suitable substrate exists.

Habitat. Adults live in sea caught down to depths of 600–700 m. Spawns in sections of rivers and streams with strong currents. Ammocoetes live in detritus-rich sand or clay sediments, selecting and changing habitats according to sediment size as they grow.

Biology. Anadromous; upstream migration of adults in autumn and spring. In Kura and Iran, autumn run begins in late October and peaks in December and January; spring run begins in mid-March to late May, with peak spawning migration in April (at 16°C) (Iran). In shorter rivers, most individuals migrate in spring. Migration occurs at dusk

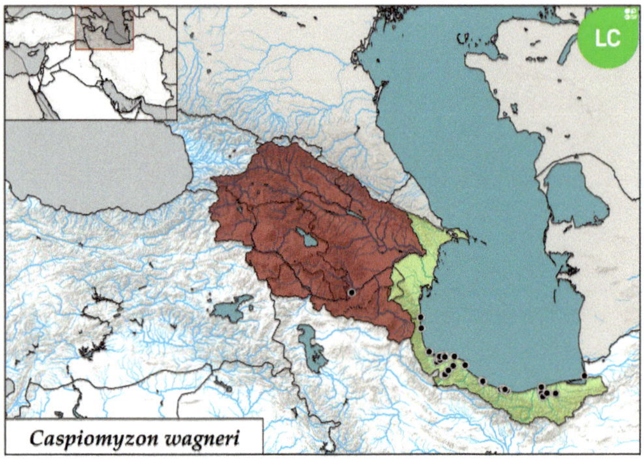

and night, peaking about 2 hours after sunset and during daylight hours when water temperature rises to 13–14°C. Migrate 2–16 km per night. Spawn in pairs in March–June (rarely in July) when temperature reaches 13–23°C. Eggs laid in crescent-shaped nests made of gravel and small stones. Ammocoetes hatch in 8–10 days. Ammocoete stage lasts 2–4 years. Feeds on detritus and microorganisms. Metamorphosis begins in mid-July and lasts until October. After metamorphosis, juveniles migrate to sea until December of same year. Adults stay at sea for at least 17 months. In lower Volga, occurrence of small (about 190–310 mm TL) and large (370–550 mm TL) mature adults suggests that adults may feed one or two summers before spawning. Feeding behaviour unknown. Due to their fat content and abundance, dried *C. wagneri* were formerly used as candles in Russia.

Conservation status. LC; historically found in all Caspian rivers. Limited by dams to lower part of rivers. Discharged when caught in Iran due to religious reasons. Rare, with only a few remaining spawning sites in rivers draining from eastern Caucasus, where it might have stabilised at a low level. In Volga, a sharp decline was observed during the 20th century. Potentially still abundant in Ural, but more information is needed at present. The status of landlocked populations in reservoirs in lower Volga must be confirmed.

Further reading. Berg 1949b, Holčík 1986 (biology); Renaud 2011 (morphology, biology); Coad 2016c (review); Abdoli et al. 2017 (migration pattern); Nazari et al. 2017 (biology).

Lampetra

There are 13 species in the genus, 4 of which are predatory and 9 nonpredatory brook lampreys. Most are known from Europe and the Asian part of the Black Sea basin. Several North American species are still placed in *Lampetra*, as a global phylogeny of lampreys shows that the genus *Lampetra* is not restricted to the Western Palaearctic. The evolution of brook lampreys has been much discussed in *Lampetra*. Brook lampreys are superficially similar to migratory lampreys of the same genus but do not smoltify and migrate downstream after metamorphosis. Instead, they remain in the streams and spawn in the approaching spring without feeding. As brook lampreys do not feed after metamorphosis, they do not grow (but shrink). Both predatory and brook lampreys may spawn simultaneously at the same sites, but they are reproductively isolated due to their size-related mating behaviour. However, it has always been believed that brook lampreys evolve from migratory lampreys, and ecological factors during their larval phase have been discussed to trigger whether or not they become migratory. Several studies have attempted to determine whether their nucleotides support the independent occurrence of predatory and nonpredatory lampreys and their recognition as separate species. The question is whether the two constitute different life-history strategies of a single but phenotypically highly bipolar species.

Analyses of mitochondrial markers from a geographically broad set of Central and Southwest European *Lampetra* populations definitively demonstrated nonmonophyly in the widespread Central European nonpredatory species *L. planeri* compared to the predatory *L. fluviatilis*. However, it remains to be seen if this is due to ongoing gene flow or incomplete lineage sorting.

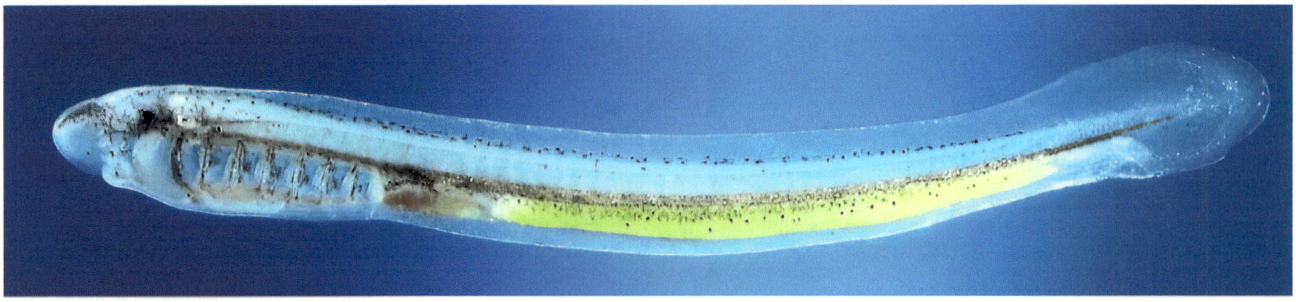

Lampetra planeri; Danube drainage, Germany; newly hatched ammocoete, ~10 mm TL. © A. Hartl.

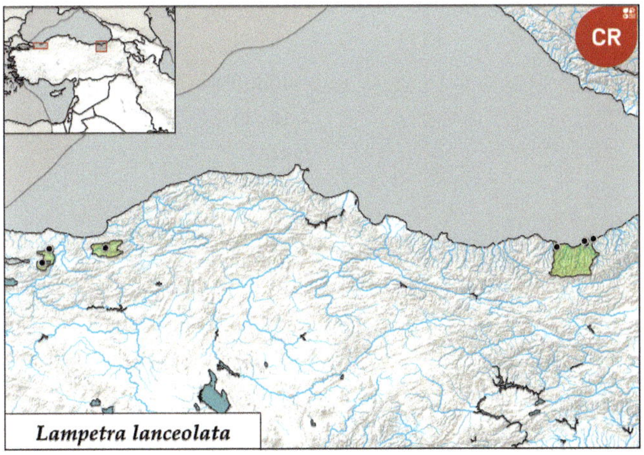

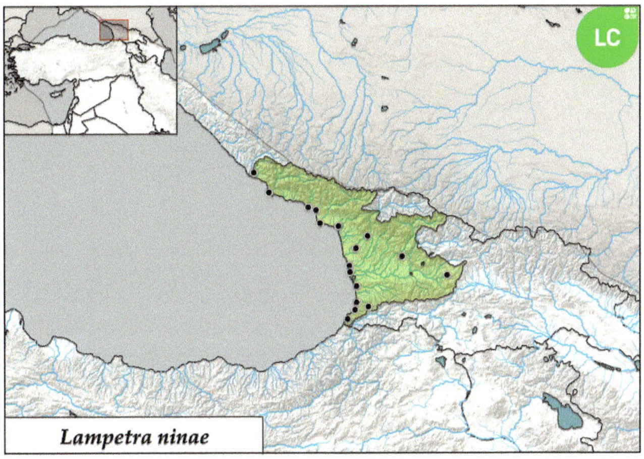

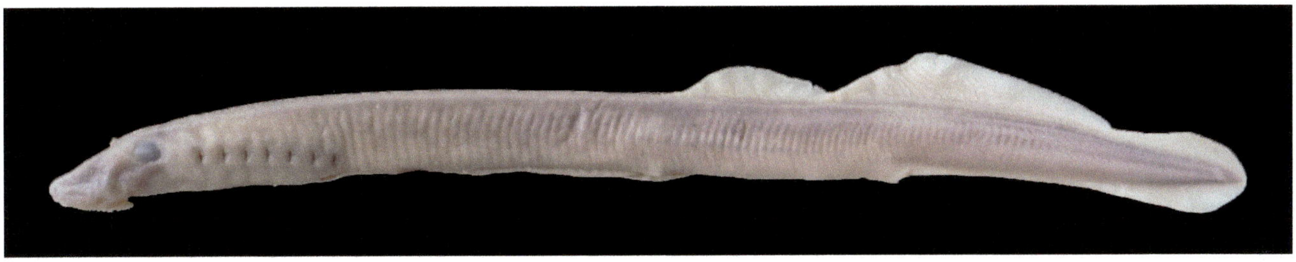

Lampetra lanceolata; İyidere, Türkiye; 129 mm TL.

A comprehensive genome-wide comparison of a predatory/nonpredatory species pair from the same location (*L. fluviatilis*/*L. planeri*) revealed strong genetic differentiation despite shared mitochondrial haplotypes in these two populations. This genome-wide strong differentiation is likely the rule, not an exception. After decades of research, we still need to learn how brook lampreys evolved from predatory lampreys. This is an area that will undoubtedly attract further research in the future.

All non-feeding species of *Lampetra* have likely evolved from *L. fluviatilis* and its ancestors. Furthermore, all brook lampreys have been isolated from this predatory lamprey at different times. Finally, an additional migratory species was widespread in the northern Black Sea basin, where *L. fluviatilis* is not found. This species was previously referred to as *Eudontomyzon* sp. "Migratory" until, after more than 100 years without records, such an animal was found in the Sea of Azov. This individual was identified as a migratory species of *Lampetra*. Due to the political unrest in Eastern Ukraine, no new data could be collected on this very rare lamprey, and thus, no molecular data are available for this species. This migratory *Lampetra* species from the Sea of Azov is likely related to *L. lanceolata* and *L. ninae*. **Further reading.** Naseka & Diripasko 2008 (rediscovery of Black Sea lamprey); Lang et al. 2009 (phylogeny); Mateus et al. 2013, Espanhol et al. 2007, Pereira et al. 2010 (book lampreys).

Lampetra lanceolata

Common name. Anatolian brook lamprey.

Diagnosis. Distinguished from *L. ninae* by: • posterial teeth absent or only two unicuspid teeth present. Size up to 140 mm TL in adults, 173 mm TL in ammocoetes.

Distribution. Türkiye: İyidere and Yanbolu drainages, coastal streams flowing into Black Sea between Rize and Trabzon, also in upper Melen, and streams flowing into Lake Sapanca.

Habitat. Lower parts of rivers with slow currents and bottoms rich in organic material. Ammocoetes live in detritus-rich sand or clay sediments. Spawns on gravel substrate.

Biology. Nonpredatory, freshwater resident. Larvae begin metamorphosis after third summer and reach a total length of 130 mm at end of summer. Likely to spawn in April and May at water temperatures between 13 and 15°C.

Conservation status. CR; distributed up to 15 km inland from mouth of İyidere, where it has declined sharply and is now very rare. There are now six hydroelectric dams in the İyidere, which caused massive habitat change due to pulsed flooding. Status and distribution in other streams are unknown. Very rare in Sapanca Lake basin.

Further reading. Kux & Steiner 1972 (description); Lang et al. 2009 (generic position); Gözler et al. 2011 (ecology); Renaud 2011 (description).

Lampetra ninae; Shakhe, Russia; 164 mm TL. © S. Tuniev.

Lampetra ninae

Common name. Caucasian brook lamprey.

Diagnosis. Distinguished from *L. lanceolata* by: • posterial teeth usually present in interrupted or complete rows, 4–23 unicuspid and bicuspid teeth. Size up to 166 mm TL in adults, slightly longer in ammocoetes.

Distribution. Caucasian Black Sea basin of Russia and Georgia: Shakhe, Mzymta, and Chakhtsutsyr drainages south to Bzyb' and Mokva in Abkhazia (Georgia) south to the border between Georgia and Türkiye.

Habitat. Foothills zones in clear, well-oxygenated brooks. Ammocoetes live in detritus-rich sands or clay sediments.

Biology. Nonpredatory, freshwater resident. Metamorphose in autumn. Spawning biology unknown but suspected to be identical to *L. lanceolata*, which overwinters and spawns in spring.

Conservation status. LC; still common and widespread within its small range.

Remarks. Described as a species of the Arctic genus *Lethenteron*. Molecular studies indicate that it is very closely

Petromyzon marinus; Rhine drainage; Germany; ammocoete, ~100 mm TL.

Petromyzon marinus; Rhine drainage; Germany; juvenile, ~120 mm TL.

Petromyzon marinus; Rhine drainage; Germany; adult, ~850 mm TL. © A Hartl.

related to *L. lanceolata*. Both may be conspecific, and further research is needed.

Further reading. Lang et al. 2009 (generic position); Naseka et al. 2009 (description); Naseka et al. 2016 (description, distribution).

Petromyzon

The monotypic genus occurs in North America, North Africa, Europe, and occasionally in the Eastern Mediterranean. It is one of the largest lampreys in the world and the most well-known and best studied, especially due to its invasion and tremendous effects in the North American Great Lakes. There is no record of larvae or an established population of *Petromyzon* in West Asia. Therefore, the solitary individuals found are likely stayers from the western Mediterranean population. On the other hand, larvae are difficult to find, and more research is needed.

Petromyzon marinus

Common name. Sea lamprey.

Diagnosis. Distinguished from other species of family in West Asia by: ● adults grey to yellow with dark-brown blotches, no blotches in juveniles / ○ supraoral lamina short, with two adjacent teeth / ● infraoral lamina with 7–8 large, sharp teeth / ○ oral disc wider than body width / ○ many small teeth on oral disc, in regular backward-curved radiating rows / ○ all teeth sharp / ● 66–75 trunk myomeres / ○ ammocoetes with black pigment on caudal. Size up to 1200 mm TL.

Distribution. Occurs very irregularly in Mediterranean rivers in Türkiye. Widespread on both sides of North Atlantic, north to Iceland, and along Norwegian coast to Barents Sea (Ura, Kola Peninsula). North Sea, Baltic, and western and central Mediterranean basins very rare in eastern Mediterranean and Baltic basins. Several land-locked populations in North America.

Habitat. Adults at sea offshore. Spawns in strong-current habitats of rivers and streams. Ammocoetes in detritus-rich sand or clay sediments.

Biology. Anadromous, parasitic. Adults migrate to rivers in autumn and winter. Spawns in pairs, April–July, mostly in May and early June, when temperatures reach at least 15°C. Spawning individuals stop their normal daylight avoidance response and breed on sunny days. Male dig a shallow nest in areas with strong currents. Ammocoete stage lasts 5½–7½ years. Feeds on diatoms and detritus, metamorphoses at 130–150 mm TL in late summer, and migrates to sea. Adults parasitise various fish species at sea, even whales and other cetaceans. It usually does not kill its hosts but feeds on small amounts of blood and body fluid from a single host for several days. Adults feed for about 3 years before migrating to spawning grounds.

Conservation status. LC; rare but widespread. Populations in Europe, which had declined due to pollution, have recovered since the 1980s but appear to be declining again in recent years.

Remarks. In 1921, 90 years after opening of the Welland Canal, the Lake Ontario population entered other Great Lakes of North America. Combined with other factors, this caused a sharp decline in many native species and the extinction of three endemic coregonids.

Further reading. Potter & Osborne 1975 (identification); Hardisty 1986b (biology); Rodríguez-Muñoz et al. 2004 (genetics); Çevik et al. 2010 (record in Türkiye).

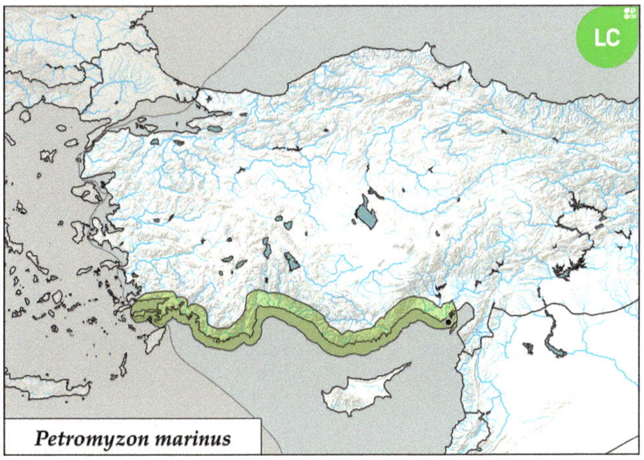

Family Carcharhinidae

Requiem sharks

The family of about 12 genera and 58 species of large sharks is found worldwide in tropical and subtropical waters. It is distinguished from all other families of sharks by having an anal fin, five gill-slits, a streamlined body, several rows of triangular teeth in the jaws, two dorsals without an anterior spine, and a short head, not laterally widened to a "hammer" shape. The largest species of the family is the tiger shark *Galeocerdo cuvier*, reaching up to 7 m. However, most requiem sharks reach 3 m or less. One species, the bull shark, regularly enters freshwater habitats and penetrates up to 4000 km inland in the Amazon. It is known to breed in freshwaters and occurs in some lakes in Central America and New Guinea, where it is the victim of massive overfishing. All requiem sharks reproduce viviparously, with internal fertilization. Unlike many other sharks, they do not feed their young with unfertilised eggs in the uterus, and they are not cannibalistic. Several marine shark species have been reported from the Shatt al Arab in Iraq. To identify them, identification resources about marine sharks must be consulted. **Further reading:** Ebert et al. 2013 (diversity); Nelson et al. 2016 (diversity).

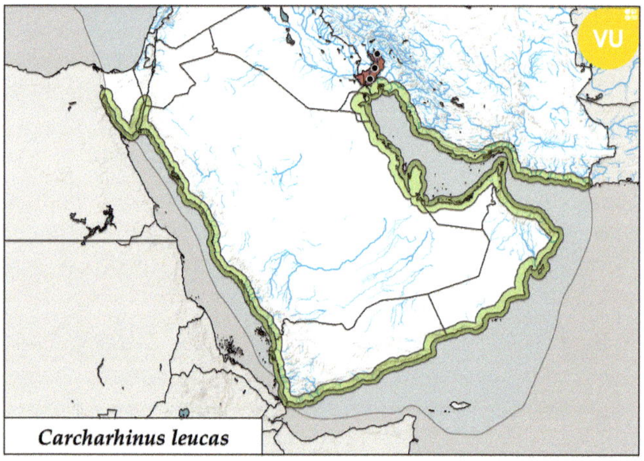

Carcharhinus leucas

Common name. Bull shark.

Diagnosis. Only shark in freshwaters in West Asia. Size up to 3400 mm TL, usually up to 2000 mm TL have been estimated in Iraqi freshwaters.

Distribution. In West Asia, in Persian Gulf basin, where it enters Euphrates-Tigris drainage. In Iran, migrates into Karun and Khowr-e Bahmanshir. In Iraq, once regularly upriver to Baghdad, but currently only up to Basra due to dams. Worldwide in warm temperate to tropical seas. Absent from Mediterranean and very rarely reported from Atlantic coast of Western Sahara. Reported to enter freshwater almost throughout its range, including Amazon, Gambia, Ganges, Mississippi, San Juan (and Lake Nicaragua), Zambezi, and many rivers in southern Africa, New Guinea, and Australia.

Habitat. Continental shelf waters to a depth of about 150 m, but usually less than 30 m. Frequently found in estuarine and freshwater areas of lower and middle reaches of rivers. Brackish and freshwater habitats are major nursery areas.

Biology. Euryhaline, moves freely between marine and freshwater. Tolerates hypersaline conditions up to 53 ‰. Life span is about 14 years. Matures at about 6 years to a length

Carcharhinus leucas; South Africa. © Shutterstock/Carlos Aguilera/133701370.

https://doi.org/10.1515/9783111677811-003

of 1600–2250 mm TL for females and 1800–2300 mm TL for males. Gestation period is 10–11 months and up to 13 embryos are reported. Usually, gives birth in estuaries. Juveniles are between 560 and 810 mm TL. Can swim up to 180 km in 24 hours. One of the most dangerous sharks. Freshwater attacks are rare but responsible for most shark attacks in shallow estuarine and coastal waters worldwide. Carnivorous, feeding near bottom on vertebrates such as fish, elasmobranchs, sea turtles, sea snakes, and large invertebrates such as crabs, sea urchins, and cephalopods. Juveniles usually take fish.

Conservation status. VU; still found in Iraqi and Iranian freshwaters. Worldwide heavily depleted with declining populations. Commonly caught in commercial and recreational fisheries. Usually not a target species but caught as by-catch or as part of a multispecies fishery. Has been commercially exploited for its skin, liver oil, and meat, with fins currently being the main product.

Further reading. Fowler et al. 2005; Coad 2015a (biology, distribution).

Freshwater shark attacks. Imagine standing knee-deep in the river, washing your fishing gear after a successful day's work. Suddenly, you hear a splash beside you, and then you fall because a shark has taken a bite out of your wade. You cannot get up because you do not have a clam, and then you see the shark coming back through the murky water. This nightmare is a bull shark attack and has happened many times. Bull sharks are often found in shallow estuarine waters, entering rivers hundreds of kilometers upstream of the estuary. They frequently hunt along shores, and their activities are likely to overlap with those of humans. Bull sharks are among the most dangerous sharks in the world, if not the most dangerous. From Iraq and Iran, 45 freshwater bull shark attacks were reported between 1938 and 1970, 19 of which were fatal. This represents 34 % of all shark attacks in freshwater worldwide, making the northern Persian Gulf one of the most dangerous freshwater areas in the world for shark attacks. Of course, we should not compare these figures with the number of fatal car accidents in the same region.
Further reading. Coad 2015a (shark attacks in the Persian Gulf).

Large lowland rivers are the freshwater habitat of *Cacharhinus leucas*, as here in lowermost Shatt al Arab in Iraq.

Family Acipenseridae

Sturgeons

Sturgeons are a small family restricted to the northern hemisphere, in Europe, northern Asia, and North America. Traditionally, all sturgeon species found in Western Asia and Europe were placed in two genera: *Acipenser* (nine species) and *Huso* (one species). Other genera in the family Acipenseridae include *Scaphirhynchus*, with three species in North America, and *Pseudoscaphirhynchus*, with 1-3 species in the Aral basin in Central Asia. The phylogeny of sturgeons has been well understood for many years, and it has long been evident that the genus *Acipenser* is paraphyletic. This has not been reflected into a generic classification, as species such as *Sterletus stellatus* have been found to occupy different positions in different analyses. Therefore, sturgeon phylogeny was considered preliminary. This was also intended to avoid renaming these highly threatened species, as it was feared that name changes would hinder sturgeon protection. Sturgeon phylogeny has recently been reanalysed and, for the first time, translated into a new generic structure. *Pseudoscaphirhynchus* and *Scaphirhynchus* remain valid genera, but *Acipenser* now comprises only *A. sturio*, *A. desotoi* and *A. oxyrinchus*. All other West Asian and European *Acipenser* species have been placed in the genus *Huso*, ignoring that *Huso* is a junior synonym of *Sterletus*. The type species of *Sterletus* is the American lake sturgeon *(S. fulvescens)*. Consequently, all sturgeon species in the Pacific basin in Asia and America, including *Huso dauricus*, have been placed in the genus *Sinosturio*. Users of species names now have three options: 1) treat the knowledge of sturgeon phylogeny still as too incomplete to risk causing confusion by changing the names, 2) place all sturgeons in the family Acipenseridae in the genus *Acipenser* (requiring few name changes), or 3) accept the proposed generic concept, as we do in this book.

Sturgeons are a very isolated group of fishes resembling sharks with a heterocercal caudal, a largely cartilaginous skeleton, and an intestine with a spiral valve. The earliest fossil sturgeon is *Protopsephurus* from the Barremian (126.5–121.4 million years ago) in the Cretaceous of China. Sturgeons are large fish with five rows of hard bony plates, no scales, four barbels, a projecting snout, and an inferior mouth forming a protrusible tube. Bony plates or scutes are more prominent in small individuals, often partly hidden under the skin in adults. Most sturgeons are anadromous, foraging in marine waters or freshened parts of seas and undertaking long spawning migrations to large rivers. Upstream migration generally starts in late summer and is less marked when temperatures fall below a threshold. Many individuals overwinter in freshwater. The main peak of upstream migration usually occurs in early spring. The individuals returning to rivers in late summer often move and spawn further upstream than those migrating in spring. All species spawn in freshwaters in spring and summer. Spawning generally takes place in fast-flowing water, on gravel bottom. Females return to foraging areas immediately after spawning, while males often remain at the spawning grounds for many weeks, waiting for other females. Females of most species do not spawn every year. Eggs and larvae are very sensitive to low oxygen concentrations. The juveniles of anadromous species usually migrate to the sea during their first summer. All sturgeons can complete their life cycles in pure freshwater, and non-anadromous, riverine individuals or populations existed in *S. nudiventris* and *S. gueldenstaedtii*.

Sterletus huso; Caspian Sea, Russia: ~3000 mm TL. © G. Zauner.

https://doi.org/10.1515/9783111677811-004

Among all Western Palaearctic fishes, sturgeons are the family most strongly impacted by human activities such as damming and overfishing. All anadromous sturgeons are now very rare in the Black Sea basin, following the impoundment of almost all spawning rivers. It is also evident that nearly all individuals found today descend from stocked juveniles produced in sturgeon farms. The last Black Sea tributaries that still support natural migrating populations are the Danube and the Rioni (Georgia). The Rioni is the key river in the Black Sea basin, home to the last populations of *S. nudiventris* (if still extant) and *S. persicus*. Despite their critical status these species are still overfished in the Rioni. The Caspian sturgeon populations are under immense pressure due to overfishing and the loss of spawning sites. It is clear that, except for *S. stellatus*, all anadromous sturgeons discussed in this book now completely depend on stocking. Overfishing at sea for meat and in rivers for caviar has resulted in their extinction. Indeed, sturgeons are a conservation success story, as most species would be completely extinct today without farming and stocking. Populations depend on stocking of juveniles from fish farms. This makes identification even more problematic as most sturgeon farms produce different hybrids. The most frequent hybrid is a cross between *S. huso* and *S. ruthenus* (called “bester”). Furthermore, non-native species such as *S. baerii* and different American species are widely produced and stocked. Fish farms in the Black and Caspian Seas used wild-caught genitors. As these are increasingly difficult to find, farms gradually shift to in-farm spawners. It is common for *S. persicus* outside their main distribution range to be confused with *S. gueldenstaedtii* by sturgeon farmers, who often mix them with this species.

Further reading. Holčík et al. 1989a (biology, systematics); Birstein & Bemis 1997 (biology, conservation); Elvira 2000 (conservation); Pikitch et al. 2005 (conservation, fisheries); Krieger et al. 2008 (molecular phylogeny); Havelka et al. 2011 (cytogenetics); Near & Thacker 2024 (fossil); Brownstein & Near 2025 (generic concept); Kottelat & Freyhof 2025 (*Sterletus* vs. *Huso*).

Figure 28. Ventral view of head of: left, *S. persicus*, 550 mm TL; right, *S. gueldenstaedtii*, 450 mm TL.

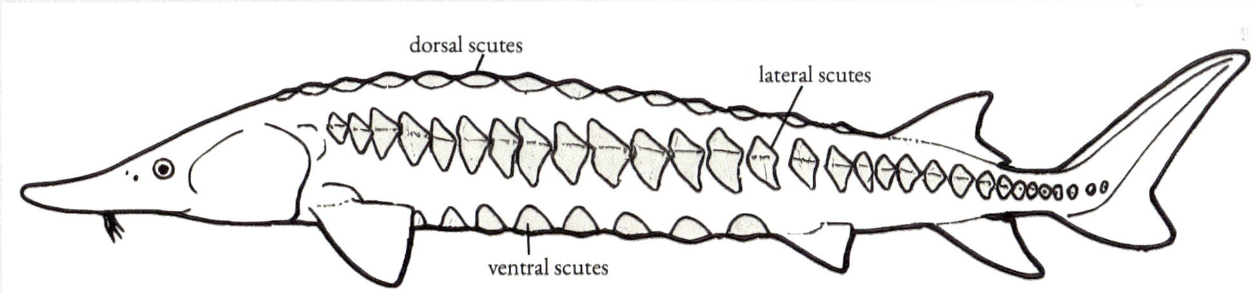

Scute terminology in Acipenseridae. A series of scutes or bony plates partially cover the body of a sturgeon. The scutes are more prominent in juveniles and become embedded and reduced as they grow. Scutes are most deeply embedded and reduced in adult *S. huso*. Scutes are arranged in five longitudinal rows, one row along the dorsal midline between the head and the dorsal origin (dorsal scutes), one midlateral row from the head to the caudal (lateral scutes), and one row between the pectoral base and the anal origin (ventral scutes). Rhombic or star-shaped bony denticles may be present at least between the dorsal and lateral scute rows; some denticles are enlarged to form plates, especially in the posterior part of the body. These denticles are very prominent in juveniles of *S. stellatus*, *S. gueldenstaedtii*, and *S. persicus* but become relatively smaller and often deeply embedded in adults. In other species, denticles are often absent or very small. In *Acipenser sturio*, the denticles are more prominent, large, and clearly visible in adults. All scutes and denticles may be lacking in some individuals of all species.

Key to species of Acipenseridae in West Asia

1a - Gill membranes joined broadly to each other and free from isthmus; mouth very large, crescent-shaped.
..................*S. huso*

1b - Gill membranes attached to isthmus and not joined to each other and attached to isthmus; mouth small and straight.
..................2

2a - Lower lip continuous, not interrupted in middle; first dorsal scute larger than following ones, fused with head, forming an obtuse angle with head profile; barbels fimbriate.
..................*S. nudiventris*

2b - Lower lip interrupted in middle; first dorsal scute not larger than following ones; barbels fimbriate or not.
..................3

3a - 20–49 fan-shaped gill rakers terminated by several tubercles; 32–62 lateral scutes.
..................*S. baerii*

3b - 15–35 rod-shaped gill rakers; 23–50 lateral scutes.
..................4

4a - Rhombic or elongated denticles between dorsal and lateral scutes; a series of plates along posterior part of anal base; 2–4 plates along lower edge of caudal peduncle.
..................*A. sturio*

4b - Star-shaped or roundish denticles between dorsal and lateral scutes; no plates along posterior part of anal base; 0–1 plate along lower edge of caudal peduncle.
..................5

5a - Base of barbels closer to mouth than to tip of snout; snout very long (59–65 % HL), narrow, depressed.
..................*S. stellatus*

5b - Base of barbels closer to tip of snout than to mouth; snout short (22–36 % HL) and blunt.
..................6

6a - Head triangular, if seen from below (Figure 28); dorsal profiles of head and body forming an obtuse angle; belly yellowish-white; dorsum golden brown; snout short, blunt and somewhat rounded, its depth at eye about equal to or smaller than its length.
..................*S. gueldenstaedtii*

6b - Head squarish, if seen from below (Figure 28); dorsal profiles of head and body continuous; belly white; dorsum greyish-blue to black; snout elongate massive and curved downward, its depth at eye smaller than its length.
..................*S. persicus*

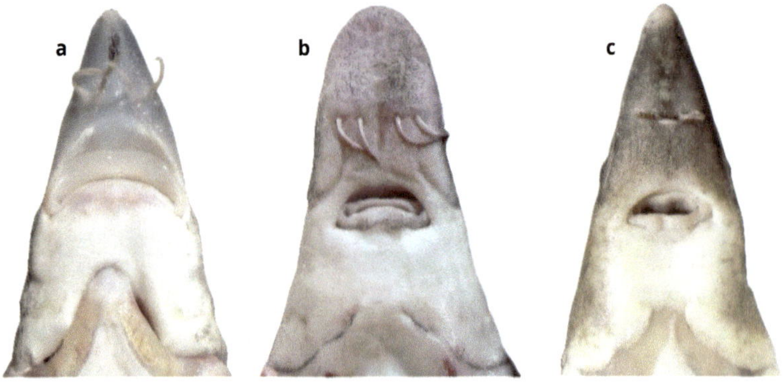

Figure 29. Ventral view of the head of **a**, *S. huso*; **b**, *S. nudiventris*; **c**, *S. ruthenus*.

Acipenser sturio; Gironde, France; ~1000 mm TL. © A. Hartl.

Acipenser sturio

Common name. Atlantic sturgeon

Diagnosis. Distinguished from species of *Sterletus* in West Asia by: ○ many rhombic or elongated denticles between dorsal and lateral scutes / ○ a series of plates along posterior part of anal base / ○ 2–4 plates along lower edge of caudal peduncle / ○ first dorsal scute fused with head / ○ 24–40 lateral scutes / ○ base of barbels midway between mouth and tip of snout or closer to mouth / ○ barbels not fimbriate / ○ 15–29 gill rakers / ○ lower lip interrupted in middle / ○ diploid. Largest documented size 6000 mm TL and 1000 kg, but today rarely exceeds 2000 mm TL.

Distribution. Only spawning river in West Asia was Rioni (Georgia). Found along all coasts of Mediterranean and Black Sea. Also, North and White Seas, and European coasts of Atlantic. Occasionally recorded from Iceland. Anadromous, in most large rivers, but not recorded from Danube upstream of delta. Currently confined to Garonne (France).

Habitat. At sea, coastal and estuarine areas on soft ground. In freshwaters, estuaries, and large rivers.

Biology. Anadromous. Males first spawn at 9–13 years, females at 11–18 years. Females spawn every other year and males every year in March–August when temperatures rise above 20°C, later in northern populations and in populations spawning further upstream. Spawning migrations peak in spring, with a secondary weak peak in late summer and autumn. Individuals migrating in late summer overwinter in rivers and spawn in spring. Spawns in strong-current habitats, in large and deep rivers, on stone to gravel bottoms. Spent individuals return directly to sea. Some juveniles migrate to sea during first summer, but most leave freshwater habitats after 2–4 years and remain at sea until maturity. At sea, feeds on a variety of molluscs, worms, crustaceans, and small fish. Atlantic populations feed benthically, while Black Sea populations feed in pelagic zone, mainly on anchovies.

Conservation status. CR; survival appears to depend only on stocking. Decline began in 19th century or possibly earlier due to massive overfishing, damming, river regulation and pollution. Last natural spawning in 1994. Last remaining population (in France) still declining. Population in Rioni (Georgia) disappeared in late 20th century. Bycatch and accidental kills are still a major problem in France. Reintroduction programmes are underway in France and Germany.

Further reading. Holčík et al. 1989b (biology); Holčík 2000 (conservation); Elvira 2000 (biology); Bacalbasa-Dobrovici & Holčík 2000 (Black Sea population); Rochard, in Keith & Allardi 2001 (status in France); Ludwig et al. 2002 (genetics).

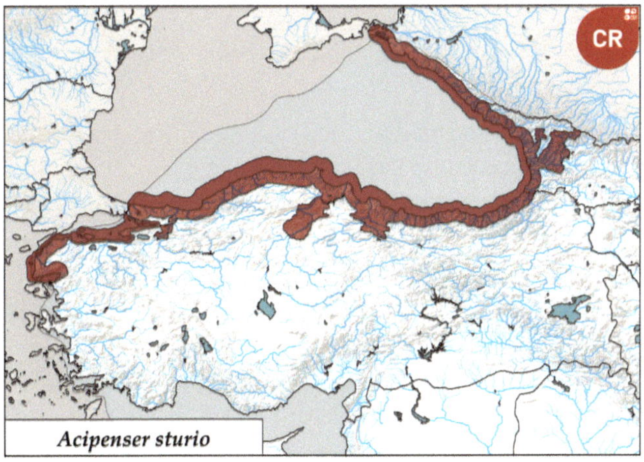

Sterletus baerii; aquaculture, Russia; ~700 mm TL. © A. Hartl.

Sterletus baerii

Common name. Siberian sturgeon

Diagnosis. Distinguished from other species of *Acipenser* and *Sterletus* in West Asia by: • 20–49 fan-shaped gill rakers terminated by several tubercles / ○ lower lip interrupted in middle / ○ 10–20 dorsal scutes / ○ 32–62 lateral scutes / ○ 7–16 ventral scutes / ○ barbels not fimbriate / ○ tetraploid. Size usually up to 2000 mm TL and 210 kg.

Distribution. Siberia (from Ob to Kolyma drainages, Lake Baikal). Aquaculture has resulted in deliberate and accidental releases, especially in Caucasus. No known established populations in West Asia.

Habitat. Deep parts of large rivers with moderate to rapid currents. Spawns in main channel over stone-gravel or gravel-sand bottom in strong current.

Biology. Non-anadromous. Males first spawn at 11–24 years, females at 20–28 years (9–10 and 11–12 years, respectively, in Lena). Females spawn every 3–5 years and males every 2–3 years in May–June at 9–18°C. In Siberian part of range, there are both sedentary and migratory parts of populations. Migratory individuals feed in estuaries and migrate upstream to spawn. Before construction of dams on Ob, they migrated about 3000 km upstream. Spawning migration begins in autumn. Sedentary individuals inhabit middle and upper reaches of rivers and do not undertake long migrations. Feeds on a variety of benthic organisms, including crustaceans and chironomid larvae.

Conservation status. Non-native and not established; released for commercial fisheries or escaped from fish farms. CR in native range.

Further reading. Sokolov & Vasil'ev 1989b (biology).

Sterletus gueldenstaedtii; Volga, Russia; ~900 mm TL. © A. Hartl.

Sterletus gueldenstaedtii

Common name. Russian sturgeon

Diagnosis. Distinguished from other species of *Acipenser* and *Sterletus* in West Asia by: ○ head triangular, if seen from below / ○ base of barbels closer to tip of snout than to mouth / ○ barbels not fimbriate / ○ lower lip interrupted in middle / ○ 15–31 gill rakers / ○ 8–19 dorsal scutes / ○ 24–50 lateral scutes / ○ 6–13 ventral scutes / ○ snout short (22–36 % HL) and blunt / ○ head length 17–19 % TL / ○ many star-shaped bony denticles of various sizes between dorsal and lateral scutes / ○ dorsum golden brown, belly yellowish white / ○ dorsal profiles of head and body forming an obtuse angle / ○ no plates along posterior part of anal base / ○ 0–1 plate along lower edge of caudal peduncle / ○ tetraploid. Size up to 2360 mm TL and 115 kg. Evidence does not support records of size up to 4000 mm TL. Subfossil specimens about 3000 mm TL.

Distribution. Caspian, Black, and Azov Sea basins.

Habitat. At sea, shallow coastal and estuarine areas. In freshwaters, deep parts of large rivers with moderate to fast currents. Spawns in strong currents (1.0–1.5 m/s) in large and deep rivers on stony or gravelly bottoms.

Biology. Anadromous and freshwater populations. A complex pattern of spawning migrations includes spring and autumn runs. Individuals migrating in spring enter freshwater just before spawning, tend to spawn in lower reaches of rivers (320–650 km in unregulated Ural). Individuals migrating in autumn overwinter in rivers and spawn further upstream in spring (900–1200 km in Ural). Males first spawn at 8–13 years, females at 10–16 years. Females spawn every 4–6 years and males every 2–3 years in April–June when temperatures rise above 10°C. Larvae drift on currents; juveniles then move to shallower habitats before migrating to sea during their first summer. Remain at sea until maturity. Feeds on a wide variety of benthic molluscs, crustaceans and small fish.

Conservation status. CR; survival appears to depend only on stocking. Very rare in Black Sea basin. Almost all spawning grounds lost due to dam construction. Heavily overfished. Caspian populations are under massive pressure from overfishing and loss of spawning habitat. Almost all migrating spawners are poached below Volgograd dam.

Further reading. Vlasenko et al. 1989a (biology).

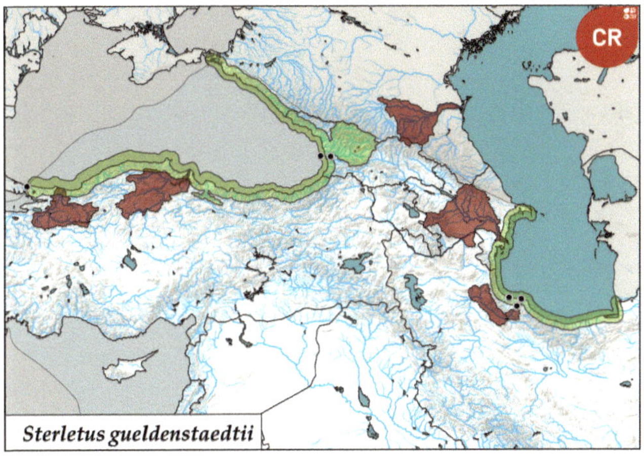

Sterletus gueldenstaedtii

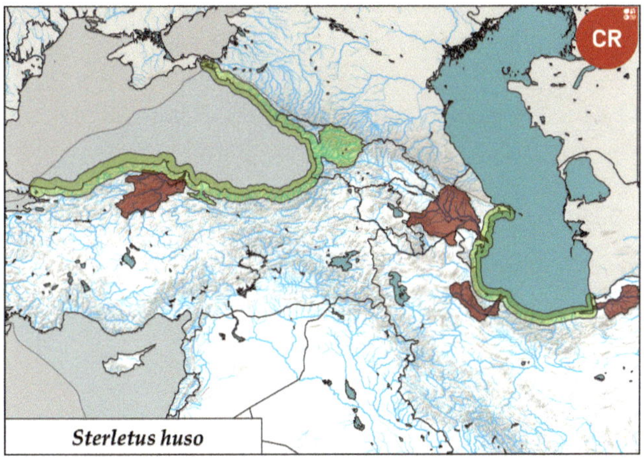

Sterletus huso

Sterletus huso; Volga, Russia; ~1000 mm TL. © A. Hartl.

Sterletus huso

Common name. Beluga

Diagnosis. Distinguished from other species of *Acipenser* and *Sterletus* in West Asia by: • gill membranes joined and free from isthmus / • mouth very large, crescent-shaped / • barbels laterally flattened / ○ 9–17 dorsal scutes / ○ 37–53 lateral scutes / ○ 7–14 ventral scutes / ○ 17–36 gill rakers. Largest documented size 8000 mm TL and 3200 kg, but generally below 5000 mm TL and 1000 kg.

Distribution. Caspian, Black, Marmara, Azov, and Adriatic Sea basins. Extirpated and actually reintroduced in Adriatic basin.

Habitat. At sea, pelagic, following food organisms. Migrates further upstream to spawn than any other sturgeon (in Danube to Germany). Spawns in strong-current habitats in main channel of large and deep rivers on rocky or gravelly bottoms. Juveniles spend first summer in shallow river habitats.

Biology. Anadromous. Males first spawn at 10–16 years, females at 14–20 years. Spawns every 3–4 years in April–June. A complicated pattern of spawning migrations includes a peak in late winter and spring and a peak in late summer and autumn. In spring, individuals migrate from sea prior to spawning. Individuals migrating in autumn remain in river until following spring. Yolk sac larvae pelagic for 7–8 days, drifting with current. Juveniles migrate to sea during their first summer and remain there until maturity. At sea, feeds on a wide variety of fish; seals and birds have also been reported as prey.

Conservation status. CR; survival appears to depend only on stocking. Extirpated from Adriatic (1970s) and Sea of Azov. Last populations in Black Sea basin migrate to Danube and Rioni (Georgia). All other populations almost extirpated due to overfishing and damming of spawning rivers. Occasional individuals are still recorded from Azov and Black Seas. No natural spawning in Iranian or Turkish waters. Most females recorded are in their first year of maturity.

Further reading. Pirogovskii et al. 1989 (biology).

Sterletus huso sold illegally at a roadside market close to Batumi in Georgia in 2014.

Sterletus nudiventris; aquaculture, Iran; ~750 mm TL. © K. Abbasi.

Sterletus nudiventris

Common name. Ship sturgeon

Diagnosis. Distinguished from other species of *Acipenser* and *Sterletus* in West Asia by: ● lower lip continuous, not interrupted in middle / ● first dorsal scute larger than following ones, fused with head, forming an obtuse angle with head profile / ○ barbels fimbriate / ○ 24–45 gill rakers / ○ 11–17 dorsal scutes / ○ 49–74 lateral scutes / ○ diploid. Size up to 2200 mm TL.

Distribution. Black, Azov, Caspian, and Aral Sea basins. Danube up to Bratislava, Volga up to Kazan, Ural up to Chkalov, unknown or very rare in others. Introduced in Lake Balkhash (Kazakhstan) from Syr-Darya (Aral basin) in 1960s.

Habitat. At sea, near coasts and estuaries. In freshwater, deep sections of large rivers. Juveniles in shallow river habitats. Spawns in strong-current habitats in mainstem of large and deep rivers on stony or gravelly bottoms.

Biology. Anadromous, with some non-migratory freshwater populations. Males first spawn at 6–15 years, females at 12–22 years. Recorded life span of 36 years, but probably much longer. Spring and autumn migrations occur in most drainages. Individuals migrating in autumn remain in river until following spring to spawn. Females spawn every 2–3 years and males every 1–2 years in March–May and at temperatures above 10°C. Most juveniles migrate to sea in their first summer and remain there until maturity. Recorded migrating 1500 km upstream in Amu Darya. Some individuals stay in freshwater for longer periods. Feeds on a wide variety of benthic fish, molluscs and crustaceans.

Conservation status. CR; survival appears to depend on stocking. Extirpated from Aral and Black Sea basins. Only occasional records from Danube, Rioni, and lower Volga. Only a very small population may remain in Ural (Russia, Kazakhstan). No natural spawning in Iran. Established in Lake Balkhash, where it forms a small population, but details are not known. Caspian *S. stellatus* introduced in Aral basin were infected by parasitic monogenean flatworm *Nitzschia sturionis*, which caused collapse of Aral population of *S. nudiventris*.

Further reading. Sokolov & Vasil'ev 1989a (biology).

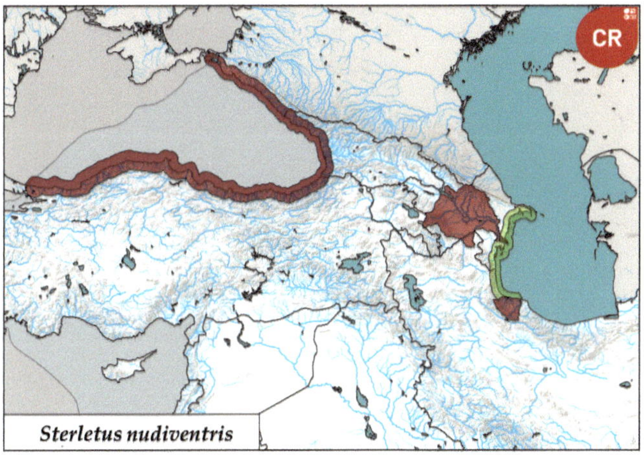

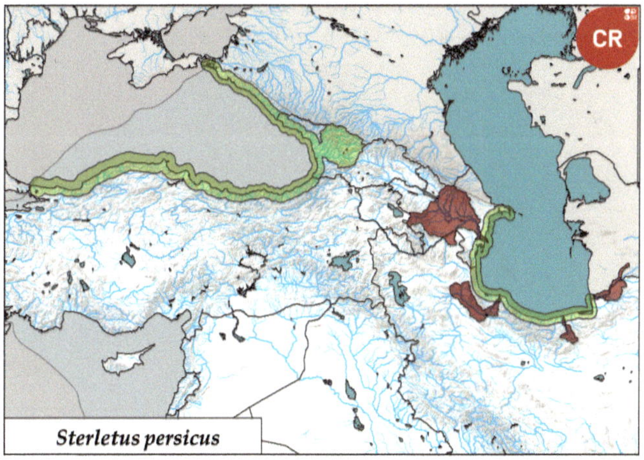

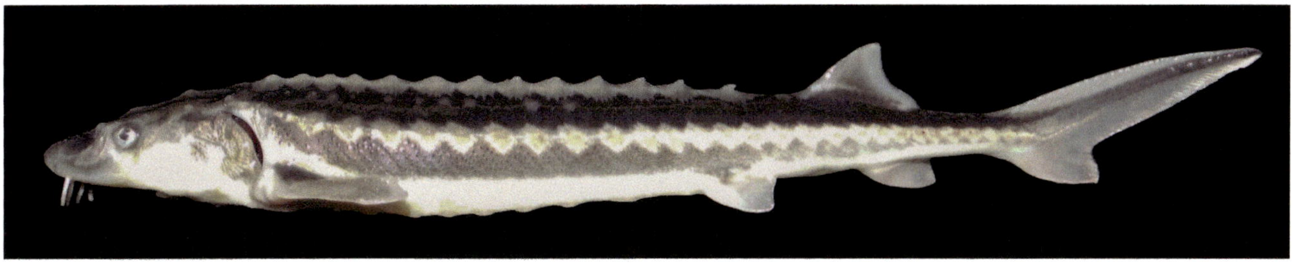

Sterletus persicus; aquaculture, Iran; ~550 mm TL.

Sterletus persicus
Common name. Persian sturgeon
Diagnosis. Distinguished from other species of *Acipenser* and *Sterletus* in West Asia by: ○ head squarish, if seen from below / ○ dorsum greyish-blue to black, belly white / ○ base of barbels closer to tip of snout than to mouth / ○ barbels not fimbriate / ○ lower lip interrupted in middle / ○ snout elongate, massive and curved downward / ○ dorsal profiles of head and body continuous / ○ no plate along posterior part of anal base / ○ no plate along posterior part of anal base / ○ 0–1 plate along lower edge of caudal peduncle / ○ 15–31 gill rakers / ○ 7–19 dorsal scutes / ○ 23–41 lateral scutes / ○ 7–13 ventral scutes. Size up to 2280 mm TL and 70 kg.
Distribution. Caspian basin, most abundant in southern part. Also, in south-eastern Black Sea basin along Georgian and Anatolian coasts where it enters Rioni and previously lower Danube.
Habitat. At sea, coastal and estuarine zones. Spawns in strong-current habitats in main channel of large and deep rivers on stony or gravel bottoms. Juveniles spend their first summer in riverine habitats.
Biology. Anadromous. Males first spawn at 8–15 years, females at 12–18 years. Does not spawn every year. Spawns June–August when temperature rises above 16°C. In southern Caspian basin, spawns April–September, but reproduction is interrupted from June to August when temperatures rise above 25°C. Most individuals migrate upstream in April–May, but some may enter rivers at other times of the year. In southern Caspian basin, there is a second run in September–October. Juveniles migrate to sea during their first summer and remain there until maturity. At sea, it feeds on a wide variety of benthic molluscs, crustaceans, and small fish.
Conservation status. CR; survival appears to depend only on stocking. Presence generally ignored; individuals often confused with *S. gueldenstaedtii*. Last spawning population migrates to Rioni in Georgia. Probably no natural reproduction anymore in Caspian basin. Has been massively stocked from artificial propagation in Iran. Recently, Iranian stocking programmes for this species have been discontinued.
Remarks. *Acipenser colchicus* has been proposed as a valid species for the Black Sea population. Molecular or other data have not supported this hypothesis, and we treat *S. colchicus* as a synonym of *S. persicus*. Similar datasets also do not support *S. persicus* as distinct from *S. gueldenstaedtii*, while these two species are always recognised as morphologically distinct, especially in the Caspian basin. Further research is needed to resolve this taxonomic issue.
Further reading. Vlasenko et al. 1989b (biology).

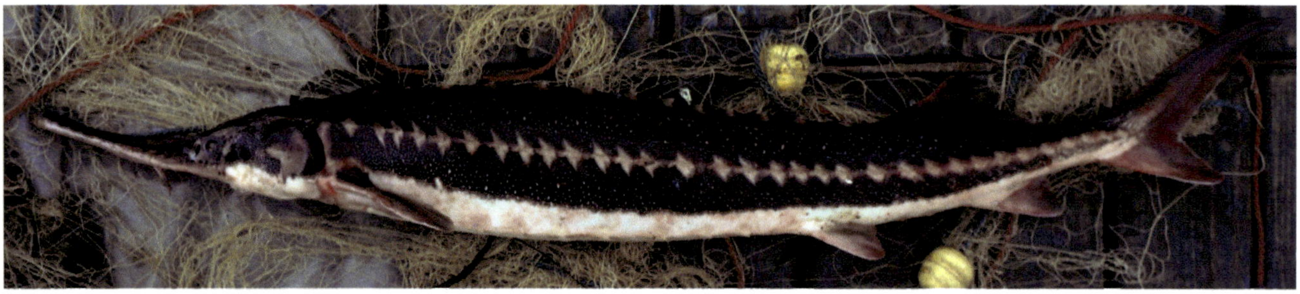

Sterletus stellatus; Danube, Hungary; ~1200 mm TL. © Z. Sallai.

Sterletus stellatus
Common name. Stellate sturgeon
Diagnosis. Distinguished from other species of *Acipenser* and *Sterletus* in West Asia by: ● snout very long (59–65 % HL), narrow, dorsoventrally compressed / ○ star-shaped bony denticles between rows of scutes / ○ no plate along posterior part of anal base / ○ 0–1 plate along lower edge of caudal peduncle / ○ 9–16 dorsal scutes / ○ 26–43 lateral scutes / ○ 9–14 ventral scutes / ○ base of barbels closer to mouth than to tip of snout / ○ lower lip interrupted in middle / ○ 24–29 gill rakers / ○ diploid. Size up to 2180 mm TL and 54 kg.
Distribution. Caspian, Black, Azov, and Aegean basins, ascending rivers to spawn. Introduced in Aral basin.
Habitat. At sea, in coastal and estuarine areas. Feed on bottom, usually on clayey sand, and intensively in middle and upper water layers. Spawns in strong currents in main channel of large and deep rivers, on stone or gravel bottoms. Also, spawns on flooded riverbanks; if gravel bottom is unavailable spawns on sand or sandy clay. Juveniles occupy shallow river habitats during first summer.
Biology. Anadromous. Males first spawn at 6–12 years, females at 8–14 years. Females spawn every 3–4 years and

males every 2–3 years in April–September. Spawns only under relatively constant hydrological conditions. Fluctuating hydrological conditions result in high egg mortality. Upstream migration with two peaks in spring and autumn. Migrates at higher temperatures and therefore later than other sturgeons. Males spend no more than 6 weeks at spawning sites, and females only 10–12 days. Spent individuals return directly to sea. Yolk sac larvae pelagic for 2–3 days, drifting with current. Juveniles migrate to sea during their first summer and remain there until maturity. At sea, feeds on a wide variety of crustaceans, molluscs and benthic and pelagic fish.

Conservation status. CR; survival appears to depend on stocking. Extirpated from Aegean Sea. In Black Sea basin, last natural populations still migrate to Danube and Rioni (Georgia), where it is heavily overfished. Very few spawners remain in rest of Black Sea basin. Large numbers were stocked from Caspian to Azov basins in 1961–1986. Caspian populations under massive pressure from overfishing and loss of spawning grounds. Stocks are rapidly declining. Almost all migrating spawners poached below Volgograd dam.

Further reading. Shubina et al. 1989 (biology); Economidis et al. 2000 (Aegean Sea).

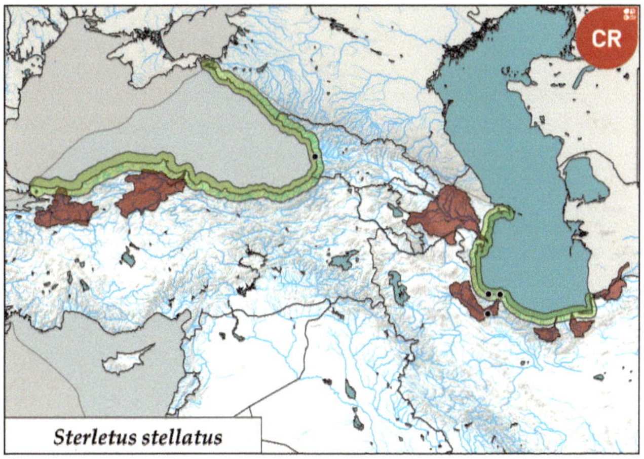

Rioni in Georgia. One of the last spawning sites of sturgeons left in West Asia.

Family Anguillidae

Freshwater eels

Anguilla belongs to Elopomorpha, a basal phylogenetic group that contains various types of fishes. These range from tarpons, ladyfishes, and bonefishes to the large diversity of true eels (Anguilliformes). All members of this group have leptocephalus larvae, unique to the Elopomorpha. *Anguilla* is unmistakably distinct from any other eel-like fishes in freshwaters of West Asia, characterised by the dorsal, caudal, and anal fins being long and confluent and the absence of dorsal spines and pelvic fins. The family is small, with about 15–20 species in a single genus (*Anguilla*). All species are catadromous (freshwater residents migrating to spawn at sea). *Anguilla* is known from all oceanic basins except the Pacific slope of the Americas. All species spawn in warm subtropical areas of deep seas. However, surprisingly little is known about their marine habitat. The European eel *Anguilla anguilla* and the Japanese eel *A. japonica* are the best-studied species.

The reproduction of *A. anguilla* has long been a mystery, given that sexually mature eels have not yet been found. Early ichthyologists were wrong to think that small gobies gave birth to young eels because they mistook parasitic nematodes for eel larvae. Another belief, which lasted until the late 19th century, was that eelpout *Zoarces viviparus* gave birth to small eels. In 1856, *Leptocephalus brevirostris* was first described from the Mediterranean. It was only in 1896 that it was recognised as the larvae of *Anguilla*. Eel larvae are called leptocephalus (plural leptocephali). It was then hypothesised that the spawning sites were in the Mediterranean, but eel larvae smaller than 30 mm TL have never been observed there. In 1904, the first leptocephali were recorded outside the Mediterranean (Faroe Islands). Since then, leptocephali smaller than 7 mm TL and yolk-sack larvae have been observed in the Sargasso Sea. In East Asia and Europe, eels are of major commercial importance, and many threats have caused drastic declines in their population. **Further reading.** Fischer & Bianchi 1984 (identification Indian Ocean); Chen et al. 2014 (Elopomorpha); Nelson et al. 2016 (diversity).

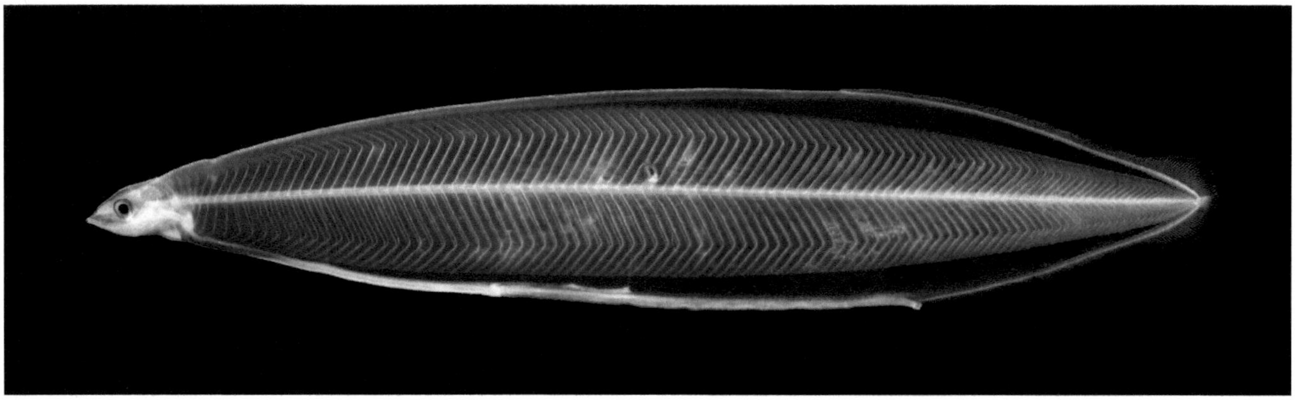

Leptocephalus larvae of an eel; Bermudas; ~100 mm SL. © S. Johnsen.

Key to species of *Anguilla* in West Asia

1a - Dorsal origin slightly in front of anus.
..................*A. bicolor*
1b - Dorsal origin clearly in front of anus.
..................2

2a - Flank plain olive yellowish to dark-brown or black, silvery grey from jaw to anus.
..................*A. anguilla*
2b - Flank mottled in individuals longer than 300 mm TL.
..................*A. bengalensis*

https://doi.org/10.1515/9783111677811-005

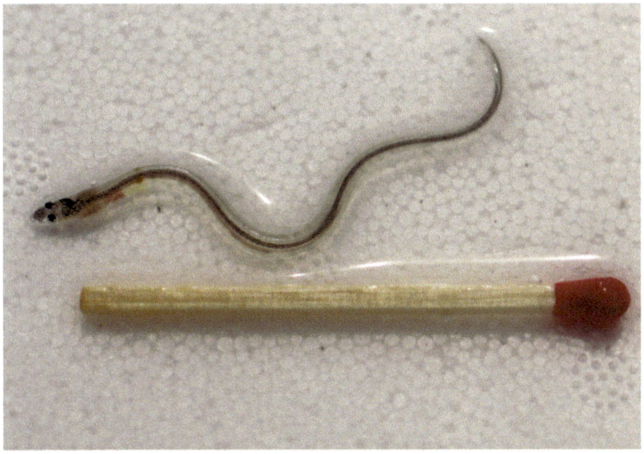

Glass eel of *A. anguilla*; Germany; © J. Simon.

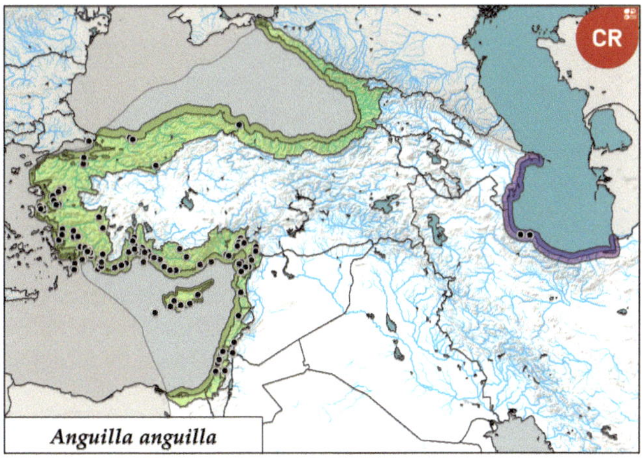

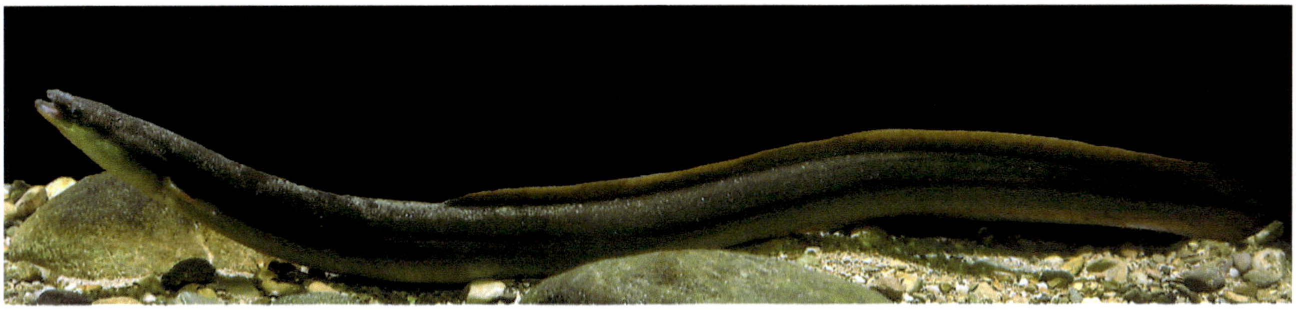

Anguilla anguilla; upper Rhine drainage, Germany; ~600 mm SL. © A. Hartl.

Anguilla anguilla

Common name. European eel

Diagnosis. Distinguished from other freshwater fishes in Mediterranean, Caspian, and Black Sea basins by: • body very elongate, slender, and cylindrical / • dorsal, caudal, and anal long and confluent / • flank plain olive yellowish to dark-brown or black, silvery grey from jaw to anus / ○ dorsal origin clearly in front of anus / ○ pelvic absent. Size usually about 400–600 mm TL, rarely larger, although individuals up to 2000 TL are reported.

Distribution. All rivers flowing to Mediterranean, Black, North, and Baltic Seas and Atlantic as far south as Canary Islands. Stocked in many inland waters and occasionally caught on Caspian coast and in rivers of Iran. Occasional individuals reach Volga drainage through canals. Very rarely in White and Barents Seas, recorded as far east as Pechora. Small numbers enter Black Sea, migrating east to Kuban drainage. Parts of population remain at sea (northwest Atlantic and Mediterranean).

Habitat. All types of benthic habitats, from small streams to banks of large rivers and lakes. Part of population remain at sea. Occurs naturally only in waters connected to sea, stocked elsewhere. Hypothesised to spawn only in Sargasso Sea, in western subtropical Atlantic (about 26°N 60°W). Spawning area is about 2000 km wide.

Biology. Catadromous. Spawning peaks in early March and continues until July. No concrete data on spawning sites, but presumably at depths of 100–200 m and at about 20°C. Adults die after spawning. Mechanisms by which Leptocephali reach European coasts are not well understood. Evidence shows that larvae swim actively and that drifting plays a minor role. A statistical re-analysis of data does not support the well-known classical scheme of larval migration in Gulf Stream. Leptocephali reach continental slope at about 70 mm and metamorphose into glass eels (almost adult appearance but transparent body), which enter estuaries. Glass eels are observed on Portuguese coast in autumn, in North Sea in winter, and in Baltic Sea in spring. Pigmentation increases during upstream migration, and juveniles are called elvers. Feeding phase lasts 5–8 years for males and up to 12 years or more for females. Males are rarely recorded more than 200 km upstream. Downstream migration begins in late summer or autumn, and adults arrive at spawning sites after spring. In male, eye diameter increases significantly before migration. Migrates in open water layers of ocean, at about 300–700 m depth. Feeds on

a wide variety of benthic organisms. Heavily infected by nematode *Anguillicoloides crassus*, which feeds on blood in swim bladder, this parasite was introduced from Japan with experimental stocks of Japanese eel *A. japonica*.
Conservation status. CR; has strongly declined. Upstream and downstream migration barriers, including mortality from hydroelectric turbines, are a major threat. Furthermore, accumulation of lipophilic chemical contaminants stored by fish and released as fat stores are broken down during migration. These have toxic effects on migrating adults. *Anguillicoloides crassus* is thought to affect ability of European eel to reach their spawning grounds. European eel is still heavily fished. Exports outside Europe are banned, all trade within Europe (for consumption, culture, and stocking) is regulated, and quotas are in place. However, poaching and illegal trade are believed to occur throughout European eel fishery and are a major wildlife crime.
Further reading. Schmidt 1925 (breeding site); Boëtius & Harding 1985 (migration); Tesch 1991, 1999 (biology); Wirth & Bernatchez 2001 (genetics); Dekker 2003 (decline); Pike et al. 2020a (biology, distribution, conservation status).

Anguilla bengalensis; India; 800 mm TL. © J. Els.

Anguilla bengalensis; India; 800 mm TL. © J. Els.

Anguilla bengalensis
Common name. Mottled Eel
Diagnosis. Distinguished from *A. bicolor* by: • dorsal origin at about midway between pectoral base and anus, clearly in front of anus / • mottled colour pattern in individuals longer than 300 mm in TL. Size usually about 800 mm, rarely up to 2000 mm total length.
Distribution. In Arabian Peninsula in Wadi Haj and Wadi Masila in Yemen. Also, in rivers on east coast of Africa, from Pongola in South Africa to Genale in south-west Ethiopia. In Asia and Indian Ocean from Pakistan east to Aceh in Indonesia and peninsular Malaysia. Known to occur on oceanic islands, including Madagascar, Mauritius, Reunion, and others. Unknown if there is one or more spawning grounds in Indian Ocean.

Habitat. Mostly found in lowland and coastal areas, but many individuals migrate far upstream. Most abundant inland eel in India. Juveniles found in habitats with coarse gravel and rocks, often in fast-flowing waters.

Biology. Catadromous. Adults reach maturity at 5.5–10.5 years and die after spawning. A seaward spawning migration in April and September (Malaysia) suggests a very long spawning season of at least 6 months. Very few leptocephali of this species have been found. Juveniles feed on insects and other aquatic invertebrates, while adults feed on fish and crabs. Considered to be an "unspoiled rare ornamental species of Himalayan drainage," often kept in temple ponds for religious purposes in India.

Conservation status. NT; although there is little information on population trends, it is believed that range and abundance are declining worldwide due to dam construction. It is also used as a food source, but despite its wide distribution, there is very little species-specific information on its use and trade, as it is usually lumped together with other eels in catch statistics. Overfishing is a major concern. In some parts of India, it contains levels of contaminants that prove toxic to humans.

Remarks. Often, two subspecies are distinguished: *Anguilla b. bengalensis* in Indian Ocean and *Anguilla b. labiata* in Eastern Africa and the Arabian Peninsula. Both subspecies are treated here as conspecific, but further research is needed.

Further reading. Talwar & Jhingran 1992 (biology, identification); Attaala & Rubaia 2005 (Record from Yemen); Crook & Nakamura 2013 (trade); Jacoby et al. 2015 (threats, conservation); Shiraishi & Crook 2015 (trade); Arai & Abdul Kadir 2017a (distribution, habitats); Arai & Arai Kadir 2017b (spawning); Chai & Arai 2018 (age at maturation); Pike et al. 2020b (biology, distribution, conservation status).

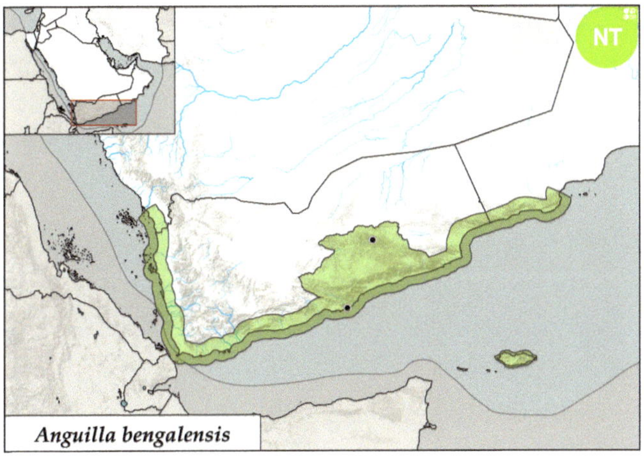

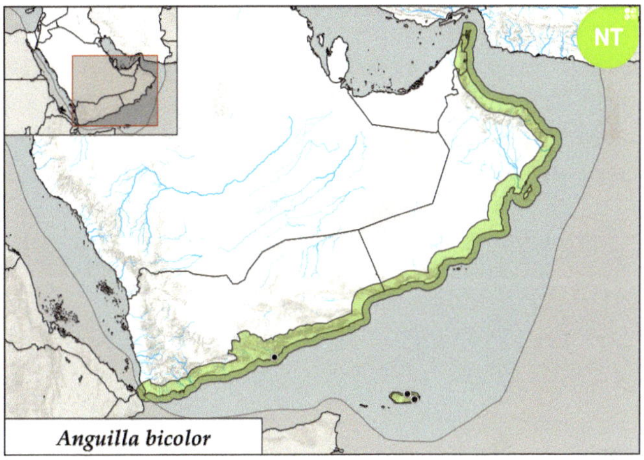

Anguilla bicolor; Thailand; ~400 mm TL. shutterstock_1595226994 dwi putra stock.

Anguilla bicolor
Common name. Shortfin Eel
Diagnosis. Distinguished from *A. bengalensis* by: • dorsal origin slightly in front of anus / • plain olive yellowish to dark-brown or black, silvery grey from jaw to anus. Size usually about 500–600 mm, rarely up to 1200 mm TL.
Distribution. In Arabian Peninsula, recorded from Yemen along Arabian Sea and Gulf of Aden but may occur elsewhere in coastal waters. Widespread in Indian Ocean from east coast of Africa to north-west Australia and Greater Sundaland.
Habitat. Mostly found in lowland and coastal areas, inhabiting estuaries, brackish waters, freshwater lakes, streams, ponds, and small rivers. In Malaysia, 75 % of adults have been found in freshwater, 20 % in coastal seawater, and 5 % in brackish water. Spawns in deep tropical and subtropical oceans.
Biology. Catadromous. Arabian eels not yet been extensively studied. Live up to 20 years, adults die after spawning. Very little known about marine phase. Leptocephali have never been found, and more information on spawning areas or early life history is needed. Believed to spawn during a broad, prolonged season (data from Sumatra). Migrates downstream in central Vietnam during storm season, October–December. Feeds on small fish, crustaceans, and molluscs.
Conservation status. NT; due to decline in availability of both *A. japonica* and *A. anguilla* (species traditionally used for aquaculture and consumption), *A. bicolor* is next preferred species for plain eels, driving up demand for this species in East Asia. Although there are very little data available to estimate population changes, it is clear that number of glass eels exported worldwide is increasing rapidly to meet demand. *Anguilla bicolor* can be purchased online from many suppliers. If catches continue to increase at current rates, exploitation is likely to pose a significant threat to populations worldwide, and shortfin eel may soon be listed as an endangered species.
Remarks. Molecular data suggest that the Indo-West Pacific *A. bicolor pacifica* (from southern China, the Philippines, and the Indonesian islands of Borneo, Sulawesi, and New Guinea) may represent its own conservation unit or an independent species.
Further reading. Talwar & Jhingran 1992 (biology, identification); Pethiyagoda 1991 (habitat); Chino & Arai 2010 (occurrence in Indonesia); Crook & Nakamura 2013 (trade); Shiraishi & Crook 2015 (trade); Pike et al. 2020c (biology, distribution, conservation status).

Wadi Hadhramaut in Yemen, is the habitat of *Anguilla bengalensis* and several endemic species.

Order Clupeiformes

Herrings

A large group of pelagic fishes living in all seas except the Antarctic. Clupeiformes is one of the most important fish orders in all open-ocean ecosystems. It plays an important role in fisheries. Most species are marine, but many enter estuaries to forage or spawn. Some are anadromous or permanent freshwater inhabitants. Clupeiformes are unmistakably distinct from other freshwater fish in the area, identified by the presence of a ventral keel with scute-like scales forming a toothed edge when viewed in profile, a large fleshy eyelid, no lateral line, and no adipose fin. The order has recently been reviewed, and a new family structure was proposed, which is followed here. Shads of the genus *Alosa* are now placed in their own family, Alosidae. *Clupeonella* is placed in Ehiravidae, *Nematalosa* and *Tenualosa* are in Dorosomatidae. The earliest fossil clupeiform is the pan-clupeoid *Cynoclupea* from the Barremian-Aptian (129.4–113.0 million years ago) in the Cretaceous of Brazil. **Further reading.** Whitehead 1985 (diversity); Hoestlandt 1991a (morphology); Nelson et al. 2016 (diversity); Wang et al. 2022 (phylogeny); Near & Thacker 2024 (earliest fossil).

Key to genera of Clupeiformes in freshwaters of West Asia

1a - Snout pig-like, protruding mouth; lower jaw very long, its articulation extending behind posterior eye margin, usually extending to a point well beyond, almost to vertical through posterior eye margin.
.................*Thrissina*
1b - Snout not protruding mouth; articulation of lower jaw not extending behind posterior eye margin.
.................2

2a - Last dorsal ray elongate, filamentous, longer than first branched dorsal ray; mouth subterminal.
.................*Nematalosa*
2b - Last dorsal ray, shorter than first branched dorsal ray; mouth terminal or superior.
.................3

3a - Last two anal rays longer than preceding rays and paddle-shaped; upper jaw without median notch; no black lateral spots; no elongated scales at base of caudal lobes.
.................*Clupeonella*
3b - Last two anal rays of same length as preceding rays; upper jaw with median notch; black lateral spots present in most individuals; elongated scales at base of caudal lobes.
.................4

4a - Scales on caudal base not elongated.
.................*Tenualosa*
4b - Two elongated scales on caudal base.
.................*Alosa*

Lake Apolyont in Türkiye was known for landlocked shads. These seem to have vanished from all lakes in West Asia.

https://doi.org/10.1515/9783111677811-006

Family Alosidae

True shads

Four genera are placed in this family: *Alosa*, *Brevoortia*, *Sardina*, and *Sardinops*. Only *Alosa* enters freshwaters in West Asia. *Sardina pilchardus* is a common species in the Mediterranean and Black Seas and is occasionally found in estuaries. It is not known to enter freshwater habitats regularly and is therefore excluded from the coverage of this book. *Alosa* are anadromous or form landlocked populations in the Atlantic, the Mediterranean, Caspian, and the Black Sea basins. Six species occur in North America. In the Caspian basin, three species, *A. braschnikowii*, *A. saposchnikowii*, and *A. sphaerocephala*, are fully marine and do not enter freshwater habitats. Caspian *A. caspia* and Black Sea *A. maeotica* regularly occur in coastal regions of Limans and river estuaries but only regionally and rarely enter freshwater regions to spawn. Only Caspian *A. kessleri* and *A. volgensis*, as well as Black Sea *A. immaculata*, are anadromous and migrate longer distances upriver. Black Sea *A. tanaica* and widespread *A. agone* regularly enter lower regions of rivers and freshened coastal lakes. *Alosa volgensis* from the northern Caspian basin may have occurred in the waters of the southern Caspian Sea. Still, this species was never identified from the region and went extinct in the 20th century.

Shads are very sensitive to environmental impacts such as impoundment, water pollution, and overfishing. Most anadromous species have sharply declined during the 20th century. Species undertaking long upriver migrations, such as *A. immaculata* and *A. kessleri*, are very vulnerable as most of their original spawning sites are no longer accessible, their spawning migrations being interrupted by dams and other alterations in river morphology. The status of species or populations spawning in the sea and lower stretches of rivers is better, as they can often find spawning habitats below dams.

The systematics of shads need critical review, and molecular data suggest that all Ponto-Caspian species are very closely related or frequently have been the victims of introgressive hybridisation. The Ponto-Caspian shads may form a recently evolved "species flock," as it is known from cichlids in East Africa. Several subspecies and "forms" have been described from the Black and Caspian Seas, and their status needs to be revised based on the basis of adequate fresh material and molecular characters. **Further reading.** Berg 1949b (diversity, biology); Whitehead 1985 (diversity); Holčík 1986 (diversity, biology, distribution).

Keys to species of *Alosa* in freshwaters of West Asia

Caspian basin

1a - 59–92 gill rakers, shorter than branchial filaments.
..................*A. kessleri*
1b - 70–180 gill rakers, about as long as or longer than branchial filaments.
..................2

2a - Head length 23–26 % SL; 99–105 gill rakers, about as long as or longer than branchial filaments, with well-developed spines.
..................*A. volgensis*
2b - Head length 25–28 % SL; 70–180 gill rakers, long and close together, 1.5–2 times longer than branchial filaments in individuals larger than 150 mm SL, without spines.
..................*A. caspia*

Mediterranean and Black Sea basin

1a - Teeth on palatine and vomer absent or poorly developed.
..................2
1b - Teeth on palatine and vomer well developed.
..................3

2a - Teeth on palatine and vomer poorly developed; 66–96 gill rakers, longer than branchial filaments.
..................*A. tanaica*
2b - No teeth on palatine; 28–50 gill rakers (rarely up to 60) about as long as branchial filaments in individuals larger than 220 mm SL.
..................*A. agone*

3a - 36–69 gill rakers, longer than branchial filaments.
..................*A. immaculata*
3b - 29–39 gill rakers, about as long as branchial filaments.
..................*A. maeotica*

https://doi.org/10.1515/9783111677811-007

Alosa agone; lower Elbe, Germany; ~200 mm SL.

Alosa agone

Common name. Twaite shad.

Diagnosis. Distinguished from other species of *Alosa* entering freshwater in Black Sea and Mediterranean basins by: ○ 28–50 gill rakers, rarely up to 60 / ○ length of gill rakers equal to length of branchial filaments in individuals longer than 220 mm SL / ○ no teeth on palatine / ○ dorsal profile curved / ○ usually a series of 4–8 black blotches behind gill opening (sometimes a single blotch). Size up to 500 mm SL.

Distribution. Mediterranean basin and rarely in northern Black Sea, occasionally east to Crimea. Southern Baltic east to Nemunas in Lithuania, North Sea north to Bergen, Atlantic coasts from Scotland and Ireland to Morocco. Before dams migrated from Drin to Lake Ohrid.

Habitat. At sea, pelagic. Juveniles stay close to shore and estuaries. Migrates from sea to rivers, spawning in main rivers often only a few kilometers above brackish water limit but used to ascend Rhône for 600 km and Nemunas for 400 km before construction of dams. In West Asia only known from lowermost rivers. Spawns in large rivers. Also reported in small rivers above gravel beds.

Biology. Anadromous. Males migrate upstream at 2–3 years, females at 3–4 years. Many individuals spawn for 3–4 seasons. Adults congregate near estuaries in April and enter rivers when temperatures reach 10–12°C, mainly in May–June. Spawning begins when temperatures reach about 15°C or higher in May–June. Spawns in large, very noisy schools near surface after midnight. Eggs sink to bottom or are pelagic. Spent fish migrate back to sea. Most juveniles migrate to estuary during their first summer and move to sea at end of second year, where most remain until maturity. Individual fish are thought to return to their natal spawning grounds. Northern distribution appears to be limited by water temperature. At sea, feeds mainly on crustaceans and small fish. Adults do not feed in freshwater. Juveniles feed on planktonic crustaceans.

Conservation status. LC; very locally distributed, a victim of pollution and damming of major rivers. Most populations declined in early 20th century but appeared to have stabilised at low levels since then.

Remarks. Up to seven subspecies of *A. fallax* have been recognised, several of which were treated as species until recently. Molecular studies using a variety of markers have identified two major groups of populations. One is widespread in northern Europe, the Atlantic, and the westernmost Mediterranean basin (*A. fallax*), and a second is found in the Ebro (Spain) across the northern Mediterranean basin to the east (*A. agone*). This is the one that is also found in West Asia. The molecular differences are very small, comparable to intraspecific differences in many other species. In addition, morphological differences are vague and focus on average gill raker numbers, which overlap widely between different "species". We treat all populations of former *A. fallax* as *A. agone*, which has priority over *A. fallax*, which becomes a junior synonym.

Further reading. Quignard & Douchement 1991 (biology); Kottelat 1997 (systematics); Bianco 2002 (biology).

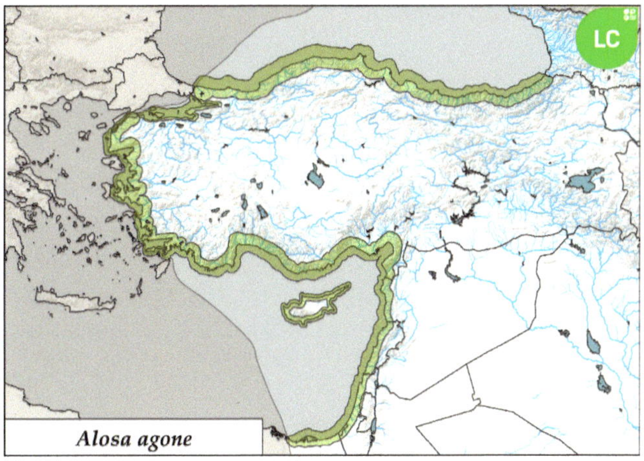

Alosa caspia; Caspian Sea, Iran; ~200 mm SL.

Alosa caspia

Common name. Caspian shad.

Diagnosis. Distinguished from other species of *Alosa* in Caspian basin by: • head length 25–28 % SL / ○ eye diameter 19–24 % HL / ○ pectoral length 16–18 % SL / ○ teeth on palatine and vomer poorly developed / ○ 70–180 gill rakers, long, and close together, 1.5–2 times longer than branchial filaments in individuals longer than 150 mm SL. Size up to 280 mm SL.

Distribution. Caspian Sea, from where adults occasionally ascend the lowermost part of rivers and migrate a short distance upstream to spawn.

Habitat. At sea, pelagic, in coastal waters with steady currents, avoiding areas of stable salinity. Migrates from sea to mouths of large rivers, spawns in fresh or slightly brackish water in shallow areas washed by large river currents.

Biology. Anadromous. Males migrate upstream at 2–3 years, females at 4–5 years. Many individuals spawn in 2–4 seasons. Two migration peaks, one in late April (mostly males) and one in early May (mostly females), enter rivers when temperature rises above 10°C. Spawns when temperature reaches 15°C or more, in May–June. Spawns usually in upper 3 m. Eggs semipelagic and demersal. Spent fish return to sea. Juveniles migrate to sea during their first summer and remain at sea until maturity. At sea feeds mainly on zooplankton such as copepods and mysids.

Conservation status. LC; widespread.

Further reading. Heckman 1991a (biology).

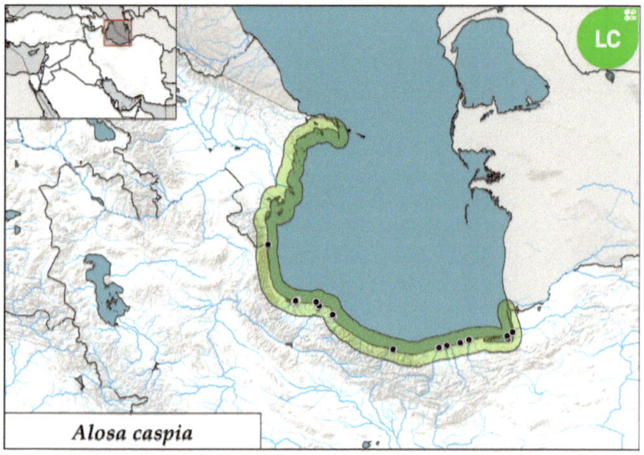

Alosa immaculata; South Bug, Ukraine; 217 mm SL.

Alosa immaculata

Common name. Large Black Sea shad.

Diagnosis. Distinguished from other species of *Alosa* entering freshwater in Black Sea basin by: ● 36–69 gill rakers, longer than branchial filaments / ○ teeth on palatine and vomer well developed. Size up to 350 mm SL.

Distribution. Black and Azov Seas, from where adults ascend rivers and migrate far upstream to spawn. Presence in Marmara basin questionable. Present in Rioni, but distribution in northern Anatolia poorly known. Used to ascend Danube for about 1600 km to Mohacs (Hungary), Don for 900 km to Oust-Medveditskaya, Dniepr to Kyiv, Dniester to Yampol, and South Bug to Voznessensks. Reported to spawn in Kuban below Krasnodar dam since 1983. A landlocked population occurs in Don.

Habitat. At sea, pelagic, in deep water. Migrates to middle reaches of large rivers, spawning where current is strongest, near surface, usually at 2–3 m depth in main channel.

Biology. Anadromous, migrating upstream to spawn at three years, rarely earlier. Few individuals spawn two seasons. Appears along coast in March–April, enters rivers when temperatures reach about 6–9°C between late March and late April. Migration usually peaks in May. Spawning begins when temperatures rise above 15°C in April–August. Spawns usually in afternoon (1–8 pm), eggs pelagic. Spent individuals return to sea to feed. Juveniles inhabit floodplain and shallow riverine habitats, migrating to sea or estuarine habitats during first summer; move to sea in autumn and remain there until maturity. At sea, it feeds on a wide variety of zooplankton (especially crustaceans) and small fish.

Conservation status. VU; damming of major rivers significantly reduced available spawning habitat and migration routes in 20th century. Decline appears to be continuing due to severe overfishing. Largest population in Danube.

Further reading. Kolarov 1991b (biology; as *A. pontica*); Kottelat 1997 (systematics).

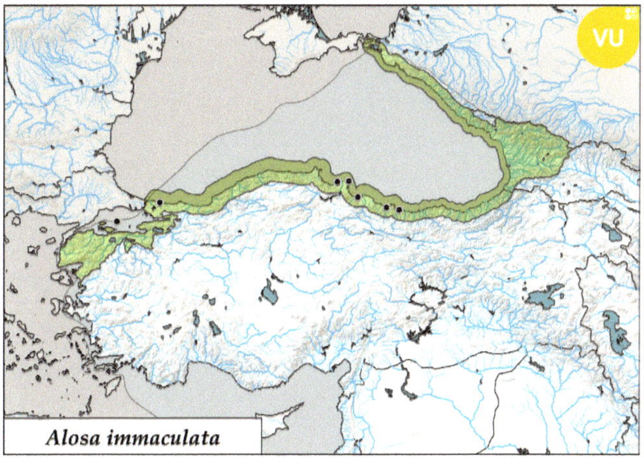

Alosa kessleri

Common name. Black-back shad.

Diagnosis. Distinguished from other species of *Alosa* in Caspian basin by: ● dorsum black in spring / ● teeth on palatine and vomer well developed / ○ 59–92 thick gill rakers, shorter than gill filaments, with conspicuous lateral spines in juveniles but lost in adults / ○ head length 22–25 % SL / ○ eye diameter 15–20 % HL / ○ pectoral length 14–16 % SL. Size up to 400 mm SL.

Distribution. Caspian Sea from where adults ascend Kura and Aras (Azerbaijan) and Sefid (Iran). Possibly also in other rivers. In European Caspian basin, in Volga, few enter Ural and Terek. Previously migrated Volga upstream to Kama and Oka systems. Migration now blocked by Volgograd Dam. There is evidence that it has formed landlocked populations in Volga reservoirs.

Alosa kessleri; Volga delta, Russia; ~300 mm SL.

Habitat. At sea, pelagic, in a wide variety of habitats. Migrates to middle reaches of large rivers, spawning near banks in main channel and in almost still waters such as river bays, eddies, and floodplains.

Biology. Anadromous. Migrates upstream to spawn at 4–5 years. Enters rivers with immature gonads. Some spawn for 2–4 seasons, but most females die after spawning. Spawners appear along coast in March–April and enter rivers in April–May when temperatures reach about 9°C, peaking at 12–15°C. Spawning run originally lasted 30–50 days. Spawning begins in May–August when temperatures rise above 15°C and continue as long as temperatures remain at 15–23°C. Spawning is most intense between 4 and 10 pm. Eggs are bathypelagic. Spent fish return to sea to feed. In autumn, move to southern part of sea to overwinter. Juveniles migrate to sea or estuaries during their first summer until maturity. At sea, feeds on a wide variety of zooplankton, crustaceans, and small fish.

Conservation status. LC; damming of major rivers in 20th century has significantly reduced available spawning habitat and migration routes and heavily fished, especially on Volga, where largest population exists. Populations in West Asia small, most abundant in Kura and Sefid.

Further reading. Heckman 1991b (biology).

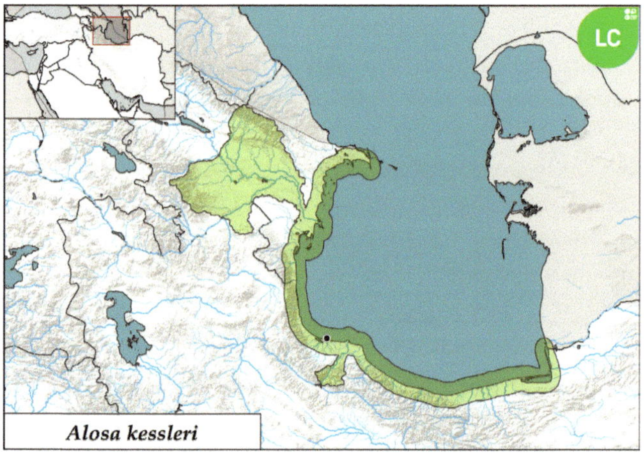

Alosa maeotica; Black Sea, Romania; ~250 mm SL.

Alosa maeotica

Common name. Azov shad.

Diagnosis. Distinguished from other species of *Alosa* entering freshwater in Black Sea basin by: ● 29–39 gill rakers, about as long as branchial filaments / ○ teeth on palatine and vomer well developed. Size up to 280 mm SL.

Distribution. Black Sea and Sea of Azov, from where adults enter coastal lagoons to spawn.

Habitat. At sea, pelagic in deep waters, entering brackish lagoons to spawn.

Biology. Spawns first time at 2 years. Many individuals spawn in 2–4 seasons. In Romania, spawners enter Lake Razelm at about 7°C in March. Spawning begins at about 15°C in late April–June. Eggs sink to bottom. Spent fish return to sea to feed. In autumn, migrates to southern Black Sea to overwinter. Juveniles migrate to sea or remain in estuaries during first summer. Feeds mainly on small fish and crustaceans.

Conservation status. LC; still common in coastal lagoons.

Further reading. Kolarov 1991a (biology).

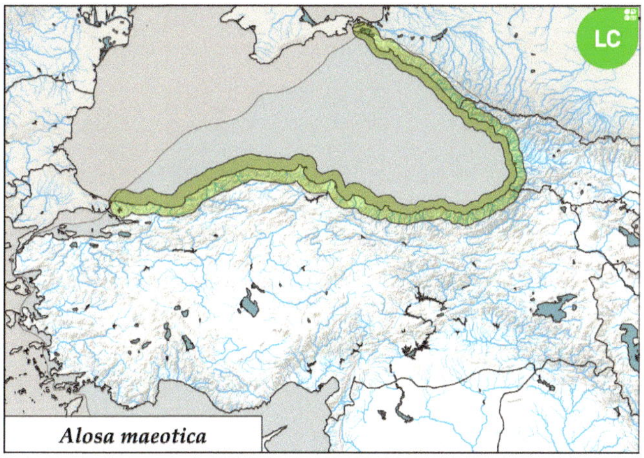

Alosa tanaica; Lake Razelm, Romania; ~160 mm SL.

Alosa tanaica

Common name. Little Black Sea shad.

Diagnosis. Distinguished from other species of *Alosa* entering freshwater in Black Sea basin by: ● 66–96 gill rakers, longer than branchial filaments / ● teeth on palatine and vomer poorly developed. Size up to 200 mm SL.

Distribution. Black and Azov Seas, from where adults ascend rivers and migrate a short distance upstream to spawn. Inland populations in Anatolian lakes of Sapanca, Apolyont, and Manyas.

Habitat. At sea, pelagic, in deeper layers (50–70 m) of coastal waters. Migrates from sea to estuaries and lower reaches of large rivers, spawns in fresh or slightly brackish water, usually close to shore, in upper 2–4 m, in almost still waters such as floodplains or lakes.

Biology. Anadromous. Migrates upstream at 1–2 years. Many individuals spawn in 2–4 seasons. Spawners appear along coast in late January–March, enter rivers when temperature reaches about 10°C, late April–May. Spawning begins when temperature reaches about 15°C or higher in May–June. Eggs are bathypelagic or sink to bottom. Fry migrate to estuaries and coastal lagoons or to sea near estuaries to feed. In autumn, move to sea to overwinter. Juveniles migrate to sea or estuaries during first summer until they reach maturity. At sea, feed on a wide variety of zooplankton (crustaceans) and small fish.

Conservation status. LC; most populations declined in early decades of 20th century due to pollution. Seems to have stabilised at moderate levels since then. Situation in West Asia needs to be better known.

Remarks. Landlocked populations in Türkiye have not been studied for many years and might have vanished. They need a critical review.
Further reading. Mogil'chenko 1991 (biology); Kottelat 1997 (systematics).

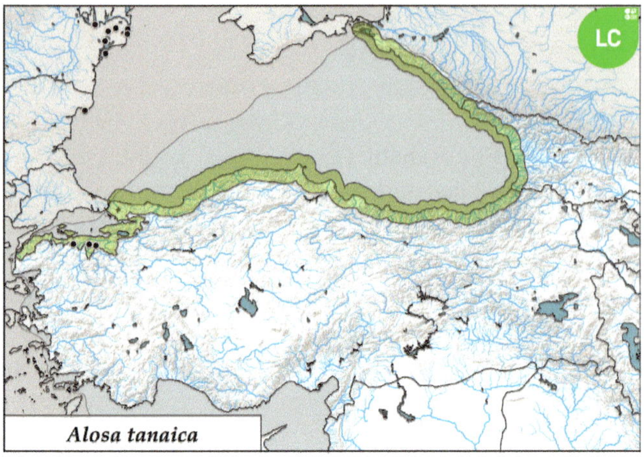

Clupeiform fishes support small scale fisheres all over their range.

Family Dorosomatidae

Gizzard shads

A large family of about 115 species in 30 genera. They are mostly tropical and subtropical shads, many of them found in coastal waters, entering freshwaters during parts of their life, or even being pure freshwater species such as African *Congothrissa*, *Limnothrissa*, and *Sierrathrissa*. Many species are of high commercial value, especially for artisanal fisheries. Two genera, *Nematalosa* and *Tenualosa*, both with a wide distribution range in the Indian Ocean, enter freshwaters in West Asia. *Nematalosa nasus* and *Tenualosa ilisha* are reported from the lower parts of the Shatt al Arab/Arvand. Only *T. ilisha* regularly enters freshwater habitats, while *N. nasus* is restricted to brackish waters. Both species are included in the identification key of Clupeiformes above.

Tenualosa comprises five species distributed in the subtropical and tropical Indo-Pacific. All species are anadromous or landlocked, with rare reports of resident marine stocks. One species, *T. thibaudeaui*, is a pure freshwater species endemic to the Mekong. All *Tenualosa* are highly valued as food fish and provide a livelihood for many, if not millions, of small-scale fisher families. Overfishing is a serious issue throughout their range. However, as this is due mostly to artisanal fisheries, catches are rarely reported accurately, and the extent of the decline of all species might be much underestimated.

The scientific and common names of *T. ilisha* require some clarification. This species is widely called *Hilsa*, which is confused with the Kelee shad, *Hilsa kelee*. Kelee and Hilsa are sympatric in the Gulf of Oman and along the southern coast of the Arabian Peninsula. Kelee enters estuaries, and both species are easily distinguished by the presence of prominent striae on the top of the head in *H. kelee* (vs. absence in *T. ilisha*). **Further reading.** Whitehead 1985 (diversity).

Nematalosa nasus; Basra, Iraq, ~150 mm SL. © A. Ali.

Tenualosa ilisha

Common name. Hilsa.

Diagnosis. Distinguished from *Nematalosa nasus* in lower reaches of rivers in Persian Gulf basin by: • last dorsal ray shorter than first branched dorsal ray / • upper jaw with median notch. Size up to 600 mm SL, usually about 360 mm SL, and up to 2.5 kg.

Distribution. From Persian Gulf along coasts of Indian subcontinent east to Irrawaddy and Salween (Myanmar). In West Asia, enters Karun and Shatt al Arab/Arvand to spawn upstream in Euphrates and Tigris. In Tigris, migrates as far north as Qal'at Salih, and in Euphrates, as far north as Yaou. Local freshwater populations in many large rivers in South Asia.

Habitat. At sea in coastal waters, migrates to lower and middle reaches of large rivers to spawn. A resident population is suspected in Kuwait Bay, which does not appear to migrate upstream and spawns in sea or in freshwater fan of Shatt al Arab/Arvand.

Biology. Anadromous, usually migrating 50–100 km upstream. Long-distance migratory populations in Pakistan (Indus), Bangladesh/India (Ganges), and Myanmar

https://doi.org/10.1515/9783111677811-008

Tenualosa ilisha; Basra, Iraq; ~250 mm SL. © A. Ali.

(Irrawaddy). In Ganges reported to migrate 1300 km upstream, but freshwater resident populations also occur in upper Ganges. Migrates upstream to spawn at 2 years and 250 mm (males) and 3 years and about 330 mm (females). Lives up to 7 years, usually 4–5 years. First spawner appears at Shatt al Arab/Arvand in February and March at high tide and spends 1–2 months feeding. Continues to migrate upstream to spawn from April–July. Does not feed during migration. Reported to travel up to 70 km in a day. Males usually ascend rivers earlier than females. Two spawning peaks reported, one in March–May and one in July–August, but data are conflicting. In South Asia, main spawning season during southwest monsoon, with a shorter season from January to February or March. Spawns on submerged vegetation in fast currents. Larvae emerge in 1 day (23°C). Larvae and juveniles enter backwaters and tributaries of rivers. Spent fish return to sea between August and November and migrate south to overwinter. Feeds on a wide variety of small phytoplankton and zooplankton.

Conservation status. VU; most important Indo-Pacific shad for commercial and artisanal fisheries, particularly in Gulf of Bengal. Also, one of the most important commercial species in Kuwait. Despite its wide distribution, Hilsa has declined sharply throughout its range, including Bangladesh, where world's largest population occurs. Overfishing, pollution, and dam construction have led to massive population declines. Often highly contaminated with pollutants. Northernmost distribution in Iraq today is Hawr al Hammar due to dam construction.

Further reading. Whitehead 1985 (biology, distribution); Coad 2010a (biology in Persian Gulf basin).

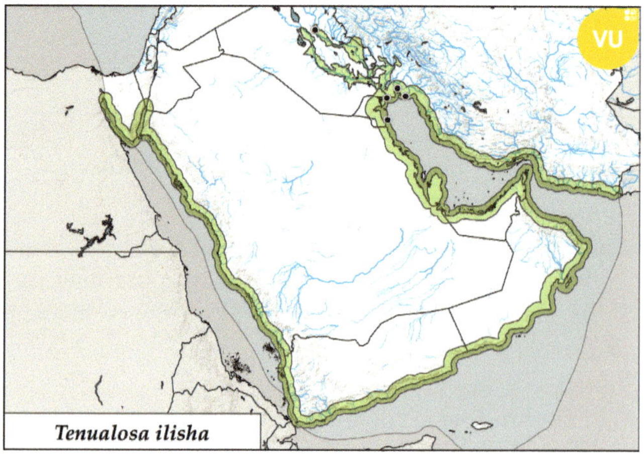

Family Ehiravidae

Dwarf sprats

A small family of about 27 species in 11 genera. They are mostly small, tropical, and subtropical herrings, mostly found in brackish and freshwaters; few are of commercial importance. Ehiravidae includes several pure freshwater species, such as the Southeast Asian genus *Sundasalanx*, paedomorphic fishes reaching up to 30 mm SL. In West Asia, the only genus entering freshwaters is *Clupeonella* (tyulkas). They are a small group of brackish water species endemic to the Marmara, Black, and Caspian Sea basins where they frequently inhabit coastal lagoons, backwaters, and the slow-flowing sections of lower rivers. One species, *C. abrau* from Lake Abrau in Russia, is endemic to a small lake close to the coast. In freshwaters of West Asia, we recognise one species, *C. cultriventris*. Two additional tyulkas, *C. engrauliformis* and *C. grimmi*, are endemic to the Caspian Sea, both strictly marine and not reported to enter freshwater habitats. *Clupeonella tscharchalensis* is a freshwater species that inhabits rivers in the northern Caspian basin. It is described from Lake Chelkar (47°50′N 59°36′E) in Kazakhstan. Until further studies are conducted on tyulkas from Lake Chelkar, we classify *C. tscharchalensis* as a valid species with fewer gill rakers than *C. cultriventris*. **Further reading.** Hoestlandt 1991a (diversity, distribution, biology); Kottelat & Freyhof 2007 (diversity).

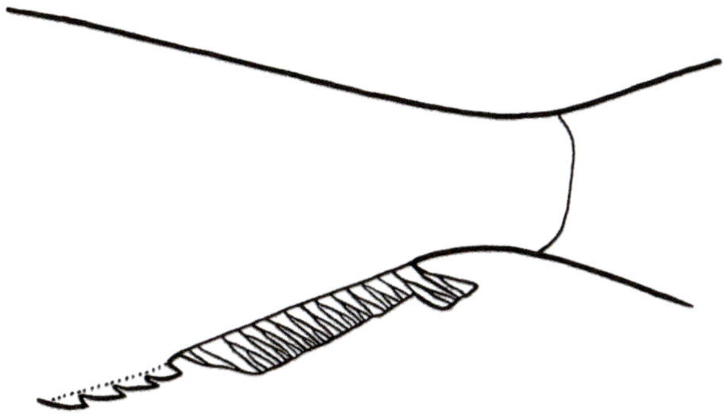

Figure 30. Paddle-shaped last two anal rays and scute-like scales in front of anal fin in *Clupeonella* (from Kottelat & Freyhof 2007, after Whitehead 1985).

Key to species of *Clupeonella* in West Asia

1a - Pectoral tip rounded; interorbital distance 13–15 % SL (42–51 gill rakers).
..................*C. grimmi*
1b - Pectoral tip pointed; interorbital distance 16–18 % SL.
..................2

2a - Ventral keel soft; body depth 16–19 % SL; 56–67 gill rakers.
..................*C. engrauliformis*
2b - Ventral keel hard; body depth 18–25 % SL; 49–62 gill rakers.
.... *C. cultriventris*

Clupeonella grimmi, Caspian Sea, Iran; 100 mm SL.

https://doi.org/10.1515/9783111677811-009

Clupeonella cultriventris; Lower South Bug, Ukraine; ~75 mm SL.

Clupeonella cultriventris

Common name. Common tyulka.

Diagnosis. Distinguished from other species of *Clupeonella* in Caspian basin by: ● 49–62 gill rakers / ● ventral keel hard, scratching human skin / ○ body depth 18–25 % SL / ○ tip of pectoral pointed / ○ interorbital distance 16–18 % SL. Size up to 120 mm SL.

Distribution. Caspian, Marmara, Azov, and Black Sea basin. Lakes Sapanca and Apolyont. Caspian Sea and Lower reaches of rivers as Volga, Ural, and possibly Terek.

Habitat. Euryhaline, in coastal waters, lagoons and lakes, estuaries, and lower reaches of large rivers with salinity up to 13 ‰. At sea, pelagic. Spawns near estuaries.

Biology. Pelagic. Marine, landlocked, or anadromous. Lives up to 6 years. Spawns in open water late in evening in April–July at 10–25°C. Eggs pelagic, usually laid in 5 m deep with at a salinity of 0.02–15 ‰. Two spawning runs in Azov Sea, one in spring and one in autumn. Individuals migrating in autumn overwinter near spawning grounds and spawn in spring. Return to sea after spawning. Juveniles migrate to sea during first summer. Feeds on zooplanktonic crustaceans (copepods, mysids, cladocerans), following their daily vertical movements in surface layers at night, descending during day.

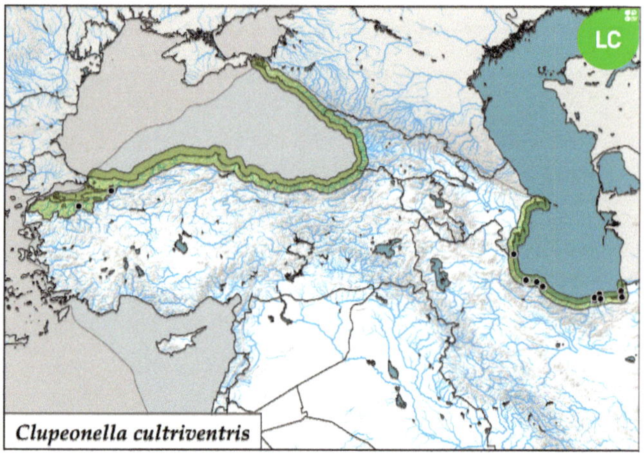

Conservation status. LC; very widespread and often abundant.

Remarks. *Clupeonella muhlisi* from Lake Apolyont in Western Anatolia and Caspian *C. caspia* were treated as separate species. We could not find diagnostic morphological or molecular characters to distinguish these populations from *C. cultriventris.* Therefore, they are treated as synonyms.

Further reading. Hoestlandt 1991b (biology); Kottelat 1997 (systematics).

Family Engraulidae

Anchovies

Anchovies are one of the most important families of marine fishes from an ecological and commercial perspective. Most are marine, but some tolerate low salinities, and several species migrate into rivers to spawn in brackish or freshwaters. Few are confined to freshwater habitats. Most anchovies are plankton filter feeders and are major prey species in the marine food chains. They are usually small (mostly 80–150 mm SL), and many species school in such numbers that they form the basis of some of the world's most important fisheries. The Peruvian anchoveta *Engraulis ringens* has yielded greater catches than any other single wild fish species worldwide, with annual harvests varying between 4.2 and 8.3 million tonnes (2008–2012), almost all used for making fishmeal. Often, catches in Mediterranean and Black Seas are dominated by anchovy *Engraulis encrasicolus* (200,000 to 393,500 tonnes in the Black Sea). This species accounts for almost all of the commercial catches. Most of the harvest is used for the fishmeal industry, primarily to feed fish in aquaculture. Superficially similar silversides (Atherinidae) have two dorsals (vs. one in anchovies). *Thrissina mystica* and *T. whiteheadi* have been recorded in the Persian Gulf from the Shatt al Arab and the Hammar Marsh. The genus *Thrissina* is restricted to the Indo-Pacific, where 38 species are actually recognised. *Thrissina* can be distinguished from other anchovies accidentally entering freshwaters and brackish estuaries by the upper pectoral ray not extending into a long filament (vs. long pectoral filament present in *Setipinna*), 31–45 branched anal rays (less than 25 in *Encrasicholina* and *Stolephorus*) and the anal isolated from the caudal (vs. joined in *Coilia*). The taxonomy of *Thrissina* is not yet settled, and the thryssas of the northern Persian Gulf are only tentatively identified. Other species might occur and should be identified using the key provided. **Further reading.** Carpenter et al. 1997 (identification, distribution); Hata et al. 2020 (T. serena); Hata et al. 2023 (as *T. malabarica*); Hata et al. 2025 (*Thrissina mystica*). Hata & Motomura 2020 (*T. serena*). Many thanks to Harutaka Hata (Japan) for reviewing this chapter.

Key to species of *Thrissina* in West Asia

1a - Maxilla short, not reaching to edge of gill cover; scutes covering abdomen less than 18.
..................*T. baelama*
1b - Maxilla long, reaching to edge of gill cover; scutes covering abdomen 20 or more.
..................2

2a - Maxilla reaching posteriorly beyond pectoral-fin insertion.
..................3
2b - Maxilla not reaching to pectoral-fin insertion.
..................4

3a - Maxillary reaching much beyond pectoral-fin base; 17–19 gill rakers on lower limb of first gill arch.
..................*T. dussumieri*
3b - Maxilla reaching to pectoral-fin base; 21–23 gill rakers on lower limb of first gill arch.
..................*T. serena*

4a - 25–26 scutes covering abdomen; 18–21 gill rakers on lower limb of first gill arch; 39–43 branched anal rays.
..................*T. whiteheadi*
4b - 26–30 scutes covering abdomen; 13–16 gill rakers on lower limb of first gill arch, 34–40 branched anal rays.
..................*T. mystica*

https://doi.org/10.1515/9783111677811-010

Thrissina mystica; Kuwait Bay; 142 mm SL. © J. Randall, Bishop Museum.

Thrissina mystica

Common name. Balam.

Diagnosis. Distinguished from *T. whiteheadi* by: • 3–16 gill rakers on lower limb of first gill arch / • 34–40 branched anal rays. Size up to 270 mm SL.

Distribution. From Persian Gulf east to south India. In Shatt al Arab/Arvand, Hammar Marsh, and up to Nasiriyah on Euphrates.

Habitat. Pelagic in coastal waters, entering freshwaters.

Biology. Spring visit to Hammar Marsh and presence of juveniles suggest spawning in brackish waters. Two spawning peaks, one in December–January and one in March–April. Feeds on a wide variety of planktonic invertebrates, fish larvae, and small fish.

Conservation status. LC.

Remark. Usually identified as *T. hamiltonii*, or as *T. malabarica. Thrissina malabarica* occurs in East Asia.

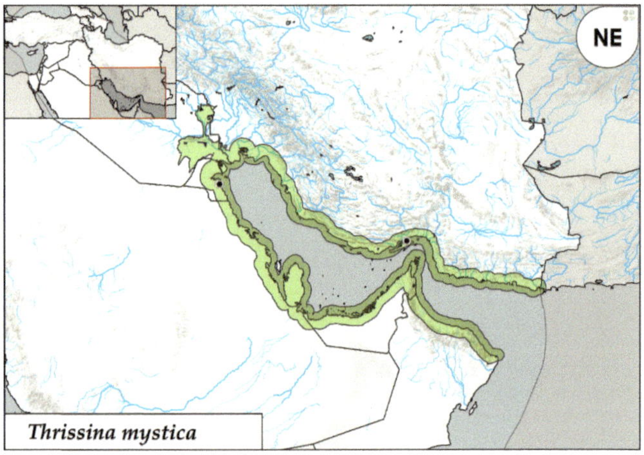

Further reading. Carpenter et al. 1997 (identification, distribution); Hata et al. 2021 (redescription). Hata et al. 2023 (*T. malabarica*); Hata et al. 2023 (as *T. malabarica*); Hata et al. 2025 (*Thrissina mystica*).

Thrissina whiteheadi; Dammam, Saudi Arabia; 205 mm SL. © J. Randall, Bishop Museum.

Thrissina whiteheadi

Common name. Persian Gulf thryssa.

Diagnosis. Distinguished from *T. mystica* by: • 18–21 gill rakers on lower limb of first gill arch / • 39–43 branched anal rays. Size up to 205 mm SL.

Distribution. Endemic to Persian Gulf. In Shatt al Arab/Arvand, Hammar Marsh and Lake Hawizah, widespread in sea.

Habitat. Pelagic in coastal waters, entering freshwaters.

Biology. No data.

Conservation status. LC

Further reading. Carpenter et al. 1997 (identification, distribution); Hata et al. 2021 (redescription).

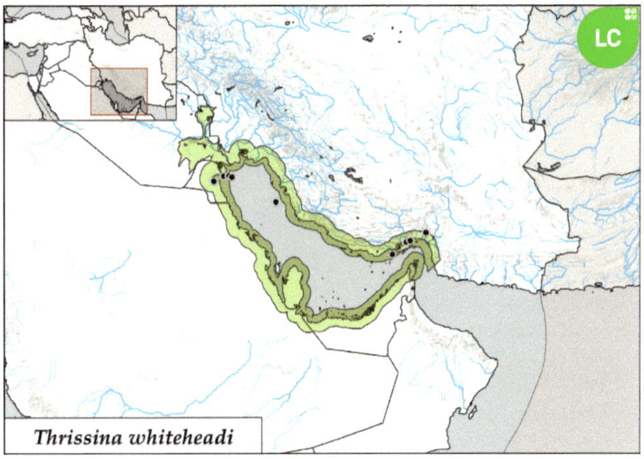

The Caspian Sea is habitat of a great diversity of Clupeiform fishes in West Asia.

Family Chanidae

Milkfish

The milkfish is a living representative of an ancient fish group (Gonorynchiformes or Anotophysi) related to Otophysi fishes, such as Characiformes, Cithariniformes, Cypriniformes, Gymnotiformes, and Siluriformes. Beside the milkfish, Gonorynchiformes only include Gonorynchidae and Kneridae. The milkfish is easily identified by its compressed and oblong body, small, toothless, terminal mouth, deeply forked caudal, complete lateral line, and small cycloid scales. Milkfishes have lateral pouches on the posterior part of the branchial chamber, forming an epibranchial organ, which allows them to breathe air. West Asia is the distribution border of the milkfish, where it occurs in the Persian Gulf.

The milkfish is one of the most important aquaculture species in the world. It is a primary consumer of plants and detritus, so it can be produced with much less energy input than most other aquaculture species. Milkfish aquaculture first began around the 12th century in the Philippines. Larvae are collected from rivers and grown in ponds into juveniles, which are marketed fresh, smoked, canned, and frozen. Since the 1980s, broodstocks have been raised and spawned in captivity to produce larvae in hatcheries. No milkfish is kept in aquaculture in West Asia, likely due to low winter temperatures. In Indonesia, Taiwan, and the Philippines, more than a quarter of a million tonnes of milkfish are harvested annually in brackish ponds, contributing roughly 60 % of the total fish production from Southeast Asian aquaculture. **Further reading.** Juario et al. 1984, Lee et al. 1986 (aquaculture); Carpenter et al. 1997 (identification, distribution); Nelson et al. 2016 (phylogenetic position).

Chanos chanos; Nha Trang, Vietnam; about 200 mm SL.

Chanos chanos

Common name. Milkfish.

Diagnosis. Distinguished from superficially similar Clupeids by having four branchiostegal rays (vs. 6–7 in Clupeiformes) and no scutes along the belly (vs. present). Size up to 1850 mm SL and 18.6 kg, usually less than 1500 mm SL and 14 kg.

Distribution. West Asia: Red Sea, Persian Gulf, and Gulf of Oman and along coasts of Arabian Peninsula. May be only a summer visitor in northern Red Sea and Persian Gulf. Widespread in tropical Indo-Pacific but absent from areas affected by cold currents. Distribution coincides with coral reef areas where water is warm (>20°C), clear and shallow. Migrates up to 100 km inland from Shatt al Arab/Arvand. Also reported from Baghu near Bandar-e Abbas, lower Mand, and Shour-Shirin or Tiyab estuaries. Likely to be found at least occasionally in all major estuaries in Persian Gulf basin.

https://doi.org/10.1515/9783111677811-011

Habitat. Marine. Juveniles enter estuaries and lower parts of rivers. Can acclimatise to freshwater. Adults occur near coasts or around islands where reefs are well developed. Spawns in marine waters in sheltered sandy bays less than 30 km from shore.

Biology. Age unknown in wild due to lack of aging method. Lives up to 12 years in ponds, but large wild adults are thought to be much older. Mature at 3–9 years in ponds. A warm water species, preferring water temperatures above 20°C, can survive up to 41°C. Dies at about 12°C. Euryhaline, can survive in fresh, marine, hypersaline waters (0–158 ‰). Very resistant to low dissolved oxygen, high ammonia, nitrite concentrations, crowding, and starvation. Spawning season long near equator becomes progressively shorter with a single peak at higher latitudes. Spawning thought to be triggered by rising spring temperatures (25–30°C) and high tides. Adults migrate offshore from coastal waters after spawning. High fecundity with very small eggs, 0.3–1.0 million eggs per kg body weight, 1.1–1.2 mm in diameter when fertilised. Hatching varies from 20 to 25 hours after fertilization at 26–32°C and 29–34 ‰ salinity. Eggs and ribbon-like larvae, pelagic up to 2–3 weeks. Eggs sink at salinities below 30 ‰. Larvae metamorphose and, at 10–17 mm TL, migrate ashore and settle in coastal wetlands (mangroves, estuaries). After about 4 weeks, most juveniles leave coastal waters. Some remain in estuaries for up to 4 years before returning to sea. Occasionally enters freshwater during juvenile stage. Grows to 200–430 mm TL in a year in wild. Larvae and small juveniles feed on copepods and diatoms. Larger juveniles are mainly herbivorous, feeding on cyanobacteria, algae, detritus, and small invertebrates grazed from substrate. Adults are opportunistic feeders, filtering, grazing, or snatching small prey and taking larger quantities of zooplankton and small fish.

Conservation status. LC; conservation status of wild populations is poorly known. Research is strongly focused on aquaculture aspects.

Further reading. Juario et al. 1984, Lee et al. 1986 (biology).

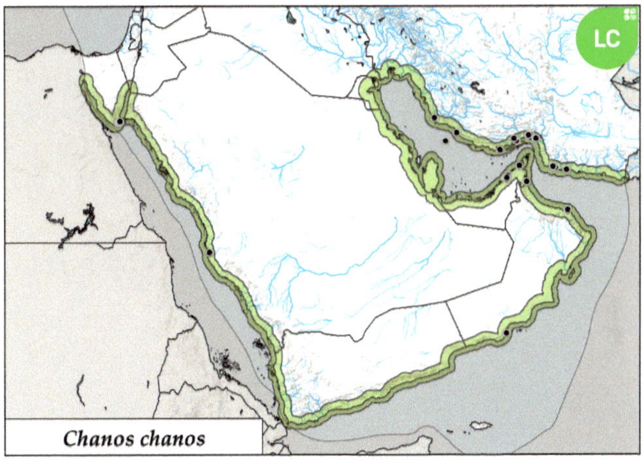

Chanos chanos at Maui, Hawaii. © A. Koharu Uyeoka.

Microbialites in the world's largest saline soda lake, Lake Van © S. Dura

Order Cypriniformes

Carps and loaches

In West Asia, 434 species, 70 % of all freshwater fishes, belong to the order Cypriniformes. Cypriniforms represent the largest group of freshwater fish worldwide, with about 4300 species recognised, though about 10,000 species exist. The earliest fossil is *Jianghanichthys* from China's early Eocene (56–48 million years ago). Bayesian relaxed molecular clock analyses of Cypriniformes result in an average posterior crown age estimate of 97 million years ago, with the credible interval ranging between 85 and 115 million years ago. The origin of Cypriniformes is found in the Late Cretaceous, similar to the evolutionary roots of modern birds, mammals, and frogs, three other species-rich groups of vertebrates. As in modern birds, mammals, and frogs, major diversification events in Cypriniformes have likely started in the Paleogene after the collapse of ecosystems at the Cretaceous–Paleogene boundary.

Until the late 20th century, only two major families were recognised: Cyprinidae for all carps, barbels, and minnows, and Cobitidae for all loaches. Just three small families were excluded: Gyrinocheilidae for three Southeast Asian suckers; Psilorhynchidae for some South Asian torrent minnows; and Catostomidae for one Chinese and many American suckers. Since Cobitidae has been split into 8 families and Cyprinidae into 12 families, not all of which are recognised by all authors, only the position of a few genera, such as *Sundadanio*, *Paedocypris*, *Tanichthys*, *Leptobarbus*, *Tinca*, and *Zacco* and some of their allies remain under discussion. Similarly, the phylogenetic relationships 'within' and 'among' the major groups have proved difficult to resolve across the entire order.

The phylogenetic position of a few genera is irrelevant compared to the benefits in communication associated with the erection of the former subfamilies of Cyprinidae into the range of families. We recognise monophyletic groups as their own families, different from Cyprinidae as Acheilognathidae, Leuciscidae, Danionidae, Gobionidae, Tincidae, and Xenocyprididae. These are the families found within the geographic coverage of this book. **Further reading.** Howes et al.1991; Cavender & Coburn 1992; Arai & Kato 2003; Saitoh et al. 2006; He et al. 2008; Mayden et al. 2008; Yang et al. 2012; Imoto et al. 2013; Liu et al. 2015 (*Jianghanichthys*); Stout et al. 2016 (phylogenetics); Feng et al. 2017 (phylogeny of frogs); Hughes et al. 2018 (molecular estimation of age); Tan & Armbruster 2018 (family placement of genera).

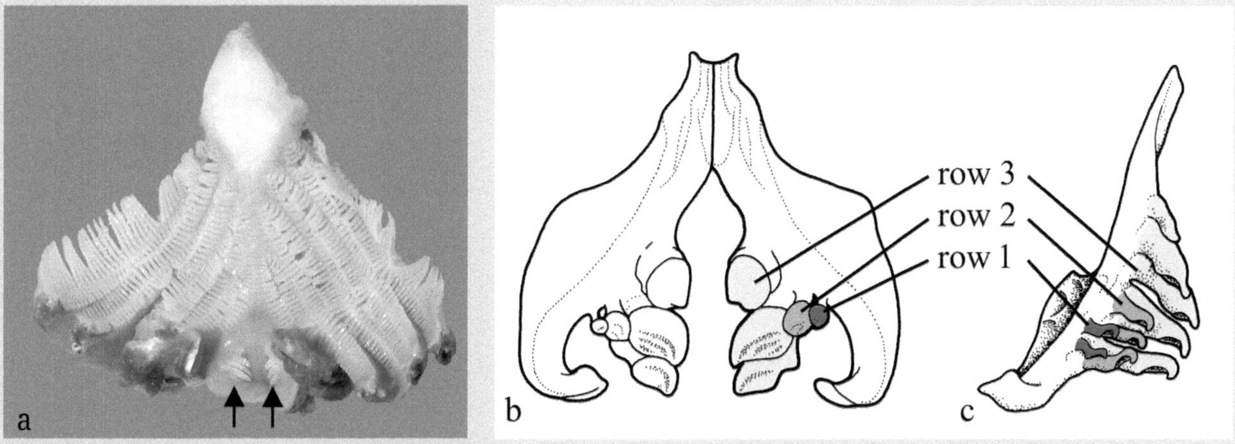

Figure 31. Dorsal view of branchial arches of *Carassius gibelio* **a,** showing four gill arches and one pharyngeal arch on each side; arrows point to pharyngeal teeth. **b,** Dorsal view of pharyngeal arches of *Cyprinus*, showing a way of counting pharyngeal teeth (1,1,3–3,1,1 teeth). **c,** Anteriordorsal view of left pharyngeal arch of *Barbus* (2,3,5–5,3,2 teeth) (from Kottelat & Freyhof 2007, after Holčík 1989).

Pharyngeal teeth in Cypriniformes. The pharyngeal teeth are located on the fifth gill arch or pharyngeal arch (gill arches 1–4 carry gill filaments and gill rakers). Their number, arrangement, and shape are useful characters to distinguish some species. To examine them, it is necessary to remove the pharyngeal arch. This dissection requires some experience and training, magnification, and good light. The standard procedure is to remove a single pharyngeal arch, usually the right one, to leave the other side of the head intact. To extract the pharyngeal arch, the opercle

https://doi.org/10.1515/9783111677811-012

is folded forward, and a small sharp tool (forceps, scalpel) is inserted between the arch and the anterior edge of the cleithrum. The tool is moved dorsally and ventrally along the cleithrum to cut the tissue surrounding the arch. The dorsal end of the bow is released first. The bow is grasped by its dorsal or ventral end and moved back and forth to release it. It should not be grasped in the middle as the tool will damage the bow or the teeth. The pharyngeal arches of small individuals are fragile, but firm handling is required to extract them. Remaining tissue is removed from the arch and between the teeth, either manually or by soaking the arch in dilute potassium hydroxide (KOH) solution (0.5–1.0 %) (this should be done carefully as teeth of small species are easily broken during handling or damaged by the chemical).

Teeth are arranged in 1–3 rows. All teeth in each row are counted, including broken and missing teeth. Some teeth may be missing due to breakage during preparation or as part of the replacement process. An empty socket indicates the position of a missing tooth. Teeth counts are given as a formula, with the left arch count given first, separated from the right by a hyphen. For each side, the tooth counts on the different rows are separated by commas, from the outermost to the innermost row of the left arch, then from the innermost to the outermost row of the right arch. For example, the formula 2,3,5–5,2,2 means that there are three rows of teeth on each arch, that the left arch has two teeth on the outer row, 3 on the middle row, and 5 on the inner row, that the right arch has two teeth on the outer row, 2 on the middle row and 5 on the inner row (Figure. 31c).

The number of teeth may differ between the two sides of an individual (usually by one unit), with the left arch generally having more teeth. Within a species, the number of pharyngeal teeth usually shows little variability.

The formula of the pharyngeal teeth, especially the number of rows, has often been used as a character to distinguish genera. The loss of tooth rows from 3 to 2 or 1 occurred independently in different lineages, especially during the evolution of small-sized species and genera, so this character cannot be used alone to define genera.

Pharyngeal teeth are regularly replaced, and empty sockets in the pharyngeal arch may result from lost teeth. New teeth develop next to the pharyngeal arch, embedded in tissue and without contact with bone. As the new tooth grows, it will attach to the part of the pharyngeal arch occupied by the lost tooth. **Further reading.** Chu 1935; Jenkins & Burkhead 1993.

Key to families of Cypriniformes based on species native and introduced to West Asia

1a - Three pairs of barbels.
..................2
1b - No, one or two pairs of barbels.
..................3

2a - An erectable suborbital spine.
..................Cobitidae
2b - No suborbital spine, lacrimal bone elevated with a suborbital flap or groove in males of some species.
..................Nemacheilidae

3a - Female with long ovipositor; a prominent dark-grey midlateral stripe on posterior part of flank.
..................Acheilognathidae
3b - Female without long ovipositor; if a stripe is present, then running from head or anterior body to caudal base.
..................4

4a - Usually 5½ branched anal rays and two pairs of barbels (rarely one pair or no barbels); if 6–7½ branched anal rays, then always two pairs of barbels (*Cyprinion, Carasobarbus*).
..................Cyprinidae

4b - 6–48½ branched anal rays; usually no or one pair of barbels.
..................5

5a - One pair of barbels (except in *Pseudorasbora*).
..................6
5b - No Barbels.
..................7

6a - 8½ branched dorsal rays; 6–8½ branched anal rays; scales deeply embedded, body compressed; belly flattened; no lateral blotches or stripe.
..................Tincidae
6b - 7½ branched dorsal rays; 6½ branched anal rays; scales not embedded on flank; in all but one species (*Pseudorasbora*), body cylindrical or slightly depressed, belly flattened; mouth inferior; a row of dark-brown or grey blotches on body; in *Pseudorasbora* mouth superior; a dark-grey stripe from eye to caudal base.
..................Gobionidae

7a - Last unbranched dorsal ray spinous, rigid; belly keeled from pectoral base to anus.
..................Xenocyprididae (*Hemiculter*, *Hypophthalmichthys*)
7b - Last unbranched dorsal ray neither spinous nor rigid; belly rounded or with a keel starting behind pelvic origin (except *Pelecus cultratus*, which has very large pectoral, reaching beyond pelvic base).
..................8

8a - Distance from anal origin to caudal base shorter than distance from anal origin to pelvic origin.
..................Xenocyprididae (*Ctenopharyngodon*, *Mylopharyngodon*)
8b - Distance from anal origin to caudal base greater than or equal to distance from anal origin to pelvic origin.
..................Leuciscidae

Despite being only distantly related, young *Ctenopharyngodon* (upper) might challenge identification as being quite similar to *Squalius* species (lower).

Family Acheilognathidae

Bitterlings

A family that includes about 75 species placed in seven genera distributed mostly in East Asia. Bitterlings are deep-bodied, compressed, and small-sized fishes. Males have plates bearing tubercles on the snout. All species have a complex reproduction cycle involving mussels. The female has a long, flexible ovipositor, which she inserts into the mussel's exhalant siphon to deposit a few eggs. The diversity of *Rhodeus* in West Asia and Europe is likely overestimated. All populations are very closely related, and external characters distinguish only *R. colchicus*. The most comprehensive phylogenetic analyses identified six different molecular groups of populations. These include *R. colchicus* (eastern Black Sea basin), *R. caspius* (southern Caspian basin), *R. meridionalis* (Vardar & Pinios in Greece and North Macedonia), *R. amarus* (Europe from east to Elbe drainage, Black Sea basin). Two additional lineages occur, one in the Danube, the Ionian Sea basin, and Central Europe west of the Elbe drainage, and a last one in the Strymon and a few adjacent rivers in the northern Aegean basin. The phylogenetic structure of these six clades is poorly supported. Therefore, *R. colchicus* might be the sister group of all other western Palearctic bitterlings. This would mean that all these populations can be identified as *R. amarus*, as they are closely related and not distinguished by morphological characters. However, a consensus on the species status of *R. colchicus* has not yet been reached. Thus, we retain *R. amarus*, *R. caspius*, and *R. colchicus* as separate species in West Asia. **Further reading.** Arai & Akai 1988 (systematics); Okazaki et al. 2001 (genetics); Liu et al. 2006 (host specificity); Bartáková et al. 2019 (phylogeny).

Rhodeus amarus; oviposition; Danube drainage, Germany; male (left) and female (right, with ovipositor), ~50 mm SL. © A. Hartl.

https://doi.org/10.1515/9783111677811-013

Rhodeus amarus; Odra drainage, Germany; female, ~45 mm SL.

Rhodeus amarus; Sakarya, Türkiye; male, ~60 mm SL.

Rhodeus amarus

Common name. Bitterling.

Diagnosis. Distinguished from *R. colchicus* by: • second infraorbital bone narrow • 35–38, usually 36–37, total vertebrae / • 34–41, usually 36–40, scales in lateral series. No morphological character known to distinguish from *R. caspius*. Size up to 95 mm SL.

Distribution. Marmara and Black Sea basins (except easternmost) and all rivers draining into Aegean south to Büyük Menderes. Very widespread and invasive in Europe, where it is native in northern Aegean basin west of Nestos. Widespread from France to Volga, in southern Russia in Don and Kuban drainages, Crimea, Great Britain, north of Italy, and possibly elsewhere. Suspected to be native only to middle and lower reaches of Black Sea tributaries and Aegean basin.

Habitat. Still or slow-flowing waters with dense aquatic vegetation and sandy-silt bottoms, such as lowland ponds, canals, streams, rivers, backwaters, and oxbows where mussels are present. Often in artificial waterbodies, rare in fast-flowing habitats.

Biology. Spawns first time at 1 year and about 30–35 mm SL. Exceptionally lives up to 5 years, but most individuals do not survive year of first reproduction, and population sizes might fluctuate greatly over years. Spawns April–August at water temperatures above 15°C. Males defend small territories around one or a few bivalves (mostly *Unio* spp.), moving with bivalves. Females select bivalves with high oxygen concentrations in exhalation siphon. Avoid bivalves that already contain high numbers of *R. amarus* larvae or are gravid with glochidia (bivalves' own larvae). Also, avoid *Anodonta cygnea*, which can expel most eggs and larvae and usually has low oxygen concentrations in its exhalation siphon. Males display pre-oviposition ejaculation when females inspect or approach mussel. When a female is ready to spawn, she extends ovipositor and muscular conical structure at its base. A batch of eggs is positioned at basal opening of ovipositor, behind which urine collects. She then pushes cone into mussel's siphon. The cone contracts, and urine, under pressure, forces eggs through ovipositor. Oviposition is a rapid process. At same time, male releases sperm into inhalant siphon. After spawning, sneaking males attempt to fertilise eggs. Females typically lay 80–250 eggs yearly, spawning up to five times per season. Eggs are oval, about 2.5–3.0 mm long, and are incubated within gill chamber of mussel. Yolk sac has lateral outgrowths, which allow larvae to attach to mussel's gills. After 20–30 days, feeding larvae leave mussel. Larvae show adaptations to very low oxygen concentrations in mussels, such as very rapid hatching (36 h) and enlarged skin respiratory system. Wing-like yolk projections hold larvae in mussel's gills. Bitterlings and mussels have a parasitic relationship. Mussels suffer a reduction in fitness when carrying *R. amarus* larvae. Larvae and mussels compete for oxygen, and larvae appear to inhibit free water circulation through mussel gills, which may damage them. Bitterlings are rarely infested by glochidia (ectoparasitic mussel larvae). Feeds mainly on algae, diatoms, and other plant material. Frequently translocated, bred, and used for human pregnancy testing before development of modern techniques, resulting in numerous introductions. Females injected with a pregnant woman's urine protrude their ovipositor. Has spread since Middle Ages and is widely introduced with carp stocked from fish farms inhabited by bitterlings. Warm-water species expanding its range due to climate change.

Conservation status. LC; abundant and expanding throughout most of its range. A major invasive species in Europe.

Further reading. Abdurakhmanov 1962 (morphology, biology); Holčík & Jedlička 1994 (morphology); Kottelat 1997 (systematics); Holčík 1999 (biology); Bogutskaya & Komlev 2001 (morphology); Smith et al. 2004 (reproduction); Bohlen et al. 2006 (genetics); Van Damme et al. 2007 (invasion); Bartáková et al. 2019 (phylogeny).

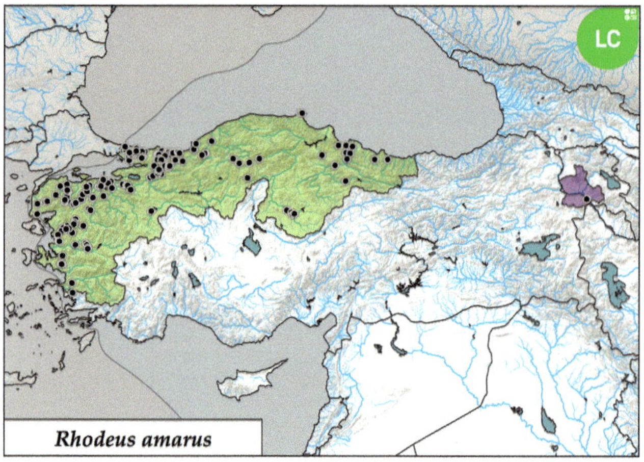

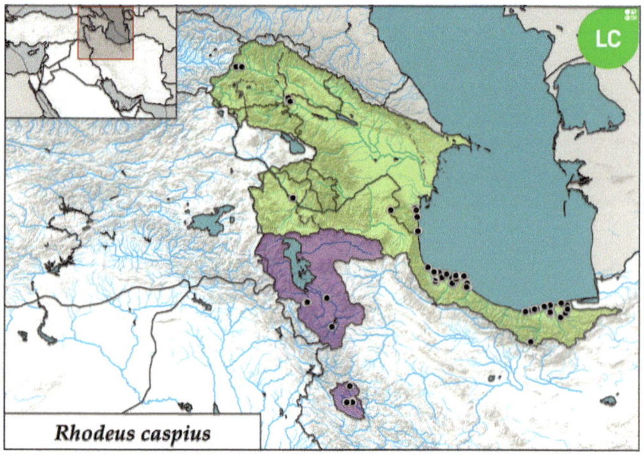

Rhodeus caspius; Lenkoran, Azerbaijan; male, ~45 mm SL.

Rhodeus caspius

Common name. Caspian bitterling.

Diagnosis. Distinguished from *R. colchicus* by: ● second infraorbital bone narrow / ○ 35–36 total vertebrae / ● 34–37 scales in lateral series. No morphological character known to distinguish from *R. amarus*. Size up to 60 mm SL, likely to grow larger.

Distribution. Caspian basin from Kura east to Gorgan (Iran). Introduced to Lake Urmia basin, upper Karkheh in Iran, and possibly elsewhere. Distribution in northern Caspian unknown.

Habitat. Slow-flowing streams and small rivers, backwaters, ponds, and lakes, usually with dense underwater vegetation and sand-silt substrate.

Biology. Spawns April–September. Mating and spawning Behaviour similar to *R. amarus*. Larvae incubated by *Unio* bivalves. Feeds on algae, benthic insect larvae, and planktonic crustaceans.

Conservation status. LC.

Remarks. Diagnostic character proposed to distinguish this species from *R. amarus* could not be confirmed.

Further reading. Bartáková et al. 2019 (phylogeny); Esmaeili et al. 2020a (description).

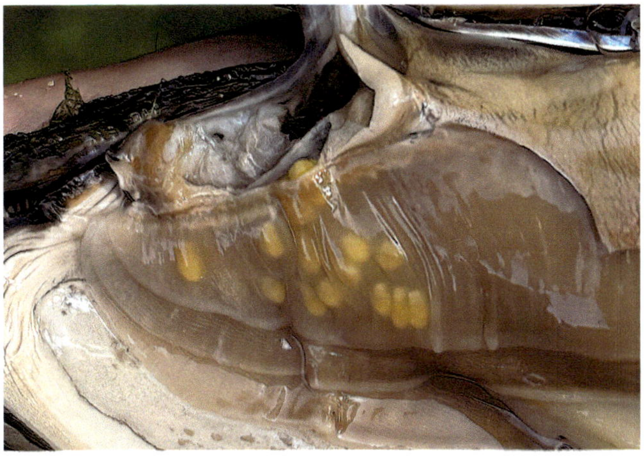

Eggs of *Rhodeus* in the gills of a *Unio* mussel. © M. Reichard.

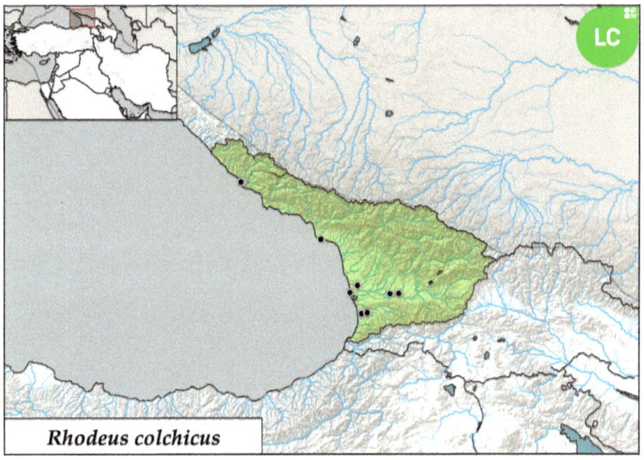

Rhodeus colchicus; Sochi, Russia; ~65 mm SL.

Rhodeus colchicus

Common name. Georgian bitterling.

Diagnosis. Distinguished from other species of *Rhodeus* in West Asia by: • second infraorbital bone wide / • 33–36, usually 34–35, total vertebrae / • 30–38, usually 33–35, scales in lateral series. Size up to 81 mm SL.

Distribution. Caucasian Black Sea basin of Russia from Kherota south to Natanebi in Georgia.

Habitat. Slow-flowing streams and small rivers, backwaters, ponds, and lakes, usually with dense underwater vegetation and sand-silt substrate.

Biology. Spawns April–September. Mating and spawning behaviour similar to *R. amarus*. Larvae incubated by *Unio* bivalves. Feeds on algae, benthic insect larvae, and planktonic crustaceans.

Conservation status. LC; frequent and widespread within its small range.

Further reading. Elanidze 1983 (morphology, biology); Bogutskaya & Komlev 2001 (description); Bohlen et al. 2006 (phylogeography); Bartáková et al. 2019 (phylogeny).

Stream in Lake Sapanca basin, Türkiye. A typical habitat of *Rhodeus amarus*.

Family Cyprinidae

Barbels and carps

Cyprinids are the most speciose fish family, with about 1800 species recognised worldwide. In our region, 124 species have been identified. In West Asia, true cyprinids come in five evolutionary lineages and a variety of body shapes and ecological groups. True barbels of the genera *Barbus*, *Luciobarbus*, *Capoeta*, and *Cyprinion* are distinguished by having a serrated posterior margin of the last branched dorsal ray. True barbels are most diverse in West Asia and Europe. Labeonines of the genera *Garra*, *Tariqilabeo*, and *Bangana* have a horny sheath on the lower jaw, often a complicated, strongly modified mouth morphology, and a smooth posterior margin of the last branched dorsal ray. Labeonines are a diverse group of fishes, being widespread and speciose, especially in tropical Asia and Africa. Yellowfishes or Torini of the genera *Arabibarbus*, *Carasobarbus*, *Mesopotamichthys*, and most likely also unstudied *Caecocypris*, form a third independent evolutionary lineage. They have large, shield-shaped scales and a smooth posterior margin of the last branched dorsal ray. Along with the African genera *Acapoeta*, *Pterocapoeta*, *Labeobarbus* (including *Varicorhinus* and *Sanagia*) and several South Asian genera, Torini forms a well-supported monophyletic group. Carps of the genera *Carassius* and *Cyprinus* have an exceptionally long dorsal fin and a serrated posterior margin of the last branched dorsal ray. Carps are moderately diverse, especially in China, and all but two species found in West Asia are non-native to the region. The last evolutionary lineage is represented by snow barbels, which are split into two phylogenetic groups that are not each other's closest relatives, both found in Iran. The Schizothoracines include the genus *Schizothorax* and several others in the high mountains of Asia, as well as the Schizopygopsines, which include *Schizopygopsis* and several other Asian mountain genera.

In Cyprinidae, the different structures around the mouth are essential for genus diagnosis. These include the rostral cap, the upper and lower lips, the upper and lower jaws, and the barbels. They vary in development or may be absent in different genera. The rostral cap is the fleshy tissue at the tip of the snout. The rostral groove separates the rostral cap from the upper lip. In several genera, the rostral cap develops into a pendulous fold that partly or completely covers the upper lip and the upper jaw. In such cases, the upper lip may be missing. The lower lip is separated from the skin of the throat by a postlabial groove. This groove is usually interrupted in the middle. In some species with fleshy lips, the median part may be hypertrophied and folded backward into a pendulous lobe. A cornified tissue with an anterior cutting edge may cover the lower jaw. In such cases, the lower jaw is often exposed, and the lower lip is restricted to the sides of the jaw. Most species of the genus *Garra* have a prominent gular disc posterior to the lower jaw. It is important to note that Labeonines have no lower lip. **Further reading.** Yang et al. 2015 (phylogeny).

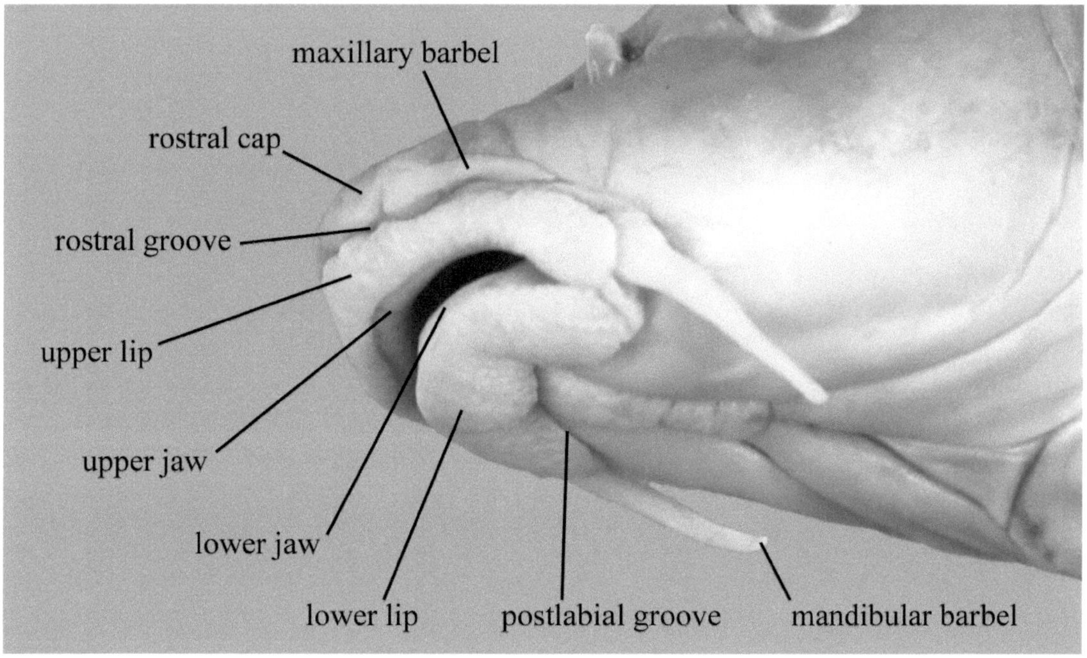

Figure 32. Mouth parts of Cyprinidae.

https://doi.org/10.1515/9783111677811-014

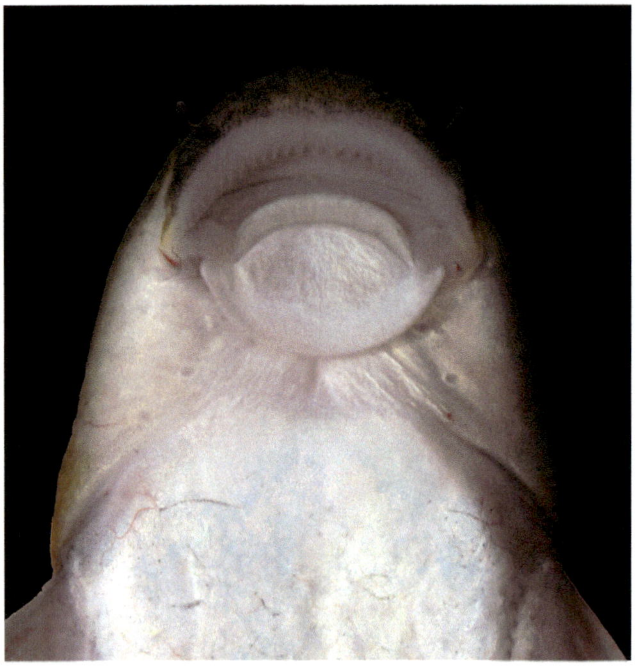

Figure 33. Mental structures and gular disk in *Garra*.

Figure 34. Lower lip with a sharp-edged horny sheath in *Capoeta*.

Key to genera of Cyprinidae in West Asia

1a - Last unbranched anal rays strongly ossified; serrated posteriorly.
..................2
1b - Last unbranched anal ray not strongly ossified, not serrated posteriorly.
..................3

2a - No barbels; pharyngeal teeth in one row.
..................*Carassius*
2b - Two pairs of barbels; pharyngeal teeth in three rows.
..................*Cyprinus*

3a - No barbels.
..................4
3b - One or two pairs of barbels.
..................8

4a - Scales restricted to lateral line, flank above pectoral, and few enlarged scales around anus and anal base.
..................*Schizopygopsis*
4b - Flank fully covered by scales.
..................5

5a - A distinctive row of enlarged, tiled scales along anal base.
..................*Schizocypris*
5b - No row of enlarged, tiled scales along anal base.
..................6

6a - An elevated gular disk behind lower jaw.
..................*Garra* (in part)
6b - Lower lip and lower jaw normally developed, without gular disk.
..................7

7a - Eyes absent or very small, often covered with skin; body whitish or pinkish without any colour patterns.
..................*Caecocypris*
7b - Eyes present and well developed; body silvery, golden or brown.
..................*Mesopotamichthys*

8a - A distinctive row of enlarged, tiled scales along anal base.
..................*Schizothorax*
8b - No row of enlarged, tiled scales along anal base.
..................9

9a - Last unbranched dorsal ray spinous, with serrae along its posterior edge.
..................10
9b - Last unbranched dorsal ray soft or spinous, without serrae along its posterior edge.
..................13

10a – 6–7½ branched anal rays; 12–17½ branched dorsal rays.
..................*Cyprinion*
10b - 5½ branched anal rays; 7–9½ branched dorsal rays.
..................11

11a - Lower lip absent or widely interrupted, lower jaw with a sharp-edged horny sheath.
..................*Capoeta*
11b - Lower lip with a fleshy lobe in middle, if interrupted, lower jaw without sharp-edged horny sheath.
..................12

12a - Body and fins covered with distinctive small black or dark-brown spots, many species with irregularly shaped brown blotches; nuptial tubercles very small, sand-like all over head.
..................*Barbus* and *Luciobarbus subquincunciatus*
12b - In adults larger than 150 mm SL, body uniform yellowish, brown or grey, without dark-brown or black blotches, sometimes with a faith lateral stripe; few, large nuptial tubercles on snout.
..................*Luciobarbus*

13a - Upper lip very narrow, almost completely reduced, hidden under rostral cap.
..................14
13b - Upper lip well developed, separated from skin of snout.
..................16

14a - A gular disc behind lower jaw, usually labrum with free margin, margins fused with skin of throat in *G. elegans*.
..................*Garra* (in part)
14b - No gular disc behind lower jaw.
..................15

15a - 29–34 total scales in lateral series, lower jaw round, without cutting edge; prominent lateral papillate pad on lower jaw.
..................*Garra* (in part)
15b - 34–38 total scales in lateral series; lower jaw almost straight, with a sharp cutting edge; no lateral papillate pad on lower jaw.
..................*Tariqilabeo*

16a - Rostral cap with a deep lateral groove (labial fold); lower lip absent.
..................*Bangana*
16b - Rostral cap without lateral groove; lower lip often with a fleshy lobe or interrupted in middle, but well developed at its margins.
..................17

17a - 6½ branched anal rays; 9–11½ branched dorsal rays; body depth usually 25–40 % SL.
..................*Carasobarbus*
17b - 5½ branched anal rays; usually 8½ branched dorsal rays; body depth usually 18–26 % SL.
..................*Arabibarbus*

Arabibarbus grypus; Tigris, Türkiye; 450 mm SL.

Arabibarbus arabicus; Wadi Maur, Yemen; 106 mm SL. © F. Krupp.

Arabibarbus

A genus of three very large (*A. grypus*) to medium-sized species endemic to West Asia. These are hexaploid barbels of the yellowfish (Torini) group, distinguished by large, shield-shaped scales with numerous parallel radii. The last unbranched dorsal ray is weakly to strongly ossified, without serrae. They have two pairs of barbels, 8½ branched dorsal rays, and 5½ branched anal rays. Their phylogenetic position has long been uncertain, but most authors categorised them in *Barbus*, as no alternative was available. One species, *A. grypus*, occurs in the Persian Gulf basin. Some authors have placed it in the Asian genus *Tor*. However, it has only recently become clear that these three species belong together, and with *Carasobarbus*, *Mesopotamichthys*, and *Ptercapoeta* form the sister group of the large African genus *Labeobarbus*. Both Arabian species are superficially similar but allopatric, and the molecular characters studied suggest that they have a long, isolated evolutionary history. **Further reading.** Borkenhagen 2014 (description, phylogeny); Yang et al. 2015; Borkenhagen 2017 (phylogeny).

Key to species of *Arabibarbus*

1a - Body cylindrical; last unbranched dorsal ray weakly ossified; posterior barbel length 4–6 % SL; 29–36, usually 30–35, total lateral-line scales.
..................*A. grypus*

1b - Body laterally compressed; last unbranched dorsal ray strongly ossified; posterior barbel length 7–12 % SL; 32–44, usually 35–41, total lateral-line scales.
..................2

2a - 29–36, usually 32–35, total lateral-line scales; head depth 14–17 % SL; pectoral length 18–22 % SL; pelvic length 15–19 % SL; head length 24–31 % SL.
..................*A. arabicus*

2b - 29–32, usually 30–31, total lateral-line scales; head depth 16–18 % SL; pectoral length 20–24 % SL; pelvic length 17–20 % SL; head length 28–33 % SL.
..................*A. hadhrami*

Arabibarbus arabicus

Common name. Arabian Shabout.

Diagnosis. Distinguished from species of *Carasobarbus* in Arabian Peninsula by: • 5½ branched anal rays / • 8½ branched dorsal rays. Distinguished from *Arabibarbus hadhrami* by: ○ 29–36, usually 32–35 total lateral-line scales (mode 33) / ○ head depth 14–17 % SL / ○ pectoral length 18–22 % SL / ○ pelvic length 15–19 % SL / ○ head length 24–31 % SL. Size up to 351 mm SL.

Distribution. Southern Red Sea and Gulf of Aden basins from southern Saudi Arabia to Yemen.

Habitat. Wadis with sections of deep pools.

Biology. No data.

Conservation status. LC; fishing might be a threat. Actual situation unknown.

Further reading. Trewavas 1941 (description); Banister & Clarke 1977 (distribution); Krupp 1983 (distribution); Borkenhagen 2014 (review).

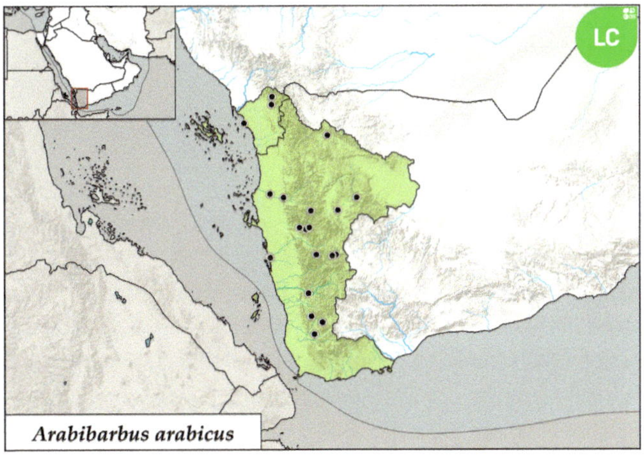

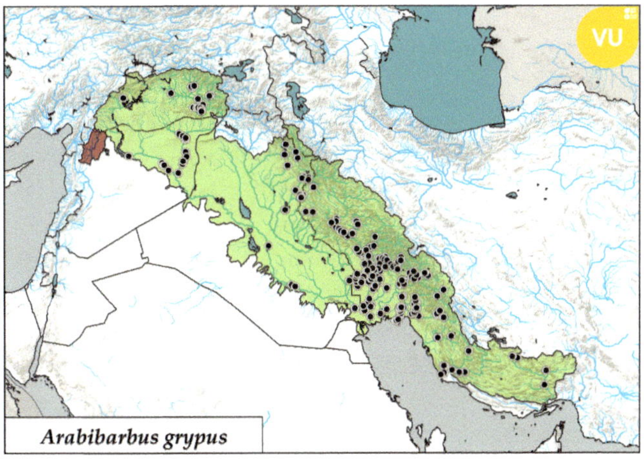

Arabibarbus grypus; Lesser Zab, Iraq; 450 mm SL.

Arabibarbus grypus; Karun, Iran; 80 mm SL. © K. Borkenhagen.

Arabibarbus grypus

Common name. Shabout.

Diagnosis. Distinguished from species of *Luciobarbus*, *Barbus*, *Carasobarbus*, and *Mesopotamichthys* in West Asia by: ○ two pairs of barbels / ○ body cylindrical / ○ last unbranched dorsal ray weakly ossified, segmented, without serration at its posterior margin / ○ dorsal origin above or in front of pelvic origin / ○ 29–36, usually 30–35, total lateral-line scales / ○ posterior barbel length 4–6 % SL / ○ median lobe on lower lip well developed, often hypertrophied. Size up to 1300 mm SL and 30 kg.

Distribution. Euphrates, Tigris, Karun, and Karkheh drainages, Qweiq (now extirpated), and rivers of Persian Gulf in Iran south to Minab.

Habitat. Large to medium-sized rivers with moderate currents. Often found in reservoirs. Spawns on sand, gravel, or submerged vegetation in fast-flowing waters.

Biology. Males mature at about 400 mm SL, females at about 450 mm SL, first spawning at 3–5 years. Lives up to 17 years, probably much longer. Spawns in Karun late April–early August and May–July in Atatürk reservoir, usually during floods when water is turbid. Migrations occur to fast-flowing stretches of rivers. Males congregate at spawning sites. Spawning migrations typically begin in April or May. Spawns only once a year. Appears in schools on spawning grounds just before dark and remains there until just before midnight, making loud noises by splashing, jumping, and chasing. Eggs are transparent and sticky and attached to stones or plants. In Iraq, reported to migrate upstream during hot, dry summers and then return downstream in September and October. Predominantly herbivorous, feeding on algae and leaves, fruits, and seeds of higher plants, with small amounts of aquatic invertebrates and fish.

Conservation status. VU; threatened mainly by overfishing, to a lesser extent by pollution, water abstraction, and dams. Extirpated from Qweiq.

Remarks. Reported to weigh up to 100 kg, but such records probably refer to *Luciobarbus esocinus*. Often reported from the Orontes drainage, but there are only two records of two individuals each from Hamah in Syria in 1881, and the species has never been recorded from the Orontes since. Likely, these two fish were later mislabelled or transported from the Euphrates or Qweiq to Hamah as food fish. Often placed in Asian genus *Tor*.

Further reading. Karaman 1971 (description, as *Tor grypus*); Coad 2010a (distribution, biology); Borkenhagen 2014 (morphology, generic position); Coad 2021a (biology, morphology).

Tigris at Hasankeyf in 2009. Large rivers with diverse structures are a typical habitat of *Arabibarbus grypus*.

Arabibarbus hadhrami; Wadi Hadhramaut drainage, Yemen; 129 mm SL. © S. Tränkner/SGN.

Arabibarbus hadhrami

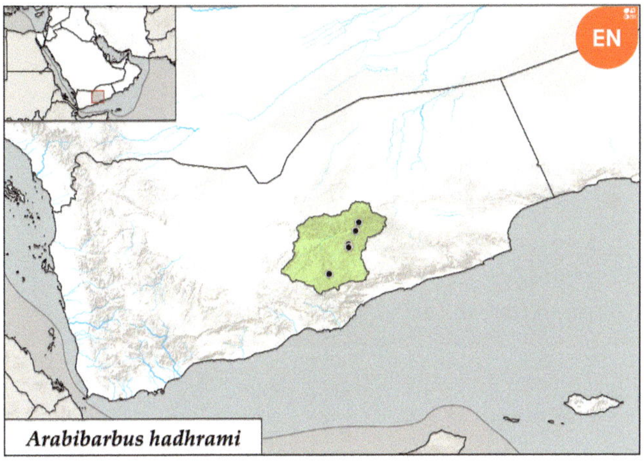

Common name. Hadhramaut Shabout.
Diagnosis. Distinguished from species of *Carasobarbus* in Arabian Peninsula by: ● 5½ branched anal rays / ● 8½ branched dorsal rays. Distinguished from *Arabibarbus arabicus* by: ○ 29–32 total lateral-line scales (mode 30) / ○ head depth 16–18 % SL / ○ pectoral length 20–24 % SL / ○ pelvic length 17–20 % SL / ○ head length 28–33 % SL. Size up to 205 mm SL.
Distribution. Yemen: Wadi Hadhramaut drainage.
Habitat. Wadis with sections of deep pools.
Biology. No data.
Conservation status. EN; appears to be declining within its very small range.
Remarks. Poorly differentiated from *A. arabicus* by morphological characters but well differentiated by molecular characters.
Further reading. Borkenhagen 2014 (description).

Bangana dero; Mashkid drainage, Iran; ~130 mm SL. © H. Mousavi-Sabet.

Bangana

Bangana comprises approximately seven large or very large species distributed across South and Southeast Asia. It is closely related to *Labeo* that is found in tropical Asia and Africa. *Bangana* are rheophilic and feed on detritus and algae. Some are territorial and defend "gardens" on stones. They are regionally important food fish, but none is used in aquaculture. One species is known from Iran, where the genus reaches its westernmost distribution.

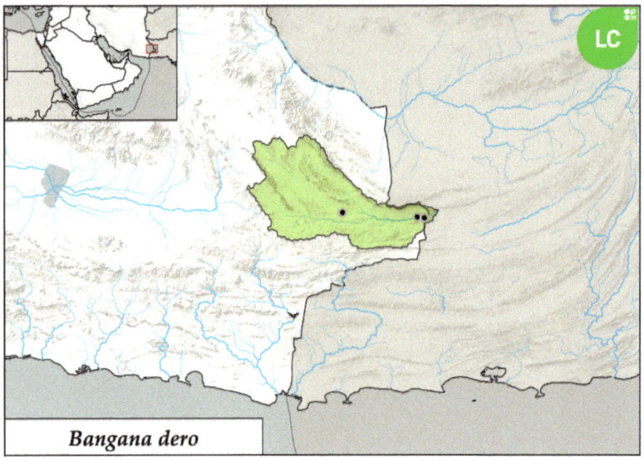

Bangana dero

Common name. Kalabans.

Diagnosis. Distinguished from other cyprinid species in southern Iran by: • transversal groove along snout / ○ mouth inferior / ○ lips thick / ○ upper lip well developed, separated from skin of snout by a groove / ○ no gular disc / ○ one pair of barbels / ○ 38–44 total lateral-line scales / ○ 5½ branched anal rays. Size up to 500 mm SL.

Distribution. Iran: Mashkid drainage on border with Pakistan. Widespread east into Pakistan, Afghanistan, and along Himalayan foothills in India, Bhutan, Nepal, Bangladesh, Myanmar, and China. Introduced in Peninsular India and Sri Lanka.

Habitat. Iran: Small streams in desert areas. Usually in torrential mountain streams, rivers, and lakes at high altitudes. Spawns on gravel in fast or moderately fast-flowing water.

Biology. Lives up to 8 years. Matures after one year and 200–300 mm SL (NE India). Spawns on rocks or vegetation in spring. Makes short spawning migrations to torrents. Spawns March–June (Garhwal Himalayas), June–July (NE India). May migrate to warmer regions of lakes and streams during winter. Herbivorous, feeding on detritus, diatoms, and algae, occasionally on invertebrates or small vertebrates. Highly valued as a food fish.

Conservation status. LC.

Remarks. *Bangana gedrosicus* is a related species reported from the Mashkid drainage in Pakistan. *Bangana dero* is distinguished from *B. gedrosicus* by having 5½ branched anal rays (vs. 6–7½) and 16–17 pectoral rays (vs. 14). *Tariqilabeo macmahoni* is another superficially similar species reported from the Mashkid basin. *Bangana dero* is distinguished from *Tariqilabeo macmahoni* in having 39–44 total lateral-line scales (vs. 35–36).

Further reading. Mohindra et al. 2005 (general information; as *Labeo dero*); Esmaeili et al. 2013a (Iran, morphology); Coad 2021a (biology, morphology).

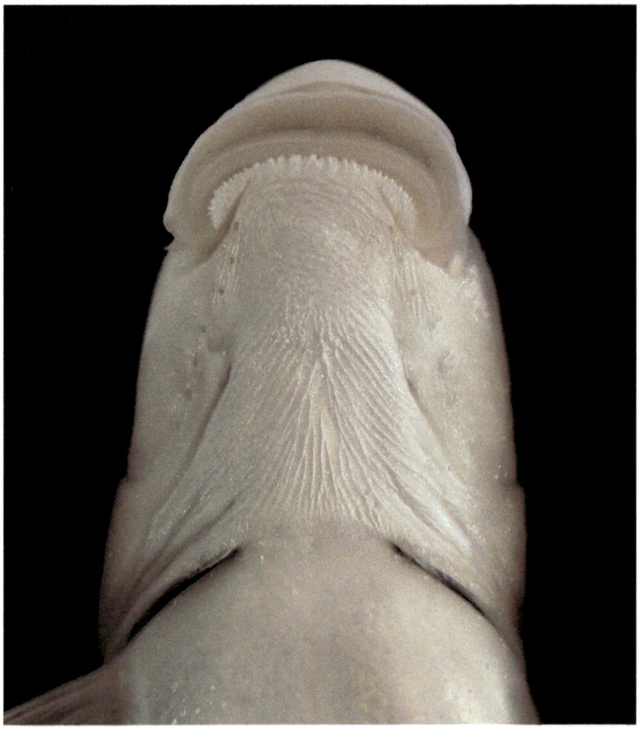

Figure 35. *Bangana dero*; Mashkid drainage, Iran; ~130 mm SL. © H. Mousavi-Sabet.

Barbus

Barbels are medium- to large-sized fishes that live near the bottom in a variety of habitats, from lakes and slow-flowing large rivers to hill streams. They are very speciose in Europe, especially in the rivers of the Mediterranean basin. Only 12 species are known from West Asia, where the genus is very widespread. The last unbranched dorsal ray may be spinous (flexible in its uppermost part only) and either rigid or not, with its posterior edge serrated on part or entire length. The extent and intensity of serration is changing with growth. In many species, females have a longer and differently shaped anal than males, which is assumed to play a role in excavating the substrate for spawning. In all species, juveniles are covered with spots and blotches. Compared to adults, the last unbranched dorsal ray is shorter, less serrated, and less rigid. Some *Barbus* species have uniform brown to grey adults without blotches, while all species have small dark-brown or black spots on their scales. Diagnosis is based solely on adults.

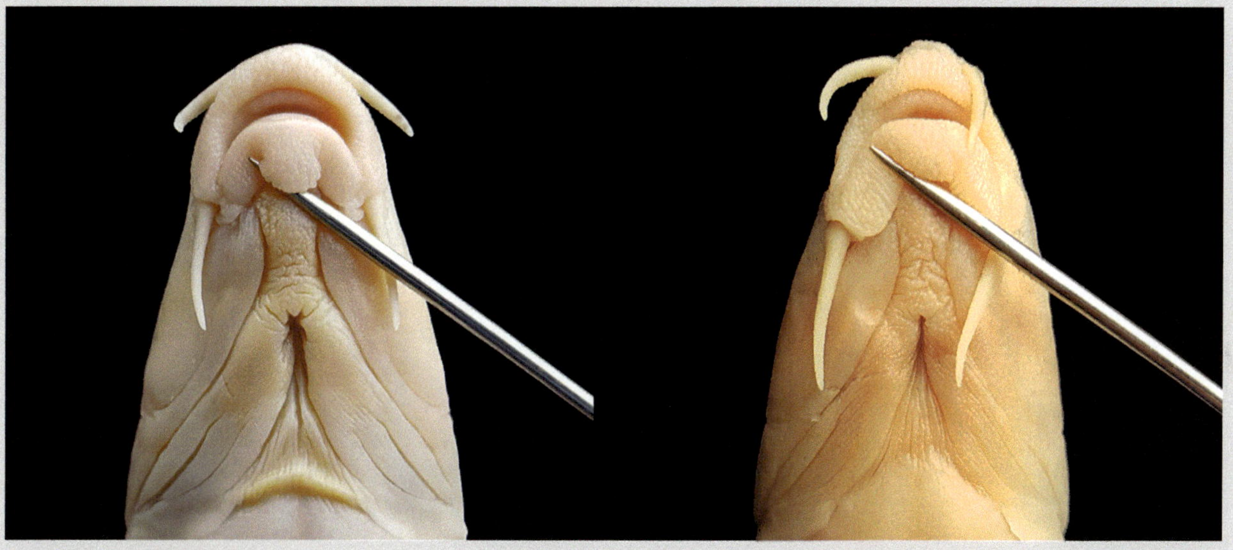

Lower lip of barbels. The terminology used in the literature to describe the shape of the lower lip of *Barbus* and *Luciobarbus* species is often inconsistent or morphologically inaccurate. It has been uncritically compiled, adapted, or translated. Here, we use only the following terms and definitions. In several species, the lower lip is thick and fleshy. In some species, the median part of the lip is produced posteriorly into a lobe (often referred to in the literature as a tongue-like lobe or three-lobed lower lip). Posteriorly, the **lobe is not continuous with the skin of the throat**; laterally, it is separated from the rest of the lip by notches. The lobe may be long, protruding, and pendulous, or square-shaped, extending posteriorly only to the corner of the mouth. In other species, the median part of the lower lip is thickened but not produced into a lobe; it appears as a swollen pad, bounded posteriorly by a more or less deep groove; the posterior extremity of the **pad is not free but continuous with the skin of the throat** (often ambiguously described in the literature as a weakly developed lobe). The difference between a lobe and a pad may be slight; we make the distinction by gently passing a needle under the lobe, which results in one of two possibilities: if it is a lobe, the needle can be inserted under the lobe (left in the above image) and moved laterally under the entire width of the lobe; or, in the case of a pad, the needle can be inserted into the groove (if any) behind the pad (right in the above image), but cannot be moved under the pad. In some species, there may be a transition from a pad to a lobe during growth; sometimes, different individuals or populations show different conditions.

Key to species of *Barbus* in West Asia

1a - 39–44 total vertebrae; small-sized fish reaching about 200 mm SL.
..................2
1b - 44–49 total vertebrae; large-sized fish reaching usually 300–500 mm SL.
..................7

2a - Lower lip with a median lobe.
..................3
2b - Lower lip with a median swollen pad.
..................5

3a - 62–72 total lateral-line scales.
..................*B. niluferensis*
3b - 50–62 total lateral-line scales.
..................4

4a - Last unbranched dorsal ray 52–72 % ossified; posterior margin of dorsal slightly convex.
..................*B. pergamonensis*
4b - Last unbranched dorsal ray 33–50 % ossified; posterior margin of dorsal straight or slightly concave.
..................*B. xanthos*

5a - Dorsal origin slightly in front of vertical of pelvic origin.
..................*B. ida*
5b - Dorsal origin behind or above vertical of pelvic origin.
..................6

6a - 71–90 total lateral-line scales.
..................*B. miliaris*
6b - 52–77 total lateral-line scales.
..................*B. cyri/B. lacerta/B. karunensis*

7a - Lower lip with median swollen pad.
..................*B. tauricus*
7b - Lower lip with median lobe.
..................8

8a - Flank plain brown or with many minute dark-brown spots in adults.
..................*B. rionicus*
8b - Flank back and head with many irregularly shaped black or brown spots, often with large, dark-brown blotches in juveniles and adults.
..................9

9a - Lateral line with 53–58 scales; anal longer in female than in male.
..................*B. escherichii*
9b - Lateral line with 58–71 scales; anal of equal length in female and male.
..................*B. anatolicus*

Barbus anatolicus; Kızılırmak, Türkiye; 236 mm SL.

Barbus anatolicus

Common name. Anatolian barbel.

Diagnosis. Distinguished from other species of *Barbus* in West Asia by: ● anal of equal length in female and in male / ○ 58–71 total lateral-line scales / ○ 9–11 gill rakers / ○ 45–49 total vertebrae / ○ lower lip with median lobe / ○ last simple dorsal ray moderately ossified, serrated along about 70–80 % of its posterior margin / ○ many small irregular shaped black or brown spots, smaller or as large as scales, often forming large, dark-brown blotches on head, back, and flank in adults and juveniles / ○ posterior dorsal margin concave. Size up to 400 mm SL.

Distribution. Türkiye: Kızılırmak and Yeşilırmak drainages.

Habitat. Medium to large rivers. Wide variety of habitats with moderate to strong currents and rocky or gravelly substrate.

Biology. Matures at about 2–4 years, males one year earlier than females. Usually a fractional spawner, some females appear to spawn only once a year. Feeds mainly on aquatic invertebrates and algae.

Conservation status. LC.

Further reading. Kotlík et al. 2004 (molecular phylogeny; as clade IV); Turan et al. 2018c (description).

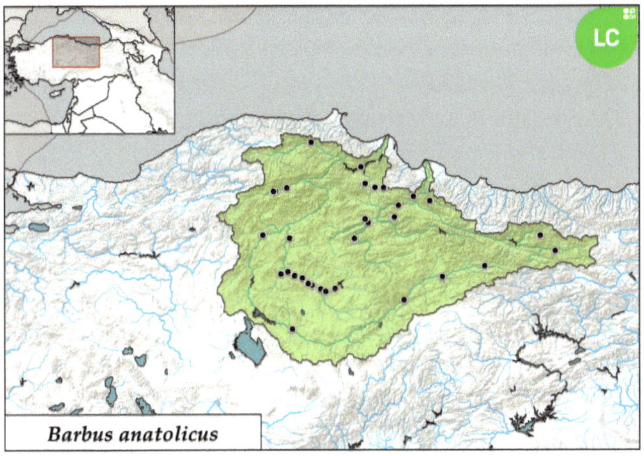

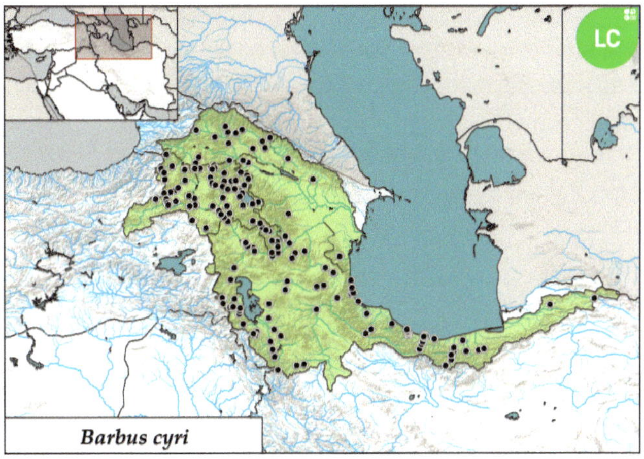

Barbus cyri; Urmia basin, Iran; female, ~120 mm SL.

Barbus cyri; Kura drainage, Türkiye; female, 130 mm SL. © M. Özuluğ.

Barbus cyri

Common name. Kura barbel.

Diagnosis. Distinguished from other species of *Barbus* in West Asia by: ○ 50–72+2–4 [52–76 total] lateral-line scales / ○ 39–44 total vertebrae / ○ postdorsal length 48–61 % SL / ○ dorsal base length 16–21 % SL / ○ anal length 6–10 % SL / ○ 28–33 scale around caudal peduncle / ○ 9–13 scales below lateral line / ○ lower lip with a median swollen pad / ○ dorsal origin at or behind vertical of pelvic origin / ○ numerous irregular dark-brown blotches on back, flanks and head / ○ anal longer in female than in male. Size up to 360 mm SL; usually smaller.

Distribution. Caspian basin: Kura and Aras west of Atrak. Also, in Lake Urmia basin.

Habitat. Mountain and foothill zones in streams and small rivers with fast, clear, well-oxygenated water and gravel substrate; also in large lakes; in steep rivers down to estuaries. In lowland areas often most common in small streams and rare in larger rivers. Spawns in streams on gravel beds.

Biology. Matures at about 2–3 years, males one year earlier than females. Usually lives 3–5 years. Spawns April–August, depending on altitude, at water temperatures of 14–20°C. Fractional spawner, spawns 2–3 times a season, some females appear to spawn only once a year. Three life history forms have been described from Lake Sevan (Armenia), differing in size, distribution, migration patterns, and spawning habitats (lacustrine, river-lacustrine, and riverine). Feeds mainly on aquatic invertebrates and algae.

Conservation status. LC.

Remarks. *Barbus goktschaicus* from Lake Sevan basin (Armenia) and *B. urmianus* from Lake Urmia basin are synonyms. *Barbus cyri* can only be distinguished from *B. lacerta* and *B. karunensis* by molecular characters. While forming their own molecular clusters, these three species are very closely related and may be treated as one species with three population groups in the future.

Further reading. Bogutskaya 2003b, c (morphology, biology); Khaefi et al. 2017a (review); Eagderi et al. 2019d (description of *B. urmianus*); Kaya et al. 2020a (distribution); Coad 2021a (biology); Jouladeh-Roudbar et al. 2023d (*B. urmianus* as synonym of *B. cyri*); Gandlin et al. 2022 (phylogeography).

Barbus cyri; Tajan, Iran; male, 150 mm SL.

Barbels challenging species recognition. We usually use two completely independent datasets to determine whether two fish populations represent one or two species. These are molecular and morphological datasets. They may or may not agree. Suppose fish from two populations can be distinguished by non-overlapping or slightly overlapping morphological characters that are not expected to be phenotypically variable (based on the observer's experience). In that case, they can be recognised as two species. If fish from the two populations also form two "well-separated" groups in molecular analyses, all is well, and they are usually accepted as two species. But what happens when two morphologically indistinguishable populations are placed in two molecular groups? The conceptual framework for resolving such cases needs to be better discussed. Some authors use more or less arbitrary thresholds for molecular markers and call all populations beyond these thresholds "cryptic species." While we agree that some species are indistinguishable by external and internal characters, the term "cryptic species" has been applied to many very different cases and should be avoided as it is not clearly defined. In West Asia, the *Barbus lacerta* group represents such a case. Three poorly differentiated molecular lineages in the Caspian and Persian Gulf basins are found in allopatry. These lineages are usually recognised as *B. cyri*, *B. karunensis*, and *B. lacerta*, but they cannot be distinguished by the morphological characters examined so far. Whether they should be treated as one species (*B. lacerta*) occurring in three distinct populations or as three poorly differentiated species is a matter of debate, as there is no clear conceptual background to resolve such situations. However, such cases are relatively rare, and few similar situations exist in *Garra* and *Oxynoemacheilus*. Thus far, nearly all species "well separated" by molecular characters can also be distinguished externally.

Barbus escherichii; Sakarya, Türkiye; 80 mm SL.

Barbus escherichii; Sakarya, Türkiye; 184 mm SL.

Barbus escherichii

Common name. Sakarya barbel.

Diagnosis. Distinguished from other species of *Barbus* in West Asia by: ● adults with numerous roundish, small, dark-brown dots, as large or larger than scales / ○ 53–58 total lateral-line scales / ○ last unbranched dorsal ray strongly ossified / ○ 7–10, usually 8–9, gill rakers / ○ head length 25–28 % SL / ○ lower lip with a median lobe / ○ juveniles with numerous irregular dark-brown blotches on back, flanks and head / ○ 45–48, usually 46–47, total vertebrae / ○ anal longer in female than in male. Size up to 300 mm SL.

Distribution. Türkiye: Sakarya drainage.

Habitat. Large rivers to mid-sized streams with gravel bottoms. Spawns in riffles and rapids. Also, in reservoirs, from which it migrates into tributaries to spawn.

Biology. Matures at about 2–4 years, males one year earlier than females. A fractional spawner. Feeds mainly on aquatic invertebrates and algae.

Conservation status. LC.

Remarks. Previously, the name *B. escherichii* was often applied to *B. tauricus* in Anatolian Black Sea basin. *Barbus escherichii* cannot be distinguished from *B. tauricus* by

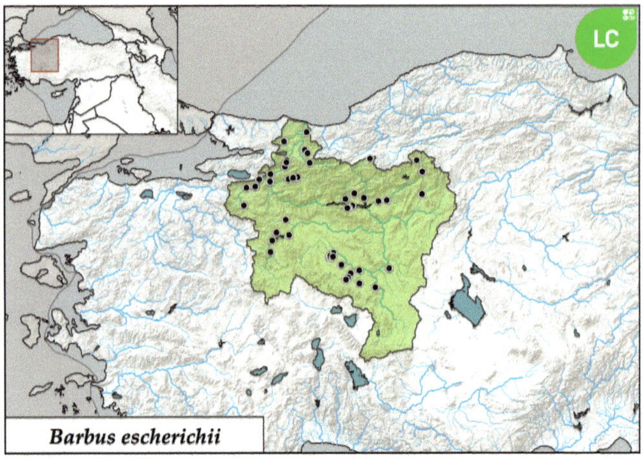

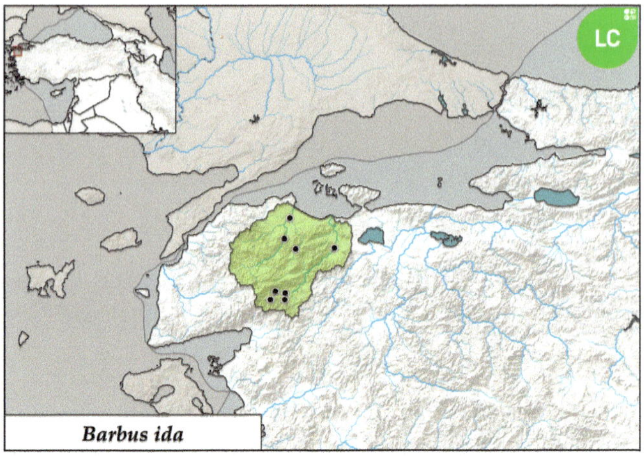

Barbus ida; Gönen, Türkiye; 150 mm SL.

its COI DNA sequences, and both species may be closely related, or *B. escherichii* may have been the victim of introgressive hybridisation. More research is needed on this species group to understand its diversity better.

Further reading. Steindachner 1897 (description); Turan et al. 2009b (comparison).

Barbus ida

Common name. Biga barbel.

Diagnosis. Distinguished from other species of *Barbus* in West Asia by: ○ 56–61 total lateral-line scales / ○ 9–10 gill rakers / ○ 39–40 total vertebrae / ○ head length 24–27 % SL / ○ lower lip with a median pad / ○ numerous irregular dark-brown blotches on back, flank, and head / ○ dorsal origin slightly in front of vertical of pelvic origin / ○ last unbranched dorsal ray 41–55 % ossified / ○ posterior margin of dorsal straight or slightly convex / ○ anal longer in female than in male. Size up to 160 mm SL.

Distribution. Türkiye: Gönen and Biga drainages on Biga Peninsula.

Habitat. Headwaters of streams and small rivers with fast, clear, well-oxygenated water and gravel substrate.

Biology. No data.

Conservation status. LC; has a small range with stable populations.

Remarks. *Barbus ida* occurs locally sympatric with *B. tauricus*. It usually inhabits small headwater streams, whereas *B. tauricus* is found in larger, slow-flowing rivers.

Further reading. Güçlü et al. 2021 (description).

Streams like this tributary of the Murat in the Euphrates drainage, Türkiye, host a diverse fish fauna, including *Barbus lacerta* and many others.

Barbus karunensis; Karun, Iran; 120 mm SL.

Barbus karunensis

Common name. Karun barbel.

Diagnosis. Distinguished from other species of *Barbus* in West Asia by: ○ 59–66+2–4 [60–70 total] lateral-line scales / ○ 39–44 total vertebrae / ○ postdorsal length 50–53 % SL / ○ dorsal base length 17–22 % SL / ○ anal length 7–9 % SL / ○ 26–29 scale around caudal peduncle / ○ 9–11 scales below lateral line / ○ 33–42 predorsal scales / ○ lower lip with a median swollen pad / ○ dorsal origin at or behind vertical of pelvic origin / ○ anal longer in female than in male. Size up to 154 mm SL.

Distribution. Iran: Sezar, Bakhtiyari, Beshar, and Khersan in upper and rarely in lower Karun drainage.

Habitat. Mountain and foothill zones, in streams and small rivers with fast, clear, well-oxygenated water and gravelly substrate.

Biology. Matures at about 2–4 years, males one year earlier than females. A fractional spawner. Feeds mainly on aquatic invertebrates and algae.

Conservation status. LC.

Remarks. *Barbus karunensis* can only be distinguished from *B. cyri* and *B. lacerta* by molecular characters.

Further reading. Khaefi et al. 2017a (description); Eagderi et al. 2019d (morphology).

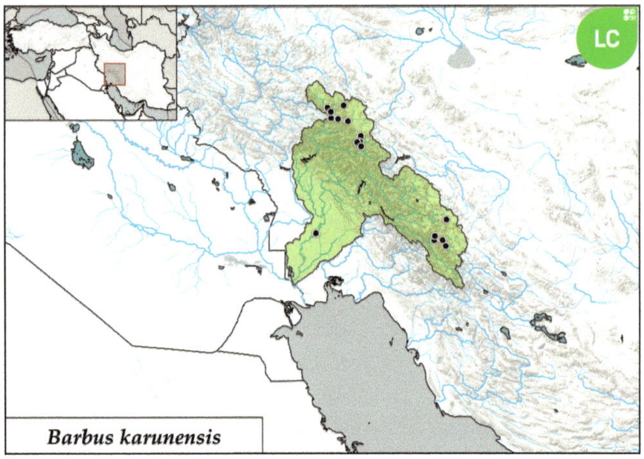

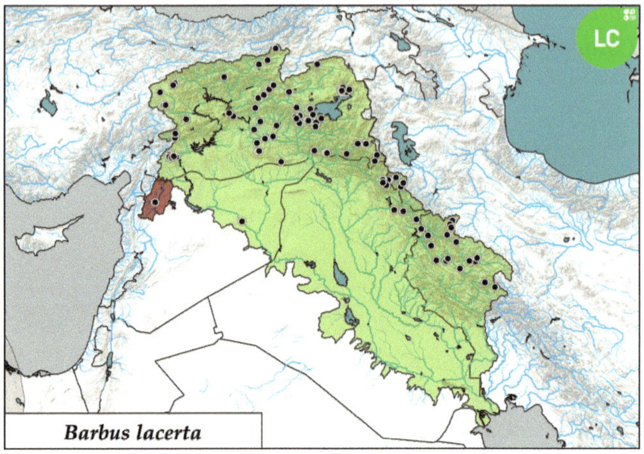

Barbus lacerta; Euphrates, Türkiye; 112 mm SL.

Barbus lacerta; Lake Nemrut, Türkiye; 110 mm SL.

Barbus lacerta

Common name. Lizard barbel.

Diagnosis. Distinguished from other species of *Barbus* in West Asia by: ○ 52–73+2–4 [56–77 total] lateral-line scales / ○ 39–44 total vertebrae / ○ postdorsal length 46–59 % SL / ○ dorsal base length 16–22 % SL / ○ anal length 6.0–9 % SL / ○ 25–32 scale around caudal peduncle / ○ 35–48 predorsal scales / ○ lower lip with a median swollen pad / ○ dorsal origin at or slightly behind vertical of pelvic origin / ○ numerous irregular dark-brown blotches on back, flanks and head / ○ anal longer in female than in male. Size up to 186 mm SL.

Distribution. Qweiq, Euphrates, Tigris, and Karkheh drainages, also in Lake Van basin.

Habitat. Mountain and foothill areas, in streams and small rivers with fast, clear, well-oxygenated water and gravel substrate. In lowlands usually most common in small streams and absent from large rivers.

Biology. Lives up to 7 years. Matures at about 2–4 years, males one year earlier than females. A fractional spawner. Feeds mainly on aquatic invertebrates and algae.

Conservation status. LC; one of the most common species in its range, extirpated from Qweiq.

Remarks. Rarely found syntopic with *Luciobarbus* species that occur in larger rivers and lowland habitats. *Barbus kosswigi*, described from the upper Greater Zab drainage in Anatolian Hakkari province, and *B. ercisianus* from the Lake Van basin are synonyms. *Barbus lacerta* can only be distinguished from *B. cyri* and *B. karunensis* based on molecular characters.

Further reading. Khaefi et al. 2017a (review); Eagderi et al. 2019d (morphology); Coad 2021a (biology, morphology).

Barbus miliaris; Jaj, Namak basin, Iran; ~110 mm SL.

Barbus miliaris

Common name. Namak barbel.

Diagnosis. Distinguished from other species of *Barbus* in West Asia by: ○ 69–87+3–5, totally 71–90, lateral-line scales / ○ 39–44 total vertebrae / ○ postdorsal length 52–59 % SL / ○ dorsal base length 18–23 % SL / ○ anal length 7–9 % SL / ○ 28–35 scale around caudal peduncle / ○ 12–15 scales below lateral line / ○ 37–45 predorsal scales / ○ lower lip with a median swollen pad / ○ dorsal origin at or slightly behind vertical of pelvic origin / ○ numerous irregular dark-brown blotches on back, flanks and head / ○ anal longer in female than in male. Size up to 240 mm SL, usually smaller.

Distribution. Iran: Qom, Qareh Chai, and Jaj drainages in Lake Namak basin. Hable and Nam drainages and Cheshme Ali spring in Western Kavir basin.

Habitat. Slow to fast-flowing rivers with gravel and rocky bottoms.

Biology. Matures at about 2–4 years, males one year earlier than females. A fractional spawner. Feeds mainly on aquatic invertebrates and algae.

Conservation status. VU; possibly extirpated from Lake Namak basin and Cheshme Ali, now found only in small numbers in 5–10 independent, mostly declining populations in Nam and Hable in Western Kavir basin.

Further reading. De Filippi 1863 (description); Karaman 1971: (morphology; as *B. mursa miliaris*); Jouladeh-Roudbar et al. 2015b (Kavir basin); Khaefi et al. 2017a (review); Eagderi et al. 2019d (morphology).

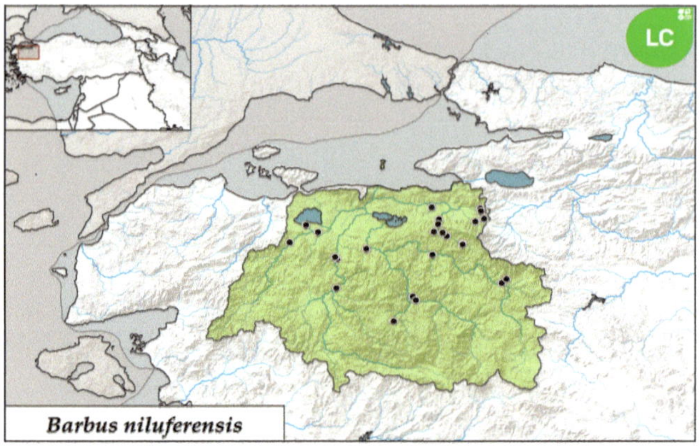

Barbus niluferensis; Susurluk drainage, Türkiye; 135 mm SL.

Barbus niluferensis

Common name. Susurluk barbel.

Diagnosis. Distinguished from other species of *Barbus* in West Asia by: ○ 62–72 total lateral-line scales / ○ last unbranched dorsal ray moderately ossified / ○ 6–8 gill rakers / ○ head length 23–26 % SL / ○ lower lip with a median lobe / ○ numerous irregular dark-brown blotches on back, flanks and head / ○ 43–44 total vertebrae / ○ anal longer in female than in male. Size up to 160 mm SL.

Distribution. Türkiye: Susurluk drainage.

Habitat. Headwater streams with fast to moderate flow, cobble, and gravel bottom.

Biology. Matures at about 2–4 years, males one year earlier than females. Usually a fractional spawner, some females appear to spawn only once a year. Feeds mainly on aquatic invertebrates and algae.

Conservation status. LC.

Remarks. *Barbus niluferensis* occurs locally sympatric with *B. tauricus*. It usually inhabits small headwater streams, whereas *B. tauricus* is found in larger, slow-flowing rivers.

Further reading. Turan et al. 2009b (description).

Barbus pergamonensis

Common name. Pergamon barbel.

Diagnosis. Distinguished from other species of *Barbus* in West Asia by: ○ 50–62 total lateral-line scales / ○ last unbranched dorsal ray ossified / ○ 8–11, usually 9, gill rakers / ○ 41–42 total vertebrae / ○ head length 24–26 % SL / ○ lower lip with a median lobe / ○ numerous irregular dark-brown blotches on back, flanks and head / ○ last unbranched dorsal ray 52–72 % ossified / ○ posterior margin of dorsal slightly convex / ○ anal longer in female than in male. Size up to 172 mm SL.

Distribution. Türkiye: Karamenderes south to Madra, Bakırçay and Gediz (Aegean basin). Also, on Lesbos Island (Greece).

Barbus pergamonensis; Bakırçay drainage, Türkiye; 120 mm SL.

Barbus cf. *pergamonensis*; Tahtalı drainage, Türkiye; 115 mm SL.

Habitat. A wide range of streams and rivers, usually in fast to moderately fast-flowing waters. Inhabit reservoirs where inflowing streams provide spawning habitat. Spawns in fast-flowing waters in riffles on gravelly substrate.

Biology. Matures at about 2–4 years, males one year earlier than females. A fractional spawner. Feeds mainly on aquatic invertebrates and algae.

Conservation status. LC; Still widespread, but range greatly reduced during 20th century due to massive pollution.

Remarks. Between the ranges of *B. pergamonensis* in north and *B. xanthos* in south, a third molecular cluster occurs in Tahtalı and Küçük Menderes drainages, which may represent an undescribed species. On the other hand, *B. pergamonensis* and *B. xanthos* are already very similar, and further studies may show that the characters distinguishing the two species are not be confirmed. They may represent a case similar to the *B. lacerta* group (see above).

Further reading. Karaman 1971 (description); Güçlü et al. 2020 (morphology, distribution).

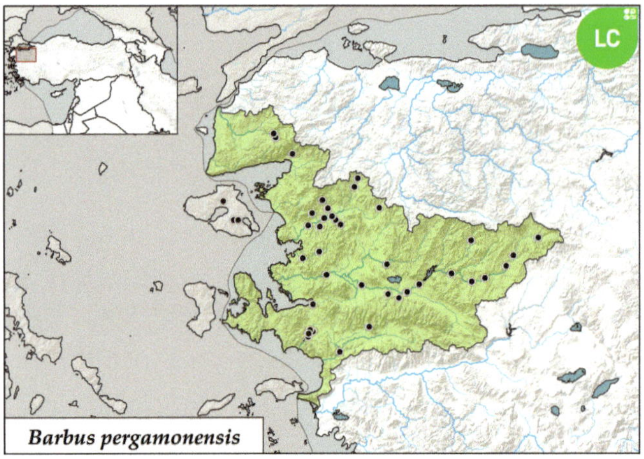

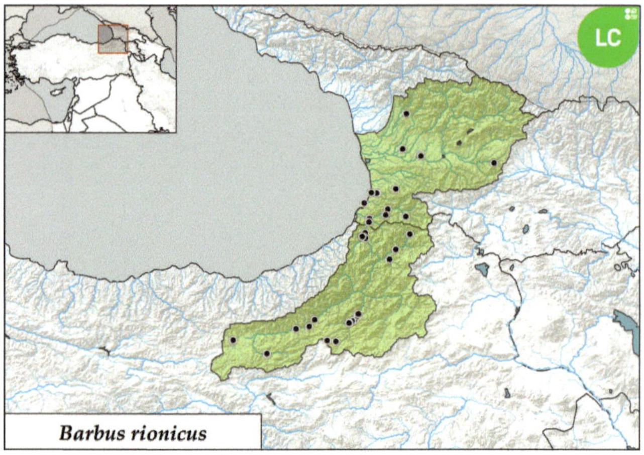

Barbus rionicus; Çoruh drainage, Türkiye; ~100 mm SL.

Barbus rionicus

Common name. Colchic barbel.

Diagnosis. Distinguished from other species of *Barbus* in West Asia by: ○ flank plain brown or grey with many minute, dark-brown spots, smaller than scales, in individuals over 100 mm SL / ○ lower lip with a median lobe / ○ 50–62, usually 54–58, total lateral-line scales / ○ last dorsal unbranched ray weakly to moderately ossified / ○ 8–12 gill rakers / ○ head length 24–28 % SL / ○ 44–47, usually 45–46, total vertebrae / ○ anal longer in female than in male. Size up to 500 mm SL, but expected to grow larger.

Distribution. Georgia and Türkiye: Eastern Black Sea basin from Kodori south to Rioni and Çoruh.

Habitat. Mountain to hill streams with gravel bottoms and moderate- to fast-flowing water. Prefers stretches with large rocks and rapids alternating with deep holes. Spawns in riffles and rapids.

Biology. Males mature at 2–3 years, 1–3 years earlier than females. Spawns May–September with a peak from May–July at water temperatures of 14–21°C. Spawns in large groups, fractional spawners. Migrates upstream to spawn. Feeds on benthic invertebrates and aquatic plants.

Conservation status. LC.

Further reading. Elanidze 1983 (morphology, biology; as *B. tauricus escherichii*); Yıldırım et al. 2001 (biology, Çoruh); Bogutskaya et al. 2003c (morphology, biology, as *B. tauricus escherichii*, partly); Kotlík et al. 2004 (molecular).

Barbus tauricus; Yenice drainage, Türkiye; ~130 mm SL.

Barbus tauricus; Biga drainage, Türkiye; 173 mm SL.

Barbus tauricus

Common name. Crimean barbel.

Diagnosis. Distinguished from other species of *Barbus* in West Asia by: ○ flank plain brown or grey, with many minute dark-brown spots, smaller than scales, in individuals over 100 mm SL / ○ lower lip with a median swollen pad / ○ 51–64, usually 53–56, total lateral-line scales / ○ last unbranched dorsal ray moderately ossified / ○ 9–12 gill rakers / ○ head length 26–30 % SL / ○ 44–48, usually 45–46, total vertebrae / ○ anal longer in female than in male. Size up to 700 mm SL.

Distribution. Southern tributaries of Marmara Sea from Lake İznik west to Karamenderes. Black Sea basin from Crimean Peninsula east along Caucasian Black Sea basin of Russia. All along Anatolian Black Sea coast, west to Thrace, and north to Kamchiya in Bulgaria. In Azov Sea basin in lower Kuban drainage, possibly hybridised with *B. kubanicus*. Absent or restricted to lowermost Sakarya, Kızılırmak, and Yeşilırmak and eastern Black Sea basin in Georgia and Çoruh (Türkiye).

Habitat. Widespread, from mountain streams with strong currents to brackish estuaries. Inhabits river reaches from coastal areas to about 600 m above sea level.

Biology. Males spawn first time at 2 years, about 65 mm SL, females at 3 years, about 100 mm SL. Moves upstream to upper reaches or near rapids to spawn on gravel or rocky bottom. Spawns May–July when water temperature reaches 15–18°C, often in large groups. Females lay 1–3 portions of eggs. Eggs are not sticky and remain between or under pebbles until hatching. In early winter moves to deeper areas with less current and stops feeding. Feeds mainly on zoobenthos.

Lower Sakarya in Türkiye. Typical habitat of rheophilic fishes.

Life cycles of rheophilic fishes. Lampreys, most sturgeons, many cyprinids, leuciscids, nemacheilids, sisorid catfishes, and salmoniforms spend at least part of their lives in moderate- to fast-flowing waters, and their life cycles show many similarities. Adults spawn in fast-flowing waters on gravel or submerged vegetation. Most species lay their eggs in the gravel or spawn above the gravel, and the eggs sink into the interstices. Predators usually eat the eggs that remain on the surface of the gravel. Eggs develop in the gravel. Yolk sac larvae usually move away from the light (they are negatively phototactic) or try to move deeper into the gravel. Larvae have been observed several meters into the substrate, but only the first few centimeters are usually suitable for eggs and larvae. When the yolk sac reserves are exhausted, the larvae become positively phototactic and leave the gravel to find suitable habitats. Larval emergence usually peaks shortly after dawn, and the larvae drift downstream. Some sturgeon and cyprinid larvae drift for several days and may find their nursery habitat several kilometers downstream of the spawning sites. This can lead to strong longitudinal variation in habitat use and extensive migrations at different life stages. Larvae of most riverine species inhabit shallow banks and bays. As they grow, they move to deeper water with faster currents. Habitat use is strongly influenced by various abiotic (e.g. available habitats, climate) and biotic (e.g. food availability, predation risk) factors.

Conservation status. LC.

Remarks. *Barbus tauricus* was previously restricted to Crimea, but new data show that it is much more widespread. *Barbus kubanicus*, *B. rionicus*, *B. anatolicus*, and *B. escherichii* occur in rivers within the range of *B. tauricus*, and it is expected that hybrids of both species and introgressed individuals are common. *Barbus tauricus* seems to inhabit rivers close to the coast, while other species occur inland. *Barbus bergi* from European Türkiye and Bulgaria and *B. oligolepis* from the Marmara basin are synonyms, as proposed differentiating characters could not be confirmed. *Barbus tauricus* locally occurs sympatric with *B. niluferensis* and *B. ida*, but it inhabits larger river sections.

Further reading. Bogutskaya et al. 2003c (morphology, biology, as *B. tauricus escherichii*, partly); Levin et al. 2015 (phylogeny); Turan et al. 2009b (description of *B. oligolepis*); Turan et al. 2018c (distribution).

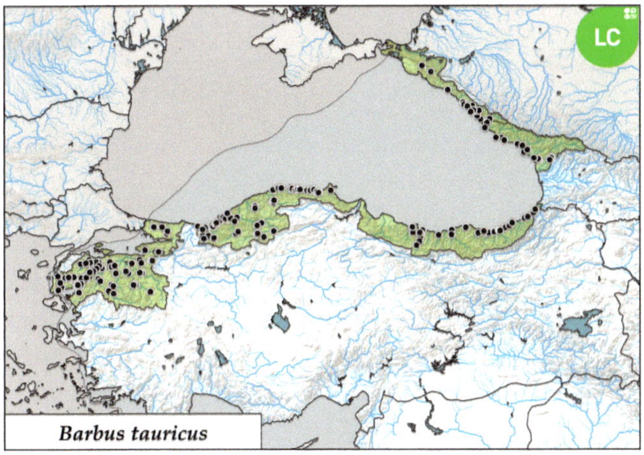

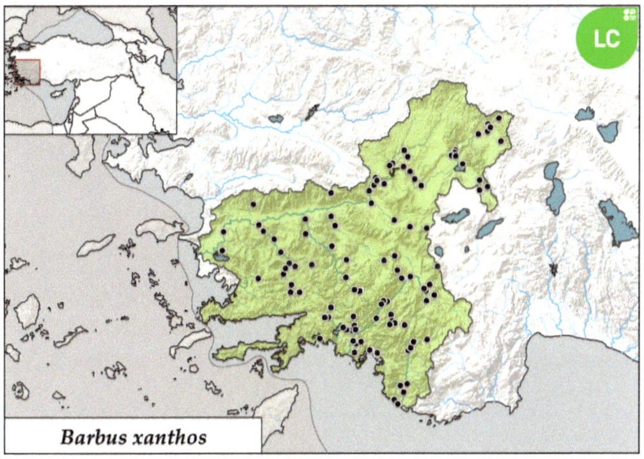

Barbus xanthos; Büyük Menderes, Türkiye; 120 mm SL.

Barbus xanthos

Common name. Menderes barbel.

Diagnosis. Distinguished from other species of *Barbus* in West Asia by: ○ 53–60 total lateral-line scales / ○ last unbranched dorsal ray weakly ossified / ○ 7–10 gill rakers / ○ 41–43 total vertebrae / ○ head length 22–27 % SL / ○ lower lip with a median lobe / ○ numerous irregular dark-brown blotches on back, flanks, and head / ○ last unbranched dorsal ray 33–50 % ossified / ○ posterior margin of dorsal straight or slightly concave / ○ anal longer in female than in male. Size up to 190 mm SL.

Distribution. Türkiye: Büyük Menderes south to Eşen (Aegean basin).

Habitat. A wide range of stream and river habitats, usually in fast to moderately fast-flowing waters. In reservoirs, where inflowing streams provide spawning habitat. Spawns in fast-flowing waters in riffles on gravel substrate.

Biology. No data. Expected to be similar to *B. pergamonensis*.

Conservation status. LC.

Further reading. Güçlü et al. 2013 (morphology, distribution); Güçlü et al. 2020 (description).

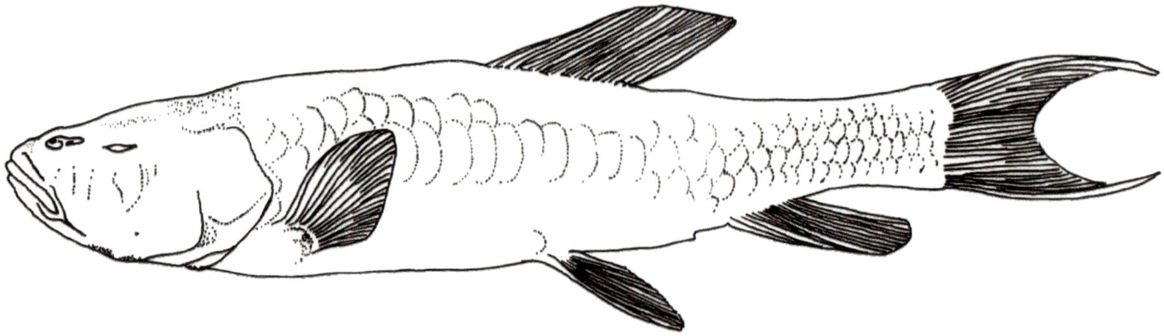

Caecocypris basimi; Haditha, Iraq, 49 mm SL. Modified from Coad 2010.

Caecocypris

This genus is represented by a single species: a small, whitish, eyeless, troglomorphic fish without barbels and with large scales. The orbit is filled with fatty tissue, and vestigial eyes are present only in juveniles. The species was only found once in 1979; virtually nothing is known about it. The large scales, body shape, and absence of barbels indicate a relationship to the geographically adjacent genus *Mesopotamichthys*.

Caecocypris basimi

Common name. Haditha cavefish.

Diagnosis. Distinguished from other subterranean fishes in West Asia by: • barbels absent / ○ no gular disc. Size up to 49 mm SL.

Distribution. Iraq: aquifer near Haditha, accessible through a well about 3 m below Shaikh Hadeed.

Habitat. Underground waters.

Biology. No data.

Conservation status. CR; possibly extinct. An attempt to locate this species in 2012 was unsuccessful.

Remarks. Coexists with *Garra widdowsoni*.

Further reading. Banister & Bunni 1980 (description).

Inside the pumping station in Shaikh Hadeed in Iraq, the only window to the habitat of *Caecocypris basimi* and *Garra widdowsoni*. © O. F. Al-Sheikhly.

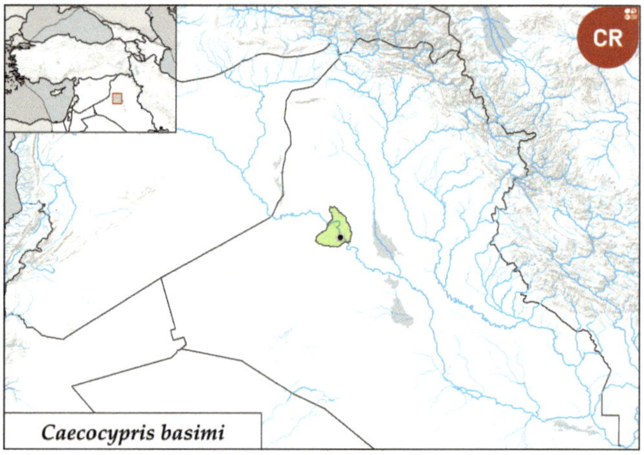

Capoeta

Capoeta are medium-sized fishes found throughout West Asia. All 27 species of the genus are present in the region, with the distribution range of *C. heratensis* extending to the Amu Darya drainage in the Aral basin. *Capoeta* are unmistakable, with their 1–2 pairs of short barbels, straight or rounded lower jaw, horny cutting sheath, and strong, serrated last unbranched dorsal ray. Some *Capoeta* have two pairs of barbels, while most have one. There is some variation in this character, with occasional individuals having one, two, or even three barbels.

The molecular diversity of *Capoeta* was the focus of several detailed studies, resulting in the description of 12 new species since the year 2000. This makes *Capoeta* the second-largest genus of Cyprinids in West Asia (after *Garra*). These studies also resulted in the description of several "molecular" species that are hardly distinguishable by external characters. Several of the populations that were proposed as species could not be confirmed by subsequent studies and are treated as synonyms. *Capoeta* frequently forms hybrids with sympatric *Luciobarbus*. Many species of *Capoeta* show remarkable variation in the shape of the mouth, especially the lower jaw, which is often wide, straight in some individuals, narrow, and arched in others. The hypothesis that these differences in mouth shape are attributed to gender was not supported when larger series were studied.

The genus *Paracapoeta* was proposed for *C. anamisensis*, *C. barroisi*, *C. erhani*, *C. mandica*, and *C. trutta*. We reject this genus as valid. This species group forms a monophyletic group, very closely related to *Capoeta*, and morphological characters poorly distinguish it. *Capoeta anamisensis* does not show any of the proposed diagnostic character states for *Paracapoeta*.

Capoeta forms an important component of many fish communities in West Asia, which are often dominated by these cyprinids by biomass. They are very abundant in many places and found in almost all kinds of water bodies, from cold mountain streams to large lowland lakes and small, brackish desert streams, as long as the fish have access to small patches of stony substrate and flowing waters. Spawning also takes place at the outflow of springs or wave-washed shores of lakes. Some *Capoeta* species are likely very powerful invasive species, but there have been few documented cases of translocations so far. **Further reading.** Levin et al. 2012 (phylogeny); Ghanavi et al. 2016 (phylogeny, Iran); Geiger et al. 2014 (phylogeny, Mediterranean); Bektaş et al. 2017 (phylogeny); Turan et al. 2022a (*Paracapoeta*).

Keys to species of *Capoeta*

Marmara and Mediterranean basins, including endorheic basins in Central Anatolia, the Damascus, and the Dead Sea basins

1a - Two pairs of barbels.
..................2
1b - One pair of barbels.
..................3

2a - Posterior edge of last unbranched dorsal ray smooth, without serrae; 49–57 total lateral-line scales.
..................*C. antalyensis*
2b - Posterior edge of last unbranched dorsal ray with serrae; 69–87 total lateral-line scales.
..................*C. tinca*

3a - Edge of lower jaw not cornified; lips fleshy; snout long and pointed.
..................*C. pestai*
3b - Edge of lower jaw cornified; lips narrow, not fleshy; snout short and blunt.
..................4

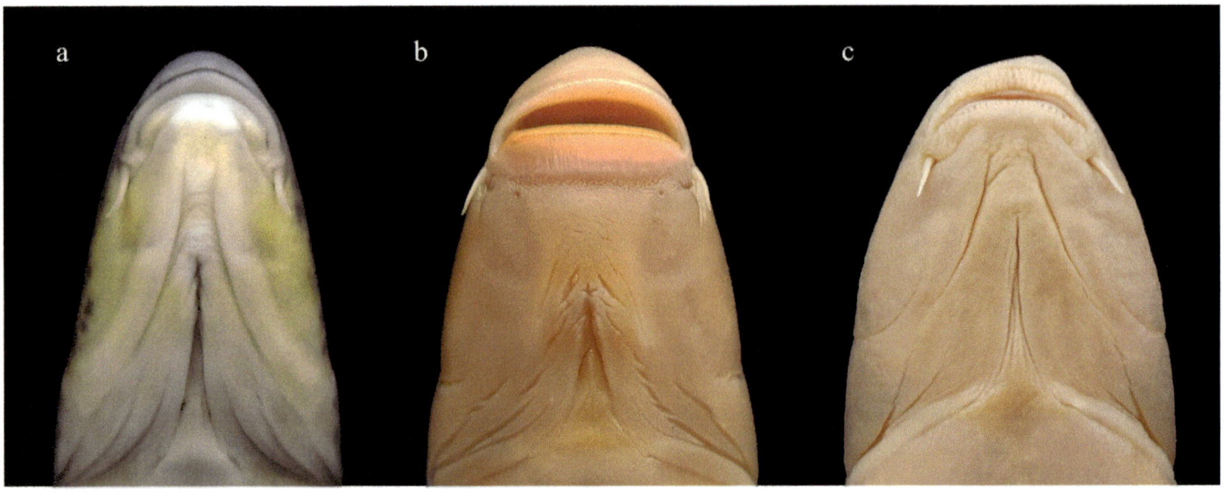

Figure 36. Shape of lower jaw: **a**, non-cornified in *C. pestai* from Lake Beyşehir; **b**, cornified in *C. damascina* from Nusaybin, Euphrates; **c**, lower lip pleated in *C. sieboldii* from Sakarya.

4a - Flank silvery or brown with many small black spots; last unbranched dorsal ray very strongly ossified, with strong serrae reaching almost to tip of ray.
..................5
4b - Flank beige or brown without small black spots; juveniles and some adults with large black blotches; last unbranched dorsal ray moderately ossified, without serrae or with small serrae, not or poorly ossified and without serrae on tip of ray.
..................6

5a - 76–83 total lateral-line scales; pelvic–hypural distance when carried forward, always falling in front of tip of snout; predorsal keel not or very slightly elevated; black spots on dorsal head smaller than on predorsal flank.
..................*C. barroisi*
5b - 63–79 total lateral-line scales; pelvic–hypural distance when carried forward, falling in between midpoint of nostril and tip of snout and center of eye; predorsal keel elevated; black spots on dorsal head about equally sized than on predorsal flank.
..................*C. erhani*

6a - Last unbranched dorsal ray weak, without serrae.
..................*C. caelestis*
6b - Last unbranched dorsal ray ossified, with serrae.
..................7

7a - Usually 68–82 total lateral-line scales; usually 9½ branched dorsal rays.
..................*C. damascina*
7b - Usually 58–71 total lateral-line scales; usually 8½ branched dorsal rays.
..................8

8a - Last unbranched dorsal ray with 21–27 small serrae; posterior margin of dorsal concave.
..................*C. bergamae*
8b - Last unbranched dorsal ray with 7–20 small serrae; posterior margin of dorsal straight.
..................*C. aydinensis*

Black Sea basin

1a - Two pairs of barbels.
..................2
1b - One pair of barbels.
..................4

2a - Last dorsal ray without serrae in individuals smaller than 100 mm SL, with 7–9 serrae in larger individuals.
..................*C. svanetica*
2b - Last dorsal ray with 12–28 serrae.
..................3

3a - 12–14 scale rows between lateral line and dorsal origin; 12–18 gill rakers; eye diameter smaller than cheek length; shape of lower jaw monomorphic, wide and straight in all individuals.
..................*C. banarescui*
3b - 14–17 scale rows between lateral line and dorsal origin; 16–22 gill rakers; eye diameter equal to or greater than cheek length; shape of lower jaw polymorphic, wide and straight or arched and narrow in some individuals.
..................*C. tinca*

4a - 7–10 gill rakers; 7½ branched dorsal rays; 68–82 total lateral-line scales; 0–8 serrae along posterior margin of last simple dorsal ray.
..................*C. oguzelii*
4b - 18–35 gill rakers; 8–9½ branched dorsal rays; 46–60 total lateral-line scales; many serrae along posterior margin of last simple dorsal ray.
..................5

5a - Lower and upper lips pleated, well developed and fleshy; edge of lower jaw not cornified; 28–35 gill rakers.
..................*C. sieboldii*
5b - Lower and upper lips smooth and narrow; edge of lower jaw cornified; 17–26 gill rakers.
..................*C. capoeta*

Caspian basin, Lake Urmia and Namak basins, Hari drainage, rivers flowing to the Persian Gulf south of Zohreh and Iranian endorheic basins

1a - Flank plain silvery or silvery with many small black spots, rarely brown with many black spots; last unbranched dorsal ray moderate to strongly ossified, with strong serrae.
..................2
1b - Flank beige, golden or brown without small black spots, juveniles and some adults with large black blotches; last unbranched dorsal ray soft or moderately ossified, with small serrae.
..................3

2a - Flank plain silvery or pale brown, often with grey stripe.
..................*C. anamisensis*
2b - Flank silvery or brown with many small black spots.
..................*C. mandica*

3a - Two pairs of barbels.
..................*C. heratensis*
3b - One pair of barbels.
..................4

4a - 63–95 total lateral-line scales.
..................5
4b - 32–63 total lateral-line scales.
..................7

5a - A deep groove in front of nares.
..................*C. coadi*
5b - No groove in front of nares.
..................6

6a - 23–28 circumpeduncular scales; 7–10 scales between lateral line and anal origin.
..................*C. saadii*
6b - 29–31 circumpeduncular scales; 10–11 scales between lateral line and anal origin.
..................*C. buhsei*

7a - Usually 7½ branched dorsal rays; 12–15 gill rakers.
..................*C. fusca*
7b - Usually 8–9½ branched dorsal rays; 15–25 gill rakers.
..................8

8a - Usually 9½ branched dorsal rays.
..................*C. kaput*
8b - Usually 8½ branched dorsal rays.
..................9

9a - 36–44 total lateral-line scales.
..................*C. aculeata*
9b - 46–59 total lateral-line scales.
..................10

10a - 44–47 total vertebrae.
..................*C. capoeta*
10b - Less than 44 total vertebrae.
..................*C. razii*

Euphrates and Tigris drainages, including Lake Van basin, excluding Karkheh, Karun, and Zohreh drainages in Iran

1a - Flank silvery with many small black spots, rarely brown with many black spots; last unbranched dorsal ray very strongly ossified, as long or longer than head, rarely shorter, with strong serrae.
..................*C. trutta*
1b - Flank beige, golden or brown without small black spots, juveniles and some adults with large black blotches; last unbranched dorsal ray soft or moderately ossified, shorter than head, with small serrae.
..................*C. damascina*

Karkheh, Karun, and Zohreh drainages in Iran

1a - Last unbranched dorsal ray as long as or longer than head, rarely shorter; adults larger than 150 mm SL, usually silvery with many small dark-brown or black blotches.
..................*C. trutta*
1b - Last unbranched dorsal ray shorter than head; adults larger 150 mm SL brown or silvery, without or with minute dark-brown or black spots.
..................2

2a - 32–44 total lateral-line scales.
..................*C. aculeata*
2b - 56–104 total lateral-line scales.
..................3

3a - A deep groove in front of nares.
..................4
3b - No groove in front of nares.
..................5

4a - 12–18 gill rakers on lower limb.
..................*C. damascina*
4b - 10–13 gill rakers on lower limb.
..................*C. coadi*

5a - Head depth at eye 90–102 % of postorbital length.
..................*C. ferdowsii*
5b - Head depth at eye 105–126 % of postorbital length.
..................*C. shajariani*

Capoeta aculeata; Lake Namak basin, Iran; ~200 mm SL.

Capoeta aculeata; Karun drainage, Iran; 183 m SL.

Capoeta aculeata

Common name. Largescale scraper.

Diagnosis. Distinguished from other species of *Capoeta* in Iranian Tigris and endorheic basins by: ● 36–44 total lateral-line scales / ○ 14–15 predorsal scales / ○ one pair of barbels / ○ flank beige, golden or brown without small black spots, juveniles and some adults with large black blotches / ○ last unbranched dorsal ray soft or moderately ossified, with small serrae / ○ usually 8½ branched dorsal rays / ○ 16–17 circumpeduncular scale rows / ○ 6–8, usually 7 scale rows between dorsal origin and lateral line / ○ 5–8, usually 6 scales between anal origin and lateral line / ○ 4–7, usually 5–6 scales between pelvic origin and lateral line / ○ 17–22 gill rakers / ○ 38–39 total vertebrae. Size up to 370 mm SL.

Distribution. Iran: Tigris drainage, endorheic Kor and Esfahan basins. Qom, Qareh Chai, Jaj, Khar, Ab-e Kamar drainages in Lake Namak basin. Hable, Nam and Bidvaz drainages in Western Kavir basin.

Habitat. Streams, rivers, and springs with moderately fast- to slow-flowing water. From reservoirs, they migrate to rivers to spawn.

Biology. Feeds on detritus, periphyton, and occasionally small invertebrates.

Conservation status. LC.

Remarks. *Capoeta aculeata* had been described from Iran without details of type locality. *Capoeta bergi*, described from the Lake Namak basin, was treated as a synonym. *Capoeta alborzensis* was described from the Lake Namak basin based on the assumption that the type locality of *C. aculeata* is in Kor basin and *C. bergi* is not an available name. No type material of *C. aculeata* was examined for that study. Later research identified the populations from the Lake Namak basin as *C. aculeata* and treated *C. gracilis* from the Esfahan basin and *C. macrolepis* from the Tigris and endorheic Kor as valid species. Again no type material was examined. We consider *C. aculeata*, *C. macrolepis*, *C. gracilis*, *C. bergi*, and *C. alborzensis* as conspecific as they are very closely related, with an COI distance < 1 %, and none of morphological characters proposed to distinguish them could be confirmed.

Further reading. Coad & Krupp 1994 (re-validation); Jouladeh-Roudbar et al. 2015b (distribution); Ghanavi et al. 2016 (phylogeny); Jouladeh-Roudbar et al. 2016a (description as *C. alborzensis*); Zareian et al. 2017 (phylogeny, morphology); Jouladeh-Roudbar et al. 2020 (distribution as *C. alborzensis*); Coad 2021a (biology, morphology).

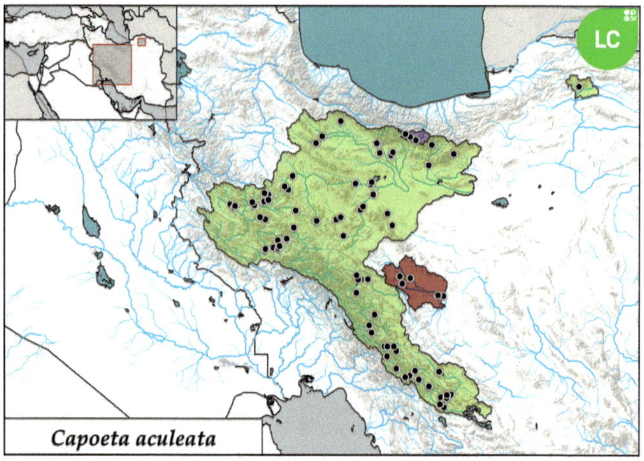

Capoeta aculeata

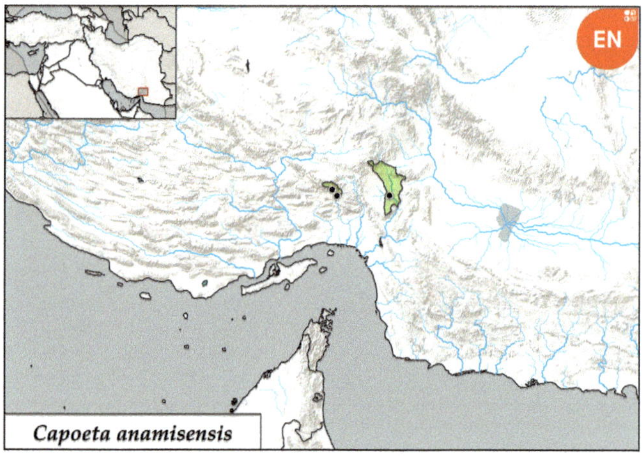

Capoeta anamisensis

Intrapopulation variation in mouth shape. In some, but not all, *Capoeta* species, there is a remarkable diversity of mouth shapes. In some individuals of the same size, the jaw is almost straight, while in others, it is horseshoe-shaped or strongly curved. These differences in mouth shape have been interpreted as sexual dimorphism. However, examination of a larger series has shown that mouth shape is unrelated to sex and that different mouth shapes occur at all ages and body sizes and reflect individual differences. The differences in mouth shape do not represent distinct morphs, and all intermediate mouth shapes are observed. Different mouth shapes may be related to ecological adaptations or feeding strategies that allow individuals within the same population to exploit a variety of food resources or habitats. However, this remains speculative and needs to be demonstrated by studying individual feeding behaviour, diet analysis and trophic signatures, such as stable isotopes.

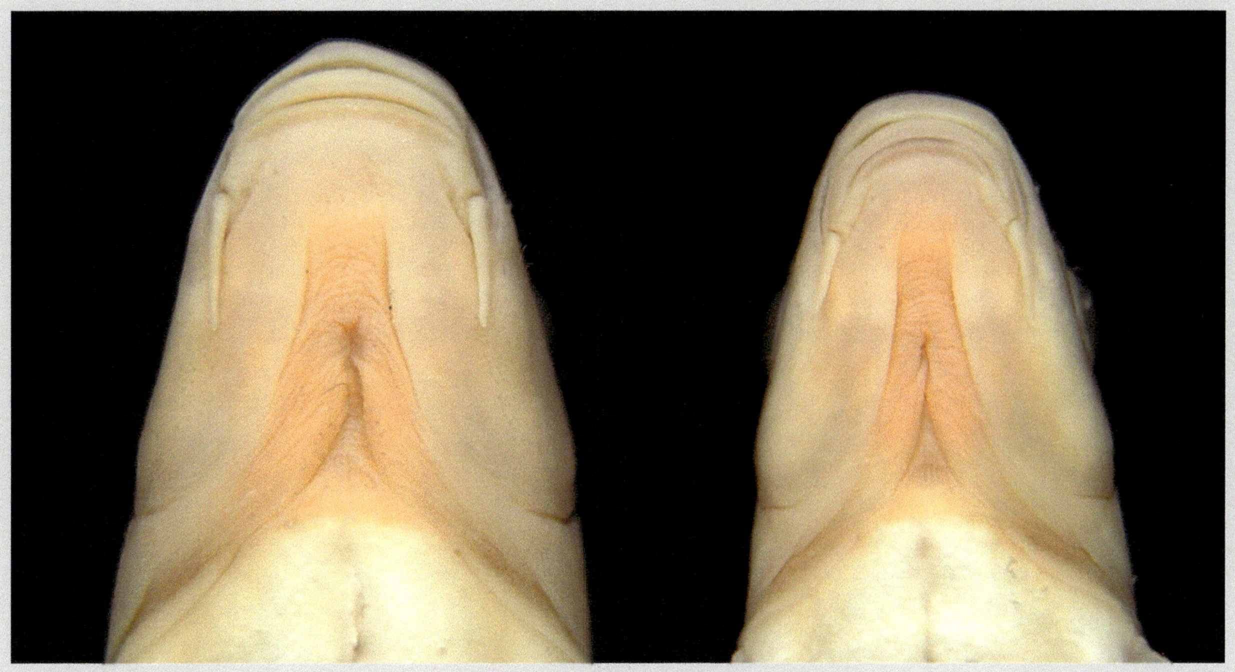

Capoeta caelestis; Göksu drainage; Türkiye; wide- and narrow-mouth forms.

Capoeta anamisensis; Minhab drainage, Iran; 120 mm SL.

Capoeta anamisensis

Common name. Hormuz scraper.

Diagnosis. Distinguished from other species of *Capoeta* in Persian Gulf basin south of Zohreh and in Iranian endorheic basins by: ○ 56–67 total lateral-line scales / ○ flank plain silvery, often with an indistinct stripe / ○ last unbranched dorsal ray very strongly ossified, with strong serrae / ○ 8½–9½ branched dorsal rays / ○ 11–12 scale rows between dorsal origin and lateral line / ○ 6–8, usually six scales between anal origin and lateral line / ○ 10–13 scales between pelvic origin and lateral line / ○ one pair of barbels / ○ 21–25 gill rakers. Size up to 230 mm SL.

Distribution. Iran: Upper Minab, Hasan Langi, and Haji Abad drainages, which flow into Strait of Hormuz.

Habitat. Streams and rivers with moderate to fast-flowing water, sand, gravel or rock substrate.

Biology. Feeds on detritus, periphyton, and occasionally small invertebrates.

Conservation status. EN; appears to be declining within its very small range.

Further reading. Zareian et al. 2016 (description).

Capoeta antalyensis; Köprüçay drainage, Türkiye; 163 mm SL.

Capoeta antalyensis

Common name. Pamphylian scraper.

Diagnosis. Distinguished from other species of *Capoeta* in Mediterranean basin and endorheic basins in Central Anatolia by: ● two pairs of barbels / ● 49–57 total lateral-line scales / ○ posterior edge of last unbranched dorsal ray smooth, without serrae / ○ 15–17 gill rakers / ○ edge of lower jaw cornified. Size up to 240 mm SL.

Distribution. Türkiye: Aksu, Köprüçay drainages and coastal streams east to Manavgat.

Habitat. Fast-flowing stretches of rivers and larger streams with gravel or rock substrate.

Biology. Feeds on detritus, periphyton, and occasionally small invertebrates.

Conservation status. VU; believed to be declining within its small range.

Remarks. In Ilıca and Manavgat, individuals with two, three and four barbels occur together. Hybrids of *C. antalyensis* and *C. caelestis* inhabit these streams. No hybrids were observed west of Gündoğdu (36.8502, 31.2900), where only *C. antalyensis* is found.

Further reading. Battalgil 1944 (description); Turan et al. 2006a (morphology); Bektaş et al. 2017 (phylogeny); Kaya et al. 2019 (hybrids).

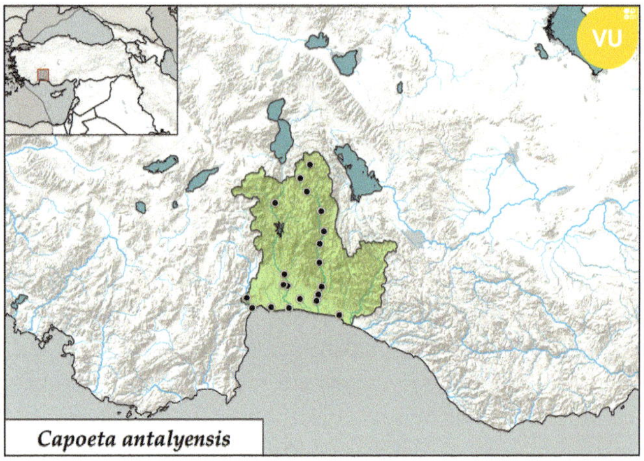

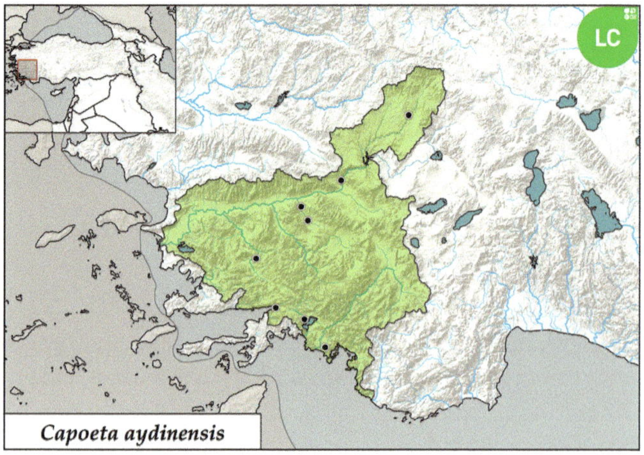

Capoeta aydinensis; Büyük Menderes drainage, Türkiye; 153 mm SL.

Capoeta aydinensis

Common name. Carian scraper.

Diagnosis. Distinguished from other species of *Capoeta* in Marmara and Mediterranean basins and Central Anatolia by: ○ 58–71 total lateral-line scales / ○ 8½ branched dorsal rays / ○ 18–24 gill rakers / ○ keel in front of dorsal origin well developed / ○ posterior dorsal margin straight / ○ flank plain brown, rarely silvery in individuals larger than 150 mm SL / ○ last simple dorsal ray weakly ossified, with 7–20 small serrae along its posterior edge origin / ○ 11–12 scale rows between dorsal origin and lateral line / ○ 7–9 scales between anal origin and lateral line / ○ one pair of barbels / ○ edge of lower jaw cornified. Size up to 370 mm SL.

Distribution. Türkiye: Büyük Menderes south to Dalaman and streams Namnam and Tersakan. Recorded from Gökova Bay and Lake Köyceğiz.

Habitat. A wide range of streams and rivers with moderate to fast-flowing water and sand or gravel and rock substrate.

Biology. No data.

Conservation status. LC.

Further reading. Bektaş et al. 2017 (phylogeny); Turan et al. 2017c (description); Akyol et al. 2020 (record from Gökova Bay).

Capoeta banarescui; Çoruh drainage, Türkiye; 118 mm SL. © M. Özuluğ.

Capoeta banarescui

Common name. Colchic scraper.

Diagnosis. Distinguished from other species of *Capoeta* in Black Sea basin by: ○ two pairs of barbels / ○ 64–77 total lateral-line scales / ○ 12–14 scale rows between lateral line and dorsal origin / ○ 8–9 scale rows between lateral line and anal origin / ○ 12–18 gill rakers / ○ 12–20 serrae on last unbranched dorsal ray / ○ eye diameter smaller than cheek length / ○ edge of lower jaw cornified / ○ lips narrow, not fleshy or pleated / ○ shape of lower jaw monomorphic, wide and straight in all individuals / ○ snout short and blunt. Size up to 350 mm SL.

Distribution. Türkiye and Georgia: Yeşilırmak drainage east to Inguri.

Habitat. A wide range of moderately to rapidly flowing streams and rivers with sand, gravel, and rock substrate. Spawns on coarse sand or gravel in flowing water.

Biology. Lives up to 12 years. Matures at 2 (male) or 3 (female) years. Spawns May–July. Feeds on detritus, algae, and invertebrates.

Conservation status. LC.

Further reading. Yıldırım & Aras 2000 (biology); Turan et al. 2006a (description); Bektaş et al. 2017 (phylogeny).

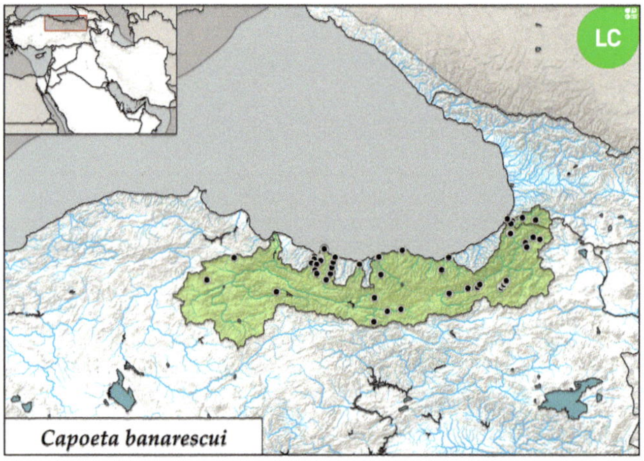

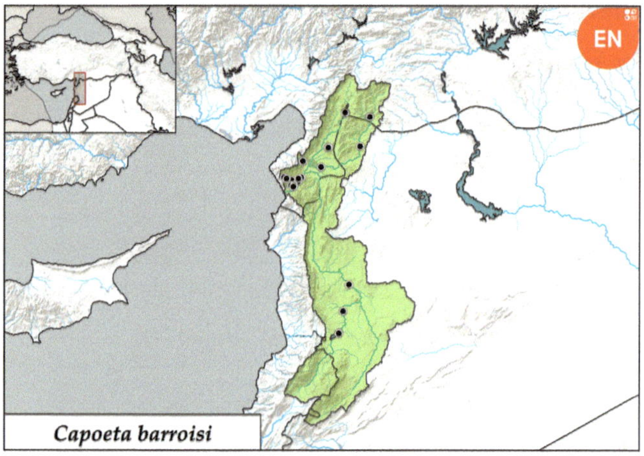

Capoeta barroisi; Orontes drainage, Türkiye; individual with long dorsal ray, 155 mm SL.

Capoeta barroisi; Orontes drainage, Syria; 200 mm SL.

Capoeta barroisi
Common name. Orontes scraper.
Diagnosis. Distinguished from other species of *Capoeta* in Mediterranean and endorheic basins in Levant by: ○ one pair of barbels / ○ 76–83 total lateral-line scales / ○ flank silvery with many small black spots / ○ black spots on dorsal head smaller than on predorsal body / ○ pelvic–hypural distance when carried forward, always falling in front of tip of snout / ○ last unbranched dorsal ray very strongly ossified, strongly serrated / ○ 26–29 gill rakers / ○ predorsal keel not or very slightly elevated. Size up to 320 mm SL.
Distribution. Türkiye and Syria: Orontes drainage.
Habitat. Lakes, reservoirs, and larger lowland rivers. Likely to migrate to inflowing rivers or streams to spawn.
Biology. Feeds on detritus, periphyton, and occasionally small invertebrates.
Conservation status. EN; appears to be declining within its small range. Restricted to a few localities in Orontes drainage such as Tahtaköprü reservoir, lower Orontes, upper Afrin and Lake Gölbaşı (Kırıkhan) in Türkiye and Qattinah reservoir in Syria. However, the exact distribution should be reviewed.
Remarks. Individuals of *C. barroisi* with the last unbranched dorsal ray as long as or longer than the head are often misidentified as *C. trutta*. *Capoeta trutta* does not occur in the Orontes.
Further reading. Turan et al. 2008b (distribution, morphology).

***Capoeta*, a genus of hybrid origin?** It has long been suspected that the relationship of *Capoeta* is with the Oriental algae scrapers of the genera *Onychostoma*, *Semiplotus*, and *Scaphiodonichthys*. Since, mitochondrial molecular markers have demonstrated that *Capoeta* is closely related to the genera *Aulopyge*, *Barbus* and *Luciobarbus*. All mitochondrial phylogenetic analyses indicate that *Capoeta* is nested within *Luciobarbus*. As is the case with the majority of Palearctic barbels, all *Luciobarbus* studied are tetraploid, with 100 chromosomes. Conversely, all *Capoeta* studied are hexaploid, with 150 chromosomes. The shift from tetraploidy to hexaploidy is a pivotal event in the evolution of *Capoeta*. Polyploidisation in fish is often associated with hybridisation, and it can be postulated that this was also the case in the origin of *Capoeta*. If a tetraploid mother (chromosomes in the egg: $n = 2$) and a diploid father (chromosome in the sperm cell: $n = 1$) (or vice versa) hybridise, the offspring is triploid ($n = 3$) and is likely to be sterile. It can be postulated that triploids cannot produce viable gametes.

It is plausible that these hybrids were sufficiently abundant in a certain situation to mate with each other and were occasionally able to produce mitotic gametes, i.e., gametes without meiotic division. These eggs and sperm should have been triploid ($n = 3$), and after fertilisation, a new fish with an even number of chromosomes, here six, would result. This hexaploid fish could then spawn normally with the other hexaploid hybrids, which might have constituted the initial step of the new lineage, the genus *Capoeta*. Support for this hybridisation event is further strengthened by nuclear DNA evidence from the RAG1 gene which, together with mitochondrial markers, suggests a biparental genomic contribution. While the mitochondrial data confirm *Luciobarbus* as the maternal source, the RAG1 tree indicates a mixed ancestry, consistent with an ancient Cyprinion-like paternal lineage, and reinforces the role of polyploidisation in the evolution of *Capoeta*. These hybridisation and polyploidisation events may have enabled the hybrids to occupy a different ecological niche than their parents, resulting in the evolution of reproductive isolation from their parents. **Further reading.** Yang et al. 2015 (hybrid origin).

Capoeta bergamae; Bakacak, Türkiye; 145 mm SL.

Capoeta bergamae

Common name. Aegean scraper.

Diagnosis. Distinguished from other species of *Capoeta* in Marmara and Mediterranean basins and Central Anatolia by: ○ 60–71 total lateral-line scales / ○ 8½, rarely 7½, branched dorsal rays / ○ 18–24 gill rakers / ○ keel in front of dorsal origin well developed / ○ posterior dorsal margin concave / ○ flank plain brown, rarely silvery in individuals larger than 150 mm SL / ○ last simple dorsal ray weakly ossified, with 21–27 small serrae along its posterior edge origin / ○ 11–14 scale rows between dorsal origin and lateral line / ○ 7–9 scales between anal origin and lateral line / ○ one pair of barbels / ○ edge of lower jaw cornified. Size up to 198 mm SL.

Distribution. Türkiye: Gediz north to Bakacak drainage (northern Biga Peninsula).

Habitat. Wide range of rivers and streams with clean and, at least seasonally, running water.

Biology. Feeds on detritus, periphyton, and occasionally small invertebrates.

Conservation status. NT; declined in recent years, but still too widespread and not declining fast enough to qualify for a threat category. Extirpated from Küçük Menderes.

Further reading. Karaman 1969 (description); Turan et al. 2017c (distribution, morphology); Bektaş et al. 2017 (phylogeny).

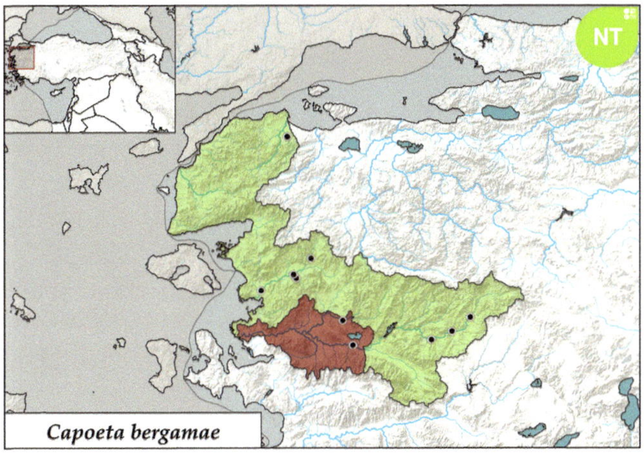

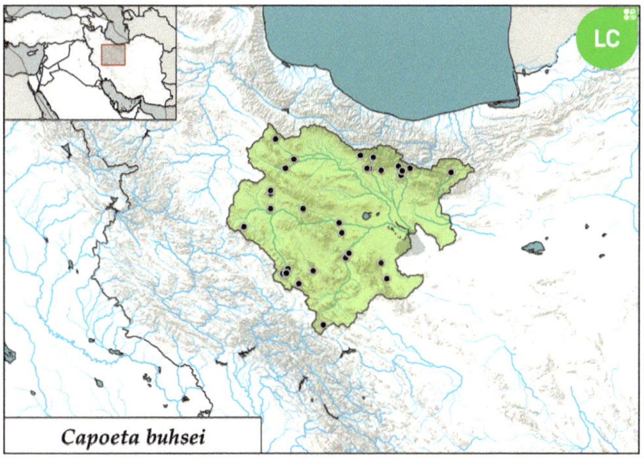

Capoeta buhsei; Lake Namak basin, Iran; ~200 mm SL.

Capoeta buhsei

Common name. Namak scraper.

Diagnosis. Distinguished from other species of *Capoeta* in Iranian endorheic basins by: ○ 82–95 total lateral-line scales / ○ usually 8½ branched dorsal rays / ○ flank beige or brown without small black spots, juveniles and some adults individuals with large black blotches / ○ no groove in front of nares / ○ one pair of barbels / ○ last unbranched dorsal ray soft or moderately ossified, serrated / ○ 14–16 scales between lateral line and dorsal origin / ○ 29–31 circumpeduncular scales / ○ 12–14 gill rakers, 7–10 on lower limb / ○ 10–11 scales between lateral line and anal origin. Size up to 230 mm SL.

Distribution. Iran: Qom, Qareh chai, Jaj, Karaj, Khar, Kan and Mazlaghan drainages in Lake Namak basin. Hable, Nam, Qolrudbar and Boneh Koh in Western Kavir basin.

Habitat. Streams with sand, mud or gravel bottom.

Biology. Feeds on periphyton.

Conservation status. LC.

Further reading. Alwan 2010 (morphology, phylogeny); Alwan et al. 2016a (morphology, phylogeny); Jouladeh-Roudbar et al. 2017b (phylogeny, morphology); Zareian & Esmaeili 2017 (description); Jouladeh-Roudbar et al. 2020 (distribution); Coad 2021a (biology, morphology).

Capoeta caelestis; Göksu, Türkiye; 137 mm SL.

Capoeta caelestis

Common name. Taurus scraper.

Diagnosis. Distinguished from other species of *Capoeta* in Mediterranean basin by: ○ one pair of barbels / ○ 57–69 total lateral-line scales / ○ last unbranched dorsal ray weak, slender, without serrae / ○ 8½ or 9½ branched dorsal rays / ○ 16–21 gill rakers / ○ 10–13 scales rows between lateral line and dorsal origin / ○ 7–8 scales between anal origin and lateral line / ○ 23–24 circumpeduncular scales / ○ flank beige, golden or brown without small black spots, juveniles and some adults individuals with large black blotches. Size up to 340 mm SL.

Distribution. Türkiye: Ilıca and Manavgat east to Göksu drainage.

Habitat. Coastal streams to mountain rivers. Often sole species in a stream. Spawns in fast-flowing water in rapids and riffles.

Biology. No data.

Conservation status. LC.

Remarks. In Ilıca and Manavgat, individuals with one pair, two pairs and three barbels occur together. These are hybrids between *C. antalyensis* and *C. caelestis*.

Further reading. Küçük 2007a (distribution); Schöter et al. 2009 (description); Alwan 2010 (morphology, phylogeny); Alwan et al. 2016b (morphology, phylogeny).

Figure 37. Gut coils in *Capoeta*, as visible in an x-ray picture, indicate the herbivory feeding behaviour of all species in this genus.

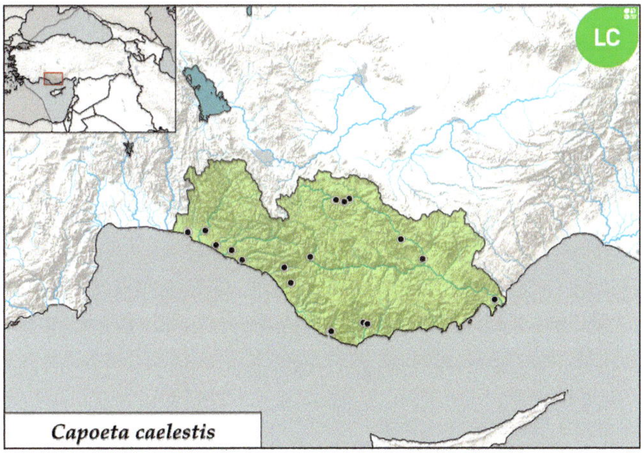

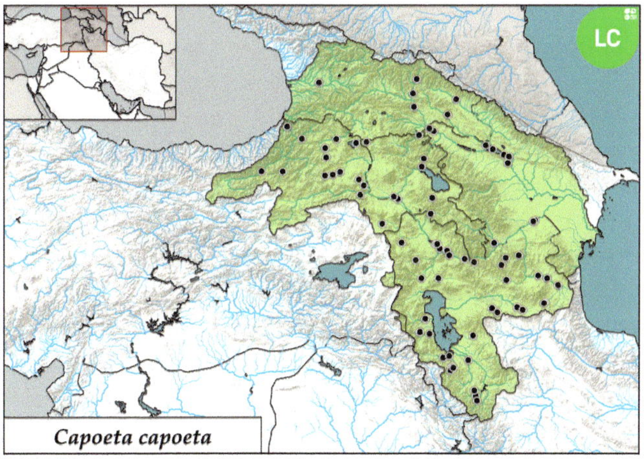

Capoeta capoeta; Urmia basin, Iran; ~140 mm SL.

Capoeta capoeta; Aras drainage, Türkiye; ~190 mm SL.

Capoeta capoeta

Common name. Caucasian scraper.

Diagnosis. Distinguished from other species of *Capoeta* in Black and Caspian Sea basins by: ○ one pair of barbels / ○ 46–59 total lateral-line scales / ○ usually 8½ branched dorsal rays / ○ last unbranched dorsal ray strong, with many serrae / ○ 18 or more predorsal scales / ○ back behind head and in front of dorsal origin strongly compressed / ○ 44–47 total vertebrae / ○ 17–26 gill rakers / ○ lower and upper lips smooth and narrow / ○ edge of lower jaw cornified. Size up to 380 mm SL.

Distribution. Caspian basin: Kura and Aras drainages (including Lake Sevan). Black Sea basin: Çoruh (Türkiye) northeast to Rioni (Georgia). Lake Urmia basin.

Habitat. Stream and river habitats from coastal to mountain streams with gravel substrates. Also in lakes and reservoirs. Spawns in fast-flowing waters in rapids and riffles on gravel or sand; also on wave-washed lake shores; migrates to lake tributaries to spawn.

Biology. Females live up to 10 years, males up to 6; usually 4–5 years. Spawns between March and September. Fractional spawner, individual females may spawn up to three times a year. Feeds mainly on detritus and periphyton.

Conservation status. LC.

Remarks. As with many barbels, the eggs are toxic and even inadequately cleaned fish can be dangerous to humans. *Capoeta ekmekciae* from the easternmost Black Sea basin (Çoruh in Türkiye northeast to Rioni in Georgia) and *C. sevangi* from Lake Sevan in Armenia are synonyms. The geographical boundaries between this species and *C. razii* south of Kura are unclear, and hybrids are expected to occur there and potentially in Aras.

Further reading. Berg 1949b (distribution, biology); Bănărescu 1999b (biology); Turan et al. 2006b (*Capoeta ekmekciae);* Zareian et al. 2017 (phylogeny, morphology); Kuljanishvili et al. 2020 (distribution, *C. ekmekciae* and *C. sevangi* as synonyms); Çiçek et al. 2022 (*C. ekmekciae* as a synonym).

Capoeta coadi; Karun drainage, Iran; 172 mm SL.

Capoeta coadi

Common name. Karun scraper.

Diagnosis: Distinguished from other species of *Capoeta* in Karkheh, Karun, Zohreh and endorheic Zayandeh drainages by: ○ usually 9½ branched dorsal rays / ○ one pair of barbels / ○ flank beige, golden or brown without small black spots, juveniles and some adults individuals with large black blotches / ○ last unbranched dorsal ray weakly ossified and serrated / ○ 14–18 gill rakers, 10–13 gill rakers on lower limb / ○ 65–84 total scales along lateral line / ○ 12–17 scales between lateral line and dorsal origin / ○ 9–11 scales between lateral line and pelvic origin / ○ 25–32 circumpeduncular scales / ○ a deep groove in front of nares. Size up to 280 mm SL.

Distribution. Iran: Karun drainage and Esfahan basin.

Habitat. Moderate or fast-flowing rivers, usually with gravel substrate and clear water.

Biology. Lives up to 10 years. Males mature at 2 years, females at 3–6 years. Spawns May–June. Feeds on periphyton.

Conservation status. LC.

Remarks. *Capoeta birunii* from the Zayandeh drainage is a synonym.

Further reading. Alwan 2010 (morphology, phylogeny); Alwan et al. 2016a (description); Jouladeh-Roudbar et al. 2016a (phylogeny); Jouladeh-Roudbar et al. 2017b (phylogeny, morphology); Zareian & Esmaeili 2017 (description of *C. birunii*); Jouladeh-Roudbar et al. 2020 (distribution, *C. birunii* as a synonym of *C. coadi*); Coad 2021a (biology, morphology); Jouladeh-Roudbar 2022 (*C. birunii* as a synonym of *C. coadi*).

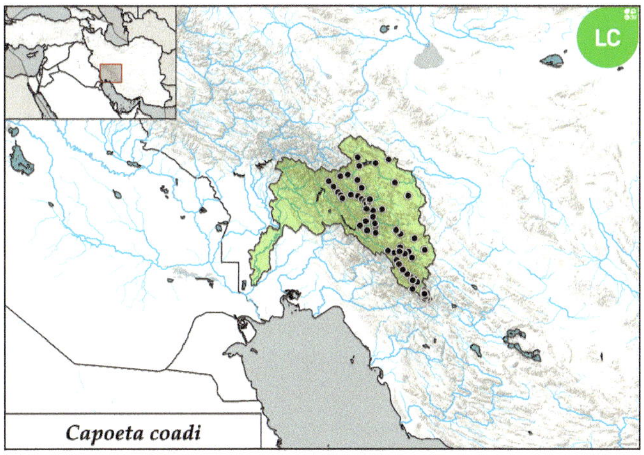

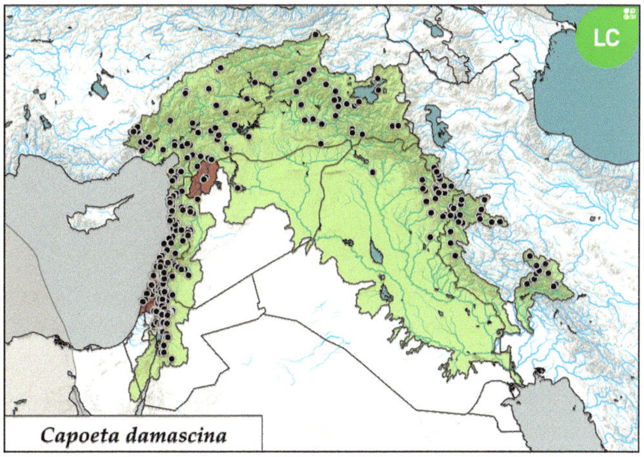

Capoeta damascina; Barada spring, Damascus basin, Syria; 163 mm SL.

Capoeta damascina; Ceyhan drainage, Türkiye; 187 mm SL.

Capoeta damascina; Dez in Karun drainage, Iran; 194 mm SL.

Capoeta damascina (*Capoeta umbla* morphotype); Tigris, Türkiye; ~320 mm SL.

Capoeta damascina
Common name. Levantine scraper.
Diagnosis. Distinguished from other species of *Capoeta* in Mediterranean, Damascus, and Dead Sea basins, Lake Van basin, Euphrates, and Tigris (except Karkheh and Karun) by: ○ flank beige, golden, or brown without small black spots, juveniles and some adults individuals with large black blotches / ○ last unbranched dorsal ray weakly to moderately ossified and serrated / ○ 8½–10½, usually 9½ branched dorsal rays / ○ usually 61–82 total lateral-line scales, many individuals have 80–104 total lateral-line scales in Tigris and Euphrates / ○ 11–20 scales between lateral line and dorsal origin / ○ 7.5–14.5 scales between lateral line and pelvic origin / ○ 23–34 circumpeduncular scales / ○ usually 18–23 gill rakers, 12–18 gill rakers on lower limb, 12–15 in Damascus basin / ○ one pair of barbels / ○ last unbranched dorsal ray shorter than head / ○ usually 18–25 scales above lateral line / ○ 11.5–15.5 scales below lateral line / ○ 31–39 circumpeduncular scales. Size up to 450 mm SL.
Distribution. Seyhan, Ceyhan, and Orontes southwards to Litani in Lebanon. Damascus and Qweiq endorheic basins, Jordan drainage, and major tributaries of Dead Sea (such as Moujib and Hasa). Euphrates and Tigris, including tributaries of Lake Van. Absent from Iranian rivers flowing into Persian Gulf south of Karun, except Dez, a tributary of Karun. Absent from other parts of Karun and Karkheh.
Habitat. A wide range of running waters, lakes, and reservoirs with (seasonally) inflowing rivers or streams. Occupies cold headwaters of mountain zones, medium-sized rivers, springs, and associated wetlands, down to small wadis and brackish desert streams. Spawns on gravel, wave-washed lake shores, and spring discharge horizons. Larvae and juveniles inhabit riparian habitats.

Biology. Lives up to 12 years. Spawns first time at 130–200 mm SL, usually matures at 2 (males) and 4 (females) years. Spawns May–June or late July (Lebanon), January–March or May (upper Jordan). Migrates regionally from lakes to rivers to spawn. In upper Jordan, migrates in December–February to spawning areas in upper reaches. Rainfall, flooding, and a drop in water temperature trigger migration. Several males usually follow a female to spawn. Eggs are sticky and are deposited in gravel excavated by female during spawning. Eggs poisonous. Adults migrate downstream after spawning. Juveniles have many dark brown blotches on a silvery or golden background. Feeds on detritus, periphyton, and invertebrates.

Conservation status. LC.

Remarks. Reported from an oasis on Sinai Peninsula, where it may have been introduced (but now extirpated). Despite its wide distribution, recent molecular studies involving fish across its range have not revealed a strong phylogeographic structure. *Capoeta kosswigi* from Lake Van basin and *C. angorae* from Seyhan and Ceyhan are synonyms. *Capoeta pyragyi*, described from the Dez, a tributary of Karun in Iran, is identical to *C. damascina* by its COI DNA sequences, and all morphological characters given in original description overlap with those of *C. damascina*. It is also treated as a synonym.

In the Tigris, high-scale count individuals (80–104, total lateral-line scales) occur in pure populations or syntopy with low-scale count individuals (61–82). In the Euphrates, no pure high-scale count populations have (yet) been reported, but low and high-scale count individuals often occur in syntopy, whereas in the Mediterranean, only low-scale count fish have been found. The high-scale count individuals are often recognised as a separate species, *C. umbla*. High and low-scale-count fish are not distinguished by their COI DNA sequences (but few individuals differ slightly in their LSU sequences), all other meristic characters overlap, and high and low-scale-count individuals are superficially very similar in all other characters. Furthermore, the distribution of scale numbers in Van Lake basin bridges the gap between *C. damascina* and "*C. umbla.*" The background of this situation has yet to be understood. A first hypothesis is that they are two species often found in sympatry; intermediate individuals could be hybrids, and the lack of COI DNA sequences could be due to introgressive hybridisation. The overall high similarity of both "species" and their syntopic occurrence challenge this view. Such superficially similar and closely related species only rarely occur in syntopy, and it is difficult to imagine that this is the case for *Capoeta*, which is known to hybridise even with sympatric other *Capoeta* and *Luciobarbus* species. Alternatively, the often bimodal scale numbers in syntopic fishes may indicate that scale number inheritance may be associated with two alternative haplotypes leading to either high- or low-scale numbers rather than intermediate scale numbers, as expected in hybrids. This hypothesis suggests a species that occurs in individuals with high- or low-scale numbers, both in mixed and pure populations. Until this hypothesis is rejected, we treat *C. umbla* and *C. damascina* as conspecifics.

Further reading. Türkmen et al. 2002 (biology); Alwan 2010 (revision of *C. damascina* group); Alwan et al. 2016b (morphology, phylogeny); Esmaeili et al. 2016a (description, molecular phylogeny); Kaya et al. 2016 (distribution); Zareian & Esmaeili 2017 (description); Jouladeh-Roudbar et al. 2017c (description); Kaya 2019 (distribution); Coad 2021a (biology, morphology).

Capoeta erhani; Seyhan drainage, Türkiye; 129 mm SL.

Capoeta erhani; Seyhan drainage, Türkiye; 220 mm SL.

Capoeta erhani
Common name. Cilician scraper.
Diagnosis. Distinguished from other species of *Capoeta* in Mediterranean basin by: ○ one pair of barbels / ○ 63–79 total scales in lateral line / ○ flank silvery or brown with many small black spots / ○ black spots on dorsal head about equally sized than on predorsal body / ○ pelvic–hypural distance when carried forward, falling in-between midpoint of nostril-the tip of snout and center of eye / ○ last unbranched dorsal ray very strongly ossified, strongly serrated / ○ 25–30 gill rakers / ○ predorsal keel elevated. Size up to 320 mm SL.
Distribution. Türkiye: Seyhan and Ceyhan drainages.
Habitat. Wide range of moderately to rapidly flowing streams and rivers with sandy, gravelly, or rocky bottoms. Common in reservoirs and large rivers.
Biology. Lives up to 5 years.
Conservation status. LC.
Remarks. *Capoeta turani* was described from Seyhan, but later studies found that the diagnostic characters overlap to a large extent. It is treated as a synonym of *C. erhani*.
Further reading. Turan et al. 2008b (description); Özuluğ & Freyhof 2008b (description as *C. turani*); Geiger et al. 2014; Bektaş et al. 2017 (phylogeny); Turan et al. 2022a (*C. turani* as a synonym of *C. erhani*).

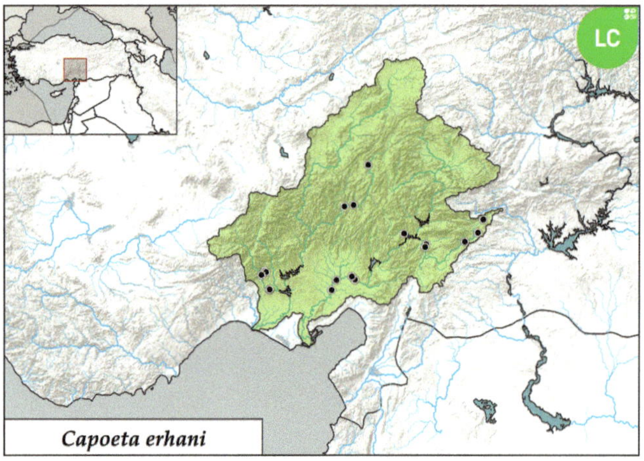

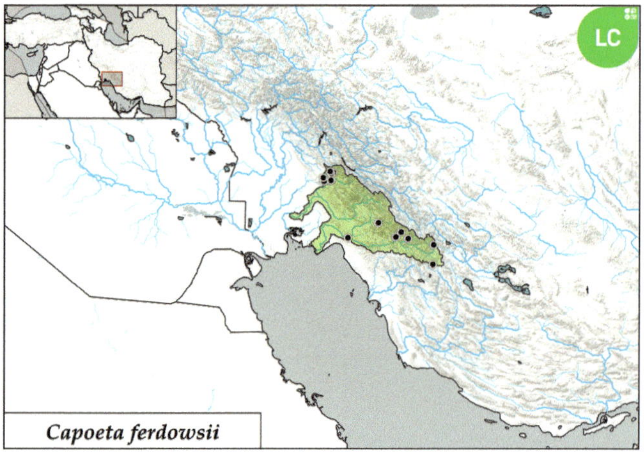

Capoeta ferdowsii; Zohreh drainage, Iran; ~ 130 mm SL.

Capoeta ferdowsii

Common name. Zohreh scraper.

Diagnosis. Distinguished from other species of *Capoeta* in Karkheh, Karun and Zohreh drainages by: ○ flank beige, golden or brown without small black spots, juveniles and some adults individuals with large black blotches / ○ usually 9½ branched dorsal rays / ○ last unbranched dorsal ray shorter than head, weakly ossified and serrated / ○ 71–80 lateral-line scales / ○ 13–17 scales between dorsal origin and lateral line / ○ 8–11 scales between anal origin and lateral line / ○ 26–30 encircling least circumference of caudal peduncle / ○ 13–15 gill rakers, 10–12 on lower limb / ○ no groove in front of nares / ○ head depth at eye 90–102 % of postorbital length. Size up to 170 mm SL, likely to grow much larger.

Distribution. Iran: Zohreh, Jarrahi, and Fahlian drainages.

Habitat. Medium to fast-flowing rivers with sand, gravel, or mud substrate.

Biology. No data.

Conservation status. LC.

Further reading. Jouladeh-Roudbar et al. 2017c (description); Zareian & Esmaeili 2017 (description).

Large, free-flowing rivers support a diverse fish fauna. The Aras at the border between Azerbaijan and Iran hosts *Capoeta capoeta* and *C. kaput*.

Capoeta fusca; Qanat-e Segonbadan, Iran; 124 mm SL.

Capoeta fusca

Common name. Qanat scraper.

Diagnosis. Distinguished from other species of *Capoeta* in Iranian endorheic basins by: ○ 7½ branched dorsal rays / ○ usually 42–50 total lateral-line scales / ○ usually 11–15 gill rakers / ○ one pair of barbels / ○ last unbranched dorsal ray soft with few fine serrae along proximal half / ○ flank beige, golden, or brown without small black spots, juveniles and some adult individuals with large black blotches. Size up to 180 mm SL, likely to grow larger.

Distribution. Sistan basin in Afghanistan. Atrak drainage, North and South Bejestan, Eastern Kavir, Gonabad, and Lut basins in Iran.

Habitat. A wide range of small to large desert streams, springs, qanats and canals. Habitats are often very small and isolated.

Biology. Lives up to 6 years. Spawns March–August, fractional spawning. Feeds on plants and filamentous algae, occasionally on invertebrates.

Conservation status. LC; although the species is widespread, its habitats are usually very small and under severe water stress.

Remarks. Scales on the belly and lower flank are deeply embedded in skin in some individuals or populations. This species is very closely related to *C. aculeata*.

Further reading. Johari et al. 2009 (biology); Patimar & Mohammadzadeh 2011 (biology); Jouladeh-Roudbar et al. 2020 (distribution); Coad 2021a (biology, morphology).

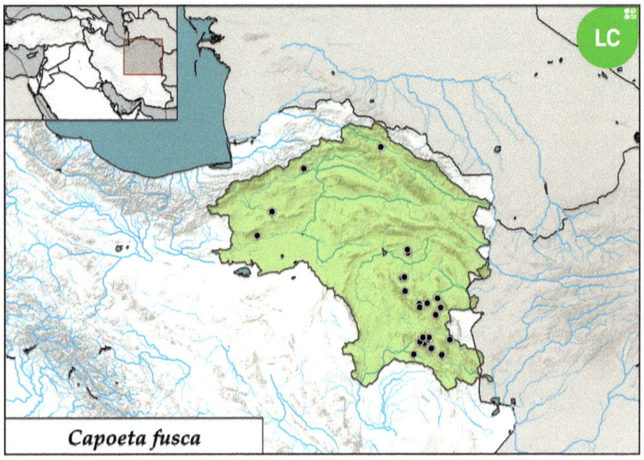

Capoeta fusca

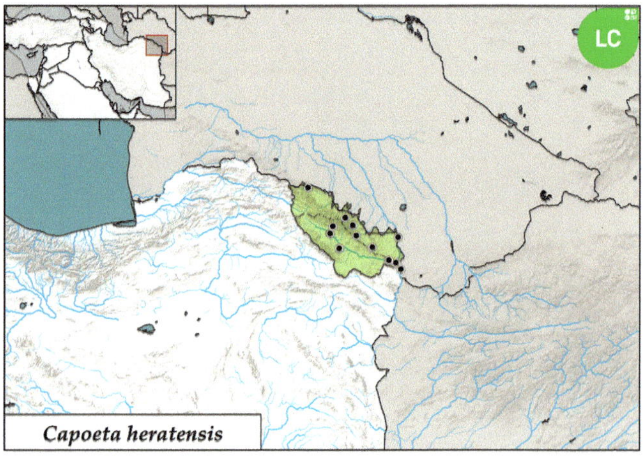

Capoeta heratensis

Capoeta heratensis; Kalat, Iran; about 150 mm SL. © H. Mousavi-Sabet.

Capoeta heratensis

Common name. Central Asian scraper.

Diagnosis. Distinguished from other species of *Capoeta* in Iranian endorheic basins by: • usually two pairs of barbels / ○ 50–61 total lateral-line scales / ○ 17–22 gill rakers / ○ usually 8½ branched dorsal rays / ○ 8–11 scales between dorsal origin and lateral line / ○ 7–8 scales between anal origin and lateral line / ○ 7–10 scales between pelvic origin and lateral line / ○ back behind head and in front of dorsal origin not or very slightly compressed / ○ last unbranched dorsal ray strong, with many serrae. Size up to 292 mm SL.

Distribution. Hari drainage in Iran, Afghanistan, and Turkmenistan. In Morghab, Karakum Canal, and Kopetdag streams in Turkmenistan, Zeravshan, and Amu Darja in Central Asia.

Habitat. A wide range of streams and rivers, including reservoirs and lakes, from which it migrates into tributaries to spawn.

Biology. Lives for 8 years, probably longer, reported to mature at 2–4 years and 100 mm (probably TL). Small spawners appear to be all males. Spawns April–August at temperatures between 16 and 23°C (Uzbekistan) on gravel bottoms. Fractional spawners, females spawn more than once in a season.

Conservation status. LC.

Remarks. Individuals with three or two barbels are common in some populations. *Capoeta steindachneri* is a synonym.

Further reading. Berg 1949b (morphology, distribution); Jouladeh-Roudbar et al. 2017b (morphology); Zareian et al. 2017 (phylogeny, morphology).

Capoeta kaput; Aras, Iran; ~300 mm SL.

Capoeta kaput

Common name. Blue scraper.

Diagnosis. Distinguished from other species of *Capoeta* in Caspian basin by: ● usually 9½ branched dorsal rays / ● flank bluish or bluish-green without small black spots, juveniles with large black blotches / ○ 54–63 total lateral-line scales / ○ 24–25 gill rakers. Size up to 900 mm SL.

Distribution. Caspian basin: Aras drainage.

Habitat. Larger rivers with moderate or fast-flowing water.

Biology. Feeds on detritus, periphyton, and occasionally small invertebrates.

Conservation status. LC.

Further reading. Levin et al. 2019 (description); Kaya 2019 (distribution); Kuljanishvili et al. 2020 (distribution).

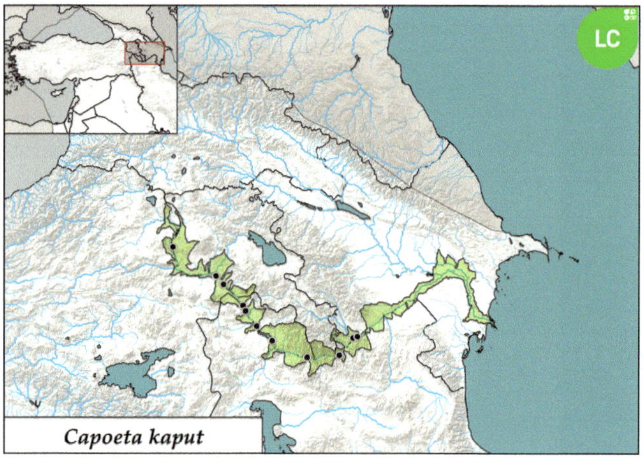

Capoeta kaput

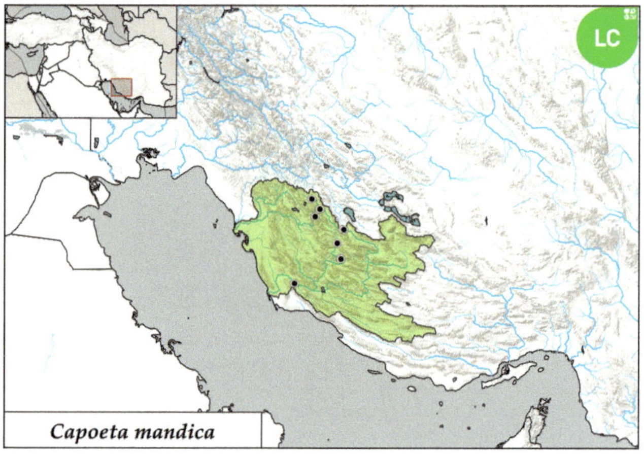

Capoeta mandica

Finding fishes in West Asia is always connected with lots of driving in full cars on bad roads, as here in Iran.

Capoeta mandica; Mond drainage, Iran; 131 mm SL.

Capoeta mandica

Common name. Mond scraper.

Diagnosis. Distinguished from other species of *Capoeta* in Persian Gulf basin south of Zohreh and in Iranian endorheic basins by: ○ flank silvery or brown with many small black spots, especially on head and back / ○ last unbranched dorsal ray very strong, shorter than head, strongly serrated / ○ 58–68 total lateral-line scales / ○ 21–27 gill rakers / ○ 12–13 scales between dorsal and lateral line / ○ 8–10 scales between lateral line and anal / ○ usually 8½ branched dorsal rays. Size up to 300 mm SL.

Distribution. Iran: Helleh and Mond drainages.

Habitat. Small rivers and streams with sand or gravel bottoms. Usually in medium or fast-flowing clear or muddy waters.

Biology. Lives up to 6 years. Spawns in spring, most likely May–June.

Conservation status. LC.

Further reading. Bianco & Bănărescu 1982 (description); Zareian et al. 2018 (description).

Capoeta oguzelii; Ezine, Türkiye; ~100 mm SL.

Capoeta oguzelii

Common name. Ezine scraper.

Diagnosis. Distinguished from other species of *Capoeta* in Black Sea basin by: ○ lower jaw strongly arched, without cornified edge / ○ 7–10 gill rakers / ○ 7½ branched anal rays / ○ one pair of barbels / ○ last unbranched dorsal ray weakly ossified, its posterior edge of with 0–8 serrae / ○ 68–82 total lateral-line scales. Size up to 139 mm SL.

Distribution. Türkiye: Upper reaches of Ezine (Kastamonu).

Habitat. Streams with moderate, slow-flowing water and a gravel, sandy or muddy bottom.

Biology. No data.

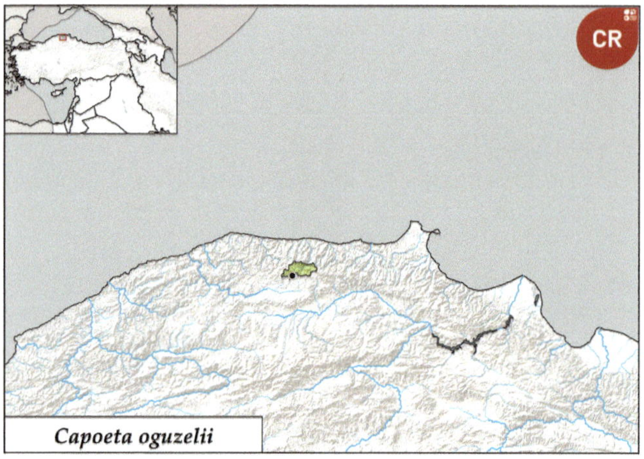

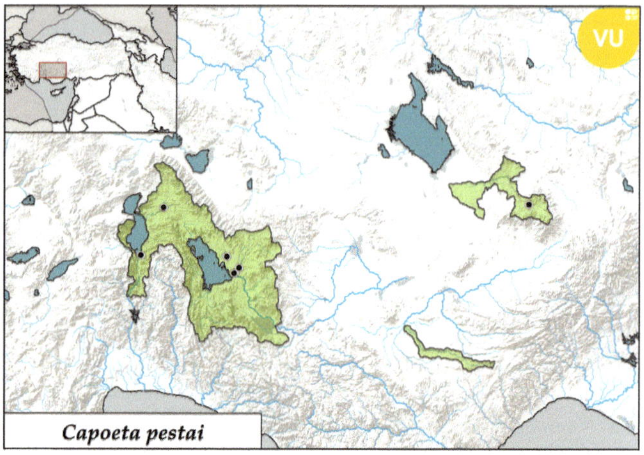

Capoeta pestai; Lake Eğirdir basin, Türkiye; 210 mm SL.

Conservation status. CR; only a few small villages around type locality. Sudden pollution or desiccation may cause immediate extinction.

Further reading. Elp et al. 2018 (description); Kaya 2019 (distribution, conservation).

Capoeta pestai

Common name. Long-snout scraper.

Diagnosis. Distinguished from other species of *Capoeta* in Central Anatolia and Mediterranean basin by: ● edge of lower jaw arched, not cornified / ● lips fleshy / ● snout long, pointed / ○ one pair of barbels / ○ last unbranched dorsal ray strongly ossified, as long or longer than head in juveniles smaller than 80 mm, shorter in adults, its posterior edge of with many serrae / ○ 83–92 total lateral-line scales / ○ 16–18 gill rakers / ○ usually 8½ branched dorsal rays. Size up to 400 mm SL.

Distribution. Türkiye: Lake Eğirdir and Beyşehir basins, Melendiz (Ihlara Valley) and Yeşildere (Karaman) drainages in Central Anatolia.

Habitat. Streams with moderately fast-flowing or almost stagnant water. Regularly entering lakes.

Biology. No data.

Conservation status. VU; remaining populations mostly in streams. Formerly a predominantly lacustrine species, migrating to lake tributaries to spawn. Since the introduction of *Sander lucioperca*, few individuals migrate to the lakes, but the situation has improved due to overfishing of *Sander*.

Remarks. Eggs are toxic. According to one report, diarrhea, vomiting and diaphoresis were reported 4.5 h after ingestion of ovaries. *Capoeta pestai* and *C. mauricii* are indistinguishable by molecular characters (COI, cytochrome B) and morphological characters proposed to distinguish the two species could not be confirmed. *Capoeta mauricii* is treated as a synonym of *C. pestai*.

Further reading. Küçük et al. 2009a (morphology, distribution, description of *C. mauricii*); Küçük et al. 2009b (conservation); Geiger et al. 2014, Bektaş et al. 2017 (phylogeny); Özeren et al. 2019 (distribution, synonyms); Bayçelebi et al. 2020a (distribution).

Hybrid between *Capoeta damascina* and *Luciobarbus schejch*; Euphrates drainage, Türkiye; 135 mm SL.

***Capoeta* x *Luciobarbus* hybrids.** Hybrids between species of *Capoeta* and *Luciobarbus* have been recorded in areas where both species occur together. These hybrids were known to early ichthyologists exploring West Asia and have been named as *Barbus* species. Hybrids are often difficult to distinguish in the field, especially if they are young. In several *Capoeta* and *Luciobarbus*, the shape of the mouth is quite variable, and hybrids may bridge the gap between the mouth and head shapes of *Capoeta* and *Luciobarbus*. All *Luciobarbus*, for which data is available, are tetraploid with 100 chromosomes, while all *Capoeta* are hexaploid with 150 chromosomes. If a tetraploid (n = 4) parent and a hexaploid (n = 6) parent hybridise, the offspring is pentaploid (n = 5) and is likely to be sterile, as all odd-numbered hybrids are. The occasional findings of mitochondrial bodies of *Luciobarbus* in *Capoeta* species demonstrate that there must be a way of hybridogenous introgression via these pentaploid hybrids. This implies that at least some hybrids must be able to produce even-chromosome eggs or sperm (n = 2 or 4). These eggs can develop and produce a new, even-chromosome-numbered hybrid, which can reproduce with the pure species and transfer the mitochondrial body of one species into the other. However, this is merely speculative; the only certainty is that it must function somehow. Further research is required to elucidate the process of introgressive hybridisation in uneven chromosome cyprinids. In areas where two *Capoeta* species co-occur, hybrids are also observed, which exhibit morphological intermediary characteristics between the parental species. In *Capoeta*, it is frequently observed that in pure fish, the mitochondrial body of a syntopic species is present. For instance, in the Orontes drainage, *C. damascina* with the mitochondrial body of *C. barroisi* and vice versa are frequently observed, while morphologically intermediate fish are notably scarce.

Capoeta razii; Tajan, Iran; 189 mm SL.

Capoeta razii

Common name. Elbrus scraper.

Diagnosis. Distinguished from other species of *Capoeta* in Caspian basin by: ○ 46–54 total lateral-line scales / ○ 18 or more predorsal scales / ○ less than 44 total vertebrae / ○ usually 8½ branched dorsal rays / ○ last unbranched dorsal ray strong, with 15–25 serrae along its posterior edge / ○ 7–9 scales between dorsal origin and lateral line / ○ 6–7 scales between anal origin and lateral line. Size up to 300 mm SL.

Distribution. Azerbaijan and Iran: Atrak west to coastal rivers south of Kura estuary. Also, (possibly introduced) in upper Gamasiab (Karkheh drainage), Nam (Western Kavir basin), and Jaj in Lake Namak basin.

Habitat. A wide range of moderately to rapidly flowing streams and rivers with sand, gravel or rock bottoms.

Biology. Lives up to 10 years. Spawns April–July. Males mature at 2 years, females at 2–3 years. Eggs are adhesive and laid in gravel. Feeds on detritus, periphyton and invertebrates.

Conservation status. LC.

Remarks. This species is usually identified as *C. gracilis*, described from the Zayandeh drainage in Iran (a synonym of *C. aculeata*). Both species are well differentiated. *Capoeta razii* occurs geographically adjacent to *C. capoeta*, and both species are almost indistinguishable by published morphological characters. Only molecular characters allow these species to be differenciated. Fish identified as *C. razii* by mtDNA has been found in Aras drainage, where only *C. capoeta* and *C. kaput* should occur. Hybrids between different *Capoeta* species are expected to occur in that region.

Further reading. Jouladeh-Roudbar et al. 2017b (description); Zareian et al. 2017 (phylogeny, morphology); Jouladeh-Roudbar et al. 2020 (distribution); Koohestan Eskandari, 2003 (biology).

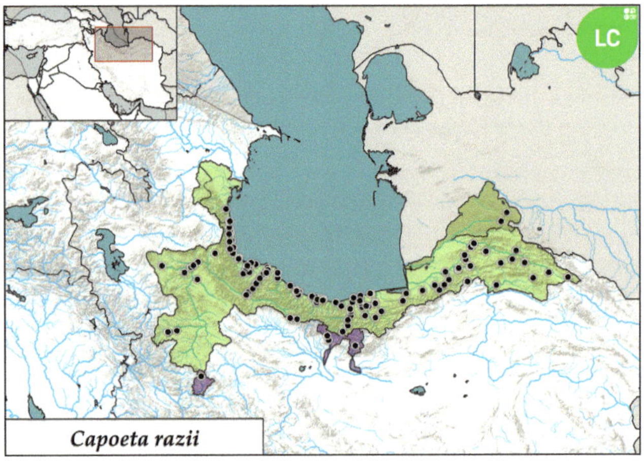

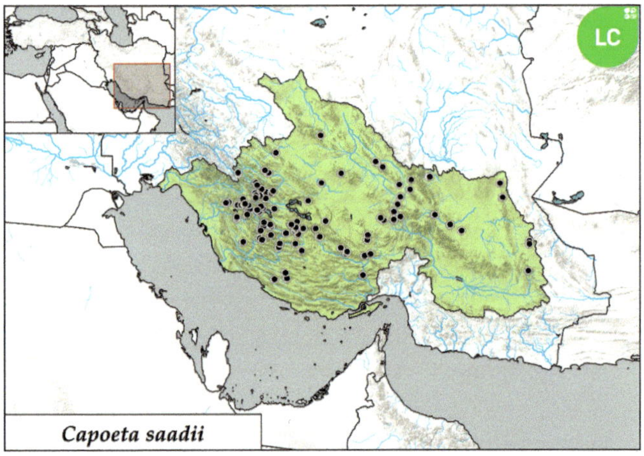

Capoeta saadii; Kol drainage, Iran; ~100 mm SL.

Capoeta saadii; Kor drainage, Iran; 192 mm SL.

Capoeta are dominant in biomass in many rivers in West Asia as here in the Kor in Iran, a habitat of *Capoeta saadii* and *C. aculeata*.

Capoeta saadii
Common name. Shiraz scraper.
Diagnosis. Distinguished from other species of *Capoeta* in Persian Gulf basin south of Zohreh and in Iranian endorheic basins by: ○ flank beige, golden or brown without small black spots, often with a lateral stripe, juveniles and some adults individuals with large black blotches / ○ 63–79 total lateral-line scales / ○ last unbranched dorsal ray soft or moderately ossified, shorter than head, with small serrae / ○ no groove in front of nares / ○ (7) 8–9½ branched dorsal rays / ○ 11–16 scales between lateral line and dorsal origin / ○ 7–10 scales between lateral line and pelvic origin / ○ 23–28 (29) circumpeduncular scales / ○ 11–17 gill rakers, usually 9–12 (15) on lower limb, 12–13 in Helleh drainage / ○ pectoral length 15–21 % SL. Size up to 384 mm SL.
Distribution. Iran: Coastal rivers from Helleh south to Kol. Also, found in endorheic basins of Kor, Esfahan and Sirjan, Jazmurian, Kerman, and Lut.
Habitat. A wide range of moderately to rapidly flowing streams and rivers with sand, gravel, and rock substrates.
Biology. No data.
Conservation status. LC.
Remarks. Analysis of COI DNA sequences places *C. raghazensis*, described from the Reghez Canyon in Kol drainage, within *C. saadii*. As morphological characters proposed to distinguish the two species could not be confirmed, *C. raghazensis* is treated as a synonym of *C. saadii*.
Further reading. Alwan 2010 (morphology, phylogeny); Alwan et al. 2016b (morphology, phylogeny); Zareian & Esmaeili 2017 (description); Jouladeh-Roudbar et al. 2020 (distribution); Coad 2021a (biology, morphology); Eagderi & Mousavi-Sabet 2021 (description of *C. raghazensis*).

Capoeta shajariani; Gamasiab, Karkheh drainage, Iran; 149 mm SL.

Capoeta shajariani
Common name. Karkheh scraper.
Diagnosis: Distinguished from other species of *Capoeta* in Karkheh, Karun and Zohreh drainages by: ○ flank beige, golden or brown without small black spots, juveniles and some adults individuals with large black blotches / ○ 70–80 scales in lateral-line series / ○ last unbranched dorsal ray moderately ossified, shorter than head, serrated / ○ no groove in front of nares / ○ usually 9½ branched dorsal rays / ○ 13–15 scales between lateral line and dorsal origin / ○ 9–11 scales between anal origin and lateral line / ○ 25–31 circumpedunclular scales / ○ 15–19 gill rakers, 11–14 on lower limb / ○ head depth at eye 105–126 % of postorbital length. Size up to 450 mm SL.
Distribution. Iran: Gamasiab, Dinevar, Razavar, Kashkan, Kahman, and Seymareh in Karkheh drainage.
Habitat. Slow to moderate flowing rivers with gravel beds.
Biology. Feeds on detritus, periphyton, and occasionally small invertebrates.
Conservation status. LC.
Further reading. Jouladeh-Roudbar et al. 2017c (description); Zareian & Esmaeili 2017 (description).

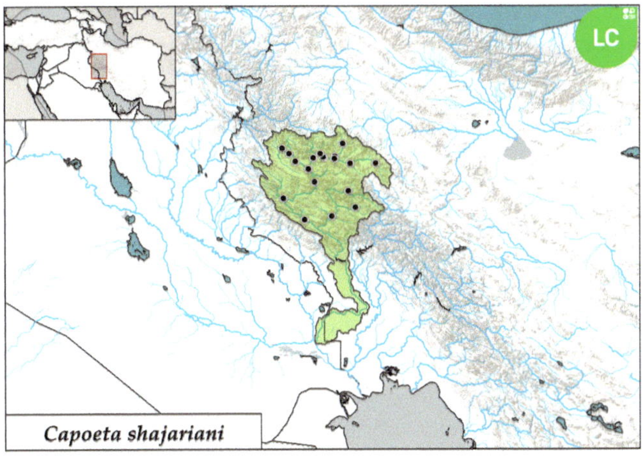

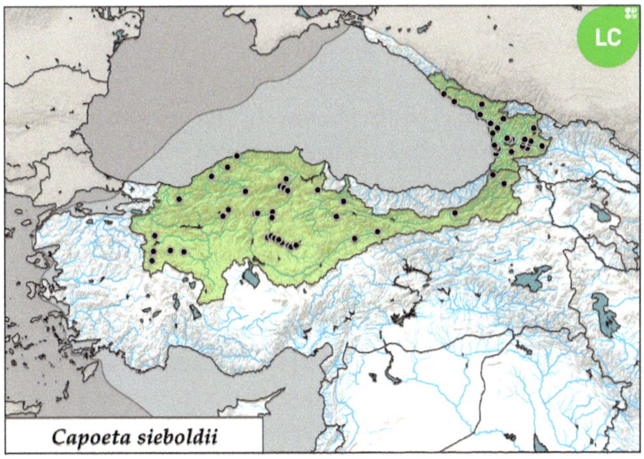

Capoeta sieboldii; Sakarya drainage, Türkiye; 221 mm SL.

Capoeta sieboldii

Common name. Nipple-lip scraper.

Diagnosis. Distinguished from other species of *Capoeta* in Black Sea basin by: ● lower and upper lips pleated, well developed and fleshy / ○ lower jaw without cornified edge / ○ one pair of barbels / ○ 52–60 total lateral-line scales / ● 28–35 gill rakers / ○ last unbranched dorsal ray ossified, its posterior edge of with many serrae / ○ usually 8½ branched dorsal rays / ○ 9–11 scales between lateral line and dorsal origin / ○ 8–10 scales between lateral line and pelvic origin. Size up to 400 mm SL.

Distribution. Türkiye and Georgia: Sakarya drainage east to northern coastal streams in Georgia.

Capoeta svanetica; Rioni, Georgia; ~100 mm SL. © G. Epitashvili.

Habitat. A wide range of streams and rivers with moderate to fast-flowing water and sand, gravel and rock substrate. In reservoirs, it migrates upstream to spawn.
Biology. Lives up to 8–10 years. Feeds on detritus, periphyton, and occasionally small invertebrates.
Conservation status. LC.
Further reading. Bektaş et al. 2017 (phylogeny).

Capoeta svanetica
Common name. Luchunis scraper.
Diagnosis. Distinguished from other species of *Capoeta* in Black Sea basin by: ● last unbranched dorsal ray not serrated or with 7–9 serrae / ○ two pairs of barbels / ○ 70–74 total lateral-line scales / ○ 10–12 scale rows between lateral line and dorsal origin / ○ 7–8 scale rows between lateral line and anal origin / ○ 12–15 gill rakers / ○ edge of lower jaw cornified / ○ lips narrow, not fleshy or pleated / ○ lower jaw monomorphic, wide and straight / ○ snout short and blunt. Size up to 136 mm SL, likely to grow larger.
Distribution. Georgia: Rioni drainage.
Habitat. A wide range of streams and rivers with moderate to fast-flowing water and sand, gravel and rock substrate. In reservoirs, it migrates upstream to spawn.
Biology. No data.
Conservation status. NE; due to very limited knowledge of this species.
Remarks. *Capoeta svanetica* is superficially very similar to the hybrid between *C. capoeta* and *Barbus rionicus*, and this possibility has not been rejected in its original description.
Further reading. Roman et al. 2022 (description).

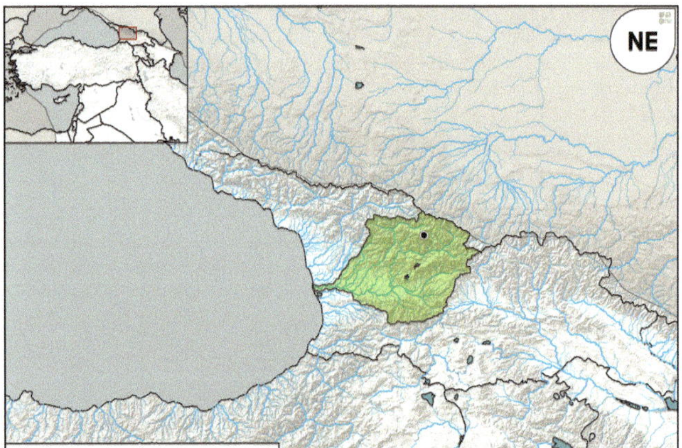

Capoeta svanetica

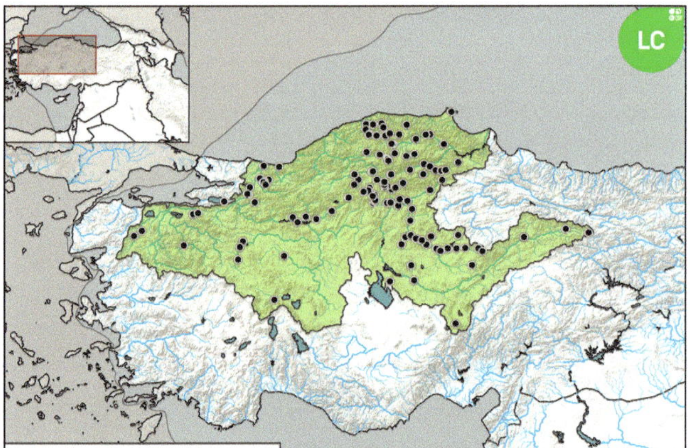

Capoeta tinca

Capoeta tinca; Sakarya drainage, Türkiye; 170 mm SL.

Capoeta tinca; Susurluk drainage, Türkiye; 108 mm SL.

Capoeta tinca

Common name. Fourbarbel scraper.

Diagnosis. Distinguished from other species of *Capoeta* in Marmara and Black Sea basin and Central Anatolia by: ○ two pairs of barbels / ○ 69–87 total lateral-line scales / ○ 14–17 scale rows between lateral line and dorsal origin / ○ 9–11 scale rows between lateral line and anal origin / ○ 16–22 gill rakers / ○ 17–28 serrae on last unbranched dorsal ray / ○ edge of lower jaw cornified / ○ lips narrow, not fleshy or pleated / ○ shape of lower jaw polymorphic, wide and straight or arched and narrow in parts of individuals / ○ snout short and blunt. Size up to 220 mm SL, likely to grow larger.

Distribution. Türkiye: Lake İznik, Eber and Akşehir basin and Sursuluk (Marmara basin) east to Kızılırmak drainage.

Habitat. Ubiquitous, inhabits a wide range of water bodies at least temporarily connected to running waters such as springs, streams, rivers, and cold headwaters. Found in lakes, reservoirs, and large rivers.

Biology. Lives up to 10 years. Matures at 2 (male) or 3 years (female). Spawns May and June (Central Anatolia).

Conservation status. LC.

Remarks. Populations from the Black Sea basin were recognised as *C. baliki*, described from Sakarya. As more data became available, they were found to be conspecific with *C. tinca*. No characters could be found to distinguish the two species. Populations from Kızılırmak are very closely related to *C. tinca*, with only minor molecular differences.

Further reading. Ekmekçi 1996 (biology); Ekmekçi & Özeren 2003 (biology); Turan et al. 2006a (description as *C. baliki*); Bektaş et al. 2017 (phylogeny); Yoğurtçuoğlu et al. 2020a (distribution).

Capoeta trutta; Tigris, Türkiye; ~250 mm SL.

Capoeta trutta; Euphrates drainage, Türkiye; individual with long dorsal ray, 180 mm SL.

Capoeta trutta; Euphrates drainage, Türkiye; individual with short dorsal ray, 100 mm SL.

Capoeta trutta

Common name. Spotted scraper.

Diagnosis. Distinguished from other species of *Capoeta* in Qweiq, Euphrates, and Tigris drainages by: ● last unbranched dorsal ray very strong, longer than head, strongly serrated / ○ one pair of barbels / ○ flank silvery with many small black spots, spots often only on back in large individuals, rarely without spots / ○ 68–90 total lateral-line scales / ○ 15–17 scale rows between lateral line and dorsal origin / ○ 9–12 scale rows between lateral line and anal origin / ○ 24–29 gill rakers / ○ usually 8½ branched dorsal rays. Size up to 450 mm SL.

Distribution. Qweiq, Euphrates, Tigris, and Karun drainages.

Habitat. Moderately fast-flowing to standing waters, from mid-sized mountain streams to large lowland rivers. Often abundant in reservoirs and lakes, from where they migrate to inflowing streams and rivers to spawn.

Biology. Lives up to 10 years. First spawns at 2 (males) and 3 (females) years. Spawns March–July in Tigris in Iraq and May–June in Tigris in Türkiye. Feeds on epilithic algae and detritus. Often grazes top mud layers in reservoirs.

Conservation status. LC; extirpated from Qweiq.

Remarks. Fish with the last unbranched dorsal ray shorter than the head are often identified as *C. barroisi*, a species endemic to the Orontes drainage.

Further reading. Coad 2010a (description, biology); Kaya 2019 (distribution); Jouladeh-Roudbar et al. 2020 (distribution); Coad 2021a (biology, morphology).

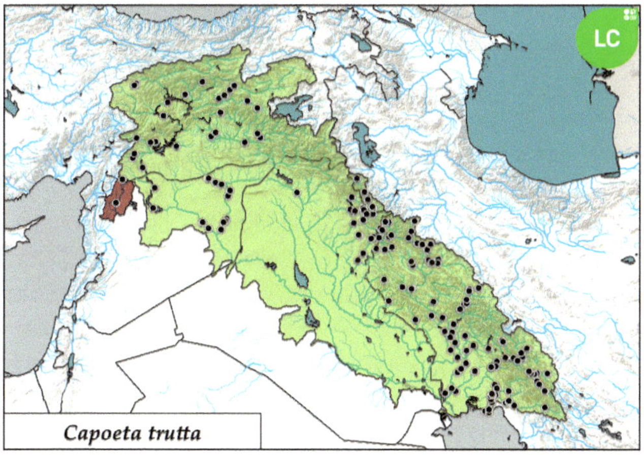

Local fish markets are an important source of fishes for scientists also.

Capoeta trutta often establishes large populations in reservoirs, such as Atatürk reservoir in Türkiye.

How fishes moved between the Euphrates and the Levant. Fish movement between the Euphrates and the Levant is a subject of interest to those studying the region's biogeography. During most of the Miocene and Pliocene periods, the waters of the Levant flowed either west to the Mediterranean or east to the Euphrates. The first region to become isolated from the Euphrates was the Damascus drainage. Until the Upper Pliocene, the rivers of the present-day Damascus basin flowed through the current oasis of Palmyra to the Euphrates. Following the subsidence of the Damascene Plain and the upheaval of the southeastern Syrian highlands during the Upper Pliocene (approximately 3 million years ago), the rivers of the Damascene Plain became isolated from the Euphrates. At roughly the same time, these rivers began to flow to the upper Jordan drainage. The present endorheic basin of Damascus came into existence only after the basaltic eruptions of the Hauran and Gabal ad-Duruz during the late Pleistocene, which severed its connection with the Jordan. Furthermore, the present-day upper and middle parts of the Orontes flowed independently to the Euphrates. Up to the very early Pleistocene, a connection existed between the Orontes and Jordan–Litani system caused by the movement of the watershed. During that time, fishes could migrate from the Euphrates through the upper Orontes to the Litani and the Jordan. During the lower Pleistocene, the al-Ghab Valley subsided, resulting in the uplifting of Gabal az-Zawiyah Mountain. The uplifting of Gabal az-Zawiyah, which extended toward the Palmyra fold belt, severed the connections between the Proto-Orontes and the Euphrates. This subsequently led to a gradual separation. Currently, except *Capoeta damascina*, all fish species of the Damascus, Jordan, and Orontes basins are distinguishable from those of the Euphrates. However, the Jordan still shares several species with the Damascus basin (as *Oxynoemacheilus insignis*, *Garra nanus*, *Tristramella simonis* and the very closely related species pair *Pseudophoxinus syriacus* and *P. drusensis*).

Carasobarbus

Carasobarbus are a medium-sized group of barbels known from West Asia and Morocco. They are hexaploid and distinguished by large, shield-shaped scales with numerous parallel radii. The last unbranched dorsal ray is weakly to strongly ossified, without serrae. They have one or two pairs of barbels, 9–11½ branched dorsal rays, and 6½ branched anal rays. *Carasobarbus kosswigi* and *C. sublimus* were previously classified in a separate genus, *Kosswigobarbus*, which was recently synonymised with *Carasobarbus*. They are most species-rich in the Tigris, where six species are found. In the Arabian Peninsula, two species are known which have close relationships with Mesopotamian *Carasobarbus*. *Carasobarbus* are inhabitants of slow-flowing or stagnant waters, often with much vegetation, while former *Kosswigobarbus* inhabit large, free-flowing, deep rivers. Some species are large fish that are commercially exploited in many areas. As in many other barbels, their eggs are poisonous. Some people have reported feeling dizzy, having abdominal pain, vomiting, diarrhea, dry mouth, and faintness after eating a kebab made with about ¼ of an ovary of *C. luteus*. **Further reading.** Karaman 1971 (description); Coad 2010a (toxic eggs); Borkenhagen & Krupp 2013 (description); Borkenhagen 2014 (phylogeny); Yang et al. 2015; Borkenhagen 2017 (phylogeny, in tribe Torini).

Keys to species of *Carasobarbus* in West Asia

Contributing author: Kai Borkenhagen (Büsum) contributed to the chapter on this genus.

Arabian Peninsula

1a - One pair of barbels; usually 10½ branched dorsal rays.
..................*C. apoensis*
1b - Two pairs of barbels; usually 9½ branched dorsal rays.
..................*C. exulatus*

Mediterranean and Persian Gulf basins

1a - Lower lip without median lobe.
..................2
1b - Lower lip with median lobe.
..................4

2a - Usually one pair of barbels; 25–30 total lateral-line scales.
..................*C. luteus*
2b - Usually two pairs of barbels; 29–38 total lateral-line scales.
..................3

3a - Usually 12 circumpeduncular scales; usually 7–9 gill rakers; head length 24–29 % SL; adults larger than 100 mm SL head grey and caudal peduncle and caudal yellow.
..................*C. canis*
3b - Usually 14–16 circumpeduncular scales; usually 9–12 gill rakers; head length 20–24 % SL; adults larger than 100 mm SL whole body golden or greenish grey.
..................*C. chantrei*

4a - 24–29 total lateral-line scales.
..................*C. sublimus*
4b - 32–44 total lateral-line scales.
..................5

5a - 32–37 total lateral-line scales.
..................6
5b - 38–44 total lateral-line scales.
..................7

6a - 32–37 (mode 36) total lateral-line scales; head length 25–27 % SL; posterior barbel 13–20 % HL; snout length 36–44 % HL.
..................*C. kosswigi*
6b - 32–34 (mode 33–34) total lateral-line scales; head length 20–24 % SL; posterior barbel 21–38 % HL; snout length 25–31 % HL.
..................*C. hajhosseini*

7a - A prominent black blotch on posterior caudal peduncle in individuals smaller than 85 mm SL; head length 22–25 % SL; dorsal fin height 19–26 % SL; distance between pelvic and anal fin origins 24–25 % SL.
..................*C. doadrioi*
7b - No black blotch on posterior caudal peduncle; head length 19–20 % SL; dorsal fin height 26–30 % SL; distance between pelvic and anal fin origins 26–28 % SL.
..................*C. saadatii*

Carasobarbus apoensis; Saudi Arabia; 212 mm SL. © J. Els.

Carasobarbus apoensis

Common name. Arabian himri.

Diagnosis. Distinguished from *Carasobarbus exulatus* and superficially similar *Arabibarbus arabicus* in Arabian Peninsula by: ● one pair of barbels / ● usually 10½ branched dorsal rays / ○ 6½ anal rays. Size up to 288 mm SL.

Distribution. Saudi Arabia: Hijaz mountain range, in wadis, draining both inland and toward Red Sea.

Habitat. Mainly upper reaches of wadis with large seasonal variations in flow.

Biology. Lives up to 19 years, possibly longer in captivity. Spawns in early spring. Feeds on both aquatic and terrestrial invertebrates and plants.

Conservation status. EN; appears to be declining within its very small range.

Remarks. Very closely related to the Mesopotamian *C. luteus*. The two species may only have diverged in post-glacial times when a river flowing from the mountains of western Saudi Arabia to the southernmost Shatt al-Arab dried up.

Further reading. Banister & Clarke 1977 (description); Krupp 1983 (morphology, distribution); Borkenhagen & Krupp 2013 (description, distribution).

Carasobarbus canis

Common name. Jordan himri.

Diagnosis. Distinguished from other species of *Carasobarbus* in Persian Gulf and Mediterranean basins by: ● head grey and caudal peduncle and caudal yellow in adults larger than 100 mm SL / ○ usually 12 circumpeduncular scales / ○ lower lip without median lobe / ○ last unbranched dorsal ray usually markedly shorter than head / ○ usually 7–9 gill rakers / ○ head length 24–29 % SL / ○ two pairs of barbels / ○ 29–35 total scales along lateral line. Size up to 550 mm SL.

Distribution. Jordan drainage. Introduced to Azraq Oasis in Jordan and coastal rivers Na'aman and Yarqon in Israel, now all extirpated.

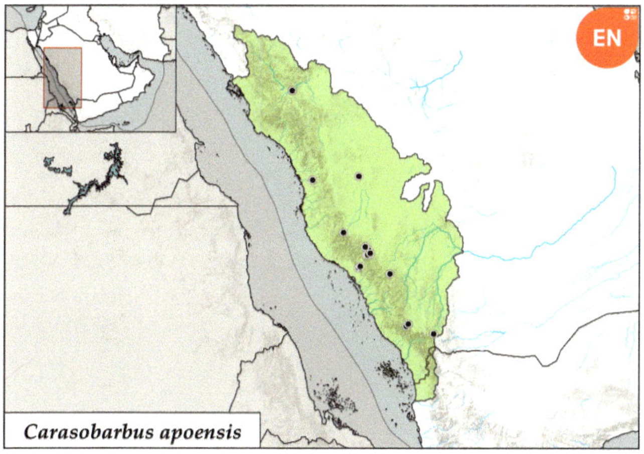

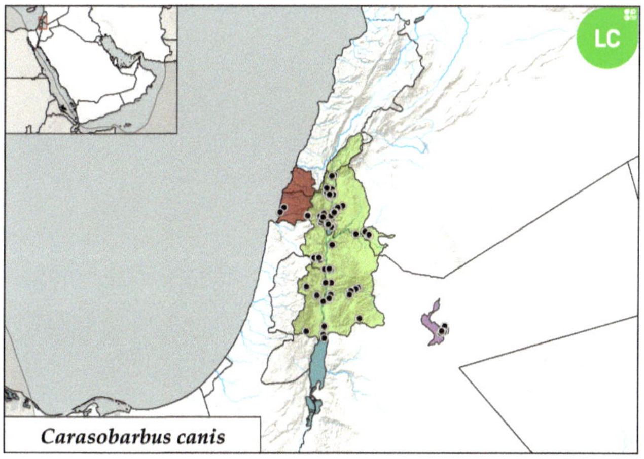

Habitat. A wide range of rivers, lakes, and marshes. Spawns along banks, usually on hard bottoms.
Biology. Spawns December–January (Lake Tiberias). Eggs adhere to substrate. Diet consists of fish, invertebrates, algae, and detritus. Relative proportion of fish in diet increases with body length. *Mirogrex* are their main prey in Lake Tiberias.
Conservation status. LC.
Remarks. Records from Euphrates drainage are based on misidentifications.
Further reading. Ben-Tuvia 1978 (biology); Spataru & Gophen (1985b) (feeding); Fishelson et al. 1996 (reproduction); Tsigenopoulos et al. 2010 (phylogeny); Borkenhagen et al. 2011 (phylogeny); Borkenhagen & Krupp 2013 (description, distribution).

Carasobarbus canis; Jordan drainage, Syria; 180 mm SL.

Reservoir of Yarmuk in Jordan. *Carasobarbus* can establish large populations in artificial waterbodies.

Carasobarbus chantrei; Orontes drainage, Türkiye; 155 mm SL.

Carasobarbus chantrei

Common name. Orontes himri.

Diagnosis. Distinguished from other species of *Carasobarbus* in Persian Gulf and Mediterranean basins by: ○ head and flank golden or greenish grey in adults larger than 100 mm SL / ○ usually 14–16 circumpeduncular scales / ○ lower lip without median lobe / ○ last unbranched dorsal ray usually shorter or as long as head, moderately ossified / ○ usually 9–12 gill rakers / ○ head length 20–24 % SL / ○ two pairs of barbels / ○ 31–38 total lateral-line scales. Size up to 385 mm SL.

Distribution. Orontes drainage in Türkiye and Syria and Nahr Marqīyah, a coastal stream in Syria.

Habitat. Springs, lakes, reservoirs, and moderately fast-flowing rivers.

Biology. No data.

Conservation status. LC. Still present in lower Orontes in Türkiye and widespread in Syria. However, the situation in Syria needs some actual surveys.

Further reading. Krupp 1985b (description); Ekmekçi & Bănărescu 1989 (diversity); Tsigenopoulos et al. 2010 (phylogeny); Borkenhagen et al. 2011 (phylogeny); Borkenhagen & Krupp 2013 (description, distribution); Bayçelebi 2020 (distribution).

Carasobarbus doadrioi

Common name. Karun kiss-lip himri.

Diagnosis. Distinguished from other species of *Carasobarbus* in Persian Gulf and Mediterranean basins by: ● 40–44 total lateral-line scales / ○ well-developed median lobe on lower lip / ● prominent black blotch on posterior caudal peduncle in individuals smaller than 85 mm SL / ○ two pairs of barbels / ○ head length 22–25 % SL / ○ dorsal fin height 19–26 % SL / ○ distance between pelvic and anal fins 24–25 % SL. Size up to 200 mm SL.

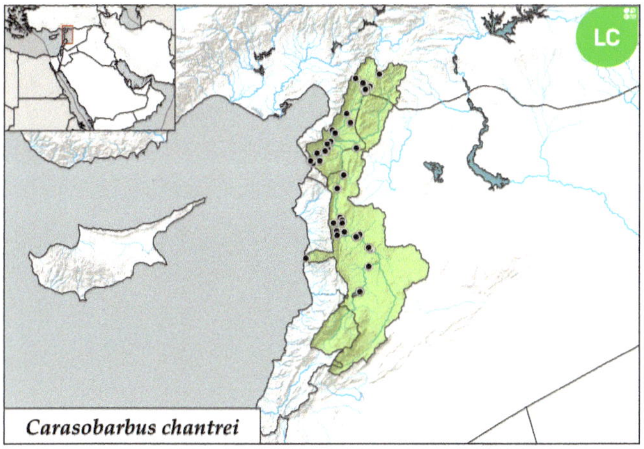

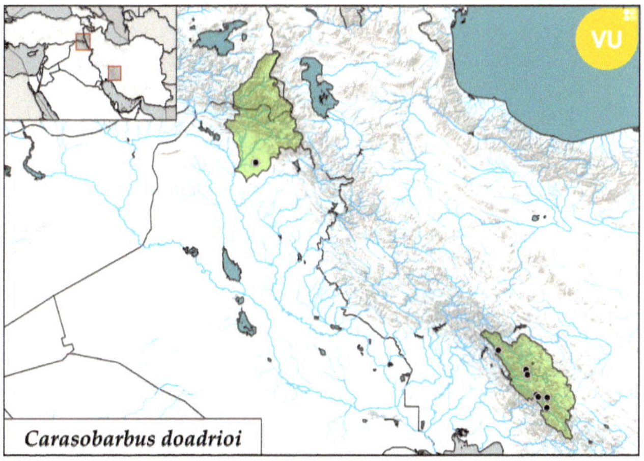

Distribution. Iran and Iraq: Upper Tigris and Karun.
Habitat. Poorly known; seem to inhabit deep, summer-warm mountain river stretches with fast-flowing water and gravel bottoms.
Biology. No data.
Conservation status. VU.
Further reading. Jouladeh-Roudbar et al. 2024 (description).

Carasobarbus doadrioi; Karun, Iran; ~150 mm SL.

Carasobarbus exulatus; Wadi al Khun, Yemen; 69 mm SL. © F. Krupp.

Carasobarbus exulatus; Wadi Hadhramaut, Yemen; 212 mm SL. © F. Krupp.

Carasobarbus exulatus

Common name. Hadhramaut himri.

Diagnosis. Distinguished from *Carasobarbus apoensis* and sympatric *Arabibarbus hadhrami* in Arabian Peninsula by: ○ two pairs of barbels / ○ usually 9½ branched dorsal rays / ○ 6½ anal rays / ○ maximum body depth 26–36 % SL / ○ head shorter than maximum body depth. Size up to 290 mm SL, likely to grow larger.

Distribution. Yemen: Wadi Hajr and upper Wadi Hadhramaut drainages.

Habitat. Wadis with high seasonal variation and sections of deep pools over gravel, bare rock or soft substrate. Large individuals usually in deep pools.

Biology. No data.

Conservation status. EN; appears to be declining within its very small range. Known from two drainage systems threatened by water abstraction, also fished for human consumption.

Further reading. Banister & Clarke 1977 (description); Krupp 1983 (description, distribution); Borkenhagen & Krupp 2013 (description, distribution).

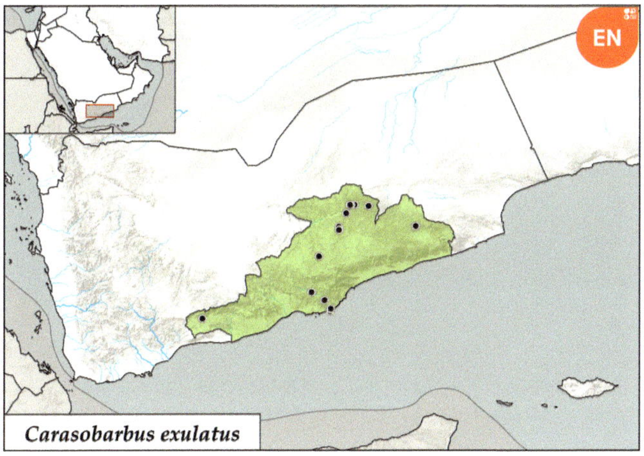

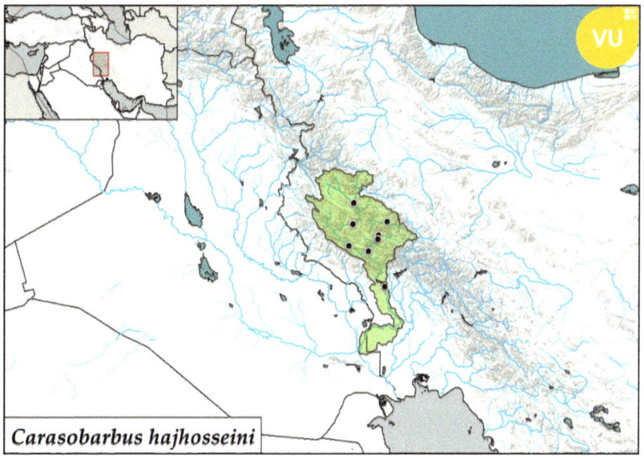

Carasobarbus hajhosseini; Karkheh, Iran; 109 mm SL.

Carasobarbus hajhosseini; Karkheh drainage, Iran; ~65 mm SL. © K. Borkenhagen.

Carasobarbus hajhosseini

Common name. Karkheh kiss-lip himri.

Diagnosis. Distinguished from other species of *Carasobarbus* in Persian Gulf and Mediterranean basins by: ○ 32–34 total lateral-line scales / ○ poorly-developed median lobe on lower lip / ○ no black blotch on posterior caudal peduncle in adults / ○ two pairs of barbels / ○ head length 20–24 % SL / ○ posterior barbel 21–38 % HL / ○ snout length 25–31 % HL. Size up to 150 mm SL.

Distribution. Iran: Gamasiab, Kahman, Kashkan, and Seymareh, all in Karkheh drainage.

Habitat. Poorly known; seem to inhabit deep, summer-warm mountain river stretches with fast-flowing water and gravel bottoms.

Biology. No data.

Conservation status. VU.

Further reading. Jouladeh-Roudbar et al. 2024 (description).

Habitat of *Carasobarbus kosswigi*, free-flowing upper reaches of Lesser Zab in Iraq.

Carasobarbus kosswigi; Lesser Zab, Iraq; 137 mm SL.

Carasobarbus kosswigi

Common name. Kiss-lip himri.

Diagnosis. Distinguished from other species of *Carasobarbus* in Persian Gulf and Mediterranean basins by: ○ 32–37 total lateral-line scales / ○ well-developed median lobe on lower lip / ● last unbranched dorsal ray longer than head, strongly ossified / ○ head length 25–27 % SL / ○ posterior barbel 13–20 % HL / ○ snout length 36–44 % HL / ○ two pairs of barbels / ○ no black blotch on posterior caudal peduncle / ○ 14–16 circumpeduncular scales. Size up to 180 mm SL.

Distribution. Euphrates and Tigris drainages.

Habitat. Poorly known; seem to inhabit deep, summer-warm mountain river stretches with fast-flowing water and gravel bottoms.

Biology. Feeds on benthic invertebrates.

Conservation status. LC.

Further reading. Ladiges 1960 (description); Tsigenopoulos et al. 2010 (phylogeny); Borkenhagen et al. 2011 (phylogeny); Borkenhagen & Krupp 2013 (description, distribution); Jouladeh-Roudbar et al. 2024 (distribution).

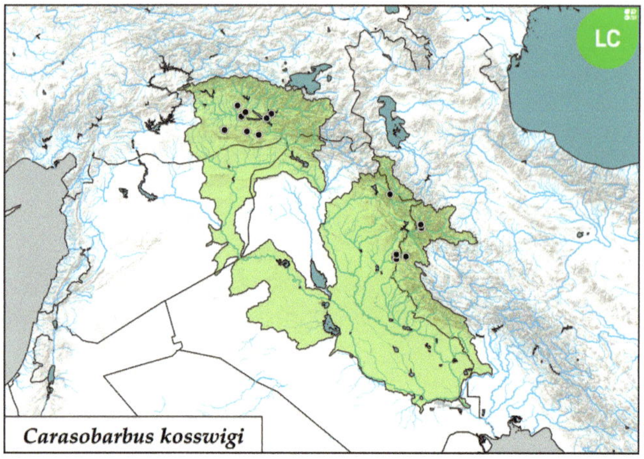

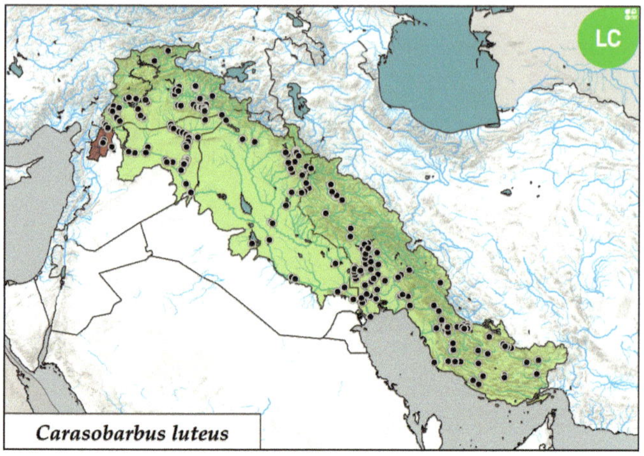

Carasobarbus luteus; Golabi spring, Iran; 126 mm SL.

Carasobarbus luteus

Common name. Mesopotamian himri.

Diagnosis. Distinguished from other species of *Carasobarbus* in Persian Gulf and Mediterranean basins by: ● usually one pair of barbels / ○ fleshy lobe on lower lip absent / ○ 25–30 total lateral-line scales / ○ usually 12 circumpeduncular scales / ○ last unbranched dorsal ray about as long as head or slightly shorter, soft, and weakly ossified. Size up to 370 mm SL.

Distribution. Qweiq, Euphrates, Tigris, and Karun drainages. Persian Gulf south to Kol drainage. Also, in Kor endorheic basin (Iran).

Habitat. Lowland rivers, backwaters, lakes, reservoirs, springs, and ponds, rarely in fast-flowing waters. Tolerant of high salinity and reported from fully marine waters. Spawns on submerged vegetation.

Biology. Lives up to 8 years. Spawns first time at 1–2 years and about 90–140 mm SL; in Atatürk reservoir at 3 years. Spawns April–June (in Hammar Marsh) or May–July (in Shatt al-Arab/Arvand). During spawning, male become reddish-brown on anterior part of body and greenish on caudal peduncle, while female is less colourful. Males can produce a series of sharp clicking sounds. Eggs are yellow to orange and sticky. Growth is reported to be slower in brackish water than in freshwater. Mainly herbivorous, taking detritus and small invertebrates, especially when young.

Conservation status. LC. Extirpated in Syrian reaches of Qweiq.

Remarks. Individuals with two pairs of barbels are very rare. In the Naband drainage in Iran, all individuals have two pairs of barbels, and this population may represent an undescribed species. Records from Orontes drainage and Damascus basin appear to have been mislabelled or taken from markets where fish from the Euphrates are sold. In some places, it is kept as a "sacred" fish in parks and holy places where it can be abundant. It adapts well to visitors, and people enjoy seeing it begging for food.

Further reading. Ahmed et al. 1984 (reproductive biology); Naama & Mushen 1986 (feeding); Mohamed et al. 1993 (occurrence in marine waters); Coad 2010a (biology); Tsigenopoulos et al. 2010 (phylogeny); Borkenhagen et al. 2011 (phylogeny); Borkenhagen & Krupp 2013 (description, distribution); Kaya et al. 2016 (distribution); Coad 2021a (biology, morphology).

Carasobarbus luteus; Tigris, Türkiye; 130 mm SL.

Large rivers as the Botan are habitat of *Carasobarbus kosswigi*.

Balıklıgöl: From ancient myths to saved fish. West Asia is a spiritual center for major religions and beliefs and one of the places where the world's earliest civilisations originated. As a result, a plethora of myths, legends, and folktales based on both pagan and Islamic beliefs exist, many of which continue to shape modern-day lives at the local level. Balıklıgöl (or Halil-Ür Rahman Lake), a pool in the southwest of the city center of Şanlıurfa, is one of the most attractive places in this regard. It also provides an illustrative example of how verbal cultural heritages can potentially become a powerful tool in fish conservation. One of the myths associated with the lake dates back to approximately 8000 BC and is linked with Atargatis, the great goddess of northern Syria. The prevailing belief was that the fish in city ponds and canals associated with the goddess's reign were either a transformed form of herself or entities created by her activities, such as bathing in these waters. Another prominent and well-established belief is rooted in early Islam and is linked to the story of the conflict between Prophet Abraham and King Nimrod. In this narrative, Nimrod is angry with Abraham for rejecting his idols. Nimrod then throws Abraham into a raging fire with an arbalest, but God transforms the flames into water (the pool of Balıklıgöl) and the logs into fish. While there have been a few other similar myths, as was believed by the older societies, the pool and the fishes inside are still considered sacred by all means, and catching or eating the fishes is strictly forbidden, which makes this a case of one of the oldest practises of fish conservation. The Balıklıgöl pool is connected to the stream Karakoyun, a tributary of the upper Euphrates, which is fed by a large karstic aquifer. Four fish species occur in the pools: *Alburnus caeruleus*, *Carasobarbus luteus*, *Capoeta damascina*, and *Garra rufa*. In her pilgrimage account, written around 380 BC, Etheria already observed fountains full of fish of size, brightness, and flavor that she had never seen before.

Further reading. McClure & Feltoe 1919 (The pilgrimage of Etheria); Arslan 2015 (history); Ünlü 2021 (species).

Carasobarbus saadatii; Karun, Iran; 175 mm SL.

Carasobarbus saadatii

Common name. Zagros kiss-lip himri.

Diagnosis. Distinguished from other species of *Carasobarbus* in Persian Gulf and Mediterranean basins by: ○ 38–40 total lateral-line scales / ○ well-developed median lobe on lower lip / ○ no black blotch on posterior caudal peduncle in individuals larger than 85 mm SL / ○ two pairs of barbels / ○ head length 19–20 % SL / ○ dorsal fin height 26–30 % SL / ○ distance between pelvic and anal fins 26–28 % SL. Size up to 250 mm SL.

Distribution. Iran: lower Karun. Possibly more widespread.

Habitat. Poorly known; seem to inhabit deep, summer-warm mountain river stretches with fast-flowing water and gravel bottoms.

Biology. No data.

Conservation status. VU.

Further reading. Jouladeh-Roudbar et al. 2024 (description).

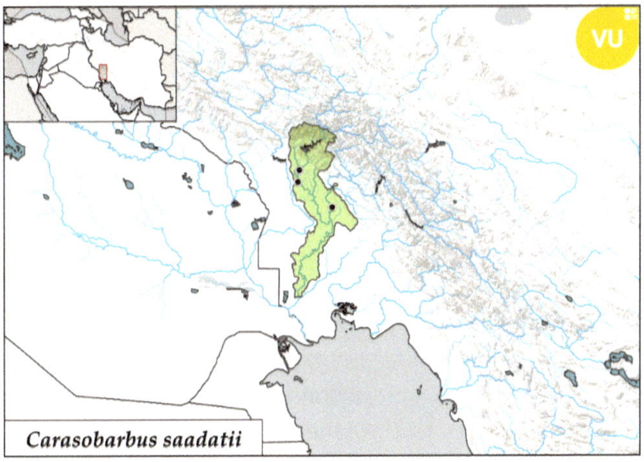

Carasobarbus sublimus; Zohreh, Iran; ~145 mm SL.

Carasobarbus sublimus

Common name. Persian kiss-lip himri.

Diagnosis. Distinguished from other species of *Carasobarbus* in Persian Gulf and Mediterranean basins by: ● 24–29 total lateral-line scales / ○ a fleshy median lobe on lower lip / ○ last unbranched dorsal ray about as long as head, strongly ossified / ○ two pairs of barbels / 12 circumpeduncular scales. Size up to 145 mm SL.

Distribution. Iran: Jarrahi and Zohreh drainages.

Habitat. Streams with medium to fast currents and hard substrate.

Biology. No data.

Conservation status. EN.

Further reading. Coad & Najafpour 1997 (description); Borkenhagen et al. 2011 (phylogeny); Borkenhagen & Krupp 2013 (description, distribution); Jouladeh-Roudbar et al. 2024 (distribution).

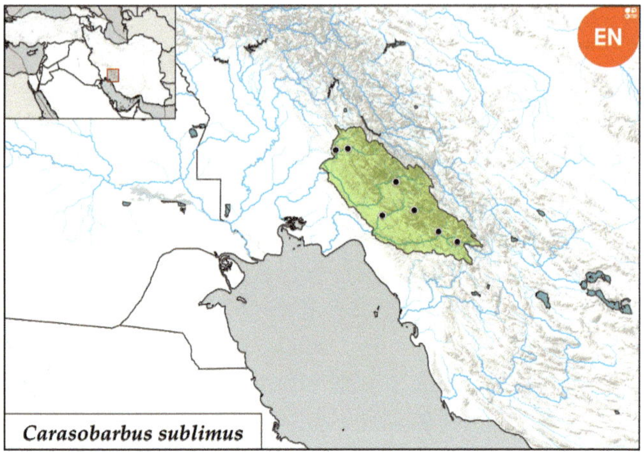

Carassius

Carps of the genus *Carassius* are immediately distinguishable from all other cyprinids in the region by having a long dorsal fin, a terminal mouth without barbels, and the last unbranched anal and dorsal rays strongly serrated. Crucian carp *C. carassius* is the only species native to the area. Three different species are known to be involved in the so-called *C. auratus* complex. These are (1) *C. auratus*, a widespread ornamental species that has become feral in many areas; (2) *C. gibelio*, an Asian species that may have been introduced with stocked fishes; (3) *C. langsdorfii* believed to be endemic to Japan and introduced as a pest with ornamental koi. An additional unnamed lineage potentially originating from Central Asia was introduced mostly within countries of the former Soviet Union and may be found in West Asia in the future.

All three mitochondrial lineages of non-native *Carassius* in West Asia are diploid and reproduce normally. This means that haploid male and female gametes join and produce a diploid offspring. In *C. auratus*, this is the only mode of reproduction. In the other lineages, triploid, all-female fish are common (but not found everywhere), and these reproduce by gynogenesis. Here, the sperm only triggers the triploid egg, and the offspring are clonal, triploid copies of the mother. If the sperm is included in the genome of the egg cell, then tetraploid individuals hatch. This might occasionally be the case in all populations with triploid fish. Triploid individuals are common, often even dominant, in many populations of the mitochondrial lineages called *C. gibelio* and *C. langsdorfii*. These lineages can switch between a recombinant and a clonal mode of reproduction.

The assignment of individual *Carassius* to *C. auratus*, *C. gibelio*, or *C. langsdorfii* by their mitochondrial DNA is not an accurate reflection of their species identities. Mitochondrial DNA lacks recombination, which is not useful to identify hybrids. In particular, triploid individuals frequently possess genomes from two (or all three) different "species". There is strong evidence that *C. auratus*, *C. gibelio*, and *C. langsdorfii* form large and highly diverse hybrid swarms in their non-native range. As all three "species" could reproduce by diploid sexual reproduction, they appear to have hybridised extensively, and their genomes are largely intermingled. We are confident that future nuclear DNA studies will not support three distinct, reproductively isolated non-native species in West Asia or Europe. However, some local populations may be pure *C. auratus*, *C. gibelio*, or *C. langsdorfii*. Identifying non-native *Carassius* to the species level is impossible because of the lack of morphological and molecular characters to identify pure species. We do not even know whether pure populations of non-native *Carassius* exist in West Asia, except for locally stocked *C. auratus*. We, therefore, refer to all populations as "*C. auratus* complex", pending future studies.

It should not be forgotten that *Carassius* hybridise with *Cyprinus carpio*. Their hybrids are fertile and may form populations independent of their parental species. Such hybrid populations are known from Europe but have yet to be discovered in West Asia. *Carassius* x *Cyprinus* hybrids can be recognised by their short barbels. **Further reading.** Murakami et al. 2001 (polyploidy systems in Japan); Szczerbowski 2002 (biology); Hänfling et al. 2005 (genetics); Gao et al. 2012 (diversity in Asia); Kalous et al. 2013 (mtDNA records of *C. langsdorfii* in Europe); Rylková et al. 2013 (non-native *Carassius* in Europe); Halas et al. 2018 (diversity in North America); Khosravi et al. 2020 (Iran).

Individual with mtDNA of *C. auratus*; Sirvan drainage, Iraq; ~130 mm SL.

Carassius auratus

Common name. Goldfish.

Diagnosis. Distinguished from *C. carassius* in West Asia by: • flank silvery, golden-green or reddish / • last simple anal and dorsal rays strongly serrated / • 38–52 gill rakers / • 26–33 total lateral-line scales / • free edge of dorsal concave or straight / • usually 5½ branched anal rays / • peritoneum black. Size up to 350 mm SL.

Distribution. Widespread in West Asia. Most non-native *Carassius* in West Asia have mitochondria belonging to this species. Reportedly native to East Asia, from Amur to Xi Jiang drainages in Russia, China, Korea, and Japan, several species appear to be confused under a single name, as evidenced by available data on Japanese populations. Domesticated in China more than 1000 years ago, introduced to Japan in 16th century and imported from Japan to Europe in 1611 (Portugal), 1691 (England), and 1755 (France) and subsequently introduced throughout West Asia, Europe, and most of the world.

Habitat. A wide variety of still waters and moderately slow-flowing rivers. Very tolerant of low oxygen and pollution. Usually associated with submerged vegetation or periodic flooding. Spawns on shallow, warm banks, on submerged vegetation or roots.

Biology. Only diploid populations known. Spawns first time at 1–2 years. Spawns June–July when temperatures reach 15–20°C. Juveniles need high temperatures to grow. Single females spawn with a few males in dense vegetation. Eggs are sticky and attached to water plants or other submerged objects. Usually lives for about 20 years in artificial conditions. Omnivorous, feeds mainly on plankton, benthic invertebrates, plant material, and detritus.

Conservation status. Non-native; introduced as a weed with stocked carp.

Further reading. Kottelat 1997 (systematics); Szczerbowski 2002 (biology); Kottelat & Freyhof 2007 (distribution, biology).

Carassius carassius

Common name. Crucian carp.

Diagnosis. Distinguished from non-native species of *Carassius* in West Asia by: ○ flank golden-green / • last simple anal and dorsal rays weakly serrated / • 23–33 gill rakers / • 31–36 total lateral-line scales / • free edge of dorsal convex / • usually 6½ branched anal rays / • peritoneum whitish. Size up to 550 mm SL.

Distribution. Locally in lower Sakarya (Akgöl), where it might be native, and in the upper Aras drainage (Lake Çıldır), where it may be non-native. Native to North, Baltic, White, Barents, northern Black, and Caspian Sea basins; Aegean Sea basin only in Maritza drainage; eastward to Kolyma drainage (Siberia); westward to Rhine and eastern drainages of England. Absent from North Sea basin in Sweden and Norway. In Baltic basin, north to about 66°N. Widely introduced to Italy, England, and France but possibly often confused with *C. auratus*.

Habitat. Usually restricted to densely vegetated backwaters and oxbows of lowland rivers. Also, in small, well-vegetated lakes and channels. Tolerates high summer temperatures and very low oxygen concentrations in summer and under ice cover. Able to survive in almost completely frozen water or almost-dry habitats by burying itself in mud. Spawns in dense submerged vegetation.

Carassius carassius; Danube delta, Romania; ~100 mm SL.

Individual with mtDNA of *Carassius gibelio*; Lower Danube, Romania; ~115 mm SL.

Biology. Lives about 10 years. Males reproduce for the first time at 3 years, females at 4 years in central and eastern Europe, at 2 years in southern Europe. Spawns in May–July at temperatures above 18°C. Individual females spawn with several males. Males follow ripe females, often with much splashing. Females spawn 3–5 times during a season. Eggs are sticky and are attached to water plants. Omnivorous, feeds all day but mostly at night on plankton, benthic invertebrates, plant material, and detritus. Seems to be a weak competitor, usually absent from waters with rich ichthyofauna and abundant predatory species. Very abundant in the absence of other fish species. High-bodied, fast-growing individuals in habitats with predatory fish more elongate in habitats without predatory fish.
Conservation status. LC.
Further reading. Szczerbowski & Szczerbowski, in Bănărescu & Paepke, 2002 (biology).

Carassius gibelio
Common name. Prussian carp.
Diagnosis. Distinguished from *C. carassius* in West Asia by: • flank silvery / • last simple anal and dorsal rays strongly serrated / • 38–52 gill rakers / ○ 26–33 total lateral-line scales / • free edge of dorsal concave or straight / • usually 5½ branched anal rays / • peritoneum black. Size up to 350 mm SL.
Distribution. Mitochondria of this species are found throughout West Asia. Often considered native from Central Europe to Siberia. As there are no historical records from Europe or West Asia, considered non-native to region. Native to Amur drainage in Russia and China and introduced in Europe with fish imported from Amur. Introduced from Europe to North Africa, West Asia, and elsewhere.
Habitat. Identical to *C. auratus*.
Biology. Usually, diploid and tetraploid individuals of both sexes, often together with triploid females. Also, all-female

populations where all individuals are triploids. Triploids reproduce by gynogenesis as sperm parasites of other species, such as *Cyprinus carpio* and various leuciscids. Lives about 10 years. Spawns first time at 3–4 years in central and eastern Europe, at 1–2 years in West Asia. Spawns May–July at temperatures above 14°C, with spawning most intensive in early morning. Older individuals spawn earlier in season than younger ones. Males move to spawning sites before females. Single female spawns with several males. Males follow mature female, often with much splashing. Female spawn up to three times per season. Sticky eggs are attached to water plants or other submerged objects. Feeding larvae and juveniles inhabit highly complex habitats such as reed beds. Omnivorous, feeding on plankton, benthic invertebrates, plant material, and detritus.

Conservation status. Non-native; introduced as a weed with stocked carp.

Further reading. Kottelat 1997 (systematics); Vasil'ev & Vasil'eva 2000 (genetics); Szczerbowski 2002 (biology; as *C. auratus*, in part); Hänfling et al. 2005 (genetics); Kottelat & Freyhof 2007 (distribution, biology); Yerli et al. 2014 (distribution).

Many thousands of small reservoirs in West Asia, as in upper Euphrates drainage, have been stocked with *Carassius*, making them the most widespread freshwater fish in the region.

Individual with mtDNA of *Carassius langsdorfii*; Biskra, Algeria; ~160 mm SL.

Carassius langsdorfii

Common name. Ginbuna.

Diagnosis. Distinguished from *C. carassius* in West Asia by: • flank silvery / • last simple anal and dorsal rays strongly serrated / • 38–52 gill rakers / ○ 26–33 total lateral-line scales / • free edge of dorsal concave or straight / • usually 5½ branched anal rays / • peritoneum black. Size up to 350 mm SL.

Distribution. Native to Japan. Mitochondria of this species are reported from Lar National Park (Iran) but are considered widespread in West Asia. Already found in Czech Republic, Greece, Germany, Ukraine, Italy and Bosnia-Herzegovina (Europe), from Lake Tahoe (USA), and British Columbia (Canada).

Habitat. Identical to *C. auratus*.

Biology. In its native range, it occurs as a diploid with normal sexual reproduction, as a triploid with gynogenetic reproduction, and rarely as a tetraploid. Only triploids have been recorded in Europe. Species may shift ploidy levels, diploids and tetraploids are expected to occur. Triploid Ginbuna has their evolutionary origin in hybridisation of diploid *C. langsdorfii*, *C. gibelio,* and *C. auratus*, as these maternal lineages have been detected in Japanese triploid *C. langsdorfii*.

Conservation status. Non-native.

Remarks. Likely introduced as a weed with ornamental koi from Japan but expected to spread as a weed in carp farms from where it escapes into the wild.

Further reading. Murakami et al. 2001 (polyploidy systems in Japan); Kalous et al. 2013; Rylková et al. 2013 (Europe); Halas et al. 2018 (North America); Khosravi et al. 2020 (Iran).

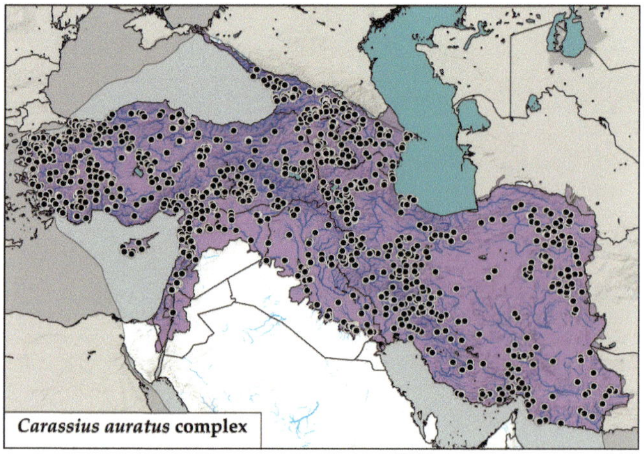

Cyprinion

Lotaks are a small genus known from Arabia, the Persian Gulf basin, adjacent Iran and Pakistan, east to the Indus drainage. They are medium-sized fish that inhabit a wide range of running waters, from fast-flowing mountain torrents to brackish desert and coastal streams. Lotaks can be identified by their usually 7½, rarely 6½, branched anal rays (compared with 5–6½ in other barbels), laterally compressed body, the sharp edge on the lower jaw (except in *C. milesi*), one pair of barbels, the serrated posterior margin of the last unbranched dorsal ray, and the 12½–17½ branched dorsal rays. They are related to South and Southeast Asian genera, such as *Semiplotus* and *Scaphiodontichthys*, but the relationships in this group of fishes are poorly understood.

To the extent that they have been studied, lotaks are diploid, with $2n = 50$ chromosomes. Molecular data suggest that *Cyprinion* hybridised with the ancestors of *Capoeta* and *Labeobarbus* and was thus involved in the allo-polyploidisation event that led to the formation of these genera. All but one of the lotaks occur in West Asia. Only *C. watsoni* is described from outside the region, from the rivers of the Sind Hills and the Salt Range in the Indus drainage of Pakistan.

While *Cyprinion* is relatively tolerant of water pollution and salinisation, it is sensitive to low temperatures and absent from the mountainous headwaters of the Euphrates in Eastern Anatolia. They feed on a wide variety of food, usually periphyton, but will take whatever is available. They spawn in flowing water, laying their eggs between gravel and rocks. Nuptial male lotaks can be territorial and aggressively defend spawning sites against other males. **Further reading.** Banister & Clarke 1977; Krupp 1983 (Arabian species, identification); Yang et al. 2015 (phylogenetic position).

Keys to *Cyprinion* in West Asia

Arabian Peninsula

1a - Last unbranched dorsal ray short, thin, serrated along proximal third of ray in large individuals; 14–16 circumpeduncular scales.
..................*C. muscatense*
1b - Last unbranched dorsal ray long, thick, serrated along almost entire length in large individuals; 16–22 circumpeduncular scales.
..................2

2a - 19–22 circumpeduncular scales; 8–9 scales between lateral line and dorsal origin.
..................*C. mhalense*
2b - 16–18 circumpeduncular scales; 6–7 scales between lateral line and dorsal origin.
..................*C. acinaces*

Persian Gulf and Iranian endorheic basin

1a - Mouth subterminal, oblique in lateral view, clearly U-shaped.
..................*C. milesi*
1b - Mouth inferior, variably arched to transverse.
..................2

2a - Mouth semi-circular, lower jaw finger-like with papillious median pad on lower lip in adults larger than 100 mm SL.
..................*C. kais*
2b - Mouth straight, with sharp cutting edge on lower jaw, without papillious central pad on lower lip; no large lateral lobes on lower lip.
..................3

3a - 9–12½, usually 10–11½, branched dorsal rays; orange spots along lateral line.
..................*C. microphthalmum*
3b - 11–17½, usually 12–15½, branched dorsal rays; no orange spots along lateral line.
..................4

4a - Posterior dorsal margin moderately to markedly concave; 39–45, usually 40–44, total lateral-line scales. ....*C. macrostomum*
4b - Posterior dorsal margin about straight or slightly concave; 32–39, usually 34–38, total lateral-line scales. ....*C. tenuiradius*

Cyprinion acinaces; Wadi Dauan, Yemen; 88 mm SL. © F. Krupp

Cyprinion acinaces

Common name. Arabian lotak.

Diagnosis. Distinguished from other species of *Cyprinion* in Arabian Peninsula by: ● 16–18 circumpeduncular scales / ○ last unbranched dorsal ray long, thick, serrated along almost entire length in large individuals / ○ 37–41, usually 38–40, total lateral-line scales / ○ 6–7 scales between lateral line and dorsal origin. Size up to 110 mm SL.

Distribution. Yemen: Wadi Hadhramaut drainage. Saudi Arabia: Coastal wadis of Hijaz mountains such as Hediyah, Sulayma, and wadis near Jeddah.

Habitat. Wadis with high seasonal fluctuations in discharge, usually over bare rock or gravel, rarely over soft substrate.

Biology. Feeds on periphyton.

Conservation status. LC.

Remarks. *Cyprinion acinaces hijazi* from northern Saudi Arabia is treated as a synonym, but further study is encouraged to determine whether it may be a separate species.

Further reading. Banister & Clarke 1977 (description); Krupp 1983 (subspecies); Freyhof et al. 2020 (distribution, identification).

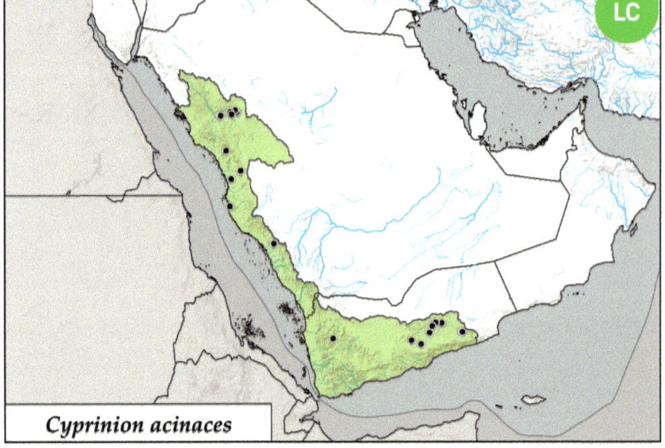

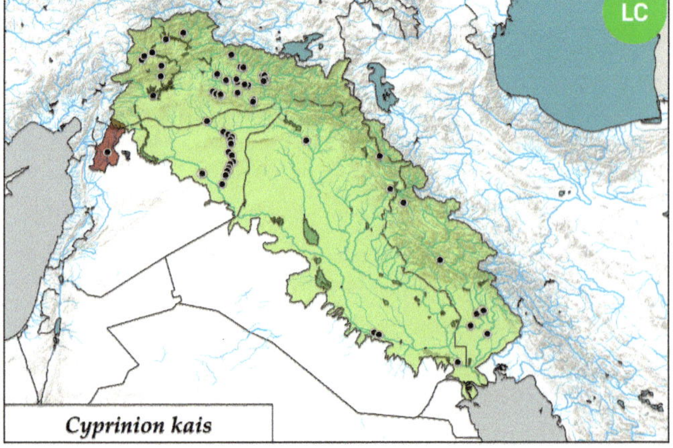

Cyprinion kais; Lesser Zab drainage, Iraq; male, ~100 mm SL.

Cyprinion kais; Tigris, Türkiye; male, ~110 mm SL.

Cyprinion kais

Common name. Smallmouth lotak.

Diagnosis. Distinguished from other species of *Cyprinion* in Persian Gulf and Iranian endorheic basins by: ● mouth semi-circular with two large lateral lobes on lower lip / ○ pelvic origin below or in front of dorsal origin / ● finger-like lower jaw with a papillious central pad / ○ 10–15 gill rakers on lower branch of first arch / ○ pelvic origin below dorsal origin / ○ no black spots or speckles on gill cover. Size up to 210 mm SL.

Distribution. Qweiq, Euphrates, Tigris, and Karun drainages.

Habitat. Usually rheophilic in fast-flowing waters in rapids and riffles. Less common in slow-flowing rivers and reservoirs. Rare in small streams and absent from cold headwaters.

Biology. No data.

Conservation status. LC; extirpated from Qweiq.

Remarks. Usually syntopic with *C. macrostomum*. Juveniles of both species are quite similar. Mitochondrial data suggest that *C. kais* is completely introgressed by *C. macrostomum*, having lost its own mitochondrial bodies. It has even been interpreted as a trophic, neotenic morphotype of *C. macrostomum*. However, the position of fins, the nuptial tubercles, and the colour pattern are so different that there is no doubt that they are two separate species.

Further reading. Kafuku 1969 (morphology); Bănărescu & Herzig-Straschil 1995 (morphology, distribution); Coad 2010a (overlap with *C. macrostomum*); Daştan et al. 2012 (genetic diversity); Kaya et al. 2016 (distribution); Keivany et al. 2016 (distribution); Nasri et al. 2018 (morphology); Coad 2021a (biology, morphology).

Cyprinion macrostomum; Tigris drainage, Türkiye; ~100 mm SL.

Cyprinion macrostomum; Lesser Zab drainage, Iraq; 150 mm SL.

Cyprinion macrostomum

Common name. Largemouth lotak.

Diagnosis. Distinguished from other species of *Cyprinion* in Persian Gulf and Iranian endorheic basins by: ○ mouth wide, with sharp cutting edge on lower jaw, without papillious central pad on lower lip / ○ pelvic origin behind dorsal origin / ○ no large lateral lobes on lower lip / ○ no orange spots along lateral line / ○ 13–21 gill rakers / ○ 39–45, usually 40–44, total lateral-line scales / ○ last unbranched dorsal ray long, thick, flexible only on very tip, usually serrated along almost entire length / ○ posterior dorsal margin moderately to markedly concave / ○ 13–16½, rarely 17½, branched dorsal rays. Size up to 300 mm SL.

Distribution. Qweiq, Euphrates, Tigris and Karun drainages. Also, in Gulf basin south to Zohreh.

Habitat. Ubiquitous species found in almost all types of habitats, from very small streams to large rivers, including backwaters and reservoirs, from where it migrates into tributaries to spawn in rapids and riffles. Often very abundant in small streams, especially if moderately polluted.

Biology. Lives up to 7 years, spawns first time at second or third year at about 100 mm SL. Spawns May–June, some into July. Eggs are laid in gravel in riffles. Feeds on epilithic algae, detritus, and plant material scraped from hard surfaces such as stones, wood, or plants.

Conservation status. LC; extirpated from Qweiq.

Remarks. Record from Orontes based on one individual reportedly collected from Lake Amik in 1881. With no further records of this hardy and ubiquitous species, it is very likely that this individual was either mislabelled or the fish was transported from the Euphrates to a fish market at Lake Amik.

Further reading. Kafuku 1969 (morphology); Bănărescu & Herzig-Straschil 1995 (morphology, distribution); Daştan et al. 2012 (genetic diversity); Kaya et al. 2016 (distribution); Keivany et al. 2016 (distribution); Nasri et al. 2018 (morphology); Coad 2021a (biology, morphology).

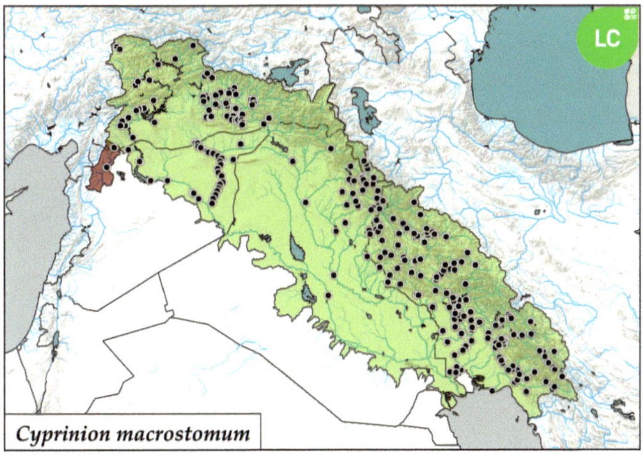

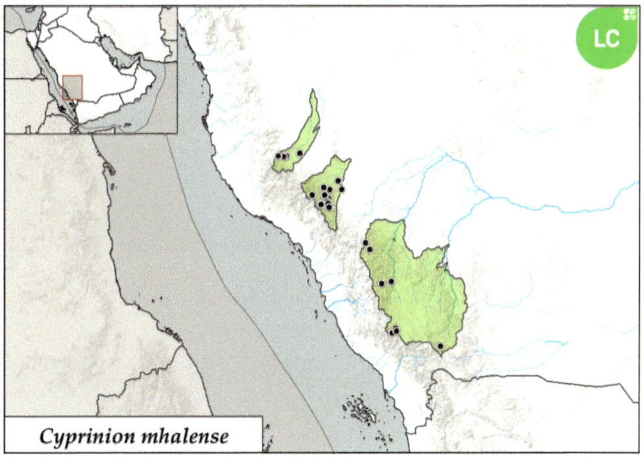

Cyprinion mhalense; Wadi Turabah, Saudi Arabia; 49 mm SL. © F. Krupp.

Cyprinion mhalense

Common name. Dawasir lotak.

Diagnosis. Distinguished from other species of *Cyprinion* in Arabian Peninsula by: ● usually 40–44 total lateral-line scales / ● 19–22 circumpeduncular scales / ● 8–9 scales between lateral line and dorsal origin / ○ last unbranched dorsal ray long, thick, serrated along almost entire length in large individuals. Size up to 160 mm SL.

Distribution. Saudi Arabia: Pleistocene and extant tributaries of Wadi ad-Dawasir, which flows to interior of Peninsula, including wadis Turabah, Afrak, Shuqub, Buwah, Shumrukh, Noaman, and Habayaba, between 1390 and 1830 m above sea level.

Habitat. Wadis with large seasonal variations in flow.

Biology. No data.

Conservation status. LC.

Further reading. Alkahem & Behnke 1983 (description); Krupp 1983 (description); Freyhof et al. 2020 (distribution, identification).

Cyprinion microphthalmum; Makran, Iran; 140 mm SL.

Cyprinion microphthalmum; Minhab, Iran, about 100 mm SL.

Cyprinion microphthalmum

Common name. Orange stripe lotak.

Diagnosis. Distinguished from other species of *Cyprinion* in Persian Gulf and Iranian endorheic basins by: ○ mouth wide, variably arched to almost straight with cutting edge in adults / ○ orange spots along lateral line / ○ no lateral lobes or papillious central pad on lower lip / ○ 8–18 gill rakers / ○ 31–43, usually 35–40, total lateral-line scales / ○ last unbranched dorsal ray thin, flexible along at least upper third, usually serrated from half to two-thirds of length / ○ dorsal margin usually from slightly concave to slightly convex / ○ 9–12½, usually 10–11½, branched dorsal rays / ○ back in front of dorsal with naked median area of variable length and about one scale wide / ○ scales on belly and upper anterior flank variably imbricated and often embedded in skin or absent. Size up to 200 mm SL.

Distribution. Shur (Iran) east to Makran region of Pakistan, where it occurs east of Bahukalat. Also, in endorheic basins of Jazmurian and Mashkid (Iran and Pakistan), Gonabad, and Lut (Iran and Pakistan).

Habitat. Ubiquitous. Inhabits a wide range of streams and rivers. Can be found in almost any freshwater habitat within its range. Spawns on gravel, usually in riffles.

Biology. Fractional spawner, individual females spawn several times during spring and summer. Males defend territories and have a dark-grey nuptial colour. Feeds on epilithic algae, detritus, and plant material scraped from hard surfaces such as stones, wood, or plants.

Conservation status. LC.

Remarks. The shape of mouth varies from almost terminal and moderately oblique in lateral view (similar to *C. milesi*) to inferior and transverse, with no real gap visible in lateral view. The thickness and degree of serration of the last unbranched dorsal ray also appear variable. *Cyprinion kirmanense*, from southeastern Iran, is usually treated as a synonym of *C. microphthalmum* but may be a valid species. *Cyprinion microphthalmum* has often been considered a synonym of *C. watsoni* from Pakistan.

Further reading. Bianco & Bănărescu 1982 (morphology, Kol drainage, as *C. watsoni*); Howes 1982 (anatomy; revalidation of *C. microphthalmum*); Nasri et al. 2013 (morphometrics, as *C. watsoni*); Nasri et al. 2018 (morphology).

Cyprinion microphthalmum inhabits a wide range of waters, including qanats.

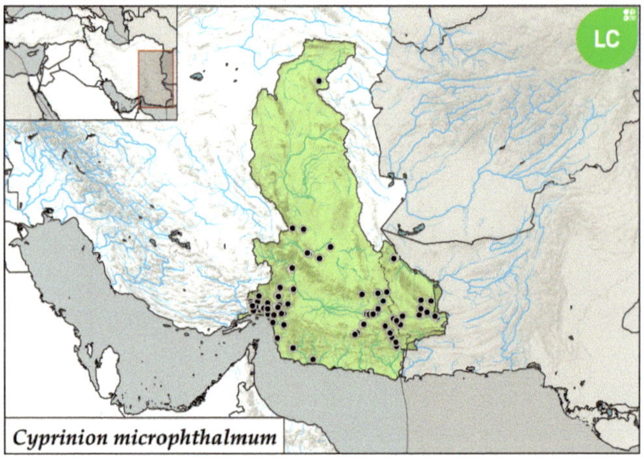

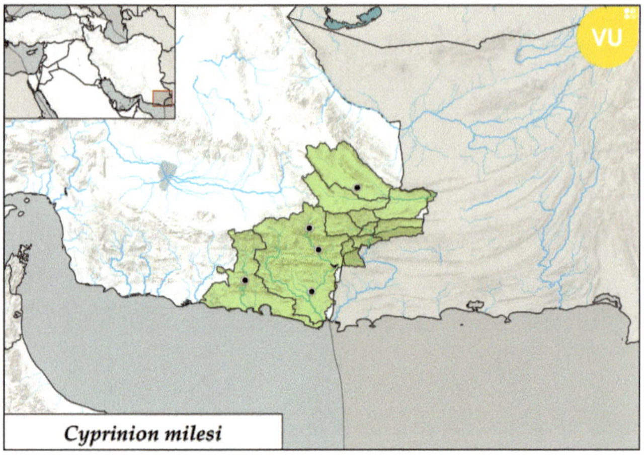

Cyprinion milesi; Sarbaz drainage, Iran; ~100 mm SL. © H. Mousavi-Sabet.

Cyprinion milesi

Common name. Bighead lotak.

Diagnosis. Distinguished from other species of *Cyprinion* in Persian Gulf and Iranian endorheic basins by: ● mouth subterminal, oblique in lateral view, clearly U-shaped / ○ no lateral lobes or pad on lower lip / ○ 11–12 gill rakers / ○ 34–39 total lateral-line scales / ○ last unbranched dorsal ray variably thick, weakly to strongly ossified and variably serrated / ○ dorsal margin straight / ○ 10–13½ branched dorsal rays / ○ back in front of dorsal usually without naked median area / ○ scales present on belly. Size up to 190 mm SL.

Distribution. Iran: Tang-e Sarhe toward Sarbaz and Mashkid, reported from Akar Kaur (Gwadar/Pakistan), both adjacent in Makran region. Also, in endorheic basins in Pakistan such as Kalat and Quetta, in Bampur and near Guadjik in Baschakird mountains, and in Panjgur basin.

Habitat. Streams and pools with fresh or brackish waters.

Biology. No data.

Conservation status. VU; thought to be declining in its small range, mainly due to desiccation of streams.

Remarks. A subterminal and moderately oblique mouth (in lateral view) is also found in individuals identical to *C. microphthalmum* in their mitochondrial DNA. The taxonomic status of *C. milesi*-like fishes in Iran requires further study. *Cyprinion milesi* is very similar to *C. watsoni* from the Indus, and it needs to be clarified if both are distinct.

Further reading. Mirza 1972 (distribution); Howes 1982 (anatomy); Nasri et al. 2013 (morphometrics); Nasri et al. 2016 (morphology); Coad 2021a (biology, morphology).

Cyprinion cf. *microphthalmum*; Minhab, Iran; long head form, ~110 mm SL.

Cyprinion milesi **in Iran.** *Cyprinion milesi* is readily distinguishable from other *Cyprinion* species in West Asia by its subterminal, clearly U-shaped mouth, which lacks the sharp cutting-edge characteristic of most *Cyprinion*. There are numerous records of *C. milesi* from the southern Persian Gulf basin in Iran, all based on *Cyprinion* with a subterminal, clearly U-shaped mouth, without a cutting edge on the lower jaw. These *C. milesi* are often found at the same localities as *C. microphthalmum*. It has been proposed that *C. milesi* is merely a morph of the latter species, retaining juvenile features in the mouth's structure. During the fieldwork for this book, we examined such individuals. Except for the head and mouth shape, these fish are indistinguishable from the syntopic *C. microphthalmum*. Mitochondrial molecular data support the view that these are aberrant individuals of *C. microphthalmum*. However, it cannot be ruled out that these are *C. milesi* carrying mtDNA of *C. microphthalmum* due to past introgressive hybridisation. *Cyprinion milesi* carrying its own mtDNA has been found in a few rivers, e.g., Bahukalat, Tang-e Sarhe, and Iranian Sarbarz. However, it is expected to be more common in adjacent Pakistan. In the future, it would be beneficial to conduct further research into the evolutionary and ontogenetic background of "aberrant" individuals of *C. microphthalmum*.

Cyprinion muscatense; Wadi Bani Khalid, Oman; ~90 mm SL.

Cyprinion muscatense

Common name. Hajar lotak.

Diagnosis. Distinguished from other species of *Cyprinion* in Arabian Peninsula by: ● last unbranched dorsal ray short, thin, and serrated along only proximal third of ray in large individuals / ● 14–16 circumpeduncular scales / ○ 37–41 total lateral-line scales / ○ 6–7 scales between lateral line and dorsal origin. Size up to 134 mm SL.

Distribution. Oman and United Arab Emirates: Wadis of Hajar mountains.

Habitat. Wadis with large seasonal variations in flow. Adults occupy larger pools than juveniles. Spawns in flowing water on gravel beds.

Biology. First spawns at 8–13 months. Fractional spawner, individual females spawn several times in a season. Survives more than 20 years in captivity; no data available from wild, but life expectancy in wild is likely to be much shorter. Fast-moving, active, mid-water dweller, usually found in small schools. In isolated pools, adults are generally very wary of human presence and initially remain hidden or in deeper areas close to hiding places; this may be an adaptation to predation by piscivorous birds. During spawning season, adults develop a distinctive pale blue colouration on anterior part of body, particularly lips and pectoral fins. Colour changes from mottled golden juveniles to silvery-grey adults. Juveniles are also proportionally thinner, i.e., body is less deep. Feeds on periphyton, opportunistically on aquatic and terrestrial invertebrates, and occasionally on plant material.

Conservation status. LC.

Remarks. This species is widespread in Hajar mountains. While the sympatric *Garra* are separated into different species, all populations of *C. muscatense* are very closely related.

Further reading. Alkahem & Behnke 1983; Krupp 1983 (description); Freyhof et al. 2020 (distribution, identification).

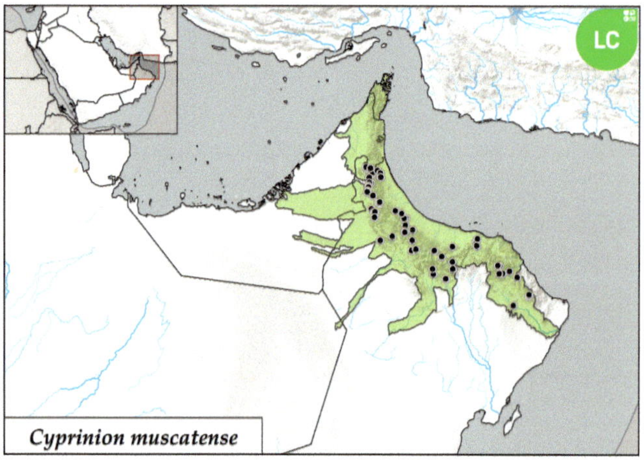

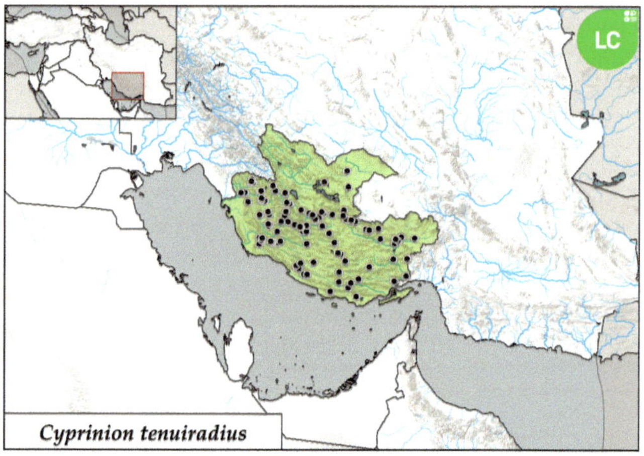

Cyprinion tenuiradius; Mond, Iran; ~90 mm SL. © H.R. Esmaeili.

Cyprinion tenuiradius; Golabi spring, Iran; ~100 mm SL.

Cyprinion tenuiradius

Common name. Persian lotak.

Diagnosis. Distinguished from other species of *Cyprinion* in Persian Gulf and Iranian endorheic basins by: ○ mouth wide, variably arched to straight / ○ no lateral lobes or pad on lower lip or small lateral lobes present / ○ 10–21 gill rakers / ○ 32–39, usually 34–38, total lateral-line scales / ○ last unbranched dorsal ray thin, flexible along at least upper third, usually serrated from half to two-thirds of length / ○ posterior dorsal margin about straight or slightly concave / ○ with or without orange spots along lateral line / ○ 11–15½, usually 12–14½, branched dorsal rays / ○ back in front of dorsal with naked median strip of variable length and about one scale wide / ○ scales on belly and upper anterior flank variably imbricated and often small and embedded in skin or absent. Size up to 130 mm SL.

Distribution. Iran: Persian Gulf basin from Helleh to Kol drainages, and Kor endorheic basin.

Habitat. A wide range of streams and rivers.

Biology. No data.

Conservation status. LC.

Remarks. *Cyprinion tenuiradius* occurs between the range of *C. macrostomum* in the north and *C. microphthalmum* in the south. It is also morphologically intermediate between these two species, with populations more or less resembling *C. macrostomum* or *C. microphthalmum*, often making identification difficult. Furthermore, *C. tenuiradius* shares its mtDNA with *C. macrostomum* or *C. microphthalmum*, and no species-specific mtDNA characters have been found so far. It is expected that *C. tenuiradius* has its evolutionary origin in hybridisation of *C. macrostomum* and *C. microphthalmum*, and the species' distribution area is a wide hybrid zone between two parental species. We accept it as a species of hybrid origin, a problematic position that needs to be studied in more detail.

Further reading. Bănărescu & Herzig-Straschil 1995 (description); Teimori et al. 2010 (distribution); Nasri et al. 2018 (morphology).

Cyprinus

A genus of about 24 species distributed mainly in southern China and northern Vietnam. Several species were endemic to lakes and may have formed adaptive radiations and species flocks. All these radiations are extinct, and only a few of the species involved remain. A few locally and regionally endemic carps are known from East Asia, all potentially under massive pressure from introduced non-native carps. In West Asia, only *Cyprinus carpio* is native. In Europe, carp appears to have been domesticated since the Middle Ages, and cultivated stocks are domesticated from the Danube wild population. Wild stocks occur naturally only in rivers draining the Black, Caspian, and Aral Seas. Carps are now found almost everywhere in West Asia. Since the 1960s, they have spread massively into all reservoirs and other large bodies of water. Interestingly, carp are relatively rare in most natural bodies of water and seem to have established populations primarily in large rivers and reservoirs. Japanese ornamental varieties (koi) are thought to be derived from *C. rubrofuscus* or to be hybrids with *C. carpio*. However, molecular data do not support this hypothesis, suggesting instead that the native range of *C. carpio* extends deep into China, with many native Chinese populations belonging to this species. The situation may be complicated by the widespread introduction of European carp into the native range of *C. rubrofuscus*.

Cyprinus carpio is widely cultivated around the world and is one of the most important species for aquaculture and inland fisheries in temperate regions. Although it is considered a pest in some countries, such as the USA, it is generally regarded as a valuable addition to fishing opportunities. In West Asia, where it is non-native to most catchment areas, it has been and still is stocked widely. The presence of carp outside of its native range is perceived as normal. If you ask someone at a reservoir which fish live there, the answer is almost always carp. Carp are especially recognised and appreciated because they grow big. Humans are fascinated by big fish, whether for food, recreational fishing or just to see them. Catching big fish is an ancient, deeply ingrained human fascination, and the person who catches a big fish is highly regarded by their social circle. Carp is one of the few large species that is ecologically well-suited to being stocked in artificial water bodies. *Silurus glanis* is another. The production of carp is big business, with all kinds of equipment available, ranging from bait and fishing tackle to fish finders, tents, video games, films and clothing. It is just another multi-million-pound industry. Carp angling is also a social event with international competitions and tourist travels. The internet is full of offers of what one needs and can do to catch 'the big carp'. As with all businesses, the aim is to expand and create new markets. West Asian countries are a new market for all carp fishing products. All that is needed is enough advertising to convince people to buy them. Reservoirs and carp already exist everywhere and are perceived as normal and native, even though they are only native to a very small part of the area.

Cyprinus carpio; Regreg drainage, Morocco; ~200 mm SL.

Cyprinus carpio

Common name. Common carp.

Diagnosis. Distinguished from other species of Cyprinidae in West Asia by: ○ two pairs of barbels / ○ 15–20½ branched dorsal rays / ○ last unbranched anal ray strongly ossified and serrated at its posterior margin / ○ caudal deeply emarginate. Size up to 1100 mm SL and 50 kg, usually less than 400 mm SL.

Distribution. Black, Caspian, and Aral basins. Introduced worldwide, including almost all drainages in West Asia.

Habitat. Warm, deep, slow-flowing, still waters like lowland rivers and large, well-vegetated lakes. Often semi-anadromous in Azov and Caspian basins. Introduced in all types of waters, especially reservoirs. Tolerates permanent salinities up to 8 ‰, temporarily up to 18.6 ‰. Spawns along banks or in backwaters. Successful survival of larvae only in warm water, under shallow submerged vegetation. Eggs hatch in water up to 10 ‰, with good results down to 6.6 ‰.

Biology. Males first spawn at 1–5 years, females at 1–6 years. Usually mature at 1–2 years in warm waters such as southern Iraq. Lives to 50 years and usually spawns annually. Age of maturity varies with latitude and altitude. Spawns late February–late April in Iraq and May–June in northern part of range when temperatures are above 18°C. Reported to spawn again in October–November in southern Iraq. Adults often undertake extensive spawning migrations to suitable backwaters and flooded meadows. Individual female spawn with few males in dense vegetation. Sticky eggs are attached to aquatic plants or other submerged objects. Larvae and juveniles inhabit warm and shallow flooded river margins or backwaters, feeding mainly on very small zooplankton (rotifers). Juveniles and adults feed on various benthic organisms and plant material. Most active at dusk and dawn. Very tolerant of low oxygen levels.

Conservation status. LC in native range. Wild populations have declined sharply since the 18th century due to river regulation and wetland draining. Cultivated in large quantities for human consumption and stocked for recreational fishing. Wild populations are slowly but steadily declining due to hybridisation with domesticated stocks. Probably very few populations remain "genetically unpolluted" due to this long process. The largest "wild" populations in West Asia appear to exist in lowland wetlands of the southern Caspian basin.

Further reading. Steffens 1958 (biology); Balon 1974, 1995 (domestication); Kottelat 1997 (systematics); Kottelat 2001a (Asian carps); Chu & Chen 1989 (Asian carps); Baruš et al. 2002 (biology); Coad 2010a (biology in Iraq, salinity tolerance).

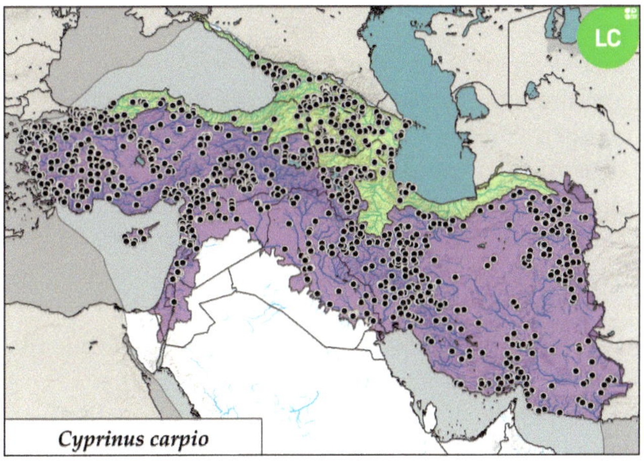

Cyprinus carpio; aquaculture variety, mirror carp, Germany; 450 mm SL. © A. Hartl.

Garra

The genus *Garra* comprises small fishes widely distributed in West South and Southeast Asia, and tropical Africa. The genus is highly diverse, with approximately 190 species currently recognised, while many others await formal description. It is the most speciose genus of Cyprinids in West Asia, with 42 species. The diversity of *Garra* in West Asia has long been significantly underestimated, as several species exhibit highly similar meristic character states and even the complex gular discs behind the mouth exhibit minimal consistent interspecific variation. Recent molecular studies have greatly contributed to the understanding of the diversity of this genus.

Most *Garra* species can be readily distinguished from other cyprinids by a prominent, uniquely shaped gular disc on the skin behind the lower jaw. This gular disc is structured into an anteromedian fold and a central callus. *Garra* species are often adapted to swift-flowing mountain streams, where they are aided in maintaining their position in the current by the gular disc and the lateral position of their large, paired fins. Furthermore, the air bladder is often reduced to reduce buoyancy. Rheophilic, current-loving species usually have a very large gular disc, but the gular disc can be reduced or even absent in some species that do not reside in fast currents.

The presence of a gular disc has long been regarded as the most important diagnostic character for *Garra*. Recent phylogenetic studies suggested that its presence is a plesiomorphic condition (an ancestral character state) in a range of labeonine cyprinids. This is evidenced by the fact that several labeonine lineages that lack the gular disc are nested within lineages that have a gular disc. Historically, species without a gular disc have been excluded from *Garra*. In West Asia, such species were previously placed in *Hemigrammocapoeta* and *Iranocypris*, but looking more closely, all of these still have remnants of some elements of the gular disc. It has long been suggested that they are closely related to *Garra*, and the "absence" of the gular disc is a secondary, adaptive character state in these fishes. This was recently supported by molecular data, which indicated that *Hemigrammocapoeta* species do not form a monophyletic lineage, but the species are related to three different groups of *Garra* species. All former *Hemigrammocapoeta*, *Iranocypris*, *Tylognathus*, and *Typhlogarra* from West Asia are included here within the genus *Garra*.

All *Garra* species feed on periphyton, plants, and detritus but are opportunistic feeders when other food sources are readily available. In streams where *Garra* are common, they have a substantial effect on the stream ecosystem by permanently grazing almost all surfaces. In West Asia, most species of *Garra* are ubiquitous inhabitants of streams and rivers. Still, they also occur in springs, lakes with stony shores, and even in marshes with dense vegetation. In southern Iran and the Arabian Peninsula, *Garra* are typically found in isolated pools in wadis that are otherwise dry, except when rain floods the river courses. They also occur in subterranean habitats.

Many species of *Garra* display prominent tubercles on the snout. The function of these tubercles has been discussed extensively, and it is generally believed that they function in territoriality and social interaction. In several species of *Garra*, the outer tubercles on the snout are enlarged and serve a social function. In most species, only some individuals develop prominent tubercles, and this character state is not associated with the sex of individuals. It has been proposed that individuals holding a territory develop tubercles, and those without territory do not. It has also been suggested that the snout tubercles may reduce the impact of rushing waters. However, these hypotheses remain untested. **Further reading.** Menon 1964 (diversity and biogeography); Krupp 1985d (phylogenetic position of *Hemigrammocapoeta*); Stiassny & Getahun 2007 (diversity in Africa); Yang et al. 2012 (phylogeny of Labeonines); Hamidan et al. 2014 (*Iranocypris*); Behrens-Chapuis et al. 2015 (molecular phylogeny); Esmaeili et al. 2016c; Zamani-Faradonbe et al. 2021b (Iran); Hashemzadeh Segherloo et al. 2016a (convergent reduction of gular disc; phylogeny).

Garra mamshuqa; Wadi al-Khun, Yemen; 52 mm SL. © F. Krupp.

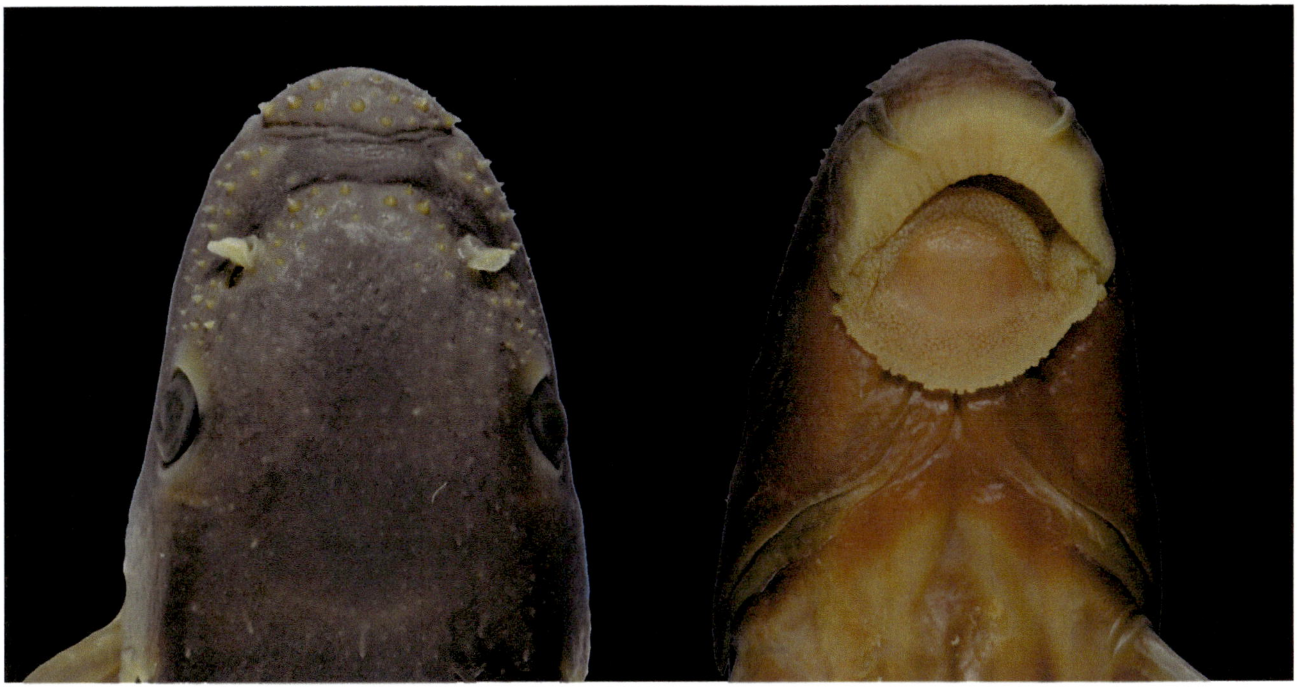

Figure 38. Head of *Garra jordanica* showing the generalised shape and structures of the gular disc and head in *Garra*.

Keys to species *Garra* in West Asia

Please note that the gular disc is present in all species; it is only incompletely developed in some.

Central Anatolia, Mediterranean, and Dead Sea basins

1a - Gular disc incompletely developed.
..................2
1b - Gular disc fully developed.
..................7

2a - No barbels; a papillate and elevated mental pad behind lower jaw.
..................3
2b - One or two pairs of barbels; no mental pad behind lower jaw.
..................4

3a - 9–15 gill rakers; few papillae on mental pad.
..................*G. kemali*
3b - 15–17 gill rakers; many papillae on mental pad.
..................*G. klatti*

4a - 8½ branched dorsal rays; two pairs of barbels; lateral line complete.
..................*G. nanus*
4b - 7½ branched dorsal rays; one pair of barbels; lateral line incomplete.
..................5

5a - Skin behind lower jaw continuous, with a shallow depression in middle; chin behind lateral lobes rounded anteriorly, not reaching between lateral lobes.
..................*G. festai*
5b - Skin behind lower jaw with a wide lateral lobe, very narrow, almost split in middle; chin between lateral lobes strongly pointed anteriorly, reaching deeply between lateral lobes.
..................6

6a - Skin behind lower jaw with lateral lobe strongly papillate on entire surface; scale in rows 2–3 above and below lateral line densely pigmented in middle, forming indistinct stripes; wide orange or reddish lateral stripe from eye to caudal base in live nuptial male.
..................*G. culiciphaga*
6b - Skin behind lower jaw with lateral lobe smooth or with indistinct furrows, papillate in middle in some individuals; scale in rows 2–3 above and below lateral line not more densely pigmented in middle, without stripes; live nuptial male with plain golden flank.
..................*G. caudomaculata*

7a - One pair of barbels.
..................*G. variabilis*
7b - Two pairs of barbels.
..................8

8a - 7½ branched dorsal rays; a small field of tubercles behind upper, posterior eye margin; 31–32+1–2 lateral-line scales; no black blotch on flank at middle of posterior caudal peduncle.
..................*G. ghorensis*
8b - 8½ branched dorsal rays; no tubercles behind upper, posterior eye margin; 33–36+1–2 lateral-line scales; a black blotch on flank at middle of posterior caudal peduncle at least visible in juveniles.
..................9

9a - Snout blunt; body depth at dorsal origin 18–21 % SL.
..................*G. turcica*
9b - Snout slightly pointed; body depth at dorsal origin 21–26 % SL.
..................10

10a - 12–17 gill rakers on lower limb of first gill arch; 5–6, rarely 4 scales between tip of pectoral and pelvic origin; maximum head width 64–72 % HL.
..................*G. jordanica*
10b - 17–21 gill rakers on lower limb of first gill arch; 2–3, rarely 4 scales between tip of pectoral and pelvic origin; maximum head width 73–82 % HL.
..................*G. orontesi*

Persian Gulf basin and endorheic basins in Iran, including Hari drainage

1a - Subterranean, body whitish, orange or pink; eye reduced and invisible externally.
..................2
1b - Epigean, body brown or grey, usually mottled; eye fully developed.
..................5

2a - Gular disc incompletely developed.
..................*G. typhlops*
2b - Gular disc completely developed.
..................3

3a - Lateral line complete or almost complete, with 28–35 scales.
..................*G. widdowsoni*
3b - Body completely naked or few scales present at pelvic base and on flank.
..................4

4a - Scales present at pelvic base, often isolated scales on flank; lateral line reduced, no visible pores.
..................*G. tashanensis*
4b - Body completely naked; lateral line on body well developed with 28–35 pores.
..................*G. lorestanensis*

5a - Lateral and distal margin of labrum of gular disc fused with skin of chin.
..................*G. elegans*
5b - Lateral and distal margin of labrum of gular disc free from skin of chin.
..................6

6a - Barbels absent.
..................*G. roseae*
6b - One or two pairs of barbels present.
..................7

7a - One pair of barbels; if two pairs, rostral barbel very small, sometimes just a tubercle.
..................8
7b - Two pairs of normally developed barbels.
..................10

8a - 20–26 gill rakers.
..................*G. variabilis*
8b - 10–16 gill rakers.
..................9

9a - Predorsal midline, chest, and belly almost without scales.
..................*G. nudiventris*
9b - Predorsal midline, chest, and belly fully covered by scales.
..................*G. rossica*

10a - 8+8 branched caudal rays.
..................*G. persica*
10b - 9+8 branched caudal rays.
..................11

11a - Usually 8½ branched dorsal rays.
..................12
11b - Usually 7½ branched dorsal rays.
..................13

12a - 32–38 total lateral-line scales.
..................*G. gymnothorax*/*G. rufa*
12b - 30–33 total lateral-line scales.
..................*G. tiam*

13a - 11–16 total gill rakers.
..................*G. rezai*
13b - 16–24 total gill rakers.
..................14

14a - Usually two scales between anus and anal origin.
..................*G. amirhosseini*
14b - Usually three scales between anus and anal origin.
..................15

15a - Posterior margin of scales on chest not embedded in skin.
..................*G. hormuzensis*
15b - Posterior margin of scales on chest embedded in skin.
..................*G. meymehensis*

Hajar Mountains of United Arab Emirates and northern Oman

1a - Subterranean, whitish or pink, eye reduced and invisible externally.
..................*G. cavernicola*
1b - Epigean, brown or grey eye fully developed.
..................2

2a - Tip of dorsal white.
..................3
2b - Tip of dorsal similar to middle and proximal part of fin.
..................4

3a - A pale- or deeply orange spot at upper opercle; middle caudal rays bold black, membranes white; 15–18 gill rakers on lower limb of first gill arch.
..................*G. barreimiae*
3b - No orange spot at upper opercle; middle caudal rays and membranes same colour as other caudal rays and membranes or slightly darker; 10–14 gill rakers on lower limb of first gill arch.
..................*G. shamal*

4a - Flank strongly mottled, with individual or series of orange scales.
..................*G. sharq*
4b - Flank usually plain greenish, brown, black, rarely with a midlateral stripe or faint mottling, never with orange midlateral scales.
..................5

5a - Flank scales with bold, dark-brown or grey margins.
..................*G. longipinnis*
5b - Flank scales same colour as centers of scales or with very indistinct dark-brown or grey margins.
..................*G. gallagheri*

Saudi Arabia, Yemen and Dhofar (Oman)

1a - Subterranean, whitish or pink; eye small.
..................*G. dunsirei*
1b - Epigean, brown, grey, often mottled; eye fully developed.
..................2

2a - A deep groove between tip of snout and nostrils; 12 circumpeduncular scales.
..................*G. mamshuqa*
2b - A shallow groove between tip of snout and nostrils or groove absent; 12–20 circumpeduncular scales.
..................3

3a - 36–39 total lateral-line scales; 20, very rarely 18, circumpeduncular scales.
..................*G. buettikeri*
3b - 30–36 total lateral-line scales; 12–18 circumpeduncular scales.
..................4

4a - Dorsal dark-grey or black; 8½ branched dorsal rays.
..................*G. sahilia*
4b - Dorsal hyaline or pale-brown with black spots at bases of branched rays; 7½ branched dorsal rays.
..................5

5a - Chest and belly naked.
..................*G. lautior*
5b - Chest and belly covered by scales at least to level of pectoral base.
..................6

6a - Usually 16–18 circumpeduncular scales; tubercles on snout usually present.
..................*G. tibanica*
6b - Usually 14–16 circumpeduncular scales; tubercles on snout absent.
..................7

7a - 8–13 gill rakers on lower limb of first gill arch; gular disc wider than long; prominent hump on back in some individuals.
..................*G. sindhae*
7b - 12–15 gill rakers on lower limb of first gill arch (10–13 in Laggashaylon); gular disc longer than wide; without hump on back.
..................*G. smartae*

Garra amirhosseini; Sartang-e-Bijar hot spring, Iran; 49 mm SL.

Garra amirhosseini
Common name. Ilam garra.
Diagnosis. Distinguished from other species of *Garra* in Persian Gulf and Iranian endorheic basins by: ○ chest and belly with very small, deeply embedded scales / ○ 17–19 scales on predorsal midline, often deeply embedded / ○ usually 7½ branched dorsal rays / ○ 9+8 branched caudal rays / ○ usually 2 scales between anus and anal origin / ○ 16–20 total gill rakers on first branchial arch / ○ 33–36 total scales along lateral line / ○ two pairs of normally developed barbels / ○ gular disc present / ○ lateral and distal margin of labrum of gular disc free from skin of chin / ○ body brown or grey, usually mottled / ○ eye fully developed. Size up to 105 mm SL.
Distribution. Iran: Sartang-e-Bijar sulfuric spring at Mehran, Meymeh, Godar khosh, Konjan cham, Zeynegan, and Kangir in Tigris drainage.
Habitat. Springs and spring-fed streams.
Biology. No data.
Conservation status. VU; known from less than 10 populations, and habitats are declining due to multiple threats, including water abstraction and climate change.
Remarks. Occurs in sympatry with *G. rufa*. *Garra amirhosseini* has COI DNA sequences very similar to *G. elegans*, while both species are clearly separated by morphological characters.
Further reading. Esmaeili et al. 2016c (description, molecular phylogeny).

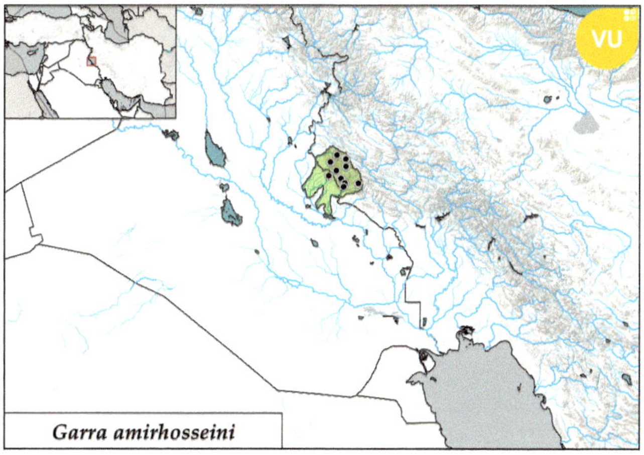

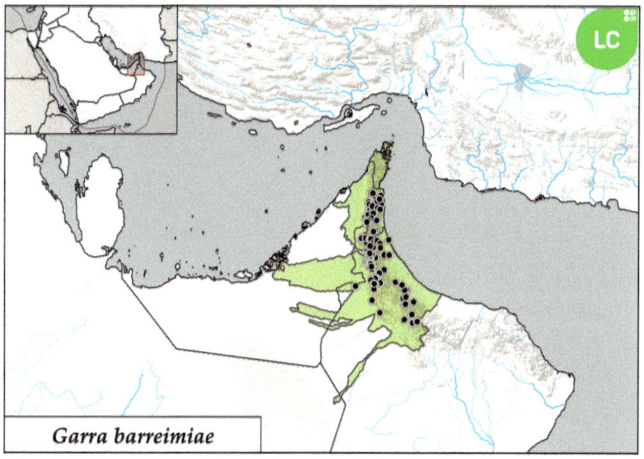

Garra barreimiae; Al-Juwayf, Oman; 49 mm SL.

Garra barreimiae

Common name. Orange-ear garra.

Diagnosis. Distinguished from other species of *Garra* in Hajar Mountains by: • 15–18 gill rakers on lower limb of first gill arch / • a pale- or deeply orange spot at upper opercle / ○ strongly mottled flank pattern with individual or series of orange midlateral scales / ○ middle caudal rays bold black, membranes white / ○ dorsal tip white / ○ anal and paired fins usually orange. Size up to 70 mm SL.

Distribution. United Arab Emirates and Oman: Wadis on both flanks of northern Hajar Mountains, extending south from southern edge of Ru'us al-Jibal range (Musandam Peninsula) to at least Wadi Hawasina on coast of Gulf of Oman and Wadi al-Kabir and its tributaries on inner flank.

Habitat. Streams, falaj systems, and springs with fresh or brackish water and gravel or rock bottoms. Typically found in small rock or gravel pools, shallow sections of larger slow-flowing pools, springs, and fast-flowing perennial sections of wadis. In small pools, juveniles occupy shallow sections, while larger adults are usually found at greater depths near hiding places.

Biology. Believed to spawn in anticipation of wadi flooding, but exact triggers remain unknown. Often explores upstream against current, dispersing when wadis flow. Can travel short distances out of water, including climbing steep, damp rock surfaces (e.g., near waterfalls or during rain). Can survive almost complete disappearance of surface water from their habitat; it has been speculated that juveniles may be able to survive in fossorial aquatic environments within wadi gravels. Accumulations of dead garra have been found in gravel under boulders at low points in wadis, apparently retreated as water levels have dropped below surface. In shallow pools, movements may be frantic on first encounter. Adults swim and feed close to substrate. Feeds mainly on periphyton but opportunistically on other food if available, e.g., dead insects floating on surface or decaying fruit or plant material. Seen to defend small feeding territories and "gardening" of algae fields might occur.

Garra buettikeri; Wadi Shumrukh, Saudi Arabia; 170 mm SL. © N.A. Hamidan.

Conservation status. LC.
Further reading. Fowler & Steinitz 1956 (description); Banister & Clarke 1977 (subspecies); Krupp 1983 (morphology); Feulner 1998 (distribution, biology); Khalaf 2009 (*G. b. wurayahi*); Kruckenhauser et al. 2011 (molecular data; as *G. barreimiae* clade 1); Hamidan et al. 2014 (molecular data); Lyon et al. 2016 (phylogeny); Kirchner et al. 2020 (distribution, identification).

Garra buettikeri

Common name. Asir garra.
Diagnosis. Distinguished from other species of *Garra* in Saudi Arabia, Yemen, and Dhofar (Oman) by: • 36–39 total lateral-line scales / • 20, very rarely 18 circumpeduncular scales / ○ 7½ branched dorsal rays / ○ 10–14 gill rakers on lower limb of first gill arch / ○ dorsal hyaline or pale-brown with black spots at bases of branched rays / ○ groove between tip of snout and nostrils shallow or absent. Size up to 180 mm SL, likely to grow larger.
Distribution. Saudi Arabia: Wadis of Asir Mountains that drain inland toward Wadi ad-Dawasir, such as Wadi Turabah, between 1400–2400 m above sea level.
Habitat. Wide variety of habitats, in fast-flowing to almost stagnant water over bare rock or soft sediment with gravel.
Biology. Feeds mainly on periphyton but opportunistically also on other food items.
Conservation status. VU; known from fewer than 10 populations, and habitat is declining due to multiple threats, including water abstraction and climate change.
Remarks. In Abala (2390 m), fish have 6–8 rakers on the lower limb of first gill arch and may represent an own species. *Garra buettikeri* is closely related to *G. tibanica* from western Arabia and *G. dunsirei*, *G. smartae*, and *G. sindhae* from Oman.
Further reading. Krupp 1983 (description); Hamidan et al. 2014 (phylogeny); Lyon et al. 2016 (phylogeny); Freyhof et al. 2020 (distribution, identification).

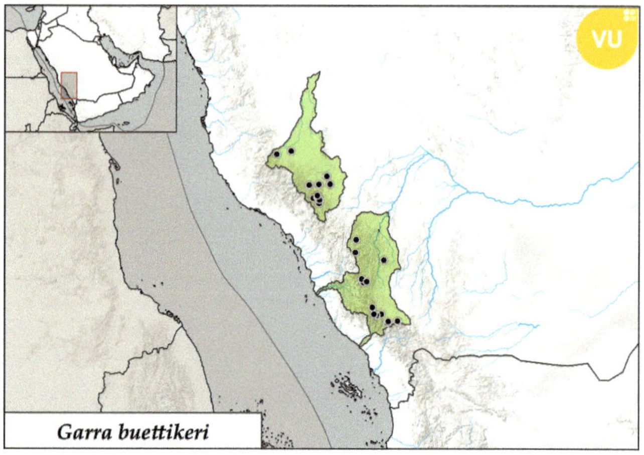

Garra buettikeri

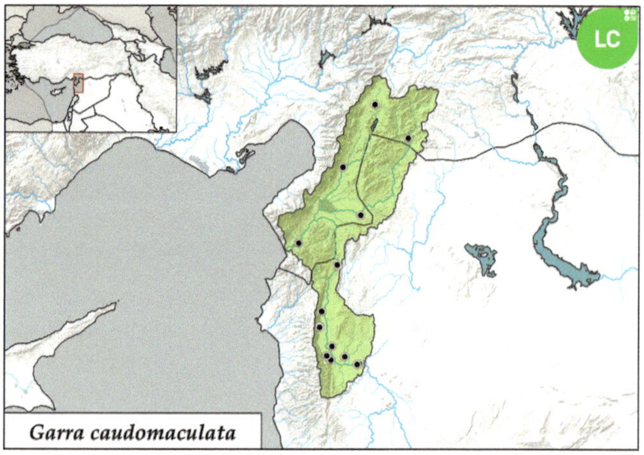

Garra caudomaculata

Garra caudomaculata; Orontes drainage, Syria; female, 47 mm SL.

Garra caudomaculata; Orontes drainage, Syria; male, 42 mm SL.

Garra caudomaculata
Common name. Orontes golden barb.
Diagnosis. Distinguished from other species of *Garra* in Central Anatolia, Mediterranean, and Dead Sea basins by: ○ gular disc incomplete / ○ skin behind lower jaw with a wide lateral lobe, very narrow, almost split in middle / ○ lateral lobe on skin behind lower jaw smooth or with indistinct furrows, papillate in middle in some individuals / ○ chin between lateral lobes strongly pointed anteriorly, reaching deeply between lateral lobes / ○ a black spot at posteriormost caudal peduncle in juveniles / ○ 7½ branched dorsal rays / ○ 31–34 scales in total lateral series / ○ lateral line incomplete, with 8–11 pored scales / ○ one pair of very small barbels / ○ scale in rows 2–3 above and below lateral line not more densely pigmented in middle, without stripes / ○ live nuptial male with plain golden flank. Size up to 60 mm SL.
Distribution. Türkiye and Syria: Orontes drainage and Nahr al-Kabir at border between Syria and Lebanon.
Habitat. Slow-flowing rivers and streams, densely vegetated springs, and wetlands.
Biology. Spawns after first winter. Fractional spawner.
Conservation status. LC.

Remarks. Earlier placed in *Hemigrammocapoeta* and often considered as a synonym of *G. culiciphaga. Garra caudomaculata*, *G. culiciphaga*, and *G. nanus* form a monophyletic lineage.

Further reading. Battalgil 1942 (description); Geiger et al. 2014 (molecular phylogeny, placement in *Garra*); Hamidan et al. 2014 (revalidation); Behrens-Chapuis et al. 2015 (molecular phylogeny); Bayçelebi 2020 (distribution).

Garra cavernicola; Al-Hoota cave, Oman; ~50 mm SL.

Orontes in Syria, just south of border with Türkiye. Habitat of a rich fish fauna including *Garra caudomaculata* and *G. orontesi*.

Garra cavernicola

Common name: Al-Hoota garra.

Diagnosis. Distinguished from other species of *Garra* in Hajar Mountains by: ○ eye reduced and invisible externally / ○ whitish or pink, without colour pattern. Size up to 80 mm SL.

Distribution. Oman: Al-Hoota cave system.

Habitat. Subterranean waters.

Biology. No data.

Conservation status. VU; known from a very small area. The population is affected by climate change.

Remarks. This species is genetically very closely related to neighboring surface-dwelling populations of *G. longipinnis* and has been treated as a 'form' of *G. longipinnis*. Despite being very young, this evolutionary lineage shows strong and distinct evolutionary trajectories and represents a different species under the Evolutionary Species Concept. A second subterranean population may occur near the town of Al Hamra and may represent an undescribed species.

Further reading. Banister 1984 (finding of cave *Garra*); Kruckenhauser et al. 2011; Kirchner et al. 2020 (Omani cave *Garra*); Hamidan et al. 2014 (phylogeny; as *G.* cf. *longipinnis*); Freyhof et al. 2020 (distribution, identification); Freyhof 2025 (description).

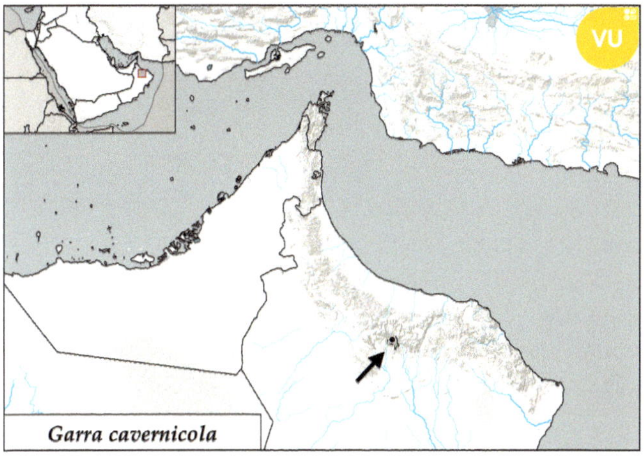

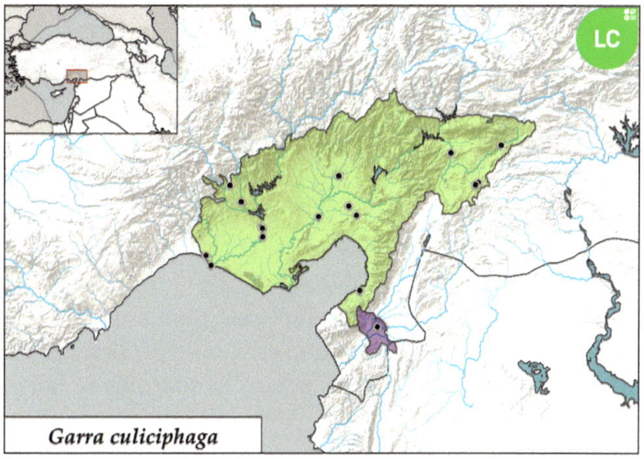

Garra culiciphaga; Ceyhan drainage, Türkiye; female, 48 mm SL.

Garra culiciphaga; Seyhan drainage, Türkiye; male, 50 mm SL.

Garra culiciphaga

Common name. Red stripe barb.

Diagnosis. Distinguished from other species of *Garra* in Central Anatolia, Mediterranean, and Dead Sea basins by: • wide, orange or reddish lateral stripe from eye to caudal base in nuptial male / ○ gular disc incomplete / ○ skin behind lower jaw with a wide lateral lobe, very narrow, almost split in middle / ○ lateral lobe on skin behind lower jaw strongly papillate on entire surface / ○ chin between lateral lobes strongly pointed anteriorly, reaching deeply between lateral lobes / ○ a black spot at posteriormost caudal peduncle in juveniles / ○ 7½ branched dorsal rays / ○ 29–34 scales in total lateral series / ○ lateral line incomplete, with 7–11 pored scales / ○ one pair of very small barbels / ○ scale in rows 2–3 above and below lateral line densely pigmented in middle, forming indistinct stripes. Size up to 70 mm SL.

Distribution. Türkiye: Lower Tarsus, Seyhan and Ceyhan drainages, also in a coastal stream at İskenderun. Likely introduced to Lake Balık (Gölbaşı-Hatay).

Habitat. Slow-flowing rivers and streams, densely vegetated springs, and wetlands. Spawns in dense vegetation of submerged plants, roots, or algae.

Biology. Usually solitary. Lives 3–4 years. Spawns first time after first winter. Spawns April–September, fractional spawner. Several males follow female into dense vegetation and spread small, transparent, and sticky eggs widely in vegetation. First-time spawners spawn later in spring than older spawners. Larvae solitary, inhabit dense vegetation, feed on plankton and detritus.

Conservation status. LC.

Remarks. Earlier placed in *Hemigrammocapoeta*.

Further reading. Pellegrin, 1927 (description); Geiger et al. 2014 (molecular phylogeny, placement in *Garra*); Hamidan et al. 2014, Behrens-Chapuis et al. 2015 (molecular phylogeny); Bayçelebi 2020 (distribution).

***Garra culiciphaga*: a misnamed fish?** The field of etymology is replete with misconceptions, and at times, consulting the original descriptions or older literature can reveal errors even in well-known names that have been in use for decades. For instance, the name "*culiciphaga*," ascribed by Jacques Pellegrin to the species, is understood to refer to its efficacy in the destruction of mosquito larvae (*Culex*, mosquito; *phaga*, to eat). Pellegrin had received the type specimens from D.E. Tok, the head doctor of the antimalarial service in the Adana region. In his own words, Pellegrin stated that "according to the experiments of the latter (D.E. Tok), these fish are excellent destroyers of mosquito larvae, hence the name given to the species." In the original description, he stated that he had received the material with a postscript attached, denoting the fish's mosquito-eating behaviour, which D.E. Tok had sent. However, in the batch of samples he had received, there was another species, which he noted as "*Aphanius fasciatus*," known as an "excellent mosquito destroyer." D.E. Tok had likely referred to *A. fasciatus* as "d'excellents destructeurs," a case that was misinterpreted by Pellegrin. According to the rules of zoological nomenclature, the name *culiciphaga* cannot be changed, even if based on misinformation. **Further reading.** Pellegrin 1927 (description).

The Sinkhole at Tawi Attair in Oman is the only habitat of *Garra dunsirei*. © J. Els.

Garra dunsirei; Tawi Attair, Oman, 60 mm SL.

Garra dunsirei

Common name. Tawi Attair garra.

Diagnosis. Distinguished from other species of *Garra* in Saudi Arabia, Yemen, and Dhofar (Oman) by: ● neural spines very short and not penetrating between dorsal pterygiophores / ○ subterranean, body whitish or pink / ○ eye small / ○ scales on chest and belly reduced or absent. Size up to 81 mm SL.

Distribution. Oman: Sinkhole at Tawi Attair in Jabal Qara Mountains in Dhofar region.

Habitat. Subterranean species. Found in a large side passage about 200 m below surface, which may be a window into a much larger underground cave system.

Biology. No data. Captive individuals have been reported to live up to 16 years; no data are available from wild, but life expectancy in wild is likely to be much shorter.

Conservation status. CR; restricted to a single known locality where it has not been found in recent years. Fewer than 250 mature individuals are believed to exist, although this may be underestimated due to the difficulty of thoroughly exploring the sinkhole.

Remarks. Its closest relatives are *G. smartae* and *G. sindhae*. Despite being a young species, it shows very strong and distinct evolutionary trajectories and represents a different species under the Evolutionary Species Concept.

Further reading. Banister 1987 (description); Hamidan et al. 2014 (phylogeny); Freyhof et al. 2020 (distribution, identification); Sayyadzadeh et al. 2023 (phylogeny).

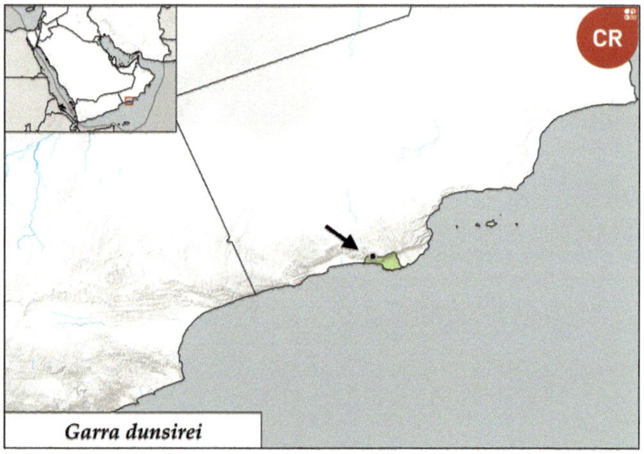

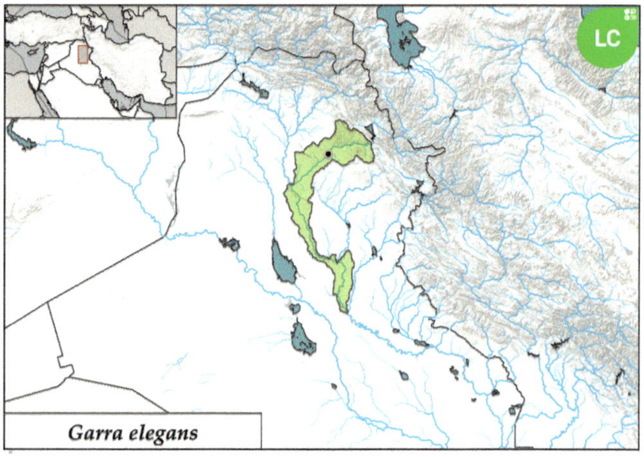

Garra elegans; Lesser Zab, Iraq; 54 mm SL.

Garra elegans

Common name. Mesopotamian garra.

Diagnosis. Distinguished from *Garra* in Persian Gulf and endorheic basins in Iran by: ● lateral and distal margin of labrum fused with chin skin / ○ chest and anterior belly without scales / ○ mid-dorsal area in front of dorsal origin covered by scales / ○ 9+8 branched caudal rays / ○ 17–20 gill rakers / ○ two pairs of normally developed barbels / ○ body brown or grey, usually mottled / ○ eye fully developed. Size up to 87 mm SL.

Distribution. Iraq: Lower Lesser Zab and main stem of Tigris.

Habitat. Large lowland rivers.

Biology. No data.

Conservation status. LC; very poorly known.

Remarks. Distribution very poorly known but thought to occur mainly in lower parts of large rivers. Previously placed in *Tylognathus*, *Hemigarra*, or *Hemigrammocapoeta*. Molecular markers place it close to *G. amirhosseini*. *Garra elegans* is usually described as lacking a gular disc, but the disc is fully attached to the chin.

Further reading. Behrens-Chapuis et al. 2015 (molecular phylogeny); Freyhof 2016b (description).

Unlike other *Garra* species in West Asia, *G. elegans* seems specialised to large lowland rivers. It was found here, in the middle of Lesser Zab in Iraq.

Garra festai; Ammiq marsh, Lebanon; 45 mm SL.

Garra festai

Common name. Ammiq barb.

Diagnosis. Distinguished from other species *Garra* in Central Anatolia, Mediterranean, and Dead Sea basins by: ○ gular disc incomplete / ○ 7½ branched dorsal rays / ○ one pair of barbels / ○ lateral line incomplete / ○ skin behind lower jaw continuous, with a shallow depression in middle / ○ chin behind lateral lobes rounded anteriorly, not reaching between lateral lobes. Size up to 90 mm SL.

Distribution. Lebanon: Ammiq marshes in upper Litani drainage.

Habitat. A small wetland with dense reeds and springs. Spawns in dense vegetation.

Biology. Lives for 5 years and spawns first time after first winter. Fractional spawner.

Conservation status. CR; limited to a small wetland that partially dries out in summer. Ammiq marshes (253 ha) are remnants of much more extensive marshes and lakes that once existed in Bekaa Valley, and since 1970, 90 % of Ammiq marshes have been drained. It is an Important Bird Area, listed in Directory of Wetlands, a Ramsar site; part of it is a UNESCO Biosphere Reserve.

Remarks. Earlier placed in *Hemigrammocapoeta*. *Garra festai* seems not to be closely related to *G. caudomaculata*, *G. culiciphaga*, and *G. nanus*.

Further reading. Tortonese 1938 (description); Krupp 1985d (distribution); Geiger et al. 2014 (molecular phylogeny, placement in *Garra*); Behrens-Chapuis et al. 2015 (molecular phylogeny); Hashemzadeh Segherloo et al. 2016a (molecular phylogeny, reduction of gular disc).

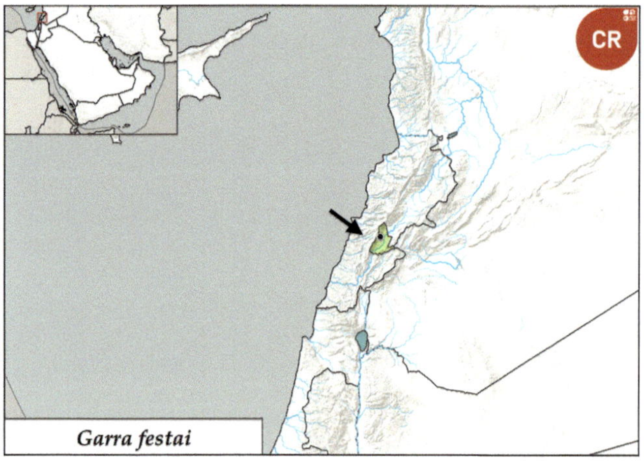

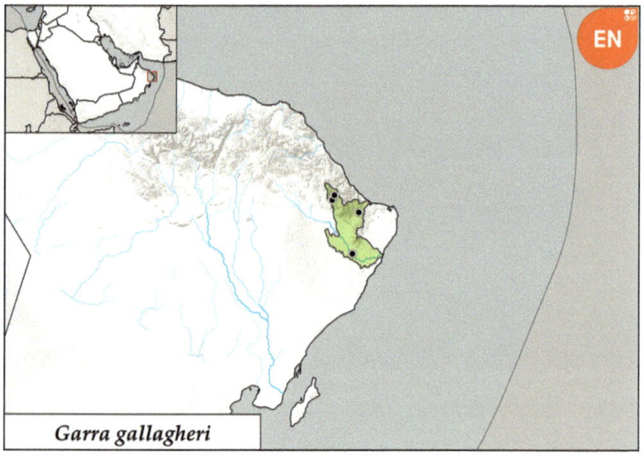

Garra gallagheri; Wadi Bani Khalid, Oman; 55 mm SL.

Garra gallagheri

Common name. Black garra.

Diagnosis. Distinguished from other species of *Garra* in Hajar Mountains by: ○ scales without bold dark-brown or grey margins or scale pockets / ○ flank plain or almost plain, golden-greenish, brown, grey, or black without or with little mottling, a faint dark-green lateral stripe present in some individuals / ○ no orange midlateral scales on flank / ○ no orange spot at upper opercle / ○ tip of dorsal not white / ○ 12–14 gill rakers on lower limb of first gill arch. Size up to 75 mm SL.

Distribution. Oman: Wadi Bani Khalid and downstream falaj systems, as in towns of Bani Bu Ali and Al Mudayrib.

Habitat. Ponds, falaj systems, and springs with fresh or brackish water and rocky bottoms. Usually in stagnant or slow-flowing water. In small pools, juveniles occupy shallow areas, while larger adults are typically found at greater depths near hiding places.

Biology. Feeds on periphyton and detritus but can take other food if readily available.

Conservation status. EN; appears to be declining within its very small range. Known from only three populations, two of which are very small and vulnerable to renovation of falaj systems. The main population in Wadi Bani Khalid has experienced a large decline due to restriction of wadi,

Ein Al-Haditha in Jordan is a habitat of *Garra ghorensis*.

but is thought to be relatively secure now. Climate change could have a significant impact on this population.

Remarks. Closely related to *G. longipinnis*, from which molecular characters well separate it, but differences in colour shades are vague and may not always be sufficient to distinguish the two species. Quite variable in colour within Wadi Bani Khalid, but all colour patterns represent the same population.

Further reading. Krupp 1988 (description); Kruckenhauser et al. 2011 (molecular data; as *G. barreimiae* clade 2); Lyon et al. 2016 (phylogeny); Kirchner et al. 2020 (distribution, identification).

Garra ghorensis

Common name. Dead Sea garra.

Diagnosis. Distinguished from other species of *Garra* in Mediterranean and Dead Sea basins by: ● tubercles present on upper posterior eye margin / ○ gular disc fully developed / ○ two pairs of normally developed barbels / ○ 31–32+1–2 lateral-line scales / ○ 12–16, usually 14 circumpeduncular scales / ○ slender body / ○ 7½ branched dorsal rays / ○ 9–11 gill rakers on lower limb of first gill arch / ○ mid-dorsal area in front of dorsal origin scaled / ○ chest naked or one to several scales forming one row between pectoral bases / ○ black blotch at posterior caudal peduncle absent. Size up to 101 mm SL.

Distribution. Jordan and Israel: Southern tributaries of Dead Sea: Lower part of Wadi Ibn Hammad, Ein Al-Haditha, Wadi Al-Hassa (including Wadi al-Burbaitah), Wadi Fifa, and Wadi Khneizerah.

Habitat. Fresh and brackish streams and springs with gravel bottoms. Usually in moderate currents.

Biology. Feeds almost exclusively on algae. Tolerates high water temperatures. Spawns in groups in late spring and early summer.

Conservation status. EN; extirpated from Israel. Population in Ain al-Haditha is now restricted to a very short stream, perhaps 300 m long, which is affected by non-native *Oreochromis aureus*. Downstream, spring water is entirely used for irrigation.

Further reading. Krupp 1982 (description); Krupp & Schneider 1989 (description); Geiger et al. 2014 (molecular phylogeny); Hamidan et al. 2014 (molecular phylogeny, morphological comparisons).

Garra ghorensis; spring at Ghor al Hadithah, Jordan; 101 mm SL.

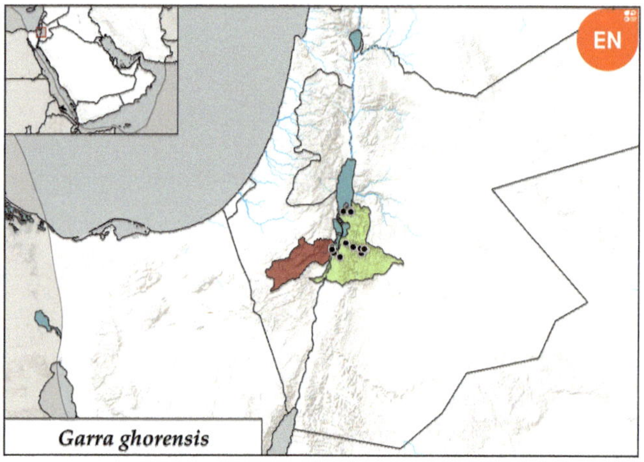

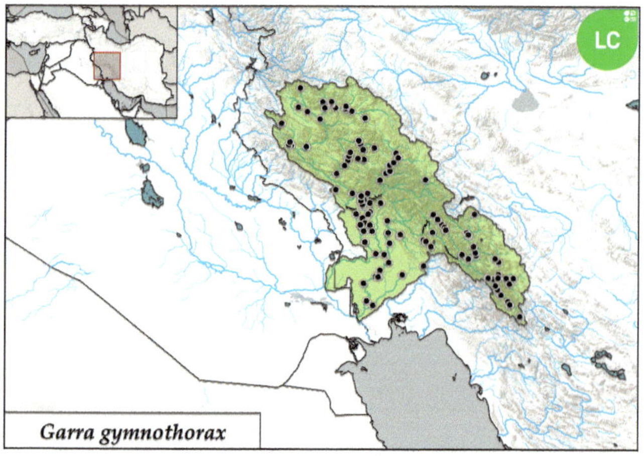

Garra gymnothorax; Gamasiab in Karkheh drainage, Iran; 120 mm SL.

Garra gymnothorax

Common name. Karun garra.

Diagnosis. Distinguished from other species of *Garra* in Persian Gulf and endorheic basins in Iran by: ○ chest without or with normally sized, often embedded scales / ○ mid-dorsal area in front of dorsal origin covered by embedded scales / ○ usually 8½ branched dorsal rays / ○ 9+8 branched caudal rays / ○ 17–23 gill rakers / ○ 32–37 total scales along lateral line / 13–17 circumpeduncular scale rows / ○ two pairs of normally developed barbels / ○ gular disc fully developed / ○ lateral and distal margin of labrum of gular disc free from skin of chin / ○ body brown or grey, usually mottled / ○ eye fully developed. Size up to 151 mm SL.

Distribution. Iran: Karun and Karkheh drainages. Most populations in Karun are hybrids with *G. rufa*.

Habitat. Streams and rivers of all kinds, as long as there are at least small areas of rock or gravel.

Biology. Lives up to 4 years. Spawns March–September with a peak in May–June. Usually total spawners. Benthic grazer feeding on algae, cyanobacteria, and small invertebrates scraped from hard surfaces such as stones, roots, and plants.

Conservation status. LC.

Remarks. *Garra gymnothorax* was described based on fish from the Karun having a naked chest, whereas most *Garra* from the Euphrates, Tigris, Karun (and Karkheh) have a scaled chest. In the Karun and Karkheh, there is an endemic mtDNA lineage of *Garra* that is well differentiated from *G. rufa*. This endemic mtDNA lineage is found in fish with naked as well as with scaled chests. Fish containing mitochondria with this endemic sequence have been identified as *G. gymnothorax*. No known morphological characters distinguish fish with mtDNA of "*G. gymnothorax*" from those with mtDNA of *G. rufa*. While all fish in Karun and Karkheh have *G. gymnothorax* mtDNA, genome-wide SNP data show that many populations in Karun (including fish at the type locality of *G. gymnothorax*) are hybrids between "*G. gymnothorax*" and *G. rufa*. Both *G. gymnothorax* and *G. rufa* mtDNA are found in Karun, suggesting extensive hybridisation between the two species.

Further reading. Berg 1949a (description); Abedi et al. 2011 (biology, as *G. rufa*); Hamidan et al. 2014, Behrens-Chapuis et al. 2015 (molecular data); Keivany et al. 2015 (morphological diversity; similarity with *G. rufa*); Sayyadzadeh et al. 2015b (discussion on taxonomic status); Esmaeili et al. 2016c (molecular phylogeny, description, revalidation); Hashemzadeh Segherloo et al. 2016a (molecular phylogeny, gular disc morphology).

Garra hormuzensis; Golabii spring, Iran; 70 mm SL.

Garra hormuzensis

Common name. Kol garra.

Diagnosis. Distinguished from other species of *Garra* in Persian Gulf and endorheic basins in Iran by: ○ mid-dorsal area in front of dorsal origin covered by scales / ○ posterior scale-margin on chest not embedded in skin / ○ chest and belly covered by scales / ○ usually 7½ branched dorsal rays / ○ 9+8 branched caudal rays / ○ usually 3 scales between anus and anal origin / ○ 16–21 gill rakers / ○ two pairs of normally developed barbels / ○ gular disc fully developed / ○ lateral and distal margin of labrum of gular disc free from skin of chin / ○ body brown or grey, usually mottled / ○ eye fully developed. Size up to 70 mm SL, likely to grow larger.

Distribution. Iran: Golabii spring, Haji Abad, Roudbal, and Fadami in Kol drainage.

Habitat. Moderately warm to very warm springs and muddy streams to moderately fast-flowing cold waters.

Biology. No data.

Conservation status. VU; known from fewer than 10 populations and declining habitat due to water abstraction and climate change.

Remarks. The mtDNA of this species is very similar to the mtDNA of "*G. mondica*." Also, occurs in sympatry, but not syntopy, with *G. persica*.

Further reading. Zamani-Faradonbe et al. 2021a (description).

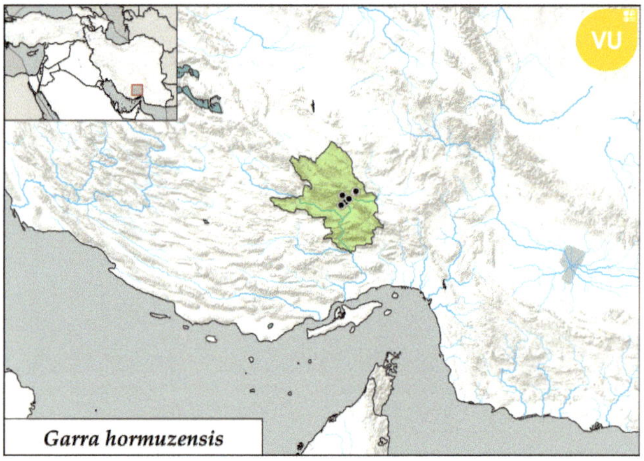

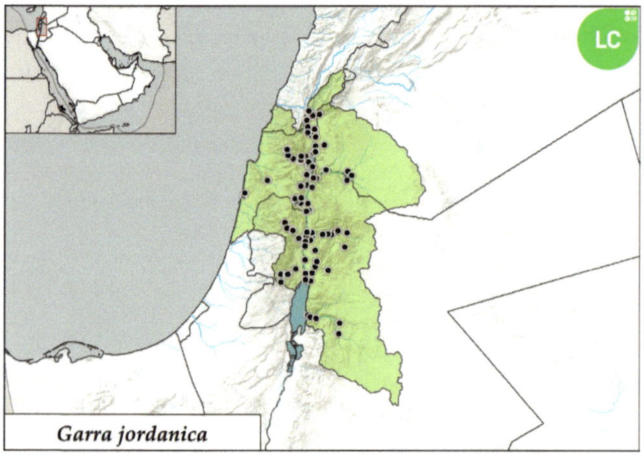

Garra jordanica; Jordan drainage, Syria; 96 mm SL.

Garra jordanica

Common name. Jordan garra.

Diagnosis. Distinguished from other species of *Garra* in Mediterranean and Dead Sea basins by: ○ gular disc fully developed / ○ tubercles absent on upper posterior eye margin / ○ two pairs of normally developed barbels / ○ 8½ branched dorsal rays / ○ 35–37 total lateral-line scales / ○ 16 circumpeduncular scales / ○ 12–17, usually, 16, gill rakers on lower limb of first gill arch / ○ 5–6, rarely 4 scales between tip of pectoral and pelvic origin / ○ maximum head width 64–72 % HL / ○ body depth at dorsal origin 20–25 % SL / ○ snout slightly pointed / ○ chest and belly scaled, 4–7 rows of deeply embedded scales on chest between pectoral bases / ○ mid-dorsal area in front of dorsal origin covered by scales / ○ dark-brown or grey blotch at posterior caudal peduncle present. Size up to 137 mm SL.

Distribution. Jordan drainage and lower Wadi Moujib, at east side of northern Dead Sea in Jordan.

Habitat. Springs, rivers, and streams with moderate to fast-flowing waters. Also in reservoirs and lakes.

Biology. No data.

Conservation status. LC.

Further reading. Goren 1974 (morphology, distribution; as *Garra rufa*); Geiger et al. 2014 (phylogeny, as *G.* cf. *ghorensis*); Hamidan et al. 2014 (description, molecular phylogeny).

Meyil Lake in Türkiye is a sinkhole in the Pleistocene Lake Tuz basin and one of the few habitats of *Garra kemali*.

Garra kemali; Lake Beyşehir basin, Türkiye; female, 74 mm SL.

Garra kemali; Lake Meyil, Türkiye; male, 50 mm SL.

Garra kemali

Common name. Tuz golden barb.

Diagnosis. Distinguished from other species of *Garra* in Central Anatolia, Mediterranean, and Dead Sea basins by: ○ no barbels / ○ gular disc incomplete / ○ 9–15 gill rakers / ○ few papillae on mental pad / ○ 36–45+1–2 scales in lateral series / ○ lateral line incomplete, with 5–14 pored scales. Size up to 74 mm SL.

Distribution. Türkiye: Ereğli marshes and Lake Meyil in Lake Tuz basin. Lakes Beyşehir and Seydişehir basins, including Çarşamba Canal. Also, in Hirfanlı reservoir (Kızılırmak).

Habitat. Densely vegetated springs, lakes, reservoirs, and slow-flowing streams.

Biology. Matures within a year. Males with bright golden nuptial colouration. Spawns on vegetation. Comes close to shore in spring but stays deep and far from shore in Hirfanlı reservoir in late summer and winter.

Conservation status. VU; occurs in less than 10 populations. Extirpated from Ereğli marshes due to draining of wetland. Other populations in decline.

Remarks. Very abundant when discovered in Hirfanlı reservoir in 2017. Despite regular monitoring efforts, not seen again until 2024, when it was found again. Previously placed in *Hemigrammocapoeta*.

Further reading. Hankó 1925 (description); Geiger et al. 2014 (molecular phylogeny, placement in *Garra*); Behrens-Chapuis et al. 2015 (molecular phylogeny); Küçük et al. 2015 (morphology, in *Hemigrammocapoeta*); Yoğurtçuoğlu et al. 2018a (distribution).

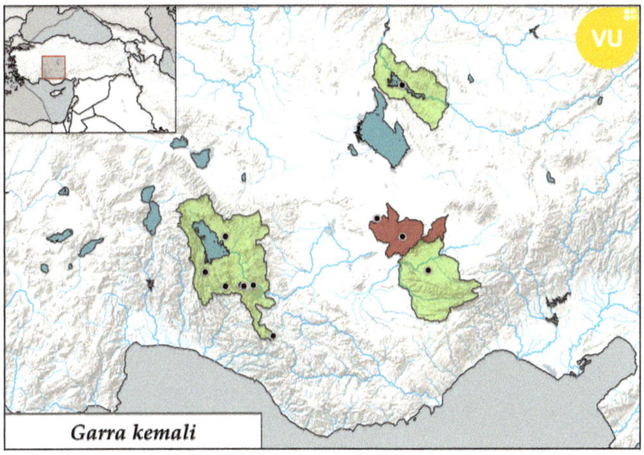

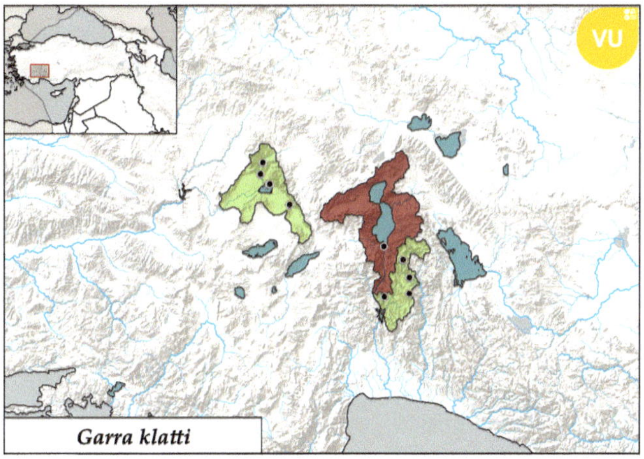

Garra klatti; Köprüçay drainage, Türkiye; female, ~70 mm SL.

Garra klatti; Lake Işıklı basin, Türkiye; male, 49 mm SL.

Garra klatti

Common name. Anatolian golden barb.

Diagnosis. Distinguished from other species of *Garra* in Central Anatolia, Mediterranean, and Dead Sea basins by: ○ no barbels / ○ gular disc incomplete / ○ 15–17 gill rakers / ○ many papillae on mental pad / ○ 36–41+1–2 scales in lateral series / ○ lateral line incomplete, with 6–32 pored scales. Size up to 75 mm SL.

Distribution. Türkiye: Lake Eğirdir basin, Lake Işıklı basin in upper Büyük Menderes drainage, Lake Gölcük (Isparta), and upper Köprüçay drainage.

Habitat. Springs, lakes, and slow-flowing streams.

Biology. Lives up to 4 years. Spawns in second year.

Conservation status. VU; extirpated from Gölcük and Eğirdir basins due to invasion of non-native species. Other populations in decline.

Remarks. Earlier placed in *Hemigrammocapoeta* or *Crossocheilus* and considered as a synonym of *G. kemali* by some authors. *Hemigrammocapoeta menderesensis* is a synonym.

Further reading. Kosswig 1950 (description); Karaman 1971 (morphology); Sarı et al. 2008 (biology); Küçük et al. 2009b (conservation); Geiger et al. 2014 (molecular phylogeny, placement in *Garra*); Hamidan et al. 2014 (molecular phylogeny); Behrens-Chapuis et al. 2015 (molecular phylogeny); Küçük et al. 2015 (morphology, in *Hemigrammocapoeta*).

The Levantine Rift System. The Levantine Rift System represents the northernmost part of the East African Rift System. This vast geological formation is currently undergoing a process of rifting, whereby the African plate is splitting apart to create a new ocean. Simply put, a rift is a fracture in the Earth's surface that widens over time. In the East African Rift System, the Somalian Plate is pulling apart from the African plate. The African Plate moves to the north, while the Somali Plate moves to the east. The African and the Arabian Plates move in a north-northeast direction, but the Arabian Plate moves faster, which leads to the rifting process. Between the Sinai and East Anatolia, the rift is subdivided into several elements, which have formed the complex geological situation observed today. The Dead Sea basin and most of the Litani and Orontes drainages are situated in the rift structures of the northern part of this rift system, the Levantine Rift Valley. Notably, the rifting and associated uplifting of the lateral mountains are relatively recent processes. In the Pliocene epoch (~3–2 million years ago), tectonic uplifting of the current-day land between the Mediterranean and the rift finally disconnected the rift valley from the Mediterranean, which flooded the area at high water levels. The last connection, through the Jezreel Valley, was disconnected only ~2 million years ago. The uplifting of the mountains of the Levant, which are also a consequence of the rifting process, resulted in the interruption of the majority of rivers in this region, which previously flowed into the Mediterranean Sea. Furthermore, it severed the westernmost tributaries of the Euphrates from this drainage system, redirecting their watercourse to the Levant. This has resulted in a complex river history for the region, with several biogeographic puzzles that help explain the Levant's fish communities, which comprise both Mesopotamian and Mediterranean elements. **Further reading.** Alwan 2013 (biogeography).

Garra lautior; Wadi Sah, Yemen; 110 mm SL. © S. Tränkner/SGN.

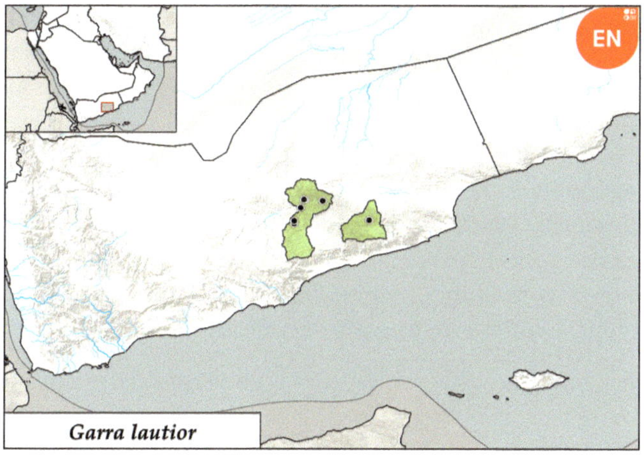

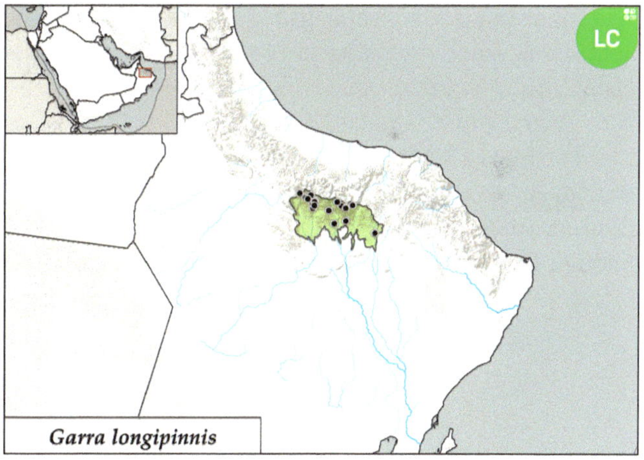

Garra lautior

Common name. Smooth Hadhramaut garra.

Diagnosis. Distinguished from other species of *Garra* in Saudi Arabia, Yemen, and Dhofar (Oman) by: ○ 34–36 total lateral-line scales / ○ 14–16 circumpeduncular scales / ○ 7½ branched dorsal rays / ○ 11–17 gill rakers on lower limb of first gill arch / ○ chest and belly naked / ○ no tubercles on snout / ○ dorsal hyaline or pale-brown with black spots at bases of branched rays / ○ groove between tip of snout and nostrils shallow or absent. Size up to 120 mm SL.

Distribution. Yemen: Wadi Hadhramaut drainage.

Habitat. Wadis with large seasonal variations in flow.

Biology. No data.

Conservation status. EN; appears to be declining within its very small range. Threatened by water abstraction, but threats and distribution poorly known.

Further reading. Krupp 1983 (morphology, as *Garra* incertae sedis); Banister 1987 (description); Freyhof et al. 2020 (distribution, identification).

Garra longipinnis; Saiq, Oman; 57 mm SL.

Garra longipinnis

Common name: Jabal Akhdar garra.

Diagnosis. Distinguished from other species of *Garra* in Hajar Mountains by: ○ scales with dark-brown or grey, bold scale pockets on back and dark-brown or grey scale margins on flank / ○ flank plain greenish, brown, or grey with little or no mottling, a faint dark-green lateral stripe in some individuals / ○ no orange midlateral scales on flank / ○ no orange spot at upper opercle / ○ tip of dorsal not white / ○ 12–14 gill rakers on lower limb of first gill arch. Size up to 80 mm SL.

Distribution. Oman: Mountain wadis and associated falaj systems draining southwards from Jebel Akhdar range and Semail Gap, from Jebel Kawr and Jebel Shams eastwards to about Lizq and nearby tributaries of Wadi ‘Andam.

Habitat. Springs, streams, and falaj systems with gravel or rocky bottoms. Usually in stagnant or slow-flowing water.

Biology. Feeds on periphyton and detritus.

Conservation status. LC.

Further reading. Banister & Clarke 1977 (description); Krupp 1983 (description); Hamidan et al. 2014 (phylogeny; as *G.* cf. *longipinnis*); Kirchner et al. 2020 (distribution, identification).

Garra lorestanensis

Common name. Lorestan cave garra.

Diagnosis. Distinguished from other species of *Garra* in Persian Gulf and endorheic basins in Iran by: ○ eye reduced and invisible externally / ○ whitish or pink, without colour pattern / ○ body without scales / ○ lateral line on body well developed, with 6–31 pores / ○ 9+8 branched caudal rays / ○ 7–8½ branched dorsal rays / ○ 10–12 gill rakers / ○ rostral cap poorly developed, partly covering upper lip / ○ gular disc fully developed longer than wide / ○ two pairs of normally developed barbels. Size up to 74 mm SL.

Distribution. Iran: Loven spring in Ab-e Sirum (or Ab-e Serum) valley near Tang-e Haft, and springs around Tuveh in Dez system of Tigris drainage. Probably more widespread.

Habitat. Underground waters.

Biology. No data.

Conservation status. LC; as with all subterranean species, distribution and threats are difficult to understand.

Remarks. Coexists with *Garra typhlops* and the cave loach *Eidinemacheilus smithi.*

Further reading. Hashemzadeh Segherloo et al. 2012 (genetic difference of sympatric *Garra*); Mousavi-Sabet & Eagderi 2016b (description).

Garra lorestanensis; Tuveh, Iran; 56 mm SL.

Loven spring in Iran is the outflow of an aquifer inhabited by *Garra lorestanensis*, *G. typhlops*, and *Eidinemacheilus smithi*.

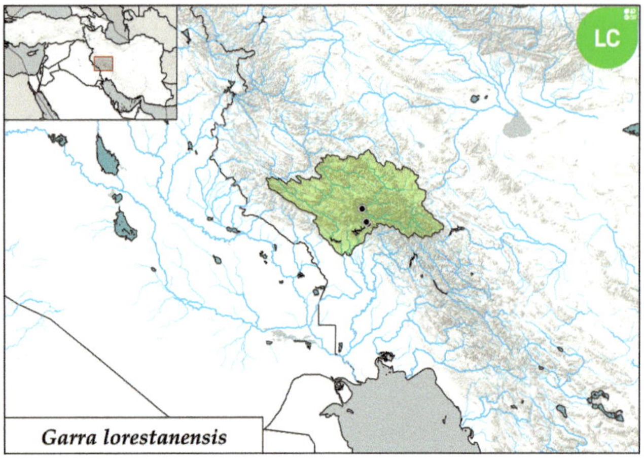

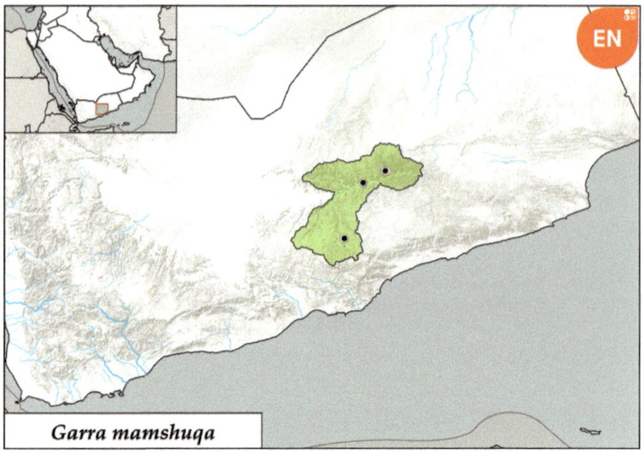

Garra mamshuqa; Wadi al-Khun, Yemen; 52 mm SL. © F. Krupp.

Garra mamshuqa

Common name. Hadhramaut garra.

Diagnosis. Distinguished from other species of *Garra* in Saudi Arabia, Yemen, and Dhofar (Oman) by: ● groove between tip of snout and nostrils deep / ○ usually 8½, rarely 7½, branched dorsal rays / ○ 34–36 total lateral-line scales / ○ usually 12 circumpeduncular scales / ○ 15–17 gill rakers on lower limb of first gill arch / ○ tubercles on snout present / ○ gular disc as long as wide or longer than wide / ○ dorsal hyaline or pale-brown with black spots at bases of branched rays. Size up to 74 mm SL, likely to grow larger.

Distribution. Yemen: Wadi Hadhramaut–Wadi Masila drainage.

Habitat. Wadis with high seasonal fluctuations in discharge over soft substrate, gravel, or bare rock.

Biology. Feeds mainly on periphyton.

Conservation status. EN; appears to be declining within its very small range due to water abstraction and climate change.

Further reading. Krupp 1983 (description); Banister 1987 (comparison with *G. lautior*); Freyhof et al. 2020 (distribution, identification).

Garra meymehensis; Meymeh, Iran; ~60 mm SL.

Garra meymehensis

Common name. Meymeh garra.

Diagnosis. Distinguished from other species of *Garra* in Persian Gulf and endorheic basins in Iran by: ○ chest with normal-sized, posterior margin of scales embedded in skin / ○ mid-dorsal area in front of dorsal origin covered by embedded scales / ○ usually 3 scales between anus and anal origin / ○ usually 7½ branched dorsal rays / ○ 9+8 branched caudal rays / ○ 16–20 gill rakers / ○ 33–37 total scales along lateral line / ○ 13–16 circumpeduncular scale rows / ○ lateral and distal margin of labrum of gular disc free from skin of chin / ○ two pairs of normally developed barbels / ○ gular disc present / ○ body brown or grey, usually mottled / ○ eye fully developed. Size up to 73 mm SL, likely to grow larger.

Distribution. Iran: Meymeh, Doiraj, and Changoleh (Tigris drainage). Iraq: Nahre Al-Tayeb.

Habitat. Springs and spring-fed streams.

Biology. Feeds mainly on periphyton.

Conservation status. VU; with small, declining populations.

Remarks. Occurs in sympatry with *G. rufa* and *G. amirhosseini.*

Further reading. Zamani-Faradonbe et al. 2021b (description); Jouladeh-Roudbar et al. 2020 (distribution).

Garra "mondica"; Konar Siyah spring, Iran; 73 mm SL.

"*Garra mondica*," a ghost of evolutionary past. Iranian "*G. mondica*" was considered endemic to the springs of Tang-e-Mohr, Konar Siyah, and Firozabad in the Mond drainage. In fact, "*G. mondica*" is a long-extinct species, and only its mitochondrial genome has survived in a few populations of *G. rufa* as a remnant of its evolutionary past. "*Garra mondica*" was described as a separate species based on mtDNA alone; none of the morphological characters proposed as diagnostic could be confirmed. Genome-wide SNP data examined for "*G. mondica*" show that its nuclear DNA, which codes for all its morphology, ecology, and behaviour, is identical to that of *G. rufa*. Only the mitochondrial genome of *G. mondica* survived the intense introgression of *G. rufa* after it invaded the Mond drainage. As "*G. mondica*" was described based on *G. rufa* individuals containing mitochondrial DNA of the extinct "*G. mondica*," it is treated as a synonym of *G. rufa*. Similar discordances between nuclear DNA and mtDNA have been found in *G. gymnothorax*, and the situation should be carefully investigated in *G. amirhosseini*, *G. hormuzensis*, *G. meymehensis*, and *G. tiam*. All these species occur in sympatry or proximity to *G. rufa*. **Further reading.** Sayyadzadeh et al. 2015b (description); Zamani-Faradonbe et al. 2021b (characters, distribution).

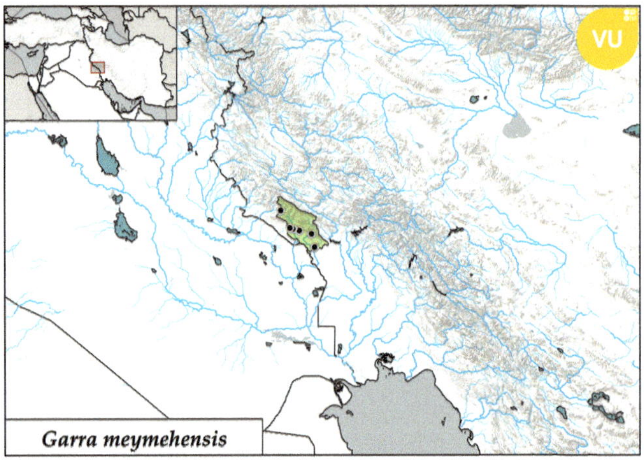

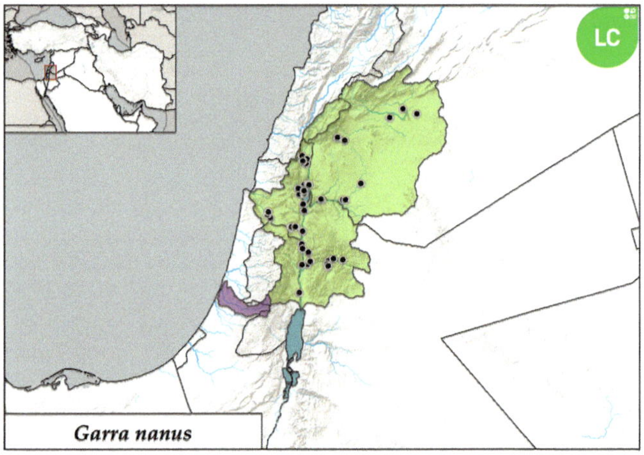

Garra nanus; Al Tammasiyyar, Syria; 61 mm SL.

Garra nanus; Jordan drainage, Syria; 58 mm SL.

Garra nanus

Common name. Damascus barb.

Diagnosis. Distinguished from other species of *Garra* in Central Anatolia, Mediterranean, and Dead Sea basins by: ○ gular disc absent / ○ two pairs of barbels / ○ 8½ branched dorsal rays / ○ a complete total lateral-line with 31–37 scales / ○ 9–15 gill rakers. Size up to 100 mm SL.

Distribution. Syria: al-Awaj in Damascus basin. Qishon in northern Israel and Jordan drainage including Lake Tiberias. Introduced to a coastal stream in southwest Israel.

Habitat. Slow-flowing or stagnant waters of streams, lakes, and springs. Usually in warm, muddy, and even slightly polluted waters.

Biology. Spawns in winter. Feeds mainly on plants, periphyton, and detritus.

Conservation status. LC.

Remarks. Earlier placed in *Hemigrammocapoeta. Garra nanus*, *G. caudomaculata*, and *G. culiciphaga*, form a monophyletic lineage. *Garra sauvagei* is a synonym, but molecular differences between populations are considerable, and the species needs revision.

Further reading. Goren 1974 (distribution, habitat; as *Tylognathus steinitziorum*); Krupp & Schneider 1989 (morphology); Geiger et al. 2014 (molecular phylogeny, placement in *Garra*); Behrens-Chapuis et al. 2015 (molecular phylogeny); Hashemzadeh Segherloo et al. 2016a (molecular phylogeny, reduction of gular disc).

Garra nudiventris; Birjand, Iran; ~55 mm SL.

Garra nudiventris

Common name. Lut garra.

Diagnosis. Distinguished from other species of *Garra* in Persian Gulf and endorheic basins in Iran by: ○ one or two pairs of barbels; if two pairs, rostral barbel very small sometimes just a tubercle / ○ gular disc small / ○ chest, belly, and predorsal midline naked / ○ 5½–6½ transverse scale rows between lateral line and dorsal origin / ○ 4½–5½ transverse scale rows between lateral line and pelvic origin / ○ 10–11 total gill rakers / ○ 7½ branched dorsal rays / ○ lateral and distal margin of labrum of gular disc free from skin of chin / ○ body brown or grey, usually mottled / ○ eye fully developed. Size up to 81 mm SL.

Distribution. Iran: Qanats and streams in Lut and Bejestan basins.

Habitat. Springs, canals, and slow-flowing streams.

Biology. No data

Conservation status. LC.

Remarks *Garra nudiventris* and *G. rossica* are very closely related and may be recognised as conspecifics in future.

Further reading. Esmaeili et al. 2016c (discussion on *G. nudiventris*); Mousavi-Sabet et al. 2019a (distribution).

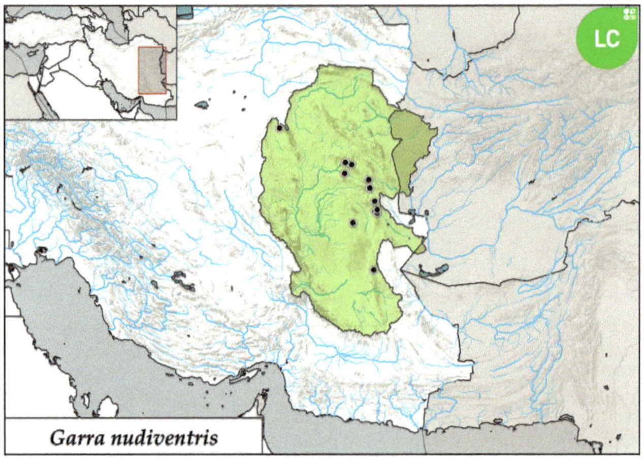

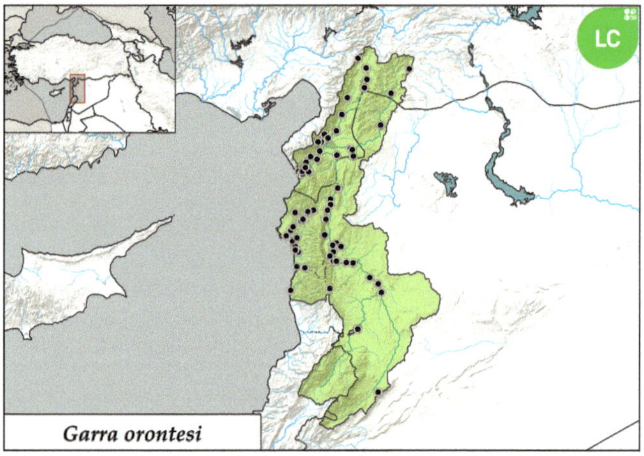

Garra orontesi; Orontes drainage, Türkiye; 117 mm SL.

Garra orontesi; Orontes drainage, Türkiye; ~90 mm SL.

Garra orontesi

Common name. Orontes garra.

Diagnosis. Distinguished from other species of *Garra* in Central Anatolia, Mediterranean, and Dead Sea basins by: ○ two pairs of normally developed barbels / ○ gular disc fully developed / ○ usually 8½ branched dorsal / ○ 35–38 total lateral-line scales / ○ 17–21 gill rakers on lower limb of first gill arch / ○ no tubercles on upper posterior eye margin / ○ maximum head width 73–82 % HL / ○ 2–3, rarely 4 scales between tip of pectoral and pelvic origin / ○ a slightly pointed snout / ○ body depth at dorsal origin 22–25 % SL / ○ caudal peduncle 13–14 % SL. Size up to 130 mm SL.

Distribution. Orontes and Nahr al-Kabir North drainages in Türkiye and Syria.

Habitat. A wide range of lakes, reservoirs, canals, streams, and rivers with at least small rock or gravel areas. Springs without much current, but usually in moderate to fast-flowing waters. Resistant to moderate pollution and can reach very high abundance in shallow, polluted streams.

Biology. Feeds on periphyton.

Conservation status. LC.

Further reading. Yalçin-Özdilek & Ekmekçi 2006 (food); Hamidan et al. 2014 (molecular phylogeny); Hashemzadeh Segherloo et al. 2016a (molecular phylogeny); Bayçelebi et al. 2021a (description).

Garra persica; Rudan drainage, Iran; 50 mm SL.

Garra persica

Common name. Persian garra.

Diagnosis. Distinguished from other species of *Garra* in Persian Gulf and endorheic basins in Iran by: ● 8+8 branched caudal rays / ○ chest and belly covered by normally developed scales, absent at isthmus in some individuals / ○ mid-dorsal area in front of dorsal origin covered by scales / ○ usually 7½ branched dorsal rays / ○ 15–21 gill rakers / ○ lateral and distal margin of labrum of gular disc free from skin of chin / ○ gular disc usually present / ○ two pairs of normally developed barbels / ○ body brown or grey, usually mottled / ○ eye fully developed. Size up to 85 mm SL.

Distribution. Iran: Mehran drainage east to Makran region in northern Gulf of Oman. Also, in Jazmurian basin.

Habitat. A wide range of rivers and streams with gravel bottoms and running water.

Biology. No data.

Conservation status. LC.

Further reading. Berg 1914 (description); Coad 1982a (description, revalidation); Esmaeili et al. 2016c (molecular phylogeny, description); Hashemzadeh Segherloo et al. 2016a (molecular phylogeny); Coad 2021a (biology, morphology).

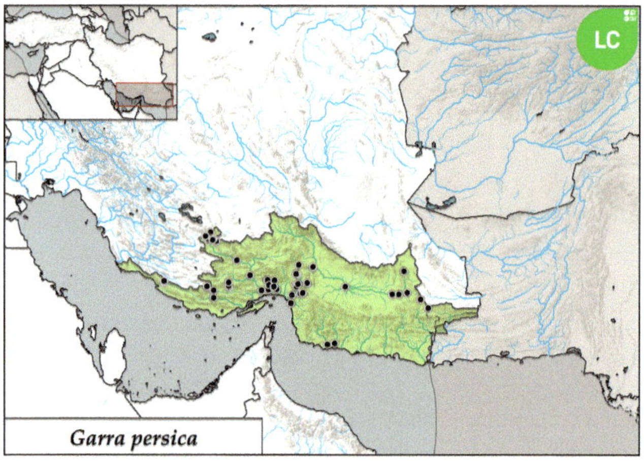

Garra persica

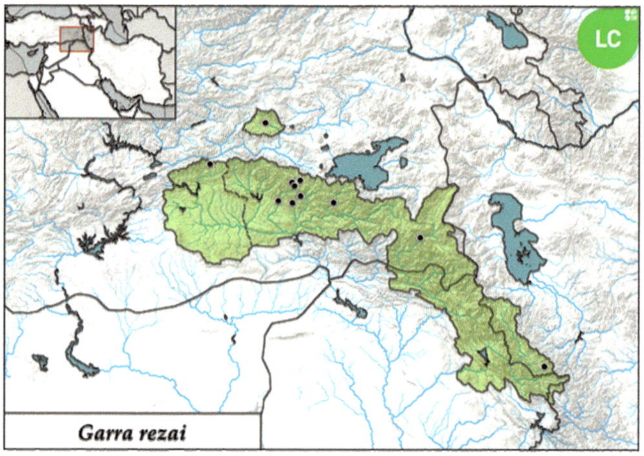

Garra rezai

Dollo's law and the lower lip in *Garra*. In evolution, "Dollo's Law" states that once a complex trait has been lost, it cannot evolve again in the same way. Although the law has fallen into disrepute, it is still one of the most important rules in evolution. This is interesting because species in the former genus *Hemigrammocapoeta* (*G. culiciphaga*, *G. caudomaculata*, *G. festai*, *G. nanus*) have a well-developed "lower lip". Molecular characters place these in *Garra* and treat *Hemigrammocapoeta* as a synonym. This hypothesis has been rejected because garras have only a very short remnant of the lower lip in the corner of the mouth, usually called the labellum. If *Hemigrammocapoeta* were part of *Garra*, the lower lip would have re-evolved in these species, breaking Dollo's law. The gular disc on the ventral side of the head of garras has often been regarded as a modified lower lip. But the gular disc is a structure completely independent of the lips or mouth. It develops from tissues of the gular region, an area behind the mouth, between the lower jaws, often extending to the end of the ventral head. In *Garra*, as in many other algae-eaters (*Capoeta*, *Chondrostoma*), the lower jaw is covered by a keratinous sheath. In contrast to *Capoeta* or *Chondrostoma*, there is a skin fold behind the lower jaw, separated from it by a groove, which some authors have called the lower lip. This skin fold is not homologous to the lower lip and is called the subtoral fold. Species in the former genus *Hemigrammocapoeta* also have a complete subtotal fold behind the lower jaw, so their "lower lip" cannot be homologous to the subtotal fold. *Hemigrammocapoeta* has a well-developed "lower lip" separated in the middle by a short, pointed papillate structure. This structure has a small transverse split in the middle and may represent the reduced torus and labrum of the gular disc. As the structures of the gular disc in *Garra* are not split in the middle, the "lower lip" of *Hemigrammocapoeta* is most likely not homologous with any of them. Their "lower lip" is probably homologous to the labellum, which is usually restricted to the corner of the mouth in *Garra*. Like other Cyprinidae, the enlarged labellum/lower lip is continuous with the upper lip. As the labellum in *Garra* is the remnant of the lower lip, the "lower lip" of *Hemigrammocapoeta* represents a true lower lip. Do these species violate Dollo's law by re-evolving the lower lip? Not in our opinion, because Dollo's law is about the loss of complex characters, and (1) the lower lip was not completely lost, and (2) a shorter or longer lower lip is a simple change of a character state, not a complex one. **Further reading.** Kottelat 2020 (gular disc).

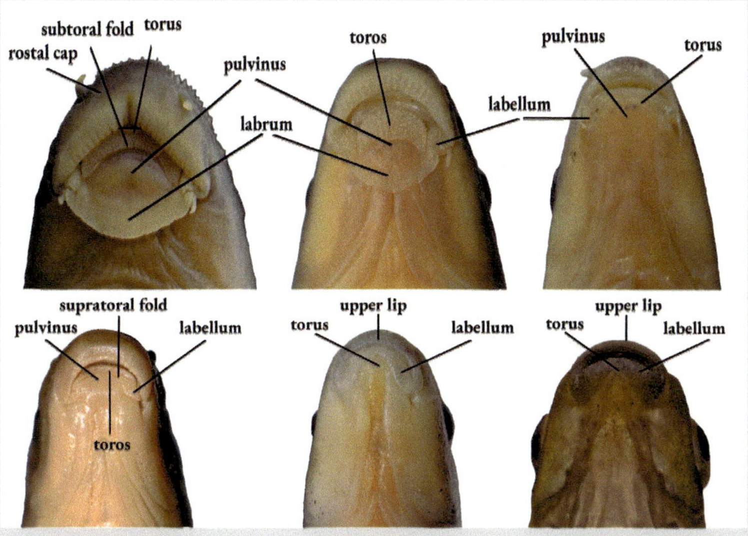

From left: *G. rufa*, *G. variabilis*, *G. elegans*, lower row: *G. klatti*, *G. festai*, *G. culiciphaga*.

Garra rezai; Yanarsu in Tigris drainage, Türkiye; ~100 mm SL.

Garra rezai

Common name. Tigris garra.

Diagnosis. Distinguished from other species of *Garra* in Persian Gulf and endorheic basins in Iran by: ○ two pairs of barbels / ○ chest with embedded scales / ○ belly and predorsal midline covered by scales, 15–19 scales along predorsal midline / ○ gular disc short and wide / ○ 35–40 total lateral-line scales / ○ 5½ (rarely 4½ and 6½) transverse scale rows between lateral line and dorsal origin / ○ 3½–4½ transverse scale rows between lateral line and pelvic origin / ○ 15–18 circumpeduncular scales / ○ 9+8 branched caudal rays / ○ axillary scale at pelvic origin large / ○ 11–16 total gill rakers / ○ lateral and distal margin of labrum of gular disc free from skin of chin / ○ usually 7½ branched dorsal rays / ○ body brown, usually mottled / ○ eye fully developed. Size up to 127 mm SL.

Distribution. Türkiye: Kaynarca (Murat drainage) in Euphrates and Yanarsu, Botan, Menfez, Hezil and Batman drainages, as well as from Aktoprak in Tigris. Iran: Boein-e Olia in Choman drainage, a tributary of Lesser Zab.

Habitat. Springs and slow-flowing streams.

Biology. No data

Conservation status. LC.

Remarks. The known distribution of this species is highly disjunct, and it is expected that it will be found in other streams in the region.

Further reading. Mousavi-Sabet et al. 2022 (description); Kaya et al. 2024a (distribution in Türkiye).

Garra roseae; stream Tang-e-Sarhe, Iran; 45 mm SL.

Garra roseae

Common name. Tang-e Sarhe garra.

Diagnosis. Distinguished from other species of *Garra* in Persian Gulf and endorheic basins in Iran by: ● barbels absent / ○ chest without scales / ○ belly and predorsal midline covered by scales / ○ gular disc small / ○ 7½–8½ transverse scale rows between lateral line and dorsal origin / ○ 6½ transverse scale rows between lateral line and pelvic origin / ○ no axillary scale at pelvic origin / ○ 11–13 total gill rakers / ○ 7½ branched dorsal rays / ○ lateral and distal margin of labrum of gular disc free from skin of chin / ○ body brown or grey usually mottled / ○ eye fully developed. Size up to 51 mm SL.

Distribution. Iran: Tang-e Sarhe in Makran region.

Habitat. Springs and slow-flowing streams.

Biology. No data

Conservation status. EN; appears to be declining within its very small range.

Further reading. Mousavi-Sabet et al. 2019a (description).

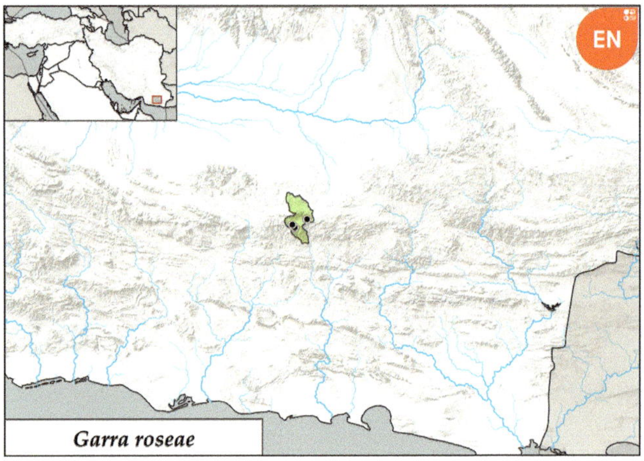

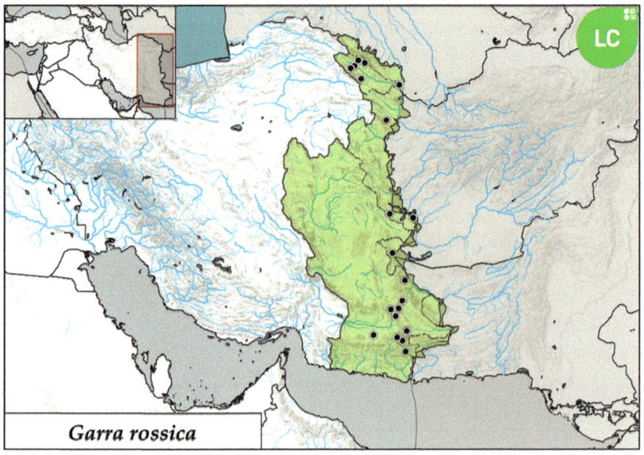

Garra rossica; Nobahar, Hari drainage, Iran; 65 mm SL.

Garra rossica

Common name. Afghan garra.

Diagnosis. Distinguished from other species of *Garra* in Persian Gulf and endorheic basins in Iran by: ○ one or two pairs of barbels; if two pairs, rostral barbel very small, sometimes just tubercle / ○ gular disc small / ○ chest, belly, and predorsal midline fully covered by scales / ○ 5½–6½ transverse scale rows between lateral line and dorsal origin / ○ 4½–5½ transverse scale rows between lateral line and pelvic origin / ○ an axillary scale at pelvic origin / ○ 13–16 gill rakers / ○ 7½ branched dorsal rays / ○ lateral and distal margin of labrum of gular disc free from skin of chin / ○ body brown or grey, usually mottled / ○ eye fully developed. Size up to 95 mm SL.

Distribution. Hari, Morghab, Bejestan, Sistan, Mashkid, and Jazmurian basins and Makran region in Iran, Turkmenistan, and Afghanistan. North-western Baluchistan west of Central Brahui, Mashkid basin, and Kharan-Makran areas in Pakistan.

Habitat. Springs, slow-flowing streams, canals, and various artificial habitats.

Biology. Mature at 2–3 years. Feeds on algae and detritus.

Conservation status. LC.

Remarks. *Garra phryne* from Sistan in Iran is a synonym. *Garra rossica* and *G. nudiventris* are very closely related and may be accepted as conspecifics in future.

Further reading. Nikolskii 1900 (description); Esmaeili et al. 2016c (discussion on *G. nudiventris*); Mousavi-Sabet et al. 2019a (distribution); Coad 2021a (biology, morphology).

Kashaf in Hari drainage, a habitat of *Garra rossica*.

Garra rufa; Sirvan drainage, Iraq; ~100 mm SL.

Garra rufa; Great Zab drainage, Iraq; ~110 mm SL.

Garra rufa

Common name. Red garra.

Diagnosis. Distinguished from other species of *Garra* in Persian Gulf and endorheic basins in Iran by: ○ chest and belly covered by scales, scales embedded in skin, rarely absent / ○ mid-dorsal area in front of dorsal origin covered by scales / ○ usually 8½ branched dorsal rays / ○ 20–24 gill rakers / ○ 32–38 total scales along lateral line / ○ 11–13 circumpeduncular scale rows / ○ 9+8 branched caudal rays / ○ gular disc fully developed / ○ two pairs of normally developed barbels / ○ lateral and distal margin of labrum of gular disc free from skin of chin / ○ body brown or grey, usually mottled / ○ eye fully developed. Size up to 200 mm SL.

Distribution. Qweiq, Euphrates, and Tigris drainages. Also, in Iranian Persian Gulf basin from Zohreh south to Mond, including endorheic Kor basin. Absent from Karkheh drainage. In Karun, most *Garra* are hybrids of this species and *G. gymnothorax*.

Habitat. Moderately warm to very warm streams, rivers, lakes, reservoirs, and artificial habitats of all types, provided there are at least small rock or gravel areas. Usually in moderately to rapidly flowing waters. Resistant to moderate pollution and may reach very high abundance in shallow, polluted streams.

Biology. Lives up to 7 years. Usually spawns after one or two years, depending on water temperature, hatching month,

and food availability. Spawns in spring and summer, March–July, depending on altitude and latitude. Fractional spawner. Spawns in groups over gravel substrate in open water column. Eggs are not sticky and fall into crevices. Larvae pelagic and increasingly lithophilic as they grow. Benthic grazer feeding on detritus, algae, cyanobacteria, and small invertebrates scraped from hard surfaces such as stones, roots, and plants.
Conservation status. LC; extirpated from Qweiq.
Remarks. This is the original Kangal doctor fish. Replaced by *G. gymnothorax* in Iranian Karkheh drainage. Both species are indistinguishable by external characters, although they belong to different phylogenetic clades.
Further reading. Patimar et al. 2010b (biology); Hamidan et al. 2014 (molecular data, restriction to Persian Gulf basin); Keivany et al. 2015 (morphological diversity); Esmaeili et al. 2016c (molecular phylogeny, description); Hashemzadeh Segherloo et al. 2016a (molecular phylogeny, gular disc morphology); Coad 2021a (biology, morphology); Kaya et al. 2024 (distribution in Türkiye).

Balıklı Kaplıca: from a remote spring to a city spa. The Anatolian village of Kangal in the upper Euphrates drainage is well known internationally, not for fish, but for the famous Kangal dogs, a large breed originally used to protect herds from wolves in the Anatolian plains. Less well known is the Balıklı Kaplıca bath and the Kangal fish, also known as the doctor fish, which became famous in the late 1990s. The Balıklı Kaplıca hot springs have been a popular bathing site for centuries. Additionally, the fish found in the springs are renowned for gently nibbling on and removing dead skin from bathers. These fish are believed to be beneficial for people who have psoriasis. There are two species of fish known for their scraping behaviour in Kangal: *Garra rufa* and *Cyprinion macrostomum*. Both are active in the spring. The therapeutic properties are attributed to the elevated levels of selenium in the water, which has been demonstrated to be beneficial in treating certain skin conditions. The fish facilitate the action of selenium by softening and cleaning the psoriatic plaque and scales, thereby exposing the lesions to the water. The treatment of psoriasis by *G. rufa* from Kangal has developed into a global health industry. In the 1950s, accommodation in Balıklı Kaplıca was provided in earthen buildings, with natural pools surrounded by timbers. The first concrete buildings, with summer opening hours, were constructed in the 1960s. Balıklı Kaplıca commenced operation under the modern system in the 1980s. Indoor pools serve the health industry both in winter and summer. Existing buildings were renovated, and the total bed capacity was increased to 270 after a new hotel with a 100-bed capacity was built. Today, it is a large and highly developed spa complex. *Garra rufa* from Kangal was taken illegally, most likely in the 1990s, and the species was eventually bred in Europe in large quantities. While initially perceived as a mere novelty, bathing with Kangal fish has become commercially profitable. Fish spas have increased globally, with numerous establishments operating in multiple cities, particularly holiday resorts. These spas offer a range of fish species, including *G. rufa* and other *Garra* species, and *Oreochromis* cichlids, which are frequently used. Recent molecular studies have shown that many hotels in southwest Türkiye are actually using *Garra turcica* instead of *G. rufa*. Concurrently, wild *Garra* species are distributed throughout their range and may be locally abundant. They come and nibble on you if you stand quietly in a stream.

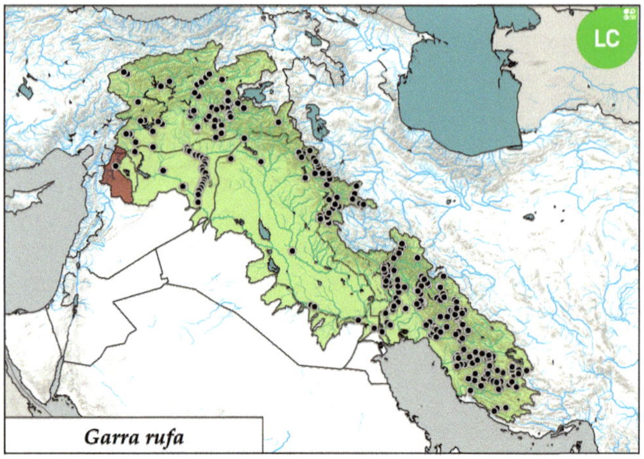

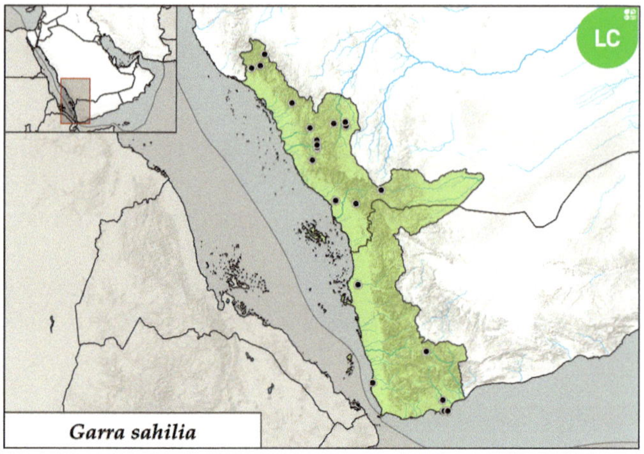

Garra sahilia; Wadi Maur, Yemen; 129 mm SL. © F. Krupp.

Garra sahilia

Common name. Coastal garra.

Diagnosis. Distinguished from other species of *Garra* in Saudi Arabia, Yemen, and Dhofar (Oman) by: ● dorsal membranes dark-grey or black / ○ 32–36 total lateral-line scales / ○ 16–17, rarely 15 or 18, circumpeduncular scales / ○ usually 8½ branched dorsal rays / ○ 11–16 gill rakers on lower limb of first gill arch, 6–10 in Saudi Arabian populations / ○ groove between tip of snout and nostrils shallow or absent / ○ eye fully developed. Size up to 129 mm SL.

Distribution. Yemen and Saudi Arabia: streams flowing to Gulf of Aden and southern Red Sea coast. Yemen: Wadis Bana, al Kabir, Lahej, Tiban, Murlwani, Maur, and others. Found up to 1130 m above sea level. Saudi Arabia: Wadis Minshah, Daga, Gaanah, north of Jizan, Sharfa, from 100–300 m above sea level.

Habitat. Wadis with high seasonal fluctuations in discharge.

Biology. No data.

Conservation status. LC.

Remarks. Molecular data suggest that this species is more closely related to the Anatolian and Mesopotamian *Garra* than to other species from the western Arabian Peninsula. The Saudi Arabian populations are described as *Garra sahilia gharbia*. They may represent a separate species, distinguished by having 6–10 gill rakers on the lower limb of first gill arch, whereas *Garra s. sahilia* has 11–16 gill rakers.

Further reading. Krupp 1983 (description); Hamidan et al. 2014 (phylogeny); Freyhof et al. 2020 (distribution, identification).

Garra shamal; Wadi Bani Kharus, Oman; 70 mm SL.

Garra shamal

Common name. Muscat garra.

Diagnosis. Distinguished from other species of *Garra* in Hajar Mountains by: ● middle caudal rays' with same colour as other rays or slightly darker, not bold black / ● membranes between these rays brown or hyaline / ○ strongly mottled flank pattern usually lacking orange scales on flank / ○ white dorsal tip / ○ no orange spot at upper opercle / ○ 10–14 gill rakers on lower limb of first gill arch / ○ eye fully developed. Size up to 75 mm SL.

Distribution. Oman: Wadis draining from Hajar Mountains to coast of Gulf of Oman, east and west of Muscat, including northern flank of Jebel Akhdar, extending from Wadi Hoqain (near Rustaq) in west to Wadi Mayh (Wadi Yiti), east of Muscat.

Habitat. Headwater streams and falaj systems draining mountain front, springs, streams, and falaj systems with fresh or brackish water, gravel, or bedrock bottoms. Usually in stagnant or slow-flowing water.

Biology. Feeds on periphyton and detritus.

Conservation status. VU; relatively small area with fewer than 10 independent populations. Has experienced loss of several populations, and many are vulnerable to overuse of water, pollution, and climate change.

Further reading. Kruckenhauser et al. 2011 (molecular phylogeny); Pichler et al. 2018 (morphology, as north clade); Kirchner et al. 2020 (description); Kirchner et al. 2020 (distribution, identification).

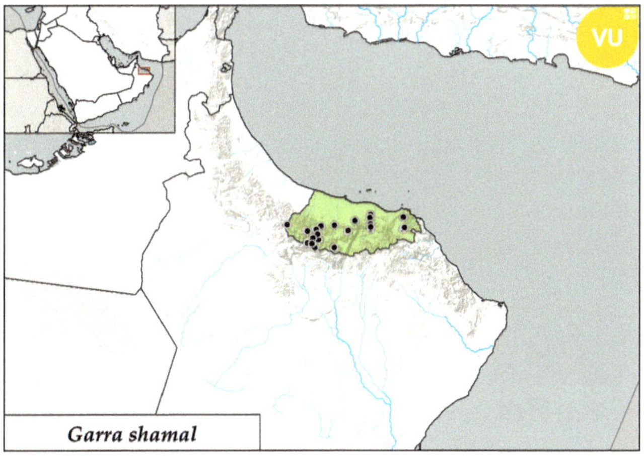

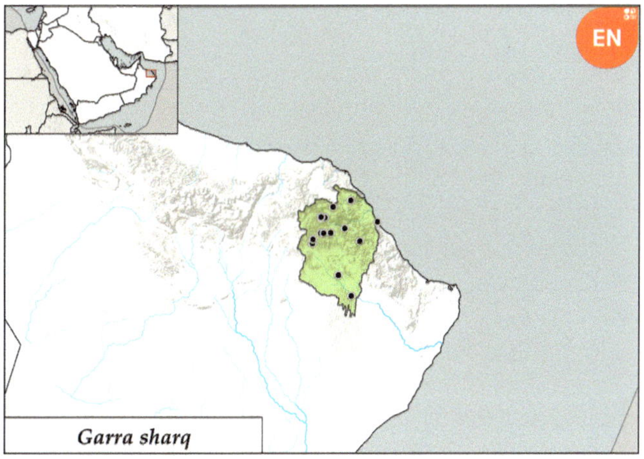

Garra sharq; Wadi Khabbah, Oman; 49 mm SL.

Garra sharq

Common name. Dayqah garra.

Diagnosis. Distinguished from other species of *Garra* in Hajar Mountains by: ○ strongly mottled flank pattern with individual or series of orange midlateral scales / ○ no orange spot at upper opercle / ○ tip of dorsal not white / ○ membranes between central caudal rays black, not silvery / ○ 10–14 gill rakers on lower limb of first gill arch / ○ eye fully developed. Size up to 50 mm SL, likely to grow larger.

Distribution. Oman: Wadi Dayqah drainage, including wadis Tayyin, Deema, and Khabbah, flowing to coast entering sea at Daghmar. Also, in isolated falaj systems and springs in endorheic basin north of Wahiba Sands.

Habitat. Mountain wadis, streams, falaj systems, and springs with freshwater, gravel, or bedrock bottoms. Usually in slow-flowing water.

Biology. Feeds on periphyton and detritus.

Conservation status. EN; significant decline due to water abstraction and expected further decline due to climate change.

Remarks. The distribution of this species in Wadi Dayqah drainage and adjacent endorheic basins suggests recent connections between two watersheds, allowing the invasion of *G. sharq*. While *G. sharq* occurs in upper Wadi Dayn and in falaj systems of town of Al-Mintirib, other falaj systems in area are occupied by *G. gallagheri*. As the *Garra* from Wadi Dayn and Al-Mintirib have only been examined for molecular characters, we cannot exclude the possibility that both species are sympatric or that these are hybrid populations between the two species.

Further reading. Pichler et al. 2018 (morphology); Kirchner et al. 2020 (description); Freyhof et al. 2020 (distribution, identification).

Habitat of *Garra sharq* in Oman.

Wadis in Hajar Mountains are usually small and have little water except few days when all stream bed is flushed.

Garra sindhae; Wadi Andhur, Oman; ~45 mm SL. © G. Lyon.

Garra sindhae

Common name. Andhur garra.

Diagnosis. Distinguished from other species of *Garra* in Saudi Arabia, Yemen, and Dhofar (Oman) by: ● prominent hump on back in some individuals / ○ 33–36 total lateral-line scales / ○ 14(–16) circumpeduncular scales / ○ 7½ branched dorsal rays / ○ 8–13 gill rakers on lower limb of first gill arch / ○ chest and belly behind pectoral origin fully covered by deeply embedded scales / ○ gular disc usually wider than long / ○ tubercles on snout absent / ○ dorsal hyaline or pale-brown with black spots at bases of branched rays / ○ groove between tip of snout and nostrils shallow or absent / ○ eye fully developed. Size up to 82 mm SL.

Distribution. Oman: Wadi Andhur in Dhofar region.

Habitat. Wadis with high seasonal fluctuations in discharge.

Biology. No data.

Conservation status. CR; restricted to a short stretch of Wadi Andhur. There appears to be no direct threat from human activity, but reduced rainfall due to climate change could threaten this species.

Remarks. Mistakenly treated as a synonym of *G. dunsirei*, which is quite distinct and, as a subterranean species, has its own distinct evolutionary history compared to *G. sindhae*. Described as *G. sindhi*, but the species name is dedicated to a female and must, therefore, be declinate to *G. sindhae*.

Further reading. Lyon et al. 2016 (description, phylogeny); Freyhof et al. 2020 (distribution, identification); Sayyadzadeh et al. 2023 (phylogeny).

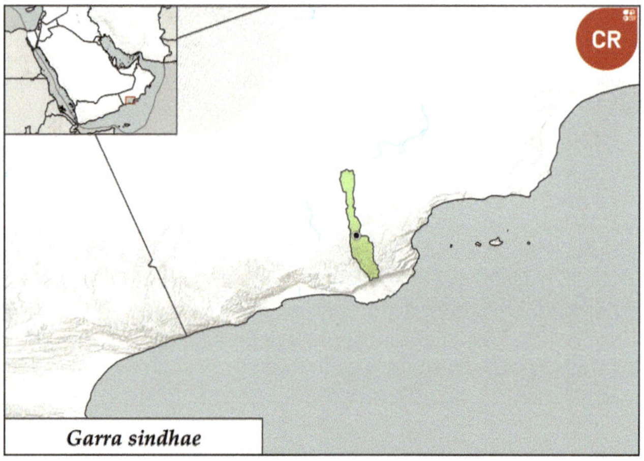

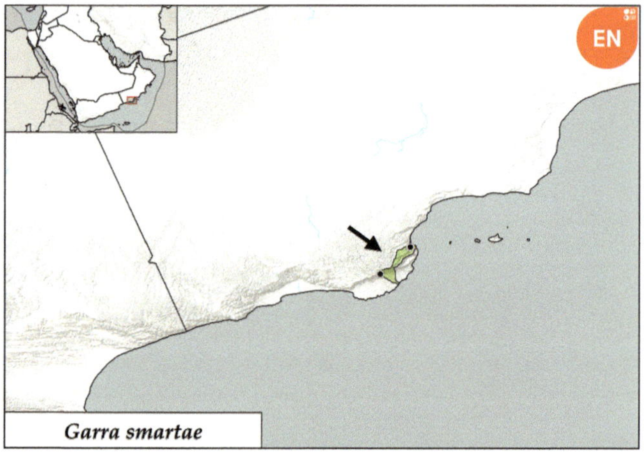

Garra smartae; Wadi Hasik, Oman; 58 mm SL. © J. Els.

Garra smartae

Common name. Hasik garra.

Diagnosis. Distinguished from other species of *Garra* in Saudi Arabia, Yemen, and Dhofar (Oman) by: ● no prominent hump on back / ○ 33–35 total lateral-line scales / ○ 14(–16) circumpeduncular scales / ○ 7½ branched dorsal rays / ○ 12–15 gill rakers on lower limb of first gill arch, 10–13 in Laggashaylon / ○ chest and belly behind pectoral origin fully covered by scales but scales on ventral midline and in front of pectoral base deeply embedded and small or absent / ○ gular disc usually longer than wide / ○ tubercles on snout absent / ○ dorsal hyaline or pale-brown with black spots at bases of branched rays / ○ groove between tip of snout and nostrils shallow or absent / ○ eye fully developed. Size up to 76 mm SL.

Distribution. Oman: Wadis Hasik and Laggashaylon in Dhofar region.

Habitat. Wadis with high seasonal fluctuations in discharge.

Biology. Feeds on periphyton and detritus. In captivity, individuals reach sexual maturity at five to six months.

Conservation status. EN; appears to be declining within its very small range.

Remarks. The Laggashaylon population appears to bridge the morphological gap between *G. smartae* and *G. sindhae*, and published molecular data suggest that this population may represent an undescribed species. *Garra smartae* has been erroneously treated as a synonym of *G. dunsirei*, which is well differentiated and, as a subterranean species, has its own distinct evolutionary trajectory when compared to *G. smartae*. Described as *Garra smarti*, but the species name is dedicated to a female and must, therefore, be declinate to *G. smartae*.

Further reading. Krupp & Budd 2009 (description); Lyon et al. 2016 (phylogeny); Freyhof et al. 2020 (distribution, identification); Sayyadzadeh et al. 2023 (phylogeny).

Garra tashanensis; Tashan Cave, Iran; 27 mm SL.

Garra tashanensis

Common name. Tashan cave garra.

Diagnosis. Distinguished from other species of *Garra* in Persian Gulf and endorheic basins in Iran by: ● lateral line on body reduced, visible pores absent / ○ eye reduced and invisible externally / ○ whitish or pink, without colour pattern / ○ body almost naked, few scales at pelvic base only / ○ 9+8 branched caudal rays / ○ 7½ branched dorsal rays / ○ rostral cap well developed, completely covering upper lip / ○ gular disc fully developed as long as wide / ○ two pairs of normally developed barbels. Size up to 35 mm SL.

Distribution. Iran: Tashan Cave at Sarjusher and Tang-e Ban spring in Jarahi drainage.

Habitat. Underground waters.

Biology. No data.

Conservation status. EN; appears to be declining within its very small range. Only two locations are known: Tashan cave and Tang-e Ban spring. Tashan cave spans approximately 900 m in length and reaches a depth of about 50 m. Tang-e Ban is a seasonal karstic spring that only flows during winter and spring, during which fish are washed from the cave to the surface.

Further reading. Mousavi-Sabet et al. 2016c (description).

Did freshwater fishes cross the Bab-al-Mandeb? Numerous studies concern species movements between Africa and Eurasia, including the migrations of early humans 'out of Africa'. A frequently cited dispersal route is across a land bridge in the southern Red Sea, which is suggested to have emerged during glacial sea-level low stands. The existence of a land bridge between Africa and the Arabian Peninsula is supported by numerous biogeographical similarities between the two regions, as evidenced by the distribution of different animal groups. This is the land bridge over the Bab-al-Mandeb Strait between Yemen and Djibouti. However, the precise timing of the final opening of the Red Sea to the Gulf of Aden remains a matter of contention, and the exact manner and timing of the entry of Asian freshwater fishes into Africa remain unclear. Geological data prove that Africa and Arabia were the last to be connected at the Bab-al-Mandeb-strait during the Miocene (10–5 million years ago). Postulated land bridge periods in younger, post-Miocene times are not supported. This indicates that no early humans or other biota, such as freshwater fishes, have walked on dry land or swam in freshwater from Djibouti to Yemen or vice versa. Following the Miocene, freshwater fishes could no longer cross the Bab-al-Mandeb Strait. However, they could still reach Africa from Asia via the Levant and the northern Arabian Peninsula and possibly through the Nile. It is evident that most freshwater fishes migrated from Arabia to Africa before the end of the Miocene, yet the specific dispersal routes remain uncertain.

Today, all native Arabian freshwater fish species are closely related to species in the Persian Gulf basin and not to African fishes. The only group of Arabian freshwater fishes that are not closely related to other West Asian species are the *Garra* of the *G. tibanica* group (*G. tibanica*, *G. buettikeri*, *G. dunsirei*, *G. smartae*, and *G. sindhae*). Indeed, Ethiopian *G. makiensis* is also related to the *G. tibanica* group. This supports the hypotheses of dispersal events and vicariance around the southern Red Sea area. However, *G. makiensis* is not very closely related to the species of the *G. tibanica* group, indicating that the last common ancestor lived much earlier than the late Miocene. The last common ancestor of all the other African *Garra* species is related to the last common ancestor of the many species of the *G. rufa* group. It has invaded Africa independently, probably through the Levant and the Nile. **Further reading.** Bosworth et al. 2005, Fernandes et al. 2006, Autin et al. 2010 (Bab-al-Mandeb-strait); Stewart & Murray 2017 (fossils of Awash); Englmaier et al. 2020 (relationship of Arabian and Ethiopian *Garra*).

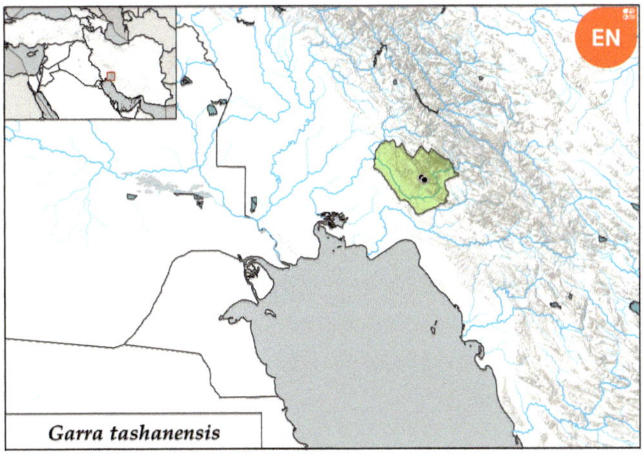

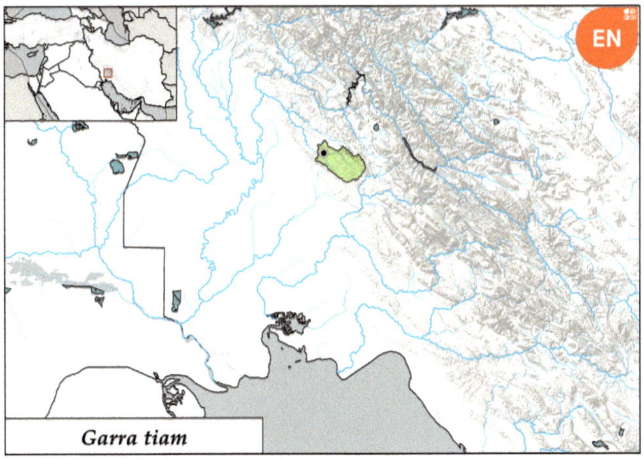

Garra tiam; Abshur drainage, Iran; ~60 mm SL.

Garra tiam

Common name. Abshur garra.

Diagnosis. Distinguished from other species of *Garra* in Persian Gulf and endorheic basins in Iran by: ○ chest with normal-sized, often embedded scales / ○ mid-dorsal area in front of dorsal origin with 9–11 scales / ○ usually 8½ branched dorsal rays / ○ 9+8 branched caudal rays / ○ 18–22 gill rakers / ○ 30–33 total scales along lateral line / ○ 30–33 total lateral-line scales / ○ 12–13 circumpeduncular scales / ○ two pairs of normally developed barbels / ○ lateral and distal margin of labrum of gular disc free from skin of chin / ○ gular disc fully developed / ○ body brown or grey usually mottled / ○ eye fully developed. Size up to 63 mm SL, likely to grow larger.

Distribution. Iran: Abshur drainage, a tributary of Karun.

Habitat. Streams and springs with rocks or gravel bottom.

Biology. Feeds mainly on periphyton.

Conservation status. EN; found in a very small area where climate change and desiccation of streams is ongoing.

Further reading. Zamani-Faradonbe et al. 2021b (description).

Garra tibanica; Wadi Dhila, Saudi Arabia; 65 mm SL. © F. Krupp.

Garra tibanica

Common name. Hijaz garra.

Diagnosis. Distinguished from other species of *Garra* in Saudi Arabia, Yemen, and Dhofar (Oman) by: ○ 34–36, very rarely 32, 33 or 37, 38 total lateral-line scales / ○ 16–18, very rarely 14, 15 or 20 circumpeduncular scales / ○ 6–8½, usually 7½, branched dorsal rays / ○ 6–17, usually 9–15, gill rakers on lower limb of first gill arch / ○ gular disc of considerably variable shape and size, usually wider than long / ○ chest and belly usually covered by scales, scales may be embedded and barely overlapping or partly absent leaving some naked areas / ○ tubercles on snout usually present / ○ dorsal membrane hyaline or pale-brown with black spots at bases of branched rays / ○ groove between tip of snout and nostrils shallow or absent / ○ eye fully developed. Size up to 131 mm SL.

Distribution. Wadi Hajr in Yemen to Khaibar area in Saudi Arabia.

Habitat. Wadis with high seasonal fluctuations in discharge.

Biology. Observed moving over short distances on land, climbing steep, moist rocky surfaces, especially near waterfalls, to reach higher pools.

Conservation status. LC.

Remarks. *Garra tibanica* was synonymised with *G. quadrimaculata* from Ethiopia, a view not supported by later studies. *Garra brittoni, Garra tibanica dhamarica, Garra tibanica elbahrica, Garra tibanica kasaba, Garra tibanica multaka, Garra tibanica nakalani, Garra tibanica scorteccii,* and *Garra tibanica yemenica*, which are all from Yemen, are treated as synonyms of *Garra tibanica*. New fieldworks should test this hypothesis.

Further reading. Trewavas 1941 (description); Krupp 1983 (revised diagnosis); Balletto & Spanò 1977 (description, subspecies); Hamidan et al. 2014 (phylogeny); Freyhof et al. 2020 (distribution, identification).

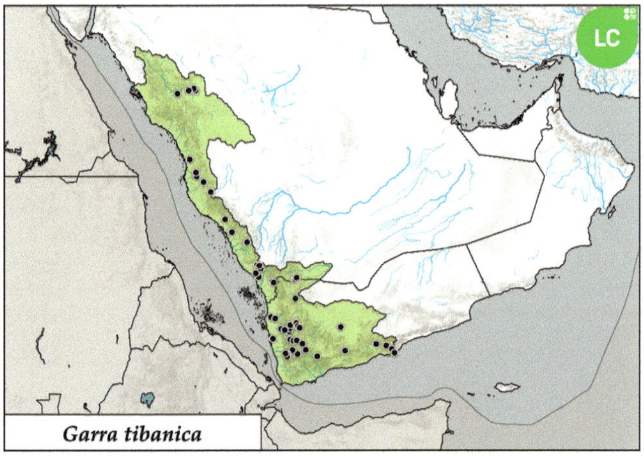

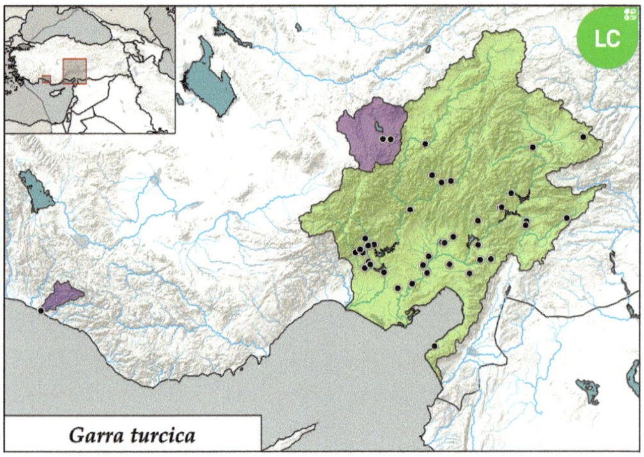

Garra turcica; Seyhan drainage, Türkiye; 90 mm SL.

Garra turcica; Ceyhan drainage, Türkiye; 103 mm SL.

Garra turcica

Common name. Cilician garra.

Diagnosis. Distinguished from other species of *Garra* in Central Anatolia, Mediterranean, and Dead Sea basins by: ○ two pairs of normally developed barbels / ○ gular disc fully developed / ○ tubercles absent on upper posterior eye margin / ○ 33–37 total lateral-line scales / ○ 8½ branched dorsal rays / ○ snout blunt / ○ body depth at dorsal origin 18–21 % SL / ○ chest and belly scaled / ○ mid-dorsal area in front of dorsal origin covered by scales / ○ dark-grey blotch at posterior caudal peduncle present. Size up to 115 mm SL.

Distribution. Türkiye: Seyhan and Ceyhan drainages and small rivers south to Arsuz. Introduced to Ilıca in Manavgat, also to the Sultan Marshes via Zamantı Tunnel.

Habitat. Streams and rivers with rocks or gravel bottoms.

Biology. No data.

Conservation status. LC.

Remarks. Molecular studies demonstrate that the majority of producers claiming to supply *G. rufa* to fish spas in Antalya (Türkiye) are in fact supplying *G. turcica*.

Further reading. Karaman 1971 (description); Kara et al. 2010: (morphology, distribution); Hashemzadeh Segherloo et al. 2016a (molecular phylogeny).

Garra typhlops; Tuveh, Iran, ~40 mm SL. © S. Vatandoust.

Garra typhlops; Tuveh, Iran, ~45 mm SL. © S. Vatandoust.

Garra typhlops

Common name. Zagros cave garra.

Diagnosis. Distinguished from other species of *Garra* in Persian Gulf and endorheic basins in Iran by: ○ eye reduced and invisible externally / ○ whitish, pink, orange, without colour pattern / ○ body without scales / ○ 9+8 branched caudal rays / ○ 8½ branched dorsal rays / ○ 10–13 gill rakers / ○ gular disc incomplete / ○ two pairs of normally developed barbels. Size up to 64 mm SL.

Distribution. Iran: Loven spring in Ab-e Sirum (or Ab-e Serum) valley near Tang-e Haft, and springs around Tuveh in Dez system of Karun drainage. Also, in a tunnel of Simareh Dam (now encased). Probably more widespread.

Habitat. Underground waters.

Biology. No data.

Conservation status. LC; as with all subterranean species, its distribution is difficult to understand, and threat levels may be low.

Remarks. A record from Simareh Dam is likely to belong to an undescribed species. Previously placed in a separate genus, *Iranocypris*. Coexists with the *Garra lorestanensis* and the cave loach *Eidinemacheilus smithi*.

Further reading. Bruun & Kaiser 1944 (description); Sargeran et al. 2008 (morphology); Hashemzadeh Segherloo et al. 2012 (molecular data); Farashi et al. 2014 (placement in *Garra*); Hamidan et al. 2014; Mousavi-Sabet & Eagderi 2016b (taxonomy); Coad 2021a (biology, morphology).

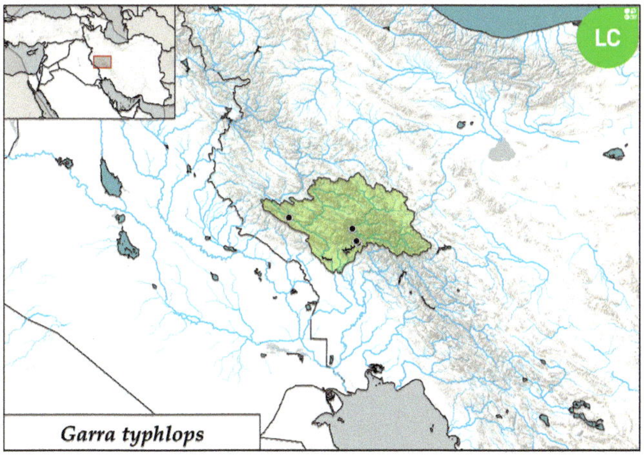

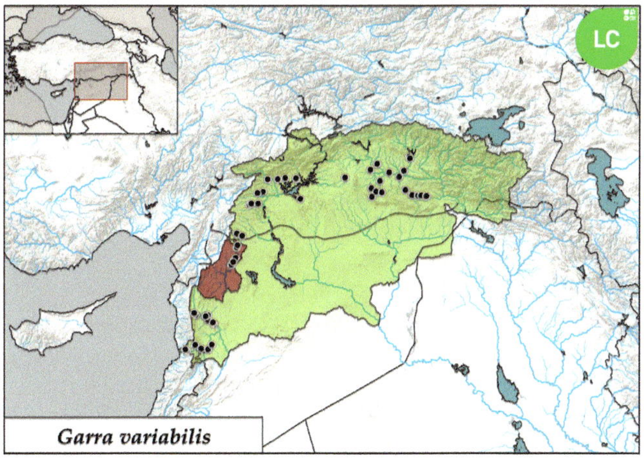

Garra variabilis, Tigris drainage, Türkiye; ~120 mm SL.

Garra variabilis

Common name. Small mouth garra.

Diagnosis. Distinguished from other species of *Garra* in Mediterranean and Gulf basins by: ○ one pair of barbels / ○ 20–26 gill rakers / ● gular disc small and narrow / ○ usually 7½ branched dorsal rays / ○ tubercles absent on upper posterior eye margin / ○ body brown or grey, usually mottled / ○ lateral and distal margin of labrum of gular disc free from skin of chin / ○ eye fully developed / ○ 32–38 lateral-line scales. Size up to 140 mm SL.

Distribution. Orontes in Syria and Lebanon and Syrian Nahr al-Kabir (South) in the Mediterranean basin. Also, Upper Euphrates and Tigris in Türkiye, Syria, and Iraq.

Habitat. Streams, springs, and rivers with slow-flowing or stagnant water, often with dense vegetation or submerged wood.

Biology. No data.

Conservation status. LC; extirpated from Qweiq.

Remarks. Not recorded from the Iranian part of Tigris drainage. While there are records from the Tigris in Iraq, it could not be found in Greater and Lesser Zab and the Sirvan, three tributaries of Tigris in Iraq.

Further reading. Geiger et al. 2014 (molecular phylogeny, placement in *Garra*); Hamidan et al. 2014 (molecular phylogeny); Behrens-Chapuis et al. 2015 (molecular phylogeny); Esmaeili et al. 2016c (molecular phylogeny); Kaya et al. 2016 (distribution).

Garra widdowsoni; Haditha, Iraq; ~50 mm SL.

Garra widdowsoni

Common name. Haditha garra.

Diagnosis. Distinguished from other species of *Garra* in Persian Gulf and endorheic basins in Iran by: ● scales restricted to lateral midline, rarely fully scaled / ○ lateral line complete, with 28–35 scales in lateral series / ○ eye reduced and invisible externally / ○ cream white or pink, without colour pattern / ○ usually 8½ branched dorsal rays / ○ gular disc fully developed / ○ two pairs of normally developed barbels. Size up to 61 mm SL.

Distribution. Iraq: groundwater table near Haditha, from three wells; one about 3 m below Shaikh Hadeed, a second 9 m deep, 135 m south-west of Shaikh Hadeed, and a third called "Pigeon Hole," 12 km south of Sheik Hadid shrine, all in Euphrates drainage.

Habitat. Underground waters.

Biology. Appears to spawn all year round as all sizes of fish are present. Grazes epilithic layers of bacteria and ciliates from rock surfaces. Also, takes food from water surface by upside-down grazing.

Conservation status. CR; three sites indicate a wider distribution in Haditha Karst. Water table in karst is now much lower than in past, and fish can only be found at Shaikh Hadeed. All other springs are dry. While this species was once very abundant, and hundreds were sold as ornamental fish in Baghdad, it is now rare in accessible parts of groundwater. Water is abstracted from wells, and the construction of a reservoir on the Euphrates upstream of Haditha has lowered the water table. The species may be on the verge of extinction. However, it is impossible to study which part of population lives in inaccessible parts of karst and which part lives in inaccessible parts.

Remarks. Often placed in a separate genus *Typhlogarra* but closely related to *Garra rufa*. Coexists with *Caecocypris basimi*.

Further reading. Trewavas 1955 (description); Hamidan et al. 2014 (molecular phylogeny, placement in *Garra*).

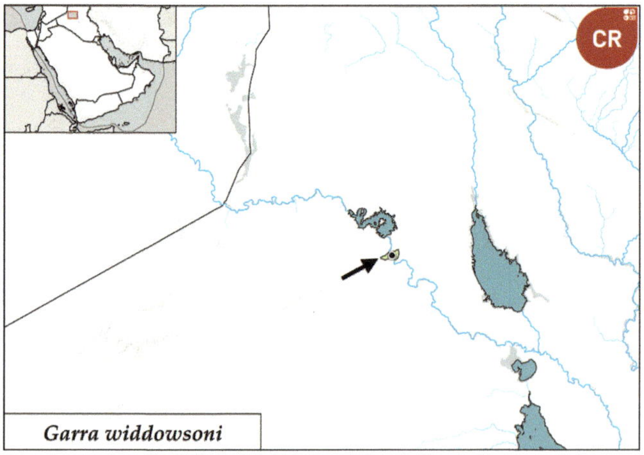

Springs at the edge of mountain slopes in the Orontes drainage, such as this one south of Qala'at in Syria, are the habitat of *Garra variabilis*.

Cave Garras in West Asia. Since their discovery, the troglomorphic animals, adapted to live in subterranean habitats, have fascinated humans. The most striking features of many troglobionts (cave-dwelling species) are the reduction of eyes and pigmentation. The discussion about the forces driving convergent evolution of reduced eyes and loss of pigmentation in typical cave-living animals was initiated very early in the history of evolutionary theory. It is usually attributed to subterranean animals' access to very limited resources and their need to save energy. The loss of pigmentation and eyes provides an advantage by conserving energy. In West Asia, six species of *Garra* have entered cave and underground waters, while only three other fish species did so (the cyprinid *Caecocypris basimi* and the loaches *Eidinemacheilus smithi* and *E. proudlovi*). It can be hypothesised that species of *Garra* are preadapted to subterranean life. A study on the eye morphology of *G. rufa* suggests that it has limited eyesight. Furthermore, *Garra* species are already adapted to grazing algae and other biofilms from surfaces and possess a long and coiled gut, which allows them to digest food effectively. In underground habitats, food resources are likely to be very restricted, and fishes that can graze bacteria- and ciliate-dominated biofilms from surfaces might not encounter the same challenges as other fish species may face regarding food scarcity. In light of this advantage, cave *Garra* may have adapted relatively quickly to the new environment by reducing their eye, colour, and scales, which are not as necessary in such a predator-free environment. **Further reading.** Romero & Paulson 2001 (subterranean fishes); Kruckenhauser et al. 2011; Kirchner et al. 2017 (molecular studies on Omani cave *Garra*).

Luciobarbus

Luciobarbus are medium to large fishes that often co-occur with superficially similar *Barbus* and *Capoeta* species. *Luciobarbus* are less rheophilic and prefer more stagnant habitats than sympatric *Barbus*. In West Asia, all but one *Luciobarbus* species can be readily distinguished from *Barbus* species by their uniform yellowish, brown, or grey colouration, sometimes with an indistinct lateral stripe (vs. body and fins covered with distinctive, irregularly shaped spots and blotches), and by the presence of large nuptial tubercles on the snout (vs. very small nuptial tubercles throughout the head). Juveniles of both genera show a pattern of dark-brown blotches on the body and are difficult to distinguish at first sight. *Luciobarbus* can also be immediately distinguished from *Capoeta* by the absence of a keratinised lower jaw (vs. present in most species), a long, pointed head (vs. short and roundish), and the presence of two pairs of barbels (vs. usually one pair). *Luciobarbus* and *Capoeta* often hybridise. Some *Luciobarbus,* such as *L. pectoralis* and *L. schejch,* show a polymorphic lower lip, with a pendulous central lobe and fleshy lips in some individuals, and an interrupted lower lip with a hard and exposed lower jaw in other individuals, including all intermediate lip shapes. This polymorphism has been misinterpreted as species diagnostic, and individuals with a fleshy lobe on the lower lip have often been identified as *L. barbulus* (= *L. schejch*) throughout West Asia. In contrast, individuals without such a lobe have been identified as *L. pectoralis*. Both species are strictly allopatric, and both lip forms are found in both species. The lip shape is not variable in other *Luciobarbus*, such as *L. esocinus* and *L. subquincunciatus*. Furthermore, *Luciobarbus* species exhibit significant allometric growth in body shape and the length and strength of the last unbranched dorsal ray. This has led to some taxonomic confusion, particularly in the Persian Gulf basin.

The identification of *Luciobarbus* species in the large rivers of the Euphrates and Tigris drainages has only recently been resolved. *Luciobarbus esocinus* and *L. subquincunciatus* are two well-differentiated species, *L. schejch* often hybridises with *L. esocinus*. Individuals cannot be identified by their COI DNA sequence alone, as they may have their own or the mtDNA of *L. esocinus*. Fish commonly identified as *L. barbulus* or *L. xanthopterus* all belong to *L. schejch*. The syntypes of *L. xanthopterus* have been identified as hybrids of *L. esocinus* and *L. schejch,* and this name is not available for any of the species. **Further reading.** Freyhof et al. 2025 (revision).

Luciobarbus esocinus; Tigris, Türkiye; ~1300 mm SL.

Potential hybrid between *Luciobarbus esocinus* and *L. schejch*; Iraq: Shatt-al Arab at Basra. © A. Ali.

Potential hybrid between *Luciobarbus esocinus* and *L. schejch*; Iraq: Baghdad. © O. F. Al-Sheikhly.

Keys to species of *Luciobarbus* in West Asia

Caspian basin and Hari drainage

Differences between *L. capito* (Caspian basin) and *L. conocephalus* (Hari & Aral basin) need to be better understood, and future research will require more effort. Molecular data support both as distinct species.

1a - Predorsal length shorter than postdorsal length; 7½ branched dorsal rays; 16–25 gill rakers.
..................*L. brachycephalus*
1b - Predorsal length equal to or longer than postdorsal length; 8½ branched dorsal rays; 10–19 gill rakers.
..................2

2a - Lips very thick, lower lip with a median lobe or pad; 83–103 total lateral-line scales; back not or slightly compressed in front of dorsal origin; colouration of back and belly not sharply contrasted.
..................*L. mursa*
2b - Lower lip very thin, interrupted in middle, without fleshy middle lobe; 52–72 total lateral-line scales; back between head and dorsal origin laterally compressed, forming a keel; sharp contrast between dark-brown or yellowish back and whitish or yellow belly in large individuals.
..................*L. capito* / *L. conocephalus*

Mediterranean basin

1a - 43–48 total lateral-line scales.
………………*L. graecus*
1b - 48–63 total lateral-line scales.
………………2

2a - 48–58 total lateral-line scales.
………………*L. pectoralis*
2b - 55–63 total lateral-line scales.
………………*L. longiceps*

Persian Gulf basin

1a - Numerous large, roundish, dark-brown, or black blotches on head, body, and fins arranged in an almost quincunx pattern; 75–88 total lateral-line scales.
………………*L. subquincunciatus*
1b - Body plain yellow, grey; or silvery or back and upper part of body brown and lower body yellow or silvery; irregularly shaped, small pale-brown spots and blotches present in juveniles 45–73 (78) total lateral-line scales.
………………2

2a - Head pike-like, strongly pointed, depressed; mouth almost terminal or terminal; snout length 2.0–2.9 times in postorbital length in individuals larger than 150 mm SL; 8–11 total gill rakes; 60–73(–78) total lateral-line scales; juveniles often with many, small dark-brown spots on upper flank.
………………*L. esocinus*
2b - Head barbel-like, not strongly pointed or depressed; mouth inferior in small individuals, inferior or subterminal in large individuals; snout length 1.1–1.9 times in postorbital length; 11–24 total gill rakes; 50–62 total lateral-line scales; juveniles without small dark-brown spots on upper flank.
………………*L. schejch*

Luciobarbus are the target of local fisheries all over their range as here in the upper Tigris in Türkiye.

Luciobarbus brachycephalus; Caspian Sea, Iran; ~250 mm SL.

Luciobarbus brachycephalus

Common name. Shorthead barbel.

Diagnosis. Distinguished from other species of *Luciobarbus* in Caspian basin and Hari drainage by: • predorsal length shorter than postdorsal length / • usually 7½ branched dorsal rays / ○ 16–25 gill rakers / • back between head and dorsal origin laterally compressed, forming a keel / ○ 62–90 (usually 65–77) total lateral-line scales / ○ lower lip thin, without median lobe or pad. Size up to about 1000 mm SL and 21 kg.

Distribution. Southern and western Caspian basin. Land-locked population in Mingachevir reservoir (Azerbaijan). Migrates up larger tributaries of western and southern coasts to spawn: Terek, Samur, Kura, and lower Aras. Rarely in lower Volga (to Volgograd) and Ural. Recorded from Karakum Canal and Kopetdag reservoir in Turkmenistan. Widespread in Aral basin (Central Asia).

Habitat. At sea, to 25 m depth. In rivers, in deep stretches with gravel or stone bottoms. Spawns in fast-flowing water in hard bottom areas, 1–2 m deep.

Biology. Mostly semi-anadromous, rarely land-locked in Caspian basin, riverine in Aral basin. Lives up to 10 years (males) and 13 years (females). Spawns first time at about 500–700 mm SL, 5–7 years, females later than males. Spawns April to August, peaking at 23–27°C. Some individuals begin spawning migration in August–September and spawn next spring after overwintering in river. Some enter rivers in early spring (March) and spawn in same year. Spring run begins at water temperatures between 7–11°C and is most intense in early summer at 25–27°C. Females lay 100,000–1,250,000 bright yellow eggs in 2–3 portions during a single season. Eggs are semi-pelagic and hatch after at least 2 days at 25°C while drifting downstream. Larvae settle in areas of slow current; after 2–12 months, juveniles drift downstream to sea (or reservoir if river is dammed). In the low-salinity zones of the sea, it feeds mainly on benthic crustaceans. Does not feed during upstream migration. Starts feeding again at spawning grounds, mainly on insects, juveniles of other fish, and rarely on algae, seeds, and other plant material.

Conservation status. EN; stocks in Caspian basin declined sharply due to massive habitat alteration, particularly overfishing and dam construction. Currently very rare in Caspian basin, but still reported from coastal rivers in Azerbaijan, the Iranian Lenkoran and Terek in Dagestan. It is regularly found in middle Amu Darya drainage, where it is expected to decline sharply due to increased hydroelectric development, overfishing, and water diversion from rivers for irrigation, especially in Turkmenistan and Afghanistan (Qosh Tepa Canal).

Further reading. Abdurakhmanov 1962 (biology); Magomedov 1981 (biology); Mitrofanov et al. 1988 (description, biology); Bogutskaya 2003a (biology).

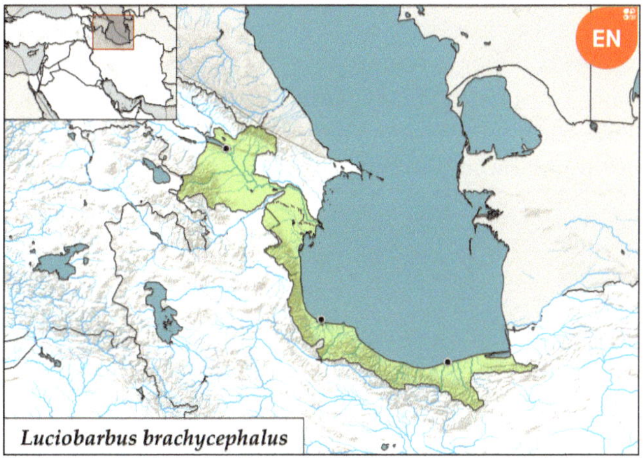

Luciobarbus capito; Aras, Armenia; ~300 mm SL. © S. Pipoyan.

Luciobarbus capito

Common name. Bulatmai barbel.

Diagnosis. Distinguished from other species of *Luciobarbus* in Caspian basin and Hari drainage by: ○ 52–72 (usually 60–66) total lateral-line scales / ○ 13–15 gill rakers / ○ upper part of body dark-brown, lower part yellow in life in large individuals / ○ lower lip thin, without median lobe or pad / ○ 8½ branched dorsal rays / ○ back between head and dorsal origin laterally compressed, forming a keel / ○ predorsal length longer than postdorsal length / ○ dorsal moderately long, 13–14 % of body depth in individuals of about 400 mm SL. Size up to about 920 mm SL and 16 kg.

Distribution. Caspian basin: tributaries of western and southern coasts, from Volga south to Atrak (Iran). Always very rare in Volga. Introduced in Jaj reservoir in Namak Lake basin (Iran).

Habitat. Resident populations mostly in large to medium-sized rivers, often also in reservoirs. At sea, usually near coast. Forages in estuaries. Spawns in lowland streams and rivers on sand-gravel bottoms, usually in strong currents.

Biology. Semi-anadromous and resident individuals and populations. Spawns first time at 3–7 years, females later than males, above 200 mm SL. Spawns April–August. Males spawn every year, some females annually, others apparently every 2–3 years. Semi-anadromous populations begin migration in late summer-autumn and spawn in next spring, but some enter rivers in early spring and spawn in same year. Migrates to uppermost tributaries of rivers. Non-anadromous populations migrate to upper reaches of tributaries just before spawning or spawn in lakes and reservoirs on sandy to muddy bottoms. Larvae feed on zooplankton and small benthic invertebrates. Juveniles and adults feed on invertebrates, algae, detritus, plant material, and small fish.

Conservation status. VU; resident populations still widespread, but abundance greatly reduced due to overfishing. Semi-anadromous populations are now very rare due to dam constructions. Terek and Kura were major spawning rivers for anadromous fish. Very rare in Terek today. Decline expected to continue due to increasing hydropower development, overfishing in estuaries, and many strong ecological impacts on the Caspian basin.

Further reading. Kazancheyev 1981 (biology); Bogutskaya et al. 2003a (description, biology); Kaya et al. 2020a (distribution); Coad 2021a (biology, morphology).

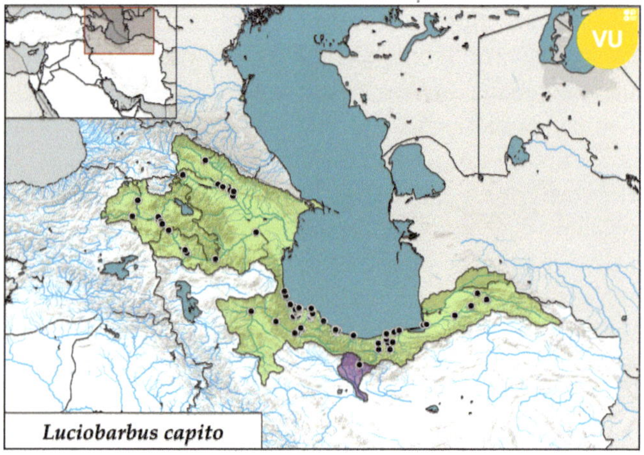

Luciobarbus conocephalus; Hari, Iran; ~250 mm SL.

Luciobarbus conocephalus

Common name. Conehead barbel.

Diagnosis. Distinguished from other species of *Luciobarbus* in Caspian basin and Hari drainage by: ○ 58–70 (usually 61–68) total lateral-line scales / ○ 13–15 gill rakers / ○ upper part of body dark-brown, lower part yellow in life in large individuals / ○ lower lip thin, without median lobe or pad / ○ 8½ branched dorsal rays / ○ back between head and dorsal origin laterally compressed, forming a keel / ○ predorsal length longer than postdorsal length / ○ dorsal long, 15–16 % of body depth in individuals of about 400 mm SL. Size up to about 664 mm SL.

Distribution. Iran: Hari drainage. Morghab, Aral basins, and Chu drainage in Central Asia. In Karakum Canal, Kopetdag, and Uzboi lakes in Turkmenistan.

Habitat. Large rivers and reservoirs, from which fish migrate into tributaries to spawn.

Biology. Poorly known, mostly a freshwater species that does not undertake long migrations.

Conservation status. LC.

Remarks. This species is distinguished from *L. capito* by molecular characters, but no material was available to test whether morphological characters can also distinguish the two. It might be an old species and by this a synonym of *L. capito*.

Further reading. Berg 1949b (diagnosis, distribution); Bogutskaya et al., in Bănărescu & Bogutskaya 2003a (distribution).

Luciobarbus esocinus

Common name. Pike barbel.

Diagnosis. Distinguished from other species of *Luciobarbus* in Persian Gulf basin by: ● mouth subterminal in small, almost terminal or terminal in large individuals / ● head pike-like, strongly pointed, depressed / ● snout length 2.0–2.9 times in postorbital length in individuals larger than 150 mm SL / ● dorsal head profile straight or concave / ○ 8–11 gill rakers / ○ 60–73(–78) total lateral-line scales /

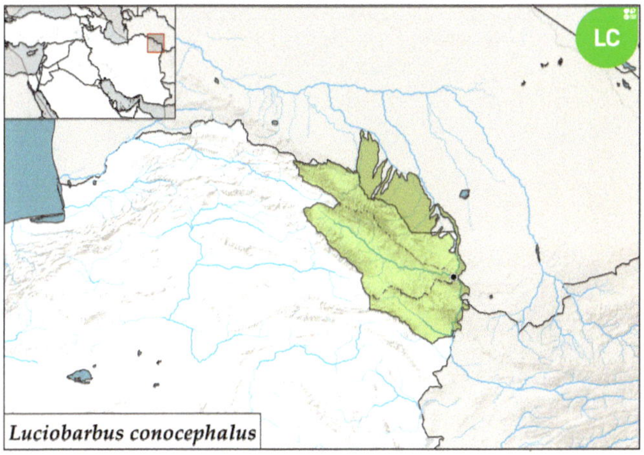

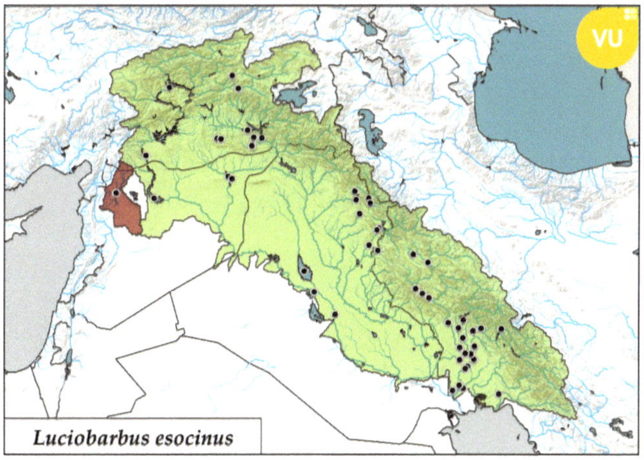

Luciobarbus esocinus; Lesser Zab, Iraq; ~230 mm SL.

• juveniles often with many, small dark-brown spots on upper flank. Size up to 2000 mm SL and 198 kg, regularly larger than 50 kg.
Distribution. Euphrates, Tigris, Karun and Jarahi drainages. Extirpated in Qweiq.
Habitat. Large rivers and reservoirs, from which fish migrate into tributaries to spawn.
Biology. Biology of this enigmatic species virtually unknown. Lives up to 17 years, probably longer. Spawns first time at a size greater than 1000 mm SL, probably at an age of about 10 years, in March–April. Predatory, feeding mainly on fish.
Conservation status. VU; builds up large populations in reservoirs, from where it migrate into tributaries to spawn. An important commercial species that has declined due to overfishing. Affected by pollution, draining of marshes, damming, and water abstraction.
Remarks. Growth hormones from this species have been cloned and used to increase growth in *Cyprinus carpio*. Often hybridises with sympatric *L. schejch*, and most hybrids seem to have had *L. esocinus* as their mother. Type material of *L. xanthopterus* has been identified as such hybrids.
Further reading. Coad 2010a (biology, distribution); Coad 2021a (biology, morphology); Freyhof et al. 2025 (revision).

Luciobarbus graecus
Common name. Greek barbel.
Diagnosis. Distinguished from other species of *Luciobarbus* in Mediterranean basin by: • 43–48 total lateral line scales. Size up to 400 mm SL.
Distribution. Türkiye: Bakacak on Biga Peninsula west to Çanakkale and south to Büyük Menderes. Also, lakes Yliki and Paralimni plus Sperchios drainage in southwestern Greece. Introduced to Beletsi and Marathon reservoirs and Kifissos drainage.
Habitat. Large to medium-sized streams and rivers with moderate currents. Also, in reservoirs.
Biology. Spawns April–July.
Conservation status. LC.
Remarks. *Luciobarbus kottelati*, and *L. lydianus* are synonyms. The biogeography of *L. graecus* is a mystery, and it is possible that *L. graecus* was introduced to Greece before the time of ichthyological exploration.

Luciobarbus graecus; Büyük Menderes, Türkiye; 170 mm SL.

Luciobarbus graecus; Biga Peninsula, Türkiye; 195 mm SL.

Further reading. Turan et al. 2008a (description as *L. kottelati*); Güçlü et al. 2013, İlhan et al. 2020, 2021 (distribution in Anatolia); Freyhof & Yoğurtçuoğlu 2024 (synonymy of *L. kottelati* and *L. lydianus*)

Luciobarbus longiceps
Common name. Jordan barbel.
Diagnosis. Distinguished from other species of *Luciobarbus* in Mediterranean basin by: ● 55–63 total lateral-line scales / ○ 10–13 gill rakers on lower arch / ○ pelvic origin behind or below dorsal origin. Size up to 700 mm SL.
Distribution. Jordan drainage, as well as Naaman at coast of Israel.
Habitat. Large to medium-sized streams and rivers with moderate currents. Also in reservoirs and lakes.
Biology. Matures at about 200 mm SL (males) and 250 mm SL (females). Migrates upstream in winter to spawn in small streams in January–March. Feeds on invertebrates, especially molluscs.
Conservation status. EN; declined sharply due to overfishing and desiccation of rivers and streams.
Further reading. Krupp & Schneider 1989 (diagnosis, distribution).

Luciobarbus longiceps; Jordan drainage, Syria; 150 mm SL.

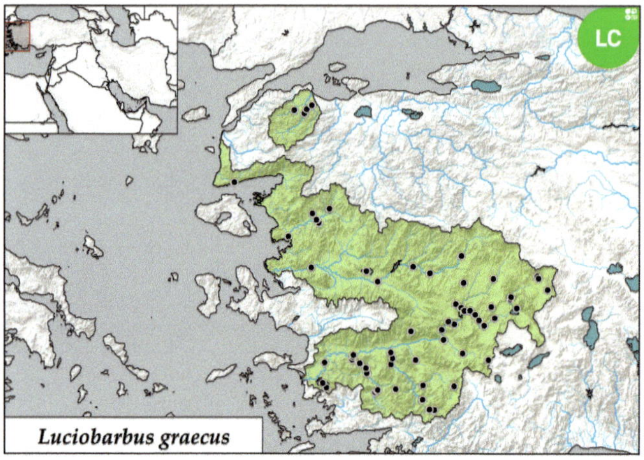

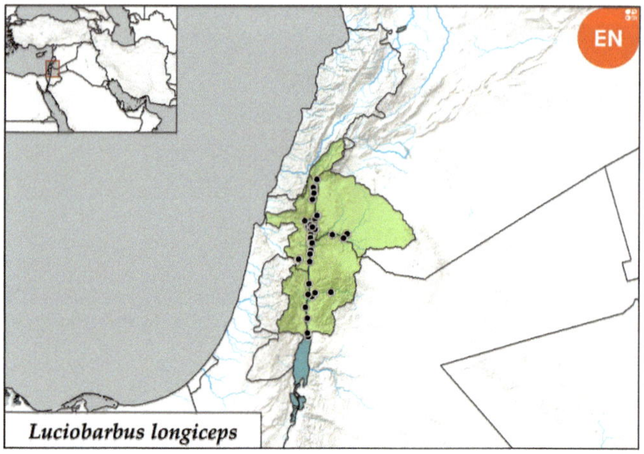

Luciobarbus mursa; Tajan, Iran; 198 mm SL.

Luciobarbus mursa

Common name. Mursa.

Diagnosis. Distinguished from other species of *Luciobarbus* in Caspian basin and Hari drainage by: ● lower lip thick and fleshy, with median lobe or pad / ● 83–103 total lateral-line scales / ○ 8½ branched dorsal rays / ● back between head and dorsal origin not or slightly compressed / ○ colouration of back and belly not sharply contrasted / ○ predorsal length equal or longer than postdorsal length / ○ body plain brown in life. Size up to about 400 mm SL.

Distribution. Kura and Aras drainages east to Gorgan.

Habitat. Wide range of streams and rivers with fast to moderately fast-flowing water. Also, in lakes and reservoirs, from where fish migrate to rivers and streams to spawn.

Biology. Spawns first time at 2–3 years, females later than males, between April and August. Migrates to uppermost tributaries of rivers to spawn in fast-flowing water on gravel. Feeds on invertebrates, algae, detritus, plant material, and small fish.

Conservation status. VU; important commercial species, particularly prized in Georgia and severely overfished, especially in Kura and Aras.

Remarks. Reports from Urmia basin (Iran) have yet to be confirmed and are probably mistaken.

Further reading. Berg 1949b (identification, distribution); Bogutskaya et al. 2003b (biology); Kaya et al. 2020a (distribution); Coad 2021a (biology, morphology).

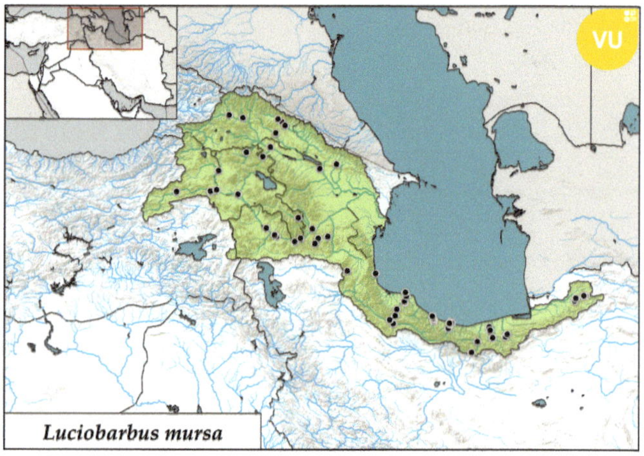

Luciobarbus pectoralis; Orontes drainage, Syria; 190 mm SL.

Luciobarbus pectoralis

Common name. Levantine barbel.

Diagnosis. Distinguished from other species of *Luciobarbus* in Mediterranean basin by: ○ 48–58 total lateral-line scales / ○ 8–13 gill rakers on lower arch / ○ 10–11 scale rows between dorsal origin and lateral line / ○ 7–8 scale row between anal origin and lateral line. Size up to 470 mm SL.

Distribution. Türkiye and Syria: Göksu east to Orontes drainage. Absent in between Göksu and Seyhan drainages.

Habitat. Medium to large streams and rivers. Also, in reservoirs from which it migrates to rivers or streams to spawn. Absent from cold mountain streams and springs. Often abundant at edge of rapids or riffles.

Biology. No data.

Conservation status. LC.

Remarks. Often reported to occur in Euphrates and Tigris drainages, but none of these reports have been confirmed. Most likely, all records are based on misidentified specimens of *L. schejch*. *Luciobarbus lorteti*, described from Lake Amik in Orontes drainage, is treated as a synonym of *L. pectoralis*. *Luciobarbus lorteti* was distinguished from *L. pectoralis* by the presence of a molariform (vs. normally developed) 4th pharyngeal tooth. Such individuals are still found in the wild. Fish with a molariform 4th pharyngeal tooth have no other distinguishing characters, and molecular data do not allow them to be distinguished from *L. pectoralis*. The shape of the 4th pharyngeal tooth may be related to the size of individual and likely to the availability and use of hard-shelled prey such as molluscs.

Further reading. Krupp 1985c (*B. lorteti*); Krupp 1985d (distribution, description); Bayçelebi 2020 (distribution).

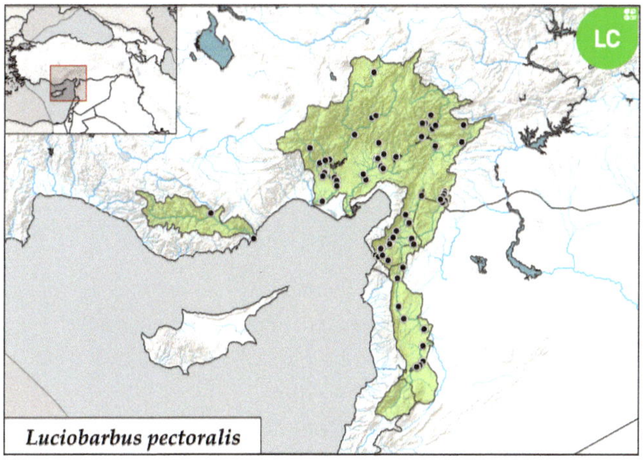

Luciobarbus schejch; Tigris, Türkiye; 195 mm SL.

Luciobarbus schejch; Lesser Zab, Iraq; 143 mm SL.

Luciobarbus schejch; Sosangerd, Iran; 331 mm SL; 229 mm SL.

Luciobarbus schejch; Tigris, Iraq; 259 mm SL.

Luciobarbus schejch

Common name. Gattan.

Diagnosis. Distinguished from other species of *Luciobarbus* in Persian Gulf basin by: ● back and upper flank yellowish or brown, lower flank yellowish or cream white in life, blackish or grey in large individuals / ○ head not strongly pointed, not depressed, barbel-like / ○ dorsal head profile convex / ○ mouth inferior or subterminal in very large individuals / ○ snout length 1.1–1.9 times in postorbital length / ○ 50–62 total lateral-line scales / ○ 11–24 gill rakers / ○ lips fleshy, usually with median pad, lips hypertrophied in some individuals / ○ juveniles without small dark-brown spots on upper flank. Size up to 800 mm SL and 30 kg (such large individuals might be hybrids with *L. esocinus*).

Distribution. Euphrates, Tigris, Karun drainages, and coastal rivers of Persian Gulf south to Kol.

Habitat. Large streams, mountain and lowland rivers, lakes, reservoirs, and marshes, from which it migrates into tributaries to spawn. Spawns on gravel substrate in shallow areas of large rivers in 30–150 cm water depths.

Biology. Lives up to 11 years, probably longer. Maturity of males at 3 years and about 330 mm SL, females at 6 years and about 600 mm SL. Spawns June–August (Mosul), at dusk until just before midnight, with loud splashing, jumping, and chasing. Exclaims large areas of gravel to spawn in "nests." Spawns once a year. Tolerant of low oxygen and moderate salinity. Feeds on a wide variety of aquatic invertebrates, small fish, detritus, and plants. Locally, molluscs are an important prey item.

Conservation status. LC; extirpated from Qweiq. Establishes large populations in reservoirs. Highly overexploited in large rivers.

Remarks. Many, but not all, individuals of *L. schejch* carry mitochondria from *L. esocinus*, as hybridisation between the two species is common. Usually, male *L. schejch* spawns with female *L. esocinus*. *Luciobarbus schejch* with mitochondria of *L. schejch* (not *L. esocinus*) have been identified as *L. kersin*. *Luciobarbus schejch* with its own mitochondria and those with mitochondria of *L. escocinus* are widespread, occur in sympatry, and are indistinguishable by morphological characters. *Luciobarbus kersin* is a synonym of *L. schejch*. *Luciobarbus schejch* is often known as *L. barbulus* or, in parts of its range, as *L. mystaceus*. *Luciobarbus barbulus* is a junior synonym of *L. schejch*. *Luciobarbus mystaceus* has been described from the Kura in Georgia, and this species is a synonym of *L. capito*. *Barbus rajanorum*, described from Aleppo, is a hybrid of *L. schejch* and *Capoeta damascina*.

Further reading. Khaefi et al. 2017b (as *L. barbulus*); Coad 2010a (biology, distribution as *L. pectoralis*, *L. barbulus*, and *L. xanthopterus*); Coad 2021a (biology, morphology, as *L. barbulus*); Freyhof et al. 2025a (taxonomic revision).

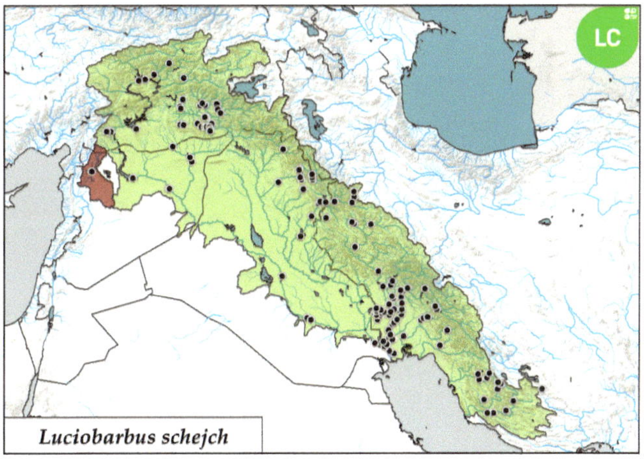

Luciobarbus subquincunciatus; Tigris, Türkiye; 500 mm SL. © M. Oral

Luciobarbus subquincunciatus

Common name. Leopard barbel.

Diagnosis. Distinguished from other species of *Luciobarbus* in West Asia by: • numerous large roundish dark-brown or black blotches on head, body, and fins arranged in an almost quincunx pattern / • 3rd inner pharyngeal tooth enlarged, molariform / ○ 75–88 lateral-line scales on body / ○ no median lobe in lower lip. Size up to 600 mm SL, but likely to grow much larger.

Distribution. Euphrates and Tigris drainages from southern Anatolia to Shatt-al Arab/Arvand and Hammar marshes, including Karun.

Habitat. Large lowland rivers. Rarely reported from reservoirs and marshes.

Biology. Feeds mainly on molluscs but also takes considerable amounts of algae and detritus.

Conservation status. CR; very rare throughout its range. Still in very small numbers in Tigris around Cizre, Botan, a tributary of Tigris, Murat, a major tributary of Euphrates, and Karun in Iran. Very rare in Tigris below Diyarbakır and adjacent waters in Iraq. Probably almost extinct in Khabur, a tributary of Euphrates, but situation in main stem of Tigris is unknown.

Remarks. Placed in *Bertinius* by some authors, a genus based on enlarged, molariform pharyngeal teeth, a character state found in several unrelated *Luciobarbus* species.

Further reading. Coad 2010a (biology, distribution); Coad 2021a (biology, morphology).

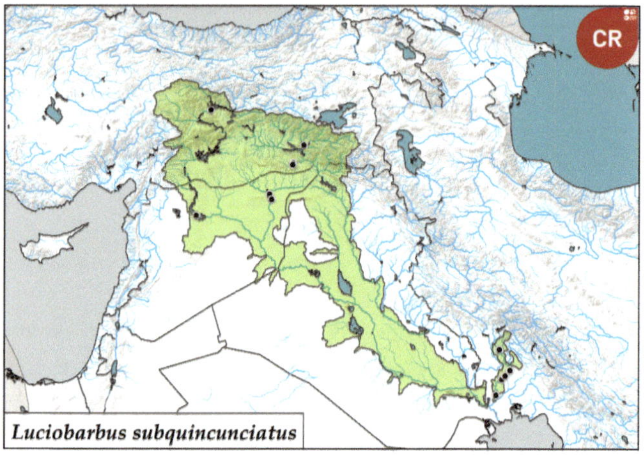

Luciobarbus and *Mesopotamichthys* are of major commercial importance in the marshes in Mesopotamia.

Mesopotamichthys sharpeyi; Shatt al-Arab drainage, Iraq; ~250 mm SL.

Mesopotamichthys

A monospecific cyprinid genus with large scales and no barbels. The only species within the genus, *Mesopotamichthys sharpeyi* forms a well-distinct clade with *Carasobarbus*, *Arabibarbus*, Moroccan *Ptercapoeta*, and *Atlantor reinii* and is most likely related to the unstudied subterranean *Caecocypris basimi* from Iraq. They all belong to the Torini group, which is also widespread in tropical Asia and Africa. **Further reading.** Borkenhagen 2014 (comparison; phylogeny).

Mesopotamichthys sharpeyi

Common name. Binni.

Diagnosis. Distinguished from superficially similar *Carasobarbus* and *Arabibarbus* by: ● no barbels / ○ 7–9½, usually 8½, branched dorsal rays / ○ 5½ branched anal rays / ○ lower lip without median lobe or pad / ○ body depth 28–35 % SL. Size up to 470 mm SL and 4 kg.

Distribution. Southern part of Euphrates and Tigris drainage. Iran: Hor Al Azim and Shadegan marshes and lower parts of Zohreh, Karkheh, and Karun. Syria: Asad reservoir and other large reservoirs (probably introduced). Iraq: Shatt-al Arab/Arvand and lower part of Euphrates, Tigris, Sirvan, including Al-Hammar Marsh and Huwazah Marsh, Lakes Saniyah, Habbaniyah, Tharthar, and Razzazah, and reservoirs Al Qadisiyah and Dukan (some probably introduced).

Habitat. Large rivers, lakes, and marshes, usually in stagnant water with dense vegetation in salinities up to 19 ‰. Spawns in stagnant water, usually on vegetation.

Biology. Lives up to 9 years. Usually spawns first at 2–4 years and 210–300 mm SL, on plants in February–April. Comes to spawning sites at dusk and dawn. Eggs are yellow and sticky. Mostly herbivorous, feeding on algae and plants.

Conservation status. VU; declined in recent years in all Iraqi and Iranian marshes and now virtually absent from rivers. Large populations restricted to Huwazah Marsh in Iraq and Shadegan Marsh in Iran. Few populations introduced in reservoirs in Syria.

Remarks. Next to *Tenualosa ilisha*, it is considered the tastiest fish from the Iraqi marshes.

Further reading. Karaman 1971 (generic description); Ekmekçi & Bănărescu 1998 (description); Coad 2010a (biology); Borkenhagen 2014 (phylogeny); Yang et al. 2015 (phylogeny); Coad 2021a (biology, morphology).

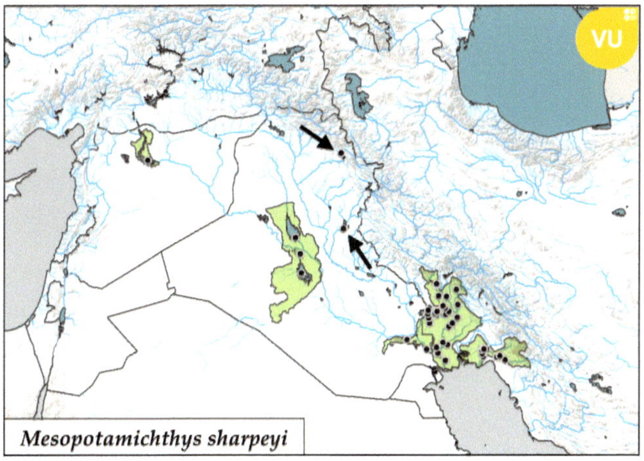

Mesopotamichthys sharpeyi

Schizocypris*, *Schizopygopsis* & *Schizothorax

Snow barbel or snow trouts are common names for two large, monophyletic radiations of Asian high-altitude cyprinids. All but one genus this book covers belong to Schizothoracinae, which includes four genera. Only *Schizopygopsis* is placed in the subfamily Schizopygopsinae, consisting of nine snow barbels genera. Phylogenetic hypotheses derived from several studies suggest that these groups may have originated repeatedly and ambiguously through hybridisation and polyploidisation within the Cyprinidae family. This may be responsible for the controversial classification among subfamilies. The history of polyploidisation in Cyprinidae remains uncertain, presenting significant challenges for phylogenetic systematics. Preliminary results indicate that Schizothoracinae and Schizopygopsinae are the maternal siblings of Cyprininae and paternal siblings of Torinae.

This hypothesis suggests that hybridisation between early carps and yellowfish may have played a pivotal role in the evolution of snow barbels.

Snow barbels are speciose and abundant in the mountains east of the Zagros and Elbrus. They are often the most common fish in these regions, dominating the fish communities by biomass. Snow barbels are highly specialised and represent the ecological pendants of barbels and scrapers. Several species show high individual variation in head and mouth shape and the development of lips. Snow barbels are a large group of fishes (more than 150 species) but only occur marginally in the area covered by this book. Their taxonomy and species diversity have never been comprehensively studied, and much remains to be done. Two species of *Schizothorax* and one species of *Schizocypris* occur in Eastern Iran. *Schizopygopsis stoliczkai* is occasionally reported from the fauna of Iran (Sistan basin), and *Schizothorax intermedius* is listed for the Hari. These records are based on misidentifications, and our area has no documented records of these species. *Schizothorax intermedius* is likely restricted to the upper Indus drainage, while *S. stoliczkai* is widely distributed in mountain areas of Afghanistan, Pakistan, India, and western China.

As is the case with barbel eggs, the eggs of these fish are poisonous to humans. Deaths have been reported as a result of eating the eggs of *Schizothorax*. Severe cases present with muscular cramps, paralysis, convulsions, and coma. Victims typically recover within 3–5 days after stomach evacuation, although this may take longer. It is advisable to exercise caution when consuming fish during its breeding season, as it is important to ensure that all traces of the eggs are removed to avoid contamination of the flesh. It should be noted that cooking does not destroy the toxin. **Further reading.** Yang et al. 2015 (phylogeny of Cyprinidae); Tan & Armbruster 2018 (subfamilies in Cyprinidae).

Schizocypris altidorsalis; Sistan, Iran; ~100 mm SL. © H. Mousavi-Sabet.

Schizocypris altidorsalis
Common name. Gorgak.
Diagnosis. Distinguished from other species of snow barbels in Iran by: • barbels absent / ○ body fully covered by scales, 81–96 scales along lateral line / ○ last unbranched dorsal ray longer than head, spinous, strongly serrated. Size up to 351 mm SL.
Distribution. Afghanistan and Iran: Sistan basin, including Helmand drainage, Chahnime reservoirs, and Hamun Lakes.
Habitat. All kinds of permanent waterbodies, migrate to seasonal rivers and streams at high water levels to spawn.
Biology. Feed mostly on aufwuchs and detritus.
Conservation status. LC; situation in Helmand is unknown.
Further reading. Bianco & Bănărescu 1982 (description); Coad & Keyzer-de Ville 2005 (identification); Coad 2021a (identification, distribution).

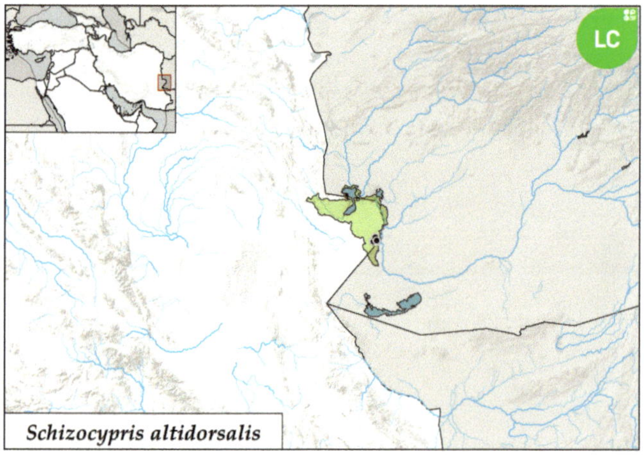

Key to species of snow barbels in Iran.

1a - Barbels absent.
............*Schizocypris altidorsalis*
1b - Two pairs of barbels.
............2
2a - 9–18 gill rakers.
............*Schizothorax pelzami*
2b - 21–41 gill rakers.
............*Schizothorax zarudnyi*

Schizothorax pelzami; Cheshmeh Ali Spring, Western Kavir basin, Iran; ~170 mm SL. © H. Mousavi-Sabet.

Schizothorax pelzami; Hari drainage, Iran; ~170 mm SL. © Mousavi-Sabet.

Schizothorax pelzami

Common name. Kavir snow barbel.

Diagnosis. Distinguished from other species of snow barbels in Iran by: ○ two pairs of barbels / ○ 9–18 gill rakers. Size up to 480 mm SL and 3 kg.

Distribution. Iran: Jam, Sharak, Akhlamad, Kashaf, and small water bodies in Khorasan, upper Kal Shur, Jajarm, and Jovein in Eastern and Western Kavir basins. Also, in Cheshmeh Ali (Iran) and Hari drainage in Afghanistan, Iran, and Turkmenistan, as well as in Morghab drainage in Afghanistan and Turkmenistan.

Habitat. All types of permanent waterbodies such as springs, streams, rivers, and qanats. Migrates to seasonal rivers and streams at high water to spawn.

Biology. Lives more than 7 years. Feeds on invertebrates, larger individuals mainly on small fish.

Conservation status. LC; situation outside Iran unknown.

Remarks. Occasionally hybridised with *Capoeta heratensis*. This is the westernmost species of snow barbel, and the Cheshmeh Ali spring is its westernmost record. *Schizothorax iranicus* may be a separate species based on molecular data, but no morphological diagnosis has been made.

Further reading. Aliev et al. 1988 (distribution); Abdoli et al. 2007 (ecology); Jouladeh-Roudbar et al. 2020 (distribution); Coad 2021a (biology, morphology).

Schizothorax zarudnyi; Sistan basin, Iran; ~250 mm SL.

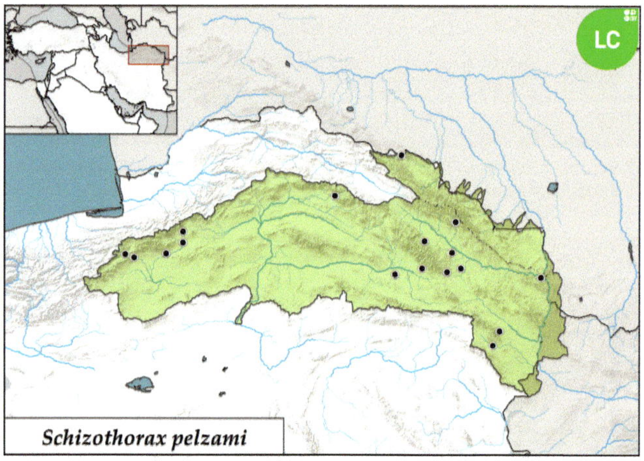

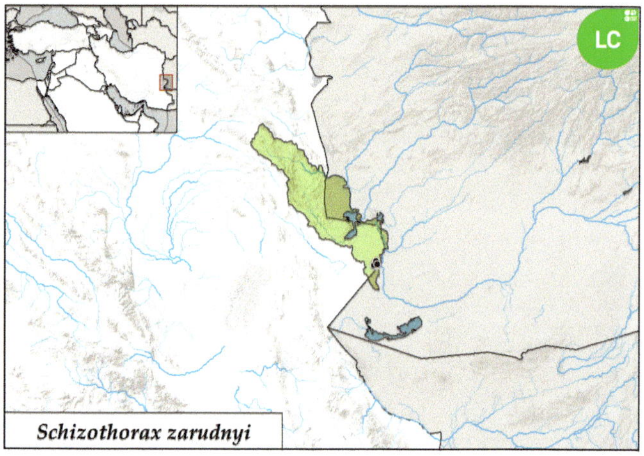

Schizothorax zarudnyi

Common name. Hamoun snow barbel.

Diagnosis. Distinguished from other species of snow barbels in Iran by: ○ two pairs of barbels / ○ 21–41 gill rakers. Size up to 530 mm SL, but maybe much larger. Up to 12 kg.

Distribution. Afghanistan and Iran: Sistan basin, including Helmand drainage. Iran: Zahak, Chahnime reservoirs, and Hamun Lakes.

Habitat. Pelagic in lakes, reed beds, and reservoirs. Appears to spawn in running waters. Expected to occur also in large rivers.

Biology. Mature after 4 years and at about 290–310 mm (male) and 380–400 cm (female, probably TL). Spawns March–April at water temperatures above 18°C. Reported to feed mainly on fish.

Conservation status. LC; situation in Helmand is unknown. A commercial species artificially propagated and stocked in Iran. Once widespread in Sistan but now reportedly restricted to the Chahnime reservoir near Zahak as other major wetlands have dried up and severely affected by invasion of non-native species such as *Cyprinus carpio*, *Carassius auratus*, *Ctenopharyngodon idella*, and *Hypophthalmichthys molitrix*.

Remarks. Placed in *Oreinus* by some authors, making the genus *Schizothorax* polyphyletic. We retain *S. zarudnyi* in *Schizothorax* until the case is better resolved.

Further reading. Gharaei 2012 (distribution); Coad 2014 (identification, distribution); Coad 2021a (biology, distribution).

Polyploid evolution in Cyprinidae. The achievement of evolutionary novelty is challenging due to the multitude of evolutionary pressures and the lengthy process of mutation and selection. On rare occasions, species hybridisation may result in fertile hybrids that exhibit an ecological benefit from the novel combination of genes and the subsequent morphological, physiological, and behavioural characteristics. However, hybrids of very different species may encounter difficulties due to the distinctiveness of their genomes, which may prevent the normal meiotic process essential for the formation of gametes (egg and sperm). Such hybrids are often sterile and unable to transmit their novel character combination to the next generation. Numerous hybrids between different species in Cyprinidae and Leuciscidae are known, and it is assumed that hybridisation, even between distantly related fish species, has occurred frequently for millions of years.

In exceptional cases, a hybrid may exhibit a doubling of chromosomes inherited from each parent, restoring meiotic pairing and fertility. This process, known as polyploidisation, results in a transition from a diploid, sterile hybrid to a tetraploid, fertile hybrid. Additionally, fish can duplicate their genome during their ontogenetic development, resulting in a doubling of the number of chromosomes inherited from the same parental species. This phenomenon, known as **auto-polyploidisation**, occurs independently of hybridisation. In a further pathway, a hybrid between species produces clonal, full-genomic gametes with two sets of chromosomes, as in all other body cells. This is an accidental occurrence in the gamete-tissue. In such a diploid, the clonal egg is fertilised by a haploid sperm, resulting in triploid offspring. Triploid individuals, which are usually females, experience difficulties in producing gametes via meiosis and are once again sterile. Except in the case in which the female may again produce clonal eggs, now with three sets of chromosomes (as in many *Carassius*), the offspring will be tetraploid (with four copies of each chromosome) if a haploid sperm of the parental species fertilise these. These hybrids are viable and fertile and may have a double-size genome, with four sets of chromosomes from each parental species. This represents one pathway in the process of **allo-polyploidisation**. Other pathways include F1 hybrids producing clonal, diploid eggs and sperm. In all cases, the fish overcomes the sterility associated with hybrids as soon as an even ($4n$, $6n$) chromosome number has been re-established with even numbers of chromosomes from each parental species.

In evolutionary terms, polyploidisation is rare in vertebrates, occurring only in a few fish groups, including sturgeons and some cypriniforms. Many of the polyploidisation events in Cyprinidae are ancient, having happened at the base of large radiations that are recognised at the generic or subfamily level today. One example is an allo-tetraploidisation event at the root of all barbels (Barbini), which are all tetraploid. The allo-tetraploidisation event at the root of Cyprinini (carps and goldfish) is believed to have occurred approximately 12 million years ago, at the end of the middle Miocene. The hybridisation event leading to the hexaploid genus *Capoeta* occurred more recently, during the late Miocene. As previously discussed in the context of *Capoeta*, identifying the parental species associated with allo-polyploidisation events is of interest. In the case of Cyprinini, one species of Barbini or Acrossocheilini was potentially the maternal contributor. In contrast, a species closely related to or included within Labeonini was potentially the paternal source. These can be identified as each polyploid fish possesses each gene four times (if tetraploid), with two copies originating from the mother and two from the father. As both parental species may have been only distantly related, the phylogeny of the polyploid fish has two roots. It does not follow the "tree-like" evolutionary pathway often seen in textbooks. This phenomenon makes it difficult to reconstruct the phylogeny of cyprinid fishes, which is why we still have a poor understanding of the relationships of different groups of Cyprinids. It is noteworthy that maternal subgenomes frequently predominate in the genomes, a phenomenon that introduces an additional layer of complexity.

One might inquire why polyploidy is so uncommon in fishes compared to plants. Polyploid animals appear to encounter significant challenges in coping with genomic and developmental chaos from merging two genomes. A fish cannot organise its functions from its new genome, as conflicting recipes exist for the same transcriptional challenges. Fish employ diverse strategies to balance dynamic subgenomic diversification during continuous re-diploidisation. They regain functional diploidy over time by eliminating certain parts of their genome, potentially those that disrupt transcription. Originally, polyploid fishes became functional diploids again through this process. This phenomenon is observed in numerous species within the Cyprinidae family. **Further reading.** Otto & Whitton 2000 (evolution); Mable 2013 (ecological benefits of polyploids and hybrids); Yang et al. 2015 (phylogeny of Cyprinidae); Ma et al. 2014 (timing of polyploidisations in Cyprinini); Luo et al. 2020 (diploidisation); Xu et al. 2023 (subgenomic differentiation).

Tariqilabeo

Tariqilabeo comprises small labeonines related to *Garra*. They share several characteristics with the latter, including a rounded snout, a mouth usually inferior and crescent-shaped, a horny edge on the lower jaw, a rostral cap usually fringed, and non-thickened unbranched dorsal rays. *Tariqilabeo* can be readily distinguished from superficially similar species of *Garra* lacking a gular disc, and the lower jaw and area behind it are heavily tuberculated with glandular openings and irregularly arranged hard ridges. Eight species are known. Their systematics and diversity have not been well studied, and numerous changes in taxonomy are expected in the future. Previously, species were included in the Southeast Asian genus *Crossocheilus* and *Gonorhynchus*. *Crossocheilus* is now restricted to a few species related to *C. siamensis* and is only found from Myanmar to the east. **Further reading.** Kottelat 2016 (generic name).

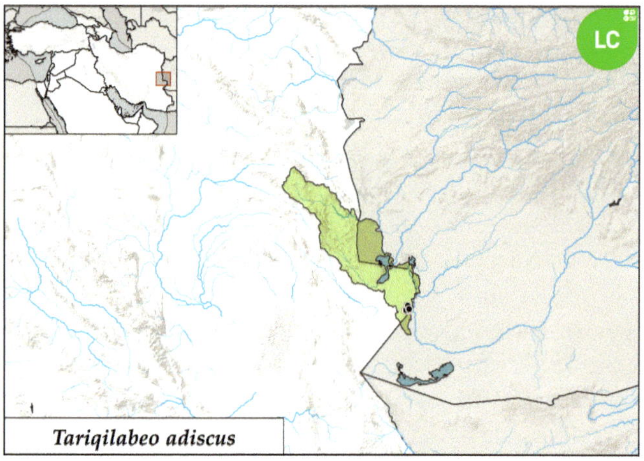

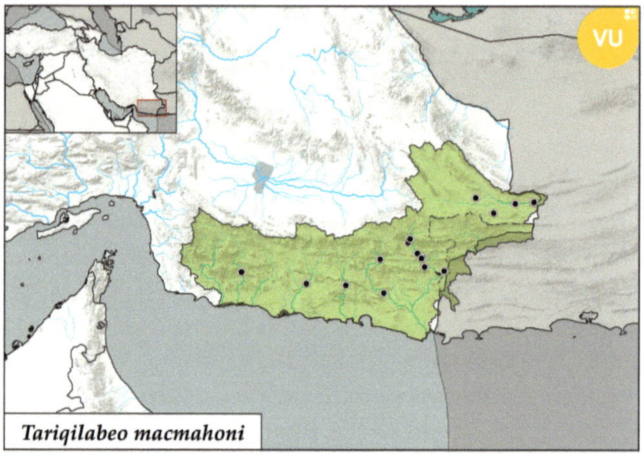

Tariqilabeo adiscus; Chahnime reservoirs, Sistan basin, Iran; ~70 mm SL.

Tariqilabeo adiscus

Common name. Sistan latia.

Diagnosis. Distinguished from *T. macmahoni* by: • two pairs of relatively long barbels / • 3–5, usually 4, scales between anus and anal origin / • 19–22 total gill rakers / • maximum body depth 17–22 % SL. Size up to 93 mm SL.

Distribution. Afghanistan and Iran: Sistan basin, including Helmand drainage. Iran: Zehak, Chahnime reservoirs, and Hamun Lakes.

Habitat. Wide range of slow-flowing and stagnant-water habitats. Common in small streams with rocky or muddy bottoms, irrigation ditches, reedbed channels, and pools.

Biology. Bottom feeders, feeding on detritus and algae.

Conservation status. LC; situation in Helmand unknown.

Remarks. Despite conflicting opinions, we maintain this species as valid and that fish from the Helmand drainage reportedly have 20–26 gill rakers. More material needs to be examined better to understand the diversity of *Tariqilabeo*

Tariqilabeo macmahoni; Sarbaz, Iran; ~75 mm SL.

in West Asia. Often identified as *Crossocheilus latius*, a species of *Tariqilabeo* from India from which it differs in having 19–22 gill rakers (vs. 37–39 in *T. latius*) and 34–38 total lateral-line scales (vs. 39–41).

Further reading. Annandale 1919 (description); Sayyadzadeh et al. 2015a (description, as *Gonorhynchus adiscus*); Behrens-Chapuis et al. 2015 (molecular phylogeny, as *Gonorhynchus adiscus*); Coad 2021a (biology, distribution).

Tariqilabeo macmahoni

Common name. Mashkid latia.

Diagnosis. Distinguished from *T. adiscus* by: • usually one pair of rostral barbels; if present, maxillary barbels very short / • 2–3 scales between anus and anal origin / • 22–25 total gill rakers / • maximum body depth 22–30 % SL. Size up to 95 mm SL.

Distribution. Iran: Bahukalat and Sarbaz in Makran region, Mashkid basin in Iran and Pakistan and Makran region in Pakistan.

Habitat. Wide range of running water habitats, including qanats and small coastal streams.

Biology. No data.

Conservation status. VU; The 2020 Red List assessment classified *Tariqilabeo macmahoni* as VU in Pakistan; however, the species' occurrence in Iran was overlooked, and including this range would likely result in a LC status.

Remarks. Often identified as *T. diplochilus*, a species that has been described from the Indus drainage. Mashkid latia differs from *T. diplochilus* by having a very short or absent maxillary barb (vs. very long), a long rostral barb (vs. short), and 3–5 scales between anus and anal origin (vs. 2–3). *Tariqilabeo iranicus* has been recently described from the Makran region of Iran, based on literature, which erroneously states that *T. macmahoni* lacks fringes on the rostral fold. This is clearly not the case and therefore *T. iranicus* is a synonym of *T. macmahoni*.

Further reading. Behrens-Chapuis et al. 2015 (molecular phylogeny); Sayyadzadeh et al. 2015a (description); Ciccotto & Page 2016 (Indus and Iran); Coad 2021a (biology, distribution). Esmaeili et al. 2025 (*T. iranicus*).

In the Helmand river in Iran, the flow of water downstream is only permitted after significant rainfall. This environment is home to a diverse community of fish species, including *Tariqilabeo adiscus* and snow barbels.

Family Danionidae

Zebrafish and allies

Danionidae is a large family of cypriniform fishes previously treated as a subfamily of Cyprinidae. The exact composition of this family is still under discussion, and the question of which genera should be included has not yet been fully resolved. Further work is needed to settle this unsatisfactory situation. The family is speciose, with approximately 300 species and 50 genera. It is particularly diverse in sub-Saharan Africa and subtropical and tropical Asia, including numerous small rainforest species. Many of these are popular ornamental fish. The vertebrate model fish *Danio rerio* and the miniature fish *Paedocypris progenetica*, one of the smallest known vertebrates, are included in this family. All species are distinctive, small, herring-like fishes characterised by the lateral line on the lower part of the caudal peduncle. **Further reading.** Tang et al. 2010 (phylogeny).

Barilius mesopotamicus; Batman drainage, Türkiye; 45 mm SL.

Barilius mesopotamicus

Common name. Mesopotamian barilius.

Diagnosis. Distinguished from superficially similar species of *Alburnus*, juvenile clupeids, and *Cabdio occidentalis* by: ● one pair of short barbels / ○ 6–11 roundish, dark, grey blotches on flank. Size up to 65 mm SL.

Distribution. Euphrates and Tigris drainages in southern Türkiye, Syria, Iraq, and Iran. In the lower part of Karun, Zohreh, Helleh, and Mond all flowing into the Persian Gulf. Also, in Dalaki, an endorheic basin nested within Tigris drainage.

Habitat. Upper water column near surface of small to large rivers with slow to moderate currents.

Biology. A wide range of fish of all sizes caught in late summer suggests multiple spawning and a 1-year maturity with a short life span. Absence from cold headwater streams indicates limited tolerance to low temperatures. Feeds on floating aquatic and terrestrial invertebrates.

Conservation status. LC.

Further reading. Coad & Krupp 1983 (description); Liao et al. 2011 (molecular phylogeny).

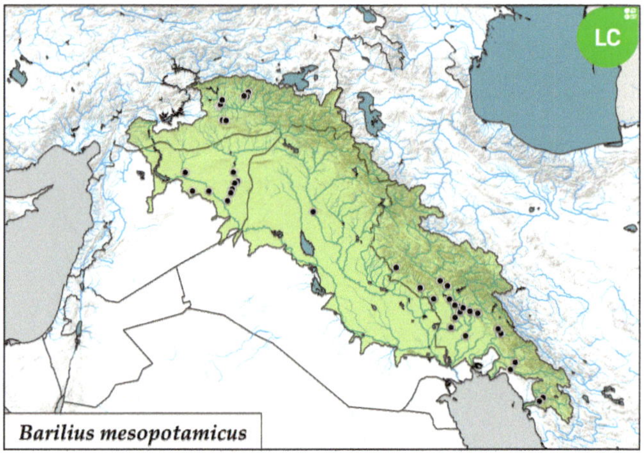

https://doi.org/10.1515/9783111677811-015

Cabdio occidentalis; Mashkid basin, Iran; ~65 mm SL. © H. Mousavi-Sabet.

Cabdio occidentalis

Common name. Persian morar.

Diagnosis. Distinguished from superficially similar species of *Alburnus* and juvenile clupeids by: • mouth subterminal; and from *Barilius mesopotamicus* by: • no barbels / • infraorbital bones large, almost as deep as eye diameter. Size up to 150 mm SL.

Distribution. Iran and Pakistan: Mashkid basin and Garo toward Sarbaz and Bahukalat in Makran region east to Indus.

Habitat. A wide variety of small streams to large, slow-flowing rivers. Also found in reservoirs and lakes.

Biology. Omnivorous, feeding mainly on phytoplankton and zooplankton. Spawns in February–May.

Conservation status. LC.

Remarks. Populations in Indus drainage may belong to this species. Commonly identified as *Cabdio morar*. Formerly known as *Aspidoparia*, which is a synonym of *Cabdio*. Molecular data place it in *Salmostoma*, from which it is distinguished by a subterminal (as opposed to terminal or superior) mouth.

Further reading. Berg 1949a (description of Iranian material); Tang et al. 2010 (molecular phylogeny); Hossain et al. 2015 (feeding); Lalramliana et al. 2019 (diversity, molecular data); Jouladeh-Roudbar et al. 2023a (description).

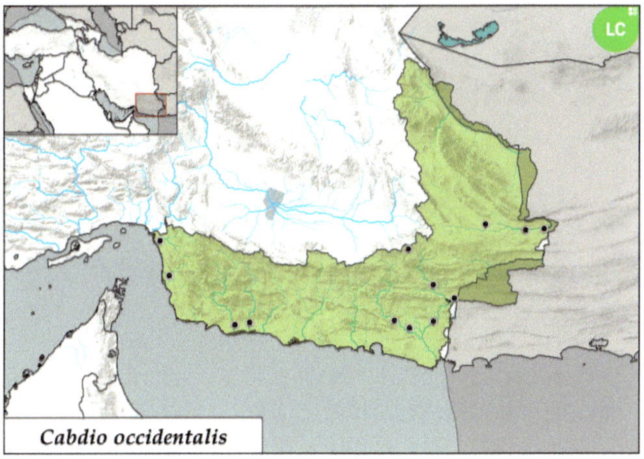

Family Gobionidae

Gudgeons

Gudgeons are a family of Cypriniformes that are most diverse in East Asia, from the Amur drainage southward to Central Vietnam; some 28 genera and 231 species are known. In West Asia, there are 17 species, all of which, except *Pseudorasbora*, being benthic fishes with a flat belly, laterally spread pectoral fins, a single pair of maxillary barbels, and 6½ branched anal rays (vs. 5½ in most Cyprinidae). Some Asian genera exhibit a markedly different morphology from that of true gudgeons. They have a more compressed body, the mouth is inferior to subterminal (even superior in *Pseudorasbora* and a few others), and they are distinguished from many other East Asian Cypriniformes by having 6½ branched anal rays. Some species of *Sarcocheilichthys* are noteworthy for spawning in the gill cavity of unionid freshwater mussels, similar to bitterlings; they, too, have developed an ovipositor. Interestingly, gudgeons and bitterlings are closely related. One East Asian species, *Pseudorasbora parva*, was accidentally introduced to Europe and Central Asia. From both directions, it made its way to West Asia, where it is now widely established. Except for *Pseudorasbora*, all gudgeons are superficially similar, bottom-dwelling fishes with a midlateral row of dark-brown or grey blotches on the flank and spotted bodies and fins. **Further reading.** Bănărescu & Nalbant 1973 (systematics); Turan et al. 2012a (identification of Anatolian species).

Key to genera of Gobionidae in West Asia

1a - Barbels absent; mouth superior.
..................*Pseudorasbora*
1b - Barbels present; mouth inferior.
..................2

2a - Caudal peduncle 3.1–3.8 times its depth; epithelial crests on scales on back.
..................*Romanogobio*
2b - Caudal peduncle 1.5–2.5 times its depth; no epithelial crests on scales on back.
..................*Gobio*

Key to species of *Gobio* in West Asia, excluding *G. nigrescens* from Hari drainage

1a - 42–63 total lateral-line scales.
..................2
1b - 39–45 total lateral-line scales.
..................5

2a - Scales extend forward to level of pectoral origin or to half distance of pectoral origin and isthmus.
..................*G. microlepidotus*
2b - Scales on chest almost extend to isthmus.
..................3

3a - Head length 24–27 % SL.
..................*G. hettitorum*
3b - Head length 27–29 % SL.
..................4

https://doi.org/10.1515/9783111677811-016

4a - Distance between anus and anal origin 5–7 % SL; caudal-peduncle length 1.7–2.2 times in caudal-peduncle depth.
..................*G. intermedius*
4b - Distance between anus and anal origin 7–8 % SL; caudal-peduncle length 1.4–1.7 times in caudal-peduncle depth.
..................*G. maeandricus*

5a - 8–10 scales between posterior extremity of pelvic base and anus.
..................6
5b - 4–8 scales between posterior extremity of pelvic base and anus.
..................7

6a - Diameter of scales on belly approximately equal to pupil diameter.
..................*G. insuyanus*
6b - Diameter of scales on belly clearly smaller than pupil diameter.
..................*G. kizilirmakensis*

7a - Interorbital distance 1.6–2.2 times in eye diameter.
..................8
7b - Interorbital distance 1.0–1.6 times in eye diameter.
..................10

8a - No black spots on head.
..................*G. gymnostethus*
8b - Many small black spots on head.
..................9

9a - Distance between anus and anal origin 5–7 % SL; body depth at dorsal origin 22–27 % SL.
..................*G. fahrettini*
9b - Distance between anus and anal origin 7–10 % SL; body depth at dorsal origin 18–22 % SL.
..................*G. baliki*

10a - Breast completely scaled, scales extending to isthmus; snout rounded, distinctly concave anterior to nostril.
..................11
10b - Breast naked or incompletely scaled, if scales on chest present, scales extending forward to level of pectoral origin; snout pointed, not or slightly concave anterior to nostril.
..................12

11a - Pectoral reaching to or 1 scale in front of pelvic origin in mature male.
..................*G. caucasicus*
11b - Pectoral reaching 3–4 scales in front of pelvic origin in mature male.
..................*G. sakaryaensis*

12a - Head length 28–30 % SL.
..................*G. kovatschevi*
12b - Head length 24–26 % SL.
..................*G. artvinicus*

Gobio

Gobio comprises approximately 50 species distributed across a broad range of geographical locations, from Portugal to Korea. Most species (36) are found in the West Palearctic, with 14 in West Asia. *Gobio* are common in the Caucasus and Central Anatolia, where their species numbers require critical revision. The species of Lake Tuz basin (*G. hettitorum*, *G. gymnostethus*, *G. insuyanus*, *G. microlepidotus*) are very closely related, and it would be beneficial to test the morphological characters proposed to distinguish these species, which may be conspecific. A similar situation exists for all Caucasian species, where *G. artvinicus* may be conspecific with *G. caucasicus*. European species, including *G. delyamurei*, *G. holurus*, *G. krymensis*, *G. brevicirris*, and Asian *G. caucasicus*, form a cluster of populations that also require taxonomic revision. The biology of West Asian gudgeons is poorly understood, but it is thought that they behave similarly to the few European species studied. They spawn at night in the open water column of lakeshores or fast-flowing sections of streams and rivers. The sticky eggs drift a short distance with the current. Larvae are benthic, and females are fractional spawners that reproduce over a long period in spring and summer. They reach maturity after the first or second winter and live for about 3–8 years, feeding on a wide range of benthic invertebrates digging out from fine substrates as mud and sand.

Gobio artvinicus; Çoruh, Türkiye; 50 mm SL. © M. Özuluğ.

Gobio artvinicus

Common name. Colchic gudgeon.

Diagnosis. Distinguished from other species of *Gobio* in West Asia by: ○ 39–42 total lateral-line scales / ○ chest between pectorals naked or covered by scales / ○ scales on belly approximately equal or slightly larger than pupil diameter / ○ scales on base of pectoral usually embedded / ○ pectoral reaching 3–4 scales in front of pelvic origin in mature male / ○ head length 24–26 % SL / ○ head width at anterior eye-margin 11–12 % SL / ○ distance between pelvic origin and anal origin 2.8–3.3 times in distance between anus and anal origin / ○ 6 scales rows between lateral line and dorsal origin / ○ 4–5 scales between lateral line and anal origin / ○ 4–6 scales between anus and anal origin / ○ 5–6 scales between posterior extremity of pelvic base and anus / ○ interorbital distance 1.1–1.6 times in eye diameter / ○ snout pointed, not or slightly concave anterior to nostril / ○ 5–8 mid-lateral blotches. Size up to 107 mm SL.

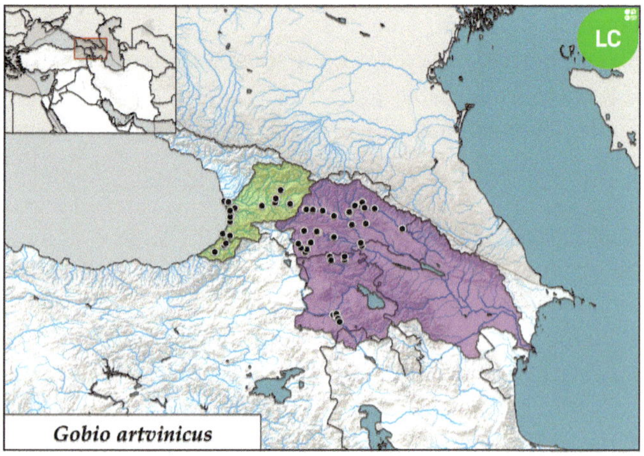

Distribution. Eastern Black Sea basin from Rioni (Georgia) south to Çoruh (Türkiye). Non-native and spreading in Kura and Aras drainages.
Habitat. A wide range of lowland and foothill streams and rivers, usually in slow to moderately fast-flowing waters on sand or gravel bottoms. Absent from high altitude streams with fast currents and low water temperatures.
Biology. No data.
Conservation status. LC.
Remarks. This species is much more variable than originally described, and the diagnostic characters need to be re-examined in the future. While fish from the Çoruh have a bare or incompletely scaled chest, the scales of populations from the Rioni and the Caspian basin cover the chest. There is some confusion about this species with *G. caucasicus*, which occurs in the adjacent Caucasian Black Sea basin and Caspian basin of Russia, and *G. artvinicus* is likely a synonym of *G. caucasicus*.
Further reading. Turan et al. 2016a (description); Epitashvili et al. 2020 (distribution); Kaya et al. 2020a (distribution); Yeşilçiçek 2022a (size).

Gobio baliki; Büyük Melen drainage, Türkiye; 86 mm SL.

Gobio baliki
Common name. Melen gudgeon.
Diagnosis. Distinguished from other species of *Gobio* in Anatolia by: ○ 40–42 total lateral-line scales / ○ chest completely scaled, scales extending forward to isthmus / ○ head length 24–28 % SL / ○ body depth at dorsal origin 18–22 % SL / ○ distance between anus and anal origin 7–10 % SL / ○ numerous small black spots on head / ○ depth of anal approximately equal to length of pelvic / ○ snout pointed / ○ 6–7 scales rows between lateral line and dorsal origin / ○ 4–5 scales between lateral line and anal origin / ○ 5–6 scales between anus and anal origin / ○ 5–6 scales between posterior extremity of pelvic base and anus / ○ 8–9 midlateral blotches. Size up to 93 mm SL.
Distribution. Türkiye: Büyük Melen drainage.
Habitat. Streams and small rivers with fast-flowing water on sandy or gravelly bottoms.
Biology. No data.
Conservation status. VU; seems to be in decline due to many threats.
Further reading. Turan et al. 2017a (description).

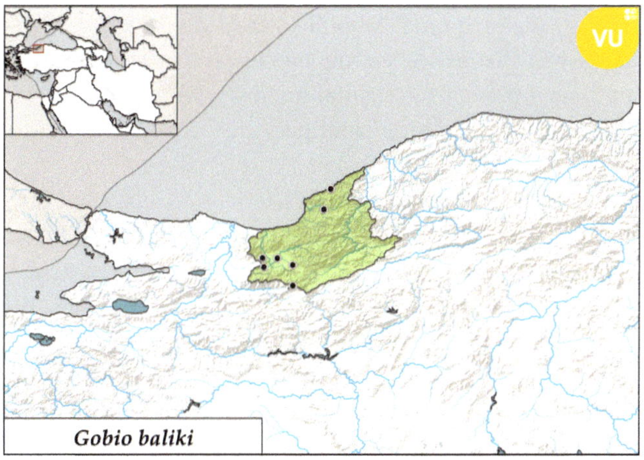

Gobio caucasicus; Sochi, Russia; ~70 mm SL.

Gobio caucasicus

Common name. Caucasian gudgeon.

Diagnosis. Distinguished from *Gobio* species in Caucasus by: ● pectoral reaching to or one scale in front of pelvic origin in mature male / ○ chest scaled between pectorals, scales extending forward beyond pectoral base (at least in some individuals of any population) / ○ head width at anterior eye-margin 9–10 % SL / ○ distance between pelvic origin and anal origin 3.3–4.1 times in distance between anus and anal origin / ○ head length 26–28 % SL / ○ head depth at nape 16–17 % SL, at eye 48–62 % HL / ○ head width 15–17 % SL, 58–62 % HL / ○ eye diameter 4–6 % SL, 3.1–3.7 times in head depth, 1.4–1.5 times in interorbital distance (at 50–60 mm SL) / ○ depth of caudal peduncle 33–36 % HL / ○ snout rounded, distinctly concave anterior to nostril. Size up to 100 mm SL.

Distribution. Russia: Caucasian Black Sea basin. In Kuma, Terek, and Sulak drainages in Caspian basin.

Habitat. Lowland and foothill streams and rivers, usually in slow to moderate-flowing waters on sandy or gravelly bottoms. Absent from high altitude streams with fast currents and low water temperatures.

Biology. No data.

Conservation status. LC.

Remarks. The gudgeons of the Caucasian Black Sea basin of Russia are usually identified as *G. caucasicus*. This species has been described based on syntypes from the European Caspian basin (often identified as *G. holurus*) and syntypes from the Rioni in Georgian Black Sea basin (conspecific with *G. artvinicus* based on molecular data). Molecular data (COI) show that some populations of "*G. caucasicus*" from the Black Sea basin are closely related to the Caspian *G. holurus*; others are closely related to *G. kubanicus*, *G. brevicirris* and *G. krymensis* from the European Black Sea basin. Finally, some populations represent a distinct molecular lineage, but the situation needs to be better understood. As different molecular lineages are found in sympatry, introgressive hybridisation is likely to have played a major role. This makes it difficult to classify species based on mtDNA alone. The identity of *G. caucasicus* still needs to be solved, as no lectotype has been identified. For now, we identify *G. holurus* as conspecific with *G. caucasicus*, as two of three syntype localities are within the range of this species.

Further reading. Berg 1949b (description); Kottelat & Freyhof 2007 (diagnosis as *G. holurus*); Turan et al. 2016a (characters as *G.* cf. *caucasicus*).

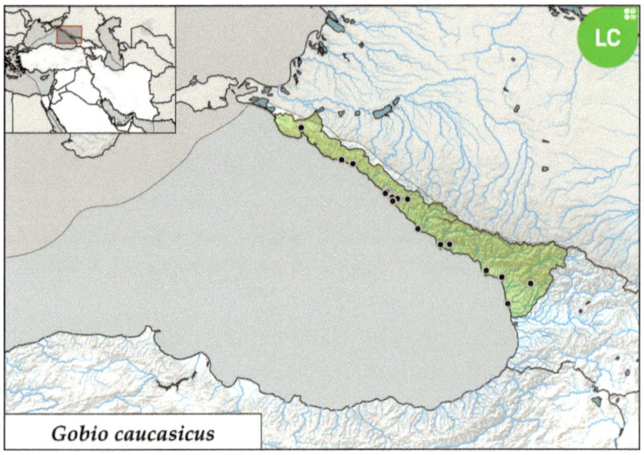

Gobio fahrettini; Ilgın basin, Türkiye; ~90 mm SL.

Gobio fahrettini

Common name. Ilgın gudgeon.

Diagnosis. Distinguished from other species of *Gobio* in Anatolia by: ○ 39–42 total lateral-line scales / ○ chest completely scaled, scales extending forward to or almost to isthmus / ○ head length 26–30 % SL / ○ distance between anus and anal origin 5–7 % SL / ○ body depth at dorsal origin 22–27 % SL / ○ numerous small black spots on head / ○ snout long, with rounded tip, 11–14 % SL / ○ caudal peduncle length 1.6–2.2 times its depth / ○ 6–7 scales rows between lateral line and dorsal origin / ○ 6–8 scales between posterior extremity of pelvic base and anus / ○ 6–8 mid-lateral blotches. Size up to 122 mm SL.

Distribution. Türkiye: Lake Ilgın basin.

Habitat. Slow to moderate-flowing streams with sandy or gravelly bottoms.

Biology. No data.

Conservation status. EN; appears to be declining within its very small range.

Remarks. Some individuals of *G. fahrettini* share their mitochondrial DNA with *G. microlepidotus*, and the two species may have hybridised in the past. The reasons for this hybridisation are not known, but the artificial introduction of *G. microlepidotus* into the Ilgın basin cannot be excluded.

Further reading. Turan et al. 2018b (description).

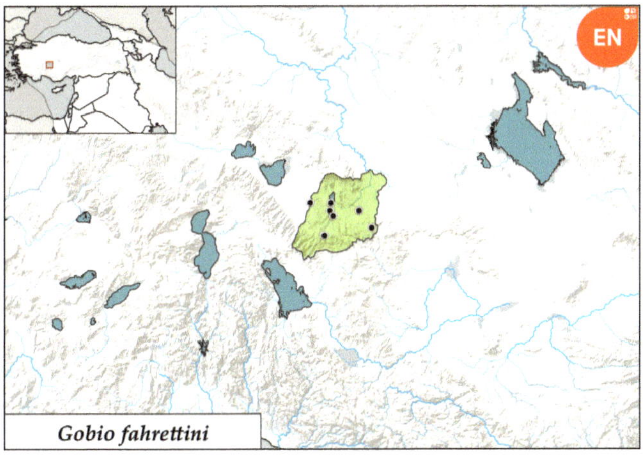

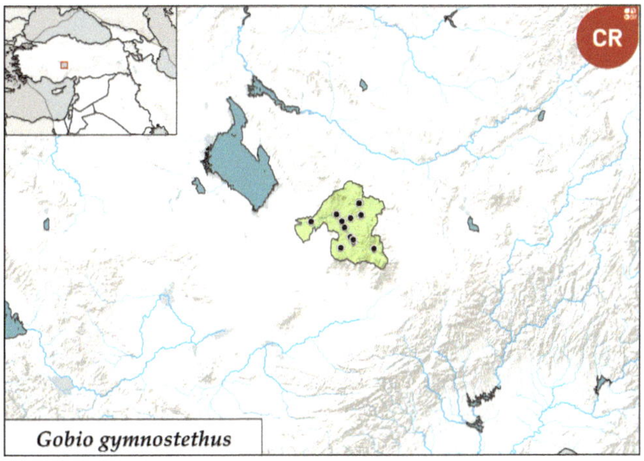

Gobio gymnostethus; Melendiz drainage, Türkiye; ~120 mm SL.

Gobio gymnostethus

Common name. Cappadocian gudgeon.

Diagnosis. Distinguished from other species of *Gobio* in Anatolia by: ○ 41–44 total lateral-line scales / ○ chest completely scaled, scales extending forward to or almost to isthmus / ○ head length 25–28 % SL / ○ 5–6 scales between posterior extremity of pelvic base and anus / ○ interorbital distance 1.6–1.9 times in eye diameter / ○ no black spots on head / ○ 6 scale rows between dorsal origin and lateral line / ○ pelvic not reaching behind anus in female / ○ interorbital distance 1.6–2.2 times in eye diameter / ○ head profile behind nostrils markedly convex. Size up to 140 mm SL.

Distribution. Türkiye: Melendiz drainage.

Habitat. Slow to moderate-flowing streams with sandy or gravelly bottoms.

Biology. No data.

Conservation status. CR; abundant but declining within its very small range, which is heavily impacted by water extraction and dams.

Further reading. Ladiges 1960 (description); Naseka et al. 2006 (identification).

Melendiz, here in Ihala valley, is the largest remaining tributary of Lake Tuz and the habitat of a rich fish fauna.

Gobio hettitorum; Karaman; Türkiye; ~120 mm SL.

Gobio hettitorum

Common name. Taurus gudgeon.

Diagnosis. Distinguished from other species of *Gobio* in Anatolia by: ○ 44–46 total lateral-line scales / ○ chest completely scaled, scales extending forward to isthmus / ○ head length 24–27 % SL / ○ 10–14 scales between posterior extremity of pelvic base and anus / ○ interorbital distance approximately 1.6–1.8 times in eye diameter / ○ 8–10 scale rows between dorsal origin and lateral line / ○ caudal length 17–20 % SL / ○ anal length 15–17 % SL. Size up to 124 mm SL.

Distribution. Türkiye: Yeşildere, Deliçay, Gödet, and Ayrancı drainages (Karaman), and Ereğli marshes.

Habitat. Small streams with clear water and sand and gravel bottoms.

Biology. No data.

Conservation status. VU; appears to be declining in its small range due to the desiccation of streams. Extirpated from Ereğli around 2010.

Further reading. Ladiges 1960 (description); Naseka et al. 2006 (identification).

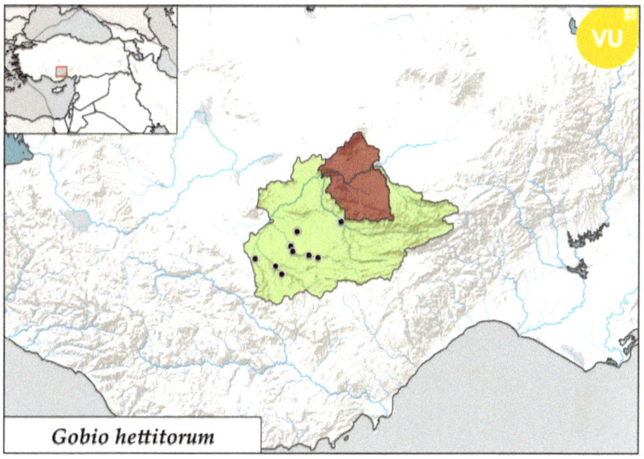

Gobio insuyanus; Insuyu, Türkiye; ~130 mm SL.

Gobio insuyanus

Common name. Cihanbeyli gudgeon.

Diagnosis. Distinguished from other species of *Gobio* in Anatolia by: ○ 39–45 total lateral-line scales / ○ head length 25–30 % SL / ○ chest completely scaled, scales extending forward to or almost to isthmus / ○ 8–10 scales between posterior extremity of pelvic base and anus / ○ scales on belly approximately equal to pupil diameter / ○ 8–9 scale rows between dorsal origin and lateral line / ○ predorsal length 48–52 % SL. Size up to 135 mm SL.

Distribution. Türkiye: Insuyu (Cihanbeyli) drainage in western Lake Tuz basin.

Habitat. Springs and spring-associated streams with slow to moderately fast-flowing waters on sandy and gravelly bottoms, often among very dense aquatic vegetation.

Biology. Lives up to 7 years. Omnivorous, feeding mainly on detritus and macroinvertebrates such as gammarids and dipteran larvae.

Conservation status CR; restricted to one site and in serious decline due to desiccations of its habitat.

Further reading. Ladiges 1960 (description); Naseka et al. 2006 (identification); Johnson et al. 2025 (biology)

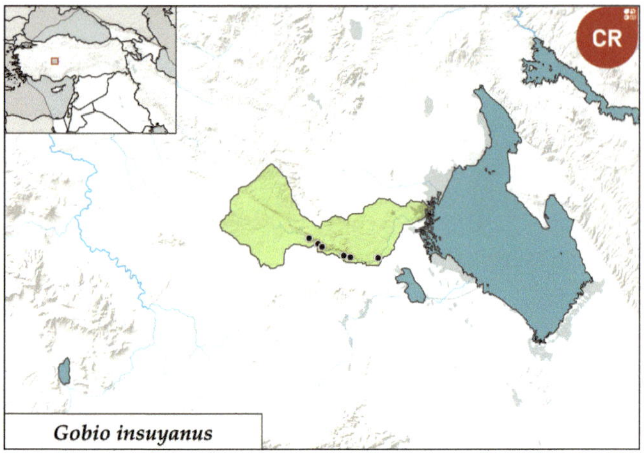

Insuyu is increasingly under pressure due to water extraction.

Gudgeons and the history of Lake Tuz basin. The present-day Lake Tuz (Türkiye) is an inhospitable environment for freshwater species. It is the second-largest lake in Türkiye, with a surface area of 1,665 km^2. It is also one of the largest hypersaline lakes in the world. During the last glacial pluvial period, Central Anatolia received significantly more rainfall than it does today, and Lake Tuz was a vast freshwater lake. Two high lake-level stages have been identified, dated approximately 20,000–17,000 years ago during the glacial maximum and again 13,000 years ago, during the Younger Dryas period. At the high lake stages, the water level rose by more than 15 m above the present lake level. This resulted in the lake area spreading widely to the southern edge of the basin, encompassing the cities of Ereğli and Karaman. In the Holocene, the lake level fell, and extensive lake terraces were formed in the southeast part of the basin. In the wetter periods, Lake Tuz was most likely inhabited by freshwater fishes. It can be speculated that these were the species surviving today within its former catchment area, including *Cobitis turcica*, *Oxynoemacheilus eregliensis*, *O. axylos*, *Seminemacheilus ekmekciae*, *Anatolichthys anatoliae*, *Garra kemali*, *Pseudophoxinus*, and *Gobio* species. The large Lake Tuz basin has dried out since and become so hypersaline that even exceptional floods cannot support freshwater fish's survival in its waters. Fishes survived this postglacial desiccation event in the streams and springs that once flowed to Lake Tuz. There, they became isolated, and some evolved into the currently recognised species. Furthermore, the system has been employed as a model for the evolution of new species in allopatry in populations that are small, isolated, and exhibit stream-specific characters. This is particularly evident in the genus *Gobio*. It is obvious that low-resolution molecular markers, such as COI or Cytochrome b, are not suitable for differentiating the recently separated species within the Lake Tuz gudgeon complex, which includes *G. gymnostethus*, *G. hettitorum*, *G. insuyanus*, and *G. microlepidotus*. **Further reading.** Kashima 2010 (Lake levels of Lake Tuz).

Gobio intermedius; Lake Eber basin, Türkiye; ~120 mm SL.

Gobio intermedius

Common name. Eber gudgeon.
Diagnosis. Distinguished from other species of *Gobio* in Anatolia by: ○ 42–49 total lateral-line scales / ○ chest completely scaled, scales extending forward to isthmus / ○ head length 27–29 % SL / ○ distance between anus and anal origin 5–7 % SL / ○ caudal-peduncle length 1.7–2.2 times in caudal-peduncle depth / ○ 8–10 scales between posterior extremity of pelvic base and anus / ○ interorbital distance approximately 1.4–1.6 times in eye diameter / ○ 8–9 scale rows between dorsal origin and lateral line / ○ caudal length 12–17 % SL anal length 20–24 % SL. Size up to 127 mm SL.
Distribution. Türkiye: Lakes Eber and Akşehir basins.
Habitat. Streams with moderate current on gravel to sandy bottoms. Also found in lakes.
Biology. No data.
Conservation status. EN; Lakes Akşehir and Eber dried up. Both occasionally have water, but it is highly polluted, and water level is less than 1 m in summer and almost anoxic. Fish are now largely confined to a few headwaters of Akarçay, which are heavily polluted, dammed and seasonally dry.
Further reading. Battalgil 1944 (description); Naseka et al. 2006 (diagnosis).

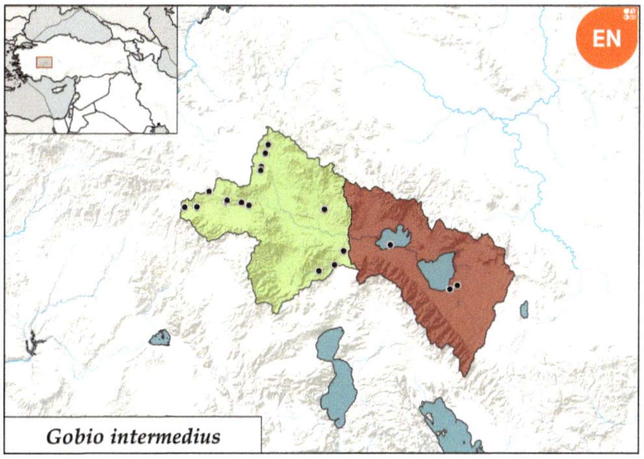

Gobio kizilirmakensis; Filyos drainage, Türkiye; ~110 mm SL.

Gobio kizilirmakensis

Common name. Filyos gudgeon.

Diagnosis. Distinguished from other species of *Gobio* in Anatolia by: ○ 40–43 total lateral-line scales / ○ chest completely scaled, scales extending forward to or almost to isthmus / ○ head length 26–29 % SL / ○ scales on belly smaller than pupil / ○ caudal-peduncle length 2.1–2.7 times its depth / ○ 5–6 scale rows between anus and anal origin / ○ 8–10 scale rows between posterior extremity of pelvic base and anus / ○ 8–9 mid-lateral blotches. Size up to 111 mm SL.

Distribution. Türkiye: Ulusu in Filyos drainage.

Habitat. Moderately fast-flowing streams and small rivers with sandy or gravelly bottoms.

Biology. No data.

Conservation status. EN; appears to be declining within its very small range. Some food factories around Çerkes heavily pollute the Ulusu.

Remarks. Despite its name, this species does not occur in Kızılırmak drainage. The Ulusu was considered a tributary of Kızılırmak, but it belongs to Filyos.

Further reading. Turan et al. 2016a (description); Yoğurtçuoğlu et al. 2020a (distribution).

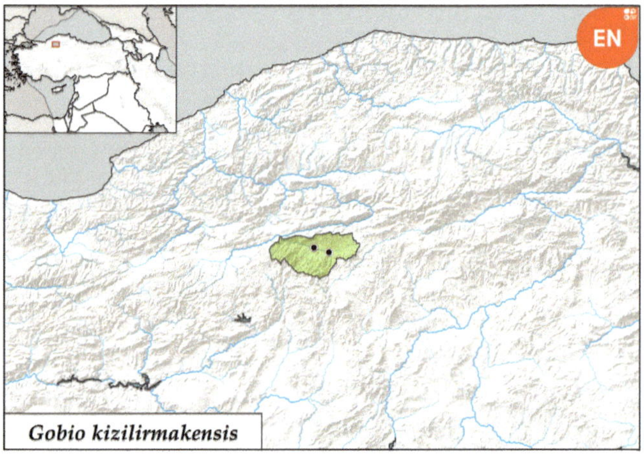

Gobio kovatschevi; Bakacak, Biga Peninsula, Türkiye; 100 mm SL.

Gobio kovatschevi

Common name. Thracian gudgeon.

Diagnosis. Distinguished from other species of *Gobio* in Anatolia by: ○ 39–44 total lateral-line scales / ○ chest naked, incompletely or completely scaled / ○ head length 28–30 % SL / ○ 4–5 scales between posterior extremity of pelvic base and anus / ○ interorbital distance 1.4–1.6 times in eye diameter / ○ 5–6 scale rows between dorsal origin and lateral line / ○ pelvic reaching behind anus in female / ○ head profile behind nostrils straight or slightly convex / ○ snout pointed, not or slightly concave anterior to nostril. Size up to 102 mm SL.

Distribution. Türkiye and Bulgaria: Black Sea basin from Provadiskaya (near Varna, Bulgaria) south to İstanbul and Biga Peninsula south to Tuzla. Also, in Büyük Melen drainage in Central Black Sea basin.

Habitat. Upper and middle reaches of a small, slow-flowing stream with a sandy or gravelly bottom.

Biology. No data.

Conservation status. LC.

Remarks. In Europe, *G. kovatschevi* appears to have had some contact with *Gobio* from the Danube, as two different types of mtDNA are found in this species, indicating past introgressive hybridisation.

Further reading. Chichkoff 1937 (description); Kottelat & Freyhof 2007 (diagnosis).

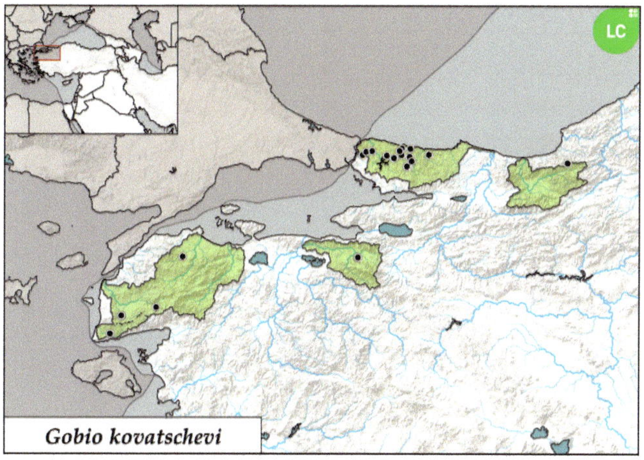

Gobio maeandricus; Işıklı spring, Türkiye; 130 mm SL.

Gobio maeandricus

Common name. Işıklı gudgeon.

Diagnosis. Distinguished from other species of *Gobio* in Anatolia by: ● 53–56 total lateral-line scales / ○ chest completely scaled, scales extending forward to, or almost to isthmus / ○ head length 27–29 % SL / ○ distance between anus and anal origin 7–8 % SL / ○ caudal-peduncle length 1.4–1.7 times in caudal-peduncle depth / ○ 10–12 scales between posterior extremity of pelvic base and anus / ○ 9–10 scale rows between dorsal origin and lateral line. Size up to 138 mm SL.

Distribution. Türkiye: Lake Işıklı basin in upper Büyük Menderes drainage and Karadirek near Sandıklı, an isolated basin that flows through underground to Lake Işıklı.

Habitat. Slow to moderate-flowing streams with sandy and gravelly bottoms, often with very dense aquatic vegetation.

Biology. No data.

Conservation status. EN; appears to be declining within its very small range.

Remarks. Extirpated *Gobio* populations from Lake Burdur basin may have belonged to this species.

Further reading. Naseka et al. 2006 (description); Güçlü et al. 2013 (distribution).

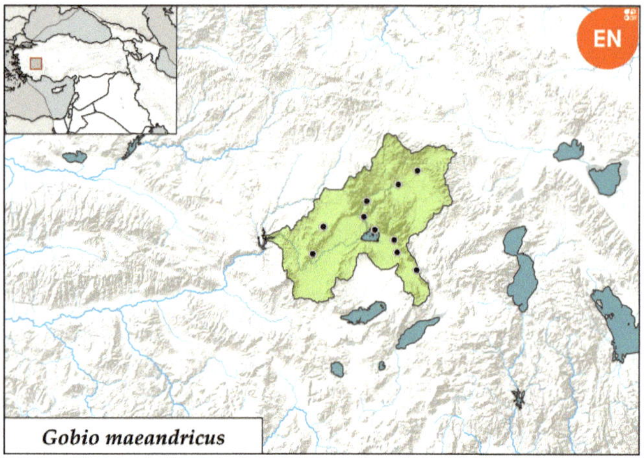

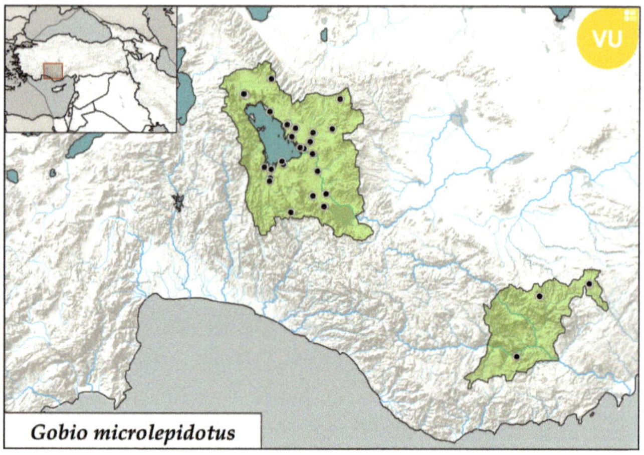

Pamukkale: Mass tourism is a significant feature of Türkiye's tourism industry, contributing heavily to its economy.

Gobio microlepidotus; Lake Beyşehir basin, Türkiye; 150 mm SL.

Gobio microlepidotus
Common name. Beyşehir gudgeon.
Diagnosis. Distinguished from other species of *Gobio* in Anatolia by: ○ 43–48 total lateral-line scales / ○ scales on chest extending forward to level of pectoral origin or to half distance of pectoral origin and isthmus / ○ head length 24–28 % SL / ○ 9–11 scales between posterior extremity of pelvic base and anus / ○ 8–10 scale rows between dorsal origin and lateral line / ○ scales on pectoral base often deeply embedded in skin. Size up to 171 mm SL.
Distribution. Türkiye: Lake Beyşehir basin and upper Göksu and Limon drainages in Mediterranean basin.
Habitat. Slow to moderate-flowing streams with sandy and gravelly bottoms, often with very dense aquatic vegetation.
Biology. No data.
Conservation status. VU; appears to be in decline within its small range.
Remarks. *Gobio battalgilae* is a synonym.
Further reading. Battalgil 1942 (description); Naseka et al. 2006 (description as *G. battalgilae*); Turan et al. 2018b (identification); Turan & Bayçelebi 2019 (Mediterranean records).

Gobio nigrescens; Hari drainage, Iran; 95 mm SL.

Gobio nigrescens
Common name. Hari gudgeon.
Diagnosis. Distinguished from *G. caucasicus*, the geographically most adjacent species in Caspian basin by: ○ pectoral reaching 3–5 scales in front of pelvic origin in mature male / ○ no scales on chest and between pectorals. Size up to 95 mm SL.
Distribution. Hari drainage in Afghanistan, Iran, and Turkmenistan. Amu Darya and Zeravshan in Central Asia (recorded from Uzbekistan). Also, in Kashaf (Iran).
Habitat. Slow to moderate-flowing streams and rivers with sandy or gravelly bottoms.
Biology. No data.
Conservation status. LC; extirpated from Kashaf drainage.

Remarks. This species is not related to other West Asian gudgeons but belongs to a group of Central Asian and Siberian species. It is sometimes identified as *G. lepidolaemus* from the adjacent Aral basin, from which it differs in chest without scale (vs. covered with scales), more scale rows on the caudal peduncle (½3/1/3½ vs. ½2/1/2½) and 16 circumpeduncular scales (vs. 12–14). *Gobio* populations found in Morghab and rivers of Kopek Dag in Turkmenistan are awaiting identification and may belong to this species.
Further reading. Mousavi-Sabet et al. 2016a (description); Sheraliev & Peng 2021 (distribution in Uzbekistan).

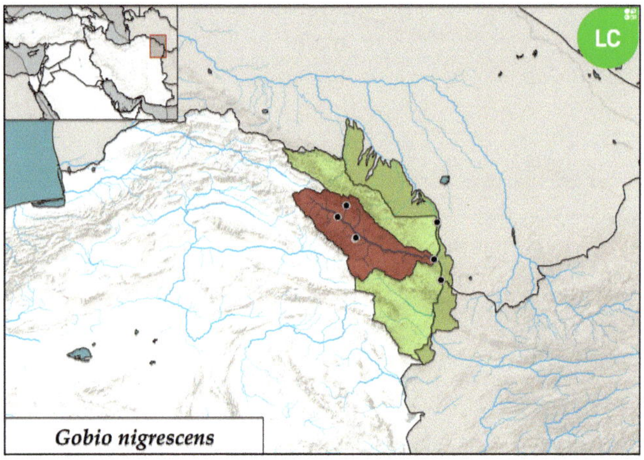

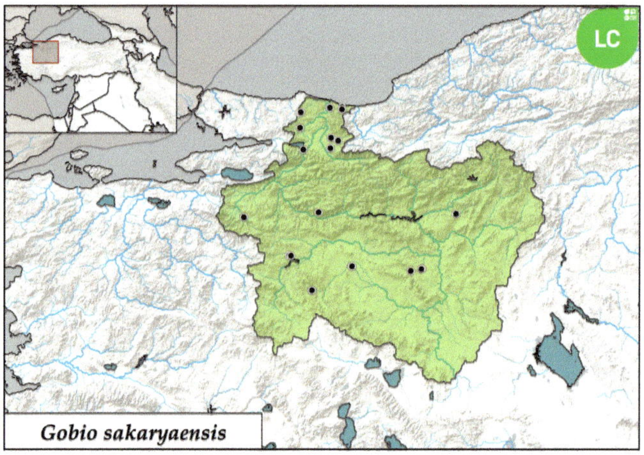

Gobio sakaryaensis; Sakarya drainage, Türkiye; ~100 mm SL.

Gobio sakaryaensis
Common name. Sakarya gudgeon.
Diagnosis. Distinguished from other species of *Gobio* in Anatolia by: ○ 39–42 total lateral-line scales / ○ chest completely scaled, scales extending forward to isthmus / ○ head length 27–30 % SL / ○ pectoral reaching 3–4 scales in front of pelvic origin in mature male / ○ 6–8 scales between posterior extremity of pelvic base and anus / ○ interorbital distance 1.4–1.6 times in eye diameter / ○ 6–7 scale rows between dorsal origin and lateral line / ○ pelvic reaching behind anus in female / ○ head profile behind nostrils straight or slightly convex / ○ snout rounded, distinctly concave anterior to nostril. Size up to 120 mm SL.
Distribution. Türkiye: Sakarya drainage.
Habitat. Small to medium-sized lowland and foothill streams with moderately fast-flowing water, sand, and gravel bottoms, and pool-riffle microhabitats.
Biology. No data.
Conservation status. LC.
Further reading. Turan et al. 2012a (description).

Pseudorasbora

A genus of six small species found in East Asia. *Pseudorasbora parva* is one of the world's most successful invasive fish species. It was first introduced from China to Europe in the early 1960s as a by-product of Chinese carp aquaculture. Following the initial introduction, the establishment of *P. parva* populations in European water bodies happened quickly and reached Anatolia and the Caucasus. Consequently, in less than six decades, *P. parva* has become widespread in central and western Europe, Türkiye, and Central Asia. Genetic analysis of the European *P. parva* invasion indicates that two Chinese lineages from southern and northern China constitute the gene pool of all European and Anatolian populations. These two lineages exhibit a large hybrid zone in Central China, suggesting that fish from that area may have been originally introduced to Europe. The initial introduction of *P. parva* into the Iranian inland waters (first recorded in 1991) was distinct from other Eurasian countries, as it originated from Japan. This is likely associated with commercial links between Japan and Iran in the 1980s when Iran was politically isolated from Europe. Since then, the Japanese lineage of *P. parva* has also spread to eastern Türkiye and has found contacts with the Chinese *P. parva* in the border area of Iran and Azerbaijan. **Further reading.** Hardouin et al. 2018 (phylogeography); Baltazar-Soares et al. 2020 (Dispersal in Europe); Ganjali et al. 2020 (invasion of Iran).

Pseudorasbora parva

Common name. Stone moroko.

Diagnosis. Distinguished from other species of Gobionidae in West Asia by: ● mouth small, superior / ● barbels absent / ○ dark-grey midlateral stripe outside spawning time / ○ 33–38 total lateral-line scales / ○ 7½ branched dorsal rays / ○ no dorsal or ventral keels. Size up to 95 mm SL.

Distribution. Very widespread in West Asia except Arabian Peninsula. In Anatolia mostly in western and central parts but increasing in Eastern Anatolia. Still rare in Iraq. Iran, including Hari drainage and Sistan and Mashkid basins. Also in Central Asia, Afghanistan, and Pakistan. Native to Japan and Amur to Zhujiang (Pearl River) drainages (Siberia, Korea, China).

Habitat. Ubiquitous, in a large variety of habitats. Most common in well-vegetated small channels, ponds, and small lakes. Usually spawns in still or very slow-flowing water.

Biology. Lives up to 3, rarely 5 years. First spawns at 1 year if larger than 35 mm SL. Male larger than female, with deeper body and darker colour. Male with bluish-grey breeding colour and a few very large nuptial tubercles on snout. Spawns March–June until October in Lake Eğirdir. Female spawns usually 3–4 times in a season. Male clear surface of a spawning site on stones or plants. Eggs are attached to substrate and guarded by male until larvae hatch. Feeds on various small crustaceans, insects, and plant material.

Conservation status. Non-native; introduced as a weed with stocked carp. Usually not abundant in fast-running waters. Proliferates locally in ponds and other artificial habitats, quickly becoming numerically dominant and a serious food competitor for native species. Also associated as a vector of *Sphaerothecum destruens*, a generalist pathogen.

Remarks. First introduced in Romania in 1961 with fry of *Ctenopharyngodon idella* from middle Changjiang [Yangtze], and from Amur into Russia and Ukraine, from where it colonised most of West Asia, North Africa, and Europe, either by active invasion, as stocked or released bait, or accidentally mixed with fry of other species.

Further reading. Bănărescu 1999a (biology); Britton et al. 2007 (effects); Hardouin et al. 2018 (phylogeography); Ganjali et al. 2020 (Iran); Küçük et al. 2024b (biology).

Pseudorasbora parva; Büyük Menderes drainage, Türkiye; ~80 mm SL.

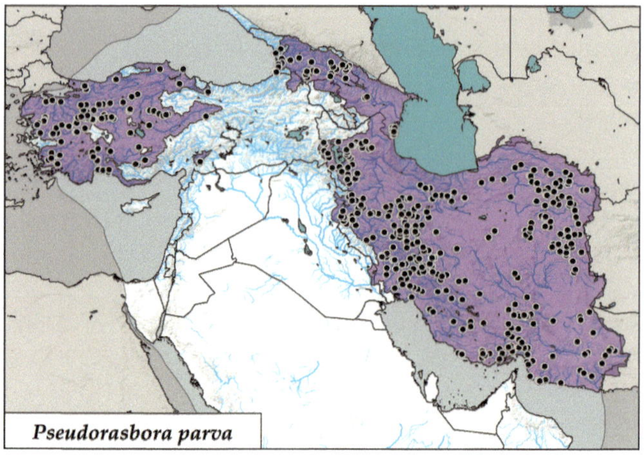

Nuptial tubercles on the head of a male *Pseudorasbora parva*.

Romanogobio macropterus; Kura drainage, Türkiye; ~100 mm SL. © M. Özuluğ.

Romanogobio

The genus comprises 17 species native to the West Palearctic, with three species occurring in East Asia. Both species in West Asia and numerous European species can be distinguished from *Gobio* species by the presence of epithelial crests on the scales of the dorsal half of the body (vs. absence). In contrast, *Romanogobio* species exhibit a slenderer body than the *Gobio* species. *Romanogobio persus* and *R. macropterus* have long been treated as conspecifics, but morphological and molecular characters distinguish them well. They are related to European species found in adjacent rivers in the north, including *R. ciscaucasicus* from Dagestan, *R. parvus* from the Kuban, *R. albipinnatus* from the Volga, and *R. tanaiticus* from the Don. All but one species (*R. benacensis*) are a highly rheophilic inhabitant of large to very large rivers with fast-flowing waters. They are nocturnal bottom dwellers, often restricted to the rivers' middle and deeper parts.

Romanogobio macropterus

Common name. Caucasian stone gudgeon.

Diagnosis. Distinguished from *R. persus* by: • 41–45, usually 42–43 total lateral-line scales / • 38–42, usually 40–41 total vertebrae / • supraorbital and infraorbital canals usually connected. Size up to 110 mm SL.

Distribution. Kura and Aras drainages.

Habitat. Fast-flowing stretches of rivers and streams with gravel and rocky substrate.

Biology. Lives up to 3 years and usually matures at 2 years. Spawns late April–June at water temperatures between 12–18°C. Fractional spawner, female usually spawn three times in a season. Feeds on aquatic invertebrates.

Conservation status. VU; has declined sharply due to hydropower development and pollution.

Further reading. Naseka et al. 1999 (morphology, biology); Kaya et al. 2020a (distribution).

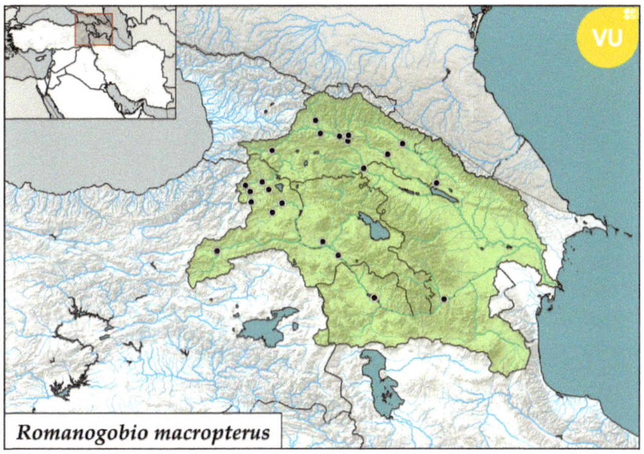

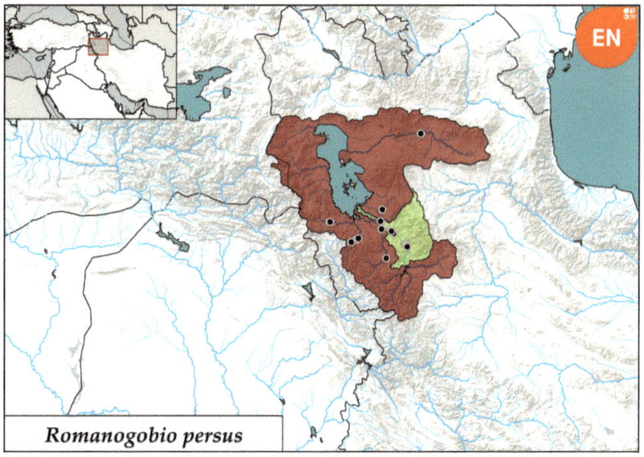

Romanogobio persus; Lake Urmia basin, Iran; 73 mm SL.

Romanogobio persus

Common name. Persian gudgeon.

Diagnosis. Distinguished from *R. macropterus* by: • 40–42, usually 40–41 total lateral-line scales / • 37–40, usually 38–39 total vertebrae / • usually no connection between supraorbital and infraorbital canals. Size up to 73 mm SL.

Distribution. Iran: Lake Urmia basin, in southern and eastern tributaries.

Habitat. Fast-flowing sections of streams and rivers with gravel or rock bottoms.

Biology. Feeds on aquatic invertebrates.

Conservation status. EN; appears to decline within its very small range. Extirpated from Mahabad, Gedar and Talkheh drainages and seems to occur now only in middle Zarrineh.

Further reading. Naseka et al. 1999 (morphology, distribution).

No water was visible when driving around Lake Urmia in autumn 2018.

Copernicus Service information 2024.

Lake Urmia. Lake Urmia (or Orumiyeh) is situated in northwestern Iran at an altitude of 1,250 m. With a total surface area ranging from 4,750 to 6,100 km^2 and a maximum depth of 16 m, it was once the largest lake in West Asia and one of the largest permanent hypersaline lakes in the world. Lake Urmia is a tectonic lake formed by the folding caused by the collision of the Arabian plate with the Eurasian plate. It may be only 0.5 million years old, and its biogeographic affinities are with the Aras adjacent to the north. During periods of higher humidity in the past, Urmia was a freshwater lake of considerable size, allowing fish to migrate between tributaries. As the lake dried out and became a 'salt lake', it was once one of the largest natural habitats for the crustacean *Artemia urmiana*, which was of considerable economic importance. However, in the latter decades of the 20th century and the early years of the 21st century, the environmental conditions around Lake Urmia deteriorated rapidly. In 2013, *A. urmiana* was declared extinct in the lake due to the extremely high salinity levels. The lake's salinity has risen to more than 300 g/L due to a combination of drought, the construction of dams, and increased groundwater pumping from the surrounding area. This has led to the desiccation of large areas of the lake. The inflowing streams and rivers have been blocked by dams, which have choked the water supply from the mountains that tower on the two sides of the lake. By the end of 2017, the lake had shrunk to 10 % of its former size (and 1/60 of the water volume in 1998). Plans for the recovery of the lake have been in place for several years, including the drainage of large amounts of water from the Aras to Lake Urmia. However, despite these efforts, the situation has stayed the same, and the lake has dried out. **Further reading.** National Geography Organization of Iran 2003 (Lake Urmia).

Family Leuciscidae

Minnows

A total of 137 species of Leuciscids have been identified in West Asia. These fishes are characterised by a short dorsal fin whose last unbranched ray is never serrated, the absence of barbels, pharyngeal teeth usually in one or two rows, and several osteological characters. They are found in Europe, North Africa, western and northern Asia, and North America but, are absent from tropical and southern Africa and tropical Asia. Leuciscidae is divided into two main groups, the Leuciscines and the Phoxinines. Both are present in West Asia. The Phoxinines are known only from the genus *Phoxinus* in West Asia. They are numerous and highly diverse in East Asia and especially in North America, where they dominate many fish faunas in terms of species and biomass. In contrast, leuciscines are most speciose in the Western Palaearctic, where most genera are endemic and only a few are distributed east of Central and East Asia. Only one genus of leuciscines is found in North America (*Notemigonus*).

The generic systematics of Leuciscidae have changed considerably in recent decades, as morphological characters poorly diagnose most genera. In many instances, the traditionally recognised genera have no congruence by morphological and molecular data. This is partly due to the earlier practise of classifying morphologically specialised species in their own, often monotypic genera (predatory species such as *Aspius*). Furthermore, some small-sized species exhibit characters that are partly interpreted as reductive. These have often been classified in the genera *Pseudophoxinus* and *Petroleuciscus* but have been transferred to other genera mostly based on molecular characters. Intergeneric hybridisation and mitochondrial DNA (mtDNA) introgression have led to the misclassification of some species based on mitochondrial molecular data alone. In such instances, we have adopted the most recent generic-level reorganisations, which have been informed by both morphological and molecular data. **Further reading.** Bogutskaya 1986, 1988a, b (osteology); Zardoya & Doadrio 1999, Rüber et al. 2007 (phylogeny).

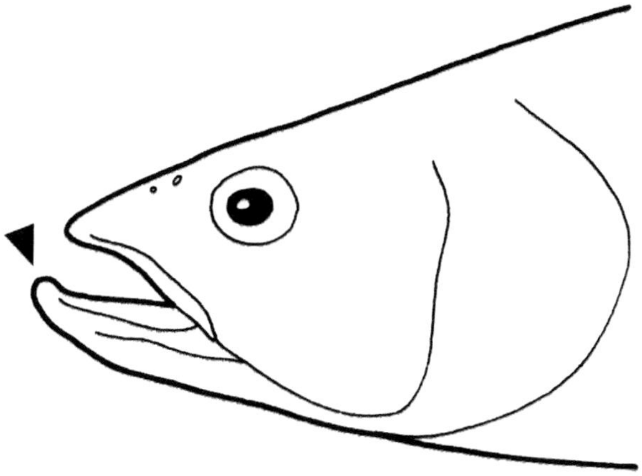

Figure 39: Symphysis of the lower jaw with a conspicuous knob (arrow) fitting in a notch in the upper jaw in *Leuciscus.*

Key to genera of Leuciscidae in West Asia

1a - Lateral line undulated, complete; belly with a sharp keel between throat and anal origin.
..................*Pelecus*

1b - Lateral line running straight, curved, complete, incomplete, or absent; belly without a sharp keel between throat and pelvic origin.
..................2

2a - Lower lip with a cornified sheath; mouth subterminal or inferior, straight or arched; mouth inferior with strongly projecting rostral cap (9–12½ branched anal rays).
..................*Chondrostoma*

2b - Lower lip without a cornified sheath; mouth superior, terminal or inferior, arched; if inferior, then 19–44½ branched anal rays.
..................3

3a - Unbranched part of pectoral rays 1–3 thicker and stiffer in male than in female; pectoral rays 2–3 slightly slanted upward, with a conspicuous downward angle at distal end of thickened part.
..................*Egirdira*

3b - Pectoral of male not modified, identical to that of female in shape, often longer in male than in female.
..................4

4a - Mouth large, cleft reaching beyond front margin of eye, superior; symphysis of lower jaw with a conspicuous knob fitting in a notch in upper jaw. (Fig. 39).
..................*Leuciscus*

4b - Mouth cleft not reaching vertical of eye, superior, terminal, or subterminal; symphysis of lower jaw without conspicuous knob; upper jaw without median notch.
..................5

https://doi.org/10.1515/9783111677811-017

5a - A midventral keel in front of anus, exposed (scaleless) at least immediately in front of anus, usually on at least half of distance between pelvic base and anus (keel weakly developed and not exposed in *Acanthobrama persidis*, some *Alburnus doriae,* and some *A. kurui*).
..................6
5b - No scaleless midventral keel in front of anus.
..................17

6a - 23–48½ branched anal rays.
..................7
6b - 7–23½ branched anal rays.
..................8

7a - 23–30½ branched anal rays.
..................*Abramis*
7b - 33–48½ branched anal rays.
..................*Ballerus*

8a - Lateral line incomplete, with 8–12 pored scales.
..................*Leucaspius*
8b - Lateral line complete.
..................9

9a - Lateral line strongly curved downward; mouth superior or terminal; size up 400 mm SL, most species less than 130 mm SL.
..................10
9b - Lateral line more or less along middle of flank; mouth subterminal to subinferior; size up to 700 mm SL.
..................12

10a - Last unbranched dorsal ray strongly ossified, often thickened.
..................*Mirogrex*
10b - Last unbranched dorsal ray as much ossified as other rays, not thickened.
..................11

11a - A small black spot on each side of each lateral line pore; usually a dark-grey midlateral stripe and several longitudinal rows of spots on flank; 6–10 gill rakers, slightly longer than distance between two rakers.
..................*Alburnoides*
11b - Body silvery, without dark-grey marks on flank, often with a faint midlateral stripe; 13–38 gill rakers, densely set, several times longer than distance between two rakers.
..................*Alburnus*

12a - 7–9½ branched anal rays.
..................13
12b - 10–23½ branched anal rays.
..................15

13a - 65–72 total lateral-line scales, body cylindrical, only slightly compressed.
..................*Leucalburnus*
13b - 35–52 total lateral-line scales; body clearly compressed.
..................14

14a - No brown stripe along flank. 35–43 total scales along lateral line; pharyngeal teeth in two rows.
..................*Acanthobrama* (*A. persidis*)
14b - A brown stripe along flank. 45–52 total scales along lateral line; pharyngeal teeth in one row.
..................*Turcichondrostoma*

15a - 10–19½ branched anal rays; last unbranched dorsal ray strongly ossified, often thickened.
..................*Acanthobrama*
15b - 19–23½ branched anal rays; last unbranched dorsal ray as much ossified as other rays, not thickened.
..................16

16a - 43–46+2–3 lateral-line scales; a scaleless dorsal groove behind nape (in adults).
..................*Blicca*
16b - 46–58+2–3 lateral-line scales; no scaleless dorsal groove in front of dorsal.
..................*Vimba*

17a - Flank with a row of dark, squarish to vertically elongated blotches, sometimes fused into an irregular midlateral stripe, followed by a separate dark-grey spot at caudal base.
..................*Phoxinus*
17b - Body uniform silvery or yellowish brown, sometimes with a broad, faint, midlateral stripe or with irregular dotted or spotted pattern, no dark-grey or brown, squarish to vertically elongated blotches, and no conspicuous dark-grey or brown spot at caudal base.
..................18

18a - Infraorbital canal not connected to preoperculo-mandibular sensory canal; lateral line incomplete in many species. (Fig. 40).
..................*Pseudophoxinus*

18b - Infraorbital canal connected to preoperculo-mandibular sensory canal; lateral line complete (except in *Ladigesocypris*).
..................19

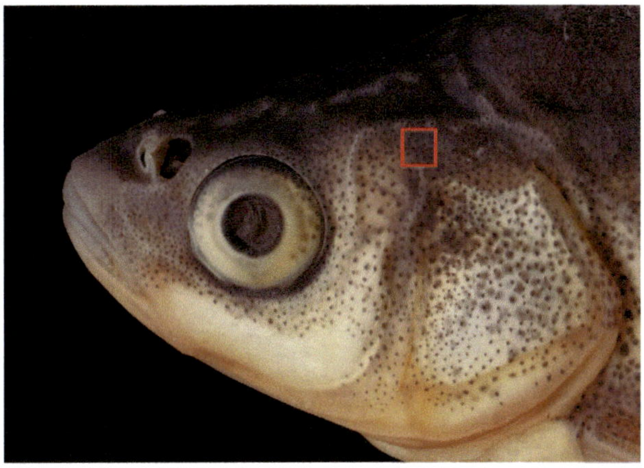

Figure 40: *Pseudophoxinus maeandricus*; red square shows disconnected infraorbital canal (left) and preoperculo-mandibular sensory canal (right).

19a - Lateral line incomplete.
..................*Ladigesocypris*
19b - Lateral line complete.
..................20

20a - Dorsal origin behind base of last pelvic ray; mouth superior or terminal.
..................21
20b - Dorsal origin above pelvic base; mouth terminal to inferior.
..................22

21a - Mouth superior; posterior margin of anal concave; keel between pelvic base and anus usually present.
..................*Scardinius*
21b - Mouth terminal; posterior margin of anal convex or almost straight; no keel between pelvic base and anus.
..................*Petroleuciscus*

22a - Posterior margin of anal concave; 4–5 simple dorsal rays (posterior 3 usually distinct, anterior 1–2 very short and often hidden in skin); pharyngeal teeth in one row.
..................*Rutilus*
22b - Posterior margin of anal convex or almost straight; 3 simple dorsal rays (posterior 2 usually distinct, anterior one very short and often hidden in skin); pharyngeal teeth in two rows.
..................*Squalius*

Leuciscid fishes are important elements in most waterbodies of West Asia, as in Lake Hazar, Türkiye.

Abramis brama; Odra, Germany; ~400 mm SL.

Abramis brama

Common name. Bream.

Diagnosis. Distinguished from species of *Ballerus*, *Blicca*, and *Vimba* by: • sub-inferior mouth, which can be extended as a tube / • 51–60 total lateral-line scales / ○ 23–30½ branched anal rays / ○ eye small, diameter about ⅔ of snout length in individuals larger than 100 mm SL / ○ pharyngeal teeth 5–5 / ○ base of paired fins hyaline or grey. Size up to 700 mm SL.

Distribution. West Asia: Marmara, Black, and Caspian Sea basins. Locally introduced in Tigris and Lake Urmia basins (both Iran). Native to most European drainages from Adour (France) to Pechora (White Sea basin); Aegean basin, in Lake Volvi and Struma and Maritza drainages east to Aral basin. Introduced in Lake Baikal and upper Ob and Yenisei drainages. Naturally absent from Iberian Peninsula, Adriatic basin, Italy, Scotland, Scandinavia north of Bergen (Norway), and 67°N (Finland). Locally introduced in Ireland, Spain, Neretva (Croatia), Italy, and likely elsewhere.

Habitat. Lakes, reservoirs, and large- to medium-sized rivers. Most common in backwaters, lower reaches of slow-flowing rivers, brackish estuaries, and warm, shallow lakes. Semi-anadromous individuals enter fresher parts of seas to feed. Usually spawns in densely vegetated backwaters, floodplains, or lake shores. Sometimes far from shore. Almost any surface can be used for spawning.

Biology. Gregarious. Forms large schools in winter, often with other fish. Lives about 10–20 years. Spawns first time at 3–4 years. Some females do not spawn every year. Spawns in May–June at temperatures above 15°C. In many populations, spawning migration begins in autumn (especially in semi-anadromous individuals), slows during winter, and continues in spring. Migrate far upstream

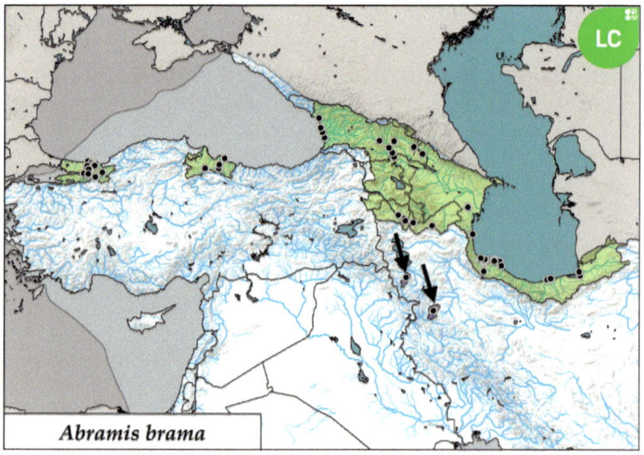

(100 km in Dniepr) to spawn. Males with nuptial tubercles on head and body. Males often defend spawning grounds along banks. Females spawn once a year for a few days (Rhine) to 1–3 times a year, 7–14 days apart (Ilmensee). Eggs are sticky and increase in size with age. Larvae and juveniles inhabit still waters and feed on plankton. Juvenile survival is high in backwaters and low in main channel of large rivers. Growth is faster in main rivers than in backwaters. Juveniles of 1–2 years move from backwaters to river for feeding. If juveniles do not have opportunity to leave backwaters, they adapt by having slower growth and reaching maturity at a smaller size (stunted populations). In lower reaches of large rivers, juveniles drift to brackish estuaries to forage when water level in flooded areas drops. Juveniles foraging in brackish water enter lower reaches of rivers to overwinter in freshwater. Feeds on benthic invertebrates excavated from fine bottom sediments, often on molluscs. May switch to particulate or even filter feeding when zooplankton abundance is high. Juveniles feed mainly on zooplankton. Often forms fertile hybrids with *Rutilus* species.

Conservation status. LC.

Further reading. Berg 1949b (biology); Molls 1999 (biology); Brylinska & Boron 2004 (biology).

Acanthobrama

Acanthobrama comprises medium-sized fish endemic to West Asia. They are distinguished by a long anal fin, an often-thickened last unbranched dorsal ray, and a usually scaleless ventral keel between the anus and the pelvic base. All species prefer lacustrine habitats as well as larger rivers and streams. Several species are of major conservation concern. *Acanthobrama centisquama* and *A. tricolor* have not been found for decades. *Acanthobrama telavivensis* was saved at the last moment and was even extinct in the wild for some years. *Acanthobrama* breams form one monophyletic lineage together with breams of the genera *Abramis*, *Ballerus*, *Blicca*, *Mirogrex*, and *Vimba*, and there had been several discussions to place these all into one genus, *Abramis*. The Aegean *Acanthobrama mirabilis* is a synonym of *Vimba vimba*, the Caucasian genus *Acanthalburnus* is a synonym of *Acanthobrama*, the southern Levantine genus *Mirogrex* has been revalidated, and the Iranian *Leuciscus persidis* has been placed in *Acanthobrama*. The phylogenetic position of *A. tricolor* has yet to be studied. *Capoetobrama kuschakewitschi*, a bream-like fish from the Aral basin, represents a distinct genus of breams. **Further reading.** Freyhof & Özuluğ 2014 (species); Perea et al. 2010 (phylogeny); Sheraliev & Peng 2021 (*Capoetobrama*).

Key to species of *Acanthobrama*

1a - Pharyngeal teeth in two rows.
..................2
1b - Pharyngeal teeth in one row.
..................4

2a - 7–9½ branched anal rays; 35–43 total lateral-line scales.
..................*A. persidis*
2b - 10–19½ branched anal rays; 50–87 total lateral-line scales.
..................3

3a - 12–19½, usually 15–17½, branched anal rays; 60–87 total lateral-line scales; 43–46, usually 44–45 total vertebrae.
..................*A. punctulata*
3b - 10–13½ branched anal rays; 50–68 total lateral-line scales; 42–43, usually 42, total vertebrae.
..................*A. urmiana*

4a - Last unbranched dorsal ray stiff and very strong, as long as or longer than body depth at dorsal origin.
..................5
4b - Last unbranched dorsal ray stiff or flexible, always shorter than body depth at dorsal origin.
..................6

5a - 82–100 total lateral-line scales.
..................*A. centisquama*
5b - 54–58 total lateral-line scales.
..................*A. hadiyahensis*

6a - Usually 8½ branched dorsal rays.
..................7

6b - Usually 7½ branched dorsal rays.
..................9

7a - 76–89 total lateral-line scales.
..................*A. thisbeae*
7b - 51–64, usually 54–62, total lateral-line scales.
..................8

8a - Snout length 27–31 % HL; tip of upper lip usually below middle of eye; usually 59–73 total lateral-line scales.
..................*A. marmid*
8b - Snout length 25–27 % HL; tip of upper lip at middle of eye or above; 51–64, usually 54–62 total lateral-line scales.
..................*A. orontis*

9a - Last unbranched dorsal ray soft and entirely segmented; 7–11½ branched anal rays; ventral keel poorly developed, covered with scales in its anterior part or fully covered with scales.
..................*A. tricolor*
9b - Last unbranched dorsal ray partly stiffened and unsegmented; 10–14½ branched anal rays; ventral keel not covered by scales along entire length.
..................10

10a - 59–69 total lateral-line scales; rigid part of last unbranched dorsal ray about 70 % of ray.
..................*A. lissneri*
10b - 51–60 total lateral-line scales; rigid part of last unbranched dorsal ray about 50–65 % of ray.
..................*A. telavivensis*

Acanthobrama centisquama; Lake Amik, Türkiye; 147 mm SL.

Acanthobrama centisquama
Common name. Long-spine bream.
Diagnosis. Distinguished from other species of *Acanthobrama* by: ○ last unbranched dorsal ray considerably thickened and very strong / ○ last unbranched dorsal ray is 85–104 % of body depth at dorsal origin, 24–28 % SL / ○ 18–21½ branched anal rays / ○ pharyngeal teeth in one row / ○ posterior margin of flank scales crenulated / ○ 82–100 total lateral-line scales / ○ 19 gill rakers / ○ 8½ branched dorsal rays. Size up to 200 mm SL.
Distribution. Lake Amik in Türkiye and Lake al-Ghab in Syria, both in lower Orontes drainage.
Habitat. Lacustrine.
Biology. No data.
Conservation status. Extinct; in modern times, only recorded from Lake Amik (Türkiye). This lake was drained in stages between the 1940s and the 1970s. Scales and pharyngeal teeth are also known from Lake al-Gab in Syrian Orontes drainage, which was drained much earlier. There is no evidence that it has survived in Lake Gölbaşı (Kırıkhan), the only remaining relic of Lake Amik. Several detailed surveys have been undertaken to find the species.
Further reading. Coad 1984 (description); Freyhof & Özuluğ 2014 (comparison).

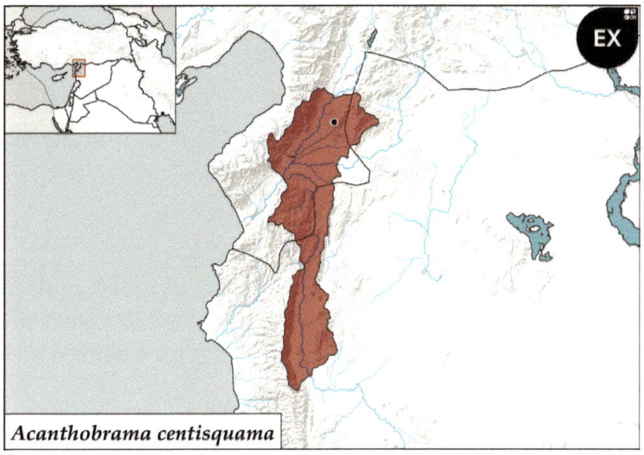

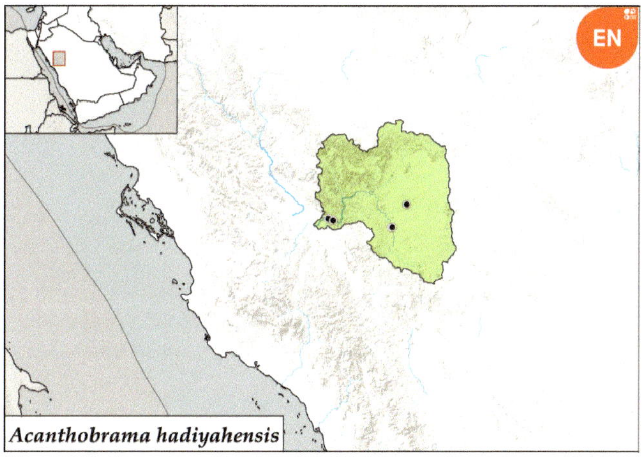

Acanthobrama hadiyahensis; Qusaiba'a reservoir, Saudi Arabia; 80 mm SL. © N.A. Hamidan.

Acanthobrama hadiyahensis

Common name. Arabian bream.

Diagnosis. Distinguished from superficially similar species of *Carasobarbus* and *Cyprinion* in Arabian Peninsula by: ○ last unbranched dorsal ray without serrae, thickened and very strong / ● no barbels / ● pharyngeal teeth in one row / ● 14–17½ branched anal rays / ○ 54–58 total lateral-line scales / ○ 17–19 gill rakers. Size up to 80 mm SL.

Distribution. Saudi Arabia: springs of Wadi Hadiyah in Khaibar area and Qusaiba'a reservoir in northern Hijaz.

Habitat. Springs and river, also reported from a reservoir.

Biology. No data.

Conservation status. EN; appears to be declining within its very small range. Known from three localities, only one been confirmed in last 30 years. Threatened by water abstraction and construction of dams and reservoirs.

Remarks. *Acanthobrama hadiyahensis* is closely related to *A. marmid*. The two species may only have diverged in post-glacial times when a river flowing from the mountains of western Saudi Arabia to the southernmost Shatt al-Arab dried up.

Further reading. Coad et al. 1983 (description); Coad 2010b (conservation); Hamidan & Aloufi 2013 (new record).

Acanthobrama lissneri; Nahr al-Yarmuk, Syria; ~70 mm SL.

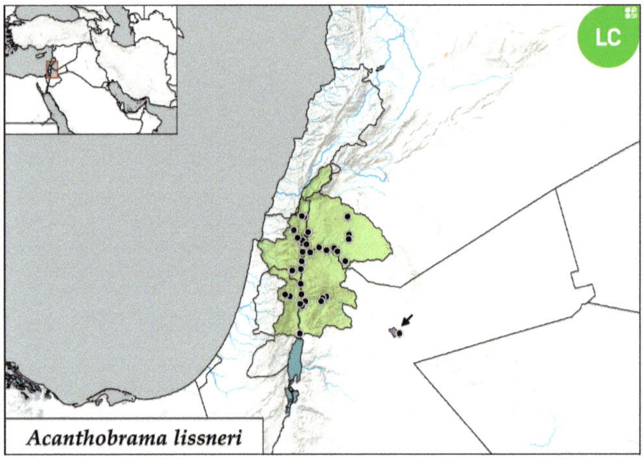

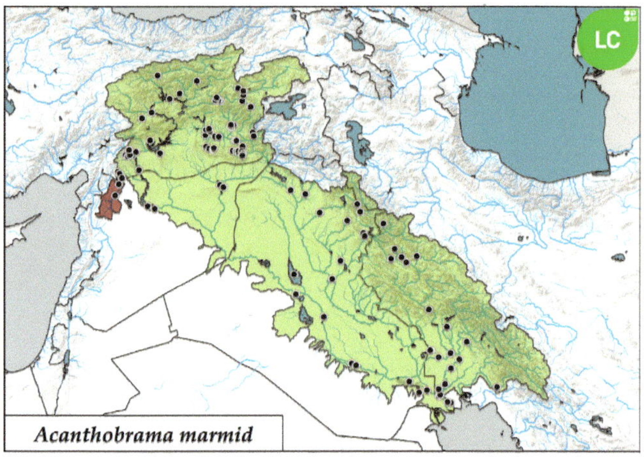

Acanthobrama lissneri

Common name. Jordan bream.

Diagnosis. Distinguished from other species of *Acanthobrama* by: ○ pharyngeal teeth in one row / ○ 59–69 total lateral-line scales / ○ rigid part of last unbranched dorsal ray about 70 % of ray / ○ 10–13½ branched anal rays / ○ ventral keel with free skinfold not covered by scales along full length / ○ 12–14 gill rakers / ○ 7½, rarely 8½, branched dorsal rays. Size up to 113 mm SL.

Distribution. Jordan drainage and Kishon, a coastal river in Israel. Introduced in Azraq oasis in Jordan.

Habitat. Streams, rivers, reservoirs, and lakes with stagnant or slow-flowing water. Feeds on zooplankton, anflug and benthic invertebrates.

Biology. Spawns January–April.

Conservation status. LC.

Further reading. Tortonese 1952 (description); Goren et al. 1973 (review, comparison); Krupp & Schneider 1989 (distribution).

Acanthobrama marmid; Tigris, Türkiye; ~110 mm SL.

Acanthobrama marmid; Euphrates, Türkiye; ~160 mm SL.

Acanthobrama marmid

Common name. Mesopotamian bream.

Diagnosis. Distinguished from other species of *Acanthobrama* by: ○ pharyngeal teeth in one row / ○ usually 59–73 total lateral-line scales / ○ tip of upper lip usually below level of middle of eye (but see fish on second picture) / ○ 13–19½, usually 14–17½, branched anal rays / ○ 9–14 gill rakers / ○ 8½, rarely 7½ or 9½, branched dorsal rays / ○ snout length 27–31 % HL. Size up to 220 mm SL.

Distribution. Qweiq, Euphrates, Tigris, and Karun drainages.

Habitat. Ubiquitous species inhabiting all types of lowland standing or slow-flowing waters, such as larger streams, rivers, springs, marshes, reservoirs, lakes, and even moderately polluted waters. Usually absent from fast-flowing and cold mountain streams.

Biology. Matures at 2–3 years and about 90 mm SL, lives 9 years, females grow faster than males. Spawns May–August. Suspected to be a fractional spawner. Feeds on aquatic invertebrates, algae, and detritus. Some populations migrate from large reservoirs to tributaries for spawning.

Conservation status. LC; in Qweiq drainage, currently restricted to Türkiye. Orontes records are misidentified *A. orontis*.

Further reading. Goren et al. 1973 (morphology, synonymy); Dağlı & Erdemli 2009 (description); Coad 2010a (biology); Coban & Yüksel 2013 (size, growth); Uçkun & Gökçe 2014 (age, growth, reproduction); Kaya et al. 2016 (distribution); Coad 2021b (biology, distribution).

Acanthobrama orontis; lower Göksu, Türkiye; ~130 mm SL

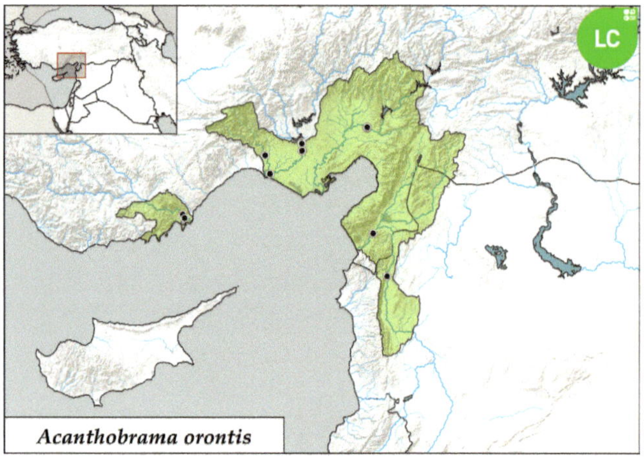

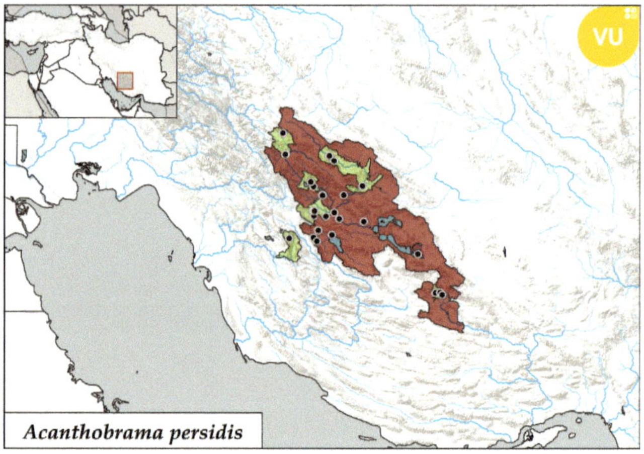

Acanthobrama orontis

Common name. Cilician bream.

Diagnosis. Distinguished from other species of *Acanthobrama* by: ○ pharyngeal teeth in one row / ○ 51–64, usually 54–62, total lateral-line scales / ○ tip of upper lip at or above level of middle of eye / ○ 8½ branched dorsal rays / ○ snout length 25–27 % HL. Size up to 156 mm SL.

Distribution. Türkiye and Syria: Göksu east to Orontes. Absent between Göksu and Seyhan.

Habitat. Lowland rivers and reservoirs. Also common in irrigation channels and artificial waters.

Biology. No data.

Conservation status. LC; possibly extirpated from Orontes, as no individuals have been seen for many years.

Further reading. Berg 1949b (description); Freyhof & Özuluğ 2014 (discussion); Bayçelebi 2020 (distribution).

Acanthobrama persidis

Common name. Persian bream.

Diagnosis. Distinguished from other species of *Acanthobrama* by: ● 7–9½, usually 8½, branched anal rays / ○ ventral keel without free skin fold in front of anus / ○ pharyngeal teeth in two rows / ○ last unbranched dorsal ray soft, not stiffened / ○ 35–43 total lateral-line scales / ○ 10–14 gill rakers / ○ 7–9½ branched anal rays / ○ 7½, rarely 6½, branched dorsal rays. Size up to 113 mm SL.

Distribution. Iran: Upper Kor basin (Khosroshin, Ghadamgah, Malosjan, and Beyza springs), Helleh, Mond, and rarely in Kol drainages (Golabi spring).

Habitat. Slow-flowing or stagnant waters. Often found in springs, but also known from large rivers, streams, irrigation ditches, and qanats, usually with medium to slow currents.

Biology. Seems to spawn in early spring. Feeds on plants and invertebrates.

Conservation status. VU; has lost many habitats due to the desiccation of springs and streams.

Further reading. Coad 1981b (description); Perea et al. 2010 (molecular phylogeny); Freyhof & Özuluğ 2014 (generic position); Teimori et al. 2015 (distribution); Coad 2021b (biology, distribution).

Acanthobrama persidis; Pirbarnoo spring, Iran; ~100 mm SL.

Acanthobrama punctulata; Kura drainage, Türkiye; ~135 mm SL.

Acanthobrama punctulata
Common name. Caucasian bream.
Diagnosis. Distinguished from other species of *Acanthobrama* by: ○ pharyngeal teeth in two rows / ○ 60–87 total lateral-line scales / ○ 12–19½, usually 15–17½, branched anal rays / ○ 12–17 gill rakers / ○ usually 44–45 total vertebrae / ○ 7–9½, usually 8½, branched dorsal rays. Size up to 230 mm SL.
Distribution. Sefid (Iran) and coastal rivers north to Kura.
Habitat. All types of standing or slow to fast-flowing waters, such as larger streams, rivers, reservoirs, and lakes, even in moderately polluted waters.
Biology. Lives up to 7 years. Matures at 1–2 years (males) and 2–3 years (females) and 80–120 mm TL. Spawns May–August in shallow, gravel-bottomed streams at a water temperature of 14–27°C.
Conservation status. LC.
Remarks. Often placed in *Acanthalburnus*, but molecular data showed that this genus should be synonymised with *Acanthobrama*. Still known as *Acanthobrama microlepis*, a name not available. *Abramis microlepis* is a junior secondary homonym of *Alburnus microlepis* Heckel, 1843, when placed in *Alburnus* by Kamensky (1901). It was replaced by *Alburnus punctulatus*, a former junior synonym, by Berg (1916) and is permanently invalid because junior homonyms and substitute names are still treated as synonyms.

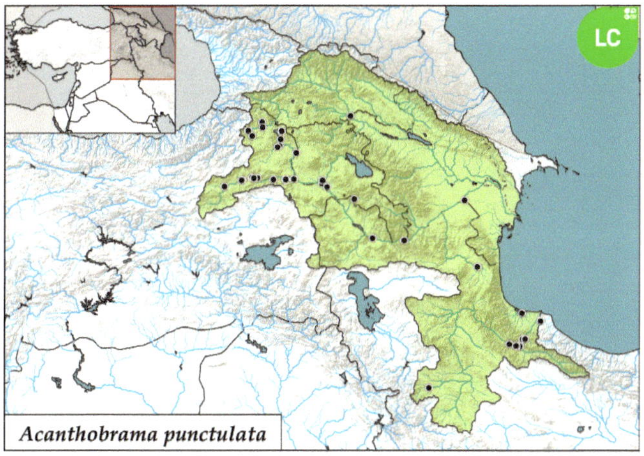

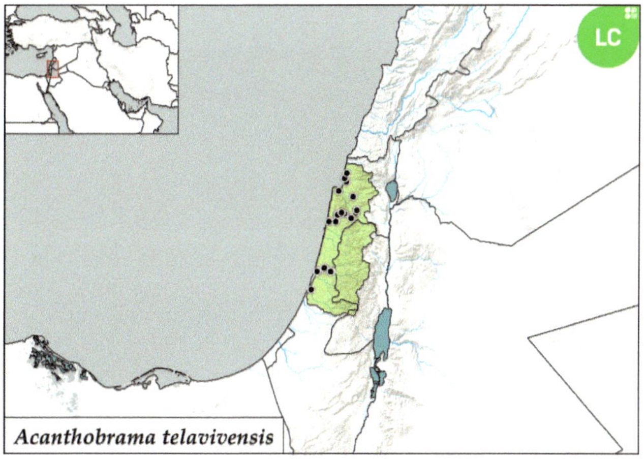

Acanthobrama telavivensis; Israel; ~75 mm SL. © M. Ford.

Further reading. Kamensky 1901; Berg 1916 (generic placement); Abdurakhmanov 1962, Elanidze 1983 (review, biology); Türkmen et al. 2001 (growth, reproduction); Perea et al. 2010 (phylogeny); Küçük et al. 2014 (generic position); Coad 2021b (biology, distribution).

Acanthobrama telavivensis

Common name. Yarkon bream.

Diagnosis. Distinguished from other species of *Acanthobrama* by: ○ pharyngeal teeth in one row / ○ 51–60 total lateral-line scales / ○ rigid part of last unbranched dorsal ray about 50–65 % of ray / ○ 12–17 gill rakers / ○ 7½, rarely 8½, branched dorsal rays / ○ 11–14½ branched anal rays / ○ ventral keel with a free skinfold not covered by scales along entire length. Size up to 95 mm SL.

Distribution. Israel: coastal streams Soreq, Yarkon, Tanninim, Daliya, and Na'aman, potentially also in Alexander, Poleg, and Hadera.

Habitat. Streams and small lakes. Spawns on gravel or rocky substrates and submerged vegetation along banks in flowing and stagnant water.

Biology. Mature after 1 year if larger than 40 mm SL, usually live 4–5 years. Spawns February–April at water temperatures between 14 and 20°C. Individual female spawn twice during season. Young hatch after 7–10 days at 13–17°C, after 5–8 days at 18–23°C. Feeds on aquatic invertebrates and detritus.

Conservation status. LC; declined since 1950s due to pollution and water abstraction. A few days before last two streams dried up in 1999, about 150 fish were captured just before the last water disappeared. These individuals were bred in captivity, and reintroduction began in 2002. Following habitat modification, reintroductions have been successful since 2006, and several new populations have been established.

Further reading. Goren et al. 1973 (description); Elron et al. 2006 (biology); Goren 2009 (extinction, reintroduction).

Acanthobrama thisbeae; lower Ceyhan drainage, Türkiye; 174 mm SL.

Acanthobrama thisbeae

Common name. Ceyhan bream.

Diagnosis. Distinguished from other species of *Acanthobrama* by: ○ pharyngeal teeth in one row / ○ 76–89 total lateral-line scales / ○ 8½ branched dorsal rays / ○ flank scales with crenulated posterior margin / ○ 14–16½ branched anal rays / ○ last unbranched dorsal ray slender and short / ○ length of last unbranched dorsal ray 62–75 % of body depth at dorsal origin. Size up to 174 mm SL.

Distribution. Türkiye: Ceyhan and Orontes drainages.

Habitat. Large lowland rivers and reservoirs.

Biology. No data.

Conservation status. LC; not seen since 1997 in Orontes.

Further reading. Freyhof & Özuluğ 2014 (description).

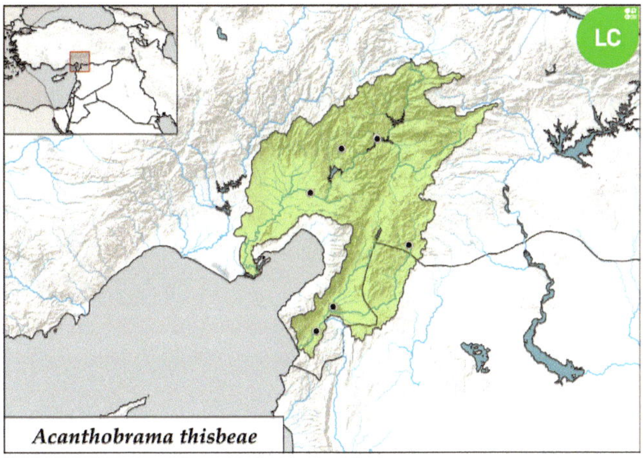

Lower parts of large rivers such as Ceyhan in Türkiye, are the major habitat for some *Acanthobrama* species.

Acanthobrama tricolor; Damascus basin, Syria; ~100 mm SL. © H. Esterbauer.

Acanthobrama tricolor

Common name. Damascus bream.

Diagnosis. Distinguished from other species of *Acanthobrama* by: ○ ventral keel poorly developed, covered with scales in its anterior part or fully covered with scales / ○ last unbranched dorsal ray soft and entirely segmented / ○ pharyngeal teeth in one row / ○ 49–60 total lateral-line scales / ○ 7–11½ branched anal rays / ○ 7½ branched dorsal rays. Size up to 127 mm SL.

Distribution. Syria: Barada and Masil al Fawwar in Al-Awaj drainage, in Golan Area of Separation.

Habitat. Springs and spring-fed streams. Historically, also in lakes.

Biology. No data.

Conservation status. CR, possibly extinct; known from lower Barada, where it has not been found since 1908. May still exist in Syria, at Masil al Fawwar in Golan Separation Area, where it was last seen in 1990.

Remarks. *Alburnus vignoni*, described from the Damascus basin, is a synonym.

Further reading. Lortet 1883 (description); Karaman 1972 (in *Acanthobrama*).

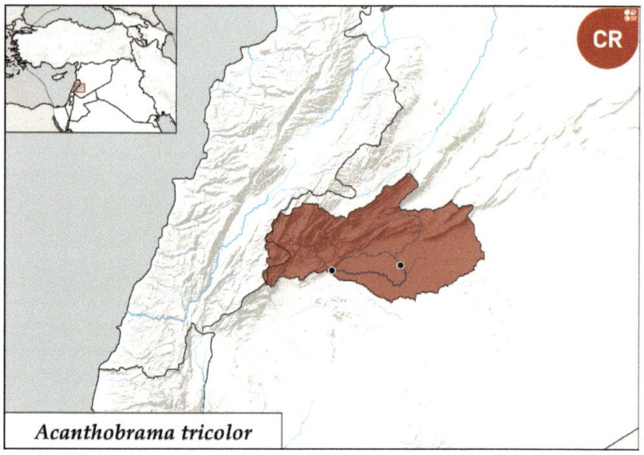

Acanthobrama urmiana; Urmia basin, Iran; ~50 mm SL.

Acanthobrama urmiana

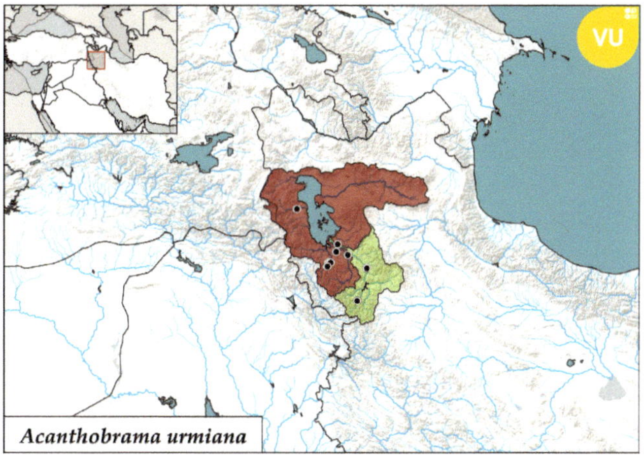

Acanthobrama urmiana

Common name. Urmia bream.

Diagnosis. Distinguished from other species of *Acanthobrama* by: ○ pharyngeal teeth in two rows / ○ 10–13½ branched anal rays / ○ 50–68 total lateral-line scales / ○ 42–43, usually 42, total vertebrae / ○ 10–14 gill rakers / ○ usually 8½ branched dorsal rays. Size up to 156 mm SL.

Distribution. Iran: Simineh, Zarrineh, and Shahr chay in Lake Urmia basin.

Habitat. Larger streams and rivers.

Biology. Lives up to 7 years. Matures at about 90–110 mm SL. Feeds on plants and invertebrates.

Conservation status. VU; has declined due to hydropower development and pollution.

Remarks. Very closely related to *A. punctulata*.

Further reading. Günther 1899 (description); Abdoli et al. 2008 (biology); Abbasi et al. 2004c (food); Coad 2021b (biology, distribution).

Alburnoides

Spirlins are small fishes that inhabit hill-stream habitats throughout West and Central Asia, extending east to the Amu Darya drainage. They also occur throughout Europe, west to France. Spirlins are superficially similar to some *Alburnus* species, particularly in turbid waters, where the lateral stripe and the dark-grey marks along the lateral line are poorly expressed. The orange bases of the paired fins are a useful indicator for distinguishing *Alburnoides* from *Alburnus* (which usually lack orange at the paired fin bases) in the field. Despite superficial similarities, both genera appear to be distantly related. All spirlin species are superficially similar, and only three species were classified as valid before the year 2000: the widespread *A. bipunctatus*, as well as *A. oblongus* and *A. taeniatus* (both from Central Asia). Since *A. oblongus* and *A. taeniatus* were transferred to *Alburnus*, 22 new *Alburnoides* species have been described, and 11 species revalidated from the list of synonyms. Currently, 34 species are recognised, 22 of which are found in West Asia. *Alburnoides varentsovi*, found on the northern slope of the Kopetdag Mountains in Turkmenistan, is an additional species in the region. It is not considered in this book. It can be challenging to distinguish between species of spirlins, and their distribution can assist in the identification, as they are all strictly allopatric. **Further reading.** Matveyev et al. 2017; Sheraliev & Peng 2021 (placement of *A. taeniatus* and *A. oblongus* to *Alburnus*).

Keys to species of *Alburnoides* in West Asia

Contributing authors: Nina Bogutskaya (Dolsko) and Davut Turan (Rize) contributed to the chapter on the genus *Alburnoides*.

Black, Mediterranean, and Marmara Sea basins

1a - Ventral keel scaleless or rarely covered by 1–3 scales on its anterior part.
..................2

1b - Ventral keel covered by scales, at least by (2) 3–7 scales on its anterior part.
..................5

2a - Usually 7–9 gill rakers.
..................3

2b - Usually 5–7 gill rakers.
..................4

3a - Mouth slightly subterminal, upper lip projecting beyond lower lip.
..................*A. kurui*

3b - Mouth terminal, upper and lower lips equally projecting.
..................*A. turani*

4a - 11–14½ branched anal rays.
..................*A. kosswigi*

4b - 14–16½ branched anal rays.
..................*A. freyhofi*

5a - 10–12½ branched anal rays.
..................6
5b - 12–16½ branched anal rays.
..................7

6a - Caudal peduncle depth 9–11 % SL; pigmentation of lateral line slightly distinct in anterior part of but indistinct in posterior part; snout length is markedly shorter than interorbital distance; 6–8, usually 7, gill rakers.
..................*A. tzanevi*
6b - Caudal peduncle depth 11–12 % SL; lateral line clearly distinct in both, anterior and posterior parts; snout length about equal to interorbital distance; usually 8–10 gill rakers.
..................*A. manyasensis*

7a - Interorbital width 7–8 % SL; caudal peduncle depth 10–11 % SL.
..................*A. coskuncelebii*
7b - Interorbital width 8–10 % SL; caudal peduncle depth 11–12 % SL.
..................8

8a - Caudal peduncle length 17–21 % SL; head width at anterior eye margin 10–11 % HL.
..................*A. fasciatus*
8b - Caudal peduncle length 21–26 % SL; head width at anterior eye margin 9–10 % HL.
..................*A. smyrnae*

Caspian basin and Hari drainage
1a - Horizontal eye diameter smaller than interorbital distance.
..................2
1b - Horizontal eye diameter equal to or larger than interorbital distance.
..................3

2a - 11–13, usually 12, predorsal vertebrae.
..................*A. tabarestanensis*
2b - 13–15, usually 14, predorsal vertebrae.
..................*A. eichwaldii*

3a - Ventral keel usually completely covered by scales, or at least anterior half of its length covered by scales.
..................*A. samiii*
3b - Ventral keel completely or almost completely scaleless.
..................*A. holciki*

Euphrates and Tigris drainages
1a - Horizontal eye diameter equal to or larger than interorbital distance.
..................*A. emineae*
1b - Horizontal eye diameter smaller than interorbital distance.
..................2

2a - Mouth terminal with pronounced chin; tip of mouth cleft situated at a horizontal line with middle of eye or lower margin of pupil; caudal lobes pointed.
..................*A. diclensis*
2b - Mouth terminal to subterminal without pronounced chin; tip of mouth cleft situated at a horizontal line with lower margin of pupil or below; caudal lobes rounded.
..................3

3a - 8–11½, usually 9–10½, branched anal rays; usually 7½ branched dorsal rays.
..................*A. nicolausi*
3b - 9–15½, usually 11–14½, branched anal rays; usually 8½ branched dorsal rays (often 7½ in *A. idignensis*).
..................4

4a - Outer anal margin straight or slightly convex; ventral keel usually partly to completely scaleless.
..................*A. idignensis*
4b - Outer anal margin straight or clearly concave; ventral keel usually completely covered by scales.
..................*A. velioglui*

Iranian endorheic basins: Western Kavir, Urmia, Namak, and Kor
1a - Horizontal eye diameter equal to or larger than interorbital distance.
..................*A. qanati*
1b - Horizontal eye diameter smaller than interorbital distance.
..................2

2a - 8–11½, usually 9–10½, branched anal rays; usually 7½ branched dorsal rays.
..................*A. petrubanarescui*
2b - 10–15½, usually 11–14½, branched anal rays; usually 8½ branched dorsal rays.
..................3

3a - Snout rounded; mouth subterminal; tip of mouth cleft situated at a horizontal line with lower margin of eye or below.
..................*A. namaki*
3b - Snout slightly pointed; mouth terminal; tip of mouth cleft situated at a horizontal line with lower half of eye.
..................*A. damghani*

Alburnoides coskuncelebii; Büyük Melen drainage, Türkiye; ~75 mm SL.

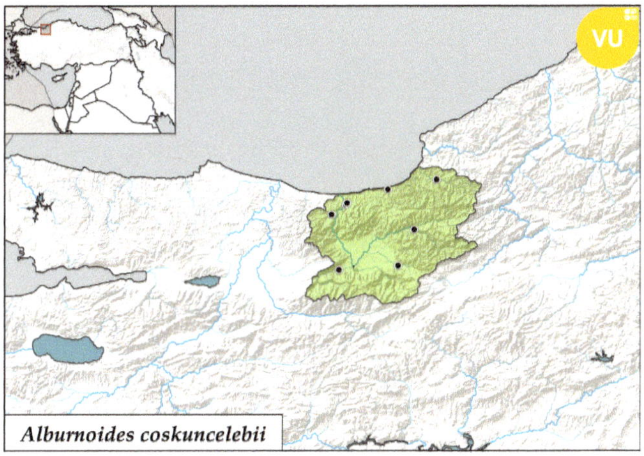

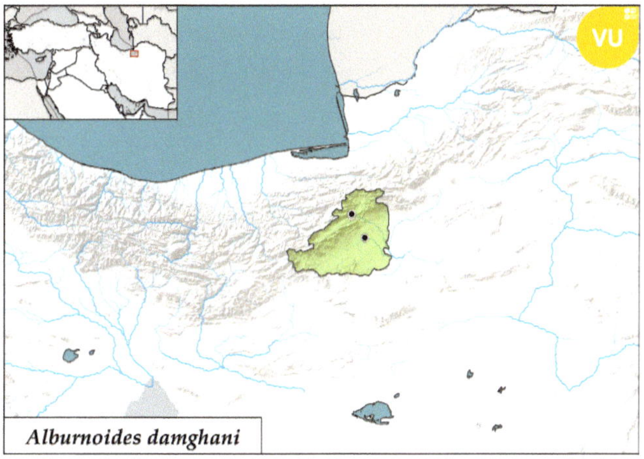

Alburnoides coskuncelebii

Common name. Büyük Melen spirlin.
Diagnosis. Distinguished from other species of *Alburnoides* in Mediterranean, Marmara, and Black Sea basins by: ○ tip of upper lip projecting lower lip / ○ 13–16½ branched anal rays / ○ 47–54+2–3 lateral line scales / ○ 6–9 gill rakers / ○ ventral keel covered by (2) 3–5 scales on its anterior part / ○ proximal 50–70 % of last unbranched dorsal ray ossified / ○ proximal 40–60 % of last unbranched anal ray ossified / ○ interorbital width 7–8 % SL / ○ caudal peduncle depth 10–11 % SL. Size up to 94 mm SL.
Distribution. Türkiye: Büyük Melen, Çayağzı, and Alaplı drainages.
Habitat. Streams and rivers with fast-flowing water, usually with cobble and gravel substrate.
Biology. No data.
Conservation status. VU; known from a small area where it is believed to be declining due to multiple threats (habitat modification, dams, drought, pollution).
Further reading. Turan et al. 2019b (description); Kaya 2020 (distribution); Canoğlu et al. 2023 (barcoding).

Alburnoides damghani

Common name. Damghan spirlin.
Diagnosis. Distinguished from other species of *Alburnoides* in Iranian endorheic basins by: ○ horizontal eye diameter smaller than interorbital distance / ○ 8½ branched dorsal rays / ○ 10–13½ branched anal rays / ○ ventral keel completely without scales or covered by scales for up to 3/4 of its length / ○ mouth terminal, upper and lower lips equally projecting / ○ snout slightly pointed / ○ tip of mouth cleft situated at a horizontal line with lower margin of eye / ○ 40–46+2–3 lateral-line scales / ○ 6–8 gill rakers. Size up to 90 mm SL.
Distribution. Iran: Cheshmeh Ali spring in Western Kavir basin.
Habitat. Spring with coarse gravel and boulders, dense riparian vegetation, and clear, well-oxygenated, fast-flowing water.

Alburnoides damghani; Cheshmeh Ali spring, Iran; ~70 mm SL.

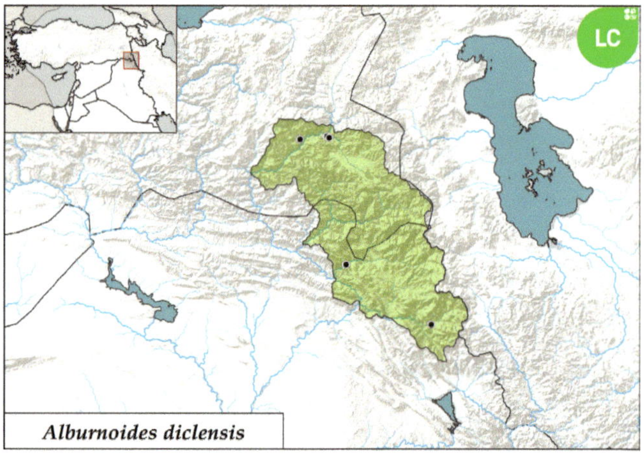

Biology. No data.
Conservation status. VU; drought and introduction of non-native species (e.g. *Oncorhynchus mykiss*) are threats.
Further reading. Jouladeh-Roudbar et al. 2016b (description).

Alburnoides diclensis

Common name. Great Zab spirlin.
Diagnosis. Distinguished from other species of *Alburnoides* in Euphrates and Tigris drainages by: ○ horizontal eye diameter smaller than interorbital distance / ○ 7–9½, usually 8½, branched dorsal rays / ○ ventral keel completely or almost completely covered by scales / ○ snout slightly rounded /

Alburnoides diclensis; Greater Zab, Iraq; ~70 mm SL.

○ mouth terminal, with pronounced chin / ○ tip of mouth cleft situated at a horizontal line with middle of eye to lower margin of pupil / ○ caudal lobes pointed / ○ 48–53+2–3 lateral-line scales / ○ 12–15½, usually 12–13½, branched anal rays / ○ 6–7 gill rakers / ○ 41–42 total vertebrae. Size up to 81 mm SL.
Distribution. Türkiye and Iraq: Greater Zab drainage, a tributary of Tigris.
Habitat. Streams and rivers with fast-flowing water, usually with cobble and gravel substrate.
Biology. No data.
Conservation status. LC.
Further reading. Turan et al. 2016b (description); Bektaş et al. 2019 (phylogeny); Canoğlu et al. 2023 (barcoding).

Alburnoides eichwaldii; Kura drainage, Armenia; ~100 mm SL. © B. Lewin.

Alburnoides eichwaldii; Aras drainage, Türkiye; ~90 mm SL.

Alburnoides eichwaldii
Common name. Kura spirlin.
Diagnosis. Distinguished from other species of *Alburnoides* in Caspian basin and Hari drainage by: ○ horizontal eye diameter smaller than interorbital distance / ○ 13–15, usually 14, predorsal vertebrae / ○ body depth at dorsal origin usually greater than head length / ○ ventral keel scaleless, partly or completely covered by scales, usually completely or almost completely scaleless / ○ snout slightly to markedly rounded / ○ upper lip slightly projecting lower lip / ○ tip of mouth cleft situated at a horizontal line slightly to markedly below middle of eye / ○ 45–54+2–3 lateral-line scales / ○ 10–14½, usually 11–13½, branched anal rays / ○ 5–9, usually 7–8, gill rakers / ○ 13–15, usually 14, predorsal vertebrae. Size up to 142 mm SL.
Distribution. Caspian basin: Kura and Aras drainages.

Habitat. Foothill streams and rivers with well-oxygenated, fast-flowing water. Also, lakes and reservoirs. Spawns in streams on gravel. Lake populations migrate to inflowing streams and rivers to spawn.

Biology. Lives up to 7 years. Females larger than males. Matures at 1–3 years. Spawns April–June at water temperatures between 13 and 18°C, over gravel beds in fast currents. Females lay 2–3 portions of eggs per season. Eggs are sticky and adhere to substrate. Feeds mainly on benthic invertebrates and terrestrial organisms, falling on water, and algae.

Conservation status. LC.

Remarks. Molecular studies suggest that *A. eichwaldii* hybridises with *A. samiii* in small coastal rivers south of Kura.

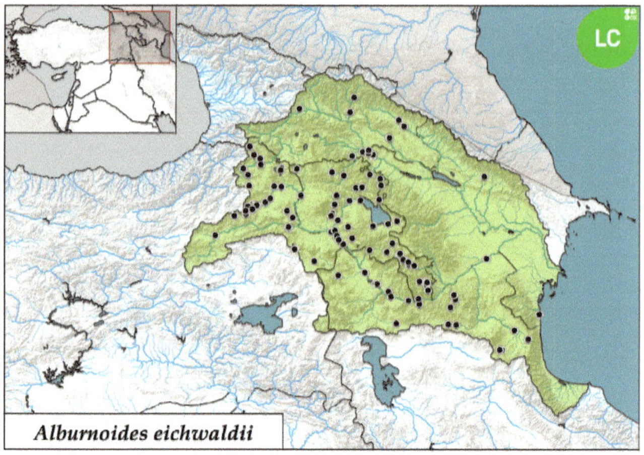

Further reading. Abdurakhmanov 1962 (biology); Bogutskaya & Coad 2009 (description); Mousavi-Sabet et al. 2015c (review); Levin et al. 2016 (barcoding, Lenkoran); Bektaş et al. 2019 (phylogeny); Kaya et al. 2020 (distribution); Canoğlu et al. 2023 (barcoding).

Hillstreams, such as the Karaj in Lake Namak basin, Iran, are the typical habitat of *Alburnoides* species.

Alburnoides emineae; Euphrates drainage, Türkiye; ~65 mm SL.

Alburnoides emineae

Common name. Khabur spirlin.

Diagnosis. Distinguished from other species of *Alburnoides* in Euphrates and Tigris drainages by: • horizontal eye diameter equal or larger than interorbital distance / ○ 8½, rarely 9½, branched dorsal rays / ○ ventral keel completely scaleless, rarely covered by few scales at its anterior part / ○ snout slightly pointed / ○ tip of mouth cleft situated at a horizontal line with middle of pupil or slightly below / ○ 39–45+2–3, usually 42–45+2–3, lateral-line scales / ○ 13–15½, usually 13–14½, branched anal rays / ○ 6–7 gill rakers / ○ 41–42 total vertebrae. Size up to 76 mm SL.

Distribution. Türkiye and most likely adjacent Syria: Beyazsu north of Nusaybin, a headwater stream of Jagh Jagh that flows to Khabur.

Habitat. Streams and rivers with fast-flowing water, usually with cobble and gravel substrate.

Biology. No data.

Conservation status. EN; appears to be declining within its very small range.

Further reading. Turan et al. 2014a (description); Bektaş et al. 2019 (phylogeny); Canoğlu et al. 2023 (barcoding).

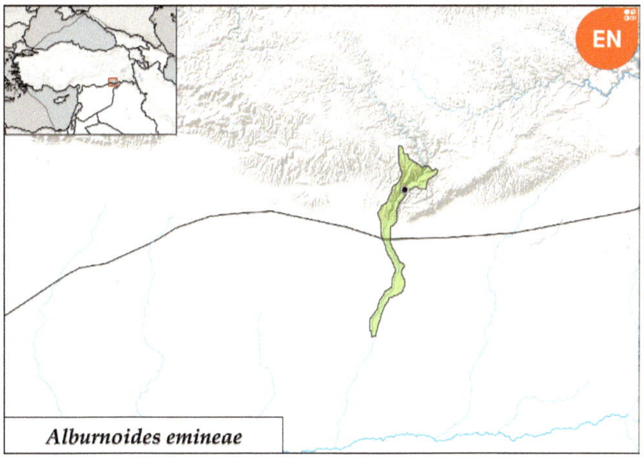

Alburnoides fasciatus; Sochi, Russia; ~90 mm SL.

Alburnoides fasciatus

Common name. Colchic spirlin.

Diagnosis. Distinguished from other species of *Alburnoides* in Mediterranean, Marmara, and Black Sea basins by: ○ ventral keel covered by (2) 3–5 scales on its anterior part / ○ 12–15½, usually 13–14½, branched anal rays / ○ 44–53+2–3 lateral-line scales / ○ 6–10 gill rakers / ○ caudal peduncle length 17–21 % SL / ○ head width at anterior eye margin 10–11 % HL / ○ proximal 30–50 % of last unbranched dorsal ray ossified / ○ interorbital width 8–10 % SL / ○ caudal peduncle depth 11–12 % SL / ○ body depth at dorsal origin usually greater than head length / ○ horizontal eye diameter smaller than interorbital distance / ○ snout pointed / ○ upper lip slightly projecting lower lip / ○ tip of mouth cleft situated at a horizontal line with middle of eye to lower margin of pupil. Size up to 127 mm SL.

Distribution. Caucasian Black Sea basin of Russia, south to Aksu drainage (Giresun) in Türkiye.

Habitat. Streams and rivers with moderate or fast-flowing water and gravel substrate; also in oxbows and lakes. Spawns in smaller streams on gravel. Lake and river populations migrate into tributaries to spawn.

Biology. Lives up to 8 years. Females larger than males. Matures at 2–3 years. Spawns April–August, at water temperatures between 13 and 19°C, over gravel bottoms in swift currents. Female lays 2–3 batches of eggs per season. Feeds mainly on benthic insect larvae, plankton, terrestrial insects, and algae.

Conservation status. LC.

Remarks. Spirlin in coastal streams east of Yeşilırmak and Aksu are likely to belong to this species.

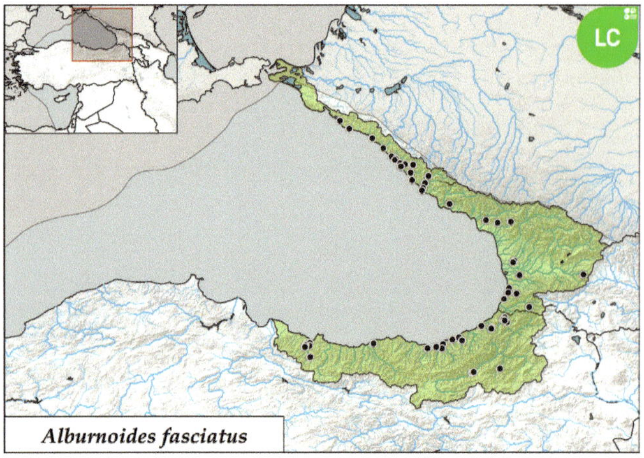

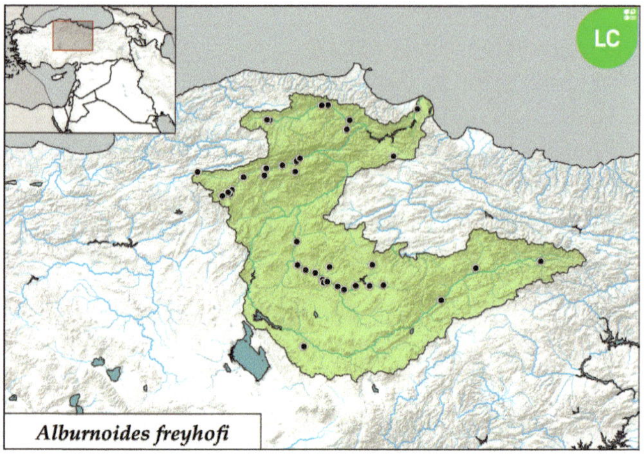

Alburnoides freyhofi; Kızılırmak drainage, Türkiye; ~75 mm SL.

Further reading. Elanidze 1983 (biology); Bogutskaya & Coad 2009 (description); Bayçelebi et al. 2015 (distribution); Turan et al. 2017b (identification); Bektaş et al. 2019 (phylogeny); Kaya 2020a (distribution, identification); Yeşilçiçek 2022b (size); Canoğlu et al. 2023 (barcoding).

Alburnoides freyhofi

Common name. Kızılırmak spirlin.
Diagnosis. Distinguished from other species of *Alburnoides* in Mediterranean, Marmara, and Black Sea basins by: ○ ventral keel usually completely scaleless / ○ body depth at dorsal origin usually greater than head length / ○ horizontal eye diameter smaller than interorbital distance / ○ snout slightly pointed / ○ mouth terminal, upper and lower lips equally projecting / ○ tip of mouth cleft situated at horizontal line with middle of eye / ○ 42–51+2–3 lateral-line scales / ○ 14–15½, rarely 16½, branched anal rays / ○ 5–7 gill rakers. Size up to 93 mm SL.
Distribution. Türkiye: Kızılırmak drainage.
Habitat. Fast-flowing streams and rivers with gravel and cobble bottoms. Also, inhabits reservoirs.
Biology. No data.
Conservation status. LC.
Further reading. Turan et al. 2017b (description); Bektaş et al. 2019 (phylogeny); Yoğurtçuoğlu et al. 2020a (distribution); Canoğlu et al. 2023 (barcoding).

Alburnoides holciki; Hari, Iran; 64 mm SL.

Alburnoides holciki; Atrak drainage, Iran; 75 mm SL.

Alburnoides holciki
Common name. Hari spirlin.
Diagnosis. Distinguished from other species of *Alburnoides* in Caspian basin and Hari drainage by: ○ horizontal eye diameter equal or slightly to clearly larger than interorbital distance / ○ ventral keel completely scaleless or almost completely scaleless / ○ 4, rarely 5, pharyngeal teeth on right 5th ceratobranchial / ○ snout pointed / ○ mouth terminal, lower lip slightly projecting upper lip / ○ tip of mouth cleft situated at a horizontal line with upper half of pupil / ○ body depth at dorsal origin greater than head length / ○ lateral line with bold dark-grey or black dots above and below pores / ○ 46–55+2–3, usually 47–51+2–3, lateral-line scales / ○ 11–16½ branched anal rays / ○ 5–9, usually 6–8, gill rakers. Size up to 93 mm SL.
Distribution. Hari and Morghab endorheic basins and Zeravshan and Amu Darya drainages (Aral basin). Also, in Atrak drainage in Iranian Caspian basin.
Habitat. Hill streams and small rivers with moderate to fast-flowing water, gravel, and silt substrate.
Biology. No data.
Conservation status. LC.

Remarks. *Alburnoides parhami* from the Atrak (Iran) is a synonym. *Alburnoides varentsovi* from the northern slopes of Kopetdag Mountains in Turkmenistan is another poorly known *Alburnoides* from Central Asia.
Further reading. Coad & Bogutskaya 2012 (description); Mousavi-Sabet et al. 2015c (description of *A. parhami*); Eagderi et al. 2019a (phylogeny).

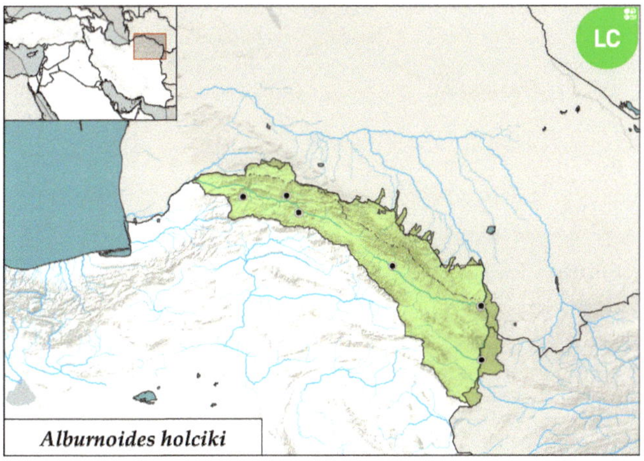

Bid Sorkh Qanat in Iran is the type locality of *Alburnoides idignensis*.

Alburnoides idignensis; Karkheh drainage, Iran; ~85 mm SL.

Alburnoides idignensis

Common name. Kermanshah spirlin.

Diagnosis. Distinguished from other species of *Alburnoides* in Euphrates and Tigris drainages by: ○ horizontal eye diameter smaller than interorbital distance / ○ no pronounced chin / ○ tip of mouth cleft situated at a horizontal line with lower margin of pupil to lower margin of eye / ○ ventral keel usually partly to completely scaleless / ○ snout stout / ○ mouth terminal or subterminal / ○ 39–49+2–3, usually 40–46+2–3, lateral-line scales / ○ 9–12½, usually 11–12½, branched anal rays / ○ usually 8½ but frequently 7½ branched dorsal rays / ○ 6–10 gill rakers / ○ 37–40, usually 39, total vertebrae / ○ caudal lobes rounded / ○ anal with straight or convex outer margin. Size up to 90 mm SL.

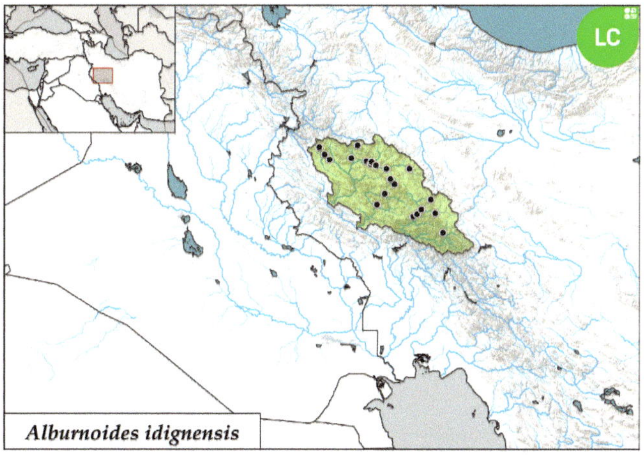

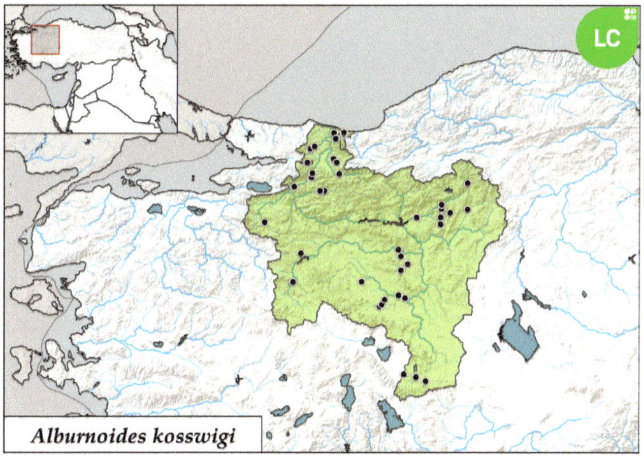

Alburnoides kosswigi; Sakarya drainage, Türkiye; ~90 mm SL.

Distribution. Iran: Karkheh and upper Dez (Karun) drainages.
Habitat. Streams and rivers with fast-flowing water, usually with cobble and gravel substrate.
Biology. No data.
Conservation status. LC.
Remarks. *Alburnoides idignensis*, *A. namaki*, and *A. nicolausi* are closely related, and future studies may not support them as independent species.
Further reading. Bogutskaya & Coad 2009 (description); Mousavi-Sabet et al. 2015c (review); Eagderi et al. 2019a (phylogeny).

Alburnoides kosswigi
Common name. Sakarya spirlin.
Diagnosis. Distinguished from other species of *Alburnoides* in Mediterranean, Marmara, and Black Sea basins by: ○ body depth at dorsal origin usually greater than head length / ○ horizontal eye diameter smaller than interorbital distance / ○ ventral keel completely scaleless or covered by 1–3 scales / ○ snout slightly pointed / ○ mouth terminal, upper lip slightly projecting lower lip / ○ tip of mouth cleft situated at horizontal line with middle of eye or slightly below / ○ 40–55+2–3 lateral-line scales / ○ 4 pharyngeal teeth in long row on right 5th ceratobranchial / ○ 11–14½, usually 12–13½, branched anal rays / ○ 5–7 gill rakers. Size up to 97 mm SL.
Distribution. Türkiye: Sakarya drainage and Lake Ilgın basin.
Habitat. Fast-flowing streams and rivers with gravel and cobbled bottoms.
Biology. No data.
Conservation status. LC.
Further reading. Turan et al. 2017b (description); Bektaş et al. 2019 (phylogeny); Canoğlu et al. 2023 (barcoding).

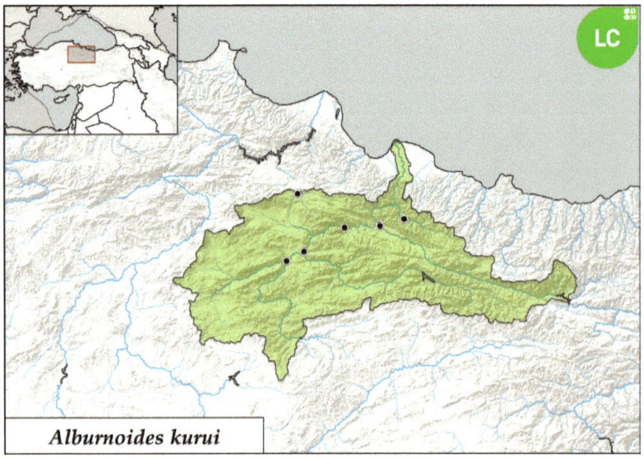

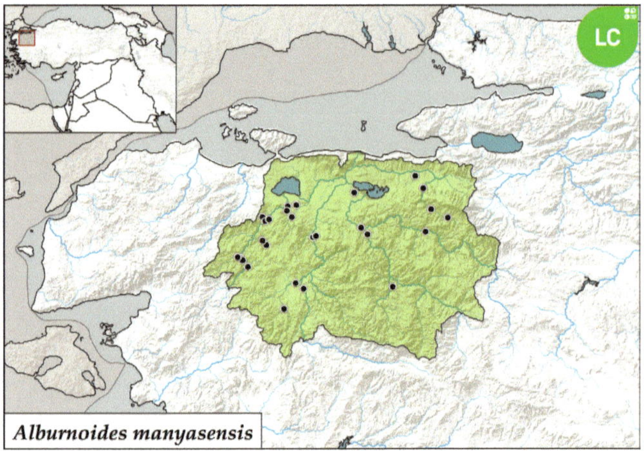

Alburnoides kurui; Yeşilırmak drainage, Türkiye; ~70 mm SL.

Alburnoides kurui
Common name. Yeşilırmak spirlin.
Diagnosis. Distinguished from other species of *Alburnoides* in Mediterranean, Marmara, and Black Sea basins by: ○ ventral keel usually completely scaleless / ○ 7–9 gill rakers / ○ body depth at dorsal origin usually greater than head length / ○ horizontal eye diameter smaller than interorbital distance / ○ snout slightly pointed / ○ mouth slightly subterminal, upper lip projecting lower lip / ○ tip of mouth cleft situated at horizontal line on or slightly below lower margin of pupil / ○ 47–56+2–3 lateral-line scales / ○ 4 pharyngeal teeth in long row on right 5th ceratobranchial / ○ 13–15½, usually 14½, branched anal rays. Size up to 90 mm SL.
Distribution. Türkiye: Yeşilırmak drainage.
Habitat. Fast-flowing streams and rivers with gravel and cobbled bottoms.
Biology. No data.
Conservation status. LC.
Further reading. Turan et al. 2017b (description); Bektaş et al. 2019 (phylogeny); Canoğlu et al. 2023 (barcoding).

Alburnoides manyasensis
Common name. Manyas spirlin.
Diagnosis. Distinguished from other species of *Alburnoides* in Mediterranean, Marmara, and Black Sea basins by: ○ ventral keel scaleless up to half its length,

Alburnoides manyasensis; Susurluk drainage, Türkiye; ~90 mm SL.

rarely almost completely or completely covered by scales / ○ body depth at dorsal origin greater than head length / ○ horizontal eye diameter usually slightly to markedly larger than interorbital distance / ○ snout short, pointed / ○ mouth terminal / ○ tip of mouth cleft situated at a horizontal line with middle of pupil or slightly below / ○ 45–52+2–3 lateral-line scales / ○ caudal peduncle depth 11–12 % SL / ○ lateral line clearly distinct in both, anterior and posterior parts / ○ snout length about equal to interorbital distance / ○ 4 pharyngeal teeth in long row on right 5th ceratobranchial / ○ 10–12½, usually 12½, branched anal rays / ○ 6–10, usually 8–10, gill rakers. Size up to 92 mm SL.

Distribution. Türkiye: Susurluk drainage.

Habitat. Streams and rivers with fast-flowing water, usually with cobble and gravel substrate. Also in lakes and reservoirs.

Biology. No data.

Conservation status. LC.

Further reading. Turan et al. 2013a (description); Turan et al. 2017b (identification); Bektaş et al. 2019 (phylogeny); Canoğlu et al. 2023 (barcoding).

Alburnoides namaki; Lake Namak basin, Iran; ~75 mm SL.

Alburnoides nicolausi; Norabad, Karkheh drainage, Iran; ~75 mm SL. Black spots are from parasites.

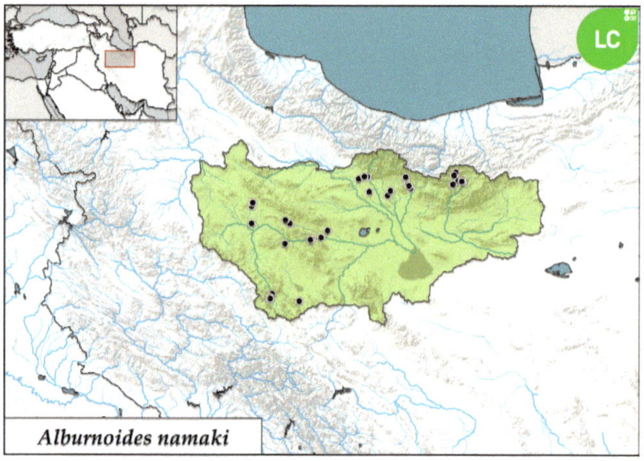

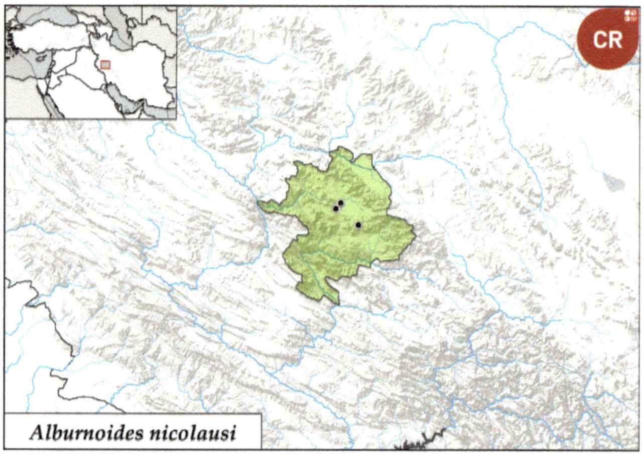

Alburnoides namaki

Common name. Namak spirlin.

Diagnosis. Distinguished from other species of *Alburnoides* in Iranian endorheic basins by: ○ horizontal eye diameter smaller than interorbital distance / ○ 8½, rarely 7½, branched dorsal rays / ○ 10–14½, usually 11–13½, branched anal rays / ○ snout rounded / ○ tip of mouth cleft situated at a horizontal line with lower margin of eye or below / ○ mouth subterminal, upper lip projecting lower lip / ○ 42–52+2–3 lateral-line scales / ○ 5–10 gill rakers. Size up to 110 mm SL.

Distribution. Iran: Lake Namak basin and Nam in Hable drainage in Western Kavir basin.

Habitat. Streams, springs, and qanats with moderate to fast-flowing water and gravel and silt substrate.

Biology. No data.

Conservation status. LC.

Remarks. *Alburnoides coadi* is a synonym. *Alburnoides idignensis*, *A. namaki*, and *A. nicolausi* are closely related, and future studies may not support them as independent species.

Further reading. Bogutskaya & Coad 2009 (description); Mousavi-Sabet et al. 2015c (review, description of *A. coadi*); Eagderi et al. 2019a (phylogeny).

Alburnoides nicolausi

Common name. Lorestan spirlin.

Diagnosis. Distinguished from other species of *Alburnoides* in Euphrates and Tigris drainages by: ● usually 7½ branched dorsal rays / ○ 8–11½, usually 9–10½, branched anal rays / ○ horizontal eye diameter smaller than interorbital distance / ○ ventral keel scaleless or covered by scales at up to ⅓ its length / ○ snout rounded / ○ mouth terminal, tip of mouth

cleft situated at about a horizontal line with lower margin of pupil / ○ no pronounced chin / ○ caudal lobes rounded / ○ 41–49+2–3, usually 42–46+2–3, lateral-line scales / ○ 5–9, usually 7–8, gill rakers. Size up to 89 mm SL.
Distribution. Iran: Nor Abad drainage in Lorestan, a tributary of Karkheh.
Habitat. Hill streams and small rivers with moderate to fast-flowing water and gravel substrate.
Biology. No data.
Conservation status. CR; known only from a small population in a very small area. Desiccation of streams is a major threat, and the species may be at the border of extinction.
Remarks. *Alburnoides nicolausi*, *A. namaki*, and *A. idignensis* are closely related, and future studies may not support them as independent species.
Further reading. Bogutskaya & Coad 2009 (description); Mousavi-Sabet et al. 2015c (review); Eagderi et al. 2019a (phylogeny).

Nor Abad in Lorestan, Iran, is the only place where *Alburnoides nicolausi* is found.

Alburnoides petrubanarescui; Lake Urmia basin, Türkiye; 73 mm SL.

Alburnoides petrubanarescui
Common name. Urmia spirlin.
Diagnosis. Distinguished from other species of *Alburnoides* in Iranian endorheic basins by: ○ 8–11½, usually 9–10½ branched anal rays / ○ usually 7½, rarely 8½ branched dorsal rays / ○ horizontal eye diameter smaller than interorbital distance / ○ mouth subterminal, tip of mouth cleft situated at a horizontal line with lower margin of eye / ○ ventral keel completely covered by scales / ○ snout stout, markedly rounded / ○ 42–49+2–3, usually 43–48+2–3, lateral-line scales / ○ 6–8 gill rakers. Size up to 109 mm SL.
Distribution. Iran and Türkiye: Simineh, Zarrineh, Mahabad, Nazlu, Talkheh, and Shahr in Lake Urmia basin.
Habitat. Fast-flowing streams with gravel and sand substrate.
Biology. No data.

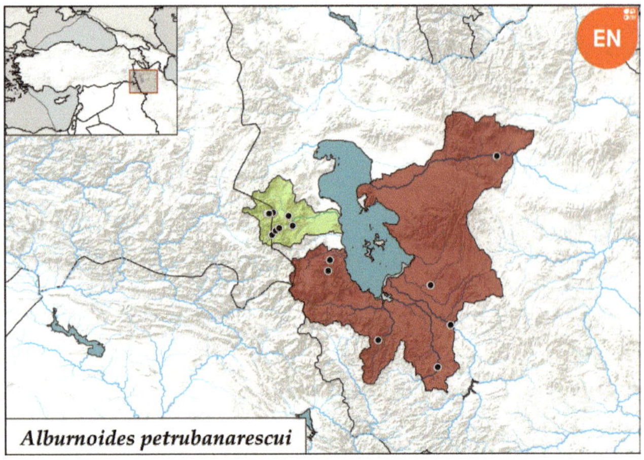

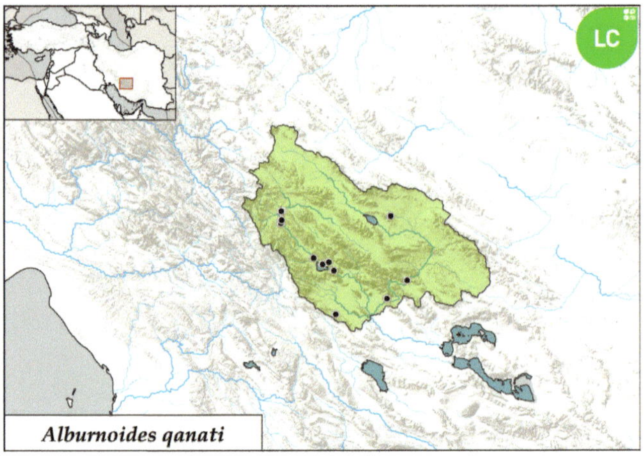

Alburnoides qanati; Kor basin, Iran; ~80 mm SL.

Conservation status. EN; seems to be declining within its very small range. Has disappeared from all tributaries in Iran except Nazlu.

Further reading. Bogutskaya & Coad 2009 (description); Mousavi-Sabet et al. 2015c (review); Kaya 2020b (records from Türkiye).

Alburnoides qanati

Common name. Qanat spirlin.

Diagnosis. Distinguished from other species of *Alburnoides* in Iranian endorheic basins by: ● horizontal eye diameter equal to or larger than interorbital distance / ○ ventral keel completely covered by scales, rarely scaleless up to ¼ its length / ○ lower lip usually projecting upper lip / ○ snout short, slightly pointed / ○ tip of mouth cleft situated at a horizontal line with middle of eye or upper margin of pupil / ○ 10–12½, usually 11½, branched anal rays / ○ 8½, rarely 7½, branched dorsal rays / ○ 41–49+2–3, usually 43–47+2–3, lateral-line scales / ○ 6–9, usually 8, gill rakers / ○ 40–41 total vertebrae. Size up to 82 mm SL.

Distribution. Iran: Kor and Harat drainages in Esfahan and Sirjan basins.

Habitat. Hill streams, spring-fed streams, small rivers with slow to moderate flow, gravel, and silt substrates.

Biology. No data.

Conservation status. LC.

Further reading. Bogutskaya & Coad 2009 (description); Mousavi-Sabet et al. 2015c (review); Eagderi et al. 2019a (phylogeny).

Alburnoides samiii; Sefid, Iran; ~70 mm SL.

Alburnoides samiii

Common name. South Caspian spirlin.

Diagnosis. Distinguished from other species of *Alburnoides* in Caspian basin and Hari drainage by: ○ horizontal eye diameter equal to or larger than interorbital distance / ○ ventral keel partly to completely covered by scales / ○ body depth at dorsal origin usually greater than head length / ○ snout short, slightly rounded / ○ mouth terminal, upper and lower lips equally projecting / ○ tip of mouth cleft situated at a horizontal line with middle of eye / ○ 47–52+2–3, usually 48–51+2–3, lateral-line scales / ○ 4 pharyngeal teeth in long row on right 5th ceratobranchial / ○ 11–15½, usually 12–14½, branched anal rays / ○ 7–10, usually 8–9, gill rakers. Size up to 85 mm SL.

Distribution. Caspian basin from Lenkoran lowlands (north of Talesh Mountains) in Azerbaijan east to Sefid in Iran.

Habitat. Streams and rivers with moderate to fast-flowing, well-oxygenated water and hard substrate. Also in lakes and reservoirs. Spawns in streams on gravel.

Biology. Females larger than males. Feeds mainly on benthic insect larvae.

Conservation status. LC.

Remarks. Molecular studies suggest that *A. eichwaldii* hybridises with *A. samiii* in small coastal rivers south of Kura.

Further reading. Coad & Bogutskaya 2009 (morphology); Mousavi-Sabet et al. 2015c (description); Levin et al. 2016 (barcoding in Lenkoran); Eagderi et al. 2019a (phylogeny).

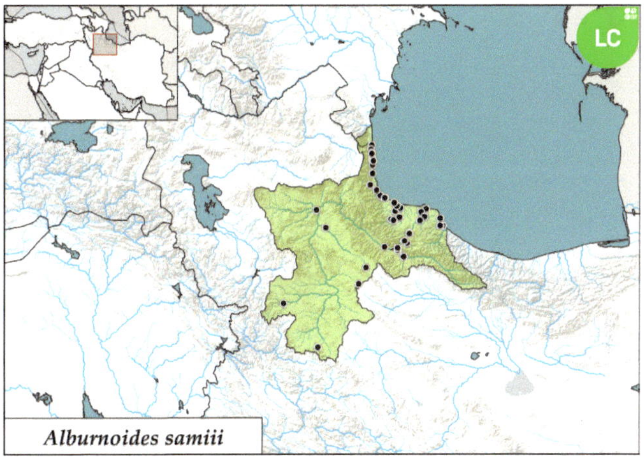

Alburnoides smyrnae; Büyük Menderes drainage, Türkiye; ~55 mm SL.

Alburnoides tabarestanensis; Tajan, Iran; ~90 mm SL.

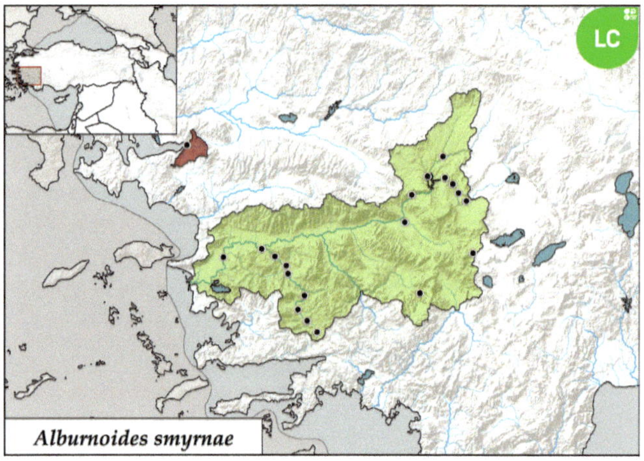

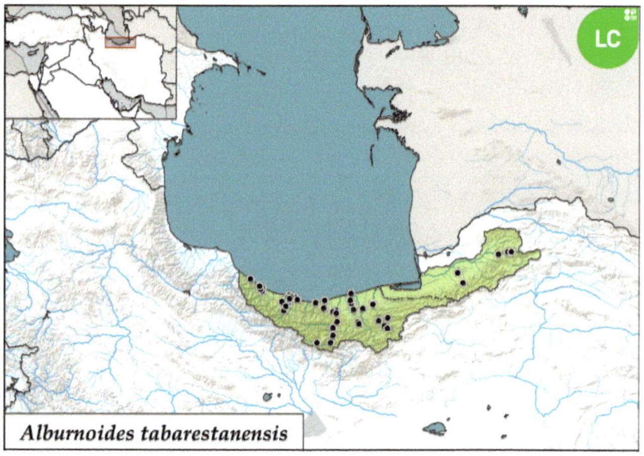

Alburnoides smyrnae

Common name. Eastern Aegean spirlin.

Diagnosis. Distinguished from other species of *Alburnoides* in Mediterranean, Marmara, and Black Sea basins by: ○ ventral keel completely or almost completely covered by scales / ○ body depth at dorsal origin greater than head length / ○ 12–15½, usually 13½, branched anal rays / ○ 40–45+2–3 lateral-line scales / ○ caudal peduncle length 21–26 % SL / ○ caudal peduncle depth 11–12 % SL / ○ head width at anterior eye margin 9–10 % HL / ○ interorbital width 8–10 % SL / ○ horizontal eye diameter equal or slightly to markedly larger than interorbital distance / ○ snout short, rounded / ○ mouth terminal to almost subterminal, upper and lower lips equally projecting / ○ tip of mouth cleft situated at a horizontal line with lower margin of pupil to lower margin of eye / ○ 4 pharyngeal teeth in long row on right 5th ceratobranchial / ○ 6–7 gill rakers. Size up to 77 mm SL.

Distribution. Türkiye: Meles (Kemer) near İzmir and Büyük Menderes drainage.

Habitat. Hill streams and small rivers with moderate to fast-flowing water and gravel substrate.

Biology. No data.

Conservation status. LC.

Remarks. Described from Meles (İzmir) but not seen in that area since. Diagnosis developed based on fish from upper Büyük Menderes is consistent with the type.

Further reading. Turan et al. 2013a (morphology, taxonomy); Turan et al. 2017b (identification); Bektaş et al. 2019 (phylogeny); Canoğlu et al. 2023 (barcoding).

Alburnoides tabarestanensis

Common name. Tabarestan spirlin.

Diagnosis. Distinguished from other species of *Alburnoides* in Caspian basin and Hari drainage by: ○ horizontal eye diameter equal or smaller than interorbital distance / ○ 11–13, usually 12, predorsal vertebrae / ○ body depth at dorsal origin usually greater than head length / ○ ventral keel usually scaleless, sometimes completely covered by scales / ○ snout slightly pointed / ○ mouth terminal, upper and lower lips equally projecting / ○ tip of mouth cleft situated at a horizontal line with middle of eye to lower margin of pupil / ○ 47–52+2–3 lateral-line scales / ○ 4 pharyngeal teeth in long row on right 5th ceratobranchial / ○ 11–14½, usually 12–13½, branched anal rays / ○ 7–10, usually 8–9, gill rakers. Size up to 95 mm SL.

Distribution. Iran: Khey to Gorgan drainages in southern and south-eastern Caspian basin.

Habitat. Streams and small rivers with moderate to fast-flowing water and gravel substrate.

Biology. No data.

Conservation status. LC.

Remarks. Spirlins from rivers between Tajan and west to Sefid and east to Atrak have yet to be identified.

Further reading. Mousavi-Sabet et al. 2015b (description); Mousavi-Sabet et al. 2015c (review); Eagderi et al. 2019a (phylogeny).

Alburnoides turani; Filyos drainage, Türkiye; ~80 mm SL.

Alburnoides turani

Common name. Filyos spirlin.

Diagnosis. Distinguished from other species of *Alburnoides* in Mediterranean, Marmara, and Black Sea basins by: ○ ventral keel usually completely scaleless, rarely with 1–3 scales / ○ 7–8 gill rakers / ○ body depth at dorsal origin usually greater than head length / ○ horizontal eye diameter smaller than interorbital distance / ○ snout slightly pointed / ○ mouth terminal, upper and lower lips equally projecting / ○ 46–53+2–3 lateral-line scales / ○ 13–14½ branched anal rays. Size up to 101 mm SL.

Distribution. Türkiye: Filyos drainage.

Habitat. Small rivers and streams with moderate to fast-flowing water and gravel substrate.

Biology. No data.

Conservation status. LC.

Further reading. Kaya 2020a (description); Canoğlu et al. 2023 (barcoding).

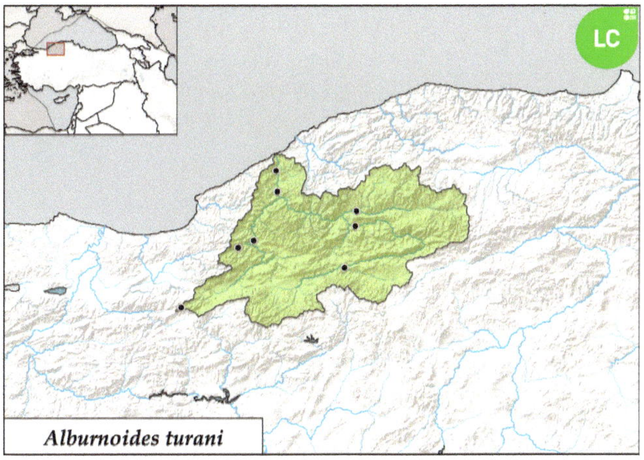

Alburnoides tzanevi; Devrekani, Türkiye; ~50 mm SL.

Alburnoides tzanevi

Common name. Thracian spirlin.

Diagnosis. Distinguished from other species of *Alburnoides* in Mediterranean, Marmara, Black, and Caspian Sea basins and Hari by: • maximum body depth usually smaller than head length / ○ horizontal eye diameter equal or larger than interorbital distance / ○ ventral keel covered by scales at about half its length / ○ caudal peduncle depth 9–11 % SL / ○ pigmentation of lateral line slightly distinct in anterior part of but indistinct in posterior part / ○ snout length is markedly shorter than interorbital distance / ○ usually 46–52+2–3 lateral-line scales / ○ 11–12½ branched anal rays / ○ 6–8, usually 7, gill rakers. Size up to 94 mm SL.

Distribution. Türkiye and Bulgaria: western Black Sea basin in Trace from Veleca south to İstanbul and east to Karasu (Sinop) drainage. Absent from Sakarya. Anatolian records need confirmation.

Habitat. Hill streams and small rivers with moderate to fast-flowing water and gravel substrate.

Biology. No data.

Conservation status. VU.

Further reading. Chichkoff 1933: 376 (description); Turan et al. 2013a (morphology).

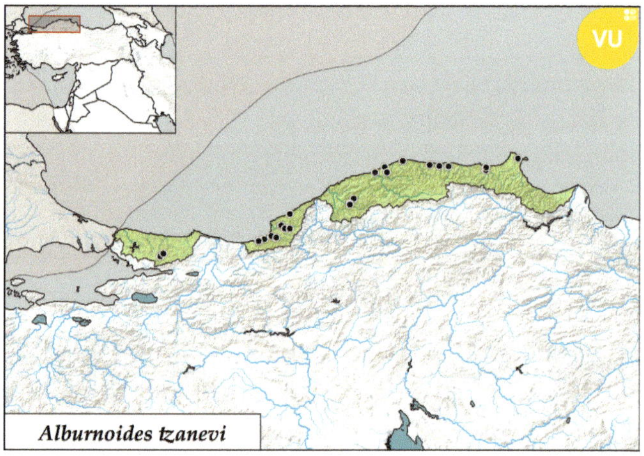

Alburnoides velioglui; Murat upper Euphrates drainage, Türkiye; ~65 mm SL.

Alburnoides velioglui; Sirvan drainage, Iraq; 46 mm SL.

Alburnoides velioglui

Common name. Mesopotamian spirlin.

Diagnosis. Distinguished from other species of *Alburnoides* in Euphrates and Tigris drainages by: ○ horizontal eye diameter smaller than interorbital distance / ○ 5–6 gill rakers / ○ ventral keel completely covered by scales / ○ no pronounced chin / ○ mouth terminal or subterminal / ○ tip of mouth situated at a horizontal line with lower margin of pupil or below up to lower margin of eye in larger individuals / ○ anal with slightly or clearly pointed tips and slightly or markedly concave outer margins / ○ 44–52+2–3, usually 46–48+2–3 lateral-line scales / ○ caudal lobes rounded / ○ 8½, rarely 9½, branched dorsal rays / ○ 11–13½ branched anal rays / ○ 41–42, usually 42, total vertebrae. Size up to 88 mm SL.

Distribution. Middle and upper Euphrates drainage in Türkiye, also found in Sirvan drainage in Iraq.

Habitat. Streams and rivers with fast-flowing water, usually with cobble and gravel substrate.

Biology. No data.

Conservation status. LC.

Further reading. Turan et al. 2014a (description); Bektaş et al. 2019 (phylogeny); Canoğlu et al. 2023 (barcoding).

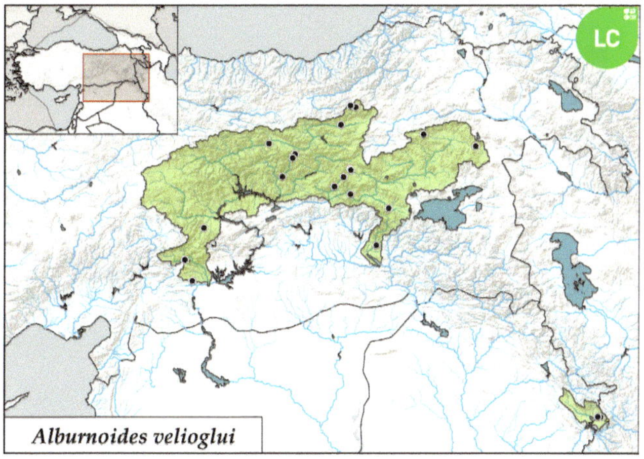

Upper Göksu in Euphrates drainage is a habitat of *Alburnoides velioglui*.

Copernicus Service information 2024

Alpaslan II Dam at the Murat, Türkiye.

Dammed: the current-day Euphrates. Climate change is worsening everything in an area in conjunction with limited rainfall and uncontrolled human population growth. In this context, a large river is seen as just the water it transports. There is a consensus that water should be used and not lost to adjacent countries or the sea. Since the early times of humanity, water from the Euphrates has been diverted for irrigation, drinking, and other purposes. The construction of the first modern water diversion structures commenced in Iraq with the erection of the Hindiya Barrage at the Iraqi Euphrates in 1913. This diverted the water of the Euphrates into irrigation canals. Another dam, the Ramadi Barrage, was constructed in 1951 in Iraq. It diverted water from the Euphrates into Lake Habbaniya to control floods. Subsequently, the Syrian Tabqa Dam (1974), the Iraqi Haditha Dam (1986), and the Syrian Tishrin Dam (1999) were constructed on the Euphrates. Following the construction of the Karakaya Dam (1987), major projects such as the Keban (1975) and Atatürk Dams (1992) were initiated as part of the Southeastern Anatolia Project in the early 1980s. Türkiye planned to incorporate 21 dams and 19 hydroelectric plants into a project designed to use the waters from the Euphrates and Tigris headwaters to vitalise economic, social, and cultural life. The construction of these dams and their reservoirs has significantly impacted the environment of the upper and middle Euphrates. The water flow of the Euphrates decreased, especially in its lower reaches, by 60–90 %, reducing the amount of water available for humans and biodiversity, and increasing its salinity. Nevertheless, more than 600 km of free-flowing river habitats remain, especially in the lower Euphrates in Iraq. This is still a significant proportion of the once large river of Mesopotamia and may be sufficient for its fish to survive. **Further reading.** Alwan 2013 (Euphrates).

Alburnus

The genus *Alburnus* (43 species) is a large group of small- to medium-sized fishes widely distributed in Europe, West, and Central Asia. In West Asia, *Alburnus*, with 24 species, is the second-largest genus of Leuciscids after *Pseudophoxinus*. Previous studies have overestimated the species diversity of *Alburnus*, as several shemayas of the *A. chalcoides* group have been recognised as separate species based on morphometric and meristic differences. All these populations are closely related, and recent research only partly confirms the differences proposed before and/or found the character states to be more overlapping than proposed. Therefore, Asian shemayas from the Aegean, Marmara, and Black Sea basins are placed in *A. derjugini*. The Central Asian *A. taeniatus* had been reported from the Hari in Iran. However, as the identification of this fish could not be confirmed, it is excluded from the coverage of this book.

Molecular characters group the species of *Alburnus* into three major species complexes, which may be recognised as separate genera in the future. True bleaks form one group, spotted bleaks belong to the second group, and Shah Kuli's belong to a third group. Sunbleaks of the genus *Leucaspius* and saramugos of the genus *Anaecypris* are closely related to, or even nested within, *Alburnus*. *Alburnus* are typically pelagic open-water dwelling fish that feed on plankton, drifting invertebrates, and small terrestrial arthropods that have fallen on the water surface. They are often common or very common and play important roles in the ecosystems they inhabit. In general, *Alburnus* are slender, silvery, herring-like fish with a superior or terminal mouth, a partly or completely open keel between the anus and the pelvic base, and a long anal fin with more than 9½ branched rays. However, several species do not fit this description, as they have fewer anal rays, no ventral keel or the keel completely

Alburnus taeniatus; Syr Darya drainage, Uzbekistan; ~70 mm SL.

Alburnus cf. *taeniatus*; Hari drainage, Iran; ~45 mm SL.

covered by scales, and a clearly terminal or even subterminal mouth. Based on anatomical studies, these species were previously classified within the genus *Petroleuciscus*. However, subsequent molecular studies led to their reclassification within the genus *Alburnus*.

Keys to species of *Alburnus* in West Asia

Caspian and Lake Urmia basins

1a - Usually 7½ branched dorsal rays.
.....................*A. filippii*
1b - Usually 8½ branched dorsal rays.
.....................2

2a - 7–10½ branched anal rays; ventral keel between posterior pelvic base and anus absent or very shallow.
.....................*A. ulanus*
2b - 9–16½ branched anal rays; ventral keel between posterior pelvic base and anus present and well developed.
.....................3

3a - 38–43+2–3 lateral line scales
.....................*A. hohenackeri*
3b - 43–65+2–5 lateral line scales
.....................4

4a - 13–15½ branched anal rays; 18–23 gill rakers; flank plain silvery without midlateral stripe in life or with very indistinct stripe; ventral keel not exposed in front of anus.
.....................*A. chalcoides*
4b - 9–12½ branched anal rays; 10–15 gill rakers; a prominent dark-brown midlateral stripe in life and preserved individuals; ventral keel exposed for 1–4 scales in front of anus.
.....................*A. atropatenae*

Persian Gulf basin, Qweiq, Lake Van, and Iranian endorheic basins, excluding Lake Urmia
1a - Anal origin below dorsal base.
.....................2
1b - Anal origin below last branched dorsal ray or behind dorsal base.
.....................3

2a - Adults with a mottled bluish-grey colour pattern on flank, just few dark-grey scales on flank in some individuals.
.....................*A. caeruleus*
2b - Adults silvery, never with dark-grey scales or blotches or with mottled bluish-grey colour pattern.
.....................*A. hohenackeri*

3a - Ventral keel between posterior pelvic base and anus absent or very shallow, completely covered by scales.
.....................4
3b - Ventral keel between posterior pelvic base and anus present, scaleless along at least 5 scales.
.....................5

4a - Usually 13–17 gill rakers.
.....................*A. doriae*
4b - 8–12 gill rakers.
.....................*A. kurui*

5a - Ventral keel completely scaleless or one scale on keel behind pelvic base; flank scales with prominent dark-grey margins; mid-lateral stripe absent or very faint.
.....................*A. tarichi*
5b - Ventral keel covered by one or more scales; flank scales without grey margins or irregular set grey pigmentation; mid-lateral stripe distinct, often bold.
.....................6

6a - Lateral line with 51–65(–72)+3 scales; head depth at nape 16–18 % SL; postorbital distance 12–15 % SL.
.....................*A. timarensis*
6b - Lateral line usually with 63–83+3 scales; head depth at nape 14–16 % SL; postorbital distance 10–12 % SL.
.....................*A. sellal*

Central Anatolia, Aegean, Marmara, and Black Sea basins
1a - 72–90 total lateral line scales.
.....................*A. akili*
1b - 38–72 total lateral line scales.
.....................2

2b - 52–69+3–5 lateral line scales; anal origin below last branched dorsal ray or behind dorsal base.
.....................*A. derjugini*
2a - 38–56+2–3 lateral line scale; anal origin below dorsal base.
.....................3

3a - 16–19½ (rarely 15½) branched anal rays.
.....................*A. alburnus*
3b - 10–15½ branched anal rays.
.....................4

4a - 18–21 gill rakers.
.....................*A. demiri*
4b - 11–19 gill rakers.
.....................5

5a - 40–47 total lateral line scales; 6–8 scale rows between lateral line and dorsal origin.
.....................*A. carianorum*
5b - 45–59 total lateral line scales; 8–10 scale rows between lateral line and dorsal origin.
.....................6

6a - In life, flank with a prominent golden upper and black lower stripe; body depth at dorsal origin 19–24 % SL; ventral keel exposed for 7–10 scales in front of anus.
.....................*A. goekhani*
6b - In life, flank without stripes or golden upper and black lower stripe faint; body depth at dorsal origin 24–30 % SL; ventral keel exposed for 0–5 scales in front of anus.
.....................*A. escherichii*

Mediterranean basin

1a - 58–89, usually 66–86 total lateral line scales.
.....................*A. sellal*
1b - 45–66 total lateral line scales.
.....................2

2a - 60–66 total lateral line scales.
.....................*A. adanensis*
2b - 45–62 total lateral line scales.
.....................3

3a - Pelvic axillary scales absent; scales on back, flank, and belly reduced, often not overlapping, absent on belly in front of pelvic base.
.....................*A. baliki*
3b - Pelvic axillary scales present; scales on back, flank, and belly well developed, overlapping.
.....................4

4a - 12–15½ branched anal rays.
.....................5
4b - 9–11½ branched anal rays.
.....................6

5a - Last unbranched dorsal ray stiff; anal origin behind vertical of last dorsal ray.
.....................*A. orontis*
5b - Last unbranched dorsal ray soft and flexible; anal origin below dorsal base.
.....................*A. escherichii*

6a - Flank plain silvery or with a distinct dark-brown mid-lateral stripe; 49–59 total lateral line scales.
.....................*A. kotschyi*
6b - Flank with many irregularly set and shaped grey, dark-brown or black blotches; 45–52 total lateral line scales.
.....................7

7a - Anal origin below 5th or 6th branched dorsal ray; 3 scale rows between lateral line and anal origin; fuzzy and usually blurred pattern of bold black or brown scales.
.....................*A. qalilus*
7b - Anal origin below or behind vertical through last dorsal ray; 4–5 scale rows between lateral line and anal origin; bold black, grey or brown scales on silvery or brown background.
.....................*A. magnificus*

Alburnus adanensis; Seyhan, Türkiye; ~67 mm SL. © M. Özuluğ.

Alburnus adanensis

Common name. Adana bleak.

Diagnosis. Distinguished from other species of *Alburnus* in Mediterranean basin by: ● 60–66 total lateral line scales / ○ 11–14 scale rows between lateral line and dorsal origin / ○ 5–6 scale rows between lateral line and anal origin / ○ 10–12½ branched anal rays. Only known individual 67 mm SL.

Distribution. Türkiye: Seyhan drainage.

Habitat. No data.

Biology. No data.

Conservation status. Extinct; not found since the 1940s. Reasons for extinction unknown.

Remarks. Has been confused with *A. kotschyi*, which also occurs in Seyhan drainage, but both species are well differentiated. It can be distinguished from *A. sellal*, recently reported to be invasive in Ceyhan, by having 60–66 total lateral line scales (as opposed to the usual 66–86).

Further reading. Battalgazi 1944 (description); Freyhof et al. 2018a (re-validation).

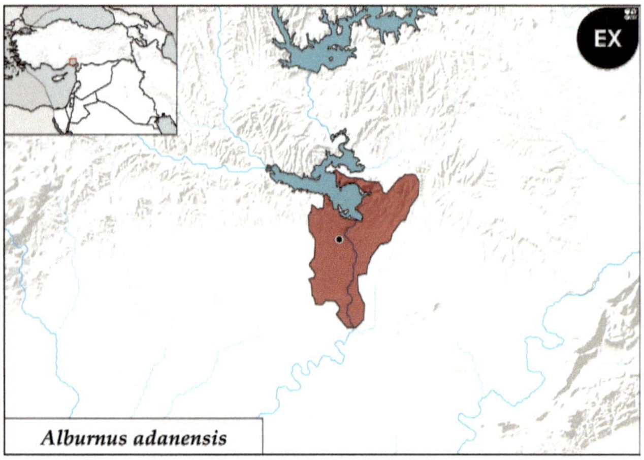

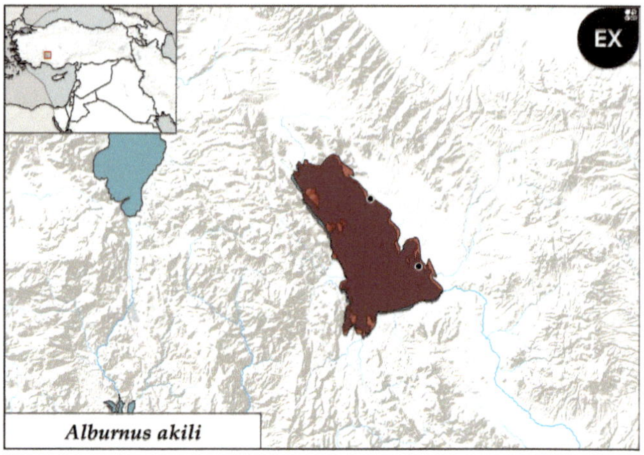

Lake Beyşehir in Türkiye was once the habitat of *Alburnus akili*. Now, it is inhabited by non-native *A. escherichii*.

Alburnus akili; Lake Beyşehir, Türkiye; 115 mm SL.

Alburnus akili

Common name. Beyşehir bleak.

Diagnosis. Distinguished from other species of *Alburnus* in Central Anatolia, Aegean, Marmara, and Black Sea basins by: ● 72–90 total lateral line scales / ○ 8½ branched dorsal rays / ○ 13½ branched anal rays. Size up to 150 mm SL.

Distribution. Türkiye: Lakes Beyşehir and Suğla (Seydişehir) basins as well as Çarşamba channel connecting both lakes.

Habitat. Lacustrine known only from lakes and slow-flowing channels.

Biology. No data.

Conservation status. Extinct; last time recorded in 1985, when population was already small. Extinction followed introduction of non-native *Sander lucioperca* in 1978 and 1980.

Remarks. In the original description, 85–88 total scales were given for this species, while a later study gives only 72–75 total scales. The correct count remains unclear, but F. Küçük counted 78–80 total lateral line scales (pers. comm.). Since the early 2000s, the non-native *A. escherichii* has become common in the former range of *A. akili*.

Further reading. Battalgil 1942 (description); Erdemli 1982 (morphology); Freyhof & Özuluğ 2010a; Küçük 2012 (extinction).

Alburnus alburnus; Susurluk drainage, Türkiye; ~90 mm SL.

Alburnus alburnus

Common name. Bleak.

Diagnosis. Distinguished from other species of *Alburnus* in Central Anatolia, Aegean, Marmara, and Black Sea basins by: ○ anal origin below branched dorsal rays 4–5 / ○ 45–48+3 lateral line scales / ○ 16–19½ branched anal rays / ○ 16–22 gill rakers / ○ ventral keel exposed from anus to pelvic base / ○ dark-brown mid-lateral stripe absent in life, faint or absent in preserved individuals / ○ mouth slightly superior. Size up to 160 mm SL.

Distribution. Black Sea basin of Russia. Türkiye: Lake Sapanca basin and Susurluk drainage, including Lakes Apolyont and Manyas (southern Marmara basin). Introduced in Cyprus. Europe north of Caucasus, Pyrenees, and Alps, eastward to Ural and Emba.

Habitat. Open waters of lakes and medium to large rivers. Forages near surface. In winter, forms large aggregations in backwaters and other still waters. Spawns in shallow riffles or on rocky lake shores, occasionally over submerged vegetation.

Biology. Gregarious. Spawns first time at 2–3 years. Usually spawns in one or two seasons, in May–August at temperatures above 12°C, 2–4 times at intervals of 1–2 weeks, early in morning. Larvae inhabit littoral zone of rivers and lakes, juveniles leave banks and occupy pelagic habitats feeding on plankton, drifting insects or invertebrates fallen on water surface. In Europe, scales were once used to make Essence d'Orient, a coating for artificial pearls.

Conservation status. LC.

Further reading. Kottelat & Freyhof 2007 (distribution, biology, diagnosis); Zogaris et al. 2012 (records from Cyprus).

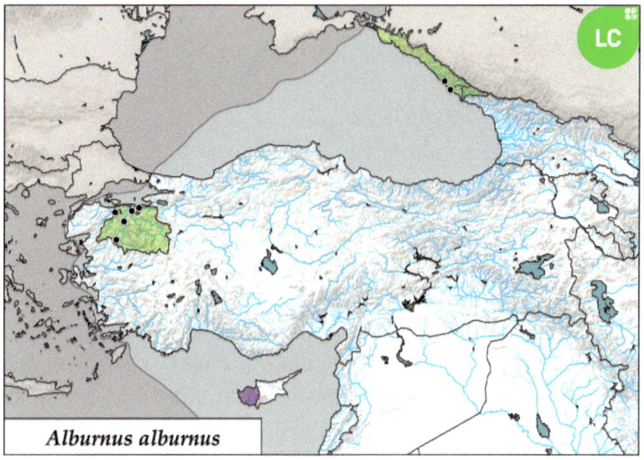

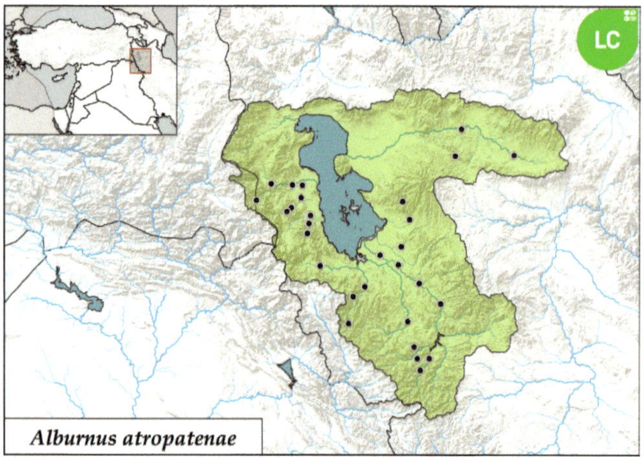

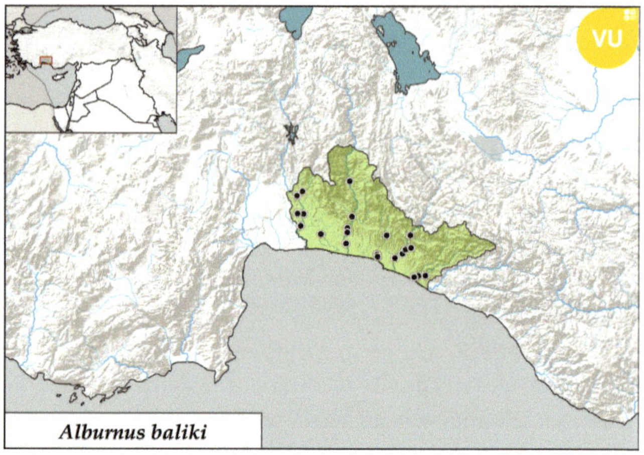

Alburnus atropatenae; Urmia basin, Türkiye; ~125 mm SL.

Alburnus atropatenae

Common name. Urmia bleak.

Diagnosis. Distinguished from other species of *Alburnus* in Caspian and Lake Urmia basins by: ○ 10–15 gill rakers / ○ 9–12½ branched anal rays / ○ 43–60+3 lateral line scales / ○ a bold, dark-brown mid-lateral stripe / ○ ventral keel exposed for 1–4 scales in front of anus / ○ usually 8½ branched dorsal rays. Size up to 130 mm SL.

Distribution. Iran and Türkiye: Lake Urmia basin. In uppermost Choman (Tigris drainage) directly at watershed to Saqez (Urmia).

Habitat. A wide range of moderately fast-flowing streams and rivers with gravel substrate.

Biology. Feeds on aquatic invertebrates.

Conservation status. LC.

Remarks. Populations from Lake Namak basin were listed as this species but were later identified as *A. doriae*. Both species are superficially similar but do not appear to be closely related. Recently, it has been considered a synonym of *A. ulanus*, as both species share the same COI DNA sequence. As both species are well differentiated morphologically, introgressive hybridisation is likely to have occurred.

Further reading. Coad & Holčík 1999 (morphology); Ghasemi et al. 2015 (distribution); Jouladeh-Roudbar et al. 2020 (as *A. ulanus*); Kaya 2020b (record from Türkiye); Coad 2021b (biology, distribution).

Alburnus baliki

Common name. Antalya spotted bleak.

Diagnosis. Distinguished from other species of *Alburnus* in Mediterranean basin by: ● pelvic axillary scales absent / ● scales on back, flank, and belly reduced, often not overlapping, absent on belly in front of pelvic base / ● nuptial male bright yellow / ○ faint grey mottling on flank including only few scales in most individuals / ○ 11–13 gill rakers / ○ 12–14½ branched anal rays / ○ 47–61 total lateral line scales / ○ ventral keel fully exposed or covered by few scales on its anterior part / ○ no dark-brown mid-lateral stripe. Size up to 60 mm SL.

Distribution. Türkiye: Aksu east to Karpuz drainages.

Habitat. Slow-flowing sections of streams and rivers, often with submerged vegetation.

Biology. No data.

Conservation status. VU; found in less than 10, mostly declining populations.

Further reading. Bogutskaya et al. 2000 (description).

Alburnus baliki; Ilıca stream, Türkiye; 45 mm SL.

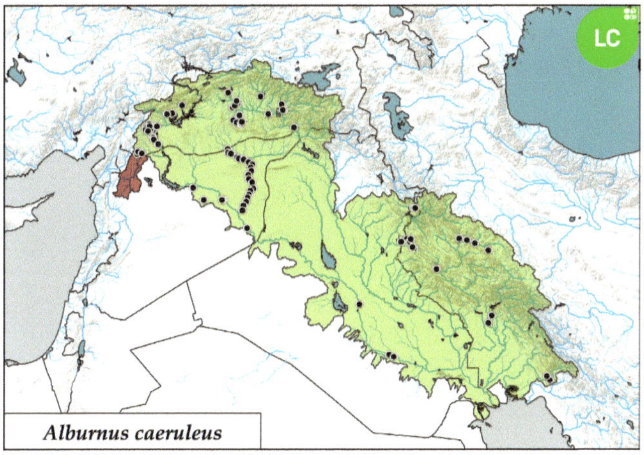

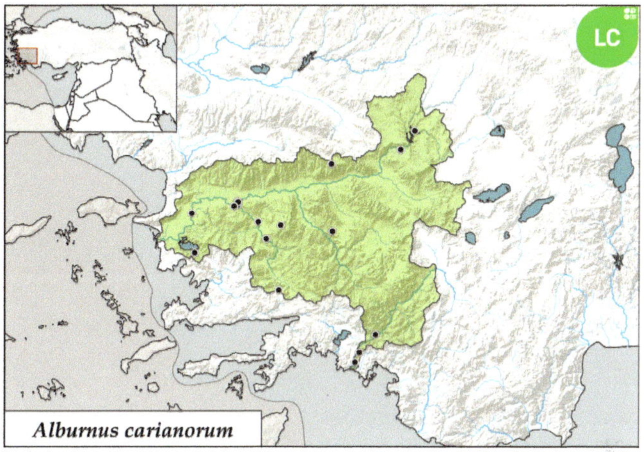

Alburnus caeruleus; Sirvan drainage, Iraq; 70 mm SL.

Alburnus caeruleus; Tigris drainage, Türkiye; ~60 mm SL.

Alburnus carianorum; Dalaman, Türkiye; ~75 mm SL.

Alburnus caeruleus

Common name. Black spotted bleak.

Diagnosis. Distinguished from other species of *Alburnus* in Persian Gulf, Qweiq, Lake Van, and Iranian endorheic basins by: ● body mottled with isolated or small groups of black pigmented scales, bluish black with silvery blotches in nuptial males / ○ anal origin below dorsal base / ○ 13–17½ branched anal rays / ○ 8–12 gill rakers / ○ 43–58 total lateral line scales / ○ ventral keel fully exposed or covered by few scales on its anterior part / ○ body depth at dorsal origin 26–34 % SL. Size up to 101 mm SL.

Distribution. Qweiq, Euphrates, and Tigris drainages, including Karun and Jarahi in Iran.

Habitat. Streams, rivers, and lakes with moderate to slow currents. Likely to be sensitive to low temperatures.

Biology. No data.

Conservation status. LC; extirpated in Syrian part of Qweiq.

Remarks. *Alburnoides recepi* is a synonym. Many populations of *A. caeruleus* share mtDNA with *Alburnus sellal*, suggesting past hybridisation and DNA introgression between the two species.

Further reading. Turan et al. 2014a (description as *A. recepi*); Birecikligil et al. 2017 (*A. recepi* as synonym); Freyhof et al. 2018b, Bektaş et al. 2020 (phylogeny, mtDNA introgression).

Alburnus carianorum

Common name. Carian bleak.

Diagnosis. Distinguished from other species of *Alburnus* in Central Anatolia, Aegean, Marmara, and Black Sea basins by: ○ 37–44+3 total scales in lateral line / ○ 11–13½ branched anal rays / ○ 6–8 scale rows between lateral line and dorsal origin / ○ 15–19 gill rakers / ○ ventral keel exposed for 2–8 scales in front of anus / ○ caudal peduncle depth 8.9–10.5 % SL. Size up to 92 mm SL.

Distribution. Türkiye: Dalaman, Tersakan, Namnam, and Büyük Menderes drainages.

Habitat. A wide range of moderately fast-flowing streams and rivers with gravel substrate.

Biology. No data.

Conservation status. LC.

Remarks. First described as A. kurui. However, as Leuciscus kurui Bogutskaya, 1995 is placed in Alburnus, A. kurui Mangit & Yerli, 2018 became a junior secondary homonym of Alburnus kurui (Bogutskaya, 1995). As both are valid species in the genus Alburnus, A carianorum is designated as a replacement name for A. kurui Mangit & Yerli, 2018.

Alburnus are a typical species of running waters in West Asia, as here in the Sirvan, a tributary of the Tigris.

Further reading. Mangit & Yerli 2018 (description as *Alburnus kurui*); Freyhof et al. 2019a (description).

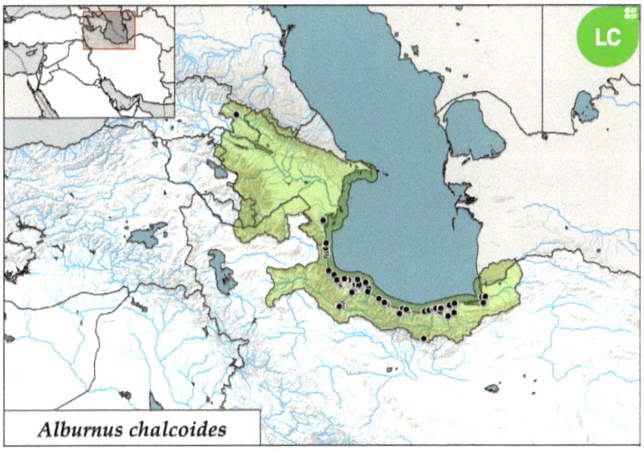

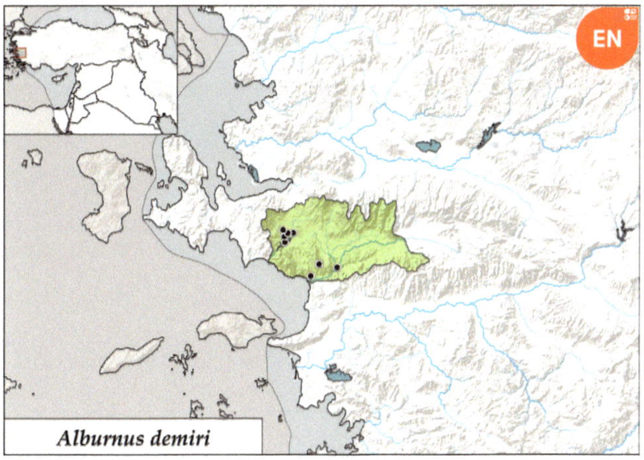

Alburnus chalcoides; Tajan, Iran; ~90 mm SL.

Alburnus chalcoides

Common name. Caspian Shemaya.

Diagnosis. Distinguished from other species of *Alburnus* in Caspian basins by: ○ 52–70+2–5 lateral line scales / ○ usually 8½ branched dorsal rays / ○ 12–17½ branched anal rays / ○ ventral keel between posterior pelvic base and anus present and well developed / ○ 18–22 gill rakers / ○ anal origin ½–2½ scales behind dorsal base, below dorsal ray 7–8 in some individuals / ○ ventral keel exposed for 3–12 scales in front of anus / ○ a faint, dark-brown mid-lateral stripe. Size up to 400 mm SL, resident fish in rivers or streams usually much smaller.

Distribution. Aral and Caspian basins, mostly western to southern coast, rarely in Ural and Volga.

Habitat. Lower parts of larger streams, rivers, lakes, estuaries, and brackish sea areas. Lives near surface, tolerating salinities up to 14 ‰. Migrates long distances upstream from sea or lakes to headwaters of streams in foothills and montane regions. Spawns in small rivers or streams with strong currents on submerged vegetation or gravel bottoms. Lacustrine and reservoir populations spawn in tributaries.

Biology. Semi-anadromous, riverine, and lacustrine populations. First spawn at 2–4 years, females 1 year later than males. Anadromous populations enter rivers in autumn (September in Kura) and migrate upstream in winter and/or spring. Spawns May–September in water 0.2–0.7 m deep, current about 1 m/s and 18–26°C, often with much splashing. Males appear to be territorial. They congregate at spawning sites and wait for mature females, which arrive later. Eggs stick to pebbles or stones. Embryonic development lasts 2–3 days; larvae initially remain between pebbles for 8–11 days, then actively migrate to shallower areas and backwaters. Many populations are riverine. In migratory populations, adults return to sea, lakes or estuaries soon after spawning to feed. Young juveniles migrate downstream in autumn of same year or following spring. Larvae and juveniles feed on zooplankton, algae, and insect larvae, adults mainly on planktonic crustaceans, terrestrial insects, and small fish. Often hybridises with sympatric *Squalius* species.

Conservation status. LC.

Further reading. Kazancheyev 1981 (Caspian); Movchan & Smirnov 1983 (morphology, biology); Kottelat & Freyhof 2007 (distribution, biology, diversity Europe); Coad 2021a (biology, morphology).

Alburnus demiri; Tahtalı drainage, Türkiye; 88 mm SL.

Alburnus demiri

Common name. Eastern Aegean bleak.

Diagnosis. Distinguished from other species of *Alburnus* in Central Anatolia, Aegean, Marmara, and Black Sea basins by: ○ 43–46+3 scales in lateral line / ○ 13–14½ branched anal rays / ○ 9 scale rows between lateral line and dorsal origin / ○ 18–21 gill rakers / ○ ventral keel exposed for 6–9 scales in front of anus / ○ caudal peduncle depth 10.5–11.1 % SL. Size up to 90 mm SL.

Distribution. Türkiye: Tahtalı and Küçük Menderes drainages.

Habitat. A wide range of moderately fast-flowing streams and rivers with gravel substrate. Reservoir populations migrate to tributaries to spawn.

Biology. Spawns May–June.

Conservation status. EN; appears to be declining within its very small range.

Further reading. Özuluğ & Freyhof 2008a (description); Mangit & Yerli 2018 (distribution).

Alburnus derjugini; Dagomys drainage, Russia; 95 mm SL.

Alburnus derjugini; Bakırçay drainage, Türkiye; 88 mm SL.

Alburnus derjugini

Common name. Georgian shemaya.

Diagnosis. Distinguished from other species of *Alburnus* in Marmara, Black Sea, and Aegean basins by: ○ 50–70+2–5 lateral line scales / ○ usually 8½ branched dorsal rays / ○ 12–17½ branched anal rays / ○ ventral keel between posterior pelvic base and anus present and well developed / ○ 18–35, usually 22–31, gill rakers / ○ anal origin ½–2½ scales behind dorsal base, below dorsal ray 7–8 in some individuals / ○ ventral keel exposed for 3–12 scales in front of anus / ○ a faint, dark-brown mid-lateral stripe. Size up to 400 mm SL, resident fish in rivers or streams usually much smaller.

Distribution. Marmara and Black Sea basin. In Turkish Aegean: Bakırçay, Güzelhisar, south to Gediz drainage.

Habitat. A wide range of streams and rivers with moderate fast-flowing water and gravel substrate.

Biology. Seem to be a resident species, no migratory populations known.

Conservation status. LC.

Remarks. Shemayas were divided into 17 species (*A. attalus*, *A. battalgilae*, *A. carinatus*, *A. chalcoides*, *A. danubicus*, *A. derjugini*, *A. istanbulensis*, *A. leobergi*, *A. mandrensis*, *A. mento*, *A. mentoides*, *A. nicaeensis*, *A. sarmaticus*, *A. sava*, *A. schischkovi*, *A. vistonicus*, *A. volviticus*) based on morphometric and meristic characters. Some species grow large, others remain very small, some are migratory or non-migratory and many are restricted to small areas. Subsequent molecular studies (COI) did not support this diversity. Molecular differences between fish from the Caspian, Black Sea, and Turkish Aegean basins are very small. In contrast, species inhabiting the Black Sea basin and the northern Aegean mostly share one haplotype, indicating a recent common evolutionary history. As more material from Türkiye became available, the proposed morphological differences could not be confirmed, especially as the number of gill rakers overlaps much more than originally described. Because of the large overlap in morphological characters and no difference in COI DNA sequences, we treat *A. attalus*, *A. battalgilae*, *A. carinatus*, *A istanbulensis*, and *A. nicaeensis* as synonyms of *A. derjugini*. It is recommended to revisit the European species also.

Further reading. Battalgil 1941 (description of Turkish species); Özuluğ & Freyhof 2007 (rediagnosis); Kottelat & Freyhof 2007 (distribution, biology, diversity Europe); Geiger et al. 2014 (phylogeny); Mangit & Yerli 2018 (phylogeny); Bayçelebi et al. 2021a (Aegean populations); Sarıahmetoğlu 2024 (Morphology of Turkish shemayas).

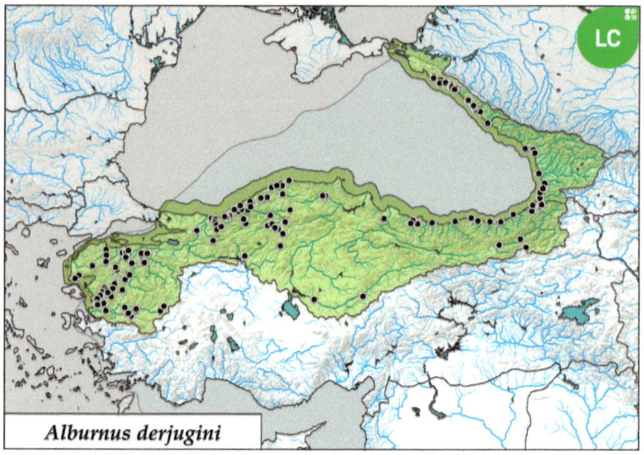

Alburnus doriae; Zayandeh drainage, Iran; ~80 mm SL.

Alburnus doriae; Karun drainage, Iran; ~130 mm SL.

Alburnus doriae

Common name. Persian shemaya.

Diagnosis. Distinguished from other species of *Alburnus* in Persian Gulf, Qweiq, Lake Van, and Iranian endorheic basins by: ○ anal origin about 1–2 scales behind dorsal base / ○ 9–12½ branched anal rays / ○ (12) 13–17 (18) gill rakers / ○ 41–53 +3 lateral line scales / ○ usually 8½ branched dorsal rays / ○ 2–5 scale rows between lateral line and pelvic origin / ○ ventral keel between pelvic base and anus often indistinct, fully covered by scales / ○ usually a bold, dark-brown mid-lateral stripe. Size up to 134 mm SL.

Distribution. Iran: Lake Namak basin, Esfahan and Sirjan basins, and few headwater streams of upper Karun.

Habitat. A wide range of moderately fast-flowing streams and rivers with gravel substrate.

Biology. No data.

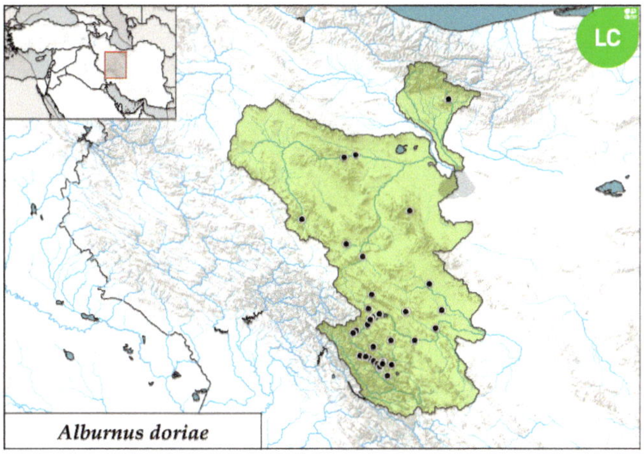

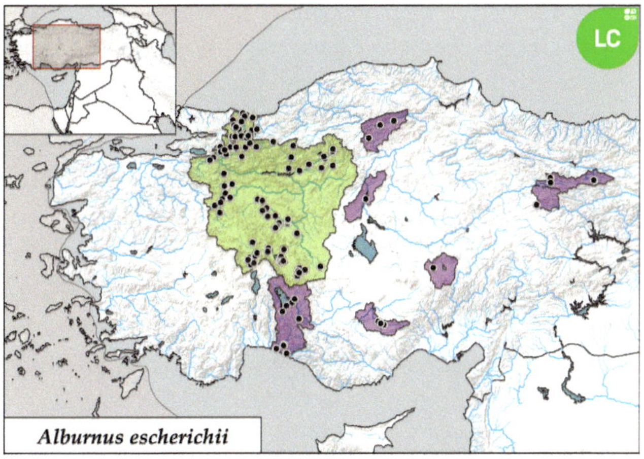

Alburnus escherichii; lower Sakarya, Türkiye; 95 mm SL.

Conservation status. LC.

Remarks. *Alburnus doriae* was a forgotten species that had not been reported for 152 years since its description in 1865. Populations of this species were later described as *Petroleuciscus esfahani* from the Zayandeh drainage and *Alburnus amirkabiri* from the Lake Namak basin; both are synonyms of *A. doriae.*

Further reading. De Filippi 1865 (description); Coad & Bogutskaya 2010 (description as *Petroleuciscus esfahani*); Mousavi-Sabet et al. 2015f (description *as A. amirkabiri*); Jouladeh-Roudbar et al. 2016c (record in Karun); Mohammadian-Kalat et al. 2017 (description, synonyms).

Alburnus escherichii

Common name. Sakarya bleak.

Diagnosis. Distinguished from other species of *Alburnus* in Central Anatolia, Mediterranean, Marmara, and Black Sea basins by: ○ 42–49+3 lateral line scales / ○ 12–16 gill rakers / ○ 12–15½ branched anal rays / ○ pelvic axillary scales present / ○ scales on back, flank, and belly well developed, overlapping / ○ in life, flank without stripes or golden upper and black lower stripe faint / ○ no mottling on flank / ○ last unbranched dorsal ray soft and flexible / ○ anal origin below dorsal base / ○ ventral keel exposed for 0–5 scales in front of anus / ○ body depth at dorsal origin 24–30 % SL / ○ 8–10 scale rows between lateral line and dorsal origin. Size up to 131 mm SL.

Distribution. Türkiye: Sakarya drainage, Lakes Eber, Akşehir, and Ilgın basins. Locally in Kızılırmak drainage (likely introduced). Introduced in Lake Beyşehir basin and Manavgat drainage and likely elsewhere.

Habitat. A wide range of moderately fast-flowing streams and rivers with gravel substrate. Spawns in fast-flowing waters on gravel beds. Lacustrine populations migrate to inflowing rivers or streams to spawn.

Biology. No data.

Conservation status. LC.

Remarks. *Alburnus nasreddini* is a synonym.

Further reading. Bogutskaya 1997b (identification); Perea et al. 2010 (phylogeny); Bayçelebi et al. 2020b (distribution, morphology, synonym); Bektaş et al. 2020 (phylogeny).

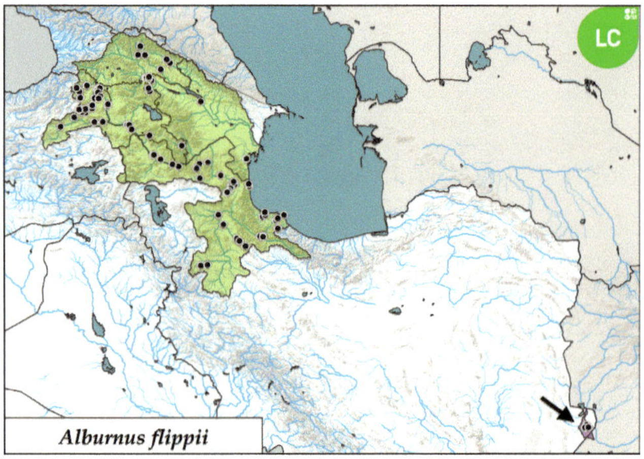

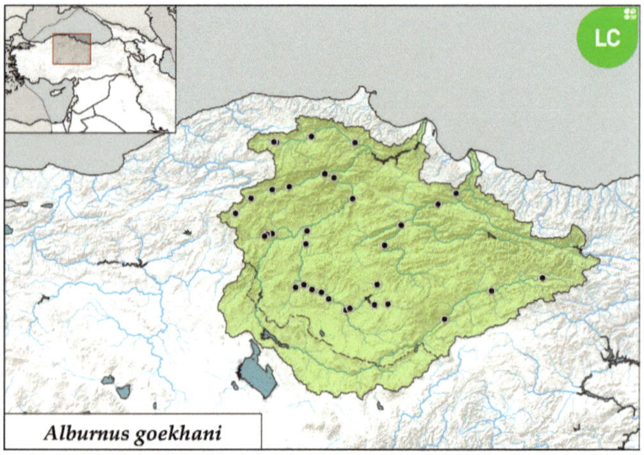

Alburnus filippii; Kura drainage, Türkiye; ~120 mm SL.

Alburnus filippii

Common name. Kura bleak.

Diagnosis. Distinguished from other species of *Alburnus* in Caspian and Lake Urmia basins by: • usually 7½ branched dorsal ray / ○ 48–61+3 lateral line scales / ○ usually 10–12½ branched anal rays / ○ 12–21 gill rakers / ○ anal origin slightly behind dorsal base / ○ a bold mid-lateral stripe / ○ ventral keel entirely naked or naked part extends more than 50 % of distance between pelvic origin and anus, rarely completely covered by scales. Size up to 125 mm SL.

Distribution. Caspian basin: Kura and Aras, east to Sefid, introduced to Chahnime lakes in Sistan basin, at altitudes up to 2000 m.

Habitat. All types of streams, lakes, and rivers except very cold mountain streams and brackish lagoons.

Biology. Life span is about 5 years, with maturity at 1 year for males and 2 years for females. Spawns May–June into July in higher altitudes.

Conservation status. LC.

Further reading. Berg 1949b (distribution, morphology); Bektaş et al. 2020 (phylogeny); Kaya et al. 2020a (distribution); Coad 2021a (biology, morphology).

Alburnus goekhani

Common name. Kızılırmak bleak.

Diagnosis. Distinguished from other species of *Alburnus* in Central Anatolia, Aegean, Marmara, and Black Sea basins by: • prominent golden upper and dark-brown or black lower stripe along flank in life / ○ body depth at dorsal origin 19–24 % SL / ○ ventral keel exposed for 7–10 scales in front of anus / ○ 48–56+2–3 lateral line scales / ○ 12–15 gill rakers / ○ 13–15½ branched anal rays / ○ anal origin situated below dorsal ray 6–8. Size up to 110 mm SL.

Distribution. Türkiye: Kızılırmak, Abdal, and Yeşilırmak drainages in southern Black Sea basin.

Habitat. A wide range of moderately fast-flowing streams and rivers with gravel substrate.

Biology. No data.

Conservation status. LC.

Further reading. Özuluğ et al. 2018 (description).

Alburnus goekhani; Yeşilırmak drainage, Türkiye; 70 mm SL.

Alburnus hohenackeri; Sefid drainage, Iran; ~70 mm SL.

Alburnus hohenackeri

Common name. Caucasian bleak.

Diagnosis. Distinguished from other species of *Alburnus* in Tigris and Caspian and Lake Urmia basins by: ○ anal origin below branched dorsal rays 4–5 / ○ 38–43+2–3 lateral line scales / ○ 11–16½ branched anal rays / ○ 16–25 gill rakers / ○ usually 8½ branched dorsal rays / ○ ventral keel well developed, entirely naked or naked part extends more than 50 % of distance between pelvic origin and anus / ○ body deep, depth 25–32 % SL, markedly laterally compressed / ○ flank silvery, without brown, grey or bluish pattern. Size up to 150 mm SL.

Distribution. Caspian basin from Kuma to Artek drainages. Widely introduced in reservoirs all over Iran.

Habitat. Slow to fast-flowing waters; prefers areas where currents are low. Common in middle and lower reaches of large rivers and their tributaries, reservoirs, and marshy streams; also in brackish water at river mouths, in estuaries and coastal lakes. Spawns in shallow water, 0.1–0.5 m deep.

Biology. Lives up to 3 years, spawns 1–2 times during life. Spawns April–July when temperatures rise above 18–23°C. Eggs sticky, laid on submerged plants, rarely on pebbles, 1.4 mm in diameter. Females lay 2–3 portions of eggs per season. Pelagic, often shoaling in upper layers of water. Feeds on plankton and other aquatic invertebrates.

Conservation status. LC.

Further reading. Berg 1949b (description; as *A. charusini*); Kottelat & Freyhof 2007 (distribution, biology, diagnosis); Kaya et al. 2020a (distribution).

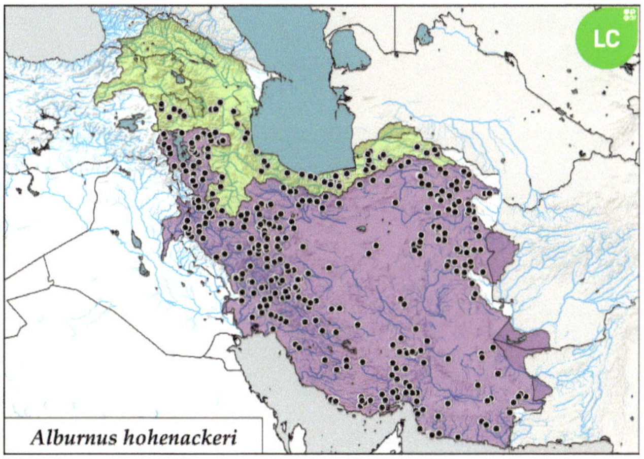

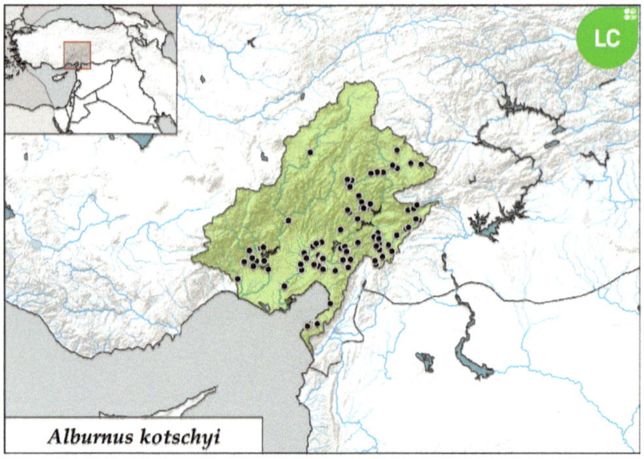

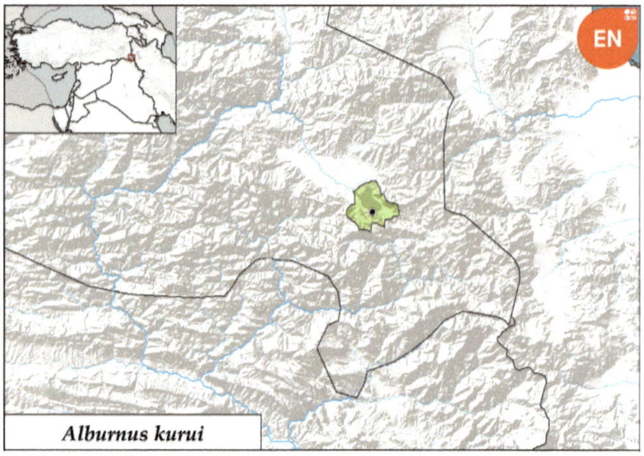

Yüksekova wetland in upper Greater Zab drainage, the habitat of endemic *Alburnus kurui*.

Alburnus kotschyi; Ceyhan drainage, Türkiye; 100 mm SL.

Alburnus kurui; Yüksekova wetland, Türkiye; 69 mm SL.

Alburnus kotschyi

Common name. Iskenderun bleak.

Diagnosis. Distinguished from other species of *Alburnus* in Mediterranean basin by: ○ prominent dark-brown mid-lateral stripe / ○ no mottling pattern on flank / ○ anal origin one scale behind last dorsal ray / ○ 9–11½ branched anal rays / ○ 49–59 total lateral line scales / ○ body depth at dorsal origin 21–26 % SL / ○ ventral keel exposed 3–10 scales in front of anus / ○ last simple dorsal ray soft / ○ pelvic axillary scales present / ○ scales on back, flank, and belly well developed, overlapping. Size up to 141 mm SL.

Distribution. Türkiye: Tarsus, Seyhan, and Ceyhan drainages and coastal streams flowing to Gulf of İskenderun north of Orontes.

Habitat. A wide range of moderately fast-flowing streams and rivers with gravel substrate.

Biology. No data.

Conservation status. LC.

Remarks. *Alburnus kotschyi* is often confused with *A. adanensis*, but both species can be easily distinguished.

Further reading. Steindachner 1863 (description); Freyhof et al. 2018a (description).

Alburnus kurui

Common name. Yüksekova bleak.

Diagnosis. Distinguished from other species of *Alburnus* in Persian Gulf, Qweiq, Lake Van, and Iranian endorheic basins by: ● 8–12 gill rakers / ○ ventral keel absent / ○ 9–11½ branched anal rays / ○ 51–58 total lateral-line scales / ○ a prominent dark-brown mid-lateral stripe / ○ body short and stout / ○ snout rounded / ○ head width at nape 15–17 % SL / ○ head depth at nape 18–20 % SL. Size up to 83 mm SL.

Distribution. Türkiye: Yüksekova wetland in upper Greater Zab drainage.

Habitat. Springs and spring-fed streams. Likely to migrate to flooded meadows in winter.

Biology. No data.

Conservation status. EN; appears to be declining within its very small range. Permanent water bodies such as small springs and streams in Yüksekova wetland are very limited and under increasing pressure from development.

Remarks. The reasons why this species is restricted to a single high-altitude wetland in Tigris drainage are not clear, but it has yet to be found elsewhere despite intensive fieldwork in the region. It was placed in *Petroleuciscus,* but it is very closely related to *Alburnus timarensis* from the Lake Van basin, although the two areas are not contiguous.

Further reading. Bogutskaya 1995 (description); Freyhof et al. 2018b (generic assignment).

Stream Serinyol in lower Orontes drainage is one of the few habitats of *Alburnus magnificus*.

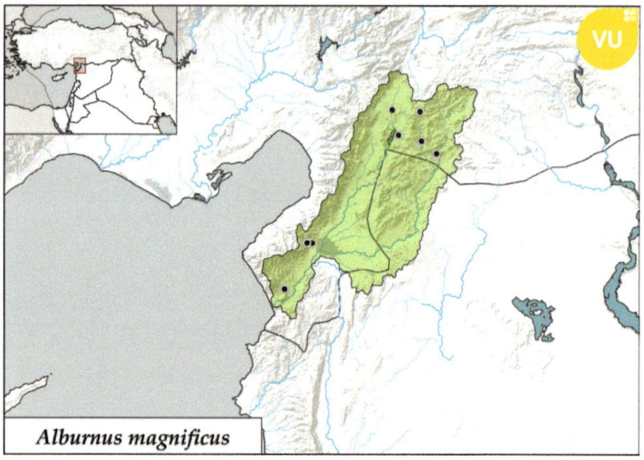

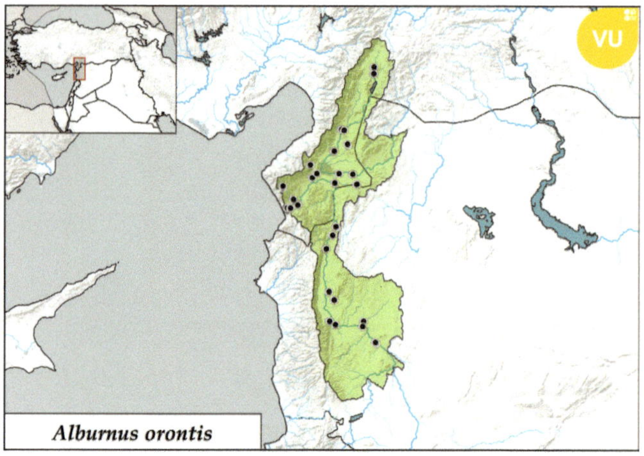

Alburnus magnificus; lower Orontes drainage, Türkiye; 75 mm SL.

Alburnus magnificus

Common name. Hatay spotted bleak.

Diagnosis. Distinguished from other species of *Alburnus* in Mediterranean basin by: ○ mottled pattern on flank, pattern clear in live individuals / ○ anal origin below or behind last dorsal ray / ○ 4–5 scale rows between lateral line and anal origin / ○ ventral keel exposed for 5–10 scales in front of anus / ○ 9–11½ branched anal rays / ○ 9–12 gill rakers / ○ 45–52 total lateral line scales / ○ 4–5 scale rows between lateral line and anal origin / ○ last simple dorsal ray soft / ○ pelvic axillary scales present / ○ scales on back, flank, and belly well developed, overlapping / ○ bold black, grey or brown scales on silvery or brown background. Size up to 82 mm SL.

Distribution. Türkiye and Syria: Orontes drainage.

Habitat. Moderately fast-flowing streams with gravel substrate.

Biology. No data.

Conservation status. VU; found in fewer than 10 declining populations. Seems to have disappeared from Syria.

Remarks. *Alburnus magnificus* is related to *A. qalilus* from coastal streams in Syria.

Further reading. Geiger et al. 2014 (phylogeny); Freyhof & Turan 2019 (description).

Alburnus orontis

Common name. Orontes spotted bleak.

Diagnosis. Distinguished from other species of *Alburnus* in Mediterranean basin by: ○ 52–59+2–3 lateral line scales / ○ 12–15½ branched anal rays / ○ ventral keel exposed 0–2 scales / ○ anal origin behind last branched dorsal ray / ○ flank plain silvery, yellowish or with fine mottled pattern / ○ last simple dorsal ray stiff / ○ pelvic axillary scales present / ○ scales on back, flank, and belly well developed, overlapping / ○ 5–7 scale rows between lateral line and anal origin. Size up to 126 mm SL.

Distribution. Türkiye and Syria: Orontes drainage.

Habitat. Moderately fast-flowing streams and rivers with gravel or sand substrate.

Biology. No data.

Conservation status. VU; occurs in less than 10 independent, mostly declining populations.

Further reading. Krupp 1985d (morphology, distribution); Geiger et al. 2014 (phylogeny); Freyhof & Turan 2019 (identification, distribution).

Alburnus orontis; lower Orontes, Syria; 90 mm SL.

Alburnus qalilus; Nahr al Sanawbar, Syria; 104 mm SL.

Alburnus qalilus

Common name. Syrian spotted bleak.

Diagnosis. Distinguished from other species of *Alburnus* in Mediterranean basin by: ○ mottled pattern on flank, pattern fuzzy, and usually blurred in live individuals / ○ anal origin below 5th or 6th branched dorsal ray / ○ 3 scale rows between lateral line and anal origin / ○ ventral keel exposed for 5–10 scales in front of anus / ○ 9–12 gill rakers / ○ 9–11½ branched anal rays / ○ 45–52 total lateral line scales / ○ 3 scale rows between lateral line and anal origin / ○ last simple dorsal ray soft / ○ pelvic axillary scales present / ○ scales on back, flank, and belly well developed, overlapping. Size up to 104 mm SL.

Distribution. Syria: Nahr al Kabir (North), Nahr al Sanawbar, and Nahr al-Hawaiz.

Habitat. A wide range of streams and rivers with moderately fast-flowing water and gravel substrate. In a reservoir at Nahr al Kabir but absent from the Nahr al Sanawbar and Nahr al-Hawaiz reservoirs.

Biology. No data.

Conservation status. EN; appears to be declining within its very small range.

Remarks. Occasionally been identified as *Leuciscus spurius*, a name given to a hybrid between *Squalius* and *Alburnus*.

Further reading. Krupp 1992b (description); Geiger et al. 2014 (phylogeny); Freyhof & Turan 2019 (distribution).

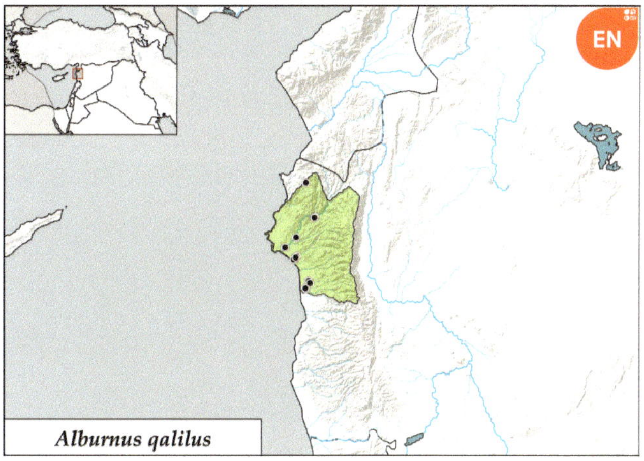

Alburnus sellal; upper Tigris, Türkiye; 90 mm SL.

Alburnus sellal

Common name. Shah kuli.

Diagnosis. Distinguished from other species of *Alburnus* in Persian Gulf, Qweiq, Lake Van, and Iranian endorheic basins by: ○ usually 63–83+3 lateral-line scales / ○ 10–25 gill rakers in riverine populations, up to 35 in lake populations / ○ 10–13½ branched anal rays / ○ 11–16 scale rows between dorsal origin and lateral line / ○ 4–7 scale rows between lateral line and pelvic origin / ○ head depth at nape 14–16 % SL / ○ postorbital distance 10–12 % SL / ○ anal origin below last branched dorsal ray or behind / ○ ventral keel covered by one or more scales / ○ flank scales without grey margins or irregular set grey pigmentation / ○ a distinct, often bold mid-lateral stripe. Size up to 180 mm SL.

Distribution. Qweiq, Euphrates, and Tigris drainages. In endorheic Kor basin in Central Iran, and rivers south to Mond and Dalaki. Introduced to Gölbaşı Lake (Ceyhan drainage) in Türkiye.

Habitat. Lakes, reservoirs, and all types of streams and rivers, from cold upland streams to subtropical rivers. Spawns in current on submerged vegetation or gravel substrate, also on wave-washed lake shores.

Biology. Gregarious. Lives up to 5–6 years, usually matures at 2. Spawns June–August. Feeds on drifting invertebrates, small animals that have fallen to surface, mainly on zooplankton in lakes, reservoirs, and large rivers.

Conservation status. LC; extirpated from Syrian part of Qweiq.

Alburnus sellal; Kor basin, Iran; 140 mm SL.

Alburnus sellal; Lake Hazar, Türkiye; 110 mm SL.

Remarks. The records from Orontes are most likely mislabelled, and this species has not been found here in 20th and 21st centuries. *Alburnus heckeli, A. mossulensis*, *A. selcuklui*, and *A. zagrosensis* are synonyms. *Alburnus heckeli* has been recognised as a separate species, differing in having more gill rakers (28–35 vs. 10–25 in *A. sellal*). Populations with high numbers of gill rakers, otherwise indistinguishable from *A. sellal*, have since been found in other lacustrine environments, including a reservoir. This suggests a rapid adaptation of this character to lentic conditions.
Further reading. Yildirim et al. 2007 (reproduction); Coad 2009b (description of *A. zagrosensis*); Coad 2010a (biology); Uçkun et al. 2014 (biology); Elp et al. 2015 (description of *A. selcuklui*); Mohammadian-Kalat et al. 2017 (*A. mossulensis*); Freyhof et al. 2018b (phylogeny, *A. selcuklui*); Eagderi et al. 2019c (*A. zagrosensis* as synonym); Bayçelebi 2020 (introduced record from Gölbaşı); Çiçek et al. 2021 (*A. heckeli*); Coad 2021a (biology, morphology).

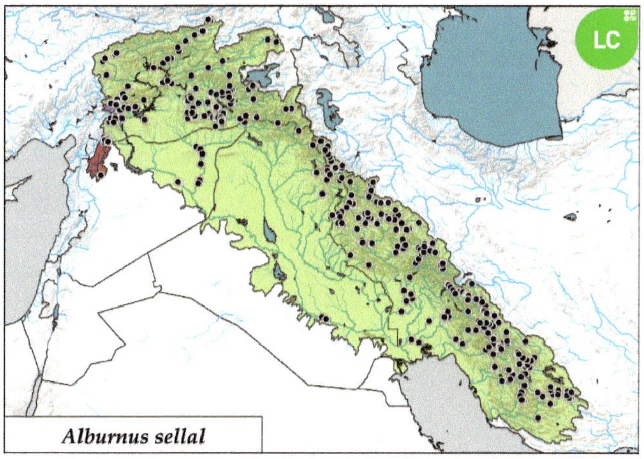

Lake Van: The largest Soda Lake in the world. Lake Van is the fourth largest endorheic lake in the world, covering a surface area of 3,755 km^2. It is situated at an altitude of 1,640 m. Lake Van is a deep lake with an average depth of 171 m and a maximum recorded depth of 451 m. Only the upper ~30 m water depth is warm and inhabited by fish, and deeper layers are too cold with rapidly decreasing oxygen concentrations. The Lake Van basin is situated within the upper Euphrates catchment. During the Pleistocene period, the lake outlet was blocked when lava from the Nemrut volcano flowed westward toward the Murat, a major tributary of the Euphrates. The lake's water level has fluctuated dramatically in the past and present. The highest recorded lake levels of 72 m above the current height were reached approximately 18,000 years ago. A significant decline to a depth of over 300 m below the present level occurred approximately 9,500 years ago, followed by an equally dramatic rise around 6,500 years ago. In recent decades, smaller fluctuations of a few metres have also been observed. Lake Van is situated in a region where the Arabian Plate from the south meets the Eurasian Plate from the north and the east. The lake lies within a tectonic depression, part of an active fault system responsible for regional volcanism, seismic activity, and hydrothermal processes. Evaporation processes, hydrothermal activity, and chemical weathering of volcanic rocks, together with a supply of alkaline tephra into the lake basin, are responsible for the extreme alkalinity of the lake water (pH 9.8, salinity 21.4 ‰). The lake is characterised by layers of sediments up to 400 m thick, which date back to approximately 500,000 years ago. These sediments have attracted the attention of climatologists and volcanologists who are interested in drilling cores. The cores demonstrated that Lake Van was initially a freshwater lake. However, the freshwater fish fauna of Lake Van has undergone a significant decline, with only two species remaining: *Alburnus tarichi*, which forages in the lake but migrates to freshwater tributaries to spawn, and *Oxynoemacheilus ercisianus*, which inhabits both the lake and its tributaries. **Further reading.** Sumita & Schmincke 2013 (Lake Van).

Alburnus tarichi; Lake Van basin, Türkiye; 155 mm SL. © M. Özuluğ.

Alburnus tarichi

Common name. Tarek.

Diagnosis. Distinguished from other species of *Alburnus* in Persian Gulf, Qweiq, Lake Van, and Iranian endorheic basins by: ○ 69–88+3 lateral-line scales / ○ 21–29 gill rakers / ○ 10–12½ branched anal rays / ○ 11–15 scale rows between dorsal origin and lateral line / ○ 5–8 scale rows between lateral line and pelvic origin / ○ anal origin 0–4 scales behind last branched dorsal ray / ○ ventral keel scaleless or one scale behind pelvic base / ○ flank scales with grey margins or irregular set grey pigmentation / ○ mid-lateral stripe absent or very faint. Size up to 220 mm SL.

Distribution. Türkiye: Lake Van basin including Lakes Nazik and Aygır. Landlocked in Koçköprü reservoir (Zilan). Introduced to Lake Erçek in 1989.

Habitat. Lacustrine, usually feeding in open water. Swims up tributaries to spawn in Van and Erçek basins. Spawns on shores of Nazik and Aygır lakes.

Biology. Lives up to 8 years. Matures at 2 years. Spawns May–June.

Conservation status. LC; fisheries for this commercial species are well managed.

Remarks. Introduced to Lake Burdur (Central Anatolia) in 1966, where it became temporarily established but disappeared before the 1980s.

Further reading. Freyhof et al. 2018b (phylogeny); Atıcı et al. 2020 (diagnosis, distribution, introductions).

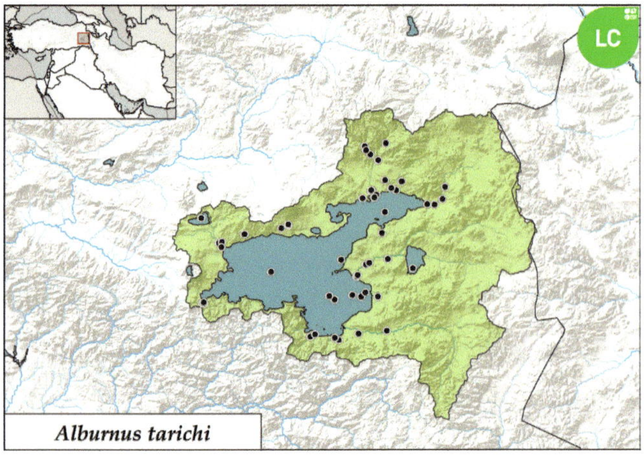

Migrating school of *Alburnus tarichi* © Saygun Dura.

Lake Van's Tarek migration as living biocultural heritage. Every year, millions of tarek make an extraordinary journey across the waters of Lake Van basin in eastern Türkiye. This incredible migration is a unique survival strategy for the species, which has adapted to the high saline and alkaline waters of Lake Van. To spawn, tarek must leave these hostile conditions and migrate to the freshwater tributaries that feed into the lake, starting a challenging migration. Migration typically begins in April, triggered by rising temperatures and increased daylight. Adult tarek swim upstream into rivers and streams in search of suitable spawning grounds. Upon reaching their spawning grounds, tarek lay their eggs in gravel beds where oxygenated water flows continuously, providing optimal conditions for egg development. Interestingly, tarek has an impressive osmoregulatory capacity, which allows it to survive in the highly alkaline waters of Lake Van. Studies have shown that it has the highest osmolarity ever recorded for a fish species, at 550 mOsm/kg. The annual tarek migration has long been a major natural event in the region, attracting both locals and tourists who come to witness the collective movement of countless iridescent fish as they travel upstream. In this sense, tarek plays a significant role in the local economy, both by creating an eco-tourism potential and by accounting for almost 30 % of Türkiye's total inland fisheries, which is the highest rate for freshwater fish catches. Historically, the importance of tarek has been documented in various records dating back to the 9th century. Local communities have relied on the fish for centuries, often salting and trading it, contributing to both the cultural and economic life of the region. The life history of the tarek is closely linked to the seasonal hydrology of its environment, which makes it highly sensitive to changes in the flow of the water and the quality of its habitat. Anthropogenic impacts such as dam construction, water pollution, habitat degradation and overfishing during vulnerable periods threaten populations and these spawning grounds. The story of the tarek migration is not only a fascinating biological phenomenon, but also a unique mass migration in West Asia. **Further reading**. Şen et al. 2015; Sarı 2008.

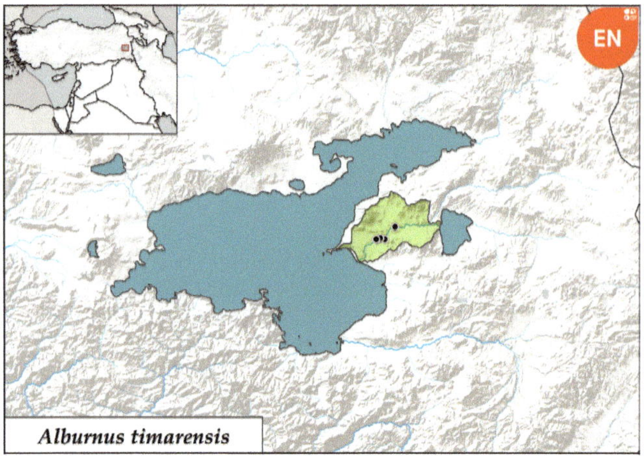

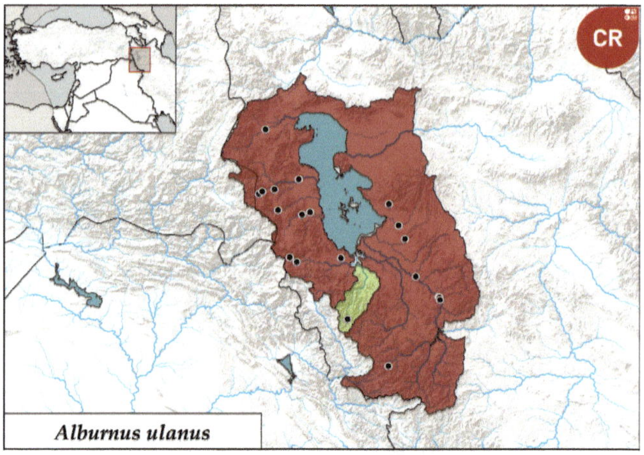

Alburnus timarensis; Karasu, Van basin, Türkiye; ~120 mm SL. © M. Özuluğ

Alburnus timarensis

Common name. Karasu tarek.

Diagnosis. Distinguished from other species of *Alburnus* in Persian Gulf, Qweiq, Lake Van, and Iranian endorheic basins by: ○ 13–17 gill rakers / ○ ventral keel present, scaleless for 5–10 scales in front of anus / ○ 9–11½ branched anal rays / ○ 51–65 (72)+3 lateral-line scales / ○ a prominent dark-brown mid-lateral stripe / ○ body slender, elongate / ○ snout pointed / ○ head width at nape 12–14 % SL / ○ postorbital distance 12–15 % SL / ○ head depth at nape 16–18 % SL / ○ flank scales without grey margins or irregular set grey pigmentation / ○ a distinct, often bold mid-lateral stripe. Size up to 113 mm SL.

Distribution. Türkiye: Karasu drainage in Lake Van basin.

Habitat. Slow to moderately fast-flowing streams.

Biology. No data.

Conservation status. EN; seems to decline within its very small range.

Remarks. A resident freshwater species found together with *A. tarichi* during the spawning season.

Further reading. Kuru 1980 (description); Elp et al. 2013 (description); Freyhof et al. 2018b (phylogeny).

Alburnus ulanus

Common name. Mahabad bleak.

Diagnosis. Distinguished from other species of *Alburnus* in Caspian and Lake Urmia basins by: • ventral keel between posterior pelvic base and anus absent or very shallow / ○ 12–16 gill rakers / ○ 7–10½ branched anal rays / ○ 33–42+3 lateral line scales / ○ a bold, dark-brown mid-lateral stripe / ○ usually 8½ branched dorsal rays. Size up to 130 mm SL.

Distribution. Iran: Lake Urmia basin.

Habitat. Lowland sections of streams and rivers with slow-flowing to standing water and submerged vegetation.

Biology. Lives for 4 years. Matures at 2 years. Spawns April–June. Feeds on aquatic invertebrates, plants, and phytoplankton.

Alburnus ulanus; Mahabad, Lake Urmia basin; ~60 mm SL.

Remarks. Formerly placed in *Petroleuciscus*. *Alburnus ulanus* shares mtDNA with *Alburnus atropatenae*, but both species are well differentiated by morphological characters, and introgressive hybridisation is probably responsible for the similar mtDNA.

Conservation status. CR; appears to be on the verge of extinction. Several field teams have failed to record it in recent years. Only one individual was found in a heavily polluted stream in 2018. It was once widespread in and around wetlands along lower reaches of tributaries of Lake Urmia. Massive habitat alteration due to construction of many dams has meant that there are now very few slow-flowing lowland rivers and streams left. These, however, have been severely affected by pollution.

Further reading. Abbasi & Sabkara 2004a (food); Abbasi & Sabkara 2004b (biology); Abbasi et al. 2006 (morphology); Ghasemi et al. 2015 (distribution); Coad 2021a (biology, morphology).

Mahabad, the last known habitat of *Alburnus ulanus*, was not in good condition in 2018.

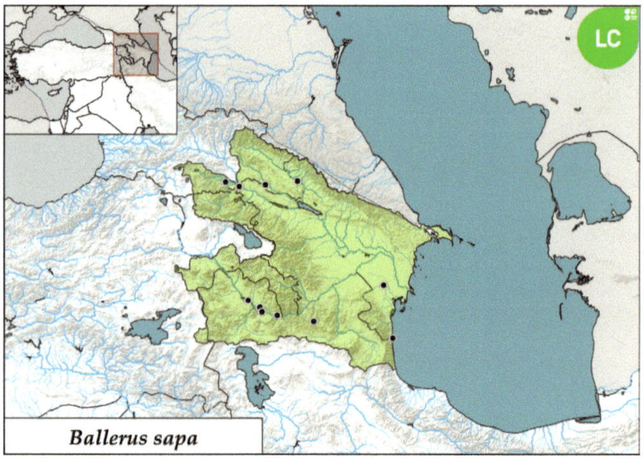

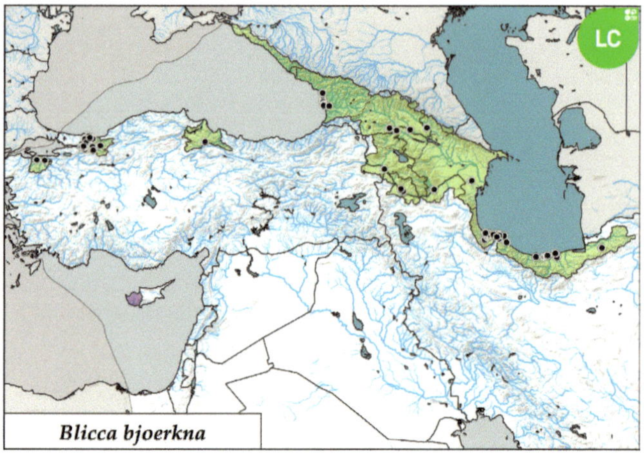

Ballerus sapa; Danube drainage, Hungary; ~150 mm SL. © Z. Sallai.

Ballerus sapa

Common name. Zobel.

Diagnosis. Distinguished from *Abramis*, *Acanthobrama*, *Blicca*, and *Vimba* by: ○ mouth small, inferior / ● 47–54 total lateral line scales / ● eye large, about equal to snout length / ● 34–41½ branched anal rays. Size up to 330 mm SL.

Distribution. Kura and Aras drainages in Azerbaijan and Georgia. Aral basin and European parts of Black, Azov, and Caspian basins. Introduced or native in northern Dvina drainage (White Sea basin), where it is spreading northwards from warmer upper reaches (Vychegda system). Non-native in Volkhov (a tributary of Lake Ladoga), Rhine, and Vistula drainages.

Habitat. Large lowland rivers and estuaries. Spawns in fast-flowing water on gravel bottom or submerged vegetation.

Biology. A rheophile, nocturnal species. First spawns at 3–4 years. Spawns April–May when temperature rises above 8°C, in large aggregations. Semi-anadromous populations feed in large brackish water habitats in estuaries around Black and Caspian Seas. Spawning migration of semi-anadromous individuals begins in November (Kura). Feeds on benthic invertebrates.

Conservation status. LC.

Remarks. Often listed for Iran but has yet to be confirmed. Very rare in Kura and Aras.

Further reading. Berg 1949b (biology); Bănărescu 1964 (biology); Kottelat & Freyhof 2007 (distribution, biology, diagnosis).

Blicca bjoerkna

Common name. Silver bream.

Diagnosis. Distinguished from *Abramis*, *Acanthobrama*, *Ballerus*, and *Vimba* by: ○ sub-inferior mouth, which cannot be extended as a tube / ○ 43–46+2–3 lateral line scales /

Blicca bjoerkna; Odra drainage, Germany; ~300 mm SL.

○ 19–23½ branched anal rays / ○ eye diameter about equal to snout length in individuals larger than 100 mm SL / ● pharyngeal teeth 2,5–2,5 / ○ orange or reddish base of paired fins in adults. Size up to 330 mm SL.

Distribution. Caspian, Marmara, and Black Sea basins. Introduced to Cyprus. In Aral, North, Baltic, White, Northern Black (south to Rioni drainage), Atlantic basins south to Adour drainage (France; possibly introduced south of Loire) and Mediterranean basins in France (Hérault and Rhône drainages). Naturally absent from Iberian Peninsula, Italy, Adriatic basin, Crimea, Great Britain (except south-east), Scandinavia north of Sundsvall (Sweden) and 65°N (Finland). Locally introduced in Spain, north-east Italy, and France, apparently introduced in small coastal drainages of Var.

Habitat. A wide variety of shallow, warm lowland lakes and slow-flowing lower reaches of large rivers and canals. Often very abundant at bottom of large sandy rivers. Spawns along banks on submerged vegetation, roots, or shallow gravel bottoms.

Biology. Predominantly nocturnal. Gregarious. Lives more than 10 years. Males first spawn at 2 years, females at 3. Most individuals spawn 2–3 times per season, about 10 days apart, and may spawn more than once. Spawns in early morning in May–July at temperatures above 15°C. Eggs are sticky, and larvae inhabit still waters. Feeds on benthic invertebrates. Often hybridises with *Vimba vimba*. Most reported hybrids with *Abramis brama* are *A. brama* with pharyngeal teeth 2.5–5.2 (as opposed to usual 5–5).

Conservation status. LC.

Further reading. Berg 1949b (biology); Bogutskaya & Naseka 2004 (generic position); Brylinska et al. 2006 (biology); Kottelat & Freyhof 2007 (distribution, biology, diagnosis); Zogaris et al. 2012 (Cyprus).

Chondrostoma

A genus of middle-sized to large Leuciscids, found from France to the Volga, and Iran. Together with the Anatolian *Turcichondrostoma*, they are related to the west European genera *Achondrostoma, Iberochondrostoma, Pseudochondrostoma, Parachondrostoma,* and *Protochondrostoma,* which all form one monophyletic lineage. All nases are also related to the European Leuciscid genera *Phoxinellus* and *Telestes* and the West Asian genus *Pseudophoxinus,* although the precise phylogenetic relationships between these groups remain unclear.

In West Asia, 11 out of 12 species of *Chondrostoma* are characterised by a well-developed rostral cap covering most of the upper lip and a variably developed cornified cutting edge on the lower jaw. *Chondrostoma* are slender fishes that inhabit fast-flowing waters near or on stone bottom. They obtain their food by scraping algae growing on stones and rocks and invertebrates living among them, which we refer to as "aufwuchs" (literally, 'growth'). Adults have an inferior mouth with a cornified sheath covering the lower lip, often with a sharp edge. In some species, the mouth is arched, as in most leuciscids. In more specialised species, the anterior edge of the lower lip (bearing the cornified sheath) is straight. In such species, the lower lip is distinctly fleshy only at the corners of the mouth, and the cornified sheath appears as a blade. The mouth morphology changes during ontogeny and all species with a straight mouth in adults have an arched mouth when they are juveniles. In West Asia, all *Chondrostoma* are difficult to distinguish, and many species are very closely related. The published diagnoses of several species do not provide an effective means of differentiating them. Consequently, the following keys are presented as a provisional guide only. Given their close phylogenetic relationship, future studies are anticipated to reveal some of the species to be conspecific.

Further reading. Elvira 1987 (revision); Robalo et al. 2007 (phylogeny); Geiger et al. 2014 (phylogeny); Çiftçi et al. 2020 (phylogeny).

Keys to species of *Chondrostoma* in West Asia

Aegean, Marmara, and Black Sea basins

1a - Usually 9, rarely 8, branched pelvic rays.
..................2
1b - Usually 8 branched pelvic rays.
..................3

2a - Usually 8½ branched dorsal rays; usually 9½ branched anal rays.
..................*C. nasus*
2b - Usually 9½ branched dorsal rays; usually 10½ branched anal rays.
..................*C. colchicum*

3a - Lateral line with 44–54 total lateral line scales.
..................*C. turnai*
3b - Lateral line with 52–69 total scales.
..................4

4a - 52–60 total lateral line scales.
..................*C. meandrense*
4b - 60–69 total lateral line scales.
..................*C. holmwoodii*

Central Anatolia and Mediterranean basin east of Antalya

1a - 32–40 gill rakers.
..................2
1b - 24–35 gill rakers.
..................3

2a - A prominent keel on back in front of dorsal.
..................*C. beysehirense*
2b - No keel on back in front of dorsal.
..................*C. kinzelbachi*

3a - Keel between anus and posterior pelvic base absent or very poorly developed; sheath covering lower lip strongly cornified.
..................*C. toros*
3b - Keel between anus and posterior pelvic base shallow but well developed; sheath covering lower lip poorly cornified.
..................*C. ceyhanense*

Caspian, Persian Gulf, and Kor endorheic basin in Iran

1a - 15–17 gill rakers; no cornified sheath covering lower lip.
..................*C. esmaeilii*
1b - (17) 25–34 gill rakers; a cornified sheath covering lower lip.
..................2

2a - Mouth cleft strongly arched; cornified sheath covering lower lip very thin.
..................*C. cyri*
2b - Mouth cleft straight or slightly arched; cornified sheath covering lower lip thick.
..................*C. regium*

Chondrostoma beysehirense; Lake Beyşehir basin, Türkiye; 140 mm SL.

Chondrostoma beysehirense

Common name. Beyşehir nase.

Diagnosis. Distinguished from other species of *Chondrostoma* in Central Anatolia and eastern Mediterranean basin by: • body considerably compressed / ○ keels covered by scales on back in front of dorsal and on belly between pectoral and pelvic / ○ posterior scale margin markedly undulated / ○ 60–71 total lateral-line scales / ○ (6–7) 7–7 pharyngeal teeth / ○ 8–9½ branched dorsal rays / ○ 8–11½, usually 9½, branched anal rays / ○ 33–40 gill rakers / ○ thick cornified sheath covering lower lip. Size up to 251 mm SL.

Distribution. Türkiye: Lake Beyşehir basin.

Habitat. Fast to moderately fast-flowing rivers and streams, often with dense submerged vegetation and gravel substrate.

Biology. Feeds on periphyton scraped from hard substrate.

Conservation status. VU; restricted to a few tributaries of the lake. Probably also inhabited lake, but currently absent due to invasion by non-native species.

Further reading. Bogutskaya 1997a (description); Güçlü et al. 2018 (morphology); Çiftçi et al. 2020 (phylogeny); Bayçelebi et al. 2020a (distribution).

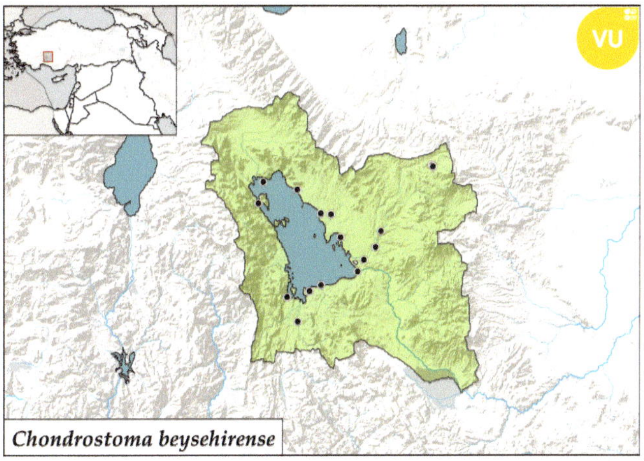

Chondrostoma ceyhanense; Ceyhan drainage, Türkiye; ~160 m SL.

Chondrostoma ceyhanense

Common name. Cilician nase.

Diagnosis. Distinguished from other species of *Chondrostoma* in Central Anatolia and eastern Mediterranean basin by: ○ 59–68 total lateral-line scales / ○ usually 6–6 pharyngeal teeth / ○ 8–9½ branched dorsal rays / ○ 9–11½, usually 9½, branched anal rays / ○ 24–29 gill rakers / ○ poorly cornified sheath covering lower lip / ○ back in front of dorsal without keel / ○ keel between anus and posterior pelvic base shallow but well-developed. Size up to 265 mm SL.

Distribution. Türkiye: Tarsus, Seyhan, and Ceyhan drainages.
Habitat. Moderate to fast-flowing large to medium-sized rivers with rocky or gravelly bottom. Common in reservoirs. Spawns in fast-flowing waters on shallow gravel beds, often in small tributaries.
Biology. Lives up to 5 years. First spawns at 2–3 years. Spawns April–June (Sır Reservoir, Ceyhan). Feeds on periphyton scraped from hard substrate.
Conservation status. LC.
Remarks. *Chondrostoma ceyhanense*, *C. kinzelbachi*, and *C. toros* are very closely related to *C. regium* and are difficult to distinguish. Future research may clarify whether they are better treated as distinct populations of *C. regium.*
Further reading. Kara & Solak 2004 (age, growth; as *C. regium*); Küçük et al. 2017 (description); Çiftçi et al. 2020 (phylogeny).

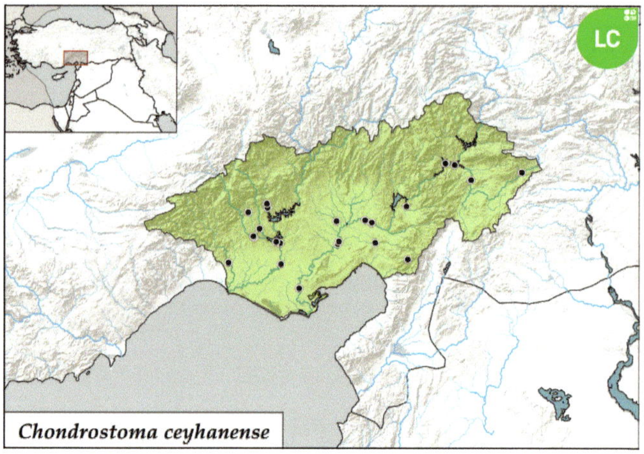

Chondrostoma ceyhanense

Chondrostoma colchicum; Çoruh drainage, Türkiye; ~140 mm SL. © M. Özuluğ.

Chondrostoma colchicum
Common name. Colchic nase.
Diagnosis. Distinguished from other species of *Chondrostoma* in Aegean, Marmara, and Black Sea basins by: ○ 55–69, usually 58–64, total lateral-line scales / ○ 6–6 or 6–5, pharyngeal teeth / ○ 8–10½, usually 9½, branched dorsal rays / ○ 8–11½, usually 9½, branched anal rays / ○ 20–30 gill rakers / usually 9, rarely 8, branched pelvic rays / ○ thick cornified sheath covering lower lip. Size up to 300 mm SL.
Distribution. Türkiye: Sakarya east to Çoruh (Türkiye and Georgia) and Black Sea basin from Çoruh north to Novorossiysk (Georgia and Russia).
Habitat. Moderate to fast-flowing, large- to medium-sized rivers with rocky or gravel bottoms, also in lakes and reservoirs. Spawns in fast-flowing water on shallow gravel beds, often in small tributaries.
Biology. Matures at 3 years. Spawning season depends on altitude and water temperature, peaking in April at 12–15°C. Females lay 2–3 batches of eggs during a spawning season. Feeds on periphyton scraped from hard substrate.
Conservation status. LC.
Remarks. *Chondrostoma angorense* from Sakarya is a synonym, and *C. colchicum* is very closely related to *C. nasus,* both may be conspecific.
Further reading. Derjugin 1899 (description); Berg 1949b (morphology); Elanidze 1983 (morphology, biology); Elvira 1987 (description); Tarkan et al. 2007 (distribution); İlhan & Balık 2008 (distribution); Bayçelebi et al. 2015 (distribution); Güçlü et al. 2018 (morphology); Çiftçi et al. 2020 (phylogeny); Küçük et al. 2023 (synonymy of *Chondrostoma angorense*).

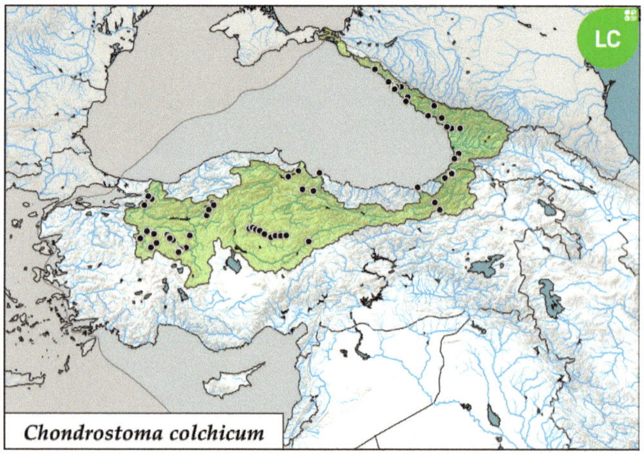

Chondrostoma colchicum

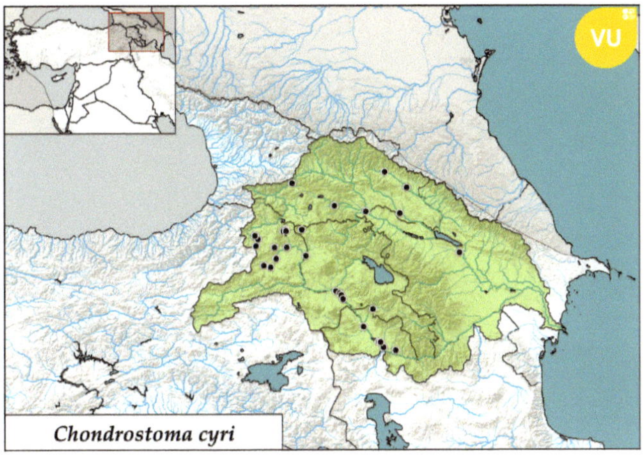

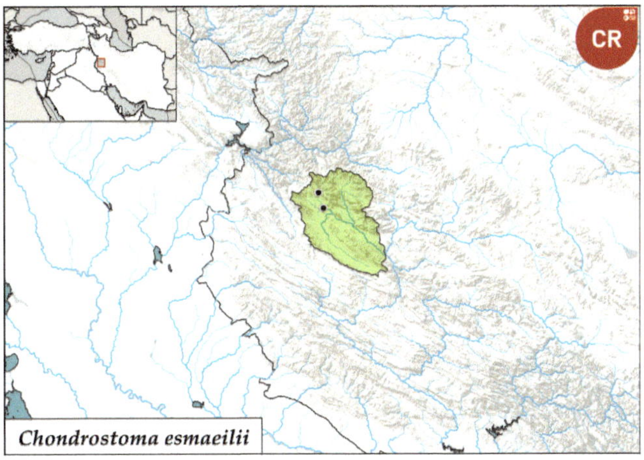

Chondrostoma cyri; Kura drainage, Türkiye; ~130 mm SL. © M. Özuluğ.

Chondrostoma cyri

Common name. Southern Caspian nase.

Diagnosis. Distinguished from other species of *Chondrostoma* in Caspian, Persian Gulf, and Kor endorheic basins in Iran by: • mouth strongly arched both in juveniles and adults / • cornified sheath covering lower lip very thin / ○ 48–73, usually 52–62, total lateral-line scales / ○ 6–5, rarely 6–6 or 5–5, pharyngeal teeth / ○ 7–9½, usually 8½, branched dorsal rays / ○ 8–10½, usually 9½, branched anal rays / ○ (17) 21–26 (32) gill rakers. Size up to 246 mm SL.

Distribution. Caspian basin: Kura and Aras drainages.

Habitat. Foothill and mountain rivers with strong to slow currents, sandy, and rocky to gravelly bottoms, also in lakes and reservoirs. Spawns on hard bottoms, migrate from lakes and reservoirs into tributaries to spawn.

Biology. Lives at least 5 years and matures at 2–3 years. Spawning season depends on altitude and water temperature, peaking in April (lower Kura), June–July (upper Kura). Females spawn one batch of eggs per year. Feeds on periphyton.

Conservation status. VU; has declined due to hydropower development and pollution.

Further reading. Elanidze 1983 (morphology, biology); Elvira 1987 (review); Güçlü et al. 2018 (morphology); Çiftçi et al. 2020 (phylogeny); Kaya et al. 2020a (distribution); Coad 2021a (biology, morphology).

Chondrostoma esmaeilii

Common name. Ravansar nase.

Diagnosis. Distinguished from other species of *Chondrostoma* in Caspian, Persian Gulf, and Kor endorheic basins in Iran by: • no cornified sheath covering lower lip / ○ mouth markedly arched both in juveniles and adults / ○ 8–9½, usually 8½, branched dorsal rays / ○ 10½ branched anal rays / ○ 51–58 total lateral-line scales / • 15–17 gill rakers / ○ 6–6 pharyngeal teeth. Size up to 120 mm SL.

Distribution. Iran: Sarab-e Ravansar, a headwater of Seymareh, a tributary of Karkheh.

Habitat. Slow to moderate-flowing streams with muddy and sandy bottoms and dense riparian vegetation.

Biology. Feeds on periphyton scraped from hard substrate. No data.

Conservation status. CR; known only from a small population in a very small area. Desiccation of the river is a major threat, and the species may be at the edge of extinction.

Remarks. Mitochondrial DNA analysis places this species in *C. regium*, which is widespread in the region. The

Chondrostoma esmaeilii; Sarab-e Ravansar, Iran; 104 mm SL.

general morphology suggests that *C. esmaeilii* may be a hybrid between *C. regium* and *Squalius verepi*. As these two species have not been found in the range of *C. esmaeilii* (despite several attempts), this species is likely a stabilised species of hybrid origin that reproduces independently of its parental species. More research is needed to clarify the situation.
Further reading. Eagderi et al. 2017c (description).

Chondrostoma holmwoodii; Bakırçay drainage, Türkiye; ~90 mm SL.

Chondrostoma holmwoodii

Common name. Eastern Aegean nase.
Diagnosis. Distinguished from other species of *Chondrostoma* in Aegean, Marmara, and Black Sea basins by: ○ 60–69 total lateral-line scales / ○ 6–6, rarely 7–6, pharyngeal teeth / ○ 8½ branched dorsal rays / ○ 8–10½, usually 9½, branched anal rays / ○ usually 8 branched pelvic rays / ○ 21–24 gill rakers / ○ thick cornified sheath covering lower lip. Size up to 230 mm SL.
Distribution. Türkiye: Gediz north to Bakırçay drainages.
Habitat. Mid-sized streams to large rivers with rocky or gravelly bottoms and moderately fast currents, also in reservoirs.
Biology. Lives up to 5 years. Matures at 2–3 years. Spawns March to early May.
Conservation status. VU; appears to be declining within its small range.
Further reading. Elvira 1987 (review); İlhan et al. 2011 (growth, biology); Güçlü et al. 2018 (morphology); Çiftçi et al. 2020 (phylogeny); İlhan et al. 2020 (distribution).

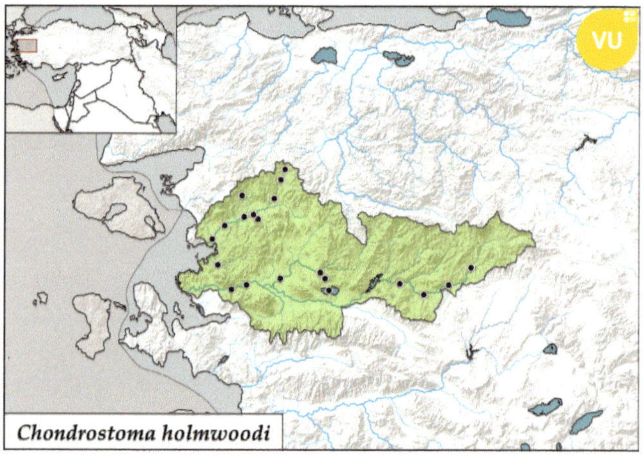

Chondrostoma kinzelbachi; Orontes drainage, Türkiye; ~180 mm SL.

Chondrostoma kinzelbachi

Common name. Levantine nase.

Diagnosis. Distinguished from other species of *Chondrostoma* in Central Anatolia and eastern Mediterranean basin by: ○ (59) 63–73 (75) total lateral-line scales / ○ usually 7–7 pharyngeal teeth / ○ 9–10½, usually 9½, branched dorsal rays / ○ 10–13½, usually 10–11½, branched anal rays / ○ 32–38 gill rakers / ○ thick sheath covering lower lip / ○ no keel on back in front of dorsal. Size up to 220 mm SL.

Distribution. Türkiye and Syria: Orontes drainage.

Habitat. Medium-sized streams to large rivers with rocky or gravelly bottoms and fast to moderately fast currents. Also, in natural lakes and large reservoirs from which it migrates into tributaries to spawn.

Biology. No data.

Conservation status. EN; appears to be declining within its very small range. Once widespread in Orontes, may now be restricted to Lake Gölbaşı and upper Karasu, including Tahtaköprü reservoir.

Remarks. *Chondrostoma kinzelbachi*, *C. ceyhanense*, and *C. toros* are all very closely related to *C. regium* and are difficult to distinguish. Future research may clarify whether they are better treated as distinct populations of *C. regium*.

Further reading. Krupp 1985a (description); Bogutskaya 1997a (identification); Güçlü et al. 2018 (morphology); Bayçelebi 2020 (distribution); Çiftçi et al. 2020 (phylogeny).

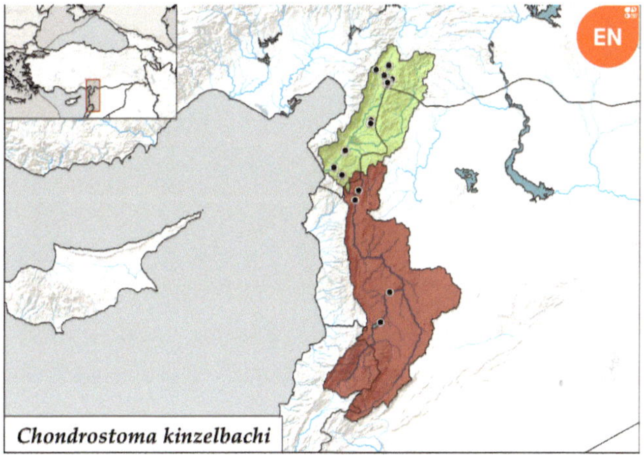

Chondrostoma meandrense; Işıklı, Türkiye; ~150 mm SL.

Chondrostoma meandrense

Common name. Işıklı nase.

Diagnosis. Distinguished from other species of *Chondrostoma* in Aegean, Marmara, and Black Sea basins by: ○ 52–60 total lateral-line scales / ○ 6–6, rarely 6–5, pharyngeal teeth / ○ 8½–9½, branched dorsal rays / ○ 9–10½ branched anal rays / ○ 23–30 gill rakers / ○ thick sheath covering lower lip / ○ usually 8 branched pelvic rays. Size up to 209 mm SL.

Distribution. Türkiye: Lake Işıklı basin in upper Büyük Menderes drainage.

Habitat. Lakes, reservoirs, medium-sized streams, and large rivers have rocky or gravelly bottoms and moderate to slow currents.

Biology. Lives up to 5 years. Matures at 1–2 years. Spawns March–May.

Conservation status. EN; appears to be declining within its very small range.

Further reading. Elvira 1987 (description); Özcan & Balık 2011 (age and growth); İlhan et al. 2011 (growth, reproductive biology; as *C. nasus*); Güçlü et al. 2013 (distribution); Güçlü et al. 2018 (morphology); Çiftçi et al. 2020 (phylogeny).

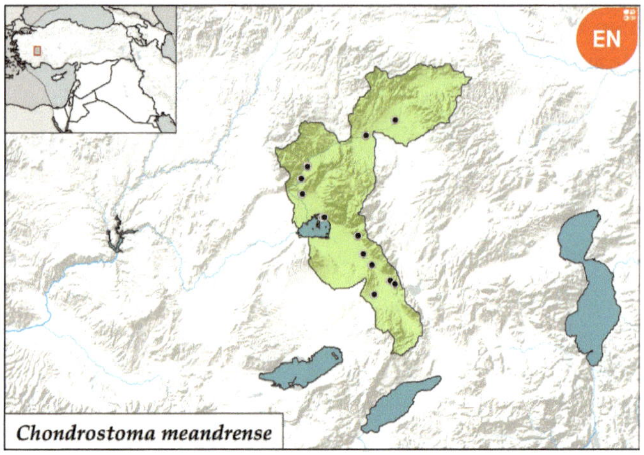

Chondrostoma nasus; Danube drainage, Romania; ~150 mm SL.

Chondrostoma nasus

Common name. Nase.

Diagnosis. Distinguished from other species of *Chondrostoma* in Aegean, Marmara, and Black Sea basins by: ○ 8–10½, usually 8½, branched dorsal rays / ○ usually 9, rarely 8 branched pelvic rays / ○ 52–72 (usually 60–63) total lateral line scales / ○ 24–31 gill rakers / ○ (9)10–11½ branched anal rays. Size up to 460 mm SL.

Distribution. Türkiye: Susurluk and Hoşap drainages on Biga peninsula. Danube, Dniestr, South Bug, and Dniepr drainages in Black Sea basin, Neman, Odra, and Vistula in southern Baltic basin and rivers of southern North Seas (westward to Meuse). Invasive or introduced in Rhône, Loire, Hérault, Seine (France), and Soca (Italy, Slovenia) drainages.

Habitat. Moderately to rapidly-flowing large to medium-sized rivers with rocky or gravelly bottoms. Spawns in fast-flowing waters on shallow gravel beds, often in small tributaries. May show a strong size-dependent longitudinal distribution in smaller rivers, with adults occupying more upper reaches.

Biology. Lives up to 12 years. First spawns at 4–5 years. May migrate tens of kilometers to spawning grounds, often in tributaries, but does not occupy these in summer. Spawns March–May when temperature reaches 12°C. Males form large aggregations, with each male defending a small territory. Females spawn only once a year and, in some populations, for a very short time (3–5 days). Females lay sticky eggs burrowed in gravel. Feeding larvae live on banks below water surface. Early juveniles are benthic and inhabit very shallow shoreline habitats. As they grow, they leave banks for faster-flowing water. Recruitment closely linked to high spring temperatures, absence of spring floods and availability of shallow-water riparian habitats. Juveniles overwinter in backwaters or in celvics along banks. Adults form dense schools in lower reaches of rivers during winter. Larvae and early juveniles with superior and later terminal mouths feed on small invertebrates. Larger juveniles and adults with inferior mouths feed on benthic diatoms, and detritus scraped from hard substrate in habitats with strong currents.

Conservation status. LC.

Remarks. Populations in Adriatic basin from Aoos (Greece, Albania) north to Drin and Skadar Lake basin constitute a separate valid species: *C. ohridanum*.

Further reading. Elvira 1987 (systematics); Nelva 1989 (biology); Maier et al. 1995 (biology); Kottelat & Freyhof 2007 (biology, distribution); Çiftçi et al. 2020 (phylogeny Susurluk population).

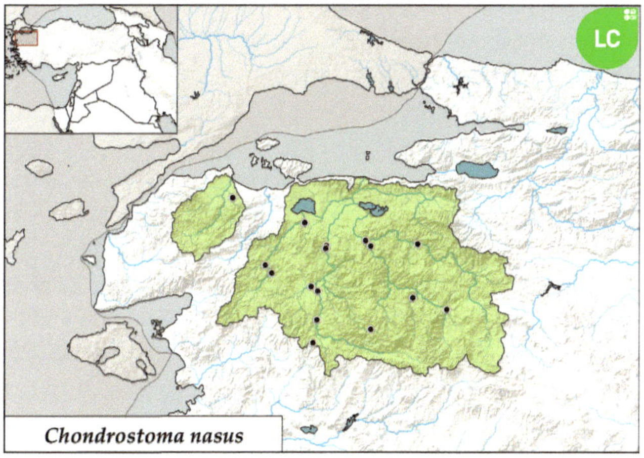

Chondrostoma regium; upper Tigris, Türkiye; ~140 mm SL.

Chondrostoma regium; Kor drainage, Iran; 124 mm SL.

Chondrostoma regium

Common name. Mesopotamian nase.

Diagnosis. Distinguished from other species of *Chondrostoma* in Caspian, Persian Gulf, and Kor endorheic basin in Iran by: ○ 47–74, usually 58–72, total lateral-line scales / ○ usually 6–6, 7–6 or 7–7 pharyngeal teeth, rarely 6–5 or 7–5 / ○ 8–9½, rarely 7½ or 10½, branched dorsal rays / ○ 8–12½, usually 9–10½, branched anal rays / ○ 22–34 gill rakers / ○ mouth cleft straight or slightly arched / ○ thick cornified sheath covering lower lip / ○ no keel on back in front of dorsal / ○ 45–48, usually 46–47 (27+20, 26+21, 26+20), total vertebrae. Size up to 400 mm SL.

Distribution. Qweiq, Euphrates, Tigris, and Karun drainages. In Zohreh drainage and Kor, Esfahan and Sirjan basins in Iran.

Habitat. Medium-sized streams and springs to large rivers with rocky or gravelly bottoms and swift to moderate currents. Spawns on gravel substrate in fast-flowing stretches of river. In reservoirs, from which they migrate to rivers or streams to spawn.

Biology. Live up to 8 years. First reproduction at 2–4 years and about 130 (males) and 150 (females) mm SL. Spawning season strongly related to altitude and latitude and triggered by water temperature of 15–25°C. Reported to spawn in January (Sirvan, Iraq), March–May (Baghdad, Iraq; Gamasiyab, Iran), May–June (Syria), and April–May in Atatürk reservoir (Türkiye). Feeds on periphyton in rivers, often taking large quantities of plankton in reservoirs.

Conservation status. LC; extirpated from Qweiq. Very rare in Kor, recorded only from Band-e Amir and Ghadamgah spring since 2007.

Remarks. There is one record from the Orontes drainage at Gis as-Sugur in Syria (caught in 1979), but this is doubtful. *Chondrostoma orientale* from the Kor in Iran is a synonym.

Further reading. Bianco & Bănărescu 1982 (description of *C. orientale*); Krupp 1985a (record from Orontes); Elvira 1987 (review); Coad 2010a (biology); Ghanbary et al. 2012 (reproductive biology); Jouladeh-Roudbar et al. 2015a (distribution in Iran); Kaya et al. 2016 (distribution); Güçlü et al. 2018 (morphology); Çiftçi et al. 2020 (phylogeny); Coad 2021a (biology, morphology).

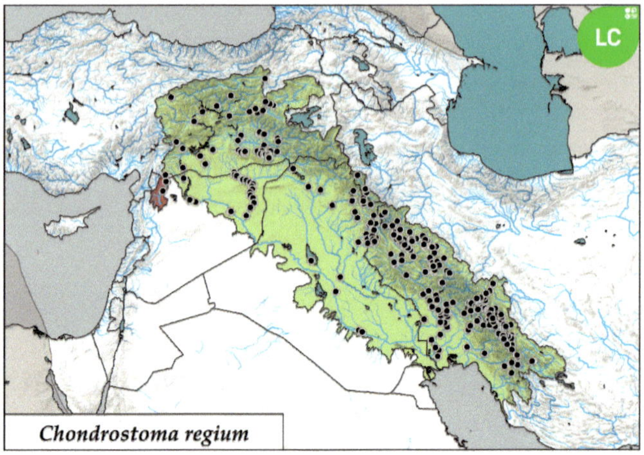

Chondrostoma regium

Chondrostoma toros; Göksu drainage, Türkiye; ~100 mm SL.

Chondrostoma toros

Common name. Taurus nase.

Diagnosis. Distinguished from other species of *Chondrostoma* in Central Anatolia and eastern Mediterranean basin by: ○ 56–64 total lateral-line scales / ○ 6–6 pharyngeal teeth / ○ 8–10½, usually 9½, branched dorsal rays / ○ 9–11½, usually 10½, branched anal rays / ○ 25–30 gill rakers / ○ thick cornified sheath covering lower lip / ○ keel between anus and posterior pelvic base absent or very poorly developed / ○ no keel on back in front of dorsal. Size up to 208 mm SL.

Distribution. Türkiye: Göksu drainage.

Habitat. Medium-sized streams to large rivers with rocky or gravelly bottoms and swift to moderate currents. Spawns on gravel substrate in swift-flowing river stretches.

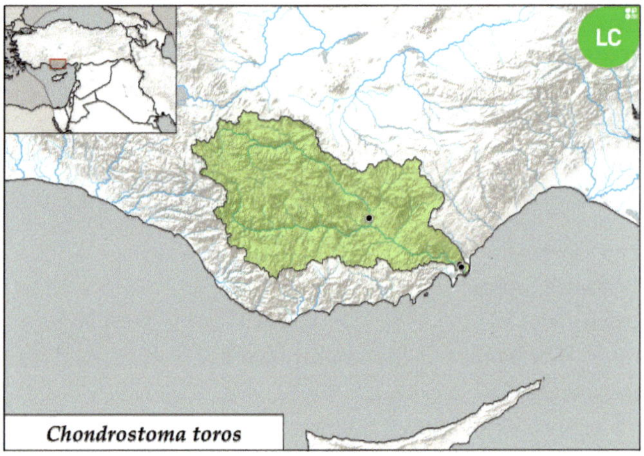

Chondrostoma toros

Chondrostoma turnai; Büyük Menderes, Türkiye; ~130 mm SL.

Biology. No data.

Conservation status. LC.

Remarks. Poorly distinguished from *C. ceyhanense,* both might be conspecific. *Chondrostoma toros*, *C. ceyhanense*, and *C. kinzelbachi* are all very closely related to *C. regium* and are difficult to distinguish. Future research may clarify whether they are better treated as distinct populations of *C. regium.*

Further reading. Bogutskaya 1997a (morphology; as *C. regium*, Göksu); Küçük et al. 2017 (description); Çiftçi et al. 2020 (phylogeny).

Chondrostoma turnai

Common name. Menderes nase.

Diagnosis. Distinguished from other species of *Chondrostoma* in Aegean, Marmara, and Black Sea basins by: ○ 44–54 total lateral-line scales / ○ 6–6, rarely 5–6 or 5–5, pharyngeal teeth / ○ 8½ branched dorsal rays / ○ 9–10½ branched anal rays / ○ 19–27 gill rakers / ○ thick cornified sheath covering lower lip / ○ usually 8 branched pelvic rays. Size up to 240 mm SL.

Distribution. Türkiye: Tahtalı, Küçük, and Büyük Menderes drainages.

Habitat. Moderate to fast-flowing large to medium-sized rivers with rocky or gravel bottoms. Spawns in fast-flowing water on shallow gravel beds, often in small tributaries.

Biology. Feed on periphyton scraped from hard substrate.

Conservation status. LC; disappeared from the Küçük Menderes drainage, which is now dry in upper and heavily polluted in lower reaches.

Remarks. Populations from Tahtalı were described as *C. smyrnae* based on very small molecular differences only. Morphological characters distinguishing *C. smyrnae* from *C. turnai* could not be confirmed based on our material, and *C. smyrnae* is treated as a synonym of *C. turnai.*

Further reading. Güçlü et al. 2018 (description); Küçük et al. 2021 (phylogeny, description of *C. smyrnae*).

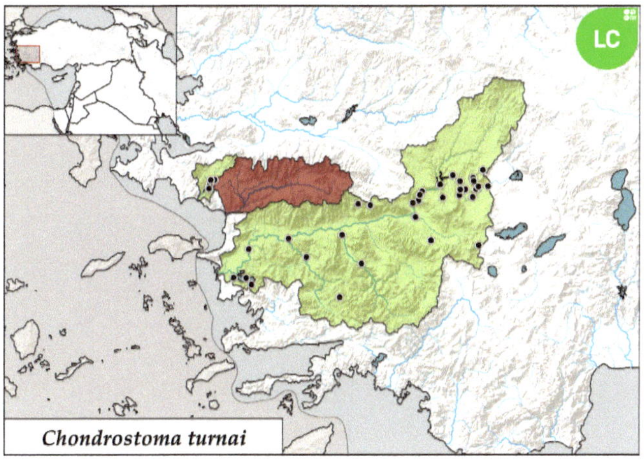

Hammar Marsh in Iraq is one of the few large wetlands left in the region. © U. Eichelmann.

Draining the Hammar Marsh. The marshes in southern Iraq were distinctive, comprising extensive wetlands once teeming with biodiversity. The construction of the Main Outfall Drain (MOD), subsequently renamed the Third River/Saddam River project, was commenced in 1953 to drain the marsh area between the Euphrates and Tigris, situated south of Baghdad. Two additional canals were constructed near the MOD: the Mother of Battles Canal and the Fidelity to the Leader Canal. The former was used to divert the flow of the Euphrates southeast to the Hammar Marsh, and the latter transported water from the MOD to the south of Basra. The drainage of the marshes was initiated with the erection of a dike to divert Central Marsh overflow away from Hammar Marsh. Drought in the area was further exacerbated by dam construction in Türkiye and Syria. The Iran–Iraq War also contributed to the deterioration of the wetlands, with 94 % of Hammar Marsh having been converted to bare ground by 2001. Following the 2003 invasion of Iraq, the Iraqi government, the United Nations, and various U.S. agencies attempted to restore the marshland. It appears that there has been some recovery of the wetland. Still, the reduced availability of water from the Euphrates and Tigris due to the construction of numerous dams, other water stressors, and climate change-driven droughts continue to pose a significant risk to the marshes. **Further reading.** Alwan 2013 (Euphrates).

Egirdira

Egirdira comprises small fish that inhabit densely vegetated streams and springs. The single species is endemic to Central Anatolia. It is distinguished from all other Leuciscids in West Asia by having the pectoral rays 2–3 thicker and slightly slanted upward, with a conspicuous downward angle at the distal end of the thickened part in the male. This character is shared with the European genera *Pelasgus* and *Delminichthys*, both related to *Egirdira*. **Further reading.** Freyhof et al 2022 (generic name, identification).

Figure 41. Modified pectoral rays in male *Egirdira nigra*.

Egirdira nigra; Lake Eğirdir basin, Türkiye; nuptial female, ~60 mm SL.

Egirdira nigra; Lake Eğirdir basin, Türkiye; nuptial male, 50 mm SL.

Egirdira nigra; Lake Eğirdir basin, Türkiye; male, ~40 mm SL.

Egirdira nigra

Common name. Egirdira.

Diagnosis. Only species in genus. Size up to 80 mm SL.

Distribution. Türkiye: Lake Eğirdir basin, Eflatunpınarı in Lake Beyşehir basin, and Yeşilyurt spring in upper Aksu drainage.

Habitat. Densely vegetated springs and streams. In Lake Eğirdir itself and in lakeshore springs. Spawns in dense vegetation.

Biology. Spawns after first winter. Lives up to 5 years. Individual females usually spawn twice a season, large females 3–4 times. Males with golden nuptial colouration, and dark brown lateral stripe, females mottled brown. Males territorial during spawning. Spawns at night.

Conservation status. VU; appears to be declining within its small range.

Remarks. Previously placed in genus *Pseudophoxinus* and usually known as *P. egirdiri*, which is a junior synonym of *P. nigra*. The Eflatunpınarı population has a minimum COI barcode distance of 1.5 % to the Eğirdir population, with no morphological differences. This population has not been observed for more than 15 years, suggesting its possible extirpation.

Further reading. Kosswig & Geldiay 1952 (description as *P. nigra*); Küçük et al. 2009b (conservation, distribution); Freyhof 2022 (generic name, identification).

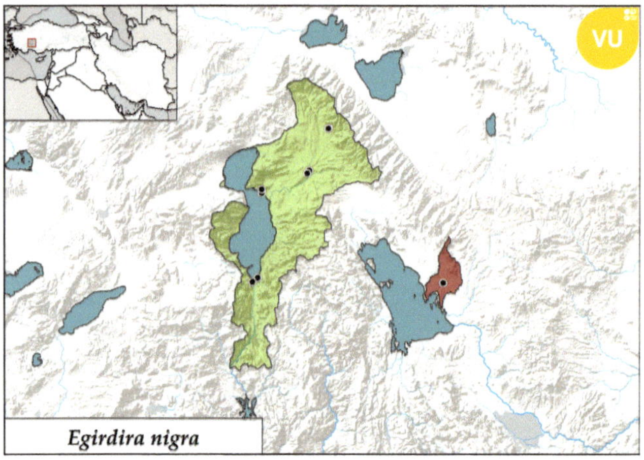

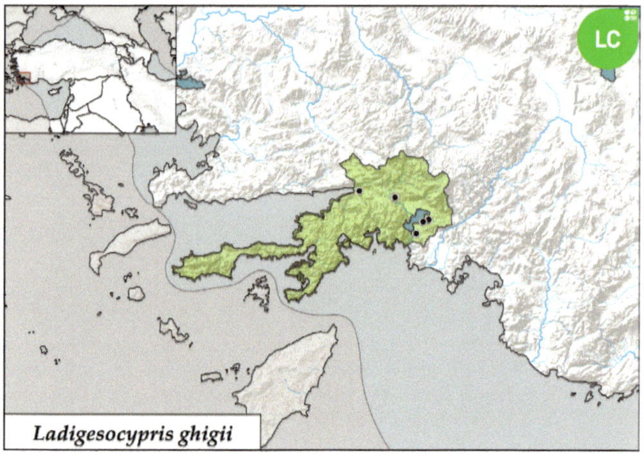

Ladigesocypris

Ladigesocypris is a small fish endemic to the southern Aegean basin. It is the sole species in the genus, closely related to *Squalius* and *Petroleuciscus*. Some authors have treated *Ladigesocypris* as a synonym of *Squalius*. This view is not supported by all molecular analyses, which are inconclusive. While some studies place *Ladigesocypris* within *Squalius*, others place the species close to *Petroleuciscus*, suggesting that *Petroleuciscus* might be a synonym of *Ladigesocypris*. The two genera are only distinguished by an incomplete lateral line in *Ladigesocypris* (vs. complete in *Petroleuciscus*). Previously, three species of *Ladigesocypris* were recognised. However, *L. irideus* from Anatolia cannot be distinguished from Greek *L. ghigii*, and *L. mermere* from Lake Marmara was found to be a synonym of *Petroleuciscus smyrnaeus*.

Ladigesocypris ghigii; Lake Köyceğiz basin, Türkiye; ~60 mm SL.

Ladigesocypris ghigii

Common name. Ghizani.

Diagnosis. Distinguished from superficially similar *Petroleuciscus smyrnaeus* in southwestern Anatolia by: • lateral line incomplete, perforating 5–11 scales / • 27–33 total scales in midlateral row / • broad grey or black midlateral stripe from head to caudal base, separating dark-brown back from white belly. Size up to 65 mm SL in male, 90 mm SL in female.

Distribution. Türkiye: Isolated populations around Marmaris, Lake Köyceğiz and in lower Dalaman drainage. Greece: Widespread on island of Rhodes, locally introduced near Athens.

Habitat. Springs and slow-flowing stretches of streams.

Biology. Lives up to 3 years. First spawns at 1 year. Spawns February–June. Fractional spawner among aquatic vegetation, especially filamentous algae. Eggs sticky, 1.3 mm in diameter, hatch in 4 days. Larvae along banks, often among vegetation. Feeds on small invertebrates and algae, in summer, mainly on plant material. Large numbers of juveniles and adults die each year during summer and autumn droughts.

Conservation status. LC.

Further reading. Bogutskaya 1996 (description); Stoumboudi et al. 2002 (reproduction, conservation); 2006 (conservation); Kottelat & Freyhof 2007 (biology, distribution).

Spring Karaot at the shore of Lake Eğirdir is the habitat of *Egirdira nigra*, *Anatolichthys iconii*, and *Capoeta pestai*.

Leucalburnus satunini; Kura drainage, Türkiye, ~110 mm SL. © M. Özuluğ.

Leucalburnus

This monotypic genus is phylogenetically isolated and may be related to *Alburnus*, although it is superficially similar to the European genus *Telestes*. *Leucalburnus* is poorly known, its ecology and distribution should be studied in greater depth, and its population trends should be monitored. Studying its ecology should not result in slaughtering hundreds of fish to calculate some useless indices, such as length-weight relationships or other trivialities. **Further reading.** Berg 1949b (distribution, morphology).

Leucalburnus satunini

Common name. Ardahan mountain dace.
Diagnosis. Distinguished from other species of Leuciscids in West Asia by: ○ scaleless keel in front of anus short, reaching about 50% of distance between vent and pelvic base / ○ body cylindrical, only slightly compressed / ○ 15–20 short, stout gill rakers / ○ 8½ branched anal rays / ○ 65–72 total lateral-line scales. Size up to 175 mm SL.
Distribution. Türkiye: Ardahan plateau in upper Kura drainage.
Habitat. Mountain streams with moderate fast-flowing water.
Biology. No data.
Conservation status. LC; no known threats, but limited range.
Further reading. Berg 1949b (distribution, morphology); Kaya et al. 2020a (distribution).

Figure 42. Gill arch of *Leucalburnus satunini*.

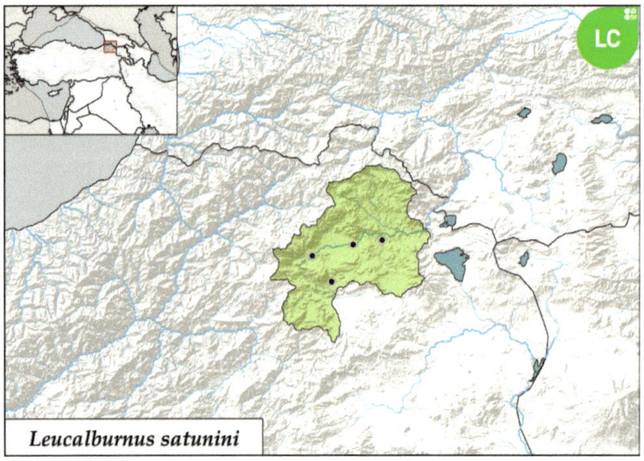

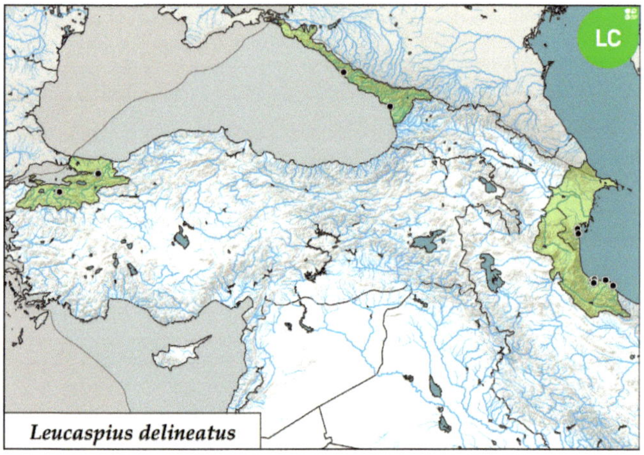

Leucaspius

Leucaspius is a group of small, slender, and silvery fishes often confused with juvenile *Alburnus*. *Leucaspius* is related to *Alburnus* and can be distinguished from it by an incomplete lateral line (vs. complete). *Leucaspius* are accidentally introduced with stocked fish from ponds, where they might be very abundant. Stocking also blurred the distribution area of the different species. Molecular data suggest that *Leucaspius* may include five species. *Leucaspius delineatus* was first described from the Danube drainage in Moravia (Czech Republic) and the adjacent region around Vienna (Austria). It is considered restricted to Central Europe, with records from the upper Danube. It has also been introduced in the Weser and Elbe drainages in Czech Republic and Germany. *Leucaspius abruptus*, described from the upper Dniestr, may be the available name for the species that is widely distributed in Eastern Europe, from the Pechora to the Rhine, including the upper Volga and the Dniestr and Dniepr in the northern Black Sea basin. Three additional molecular clades have been identified in West Asia. *Leucaspius owsianka*, described from the upper Don, is the available name for sun bleaks found in the Caucasian Black Sea basin of Russia and those from the Don and Kuban drainages. *Leucaspius thracicus*, described from the Bosporus region in Türkiye, is widely distributed in the lower Danube drainage, the lower Dniestr, Georgia, eastern Greece (Nestos), and the Marmara Sea basin. *Leucaspius dimorphus* is an available name for the populations from the western Caspian basin, including the rivers of Dagestan and the Kura and Sefid drainages. Further research is required to assess this genus fully. Currently, there is no evidence to suggest how the different molecular groups of *Leucaspius* can be distinguished externally, and their distribution remains uncertain. Therefore, we have elected to retain *L. delineatus* as the provisional designation pending further investigation. **Further reading.** Arnold & Längert 1995 (biology); Kottelat & Freyhof 2007 (distribution, biology, diagnosis).

Leucaspius delineatus (*L. thracicus*); Lake Apolyont, Türkiye; ~50 mm SL.

Leucaspius delineatus

Common name. Sun bleak.

Diagnosis. Distinguished from other Leuciscidae in Black and Marmara Sea basin by: ○ lateral line incomplete with about 8–12 pored scales / ○ keel covered by scales between pelvic origin and anus / ○ mouth superior / ○ 11–13½ branched anal rays. Size up to 90 mm SL.

Distribution. Türkiye: Izmit west to Susurluk drainage in Marmara basin, including Lakes Apolyont and Manyas (*L. thracicus*). Azov and Black Sea basin south to Rioni drainage (*L. owsianka*), northern and western Caspian basin (south to Sefid drainage in Iran) (*L. dimorphus*). In Europe, from lower Rhine and northern Germany east to Pechora drainage; Black Sea basin south to Rioni drainage, northern and western Caspian basin; Aegean Sea basin (from Maritsa to Nestos). Absent in Italy, Adriatic basin, Great Britain, and Scandinavia (except southernmost Sweden). Widely introduced in France, upper Rhine drainage, and locally in Great Britain and Switzerland.

Habitat. Lowland riverine habitats, especially oxbows and other water bodies connected to rivers only during floods. Often found in ponds, steppe lakes, and small bodies of water not connected to rivers. May occur in any habitat with few or no predators.

Biology. Gregarious. Spawns first time with one year; only for a single season. Spawns May–September when temperatures reach 16–18°C. Females may spawn about once every 3 weeks. Males are territorial; they clean spawning area and guard eggs, tied in strings around roots, reeds, aquatic vegetation, or any material floating on water surface. Feeds mainly on zooplankton and terrestrial insects. Stunted populations can form in predator-free habitats. Often rapidly invades newly created water bodies by as yet-unknown mechanisms. In Europe, scales were used to produce Essence d'Orient to coat artificial pearls.

Conservation status. LC.

Further reading. Arnold & Längert 1995 (biology); Kottelat & Freyhof 2007 (distribution, biology, diagnosis); Özuluğ & Saç 2019 (distribution).

Leucaspius delineatus (*L. owsianka*); Lower Kuban, Russia; ~55 mm SL.

Leuciscus

Daces, ides, and asps of the genus *Leuciscus* are medium-sized to large leuciscids, with the greatest diversity occurring from Western Europe to Siberia. *Leuciscus* are characterised by their silvery, elongated body, concave posterior margin of the anal, a complete lateral line, and 7–8½ branched dorsal rays. Most species inhabit large rivers and streams with clear water and moderate to swift currents. Two species, *Leuciscus aspius* and *L. vorax*, are known to occur in West Asia, where they are large predators that hunt in the open water column of lakes, marshes, and large rivers, often also in freshened parts of the sea. Both were previously classified in their own genus, *Aspius*. *Leuciscus latus*, described from the Hari in Afghanistan, has been occasionally listed for Iran. Recent records appear to be based on a misidentification of *Rutilus lacustris*, a non-native species in the Hari. The identity of *L. latus* requires further clarification, and the species has not been recorded in the 20th or 21st century. It cannot be excluded that *L. latus* is a synonym of one of the Central Asian *Leuciscus* species. Central Asian species, including *L. baicalensis*, *L. bergi*, *L. lehmanni*, *L. lindbergi*, *L. squaliusculus*, and *L. dzungaricus*, all share the same or highly similar mitochondrial DNA haplotypes (COI). This species group requires taxonomic revision. *Aspiolucius esocinus* is an asp native to the Aral basin, which appears to be closely related to *Leuciscus* and may be transferred to that genus in the future. **Further reading.** Jouladeh-Roudbar et al. 2020 (misidentification of *L. latus*).

Leuciscus aspius; Rhine drainage, Germany; ~80 mm SL.

Leuciscus aspius; Danube drainage, Germany; ~650 mm SL. © A. Hartl.

Leuciscus aspius

Common name. Asp.

Diagnosis. Distinguished from *L. vorax* by: • 61–73+3 lateral line scales / • lower lip slightly projecting upper lip. Size up to 800 mm SL.

Distribution. Widespread in Caspian basin. Sakarya and rivers west of Sakarya. In Rioni drainage (Georgia) but absent from Black Sea basin south of Danube. Large rivers draining into Aral Sea and North Sea (Weser, Elbe), Baltic Sea (southern tributaries, Norway east of Oslo, southern Sweden, Kokemären drainage in southern Finland), northern Black Sea, Sea of Azov and Caspian Sea; Aegean basin from Maritza to Lake Volvi drainages. Introduced in Rhine, Rhone, and other French rivers, Po (Italy), northern Dvina (Russia), Lake Balkhash (Central Asia), and likely elsewhere.

Habitat. Open waters of large and medium-sized lowland rivers and large lakes. Often found in freshwater parts of sea. Spawns on gravel or submerged vegetation in fast-flowing water. Lake populations migrate into tributaries.
Biology. Lives more than 10 years. First spawns at 3–5 years. Reproductive success appears to be linked to low water levels and high spring temperatures. Spawns more than one season, in March–early April (mid–May in Volga), when temperatures rise above 8°C (Rhine) or even lower (4–6°C on Terek) in places with strong currents. Spawning takes about 2 weeks, often just few days. Eggs are attached to gravel or submerged vegetation. Semi-anadromous individuals or populations (Terek) forage mainly in estuaries and freshened parts of sea, migrating to rivers only to spawn. Begins spawning migration in second half of October (Terek) and overwinters in lower part of river. Juveniles gregarious predators; adults hunt in small groups or are solitary. Juveniles and adults feed mainly on fish, especially *Atherina* and *Alburnus*. Juveniles grow rapidly, reaching about 80–100 mm SL in first year. In Europe, often hybridises with *Leuciscus idus*.
Conservation status. LC. Seems to be extirpated from Sakarya.
Remarks. Records from Lake Manyas (Türkiye) could not be confirmed. Populations from the southern Caspian and Aral basins are reported to have up to 90 total lateral line scales and often red lips. The morphology of asps has not been studied in detail, and molecular studies (COI), including asps from the Aras, Aral basin, and Europe, show that all populations are very closely related and show no geographical structure.
Further reading. Baruš & Oliva 1995 (biology); Riehl et al. 2002 (reproduction); Kottelat & Freyhof 2007 (biology, distribution); Coad 2021a (biology, morphology).

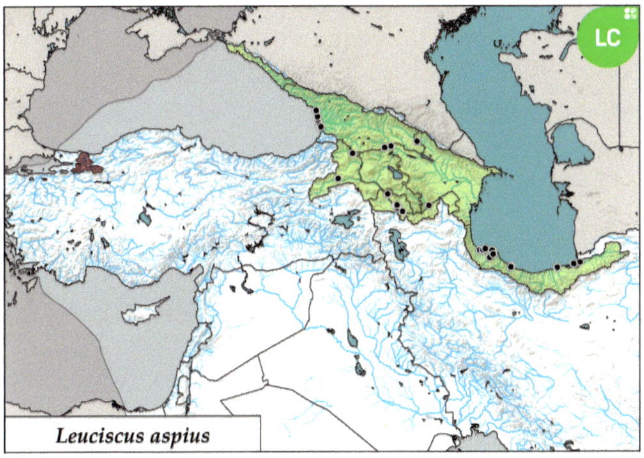

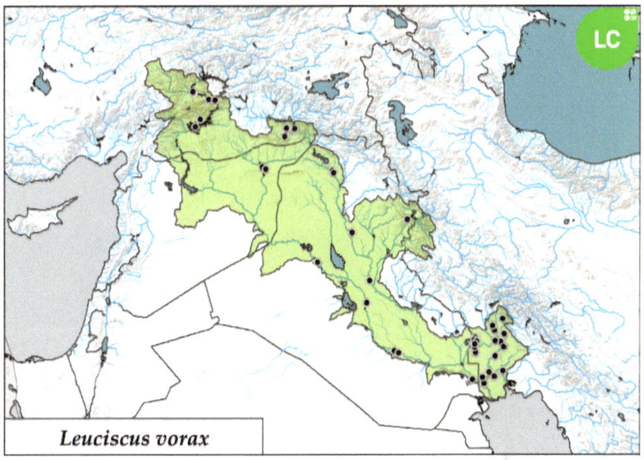

Large reservoirs in Aras drainage, like this one (Karakurt reservoir), are occupied by *Leuciscus aspius*.

Leuciscus vorax; fish market in Homs, Syria; ~300 mm SL.

Leuciscus vorax

Common name. Mesopotamian asp.

Diagnosis. Distinguished from *L. aspius* by: • 79–107 lateral line scales / • lower lip strongly projecting upper lip. Size up to 1300 mm SL and 60 kg.

Distribution. Euphrates and Tigris drainages, including lower Karun and Jarahi (Iran).

Habitat. Large rivers, lakes, and reservoirs. Spawns in current on submerged vegetation and gravel.

Biology. Lives up to 8 years, likely much longer. Matures at 3–4 years and spawns January–early March at water temperatures of 10–14°C. Juveniles prey on small fish, aquatic insects, and molluscs, while adults prey mainly on fish and frogs.

Conservation status. LC.

Remarks. Records from the Orontes drainage are doubtful and may refer to *Squalius kottelati*.

Further reading. Coad 2010a (distribution, biology); Coad 2021a (biology, morphology).

Mirogrex hulensis; Lake Hula, Israel; 120 mm SL.

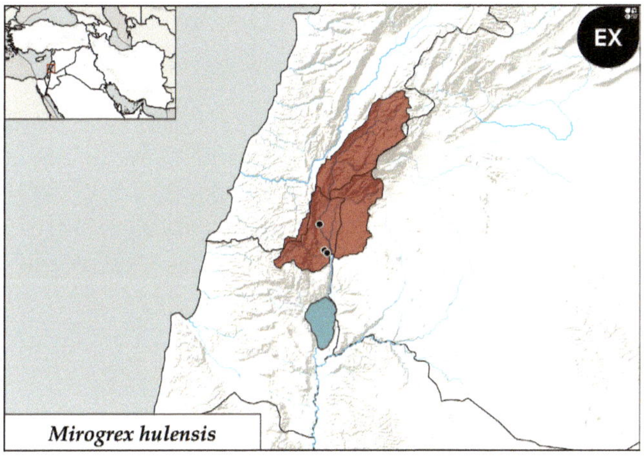

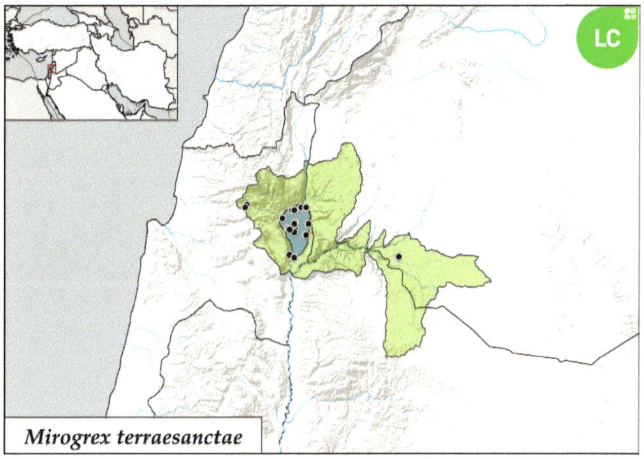

Mirogrex

Mirogrex comprises a group of pelagic fish that superficially resemble bleaks of the genus *Alburnus*. There are two species in this genus. Both are endemic to the upper Jordan drainage and are only found in three lakes, one already dry. Previously, both species were included in the genus *Acanthobrama*, but they represent an independent evolutionary lineage within the radiation of breams. Both species are closely related, and it is possible that they share a quite recent evolutionary past.

Mirogrex hulensis

Common name. Hula bream.

Diagnosis. Distinguished from *M. terraesanctae* by: • 63–73 total lateral-line scales / • 14–18, usually 16, gill rakers on lower limb of first gill arch / • body depth 22–31 % SL. Size up to 205 mm SL.

Distribution. Israel: Lake Hula and adjacent marshes.

Habitat. Lacustrine.

Biology. Pelagic, gregarious, feeding on plankton and anflug. Spawns February–April.

Conservation status. Extinct; Lake Hula and its marshes were drained in the 1950s. Last recorded in 1975. Hula marshes were revived shortly after its extinction, but it was too late for this endemic species.

Further reading. Goren et al. 1973 (description).

Mirogrex terraesanctae; Lake Tiberias, Israel; 120 mm SL.

Mirogrex terraesanctae

Common name. Tiberias bream.

Diagnosis. Distinguished from *M. hulensis* by: • 73–82 total lateral-line scales / • 16–20, usually 18–19, gill rakers on lower limb of first gill arch / • body depth 18–24 % SL. Size up to 152 mm SL.

Distribution. Lake Tiberias in Israel and Lake Muzayrib in Syria.

Habitat. Lacustrine, pelagic; spawns on wave-washed lake shores.

Biology. Pelagic, gregarious. Spawns November–May with a peak in January–February in littoral zone (0–50 cm) on rocks and gravel on wave-washed shores. Feeds on zooplankton and anflug.

Conservation status. LC; population of Lake Tiberias is large and commercially important, population of Lake Muzayrib is small and may be threatened or extirpated.

Remarks. Occasionally hybridises with *Acanthobrama lissneri*; Muzayrib Lake population appears to be at least partly introgressed by mitochondrial genes from *A. lissneri*.

Further reading. Steinitz 1952 (description); Goren et al. 1973 (review); Gafny et al. 1992 (spawning); Durand et al. 2002 (phylogeny); Perea et al. 2010 (phylogeny); Geiger et al. 2014 (phylogeny, introgression by *Acanthobrama*); Behrens-Chapuis et al. 2015 (cytochrome oxidase 1 intraspecific variability).

Lake Tiberias in Israel is the largest habitat of *M. terraesanctae*.

Pelecus cultratus; Danube, Austria; ~400 mm SL. © A. Hartl.

Pelecus cultratus

Common name. Razor fish.

Diagnosis. Distinguished from other species of Leuciscidae in West Asia by: • lateral line running in an undulated line / ○ 24–29½ branched anal rays / ○ mouth superior / ○ body very elongate and compressed / • dorsal profile almost straight / ○ sharp, scaleless keel from throat to anus / • dorsal base entirely above anal base. Size up to 500 mm SL.

Distribution. Caspian basin in Azerbaijan and Iran. Aral basin and European part of Black and Caspian basins. Absent from Anatolia and Black Sea basin south of Kuban. Baltic basin from Vistula to Neva, southern Sweden and Finland, Lakes Ladoga and Onega. Occasionally, on Baltic coast west of Vistula and Finnish coast north of 61°N.

Habitat. Surface of open waters of large rivers and lakes. Abundant in reservoirs (Don, Volga). Semi-anadromous individuals feed and spawn in pelagic zone of freshened parts of sea or lower parts of rivers, in main channels or floodplains. Lacustrine populations spawn in open waters of lakes.

Biology. Lives up to 9 years. First spawns at 3–5 years, with 200–300 mm SL. Semi-anadromous individuals enter rivers from July (Don). When rivers are ice-covered, spawning migration ceases and resumes with ice break-up in April–May. Resident individuals in rivers throughout year. Spawns May–June, earlier in southern drainages (April in Syr-Darya) at temperatures above 12°C. Eggs semi-pelagic and drift with current (in rivers). Eggs hatch after 3–4 days. Adults return to estuaries to feed immediately after spawning. Juveniles may migrate to estuaries during first summer. Feeds on zooplankton, terrestrial invertebrates, and small fish.

Conservation status. LC.

Remarks. This species represents the sole member of an isolated lineage within Leuciscidae, lacking any close relationship to other Leuciscid genera. It is unique among Western Palaearctic leuciscids, exhibiting a long, slender, highly compressed body with a sharp keel from the throat to the anus. However, this appearance is shared with some unrelated Asian species, particularly *Macrochirichthys*

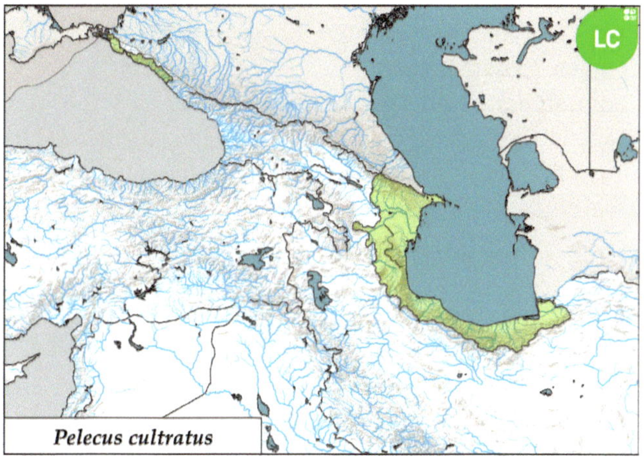

macrochirus. *Pelecus* is an important commercial species, sold dried and smoked.

Further reading. Berg 1949b (biology); Balon 1956 (biology); Bogutskaya 1988b (osteology); Kottelat & Freyhof 2007 (biology, distribution).

Petroleuciscus

The genus *Petroleuciscus* is currently restricted to four species: *P. aphipsi*, *P. borysthenicus*, *P. ninae*, and *P. smyrnaeus*. These species appear to be closely related, or even congeneric to *Ladigesocypris*. Both genera are also closely related to *Squalius*. *Petroleuciscus* is endemic to the Black, Marmara, and Aegean basins. Other species associated with this genus are now placed in different genera. *Petroleuciscus persidis*, native to Central Iran, belongs to the genus *Acanthobrama*. *Petroleuciscus kurui* and *P. ulanus*, also native to Iran, belong to *Alburnus*. *Petroleuciscus squaliusculus*, native to the Aral Sea basin, belongs to *Leuciscus*. Finally, *P. esfahani*, native to Iran, is a synonym of *Alburnus doriae*. **Further reading.** Bogutskaya 1996 (taxonomy); 2002 (systematics); Perea et al. 2010 (*P. persidis*); Mohammadian-Kalat et al. 2017 (*P. esfahani*); Freyhof et al. 2018a (*P. kurui*); Sheraliev & Peng 2021 (*P. squaliusculus*); Kalaycı 2022 (Phylogeny); Kaya et al. 2024b (*P. ninae*).

Key to species of *Petroleuciscus* in West Asia

1a - 8–9½ branched dorsal rays; 9–10½ branched anal rays; 7 scales between lateral line and dorsal origin; iris orange to red; posterior margin of anal convex or almost straight; no black pigments along free margin of each flank scale.
.....................*P. borysthenicus*
1b - 7–8½ branched dorsal rays; usually 8½ branched anal rays; 6 scales between lateral line and dorsal origin; iris whitish; posterior margin of anal straight or slightly concave; black pigments along free margin of each flank scale.
.....................2

2a - Pigments absent or very few orange pigments on anal rays in life; body depth at dorsal origin 24–27 % SL; head width at posterior eye margin 14–16 % SL; eye diameter approximately equal to snout length.
.....................*P. smyrnaeus*
2b - Numerous black pigments on anal rays in life; body depth at dorsal origin 27–30 % SL; head width at posterior eye margin 16–19 % SL; eye diameter smaller than snout length.
.....................*P. ninae*

Petroleuciscus borysthenicus; Bakacak drainage, Türkiye; ~80 mm SL.

Petroleuciscus borysthenicus
Common name. Bobyretz.
Diagnosis. Distinguished from other species of *Petroleuciscus* in West Asia by: ○ 8–9½ branched dorsal rays / ○ 9–10½ branched anal rays / ● iris orange to red / ● 7 scales between lateral line and dorsal origin / ○ 33–40 total lateral line scales / ○ posterior margin of anal convex or almost straight / ○ no black pigments along free margin of each flank scale. Size up to 185 mm SL.

Distribution. Marmara basin east to Lake Sapanca, but absent from Anatolian coast east of Sakarya, except for two records from Samsun. Caucasian Black Sea basin of Russia and Georgia. Lower parts of all tributaries of northern Black Sea and Azov basins except Don. Aegean basin from Struma drainage (Greece) eastwards.
Habitat. Lowland streams, rivers, lower reaches of mountain rivers, lakes, deltas, and backwaters with moderate to no current. Prefers warm water up to 30–32°C. Tolerant of

low oxygen concentrations, sometimes found in marshes and waters with dense vegetation. Tolerant of slightly brackish water. Prefers sand, sand-mud, or mud bottoms. Prefers shallow areas with slow currents along banks, backwaters, small lakes, and similar still waters.

Biology. Lives up to 8 years. Spawns first time at 2–3 years, males may spawn in first year. Forms spawning aggregations of up to 150 individuals. Spawns April–June at 12–28°C in 0.1–1.0 m deep water over sandy or slightly muddy bottom. Females lay 2–3 portions of adhesive eggs on or among aquatic plants. Usually found in groups of 5–30 individuals close to bottom, but sometimes close to surface, under floating leaves or other objects. Does not migrate but moves to deeper water in winter. Feeds mainly on insects, their larvae, plankton, benthic invertebrates, and algae.

Conservation status. LC; extirpated from Lake İznik basin.

Further reading. Movchan & Smirnov 1981 (description, biology); Kottelat & Freyhof 2007 (biology, distribution); Saygun et al. 2017 (distribution); Özpiçak et al. 2022 (distribution).

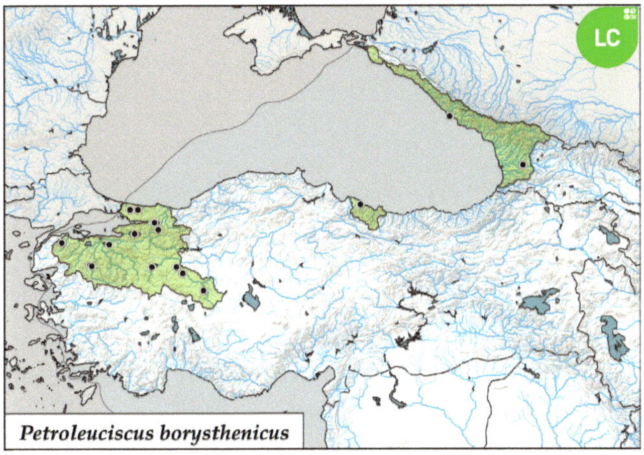

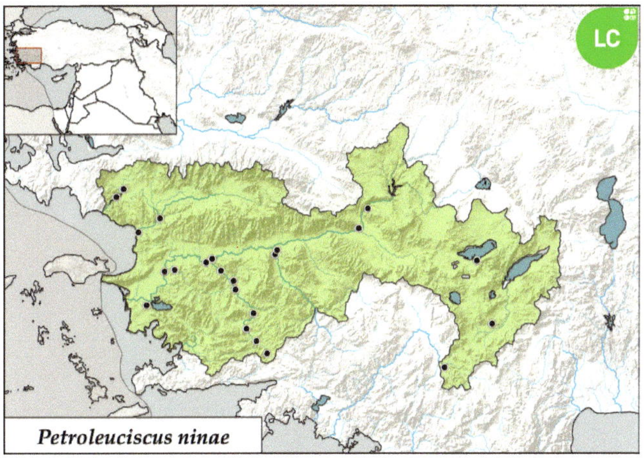

Petroleuciscus ninae; Büyük Menderes, Türkiye; ~60 mm SL.

Petroleuciscus ninae
Common name. Menderes chub.
Diagnosis. Distinguished from other species of *Petroleuciscus* in West Asia by: ○ 7–8½ branched dorsal rays / ○ 8–9½ branched anal rays / ○ iris whitish / ○ 6 scales between lateral line and dorsal origin / ○ 29–34 total lateral line scales / ○ posterior margin of anal straight or slightly concave / ○ black pigments along free margin of each flank scale / ○ numerous black pigments on anal rays in life / ○ body depth at dorsal origin 27–30 % SL / ○ head width at posterior eye margin 16–19 % SL / ○ eye diameter smaller than snout length. Size up to 95 mm SL.
Distribution. Türkiye: Büyük and Küçük Menderes, Sarıçay (Muğla) and Tahtalı drainages. Locally introduced (Acıgöl).
Habitat. Slow-flowing to standing water of rivers, reservoirs, and larger streams, often associated with aquatic vegetation.
Biology. No data.
Conservation status. LC.
Further reading. Turan et al. 2018d (description); Kaya et al. 2024b (distribution).

Petroleuciscus smyrnaeus; Bakırçay drainage, Türkiye; 72 mm SL.

Petroleuciscus smyrnaeus
Common name. Smyrna chub.
Diagnosis. Distinguished from other species of *Petroleuciscus* in West Asia by: ○ 7½ branched dorsal rays / ○ usually 8½ branched anal rays / ○ iris whitish / ○ 6 scales between lateral line and dorsal origin / ○ 32–35 total lateral line scales / ○ posterior margin of anal straight or slightly concave / ○ black pigments along free margin of each flank scale/ ○ pigments absent or very few orange pigments on anal rays in life / ○ body depth at dorsal origin 24–27 % SL / ○ head width at posterior eye margin 14–16 % SL / ○ eye diameter approximately equal to snout length. Size up to 100 mm SL.
Distribution. Türkiye: Bakırçay to Gediz drainages. Also, on Lesbos Island (Greece).
Habitat. Slow-flowing to standing water of rivers, reservoirs, and larger streams, often associated with aquatic vegetation.
Biology. No data.
Remarks. *Ladigesocypris mermere* from Lake Marmara in Gediz drainage is a synonym.
Conservation status. LC.
Further reading. Bogutskaya 1996 (description); Bogutskaya 2002 (systematics); Stoumboudi et al. 2006 (Lesbos); Kottelat & Freyhof 2007 (biology, distribution); Turan et al. 2018d (diagnosis); İlhan et al. 2020 (distribution).

Phoxinus
Phoxinus comprises small fishes readily identifiable by their black lateral stripe or series of blotches and characteristic body shape. In West Asia, *Phoxinus* is the sole representative of a large radiation of minnows, particularly diverse in North America, where almost all Leuciscids belong to phoxinine minnows. The genus *Phoxinus* is highly diverse, extending from eastern Spain to Siberia. However, this diversity has only recently received attention, and most populations are still identified as *P. phoxinus*. *Phoxinus phoxinus* is endemic to Central Europe. *Phoxinus* is only marginally distributed

in West Asia and is known from the south-eastern Black Sea basin (*P. colchicus*) and northwestern Anatolia (*P. abanticus*, *P. kottelati*). *Phoxinus* has become the victim of "molecular parataxonomists", and the description of many new species without morphological and minor molecular differences can be expected. Also, in West Asia, this might result in the recognition of many new species, all closely related and indistinguishable by external characters. **Further reading.** Artaev et al. 2024 (Caucasian *Phoxinus*).

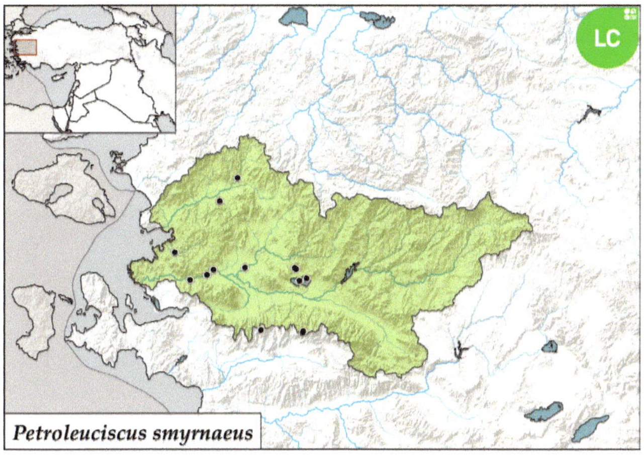

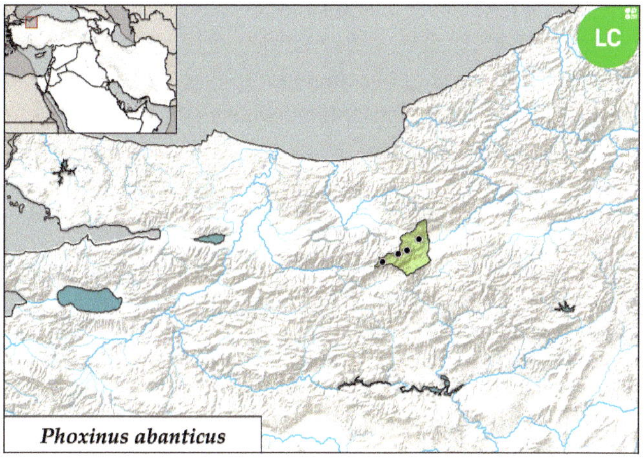

Key to species of *Phoxinus* in West Asia

1a - Scales on complete belly between pectoral base in male; a supraorbital ridge margined by a longitudinal depression extending from nostrils backwards.
....................*P. colchicus*
1b - No scales on belly between pectoral bases or belly incompletely covered by scales in male; no supraorbital ridge; no depression behind nostrils.
....................2

2a - 60–69 total lateral line scales; caudal peduncle depth: 1.8–2.3 times in length; no scales on chest in male and female.
....................*P. abanticus*
2b - 77–90 total lateral line scales; caudal peduncle depth: 2.3–2.9 times in length; middle of belly scaled but separated area by a naked area anteriorly in male.
....................*P. kottelati*

Phoxinus abanticus; Abant basin, Türkiye; ~50 mm SL. © M. Özuluğ

Phoxinus abanticus
Common name. Abant minnow.
Diagnosis. Distinguished from other species of *Phoxinus* in West Asia by: ● no scales on chest in male and female / ● 60–69 total lateral line scales / ● caudal peduncle depth 1.8–2.3 times in length / ○ no supraorbital ridge / ○ no depression behind nostrils. Size up to 63 mm SL.
Distribution. Türkiye: Lake Abant basin.
Habitat. A wide range of cold and well-oxygenated habitats, from small, fast-flowing streams to lake shores. Spawns on clean gravel areas in flowing water.
Biology. No data.
Conservation status. LC; has a small range but its population trend seems to be stable and there are no major threats.
Remarks. Closely related to *P. strandjae* and *P. kottelati* and it is expected that future research will demonstrate their conspecificity.
Further reading. Turan et al. 2023 (description).

Phoxinus colchicus; Sochi, Russia; ~75 mm SL.© S. Tuniev.

Phoxinus colchicus
Common name. Caucasian minnow.
Diagnosis. Distinguished from other species of *Phoxinus* in West Asia by: ● belly between pectoral bases entirely covered by scales in male, absent in female / ● a supraorbital ridge margined by a longitudinal depression extending from nostrils backwards / ● caudal-peduncle depth 2.1–2.9 times in its length / ○ 75–91 total lateral line scales. Size up to 82 mm SL.
Distribution. Caucasian Black Sea basin of Russia and Georgia south to Çoruh (Borçka, Türkiye). Records from Hopa (eastern Türkiye) not confirmed.
Habitat. A wide range of cold and well-oxygenated habitats, from small, fast-flowing streams to large rivers. Spawns on clean gravel beds in flowing water.
Biology. Gregarious, rheophilic. Spawns April–July, in schools, fractional spawners, females lay sticky eggs deep in clean gravel. Feeds on invertebrates, algae, and detritus.
Conservation status. LC.
Further reading. Berg 1949b (description); Kottelat 2007 (systematics); Kottelat & Freyhof 2007 (biology, distribution); Bayçelebi et al. 2015 (record from Çoruh).

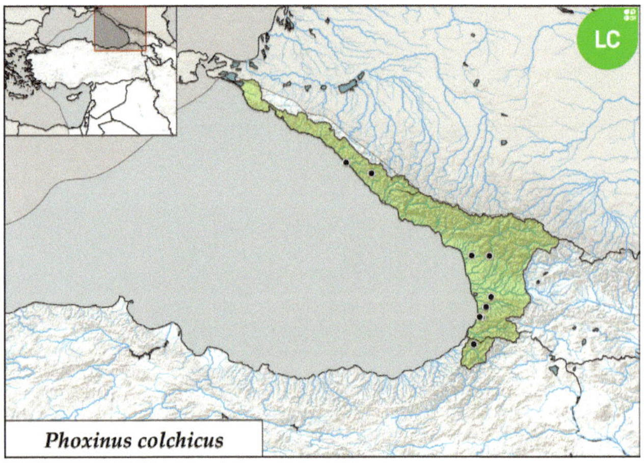

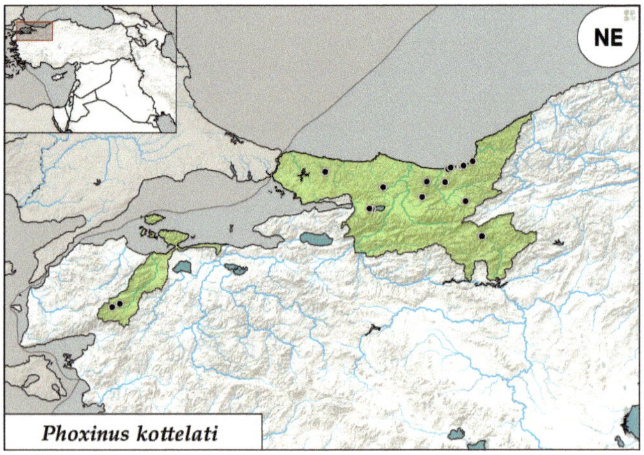

Phoxinus kottelati; Gönen drainage; female, ~65 mm SL.

Phoxinus kottelati; Gönen drainage; male, ~60 mm SL.

Phoxinus kottelati

Common name. Anatolian minnow.

Diagnosis. Distinguished from other species of *Phoxinus* in West Asia by: • 77–90 total lateral line scales / • caudal peduncle depth 2.3–2.9 times in length / • middle of belly scaled but separated by a naked area anteriorly in male, absent in female / ○ no supraorbital ridge / ○ no depression behind nostrils. Size up to 70 mm SL, likely to grow larger.

Distribution. Lake Sapanca and Biga Peninsula. Also occurs east to Zonguldak in Anatolian western Black Sea.

Habitat. Small streams with clear, well-oxygenated water, over gravel to rocky bottoms.

Biology. No data.

Conservation status. Not evaluated.

Remarks. See comments in *P. abanticus*

Further reading. Kottelat 2007 (diagnosis); Kottelat & Freyhof 2007 (biology, distribution); Özuluğ & Saç 2019 (distribution); Bayçelebi 2025 (description).

Phoxinus kottelati; Lake Sapanca basin; male, ~55 mm SL.

Pseudophoxinus

Pseudophoxinus is the largest genus of Leuciscidae in our region, comprising 26 species. These species are endemic to West Asia, with the greatest diversity occurring in Central Anatolia and the Levant. Two species are found in the western Euphrates and Qweiq drainages. *Pseudophoxinus* are small fish diagnosed by the preoperculo-mandibular sensory canal being disconnected from the infraorbital canal on the head. The disconnection of the two head canals represents an adaptive character state and has evolved several times in parallel in small Leuciscids. Several species classified in *Pseudophoxinus* due to this character state were transferred to other genera, including *Acanthobrama, Alburnus, Egirdira, Pelasgus, Rutilus,* and *Tropidophoxinellus*. *Pseudophoxinus* is closely related to the genera *Phoxinellus, Telestes,* and *Chondrostoma*, and the genus is divided into two major species groups. A notable feature of many species of *Pseudophoxinus* is the absence of a complete lateral line, which does not extend to the caudal base. This character displays considerable variability, with a complete lateral line occurring in species otherwise diagnosed by an incomplete lateral line. Conversely, individuals with a shortened lateral line may be rare in species with a complete lateral line. Nevertheless, this character is included in the key, and examining more than one individual is essential for accurate identification. Several species of *Pseudophoxinus* are of significant conservation concern due to their restricted distribution ranges and occurrence in a limited number of water bodies. Consequently, we provide comprehensive information about their distribution. All species inhabit slow-flowing stretches of streams or rivers and stagnant water with dense vegetation. Most species are restricted to or are most common in springs and spring-fed streams. Our understanding of the biology of *Pseudophoxinus* is very limited, and more research is encouraged without slaughtering hundreds of fish for useless indices.

Keys to species of *Pseudophoxinus*

Western Anatolia, including Lakes Avlan, Burdur, Işıklı, Çorak (Akgöl), Karaevli (Burdur), Salda, and Söğüt basins, and Akçay (Finike), Karadirek (Sandıklı), and Onaç drainages

1a - Lateral line complete.
.....................*P. maeandricus*
1b - Lateral line incomplete.
.....................2

2a - Mouth almost superior.
.....................*P. mehmeti*
2b - Mouth terminal or slightly subterminal.
.....................3

3a - 41–46+2–3 scales in lateral series.
......................*P. maeandri*
3b - 46–64+2–3 scales in lateral series.
......................4

4a - Pelvic with 7 branched rays.
......................*P. burduricus*
4b - Pelvic with 8–10 branched rays.
......................5

5a - 52–64+2–3 scales in lateral series; 13–16 scale rows between dorsal origin and lateral line.
......................*P. evliyai*
5b - 46–53+2–3 scales in lateral series: 10–13 scale rows between dorsal origin and lateral line.
......................*P. ninae*

Central Anatolia, including rivers and springs flowing to Bay of Antalya, wider Lake Tuz basin, Beyşehir, Eğirdir basins, Sultan marshes, and areas around Ereğli and Niğde

1a - Shallow keel between posterior end of pelvic base and anus, scaleless for about ⅓ or ¼ of its length.
......................*P. battalgilae*
1b - No keel between posterior end of pelvic base and anus.
......................2

2a - 65–109 +1–3 scales in lateral series.
......................3
2b - 26–68 +2–3 scales in lateral series.
......................7

3a - Pelvic origin below or slightly behind dorsal origin.
......................4
3b - Pelvic origin in front of pelvic origin.
......................5

4a - Last unbranched dorsal ray spinous and thickened.
......................*P. anatolicus*
4b - Last unbranched dorsal ray flexible and not thickened.
......................*P. crassus*

5a - Lower lip projecting beyond tip of upper lip.
......................*P. handlirschi*
5b - Upper lip projecting beyond tip of lower lip.
......................6

6a - Prominent epidermal black stripe from eye to caudal base; 74–85 +3–4 scales along lateral line.
......................*P. fahrettini*
6b - Body plain, without epidermal stripe; 84–94+1–2 scales along lateral line.
......................*P. hittitorum*

7a - Lateral line usually complete, reaching caudal base, rarely interrupted or shortened.
.....................*P. iconii*
7b - Lateral line usually incomplete, rarely complete.
.....................8

8a - 41–44+2–3 scales in lateral series.
.....................*P. alii*
8b - 52–68+2–3 scales in lateral series.
.....................9

9a - Pelvic axillary lobe present; eye diameter 25–29 % HL.
.....................*P. antalyae*
9b - Pelvic axillary lobe absent; eye diameter 13–20 % HL.
.....................*P. elizavetae*

Eastern Mediterranean basin from Seyhan south to the Dead Sea basin, including Qweiq and the upper Euphrates drainages

1a - Lateral line complete, reaching to caudal base.
.....................2
1b - Lateral line incomplete.
.....................4

2a - 55–62+2–3 scales in lateral series.
.....................*P. zeregi*
2b - 38–46+2–3 scales in lateral series.
.....................3

3a - No black spots below lateral line; uppermost point of mouth cleft at level of center of eye or above.
.....................*P. cilicicus*
3b - Many small black spots below lateral line; uppermost point of mouth cleft below center of eye.
.....................*P. zekayi*

4a - 26–33+2–3 scales in lateral series.
.....................*P. hasani*
4b - 36–62+2–3 scales in lateral series.
.....................5

5a - Pelvic axillary lobe present.
.....................*P. firati*
5b - Pelvic axillary lobe absent.
.....................6

6a - Pelvic short, reaching 1–5 scales in front of anus; 6–7 scales between pelvic origin and lateral line.
.....................7
6b - Pelvic long, reaching or overlapping anus; 4–5 scales between pelvic origin and lateral line.
.....................8

7a - 36–50+2–3 scales in lateral series.
......................*P. syriacus*
7b - 47–60+2–3 scales in lateral series.
......................*P. drusensis*

8a - No or a faint, brown or grey mid-lateral stripe; 7–8 gill rakers; 19–20 abdominal vertebrae.
......................*P. libani*
8b - A bold, black, or dark-grey mid-lateral stripe; 8–11 (13) gill rakers; 21–22 abdominal vertebrae.
......................*P. turani*

Onaç River, long believed to be the only habitat of *Pseudophoxinus ninae*, is usually dry below the reservoir, which also dried up. The last pumps take the last water out.

Pseudophoxinus alii; Ilıca drainage; ~45 mm SL.

Pseudophoxinus alii

Common name. Pamphylian spring minnow.

Diagnosis. Distinguished from other species of *Pseudophoxinus* in Central Anatolia by: ○ 41–44+2–3 scales in lateral series / ○ lateral line incomplete with 38–41 perforated scales / ○ prominent mid-lateral stripe / ○ no keel between posterior end of pelvic base and anus / ○ upper lip projecting beyond tip of lower lip. Size up to 98 mm SL.

Distribution. Türkiye: Ilıca and Kömürcüler, as well as middle Köprüçay and Karaöz sections of Aksu. Manavgat reservoir and Yamansaz marshes, all flowing into Bay of Antalya.

Habitat. Lowland streams with dense vegetation and sand and gravel bottoms.

Biology. No data.

Conservation status. VU; appears to be declining within its small range.

Further reading. Küçük 2007b (description).

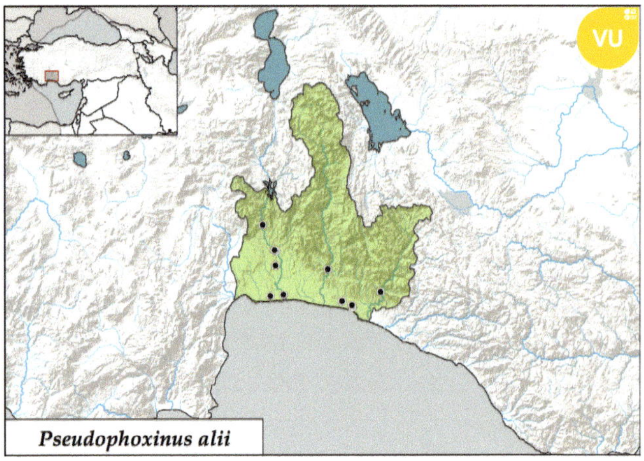

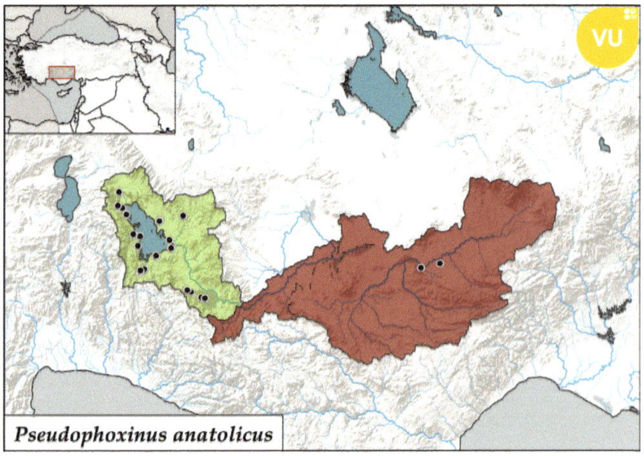

Pseudophoxinus anatolicus; Lake Beyşehir basin, Türkiye; ~100 mm SL.

Pseudophoxinus anatolicus; Lake Beyşehir basin, Türkiye; 185 mm SL.

Pseudophoxinus anatolicus

Common name. Giant spring minnow.

Diagnosis. Distinguished from other species of *Pseudophoxinus* in Central Anatolia by: ● last unbranched dorsal ray spinous and thickened / ○ 85–109+2–3 scales in lateral series / ○ lateral line complete, gradually reaching to lateral midline at caudal peduncle / ○ pelvic not reaching to anus / ○ 11–12½ scales between lateral line and pelvic origin / ○ pelvic axillary lobe absent / ○ no keel between posterior end of pelvic base and anus / ○ pelvic origin below dorsal origin / ○ upper lip projecting beyond tip of lower lip / ○ flank plain brown or grey in adults, with prominent or weak stripe in juveniles. Size up to 240 mm SL.

Distribution. Türkiye: Ereğli marshes, Lake Hotamis, Beyşehir and Suğla basins.

Habitat. Lakes and streams with dense vegetation and stagnant or slow-flowing water.

Biology. Lives up to 5 years, potentially longer.

Conservation status. VU; abundant and commercially important in Lakes Beyşehir and Suğla before the 1970s. After drainage of Lake Suğla, extensive alteration of water bodies in Ereğli marshes, and introduction of *Sander*

Pseudophoxinus antalyae; Kırkgöz spring, Türkiye; ~80 mm SL.

lucioperca into Lake Beyşehir, restricted to very few sites. Population in Ereğli-Akgöl is suspected to have been extirpated since 2018.

Remarks. *Pseudophoxinus caralis* is a synonym.

Further reading. Bogutskaya 1992 (revision, distribution); Freyhof & Özuluğ 2010c (*P. caralis*); Demirci 2016 (age).

Pseudophoxinus antalyae

Common name. Antalya spring minnow.

Diagnosis. Distinguished from other species of *Pseudophoxinus* in Central Anatolia by: ○ 52–64+2 scales in lateral series / ○ lateral line incomplete with 42–59 perforated scales / ○ no keel between posterior end of pelvic base and anus / ○ pelvic-axillary lobe present / ○ eye diameter 25–29 % HL / ○ upper lip projecting beyond tip of lower lip / ○ flank plain brown, grey or golden without prominent stripe. Size up to 97 mm SL.

Distribution. Türkiye: Düden and Kırkgöz springs and outflowing stream Yediarıklar in Bay of Antalya.

Habitat. Springs and spring-fed streams with clear water and dense vegetation.

Biology. No data.

Conservation status. EN; appears to be declining within its very small range.

Further reading. Bogutskaya 1992 (description); Küçük 2007b (distribution).

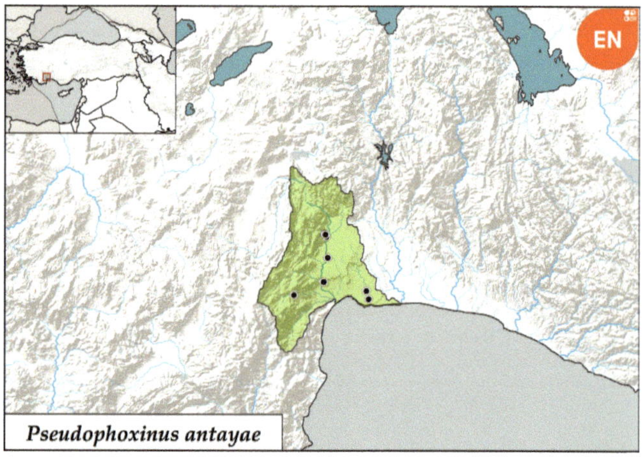

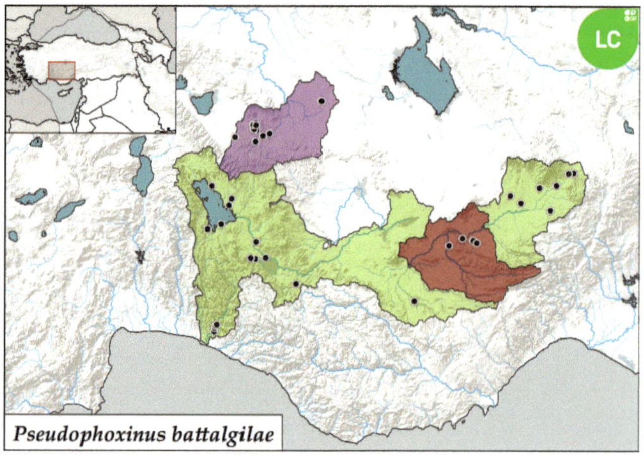

Pseudophoxinus battalgilae; Niğde, Türkiye; 63 mm SL.

Pseudophoxinus burduricus; Düğer, Lake Burdur basin, Türkiye; ~60 mm SL.

Pseudophoxinus battalgilae

Common name. Lycaonian spring minnow.

Diagnosis. Distinguished from other species of *Pseudophoxinus* in Central Anatolia by: • shallow keel between posterior end of pelvic base and anus, scaleless for about ⅓ or ¼ of its length / ○ 55–64+3 scales in lateral series / ○ lateral line complete / ○ upper and lower lip terminates at same level / ○ flank with prominent, brown or grey mid-lateral stripe. Size up to 150 mm SL.

Distribution. Türkiye: Lakes Beyşehir and Suğla basins, Akkaya reservoir south of Niğde, a reservoir at Gümüşler, Tatlıkuyu canal at Ereğli and Manavgat drainage (Mediterranean basin) including Oymapınar and Manavgat reservoirs. Introduced in Lake Ilgın basin. Expected to occur elsewhere.

Habitat. Lakes and reservoirs, springs, and slow-flowing streams. Usually in densely vegetated habitats.

Biology. No data.

Conservation status. LC; extirpated from Tatlıkuyu channel and Akkaya reservoir.

Further reading. Bogutskaya 1997b (description); Küçük et al. 2016 (diagnosis, distribution).

Pseudophoxinus burduricus

Common name. Burdur spring minnow.

Diagnosis. Distinguished from other species of *Pseudophoxinus* in Western Anatolia by: ○ 47–57+2–3 scales in lateral series / ○ lateral line incomplete with 21–39 (usually 26–37) perforated scales / ○ 7 branched pelvic rays / ○ 10–12½ scale rows between dorsal origin and lateral line / ○ mouth

slightly subterminal / ○ flank usually with indistinct, brown mid-lateral stripe. Size up to 87 mm SL.
Distribution. Türkiye: Salda and Yarışlı (Sazak spring) basins, streams Değirmendere, Karamanlı, Düğer, and Bozçay springs (a southwest tributary of Lake Burdur). Extirpated from Lake Karataş, and Karapınar spring east of Lake Salda.
Habitat. Lakes, slow-flowing streams, springs, usually with dense vegetation. In lakes also in open water.
Biology. No data.
Conservation status. EN; appears to be declining within its very small range.
Further reading. Küçük et al. 2013 (description).

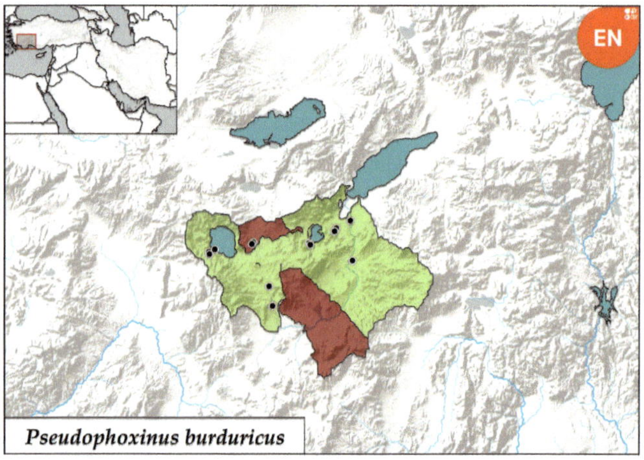

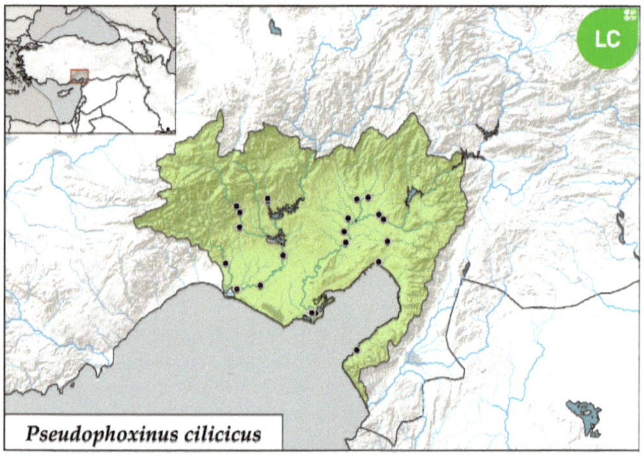

Pseudophoxinus cilicicus; lower Seyhan, Türkiye; 100 mm SL.

Pseudophoxinus cilicicus
Common name. Cilician spring minnow.
Diagnosis. Distinguished from other species of *Pseudophoxinus* in eastern Mediterranean basin by: ○ lower lip slightly projecting beyond tip of upper lip in some individuals / ○ uppermost point of cleft at level of center of eye or above / ○ 38–45+2–3 scales in lateral line / ○ no black spots below lateral line / ○ lateral line complete / ○ flank usually with indistinct, brown mid-lateral stripe. Size up to 138 mm SL.
Distribution. Türkiye: Lower Tarsus, Seyhan and Ceyhan drainages and a small stream at Arsuz.
Habitat. Lowland rivers and streams with slow-flowing or stagnant water; also in moderately polluted sections.
Biology. No data.
Conservation status. LC.
Further reading. Saç et al. 2019a (description).

Pseudophoxinus cilicicus; Arsuz, Türkiye; ~70 mm SL.

Pseudophoxinus crassus; Lake Tuz basin, Türkiye; ~85 mm SL.

Pseudophoxinus crassus
Common name. Fat spring minnow.
Diagnosis. Distinguished from other species of *Pseudophoxinus* in Central Anatolia by: ○ 65–78+2–3 scales in lateral series / ○ last unbranched dorsal ray flexible and not thickened / ○ lateral line complete / ○ lateral line gradually reaching to lateral midline at caudal peduncle / ○ pelvic not reaching to anus / ○ 8–9 gill rakers / ○ 8–10½ scales between lateral line and pelvic origin / ○ pelvic axillary lobe absent / ○ no keel between posterior end of pelvic base and anus / ○ pelvic origin below dorsal origin / ○ upper lip projecting beyond tip of lower lip / ○ flank plain brown, grey or silvery without prominent stripe. Size up to 180 mm SL.
Distribution. Türkiye: Lake Tuz basin at Cihanbeyli, Lake Kozanlı and Samsam basins, springs south of Eşmekaya and near Aksaray and Niğde.
Habitat. Springs and spring-fed streams with clear water and dense vegetation.
Biology. Lives up to 7 years and spawns first with 1 (male) and 2 (female) years in April–May. Female seem to spawn only once a season.
Conservation status. EN; appears to be declining within its very small range.
Remarks. Closely related to *P. anatolicus*.
Further reading. Ladiges 1960 (description); Bogutskaya 1992 (revision, distribution); Ekmekçi & Kırankaya 2010 (biology).

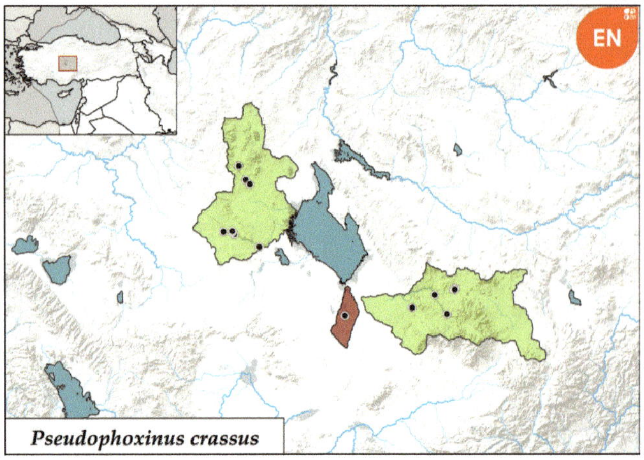

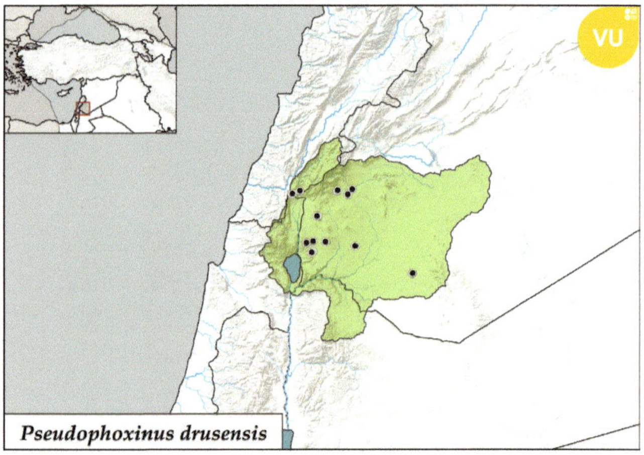

Pseudophoxinus drusensis; Al Tammasiyyar, Syria; ~80 mm SL.

Pseudophoxinus drusensis

Common name. Drusian spring minnow.

Diagnosis. Distinguished from other species of *Pseudophoxinus* in eastern Mediterranean basin by: ○ 47–60+2–3 scales in lateral series / ○ flank plain brown, grey or golden without prominent stripe / ○ pelvic short, reaching 1–5 scales in front of anus / ○ 6–7 scales between pelvic origin and lateral line / ○ lateral line incomplete / ○ pelvic axillary lobe absent. Size up to 80 mm SL, likely to grow larger.

Distribution. Israel and Syria: Northern Jordan drainage in Golan Heights, Jebel Druse, and Al Tammasiyyar in Syria.

Habitat. Springs and spring-fed streams with clear water and dense vegetation.

Biology. Spawns April–June.

Conservation status. VU; appears to be declining within its small range.

Further reading. Goren 1974 (distribution); Krupp 1985d (morphology); Krupp & Schneider 1989 (distribution).

Pseudophoxinus elizavetae; Soysallı spring, Türkiye; ~90 mm SL.

Pseudophoxinus elizavetae

Common name. Sultan minnow.

Diagnosis. Distinguished from other species of *Pseudophoxinus* in Central Anatolia by: ○ 60–68+3 scales in lateral series / ○ lateral line incomplete with 33–60 perforated scales / ○ no keel between posterior end of pelvic base and anus / ○ pelvic origin clearly behind vertical of dorsal origin / ○ flank plain brown, grey or golden without prominent stripe / ○ upper lip projecting beyond tip of lower lip / ○ pelvic-axillary lobe absent / ○ eye diameter 13–20 % HL. Size up to 100 mm SL.

Distribution. Türkiye: Sultan marshes.

Habitat. Springs and spring-fed streams with clear water and dense vegetation. Also, in muddy irrigation channels.

Biology. No data.

Conservation status. CR; Sultan marshes is threatened by desiccation. Zamantı Tunnel has been transferring water from Seyhan to the Sultan marshes since 2011. Since then, *Capoeta damascina, Garra turcica, Squalius cappadocicus,* and *Oxynoemacheilus seyhanensis* have invaded Sultan marshes. The impact of this invasion on native species is unknown.

Further reading. Bogutskaya et al. 2007 (description).

Copernicus Service information 2024.

Sultan marshes. The designation of a protected area is often insufficient for biodiversity conservation. The Ramsar site status has not prevented the drainage of the endorheic Sultan marshes since the 2000s, nor their connection to the Seyhan. This wetland, situated in the middle of the Yeşilhisar-Develi Plain, comprises a freshwater marsh covered with reeds in its southern section, the saline Lake Yay in the northern section, and the barren lands surrounding these two. The Sultan marshes are fed by streams that descend from the Ala Mountains in the south and by springs with a high output. The Akköy dam has been constructed for irrigation purposes on the Yeşilhisar, a stream that originates in the Ala Mountains. The Dündarlı drains to Sultan marshes from the south and descends into the plain west of the village Ovaçiftlik, where it joins the marsh as a delta. Furthermore, the Kovalı dam, constructed for irrigation purposes, has also dammed up the stream. Another example is the Ağaçasar dam, built for irrigation on the Yahyalı stream, which descends from the Ala Mountains. The Develi stream, which originates on Koç Mountain, disappears near the village of Sindelhöyük without reaching the lake. It has been proposed that the Soysallı spring, located north of Lake Yay and the primary habitat for the region's endemic fish species, be used for irrigation during the summer, with the remaining water diverted into the lake. Studies concerning these projects began in the 1960s and culminated in the implementation of the Great Develi Irrigation Project in the early 1970s. To address the increasing demand for water in the area, water from the Seyhan has been drained to the Sultan marshes basin since 2011. This has enabled several fish species to invade this endorheic basin. Since then, Sultan marshes have transformed from marshland to steppe habitats. All of its freshwater biodiversity is now restricted to a single spring area. All species of fish (*Anatolichthys danfordii, Pseudophoxinus elizavetae, Cobitis joergbohleni, Oxynoemacheilus ciceki* and *Seminemacheilus ahmeti*) are at risk of extinction, and future activities may result in the complete drainage of the remaining water or the introduction of non-native species that could eliminate the highly sensitive endemic fauna. **Further reading.** Çiçek & Sungur 2020 (Sultan marshes).

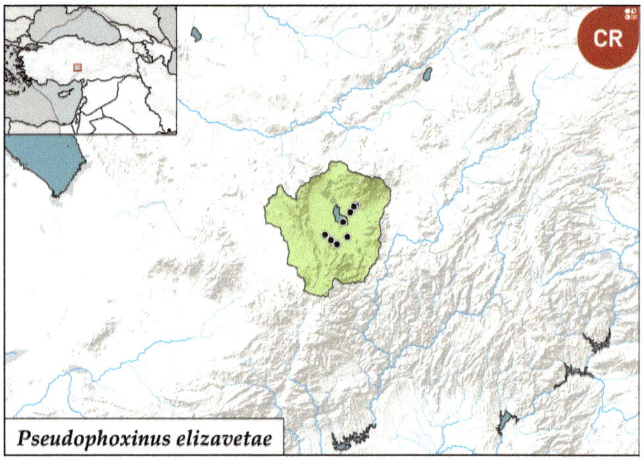

Pseudophoxinus elizavetae

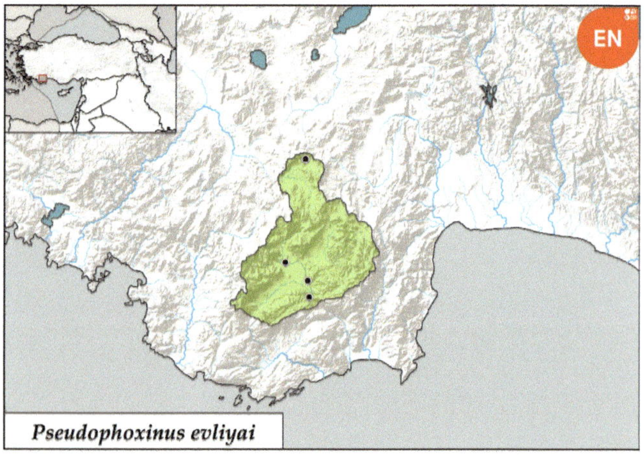

Pseudophoxinus evliyai

Pseudophoxinus evliyai; Kırkpınar, Türkiye; ~70 mm SL.

Pseudophoxinus evliyai

Common name. Lycian spring minnow.

Diagnosis. Distinguished from other species of *Pseudophoxinus* in Western Anatolia by: ○ 52–64+2–3 scales in lateral series / ○ lateral line incomplete with 16–30 perforated scales / ○ 8–10 branched pelvic rays / ○ 13–16 scale rows between dorsal origin and lateral line / ○ mouth slightly subterminal / ○ flank usually with indistinct, brown mid-lateral stripe. Size up to 87 mm SL.

Distribution. Türkiye: Kırkpınar spring, Lake Söğüt and Avlan basins, Akçay drainage (Finike).

Habitat. Lakes, springs, spring-fed streams, and canals with clear water and dense vegetation.

Biology. No data.

Conservation status. EN; appears to be declining within its very small range. Lakes Avlan and Söğüt dried up in 1990s, and the species is now restricted to irrigation canals and spring-fed streams in both basins. Both populations are very small and very restricted in their distribution.

Remarks. Originally described as *P. evliyae*, it was changed to *P. evliyai* based on gender rules.

Further reading. Freyhof & Özuluğ 2010b (description).

Pseudophoxinus fahrettini; upper Köprüçay drainage, Türkiye; ~100 mm SL.

Pseudophoxinus fahrettini

Common name. Pisidian spring minnow.

Diagnosis. Distinguished from other species of *Pseudophoxinus* in Central Anatolia by: ○ 74–85+3–4 scales in lateral series / ○ lateral line complete / ○ upper lip projecting beyond tip of lower lip / ○ 12–14 short gill rakers / ○ body depth at dorsal origin 23–29 % SL / ○ prominent black midlateral stripe / ○ lateral line slanted upwards to lateral midline at caudal peduncle / ○ 8–10½ scales between lateral line and pelvic origin / ○ last unbranched dorsal ray flexible / ○ pelvic axillary lobe present / ○ no keel between posterior end of pelvic base and anus / ○ pelvic origin clearly behind vertical of dorsal origin. Size up to 121 mm SL.

Distribution. Türkiye: upper and middle Köprüçay drainage.

Habitat. Springs and streams with clear water and dense vegetation.

Biology. No data.

Conservation status. EN; appears to be declining within its very small range.

Further reading. Freyhof & Özuluğ 2010a (description).

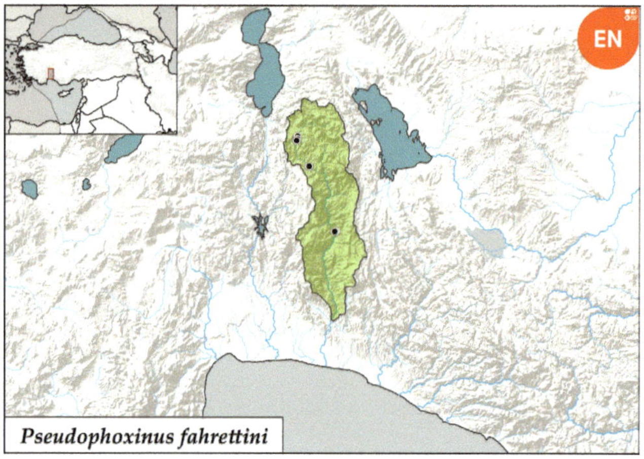

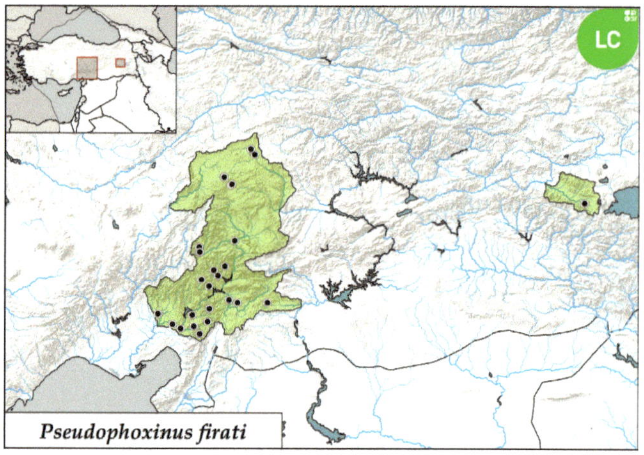

Pseudophoxinus firati; upper Euphrates drainage, Türkiye; 75 mm SL.

Pseudophoxinus firati; Güroymak Spring, upper Euphrates, Türkiye; ~90 mm SL.

Pseudophoxinus firati

Common name. Euphrates spring minnow.

Diagnosis. Distinguished from other species of *Pseudophoxinus* in eastern Mediterranean basin by: ○ 43–49+2–3 scales in lateral series / ○ lateral line incomplete, often almost complete, with 15–44 perforated scales / ○ flank with prominent brown, mid-lateral stripe / ○ pelvic axillary lobe present / ○ 6–8 gill rakers. Size up to 100 mm SL.

Distribution. Türkiye: Headwaters in Ceyhan, Seyhan (Karahalka spring) drainages. Very locally in Euphrates drainage.

Habitat. Springs and streams with clear water and dense vegetation.

Biology. No data.

Conservation status. LC.

Further reading. Bogutskaya et al. 2007 (description); Saç et al. 2019b (distribution).

Lake Eğirdir in Türkiye is the habitat of several regionally endemic species.

Pseudophoxinus handlirschi; Lake Eğirdir, Türkiye; 74 mm L.

Pseudophoxinus handlirschi
Common name. Eğirdir minnow.
Diagnosis. Distinguished from other species of *Pseudophoxinus* in Central Anatolia by: • lower lip projecting beyond tip of upper lip / ○ 76–98+3–4 scales in lateral series / ○ lateral line complete / ○ flank with brown mid-lateral stripe / ○ 15–17 long gill rakers / ○ body depth at dorsal origin 19–23 % SL / ○ pelvic origin clearly behind vertical of dorsal origin / ○ no keel between posterior end of pelvic base and anus. Size up to 104 mm SL.
Distribution. Lake Eğirdir.
Habitat. Lacustrine and pelagic.
Biology. No data.
Conservation status. Extinct; disappeared after the introduction of non-native *Sander lucioperca*.
Further reading. Küçük et al. 2009b (extinction); Freyhof & Özuluğ 2010b (diagnosis).

Pseudophoxinus hasani; Nahr al Marqīyah, Syria; ~50 mm SL.

Pseudophoxinus hasani
Common name. Marquia spring minnow.
Diagnosis. Distinguished from other species of *Pseudophoxinus* in eastern Mediterranean basin by: • 26–33+2–3 scales in lateral series / ○ lateral line incomplete with 10–20 perforated scales / ○ flank with prominent brown mid-lateral stripe. Size up to 58 mm SL.
Distribution. Syria: Nahr al Marqīyah drainage.
Habitat. Springs and spring-fed streams with clear water and dense vegetation.

Biology. No data.
Conservation status. CR; habitat quality has deteriorated severely, and stream now has little water in summer. Pollution is a serious problem in lower reaches.
Further reading. Krupp 1992b (description).

Pseudophoxinus hittitorum
Common name. Hittite spring minnow.
Diagnosis. Distinguished from other species of *Pseudophoxinus* in Central Anatolia by: ○ 84–96+1–2 lateral-line scales / ○ flank plain brown or grey without prominent stripe / ○ pelvic axillary lobe absent / ○ lateral line complete / ○ no keel between posterior end of pelvic base and anus / ○ pelvic origin clearly behind vertical of dorsal origin / ○ upper lip projecting beyond tip of lower lip / ○ 6–8 short gill rakers / ○ 26–30 scales between lateral line and dorsal origin / ○ last simple dorsal ray flexible / ○ pelvic short, not reaching to anus / ○ dorsal origin behind vertical of pelvic origin. Size up to 104 mm SL.
Distribution. Türkiye: Lake Beyşehir basin and Seydişehir canal flowing to Lake Suğla.
Habitat. Springs and spring-fed streams with clear water and dense vegetation.
Biology. No data.
Conservation status. VU; appears to be declining within its small range.
Further reading. Freyhof & Özuluğ 2010c (description); Küçük et al. 2016 (distribution).

Pseudophoxinus hittitorum; Lake Beyşehir basin, Türkiye; ~90 mm SL.

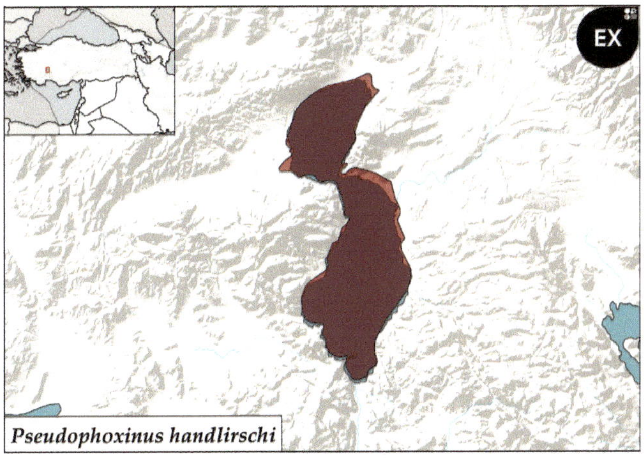

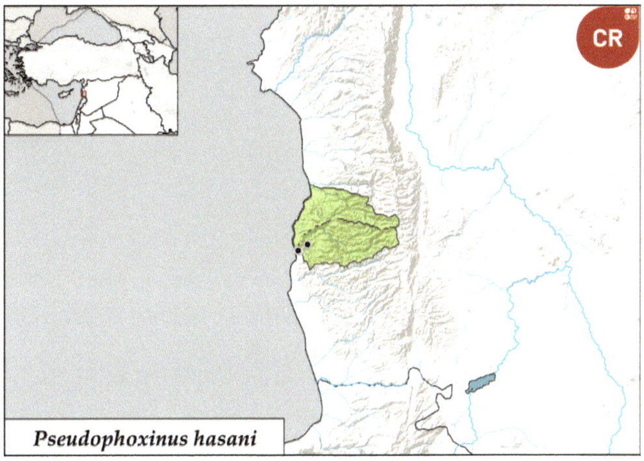

Eflatunpınarı in Türkiye is one of the habitats of *Pseudophoxinus hittitorum* and several other species native to Lake Beyşehir basin.

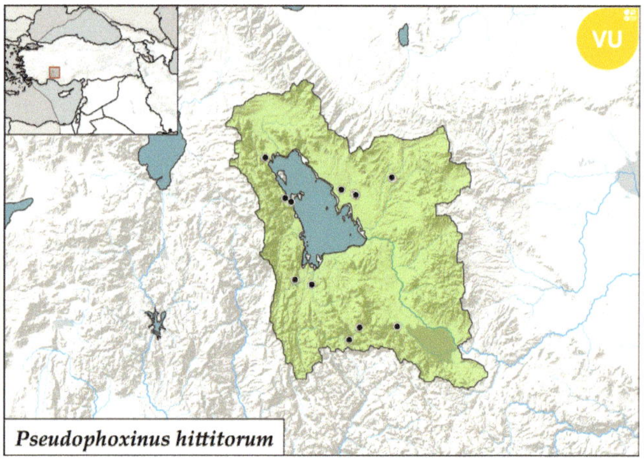

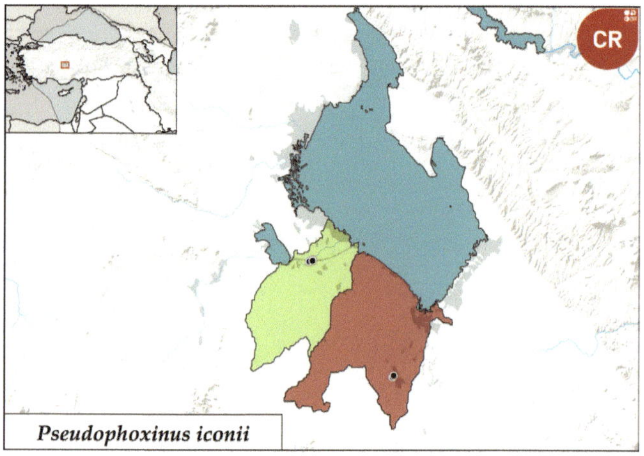

Pseudophoxinus iconii; Gölyazı spring, Lake Tuz basin, Türkiye; ~80 mm SL.

Pseudophoxinus iconii

Common name. Tuz spring minnow.

Diagnosis. Distinguished from other species of *Pseudophoxinus* in Central Anatolia by: ○ 53–63+2–3 scales in lateral series / ○ lateral line usually complete with 48–61 perforated scales / ○ no keel between posterior end of pelvic base and anus / ○ flank with brown mid-lateral stripe / ○ upper lip not or slightly projecting beyond tip of lower lip / ○ 13–15 scale rows between lateral line and dorsal origin / ○ 10–13 gill-rakers. Size up to 81 mm SL.

Distribution. Türkiye: Gölyazı, Büget, Tersakan and Bolluk drainages in Lake Tuz basin.

Habitat. Springs and slow-flowing spring-fed streams. Usually in densely vegetated habitats.

Biology. No data.

Conservation status. CR; Extirpated from Lake Büget in Eşmekaya.

Further reading. Küçük et al. 2016 (description).

Gölyazı is one of the few habitats of *Pseudophoxinus iconii* in Lake Tuz basin.

Pseudophoxinus libani; spring south of Yammoune, Lebanon; ~90 mm SL.

Pseudophoxinus libani
Common name. Lebanese minnow.
Diagnosis. Distinguished from other species of *Pseudophoxinus* in eastern Mediterranean basin by: ○ 42–57+2–3 scales in lateral series / ○ lateral line incomplete / ○ flank plain brown, grey or golden without or with faint, brown or grey mid-lateral stripe / ○ pelvic axillary lobe absent / ○ pelvic long, reaching or overlapping anus / ○ 4–5 scales between pelvic origin and lateral line / ○ 7–8 gill rakers / ○ 19–20 abdominal vertebrae. Size up to 95 mm SL.
Distribution. Orontes drainage in Türkiye and Syria except for upper Karasu, Litani drainage, and endorheic Lake El Yammoune basin in Lebanon.
Habitat. Springs, streams, and rivers, usually with clear water and dense vegetation.
Biology. Spawns between February–April.
Conservation status. LC.
Remarks. *Pseudophoxinus kervillei* is a synonym.
Further reading. Bariche & Freyhof 2016 (diagnosis, synonyms, distribution); Bayçelebi 2020 (distribution).

Pseudophoxinus libani; Orontes drainage, Syria; 72 mm SL.

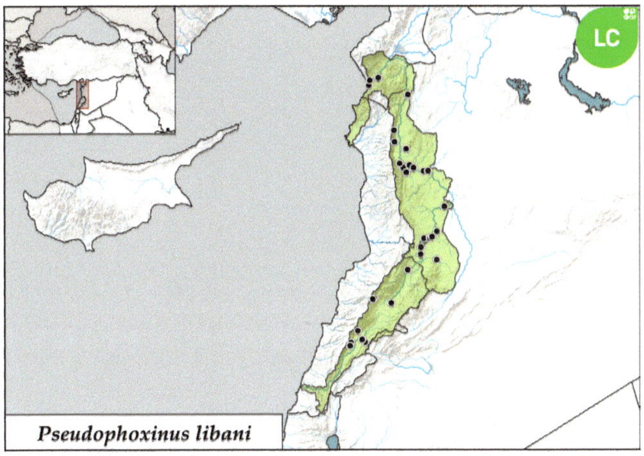

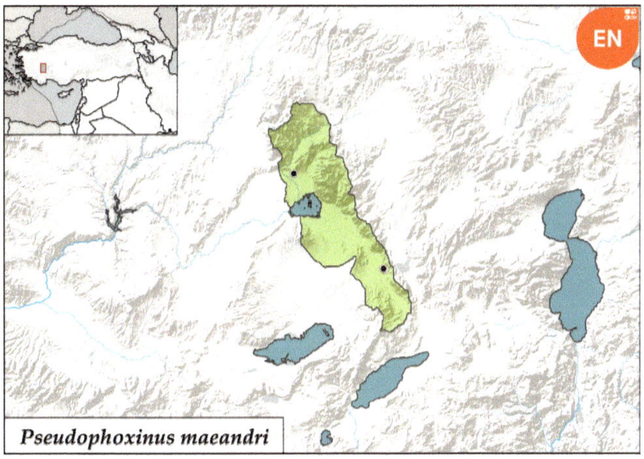

Pseudophoxinus maeandri

Common name. Apamean spring minnow.

Diagnosis. Distinguished from other species of *Pseudophoxinus* in Western Anatolia by: ○ 41–46+2–3 scales in lateral series / ○ lateral line incomplete with 19–27 perforated scales / ○ mouth terminal or slightly subterminal / ○ flank without or with indistinct, brown mid-lateral stripe. Size up to 57 mm SL.

Distribution. Türkiye: Lake Işıklı basin.

Habitat. Densely vegetated springs and lakes.

Biology. No data.

Conservation status. EN; appears to be declining within its very small range.

Further reading. Bogutskaya 1992 (revision); Freyhof & Özuluğ 2006 (description); Küçük et al. 2013 (distribution).

Pseudophoxinus maeandri; Dinar, Türkiye; ~40 mm SL.

Pseudophoxinus maeandricus; Karadirek, Türkiye; ~80 mm SL.

Pseudophoxinus maeandricus

Common name. Sandıklı spring minnow.

Diagnosis. Distinguished from other species of *Pseudophoxinus* in Western Anatolia by: • lateral line complete / ○ 60–65+2 scales in lateral series / ○ mouth slightly subterminal/ ○ flank without mid-lateral stripe. Size up to 100 mm SL.

Distribution. Türkiye: Karadirek near Sandıklı, an isolated basin flowing underground to Işıklı.

Habitat. Streams among dense vegetation.

Biology. No data.

Conservation status. CR; known only from one small stream under pressure from many threats.

Remarks. After being described from the Işıklı spring, this species was never recorded from there again. There was probably a mistake in labelling of collections.

Further reading. Bogutskaya 1992 (revision, distribution); Küçük et al. 2013 (distribution).

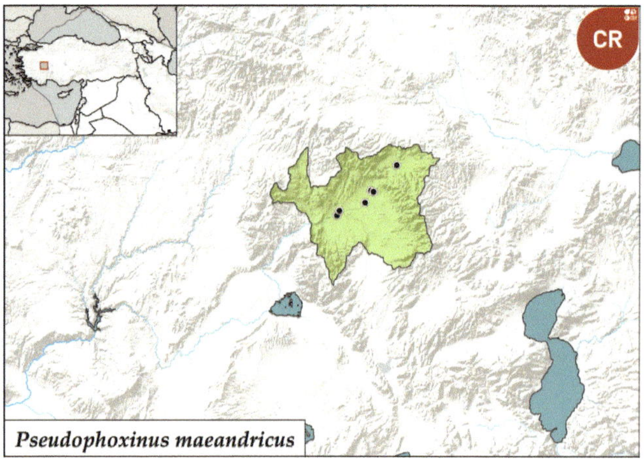

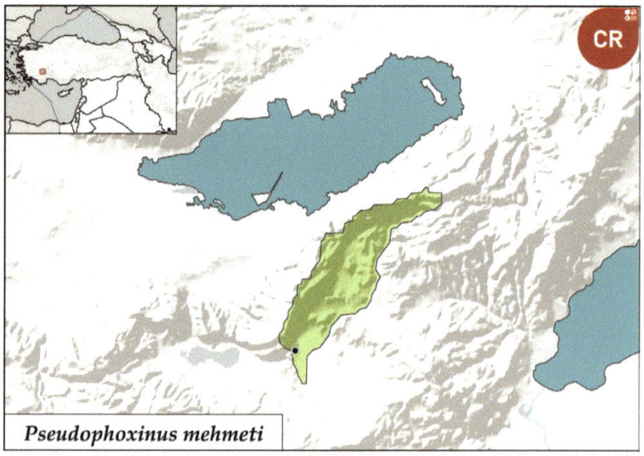

Pseudophoxinus mehmeti; Alanköy reservoir, Türkiye; ~60 mm SL.

Pseudophoxinus mehmeti; stream Alanköy, Türkiye; ~100 mm SL.

Pseudophoxinus mehmeti

Common name. Alanköy spring minnow.

Diagnosis. Distinguished from other species of *Pseudophoxinus* in Western Anatolia by: ● mouth almost superior, lower lip usually projecting slightly beyond upper lip / ○ 48–60+2 scales in lateral series / ○ lateral line incomplete with 30–50 perforated scales / ○ 11–13½ scale rows between lateral line and dorsal origin / ○ flank usually with indistinct, brown mid-lateral stripe. Size up to 100 mm SL.

Distribution. Türkiye: Lake Çorak (Akgöl) basin.

Habitat. Springs and streams, also in a reservoir.

Biology. No data.

Conservation status. CR; Alanköy reservoir has dried up, and the species is now restricted to springs and streams where it is under pressure from water extraction and stocking of *Oncorhynchus mykiss.*

Remarks. The diagnostic mouth shape of *P. mehmeti* has not been confirmed in its stream population and other distinguishing characters requires further revision.

Further reading. Ekmekçi et al. 2015 (description).

Alanköy Reservoir. A small, shallow reservoir fed by an endorheic stream during periods of high precipitation. Endemic *Pseudophoxinus mehmeti* temporarily colonises the lake from the stream.

Pseudophoxinus ninae, Onaç drainage, Türkiye; ~60 mm SL.

Pseudophoxinus ninae

Common name. Onaç spring minnow.

Diagnosis. Distinguished from other species of *Pseudophoxinus* in Western Anatolia by: ○ 46–53+2–3 scales in lateral series / ○ lateral line incomplete with 10–32 perforated scales / ○ 8–10 branched pelvic rays / ○ mouth slightly subterminal / ○ 10–13 scale rows between dorsal origin and lateral line / ○ flank usually with indistinct, brown mid-lateral stripe. Size up to 83 mm SL.

Distribution. Türkiye: Onaç drainage, Kestel marsh and springs in former Lake Karaevli.

Habitat. Reservoirs, slow-flowing streams, springs usually among vegetation.

Biology. No data.

Conservation status. EN; appears to be declining within its very small range.

Further reading. Freyhof & Özuluğ 2006 (description); Küçük et al. 2013 (distribution).

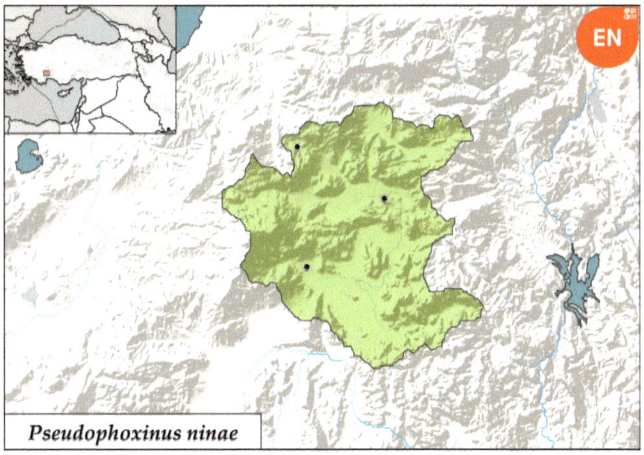

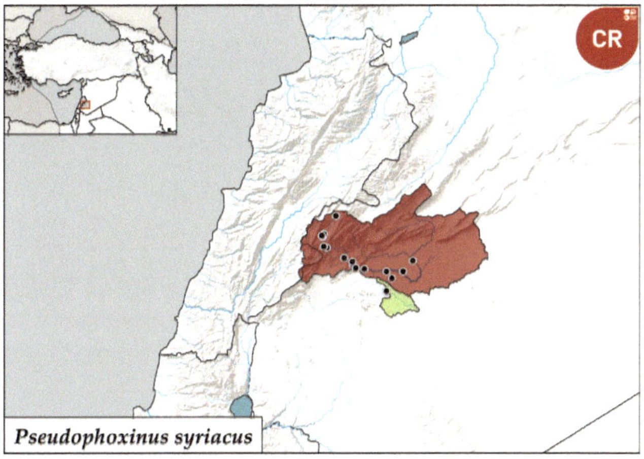

Pseudophoxinus syriacus; spring of Barada, Syria; ~65 mm SL.

Pseudophoxinus syriacus

Common name. Barada spring minnow.

Diagnosis. Distinguished from other species of *Pseudophoxinus* in eastern Mediterranean basin by: ○ 36–50+2–3 scales along lateral series / ○ lateral line incomplete / ○ flank plain brown, grey or golden without prominent stripe / ○ pelvic axillary lobe absent / ○ pelvic short, reaching 1–5 scales in front of anus / ○ 6–7 scales between pelvic origin and lateral line. Size up to 87 mm SL.

Distribution. Syria: Damascus basin.

Habitat. Springs and spring-fed streams with clear water and dense vegetation.

Biology. No data.

Conservation status. CR; known from only one site and has declined by more than 90 % due to water abstraction in early 21st century. With no new data since 2008, it cannot be excluded that it is now extinct.

Further reading. Krupp 1985d (diagnosis, distribution).

Pseudophoxinus turani; spring at Hassa, Orontes drainage, Türkiye; ~50 mm SL.

Pseudophoxinus turani

Common name. Karasu spring minnow.
Diagnosis. Distinguished from other species of *Pseudophoxinus* in eastern Mediterranean basin by: 38–46+2–3 scales in lateral series / ○ flank with bold, black or dark-grey mid-lateral stripe / ○ lateral line incomplete, with 12–25, usually 16–21, perforated scales / ○ pelvic axillary lobe absent / ○ pelvic long, reaching or overlapping anus / ○ 21–22 abdominal vertebrae / ○ 4–5 scales between pelvic origin and lateral line / ○ 8–11, rarely 13 gill rakers / ○ 10–11 scale rows between the lateral line and dorsal origin. Size up to 93 mm SL.

Distribution. Türkiye: Karasu drainage in northern Orontes catchment.
Habitat. Springs and slow-flowing streams, usually among vegetation.
Biology. No data.
Conservation status. VU; appears to be declining within its small range.
Further reading. Küçük & Güçlü 2014 (description).

Last remains of *Pseudophoxinus syriacus* (large fish) and *Paraphanius mento* in the drained En Fidje spring near Damascus, Syria.

The fate of the endorheic Damascus basin. The basin of Damascus, with its shallow Pleistocene lakes, emerged only after the outflow of the Hauran basalts that isolated it from the Jordan drainage. The streams Barada and al-Awaj flowed into large shallow lakes east of Damascus. These lakes were filled at least twice, in the middle Pleistocene and around 20,000 years BP, covering 2200 km^2 at their maximum extension. During the warmer interglacial period, the lake levels shrank, except for a humid period at approximately 4000–5000 years BP. At the end of the 19th century, evidence of the lakes still existed, indicating that they were significant hotspots for aquatic biodiversity, as observed by local inhabitants and early explorers. Several fish species, including *Pseudophoxinus syriacus*, *Acanthobrama tricolor*, and *Tristramella simonis*, were identified in the lakes. The lakes have completely dried out, and their remains are no longer visible in the Syrian desert. The Barada, one of the major tributaries of these lakes, has become a sewage drainage system, and no fish seem to remain in that small river. The renowned spring En Fidje, the source of the Barada as referenced in Islamic poetry for its beauty, has been almost entirely extracted for drinking water. The al-Awaj is the sole remaining source of the Damascus basin's distinctive fish fauna, with its headwater streams comprising a limited number of species. It may be the last surviving remnant of the entire Damascus basin.

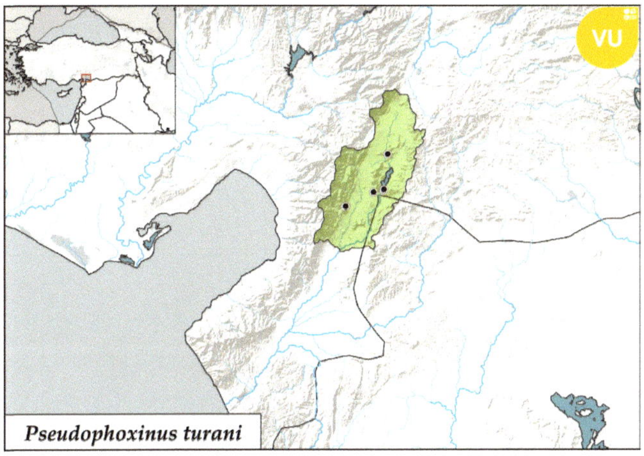

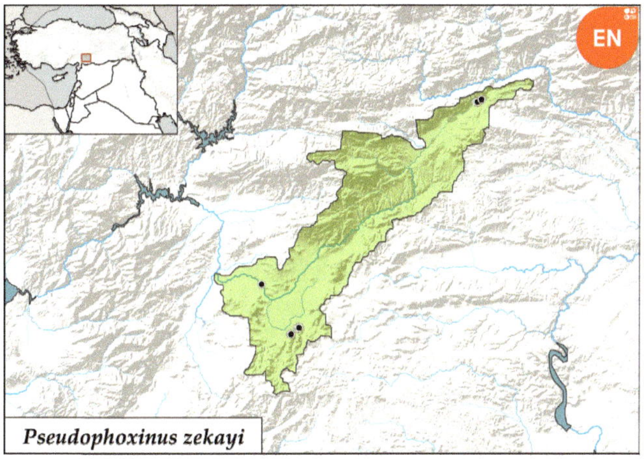

Pseudophoxinus zekayi; Çöçelli spring, Ceyhan drainage, Türkiye; ~90 mm SL.

Pseudophoxinus zekayi
Common name. Ceyhan spring minnow.
Diagnosis. Distinguished from other species of *Pseudophoxinus* in eastern Mediterranean basin by: ○ 38–46+2–3 scales in lateral series. / ○ lateral line complete / ○ flank usually with indistinct, brown mid-lateral stripe / ○ many small black spots below lateral line / ○ uppermost point of cleft below of center of eye. Size up to 113 mm SL.
Distribution. Türkiye: Eastern Ceyhan drainage in springs around Çöçelli and Lakes Gölbaşı.
Habitat. Springs, spring-fed streams, and densely vegetated lakes.
Biology. No data.
Conservation status. EN; appears to be declining within its very small range.
Further reading. Bogutskaya et al. 2007 (description); Saç et al. 2019b (distribution).

Pseudophoxinus zeregi; lower Orontes drainage, Türkiye; ~55 mm SL.

Pseudophoxinus zeregi
Common name. Levantine spring minnow.
Diagnosis. Distinguished from other species of *Pseudophoxinus* in eastern Mediterranean basin by: ○ 55–62+2–3 scales in lateral series / ○ lateral line complete / ○ flank with prominent, brown or grey mid-lateral stripe. Size up to 79 mm SL.
Distribution. Türkiye and Syria: Orontes drainage and headwaters of Qweiq. Rivers Marqīyah and Abou Noah in Syria.
Habitat. Pools, slow-flowing streams, marshes, springs usually among vegetation.
Biology. Lives up to 5 years. Spawns first time after first winter. Fractional spawner. Feeds on a wide range of small invertebrates and detritus.
Conservation status. EN; Extirpated from the Qweiq in Syria and the potentially from Orontes drainage in Türkiye. It appears to have declined sharply in recent years due to climate change, desiccation of streams, and massive pollution.
Further reading. Krupp 1985d (distribution); Geiger et al. 2014 (phylogeny).

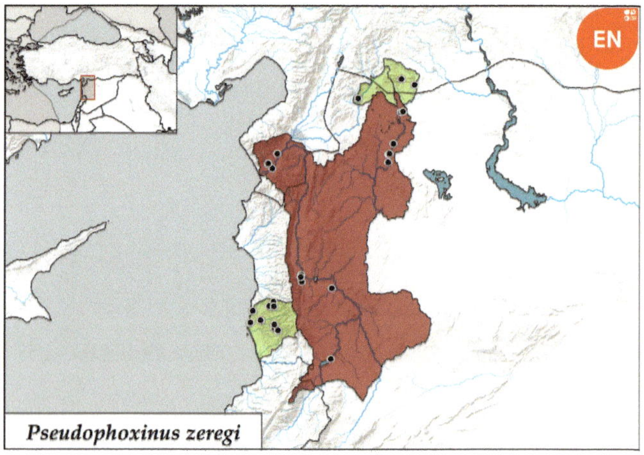

Pseudophoxinus sp. Israel, Jordan drainage, Israel; ~45 mm SL. © B. Nicca

***Pseudophoxinus* sp.**
Common name. Jordan spring minnow.
Diagnosis. The characters distinguishing this species from other *Pseudophoxinus* have yet to be worked out.
Distribution. Jordan drainage in Lebanon, Jordan and Israel.
Habitat. Springs, streams, and rivers, usually with clear water and dense vegetation. Also, found in littoral zone of Lake Kinneret.
Biology. Spawns February–April.
Conservation status. NE; due to very limited knowledge of this species.
Further reading. Goren 1974 (distribution).

Lake Karataş in Central Anatolia was once a major habitat of *Pseudophoxinus burduricus*. Recently it has dried out.

The lost River Qweiq. The Qweiq, also known as the Qweik, Quweiq, Kwaik, Kueik, Coic, Chalos, Belos, or Aleppo, was a river in the northern Levant. It had (and still has) sources in Türkiye and flowed through Aleppo, ending in a saltpan in Syria. Aleppo was an important city for early explorers and naturalists, and as a result, numerous freshwater fishes were collected for study by Russell, Linnaeus, Heckel, and others. The Qweiq was a tributary of the Euphrates, and it was not until the Holocene that this contact was broken and in the Qweiq became an endorheic basin. In the 19th century, the Qweiq was still a large river, and fish species such as *Luciobarbus esocinus* existed there. It supported the Syrian city of Aleppo for centuries with water. The Qweiq dried out successively in the 20th century. It has completely vanished from Syria, and only in Türkiye, two very small headwater streams remain, with a reservoir holding water for irrigation. Syrian farmers now pump their waters from deep layers, independent of the surface waters of the Qweiq. In addition, Aleppo constructed a canal for the Euphrates to flush out the sewage from the city. The middle Orontes was a tributary of the Qweiq, until the early Pleistocene. At that time, the Qweiq lost contact with the Orontes but continued to flow to the Euphrates. The connection between the Qweiq and the Orontes had been closed approximately 2.0–1.5 million years ago, and most freshwater fishes in the Qweiq are well differentiated from their closest relatives in the Orontes. However, some freshwater fishes in the Qweiq are closely related to those of the Orontes, suggesting that there must have been secondary connections much later. *Pseudophoxinus zeregi* and *Capoeta damascina* are widely distributed in the Orontes in the Qweiq. Similarly, *Garra variabilis* exhibits identical mitochondrial DNA haplotypes between the Qweiq and the Orontes, except those from the Euphrates, which show considerable differences. Furthermore, the mitochondrial DNA bodies of *Oxynoemacheilus namiri* from the Orontes have been identified in *O. tigris* from the Qweiq, indicating introgression. These close relationships are most likely due to Holocene contact between the middle sector of the Orontes and the Qweiq, which was interrupted by basaltic outflows south of Aleppo.

Sünnep east of Kilis, Türkiye, one of the small streams remaining from the Qweiq.

Copernicus Service information 2024.

Rutilus

Rutilus are small- to large-sized fishes endemic to Europe, North-West, and Central Asia. In West Asia, only five species occur, with two endemic to Azerbaijan: *R. atropatenus* and *R. sojuchbulagi*. Both are poorly known and have been placed in the genus *Pseudophoxinus* until recently. Many subspecies and variants of the roach, *R. rutilus*, have been described, particularly from the Caspian basin. All can be identified as *R. lacustris*. *Rutilus rutilus* has a restricted range in West Asia, while *R. lacustris* and *R. rutilus* are largely allopatric in West Asia, both species have large hybrid zones in Europe. An attempt was made to split the genus *Rutilus* into three smaller genera (*Leucos*, *Rutilus*, and *Sarmarutilus*). Finally, most ichthyologists did not accept it as all these genera are poorly distinguished and closely related. We agree with this view and maintain the current classification of all *Rutilus* species as a single genus. **Further reading.** Bianco & Ketmaier 2014 (molecular phylogeny, genera); Artaev et al. 2021 (*R. rutilus*, *R. lacustris*).

Key to species of *Rutilus* in West Asia

1a - 53–64+3 lateral-line scales.
..................*R. frisii*
1b - 34–42+3 lateral-line scales.
..................2

2a - 34–38+2–3 lateral-line scales; 6–7 scales between dorsal origin and lateral line.
..................3
2b - 39–42+2–3 lateral-line scales; 7–9 scales between dorsal origin and lateral line.
..................4

3a - Tip of snout at or below horizontal level of lower eye margin.
..................*R. sojuchbulagi*
3b - Tip of snout above horizontal level of lower eye margin.
..................*R. atropatenus*

4a - Pectoral, pelvic, and anal orange to red; iris from yellow in juveniles to deep red in adults.
..................*R. rutilus*
4b - Pectoral, pelvic, and anal hyaline, yellowish to dark-grey; iris whitish to pale yellow.
..................*R. lacustris*

Rutilus frisii, Tajan, Iran; with nuptial tubercles.

Rutilus atropatenus; Kura drainage, Azerbaijan; ~80 mm SL.

Rutilus atropatenus

Common name. Azerbaijani spring roach.

Diagnosis. Distinguished from other species of *Rutilus* in West Asia by: ○ tip of snout above horizontal level of lower eye margin / ○ 34–38+2–3 lateral-line scales / ○ 6–7 scales between dorsal origin and lateral line. Size up to 80 mm SL.

Distribution. Azerbaijan: Springs in Alazani-Agrichay valley in lower Kura drainage.

Habitat. Karstic springs, where it is the only fish species.

Biology. No data.

Conservation status. VU; appears to be declining within its small range.

Further reading. Derjavin 1937 (description); Artaev et al. 2018 (distribution); Kuljanishvili et al. 2020 (generic placement).

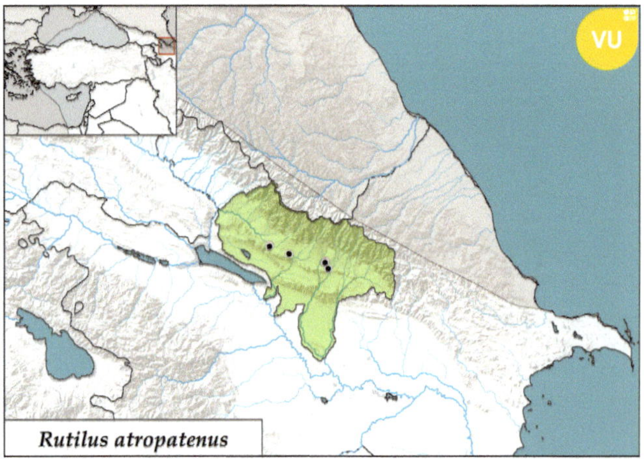

Rutilus frisii

Common name. Vyrezub.

Diagnosis. Distinguished from other species of *Rutilus* in West Asia by: ● body almost cylindrical / ● 53–64+3 lateral-line scales / ● abdomen posterior to pelvic rounded / ○ mouth subterminal / ○ iris and fins grey or slightly yellowish / ○ usually 9½ branched dorsal rays / ○ 9–10½ branched anal rays / ○ no midlateral stripe / ○ lateral line complete. Size up to 680 mm and 8 kg.

Distribution. Black and Azov Sea basins, rare in Danube (mostly restricted to few subalpine lakes), absent from Kuban; landlocked populations in reservoirs on Don; Caspian basin, common from Terek to Atrak drainages, very rare in northern Caspian basin (in Volga formerly as far as Perm and in Ural). Landlocked population in Anatolian Lake İznik.

Habitat. Coastal lakes connected to rivers, and lowland reaches of large rivers close to coast at sea. In estuaries, found in deep layers down to 20 m. Tolerates salinities up to 7–12 ‰. Inland populations inhabit large lakes or reservoirs. Semianadromous populations often enter in small rivers or streams with strong currents on gravel beds to spawn.

Biology. Riverine, lacustrine or semi-anadromous. Lives up to 12 years. Spawns first time at 4–5 years, with 430–500 mm SL. Enters rivers in second half of October (Black Sea, rarely in Caspian Sea). When rivers are ice-covered, spawning migration stops and resumes when ice breaks up or even under ice in February–April. Populations spawning in small streams enter in February–April. Spawns in April–May, somewhat earlier in southern drainages (February in Terek), when temperature reaches 8°C, peaking at 13–15°C.

Rutilus frisii; Tajan, Iran; ~300 mm SL.

Eggs adhere to rocks and gravel, rarely to submerged vegetation. Adults return to estuary immediately after spawning to feed. Eggs hatch in 10–16 days at 12–19°C. Juveniles migrate to estuaries during first summer (August). Inland populations migrate from lakes, reservoirs, or middle reaches of rivers to tributaries or headwaters in spring. Larvae and early juveniles feed on zooplankton, algae, and insect larvae. Adults feed mainly on molluscs and *Rhithropanopeus* crabs, supplemented by other benthic invertebrates. Stops feeding during migration, spawning and overwintering.

Conservation status. LC: Lake İznik population declined in the late 20th century and may now be extirpated. Migratory populations very rare in Black Sea basin. In Caspian, populations declined since 1980s and may continue to decline. Dams and large-scale overfishing have made natural spawning almost impossible in Caspian, and populations are maintained largely by massive stocking of 120–140 million juveniles annually in Iranian waters.

Further reading. Berg 1949b (biology, morphology); Kazancheyev 1981 (Caspian; biology); Magomedov 1981 (biology); Movchan & Smirnov 1981 (Black Sea; biology); Belyaeva et al. 1989 (biology); Kotlík et al. 2008 (phylogeography).

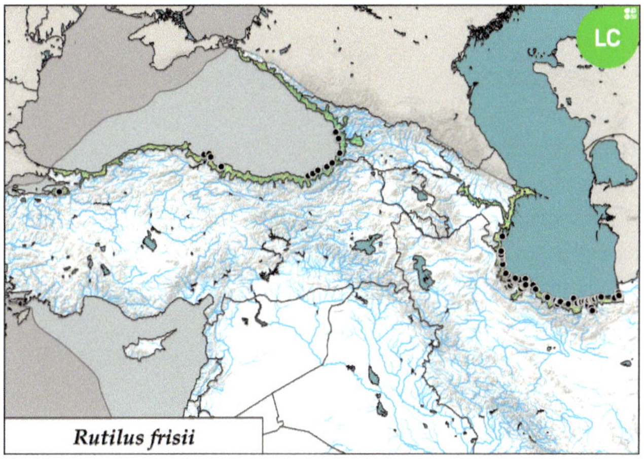

Rutilus lacustris; Aras drainage, Armenia; ~250 mm SL. © B. Lewin.

Rutilus lacustris; Kura, Azerbaijan; ~200 mm SL.

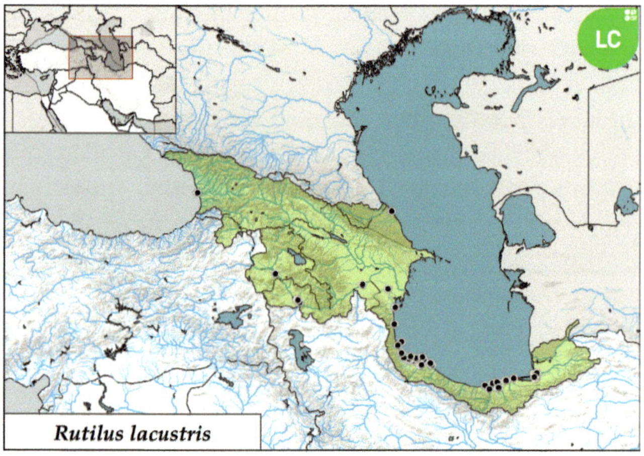

Rutilus lacustris

Common name. Taran.

Diagnosis. Distinguished from other species of *Rutilus* in West Asia by: ○ 39–42+2–3 lateral-line scales / ○ body laterally compressed / ○ abdomen posterior to pelvic compressed / ○ usually 10½ branched dorsal rays / ○ usually 9–10½ branched anal rays / ○ iris whitish to yellowish / ○ pectoral, pelvic and anal hyaline, yellowish to dark-grey / ○ lateral line complete / ○ no midlateral stripe / ○ 7–9 scales between dorsal origin and lateral line. Size up to 430 mm SL.

Distribution. Freshened areas of Caspian basin entering all tributaries. Eastern Black and Azov Seas, entering Don, Kuban, Dniepr, Dniestr, and rarely Danube drainages and lower parts of these rivers. Hari drainage (not native) but possibly native to Morghab (Turkmenistan). Also, in Aegean basin (Struma and Maritza drainages and Lake Volvi). Widespread in Aral basin and Siberia from Ob east to Lena.

Habitat. Resident populations inhabit many lowland habitats, such as rivers, lakes, reservoirs, and canals. Migratory fish spend most of the year in shallow, brackish coastal waters. At sea, adults are most abundant at depths of 2.5–4.0 m and 2–4 ‰ in summer. Enters freshwaters of estuaries, lagoons, and lower reaches of rivers to spawn.

Biology. Resident and semi-anadromous populations. Usually mature at 2–4 years and live 10–15 years. Comes to coasts in August and enters rivers in September, peaking in October. Stops migrating at end of November and overwinters in main river or estuary. Spawning migration resumes with ice break-up in March (Dniepr). Spawns in April–May. Adults return to estuaries to feed immediately after spawning. Juveniles migrate to estuaries during first summer (August). Resident populations may undertake considerable river migrations, usually spawning in flooded meadows and backwaters. Lacustrine populations usually spawn in inflowing rivers or on lake shores. Spawning usually occurs at temperatures above 10–12°C, usually on submerged vegetation, but all types of substrates will be used if no vegetation is available. Several males typically follow a single female during spawning. Feeds on various plants, detritus, and invertebrates, locally molluscs or plankton dominate diet.

Conservation status. LC.

Remarks. Molecular data suggest that all Caspian populations (and many in Europe) belong to this species. *Rutilus lacustris*, described from Siberia, has long been considered a synonym of *R. rutilus*. It had already been accepted in Europe as a valid species under *R. heckelii* (Black Sea basin) and *R. caspicus* (Caspian basin). Molecular data support the view that both are conspecific and that populations occurring east to Lena and Lake Baikal all belong to *R. lacustris*. The situation is more complicated in the Black Sea and the northern Caspian basins, as there appears to be some introgressive hybridisation between *R. rutilus* and *R. lacustris*, and many hybrid populations occur in that region.
Further reading. Berg 1949b (biology, systematics); Holčík & Skorepa 1971 (biology); Kottelat 1997 (systematics); Kottelat & Freyhof 2007 (distribution, biology as *R. heckeli* & *R. caspicus*); Larmuseau et al. 2009 (distribution); Geiger et al. 2014 (molecular data); Artaev et al. 2021 (distribution, hybrids with *R. rutilus*).

Rutilus rutilus; Elbe drainage, Germany; ~200 mm SL.

Rutilus rutilus
Common name. Roach.
Diagnosis. Distinguished from other species of *Rutilus* in West Asia by: ○ 39–42+2–3 (41–44 total) lateral-line scales / ○ body laterally compressed / ○ abdomen posterior to pelvic base compressed / ○ usually 10½ branched dorsal rays / ○ usually 10½ branched anal rays / ○ iris from yellowish in juveniles to deep red in adults / ○ pectoral, pelvic and anal orange to red / ○ no midlateral stripe / ○ lateral line complete / ○ 7–9 scales between dorsal origin and lateral line. Size up to 500 mm SL.
Distribution. In West Asia, restricted to Black Sea basin, confirmed from Lake Sapanca, Sursuluk, and Sakarya. Introduced to Cyprus. Europe north of Pyrenees and Alps, east to Volga. In Azov, Black and Caspian basins probably restricted to upper and middle reaches of rivers, replaced by *R. lacustris* and hybrids between both in lower reaches. Present in Lake Vegoritis in Greece. Naturally absent from Iberian Peninsula, western France, Adriatic Basin, Italy, UK north of 56°N, Scandinavia north of 69°N. Locally introduced and invasive in northern Italy, Spain, and Portugal. Introduced in North Africa.

Rutilus rutilus; Lake Apolyont, Türkiye; ~100 mm SL.

Habitat. Mostly in lowland areas and in lakes. Common in nutrient-rich lakes and large to medium-sized rivers and backwaters. Takes advantage of channeling, damming, and light organic pollution. Small lowland streams, brackish coastal lagoons, and fast-flowing rivers where it is restricted to sections where backwaters or shelters allow overwintering. Spawns in dense submerged vegetation in backwaters or lakes, flooded meadows, or shallow, fast-flowing river habitats on vegetated or gravel bottoms.

Biology. Lives up to 13 years. Males first spawn at 2–3 years, females 1 year later, usually at about 100 mm SL. Undertakes short spawning migrations, sometimes beginning as early as September, generally peaking at temperatures above 9°C in spring. Spawns in April–May when temperatures rise above 12°C. Usually, an entire population spawns within 5–10 days. Spawns in shoals. Eggs are sticky and hatch in about 12 days. Larvae and juveniles occupy a wide range of littoral habitats. Feeds mainly on benthic invertebrates, zooplankton, plant material, and detritus. Populations feeding predominantly on detritus are often stunted (stunted populations may also be associated with strong year classes). May shift from littoral to pelagic habitats and between benthic and zooplankton when a particular food is abundant or to avoid predation and/or competition. Decision to remain in open water or among littoral vegetation is often described as a trade-off between food intake and predator avoidance. During growth, there is an energetic need to switch from zooplankton to benthic food (chironomids, molluscs). Individuals able to feed on *Dreissena* mussels increase their growth rate but do not exploit this food source until they reach about 120 mm SL (size at which they can crush mussels). In some areas (Volga reservoirs), pelagic and benthic roach can be distinguished by life history characteristics (spawning time, spawning sites), and these may represent sympatric populations of *R. rutilus* and *R. lacustris*. Overwinters in backwaters or deep parts of lakes. Often produces fertile hybrids with *Abramis brama*.

Conservation status. LC.

Further reading. Holčík & Skorepa 1971 (morphology); Kottelat 1997 (systematics); Larmuseau et al. 2009 (distribution); Kottelat & Freyhof 2007 (distribution, biology); Zogaris et al. 2012 (Cyprus); Artaev et al. 2021 (distribution, hybrids with *R. lacustris*).

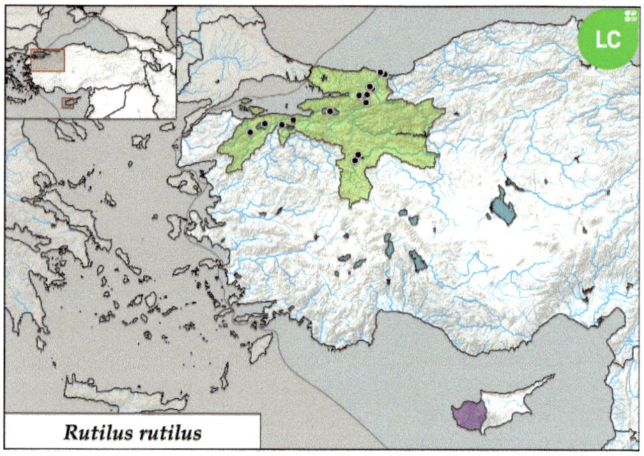

Rutilus sojuchbulagi; spring in Akstafa region, Azerbaijan; ~70 mm SL (modified from Abdurakhmanov 1950).

Rutilus sojuchbulagi

Common name. Akstafa spring roach.

Diagnosis. Distinguished from other species of *Rutilus* in West Asia by: • tip of snout at or below horizontal level of lower eye margin / ○ 35–37+2–3 scales in lateral series. Size up to 70 mm SL.

Distribution. Azerbaijan: Springs in Akstafa region in middle Kura drainage.

Habitat. Springs.

Biology. No data.

Conservation status. Extinct; has not been found since it was described in 1940s, despite several visits to its type locality and surrounding waters. Habitats in this area are heavily modified, and this species is unlikely to have survived.

Further reading. Abdurakhmanov 1950 (description); Kuljanishvili et al. 2020 (generic placement).

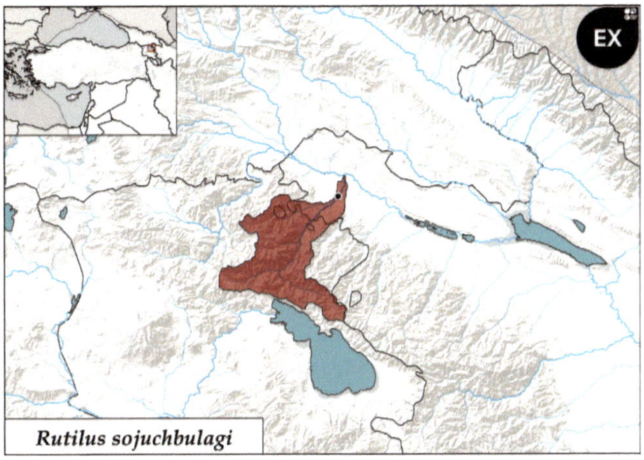

Scardinius elmaliensis; Lake Avlan, Türkiye; 74 mm SL.

Scardinius

Scardinius are small- to medium-sized fishes that inhabit various habitats, including lakes, reservoirs and slow-flowing rivers. They are typically associated with submerged vegetation. Eight species have been identified, with the genus being most speciose in the Adriatic basin and in Greece, where all but one species occur. However, some European species require critical revision, and fewer species may be recognised in the future. Two species are found in West Asia, with one, *S. elmaliensis*, being endemic.

Further reading. Kottelat & Freyhof 2007 (diversity, distribution).

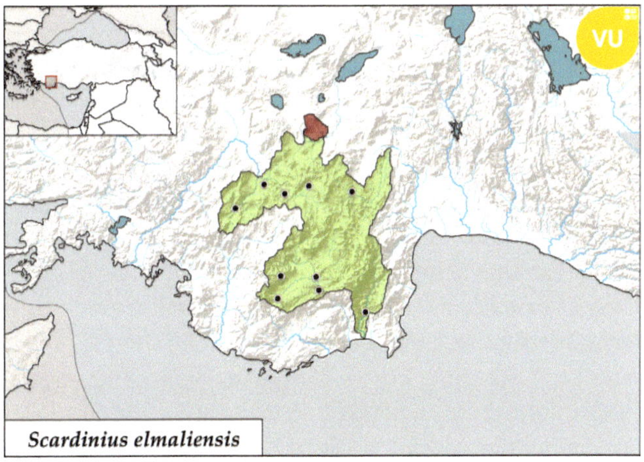

Scardinius elmaliensis

Common name. Elmalı rudd.

Diagnosis. Distinguished from *S. erythrophthalmus* in West Asia by: • 15–16 gill rakers / • 9½ branched anal rays. Size up to 186 mm SL.

Distribution. Türkiye: Akçay drainage, Lake Avlan basin and Çayboğazı reservoir. Dalaman drainage: Lake Gölhisar,

Korkuteli, Çavdır and Yapraklı reservoirs. Burdur basin: Lake Karataş and Belenli reservoir. In endorheic former Söğüt basin as in Osmankalfalar reservoir and Lake Karagöl and its canals in Elmalı.

Habitat. Reservoirs, lakes, slow-flowing streams, marshes, and springs, usually among vegetation or under bank cover.

Biology. No data.

Conservation status. VU; declining within its small range. Most populations appear to inhabit reservoirs. Extirpated from Lake Karataş.

Remarks. Molecular data place *S. elmaliensis* as a sister clade to *S. erythrophthalmus*.

Further reading. Bogutskaya 1997b (description); Geiger et al. 2014 (molecular data); Karataş et al. 2016 (molecular data, distribution).

Scardinius erythrophthalmus; Elbe drainage, Germany; ~180 mm SL.

Scardinius erythrophthalmus

Common name. Rudd.

Diagnosis. Distinguished from *S. elmaliensis* in West Asia by: • 8–13 gill rakers / • 9–13½, usually 10–12½, branched anal rays. Size up to 350 mm SL.

Distribution. In West Asia, in deltaic lakes and lower reaches of Inguri, Rioni, and Çoruh (Georgia), Marmara and Black Sea basins east to Sakarya. Also in lower Yeşilırmak and Kızılırmak (Türkiye). Absent between Kızılırmak and Coruh, as well as between Sakarya and Kızılırmak. In Caspian basin, lower reaches of Kura and Aras and lakes in Lenkoran region (Azerbaijan), Anzali and Boojagh wetlands, and Sefid drainage (Iran). Most European rivers north of Pyrenees and Alps, eastward to Ural and Eya drainages, Aral and White Sea basins; Black Sea basin in Europe. Naturally absent from Iberian Peninsula, Adriatic basin, Italy, Greece south of Pinios drainage, Great Britain north of 54°N, Ireland, and Scandinavia north of 62°N. Introduced in northern Italy, Spain, and Corsica.

Habitat. Nutrient-rich, well-vegetated lowland rivers, backwaters, oxbows, ponds, and lakes. Spawns on roots or submerged vegetation.

Biology. Lives up to 17 years. First spawns at 3–4 years. Fractional spawner. Spawns April–July when water temperature is 12–22°C. Males congregate at spawning sites and drive mature females, often with much splashing, into dense vegetation to spawn. Eggs are very sticky. Feeds mainly on plankton, terrestrial insects, and plant material. Able to adapt to adverse environmental conditions by slow growth and small size at maturity (stunted populations), e.g., in Anzali Lagoon, with a maximum total length of 146 mm, living up to 5 years and spawning at age 1+.

Conservation status. LC.

Further reading. Elanidze 1983 (distribution, biology, Georgia); Kottelat & Freyhof 2007 (distribution, biology); Uğurlu et al. 2008 (distribution in Türkiye); Patimar et al. 2010a (Anzali Lagoon).

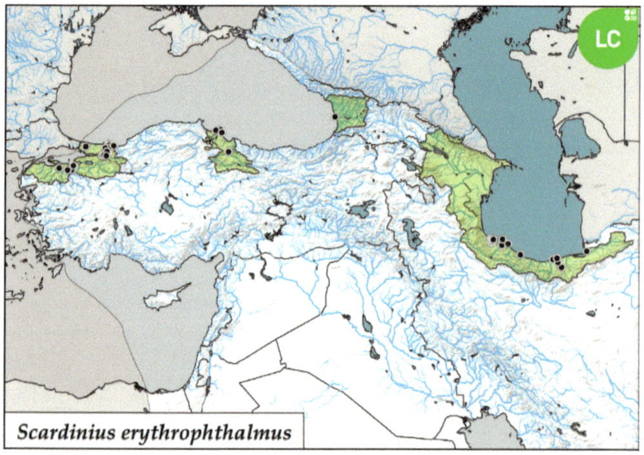

Squalius

Chubs are medium-sized to large fishes that inhabit lakes, slow-flowing large rivers, and hill streams. In this genus, 38 species are recognised, 16 of which are found in West Asia. They are particularly abundant in medium-sized streams and small rivers with gravel substrate and may constitute a significant portion of the fish biomass in such habitats. *Squalius* is characterised by scaled midventral keel in front of the anus, a complete lateral line, a slightly subterminal mouth, the dorsal origin behind the pelvic base, two rows of pharyngeal teeth, the absence of a prominent grey or brown lateral stripe (a very indistinct stripe is present in a few species), 7–10½ branched anal rays and 7–9½ branched dorsal rays. In Anatolia, mtDNA studies have only partially corroborated the morphological species hypothesis in recent years. This is due to several factors, including introgressive hybridisation and the possibility that some species may have arisen through genetic drift in small founder populations invading new habitats. Genetic drift is a potential evolutionary pathway of *S. carinus*, which has a very small range adjacent to its closely related sister species. Additionally, some species have been identified as synonyms due to underestimating their morphological variability. The species diversity of *Squalius* is still incompletely understood and future studies of nuclear DNA may change the species diversity recognised here. In Anatolia, Eastern Greece, and Bulgaria, one very widespread molecular group is distributed, having the mtDNA of *S. orpheus*. However, nuclear DNA data strongly suggest that Greek and Anatolian populations are two species: *S. orpheus* and *S. cappadocicus*. Indeed, selected populations are quite different from each other and there are plenty of intermediate populations, making separating species impossible for now. An additional complex situation is found in the Persian Gulf basin, where *S. lepidus*, *S. berak*, and *S. verepi* often share the same mtDNA sequences. They seem also to have often hybridised and such hybrids have locally stabilised, and populations are being morphologically intermediate between *S. berak/S. verepi* and *S. lepidus*, challenging their identification. In other populations, the external appearance of the fish is identical to that of one species, yet the mitochondrial DNA (mtDNA) belongs to the different species. Genomic studies will shed some light on *Squalius*' evolutionary past, hopefully soon. **Further reading.** Durand et al. 2000 (molecular diversity); Özuluğ & Freyhof 2011 (morphological diversity in Western Anatolia).

Keys to species of *Squalius* in West Asia

Contributing author: Esra Bayçelebi (Rize) contributed to the chapter on the genus *Squalius*.

Black and Marmara Sea basins

1a - Vertical black bar behind opercle conspicuous; anal rays and membranes orange or red in life in adults larger than 150 mm SL.
..................*S. orientalis*
1b - Vertical black bar behind opercle in-conspicuous; anal membranes whitish, yellowish, very pale-orange-grey or blackish, rays usually blackish in life in adults larger than 150 mm SL.
..................2

2a - 47–49 total lateral-line scales; head length 24–26 % SL.
..................*S. cii*
2b - 40–46 total lateral-line scales; head length 25–29 % SL.
..................3

3a - Dorsal origin situated behind of pelvic base; posteriormost point of anal at tip of 7th or 8th branched ray.
..................*S. pursakensis*
3b - Dorsal origin situated above pelvic base; posteriormost point of anal at tip of 3rd–5th branched ray
..................*S. cappadocicus*

Central Anatolian endorheic basins, Aegean and Mediterranean basins east to Bay of Antalya

1a - 14–15 circumpeduncular scales; scale pockets on flank with a dark-grey, crescent-shaped mark.
..................2

1b - 16–18 circumpeduncular scales; scale pockets on flank with a dark-brown or grey spot or roundish blotch on middle of flank scale.
..................4

2a - Lower lip clearly projecting beyond upper lip; usually 9½ anal rays.
..................*S. anatolicus*
2b - Upper lip clearly projecting beyond lower lip or both lips at equal level; 7–8½, very rarely 9½ anal rays.
..................3

3a - Dorsal origin situated behind of pelvic base; posteriormost point of anal at tip of 7th or 8th branched ray.
..................*S. pursakensis*
3b - Dorsal origin situated above pelvic base; posteriormost point of anal at tip of 3th–5th branched ray
..................*S. cappadocicus*

4a - A conspicuous broad, dark-brown stripe on upper part of flank between head and end of caudal peduncle; 37–40+2–3 lateral-line scales.
..................*S. kosswigi*
4b - No stripe on flank; 40–43+2–3 lateral-line scales.
..................5

5a - Dark-brown in life; lower lip projecting beyond tip of upper lip (equal to tip of upper lip in few individuals); upper head profile straight; snout conical and pointed; interorbital area straight or slightly convex.
..................*S. carinus*
5b - Silvery or golden in life; upper lip projecting beyond tip of lower lip or mouth terminal; upper head profile convex; snout rounded with obtuse tip; interorbital area convex.
..................6

6a - Mouth subterminal in adults; dorso-hypural distance reaching nostrils, rarely to anterior half of eye when reported forwards.
..................*S. fellowesii*
6b - Mouth terminal or slightly subterminal in adults; dorso-hypural distance reaching anterior half of eye when reported forwards.
..................*S. aristotelis*

Eastern Mediterranean (Seyhan, Ceyhan, Orontes)

1a - Lower lip projecting beyond upper lip; 9–10½ branched anal rays.
..................*S. kottelati*
1b - Upper lip projecting beyond lower lip or tip of both lips equal; 7–8½ branched anal rays.
..................*S. cappadocicus*

Persian Gulf, Caspian, Lake Namak, Western Kavir, and Lake Urmia basins

1a - Mouth superior in adults; 9–10½, rarely 8½ branched anal rays; posterior anal margin straight.
..................*S. lepidus*
1b - Mouth terminal or subterminal in adults; (7½) 8½, very rarely 9 ½ branched anal rays, posterior anal margin convex.
..................2

2a - Knob on lower jaw symphysis wide, thick; anal rays orange.
..................*S. namak*

2b - Knob on lower jaw symphysis narrow, thin, or absent; anal rays hyaline, grey or black, orange in some individuals and populations.
..................3

3a - Anal with numerous black pigments on rays in preserved individuals; 11–12 gill rakers; anal height 14–18 % SL.
..................*S. cappadocicus*
3b - Anal with few or without black pigments in preserved individuals; 7–11 gill rakers; anal height 17–27 % SL.
..................4

4a - Head length 27–32 % SL.
..................*S. berak / S. verepi*
4b - Head length 24–28 % SL.
..................5

5a - Posteriormost point of anal at tip of 5th or 6th branched ray.
..................*S. turcicus*
5b - Posteriormost point of anal at tip of 3rd or 5th branched ray.
..................*S. agdamicus*

Squalius agdamicus
Common name. Kura chub.
Diagnosis. Distinguished from other species of *Squalius* in Caspian basin by: ○ anal yellowish, pale-orange or hyaline in life / ○ few or without black pigments in anal in preserved individuals / ○ 43–46+1–2 lateral-line scales / ○ 7–11 gill rakers / ○ posteriormost point at tip of anal at 3rd or 5th branched ray / ○ mouth terminal or subterminal in adults / ○ head length 24–28 % SL / ○ eye diameter 18–23 % HL / ○ 8½, very rarely 9½ branched anal rays / ○ anal height 17–19 % SL. Size up to 220 mm SL, likely to grow larger.
Distribution. Caspian basin: Kura drainage.
Habitat. Fast-flowing streams with cobble and gravel bottoms.
Biology. No data.
Conservation status. LC.
Remarks. Molecular characters well differentiate chubs from the Kura and Aras. The species from the Kura is identified as *S. agdamicus*, and that from the Aras is *S. turcicus*, but their morphological characters still need to be studied better.
Further reading. Bayçelebi 2019 (description).

Squalius agdamicus; Kura drainage, Türkiye; ~160 mm SL. © M. Özuluğ.

Squalius agdamicus; Kura drainage, Azerbaijan; ~130 mm SL.

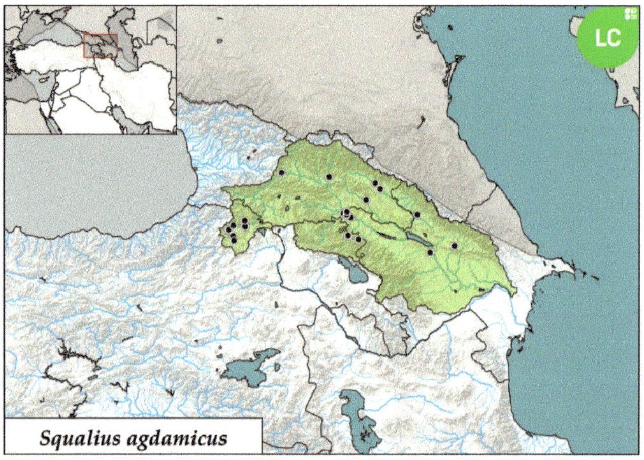

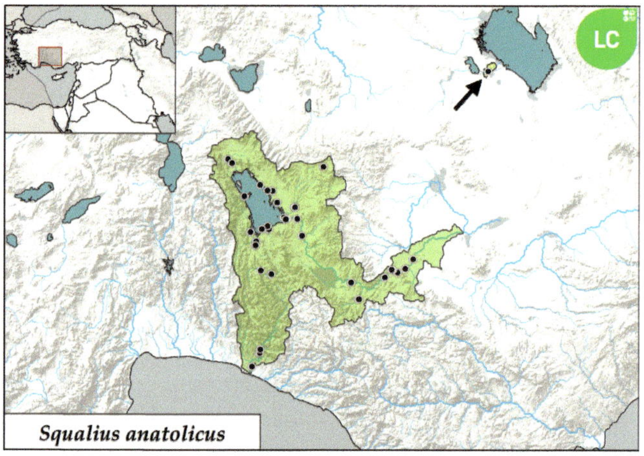

Squalius anatolicus; Lake Beyşehir basin, Türkiye; ~150 mm SL.

Squalius anatolicus

Common name. Beyşehir pike chub.

Diagnosis. Distinguished from other species of *Squalius* in Western and Central Anatolia by: ○ lower lip projecting beyond upper lip / ● head length 30–32 % SL / ○ 14 circumpeduncular scales / ○ flank silvery or golden in life / ○ scale pockets on flank with a dark-grey, crescent-shaped mark / ○ anal membranes blackish / ○ 9½ anal rays / ○ 8–10 gill rakers. Size up to 350 mm SL, likely to grow larger.

Distribution. Türkiye: Lake Beyşehir, also in Manavgat, draining to Mediterranean east of Antalya. One record also available from Lake Tuz basin.
Habitat. Lakeshores, tributaries, streams, and small rivers. Also, reservoirs and rivers up to relatively small streams.
Biology. Spawns late May–early July. Probably a fractional spawner. Feeds on invertebrates and fish as it grows.
Conservation status. LC.
Further reading. Özuluğ & Freyhof 2011 (description).

Squalius aristotelis

Common name. Tuzla chub.
Diagnosis. Distinguished from other species of *Squalius* in Western and Central Anatolia by: ● upper lip short, not or very slightly projecting beyond tip of lower lip / ○ mouth terminal or slightly subterminal in adults / ○ dorso-hypural distance reaching anterior half of eye when reported forwards / ○ silvery or golden in life / ○ no stripe on flank / ○ anal rays orange in live, blackish in preserved individuals, membranes hyaline / ○ upper lip thick, its anterior depth 3.8–5.8 times in its lateral length / ○ dorsal origin above middle of pelvic base / ○ upper head profile convex / ● snout rounded / ○ 40–45+2–3 lateral-line scales / ○ a dark-brown or black spot or blotch on flank scale pockets and at predorsal scales / ○ belly between anus and posterior extremity of pelvic base distinctly compressed / ○ prepelvic distance 53–55 % HL / ○ anal length 68–78 % HL / ○ interorbital area convex, its width 37–41 % HL / ○ head depth at nape 46–51 % HL. Size up to 183 mm SL, likely to grow larger.
Distribution. Türkiye: Tuzla drainage entering Aegean at Assos.
Habitat. Streams with Mediterranean water regime.
Biology. No data.
Conservation status. EN; appears to be declining within its very small range.
Remarks. Records from other coastal streams could not be confirmed. This species is distributed between the ranges of *S. cii* and *S. fellowesii*, and molecular data strongly suggest that *S. aristotelis* may be a species of hybrid origin. Its population includes individuals with mtDNA haplotypes of neighboring *S. cii* and haplotypes associated with the large population complex identified here as *S. cappadocicus*.
Further reading. Özuluğ & Freyhof 2011 (description).

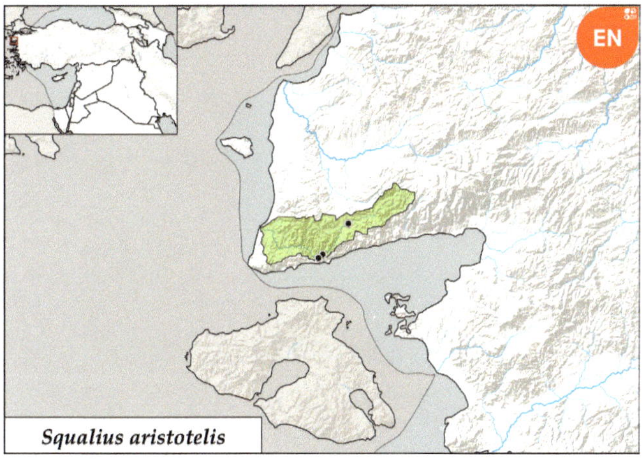

Squalius aristotelis; stream Behramkale, Türkiye; ~110 mm SL.

Squalius berak; Qweiq drainage, Türkiye; 145 mm SL.

Squalius berak; upper Euphrates drainage, Türkiye; 160 mm SL.

Squalius berak

Common name. Euphrates chub.

Diagnosis. Distinguished from *S. lepidus* in Persian Gulf basin by: • upper lip projecting beyond lower lip / • posterior anal-margin convex / • 39–46+1–2 lateral-line scales / • head blunt / • snout short / • dorsal head profile straight or slightly convex / • scale pockets along lateral midline densely covered by pigments forming a large grey or black crescent-shaped blotch, resulting in a very contrasted reticulate pattern / • anal rays orange in life / • anal with few or without black pigments in preserved individuals / • head length 26–30 % SL / • interorbital distance 37–45 % HL / • head depth 65–71 % HL / • posteriormost point of anal at tip of 3rd or 5th branched rays. Size up to 450 mm SL.

Distribution. Qweiq and Euphrates drainages.

Habitat. Small- to medium-sized streams, mostly in mountains and hilly areas.

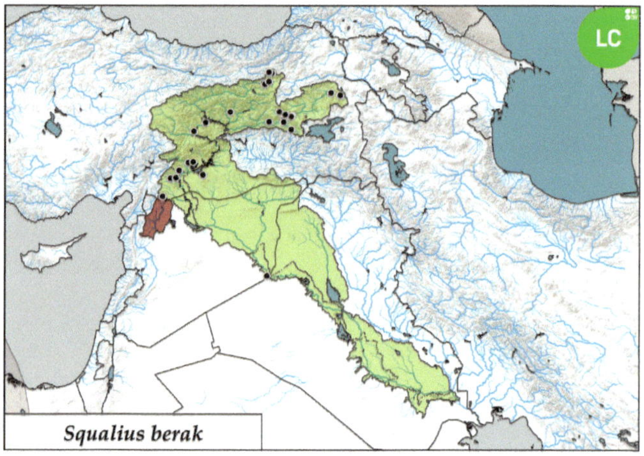

Biology. No data.

Conservation status. LC.

Remarks. In medium-sized rivers often in sympatry with *S. lepidus*, which inhabits mostly large rivers. Both species occasionally hybridise as mitochondrial DNA of *S. lepidus* is often found in *S. berak*, and there are populations with intermediate morphological character states. Individuals "difficult to identify" by characters given could be such hybrids. *Squalius semae* from the upper Euphrates is a synonym. Characters proposed to distinguish *S. verepi* from the upper Tigris from *S. berak* could also not be confirmed and both are also indistinguishable by their COI sequences. Few genomic data however indicate that *S. verepi* might be a valid species.

Further reading. Bayçelebi 2019 (description); Coad 2021a (biology, morphology); Turan et al. 2017d (description of *S. semae*); Turan 2022 (description of *S. verepi*).

Squalius cappadocicus

Common name. Anatolian chub.

Diagnosis. Distinguished from other species of Squalius in Central Anatolia, Marmara, and Black Sea basin, and upper Euphrates by: / ○ anal membranes whitish, yellowish, very pale-orange grey or blackish, rays usually blackish in life, in adults larger than 150 mm SL / ○ 38–44+2 lateral-line scales / ○ dorsal origin above pelvic base / ○ 14–15 circumpeduncular scales / ○ scale pockets on flank with a dark-brown or grey crescent-shaped mark or spot or roundish blotch on middle of flank scale / ○ mouth terminal or subterminal in adults, upper lip projecting beyond lower lip or both lips at equal level / ○ (7½) 8½, very rarely 9 ½ branched anal rays / ○ posteriormost point of anal at tip of 3rd–5th branched rays / ○ anal height 14–18 % SL / ○ head length 25–32 % SL / ○ belly between anus and posterior extremity of pelvic base rounded or very slightly compressed / ○ 11–12 gill rakers. Size up to 350 mm SL.

Distribution. Anatolian Black Sea coast west of Çoruh. Eastern tributaries of Sakarya and in Melendiz drainage (Tuz basin). Seyhan, and few tributaries in upper Euphrates including Kangal and Tohma.

Habitat. Streams and rivers stretches with moderate currents, also in reservoirs and lakes.

Biology. Lives up to 7 years. Spawns April–June in shallow gravel bottom areas. Omnivorous, feeding mainly on insects and aquatic invertebrates.

Conservation status. LC.

Remarks. The taxonomy and distribution of *S. cappadocicus* remain unresolved. It shares the COI sequence data with *S. orpheus*; both are well distinguished morphologically and supported as two species by unpublished nuclear DNA data. We restrict *S. orpheus* to Europe and preliminarily identify all Anatolian populations with COI sequences of *S. orpheus* as *S. cappadocicus*, the first name available for these fishes. COI and nDNA demonstrate *S. adanaensis* and *S. seyhanensis* to be conspecific. Both are identical to *S. cappadocicus* (and *S. orpheus*) by their COI sequences. *Squalius cappadocicus* and *S. seyhanensis* are very similar morphologically, and the morphological characters proposed to distinguish them could not be confirmed by us. We preliminarily identify *S. adanaensis*, *S. seyhanensis*, and all populations from the Black Sea basin having the COI of *S. orpheus* as *S. cappadocicus*. More research on nDNA might resolve this complicated situation in the future.

Further reading. Kottelat & Economidis 2006 (description); Özuluğ & Freyhof 2011 (description of *S. cappadocicus*); Turan et al. 2013b (description of *S. adanaensis* and *S. seyhanensis*); Özcan & Serdar 2019 (biology).

Squalius cappadocicus; lower Seyhan drainage, Türkiye; ~110 mm SL; = *S. adanaensis*).

Squalius cappadocicus; Melendiz drainage, Türkiye; ~110 mm SL; = *S. cappadocicus*).

Squalius cappadocicus; Zamantı, Seyhan drainage; ~200 mm SL; = S. *seyhanensis*).

Squalius cappadocicus; Sakarya drainage, Türkiye; ~140 mm SL.

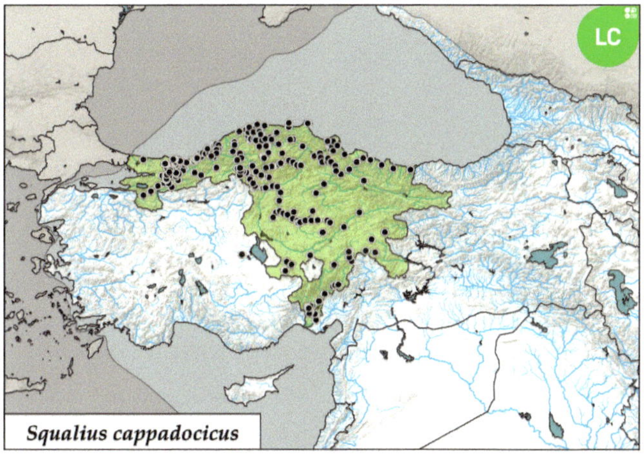

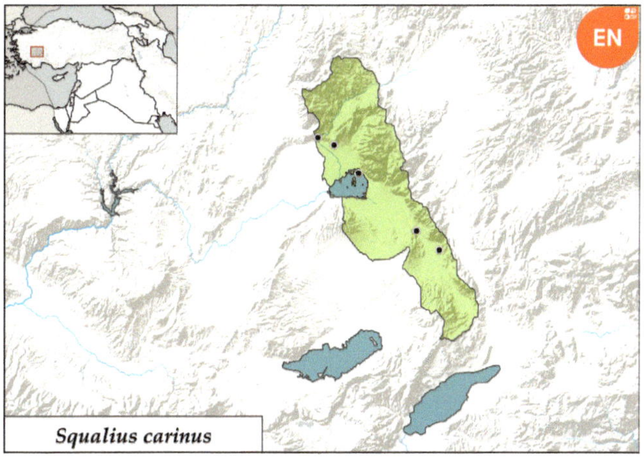

Squalius carinus; Işıklı spring, Türkiye; ~100 mm SL.

Squalius carinus

Common name. Işıklı chub.

Diagnosis. Distinguished from other species of *Squalius* in Western and Central Anatolia by: • dark-brown pigmentation in life / ○ lower lip projecting beyond tip of upper lip (equal to tip of upper lip in few individuals) / ○ uppermost point of mouth cleft about at level between middle of eye and lower margin of pupil / ○ upper head profile straight / ○ snout conical and pointed / ○ interorbital area straight or slightly convex / ○ 39–43+2–3 lateral-line scales / ○ belly between anus and posterior extremity of pelvic base distinctly compressed / ○ dorsal origin above vertical of pelvic base / ○ anal orange, membranes hyaline, rays blackish in life / ○ a dark-brown or grey spot on scale pockets on flank / ○ dark-brown stripes along upper and lower edge of exposed part of scales / ○ predorsal length 53–57 % SL / ○ head length 27–29 % SL / • head depth at eye 46–50 % HL / ○ interorbital distance 30–38 % HL. Size up to 260 mm SL.

Distribution. Türkiye: Lake Işıklı basin.

Habitat. Streams, springs, and Lake Işıklı itself, from where they migrate to streams to spawn.

Biology. Lives up to 5 years.

Conservation status. EN; appears to be declining within its very small range.

Remarks. Molecular data suggest that this species is conspecific with *S. fellowesii*. As *S. carinus* is well differentiated by morphological characters, it could be advised that it represents a young species adapted to the environment of Lake Işıklı or has been introgressed by mtDNA of *S. fellowesii*. Further research may also show that *S. carinus* is a synonym of *S. fellowesii*.

Further reading. Özuluğ & Freyhof 2011 (description).

Squalius cii; stream Bakacak, Türkiye; 200 mm SL.

Squalius cii

Common name. Marmara chub.

Diagnosis. Distinguished from other species of *Squalius* in Western and Central Anatolia by: ○ 14–16 circumpeduncular scales / ○ anal rays hyaline or pale-orange in live and preserved individuals / ○ anal membranes distinctly black / ○ upper lip clearly projecting beyond lower lip / ○ upper head profile almost straight / ○ vertical black bar behind opercle in-conspicuous / ○ 45–49 total lateral-line scales / ○ head length 24–26 % SL / ○ scale pockets on flank with a dark-grey, crescent-shaped mark / ○ belly between anus and posterior extremity of pelvic base slightly compressed / ○ head width at nape 36–41 % SL. Size up to 255 mm SL.

Distribution. Türkiye: Lake İznik basin, Sursuluk drainage, Biga Peninsula (Marmara basin) south to Karamenderes (Aegean basin).

Habitat. A wide range of lakes, streams, and rivers, including adjacent reservoirs, from which it migrates into tributaries to spawn.

Biology. No data.

Conservation status. LC.

Remarks. *Squalius cephaloides* from streams Teşvikiye and Armutlu is a synonym.

Further reading. Özuluğ & Freyhof 2011 (description); Bayçelebi et al. 2022 (synonymy of *S. cephaloides*).

Işıklı spring is an important habitat of *Squalius carinus* and other species endemic to the region.

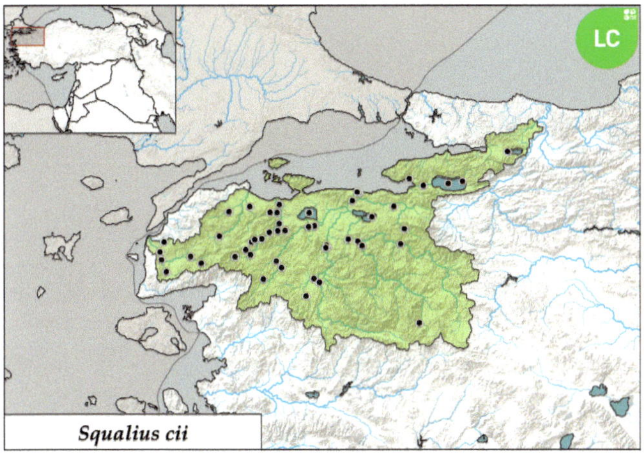

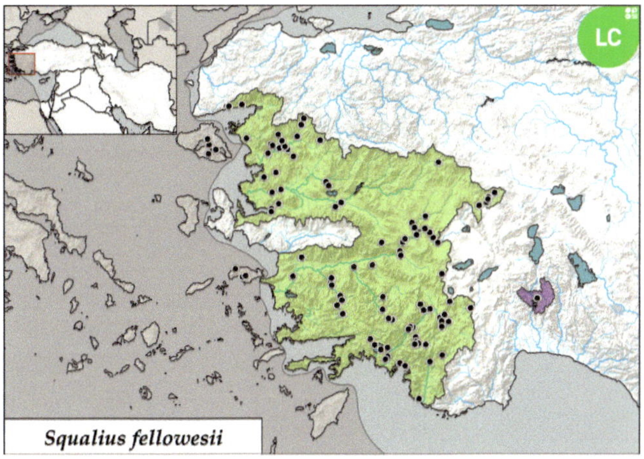

Squalius fellowesii; Dalaman drainage, Türkiye; ~130 mm SL.

Squalius fellowesii

Common name. Aegean chub.

Diagnosis. Distinguished from other species of *Squalius* in Western and Central Anatolia by: ○ 16–17 circumpeduncular scales / ○ anal rays orange in live, blackish in preserved individuals / ○ mouth subterminal in adults / ○ silvery or golden in life / ○ no stripe on flank / ○ anal membranes hyaline / ○ a dark-brown, vertically elongated or roundish spot or blotch on flank scale pockets / ○ spots or blotches on scales pockets also on predorsal back / ○ upper lip clearly projecting beyond lower / ○ upper head profile convex / ○ snout rounded with obtuse tip / ○ interorbital area convex / ○ 40–43+2–3 lateral-line scales / ○ 6–9 gill rakers / ○ abdomen between posterior extremity of pelvic base and anus distinctly compressed / ○ dorso-hypural distance reaching nostrils and anterior half of eye when reported forwards / ○ mouth subterminal in adults / ○ length of caudal peduncle 8–12 % SL. Size up to 180 mm SL.

Distribution. Türkiye: Aegean basin from Eşen north to Madra southern Biga Peninsula. Absent from Tahtalı and Küçük Menderes, as well as from Işıklı, but occurs in Karadirek. Introduced to Aksu (Mediterranean basin). Greece: Islands of Lesbos and Samos.

Habitat. Streams and rivers, including adjacent reservoirs, from which it migrates into tributaries to spawn.

Biology. Lives up to 7 years. Matures after the first winter in males and the second winter in females in most populations. Spawns between early April and late May.

Conservation status. LC.

Remarks. Molecular data suggest this species is closely related to *S. carinus* from Lake Işıklı basin.

Further reading. Özuluğ & Freyhof 2011 (description); Şaşı & Özay 2017 (biology).

Squalius fellowesii; Büyük Menderes drainage, Türkiye; ~120 mm SL.

Squalius kosswigi; stream Şaşal Türkiye; ~80 mm SL.

Squalius kosswigi

Common name. Striped chub.

Diagnosis. Distinguished from other species of *Squalius* in Western and Central Anatolia by: ● inconspicuous, broad, dark-brown stripe on upper part of flank between head and end of caudal peduncle / ● sharp contrast between dark-brown dorsal part and light-brown lateral and ventral part of body in preserved individuals / ○ snout obtuse / ○ 37–42+2–3 lateral-line scales / ○ scale pockets on flank below lateral stripe with a prominent dark-brown blotch / ○ posterior margin of flank scales usually covered with dark-brown pigments giving it a dusky pattern / ○ no dark-brown spot on scale pockets on back / ○ abdomen between posterior extremity of pelvic base and anus only very slightly compressed / ○ anal rays orange in live, blackish in preserved individuals / ○ anal membranes hyaline / ○ caudal peduncle depth 15–19 % SL. Size up to 158 mm SL.

Distribution. Türkiye: Küçük Menderes, Tahtalı drainages, and four coastal streams west to Azmak (İzmir).

Habitat. Streams with Mediterranean water regime. Can survive in very small pools in dry riverbeds in summer. Also, known from springs and reservoirs.

Biology. Lives up to 6 years. Spawns in March and April.

Conservation status. LC.

Further reading. Özuluğ & Freyhof 2011 (description).

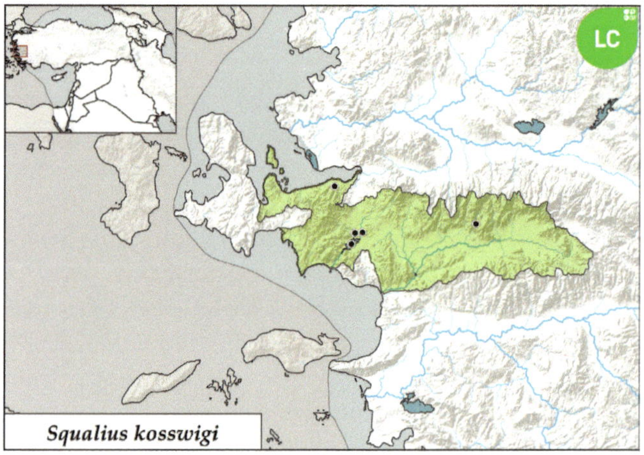

Squalius kottelati

Common name. Cilician pike chub.

Diagnosis. Distinguished from other species of *Squalius* in eastern Mediterranean basin by: • lower lip projecting beyond upper lip / • 9–10½ branched anal rays / • posterior anal margin straight / ○ 45–47 total lateral-line scales. Size up to 240 mm SL.

Distribution. Türkiye: Seyhan, Ceyhan, and Orontes drainages.

Habitat. A wide range of streams and rivers, mostly in mountain and hill areas. Also, in reservoirs from where they migrate to tributaries to spawn on gravel substrate.

Biology. No data.

Conservation status. LC. Most likely, this species also occurred in Syrian part of Orontes, but it has not been recorded in recent years.

Remarks. Very rare in Seyhan.

Further reading. Turan et al. 2009a (description); Bayçelebi 2020 (distribution).

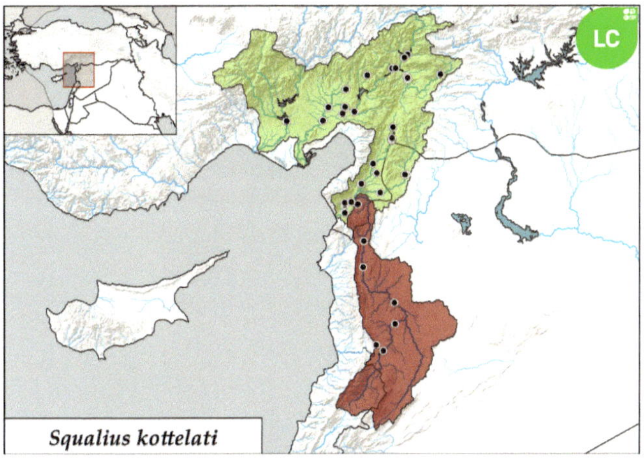

Squalius kottelati; Orontes drainage, Türkiye; 190 mm SL.

Squalius lepidus; Tigris drainage, Türkiye; 213 mm SL.

Squalius lepidus

Common name. Mesopotamian pike chub.

Diagnosis. Distinguished from *S. berak* and *S. verepi* in Persian Gulf basin by: • lower lip projecting beyond upper lip or tip of both lips equal / • posterior anal margin straight / • 9–10½, rarely 8½ branched anal rays / • 45–60+1–3 lateral-line scales / • head pointed / • snout long / • dorsal head profile straight or concave / • scale pockets on flank not or moderately covered by pigments forming a grey crescent-shaped blotch, resulting in a poorly contrasted reticulate pattern or reticulate pattern absent. Size up to 350 mm SL, likely to grow larger.

Squalius lepidus; Tigris drainage, Iran; 187 mm SL.

Distribution. Euphrates, Tigris, and Karun drainages.
Habitat. Large- to medium-sized rivers, lakes, and reservoirs from where they migrate to tributaries to spawn. Rarely in small streams.
Biology. Spawns May–June.
Conservation status. LC.
Remarks. In medium-sized rivers, it often occurs in sympatry with *S. berak* (Euphrates) or *S. verepi* (Tigris), which inhabits mostly small- to medium-sized streams. Sympatric species occasionally hybridise, as mitochondrial DNA from *S. berak* and *S. verepi* is often found in *S. lepidus*, and there are many populations similar to hybrids between these species (relatively long head). Many individuals "difficult to identify" by the characters may be such hybrids. Chubs identified as *S. lepidus* from Lake Beyşehir, Lake Akşehir, Ceyhan, and the Orontes belong to other species.

Further reading. Bogutskaya 1994 (description); Coad 2021a (biology, morphology).

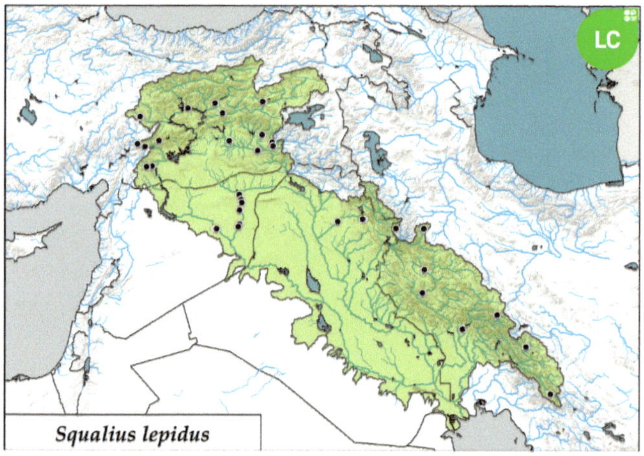

Tigris at Diyarbakır is the habitat of *Squalius lepidus* and many other large-river species.

Hybrid between *Squalius verepi* and *Alburnus sellal*; 170 mm SL; Tigris drainage, Türkiye.

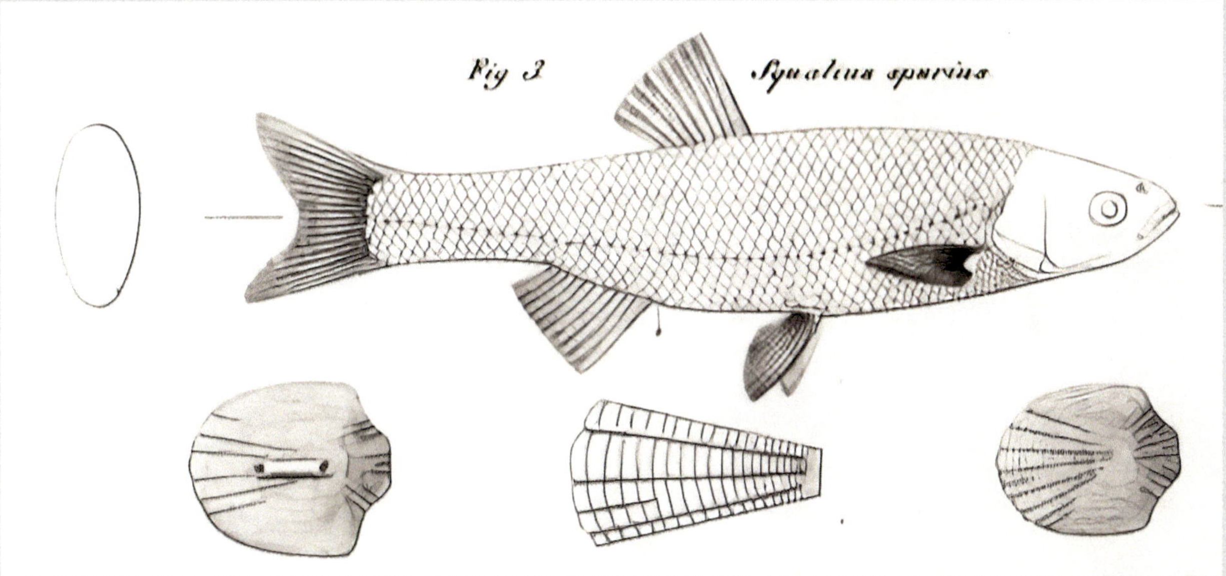

Squalius spurius; Aleppo, Syria (from Heckel 1843).

***Leuciscus spurius*, a lost mystery fish?** *Leuciscus spurius* was first described by Heckel in 1843 from the Qweiq in Aleppo, a region where the fish fauna was largely unknown to scientists at the time. It was frequently listed as occurring in the area by authors who copied information from Heckel's description. After 145 years of its description, *L. spurius* was recorded from coastal streams in Syria, but later these fish were described as *Alburnus qalilus*. The generic position of *L. spurius* was discussed on a few occasions, partly based on the sole individual from Heckel's collection and the accompanying illustration in the original description. Indeed, this species was never found again. A thorough examination of the type specimen in the fish collection of the Natural History Museum in Vienna revealed that the type of *L. spurius* is indistinguishable from the hybrids between *Squalius berak* and *Alburnus sellal*, which are often found in the Euphrates. Both parental species also occur in the Qweiq, so they are expected to hybridise there. Such hybrids exhibit a distinct phenotype that differs from their parental species. Consequently, Heckel described this individual as a different species. **Further reading.** Heckel 1843 (description); Krupp 1985d (records from Syria); Bogutskaya 1994; 1997b (discussion); Turan et al. 2009a (placement in *Squalius*).

Squalius namak; Nam, Western Kavir basin, Iran; ~280 mm SL.

Squalius namak

Common name. Namak chub.

Diagnosis. Distinguished from other species of *Squalius* in adjacent Caspian and Persian Gulf basins by: ● symphysial knob on lower jaw wide and thick / ○ a bold, grey or brown, roundish or crescent shaped blotch at tip of flank scales in preserved individuals / ○ scale pockets on flank above lateral line poorly pigmented / ○ 39–43+1–3 lateral-line scales (39–45 total) / ○ posterior anal margin convex / ○ mouth subterminal / ○ anal height 7–13% SL / ○ anal with few or without black pigments in preserved individuals / ○ 9–12 gill rakers / ○ caudal, anal and pelvic rays orange / ○ head blunt / ○ caudal peduncle length 14–20 % SL. Size up to 450 mm SL, likely to grow larger.

Distribution. Iran: Qom and Qareh Chai drainages in Lake Namak basin. Hable and Nam drainages in Western Kavir basin.

Habitat. Springs and small rivers.

Biology. No data.

Conservation status. LC.

Remarks. Individuals analysed from Cheshmeh Ali have the mtDNA of *S. turcicus*. This population could be identified as *S. turcicus* or could consist of hybrids between *S. namak* and *S. turcicus*.

Further reading. Khaefi et al. 2016 (description).

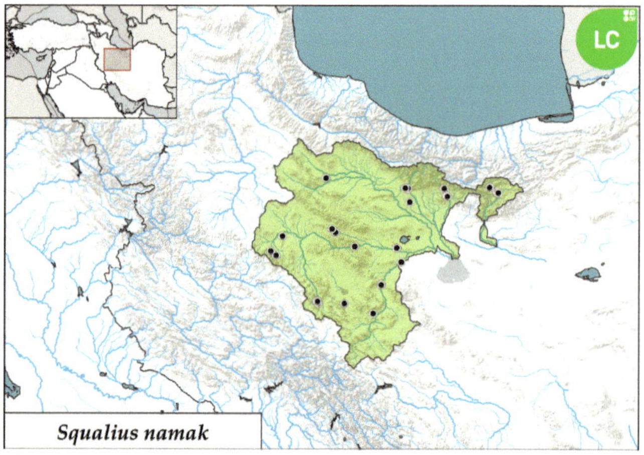

Squalius orientalis; Çoruh drainage, Türkiye; ~160 mm SL.

Squalius orientalis

Common name. Georgian chub.

Diagnosis. Distinguished from other species of *Squalius* in Black Sea basin by: ○ 9½, rarely 8½ branched anal rays / ○ anal orange to red / ○ conspicuous vertical black bar behind opercle / ○ 41–46+1–2 lateral-line scales / ○ 10–12 gill rakers / ○ caudal peduncle depth 11–12 % SL / ○ posteriormost point at tip of 3rd anal branched ray. Size up to 600 mm SL.

Distribution. Eastern Black Sea basin from Doğankent east to Çoruh and north to Ashe drainage and other small rivers south of Caucasus.

Habitat. Usually in small rivers and large streams with riffles and pools, even in very small mountain streams. On banks of slow-flowing lowland rivers and lakes, making spawning migrations to inflowing streams. Spawns in fast-flowing water over gravel banks.

Biology. Lives up to 16 years, females longer than males. Males first spawn at 2–3 years, females at 4–6 years. Juveniles gregarious, adults more solitary. Feeds on a wide variety of aquatic and terrestrial animal and plant material. Often forms hybrids with *Alburnus*.

Conservation status. LC.

Remarks. Molecular data suggest that this species is very closely related to *S. turcicus* from Aras, and both may be conspecific. However, *S. turcicus* usually has dark grey fins, whereas *S. orientalis* has orange fins, and S. *turcicus* lacks the prominent vertical black bar behind the opercle. We recognise *S. cephalus* and *S. orientalis* as two species based on molecular characters alone, as their morphology has yet to be studied. Molecular data suggest that *S. orientalis* might have a much wider distribution in the northern Black Sea basin, including all rivers east of Danube and tributaries of Baltic Sea west to Odra and eastern tributaries of Elbe (Czech Republic and Germany). Within this population group, *S. wjatkensis* from the Azov and northern Caspian Sea basins (Volga) and *S. latifrons* may be considered additional valid species in future. *Squalius latifrons* is an available name for chubs closely related to *S. orientalis* from southern Finland, Sweden (north to around Stockholm), the Baltic basin, and tributaries of the northern and western Black Sea. Fish from Bulgarian coastal rivers in Black Sea basin also belong to this molecular group, and we suspect that they are identical to those identified as *S. cephalus* in northwestern Anatolia. *Squalius cephalus* is restricted to France, Great Britain, the Danube drainage, and the North Sea basin tributaries east of Elbe.

Further reading. Bayçelebi 2019 (description); Dedeoğlu et al. 2020 (biology); Coad 2021a (biology, morphology).

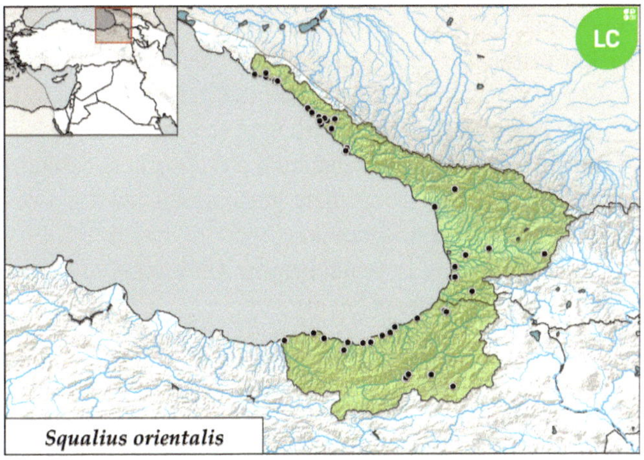

Squalius pursakensis; Porsuk subdrainage in Sakarya drainage, Türkiye; ~160 mm SL.

Squalius pursakensis

Common name. Sakarya chub.

Diagnosis. Distinguished from other species of *Squalius* in Western and Central Anatolia by: ○ dorsal origin behind vertical of last pelvic ray / ○ posteriormost point of anal at tip of 7th or 8th branched ray / ○ 14–15 circumpeduncular scales / ○ 8½, very rarely 9½ anal rays / ○ anal rays hyaline in live and preserved individuals / ○ anal membranes black, rarely orange in live, blackish in preserved individuals / ○ upper lip usually projecting beyond lower lip, lips at same level or lower lip projection in some individuals / ○ upper and ventral head profile straight in individuals larger than 100 mm SL / ○ snout conical and pointed / ○ scale pockets on flank with a dark-grey, crescent-shaped mark / ○ belly between anus and posterior extremity of pelvic base slightly or not compressed / ○ eye diameter 1.7–2.1 times in interorbital distance, 18–23 % HL / ○ head length 25–29 % SL / ○ head depth at nape 63–67 % HL / ○ 40–46 total lateral-line scales. Size up to 380 mm SL.

Distribution. Türkiye: Sakarya drainage, likely restricted to upper and middle parts, including Porsuk and Kirmir. Also, in Lakes Eber, Akşehir, and Ilgın basins.

Habitat. A wide range of streams and rivers, mostly in hilly areas. In reservoirs, from where they migrate to tributaries to spawn on gravel substrate.

Biology. Lives up to 10 years. Spawns April–June.

Conservation status. LC.

Remarks. *Squalius recurvirostris* from Eber, Akşehir, and Ilgın lakes is a synonym as it is indistinguishable from *S. pursakensis* by COI DNA sequences, and morphological characters proposed to distinguish both species could also not be confirmed. *Squalius pursakensis* and mtDNA haplotypes of *S. cappadocicus* are found in the Sakarya. It has not yet been investigated where exactly both species occur in Sakarya, possibly in different subsystems, or where exactly hybrids between *S. pursakensis* and *S. cappadocicus* occur. In addition, the morphological characters distinguishing *S. pursakensis* and *S. cappadocicus* are poorly understood and require further research.

Further reading. Özuluğ & Freyhof 2011 (re-description).

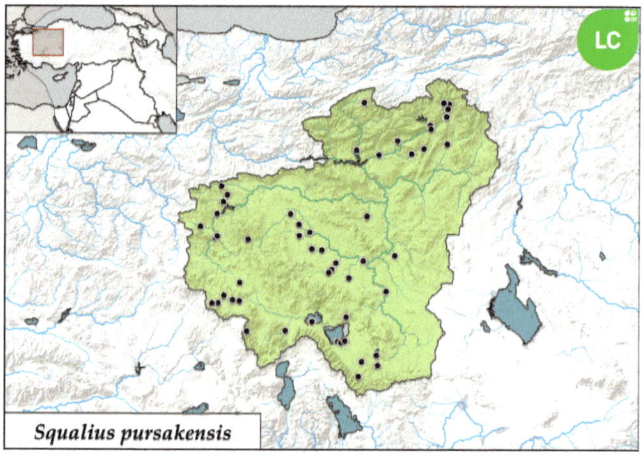

Squalius pursakensis; Lake Eber basin, Türkiye; ~190 mm SL.

Squalius turcicus; Aras drainage, Türkiye; ~190 mm SL. © M. Özuluğ.

Squalius turcicus; Cheshme Kile, Caspian basin, Iran; ~260 mm SL.

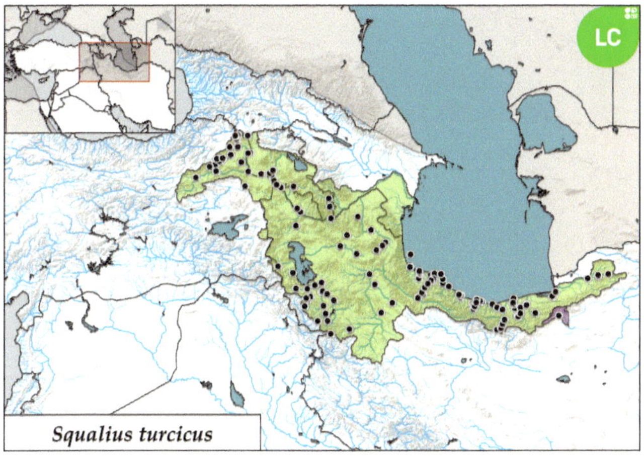

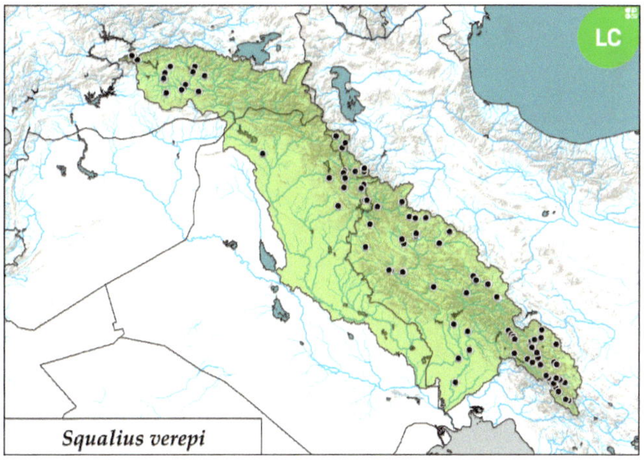

Squalius turcicus

Common name. South Caspian chub.

Diagnosis. Distinguished from other species of *Squalius* in Persian Gulf, Caspian, Lake Namak and Lake Urmia basins by: ● anal membranes and rays hyaline, greyish-white or yellow in life / ○ mouth terminal or subterminal in adults / ○ snout pointed / ○ knob on lower jaw symphysis narrow, thin or absent / ○ head length 26–28 % SL / ○ interorbital distance 35–41 % HL / ○ head width at nape 61–70 % HL / ○ anterior width 4.2–7.1 times in its lateral length / ○ (7½) 8½ (9½) branched anal rays / ○ 7–11 gill rakers / ○ anal height 18–24 % SL / ○ posterior tip of flank scales without bold blotch /

○ posteriormost point of anal at tip of 5th or 6th anal branched ray / ○ a wide band of densely-set, dark-brown or blackish melanophores along posterior margin of each flank scale. Size up to 264 mm SL.
Distribution. Caspian basin from Aras east to Atrak. Also, in Lake Urmia basin (Iran), Cheshme Ali in Western Kavir basin (Iran), Sulak drainage (Dagestan), and uppermost Choman (Tigris drainage), directly at watershed to Saqez (Urmia).
Habitat. Wide range of streams and rivers, mostly in hilly areas. In lakes, from where it migrates to tributaries to spawn. Locally spawns on wave-washed lake shores on gravel substrate.
Biology. Lives up to 8 years, possibly longer. First spawns at 2–3 years (males) and 3–4 years (females). Spawns late May–July at water temperatures of 12–18°C.
Conservation status. LC.
Remarks. Molecular data suggest that this species is closely related to *S. orientalis* from the eastern Black Sea basin, and both may be conspecific. However, *S. turcicus* usually has dark-grey fins, whereas *S. orientalis* has orange fins, and the prominent vertical black bar behind the opercle in *S. orientalis* is absent in *S. turcicus*. Further research is needed to determine whether this is one or two species. Molecular characters differentiate the Kura and Aras chubs well. The species from the Kura is identified as *S. agdamicus*, and the one from the Aras is *S. turcicus*, but their morphological characters still need to be better studied. More research is required to distinguish the two species and understand their range.
Further reading. Bayçelebi 2019 (description, distribution).

Squalius verepi; Sirvan drainage, Iraq; 150 mm SL.

Squalius verepi
Common name. Tigris chub.
Diagnosis. Distinguished from *S. lepidus* in Persian Gulf basin by: ● upper lip projecting beyond lower lip / ○ posterior anal-margin convex / ● 39–46+1–2 lateral-line scales / ○ head blunt / ○ snout short / ○ dorsal head profile straight or slightly convex / ○ scale pockets along lateral midline densely covered by pigments forming a large grey or black crescent-shaped blotch, resulting in a very contrasted reticulate pattern / ○ anal rays orange in life / ○ anal with few or without black pigments in preserved individuals / ○ head length 27–30 % SL / ○ interorbital distance 35–45 % HL / ○ head depth 61–72 % HL / ○ posteriormost point of anal at tip of 3rd or 5th branched rays. Size up to 450 mm SL.
Distribution. Tigris and Karun drainages.
Habitat. Small- to medium-sized streams, mostly in mountains and hilly areas.
Biology. No data.
Conservation status. LC.
Remarks. In medium-sized rivers often in sympatry with *S. lepidus*, which inhabits mostly large rivers. Both species occasionally hybridise as mitochondrial DNA of *S. lepidus* is often found in *S. verepi*, and there are populations with intermediate morphological character states. Individuals “difficult to identify” by characters given could be such hybrids. Morphological characters proposed to distinguish *S. verepi* from *S. berak* could not be confirmed and both are also indistinguishable by their COI sequences. Only few unpublished genomic data indicate that *S. verepi* might be a valid species.
Further reading. Turan 2022 (description).

Hybrid between *Squalius verepi* and *S. lepidus*; upper Tigris drainage, Türkiye; 120 mm SL.

Spring Karamusa near Tefenni in Türkiye is one of the few habitats of *Turcichondrostoma fahirae*. The spring is dammed and used as a reservoir for pumping.

Turcichondrostoma fahirae; Karapinar spring, Türkiye; ~80 mm SL.

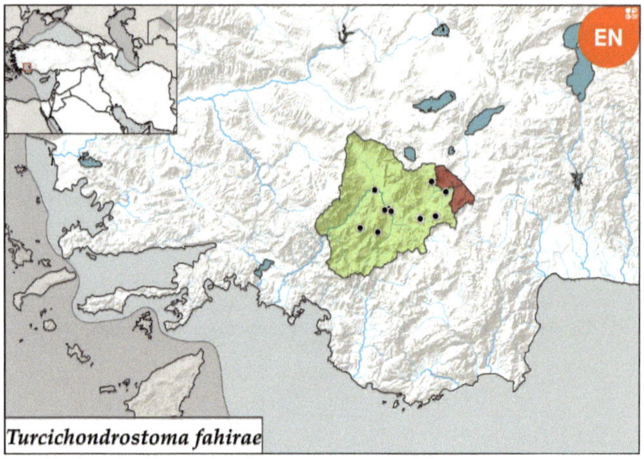

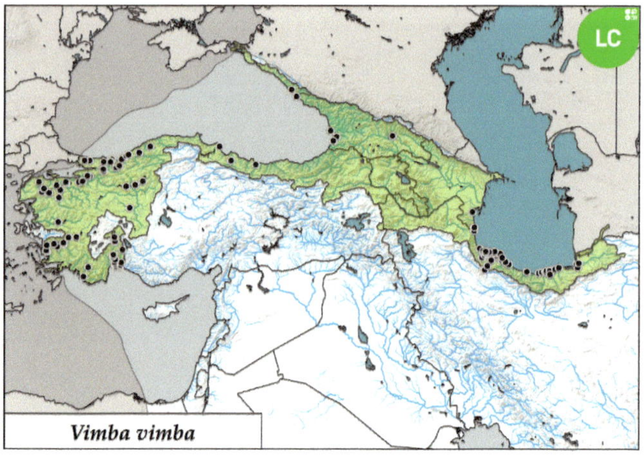

Turcichondrostoma fahirae

Common name. Tefenni nase.

Diagnosis. Distinguished from species of *Chondrostoma* in Central Anatolia, Aegean, Marmara, and Black Sea basins by: ● no cornified sheath covering lower lip / ● 12–14 gill rakers / ○ 44–54 total lateral-line scales / ○ 5–5 pharyngeal teeth / ○ 8½ branched dorsal rays / ○ usually 8½ branched anal rays. Size up to 127 mm SL.

Distribution. Türkiye: Kırkpınar (Başpınar) spring near Karamusa (Tefenni), Değirmendere flowing into Karamanlı reservoir, Lake Karatas, Elmacık reservoir and Bozçay in Burdur basin.

Habitat. Springs and streams with gravel substrate.

Biology. No data.

Conservation status. EN; appears to be declining within its very small range. Extirpated from Lake Karataş as this lake has dried out in recent years.

Remarks. The single species included in this genus had previously been classified as *Pseudophoxinus*. Molecular data indicate that this species is closely related to *Chondrostoma*. Available molecular data are contradictory, and some molecular analysis relate *Turcichondrostoma* with *Chondrostoma soetta* and *C. kneri* from the Adriatic basin in Europe. Further studies are required to resolve the generic position of *T. fahirae*.

Further reading. Ladiges 1960 (description); Bogutskaya 1992 (review; as *Pseudophoxinus fahirae*); Freyhof & Özuluğ 2010b (placement in *Chondrostoma*); Güçlü et al. 2018 (morphology); Çiftçi et al. 2020 (phylogeny); Turan et al. 2021a (description).

Vimba vimba; Gönen drainage; Türkiye; ~150 mm SL.

Vimba vimba; Lake Marmara, Türkiye; ~110 mm SL.

Vimba vimba; Danube drainage, Slovenia; ~300 mm SL.

Vimba vimba

Common name. Vimba.

Diagnosis. Distinguished from species of *Abramis*, *Acanthobrama*, *Ballerus*, and *Blicca* by: ○ 46–58+2–3 lateral-line scales / ○ 16–22½ branched anal rays / ○ no scaleless groove in front of dorsal / ○ last unbranched dorsal ray as much or slighly more ossified as other rays, thickened or not thickened / ○ sub-inferior mouth, which cannot be extended as a tube / ○ back keeled behind dorsal base / ○ pharyngeal teeth 5–5 / ○ orange or reddish base of paired fins in adults. Size up to 400 mm SL (Caspian anadromous form), many populations much smaller.

Distribution. Caspian, Black, Marmara, and Baltic Sea basins (Sweden and Finland north to 63°N), North Sea basin from Elbe to Ems drainages. In Anatolia: in Aegean basin (from Gediz to Eşen), southern Marmara basin including lakes Apolyont, Manyas, İznik, and Sapanca, and in Black Sea tributaries. Lake Eğirdir (Central Anatolia) and Aksu and Köprüçay which flow to Gulf of Antalya. Semi-anadromous in Caspian basin, entering almost all, even small, tributaries to spawn. Widespread in Europe from Elbe east to Volga and all rivers in Baltic, Black, and Caspian basin. Non-native to Rhine and northern Italy.

Habitat. Brackish estuaries, large to medium-sized rivers, and large lakes. Sedentary populations also occur in small rivers. Spawns on gravel in riffles in shallow, fast-flowing streams and rivers.

Biology. Lives up to 7–15 years. Semi-anadromous populations forage in freshened parts of sea. Caspian population undertakes long-distance migrations to overwinter and feed at sea. To spawn, enters rivers in winter (Kura) or March–April (smaller rivers) and undertakes long upstream spawning migrations. Spawns first time at 2–3 years of age. Spawns late April–late July, depending on water temperature. Commences spawning when temperature rises above 15°C. Lacustrine populations migrate to fast-flowing tributaries. Fractional spawners, individual females spawn several times over 2–3 months.

Individuals spawn only once per season, reported from upper Danube and Caspian basin, and spent adults return to feeding areas. Feeds mainly on small molluscs and insect larvae.

Conservation status. LC.

Remarks. The genus *Vimba* is closely related to other breams, such as *Acanthobrama*, *Abramis*, *Ballerus*, *Blicca*, and *Mirogrex*. Typically, four species are recognised within the genus. *Vimba melanops* is endemic to the European part of the Aegean, *V. persa* is found in the Caspian basin, *V. mirabilis* is found in the eastern Aegean basin (from Gediz south to Eşen in Türkiye), and *V. vimba* is found in the Black Sea basin. However, all *Vimba* are very closely related, and morphological characters do not allow for the distinction of different species within that genus. Therefore, we treat all *Vimba* as a single species.

Further reading. Ladiges 1960 (description of *V. mirabilis*); Abdurakhmanov 1962 (morphology, biology; Caspian basin); Bănărescu et al. 1970 (systematics); Elanidze 1983 (morphology, biology, Black Sea); Geldiay & Balık 2007 (distribution in Türkiye); Kottelat & Freyhof 2007 (summary of distribution and biology); Hänfling et al. 2009 (phylogeography); Özcan & Balık 2009 (biology, as *V. mirabilis*); Perea et al. 2010 (phylogeny); Chaichi et al. 2011 (reproduction, Caspian); Bogutskaya et al. 2013 (review; Caspian, as *Vimba persa*); Okgerman et al. 2013 (feeding).

Leuciscid fishes have a poor taste and many intromuscular bones and are of low commercial value in West Asia.

Family Tincidae

Tench

The phylogenetic position of the tench has been debated for decades, with the species currently classified as belonging to a distinct family within the order Cypriniformes. Following the publication of several molecular studies in the 21st century, *Tinca* was occasionally grouped with a diverse range of East Asian fishes, including *Zacco*, *Tanichthys*, and bitterlings belonging to the family Acheilognathidae. In this context, Tincidae was proposed as a distinct family, with the genus *Tinca* representing a monotypic taxon. The fossil record of tincids is extensive, with remains dating back to the early Oligocene (30 million years ago) in Europe, Central Asia, and Siberia. *Palaeotinca* and *Tarsichthys* were the first known genera in Tincidae and among Central Europe's earliest representatives of Cypriniformes. It is possible that they reached this area together with a small number of other Asian invaders. *Tinca* was first recorded in the Middle Miocene and has since been widespread in Europe and Central Asia. Despite the long history of Tincidae, there is only one extant species, indicating that it has very strong dispersal abilities. This may be due to the species being able to guarantee gene flow over large distances, or it may be due to a massive extinction event in the past, which reduced the species' range. The range of *Tinca* is subdivided between two slightly differentiated molecular groups of populations. The first is from the British Isles to Poland, while the second is from Germany throughout Asia to China. Both groups show a large overlap in their range within Central Europe (Germany, Poland, and the Danube), which might originate from postglacial expansion and stocking. These two population groups indicate that there might have been two glacial refuge areas. **Further reading.** Schulz-Mirbach & Reichenbacher 2006 (*Palaeotinca*); Gaudant 2007 (*Tarsichthys*); Lajbner et al. 2010 (phylogeography); Hirt et al. 2017 (phylogeny); Tan & Armbruster 2018 (Tincidae).

Tinca tinca

Common name. Tench.
Diagnosis. Distinguished from species of Cyprinidae in West Asia by: ○ golden greenish brown body / ○ orange or red iris / ○ one pair of barbels (maxillary) / ○ 96–115 total lateral-line scales, small and deeply embedded / ○ 6–9½ branched anal rays / ○ 8–9½ branched dorsal rays. Size up to 600 mm SL.
Distribution. Centuries of stocking in West Asia and Europe have blurred the original distribution pattern. In West Asia, thought to be native to Black and Caspian Sea basins. Now widely introduced in Anatolia. Presumed native to most of Europe, naturally absent only in Ireland, Scandinavia north of 61°30'N, eastern Adriatic basin, and western and southern Greece, where it is now introduced. In northern Asia, native east to Yenisei drainage south of 60°N. Introduced in North and South Africa, Tasmania, Australia, New Zealand, India, North America, Chile, and probably elsewhere.

Tinca tinca; Danube drainage, Germany; ~350 mm SL. © A. Hartl.

https://doi.org/10.1515/9783111677811-018

Habitat. Typically, in shallow, densely vegetated lakes and backwaters, but introduced often to all kinds of lakes and reservoirs. Often overwinters buried in mud. Spawns in still water among dense vegetation.

Biology. Lives up to 20 years. First spawns at 2–6 years and 70–250 mm SL, females usually a year later than males. Pelvic rays more robust, longer, and extending beyond anus in males. Spawns May–October in Central Europe, usually June–July, at temperatures above 19°C, usually 22–24°C. Several males follow a female who lays eggs in several parts above vegetation. Females may spawn 1–9 times a year, every 11–15 days, if weather is warm enough. High embryo mortality has been observed when temperatures fluctuate greatly. Larvae and juveniles restricted to dense vegetation. Tolerant of low oxygen and salinity up to 12 ‰. Feeds on detritus, benthic animals, and plant material. Adults often feed mainly on molluscs.

Conservation status. LC.

Further reading. Bogutskaya 1986, 1988b (osteology, phylogeny); Brylińska et al. 1999a, b (biology); Kottelat & Freyhof 2007 (summary of distribution and biology).

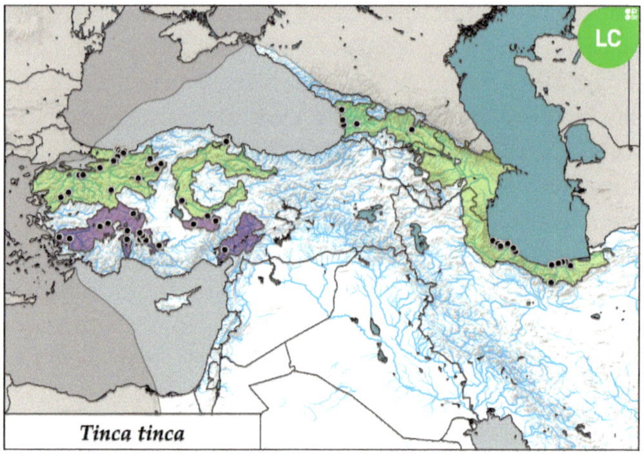

Tinca tinca inhabits mostly stagnant and densely vegetated waterbodies and occasionally establishes in such waters outside its native range (Lake Gölhisar, Türkiye).

Family Xenocyprididae

Chinese carps

The aggregation of non-native East Asian carps has been placed in several cyprinid subfamilies, depending on which molecular characters were studied. The genera *Squaliobarbus*, *Elopichthys*, *Ctenopharyngodon*, and *Mylopharyngodon* were previously classified within the Cyprinid subfamily Squaliobarbinae. *Hypophthalmichthys* has been placed in the family Leuciscidae. The results of recent, comprehensive molecular studies place these fishes in Xenocyprididae. Xenocyprididae is a diverse group that encompasses species native to Southeast Asia (e.g. *Macrochirichthys*, and *Parachela*) and East Asia (e.g., *Culter*, *Chanodichthys*, and *Ochetobius*). They represent an independent adaptive radiation within the order Cypriniformes and exhibit remarkable morphological and ecological diversity. In East Asia, they form similar ecological niches to those of Leuciscids in the West, and some are even superficially similar to *Alburnus*, *Leuciscus*, *Chondrostoma*, and others. Some species attain considerable size and exhibit rapid growth; four have been cultivated and widely introduced. Several other species are small pond dwellers, including *Aphyocypris*, *Hemigrammocypris*, and *Metzia*. Xenocypridis are particularly abundant in lowland freshwater ecosystems, with some species being ubiquitous. Some are also resistant to the majority of human impacts. One species, *Hemiculter leucisculus*, was introduced to the Aral basin and has become invasive along the southern and western coast of the Caspian Sea. Subsequently, it was introduced to the Euphrates and Tigris drainages, where it is expanding its range. *Ctenopharyngodon*, *Mylopharyngodon*, and *Hypophthalmichthys* have been stocked throughout West Asia and are typically sourced from European suppliers. However, these activities have declined due to the perceived inferiority of the flesh of these species and their subsequent low market value. There is no documented evidence of any established population of these species in West Asia. Nevertheless, their potential to become established in the future as a consequence of warming due to climate change cannot be discounted. **Further reading.** Howes 1981 (anatomy); Chen 1987 (phylogeny); Cunha et al. 2002 (genetics); Kottelat & Freyhof 2007 (summary of distribution and biology); DeVaney et al. 2009 (invasion potential).

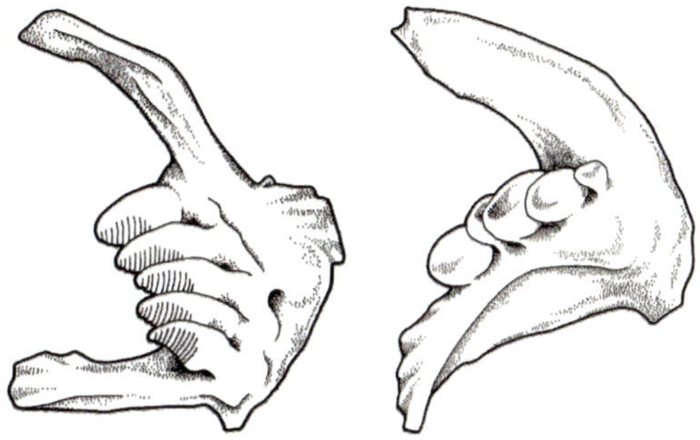

Figure 43. Frontal view of left pharyngeal arch of *Ctenopharyngodon idella* (the two small teeth on row 1 are not visible in frontal view) and left pharyngeal arch of *Mylopharyngodon piceus* (from Kottelat & Freyhof 2007, after Nico et al. 2005).

Ctenopharyngodon idella; aquaculture, Germany; ~600 mm SL. © A. Hartl.

Ctenopharyngodon idella

Common name. Grass carp.

Diagnosis. Distinguished from similar *Mylopharyngodon piceus* by: • body olive to brassy green on back, silvery white to yellow on flank / ○ body cylindrical / • pharyngeal teeth laterally compressed, serrated, with a groove along grinding surface (Figure 43), usually in two rows, 2,5–4,2. Size up to 1200 mm SL and 32 kg.

Distribution. Rarely introduced for weed control or escaped from aquaculture facilities. No established populations in West Asia. Native to East Asia, in most major Pacific drainages from Amur to Xi Jiang. Used in aquaculture in Asia, Europe, and North America; released in most major drainages. Established non-native populations in North America and in Aral Sea basin (Central Asia).

https://doi.org/10.1515/9783111677811-019

Habitat. Stocked in large rivers and almost all still waters such as lakes, reservoirs, and ponds. Often escapes from fish farms. In its native range, inhabits rivers with pronounced water level fluctuations caused by floods. Spawning and overwintering in middle and lower reaches of large rivers. In lakes, reservoirs, and backwaters during feeding season, preferably in warm, clear water with high oxygen concentrations.
Biology. In natural habitat, first spawns at 7–10 years and about 600–800 mm SL, females 1–2 years later than males. Spawns April–August. Migrates upstream and spawns during high water, in upper water layer, or on surface in sections with strong currents. Females lay one or more portions of eggs depending on length of flood. Average females lay about 1.5 million pelagic eggs, hatching while drifting downstream in 2–3 days; if river is blocked or available river reaches are too short, eggs cannot drift long enough and fail to develop. After spawning, adults leave river and migrate to flooded areas, lakes, or backwaters with dense vegetation. They return to river in autumn–winter and spend cold season in deep places in lower reaches. Larvae settle in flooded lakes and channels with little or no current after a downstream migration of about 1000 km. Larvae feed on phyto- and zooplankton and then, from about 25 to 50 mm SL, on aquatic macrophytes. Larger juveniles and adults feed mainly on macrophytes, including terrestrial macrophytes, during summer floods. Does not feed in winter. There are reliable data on natural spawning in rivers in southern Russia, but success of these reproductive events is not known.
Conservation status. Non-native; stocked for commercial fisheries. In West Asia very rarely stocked. Native stocks in eastern Russia and China are declining. Its introduction into Czech Republic was accompanied by introduction of tapeworm *Bothriocephalus gowkongensis*, which subsequently caused severe losses in *Cyprinus carpio* farmed stocks.
Further reading. Bíró 1999a (biology); Nico et al. 2001 (comparison with *Mylopharyngodon*); Kottelat & Freyhof 2007 (summary of distribution and biology).

Hemiculter leucisculus; Sirvan drainage, Iraq; ~90 mm SL.

Hemiculter leucisculus
Common name. Sharpbelly.
Diagnosis. Distinguished from superficially similar species of *Alburnus* by: • sharp keel on ventral midline from chest to pelvic base / • last unbranched dorsal ray unsegmented, stiff and spinous with flexible tip / ○ anal origin far behind vertical of last dorsal ray / ○ lateral line curves downward from its origin to end of pectoral fin. Size up to 200 mm SL.
Distribution. Non-native to southern Caspian basin in Iran, Azerbaijan, and southern Russia. Locally introduced in Tigris drainage. Native from Amur in Russia, south to China, North and South Korea, Japan, and northern Vietnam.
Habitat. Streams and rivers, reservoirs and lakes. Also in coastal lagoons in Caspian Sea.
Biology. Lives up to 6 years in Iran, begins spawning in second year. Spawns March–August, most likely fractional spawners. Feeds on algae, plant material, terrestrial and aquatic invertebrates, and small fish.
Conservation status. Non-native; introduced as a weed with stocked Common carp (*Cyprinus carpio*) or/and Chinese carps (*Ctenopharyngodon*, *Hyophthalmichthys*).
Remarks. An invasive species introduced with Chinese carp from Central Asia, invading the Caspian and Tigris basins. Could appear wherever carp are stocked as it is transported long distances with carp fingerlings. The current source of this species appears to be Caspian hatcheries in Azerbaijan

and Iran. Still, it is only a matter of time before it becomes established in hatcheries producing *Cyprinus carpio* elsewhere. This species will invade all of West Asia, Europe, and North Africa, following the path of *Pseudorasbora parva*. Several species are confused under this name. The identity of the species introduced in the Western Palaearctic has yet to be investigated.

Further reading. Coad 2010a (biology, distribution); Vasil'eva et al. 2022 (East Asian diversity).

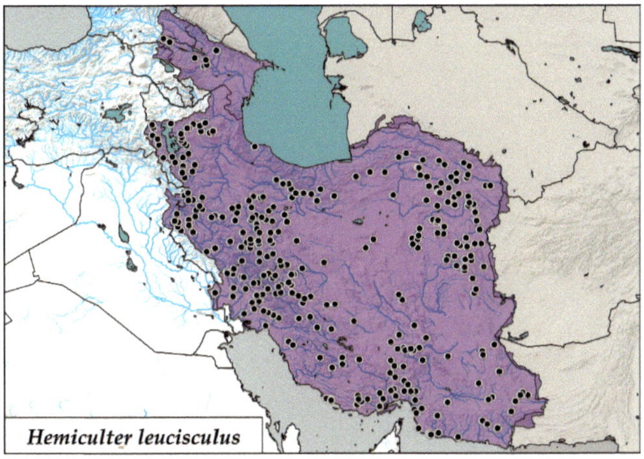
Hemiculter leucisculus

Sharpbelly, the next generation invader to the Palaearctic. What differentiates an invasive species from a non-invasive one? Several studies have analysed the ecological traits associated with the invasive power of a species. A potent invader exhibits a high tolerance and adaptability to a wide range of temperatures, oxygen levels, and pollutants. It has no special requirements to achieve its high reproductive potential and no specialisation for feeding. It has a moderate size to enable it to exist in "normal"-sized fish communities but not to become the victim of fisheries. It requires a vector to spread, so humans should carry it around and, despite being illegal almost everywhere, transfer it from one drainage system to another. Being a weed in carp ponds seems to be the most effective way. Carps are stocked everywhere, traded internationally without strong limitations, and cultured in relaxed ways regarding biosafety. Several examples of weed species in carp ponds are quickly conquering large areas. In their native range, such potential invaders are typically found in pond areas, polluted canals, and various lake and river habitats. These characteristics indicate the most successful invasive fish species of the 20th century, including carps such as *Carassius*, mosquitofish *Gambusia holbrooki* and the topmouth gudgeon *Pseudorasbora parva*. Similarly, sharpbelly, *Hemiculter leucisculus*, is also spreading. It can be confidently predicted that this species will subsequently conquer all of West Asia, Europe, and North Africa. As it occurs north to the Amur in its native range, it is difficult to estimate how far north it might spread in Russia and Central Europe. It can be predicted that sharpbelly will become established in any location where carps are stocked. It will be an abundant and ecologically influential invader, competing with *Alburnus* species. The potential consequences of this phenomenon can be observed in Central Asia, where the fish communities of large rivers and artificial waterbodies are dominated by East Asian species, with native species being almost entirely restricted to small mountain rivers. **Further readings.** Alcaraz et al. 2005; Devin & Beisel 2007; Van Kleunen et al. 2010; Matzek 2012 (invasive species traits).

Hypophthalmichthys

Two of the three species of this genus have been introduced to Europe, from where they were subsequently spread to West Asia. Both have been hybridised in aquaculture, and fish found in West Asia and Europe do not represent pure species. As far as is currently known, they did not establish self-sustaining populations except (perhaps) in the Danube, Maritza (Europe), and Terek (western Caspian basin), where data at least allow us to suspect that *H. molitix* might have established. *Hypophthalmichthys* have also been introduced to Central Asia, particularly in the Aral basin. They were introduced from the Amur (Siberia) to European Russia and from the Yangtze (China) to Romania. From these two areas, fish were transported to various locations throughout North Africa, West Asia, and Europe for aquaculture purposes. It is estimated that most major rivers were stocked, particularly in reservoirs. Many fisheries biologists have postulated that these fish can transform algal blooms from pollution into edible fish protein. The first importation of these fish to Romania is a prime example of how careless fish introductions have often been handled. In 1961, unidentified fish larvae were caught in the Yangtze near the city of Wuhan and released in an aquaculture station near Bucharest. Upon draining the ponds in the autumn, 22 Chinese fish species were recorded, including *Hypophthalmichthys* and *Pseudorasbora*. While the other introduced species could be successfully eliminated, *Pseudorasbora* escaped from the ponds and is still expanding its range. *Hypophthalmichthys* and other introduced Chinese carps also hosted several parasites that conquered West Asia and Europe. The most spectacular of these is the glochidium of large freshwater mussel *Sinanodonta woodiana*.

Hypophthalmichthys molitrix; Volga delta, Russia; ~500 mm SL.

Hypophthalmichthys molitrix

Common name. Silver carp.

Diagnosis. Distinguished from *H. nobilis* by: • sharp scaleless keel from pectoral region to anal origin / • pectoral short, not reaching beyond pelvic origin / • 650–820 long, slender gill rakers / • head length 24–29 % SL / • plain silvery colouration, greenish-grey above, whitish below. Size up to 1000 mm SL and 50 kg.

Distribution. Introduced in most drainages in West Asia. Established reproductive populations in North America, in Karakum Canal in Turkmenistan, Central Asia, potentially also in Danube (Europe). No established population in Aral Sea basin. Native to most major Pacific drainages in East Asia from Amur to Xi Jiang. Used in aquaculture almost worldwide outside of tropics.

Habitat. Stocked in large rivers and almost all still waters such as lakes, reservoirs, and ponds. Often escapes from fish farms. Aquaculture tolerates brackish water (up to 7 ‰) when stocked in estuaries and coastal lakes. In its native range, inhabits rivers with marked fluctuations in water level, overwintering in middle and lower reaches. During feeding season in shallow (0.5–1.0 m deep) and warm (above 21°C) backwaters, lakes, and flooded areas with slow currents. Spawns in rivers or tributaries above shallow rapids with gravel or sand bottoms, strong current (0.5–1.7 m/s), turbid water, temperatures above 15°C (usually 18–26°C), and high oxygen concentrations.

Biology. First spawns at 5–7 years. Juveniles and adults form large schools during spawning and overwintering. Migrates long distances upstream at onset of rapid floods and water level rise (April–August depending on location; late April–early June in Terek), able to jump over obstacles up to 1 m. Spawns during floods when water level rises 50–120 cm above normal, usually at 18–26°C. Stops spawning when conditions change (particularly sensitive to falling water levels, reduced turbidity and increased current velocity) and resumes spawning when water levels rise. Spawns in upper layer of water or even on surface. Females lay about 500,000 eggs in one or more portions, depending on duration of flood. Eggs float, are transparent, and hatch after about 2 days at about 25°C while drifting downstream. If river is blocked or if available river reaches are too short, eggs cannot drift long enough and fail to develop. After spawning, adults leave river and migrate to feeding areas. In autumn–winter they migrate to deeper places in mainstream, where they remain without feeding. Larvae drift downstream and settle in floodplain lakes, shallow banks, and backwaters with little or no current. Feeds from about 15 mm SL on phytoplankton only. Very sensitive to low temperature (below 5°C) and oxygen depletion. There is reliable data on natural spawning in rivers of southern Russia (Terek) and lower Danube. Still, it is not known whether eggs and larvae could complete development. There is no evidence that this species has become established in West Asia or North Africa.

Conservation status. Non-native; stocked for commercial fisheries. Survives in West Asia and Europe only through stocking. Potentially reproductive in Danube. Native wild stocks in Russia and China are declining.

Further reading. Holčík 1976 (introduction, Danube); Movchan & Smirnov 1983 (morphology, biology); Howes 1981 (anatomy); Kottelat & Freyhof 2007 (summary of distribution and biology).

Hypophthalmichthys nobilis; Shanghai Aquarium, ~350 mm SL. © Zhou Hang.

Hypophthalmichthys nobilis

Common name. Bighead carp.

Diagnosis. Distinguished from *H. molitrix* by: • scaled keel from pelvic to anal / • pectoral long reaching beyond pelvic origin / • 240–300 long gill rakers / • head length 27–35 % SL / • dark-grey overall colouration, flank with dark, large, very irregularly shaped blotches. Size up to 1460 mm SL and 40 kg.

Distribution. Used in aquaculture in Europe, Asia, and North America. No reliable data on natural spawning in West Asia, spawning reported from Europe (Terek). This species is rarely used in aquaculture and may have already vanished from West Asia. Established reproductive populations in North America and Karakum Canal in Turkmenistan, Central Asia. No established population in Aral Sea basin. Native to Central and South China.

Habitat. Stocked in large rivers and almost all still waters such as lakes and ponds. Often escape from fish farms. In aquaculture, adults tolerate brackish water (up to 7 ‰) when stocked in estuaries and coastal lakes. In native range, in rivers with marked water level fluctuations, overwintering in middle and lower reaches. Foraging in shallow (0.5–1.5 m deep) and warm (above 24°C) backwaters, lakes, and flooded areas with slow currents. Spawns in deep, turbid, and warm water above 18°C (usually 22–30°C), with strong currents (1.1–1.9 m/s) and high oxygen concentrations.

Biology. Lives up to 20 years. Spawns first time at 5–6 years, 550–700 mm SL, and 5–10 kg (Russian Pacific basin), or even at 2–4 years (males) and 3–5 years (females) in Turkmenistan. Migrates long distances upstream at onset of rapid flooding and water level rise (April–July, depending on location). Spawning migration begins in mid–May at temperatures of 18–19°C and lasts until end of June (Terek). Spawns during floods. Stops spawning when conditions change and resumes when water level rises. Spawns in upper water layer or even on surface in June (Terek). Females lay up to 1.1 million eggs in 1–3 portions, depending on flood duration. Eggs are yellowish, transparent, and hatch after about 2 days at around 25°C while drifting downstream in a deep open water layer. If river is blocked or available river reaches are too short, eggs cannot drift long enough and fail to develop. After spawning, adults leave river and migrate to feeding areas. Larvae drift downstream and settle in floodplain lakes, shallow banks, and backwaters with little or no current. In autumn/winter, when temperatures drop to 10°C, juveniles and adults form separate large schools and move downstream to deeper areas in main river to overwinter. Feeds mainly on zooplankton and also on algae.

Conservation status. Non-native; stocked for commercial fisheries. There are reliable data on natural spawning in rivers of southern Russia (Terek), but whether eggs and larvae could complete development is unknown. Native wild stocks in Russia and China are declining.

Further reading. Holčík 1976 (introduction, Danube); Movchan & Smirnov 1983 (morphology, biology); Kottelat & Freyhof 2007 (summary of distribution and biology).

Mylopharyngodon piceus; Shanghai Aquarium; China; ~800 mm SL. © Zhou Hang.

Mylopharyngodon piceus

Common name. Black carp.

Diagnosis. Distinguished from similar *Ctenopharyngodon idella* by: ● body black, blue-grey or dark-brown / ○ body slightly compressed / ● pharyngeal teeth molar-like, massive, smooth, very strong (Figure 43). Size up to 1800 mm SL and more than 60 kg.

Distribution. Used in aquaculture; stocked in Rioni (Georgia), Danube, Dniestr, Dniepr, Don, Kuban, and Volga drainages. Reproductive populations established only in Amu Darya (Central Asia) and possibly in Tone drainages (Japan), not in West Asia. Native to most major Pacific drainages, from Amur to Xi Jiang.

Habitat. Stocked to reservoirs. In native range, large lowland rivers and lakes usually with clear water and high oxygen concentrations.

Biology. Spawns first time at 6–11 years, females later than males, at about 1000 mm SL and 15 kg, males at 900 mm and 11 kg. Migrates upstream and spawns in open waters from March to July. Spawns when water level rises, at 19–30°C. Eggs pelagic or semi-pelagic, hatching while drifting downstream. If river is blocked or available river reaches are too short, eggs cannot drift long enough to develop. Larvae settle in floodplain lakes and channels with little or no current. Larvae feed on zooplankton, then ostracods and aquatic insects. At about 120 mm SL, juveniles begin to feed on small snails and clams; larger juveniles and adults feed almost exclusively on molluscs.

Conservation status. Non-native; stocked to control bivalves. In West Asia, persists only in Rioni (Georgia) through stocking. Native stocks in Russia and China have declined sharply. Expected to disappear if stocking is stopped.

Remarks. Introduced to control of populations of molluscan vectors of fish and human parasites. Commonly used to remove *Dreissena* mussels that clog hydroelectric plants.

Further reading. Evtushenko et al. 1994 (biology); Bíró 1999b (biology); Nico et al. 2001 (biology, introductions); Nico et al. 2005 (synopsis).

Çöçelli spring is the habitat of *Pseudophoxinus zekayi*, *Paraphanius alexandri*, and *Garra culiciphaga*.

Family Cobitidae

Spined loaches

A family of small, slender fishes found in Morocco, Europe, and Asia. It is most diverse in Southeast and East Asia, with approximately 240 species in 18 genera. Two genera are found in West Asia. They are distinguished by an erectile, movable, usually bifid, suborbital spine, hidden under the skin in some species. Many spined loaches exhibit external sexual dimorphism. In *Cobitis*, the male has a modified pectoral, which is usually much larger than the female's pectoral, and the male is smaller than the female. In *Sabanejewia*, there are no size differences between the sexes, but the male exhibits vertical swellings of the flanks. All Cobitidae are benthic, living on the substrate, and all prey on small invertebrates. The pigmentation of spined loaches is a useful diagnostic feature. Its pattern is organised in one series of blotches on the back and up to four pigmentation zones on the flank, described as Gambetta's longitudinal pigmentation zones. These zones are abbreviated as Z1 to Z4, starting with Z1 as the narrow line immediately below the mid-dorsal row of blotches and Z4 as the midlateral row of blotches. Furthermore, most species exhibit one or two black spots on the caudal base, which may be fused to form a bar in *Sabanejewia*. In *Sabanejewia* and several *Cobitis* species, the Gambetta zones are either incompletely developed or dissociated.

Three East Asian species of weatherfish (*Misgurnus anguillicaudatus*, *M. bipartitus*, *Paramisgurnus dabryanus*) are non-native in Europe and are anticipated to be found in West Asia in the near future. Their spread is facilitated by the use of non-biosave stocking materials of carp and other aquaculture fishes and through the pond and aquarium trade. East Asian weatherfish are pale-brownish with numerous dark-brown spots and small blotches, forward-directed barbels, and a mental lobe that is prolonged into two long, barbel-like projections (vs. mental lobe absent or with one very short projection in *Cobitis* and *Sabanejewia*). **Further reading.** Gambetta 1934 (pigmentation pattern); Nalbant 1963 (genera); Sawada 1982 (morphology); Ludwig et al. 2001 (molecular phylogeny); Kottelat 2012 (diversity); Geiger et al. 2014 (molecular diversity in Mediterranean); Zangl et al. 2020 (*Misgurnus* invasion).

Key to genera of Cobitidae in West Asia

1a - Body pigmentation usually organised in 1 dorsal and 2 lateral longitudinal zones; 12 branched caudal rays. ....*Sabanejewia*

1b - Body pigmentation usually organised in 1 dorsal and 4 lateral longitudinal zones; 14 branched caudal rays. ....*Cobitis*

Misgurnus anguillicaudatus; Germany; 150 mm SL.

https://doi.org/10.1515/9783111677811-020

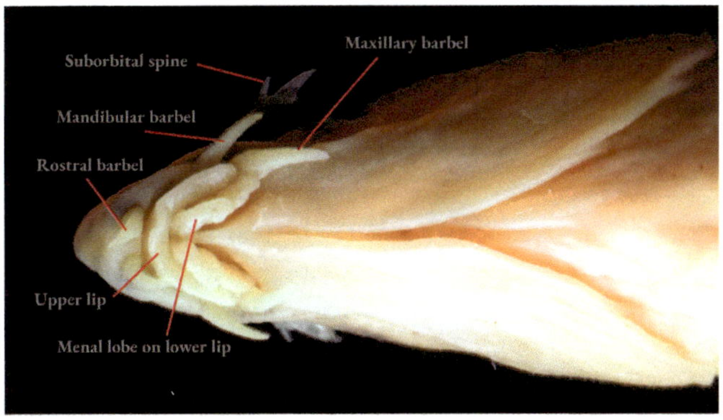

Figure 44: Head of Cobitidae.

Cobitis

Cobitis comprises approximately 126 species, distributed between Portugal and Morocco in the west and Korea and Japan in the east. In West Asia, 31 species have been documented, most of which are found in Anatolia. In conjunction with Greece, this region represents a significant diversity hotspot of the genus. Male *Cobitis* have one or two laminar projections on the dorsal surface of the anterior pectoral rays (lamina circularis, plural: laminae circularis; also known as Canestrini scales). Examining the lamina circularis in *Cobitis* requires preparation, as it is covered by skin. In species with two laminae, these may be very small, particularly the outer lamina, which may be a small lateral enlargement or a lateral crest on the fin ray. It should be noted that the inner lamina is not necessarily scale-shaped.

Cobitis is notable for exhibiting minimal morphological variation. Most species exhibit superficial similarities, share identical meristic characters, and display comparable structural characteristics, including fin position, mouth shape and structure, nares, lateral-line length, and squamation pattern. Species differentiation is primarily achieved through colour patterns, organised by the Gambetta zones relatively conservatively. Several species have a notable divergence from the typical configuration of the Gambetta zones. This is evidenced by the fragmentation of the blotches within these zones, the blurring of the borders between them, and the gradual convergence of different zones. Notably, this process of dissociating the colour pattern of the Gambetta zones is ontogenetic. In young fish, the colour pattern is often well organised in the Gambetta zones, whereas in adults, particularly in adult females, a pattern of dissociated blotches is observed. The degree to which the Gambetta zones' colour patterns are dissociated depends on the species and particular populations. For instance, considerable individual variation is observed in the colour pattern of *C. battalgilae*, which spans almost the entire spectrum of variation observed in *Cobitis*. Additionally, considerable variation is observed among different populations of *C. strumicae*, *C. phrygica*, *C. simplicispina* and *C. turcica*, with almost every population displaying distinct characteristics. In *C. dorademiri* and *C. sipahilerae*, only a small proportion of individuals exhibit a dissociated colour pattern, with most individuals displaying a highly similar phenotype.

Cobitis are typically associated with fine substrates (sand, mud) from which they feed. They burrow in the substrate to hide. A specialised filter-feeding mechanism enables *Cobitis* to separate fine organic particles and small benthic organisms (e.g., copepods, worms, insect larvae) from the substrate. *Cobitis* can swallow air, with oxygen being absorbed through the gut walls. The spawning behaviour is consistent in the few species of *Cobitis* in which it has been observed. The male follows the female, and after both enter dense vegetation (e.g., filamentous algae), the male

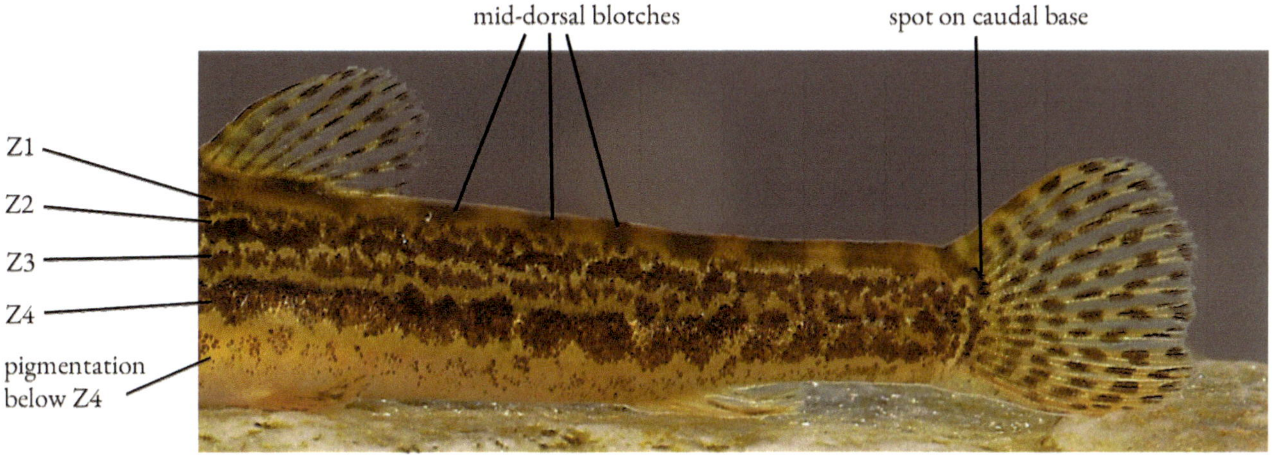

Figure 45: Colour marks on the body of *Cobitis* species. Gambetta's longitudinal pigmentation zones Z1–4.

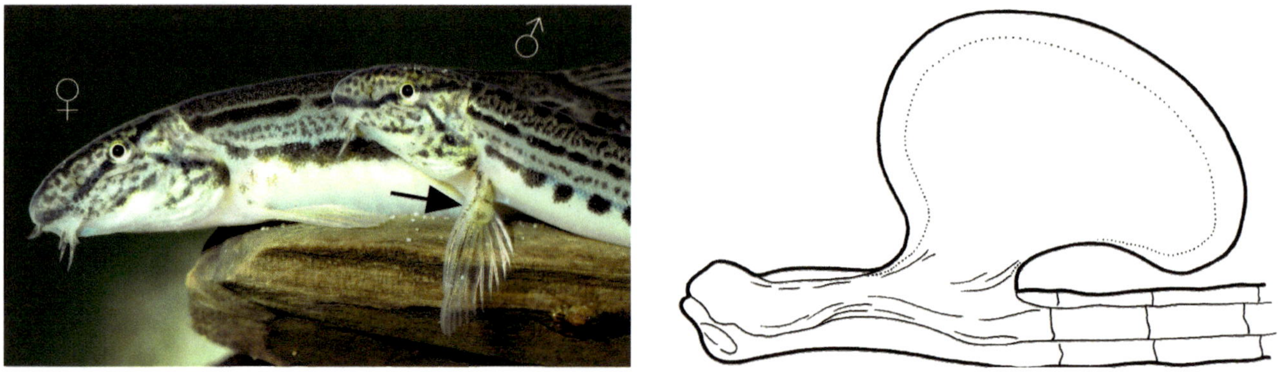

Figure 46: Sexual dimorphism in *Cobitis*; **left**, male and female *Cobitis* showing the position of the lamina circularis (arrow) on the dorsal surface of the anterior pectoral ray of male; **right**, lamina circularis, dorsal view (from Kottelat & Freyhof 2007, a © A. Hartl; b after Nalbant 1963).

forms a complete ring around the female's body behind the dorsal as the female releases the eggs. The eggs of most species are not sticky but swell to 2.5–3.5 mm in diameter and, due to this large size, are retained in the vegetation. The free larvae have large external gill filaments. The biology of almost all *Cobitis* species in West Asia remains unstudied, and there is much work to be done in the future.

Some *Cobitis* species of Central and Eastern Europe occur as a multitude of hybridogenous lineages. These hybrid lineages, made up of nearly only female individuals, reproduce by gynogenesis, often including polyploid individuals. They are described as sperm-parasites, as they are dependent on males of bisexually reproducing lineages that serve as sperm donors, with the sperm serving only to induce the development of their eggs. Such hybridogenous lineages have yet to be reported from West Asia, but their occurrence cannot be excluded. The presence of highly biased sex ratios may serve as an indicator.

Further reading. Gambetta 1934 (pigmentation pattern); Bohlen 2003a, b (reproduction); Ludwig et al. 2001 (molecular phylogeny); Bohlen et al. 2006 (molecular phylogeny); Geiger et al. 2014 (molecular diversity in the Mediterranean); Perdices et al. 2018 (molecular diversity); Freyhof et al. 2018c (species diversity in West Asia).

Keys to species of *Cobitis* in West Asia

Persian Gulf basin

1a - Male with two laminae circularis.
..................2
1b - Male with one lamina circularis.
..................3

2a - Blotches in Z4 between vertical of pectoral origin and dorsal base squarish, irregularly shaped, roundish, and usually widely separated; blotches in Z4 posterior to dorsal base not fused into stripes, restricted to one row; pelvic axillary lobe present.
..................*C. elazigensis*
2b - Blotches in Z4 between vertical of pectoral origin until caudal–peduncle comma-shaped, densely set, often fused into short stripes or dissociated into a wide band of several rows of blotches; axillary pelvic lobe absent.
..................*C. linea*

3a - Large, usually horizontally elongated, dark-brown blotches in anterior part of Z4, roundish or vertically elongated on caudal–peduncle.
..................*C. avicennae*
3b - Minute, roundish, dark-grey dots in Z4.
..................*C. kellei*

Caspian basin

1a - Subdorsal scales with a large focal zone (about ½ of maximum scale diameter).
..................*C. amphilekta*
1b - Subdorsal scales with a small focal zone (about or less than ⅒ of maximum scale diameter).
..................2

2a - Plate of lamina circularis not reaching end of third segment of attached pectoral ray; dark-brown spot at upper caudal base poorly developed.
..................*C. derzhavini*
2b - Plate of lamina circularis reaching to or beyond end of third segment of attached pectoral ray; dark-brown spot at upper caudal base well developed.
..................3

3a - Base of lamina circularis widely connected to pectoral ray; cheek between eye and opercle strongly pigmented.
..................*C. saniae*
3b - Base of lamina circularis with a narrow connection to pectoral ray; cheek between eye and opercle poorly pigmented or unpigmented.
..................*C. faridpaki*

Southern Black Sea basin and endorheic Anatolian basins of Lakes Akşehir, Eber, and Ilgın

1a - Male with 2 laminae circularis.
..................*C. simplicispina*
1b - Male with 1 lamina circularis.
..................2

2a - Pigmentation in Z3 reduced to about head length in male or pigmentation in Z3 as wide as or narrower than pigmentation in Z2.
..................*C. splendens*
2b - Pigmentation in Z3 usually complete, reaching much behind dorsal base and or pigmentation in Z3 wider than pigmentation in Z2.
..................*C. satunini*

Southern Marmara and eastern Aegean basins south to Eşen

1a - Male with 2 laminae circularis.
..................2
1b - Male with 1 lamina circularis.
..................6

2a - Two brown or black spots at upper and lower caudal base.
..................*C. strumicae*
2b - One or no black spot at upper and lower caudal base.
..................3

3a - External part of suborbital spine simple; no black spot at upper caudal base; colour pattern on flank often dissociated, not following Gambetta zones in most, but not all, individuals and populations.
..................*C. phrygica*
3b - External part of suborbital spine bifurcate; back spot at upper caudal base present, often very small, rarely absent; colour pattern on flank following Gambetta zones in all populations except in few individuals.
..................4

4a - Pigmentation below Z4 absent except some small spots at caudal–peduncle, also above anal base in few individuals; caudal–peduncle depth 1.2–1.7 times in caudal–peduncle length.
..................*C. indus*
4b - Pigmentation below Z4 present, if restricted to anal base, then caudal–peduncle depth 1.0–1.2 times in caudal–peduncle length.
..................5

5a - Pigmentation below Z4 restricted to a short line or few dots below pectoral and a second short line on lower caudal–peduncle.
..................*C. dorademiri*
5b - Pigmentation below Z4 usually present along complete flank.
..................*C. afifeae*

6a - Blotches in Z2 and Z4 very small, indistinct or confluent; flank pigmented below zone Z4; body brownish.
..................*C. puncticulata*
6b - Blotches in Z2 and Z4 well separated, fused into a bold stripe in some individuals; flank without pigmentation below zone Z4; body pale-yellowish or whitish.
..................7

7a - Focal zone of subdorsal scale large, about ⅓–½ of vertical scale diameter.
..................*C. satunini*
7b - Focal zone of subdorsal scale small, about ⅒–⅓ of vertical scale diameter.
..................8

8a - 25–36, small, comma-shaped blotches in Z4 along anterior and middle part of flank blotches squarish or irregularly shaped on posterior part of flank, rarely fused into short stripes.
..................*C. troasensis*
8b - 12–18 roundish, triangular, squarish, or horizontally elongate blotches in Z4, often densely set and fused into short stripes along anterior and middle part of flank, blotches vertically elongate on posterior part of flank.
..................9

9a - Black spot at upper caudal base roundish or ovoid, larger than eye diameter in male; Z3 fully covered by very small spots forming a sand-like pattern, especially in female.
..................*C. emrei*
9b - Black spot at upper caudal base absent or comma-shaped, smaller than eye diameter in male and female; Z3 with a row or a narrow band of small spots not forming a sand-like pattern.
..................*C. fahireae*

Mediterranean basin east of Eşen and endorheic basins in Central Anatolia (excluding basins of Lakes Akşehir, Eber, and Ilgın)

1a - Male with 1 lamina circularis.
..................*C. bilseli*
1b - Male with 2 laminae circularis.
..................2

2a - No black spot at upper caudal base.
..................3
2b - Black spot present at upper caudal base, often very small and best seen in live individuals.
..................4

3a - A bold-brown stripe in Z4 between head and vertical of anal origin; flank below Z4 unpigmented; external part of suborbital spine bifurcate.
..................*C. evreni*
3b - A series of midlateral blotches in Z4 or Gambetta zones dissociated into a pattern of many irregularly shaped small or medium-sized spots and blotches; flank below Z4 usually pigmented; external part of suborbital spine simple.
..................*C. phrygica*

4a - Back plain cream-brown, few individuals with 1–3 narrow, squarish, brown bars.
..................*C. sipahilerae*
4b - Back with 3–8 large, roundish, dark-brown blotches or blotches fused into a dark-brown stripe on back or blotches dissociated into a marbled pattern.
..................5

5a - Flank pattern not organised in Gambetta zones, completely disorganised into a pattern of many small spots and blotches.
..................6
5b - Flank pattern well organised in four Gambetta zones, blotches in Z4 organised in a wide band or two rows of spots, and blotches in some individuals or populations.
..................7

6a - Mental lobe long, reaching to or beyond lower lip; caudal truncate; prepelvic length 53–56 % SL; preanal distance 75–78 % SL in female.
..................*C. turcica*
6b - Mental lobe short, not reaching beyond lower lip; caudal slightly rounded; prepelvic length 58–59 % SL; preanal distance 81–82 % SL in female.
..................*C. joergbohleni*

7a - External part of suborbital spine simple; two distinct rows of small blotches in Z4, one along lateral midline and one distinctly below.
..................*C. pirii*
7b - External part of suborbital spine bifurcate, simple in some individuals of *C. battalgilae*, one row of large blotches in Z4 or blotches in Z4 dissociated into a band of spots and blotches or spots and blotches of Z4 completely dissociated, not forming two rows.
..................8

8a - A series of large, horizontally elongate blotches in Z4, adjacent blotches often fused to each other; blotches in Z4 not vertically or horizontally dissociated.
..................*C. erkakanae*
8b - A series of roundish, squarish, comma-shaped, or vertically elongated blotches in Z4, rarely blotches horizontally elongate or fused into short stripes in some individuals (in *C. levantina*); blotches in Z4 often vertically and horizontally dissociated into several rows or a wide band of small blotches and spots.
..................9

9a - Blotches in Z4 usually vertically elongate or squarish, often dissociated into a band of small irregularly shaped blotches and spots; interorbital distance 14–16 % HL in male.
..................*C. battalgilae*
9b - Blotches in Z4 in a single row, roundish, irregularly shaped, horizontally elongate or squarish anterior to dorsal origin, often fused into a stripe, not dissociated into a band of small spots or small blotches; interorbital distance 7–11 % HL in male.
..................10

10a - Blotches in Z2 irregularly shaped, often ½ of size or as large as in Z3, often blotches in Z2 and Z3 about size of eye or pupil diameter, Z2, and Z3 usually forming one marbled pattern; many small, dark-brown spots or blotches on lower part of caudal–peduncle, reaching forward to pectoral base in some individuals.
..................*C. levantina*
10b - Z2 with small brown spots, always much smaller than blotches in Z3, much smaller than eye or pupil diameter, Z2 and Z3 well separated; no pigmentation or very few isolated brown spots below Z4 and on lower caudal–peduncle.
..................11

11a - Anterior to dorsal origin, blotches in Z2 and Z4 almost always well separated from each other; Z1 and Z2 always well distinguished; prepelvic length 56–60 % SL in female; caudal–peduncle length 11–14 % SL in female.
..................*C. aliyeae*
11b - Anterior to dorsal origin, blotches in Z2 usually fused into short, irregularly shaped stripes or forming a mottled band of irregularly shaped, small blotches and spots, Z4 fused into a stripe in few individuals; Z1 and Z2 usually fused into a mottled band; prepelvic length 53–55 % SL in female; caudal–peduncle length 15–19 % SL in female.
..................*C. anabelae*

Cobitis afifeae; Büyük Menderes drainage, Türkiye; male, 71 mm SL.

Cobitis afifeae; stream Karadirek, Türkiye; male, 47 mm SL.

Cobitis afifeae

Common name. Menderes spined loach.

Diagnosis. Distinguished from other species of *Cobitis* in eastern Aegean basin by ○ pigmentation below Z4 usually present along complete flank / ○ colour pattern on flank often dissociated, but following Gambetta zones in all populations except in few individuals / ○ external part of suborbital spine bifurcate / ○ back spot at upper caudal base present, often very small / ○ caudal–peduncle depth 1.9–1.1 times in caudal–peduncle length / ○ two laminae circularis in male. Size up to 90 mm SL.

Distribution. Türkiye: Büyük Menderes and Küçük Menderes drainages, including Lake Işıklı basin and Karadirek, which flows underground to Lake Işıklı.

Habitat. Streams, rivers, lakes, and reservoirs, usually on sand or fine gravel in slow or moderate fast-flowing waters.
Biology. No data.
Conservation status. LC; extirpated from Küçük Menderes.
Remarks. Often occurs in sympatry with *C. fahireae*.
Further reading. Freyhof et al. 2018c (description).

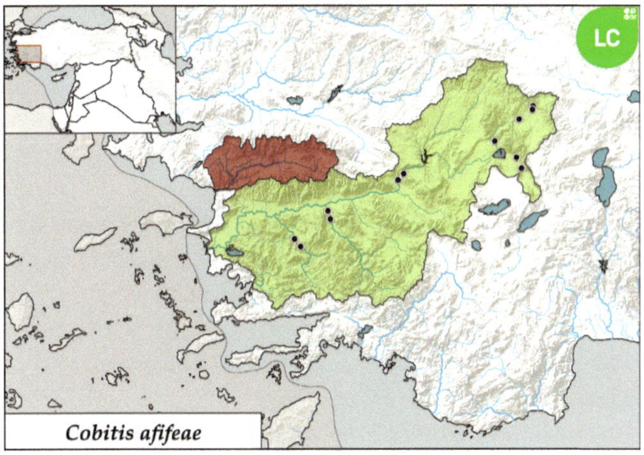

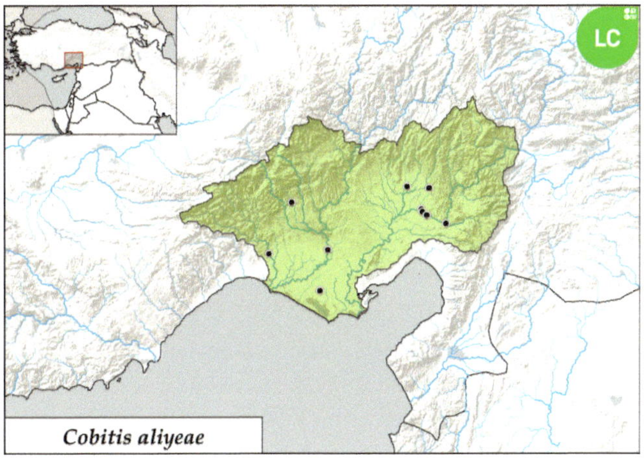

Cobitis aliyeae; lower Seyhan drainage, Türkiye; female, 65 mm SL.

Cobitis aliyeae

Common name. Cilician spined loach.
Diagnosis. Distinguished from other species of *Cobitis* in Mediterranean basin east of Eşen by: ○ flank pattern well organised in four Gambetta zones / ○ anterior to dorsal origin, blotches in Z2 and Z4 almost always well separated from each other / ○ Z1 and Z2 always well distinguished / ○ Z2 with small brown spots, always much smaller than blotches in Z3, much smaller than eye or pupil diameter / ○ Z2 and Z3 well separated / ○ Z4 formed by one series of distinct blotches / ○ no pigmentation or very few isolated brown spots below Z4 and on lower caudal–peduncle / ○ mid-dorsal pigmentation consisting in a series of 10–14 dark-brown, roundish blotches / ○ prepelvic length 56–60 % SL in female / ○ interorbital distance 7–11 % HL in male / ○ caudal–peduncle length 11–14 % SL in female. Size up to 65 mm SL.
Distribution. Türkiye: Lower Tarsus, Seyhan and Ceyhan drainages.
Habitat. Streams, rivers, lakes, and reservoirs, usually on sand or fine gravel in slow or moderate fast-flowing waters.
Biology. No data.
Conservation status. LC.
Remarks. *Cobitis aliyeae*, *C. evreni*, and *C. erkakanae* are three *Cobitis* species found in the Ceyhan drainage. All three species occur in allopatry within the drainage, *C. aliyeae* in lower Ceyhan and *C. evreni* in upper reaches of Ceyhan.
Further reading. Freyhof et al. 2018c (description).

Cobitis amphilekta; Caspian Sea near Port Il'icha, Azerbaijan; male, 41 mm SL. © E. Vasil'eva.

Cobitis amphilekta

Common name. Azerbaijani spined loach.

Diagnosis. Distinguished from other species of *Cobitis* in Caspian basin by: ● subdorsal scales with a large focal zone, about ½ of maximum scale diameter / ○ Z4 consisting of distinct, large dark-brown blotches, usually horizontally elongated on anterior part of flank, roundish or vertically elongated on caudal–peduncle / ○ spot in upper part of caudal base very small and faith. Size up to 65 mm SL.

Distribution. Kyzylagach Bay, south of Kura estuary, lower Kumbashi, about 150 m from Azerbaijan's coast, desalinated Caspian Sea area off Durnev Islands and Prorva in Kazakhstan. Expected to occur elsewhere in lower parts of rivers, coastal lagoons, and freshened parts of sea.

Habitat. Coastal lagoons, lower parts of rivers, and freshwater parts of sea with gravel, sand, or mud bottoms.

Biology. No data.

Conservation status. Extinct; not recorded since 1937. Intensive fieldwork in southern Azerbaijan revealed only *C. saniae*, even in stream sections very close to the sea.

Further reading. Vasil'eva & Vasil'ev 2012 (description).

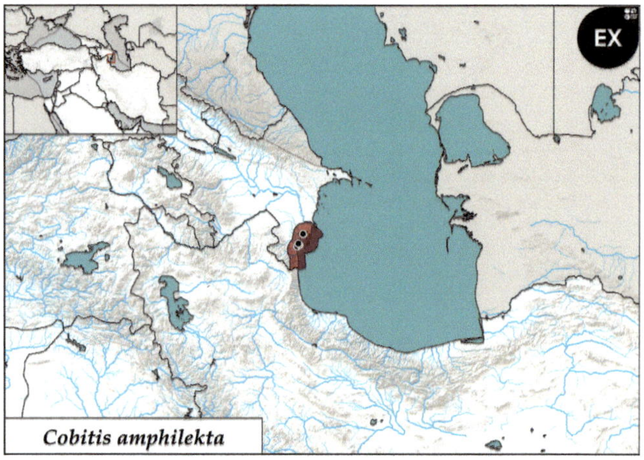

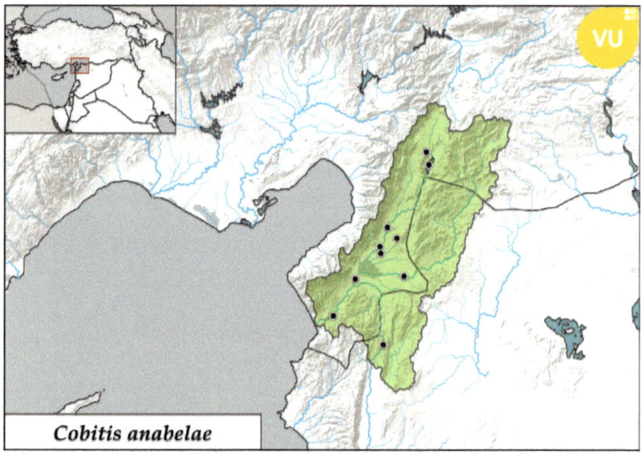

Cobitis anabelae; Karasu in Orontes drainage, Türkiye; male, 53 mm SL.

Cobitis anabelae; Gölbaşı lake in Orontes drainage, Türkiye; female, 70 mm SL.

Cobitis anabelae

Common name. Hatay spined loach.

Diagnosis. Distinguished from other species of *Cobitis* in Mediterranean basin east of Eşen by: ○ flank pattern well organised in four Gambetta zones / ○ anterior to dorsal origin, blotches in Z2 usually fused into short, irregularly shaped stripes, or forming a mottled band of irregularly shaped, small blotches and spots / ○ Z1 and Z2 usually fused into a mottled band / ○ Z2 with small brown spots, always much smaller than blotches in Z3, much smaller than eye or pupil diameter / ○ Z2 and Z3 well separated / ○ Z4 formed by one series of distinct blotches, fused into a stripe in few individuals / ○ mid-dorsal pigmentation consisting in a series of 7–11 dark-brown, roundish blotches / ○ no pigmentation or very few isolated brown spots below Z4 and on lower caudal–peduncle / ○ interorbital distance 7–11 % HL in male / ○ prepelvic length 53–55 % SL in female / ○ caudal–peduncle length 15–19 % SL in female. Size up to 82 mm SL.

Distribution. Türkiye and Syria: Lower Orontes drainage.

Habitat. Streams, rivers, lakes, and reservoirs, usually on sand or fine gravel in slow or moderate fast-flowing waters.

Biology. No data.

Conservation status. VU; appears to be declining within its small range.

Remarks. While *C. anabelae* is found in northern, lower part of Orontes drainage, the upper Orontes is inhabited by the closely related *C. levantina*. The exact distribution of both species is unknown, and it cannot be excluded that they might form a hybrid zone at the border of their ranges.

Further reading. Freyhof et al. 2018c (description).

Cobitis avicennae; Karkheh drainage, Iran; ~90 mm SL.

Cobitis avicennae

Common name. Zagros spined loach.

Diagnosis. Distinguished from other species of *Cobitis* in Persian Gulf basin by: ○ one lamina circularis in male / ○ 12–17 distinct, large, dark-brown blotches in Z4, usually horizontally elongated along anterior part of flank, roundish or vertically elongated on caudal–peduncle. Size up to 98 mm SL.

Distribution. Iran: Dinevar, Raazavar, and Gamasiab in upper Karkheh drainage.

Habitat. Streams with sandy or muddy bottoms.

Biology. No data.

Conservation status. EN; appears to be declining within its very small range.

Remarks. Records from Sirvan (Iran) and Dukan reservoir in Lesser Zab drainage (Iraq) might be identified as this species, but these identifications must be confirmed.

Further reading. Coad 2010a (record from Iraq); Mousavi-Sabet et al. 2015e (description).

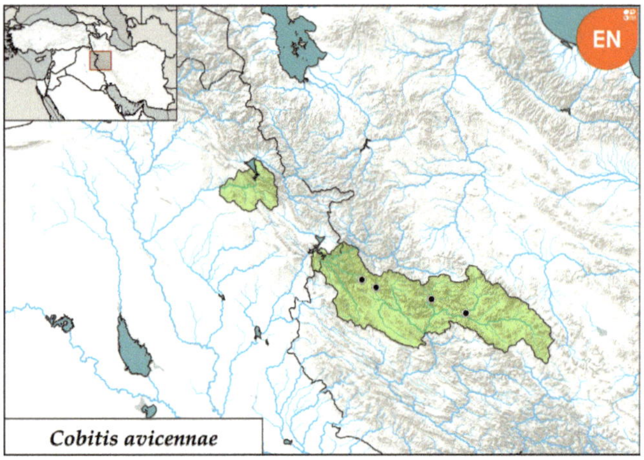

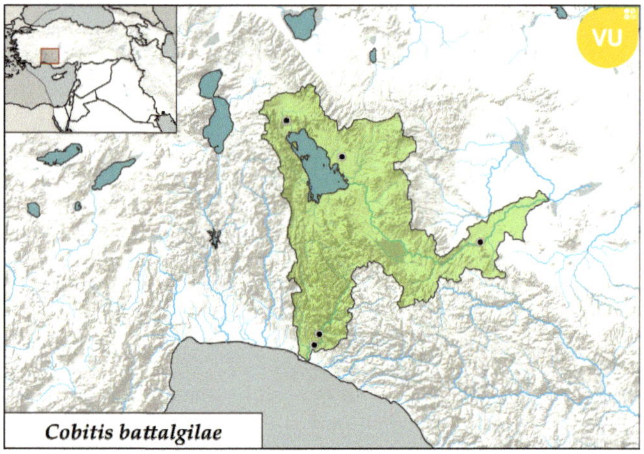

Cobitis battalgilae; spring Eflatun Pınarı, Türkiye; male, 68 mm SL.

Cobitis battalgilae
Common name. Beyşehir spined loach.
Diagnosis. Distinguished from other species of *Cobitis* in Mediterranean basin east of Eşen and endorheic basins in Central Anatolia by: ○ two laminae circularis in male / ○ pigmentation organised in Gambetta zones / ○ blotches in Z4 being usually dissociated, usually vertically elongate, often forming a band or an open field of small, irregularly shaped blotches and spots in female larger than 90 mm SL / ○ blotches in Z4 often vertically and horizontally dissociated into several rows or a wide band of small blotches and spots / ○ blotches in Z4 in not forming stripes, usually much larger than in Z2 / ○ mid-dorsal pigmentation consisting in a series of blotches, often very closely set or slightly fused / ○ a black spot at upper caudal base, often very small and comma-shaped / ○ anterior nare reaching to hint margin of posterior nare when folded down / ○ external part of suborbital spine bifurcate / ○ snout as long or shorter than distance between tip of anal and end of hypural complex / ○ caudal peduncle 14–16 % SL / ○ interorbital distance 14–16 % HL in male / ○ eye diameter in male 16–20 % HL. Size up to 137 mm SL.
Distribution. Türkiye: Lake Beyşehir basin in Çeltek in very north and in spring Eflatunpınarı in east of lake (and maybe elsewhere). Also, in Apa reservoir and in river that flows out of Lake Suğla and ends in Konya plains, as well as in Manavgat that flows to Gulf of Antalya.
Habitat. Running waters with gravel, sand, or silt bottoms, often with dense submerged vegetation.
Biology. No data.
Conservation status. VU; less than 10 populations are known; some declining but may be more widespread.
Remarks. Closely related to *C. turcica*, and few individuals might be difficult to distinguish. *Cobitis battalgilae* grows

Cobitis battalgilae; spring Eflatun Pınarı, Türkiye; female, 135 mm SL.

much larger than *C. turcica*. The distribution area of *C. battalgilae* reflects the hydrological connections of Lake Beyşehir, which seasonally flows over to Lake Suğla, from which some parts of water flow to the Konya plain, and other parts of water flow into the karstic rock masses supporting springs in Manavgat drainage through underground connections.

The colour pattern in *C. battalgilae* is especially variable within and between populations. Some individuals, usually males smaller than 70 mm SL, lack the pigmentation in Z3, and there is one series of squarish blotches in Z4. Other individuals have the Z3 partly or fully pigmented, and the blotches in Z4 are vertically elongate, often small and densely set, or some or all blotches in Z4 are divided into 2 or more elements, resulting in a colour pattern of two rows of blotches (similar to *C. pirii*, see below). In other individuals, Z4 is dissociated into a band of small, irregularly shaped blotches and spots, or the spots and blotches are dispersed on the flank, forming an open field of blotches and spots. This large variability in colour pattern in *C. battalgilae* makes it very difficult to identify this species, especially if only few individuals are available.

Further reading. Băcescu 1962 (description); Freyhof et al. 2018c (description).

Cobitis bilseli; Beyşehir canal, Türkiye; ~190 mm SL.

Cobitis bilseli

Common name. Great Beyşehir spined loach.

Diagnosis. Distinguished from other species of *Cobitis* in Mediterranean basin east of Eşen and endorheic basins in Central Anatolia by: • one lamina circularis on pectoral in male, base of pectoral ray adjacent to lamina circularis swollen / • a short skin-flap before genital papilla. Size up to 195 mm SL.

Distribution. Türkiye: Lake Beyşehir basin, in streams Sarıöz (Eylikler) and Sarıçay north of Beyşehir (and possibly others). Also, in river flowing from Lake Beyşehir to Lake Suğla and seasonally in outflow of Lake Beyşehir, possibly more widespread.

Habitat. Running waters with gravel or sand bottoms, often with dense submerged vegetation.

Biology. No data.

Conservation status. EN; appears to be declining within its very small range.

Remarks. This is the largest spined loach in the world. It was placed in a separate subgenus, *Beyshehiria*, but belongs to a group of species around *C. simplicispina*. It is related to the geographically adjacent *C. pirii* from Lake Eğirdir basin. It is the only species of this species group having one lamina circularis, while all other species have two.

Further reading. Battalgil 1942 (description); Freyhof et al. 2018c (description).

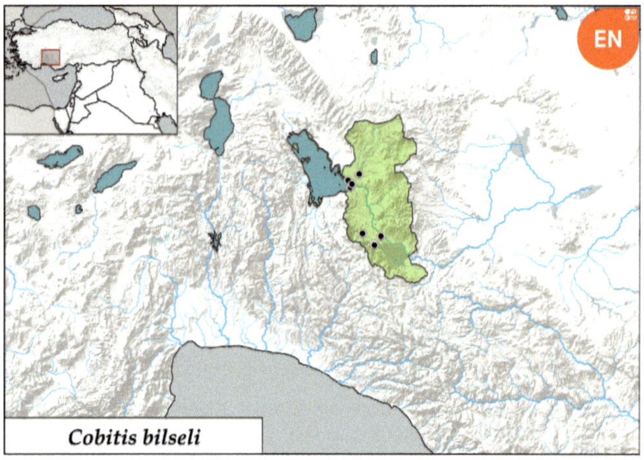

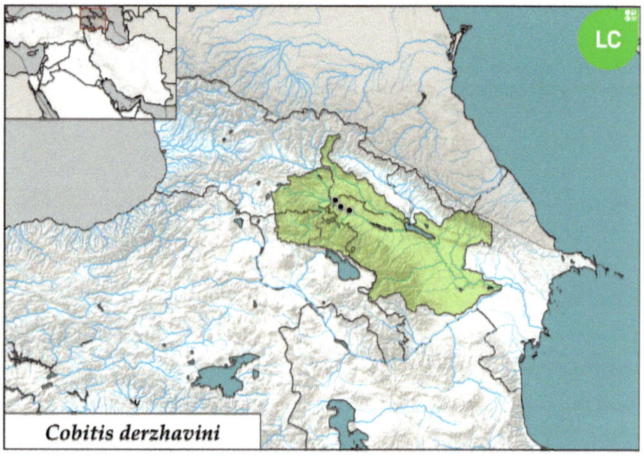

Doors regulating the water level of Lake Beyşehir, a seasonal habitat of *Cobitis bilseli*.

Cobitis derzhavini; Kara-Su in Kura drainage, Azerbaijan; female, 73 mm SL.

Cobitis derzhavini; Kara-Su in Kura drainage, Azerbaijan; male, 66 mm SL.

Cobitis derzhavini

Common name. Kura spined loach.

Diagnosis. Distinguished from other species of *Cobitis* in southern Caspian basin by: • plate of lamina circularis not reaching end of third segment of attached pectoral ray / • dark-brown spot at upper caudal base absent or poorly developed / ○ subdorsal scales with a small focal zone (about or less than 1/10 of maximum scale diameter) / ○ one lamina circularis in male. Size up to 80 mm SL.

Distribution. Azerbaijan and Georgia: Lower Kura drainage: Kara-Su and its tributaries, Agstafa, Mingechaur reservoir basin, and Katekh in Balaken region.

Habitat. Wetlands, streams, rivers, and reservoirs, usually on sand or fine gravel in slow or moderate fast-flowing waters.

Biology. No data.

Conservation status. LC.

Remarks. Sympatric with *C. saniae* in Agstafa drainage and potentially in other areas. Both species are very closely related, and their mode of reproductive isolation is interesting to research.

Further reading. Vasil'eva et al. 2020 (description); Vasil'eva & Vasil'ev. 2020 (osteology).

Cobitis dorademiri; Lake Köyceğiz basin, Türkiye; male, 87 mm SL.

Cobitis dorademiri

Common name. Köyceğiz spined loach.

Diagnosis. Distinguished from other *Cobitis* species in eastern Aegean basin by: ○ very deep caudal-peduncle, depth 1.0–1.2 times in its length / ○ two laminae circularis in male / ○ back spot at upper caudal base present, often very small / ○ usually, a very regular colour pattern organised in four Gambetta zones / ○ Z4 often dissociated into a wide band or field of small blotches and spots / ○ blotches in Z4 much larger than in Z2 / ○ pigmentation below Z4 present, often restricted to anal base / ○ external part of suborbital spine bifurcate. Size up to 90 mm SL.

Distribution. Türkiye: Lake Köyceğiz basin, lower Dalaman drainage, and some small coastal streams in between.

Habitat. Streams and rivers with sandy or gravelly bottoms.

Biology. No data.

Conservation status. EN; appears to be declining within its very small range.

Remarks. Four different *Cobitis* species in Dalaman drainage make this small river a hotspot in *Cobitis* species diversity. *Cobitis dorademiri*, *C. phrygica,* and *C. indus* have two laminae circularis. These three species seem not to occur in sympatry as *C. dorademiri* is only known from the lower, *C. indus* from the middle, and *C. phrygica* from upper Dalaman drainage, where it occurs adjacent to and potentially sympatric with *C. fahireae*. The distribution of different species in Dalaman should be better explored. It is expected that *C. dorademiri* and *C. indus*, but potentially all species, might occur in sympatry.

Further reading. Erk'akan et al. 2017 (description); Freyhof et al. 2018c (description).

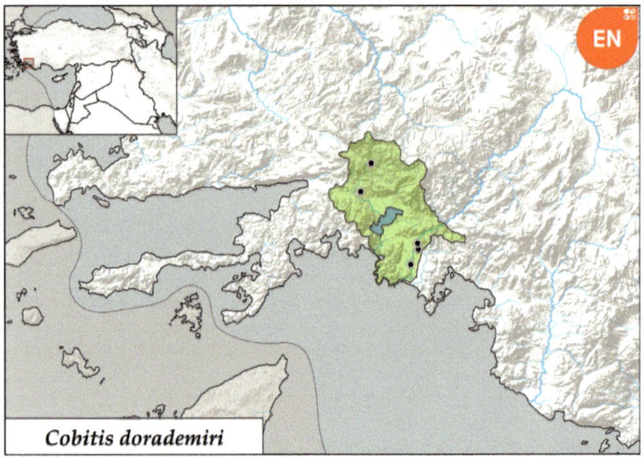

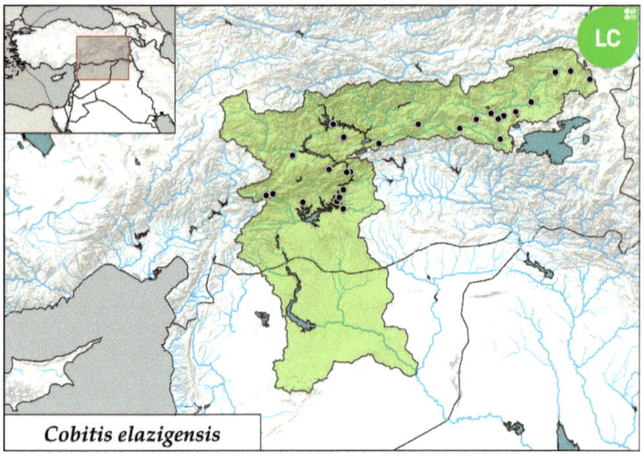

Cobitis elazigensis; upper Euphrates drainage, Türkiye; female, 122 mm SL.

Cobitis elazigensis

Common name. Euphrates spined loach.

Diagnosis. Distinguished from other species of *Cobitis* in Persian Gulf basin by: ○ two laminae circularis in male / ○ Z4 between vertical of pectoral origin and dorsal base squarish, irregularly shaped, roundish, and usually widely separated / ○ pelvic axillary lobe present. Size up to 143 mm SL.

Distribution. Türkiye: Upper Euphrates south to Atatürk reservoir. Expected to be more widespread in Euphrates, and records of *Cobitis* from Syrian Euphrates may belong to this species.

Habitat. Streams, rivers, lakes, and reservoirs, usually on sand or fine gravel in slow or moderate fast-flowing waters.

Biology. No data.

Conservation status. LC.

Further reading. Coad & Sarıeyyüpoğlu 1988 (description); Erk'akan et al. 1998 (discussion); Bohlen et al. 2006 (phylogeny); Freyhof et al. 2018c (description).

Cobitis emrei; Lake Sapanca basin, Türkiye; male, 55 mm SL.

Cobitis emrei

Common name. Sapanca spined loach.

Diagnosis. Distinguished from other species of *Cobitis* in Marmara and Black Sea basins by: ● Z3 fully covered by very small spots forming a sand-like pattern, especially in female / ● black spot at upper caudal base roundish or ovoid, larger than eye diameter in male / ○ blotches in Z2 and Z4 well separated, fused into a bold stripe in some individuals / ○ flank without pigmentation below Z4 / ○ 12–18 roundish, triangular, squarish or horizontally elongate blotches in Z4, blotches vertically elongate on posterior part of flank / ○ one lamina circularis in male / ○ focal zone of subdorsal scale small, about ⅓–⅒ of vertical scale diameter. Size up to 77 mm SL.

Distribution. Türkiye: Lake Sapanca basin. *Cobitis* in lower Sakarya drainage are expected to belong to this species.

Habitat. Lakeshores with fine sediments such as sand and silt, and inflowing streams.

Biology. No data.

Conservation status. VU; appears to be declining within its small range.

Further reading. Erk'akan et al. 1998 (distribution as *C. vardarensis*); Freyhof et al. 2018c (description).

Cobitis erkakanae; River connecting Lakes Gölbaşı and Azaplı, Türkiye; female, ~65 mm SL.

Cobitis erkakanae

Common name. Gölbaşı spined loach.

Diagnosis. Distinguished from other species of *Cobitis* in Mediterranean basin east of Eşen and endorheic basins in Central Anatolia by: ○ flank pattern well organised in four Gambetta zones / ○ a series of large, horizontally elongate blotches in Z4, adjacent blotches often fused / ○ blotches in Z4 not vertically or horizontally dissociated / ○ external part of suborbital spine bifurcate / ○ black spot present at upper caudal base / ○ two laminae circularis in male / ○ mid-dorsal pigmentation consisting in a series of 9–12, elongate blotches, often very closely set or slightly fused. Size up to 65 mm SL.

Distribution. Türkiye: Lakes Gölbaşı (Adıyaman) basin in Ceyhan drainage.

Habitat. Streams and lakes, usually on sand or fine gravel.

Biology. No data.

Conservation status. EN; appears to be declining within its very small range.

Remarks. Together with *C. aliyeae* and *C. evreni*, *C. erkakanae* is the third *Cobitis* species found in Ceyhan drainage. All three species occur in allopatry within drainage, but we are unaware of any biogeographical background explaining why the Gölbaşı lakes have their own fish fauna isolated from the Ceyhan. Not only is *C. erkakanae* endemic to the Gölbaşı lakes, but molecular data indicates that the populations of *Paraphanius* (*P. boulengeri*) and *Pseudophoxinus zekayi* are isolated from those in Ceyhan. While the water from the Gölbaşı lakes flows to the Ceyhan today, we postulate that this was an endorheic basin connected only recently to the Ceyhan.

Further reading. Freyhof et al. 2018c (description).

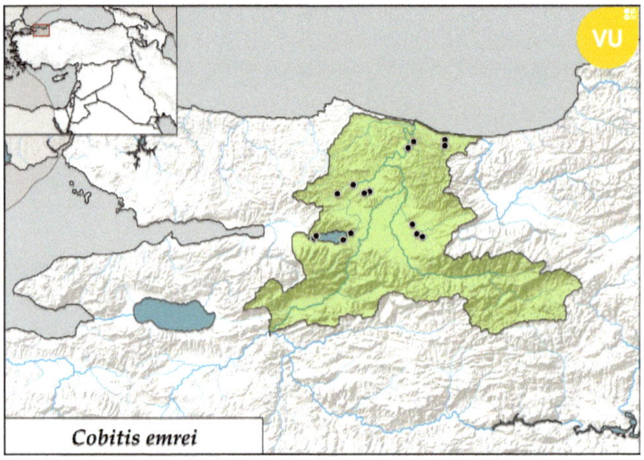

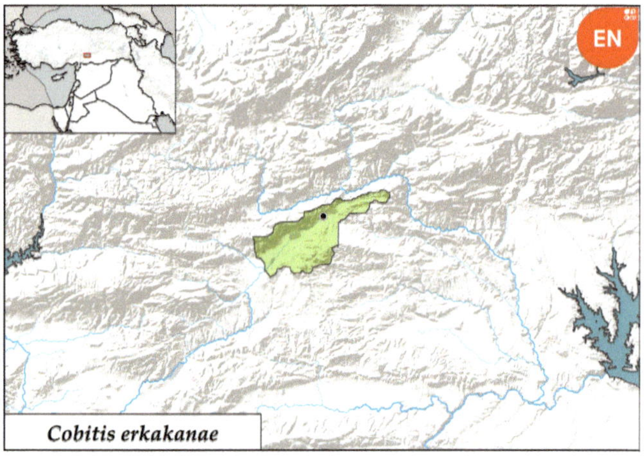

Cobitis evreni; Ceyhan drainage, Türkiye; upper: 78 mm SL; lower: 75 mm SL.

Cobitis evreni

Common name. Ceyhan spined loach.

Diagnosis. Distinguished from other species of *Cobitis* in Mediterranean basin east of Eşen and endorheic basins in Central Anatolia by: ● a bold-brown stripe in Z4 between head and vertical of anal origin / ○ flank below Z4 unpigmented / ○ external part of suborbital spine bifurcate / ○ no black spot at upper caudal base / ○ two laminae circularis in male. Size up to 80 mm SL.

Distribution. Türkiye: Göksun and Savrun in middle Ceyhan drainage.

Habitat. Running waters with gravel or sand bottoms, also on lakeshores.

Biology. No data.

Conservation status. EN; appears to be declining within its very small range.

Remarks. *Cobitis evreni*, *C. aliyeae*, and *C. erkakanae* are the three *Cobitis* species found in Ceyhan drainage. All three species occur in allopatry within Ceyhan drainage, *C. evreni* inhabiting hillstream habitats in middle and upper Ceyhan drainage.

Further reading. Erk'akan et al. 2008a (description); Freyhof et al. 2018c (description).

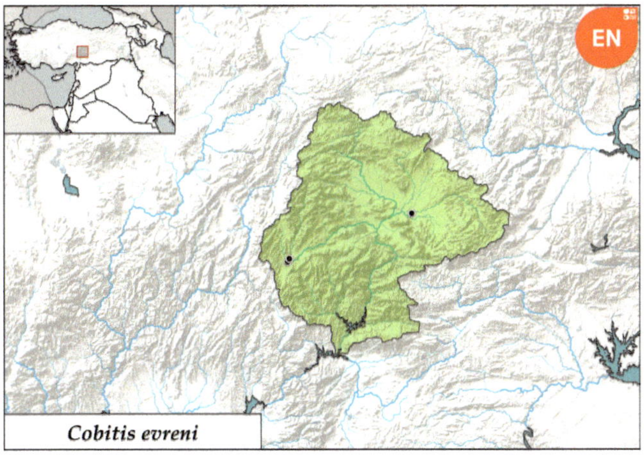

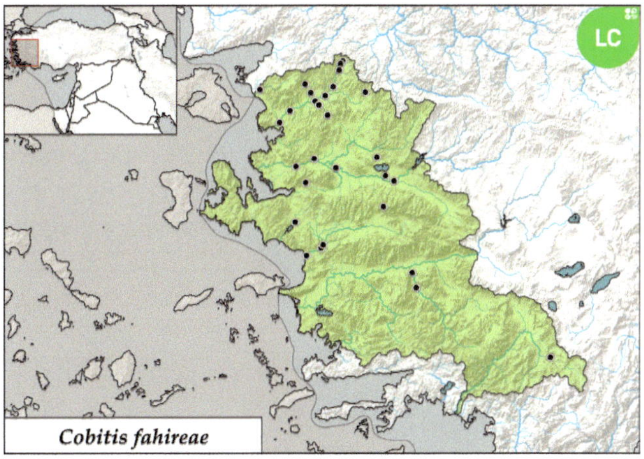

Cobitis fahireae; Büyük Menderes drainage, Türkiye; 51 mm SL.

Cobitis fahireae; Bakırçay drainage, Türkiye; male, 46 mm SL.

Cobitis fahireae

Common name. Aegean spined loach.

Diagnosis. Distinguished from other species of *Cobitis* in eastern Aegean basin by: • 12–18 roundish, triangular, squarish, or horizontally elongate blotches in Z4, often densely set and fused into short stripes along anterior and middle part of flank / • blotches vertically elongate on posterior part of flank / ○ black spot at upper caudal base absent or comma-shaped, smaller than eye diameter / ○ Z3 with a row or a narrow band of small spots not forming a sand-like pattern / ○ focal zone of subdorsal scale small / ○ blotches in Z2 and Z4 well separated, fused into a bold stripe in some individuals / ○ flank without pigmentation below zone Z4 / ○ one lamina circularis in male. Size up to 89 mm SL.

Distribution. Türkiye: Upper Dalaman drainage (around Gölhisar) north to Madra drainage in Aegean basin.

Habitat. Streams, rivers, lakes, and reservoirs, usually on sand or fine gravel in slow or moderate fast-flowing waters.
Biology. No data.
Conservation status. LC.
Remarks. *Cobitis kurui*, described from the Küçük Menderes, and *C. damlae* from the upper Dalaman are synonyms. *Cobitis damlae* was described for an albinotic individual miss-perceived as a cave fish. Molecular data indicate a moderate diversity of different populations identified as *C. fahireae* within its large range, and we cannot fully exclude that more than one species is involved. Initial examination of several populations did not reveal any characters allowing to distinguish different populations as distinct species.
Further reading. Erk'akan et al. 1998 (description); Erk'akan & Özdemir 2014 (*Cobitis damlae*); Freyhof et al. 2018c (diagnosis, synonyms).

Cobitis faridpaki; Tajan, Iran; male, 50 mm SL.

Cobitis faridpaki
Common name. Siahrud spined loach.
Diagnosis. Distinguished from other species of *Cobitis* in Caspian and Lake Namak basins by: ○ one lamina circularis in male / ○ base of lamina circularis with a narrow connection to pectoral ray / ○ cheek between eye and opercle poorly pigmented or unpigmented / ○ plate of lamina circularis reaching to or beyond end of third segment of attached pectoral ray / ○ dark-brown spot at upper caudal base well developed / ○ subdorsal scales with a small focal zone, about or less than 1/10 of maximum scale diameter. Size up to 90 mm SL.
Distribution. Iran: Eastern Caspian basin from Khey east to Gorgan, Karaj in Lake Namak basin, and Taar and Havir in upper Nam drainage in Western Kavir basin.
Habitat. Running and stagnant waters with sandy bottoms, often with dense submerged vegetation.
Biology. No data.
Conservation status. LC.
Remarks. *Cobitis keyvani* from eastern Caspian basin is a synonym. Most probably Western Kavir population was translocated to upper Nam drainage.
Further reading. Mousavi-Sabet et al. 2011 (description); Mousavi-Sabet et al. 2012 (description of *C. keyvani*); Eagderi et al. 2017a (Lake Namak); Jouladeh-Roudbar et al. 2017a (*C. keyvani* as a synonym of *C. faridpaki*).

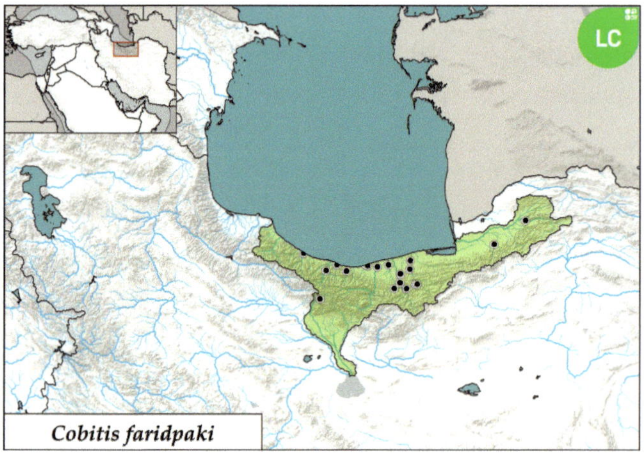

Cobitis indus; Dalaman, Türkiye; female (upper) and male (lower), ~50 mm SL. © S. Aksu.

Cobitis indus

Common name. Dalaman spined loach.

Diagnosis. Distinguished from other *Cobitis* species in eastern Aegean basin by: ○ caudal–peduncle depth 1.2–1.7 times in caudal–peduncle length / ○ two laminae circularis in male / ○ back spot at upper caudal base present, often very small / ○ usually a very regular colour pattern organised in four Gambetta zones / ○ Z4 rarely dissociated into a wide band or field of small blotches and spots / ○ blotches in Z4 slightly larger than in Z2 / ○ pigmentation below Z4 absent or restricted to a short line or few dots on caudal–peduncle / ○ external part of suborbital spine bifurcate. Size up to 69 mm SL.

Distribution. Türkiye: Middle Dalaman drainage.

Habitat. Rivers, usually on sand or fine gravel in slow or moderate fast-flowing waters.

Biology. No data.

Conservation status. DD; due to the very poor knowledge of this species.

Remarks. See remarks under *C. dorademiri.*

Further reading. Eagderi et al. 2022 (description).

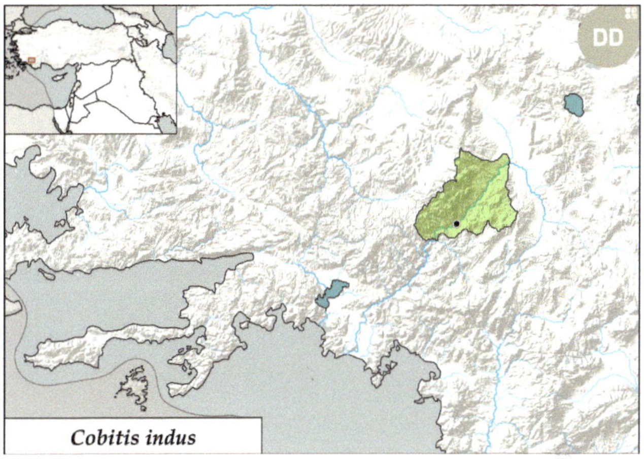

Cobitis joergbohleni; spring Soysallı, Türkiye; female, 70 mm SL.

Cobitis joergbohleni

Common name. Sultan spined loach.

Diagnosis. Distinguished from other species of *Cobitis* in Mediterranean basin east of Eşen and endorheic basins in Central Anatolia by: ○ flank pattern not organised in Gambetta zones, completely disorganised into a pattern of many small spots and blotches / ○ two laminae circularis in male / ○ black spot present at upper caudal base, often very small and best seen in life individuals / ○ 5–8 dark-brown, large and roundish blotches on back anterior to dorsal origin / ○ mental lobe short, not reaching beyond lower lip; caudal slightly rounded / ○ prepelvic length 58–59 % SL / ○ preanal distance 81–82 % SL in female. Size up to 70 mm SL.

Distribution. Türkiye: Sultan marshes in Develi depression.

Habitat. Springs, streams, and canals, usually on sand or fine gravel in slow or moderate fast-flowing waters.

Biology. No data.

Conservation status. CR; see comments under *Pseudophoxinus elizavetae.*

Remarks. *Cobitis fusunae* is a synonym.

Further reading. Freyhof et al. 2018c (description); Özdemir 2019 (*C. fusunae*).

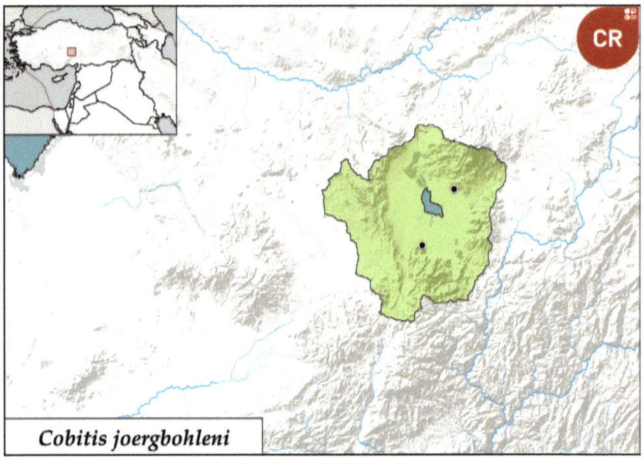

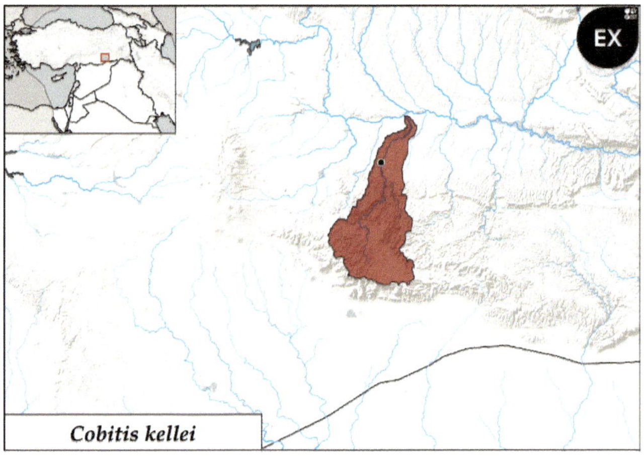

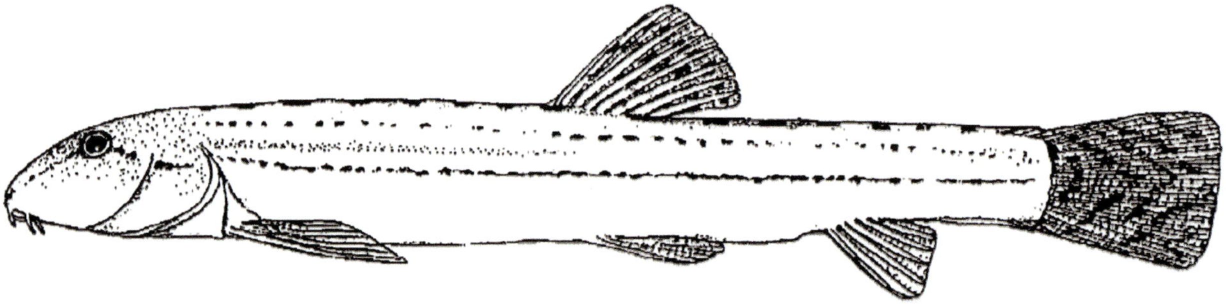

Cobitis kellei; Tigris drainage; Türkiye; male, 46 mm SL. © T.N. Nalbant.

Cobitis kellei

Common name. Diyarbakır spined loach.

Diagnosis. Distinguished from other species of *Cobitis* in Persian Gulf basin by: ○ one lamina circularis in male / ○ Z4 consisting of indistinct minute, dark-grey, roundish dots. Size up to 58 mm SL, likely to grow larger.

Distribution. Türkiye: Göksu in upper Tigris drainage.

Habitat. Described from a small lowland stream.

Biology. No data.

Conservation status. Extinct; only known occurrence of this species was in lower Göksu. It has not been found again in the region despite intensive fieldwork. Habitat conditions in Göksu had been good and it remains unclear why it might have disappeared. In recent years, however, Göksu has almost completely dried out.

Further reading. Erk'akan et al. 1998 (description); Freyhof et al. 2018c (identification, extinction).

Cobitis levantina; Litani drainage; Lebanon; female, 65 mm SL. © M. Bariche.

Cobitis levantina

Common name. Orontes spined loach.

Diagnosis. Distinguished from other species of *Cobitis* in Mediterranean basin east of Eşen by: ○ flank pattern well organised in four Gambetta zones / ○ Z4 formed by one series of distinct blotches, partly or completely fused into a stripe in some individuals / ○ black spot present at upper caudal base / ○ blotches in Z2 irregularly shaped, often ½ of size or as large as in Z3 / ○ often blotches in Z2 and Z3 about size of eye or pupil diameter / ○ Z2 and Z3 usually forming one marbled pattern / ○ many small, dark-brown spots or blotches on lower part of caudal–peduncle, reaching forward to pectoral base in some individuals / ○ interorbital distance 7–11 % HL in male / ○ external part of suborbital spine bifurcate / ○ mid-dorsal pigmentation consisting in a series of 7–12 dark-brown, roundish blotches / ○ two laminae circularis in male. Size up to 82 mm SL.

Distribution. Upper Orontes drainage in Syria and Lebanon, and Litani drainage in Lebanon.

Habitat. Running and standing waters with sandy or muddy bottoms, often with dense submerged vegetation.

Biology. No data.

Conservation status. EN; appears to be declining within its very small range.

Remarks. *Cobitis levantina* is found in southern, upper Orontes, but closely related *C. anabelae* inhabits the northern, lower part of Orontes drainage. The exact distribution of both species is unknown, and it cannot be excluded that they might form a hybrid zone at the border of their ranges.

Further reading. Krupp & Moubayed 1992 (description); Freyhof et al. 2018c (diagnosis, distribution).

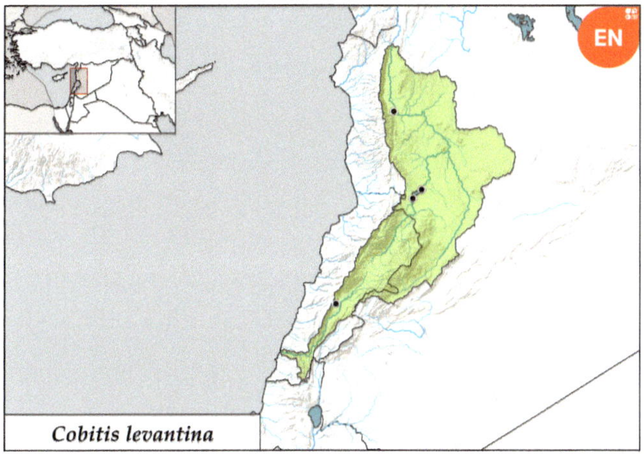

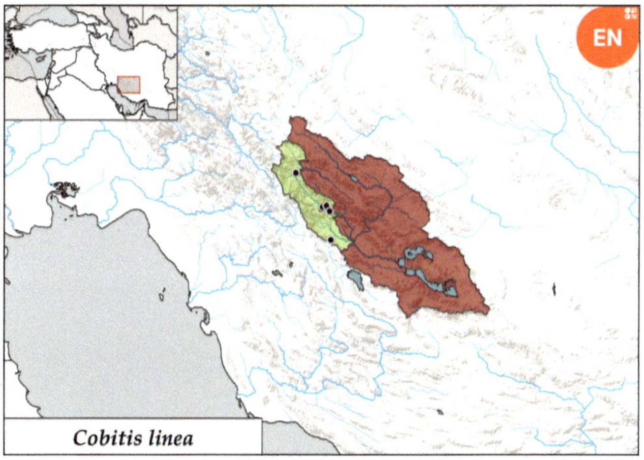

Cobitis linea; Kor, Iran; female, 89 mm SL.

Cobitis linea

Common name. Shiraz spined loach.

Diagnosis. Distinguished from species of *Cobitis* in Persian Gulf basin by: ○ two laminae circularis in male / ○ blotches in Z4 between vertical of pectoral origin until caudal-peduncle comma-shaped, densely set, often fused into short stripes or dissociated into a wide band of several rows of blotches / ○ axillary pelvic lobe absent. Size up to 100 mm SL.

Distribution. Iran: Kor drainage.

Habitat. Slow-flowing and stagnant waters with muddy or sandy bottoms.

Biology. Spawns June–July.

Conservation status. EN; appears to be declining within its very small range.

Remarks. Occasionally recorded from upper Kol, but this must be incorrect. Superficially similar loaches also occur in Iranian Karun drainage, and it cannot be excluded that these might be identified as *C. linea* in the future.

Further reading. Bianco & Nalbant 1980 (redescription); Esmaeili et al. 2008 (conservation status); Freyhof et al. 2018c (diagnosis, distribution).

Spring Ghadamgah in Kor drainage, Iran, is one of the few habitats of *Cobitis linea*.

Cobitis phrygica; spring east of Gemiş, Türkiye; female, 82 mm SL.

Cobitis phrygica; spring Kırkpınar, Türkiye; female, 79 mm SL.

Cobitis phrygica

Common name. Phrygian spined loach.

Diagnosis. Distinguished from other species of *Cobitis* in Central Anatolia and southern Aegean basin by: ○ a series of midlateral blotches in Z4 or Gambetta zones dissociated into a pattern of many irregularly shaped small or medium-sized spots and blotches / ○ flank below Z4, usually pigmented / ○ external part of suborbital spine simple / ○ no black spot at upper caudal base / ○ two laminae circularis in male. Size up to 120 mm SL.

Distribution. Türkiye: Salda basin, upper Eşen, and Dalaman drainages. In Aksu, north of Bucak, and in one spring area east of Gemiş in Acıgöl basin.

Habitat. Flowing and standing waters of springs and streams with sandy or muddy bottoms, often with dense submerged vegetation.

Biology. No data.

Conservation status. EN; largest populations in Dalaman and Eşen. Extirpated from Acıgöl basin. Other populations are confined to small areas.

Remarks. Often treated as a synonym of *C. simplicispina*, but both species are not closely related. The colour pattern in *C. phrygica* is very variable within and between populations, making it very difficult to identify. Some individuals have the flank pattern well organised in Gambetta zones; in others, the flank is covered by small spots. Also, all kinds of intermediate colour patterns are found, ranging from individuals with well-organised Gambetta zones to others with Gambetta zones completely dissociated.

Further reading. Battalgazi 1944 (description); Erk'akan et al. 1999 (diagnosis); Bohlen et al. 2006 (molecular phylogeny); Freyhof et al. 2018c (diagnosis, distribution).

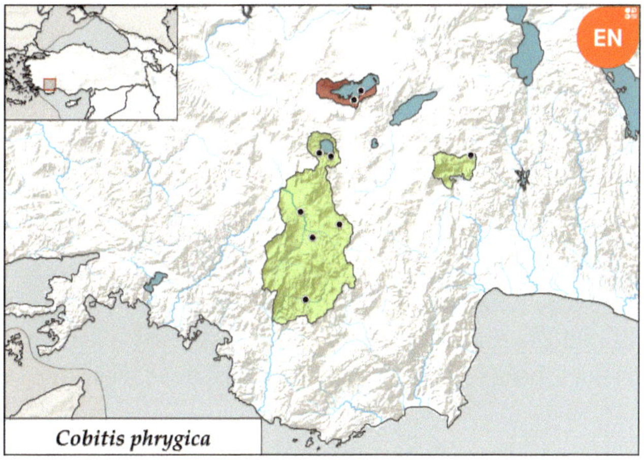

Cobitis pirii; Lake Eğirdir basin, Türkiye; female, 74 mm SL.

Cobitis pirii

Common name. Pisidian spined loach.

Diagnosis. Distinguished from other species of *Cobitis* in Mediterranean basin east of Eşen and endorheic basins in Central Anatolia by: ○ flank pattern well organised in four Gambetta zones, blotches in Z4 organised in a wide band or two rows of spots, and blotches in some individuals or populations / ○ external part of suborbital spine simple / ○ black spot present at upper caudal base / ○ back with 13–20 dark-brown, roundish blotches, often irregularly shaped or fused to each other / ○ two laminae circularis in male. Size up to 97 mm SL.

Distribution. Türkiye: Lake Eğirdir basin, upper Aksu and Köprüçay drainages.

Habitat. Streams, rivers, lakes, and reservoirs, usually on sand or fine gravel in slow or moderate fast-flowing waters.

Biology. No data.

Conservation status. VU; appears to be declining within its small range.

Further reading. Freyhof et al. 2018c (description).

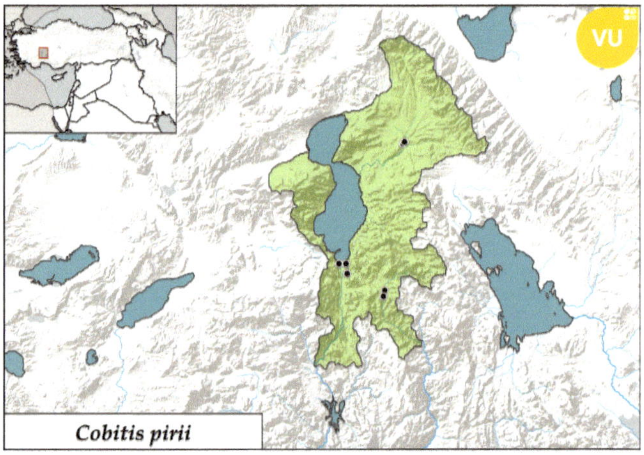

Cobitis puncticulata; Lake Ulubat basin, Türkiye; male, 64 mm SL.

Cobitis puncticulata

Common name. Brown spined loach.

Diagnosis. Distinguished from other species of *Cobitis* in Marmara basin by: ● body brownish / ● blotches of zone Z2 and Z4 very small, indistinct, or confluent / ○ flank pigmented below zone Z4 / ○ one lamina circularis in male. Size up to 78 mm SL.

Distribution. Türkiye: Lakes Manyas and Ulubat basins; Greece and Türkiye: lower Maritza drainage.

Habitat. Still to moderately flowing streams with muddy bottoms.

Biology. No data.

Conservation status. EN; appears to be declining within its very small range.

Further reading. Erk'akan et al. 1998 (description); Freyhof et al. 2008 (record from Greece); Freyhof et al. 2018c (description, distribution).

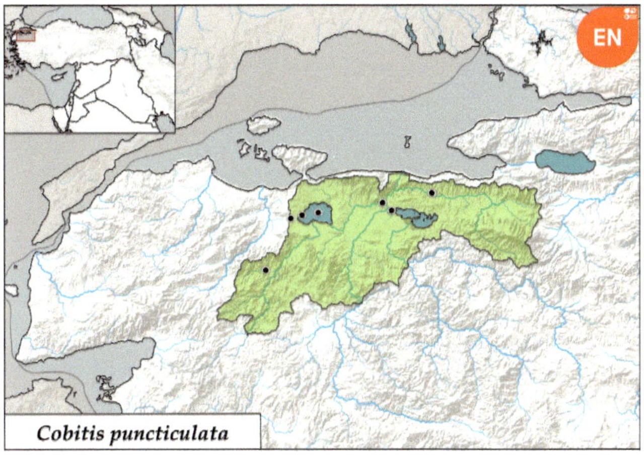

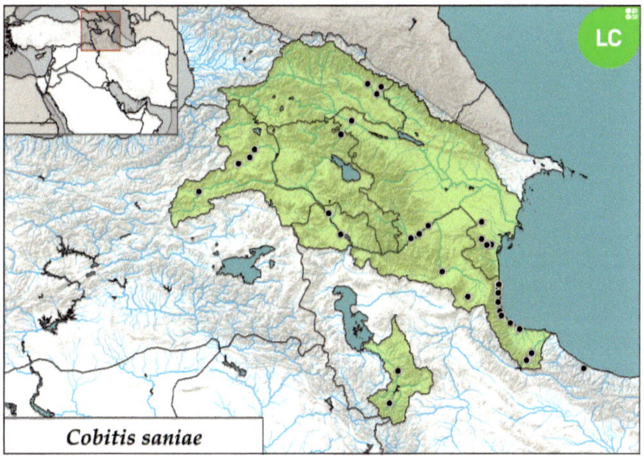

Cobitis saniae; Sefid drainage, Iran; female, 85 mm SL.

Cobitis saniae; Lenkoran, Azerbaijan; female, 80 mm SL.

Cobitis saniae

Common name. Western Caspian spined loach.

Diagnosis. Distinguished from other species of *Cobitis* in Caspian and Black Sea basin by: ○ subdorsal scales with a small focal zone, about or less than ⅒ of maximum scale diameter / ○ blotches in Z4 horizontally elongate, often very densely set and fused into short or longer stripes anterior to dorsal origin / ○ dark-brown spot at upper caudal base well developed / ○ one lamina circularis in male / ○ plate of lamina circularis reaching to or beyond end of third segment of attached pectoral ray / ○ base of lamina circularis widely connected to pectoral ray / ○ cheek between eye and opercle strongly pigmented. Size up to 85 mm SL.

Cobitis saniae; Lenkoran, Azerbaijan; male, 55 mm SL.

Distribution. Sefid (Iran) north-west to Kura and Aras in Azerbaijan, Armenia, and Türkiye and Zarrineh drainage in Lake Urmia basin.
Habitat. Streams, rivers, lakes, and reservoirs, usually on sand or fine gravel in slow or moderate fast-flowing waters.
Biology. No data.
Conservation status. LC.
Remarks. A record from middle Rioni in Georgian Black Sea basin may be a lapsus.
Further reading. Eagderi et al. 2017b (description); Vasil'eva et al. 2020 (distribution in Caucasus).

Cobitis satunini; Kintrishi, Georgia; female, ~90 mm SL. © M. Geiger.

Cobitis satunini; Tuzla drainage, Türkiye; male, ~50 mm SL.

Cobitis satunini

Common name. Colchic spined loach.

Diagnosis. Distinguished from other species of *Cobitis* in Black Sea basin by: ○ pigmentation in Z3 usually complete, reaching much behind dorsal base / ○ pigmentation in Z3 usually wider than pigmentation in Z2 / ○ blotches in Z4 irregularly roundish or squarish anterior to dorsal origin / ○ focal zone of subdorsal scale small, about ⅓–⅒ of vertical scale diameter / ○ one lamina circularis in male / ○ base of lamina circularis with a narrow connection to pectoral ray. Size up to 90 mm SL.

Distribution. From Kintrich in Georgia [entering Black Sea at 41.80388, 41.76916] clockwise to İyidere, Filyos and Bartın drainages in southwestern Black Sea basin and from Tuzla (northernmost Aegean) east to Lake İznik basin, including Sursuluk drainage.

Habitat. Still to moderately flowing clear water streams with mud or sand bottoms.

Biology. No data.

Conservation status. LC.

Remarks. *Cobitis satunini* is closely related to European *C. taenia*. *Cobitis taenia* is distinguished by having 48 chromosomes (vs. 50). Records of *Cobitis* along the southern Black Sea coast are very sparse, and we are not aware of findings of *Cobitis* (except *C. simplicispina*) between Trabzon in east and Bartın in west. *Cobitis* from Western Black Sea basin, Marmara and northern Aegean had been identified as *C. taenia*. But these fish have 50 chromosomes and are also close to *C. satunini* by morphological and molecular characters. *Cobitis osurgeticus* has recently been treated as a valid species for most of the populations identified here as *C. satunini*. It is distinguished from *C. satunini* by very small differences in chromosomes and morphometric characters (both populations are probably conspecific). *Cobitis osurgeticus* is a *numen nudum*, as its description is not valid, despite later indications, and the name is not available.

Further reading. Gladkov 1935 (description); Freyhof et al. 2018c (description, distribution); Vasil'eva et al. 2024 (*C. osurgeticus*)

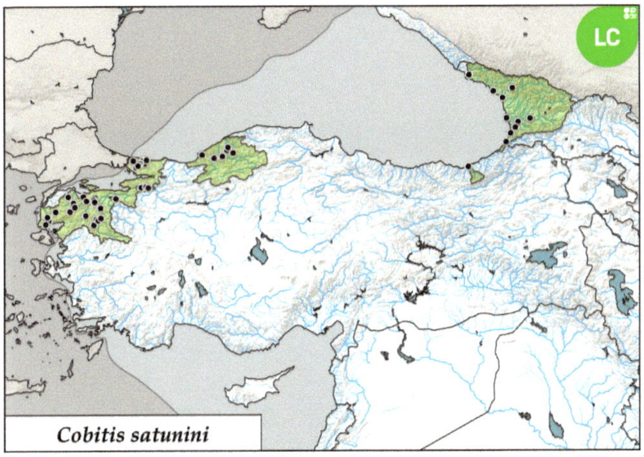

Cobitis satunini

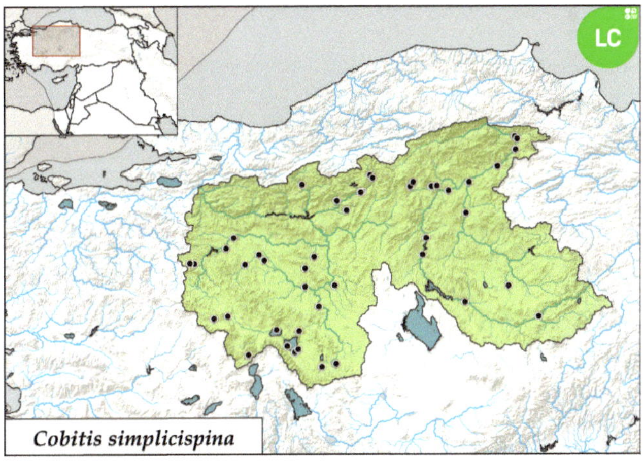

Cobitis simplicispina

Cobitis simplicispina; Sakarya drainage, Türkiye; male, 73 mm SL.

Cobitis simplicispina; Sakarya drainage, Türkiye; female, ~100 mm SL.

Cobitis simplicispina; Lake Akşehir drainage, Türkiye; male, 65 mm SL.

Cobitis simplicispina; Lake Akşehir drainage, Türkiye; female, 129 mm SL.

Cobitis simplicispina

Common name. Galatian spined loach.

Diagnosis. Distinguished from other species of *Cobitis* in Black Sea basin and endorheic Anatolian basins of Lakes Akşehir, Eber, and Ilgın by: • two laminae circularis in male. Size up to 130 mm SL.

Distribution. Türkiye: Sakarya and Kızılırmak drainages and endorheic basins of Lakes Eber, Akşehir and Ilgın.

Habitat. Streams, rivers, lakes, and reservoirs, usually on sand or fine gravel in slow or moderate fast-flowing waters.

Biology. No data.

Conservation status. LC.

Remarks. There is a doubtful record of *C. simplicispina* from Lake İznik, which has not been validated yet. Only *C. satunini* was so far found in Lake İznik. *Cobitis simplicispina* is the only species of *Cobitis* found in Black Sea basin having two laminae circularis, and it is immediately distinguished by its large size and the blotches in Z4 organised in one series, often fused into short or longer stripes (vs. blotches in Z4 well separated in other species). Molecular data place

C. simplicispina in a group of species from Central Anatolia and the Mediterranean basin, indicating a past biogeographic connection between these regions.

Further reading. Erk'akan et al. 1999 (diagnosis, synonyms); Bohlen et al. 2006 (molecular phylogeny); Freyhof et al. 2018c (description, distribution).

Cobitis sipahilerae; stream at Döşemealtı, Türkiye; female, 69 mm SL.

Cobitis sipahilerae

Common name. Kırkgöz spined loach.

Diagnosis. Distinguished from other species of *Cobitis* in Mediterranean basin east of Eşen and endorheic basins in Central Anatolia by: ● back plain cream-brown, few individuals with 1–3 narrow, squarish, brown bars / ○ two laminae circularis in male / ○ black spot present at upper caudal base, often very small and best seen in life individuals. Size up to 90 mm SL.

Distribution. Türkiye: Düden and Kırkgöz springs and outflowing stream Yediarıklar.

Habitat. Springs and spring-fed streams, usually on sand or fine gravel in slow or moderate fast-flowing waters.

Biology. No data.

Conservation status. EN; appears to be declining within its very small range.

Remarks. *Cobitis sipahilerae* is one of four freshwater fish species native to the Kırkgöz-Yediarıklar drainage and the adjacent Düden spring in Antalya. Others are *Pseudophoxinus antalyae*, *Paraphanius mentoides*, and *Seminemacheilus attalicus*. All four species are endemic to this small drainage system. The occurrence of four endemic species (the native fish fauna is comprised exclusively of strict endemic species) makes the Düden-Kırkgöz-Yediarıklar drainage a site of major importance for conservation.

Further reading. Erk'akan et al. 2017 (description); Freyhof et al. 2018c (description).

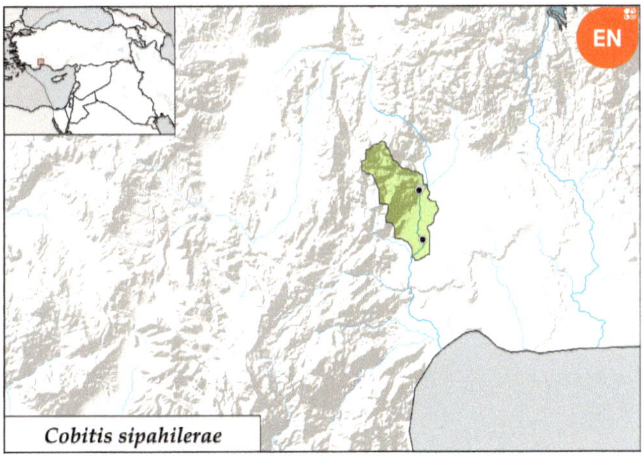

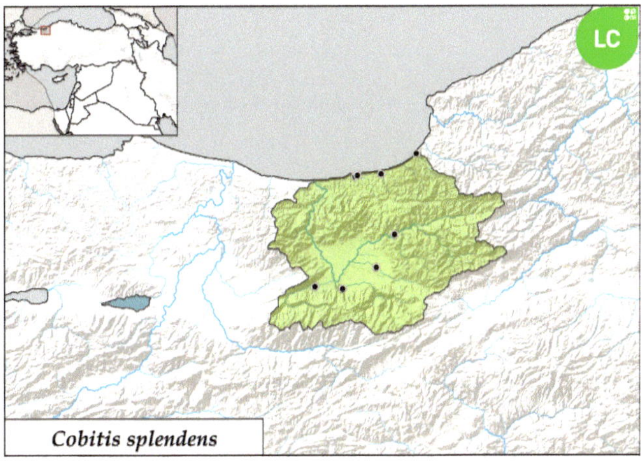

Cobitis splendens; stream Kovukkavla, Türkiye; male, ~49 mm SL.

Cobitis splendens

Common name. Bithynian spined loach.

Diagnosis. Distinguished from other species of *Cobitis* in Black Sea basin and endorheic Anatolian basins of Lakes Akşehir, Eber and Ilgın by: ○ pigmentation in Z3 reduced to about head length in male or pigmentation in Z3 as wide as or narrower than pigmentation in Z2 / ○ blotches in Z4 irregularly roundish or squarish anterior to dorsal origin / ○ focal zone of subdorsal scale small, about $\frac{1}{3}$–$\frac{1}{10}$ of vertical scale diameter / ○ one lamina circularis in male / ○ base of lamina circularis with a narrow connection to pectoral ray. Size up to 86 mm SL.

Distribution. Türkiye: Small coastal streams around Akçakoca and Büyük Melen (Black Sea).

Habitat. Lower and middle reaches of still to moderately fast-flowing, clear streams with mud or sand bottoms.

Biology. No data.

Conservation status. LC; Population may be stable, but more data is needed.

Remarks. Doubtful records of this species from Aegean and southern Marmara basins could not be confirmed and seem to be based on misidentifications of *C. satunini* and *C. fahireae*. A reduced pigmentation in Z3 in male is very characteristic to *C. splendens*. Still, this character state is not unique to *C. splendens*, and individual males with very much reduced or even no pigmentation in Z3 are found in many other *Cobitis* species. *Cobitis* from the Büyük Melen [entering Black Sea at 41.073, 30.967] are identified as *C. splendens* based on molecular characters. Still, in this population, the pigmentation in Z3 is not reduced, distinguishing it from *C. splendens*. The *Cobitis* of Anatolian Black Sea basin needs an in-depth revision.

Further reading. Erk'akan et al. 1998 (description); Erk'akan et al. 1999 (distribution); Freyhof et al. 2018c (description, distribution, Büyük Melen).

Cobitis strumicae; Provadiskaya, Bulgaria; ~70 mm SL.

Cobitis strumicae; stream Koca, Türkiye; female, 85 mm SL.

Cobitis strumicae

Common name. Struma spined loach.

Diagnosis. Distinguished from other species of *Cobitis* in Marmara basin by: ● two laminae circularis in male / ○ usually two small spots at caudal base, often very small or absent / ○ zone Z4 with 12–21 roundish or squarish blotches, sometimes indistinct or fused two by two / ○ flank not pigmented below zone Z4. Size up to 110 mm SL.

Distribution. Türkiye: Kocaçay in Çanakkale province and a doubtful record from Terme (Samsun, Black Sea basin). European Aegean and Black Sea basins: from Struma drainage to Varna area (Greece, Bulgaria, Türkiye), Lakes Volvi and Koronia (Greece); Danube basin in Bulgaria. Expected to be more widespread.

Habitat. Still to flowing waters with sandy or muddy bottoms.

Biology. No data.

Conservation status. LC.

Further reading. Economidis & Nalbant 1997 (description); Freyhof et al. 2018c (record from Anatolia); Özpiçak et al. 2023 (record from Samsun).

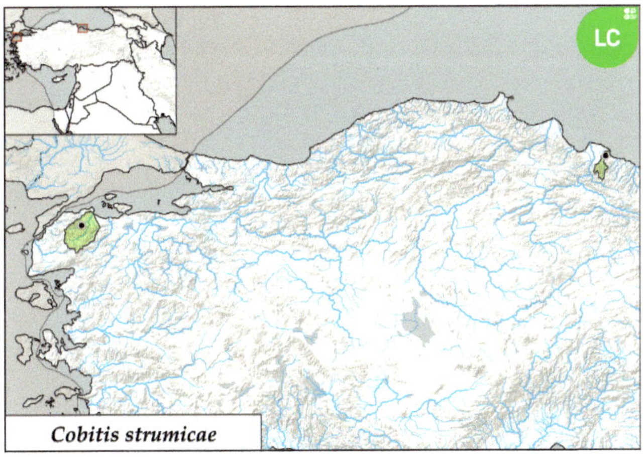

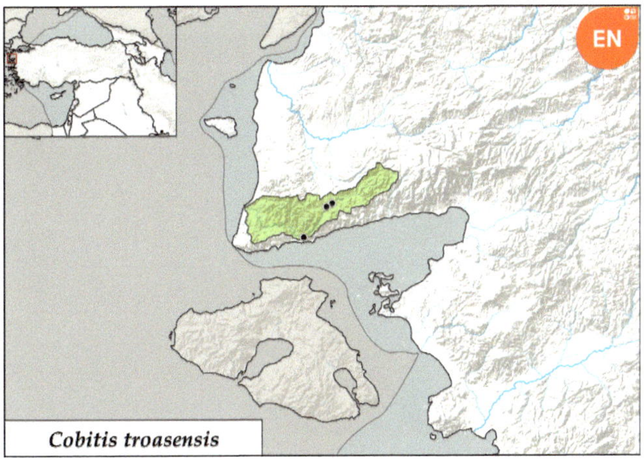

Cobitis troasensis; stream Tuzla, Türkiye; male, 60 mm SL.

Cobitis troasensis

Common name. Dardanelles spined loach.

Diagnosis. Distinguished from other species of *Cobitis* in eastern Aegean basin by: ● 25–36, small, comma-shaped blotches in Z4 along anterior and middle part of flank / ○ focal zone of subdorsal scale small, about ⅓–1⁄10 of vertical scale diameter / ○ flank without pigmentation below zone Z4 / ○ body pale-yellowish or whitish / ○ blotches squarish or irregular shaped on posterior part of flank, rarely fused into short stripes / ○ one lamina circularis in male. Size up to 76 mm SL.

Distribution. Türkiye: Tuzla drainage.

Habitat. Streams and rivers, usually on sand or fine gravel in slow or moderate fast-flowing waters.

Biology. No data.

Conservation status. EN; appears to be declining within its very small range.

Remarks. *Cobitis troasensis* is found in sympatry with *C. satunini*, making it one of very few cases in which two species of *Cobitis* with one lamina circularis occur together. Hybridisation between both should be expected.

Further reading. Freyhof et al. 2018c (description).

Cobitis turcica; Lake Tuz basin, Türkiye; female, 76 mm SL.

Cobitis turcica; Lake Tuz basin, Türkiye; female, ~70 mm SL.

Cobitis turcica

Common name. Central Anatolian spined loach.

Diagnosis. Distinguished from other species of *Cobitis* in Mediterranean basin east of Eşen and endorheic basins in Central Anatolia by: ○ flank pattern not organised in Gambetta zones, completely disorganised into a pattern of many small spots and blotches / ○ back with dark-brown, large, and roundish blotches / ○ black spot present at upper caudal base, often very small and best seen in life individuals / ○ mental lobe long, reaching to or beyond lower lip / ○ caudal truncate / ○ prepelvic length 53–56 % SL / ○ preanal distance 75–78 % SL in female / ○ two laminae circularis in male. Size up to 100 mm SL.

Distribution. Türkiye: Wider Lake Tuz basin: Ereğli, Melendiz (Ihlara Valley), springs in Sultanhanı north of Sarı Yayla and Gölyazı.

Habitat. Springs and streams with still to moderately flowing clear water and muddy or sandy bottoms. Also, lakes and reservoirs.
Biology. No data.
Conservation status. EN; appears to be declining within its very small range. Extirpated from Ereğli.
Remarks. *Cobitis turcica* and *C. battalgilae* are closely related, but almost all individuals can be well distinguished, and therefore they are treated as two species.
Further reading. Hankó 1925 (description); Erk'akan et al. 1998 (description, synonyms); Bohlen et al. 2006 (molecular phylogeny); Freyhof et al. 2018c (description, distribution).

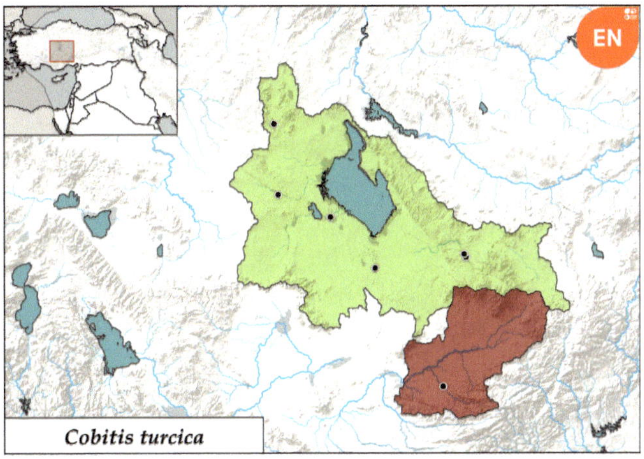

Upper Aras in Türkiye is the habitat of *Sabanejewia aurata* and many other rheophilic species.

Sabanejewia

Golden loaches are widely distributed in the northern Aegean, Black, Baltic, Caspian, and Aral basins. There are approximately 10 species of *Sabanejewia*. *Sabanejewia aralensis* from the Aral Sea basin has been occasionally reported from the Hari drainage, although its presence here has yet to be confirmed. While superficially similar to *Cobitis*, *Sabanejewia* exhibits a markedly different sexual dimorphism. Male *Sabanejewia* reach the same size as females, lack modifications of the pectoral fins, but exhibit vertical swellings of the flanks in males. The biogeography of *Sabanejewia* needs to be better understood. They are related to *Cobitis*, *Misgurnus*, *Microcobitis*, and *Bibarba*, all found in East Asia. Only *Sabanejewia* is absent from East Asia, reaching no further east than the Aral basin. The family Cobitidae probably originated in East and Southeast Asia. Consequently, it is anticipated that *Sabanejewia* may have colonised western Eurasia at an early stage, potentially during the Oligocene, in conjunction with the first Cyprinids. The genus subsequently disappeared from East Asia. **Further reading.** Sayyadzadeh et al. 2018a (phylogeny); Bohlen et al. 2020 (phylogeny).

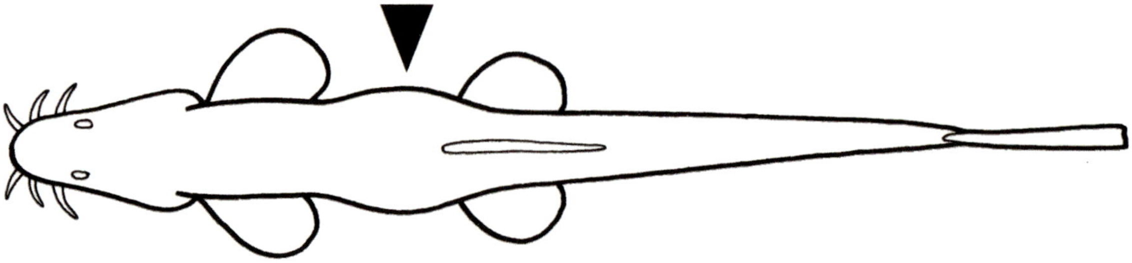

Figure 47: Vertical swelling (arrow) of male *Sabanejewia*.

Sabanejewia aurata; Kura drainage, Türkiye; ~70 mm SL. © M. Özuluğ.

Sabanejewia aurata

Common name. Caspian golden loach.

Diagnosis. Distinguished from *Sabanejewia caspia* by: • 10–15 large, distinct blotches along lateral midline • no irregularly shaped band of brown pigments below lateral blotches. Size up to 100 mm SL.

Distribution. Southern Caspian basin: Kura and Aras drainages east to Sefid.

Habitat. Hill streams with clear water and sandy or gravelly bottoms. In moderate current.

Biology. No data.

Conservation status. LC.

Remarks. Occasionally, reported to occur in all rivers of Iranian Caspian basin, but no records east of Sefid was confirmed in last decades. The identity of Hari and Morghab populations is questionable, and these fishes might belong to *S. aralensis*, a species widespread in Zeravshan, Amu, and Syr Darya in Central Asia.

Further reading. Berg 1949b (description); Sayyadzadeh et al. 2018a (description).

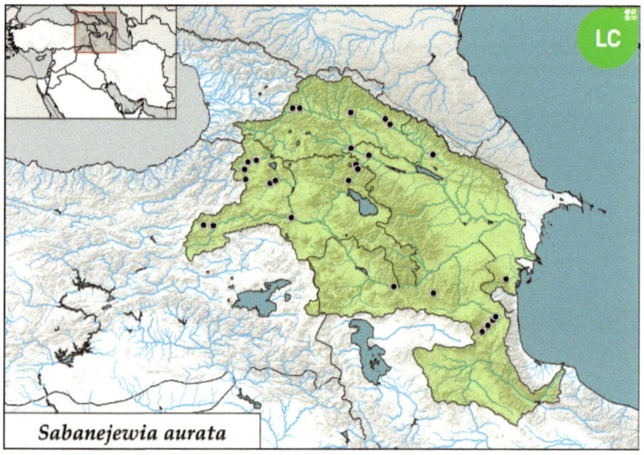

Sabanejewia caspia; Sefid drainage, Iran; ~50 mm SL.

Sabanejewia caspia
Common name. Liman golden loach.
Diagnosis. Distinguished from *Sabanejewia aurata* by: • a narrow, irregularly shaped stripe along lateral midline often overlaid by small blotches or short bars / • an irregularly shaped band of brown pigments below lateral stripe. Size up to 92 mm SL.
Distribution. Caspian basin: Lowermost Kura in Azerbaijan, Kyzylagach Bay south of Kura estuary, brackish lagoons in Lenkoran region, Anzali wetland, and lowermost Sefid in Iran. Caspian Sea at Durnev Islands and estuaries of Ural and Emba in Kazakhstan.
Habitat. Lowermost part of rivers and coastal waters. In still or very slow-flowing, often brackish waters.
Biology. Spawns in April in brackish lagoons of Lenkoran. Feeds on detritus and benthic invertebrates.
Conservation status. EN; appears to be declining within its very small range.
Further reading. Berg 1949b (description); Vasil'eva & Vasil'ev 2012 (distribution); Sayyadzadeh et al. 2018a (description).

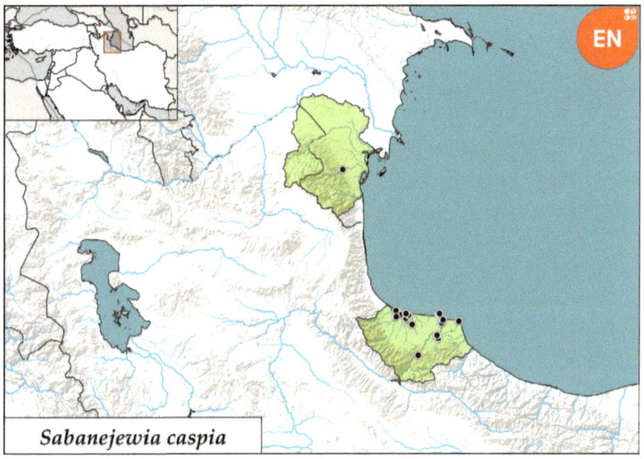

Kırkgöz is an endemicity hotspot in Türkiye. Yediarıklar, the stream flowing out of Kırkgöz, is the habitat of *Cobitis sipahilerae*, *Pseudophoxinus antalyae*, *Paraphanius mentoides*, and *Seminemacheilus attalicus*.

Family Nemacheilidae

Stone loaches

A large family with 49 genera and more than 840 known species (more than 2000 may exist) from the rivers of Asia, Europe, and Ethiopia. Nemacheilid loaches are the second largest family of cypriniform fishes and the least known large group within freshwaters in Asia. Actually, 112 known species of nemacheilid loaches are recognised in West Asia. However, further species will be discovered in the future. Stone loaches are occasionally confused with spined loaches of the family Cobitidae, which also possess three pairs of barbels. In contrast to stone loaches, spined loaches have an erectable suborbital spine that they use for defense. Such a spine is absent in stone loaches; both families have very different general appearances. In many *Oxynoemacheilus* and *Paraschistura* species, males exhibit an exposed lachrymal bone called a suborbital flap or groove. In contrast, other species lack this character. The shape of this structure can be used for identification purposes. There have been speculations that the suborbital flap or groove exposure is a seasonal character. However, this still needs to be demonstrated, and further research is required.

Figure 48. Suborbital groove in front of the eye in a male of *O. angorae.*

The generic structure of nemacheilid loaches from West Asia has been confused for a long time. Species have been placed in *Nemacheilus* (an East Asian genus), *Barbatula* (a European and North Asian genus), *Orthrias* (a synonym of *Barbatula*), *Nun* (a synonym of *Oxynoemacheilus*), *Ilamnemacheilus* (a synonym of *Oxynoemacheilus*), and *Metaschistura* (a synonym of *Paraschistura*). Molecular studies have demonstrated that eight genera can be recognised in West Asia, with *Oxynoemacheilus* being the most speciose and widespread. However, the situation still needs to be fully settled, as *Seminemacheilus* turns out to be nested within *Oxynoemacheilus* in some phylogenetic analyses, while in others, it represents its own genus. Indeed, both genera are similar in many characters and poorly distinguished. *Triplophysa stoliczkai* has been recorded from the Sistan basin in Iran, subsequently identified as *Paraschistura alta.*

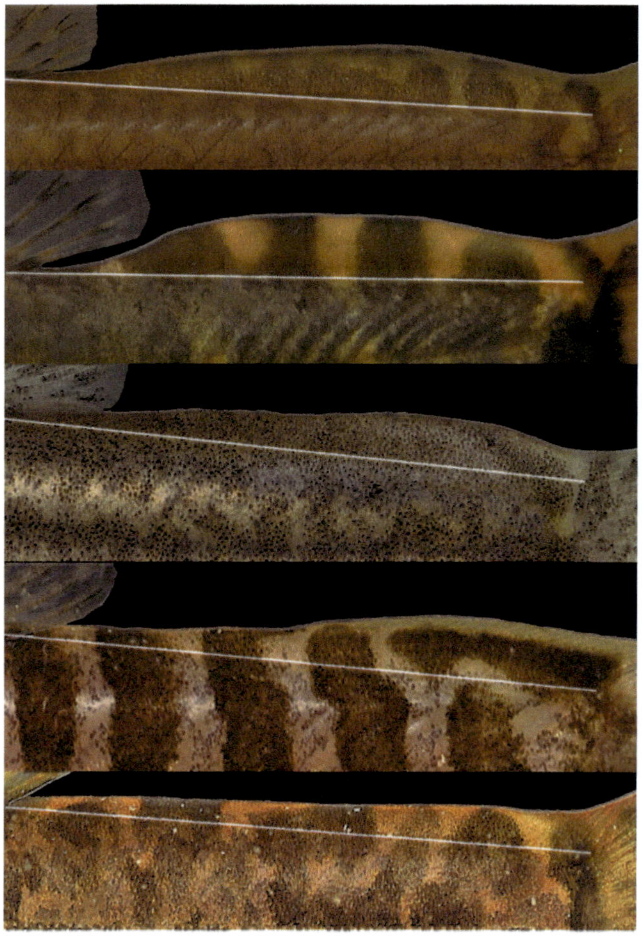

Figure 49. Adipose crests and adipose ridge (lowermost figure) on caudal peduncle in Nemacheilidae. The crest is convex, while the ridge is straight. The white line shows the proximal base of the crest and ridge.

https://doi.org/10.1515/9783111677811-021

Nemacheilid loaches inhabit a wide range of water bodies, typically found in swift-flowing water, where they feed on small invertebrates. Up to five nemacheilid species and up to three species of *Oxynoemacheilus* have been found in syntopy, making identifying species in the field challenging. Syntopic species are usually found in different microhabitats, often in different current velocities. However, the ecological niche separation in different loaches has yet to be studied. All species appear to spawn for the first time at 1 year; most individuals seem to be multiple spawners but spawn only one to two seasons. Biological details are unknown for almost all species. **Further reading.** Tang et al. 2006 (phylogeny); Šlechtová et al. 2007 (phylogeny); Prokofiev 2009 (genera); Freyhof et al. 2011 (diversity); Kottelat 2012 (diversity); Jouladeh-Roudbar et al. 2015c (*Paraschistura alta* record from Iran); Hashemzadeh Segherloo et al. 2016b (West Asian genera).

Key to genera of *Nemacheilidae* in West Asia

1a - Subterranean species, body whitish or pink; eye reduced, externally invisible.
..................*Eidinemacheilus*

1b - Epigean species, body with large brown blotches, bars, a stripe, marbled or mottled colour pattern; eye fully developed.
..................2

2a - Dorsal origin clearly behind vertical of pelvic origin; anus situated more than two eye diameters in front of anal origin.
..................*Turcinoemacheilus*

2b - Dorsal origin in front, above, or slightly behind vertical of pelvic origin; anus situated about one eye diameter or less in front of anal origin.
..................3

3a - Pectoral in male remarkably longer than in female; male with many small nuptial tubercles on head, especially on cheek and on upper pectoral (absent in *Seminemacheilus*).
..................4

3b - Pectoral in male not longer than in female; male without small nuptial tubercles on head or on pectoral.
..................5

4a - Suborbital flap or groove always absent in male; caudal very slightly emarginate, truncate, or slightly rounded; pectoral in male reaching beyond pelvic origin when adpressed; epural bone in caudal skeleton absent.
..................*Seminemacheilus*

4b - Suborbital groove present or absent in male; caudal deeply emarginate or truncate; if suborbital groove absent and caudal slightly emarginate or truncate, then pectoral in male not reaching beyond pelvic origin when adpressed; epural bone in caudal skeleton present.
..................*Oxynoemacheilus*

5a - A prominent dorsal adipose crest on caudal peduncle, not supported by procurrent caudal rays; flank marbled, mottled, or with an irregularly shaped midlateral stripe.
..................*Paracobitis*

5b - No dorsal adipose crest on caudal peduncle, or, if dorsal adipose crest present, supported by procurrent caudal rays and caudal peduncle, usually with regularly shaped bars.
..................6

6a - Colour pattern on flank with regularly- or irregularly set bars, often dissociated into irregularly shaped blotches, especially on flank in front of dorsal origin; most species with prominent black spot at base of unbranched dorsal rays and on first and sometimes second branched rays; a dark-brown or black bar or 1–2 black blotches at caudal base; no whitish or yellowish triangular patch on posteriormost upper and lower caudal peduncle; caudal deeply emarginate or forked.
.................*Paraschistura* + *Schistura*
6b - Colour pattern on flank with a marbled or mottled pattern, usually with a dark-brown or grey irregularly shaped, midlateral stripe, often restricted to caudal peduncle; no prominent black spot at base of unbranched dorsal rays or on first or second branched rays; no dark-brown or black bar or blotch at caudal base; a prominent whitish or yellowish triangular patch on posteriormost upper and lower caudal peduncle; caudal slightly emarginate.
.................*Sasanidus*

Eidinemacheilus

The two species are troglomorphic inhabitants of underground waters in the Zagros Mountains. They are superficially quite different, and molecular data suggest that both are not closely related. The phylogenetic relationships of *Eidinemacheilus* and other genera remain unresolved. *Eidinemacheilus* may be a relict genus in the Zagros, but the diversity of nemacheilid loaches in South Asia needs to be better understood. Therefore, epigean *Eidinemacheilus* species may be found there in the future. As troglomorphic fishes lack colour patterns and face very different evolutionary challenges, potential epigean *Eidinemacheilus* should be identified by molecular characters. **Further reading.** Hashemzadeh Segherloo et al. 2016b (description & phylogeny).

Eidinemacheilus proudlovei

Common name. Tabin blind loach.

Diagnosis. Distinguished from *E. smithi* by: • dorsal adipose crest absent / • 8+8 or 8+7 branched caudal rays / • jaws enlarged / • head canal system fully developed. Size up to 54 mm SL.

Distribution. Iraq: An aquifer near Kanishok in Tabin drainage. Tabin flows into Lesser Zab.

Habitat. Underground waters.

Biology. No data.

Conservation status. LC; as with all subterranean species, its distribution is difficult to understand, and threat levels may be low.

Further reading. Freyhof et al. 2016b (description).

Eidinemacheilus proudlovei; ephemeral spring in Tabin Gorge; Iraq; ~50 mm SL.

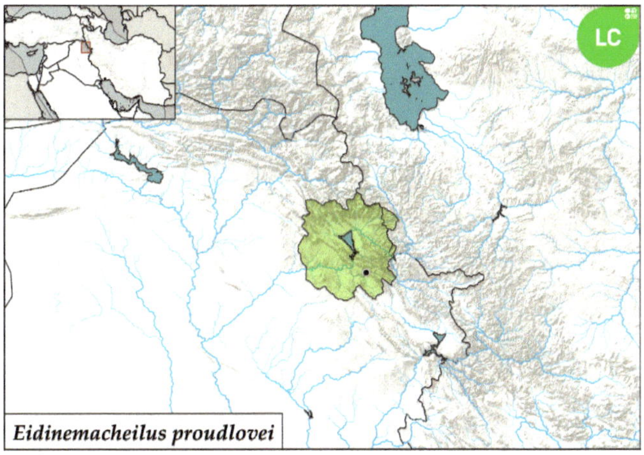

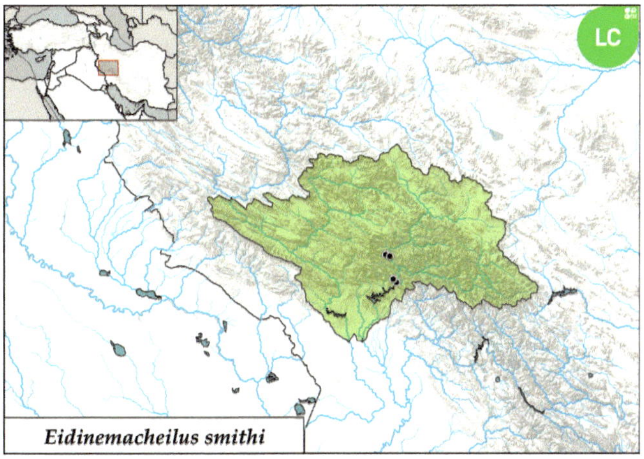

Eidinemacheilus smithi; Loven, Iran; 28 mm SL.

Eidinemacheilus smithi

Common name. Zagros blind loach.

Diagnosis. Distinguished from *E. proudlovei* by: • dorsal adipose crest present / • 7+7 branched caudal rays / • jaws normally developed / • head canal system reduced. Size up to 53 mm SL.

Distribution. Iran: Loven spring in Ab-e Sirum (or Ab-e Serum) valley near Tang-e Haft, and springs around Tuveh in Karun drainage. Probably more widespread.

Habitat. Underground waters.

Biology. No data.

Conservation status. LC; as with all subterranean species, its distribution is difficult to understand, and threat levels may be low.

Further reading. Greenwood 1976 (description); Freyhof et al. 2014b (diagnosis).

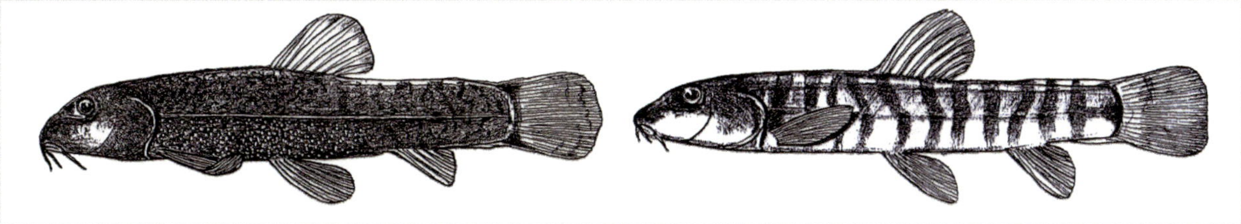

Figure 50. *Oxynoemacheilus* is the most species-rich genus of freshwater fish in West Asia. Many have small ranges, and several new species await their discovery and description. These two undescribed species from Türkiye have not been found since the 1970s. From left: Stream Nif in Gediz drainage, stream Melendiz at Ihlara. © T. Nalbant.

Reductive evolution in subterranean fishes? In most habitats, the initial observation of most organisms is one of adaptation. *Esox* are adapted for predation, while *Alburnus* are adapted for fast swimming. In contrast, subterranean animals are characterised by losses rather than gains. These organisms have much-reduced eyes and pigments and often have no or few scales. This has long fascinated biologists, dating from the time of Lamarck in the late 18th century. In no other group of biota has Lamarck's view of use and disuse persisted so long to influence modern biology as in troglobionts. Even Darwin's explanation of eyeless subterranean animals was a straightforward Lamarckian one and did not involve adaptation and the struggle for existence. In his 1859 work, Darwin wrote, "Their [eyes] loss may be attributed to disuse" (Darwin, 1859). Subsequently, authors proposed neutralistic explanations instead of natural selection to account for the phenomenon of "reductive evolution." However, this view shifted in the late 20th century, with troglomorphic features now regarded as highly adaptive to subterranean environments and selection identified as a significant driver of species adaptation to subterranean habitats. Compared to their surface-dwelling counterparts, subterranean species exhibit many constructive biological changes yet display regressive alterations, exemplified by the reduction of eyes. Indeed, the reduction of the eye is an adaptation that conserves energy. There is evidence to suggest that different changes may be connected by pleiotropy. Pleiotropy occurs when a single gene influences multiple phenotypic traits. Consequently, a mutation in a pleiotropic gene may affect some or all traits simultaneously. This contemporary perspective on the distinctive characteristics of troglomorphic species offers a novel and significant opportunity to advance our comprehension of the phenomenon of "reductive" evolution. **Further reading.** Darwin 1859 (origin of species); Cluver & Pipan 2009 (adaptation in cavefish).

Oxynoemacheilus

This is the most speciose genus of freshwater fishes in the Western Palaearctic, with 68 known species, and the largest in West Asia, with 64 species. It is distributed throughout West Asia and one species is found in Central Asia (*O. oxianus*). Three species are found in Europe in Albania and Greece (*O. bureschi, O. pindus*), and the northern Caucasus (*O. merga*). *Oxynoemacheilus* are found virtually everywhere, with one or more species in almost all habitats. They belong to the standard repertoire of nearly every river, stream, and spring. However, this genus is absent in Iran from the Caspian basin, east of the Sefid, and from many of the endorheic basins in southern Iran. It is also absent from the rivers south of the Mond, where mostly loaches of the genus *Paraschistura* are found, and from all the Arabian Peninsula.

Mature males are typically smaller than females, with upward-curled pectorals that are longer than those of females. These rays are stiffer, wider, and covered by numerous rows of minute tubercles. Outside the spawning season, it can be challenging to distinguish between sexes. In many species, adult males have a suborbital slit or groove below the modified lachrymal bone. However, this slit or groove is absent in many other species. In some species, a small groove may be present in the posterior outline of the lachrymal bone. In other species, the lachrymal bone may be visible but completely covered by skin.

Oxynoemacheilus bergianus x *Oxynoemacheilus euphraticus*; Great Zab drainage, Iraq; 70 mm SL. Individuals in this population have one or the other type of mtDNA indicating hybridisation.

Identifying *Oxynoemacheilus* species is often challenging, particularly for slender species with a deeply emarginate caudal and a suborbital groove in the male. Such species are usually widely distributed and may exhibit variations in colour patterns within and between populations. Unfortunately, in slender *Oxynoemacheilus*, several populations without morphological and only small differences in mtDNA had been recognised as separate species, indicating a need for a conceptual framework of boundaries between populations and species. Some of these are not accepted here. Many other species have a stout, massively built body, a slightly emarginate or even truncate caudal, and no suborbital groove in male. These species are often locally endemic. In sympatry, the caudal fin's shape and the caudal peduncle's depth can be useful to distinguish between most species. It has been demonstrated that the diversity in body shapes observed in *Oxynoemacheilus* does not fully correspond to the phylogeny of the different species, as slender and stout species may be closely related. However, nuclear DNA characters may lead to very different phylogenetic patterns if analysed in the future. Some stout species form monophyletic groups in *Oxynoemacheilus*, but slender species are a highly polyphyletic group. Notably, there are a variety of intermediate morphotypes within this large genus.

We know that the diversity of *Oxynoemacheilus* presented in this book sometimes not conforms to simplistic, tree-based patterns, as introgressive hybridisation between species has occurred. In several species, isolated populations differ only by small molecular differences without morphological differences, and we do not accept these as species. Of course, we observe the species diversity of *Oxynoemacheilus* (and other genera) to be streamlined with clusters of minimal COI differences and new species to be described with only 'invented' morphological differences. This process has already begun. Here, the concept of "old species" should be applied to avoid false species decriptions. **Further reading.** Prokofiev 2009 (genera); Freyhof et al. 2011 (diversity); Kottelat 2012 (diversity).

Variability in colour pattern of *Oxynoemacheilus cilicicus* left from top: lower Seyhan drainage; 50 mm SL; lower Ceyhan drainage; 55 mm SL; river connecting Lakes Gölbaşı and Azaplı; 53 mm SL; spring Çöçelli, Ceyhan drainage; 50 mm SL; right from top: Göksu delta, 55 mm SL; 55 mm SL; 55 mm SL; 52 mm SL.

Keys to species of *Oxynoemacheilus* in West Asia

Caspian basin, including Lake Urmia and Namak basins

Oxynoemacheilus bergi is a valid species from the Kura drainage. However, it needs to be better known to be included in the key. It may be a senior synonym of *O. bergianus*, but this hypothesis awaits confirmation. *Oxynoemacheilus merga* from the European Caspian basin in Azerbaijan, north of Baku, is excluded from the coverage of this book. Most *Oxynoemacheilus* species have individuals with a completely or almost complete plain brown colour pattern. The identification of species by colour pattern requires the examination of more than one or a few individuals.

1a - Suborbital groove absent in adult male; flank with 12–17 distinct, regularly shaped, and set bars; caudal slightly emarginate or truncate.
..................*O. cyri*
1b - Suborbital groove present in adult male; flank without bars or bars very irregularly shaped, and set; caudal emarginate or furcate.
..................2

2a - Caudal slightly emarginate; flank mottled, with a midlateral series of indistinct blotches, often forming an irregularly shaped midlateral stripe, without distinct bars or large blotches; caudal–peduncle depth 1.3–1.6 times in its length.
..................*O. veyselorum*
2b - Caudal deeply emarginate or furcate; flank pattern with distinct bars or large blotches, not forming a midlateral stripe; caudal–peduncle depth 2.0–3.2 times in its length.
..................3

3a - Flank mottled or with large, irregularly shaped, squarish, or vertically elongate blotches, usually as wide or narrower than interspaces; lateral blotches on caudal peduncle usually interrupted in size and/or shape at or above lateral midline; one central or no black, grey or brown blotch or short bar on caudal base, its colour identical to blotches on caudal peduncle.
..................*O. bergianus*
3b - Flank with wide, irregularly shaped bars, much wider than interspaces; bars on caudal peduncle not interrupted in size and shape at lateral midline; an upper and a lower bold, black or dark-brown blotch on caudal base, often fused into a bar, its colour distinctly darker than bars on caudal peduncle.
..................4

4a - Caudal emarginate, with many small brown blotches on rays arranged in 3–6, narrow, pale-brown bands; flank usually with irregularly shaped and spaced, dark-brown bars, often split in middle, irregularly shaped and narrower than interspaces, or flank almost plain brown.
..................*O. elsae*
4b - Caudal forked, with 1–3 wide and very bold bands; flank with brown bars or vertically elongate blotches, much wider than interspaces, usually regularly set.
..................*O. brandtii*

Black and Marmara Sea basins

No external characters are known to distinguish *O. banarescui* (Devrekani, Filyos), *O. bergianus* (Kızılırmak), *O. fatsaensis* (Elekçi, Yeşilırmak), and *O. simavicus* (Sursuluk, Sakarya, Büyük Melen). These four species are related and can only be distinguished by molecular characters.

1a - Suborbital groove absent in adult males.
..................*O. seyhanensis*
1b - Suborbital groove present in adult males.
..................2

2a - Caudal–peduncle depth 6–10 % SL.
..................*O. banarescui, O. bergianus, O. fatsaensis, O. simavicus*
2b - Caudal–peduncle depth 10–13 % SL.
..................3

3a - A median incision in upper lip in most individuals.
..................*O. cemali*
3b - No median incision in upper lip.
..................4

4a – Caudal with a wide, bold band in middle or posterior half.
..................*O. phasicus*
4b – Caudal with 1–3 narrow, pale bands.
..................5

5a - Distance between pectoral and pelvic origins 25–28 % SL; caudal slightly emarginate.
..................*O. angorae*
5b - Distance between pectoral and pelvic origins 29–34 % SL; caudal moderately emarginate.
..................*O. marmaraensis*

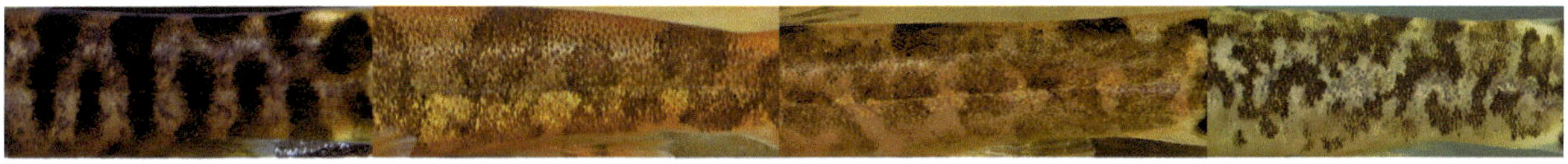

Figure 51. Colour pattern on caudal peduncle of *Oxynoemacheilus*: From left: barred, marmorate, marbled, and mottled. In marmorate patterns, blotches are larger and clearer with wider interspaces than marbled ones.

Aegean basin
1a - Belly without scales; caudal slightly emarginate or almost truncate, middle caudal ray 84–98 % of longest upper caudal ray length.
..................*O. anatolicus*
1b - Belly with few, small, embedded scales; caudal emarginate, middle caudal ray 71–91 % of longest upper caudal ray length.
..................2

2a - Body depth almost uniform between dorsal and caudal base (i.e., ratio of body depth at posterior of dorsal base to body depth at caudal base is equal or almost equal to 1).
..................3
2b - Body depth decreasing between dorsal and caudal base (i.e., ratio of body depth at posterior of dorsal base to body depth at caudal base is greater than 1).
..................4

3a - Body depth at dorsal origin 17–18 % SL; caudal peduncle 12–13 % SL, length of middle caudal ray 19–23 % SL; dorsal and ventral adipose crests present on caudal peduncle behind vertical of posterior anal base.
..................*O. theophilii*
3b - Body depth at dorsal origin 15–17 % SL; caudal peduncle 10–12 % SL; length of middle caudal ray 16–19 % SL; no dorsal and ventral adipose crests on caudal peduncle behind vertical of posterior anal base.
..................*O. fatmae*

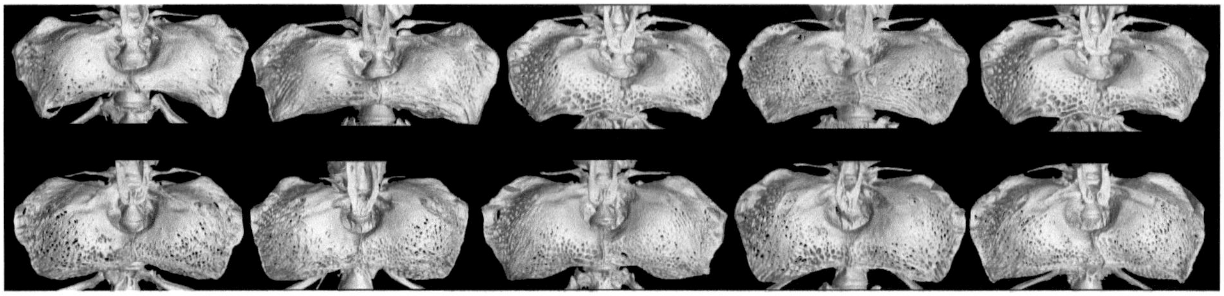

Figure 52. In Nemacheilidae, the swim bladder is situated in a bony capsule (ventral view in 10 individuals of *Oxynoemacheilus germencicus*). The structure and shape of this capsule has often been used in identification. We find the preparation of this capsule too difficult, and the variability of its characters needs to be better understood to use it as a standard character for identification.

4a - Caudal–peduncle depth 1.0–1.3 times in caudal–peduncle length; caudal peduncle 13–16 % SL; flank with coarse mottling, without bars or blotches.
..................*O. kottelati*
4b - Caudal–peduncle depth 1.3–2.2 times in caudal–peduncle length; caudal peduncle 16–22 % SL; flank with bars and/or blotches, rarely with fine mottling.
..................5

5a - Caudal–peduncle depth 1.3–1.5 times in caudal–peduncle length; scales on belly present from pelvic base to midline between pectoral- and pelvic origins.
..................*O. eliasi*
5b - Caudal–peduncle depth 1.5–2.2 times in caudal–peduncle length; scales on belly restricted between pelvic bases, rarely found more anteriorly.
..................*O. germencicus*

Central Anatolia and rivers flowing to the Bay of Antalya
1a - Flank without scales, with many roundish blotches or small spots, without bars or elongated blotches.
..................*O. ciceki*
1b - Flank with scales, mottled or with bars and / or elongated blotches.
..................2

2a - Caudal–peduncle depth 2.2–2.6 times in caudal–peduncle length.
..................*O. isauricus*
2b - Caudal–peduncle depth 1.2–2.1 times in caudal–peduncle length.
..................3

3a - Caudal deeply emarginate, middle caudal ray 65–76 % of longest upper caudal ray length.
..................*O. mediterraneus*
3b - Caudal slightly emarginate or almost truncate, middle caudal ray 71–98 % of longest upper caudal ray length.
..................4

4a - Belly with few, small, embedded scales; caudal emarginate, middle caudal ray 76–91 % of longest upper caudal ray length.
..................*O. nasreddini*
4b - Belly without scales; caudal slightly emarginate or almost truncate, middle caudal ray 84–98 % of longest upper caudal ray length.
..................5

5a - Flank with a series of dark-brown midlateral blotches usually fused into a wide, irregularly shaped midlateral stripe, rarely a mottled pattern.
..................*O. angorae*
5b - Flank with irregularly set and shaped distinct or confluent blotches and spots forming a marbled or mottled pattern, with large, roundish brown blotches in some individuals.
..................6

6a - Pre-dorsal back usually with 3–4 saddles; flank completely covered by scales.
..................*O. anatolicus*
6b - Pre-dorsal back with dark-brown fine mottled pattern, without saddles; few isolated scales on flank in front of dorsal origin.
..................7

7a - Dorsal and ventral adipose crests not elevated from dorsal profile, straight, rarely convex, caudal peduncle highest at end of hypural complex; dorsal adipose crest reaching behind vertical of posterior anal base, absent in some individuals; caudal–peduncle depth 1.5–1.9 times in its length; shortest middle caudal ray is 83–91 % of longest ray of upper caudal lobe; usually flank with irregularly set and shaped, distinct, large blotches forming a marbled pattern.
..................*O. eregliensis*
7b - Dorsal and ventral adipose crests elevated from dorsal profile, usually with convex margin, caudal peduncle shallower at end of hypural complex and at maximum height of dorsal adipose crest; dorsal adipose crest reaching to or exceeding vertical of anal base; caudal–peduncle depth 1.2–1.6 times in its length; shortest middle caudal ray is 88–98 % of longest ray of upper caudal lobe; usually flank with irregularly set and shaped confluent small blotches and spots forming a mottled pattern, rarely with larger blotches forming a marbled pattern.
..................*O. axylos*

Göksu, Seyhan, Ceyhan, and Orontes rivers, plus coastal drainages within their bounds
1a - Suborbital groove absent in male.
.....................2
1b - Suborbital groove present in male.
.....................6

2a - Caudal slightly emarginate, almost truncate.
.....................*O. seyhanensis*
2b - Caudal deeply emarginate or forked.
.....................3

3a - Lateral line terminating above anal base or on hypural complex; many isolated and embedded scales on back and flank in front of dorsal origin.
.....................*O. ceyhanensis*
3b - Lateral line terminating anterior to or below dorsal base, rarely above anus or anal base; no or very few (*O. namiri*) isolated and embedded scales on back and flank in front of dorsal origin.
.....................4

4a - Flank pattern, usually with 6–17, very distinct and regularly shaped and set bars, usually all or most flank-bars, at least behind dorsal base, extending to middorsal saddles and usually meeting contralateral.
.....................*O. namiri*
4b - Flank pattern mottled or with many, very narrow, irregularly shaped and set, pale-brown bars, usually flank-bars separated from middorsal saddles and not meeting contralaterals.
.....................5

5a - Lateral line with 23–45 pores; infraorbital canal with 10–13 pores; body depth at dorsal origin 1.2–1.4 times in HL.
..................*O. amanos*
5b - Lateral line with 13–23 pores; infraorbital canal with 8–10 pores; body depth at dorsal origin 1.0–1.2 times in HL.
..................*O. cilicicus*

6a - Caudal pattern very indistinct, almost or completely absent.
..................*O. seyhanicola*
6b - Caudal with dark-brown bands of elongated blotches.
..................7

7a - A bold, black blotch or spot at upper and lower caudal base.
..................8
7b - One central, bold-brown blotch or no blotch on caudal base, often an irregularly shaped black bar at caudal base.
..................10

8a - Flank and caudal peduncle with marbled colour pattern, without series of midlateral blotches or bars.
..................*O. argyrogramma*
8b – A series of midlateral blotches on caudal peduncle, in some individuals, flank with bars.
..................9

9a - 5–7 flank bars, irregularly shaped and set, extending to middorsal saddles and meeting contralateral; caudal peduncle 1.8–2.4 times longer than deep.
...............*O. shehabi*
9b - A midlateral series of large blotches, disconnected from saddles on back; caudal peduncle 1.4–1.8 times longer than deep.
...............*O. sarus*

10a - Caudal–peduncle depth 1.4–1.9 times in its length.
...............11
10b - Caudal–peduncle depth 1.8–3.5 times in its length.
...............13

11a - Middle caudal ray 64–74 % of length of longest ray in upper caudal lobe.
...............*O. hamwii*
11b - Middle caudal ray 83–92 % of length of longest ray in upper caudal lobe.

12a - Flank with irregularly set and shaped confluent small blotches and spots forming a mottled pattern, rarely with larger blotches forming a marbled pattern; pre-dorsal back with blotches or a dark-brown, fine, mottled pattern, without saddles; few isolated scales on the flank in front of the dorsal origin.
...............*O. eregliensis*
12b - Flank with a series of dark-brown midlateral blotches usually fused into a wide, irregularly shaped midlateral stripe, rarely a mottled pattern; pre-dorsal back usually with 3–4 saddles; flank completely covered by scales.
...............*O. angorae*

13a - A yellowish triangle at upper and lower posteriormost portions of caudal peduncle; caudal–peduncle depth 1.8–2.4 times in its length; caudal emarginate, middle caudal ray 80–91 % of length of longest ray in upper caudal lobe.
...............*O. evreni*
13b - No yellowish triangles at posteriormost portion of caudal peduncle; caudal–peduncle depth usually 2.0–3.0 times in its length; caudal deeply emarginate, middle caudal ray 70–84 % of length of longest ray in upper caudal lobe.
...............*O. bergianus*

Southern Levant (Damascus basin, Litani, and Jordan drainages)

1a - Suborbital groove absent in adult male; body scaleless or scales restricted to flank behind dorsal base, with few isolated scales on anterior flank in some individuals.
..................2
1b - Suborbital groove present in adult male; back and flank covered by scales.
..................4

2a - Body completely scaleless; 9–11½ branched dorsal rays.
..................*O. galilaeus*
2b - Anterior flank scaleless or with few isolated scales, caudal peduncle with densely set scales; 8–9½ branched dorsal rays.
..................3

3a - Lateral line terminating before vertical of dorsal origin; body depth at dorsal origin 20–25 % SL.
..................*O. panthera*
3b - Lateral line terminating about at vertical of dorsal origin; body depth at dorsal origin 16–20 % SL.
..................*O. leontinae*

4a - Lateral line incomplete, terminating between vertical or dorsal and anal base, usually to anal base.
..................*O. pantheroides*
4b - Lateral line complete, terminating at or almost at caudal base.
..................*O. insignis*

Euphrates, Qweiq drainages, and Lake Van basin

1a - Lateral line incomplete, terminating anterior to dorsal origin or above anal base.
..................2
1b - Lateral line complete, reaching to caudal base.
..................5

2a - Flank with distinct bars; an upper and a lower black spot at caudal base; caudal peduncle with a long, high dorsal adipose crest; crest with yellowish or whitish margin, reaching beyond vertical through anal origin, usually below last dorsal rays when folded down; flank with isolated and deeply embedded scales.
..................*O. tigris*
2b - Flank with a mottled, marbled, or marmorated pattern, often with narrow, indistinct bars on caudal peduncle or a series of dark-brown blotches; colour pattern on caudal base not differentiated from flank pattern, with or without a black bar, never with two black spots; caudal peduncle without or with a short dorsal adipose crest, crest without white margin, reaching to middle of caudal peduncle or to vertical through anal base; flank without scales.
..................3

3a - A central pore in supratemporal canal; a bold bar on caudal base, usually M-shaped, fused with a roundish or vertically elongate, dark-brown blotch at vertical midline; caudal usually with wide, bold, dark-brown bands.
..................*O. arsaniasus*
3b - No central pore in supratemporal canal; a mottled or marmorated pattern on caudal base, without bold bar; caudal with narrow, indistinct, pale-brown bands.
..................4

4a - Caudal–peduncle length 1.0–1.3 times its depth; body width at dorsal origin 13–17 % SL; dorsal adipose crest on caudal peduncle high, not reaching to vertical through last anal ray base in individuals larger than 50 mm SL.
..................*O. ercisianus*

4b - Caudal–peduncle length 1.3–1.6 times its depth; body width at dorsal origin 11–13 % SL; dorsal adipose crest on caudal peduncle shallow, reaching to vertical through last anal ray base in individuals larger than 50 mm SL.
..................*O. muefiti*

5a - Suborbital groove or flap absent in male.
..................6
5b - Suborbital groove present in male.
..................7

6a - Caudal peduncle without adipose crest or ridge; flank with a distinct midlateral series of large, roundish, or horizontally elongated brown blotches and a row of small brown blotches between upper pelvic base and lowermost caudal base.
..................*O. paucilepis*
6b - Caudal peduncle with a shallow, wide, often long adipose ridge; flank with a brown mottled or marmorated pattern, with irregularly shaped brown bars, especially on flank behind dorsal origin in some individuals.
..................*O. kaynaki*

7a - Caudal slightly emarginate.
..................*O. araxensis*
7b - Caudal deeply emarginate or slightly forked.
..................8

8a - One central or no black, grey or brown blotch or short bar on caudal base, its colour identical to blotches on caudal peduncle; upper lip without median incision.
..................*O. bergianus*
8b - A bold upper and a lower black blotch or spot on caudal base, often overlaid by a dark-brown or black chevron-shaped bar; upper lip with a median incision.
..................9

9a - Flank and caudal peduncle with marbled colour pattern, without bars; caudal–peduncle depth 1.4–1.8 times in caudal–peduncle length.
..................*O. argyrogramma*
9b - Caudal peduncle, and in some individuals also flank, with bars; caudal–peduncle depth 2.0–2.8 times in caudal–peduncle length.
..................*O. euphraticus*

Kor basin
1a - Caudal deeply emarginate; no dorsal crest on caudal peduncle.
..................*O. persa*
1b - Caudal slightly emarginate or truncate; shallow dorsal crest on caudal peduncle.
..................*O. tongiorgii*

Tigris drainage
The Tigris is the biodiversity hotspot for *Oxynoemacheilus*, but not all species have been described, especially from Iraq and Iran. Many species have small ranges, and their distribution might help identify them.

1a - Suborbital groove absent in male.
..................2
1b - Suborbital groove present in male.
..................8

2a - Lateral line incomplete, terminating in front of or above anal base.
..................3
2b - Lateral line complete, terminating behind vertical of anal base or at caudal base.
..................5

3a - Scales present on back and flank in front of anus; central pore in supratemporal canal present.
..................*O. frenatus*
3b - Scales absent on back and flank in front of anus; central pore in supratemporal canal absent.
..................4

4a - Lateral line very short, terminating slightly behind pectoral base, not reaching vertical through dorsal origin; a midlateral series of small, horizontally elongated, dark-brown blotches often fused into an irregularly shaped midlateral stripe, one additional stripe above and below midlateral stripe in many individuals.
..................*O. gyndes*
4b - Lateral line long, terminating under dorsal base or above anal base; flank with vertically elongated, irregularly shaped blotches or narrow bars.
..................*O. kiabii*

5a - Posterior process of bony air-bladder capsule directed posteriorly.
..................*O. zagrosensis*
5b - Posterior process of bony air-bladder capsule directed laterally.
..................6

6a - 8+8 branched caudal rays; caudal–peduncle depth 1.0–1.2 times in caudal–peduncle length; interorbital distance 1.1–1.4 times in snout length; colour pattern on flank behind dorsal base mottled or marmorated.
..................*O. zarzianus*
6b - 10+9, 9+9, or 9+8 branched caudal rays; caudal–peduncle depth 1.3–1.8 times in caudal–peduncle length; interorbital distance 1.4–1.9 times in snout length; colour pattern on flank behind dorsal base with bars or vertically elongated blotches.
..................7

7a - Very indistinct, fuzzy bars or vertically elongated blotches on caudal peduncle; maxillary barbel reaching to anterior eye margin or middle of eye; interorbital distance 1.4–1.5 times in snout length; caudal–peduncle depth 1.3–1.4 times in caudal–peduncle length.
..................*O. chomanicus*
7b - Distinct bars or vertically elongated blotches on flank behind dorsal origin; maxillary barbel reaching beyond middle of eye, usually to posterior eye margin; interorbital distance 1.6–1.9 times in snout length; caudal–peduncle depth 1.2–1.3 times in caudal–peduncle length.
..................*O. kentritensis*

8a - Flank naked, only caudal peduncle behind anus with scales; lateral line incomplete; caudal slightly emarginate.
..................*O. hazarensis*
8b - Flank completely covered by scales; lateral line complete; caudal deeply emarginate or forked.
..................9

9a - Flank with a distinct series of midlateral blotches, fused to each other or fused into a stripe, isolated patches of blotches, or a row of small dark-brown spots below lateral series of blotches.
..................*O. hanae*
9b - Flank mottled or marmorated, lower flank without colour pattern or colour pattern of midlateral flank reaching down to lower flank.
..................10

10a - Upper lip without or with very short median incision.
..................11
10b - Upper lip with well-developed median incision.
..................13

11a - One central or no black, grey or brown blotch or short bar on caudal base, its colour identical to blotches on caudal peduncle.
..................*O. bergianus*
11b - Two bold, black blotches or spots at caudal base, distinctly darker than flank pattern, usually prominent in life and preserved fish, in some individuals overlaid by a chevron-shaped, dark-brown or black bar.
..................12

12a – Flank with very dense mottling between blotches on flank; caudal–peduncle depth 1.7–3.1 times in its length.
..................*O. karunensis*
12b – No dense mottling in interspaces between blotches on flank; caudal–peduncle depth 1.4–1.7 times in its length.
..................*O. chaboras*

13a – Last bars on caudal peduncle narrower than interspaces.
..................*O. kurdistanicus*
13b – Last bars on caudal peduncle as wide or wider than interspaces.
..................14

14a – Juveniles larger than 20 but smaller than 45 mm SL with intense mottled pattern between midlateral blotches on flank.
.... *O. marunensis*
14b – Juveniles larger than 20 but smaller than 45 mm SL with poorly developed mottling between midlateral blotches on flank.
..................*O. euphraticus*

Oxynoemacheilus amanos

Common name. Hassa loach.

Diagnosis. Distinguished from other species of *Oxynoemacheilus* in Cilicia and northern Levant by: ○ suborbital groove absent in male / ○ caudal deeply emarginate / ○ lateral line incomplete with 23–45 pores, terminating between vertical of dorsal origin and anus / ○ 10–13 pores in infraorbital canal / ○ flank pattern mottled or with many, very narrow, irregularly shaped and set, pale-brown bars, usually flank-bars separated from middorsal saddles and not meeting contralaterals / ○ flank and back posterior to dorsal base covered by scales / ○ body depth at dorsal origin 1.2–1.4 times in HL. Size up to 71 mm SL.

Distribution. Türkiye: Spring İncesu at Hassa, a northern tributary of Orontes.

Oxynoemacheilus amanos; spring İncesu at Hassa; 66 mm SL.

Many freshwater fish have small distribution areas in West Asia, but few are as small as the spring İncesu, the only habitat of *Oxynoemacheilus amanos*.

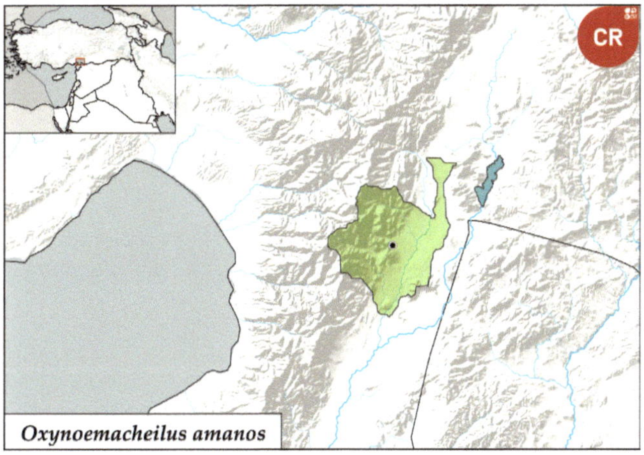

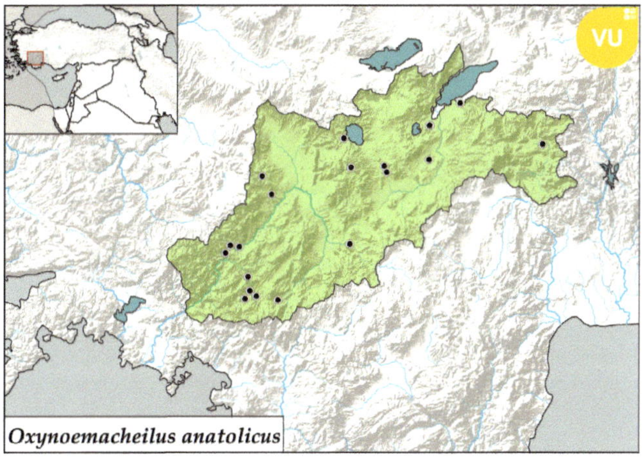

Habitat. Only known from one spring but expected to occur in nearby slow-flowing streams.
Biology. No data.
Conservation status. CR; known only from a small population in a very small, unprotected spring. Expected to decline.
Remarks. Related to *O. cilicicus* from adjacent Ceyhan drainage.
Further reading. Kaya et al. 2021 (description).

Oxynoemacheilus anatolicus

Common name. Burdur loach.
Diagnosis. Distinguished from other species of *Oxynoemacheilus* in Aegean basin and Central Anatolia by: ○ caudal emarginate, shortest middle caudal ray 84–98 % of longest ray of upper caudal lobe / ○ no prominent dorsal crest on caudal peduncle / ○ pre-dorsal back usually with 3–4 saddles / ○ flank completely covered by scales / ○ belly

Oxynoemacheilus anatolicus; Düğer, Lake Burdur basin, Türkiye; 55 mm SL.

without scales / ○ head length 24–27 % SL / ○ suborbital groove present in male / ○ caudal–peduncle depth 1.3–1.6 times in its length. Size up to 86 mm SL.
Distribution. Türkiye: Dalaman drainage and four spring-fed streams in Central Anatolia: Düger, Ereçay at Elmacık near Kemer, Karamanlı reservoir, and Pınarbaşı.
Habitat. Slow-flowing streams with dense vegetation and sand, mud, or gravel substrate.
Biology. Lives up to 5 years. Spawns after first winter. First spawn recorded in March.
Conservation status. VU; appears to be declining within its small range.
Remarks. Superficially similar loaches in Lake Salda basin may also belong to this species.
Further reading. Erk'akan et al. 2008b (description); Geiger et al. 2014 (molecular data); Innal 2019 (biology, distribution).

Oxynoemacheilus angorae; Bayındır, Sakarya drainage, Türkiye; ~60 mm SL.

Oxynoemacheilus angorae; Porsuk subdrainage, Sakarya drainage, Türkiye; ~60 mm SL

Oxynoemacheilus angorae

Common name. Angora loach.
Diagnosis. Distinguished from other species of *Oxynoemacheilus* in Marmara, and Black Sea basins, and Central Anatolia by: ○ caudal–peduncle depth 1.5–1.8 times in its length / ○ caudal slightly emarginate, shortest middle caudal ray 88–92 % of longest ray of upper caudal lobe / ○ belly without scales / ○ no prominent dorsal crest on caudal peduncle / ○ suborbital groove present in male / ○ flank marbled, often with a series of midlateral blotches fused to a wide, irregularly shaped midlateral stripe / ○ upper part of cheek mottled / ○ tip of pectoral usually not reaching

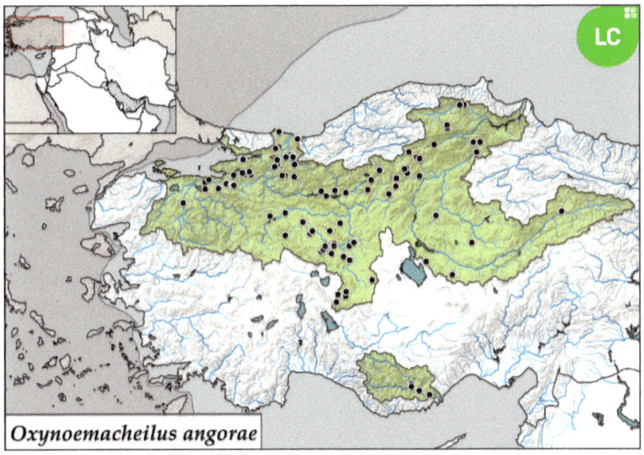

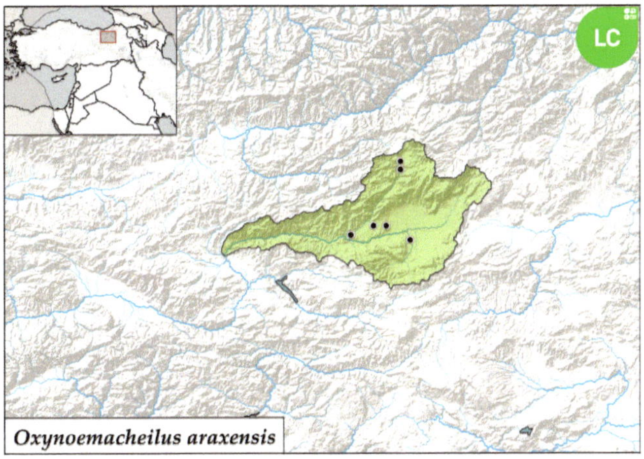

Oxynoemacheilus araxensis; upper Euphrates drainage, Türkiye; 71 mm SL. © M. Özuluğ.

pelvic origin in male / ○ distance between pectoral and pelvic origins 25–28 % SL / ○ no median incision in upper lip / ○ axillary pelvic lobe absent / ○ colour pattern on flank anterior to dorsal origin usually interrupted by pale-beige lateral line. Size up to 98 mm SL.

Distribution. Türkiye: Susurluk east to Kızılırmak drainages and endorheic Lakes İznik and Ilgın basins, also in Göksu drainge.

Habitat. A wide range of habitats, from swift mountain streams, banks of large rivers, springs to muddy lakes with dense vegetation. Moderately rheophilic, avoids very fast currents.

Biology. Spawns first time, usually at age 1; most individuals spawn for 1–2 seasons. Multiple spawner. Feeds on relatively large benthic invertebrates. Tolerant of moderate organic pollution and stream channeling.

Conservation status. LC.

Further reading. Yazıcıoğlu & Yazıcı 2016 (biology); Yoğurtçuoğlu et al. 2022 (distribution); Kaya et al. 2025a (record in Göksu).

Oxynoemacheilus araxensis

Common name. Erzurum loach.

Diagnosis. Distinguished from other species of *Oxynoemacheilus* in Euphrates drainage by: ○ caudal slightly emarginate / ○ lateral line complete / ○ suborbital groove present in male / ○ flank mottled or with 6–9 irregularly shaped dark-grey blotches, often fused, forming a lateral stripe / ○ caudal peduncle without high adipose crest or ridge / ○ flank covered by scales. Size up to 90 mm SL.

Distribution. Türkiye: Upper Karasu, a tributary of Euphrates west of Erzurum.

Habitat. Moderately fast-flowing streams.

Biology. No data.

Conservation status. LC; appears to be declining within its small range.

Remarks. Despite its name, this species is restricted to the Euphrates drainage and has not been found in the Aras (Caspian basin).

Further reading. Bănărescu et al. 1978 (description); Freyhof et al. 2019 (distribution).

Oxynoemacheilus argyrogramma; Merziman, Euphrates, Türkiye; 56 mm SL.

Oxynoemacheilus argyrogramma; Qweiq drainage, Türkiye; 50 mm SL.

Oxynoemacheilus argyrogramma

Common name. Two-spot loach.

Diagnosis Distinguished from other species of *Oxynoemacheilus* in Euphrates and Qweiq drainages by: ○ lateral line complete / ○ flank covered by scales / ○ two distinct and prominent black blotches at posteriormost caudal base, often overlaid by a dark-brown or black chevron-shaped bar / ○ caudal moderately deep emarginate / ○ flank and caudal peduncle with marbled colour pattern, without bars / ○ caudal peduncle without high adipose crest or ridge / ○ upper lip with a median incision / ○ suborbital groove present in male / ○ caudal–peduncle depth 1.4–1.8 times in caudal–peduncle length. Size up to 62 mm SL.

Distribution. Qweiq drainage in Türkiye and Syria. South-western Euphrates drainage in Türkiye, and most likely in adjacent Syrian Euphrates. In Gölbaşı lakes in uppermost Ceyhan and potentially more widespread in Ceyhan.

Habitat. Wide range of habitats in moderately fast-flowing waters, from small hill streams to banks of large rivers. Rarely found in stagnant waters as reservoirs.

Biology. No data.

Conservation status. LC.

Further reading. Heckel 1847 (description).

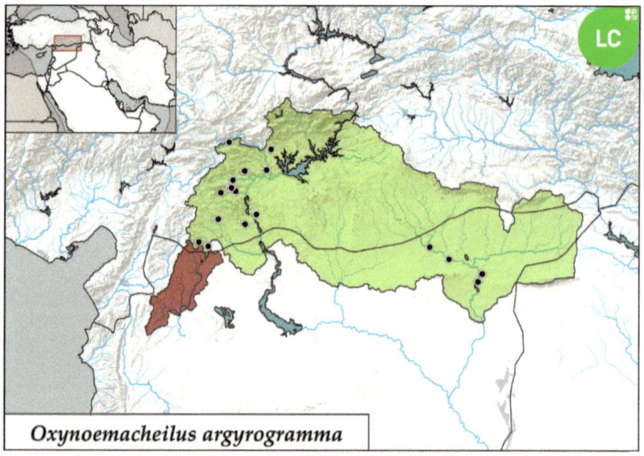

Oxynoemacheilus arsaniasus; upper Euphrates drainage, Türkiye; 90 mm SL.

Oxynoemacheilus arsaniasus

Common name. Murat loach.

Diagnosis. Distinguished from other species of *Oxynoemacheilus* in Euphrates drainage by: ○ scales absent / ○ supratemporal canal with one central pore / ○ lateral line incomplete, reaching to vertical through a point between about of one snout length in front of dorsal origin and middle of caudal peduncle / ○ suborbital groove absent in male / ○ caudal slightly emarginate / ○ pelvic axillary lobe absent / ○ no or a shallow dorsal and ventral adipose crest in individuals larger than 50 mm SL / ○ dark-brown bar or a vertically elongated blotch on caudal base or a bar fused with a roundish or vertically elongate, dark-brown blotch at vertical midline / ○ caudal usually with wide, bold, dark-brown bands / ○ caudal–peduncle length 1.3–1.6 times its depth / ○ prepelvic length 48–53 % SL. Size up to 97 mm SL.

Distribution. Türkiye: Murat drainage, a major tributary of Euphrates and in Güroymak spring near Bitlis. Potentially more widespread in Euphrates.

Habitat. Fast-flowing streams and rivers with gravel or rock substrate.

Biology. No data.

Conservation status. LC. Extirpated from Syrian part of Qweiq.

Further reading. Freyhof et al. 2019b (description).

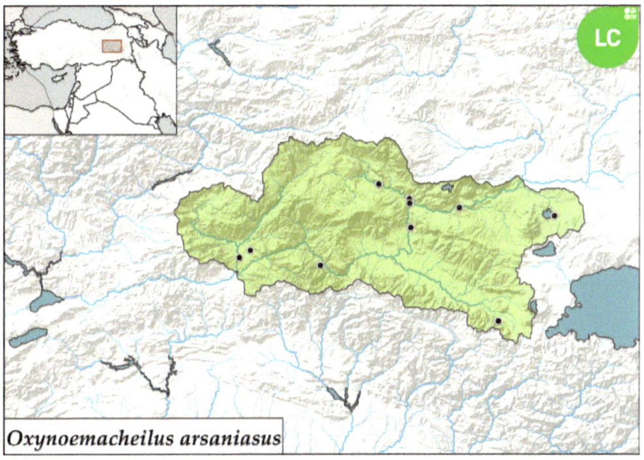

Oxynoemacheilus axylos; Sarı Yayla, Lake Tuz basin, Türkiye; 73 mm SL.

Oxynoemacheilus axylos; Gölyazı, Lake Tuz basin, Türkiye; 60 mm SL.

Oxynoemacheilus axylos

Common name. Lake Tuz loach.

Diagnosis. Distinguished from other species of *Oxynoemacheilus* in Central Anatolia and rivers flowing to Bay of Antalya by: ● elevated dorsal and ventral adipose crests, usually with convex margin and reaching to or exceeding vertical of anal base / ○ caudal–peduncle depth 1.2–1.6 times in its length / ○ caudal almost truncate, middle caudal ray 88–98 % of longest ray of upper caudal lobe / ○ head length 24–27 % SL / ○ flank usually with irregularly set and shaped, confluent, small blotches and spots, forming a mottled pattern or rarely with larger blotches forming a marbled pattern. Size up to 73 mm SL.

Distribution. Türkiye: Pınarbaşı spring in İnsuyu, Lakes Samsam and Gök in Kozanlı, and Melendiz drainage (Ihlara Valley), all in Lake Tuz basin.

Habitat. Moderately fast-flowing and standing waters of springs and streams with gravel substrate.

Biology. No data.

Conservation status. VU; appears to be declining within its small range.

Further reading. Yoğurtçuoğlu et al. 2022 (description).

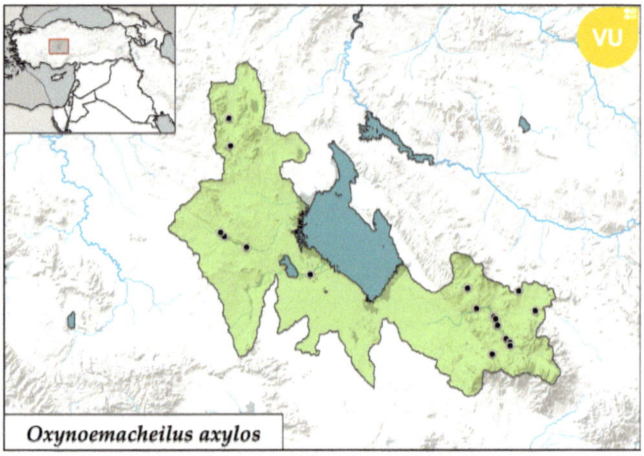

Devrekani is a major habitat of *Oxynoemacheilus banarescui* that is quite abundant in such fast-flowing stretches.

Oxynoemacheilus banarescui; Devrekani drainage, Türkiye, 70 mm SL. © M. Özuluğ.

Oxynoemacheilus banarescui

Common name. Paphlagonian sportive loach.

Diagnosis. Distinguished from other species of *Oxynoemacheilus* in the Black Sea basin by: ○ caudal–peduncle depth 7–10 % SL, usually 2.0–3.0 times in its length, very rarely 1.6–3.0, without high adipose crest or ridge / ○ caudal emarginate, middle caudal ray 80–90 % of length of longest ray in upper caudal lobe / ○ lateral line complete / ○ flank covered by scales / ○ suborbital groove present in male / ○ one central or no black, grey or brown blotch or short bar on caudal base, its colour identical to blotches on caudal peduncle / ○ no yellowish triangles at posteriormost portion of caudal peduncle / ○ pelvic very long, reaching or almost reaching to anus / ○ pectoral in male very long, reaching or almost reaching to pelvic origin / ○ flank mottled or with large, irregularly shaped, squarish, or vertically elongate blotches, usually as wide or narrower than interspaces / ○ lateral blotches on caudal peduncle usually interrupted in size and/or shape at or above lateral midline. Size up to 78 mm SL.

Distribution. Türkiye: Filyos and Devrekani drainages in Black Sea basin.

Habitat. Fast-flowing streams and rivers with gravel or rock substrate.

Biology. No data.

Conservation status. LC.

Remarks. No external characters are known to distinguish this species from *O. bergianus*, *O. fatsaensis*, and *O. simavicus*, all of which are diagnosed by molecular characters. As these four species are allopatric, the distribution area should help in identification.

Further reading. Delmastro 1982 (description); Yoğurtçuoğlu et al. 2020a (distribution); Freyhof 2022 (revision of *O. bergianus* group).

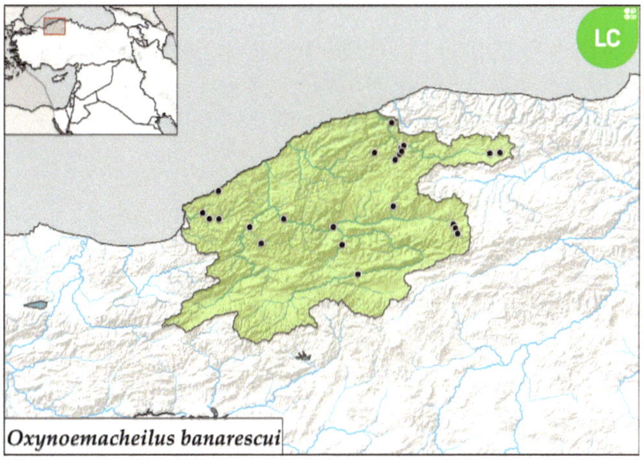

***Oxynoemacheilus bergi* and other names of Caspian loaches.** It is challenging to ascertain the number of valid species and their correct names. This often necessitates travel to type localities in remote areas. One such case is explained here. *Nemacheilus bergi* was described from the Kura drainage in Azerbaijan and was treated as a valid species without examining fish from the type locality or any other justification. In the same study, *N. alasanicus* from the upper Alazani in the Georgian Kura drainage was treated as a synonym of *N. bergi*, again without examining any materials or arguments. Early authors clearly distinguished *N. bergi* from *Oxynoemacheilus brandtii*. The account of *O. bergi* does not correspond to *O. cyri*, another species known from the Kura. *Oxynoemacheilus bergi* could be a third species of *Oxynoemacheilus* in the Kura, which has yet to be found (again). The account on *O. bergi* does not correspond to *O. veyselorum*, which has only been found in the adjacent Aras. However, it cannot be ruled out that *O. bergi* might be conspecific with the species identified as *O. bergianus*, also found in the Aras drainage. Nevertheless, there are no records of *O. bergianus* from the Kura, and fieldwork in late 2024 aimed to resolve the issue but failed to find it. If *O. bergianus* were conspecific with *O. bergi*, it would become a junior synonym of *O. bergi*. It was also suggested that two other loaches occur in the Alazani, a tributary of the Kura: *N. angorae alasanicus* and *N. brandtii gibbusnazus*. Indeed, only *O. brandtii*, very common in the Kura, has been recently found in the Alazani. *Nemacheilus angorae alasanicus* might be conspecific with *O. bergianus*, and both could be conspecific with *O. bergi*. These species should be re-examined based on materials from their type localities, and considerable efforts should be made to find the mysterious *Oxynoemacheilus bergi* and *O. alasanicus*. This is a complex issue, demonstrating that even with the publishing of this book, many questions need to be resolved. **Further reading.** Berg 1898, Gratzianov 1907 (*O. bergi*); Elanidze 1983 (Alazani loaches).

Oxynoemacheilus bergi; Kura drainage, Azerbaijan; 50 mm SL. © S. Eagderi.

Oxynoemacheilus bergi
Common name. Agstafa loach.
Diagnosis. Distinguished from other species of *Oxynoemacheilus* in Caspian basin by: ○ caudal moderately deeply emarginate / ○ caudal–peduncle depth two times in its length. Size up to 71 mm SL.
Distribution. Azerbaijan: Agstafa, a tributary of Kura.
Habitat. No data.
Biology. No data.
Conservation status. DD; due to the very poor knowledge of this species.
Remarks. This species might be conspecific with *O. bergianus*.
Further reading. Çiçek et al. 2018 (rediscovery); Freyhof et al. 2021b (discussion).

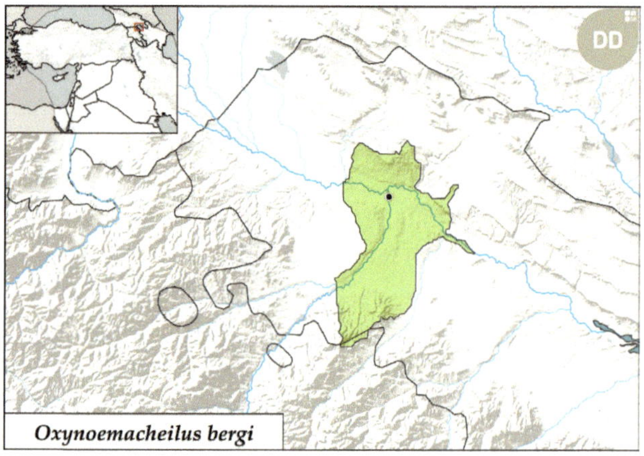

Oxynoemacheilus bergianus; upper Euphrates drainage, Türkiye; 59 mm SL.

Oxynoemacheilus bergianus

Common name. Sportive loach.

Diagnosis. Distinguished from other species of *Oxynoemacheilus* in Seyhan, Black Sea, and Caspian basins as well as Euphrates and Tigris drainages by: ○ caudal–peduncle depth 6–10 % SL, usually 2.0–3.0 times in its length, very rarely 1.8–3.5, without high adipose crest or ridge / ○ caudal deeply emarginate, middle caudal ray 70–84 % of length of longest ray in upper caudal lobe / ○ lateral line complete / ○ flank covered by scales / ○ upper lip without a median incision / ○ suborbital groove present in male / ○ one central or no black, grey, or brown blotch or short bar on caudal base, its colour identical to blotches on caudal peduncle / ○ no yellowish triangles at posteriormost portion of caudal peduncle / ○ pelvic very long, reaching or almost reaching to anus / ○ pectoral in male very long, reaching or almost reaching to pelvic origin / ○ flank mottled, marmorate, or with large, irregularly shaped, squarish, or vertically elongate blotches, usually as wide or narrower than interspaces / ○ lateral blotches on caudal peduncle usually interrupted in size and/or shape at or above lateral midline. Size up to 71 mm SL.

Distribution. Kızılırmak drainage in Anatolian Black Sea basin, upper Zamantı (Seyhan) drainage in Mediterranean basin, upper Euphrates and upper Tigris including Greater Zab, Sirvan south to Doiraj, and Meymeh. Also, in Caspian basin, from Sefid in Iran north-west to Aras and Lake Namak and Urmia basins.

Habitat. Moderate to fast-flowing waters with coarse gravel and small boulders.

Biology. No data.

Conservation status. LC.

Oxynoemacheilus bergianus; Seyhan drainage, Türkiye; 62 mm SL.

Oxynoemacheilus bergianus; upper Tigris, Türkiye; 61 mm SL.

Remarks. This is the most widespread of all *Oxynoemacheilus* species. *Oxynoemacheilus erdali* from the upper Murat drainage in Türkiye, *O. lenkoranensis* from Azerbaijan, *O. longipinnis* and *O. parvinae* from Iran, and *O. samanticus* from Seyhan in Türkiye are synonyms. No consistent morphological or colour pattern differences between these populations have been found or confirmed, and they show very small molecular differences. Recently published morphological differences are fabricated to support populations with very small molecular differences as separate species. *Oxynoemacheilus longipinnis* has been described as a separate genus, *Ilamnemacheilus*, based on a malformed individual with 10½ branched dorsal rays, several multiplied vertebrae, and a shortened caudal peduncle.

There appear to be no records of *O. bergianus* from the Kura drainage, but further fieldwork is required to confirm its presence. If this species is found to be conspecific with *O. bergi* from the Kura, then *O. bergi* would take precedence over *O. bergianus*. No external characters are known to distinguish *O. bergianus* from *O. banarescui*, *O. fatsaensis*, and *O. simavicus*, all of which are diagnosed by molecular characters. Published morphological differences are based on small sample sizes or are intentionally falsified. As these four species are allopatric, the range should aid identification.

Further reading. Derjavin 1934 (description); Abdurakhmanov 1962 (description of *Nemacheilus lenkoranensis*); Balık & Bănărescu 1978 (description of *O. samanticus*); Coad & Nalbant 2005 (description of *Ilamnemacheilus longipinnis*); Erk'akan et al. 2007 (description of *O. erdali*); Freyhof 2016a (generic position of *I. longipinnis*); Sayyadzadeh et al. 2016 (description of *O. parvinae*); Sayyadzadeh et al. 2017 (re-description of *O. longipinnis*); Kaya et al. 2016 (distribution); Freyhof 2022 (revision of *O. bergianus* group).

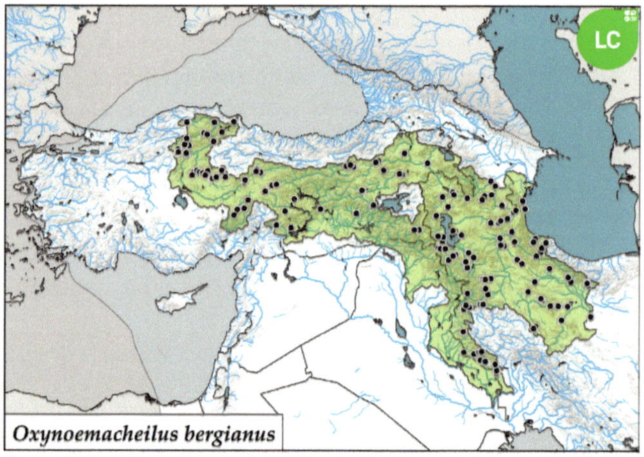

Oxynoemacheilus bergianus; upper Euphrates drainage, Türkiye; 60 mm SL.

Oxynoemacheilus brandtii; Aras drainage, Türkiye; 65 mm SL. © M. Özuluğ.

Oxynoemacheilus brandtii; Aras drainage, Türkiye; 48 mm SL. © M. Özuluğ.

Oxynoemacheilus brandtii

Common name. Caucasian loach.

Diagnosis. Distinguished from other species of *Oxynoemacheilus* in Caspian basin by: • caudal forked, usually with 1–3, wide and very bold bands / ○ an upper and a lower bold, black or dark-brown blotch on caudal base, usually fused into a bar, its colour distinctly darker than bars on caudal peduncle / ○ flank with wide, irregularly shaped bars, much wider than interspaces / ○ bars on caudal peduncle not interrupted in size and shape at or above lateral midline, rarely with a mottled pattern / ○ suborbital groove present in male / ○ caudal-peduncle depth 2.5–3.2 times in its length. Size up to 70 mm SL.

Distribution. Kura and Aras drainages.

Habitat. Fast to very fast-flowing streams and rivers with gravel or rock substrate. Usually found in riffles and rapids in middle of river.

Biology. No data.

Conservation status. LC.

Remarks. Populations studied from the Aras drainage share their mtDNA sequence with *O. elsae* but are identified as *O. brandtii* by morphological characters. *Nemacheilus brandtii gibbusnazus* from the upper Alazani in Georgian Kura drainage has been treated as a synonym, but the identity of this population may require further investigation.

Further reading. Kessler 1877 (description); Berg 1949b (description); Kaya et al. 2020a (distribution); Freyhof et al. 2021b (diagnosis, distribution).

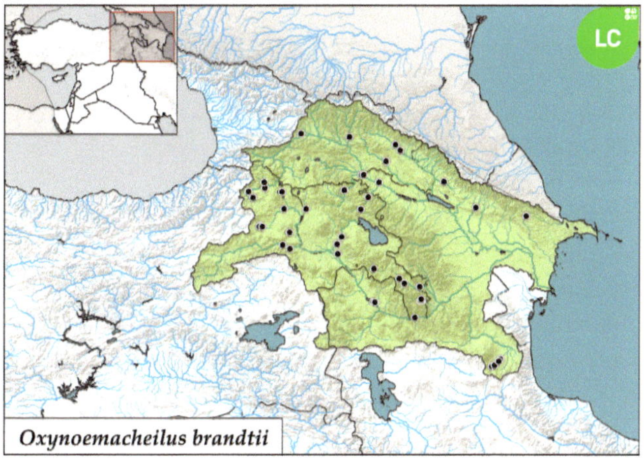

Oxynoemacheilus cemali; Çoruh drainage, Türkiye; ~65 mm SL. © M. Özuluğ.

Oxynoemacheilus cemali; Çoruh drainage, Türkiye; ~80 mm SL. © M. Özuluğ.

Oxynoemacheilus cemali

Common name. Çoruh loach.

Diagnosis. Distinguished from other species of *Oxynoemacheilus* in Black Sea basin by: ○ suborbital groove present in male / ○ axillary pelvic lobe present / ○ a median incision in upper lip in most individuals / ○ lateral line complete / ○ caudal deeply emarginate / ○ dorsal adipose crest on caudal peduncle absent / ○ tip of pectoral almost reaching vertical through dorsal origin in male / ○ 7–15 dark-grey saddles on back, 3–7 in front of dorsal origin / ○ 9–15 irregularly shaped dark-grey bars on flank posterior to dorsal origin in most individuals / ○ colour pattern on flank not interrupted by whitish lateral line / ○ caudal–peduncle length 18–21 % SL / ○ caudal–peduncle depth 10–13 % SL. Size up to 95 mm SL.

Distribution. Çoruh drainage in Türkiye and Georgia, as well as in upper part of Doğankent stream, a coastal river in Eastern Black Sea basin.

Habitat. Moderately fast-flowing waters in streams with gravel substrate.

Biology. No data.

Conservation status. LC.

Further reading. Turan et al. 2019a (description); Kuljanishvili et al. 2020 (distribution).

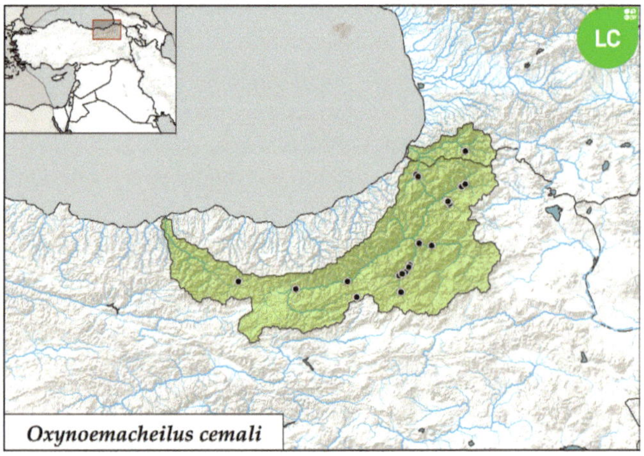

Oxynoemacheilus ceyhanensis; Ceyhan drainage, Türkiye; 50 mm SL.

Oxynoemacheilus ceyhanensis

Common name. Elbistan loach.

Diagnosis. Distinguished from other species of *Oxynoemacheilus* in Cilicia and northern Levant by: ○ caudal deeply emarginate, length of middle caudal ray 1.1–1.3 times in length of longest unbranched ray in upper caudal lobe / ○ lateral line complete, terminating at caudal base or on hypural complex / ○ suborbital groove absent in male / ○ flank and back behind dorsal base covered by many isolated and embedded scales / ○ no prominent dorsal crest on caudal peduncle / ○ caudal–peduncle depth 1.4–1.6 times in its length / ○ body depth at dorsal origin 18–20 % SL / ○ head length 25–27 % SL. Size up to 57 mm SL.

Distribution. Türkiye: Upper Ceyhan drainage.

Habitat. Moderately fast-flowing waters in streams with gravel substrate.

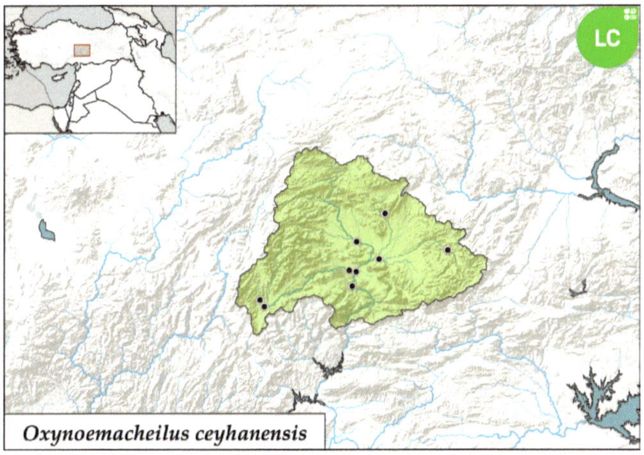

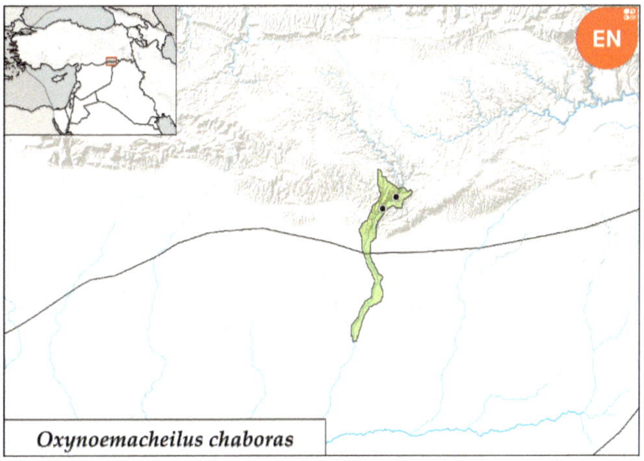

Oxynoemacheilus chaboras; stream Beyazsu, Türkiye; 51 mm SL.

Biology. No data.
Conservation status. LC.
Remarks. The distribution of this species needs to be better understood as it has usually been confused with *O. cilicicus*, a species found in the lower Ceyhan drainage.
Further reading. Erk'akan et al. 2007 (description); Kaya et al. 2020b (distribution).

Oxynoemacheilus chaboras

Common name. Khabur two-spot loach.
Diagnosis. Distinguished from other species of *Oxynoemacheilus* in Euphrates drainage by: ○ flank with a distinct series of midlateral blotches, fused to each other or fused into a stripe, below lateral series of blotches with isolated patches of blotches or a row of small dark-brown spots / ○ two distinct and prominent black blotches at posteriormost caudal base, usually overlaid by a dark-brown or black chevron-shaped bar / ○ suborbital groove present in male / ○ no or a very short median incision in upper lip / ○ lateral line complete, terminating behind vertical of anal base or at caudal base / ○ flank completely covered by scales / ○ caudal–peduncle depth 1.4–1.7 times in its length / ○ caudal deeply emarginate or forked. Size up to 60 mm SL.
Distribution. Türkiye and most likely adjacent Syria: Beyazsu north of Nusaybin in Khabur drainage, a tributary of Euphrates.
Habitat. Fast-flowing waters in streams with gravel substrate.
Biology. No data.
Conservation status. EN; restricted range and ongoing threats such as droughts.
Further reading. Kaya et al. 2024c (description).

Oxynoemacheilus chomanicus; Lesser Zab drainage, Iraq; 62 mm SL.

Oxynoemacheilus chomanicus

Common name. Choman loach.
Diagnosis. Distinguished from other species of *Oxynoemacheilus* in Tigris drainage by: ○ dorsal profile straight / ○ caudal peduncle with very indistinct, fuzzy bars or vertically elongated blotches / ○ flank completely covered by scales / ○ caudal slightly emarginate / ○ lateral line complete, terminating behind vertical of anal base or at caudal base / ○ suborbital groove absent in adult male / ○ maxillary barbel reaching to anterior eye margin or middle of eye / ○ colour pattern on anterior part of flank not interrupted by unpigmented zone along lateral line / ○ one central pore in supratemporal

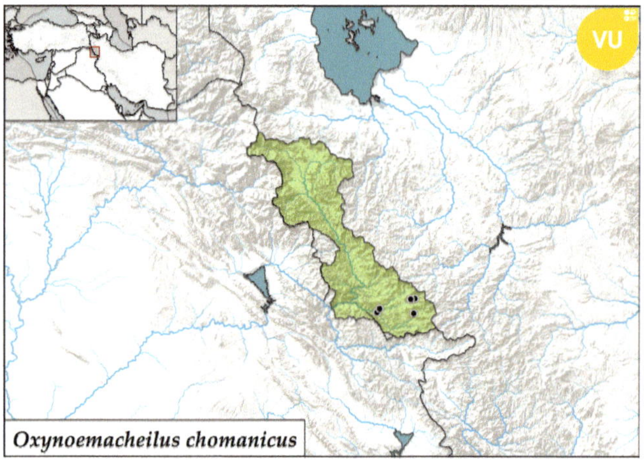

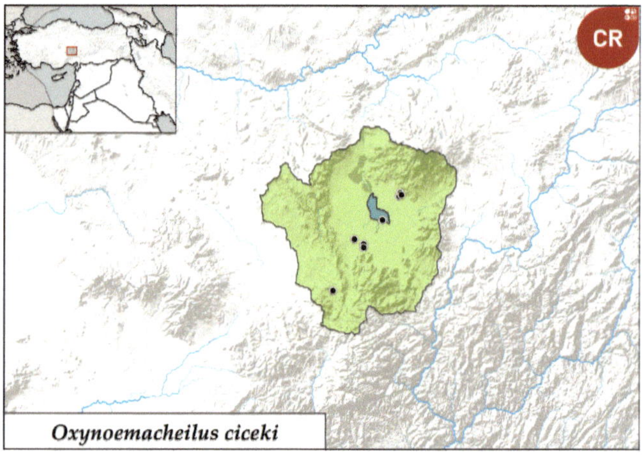

canal / ○ no median incision in upper lip / ○ interorbital distance 1.4–1.5 times in snout length / ○ caudal-peduncle depth 1.3–1.4 times in caudal-peduncle length / ○ 9+8 branched caudal rays / ○ posterior process of bony air-bladder capsule directed laterally. Size up to 77 mm SL.

Distribution. Iran and Iraq: Headwaters of Choman and Lesser Zab.

Habitat. Moderately fast-flowing waters in streams with gravel and rock substrate.

Biology. No data.

Conservation status. VU; appears to be declining within its small range.

Remarks. *Oxynoemacheilus chomanicus*, *O. zarzianus*, *O. zagrosensis*, and loaches identified as *O. frenatus* are superficially very similar and occur in Lesser Zab drainage. The distribution of these species needs to be better known, and identification may be difficult.

Further reading. Kamangar et al. 2014 (description); Sayyadzadeh & Esmaeili 2020 (discussion of Iranian and Anatolian populations of *O. frenatus*).

Oxynoemacheilus ciceki; Soysallı, Türkiye; 60 mm SL.

Oxynoemacheilus ciceki

Common name. Sultan loach.

Diagnosis. Distinguished from other species of *Oxynoemacheilus* in Central Anatolia and rivers flowing to Bay of Antalya by: ● scales absent / ○ caudal-peduncle depth 1.1–2.2 times in caudal-peduncle length / ○ caudal slightly emarginate / ○ flank with roundish blotches in variable size, without bars / ○ a small median incision in upper lip / ○ pelvic axillary lobe very small or absent / ○ lateral line complete. Size up to 60 mm SL.

Distribution. Türkiye: Sultan marshes in Develi depression.

Habitat. Standing water of springs on gravel substrate.

Biology. No data.

Conservation status. CR; see comments under *Pseudophoxinus elizavetae*.

Further reading. Sungur et al. 2018a (description); Soyubelli 2018 (biology); Çiçek & Sungur 2020 (conservation).

Oxynoemacheilus cilicicus; lower Seyhan drainage, Türkiye; 50 mm SL.

Oxynoemacheilus cilicicus

Common name. Cilician loach.

Diagnosis. Distinguished from other species of *Oxynoemacheilus* in Cilicia and northern Levant by: ○ caudal deeply emarginate, middle caudal ray 1.2–1.4 times in length of longest ray of upper caudal lobe / ○ lateral line with 13–23 pores terminating anterior to dorsal origin or slightly behind dorsal base / ○ infraorbital canal with 8–10 pores / ○ suborbital groove absent in male / ○ flank and back behind dorsal base covered by scales / ○ caudal–peduncle depth 1.0–1.4 times in its length / ○ body depth at dorsal origin 1.0–1.2 times in HL / ○ body depth at dorsal origin 20–24 % SL / ○ head length 22–25 % SL / ○ flank pattern mottled or with many, very narrow, irregularly shaped and set, pale-brown bars, usually flank-bars separated from middorsal saddles and not meeting contralateral. Size up to 58 mm SL.

Distribution. Türkiye: Lower Göksu east to Ceyhan drainage.

Habitat. Small streams to larger rivers. In slow or moderately fast-flowing stretches, usually along banks. Also along banks of reservoirs and in irrigation canals.

Biology. No data.

Conservation status. LC.

Remarks. Distribution of this species and superficially similar *O. ceyhanensis* from upper Ceyhan needs to be better explored.

Further reading. Kaya et al. 2020b (description).

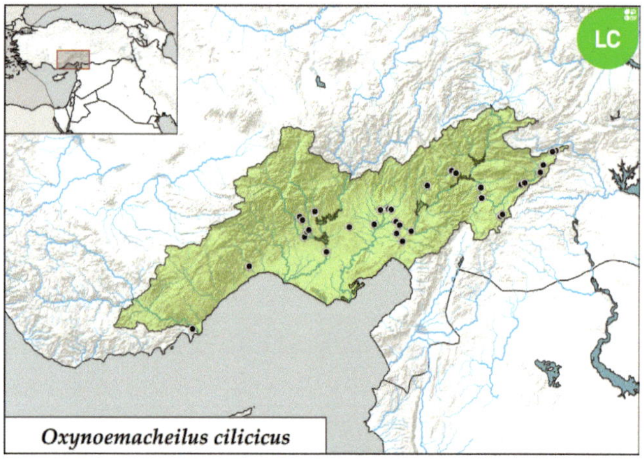

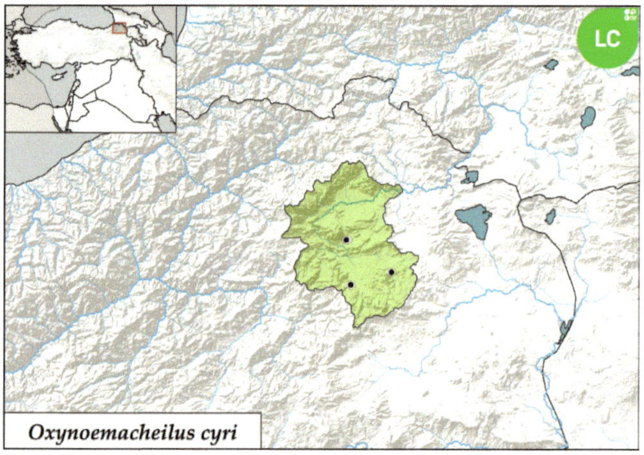

Oxynoemacheilus cyri; Kura drainage, Türkiye; ~60 mm SL. © M. Özuluğ.

Oxynoemacheilus cyri

Common name. Kura banded loach.

Diagnosis. Distinguished from other species of *Oxynoemacheilus* in Caspian basin by: • suborbital groove absent in male / • flank with 12–17 distinct, regularly shaped and set bars / • caudal very slightly emarginate or truncate. Size up to 74 mm SL.

Distribution. Türkiye: Ardahan plateau in upper Kura drainage.

Habitat. Slow or fast-flowing streams with gravel or rock bottoms and clear water.

Biology. No data.

Conservation status. LC; has a small range in upper Kura drainage; its habitat is intact, the species is common, and no threats are known.

Remarks. It is a phylogenetically isolated species, restricted to a small area at high altitudes. It appears to be endemic to Türkiye and has never been found in downstream Armenia.

Further reading. Berg 1910 (description); Kaya et al. 2020a (distribution).

Oxynoemacheilus eliasi; Şaşal, Türkiye; ~55 mm SL.

Oxynoemacheilus eliasi

Common name. Küçük Menderes loach.

Diagnosis. Distinguished from other species of *Oxynoemacheilus* in Aegean basin by: ○ inner axial stripe absent / ○ tip of pectoral usually not reaching to or reaching pelvic origin in male / ○ belly with small embedded scales / ○ body depth decreasing between dorsal and caudal base / ○ scales present from midline between pectoral- and pelvic bases to pelvic base / ○ caudal–peduncle depth 1.3–1.5 times in its length / ○ caudal emarginate, middle caudal ray 71–84 % of length of longest ray in upper caudal lobe / ○ two dark-brown blotches on caudal base, rarely connected to each other / ○ flank with bars and/or blotches. Size up to 65 mm SL.

Distribution. Türkiye: Küçük Menderes, Tahtalı and Gediz drainages.

Habitat. Standing and flowing waters of springs and streams on gravel substrate.

Biology. No data.

Conservation status. EN; appears to be declining within its small range.

Further reading. Yoğurtçuoğlu et al. 2022 (description).

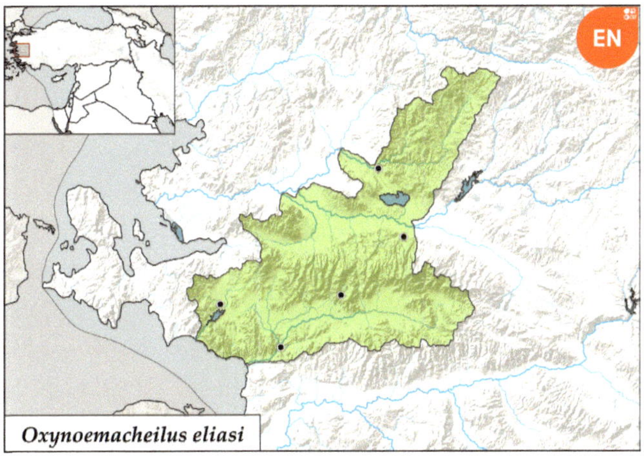

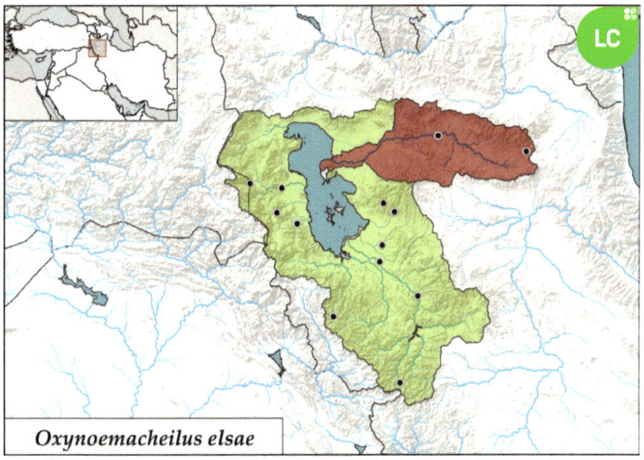

Oxynoemacheilus elsae; Lake Urmia basin, Iran; 55 mm SL.

Oxynoemacheilus elsae

Common name. Urmia loach.

Diagnosis. Distinguished from other species of *Oxynoemacheilus* in Caspian basin by: ○ caudal emarginate, usually with 3–6, narrow, pale-brown band / ○ an upper and a lower bold, black or dark-brown blotch on caudal base, often fused into a bar, it's colour distinctly darker than bars on caudal peduncle / ○ flank with irregularly shaped and spaced, dark-brown bars, often split in middle, more narrow than interspaces, or flank is almost plain brown / ○ bars on caudal peduncle not interrupted in size and shape at or above lateral midline / ○ suborbital groove present in male / ○ caudal–peduncle depth 2.3–2.9 times in its length. Size up to 68 mm SL.

Distribution. Iran and Türkiye: Zarrineh, Simineh, Mahabad, Barandoz, Nazlo, and Soufi drainages in Lake Urmia basin.

Habitat. Rivers and streams with fast-flowing water and gravel or boulder substrate.

Biology. No data.

Conservation status. LC.

Remarks. *Oxynoemacheilus brandtii* from Aras shares mitochondrial DNA with this species, but these fish can be identified as *O. brandtii* and not as *O. elsae.*

Further reading. Eagderi et al. 2018b (description); Kaya 2020b (distribution in Türkiye); Freyhof et al. 2021b (diagnosis).

Oxynoemacheilus ercisianus; Zilan, Lake Van basin, Türkiye; ~60 mm SL. © M. Özuluğ.

Oxynoemacheilus ercisianus

Common name. Van loach.

Diagnosis. Distinguished from other species of *Oxynoemacheilus* in adjacent Euphrates drainage by: ○ scales absent / ○ lateral line incomplete not reaching or reaching below dorsal base / ○ suborbital groove absent in male / ○ caudal slightly emarginate / ○ supratemporal canal without central pore / ○ pelvic axillary lobe absent / ○ dorsal adipose crest on caudal peduncle high, not reaching to vertical through last anal ray base in individuals larger than 50 mm SL / ○ caudal–peduncle length 1.0–1.3 times its depth / ○ body width at dorsal origin 13–17 % SL / ○ flank with a marmorated pattern or with large, irregularly shaped and narrowly spaced dark-brown bars or blotches, usually more prominent on flank behind dorsal base / ○ caudal base with mottled colour pattern, without bold, black bar or bold, black spots upper and lower caudal base. Size up to 82 mm SL.

Distribution. Türkiye: Lake Van and its tributaries Zilan, Deli, Bendimahi and Karasu.

Habitat. Moderately flowing streams and rivers with gravel or rock substrate. Also, in microbialites of Lake Van itself, at a maximum depth of 13 m and about 500 m offshore.

Biology. No data.

Conservation status. VU; appears to be declining within its small range.

Remarks. *Nemacheilus pulsiz* is a synonym.

Further reading. Erk'akan & Kuru 1986 (description); Krupp 1992a (description of *Nemacheilus pulsiz*); Freyhof et al. 2019b (diagnosis, distribution); Akkuş et al. 2021 (habitat in Lake Van).

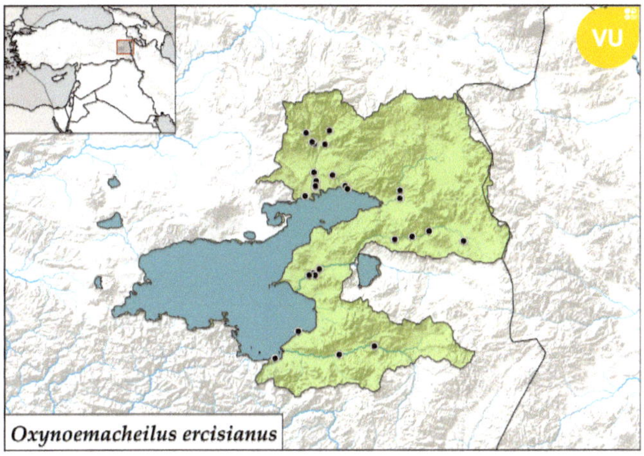

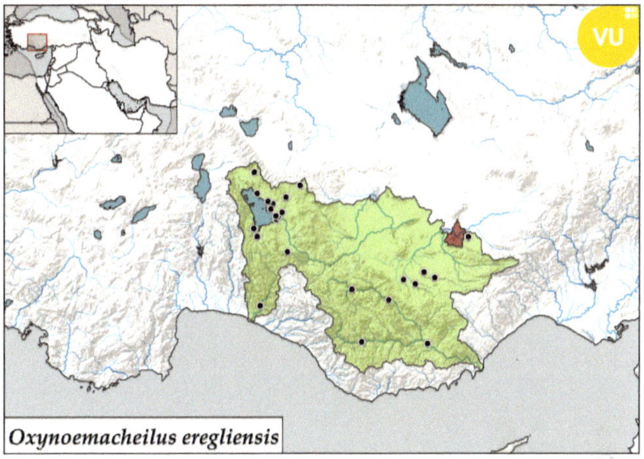

Oxynoemacheilus eregliensis; Lake Beyşehir basin, Türkiye; ~70 mm SL.

Oxynoemacheilus eregliensis

Common name. Ereğli loach.

Diagnosis. Distinguished from other species of *Oxynoemacheilus* in Central Anatolia and rivers flowing to Bay of Antalya by: ○ caudal slightly emarginate, shortest middle caudal ray 83–91 % of longest ray of upper caudal lobe / ○ belly without scales / ○ no prominent dorsal crest on caudal peduncle / ○ dorsal adipose crest reaching behind vertical of posterior anal base, absent in some individuals / ○ flank usually with irregularly set and shaped, distinct, large blotches forming a marbled pattern / ○ few isolated scales on flank in front of dorsal origin / ○ head length 24–26 % SL / ○ caudal–peduncle depth 1.5–1.9 times in its length. Size up to 90 mm SL.

Distribution. Türkiye: Lake Beyşehir and Suğla basins, Yeşildere (Karaman) in Central Anatolia, also in Göksu and Manavgat drainage in southern Anatolia.

Habitat. Small streams and springs with gravelly, sandy, or muddy bottoms and slow-flowing water.

Biology. No data.

Conservation status. VU; appears to be declining within its small range. Extirpated in Ereğli, elsewhere often in very small water bodies.

Remarks. *Oxynoemacheilus atili* is a synonym.

Further reading. Bănărescu et al. 1978 (description); Yoğurtçuoğlu et al. 2022 (distribution); Innal 2019 (size); Kaya et al. 2025a (record in Göksu).

Oxynoemacheilus euphraticus; Göksu, Euphrates drainage, Türkiye; 60 mm SL.

Oxynoemacheilus euphraticus; Great Zab drainage, Iraq; 55 mm SL.

Oxynoemacheilus euphraticus

Common name. Fırat loach.

Diagnosis. Distinguished from other species of *Oxynoemacheilus* in Euphrates and Tigris drainages by: ○ two distinct and prominent black blotches at posteriormost caudal base, often overlaid by a dark-brown or black chevron-shaped bar / ○ lateral line complete / ○ juveniles larger than 20 but smaller than 45 mm SL with poorly developed mottling between midlateral blotches on flank / ○ last bars on caudal peduncle as wide or wider than interspaces / ○ upper lip with well-developed median incision / ○ flank covered by scales / ○ caudal forked / ○ caudal peduncle without high adipose crest or ridge / ○ suborbital groove present in male / ○ caudal peduncle, and in some individuals also flank, with bars / ○ caudal–peduncle depth 2.0–2.8 times in caudal–peduncle length. Size up to 70 mm SL.

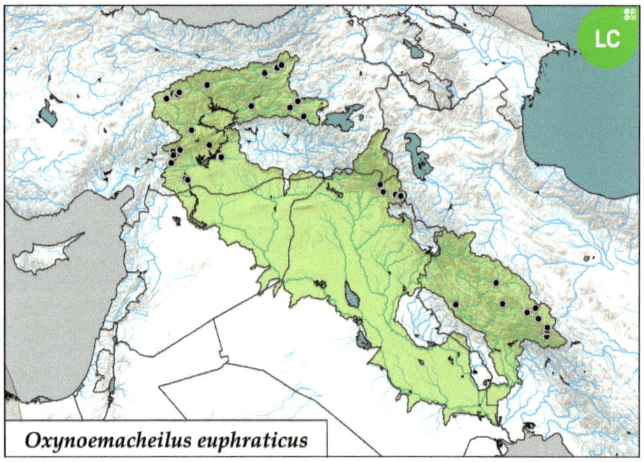

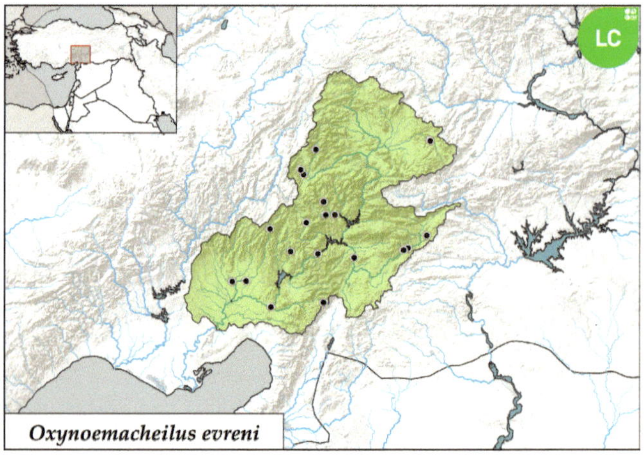

Distribution. Euphrates, Tigris, and Karun drainages.
Habitat. Streams and rivers with gravel substrate and fast to very fast-flowing water.
Biology. No data.
Conservation status. LC.
Remarks. The distribution of this species is still incompletely known and it is not recorded from Lesser Zab and Meymeh. It is superficially very similar to *O. kurdistanicus* and *O. marunensis*, and these species appear to have formed hybrid populations in some parts of the Tigris drainage. *Oxynoemacheilus freyhofi* from the Karun drainage is a synonym.
Further reading. Bănărescu & Nalbant 1964 (description); Jouladeh-Roudbar et al. 2016e (description as *O. freyhofi*); Freyhof & Özuluğ 2017 (synonyms).

Oxynoemacheilus evreni; Ceyhan drainage, Türkiye; 60 mm SL.

Oxynoemacheilus evreni
Common name. Ceyhan sportive loach.
Diagnosis. Distinguished from other species of *Oxynoemacheilus* in Cilicia and northern Levant by: ○ caudal emarginate, middle caudal ray 80–91 % of length of longest ray in upper caudal lobe / ○ caudal with dark-brown bands of elongated blotches / ○ a yellowish triangle at upper and lower posteriormost portions of caudal peduncle / ○ one central, bold-brown blotch or no blotch on caudal base, often an irregularly shaped black bar at caudal base / ○ lateral line complete / ○ flank and back behind dorsal base covered by scales / ○ suborbital groove present in male / ○ no prominent dorsal crest on caudal peduncle / ○ flank posterior to dorsal base and caudal peduncle with a series of bold, usually irregularly shaped bars reaching from dorsal midline almost to ventral midline / ○ caudal-peduncle depth 1.8–2.4 times in its length. Size up to 68 mm SL.
Distribution. Türkiye: Ceyhan drainage.
Habitat. Small streams to larger rivers. In fast-flowing stretches usually in riffles with gravel substrate.
Biology. No data.
Conservation status. LC.
Remarks. Seyhan drainage records are subject to confirmation.
Further reading. Erk'akan et al. 2007 (description); Erk'akan et al. 2014 (size); Freyhof et al. 2021a (distribution).

Oxynoemacheilus fatmae; Güzelhisar drainage, Türkiye; 70 mm SL.

Oxynoemacheilus fatmae
Common name. Güzelhisar loach.
Diagnosis. Distinguished from other species of *Oxynoemacheilus* in Aegean basin by: ○ belly with few, small, embedded scales / ○ caudal emarginate, middle caudal ray 71–91 % of length of longest upper caudal ray / ○ body depth almost uniform between dorsal and caudal base (ratio of body depth at posterior of dorsal base to body depth at caudal base is equal or almost equal to 1) / ○ body depth at dorsal origin 15–17 % SL / ○ caudal peduncle 10–12 % SL / ○ length of middle caudal ray 16–19 % SL / ○ no dorsal and ventral adipose crests on caudal peduncle behind vertical of posterior anal base. Size up to 70 mm SL.
Distribution. Türkiye: Güzelhisar drainage in northeastern Aegean basin.
Habitat. Moderate to fast-flowing streams with gravel substrate.
Biology. No data.
Conservation status. VU.
Further reading. Turan et al. 2024b (description).

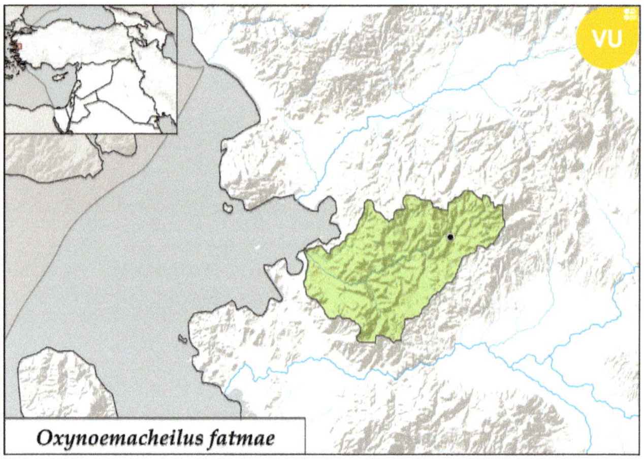

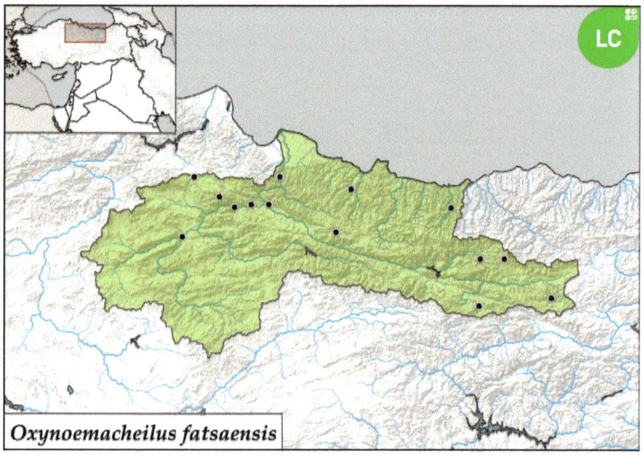

Oxynoemacheilus fatsaensis; Yeşilırmak drainage, Türkiye; 50 mm SL. © M. Özuluğ.

Oxynoemacheilus fatsaensis
Common name. Yeşilırmak sportive loach.
Diagnosis. Distinguished from other species of *Oxynoemacheilus* in Black Sea basin by: ○ caudal–peduncle depth 7–9 % SL, usually 2.0–2.6 times in its length, without high adipose crest or ridge / ○ caudal emarginate, middle caudal ray 80–90 % of length of longest ray in upper caudal lobe / ○ lateral line complete / ○ flank covered by scales / ○ suborbital groove present in male / ○ one central or no black, grey or brown blotch or short bar on caudal base, its colour identical to blotches on caudal peduncle / ○ no yellowish triangles at posteriormost portion of caudal peduncle / ○ pelvic very long, reaching or almost reaching to anus / ○ pectoral in male very long, reaching or almost reaching to pelvic origin / ○ flank mottled or with large irregularly shaped, squarish or vertically elongate blotches, usually as wide or narrower than interspaces / ○ lateral blotches on caudal peduncle usually interrupted in size and/or shape at or above lateral midline. Size up to 75 mm SL.
Distribution. Türkiye: Yeşilırmak drainage and adjacent coastal stream Elekçi.

Habitat. Fast-flowing water in streams and rivers with gravel substrate.
Biology. No data.
Conservation status. LC.
Remarks. No external characters are known to distinguish this species from *O. bergianus*, *O. banarescui*, and *O. simavicus*, and all species are diagnosed only by molecular characters. Published differences are based on small sample sizes or are spurious. As these four species are allopatric, the distribution areas should help in identification.
Further reading. Saygun et al. 2021 (description); Freyhof et al. 2022 (revision of *O. bergianus* group).

Oxynoemacheilus frenatus; Tigris at Diyarbakır, Türkiye; ~70 mm SL.

Oxynoemacheilus frenatus

Common name. Mesopotamian loach.
Diagnosis. Distinguished from other species of *Oxynoemacheilus* in Tigris drainage by: ○ lateral line incomplete, terminating in front of or above anal base / ○ caudal slightly emarginate / ○ flank mottled or marbled without bars / ○ colour pattern on anterior part of flank, often interrupted by an unpigmented zone along lateral line / ○ a shallow dorsal crest on caudal peduncle / ○ one central pore in supratemporal canal / ○ suborbital groove absent in male / ○ scales present on back and flank in front of anus. Size up to 75 mm SL.
Distribution. Mosul and upper Tigris drainage down to Batman drainage.
Habitat. Moderately fast-flowing to standing waters of springs, streams, and rivers with muddy or gravelly bottoms.
Biology. No data.
Conservation status. LC.
Remarks. Recently not recorded from rivers downstream of Batman. Superficially, very similar but genetically very different loaches occur in Iranian Lesser Zab, and these have also been identified as *O. frenatus*. It is impossible to say whether the populations from the upper Tigris are the same as those from Mosul. Only when new material from Mosul becomes available will the identity of Anatolian and Iranian populations be finally resolved.
Further reading. Heckel 1843 (description); Freyhof et al. 2017a (discussion of Anatolian and Iranian populations); Sayyadzadeh & Esmaeili 2020 (discussion of Iranian and Anatolian populations).

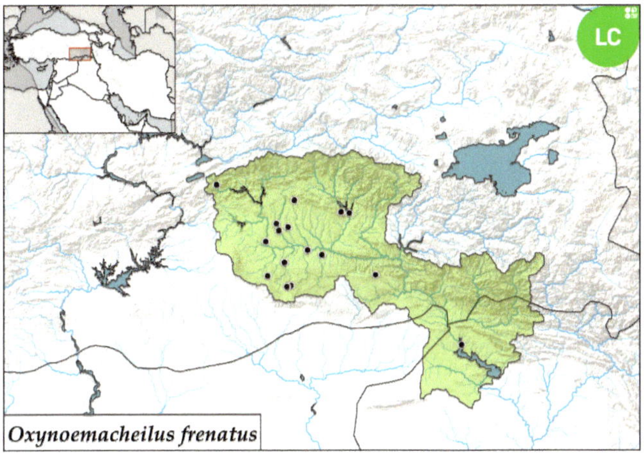

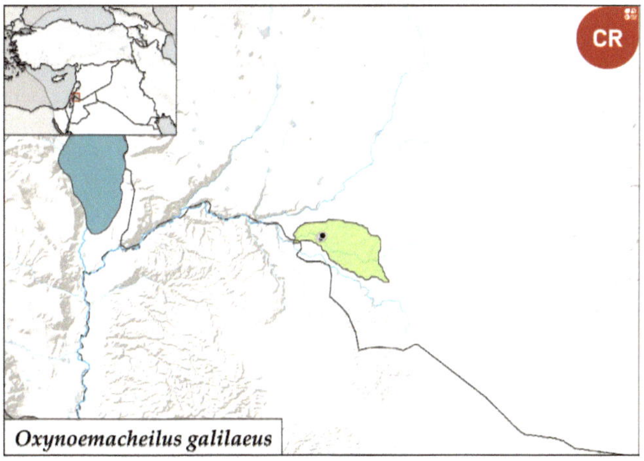

Oxynoemacheilus galilaeus; Lake Muzayrib, Syria; ~55 mm SL.

Oxynoemacheilus galilaeus
Common name. Galilean loach.
Diagnosis. Distinguished from other species of *Oxynoemacheilus* in southern Levant by: • 9–11½ branched dorsal rays / • body scaleless / ○ caudal slightly emarginate or truncate / ○ no dorsal crest on caudal peduncle / ○ suborbital groove absent in male. Size up to 64 mm SL.
Distribution. Syria: Lake Muzayrib basin.
Habitat. Lacustrine, along shores.
Biology. No data.
Conservation status. CR; Lake Muzayrib has been reported to be drying up, and the species may be extinct. No fieldwork has been carried out since 2008 to investigate the situation.
Remarks. Described from Lake Tiberias but later speculated to be from Lake Hula or Lake Muzayrib. The single type specimen corresponds well to the Lake Muzayrib population, while all fish we see from Tiberias and Hula do not. Fish from Hula and Tiberias are similar to the type of *O. galilaeus* in being scaleless and in number of anal and dorsal rays (all individuals from Hula are reported to have 6½ branched anal rays). These fish also have a very different colour pattern from *O. galilaeus* (large spots or bars on the flank vs. fine mottled pattern in *O. galilaeus*) and body shape (deep body vs. slender in *O. galilaeus*), and we identify them as *O. leontinae*. *Oxynoemacheilus galilaeus* is diagnosed by 6½ branched anal rays; this character is variable (Muzayrib), and few individuals have 6½ (including type), most have 5½ branched anal rays. It has been placed in a separate genus: *Nun*, based on fish from the Hula basin, but as *Nun* was described based on *O. leontinae*, this is its type species.
Further reading. Günther 1865 (description); Bănărescu et al. 1982 (description of *Nun* based on Hula population); Krupp & Schneider 1989 (distribution); Bănărescu & Nalbant 1995 (Figure of *Nun*); Prokofiev 2009 (osteology of Hula population); Freyhof et al. 2011 (generic placement based on Muzayrib population); Prokofiev & Golubtsov 2013 (revalidation of *Nun* based on Hula population).

Lake Muzayrib in Syria is the only habitat known for *Oxynoemacheilus galilaeus*. The lake has been reported to have dried out, and the species might now be extinct.

Oxynoemacheilus germencicus; Büyük Menderes drainage, Türkiye; ~60 mm SL.

Oxynoemacheilus germencicus

Common name. Carian loach.

Diagnosis. Distinguished from other species of *Oxynoemacheilus* in Aegean basin by: ○ caudal emarginate / ○ flank with scales / ○ caudal–peduncle depth 1.5–2.2 times in its length / ○ no prominent dorsal crest on caudal peduncle / ○ suborbital groove present in male / ○ scales on belly restricted between pelvic bases, rarely exceeding anteriorly / ○ inner axial stripe absent / ○ tip of anal reaching to or slightly beyond ventral adipose crest, if present / ○ belly with few, small, embedded scales / ○ tip of pectoral usually not reaching to or reaching pelvic origin in male / ○ body depth decreasing between dorsal and caudal base / ○ flank with bars and/or blotches, rarely plain or mottled. Size up to 83 mm SL.

Distribution. Türkiye: Büyük Menderes and Gediz drainages.

Habitat. Moderately fast-flowing to stagnant waters of springs, streams, and rivers with muddy or gravelly bottoms.

Biology. No data.

Conservation status. LC.

Remarks. *Oxynoemacheilus cinicus* and *O. mesudae*, both from Büyük Menderes, are synonyms.

Further reading. Erk'akan et al. 2007 (description of *O. germencicus* and *O. cinicus*); Erk'akan 2012 (description of *O. mesudae*); Yoğurtçuoğlu et al. 2022 (revision, distribution).

Oxynoemacheilus germencicus; Büyük Menderes drainage, Türkiye, 46 mm SL.

Oxynoemacheilus germencicus; Büyük Menderes drainage, Türkiye, 55 mm SL.

Oxynoemacheilus germencicus; Işıklı spring, Türkiye, 79 mm SL.

The Zalm in Iraqi Kurdistan is a tributary to the Sirvan. It is the habitat of *Oxynoemacheilus gyndes*, *O. hanae*, *O. marunensis*, *Paracobitis molavii*, and *Turcinoemacheilus inexpectatus*.

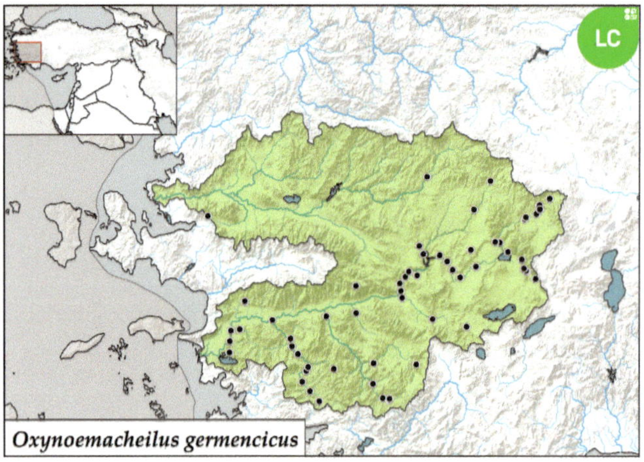

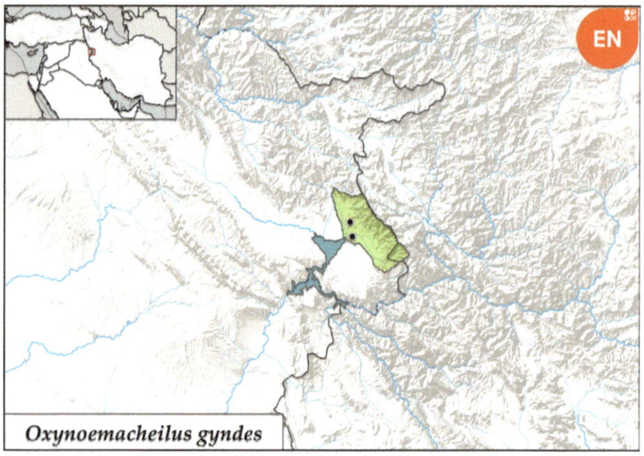

Oxynoemacheilus gyndes; Sirvan drainage, Iraq; female, 54 mm SL.

Oxynoemacheilus gyndes; Sirvan drainage, Iraq; male, 52 mm SL.

Oxynoemacheilus gyndes

Common name. Zalm loach.

Diagnosis. Distinguished from other species of *Oxynoemacheilus* in Tigris drainage by: • lateral line incomplete, terminating slightly behind pectoral base / • body naked except very dorsal part of caudal peduncle / • a midlateral series of small, horizontally elongated, dark-brown blotches often fused into an irregularly shaped midlateral stripe, one additional stripe above and below midlateral stripe in many individuals / ○ 3–4 lateral pores and no central pore in supratemporal canal / ○ caudal slightly emarginate / ○ suborbital groove absent in male / ○ upper lip without median incision. Size up to 57 mm SL.

Distribution. Iraq: Zalm, a headwater of Sirvan in Tigris drainage.

Habitat. Moderately fast-flowing and standing waters of streams, backwaters, and springs with muddy bottoms.

Biology. No data.

Conservation status. EN; appears to be declining within its very small range.

Remarks. *Oxynoemacheilus gyndes* is a very distinctive species, not closely related to other loaches in its genus.

Further reading. Freyhof & Abdullah 2017 (description).

Oxynoemacheilus hamwii; lower Orontes drainage, Türkiye; 60 mm SL.

Oxynoemacheilus hamwii

Common name. Orontes sportive loach.

Diagnosis. Distinguished from other species of *Oxynoemacheilus* in Cilicia and northern Levant by: ● yellowish triangle at upper and lower posteriormost portions of caudal peduncle / ○ one central, bold-brown blotch or no blotch on caudal base, often an irregularly shaped black bar at caudal base / ○ caudal deeply emarginate with dark-brown bands of elongated blotches / ○ middle caudal ray 64–74 % of length of longest ray in upper caudal lobe / ○ lateral line complete or almost complete terminating behind vertical of anus / ○ flank covered by scales, sparsely set on anterior, densely set on posterior flank / ○ no prominent dorsal crest on caudal peduncle / ○ suborbital groove present in male / ○ flank in juveniles with a midlateral series of horizontally elongate blotches, adults with a brown marbled pattern / ○ caudal–peduncle depth 1.4–1.9 times in its length. Size up to 64 mm SL.

Distribution. Orontes drainage in Türkiye and northern Syria (Afrin).

Habitat. Moderately fast-flowing water in streams and rivers with gravel substrate.

Biology. No data.

Conservation status. VU; appears to be extirpated in lower Orontes and is declining within its small range.

Remarks. Records of this species from the upper Orontes refer to *O. shehabi*, and records from the Euphrates and Tigris are misidentifications of other species, most likely *O. argyrogramma*, *O. chaboras*, and *O. euphraticus*.

Further reading. Krupp & Schneider 1991 (description).

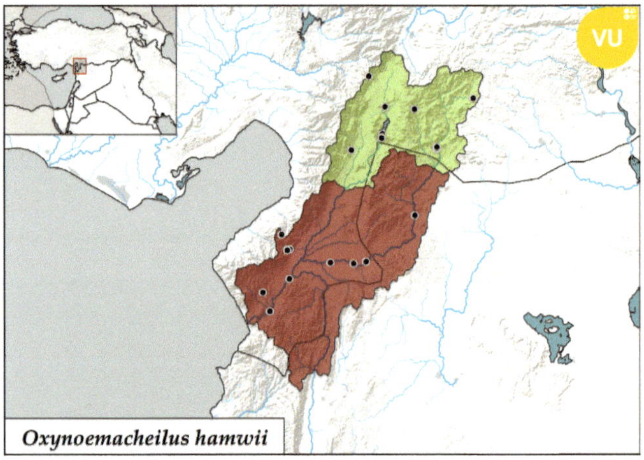

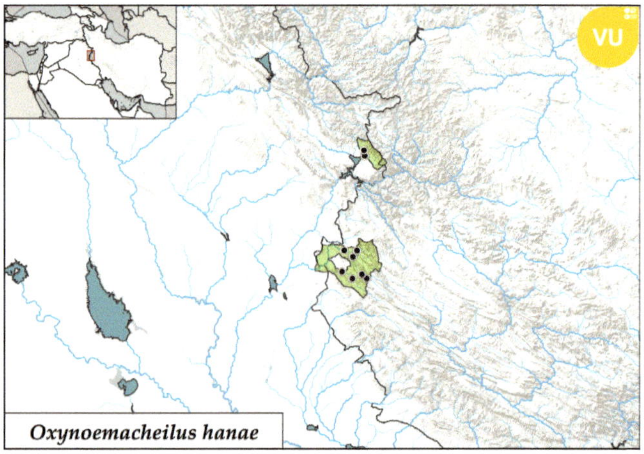

Oxynoemacheilus hanae; Sirvan drainage, Iraq; ~50 mm SL.

Oxynoemacheilus hanae

Common name. Sirvan two-spot loach.

Diagnosis. Distinguished from other species of *Oxynoemacheilus* in Tigris drainage by: ○ flank with a distinct series of midlateral blotches, fused to each other or fused into a stripe / ○ isolated patches of blotches or a row of small dark-brown spots below lateral series of blotches / ○ two distinct and prominent black blotches at posteriormost caudal base, usually overlaid by a dark-brown or black chevron-shaped bar / ○ suborbital groove present in male / ○ no or a very

short median incision in upper lip / ○ lateral line complete, terminating behind vertical of anal base or at caudal base / ○ flank completely covered by scales / ○ caudal deeply emarginate or forked. Size up to 61 mm SL.
Distribution. Iran and Iraq: Headwaters of Sirvan and Alvand drainages.
Habitat. Moderately fast-flowing streams and springs with gravel substrate.
Biology. No data.
Conservation status. VU; appears to be declining within its small range.
Remarks. The Iranian population is superficially very similar, but their COI sequence data place them close to *O. marunensis*. They may represent *O. hanae*, introgressed by *O. marunensis*. In the upper Sirvan, *O. hanae* and *O. marunensis* occur in syntopy.
Further reading. Freyhof & Abdullah 2017 (description); Jouladeh-Roudbar et al. 2020 (record from Iran).

Oxynoemacheilus hazarensis; Lake Hazar, Türkiye; ~65 mm SL.

Oxynoemacheilus hazarensis

Common name. Hazar loach.
Diagnosis. Distinguished from other species of *Oxynoemacheilus* in Tigris drainage by: ● flank naked, only caudal peduncle behind anus with scales / ○ suborbital groove present in male / ○ lateral line incomplete / ○ caudal slightly emarginate / ○ a deep incision in middle of upper lip / ○ mottled or marmorated colour pattern on flank, not interrupted by an unpigmented zone along lateral line. Size up to 67 mm SL.
Distribution. Türkiye: Lake Hazar basin.
Habitat. Lakeshores and moderately fast-flowing, stagnant waters of streams, backwaters, and springs with muddy bottoms.
Biology. No data.
Conservation status. EN; appears to be declining within its very small range.
Remarks. This species is not closely related to other species in the Tigris but belongs to a group of loaches that are mostly found in the Euphrates.
Further reading. Freyhof & Özuluğ 2017 (description); Freyhof et al. 2019b (diagnosis).

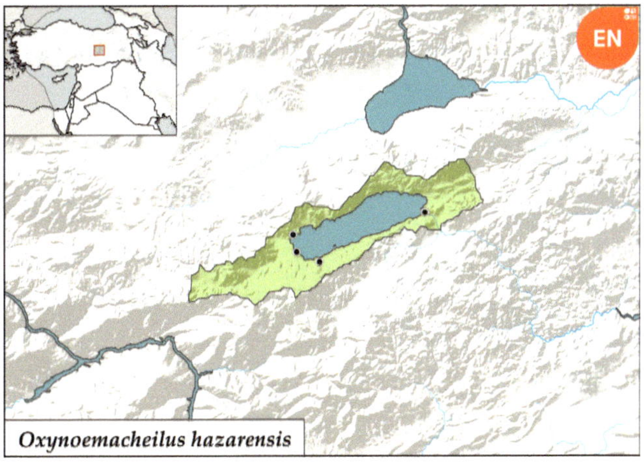

The Dead Sea basin and the Jordan. The present-day Jordan-Dead Sea basin results from tectonic activities that commenced during the Miocene/Pliocene transition. The Yizre'el Valley was formed during this period, and the Jordan Valley was lowered. During the Pliocene, the Jordan flowed from south to north and drained along with the Litani via Nahal Qishon toward the Mediterranean. During the Lower Pleistocene, the lowering of the Jordan-Wadi Araba graben continued, forming two endorheic basins: one in the Hula and the other in the Dead Sea region. At that time, the Jordan was no longer connected to the Litani but rather flowed southeastward into the Jordan Valley.

The Dead Sea with the typical salt crusts on stones along its shore (Jordan).

Between 1.50 and 1.25 Mya, the Hula basin became part of the Dead Sea drainage system by breaching the Korazim block, which had previously separated them. During the Pleistocene glacials, a series of successive lakes occupied the Jordan-Dead Sea graben. A large shallow freshwater lake (Lake Samra), covering an area from 50 km south of the Dead Sea to Lake Tiberias, existed during the Riss glacial period (425 ka–130 ka BP). At the same time, a lake entirely flooded the Hula basin, and a river connected the two lakes (Lake Samra and Hula Emeq). Between approximately 70,000 and 18,000 years ago, Lake Samra transformed into the saline Lake Lisan, which subsequently shrank and retreated to form the present-day Dead Sea. This lake was previously less saline than the present Dead Sea, as evidenced by fossil remnants of diatoms, fish, and other organisms. The most recent tectonic activity, approximately 18,000 years ago, resulted in a 300–400 m lowering of the valley to its present depth of −794 m at the bottom. This pronounced subsidence led to the deepening of Hula Emeq and the northern Dead Sea, giving rise to Lake Tiberias (14–18 ka BP). Lake Tiberias, with a surface area of approximately 170 km^2 and a depth of 43 m, is the largest natural freshwater lake in West Asia and the lowest freshwater lake on Earth, situated 209 m below sea level. The Dead Sea is the deepest hypersaline lake in the world, with a depth of almost 400 m below sea level. It is approximately 75 km long and 10–15 km wide. The lake is divided into two basins: a deep northern one (approximately 380 m deep) and a shallow southern one (approximately 2–3 m deep), separated by the Lisan Peninsula. Small springs and rivers drain into this terminal lake, but no surface watercourses ever reach it in the present era, except for Wadi al-Hasa and Wadi Moujib (both in Jordan) after heavy rain. **Further reading.** Alwan 2010 (Jordan).

Oxynoemacheilus insignis; Barada drainage, Syria; ~60 mm SL.

Oxynoemacheilus insignis

Common name. Palestine loach.

Diagnosis. Distinguished from other species of *Oxynoemacheilus* in southern Levant by: ○ caudal emarginate / ○ lateral line complete, terminating at or almost at caudal base / ○ back and flank covered by scales / ○ no prominent dorsal crest on caudal peduncle / ○ suborbital groove present in male. Size up to 100 mm SL.

Distribution. Jordan drainage, Dead Sea, and Damascus basins. In Damascus basin, now limited to Barada and Nahr al Awaj springs west of Damascus.

Habitat. Moderately fast-flowing to stagnant waters of springs, streams, and rivers with muddy or gravelly bottoms.

Biology. No data.

Conservation status. LC; likely extirpated from Barada, but still common in headwaters of Nahr al Aouaj and Jordan drainage.

Remarks. Diversity of loaches in Jordan drainage still needs to be better understood. *Oxynoemacheilus insignis* is widespread in all the eastern, left tributaries of the Jordan in Syria and Jordan, from the Yarmouk south to the Zarqa, Moujib, and tributaries of southern Dead Sea. *Noemacheilus insignis tortonesei* from the Zarqa and *N. angorae jordanicus* from Wadi Kufrinja, both in Jordan, are synonyms. *Oxynoemacheilus doriae*, *O. israeliticus* (both from the Bet-She'an Valley, Israel), and *O. pantheroides* (from the Golan Heights, Israel) are usually treated as synonyms of *O. insignis*. Molecular data strongly suggest that *O. pantheroides* is a valid species, possibly endemic to tributaries of Lake Tiberias. *Oxynoemacheilus doriae* and *O. israeliticus* should be re-examined based on fresh material from type localities, as they may also represent valid species.

Further reading. Bănărescu et al. 1982 (description of *O. dori* and *O. israeliticus*); Krupp & Schneider 1989 (synonym of *O. insignis*).

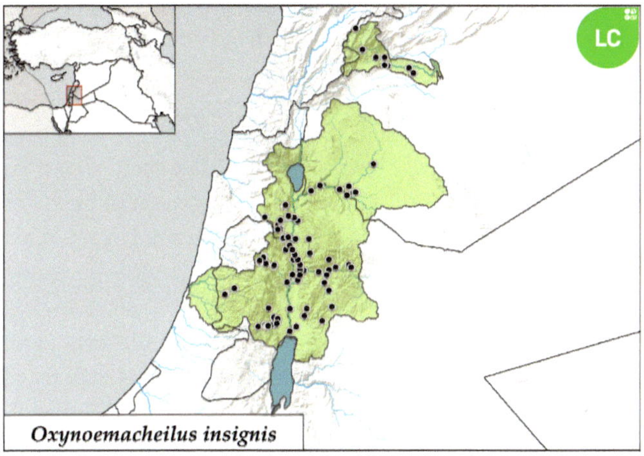

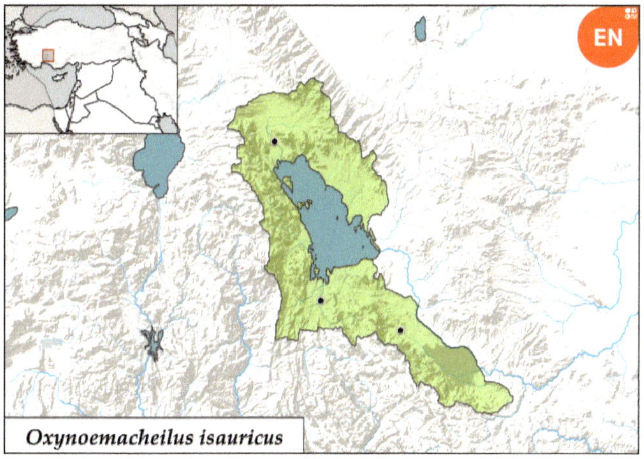

Oxynoemacheilus isauricus; Lake Beyşehir basin, Türkiye; ~50 mm SL.

Oxynoemacheilus isauricus

Common name. Beyşehir sportive loach.

Diagnosis. Distinguished from other species of *Oxynoemacheilus* in Central Anatolia and rivers flowing to Bay of Antalya by: ● caudal–peduncle depth 2.2–2.6 times in its length / ○ caudal deeply emarginate, shortest middle caudal ray 71–85 % of longest ray of upper caudal lobe / ○ preanal length 68–74 % SL / ○ head length 21–24 % SL / ○ body depth strongly decreasing between dorsal and caudal bases / ○ flank with scales / ○ flank with a series of irregularly shaped and set, vertically elongated midlateral blotches, usually disconnected from saddles on back, rarely a mottled or marbled pattern / ○ upper part of cheek without or with vermiculated dark-brown spots / ○ tip of anal not reaching ventral adipose crest / ○ prepelvic length 51–51 % SL / ○ suborbital groove present in male. Size up to 63 mm SL.

Distribution. Türkiye: Lake Beyşehir and Suğla basin. Detailed distribution is poorly known.

Habitat. Fast-flowing streams on gravel substrate.

Biology. No data.

Conservation status. EN; appears to be declining within its small range.

Further reading. Yoğurtçuoğlu et al. 2021b (description).

Oxynoemacheilus karunensis; Karkheh drainage, Iran; ~40 mm SL. © K. Borkenhagen.

Oxynoemacheilus karunensis

Common name. Karkheh two-spot loach.

Diagnosis. Distinguished from other species of *Oxynoemacheilus* in Tigris drainage by: ○ pelvic axillary lobe absent or rudimentary, shallow and knob-shaped / ○ two distinct and prominent black blotches at posteriormost caudal base, often overlaid by a dark-brown or black chevron-shaped bar / ○ suborbital groove present in male / ○ lateral line complete, terminating behind vertical of anal base or at caudal base / ○ flank completely covered by scales / ○ caudal deeply emarginate or forked / ○ no or very short incision in upper lip / ○ a series of vertically elongated blotches or short bars along lateral midline / ○ dark-brown blotches behind dorsal base narrower than interspaces / ○ caudal peduncle 1.7–3.1 times longer than deep. Size up to 58 mm SL.

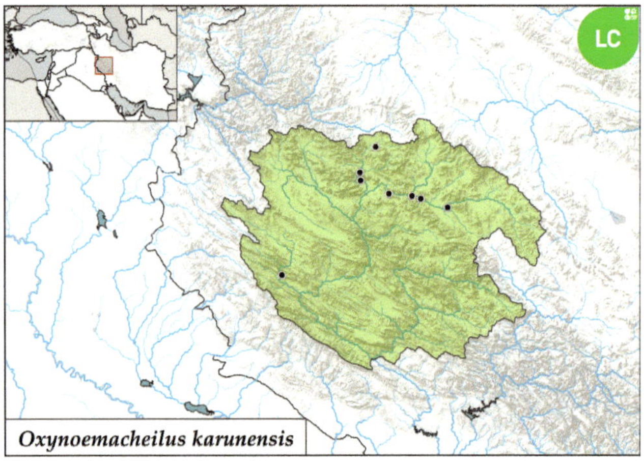

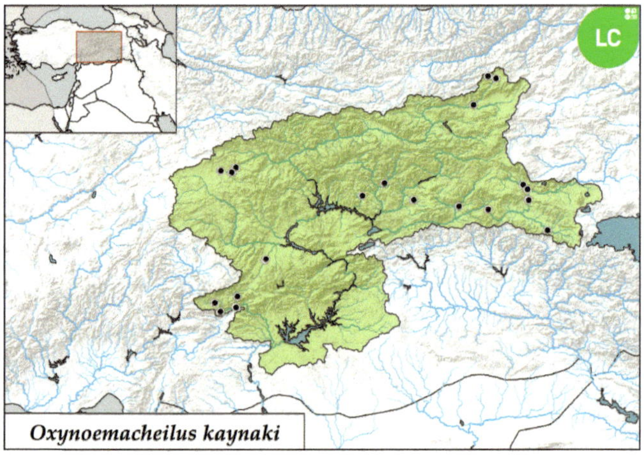

Oxynoemacheilus kaynaki; Euphrates drainage, Türkiye; 64 mm SL.

Oxynoemacheilus kaynaki; Euphrates drainage, Türkiye; 68 mm SL.

Distribution. Iran: Gamasiab, Dinevar, and upper Seymareh in Karkheh drainage.
Habitat. Moderately fast-flowing and standing waters of streams, backwaters, and springs with muddy bottoms.
Biology. No data.
Conservation status. LC.
Remarks. The type locally of this species was mistakenly thought to be in Karun drainage. It was later found to be endemic in Karkheh drainage. Fish identified as *O. karunensis* from the Jarrahi drainage had been described as *O. marunensis*.
Further reading. Freyhof 2016a (description).

Oxynoemacheilus kaynaki
Common name. Melid loach.
Diagnosis. Distinguished from other species of *Oxynoemacheilus* in Euphrates drainage by: ○ caudal peduncle with a shallow, wide, and often long dorsal adipose ridge, not elevated from dorsal profile, reaching to middle of caudal peduncle or below last dorsal ray when folded down / ○ lateral line complete / ○ suborbital groove absent in male / ○ pelvic axillary lobe present / ○ flank covered by scales / ○ caudal very slightly emarginate or truncate / ○ flank with a brown mottled or marmorated pattern, with irregularly shaped brown bars, especially on flank behind dorsal origin in some individuals / ○ caudal base with a bold, dark-brown or black bar. Size up to 68 mm SL.
Distribution. Türkiye: Upper Göksu, Kaynarca, and Gül, which are headwaters of Peri Suyu, Sultan Suyu, which flows to Karakaya reservoir, and Kangal, which flows to Keban reservoir, all in Euphrates drainage.
Habitat. Fast to moderately fast-flowing streams with gravel substrate. Usually along margins of streams with muddy substrate.
Biology. No data.
Conservation status. LC.
Further reading. Erk'akan et al. 2008b (description); Freyhof et al. 2019b (diagnosis, distribution).

Oxynoemacheilus kentritensis; Tigris drainage, Türkiye; 75 mm SL.

Oxynoemacheilus kentritensis

Common name. Botan loach.

Diagnosis. Distinguished from other species of *Oxynoemacheilus* in Tigris drainage by: ○ distinct bars or vertically elongated blotches on flank behind dorsal origin / ○ colour pattern on anterior part of flank not interrupted by unpigmented zone along lateral line / ○ 10+9, 9+9 or 9+8 branched caudal rays / ○ flank completely covered by scales / ○ caudal slightly emarginate / ○ suborbital groove absent in male / ○ one central pore in supratemporal canal / ○ no median incision in upper lip / ○ lateral line complete, terminating behind vertical of anal base or at caudal base / ○ posterior process of bony air-bladder capsule directed laterally / ○ maxillary barbel reaching beyond middle of eye, usually to posterior eye margin / ○ interorbital distance 1.6–1.9 times in snout length / ○ caudal–peduncle depth 1.2–1.3 times in caudal–peduncle length. Size up to 78 mm SL.

Distribution. Botan, as well as Hezil and Nerduş, entering Tigris in border area of Türkiye, Iraq, and Syria.

Habitat. Moderately fast-flowing and stagnant waters of gravel-bedded streams.

Biology. No data.

Conservation status. LC.

Further reading. Freyhof et al. 2017a (description).

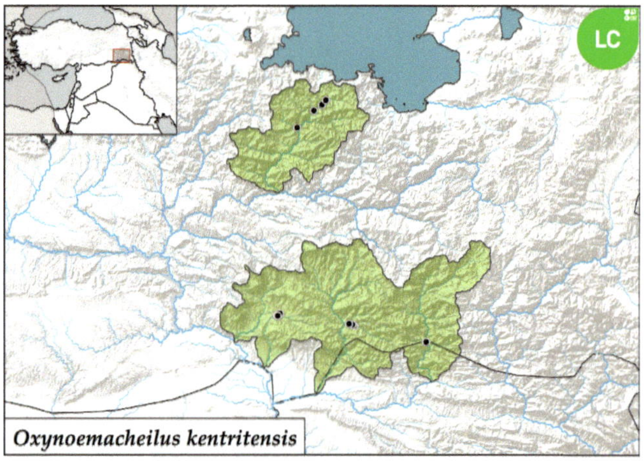

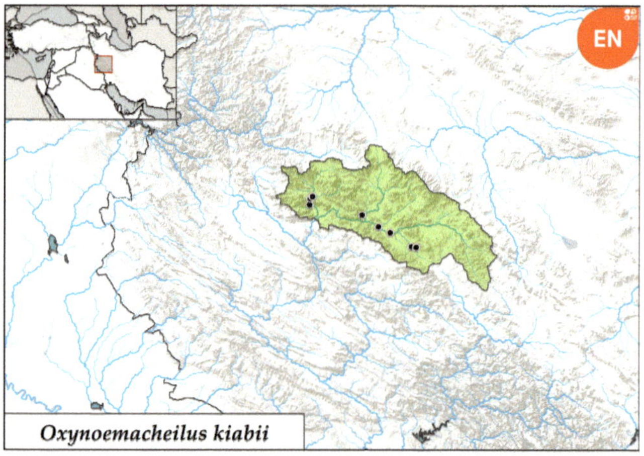

Oxynoemacheilus kiabii; Karkheh drainage, Iran; ~55 mm SL.

Oxynoemacheilus kiabii

Common name. Longhead Zagros loach.

Diagnosis. Distinguished from other species of *Oxynoemacheilus* in Tigris drainage by: ○ two, rarely one or three lateral pores in supratemporal canal and no central pore / ○ suborbital groove absent in male / ○ flank covered by embedded scales, no scales on back in front of dorsal / ○ caudal slightly emarginate / ○ lateral line incomplete, terminating under dorsal base or above anal base / ○ head length 26–30 % SL / ○ flank with vertically elongated, irregularly shaped blotches or narrow bars. Size up to 57 mm SL.

Distribution. Iran: Gamasiab and Dinevar in Karkheh drainage.

Habitat. Moderate to fast-flowing streams with gravel substrate.

Biology. No data

Conservation status. EN; appears to be declining within its very small range.

Further reading. Golzarianpour et al. 2011 (description).

Oxynoemacheilus kottelati; Havran drainage, Türkiye; ~65 mm SL.

Oxynoemacheilus kottelati

Common name. Havran loach.

Diagnosis. Distinguished from other species of *Oxynoemacheilus* in Aegean basin by: ○ no shallow dorsal crest on caudal peduncle / ○ suborbital groove present in male / ○ belly with few, small, embedded scales / ○ caudal emarginate / ○ middle caudal ray 71–91 % of length of longest upper caudal ray / ○ body depth decreasing between dorsal and caudal base / ○ caudal–peduncle depth 1.0–1.3 times in caudal–peduncle length / ○ caudal peduncle 13–16 % SL / ○ flank with coarse mottling, without bars or blotches. Size up to 65 mm SL.

Distribution. Türkiye: Havran and Karınca in northeastern Aegean basin.

Habitat. Moderate to fast-flowing streams with gravel substrate.

Biology. No data.

Conservation status. VU

Further reading. Turan et al. 2024a (description).

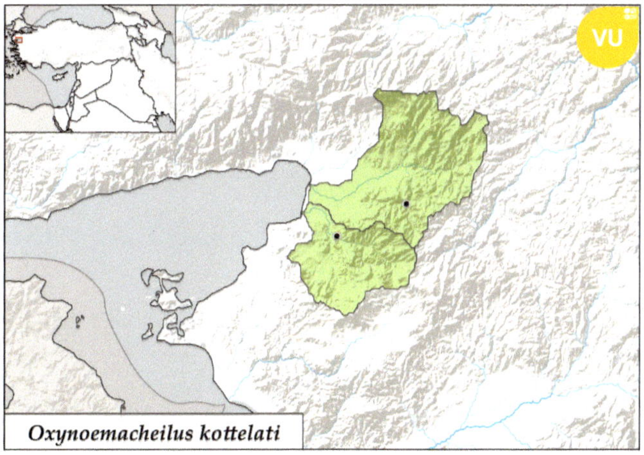

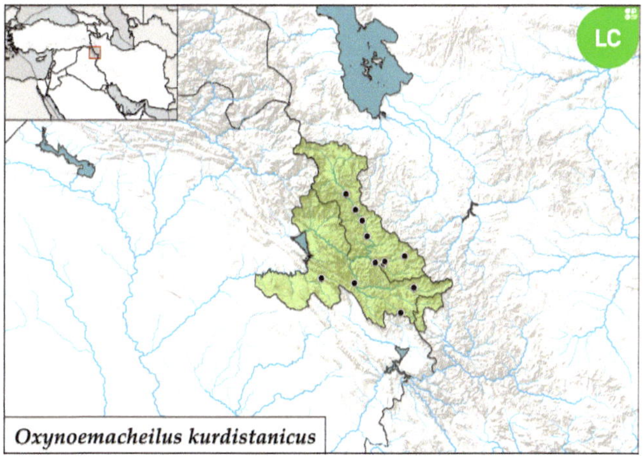

Oxynoemacheilus kurdistanicus; Lesser Zab drainage, Iraq; ~55 mm SL.

Oxynoemacheilus kurdistanicus; Lesser Zab drainage, Iraq; 59 mm SL.

Oxynoemacheilus kurdistanicus

Common name. Lesser Zab loach.

Diagnosis. Distinguished from other species of *Oxynoemacheilus* in Tigris drainage by: ○ suborbital groove present in male / ○ usually 9½ (rarely 8½) branched dorsal rays / ○ flank with a midlateral row of 6–12 dark-brown, irregularly shaped, vertically elongated bars / ○ last bars on caudal peduncle narrower than interspaces / ○ two distinct and prominent black blotches at posteriormost caudal base, often overlaid by a dark-brown or black chevron-shaped bar / ○ upper lip with a deep median incision / ○ pelvic axillary lobe absent or a very shallow pad / ○ suborbital groove present in male / ○ lateral line complete, terminating at caudal base / ○ body completely covered by scales / ○ caudal deeply emarginate or forked. Size up to 75 mm SL.

Distribution. Iran and Iraq: Lesser Zab drainage.

Habitat. Moderate to fast-flowing streams with gravel substrate.

Biology. No data.

Conservation status. LC.

Further reading. Kamangar et al. 2014 (description).

Oxynoemacheilus leontinae; Lake Tiberias, Israel; 40 mm SL.

Oxynoemacheilus leontinae

Common name. Tiberias loach.

Diagnosis. Distinguished from other species of *Oxynoemacheilus* in southern Levant by: ○ flank scaleless or with few isolated scales / ○ caudal very slightly emarginate or truncate / ○ lateral line terminating about at vertical of dorsal origin / ○ without prominent dorsal crest on caudal peduncle or with short crest on posterior caudal peduncle / ○ suborbital groove absent in male / ○ body depth at dorsal origin 16–20 % SL / ○ 8½-9½ branched dorsal rays / ○ flank with a series of irregularly shaped bars or large, roundish or vertically elongate blotches. Size up to 64 mm SL.

Distribution. Israel: Lake Hula and Tiberias basins.

Habitat. Lakeshores and small tributaries.

Biology. No data.

Conservation status. EN; appears to be declining within its very small range. Extirpated from Hula and now restricted to Tiberias basin.

Remarks. Often identified as *O. galilaeus* or *O. panthera*.

Further reading. Lortet 1883 (description); Krupp 1985d, Krupp & Schneider 1989 (distribution).

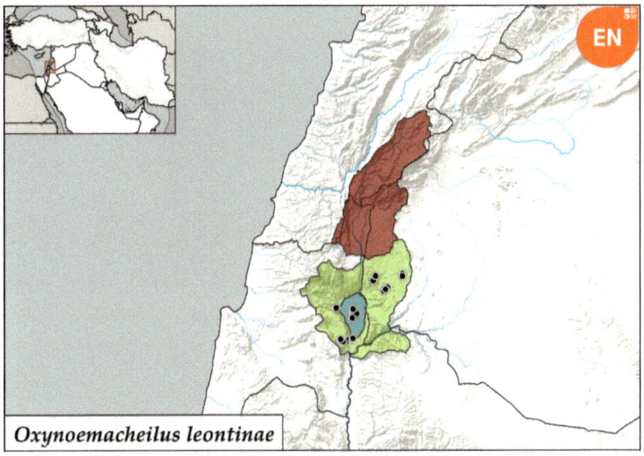

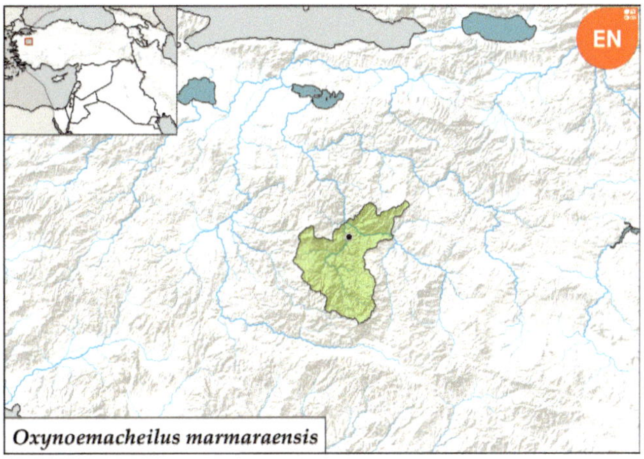

Oxynoemacheilus marmaraensis; Sursuluk drainage, Türkiye; ~75 mm SL.

Oxynoemacheilus marmaraensis

Common name. Marmara loach.

Diagnosis. Distinguished from other species of *Oxynoemacheilus* in Marmara and Black Sea basin by: ○ flank with mottled or vermiculate pattern / ○ no axillary lobe at pelvic base / ○ caudal-peduncle depth 1.2–1.6 times in its length / ○ caudal moderately emarginate, lobes slightly rounded / ○ no median incision in upper lip / ○ distance between pectoral and pelvic origins 29–34 % SL / ○ no prominent dorsal crest on caudal peduncle / ○ suborbital groove present in male. Size up to 75 mm SL.

Distribution. Türkiye: Emet drainage, a tributary of Susurluk.

Habitat. Moderately fast-flowing streams.

Biology. No data.

Conservation status. EN; poorly known and potentially more common and less threatened.

Further reading. Turan et al. 2023 (description).

Oxynoemacheilus marunensis; Jarrahi drainage, Iran; ~45 mm SL. © K. Borkenhagen.

Oxynoemacheilus marunensis; upper Tigris drainage, Türkiye; 75 mm SL.

Oxynoemacheilus marunensis

Common name. Tigris two-spot loach.

Diagnosis. Distinguished from other species of *Oxynoemacheilus* in Tigris drainage by: ○ flank with a midlateral row of 7–10 dark-brown, irregularly shaped, vertically elongated blotches, overlaid by a marmorated pattern / ○ juveniles larger than 20 but smaller than 45 mm SL with poorly developed mottling between midlateral blotches on flank / ○ two distinct and prominent black blotches at posteriormost caudal base, often overlaid by a dark-brown or black chevron-shaped bar ○ / upper lip r with a well-developed median incision / ○ pelvic axillary lobe small but well developed / ○ suborbital groove present in male / ○ lateral line complete, terminating behind vertical of anal base or at caudal base / ○ flank completely covered by scales / ○ caudal deeply emarginate or forked. Size up to 78 mm SL.

Distribution. Jarrahi drainage (Iran), Sirvan (Iran, Iraq), and upper Tigris and its tributaries (Türkiye).

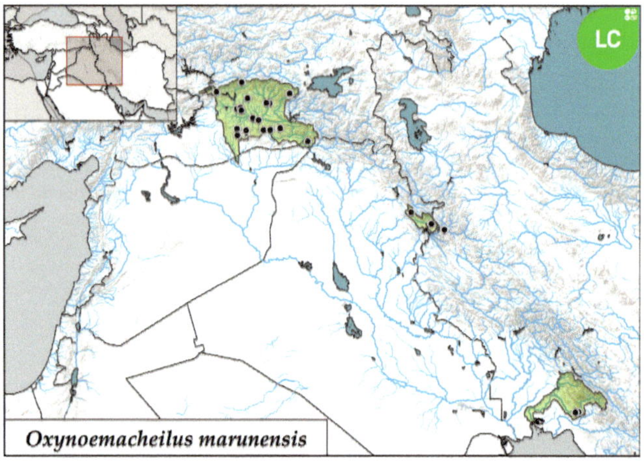

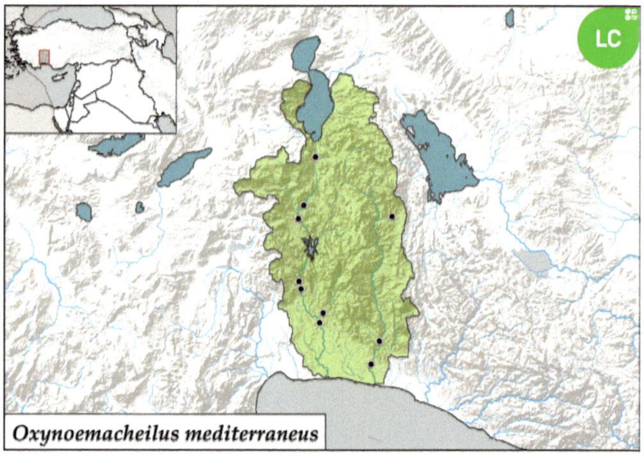

Oxynoemacheilus mediterraneus; Köprüçay drainage, Türkiye; 48 mm SL.

Habitat. Moderately fast-flowing and standing waters of streams, backwaters, and springs with muddy bottoms.
Biology. No data.
Conservation status. LC.
Remarks. This species has not (yet) been recorded from the Greater and Lesser Zab drainages, where it is replaced by *O. euphraticus* (Greater Zab) and *O. kurdistanicus* (Lesser Zab). It also appears to be absent from the Karkheh and Karun drainages. Molecular characters very slightly differentiate major populations in the Jarrahi, Sirvan, and upper Tigris and may be described as different species if clear morphological differences can be worked out.
Further reading. Sayyadzadeh & Esmaeili 2020 (description).

Oxynoemacheilus mediterraneus
Common name. Pamphylian loach.
Diagnosis. Distinguished from other species of *Oxynoemacheilus* in Central Anatolia and rivers flowing to Bay of Antalya by: ○ caudal–peduncle depth 1.2–1.5 times in its length / ● caudal deeply emarginate, shortest middle caudal ray 65–76 % of longest ray of upper caudal lobe / ○ body depth almost uniform between dorsal and caudal bases / ○ flank blotches usually connected to saddles on back / ○ flank with scales / ○ no prominent dorsal crest on caudal peduncle / ○ suborbital groove present in male. Size up to 73 mm SL.
Distribution. Türkiye: Aksu and Köprüçay drainages in Gulf of Antalya.
Habitat. Fast-flowing to almost stagnant waters of streams and rivers with mud or gravel bottoms.
Biology. No data.
Conservation status. LC.
Remarks. Loaches from Lake Eğirdir basin carry the mtDNA of *O. mediterraneus* but are identified as *O. nasreddini* by morphological characters.
Further reading. Erk'akan et al. 2007 (description); Yoğurtçuoğlu et al. 2021a (distribution).

Oxynoemacheilus muefiti; upper Euphrates drainage, Türkiye; 69 mm SL.

Oxynoemacheilus muefiti
Common name. Stout Euphrates loach.
Diagnosis. Distinguished from other species of *Oxynoemacheilus* in Euphrates drainage by: ○ scales absent / ○ lateral line incomplete, not reaching or reaching below dorsal base / ○ suborbital groove absent in male / ○ caudal slightly emarginate / ○ supratemporal canal without central pore / ○ pelvic axillary lobe absent / ○ dorsal adipose crest on caudal peduncle shallow, reaching to vertical through last anal ray base in individuals larger than 50 mm SL / ○ caudal–peduncle length 1.3–1.6 times its depth / ○ body width at dorsal origin 11–13 % SL / ○ flank with fine, brown, mottled colour pattern, usually with narrow, pale-brown bars on flank behind dorsal base, bars confluent with adjacent saddles on back / ○ caudal base with mottled colour pattern, without bold, black bar or bold, black spots upper and lower caudal base. Size up to 70 mm SL.
Distribution. Türkiye: Upper Murat drainage and Eğri, a tributary of Atatürk reservoir. Potentially more widespread in Euphrates.
Habitat. Moderately fast-flowing streams and rivers with gravel or rock substrate.
Biology. No data.
Conservation status. LC.
Remarks. This species is very closely related to *O. ercisianus* from Lake Van basin.
Further reading. Freyhof et al. 2019b (description).

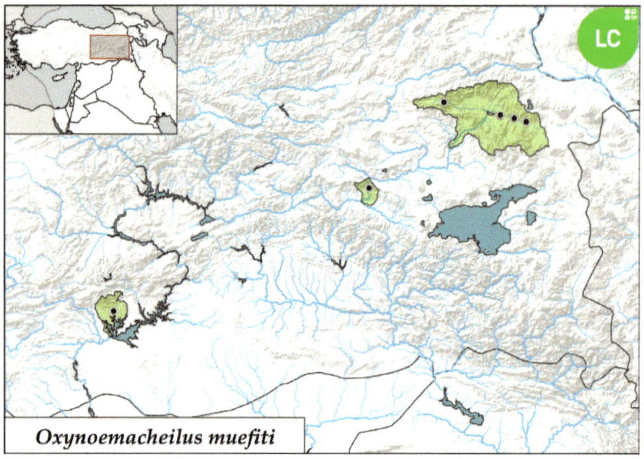

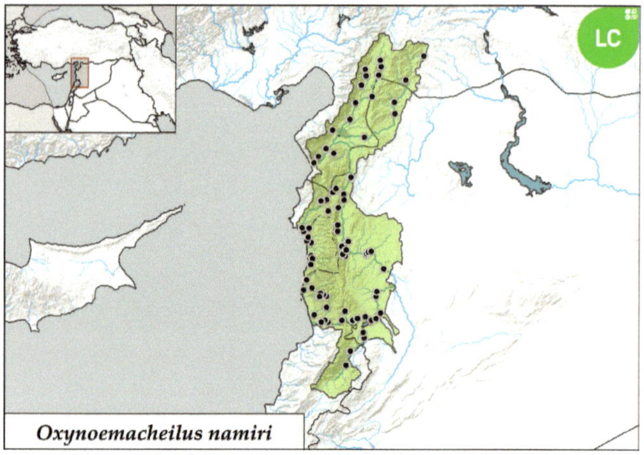

The Orontes in Al Ghab in Syria, habitat of *Oxynoemacheilus namiri*, is heavily affected by pollution.

Oxynoemacheilus namiri; Orontes drainage, Türkiye; 64 mm SL.

Oxynoemacheilus namiri; Afrin, Orontes drainage, Türkiye; 53 mm SL.

Oxynoemacheilus namiri

Common name. Levantine loach.

Diagnosis. Distinguished from other species of *Oxynoemacheilus* in Cilicia and northern Levant by: ● flank pattern usually with 6–17, very distinct and regularly shaped and set bars, usually all or most flank-bars, at least behind dorsal base, extending to middorsal saddles and usually meeting contralateral / ○ caudal deeply emarginate, length of middle caudal ray 1.3–1.5 times in length of longest unbranched ray in upper caudal lobe / ○ suborbital groove absent in male / ○ no or very shallow dorsal crest on caudal peduncle / ○ scales absent on flank in front of dorsal, sparsely set below dorsal, densely set on posterior flank, or very few isolated and embedded scales on back and flank in front of dorsal origin / ○ lateral line terminating anterior to or below dorsal base, rarely above anus or anal base, very exceptionally at caudal base / ○ head length 21–24 % SL / ○ body depth at dorsal origin 16–20 % SL / ○ caudal–peduncle depth 1.2–1.4 times in its length. Size up to 84 mm SL.

Distribution. Orontes drainage and coastal streams in Syria from Orontes south to Nahr al-Kabir on Syrian-Lebanese border.

Habitat. Moderately fast-flowing to stagnant waters of springs, streams, and rivers with muddy or gravelly bottoms. Also on banks of reservoirs and in irrigation canals.

Biology. No data.

Conservation status. LC.

Remarks. A population of *O. tigris* from the upper Qweiq in Türkiye shares mtDNA with *O. namiri*, suggesting past hybridisation between these two species and, thus, a close biogeographic link between the Orontes and the Qweiq.

Further reading. Krupp & Schneider 1991 (description).

Oxynoemacheilus nasreddini; Lake Ilgın basin, Türkiye; 75 mm SL.

Oxynoemacheilus nasreddini

Common name. Eber loach.

Diagnosis. Distinguished from other species of *Oxynoemacheilus* in Central Anatolia and rivers flowing to Bay of Antalya by: ○ a prominent inner axial stripe along midlateral flank / ○ body usually with a series of vertically elongated blotches / ○ tip of pectoral usually reaching to or slightly beyond pelvic origin in male / ○ body depth decreasing between dorsal and caudal base / ○ belly with few, small, embedded scales / ○ caudal–peduncle depth 1.5–2.1 times in its length / ○ caudal emarginate, shortest middle caudal ray 76–91 % of longest ray of upper caudal lobe / ○ prepelvic length 48–55 % SL / ○ usually a series of short bars on caudal peduncle / ○ 2–4 dark-brown bands on caudal / ○ suborbital groove present in male. Size up to 78 mm SL.

Distribution. Türkiye: Lakes Akşehir, Eber, Eğirdir and Ilgın basins.

Habitat. Slow-flowing streams with gravel or mud bottom.

Biology. No data.

Conservation status. VU; appears to be declining within its small range and has lost several habitats in recent years due to desiccation of streams and massive pollution.

Remarks. Populations from Lake Eğirdir basin share the mtDNA with *O. mediterraneus*, indicating a past introgressive hybridisation.

Further reading. Yoğurtçuoğlu et al. 2021a (description).

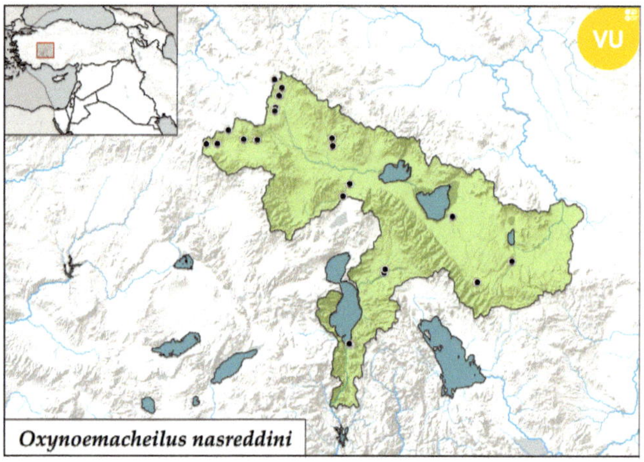

Oxynoemacheilus nasreddini

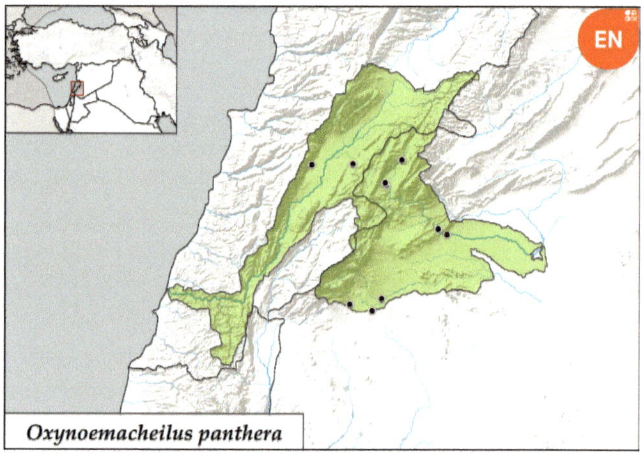

Oxynoemacheilus panthera

The use and misuse of DNA barcoding for taxonomy.

"...a classification founded on any single character, however important that may be, has always failed."
—Charles Darwin (1859)

The advent of molecular techniques has provided new tools for species identification, revolutionising the taxonomic methodology within less than two decades. One of the most widely applied methods compares nucleotide sequence similarity among organisms and attributes it to species differences. For this, sequencing a short fragment of the mitochondrial gene, i.e., cytochrome c oxidase subunit 1 (COI or DNA barcode), was hailed as a "magical tool" or "the taxonomy of the twenty-first century." Since its inception, DNA barcoding has become a widely accepted and standardised protocol for species identification worldwide. However, as with any single-character approach, it must be applied cautiously. Some believe that the barcoding gene was placed in all organisms by a higher entity to allow scientists to recognise species. This viewpoint is not as uncommon as one might suspect.

The arguments against the approach have been consistently presented. The primary criticism is based on the opposition to the assertion of the proponents of the barcoding approach, who have declared that if two organisms show a "considerable" nucleotide difference, they would be considered different species, or vice versa. The assertion is open to serious debate for three reasons. Firstly, the reduction of the complex evolutionary history of organisms to a single character: phylogenetic distance, is employed as a species concept. Secondly, no reasonable distance threshold is provided by which species can be demarcated. Thirdly, the confusion between diagnosis (assigning a fish to an already known species) and definition (deciding if a group of individuals is a species) is evident. In the real world, species are evolving entities, and phylogenetic (or COI) distance can be understood as a measure of proximity to the number of generations (time) that have passed since the last gene flow between populations.

The species concept that DNA barcoding represents ignores the existence of old and young species, where phylogenetic distance is of no consequence. This is because, while speciation does not always require a long period that leads to high phylogenetic distances, genetic divergence does not necessarily lead to speciation. Furthermore, there are several theoretical pitfalls in applying the barcoding approach, in which untrained practitioners (both taxonomists and molecular biologists) have always been prone to being trapped. For instance, at least in the case of fishes, mitochondrial capture through introgressive hybridisation is much more common than previously thought. Consequently, the sole use of COI DNA sequences to identify an individual with mitochondrial DNA introgressed by another species will result in an erroneous identification.

Oxynoemacheilus panthera; Litani drainage, Lebanon; 60 mm SL.

Oxynoemacheilus panthera

Common name. Damascus loach.

Diagnosis. Distinguished from other species of *Oxynoemacheilus* in southern Levant by: ○ caudal very slightly emarginate or truncate / ○ lateral line terminating before vertical of dorsal origin / ○ usually a prominent dorsal crest on caudal peduncle / ○ 8½–9½ branched dorsal rays / ○ anterior flank scaleless or with few isolated scales, caudal peduncle with densely set scales / ○ suborbital groove absent in male / ○ body depth at dorsal origin 20–25 % SL. Size up to 77 mm SL.

Distribution. Syria: Barada and Nahr al Awaj drainages west of Damascus; Lebanon: Litani drainage.

Habitat. Standing and slow-flowing waters of springs and streams with mud or gravel bottoms.

Biology. No data.

Conservation status. EN; appears to be declining within its very small range.

Remarks. This species is usually identified in Litani as *O. leontinae*, a poorly known loach from the Jordan drainage. The few individuals analysed from the Litani drainage host the COI sequences of *O. namiri*.

Further reading. Krupp 1985d (description, distribution).

Oxynoemacheilus panthera; spring of Barada, Syria; ~70 mm SL.

Stream al Tammasiyyar in Syria is a habitat of *Oxynoemacheilus panthera*, *O. insignis*, and *Pseudophoxinus drusensis*.

Oxynoemacheilus pantheroides; upper Jordan drainage, Israel; ~60 mm SL. © M. Ford.

Oxynoemacheilus pantheroides

Common name. Golan loach.

Diagnosis. Distinguished from other species of *Oxynoemacheilus* in southern Levant by: ○ caudal deeply emarginate / ○ no prominent dorsal crest on caudal peduncle / ○ suborbital groove present in male / ○ lateral line usually incomplete, terminating between vertical of dorsal- and anal base / ○ back and flank densely covered by scales. Size up to 60 mm SL, likely to grow larger.

Distribution. Israel: Upper Jordan drainage, possibly also in upper reaches of Jordan in Lebanon.

Habitat. Standing and slow-flowing waters of springs and streams with mud or gravel bottoms.

Biology. No data.

Conservation status. NE; due to very limited knowledge of this species.

Remarks. *Oxynoemacheilus doriae* from Israel may be an additional valid species, and further research is needed to test this hypothesis.

Further reading. Bănărescu et al. 1982 (description); Krupp & Schneider 1989 (synonym of *O. insignis*).

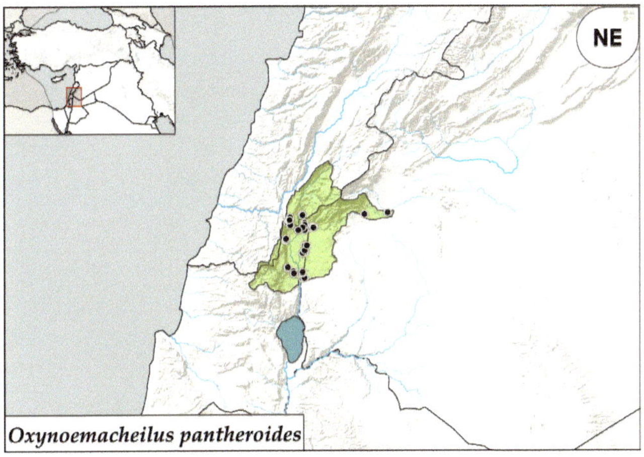

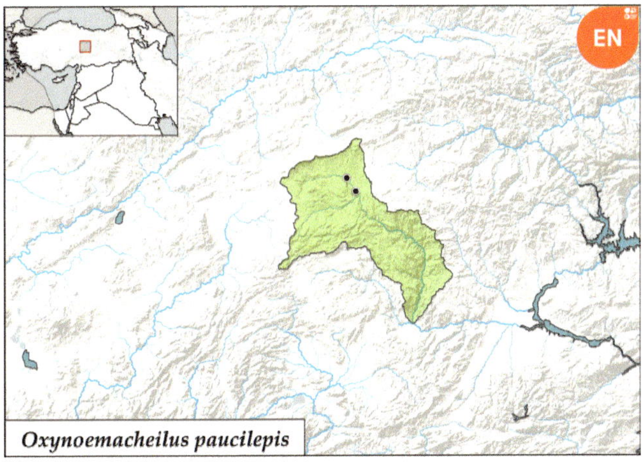

Oxynoemacheilus paucilepis; upper Euphrates drainage, Türkiye; ~35 mm SL.

Oxynoemacheilus paucilepis; upper Euphrates drainage, Türkiye; 70 mm SL.

Oxynoemacheilus paucilepis
Common name. Tohma loach.
Diagnosis. Distinguished from other species of *Oxynoemacheilus* in Euphrates drainage by: • flank with a distinct midlateral series of large, roundish or horizontally elongated brown blotches and a row of small brown blotches between upper pelvic base and lowermost caudal base / ○ lateral line complete / ○ caudal emarginate / ○ no prominent dorsal crest on caudal peduncle / ○ suborbital groove absent in male / ○ caudal peduncle without high adipose crest or ridge / ○ flank covered by scales. Size up to 76 mm SL.
Distribution. Türkiye: Mancınık, Çetinkaya, and Kalkım drainages in upper Euphrates.
Habitat. Moderately or slowly flowing waters in streams with gravel substrate.
Biology. No data.
Conservation status. EN; appears to be declining within its very small range.
Further reading. Erk'akan et al. 2007 (description).

Oxynoemacheilus persa; Kor drainage, Iran; ~60 m SL.

Oxynoemacheilus persa
Common name. Persian loach.
Diagnosis. Distinguished from *Oxynoemacheilus tongiorgii*, it is only other congeneric in Iranian Kor drainages by: • caudal deeply emarginate / • no dorsal crest on caudal peduncle. Size up to 78 mm SL.
Distribution. Iran: Kor basin, Helleh, and Mond drainages.
Habitat. Moderately fast-flowing waters in springs, streams, and rivers with gravel substrate.
Biology. Spawns May and June.
Conservation status. VU; appears to be declining within its small range.
Remarks. *Oxynoemacheilus farsicus* is a synonym.
Further reading. Nalbant & Bianco 1998 (*Oxynoemacheilus farsicus*); Mirghiyasi et al. 2016 (spawning time); Esmaeili et al. 2019 (phylogeography).

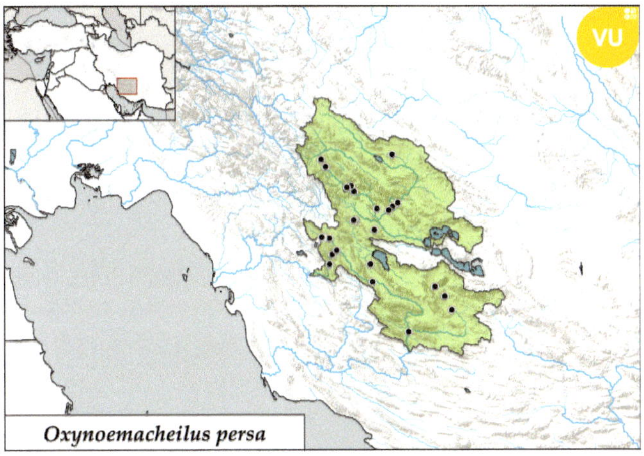

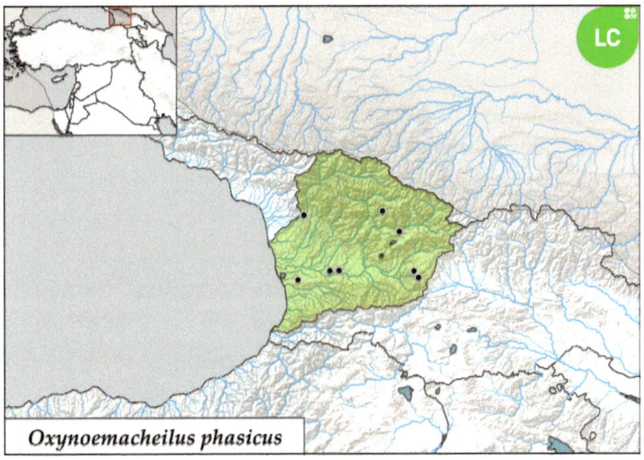

Oxynoemacheilus phasicus; Rioni drainage, Georgia; 65 mm SL © R. Lennigk.

Oxynoemacheilus phasicus

Common name. Rioni loach.

Diagnosis. Distinguished from other adjacent *Oxynoemacheilus* species by: ○ flank mottled, marbled, or with bars / ○ a moderately emarginate caudal / ○ caudal peduncle deep / ○ no median incision in upper lip / ○ pelvic not reaching to origin of genital papillae / ○ pelvic origin below first branched dorsal ray / ○ a wide, bold band at middle or posterior half of caudal / ○ suborbital groove present in male. Size up to 77 mm SL.

Distribution. Georgia: Rioni and Inguri drainages.

Habitat. Moderate or fast-flowing waters in springs, streams, and rivers with gravel substrate.

Biology. No data.

Conservation status. LC.

Further reading. Freyhof et al. 2021b (description).

Oxynoemacheilus sarus; lower Seyhan drainage, Türkiye; 47 mm SL.

Oxynoemacheilus sarus

Common name. Sarus loach.

Diagnosis. Distinguished from other *Oxynoemacheilus* species in Cilicia (including Göksu, Seyhan, and Ceyhan drainages) by: ○ a series of irregularly shaped midlateral blotches or bars disconnected from saddles on back / ○ a bold, black blotch or spot at upper and lower caudal base / ○ 3–5 dark-brown bands on caudal / ○ lateral line complete, terminating at anterior part of hypural complex / ○ caudal forked (shortest middle caudal ray is 56–70 % of longest ray of upper caudal lobe) / ○ caudal peduncle 1.4–1.8 times longer than deep, without dorsal crest / ○ a suborbital groove in male / ○ pre-dorsal length 45–52 % SL / ○ prepelvic length 48–53 % SL / ○ distance between pectoral- and pelvic origin 25–29 % SL. Size up to 54 mm SL.

Distribution. Türkiye: Ceyhan, and lower Seyhan drainages.

Habitat. Moderately fast-flowing water in streams and rivers with stones and gravel bottoms.

Biology. No data.

Conservation status. LC.

Further reading. Freyhof et al. 2021a (description).

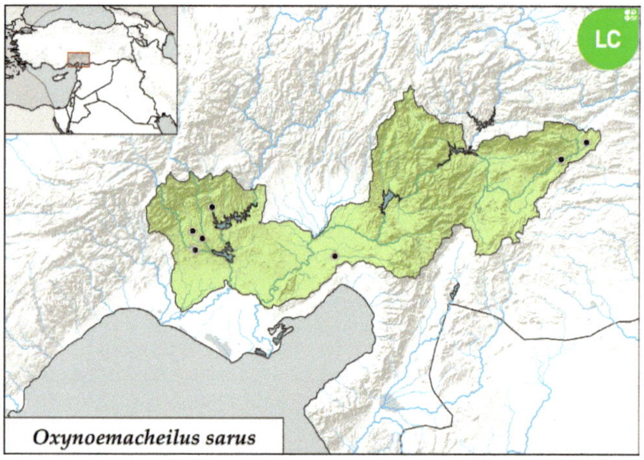

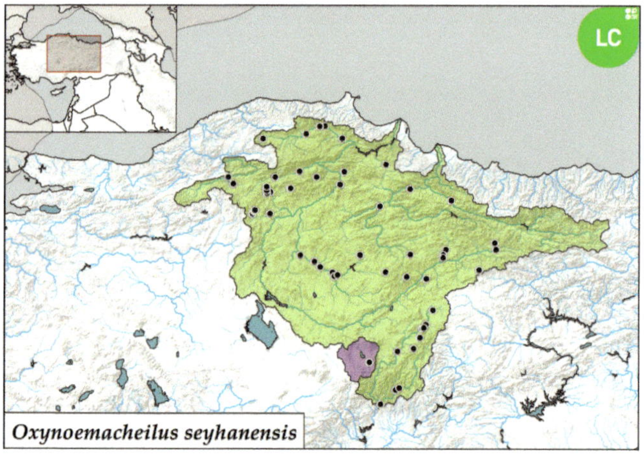

Oxynoemacheilus seyhanensis; Kızılırmak drainage, Türkiye; 68 mm SL.

Oxynoemacheilus seyhanensis; Kızılırmak drainage, Türkiye; ~75 mm SL.

Oxynoemacheilus seyhanensis; Zamantı, Seyhan drainage, Türkiye; ~65 mm SL.

Oxynoemacheilus seyhanensis
Common name. Lycaonian loach.
Diagnosis. Distinguished from other species of *Oxynoemacheilus* in Marmara and Black Sea basin as well as in Cilicia and northern Levant by: ● caudal slightly emarginate, almost truncate / ○ lateral line complete / ○ flank behind dorsal base covered by scales / ○ suborbital groove absent in male / ○ no prominent dorsal crest on caudal peduncle. Size up to 84 mm SL.
Distribution. Türkiye: Kızılırmak and Yeşilırmak drainages, also in upper and middle Seyhan, and in Çerkeş, in upper Filyos drainage. Recorded from Sultan marshes, where it entered through Zamantı tunnel.
Habitat. Moderately fast-flowing water in streams with gravel or mud substrate.
Biology. No data.
Conservation status. LC.
Remarks. *Oxynoemacheilus kosswigi* is a synonym.
Further reading. Bănărescu 1968 (description); Erk'akan & Kuru 1986 (description as *O. kosswigi*); Sungur 2020 (synonyms).

Oxynoemacheilus seyhanicola; Seyhan drainage, Türkiye; ~45 mm SL.

Oxynoemacheilus seyhanicola
Common name. Adana loach.
Diagnosis. Distinguished from other species of *Oxynoemacheilus* in Cilicia and northern Levant by: ● caudal pattern indistinct, almost or completely absent / ○ bold black band or blotches on caudal base / ○ lateral line complete, terminating at anterior part of hypural complex / ○ caudal deeply emarginate / ○ caudal peduncle without dorsal crest / ○ suborbital groove present in male / ○ caudal-peduncle depth 1.4–2.2 in its length / ○ pre-dorsal length 51–56 % SL / ○ prepelvic length 50–57 % SL / ○ distance between pectoral- and pelvic origin 27–32 % SL. Size up to 56 mm SL.
Distribution. Türkiye: Seyhan and Ceyhan drainages.
Habitat. Moderately fast-flowing water in streams and rivers with gravel substrate.
Biology. No data.
Conservation status. LC.
Further reading. Erk'akan et al. 2007 (description); Freyhof et al. 2021a (taxonomy, distribution).

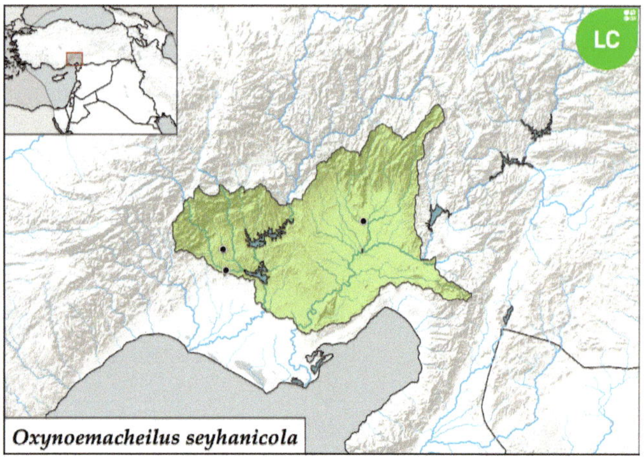

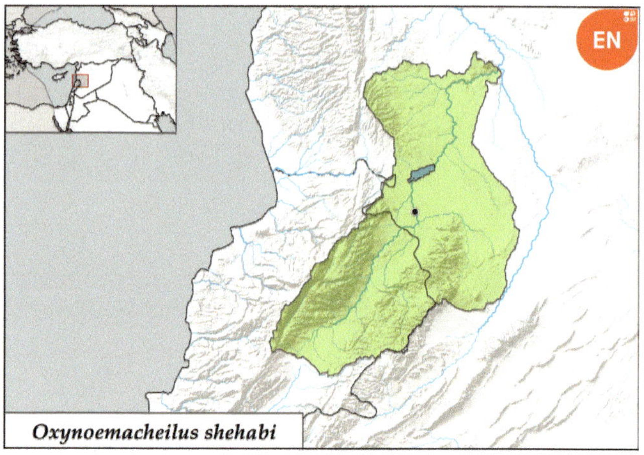

Oxynoemacheilus shehabi; upper Orontes, Syria; 46 mm SL.

Oxynoemacheilus shehabi

Common name. Syrian sportive loach.

Diagnosis. Distinguished from other species of *Oxynoemacheilus* in Cilicia and northern Levant by: ● flank with 5–7 dark-brown bars, much narrower than interspaces, irregularly shaped and set, extending to middorsal saddles and meeting contralateral / ○ a bold, black blotch or spot at upper and lower caudal base / ○ caudal with dark-brown bands of elongated blotches / ○ lateral line complete terminating at caudal base / ○ caudal deeply emarginate / ○ flank covered by scales, sparsely set on anterior, densely set on posterior flank / ○ no dorsal crest on caudal peduncle / ○ no or a very short incision in upper lip / ○ suborbital groove present in male / ○ pelvic axillary lobe well developed, fully attached to body / ○ two, usually indistinct, black blotches or spots at caudal base / ○ caudal–peduncle depth 1.8–2.4 in its length. Size up to 48 mm SL, likely to grow larger.

Oxynoemacheilus simavicus; Susurluk drainage, Türkiye; 65 mm SL.

Oxynoemacheilus simavicus; Bayındır, Sakarya drainage, Türkiye; 64 mm SL.

Oxynoemacheilus simavicus; Porsuk subdrainage, Sakarya drainage, Türkiye; 55 mm SL.

Distribution. Syria: Upper Orontes drainage upriver of Lake Qattinah.
Habitat. From small streams to larger rivers. In fast-flowing stretches.
Biology. No data.
Conservation status. EN; appears to be declining within its very small range.
Further reading. Freyhof & Geiger 2021 (description).

Oxynoemacheilus simavicus

Common name. Mysian loach.
Diagnosis. Distinguished from other species of *Oxynoemacheilus* in Marmara and Black Sea basin by: ○ caudal-peduncle depth 6–10 % SL, usually 2.0–3.0 times in its length, very rarely 1.7–3.3, without high adipose crest or ridge / ○ caudal emarginate, middle caudal ray 70–96 % of length of longest ray in upper caudal lobe / ○ lateral line complete / ○ flank covered by scales / ○ suborbital groove present in male / ○ one central or no black, grey or brown blotch or short bar on caudal base, its colour identical to blotches on caudal peduncle / ○ no yellowish triangles at posteriormost portion of caudal peduncle / ○ pelvic very long, reaching or almost reaching to anus / ○ pectoral in male very long, reaching or almost reaching to pelvic origin / ○ flank mottled or with large irregularly shaped, squarish or vertically elongate blotches, usually as wide or narrower than interspaces / ○ lateral blotches on caudal peduncle usually interrupted in size and/or shape at or above lateral midline. Size up to 74 mm SL.
Distribution. Türkiye: Susurluk, Sakarya and Büyük Melen drainages.
Habitat. Moderately fast-flowing water in streams and rivers with gravel substrate. Usually found in riffles and other fast-flowing habitats.
Biology. No data.
Conservation status. LC.
Remarks. No external characters are known to distinguish this species from *O. banarescui*, *O. bergianus*, and *O. fatsaensis*, all of which are diagnosed by molecular characters. Morphological characters and differences in colour pattern to distinguish this species from the others are based on too small sample sizes or are false. As these four species are allopatric, the range should help in identification. *Oxynoemacheilus sakaryaensis* and *O. melenicus* are synonyms of this species as they are closely related, and none of the characters proposed as diagnostic in their description could be confirmed. The diagnostic characters

distinguishing *O. simavicus*, *O. sakaryaensis,* and *O. melenicus* have been deliberately falsified to allow populations that differ slightly in molecular characters to be described as separate species.

Further reading. Balık & Bănărescu 1978, in Bănărescu, Nalbant & Balık 1978 (description); Erk'akan et al. 2014 (size); Freyhof 2022 (revision of *O. bergianus* group); Turan et al. 2023 (description of O. *sakaryaensis* and *O. melenicus*).

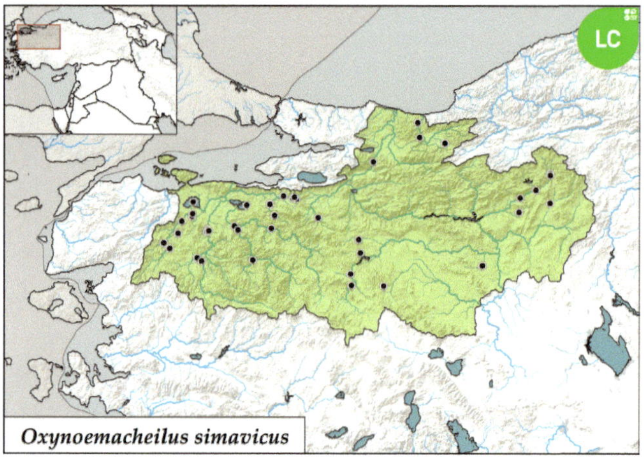

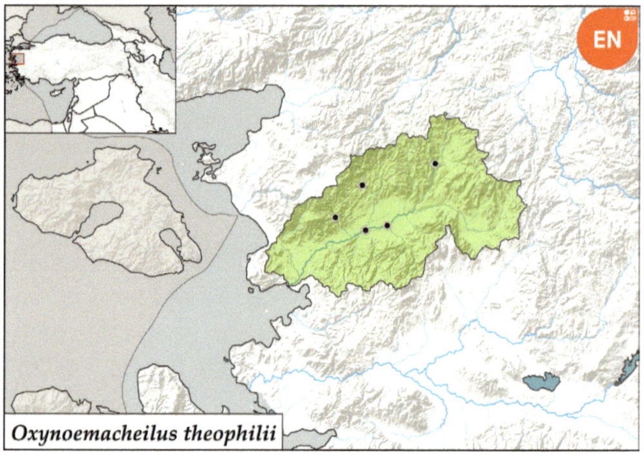

Streams in lower Seyhan drainage are habitats of *Oxynoemacheilus seyhanicola*.

Oxynoemacheilus theophilii; Bakırçay, Türkiye; 65 mm SL.

Oxynoemacheilus theophilii

Common name. Aeolian loach.

Diagnosis. Distinguished from other species of *Oxynoemacheilus* in Aegean basin by: ○ caudal deeply emarginate / ○ a shallow dorsal crest on caudal peduncle / ○ suborbital groove present in male; caudal-peduncle depth 12–13 % SL, 1.3–1.6 times in its length / ○ flank mottled / ○ body depth at dorsal origin 17–18 % SL / ○ length of middle caudal ray 19–23 % SL / ○ dorsal and ventral adipose crests present on caudal peduncle behind vertical of posterior anal base / ○ body depth between dorsal and caudal bases almost uniform / ○ caudal emarginate / ○ belly with few, small, embedded scales. Size up to 66 mm SL.

Distribution. Türkiye: Bakırçay drainage. Also, on Greek island of Lesbos.

Habitat. Moderately fast-flowing waters in streams with gravel substrate.

Biology. No data.

Conservation status. EN; appears to be declining within its very small range.

Remarks. *Oxynoemacheilus bergamensis* is a synonym.

Further reading. Stoumboudi et al. 2006 (description); Erk'akan et al. 2007 (description as *Barbatula bergamensis*); Freyhof et al. 2011 (distribution); İlhan et al. 2020 (distribution).

Oxynoemacheilus tigris; Qweiq drainage, Türkiye; ~55 mm SL; individual with mtDNA of *O. namiri*.

Oxynoemacheilus tigris

Common name. Halap loach.

Diagnosis. Distinguished from other species of *Oxynoemacheilus* in Euphrates and Qweiq drainages by: ● prominent long and high dorsal adipose crest on caudal peduncle reaching beyond vertical through anal origin, usually below last dorsal rays when folded down / ● dorsal crest usually dark-brown with yellowish or whitish margin in individuals larger than 60 mm SL / ● 10–16 bold, narrow, regularly shaped bars, very prominent on caudal peduncle / ○ caudal slightly emarginate / ○ suborbital groove absent in male / ○ lateral line incomplete / ○ many isolated, deeply embedded scales on flank / ○ two bold, black spots at caudal base in most individuals. Size up to 73 mm SL.

Oxynoemacheilus sarali; Merzimen, Euphrates drainage, Türkiye; 65 mm SL.

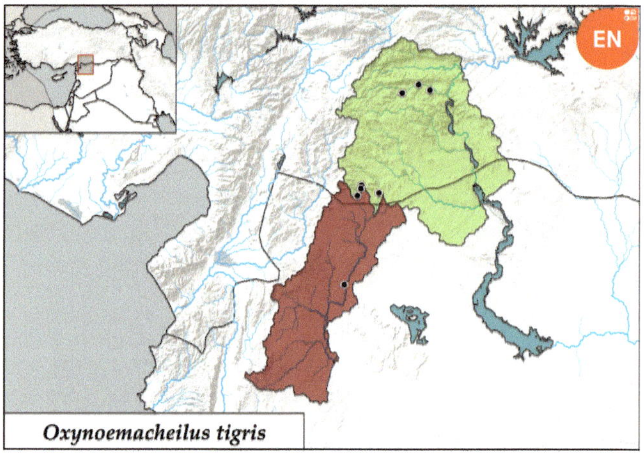

Distribution. Qweiq drainage in Türkiye and Syria, and stream Merziman, which joins Euphrates at Birecik reservoir.

Habitat. Moderately fast-flowing to standing waters of streams with gravel or mud bottoms.

Biology. No data.

Conservation status. EN; appears to be declining within its very small range. Extirpated from Syria and has a very small distribution range in Türkiye.

Remarks. A population of *O. tigris* from the upper Qweiq in Türkiye shares mtDNA with *O. namiri* from the Orontes, The population from Merziman was described as *O. sarali* in 2025, too late to be included in this book.

Further reading. Heckel 1843 (description); Freyhof et al. 2019b (diagnosis, distribution); Turan et al. 2025 (*O. sarali*).

Sünnep, a small headwater stream inhabited by *Oxynoemacheilus tigris*, is one of only two remaining tributaries of the Qweiq drainage.

The Choman, making the border between Iraq (left) and Iran (right), is a tributary of the Lesser Zab and the habitat of a diverse and rich fish community, including *Oxynoemacheilus kurdistanicus*.

Oxynoemacheilus tongiorgii; Kor drainage, Iran; female, ~60 m SL.

Oxynoemacheilus tongiorgii
Common name. Kor loach.
Diagnosis. Distinguished from *Oxynoemacheilus persa*, the only other congeneric in Iranian Kor and Mond drainages by: • caudal slightly emarginate or truncate / • shallow dorsal crest on caudal peduncle. Size up to 62 mm SL.
Distribution. Iran: Kor drainage.
Habitat. Moderately fast-flowing to standing waters of springs and streams with muddy bottoms.
Biology. No data.
Conservation status. EN; appears to be declining within its very small range.
Remarks. Previously thought to be a species of *Seminemacheilus*, but molecular data show that it belongs to *Oxynoemacheilus*. Originally described from Kol drainage, which is a mistake.
Further reading. Nalbant & Bianco 1998 (description); Freyhof et al. 2011 (generic position).

Oxynoemacheilus tongiorgii; Kor drainage, Iran; male, ~50 m SL.

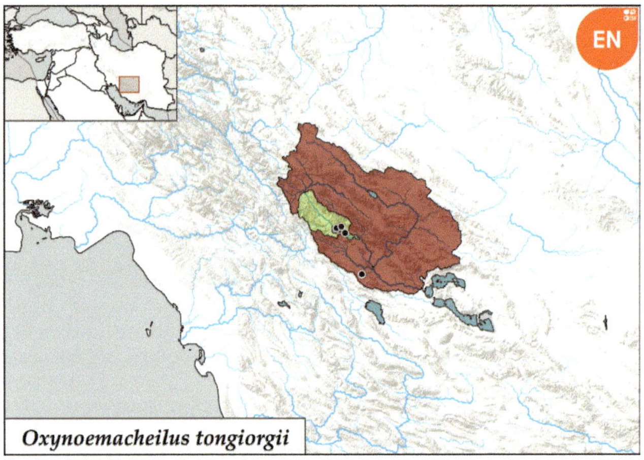

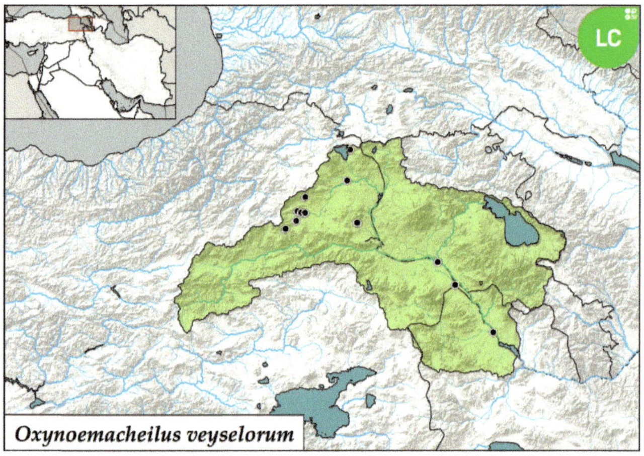

Oxynoemacheilus veyselorum; Aras drainage, Türkiye; ~80 mm SL. © M. Özuluğ.

Oxynoemacheilus veyselorum

Common name. Aras loach.

Diagnosis. Distinguished from other species of *Oxynoemacheilus* in Caspian basin by: • flank mottled, often forming an irregularly shaped midlateral stripe, without distinct bars or large blotches / ○ lateral line complete / ○ caudal slightly emarginate / ○ suborbital groove present in male / ○ caudal–peduncle depth 1.3–1.6 times in its length. Size up to 117 mm SL.

Distribution. Caspian basin: Aras drainage.

Habitat. Rivers and streams with fast-flowing water and gravel or boulder substrate.

Biology. No data.

Conservation status. LC.

Further reading. Çiçek et al. 2018 (description as *Oxynoemacheilus veyseli*); Kaya et al. 2020a; Kuljanishvili et al. 2020 (distribution).

Oxynoemacheilus zagrosensis; Shooei River, Lesser Zab drainage, Iran; ~50 mm SL.

Oxynoemacheilus zagrosensis

Common name. Mottled Choman loach.

Diagnosis. Distinguished from other species of *Oxynoemacheilus* in Tigris drainage by: • posterior process of bony air-bladder capsule directed posteriorly / ○ flank mottled / ○ caudal slightly emarginate / ○ flank completely covered by scales / ○ dorsal profile convex / ○ suborbital groove absent in male / ○ one central pore in supratemporal canal / ○ no median incision in upper lip / ○ lateral line complete, terminating behind vertical of anal base or at caudal base / ○ colour pattern on anterior part of flank not interrupted by unpigmented zone along lateral line. Size up to 61 mm SL.

Distribution. Iran: Choman drainage, a tributary of Lesser Zab.

Habitat. Moderately fast-flowing and standing waters of streams, backwaters, and springs with muddy bottoms.

Biology. No data.

Conservation status. VU; appears to be declining within its small range.

Remarks. See remarks at *O. chomanicus*.

Further reading. Kamangar et al. 2014 (description); Sayyadzadeh & Esmaeili 2020 (discussion of Iranian and Anatolian populations of *Oxynoemacheilus frenatus*).

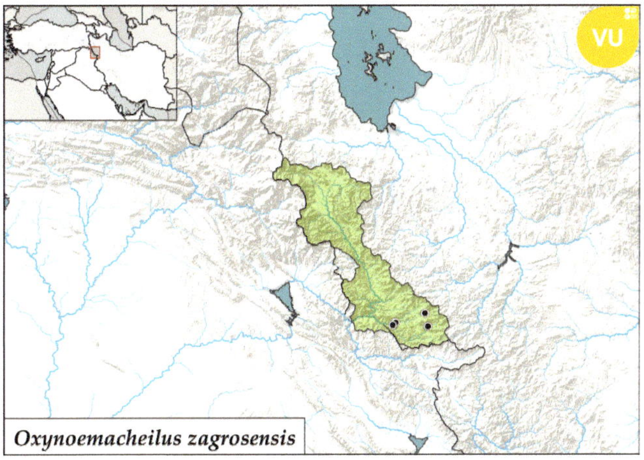

Oxynoemacheilus zarzianus; Lesser Zab drainage, Iraq; 55 mm SL.

Oxynoemacheilus zarzianus; Lesser Zab drainage, Iraq; 62 mm SL.

Oxynoemacheilus zarzianus

Common name. Zarzianian loach.

Diagnosis. Distinguished from other species of *Oxynoemacheilus* in Tigris drainage by: ● 8+8 branched caudal rays / ○ flank completely covered by scales / ○ caudal slightly emarginate / ○ dorsal profile convex / ○ colour pattern on flank behind dorsal base mottled or marmorated, without bars / ○ colour pattern on anterior part of flank not interrupted by unpigmented zone along lateral line / ○ one central and one lateral pore in supratemporal canal / ○ pelvic axillary lobe absent or a very shallow pad / ○ a median incision in upper lip / ○ suborbital groove absent in male / ○ lateral line complete, terminating behind vertical of anal base or at caudal base / ○ caudal–peduncle depth 1.0–1.2 times in caudal–peduncle length / ○ interorbital distance 1.1–1.4 times in snout length / ○ posterior process of bony air-bladder capsule directed laterally. Size up to 75 mm SL.

Distribution. Iran and Iraq: Lesser Zab, Qeshlaq, and upper Sirvan drainages.

Habitat. Moderately fast-flowing and standing waters of streams, backwaters, and springs with gravel bottoms.

Biology. No data.

Conservation status. LC.

Remarks. See remarks at *O. chomanicus*.

Further reading. Freyhof & Geiger 2017 (description); Sayyadzadeh & Esmaeili 2020 (discussion of Iranian and Anatolian populations of *O. frenatus*).

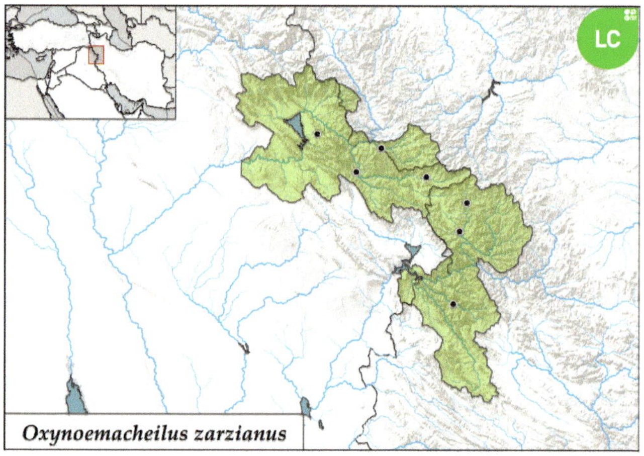

River Tajan is located in humid northern Iran and is the habitat of *Paracobitis abrishamchianorum*.

Paracobitis

A slender body and a high dorsal adipose crest on the caudal peduncle distinguish *Paracobitis*. All species lack a suborbital flap or groove, and there is no sexual dimorphism in the length of the pectoral. Additionally, other genera within the family contain crested species with prominent brown bars on the posterior part of the flank (*Paraschistura*) or a stout body with a markedly different general appearance and pronounced sexual dimorphism (male with enlarged pectoral fins in *Oxynoemacheilus* and *Seminemacheilus*), a character state that absent in *Paracobitis*. Twelve species of *Paracobitis* are currently recognised, with eight of them being described after the year 2000. All species of *Paracobitis*, except subterranean, Central Asian *P. starostini*, are found in West Asia. Slender loaches with a high adipose crest, identified as *Paracobitis*, also occur in China. They belong to a very different evolutionary lineage and are placed in the genus *Homatula*. *Paracobitis smithi* from Iran is placed in its own genus (*Eidinemacheilus*). The phylogenetic position of *Paracobitis ghazniensis* from the Afghan Helmand drainage still needs to be studied. However, the species is superficially very similar to *P. rhadinaea*; thus, we treat both as conspecific. The species diversity of *Paracobitis* in West Asia is well understood, and only one additional species might await description. **Further reading.** Bănărescu & Nalbant 1966 (*N. ghazniensis*); Kottelat 2012 (*Homatula*); Freyhof et al. 2014b (revision of *Paracobitis*); Thoni & Hart 2015 (*P. boutanensis*).

Key to species of *Paracobitis* in West Asia

1a - Posterior narial opening slit-shaped.
..................2
1b - Posterior narial opening roundish or ovoid.
..................3

2a - Back and flank densely covered by scales; flank with many, irregularly set, small blotches or bars not organised in rows.
..................*P. longicauda*
2b - Back and flank with few embedded scales or without scales; flank posteriorly with one or two rows of large, irregularly set blotches.
..................*P. rhadinaea*

3a - Body completely scaleless; isolated scales on dorsal adipose crest in few individuals.
..................4
3b - Body covered by scales; scales present at least on caudal peduncle.
..................5

4a - Caudal rays dark-brown or black with a hyaline base and a hyaline posterior margin; caudal clearly emarginate; axillary pelvic lobe well developed with free tip.
..................*P. hircanica*
4b - Caudal rays hyaline with an irregular pattern of dark-brown or black spots and elongated blotches often organised in 1–3 wide, irregularly shaped bars; caudal truncate or very slightly emarginate; axillary pelvic lobe absent or fully attached.
..................*P. atrakensis*

5a - Dorsal origin situated above, behind, or very slightly in front of vertical of pelvic origin.
..................6
5b - Dorsal origin situated clearly in front of vertical of pelvic origin.
..................9

6a - Flank cream yellowish with large, irregularly shaped brown blotches; dorsal origin situated about one eye diameter behind pelvic origin.
..................*P. basharensis*
6b - Flank brown with a fine or wide, cream-yellowish reticulate or vermiculate pattern; dorsal origin situated above or very slightly in front of vertical of pelvic origin.
..................7

7a - Axillary pelvic lobe large and well developed, usually with free tip.
..................*P. abrishamchianorum*
7b - Axillary pelvic lobe absent or very small and fully attached to body.
..................8

8a - Caudal slightly emarginate; predorsal length 49–56 % SL; prepelvic length 50–53 % DL; outer rostral barbel 16–24 % HL; maxillary barbel 19–27 % HL.
..................*P. zabgawraensis*
8b - Caudal truncate; predorsal length 56–57 %; prepelvic length 54 % SL; outer rostral barbel 24–32 % HL; maxillary barbel 27–34 % HL.
..................*P. salihae*

9a - Caudal truncate.
..................*P. molavii*
9b - Caudal emarginate.
..................10

10a - Tube of anterior nostril reaching beyond posterior tip of posterior nostril when folded back; caudal adipose crest deep, its depth at highest point 3.3–3.8 % SL in individuals of 30–50 mm SL.
..................*P. malapterura*
10b - Tube of anterior nostril not fully overlapping posterior nostril when folded back; caudal adipose crest shallow, its depth at highest point 2.1–3.2 % SL in individuals of 30–50 mm SL.
..................*P. persa*

Paracobitis abrishamchianorum; Kashpel, Iran; 95 mm SL.

Paracobitis abrishamchianorum

Common name. Caspian crested loach.

Diagnosis. Distinguished from other species of *Paracobitis* by: ○ axillary pelvic lobe large and well developed, usually with free tip / ○ scales sparsely set on flank between dorsal origin and anal base, densely set on caudal peduncle / ○ posterior narial opening roundish / ○ caudal emarginate / ○ dorsal origin usually slightly behind a vertical through pelvic origin, rarely above or in front / ○ caudal rays hyaline with an irregular pattern of dark-brown or black spots and elongated blotches often organised in 1–2 wide, irregularly shaped bars. Size up to 110 mm SL.

Distribution. Iran: Tajan, Siyah, Talar, Babol, Haraz, and Kashpal in southern Caspian basin.

Habitat. Streams and small rivers with clear water, coarse gravel, and fast currents.

Biology. No data.

Conservation status. LC.

Remarks. Originally described as *P. abrishamchiani*, the name needed to be declined, as the species was described for two persons.

Further reading. Mousavi-Sabet et al. 2019b (description).

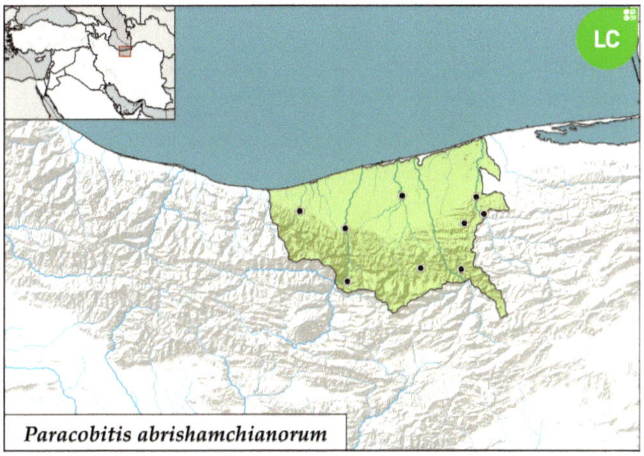

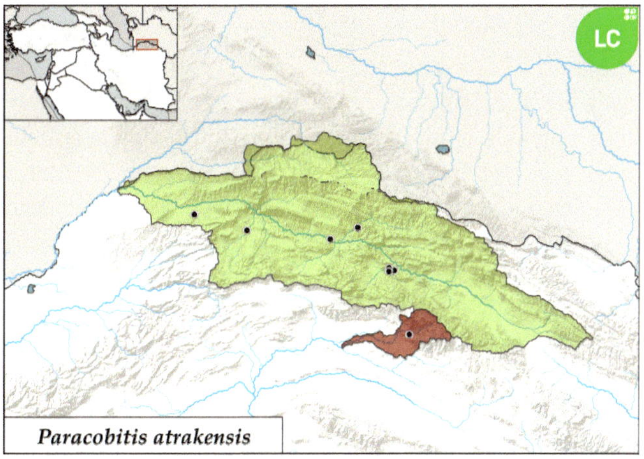

Paracobitis atrakensis; Atrak, Iran; ~125 mm SL.

Paracobitis atrakensis

Common name. Atrak crested loach.

Diagnosis. Distinguished from other species of *Paracobitis* by: ○ no scales on body, isolated scales on dorsal, adipose crest in few individuals / ○ caudal truncate or very slightly emarginate / ○ coarse mottled brown pattern on flank / ○ posterior narial opening roundish / ○ pelvic origin below vertical of last unbranched or first branched dorsal ray / ○ caudal rays hyaline with an irregular pattern of dark-brown or black spots and elongated blotches often organised in 1–3 wide, irregularly shaped bars / ○ no pelvic axillary lobe or lobe fully attached. Size up to 125 mm SL.

Distribution. Iran and Turkmenistan: Atrak drainage in southeast Caspian basin and Bidvaz drainage in Eastern Kavir basin.

Habitat. Streams with coarse gravel and moderate to fast-flowing water.

Biology. No data.

Conservation status. LC; extirpated from Bidvaz.

Further reading. Esmaeili et al. 2014d (description).

Paracobitis basharensis; Karun drainage; ~50 mm SL.

Paracobitis basharensis

Common name. Karun crested loach.

Diagnosis. Distinguished from other species of *Paracobitis* by: ● dorsal origin at about one eye diameter behind a vertical of pelvic origin / ○ many brown or yellowish, medium-sized, widely spaces and irregularly shaped blotches forming a mottled pattern on predorsal back and flank / ○ flank behind dorsal origin covered by small scales / ○ caudal almost truncate or slightly emarginate / ○ posterior narial opening roundish / ○ axillary pelvic lobe small. Size up to 61 mm SL, likely to grow larger.

Distribution. Iran: Headwaters of Bashar in upper Karun drainage.

Habitat. Streams with coarse gravel and moderate to fast-flowing water.

Biology. No data. Only four individuals have been found so far.

Conservation status. DD; due to very poor knowledge of this species. May be extinct, as it has never been found again despite intensive searches.

Further reading. Freyhof et al. 2014b (description).

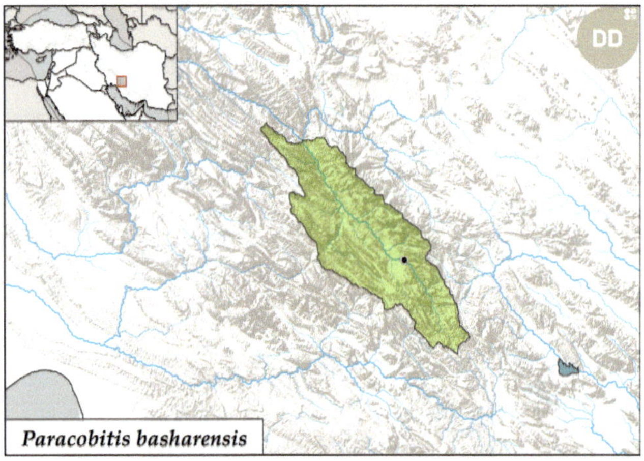

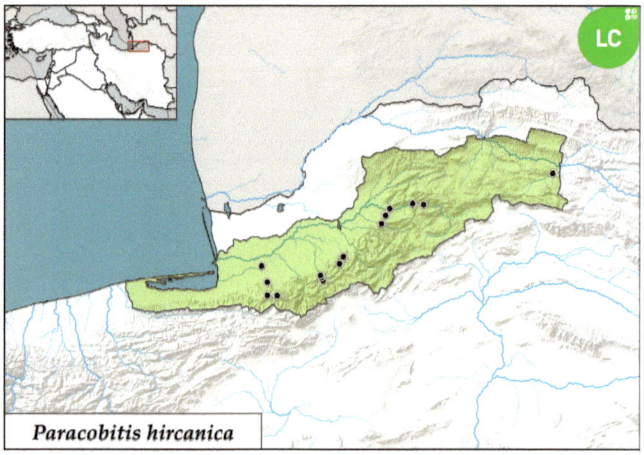

Paracobitis hircanica; Gorgan drainage, Iran; ~65 mm SL. © R. Patimar.

Paracobitis hircanica
Common name. Golestan crested loach.
Diagnosis. Distinguished from other species of *Paracobitis* by: ○ caudal rays dark-brown or black with a hyaline base and a hyaline posterior margin / ○ no scales on body / ○ caudal emarginate / ○ fine marbled or vermiculate brown pattern on flank / ○ posterior narial opening roundish / ○ axillary pelvic lobe well developed with free tip / ○ pelvic origin below vertical of last unbranched or first to second branched dorsal ray. Size up to 95 mm SL.
Distribution. Iran: Ziyarat, Zaringol, Madarsu, and Qarasu in Gorgan drainage and locally in Atrak.
Habitat. Fast to moderately fast-flowing streams.
Biology. Lives for 3 (males) to 4 years (females). Feeds on benthic invertebrates such as chironomids and stoneflies.
Conservation status. LC.
Remarks. *Paracobitis atrakensis* and *P. hircanica* occur in sympatry in Atrak and hybrids should be expected to occur.
Further reading. Patimar et al. 2009 (biology as *P. malapterura*); Mousavi-Sabet et al. 2015a (description).

Paracobitis longicauda; Hari drainage, Iran; ~80 mm SL. © H. Mousavi-Sabet.

Paracobitis longicauda; Hari drainage, Iran; ~120 mm SL.

Paracobitis longicauda
Common name. Central Asian crested loach.
Diagnosis. Distinguished from other species of *Paracobitis* by: ○ posterior narial opening slit-shaped / ○ body fully covered by scales / ○ irregularly set blotches or bars on flank behind dorsal base / ○ caudal deeply emarginate / ○ pelvic origin below second or third branched dorsal ray / ○ axillary pelvic lobe small / ○ caudal rays hyaline with an irregular pattern of dark-brown or black spots and elongated blotches often organised in 1–2 wide, irregularly shaped bars. Size up to 150 mm SL.
Distribution. Hari and Morghab in Iran, Afghanistan, and Turkmenistan. Amu Darya and Zeravshan in Aral basin.
Habitat. Streams and small rivers. From lowland habitats up to high mountain streams.
Biology. Feeds on aquatic invertebrates.
Conservation status. LC; common and widespread, especially in Amu Darya drainage.
Further reading. Freyhof et al. 2014b (diagnosis, discussion of types).

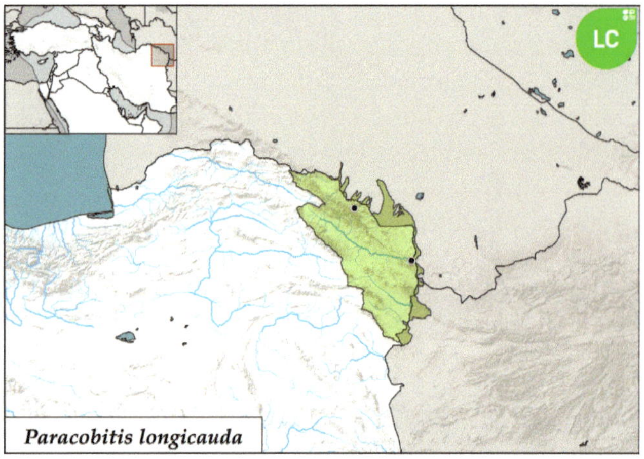

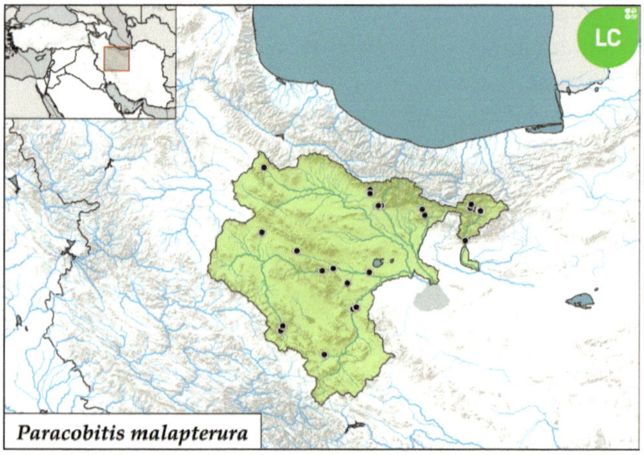

Paracobitis malapterura; Namak basin, Iran, 64 mm SL.

Paracobitis malapterura; Namak basin, Iran; ~90 mm SL.

Paracobitis malapterura

Common name. Namak crested loach.

Diagnosis. Distinguished from other species of *Paracobitis* by: ○ flank behind dorsal origin or slightly anterior to dorsal base covered by scales, anterior flank naked / ○ moderately to slightly emarginate caudal / ○ posterior narial opening roundish / ○ caudal adipose crest high, its depth at highest point 3.3–3.8 SL in individuals between 30–50 mm SL / ○ tube of anterior nostril not fully overlapping posterior nostril when folded back / ○ midlateral stripe connected to blotches and saddles on caudal adipose crest / ○ caudal rays hyaline with an irregular pattern of dark-brown or black spots and elongated blotches often organised in 1–2 wide, irregularly shaped bars / ○ pelvic origin below vertical of last unbranched or first to second branched dorsal ray / ○ axillary pelvic lobe small, often indistinct / ○ tube of anterior nostril reaching beyond posterior tip of posterior nostril when folded back. Size up to 126 mm SL.

Distribution. Iran: Qom, Qara chai, Jaj, and Karaj in Lake Namak basin and Nam and Hable in Western Kavir basin.

Habitat. Fast-flowing stream with gravel bottom.

Biology. Lives up to 5 years and spawns between March and May.

Conservation status. LC.

Remarks. There are doubtful records of this species from the Urmia basin, which need to be confirmed. *Paracobitis iranica* is a synonym.

Further reading. Freyhof et al. 2014b (diagnosis, distribution); Jamali et al. 2016 (biology).

The Hari at the border between Iran and Turkmenistan was the habitat of *Paracobitis longicauda*. Recently, the river falls often dry.

Paracobitis malapterura, MNHN 3962, B.3070, syntypes, 126 and 103 mm SL; probably collected in 1837 and "sent from Syria."

Pierre Martin Rémi Aucher-Éloy and *Paracobitis malapterura*. The provenance of a fish can occasionally be a source of considerable interest. *Cobitis malapterura* was described by Cuvier and Valenciennes (1846) based on two individuals (MNHN 3962 and B-3070) received in 1840 by the Muséum National d'Histoire Naturelle (MNHN) in Paris. Both specimens were dispatched to Paris by Rémi Aucher-Éloy, a French botanist who relocated to Constantinople (now İstanbul) to collect plants in West Asia. Between 1830 and 1838, he undertook extensive travels in West Asia. The two specimens are labelled as originating from Syria, yet Cuvier & Valenciennes (1846) note that "Mr. Aucher-Éloy (1792–1838) has sent, from Syria, a loach." The two fish were probably sent through an agent from what is now

Lebanon, as this was a common practice at the time for agents to receive material and sell it to museums and collections. The fish were obtained by MNHN in 1840, following the death of Aucher-Éloy in 1838. This suggests they were collected in the final years of Aucher-Éloy's travels. In March 1836, a significant portion of Aucher-Éloy's collections were lost in a fire in Constantinople. Consequently, the fish were probably collected after the fire.
In 1836, Aucher-Éloy undertook several expeditions to Greece and Western Anatolia (an area devoid of *Paracobitis*). In 1837, he accompanied the French zoologist M. Dufaud on a survey of northern Anatolia, traversing the shores of the north of Lake Urmia and extending as far as the Caspian Sea in Iran. In August 1837, they surveyed the western Alborz Mountain range, after which they departed the mountains on September 2. Their route took them along the lower Sefid, through Qazvin and Karaj, to Tehran, where they arrived on September 5. Dufaud died of fever in Tehran on October 21. On December 22, 1837, Aucher-Éloy continued his travels south, reaching Qom on December 24. In his account, he noted that the city was based on a small river that always had good water and never dried up. (Aucher-Éloy, 1843: 464). From Qom, he subsequently proceeded to Esfahan, and from there continued his journey southwards to Shiraz, Bandar Abbas, and later to Oman and also to Baluchistan, before returning to Esfahan, where he died.
Aucher-Éloy (1843) did not mention anything about his fish collection in his published work, but he rarely mentioned specific plants in his writings. The two loaches were the only fish sent to Paris. M. Dufaud may have collected them before he died in Tehran, as he was interested in zoological materials. Aucher-Éloy and Dufaud entered the distribution range of *Paracobitis* when they visited the Lake Namak basin after turning east from the Sefid to Tehran. Aucher-Éloy crossed rivers that were potentially inhabited by *Paracobitis* until he left Qom. It seems highly probable that Aucher-Éloy and Dufaud collected their *Paracobitis* specimens in the Namak basin and did so in the northern tributaries of the lake in the present-day Alborz province. This is because Dufaud died before they reached Qom, the next locality where they might have collected *Paracobitis*. The two syntypes, still extant at MNHN, are indistinguishable from the *Paracobitis* species found in the Lake Namak basin.
Further reading. Aucher-Éloy 1843 (travel description); Freyhof et al. 2014b (identification).

Paracobitis molavii; Sirvan drainage, Iraq; 56 mm SL.

Paracobitis molavii
Common name. Kurdistan crested loach.
Diagnosis. Distinguished from other species of *Paracobitis* by: ○ caudal truncate / ○ pelvic origin below first branched dorsal ray / ○ flank behind dorsal origin or slightly anterior to dorsal base covered by scales, anterior flank naked / ○ axillary pelvic lobe small / ○ posterior narial opening roundish. Size up to 66 mm SL, likely to grow larger.
Distribution. Iran and Iraq: Upper Zalem, Sirvan, Leyleh, and Lesser Zab drainages.
Habitat. Streams with coarse gravel and moderate to fast-flowing water.
Biology. No data.
Conservation status. VU; appears to be declining within its small range.
Further reading. Freyhof et al. 2014b (description).

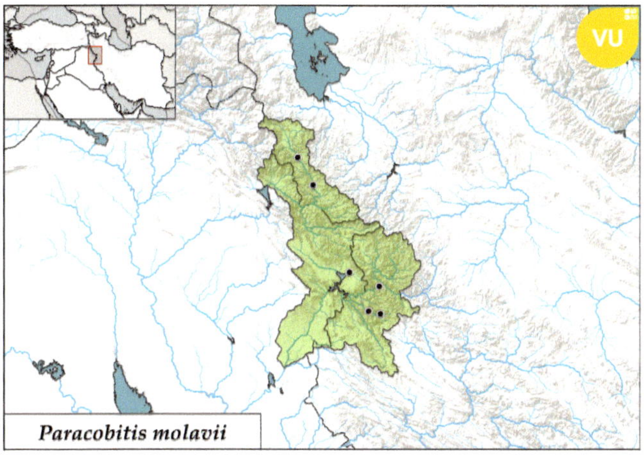

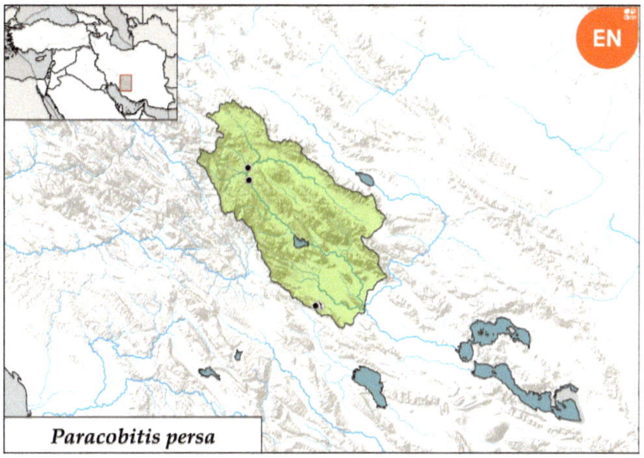

Paracobitis persa; Malsosjan spring, Iran; 75 mm SL.

Paracobitis persa

Common name. Kor crested loach.

Diagnosis. Distinguished from other species of *Paracobitis* by: ○ tube of anterior nostril not fully overlapping posterior nostril when folded back / ○ caudal adipose crest very shallow, its depth at highest point 2.1–3.2 % SL / ○ flank behind dorsal origin or slightly anterior to dorsal base covered by scales, anterior flank naked / ○ pelvic origin below last unbranched dorsal ray / ○ axillary pelvic lobe small, often indistinct / ○ moderately to slightly emarginate caudal / ○ posterior narial opening roundish / ○ midlateral stripe always disconnected from blotches and saddles on caudal adipose crest. Size up to 75 mm SL.

Distribution. Iran: Beyza, Maloosjan, and Ghadamgah springs, as well as headwaters of Kor.

Habitat. Small, slow-flowing streams and springs.

Biology. No data.

Conservation status. EN; appears to be declining within its very small range.

Further reading. Freyhof et al. 2014b (description).

Paracobitis rhadinaea; Sistan basin, Iran; ~250 mm SL.

Paracobitis rhadinaea

Common name. Giant crested loach.

Diagnosis. Distinguished from other species of *Paracobitis* by: ○ posterior narial opening slit-shaped / ○ body naked or with few, usually isolated and embedded scales / ○ usually a midlateral row of irregularly set and shaped large dark-brown blotches on flank posterior to dorsal base / ○ caudal emarginate or truncate / ○ pelvic origin below vertical of first to third branched dorsal ray / ○ axillary pelvic lobe large, well developed. Size up to 250 mm SL.

Distribution. Afghanistan and Iran: Sistan basin, including Helmand drainage, Chahnime reservoirs, and Hamun Lakes.

Habitat. Lakes and large rivers, but little-known about habitats.

Biology. Feeds on aquatic invertebrates and small fishes.

Remarks. *Paracobitis macmahoni*, *P. ghazniensis*, and *P. vignai* are synonyms for this species. *Paracobitis rhadinaea* shows remarkable variability in colour pattern, from spots to almost patternless, in scale cover, from scales on the caudal peduncle to being completely bare, and from an emarginate to a truncate or slightly rounded caudal.

Conservation status. LC; situation in Helmand is unknown. Strongly declined in Iran due to loss of wetlands.

Further reading. Regan 1906 (description); Bănărescu & Nalbant 1966 (diagnosis); Freyhof et al. 2014b (diagnosis); Mousavi-Sabet et al. 2014 (description); Mousavi-Sabet et al. 2016b (conservation); Sayyadzadeh et al. 2019a (variability, *P. vignai*).

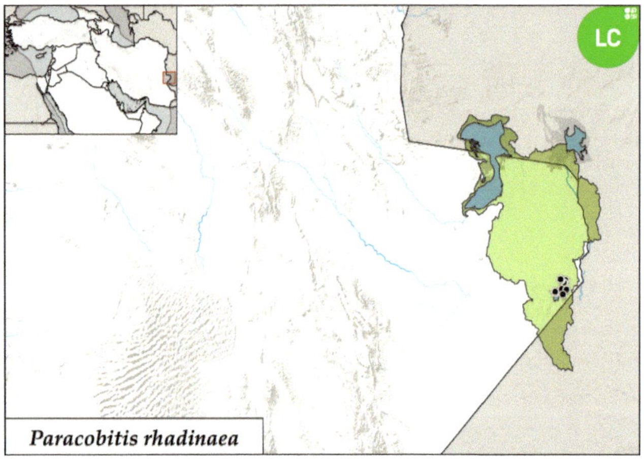

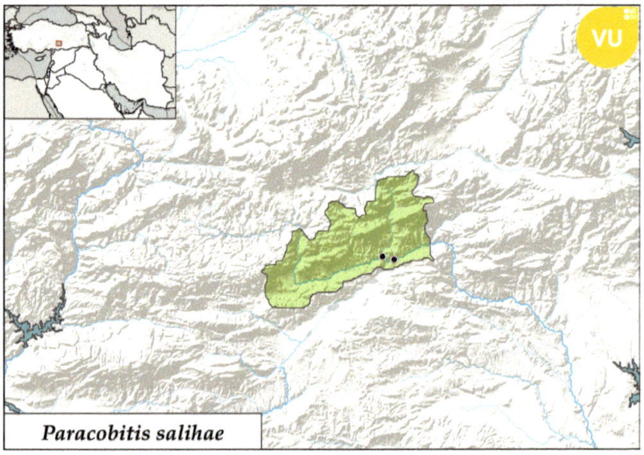

Paracobitis salihae; Euphrates drainage, Türkiye; 66 mm SL.

Paracobitis salihae

Common name. Euphrates crested loach.

Diagnosis. Distinguished from other species of *Paracobitis* by: ○ pelvic origin below or very slightly in front of dorsal origin / ○ caudal truncate / ○ no pelvic axillary lobe / ○ flank with a dark-brown vermiculate or marbled pattern / ○ flank behind vertical through adipose-crest origin covered with small, deeply embedded scales / ○ posterior narial opening roundish / ○ predorsal length 56–57 % / ○ prepelvic length 54 % SL / ○ outer rostral barbel 24–32 % HL / ○ maxillary barbel 27–34 % HL. Size up to 66 mm SL.

Distribution. Türkiye: Upper Göksu drainage.

Habitat. Streams with coarse gravel and moderate to fast-flowing water.

Biology. No data.

Conservation status. VU; appears to be declining within its small range. Its distribution is very small, and only two individuals have been found so far.

Further reading. Kaya et al. 2020c (description).

Paracobitis zabgawraensis

Common name. Greater Zab crested loach.

Diagnosis. Distinguished from other species of *Paracobitis* by: ● body almost plain brown with narrow reticulate yellowish pattern / ○ pelvic origin below or very slightly in front of dorsal origin / ○ caudal slightly emarginate / ○ no or a very small pelvic axillary lobe / ○ flank behind dorsal origin covered with small, deeply embedded scales / ○ posterior narial opening roundish. / ○ predorsal length 49–56 % SL / ○ prepelvic length 50–53 % SL / ○ outer rostral barbel 16–24 % HL / ○ maxillary barbel 19–27 % HL. Size up to 77 mm SL.

Distribution. Türkiye: Upper Yanarsu, Botan, and Nerduş drainages; Greater Zab drainage in Türkiye and Iraq.

Habitat. Streams with coarse gravel and moderate to fast-flowing water.

Biology. No data.

Conservation status. LC.

Further reading. Freyhof et al. 2014b (description); Kaya et al. 2020c (distribution).

Paracobitis zabgawraensis; headwater of Great Zab, Iraq; 73 mm SL.

The Nerduş in Türkiye is the habitat of *Paracobitis zabgawraensis*, *Turcinoemacheilus ekmekciae*, *Glyptothorax daemon*, *Oxynoemacheilus kentritensis*, *O. marunensis* and many other stream fishes.

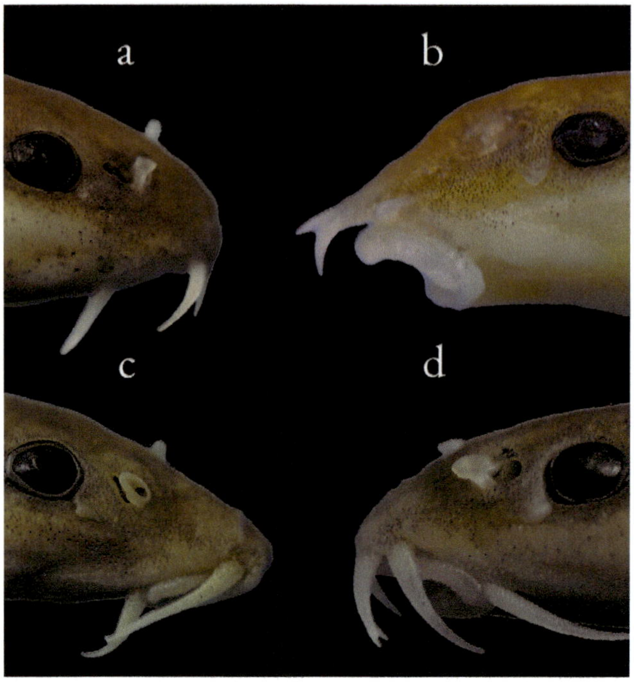

Figure 54. Suborbital flap and groove in some *Paraschistura*: a, *P. nielseni*, 46 mm SL; b, *P. ilamensis*, 38 mm SL; c, *P. hormuzensis*, 45 mm SL; b, *P. susiani*, 34 mm SL.

Paraschistura

Paraschistura is a poorly known group of medium-sized or small loaches, distributed from the Karkheh in Iran throughout southern Iran and Pakistan to the Indus and the Hari, Morghab, and Helmand in Afghanistan, Iran, Pakistan, and Turkmenistan. It is anticipated that some species of India's large loach diversity may belong to this genus. The genus *Paraschistura* was first described in 2009, and 24 species are currently recognised, 16 of which are found in West Asia. The genus *Metaschistura* is a synonym. Several osteological characters diagnose *Paraschistura*. In Iran, all species except *P. aredvii*, *P. nielseni*, *P. ilamensis*, and *P. susiani* can be readily identified by a prominent black spot at the base of the first dorsal rays. The diversity of *Paraschistura* species in Iran has been well-studied, but knowledge of diversity in Afghanistan, Pakistan, and Turkmenistan is very poor. It is anticipated that further unidentified species will be discovered in these regions. **Further reading.** Mirza et al. 1981 (Pakistan); Prokofiev 2009 (description); Freyhof et al. 2015 (revision in West Asia, *Metaschistura* synonym, phylogeny).

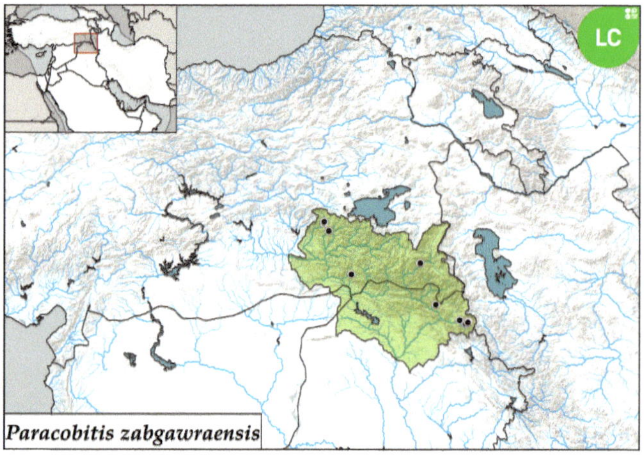

Key to species of *Paraschistura* in Iran

1a - Body completely scaleless.
..................2

1b - Scales present at least on caudal peduncle; scales on flank and caudal peduncle often isolated and deeply embedded.
..................4

2a - Flank with marbled colour pattern without distinct bars or with bars dissociated in a row of midlateral blotches and saddles on back or with 11–14 irregularly shaped and set dark-brown bars on flank, usually dissociated into two adjacent narrow bars or individual bars with a central open space.
..................*P. turcmenica*

2b - Flank with 10–12 regularly shaped and set, dark-brown bars.
..................3

3a - Body depth 11–15 % SL; pelvic origin below vertical of first branched dorsal ray; no dorsal adipose crest.
..................*P. kessleri*

3b - Body depth 14–17 % SL; pelvic origin shortly before vertical of first branched dorsal ray; a short dorsal adipose crest.
..................*P. lindbergi*

4a - Lateral line complete or almost complete, reaching to caudal base.
..................5
4b - Lateral line incomplete, reaching a point before caudal base.
..................7

5a - Caudal peduncle with a prominent dorsal adipose crest supported by 22–25 procurrent caudal rays; snout blunt.
..................*P. cristata*
5b - Caudal peduncle without dorsal, adipose crest, if with crest, then without procurrent caudal rays; snout long and pointed.
..................6

6a - Processus dentiformis absent in upper jaw.
..................*P. alta*
6b - Processus dentiformis in upper jaw small and pointed.
..................*P. makranensis*

7a - Predorsal flank scaleless, scales present on caudal peduncle, also on back in populations from Sirjan and Jazmurian.
..................8
7b - Predorsal flank and back covered by scales.
..................9

8a - Pelvic origin situated behind vertical of dorsal origin.
..................*P. abdolii*
8b - Pelvic origin situated in front of or below vertical of dorsal origin.
..................*P. kermanensis*

9a - Suborbital flap or groove absent in male.
..................10
9b - Suborbital flap or groove present in male.
..................13

10a - Dorsal origin situated in front of vertical of pelvic origin.
..................*P. aredvii*
10b - Dorsal origin situated above or very slightly behind of vertical of pelvic origin.
..................11

11a - Distance between anus and tip of pelvic 4–7 % SL.
..................*P. naumanni*
11b - Distance between anus and tip of pelvic 7–10 % SL.
..................*P. delvarii*

12a - Suborbital groove present in male.
..................*P. nielseni*
12b –Suborbital flap present in male.
..................13

13a - Suborbital flap, triangular, not overlapping horizontal slit below flap; lateral line usually reaching beyond dorsal base; bold black spot at base of first dorsal rays; snout pointed.
..................14

13b - Suborbital flap roundish or elongate, overlapping slit below flap; lateral line not reaching or reaching to dorsal base; very faint black spot at base of first dorsal rays or spot absent; snout blunt.
..................15

14a - Flank with 11–18 dark-brown bars; pelvic reaching to about 1–2 eye diameter in front of anus.
..................*P. bampurensis*
14b - Flank with 7–8 dark-brown bars; pelvic reaching to or beyond anus, to anal origin in some individuals.
..................*P. hormuzensis*

15a - Male with an elongated suborbital flap pointed downward; caudal–peduncle depth 1.6–1.8 times longer than deep.
..................*P. ilamensis*
15b - Male with a short, blunt, and roundish suborbital flap; caudal–peduncle depth 1.4–1.6 times longer than deep.
..................*P. susiani*

Paraschistura abdolii; Baft, Iran; ~65 mm SL.

Paraschistura abdolii

Common name. Sirjan loach.

Diagnosis. Distinguished from other species of *Paraschistura* in West Asia by: ○ caudal peduncle covered by deeply embedded, often isolated scales, in some populations, scales also present immediately in front of dorsal origin / ○ pelvic origin below first or second branched dorsal ray / ○ lateral line incomplete / ○ caudal emarginate / ○ 7½ branched dorsal rays / ○ 11–19 regularly or irregularly shaped pale- or dark-brown bars on flank, often dissociated into a marmorate pattern / ○ suborbital flap or groove absent in male / ○ dorsal adipose keel usually absent, if present, then without procurrent rays / ○ predorsal distance 49–53 % SL / ○ body depth at dorsal origin 10–14 % SL. Size up to 65 mm SL.

Distribution. Iran: Haji Abad and upper Minhab in Persian Gulf basin, Baft in Sirjan basin, and western tributaries of Jazmurian and Lut basins.

Habitat. Fast to slow-flowing streams with gravel substrate.

Biology. No data.

Conservation status. LC.

Remarks. Individuals from the Jazmurian and Sirjan basins lack dorsal scales, as do most individuals from the Kol drainage. In Kol drainage, there are some individuals that have scales on the dorsum, just in front of dorsal origin.

Further reading. Freyhof et al. 2015 (description).

Paraschistura abdolii; Kol drainage, Iran; ~40 mm SL. © K. Borkenhagen.

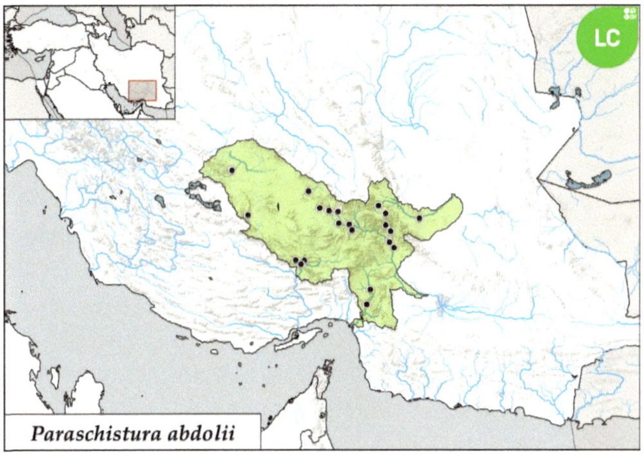

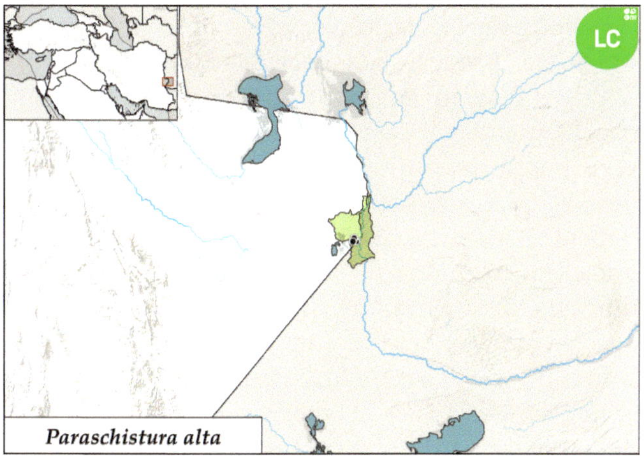

Stream Baft in Sirjan basin is the habitat of *Paraschistura abdolii* and *Capoeta saadii*.

Paraschistura alta; Sistan basin, Iran; ~80 mm SL.

Paraschistura alta

Common name. Helmand banded loach.

Diagnosis. Distinguished from other species of *Paraschistura* in West Asia by: ● processus dentiformis in upper jaw absent / ○ lateral line complete or almost complete (incomplete in Iranian fish), reaching to caudal base / ○ flank and caudal peduncle covered by scales / ○ snout long and pointed / ○ caudal forked / ○ 7½ or 8½ branched dorsal rays / ○ pelvic origin below second or third branched dorsal ray / ○ no dorsal adipose crest / ○ suborbital flap or groove absent in male / ○ a small, bold black spot at base of first dorsal rays / ○ black bar at caudal base faint / ○ 9–10 regularly set and shaped pale-brown or grey bars, absent or very faint on flank in front of dorsal origin / ○ caudal–peduncle depth 0.9–1.2 times longer than deep / ○ body depth at dorsal origin 12–20 % SL. Size up to 112 mm SL.

Distribution. Afghanistan and Iran: Sistan basin, including Helmand drainage, Chahnime reservoirs, Hamun Lakes, and Zehak.

Habitat. Large and medium-sized rivers.

Biology. No data.

Conservation status. LC; situation in Helmand unknown.

Remarks. This species was originally described as having a complete or almost complete lateral line, which is incomplete in fish recorded from Iran. Whether this species has a suborbital groove or flap in males remains unknown; it is absent in the few individuals examined.

Further reading. Nalbant & Bianco 1998 (description); Jouladeh-Roudbar et al. 2015c (record from Iran).

The Hamoun: From wetlands to wastelands. The Helmand rises in the Hindu Kush mountain range, situated west of Kabul in Afghanistan. It crosses the Dashti Margo desert before reaching the Hamoun or Sistan Lake region, located at the border between Afghanistan and Iran. The Hamoun wetlands, which form a true oasis in the middle of hundreds of kilometers of arid plains, are a testament to the river's influence. The marshes and lakes of the Hamoun wetlands supported a diverse range of wildlife and were an important stopover for migrating birds. Furthermore, the wetlands were the source of livelihood for thousands of people who depended on agriculture and fishing in the area. However, many factors, including improved irrigation practice, urban development, dams, and drought, led to a gradual decline in the wetlands since the 1970s. Since the 1990s, the Hamoun wetlands have deteriorated rapidly due to the expansion of irrigation, coupled with one of the worst droughts ever recorded. One of the major dams blocking water to the Hamoun wetlands is the Kajaki dam, located in central Afghanistan. The Taliban government closed the sluices of the dam in 1998 after a dispute with Iran. At the same time, severe droughts massively decreased the precipitation in the Hindu Kush. Until 2001, the Hamoun wetlands had completely dried out. The post-Taliban government in Kabul agreed to open the sluices again in late October, allowing water to reach the Hamoun wetlands again. The new Taliban government allows surplus water to flow during high floods. Much work still needs to be done to restore the wetlands, including ensuring adequate water flow to sustain the ecosystem. When the dam is opened, fish from Afghanistan, including the loach *Paraschistura alta*, are washed downriver. **Further reading.** Rashki et al. 2013 (Sistan Oasis)

Copernicus Service information 2024

Paraschistura aredvii; Sarab-e Bahram spring, Iran; 39 mm SL.

Paraschistura aredvii

Common name. Zohreh loach.

Diagnosis. Distinguished from other species of *Paraschistura* in West Asia by: ○ pelvic origin below first or second branched dorsal ray / ○ lateral line incomplete / ○ suborbital groove absent in male / ○ back and flank behind dorsal origin covered by scales / ○ 7–9 irregularly shaped blotches or bars on flank, wider than interspaces / ○ dorsal adipose crest absent / ○ snout blunt / ○ caudal emarginate / ○ 7½ branched dorsal rays / ○ caudal–peduncle depth 1.4–1.7 times longer than deep / ○ body depth at dorsal origin 14–17 % SL. Size up to 56 mm SL.

Distribution. Iran: Fahlian, Shiv and Zohreh drainages.

Habitat. Fast to slow-flowing streams with gravel substrate.

Biology. No data.

Conservation status. LC.

Further reading. Freyhof et al. 2015 (description).

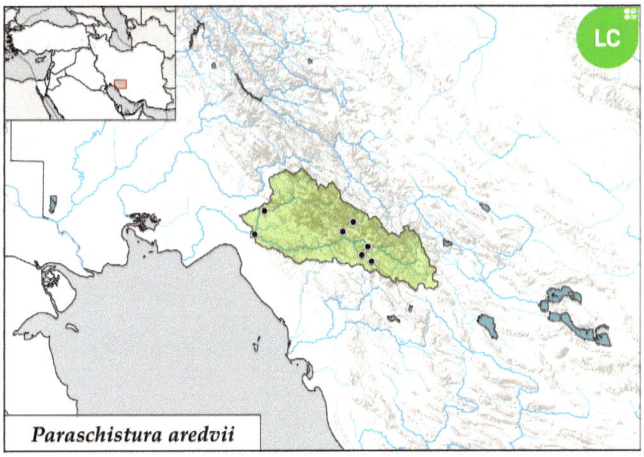

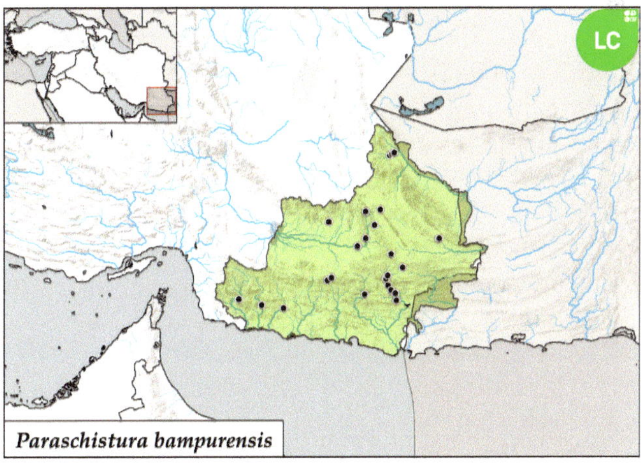

Paraschistura bampurensis; Sarbaz, Iran; ~45 mm SL.

Paraschistura bampurensis

Common name. Bampur loach.

Diagnosis. Distinguished from other species of *Paraschistura* in West Asia by: ○ snout pointed / ○ flank and back covered by scales / ○ 11–18 dark-brown bars on flank usually narrower than interspaces, especially on caudal peduncle / ○ lateral line incomplete, usually reaching beyond dorsal base, beyond anal origin in some individuals / ○ caudal forked or deeply emarginate / ○ suborbital flap present in male, triangular, not overlapping horizontal slit below flap / ○ a bold black spot at base of first dorsal rays / ○ dorsal adipose crest absent / ○ usually 8½ branched dorsal rays / ○ pelvic reaching to about 1–2 eye diameter in front of anus / ○ pelvic origin below first or second branched dorsal ray. Size up to 60 mm SL.

Distribution. Iran and Pakistan: Mashkid and Khash drainages, eastern tributaries of Jazmurian basin, Sarhe, Sarbaz, and Bahukalat drainages in Makran region.

Habitat. Moderately to rapidly flowing freshwater streams in mountain and desert landscapes.

Biology. No data.

Conservation status. LC.

Further reading. Freyhof et al. 2015 (diagnosis).

Paraschistura cristata; Hari drainage, Iran; 59 mm SL. © H. Mousavi-Sabet.

Paraschistura cristata

Common name. Khorasan loach.

Diagnosis. Distinguished from other species of *Paraschistura* in West Asia by: ● prominent dorsal adipose crest supported by 22–25 procurrent caudal rays / ○ lateral line complete or almost complete, reaching to caudal base / ○ snout blunt / ○ flank completely covered by scales / ○ pelvic reaching to or slightly in front of anus / ○ a bold black spot at base of first dorsal rays / ○ 9–13 brown bars usually faded or absent on flank in front of dorsal origin, regularly shaped and wider than interspaces or dissociated into blotches and irregularly shaped bars; rarely without bars / ○ body with 4–8 wide dark-brown bars behind dorsal origin / ○ usually a complete black bar at caudal base / ○ pelvic origin below vertical of second or third branched dorsal ray / ○ suborbital groove or flap absent in male / ○ caudal emarginate / ○ usually 8½, rarely 7½ branched dorsal rays. Size up to 70 mm SL.

Distribution. Eastern Kavir basin (Iran), Hari drainage in Afghanistan, Iran, Turkmenistan, Morghab, and Kopetdag in Karakum Desert basin (Turkmenistan). Probably more widespread in Turkmenistan.

Habitat. Fast to slow-flowing streams with gravel substrate.

Biology. Lives up to 4 years, and spawns March–June.

Conservation status. LC.

Remarks. Previously placed in the monotypic genus *Metaschistura*.

Further reading. Prokofiev 2009 (placement in *Metaschistura*); Freyhof et al. 2015 (diagnosis).

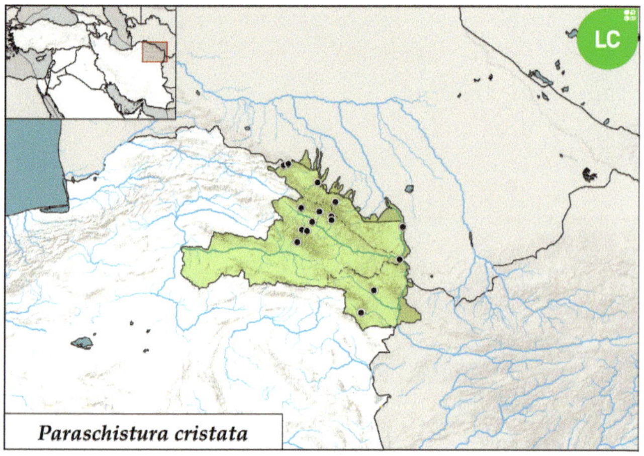

Paraschistura delvarii; Mond drainage, Iran; ~45 mm SL.

Paraschistura delvarii

Common name. Qareh Aqaj loach.

Diagnosis. Distinguished from other species of *Paraschistura* in West Asia by: ○ dorsal origin situated above, or very slightly behind vertical of pelvic origin / ○ suborbital flap or groove absent in male / ○ flank and back completely covered by scales / ○ lateral line incomplete / ○ no dorsal adipose crest / ○ caudal emarginate / ○ caudal–peduncle depth 1.2–1.9 times longer than deep / ○ snout blunt / ○ 7½ branched dorsal rays / ○ very variable colour pattern of irregularly shaped and set, often faded bars and blotches on flank / ○ a bold black or brown spot at base of first dorsal rays / ○ an irregularly shaped black bar at caudal base in some populations / ○ predorsal distance 52–56 % SL / ○ distance between tip of pelvic and anus 7–11 % SL. Size up to 51 mm SL.

Distribution. Iran: Qareh Aqaj and Firozabad in Mond drainage.

Habitat. Fast to slow-flowing streams with gravel substrate.

Biology. No data.

Conservation status. EN; appears to be declining within its very small range.

Further reading. Mousavi-Sabet & Eagderi 2015 (description); Sayyadzadeh et al. 2018b (description).

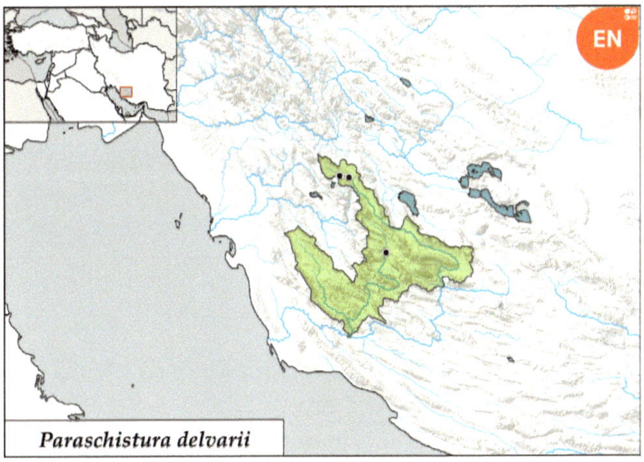

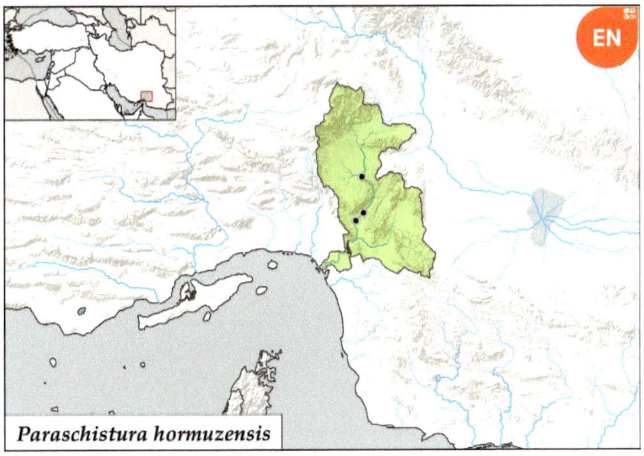

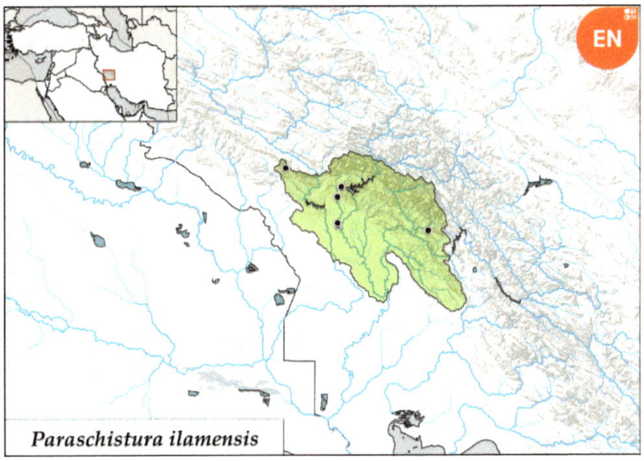

Paraschistura hormuzensis; Minhab drainage, Iran; ~45 mm SL.

Paraschistura hormuzensis
Common name. Minab loach.
Diagnosis. Distinguished from other species of *Paraschistura* in West Asia by: ○ snout pointed / ○ suborbital flap present in male, triangular, not overlapping horizontal slit below flap / ○ caudal forked / ○ lateral line incomplete / ○ 7–8 dark-brown bars on flank / ○ pelvic reaching to or beyond anus, to anal origin in some individuals / ○ body fully covered by scales / ○ body depth at dorsal origin 13–18 % SL / ○ no dorsal adipose crest / ○ a bold black spot at base of first dorsal rays / ○ usually a complete black bar at caudal base / ○ pelvic origin below first or second branched dorsal ray / ○ 7½ branched dorsal rays. Size up to 52 mm SL.
Distribution. Iran: Minab drainage.
Habitat. Fast to slow-flowing streams. Usually in shallow areas and on gravel substrate.
Biology. No data.
Conservation status. EN; appears to be declining within its very small range.
Further reading. Freyhof et al. 2015 (description).

Paraschistura ilamensis; Karkheh drainage, Iran; ~55 mm SL. © S. Vatandoust.

Paraschistura ilamensis
Common name. Ilam banded loach.
Diagnosis. Distinguished from other species of *Paraschistura* in West Asia by: ○ suborbital flap roundish or elongate, pointed downward, overlapping slit below flap / ○ snout blunt / ○ back and flank fully covered by scales / ○ pelvic origin below first branched dorsal ray / ○ 7–13 dark-brown, irregularly shaped and set, often faded bars on flank / ○ pelvic reaching to a point about 1.0–2.0 eye diameter in front of anus / ○ no dorsal adipose crest / ○ lateral line incomplete / ○ caudal emarginate / ○ caudal–peduncle depth 1.6–1.8 times longer than deep / ○ a bold or very faint black spot at base of first dorsal rays / ○ usually a complete black bar at caudal base / ○ 7½ branched dorsal rays / ○ body depth at dorsal origin 14–15 % SL. Size up to 55 mm SL.
Distribution. Iran: Bala and Cholvar are two tributaries of Dez in Karun drainage and Siah Gav in Karkheh drainage.
Habitat. Fast to slow-flowing streams with gravel substrate.
Biology. No data.
Conservation status. EN; seems to be declining within its very small range.
Remarks. *Paraschistura pasatigris* is a synonym.
Further reading. Freyhof et al. 2015 (description as *P. pasatigris*); Vatandoust & Eagderi 2015 (description); Eagderi et al. 2019b (priority).

River Rodan in Iran is the habitat of *Paraschistura hormuzensis*, *Capoeta anamisensis*, *Garra persica* and *Cyprinion microphthalmum*.

Paraschistura kermanensis; Bidkhoon drainage, Iran; 65 mm SL.

Paraschistura kermanensis

Common name. Kerman loach.

Diagnosis. Distinguished from other species of *Paraschistura* in West Asia by: ○ dorsal origin situated above, or very slightly behind vertical of pelvic origin / ○ lateral line incomplete / ○ caudal peduncle covered by scales, no scales on predorsal back and middle and anterior part of flank / ○ suborbital flap or groove absent in male / ○ very variable colour pattern of 4–21 irregularly shaped and set, often faded bars and blotches on flank / ○ a bold black spot at base of first dorsal rays / ○ an irregularly shaped black bar at caudal base in some populations / ○ no dorsal adipose crest / ○ caudal emarginate / ○ 7½ branched dorsal rays / ○ snout blunt / ○ predorsal distance 53–56 % SL / ○ caudal-peduncle depth 1.1–1.4 times longer than deep. Size up to 68 mm SL.

Distribution. Iran: Bidkhoon in Kerman basin.

Habitat. High altitude (about 2400 m), clear, fast to slow-flowing streams with gravel substrate.

Biology. No data.

Conservation status. CR; known only from a small population in a very small area. Desiccation of rivers is a major threat, and it may be at the edge of extinction.

Remarks. This species is superficially similar to *P. abdolii*, and both occur nearby.

Further reading. Sayyadzadeh et al. 2019b (description).

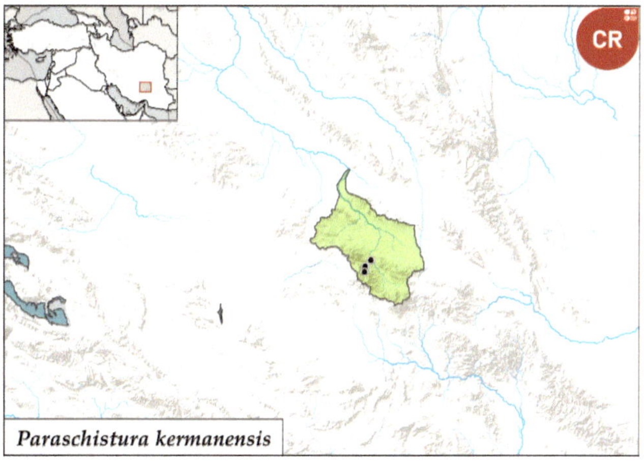

Paraschistura kessleri; Mashkid basin, Iran; ~35 mm SL. © H. Mousavi-Sabet.

Paraschistura kessleri

Common name. Pishin Lora loach.

Diagnosis. Distinguished from other species of *Paraschistura* in West Asia by: ○ 10–12 regularly shaped and set dark-brown bars on flank, as wide as interspaces / ○ scales completely absent / ○ body depth 11–15 % SL / ○ pelvic reaching to about 1–2 eye diameter in front of anus / ○ suborbital flap or groove absent in male / ○ a bold black spot at base of first dorsal rays / ○ caudal emarginate / ○ snout blunt / ○ no dorsal adipose crest / ○ pelvic origin below first branched dorsal ray / ○ usually 7½ branched dorsal rays. Size up to 45 mm SL.

Distribution. Pishin Lora drainage in Pakistan and Afghanistan and Mashkid basin in Iran and Pakistan.

Habitat. Moderately to rapidly flowing freshwater streams in mountain and desert landscapes.

Biology. No data.

Conservation status. LC.

Remarks. The records from Bodo in Pakistan probably refer to *P. lindbergi*, and both occur in sympatry in Mashkid basin. *Nemacheilus sargadensis* is a synonym of *P. kessleri*. *Paraschistura turcomana* is another species without scales. It has been described from the Kushka in Morghab drainage in Turkmenistan. Like the Hari, Morghab is an endorheic basin that originates in Afghanistan and ends in Karakum Desert, not far from the inner estuary of Hari.

Further reading. Freyhof et al. 2015 (diagnosis).

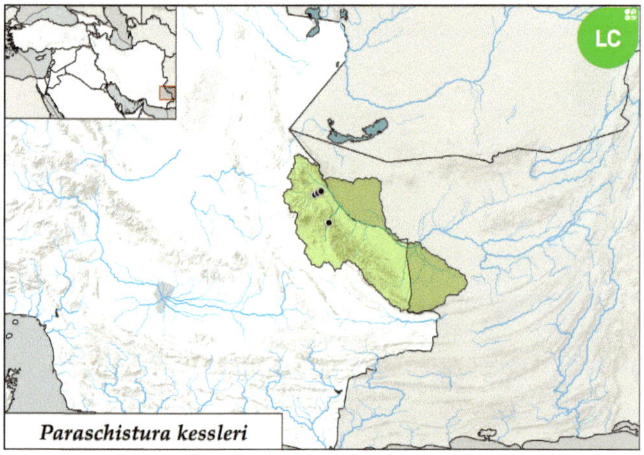

Paraschistura lindbergi; Mashkid basin, Iran; ~40 mm SL.

Paraschistura lindbergi

Common name. Farah banded loach.

Diagnosis. Distinguished from other species of *Paraschistura* in West Asia by: ○ 10–12 regularly shaped and set dark-brown bars on flank, wider or as wide as interspaces / ○ scales completely absent / ○ body depth 14–17 % SL / ○ suborbital flap or groove absent in male / ○ a bold black spot at base of first dorsal rays / ○ caudal emarginate / ○ snout blunt / ○ a short dorsal adipose crest / ○ pelvic origin shortly in front of first branched dorsal ray / ○ usually 7½ branched dorsal rays. Size up to 58 mm SL.

Distribution. Afghanistan: Farah drainage in Sistan basin; Iran and Pakistan: Mashkid basin; Pakistan: Unar and Loralai in Indus drainage.

Habitat. Streams with moderate to fast-flowing freshwater in mountainous desert landscapes.

Biology. No data.

Conservation status. NE; due to the very poor knowledge about this species.

Remarks. This species is poorly known and has not been studied since its description in 1965. The Farah, Mashkid, and Indus populations seem unlikely to be conspecific. *Paraschistura lindbergi* and *P. kessleri* occur in sympatry in Mashkid basin.

Further reading. Bănărescu & Mirza 1965 (description); Mirza et al. 1981 (distribution in Pakistan).

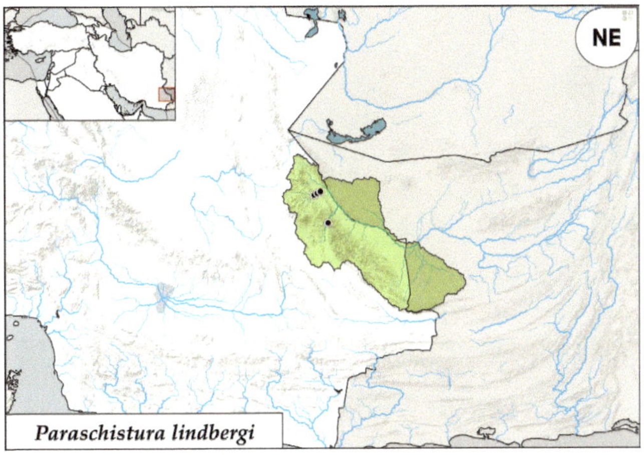

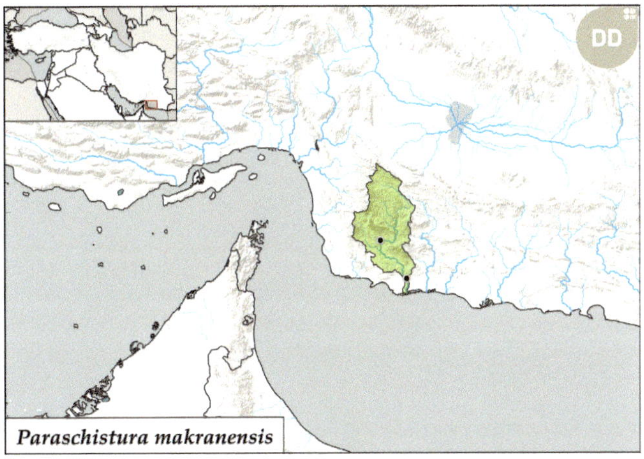

Paraschistura makranensis; Jegin drainage, Iran; 35 mm SL. © H. Mousavi-Sabet.

Paraschistura makranensis

Common name. Jegin loach.

Diagnosis. Distinguished from other species of *Paraschistura* in West Asia by: ● flank plain brown or slightly mottled / ○ lateral line complete or almost complete, reaching to caudal base / ○ caudal emarginate / ○ 7½ branched dorsal rays / ○ processus dentiformis in upper jaw small and pointed / ○ pelvic origin situated below second or third branched dorsal ray / ○ no dorsal adipose crest on caudal peduncle without procurrent rays / ○ suborbital flap present in male / ○ small, isolated, deeply embedded scales restricted to caudal peduncle and flank below and behind dorsal origin, no scales on predorsal back and flank / ○ snout pointed / ○ a bold black spot at base of first dorsal rays / ○ 2–3 black blotches at caudal base / ○ caudal–peduncle depth 1.2–1.4 times longer than deep. Size up to 40 mm SL.

Distribution. Iran: Upper Jegin drainage.

Habitat. Fast to slow-flowing streams with gravel substrate.

Biology. No data.

Conservation status. DD; due to the very poor knowledge of this species.

Further reading. Eagderi et al. 2019b (description).

Paraschistura naumanni; Kol drainage, Iran; ~40 mm SL.

Paraschistura naumanni; Golabi spring, Iran; ~45 mm SL.

Paraschistura naumanni

Common name. Persian banded loach.

Diagnosis. Distinguished from other species of *Paraschistura* in West Asia by: ○ dorsal origin situated above, or very slightly behind vertical of pelvic origin / ○ suborbital flap or groove absent in male / ○ back and flank fully covered by scales / ○ snout blunt / ○ absence of a long dorsal adipose crest / ○ lateral line incomplete / ○ caudal emarginate / ○ 7½ branched dorsal rays / ○ very variable colour pattern of irregularly shaped and set, often faded bars and blotches on flank / ○ pelvic reaching to a point about 1.0–2.0 eye diameter in front of anus / ○ no dorsal adipose crest / ○ caudal-peduncle depth 1.7–2.0 times longer than deep / ○ a bold or faint black spot at base of first dorsal rays in some populations / ○ a complete black bar at caudal base in some populations / ○ distance between anus and tip of pelvic 4–7 % SL. Size up to 51 mm SL.

Distribution. Iran: Kor basin and Kol and Mond drainages.

Habitat. Fast to slow-flowing streams and springs with gravel substrate.

Biology. No data.

Conservation status. VU; appears to be declining in its small range, where stream desiccation is a major challenge.

Remarks. In the Kol drainage, *P. naumanni* occurs in sympatry with *P. abdolii* and *P. delvarii*. Whether all three species occur in syntopy or are geographically separated within Kol drainage remains unclear. *Oxynoemacheilus persa* is another loach species found in Mond. *Paraschistura naumanni* shows a large variability in colour patterns within and between different populations, and fish from other sites may look very different.

Further reading. Freyhof et al. 2015 (description).

Spring Golabi in Iran is the habitat of *Paraschistura naumanni*, *Esmaeilius darabensis*, *Mystus cyrusi*, and other regional endemic species.

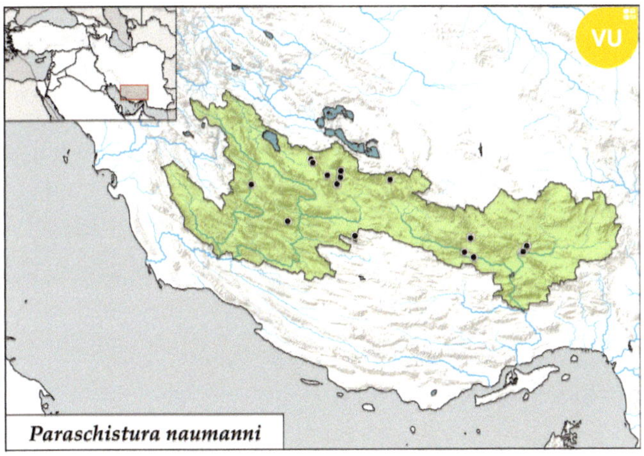

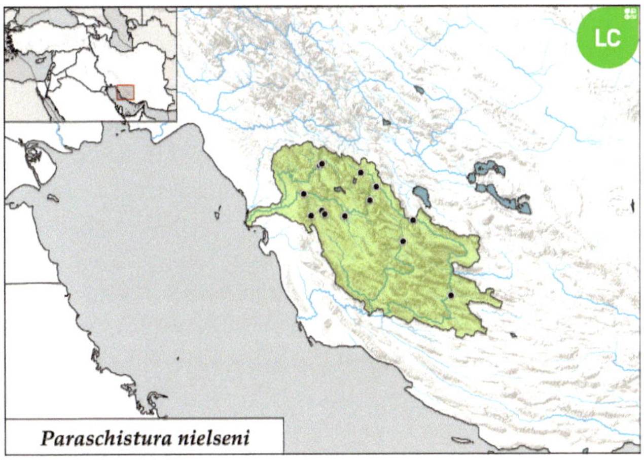

Paraschistura nielseni; Helleh drainage, Iran; ~45 mm SL.

Paraschistura nielseni

Common name. Helleh loach.

Diagnosis. Distinguished from other species of *Paraschistura* in West Asia by: ○ 6–11 regularly set and shaped dark-or pale-brown or grey bars, wider than interspaces, absent or very faint on flank in front of dorsal origin / ○ suborbital groove present in male / ○ a faint dark-grey spot at base of first dorsal rays / ○ black bar at caudal base faint or dissociated o two blotches, rarely complete / ○ snout blunt / ○ no dorsal adipose crest / ○ lateral line incomplete / ○ caudal deeply emarginate / ○ back and flank fully covered by scales / ○ body depth at dorsal origin 15–17 % SL / ○ pelvic reaching to about 1–2 eye diameters in front of anus / ○ pelvic origin below first or second branched dorsal ray / ○ 7½ branched dorsal rays. Size up to 52 mm SL.

Distribution. Iran: Helleh and Mond drainages.

Habitat. Fast to slow-flowing streams and springs with gravel substrate.

Biology. No data.

Conservation status. LC.

Further reading. Nalbant & Bianco 1998 (description); Freyhof et al. 2015 (diagnosis).

Paraschistura susiani; Marun drainage, Iran; ~50 mm SL.

Paraschistura susiani

Common name. Khuzestan banded loach.

Diagnosis. Distinguished from other species of *Paraschistura* in West Asia by: ○ suborbital present in male, short, blunt, and roundish / ○ back and flank fully covered by scales / ○ snout blunt / ○ no dorsal adipose crest / ○ lateral line incomplete / ○ caudal emarginate / ○ pelvic origin below first or second branched dorsal ray / ○ pelvic reaching to about 0.5–1 eye diameters in front of anus / ○ 7½ branched dorsal rays / ○ body depth at dorsal origin 14–16 % SL / ○ 7–10, usually regularly set and shaped, dark-or pale- brown or grey bars, wider than interspaces, absent or very faint on flank in front of dorsal origin / ○ black or brown bar at caudal base / ○ a bold or very faint dark-grey spot at base of first dorsal rays / ○ caudal peduncle 1.4–1.6 times longer than deep. Size up to 50 mm SL.

Distribution. Iran: Ab-e Ala and Zard in Jarahi drainage and Shur in Karun drainage.

Habitat. Fast to slow-flowing streams and springs with gravel substrate.

Biology. No data.

Conservation status. EN; appears to be declining within its very small range.

Further reading. Freyhof et al. 2015 (description).

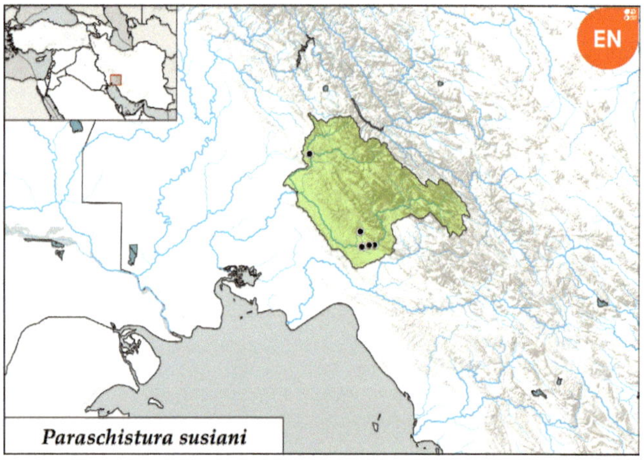

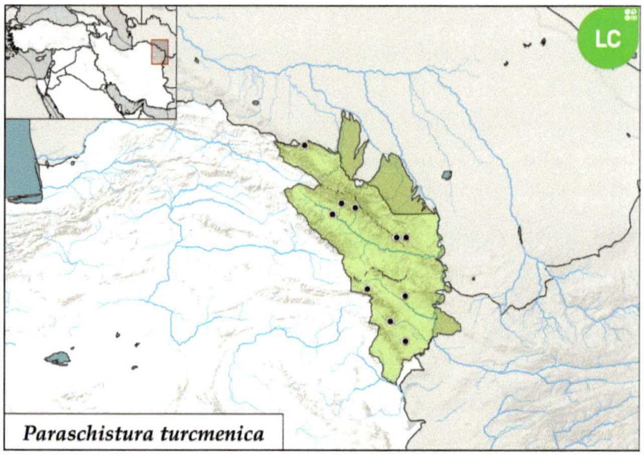

Paraschistura turcmenica; Hari drainage, Iran, ~45 mm SL.

Paraschistura turcmenica

Common name. Turkmenian loach.

Diagnosis. Distinguished from other species of *Paraschistura* in West Asia by: ○ scales completely absent / ○ flank with marbled colour pattern, or with bars dissociated in a row of midlateral blotches and saddles on back, or with 11–14 irregularly shaped and set, dark-brown bars on flank, usually dissociated into two adjacent narrow bars or individual bars with a central open space / ○ suborbital flap or groove absent in male / ○ no dorsal adipose crest / ○ caudal emarginate / ○ a bold black spot at base of first dorsal rays / ○ snout blunt / ○ usually 7½ branched dorsal rays / ○ body depth at dorsal origin 13–15 % SL / ○ pelvic reaching to about 1–2 eye diameter in front of anus / ○ pelvic origin below first or third branched dorsal ray. Size up to 53 mm SL.

Distribution. Eastern Kavir basin, Atrak drainage, western Karakum Desert, Hari in Afghanistan, Iran and Turkmenistan, Morghab in Afghanistan and Turkmenistan, and northern slope of Kopetdag in Turkmenistan.

Habitat. Streams and springs with moderate to fast-flowing freshwater in mountain and desert landscapes.
Biology. No data.
Conservation status. LC.
Remarks. *Nemacheilus kessleri turcomanus* is likely a synonym for this species. *Paraschistura turcomanus* has recently been re-validated, but while syntypes of *P. turcomana* examined for that study are typical of *P. turcmenica*, the species identified as *P. turcmenica* may be juvenile *P. cristata*.
Further reading. Freyhof et al. 2015 (diagnosis); Mousavi-Sabet et al. 2015d (*P. turcomana*).

Sasanidus

Sasanidus and *Eidinemacheilus* are two loach genera endemic to the Persian Gulf basin. Only one species of *Sasanidus* is known, and no superficially similar species have been reported from Pakistan or India. *Sasanidus* is distantly related to *Paraschistura*, and it cannot be excluded that other species of *Sasanidus* might exist on the Indian subcontinent, where a large diversity of nemacheilid loaches exist, usually identified as *Paraschistura* or *Schistura*. Furthermore, an understanding of the phylogeny of loaches is virtually lacking in this region. **Further reading.** Freyhof et al. 2016a (description, phylogeny).

Sasanidus kermanshahensis

Common name. Kermanshah loach.
Diagnosis. Distinguished from superficially similar species of *Oxynoemacheilus* in Tigris drainage by: ○ body cylindrical / • a prominent, irregular lateral stripe along lateral midline / • no dark-brown or black bar or blotch at caudal base / • a prominent whitish or yellowish triangular patch on posteriormost upper and lower caudal peduncle / • no sexual dimorphism in pectoral length / ○ caudal peduncle deep, without adipose crest / ○ caudal moderately deep emarginate / ○ lateral line incomplete, reaching to vertical between middle dorsal base and middle anal base / ○ suborbital groove present in male. Size up to 70 mm SL.

Sasanidus kermanshahensis; Karkheh drainage, Iran, ~70 mm SL.

Sasanidus kermanshahensis; Karkheh drainage, Iran, ~65 mm SL.

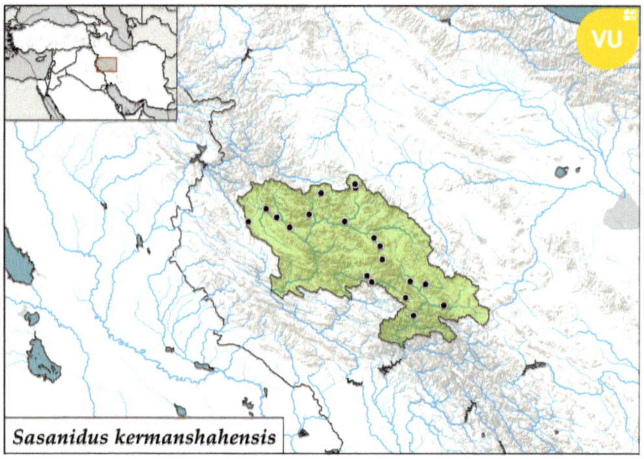

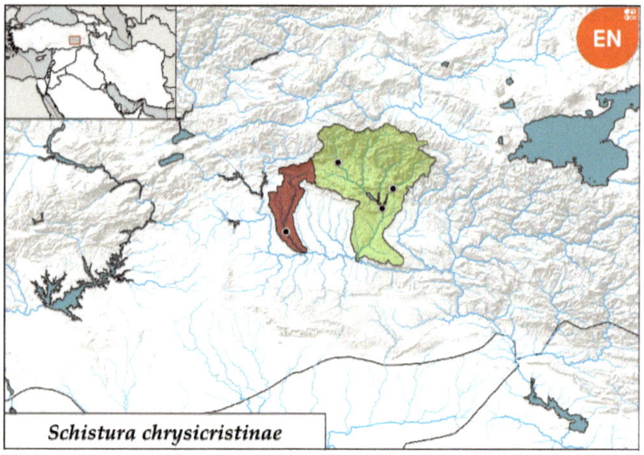

Distribution. Iran: Gamasiab, Dinevar, Razavar, and Kahman in Karkheh drainage and Sezar in Karun drainage.
Habitat. Moderately fast-flowing to standing waters of gravel-bed streams.
Biology. No data.
Conservation status. VU; appears to be declining within its small range.
Further reading. Bănărescu & Nalbant 1966 (description); Freyhof et al. 2016a (phylogeny).

Schistura

Schistura is a highly polyphyletic genus that encompasses approximately 250 species. The type species of *Schistura* is *S. rupecula*, found in northern India. The genus is thought to be widely distributed and speciose in the southern Himalayas, extending from Pakistan to China. Loaches identified as *Schistura* are widespread in India and Sri Lanka and occur all over tropical Asia and Indonesia. They are especially speciose in Indochina. Most loaches placed in *Schistura* will be recognised in different genera in the future, and the genus needs a fundamental revision. In West Asia, several loaches had been described in *Schistura*, including *S. ceyhanensis*, *S. chrysicristinae*, *S. evreni*, and *S. seyhanicola*, all from Türkiye. Except for *S. chrysicristinae*, these were later placed in *Oxynoemacheilus*. *Schistura chrysicristinae* was placed in *Paraschistura* based on the characters provided in its original description. Upon rediscovery in 2021, mitochondrial DNA data indicated a close relationship between *S. chrysicristinae* and *S. rupecula*. Consequently, *S. chrysicristinae* was reclassified within the genus *Schistura*. *Schistura chrysicristinae* is found approximately 3,000 km northeast of the genus's known range.

Schistura chrysicristinae

Common name. Batman loach.
Diagnosis. Distinguished from other species of *Nemacheilidae* in upper Tigris drainage by: ○ flank with wide, pale-brown bars or large, irregularly shaped and set blotches / ○ no black bar or bold black blotches at caudal base / ○ no suborbital flap or groove in male / pectoral in male as large as in female / ○ lateral line incomplete, reaching above anal base / ○ body covered by scales, scales densely set on caudal

Schistura chrysicristinae; Batman drainage, Türkiye; ~35 mm SL.

peduncle and flank behind dorsal origin, absent on anterior flank / ○ caudal emarginate / ○ 7–8½ branched dorsal rays. Size up to 36 mm SL, likely to grow larger.
Distribution. Türkiye: Ambar and Batman drainages.
Habitat. Fast-flowing river sections.
Biology. No data.
Conservation status. EN; appears to be declining within its very small range. Not found in Ambar in recent years.
Further reading. Nalbant 1998 (description); Kaya et al. 2025b (description, generic position).

Seminemacheilus

Seminemacheilus comprises six species endemic to Central and southern Anatolia. It is distinguished by a roundish or slightly emarginate caudal, a very stout body, a very long pectoral in male, and no or very few scales. *Seminemacheilus* is related to *Oxynoemacheilus*, and some phylogenetic studies place *Seminemacheilus* as a sister genus to *Oxynoemacheilus* or nested within the basal clades of that genus. However, the phylogenetic relationships between these taxa remain uncertain, and we treat *Seminemacheilus* as a distinct genus. *Seminemacheilus* can be distinguished from similar stout and deep-bodied *Oxynoemacheilus* species, including *O. panthera*, *O. seyhanensis*, and *O. frenatus*, by the presence of a very long pectoral fin in males that extends beyond the pelvic origin when folded back. *Seminemacheilus* are stagnophilic loaches that are rarely found in flowing waters. They typically inhabit densely vegetated backwaters, springs, or slow-flowing streams. They exhibit pronounced sexual dimorphism, with males being smaller, more slender, and having a longer pectoral fin than females.
Further reading. Yoğurtçuoğlu et al. 2020b (revision).

Key to species of *Seminemacheilus*

1a - A few deeply embedded scales on caudal peduncle; caudal–peduncle depth 1.3–1.6 times in its length; head length 21–24 % SL.
..................*S. ispartensis*
1b - Body scaleless; caudal–peduncle depth 0.9–1.3 times in its length; head length 24–28 % SL.
......................2

2a - Caudal usually slightly emarginate; one central pore in supratemporal head canal.
..................*S. attalicus*
2b - Caudal truncate or roundish; no central pore in supratemporal head canal.
......................3

3a - Belly usually with many small black or brown blotches; posterior naris reaching to anterior eye margin when folded backwards.
..................*S. ahmeti*
3b - Belly plain white or yellowish without small black or brown blotches; posterior naris not reaching to anterior eye margin when folded backwards.
......................4

4a - Caudal roundish; caudal–peduncle depth 0.9–1.1 times in its length.
..................*S. ekmekciae*
4b - Caudal truncate; caudal–peduncle depth 1.1–1.3 times in its length.
......................5

5a - 5–7 supraorbital head pores; eye diameter 13–14 % HL; snout length 38–40 % HL; prepelvic length 57–59 % SL.
..................*S. lendlii*
5b - 2–5 (6) supraorbital pores; eye diameter 14–17 % HL; snout length 35–37 % HL; prepelvic length 53–56 % SL.
..................*S. tubae*

Seminemacheilus ahmeti; spring Soysallı, Türkiye; female, 53 mm SL.

Seminemacheilus ahmeti; spring Soysallı, Türkiye; male; 51 mm SL.

Seminemacheilus ahmeti

Common name. Sultan pond loach.

Diagnosis. Distinguished from other species of *Seminemacheilus* by: • many small black or brown dots or blotches on belly / • posterior naris reaching to anterior eye margin when folded backwards / ○ body scaleless / ○ caudal truncate / ○ no central pore in supratemporal head canal / ○ 2–5 supraorbital head pores / ○ caudal–peduncle depth 1.0–1.1 times in its length / ○ head length 25–28 % SL / ○ eye diameter 13–14 % HL / ○ snout length 38–40 % HL / ○ prepelvic length 57–59 % SL / ○ flank pattern mottled, vermiculated, or with large isolated blotches. Size up to 57 mm SL.

Distribution. Türkiye: Sultan marshes.

Habitat. Springs and streams with standing or slow-flowing water and dense submerged vegetation.

Biology. No data.

Conservation status. CR; see comments under *Pseudophoxinus elizavetae*.

Further reading. Sungur et al. 2018b (description); Yoğurtçuoğlu et al. 2020b (revision).

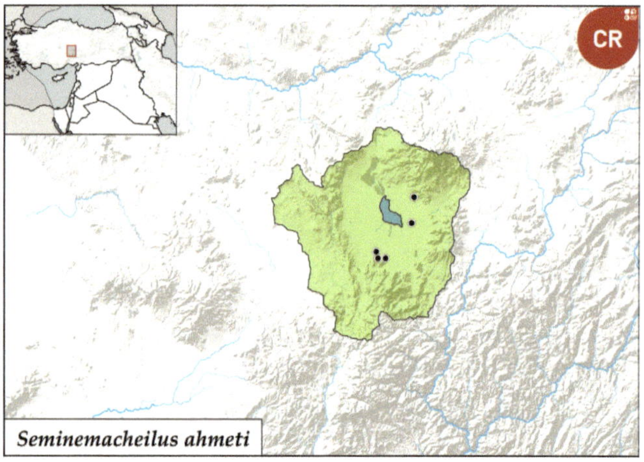

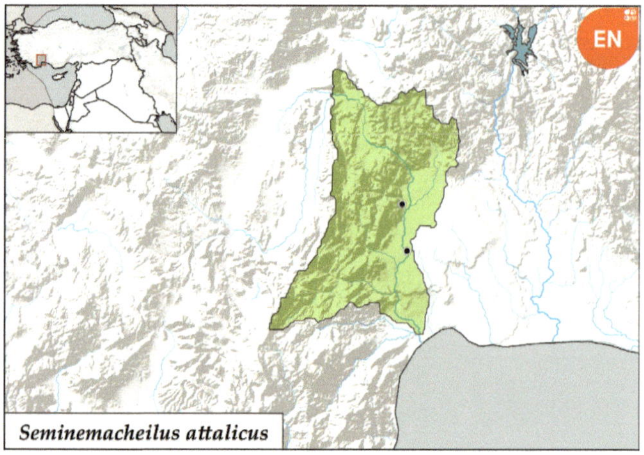

Seminemacheilus attalicus; spring Kırkgöz, Türkiye; male, 64 mm SL.

Seminemacheilus attalicus

Common name. Kırkgöz pond loach.

Diagnosis. Distinguished from other species of *Seminemacheilus* by: • caudal usually slightly emarginate / ○ flank mottled or marbled / • one central pore in supratemporal head / • 7–13 supraorbital head pores / ○ body scaleless / ○ caudal–peduncle depth 0.9–1.2 times in its length / ○ head length 24–28 % SL / ○ belly without or very rarely with black or brown blotches / ○ posterior naris not reaching to anterior eye margin when folded backwards. Size up to 86 mm SL.

Distribution. Türkiye: Spring Kırkgöz and outflowing stream Yediarıklar.

Habitat. Springs and streams with stagnant or slow-flowing water and dense submerged vegetation.

Biology. No data.

Conservation status. EN; appears to be declining within its very small range.

Further reading. Yoğurtçuoğlu et al. 2020b (description).

Seminemacheilus ekmekciae; Gölyazı, Türkiye; female, ~50 mm SL.

Seminemacheilus ekmekciae; Lake Büget, Türkiye; male, 34 mm SL.

Seminemacheilus ekmekciae

Common name. Central Anatolian pond loach.

Diagnosis. Distinguished from other species of *Seminemacheilus* by: ○ caudal rounded / ○ no central pore in supratemporal head canal / ○ 2–4 supraorbital head pores / ○ body scaleless / ○ no black or brown dots or blotches on belly / ○ posterior naris not reaching to anterior eye margin when folded backwards / ○ flank pattern mottled, vermiculated, or with large isolated blotches / ○ caudal–peduncle depth 0.9–1.1 times in its length / ○ head length 23–26 % SL. Size up to 73 mm SL.

Distribution. Türkiye: Lake Tuz basin and Ereğli marshes.

Habitat. Springs and streams with stagnant or slow-flowing water and dense submerged vegetation.

Biology. No data.

Conservation status. EN; appears to be declining within its very small range. Disappeared from Ereğli marshes as these dried up.

Further reading. Yoğurtçuoğlu et al. 2020b (description).

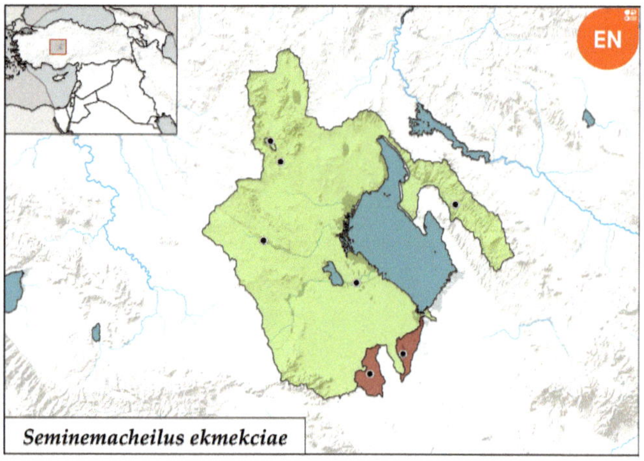

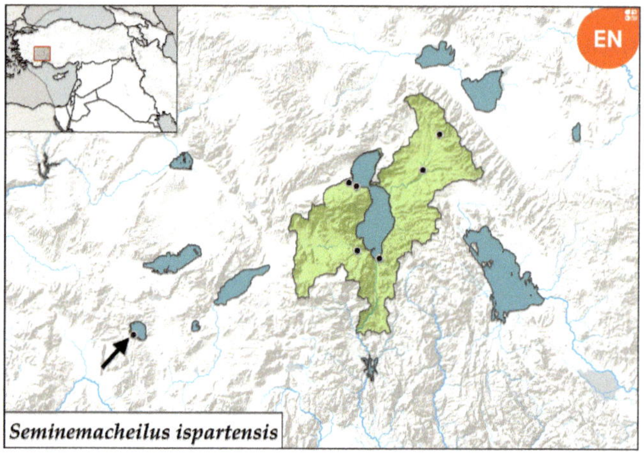

Seminemacheilus ispartensis; Lake Eğirdir basin, Türkiye; female, 72 mm SL.

Seminemacheilus ispartensis; Lake Eğirdir basin, Türkiye; male, 74 mm SL.

Seminemacheilus ispartensis

Common name. Southern pond loach.

Diagnosis. Distinguished from other species of *Seminemacheilus* by: • a few deeply embedded scales on caudal peduncle / • caudal-peduncle depth 1.3–1.6 times in its length / • head length 21–24 % SL / ○ caudal truncate / ○ no black or brown dots or blotches on belly / ○ posterior naris not reaching to anterior eye margin when folded backward / ○ no central pore in supratemporal head canal / ○ 6–10 supraorbital head pores / ○ flank pattern mottled, vermiculated, or with large isolated blotches. Size up to 91 mm SL.

Distribution. Türkiye: Lake Eğirdir and Salda basins.

Habitat. Springs, streams, and lakeshores. Usually in habitats with standing water and dense submerged vegetation. Spawning begins in March.

Biology. Lives up to 4 years.

Conservation status. EN; appears to be declining within its very small range.

Remarks. Although *Seminemacheilus ispartensis* and *S. lendlii* are almost identical in their COI sequences, *S. ispartensis* is recognised as a valid species based on clear morphological differences, with recent introgressive hybridisation suspected. *Seminemacheilus* recorded from Karapınar (Lake Yarışlı basin) may belong to this species. This population has disappeared as Karapınar has dried up.

Further reading. Bănărescu & Nalbant 1964 (distribution as *Nemacheilus lendlii*); Erk'akan et al. 2007 (description); Yoğurtçuoğlu et al. 2020b (revision).

Kırkgöz is a micro-hotspot of fish biodiversity in the Mediterranean basin inhabited by *Seminemacheilus attalicus* and three other locally endemic species.

Seminemacheilus lendlii; stream Eymir, Sakarya drainage, Türkiye; female, 68 mm SL.

Seminemacheilus lendlii; stream Eymir, Sakarya drainage, Türkiye; male, 53 mm SL.

Seminemacheilus lendlii

Common name. Northern pond loach

Diagnosis. Distinguished from other species of *Seminemacheilus* by: ○ caudal truncate / ○ posterior naris not reaching to anterior eye margin when folded backward / ○ no black or brown dots or blotches on belly / ○ flank pattern fine mottled / ○ no central pore in supratemporal canal / ○ 5–7 supraorbital head pores / ○ caudal–peduncle depth 1.1–1.2 times in caudal–peduncle length / ○ head length 25–28 % SL / ○ snout length 38–40 % HL / ○ prepelvic length 57–59 % SL / ○ eye diameter 13–14 % HL. Size up to 77 mm SL.

Distribution. Türkiye: Lake Eymir-Mogan basins (Ankara), Porsuk drainage, a tributary of upper Sakarya, and Lakes Akşehir, Eber, and Ilgın basins.

Habitat. Springs and streams with standing or slow-flowing water and dense vegetation.

Biology. No data.

Conservation status. VU; potentially vanished from Lakes Ilgin and Eymir-Mogan basins.

Further reading. Hankó 1925 (description); Yoğurtçuoğlu et al. 2020b (revision).

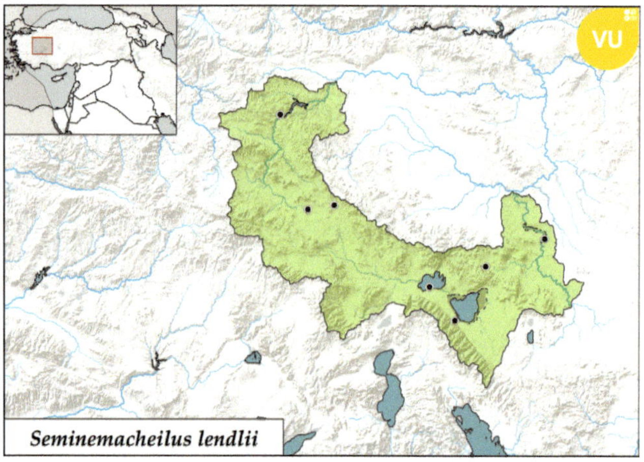

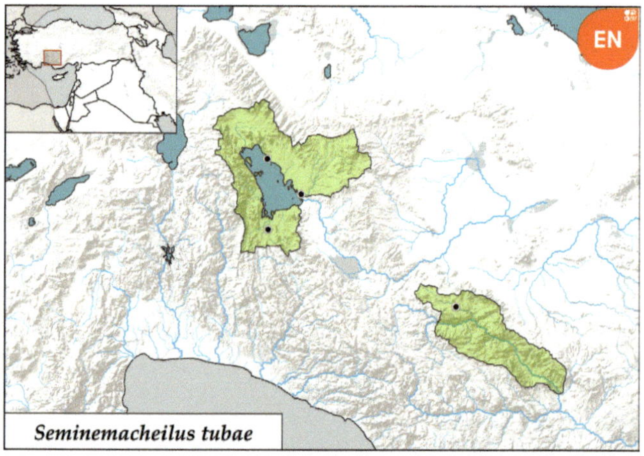

Seminemacheilus tubae; Lake Beyşehir basin, Türkiye; female, 65 mm SL.

Seminemacheilus tubae; Lake Beyşehir basin, Türkiye; male; 54 mm.

Seminemacheilus tubae

Common name. Beyşehir pond loach.

Diagnosis. Distinguished from other species of *Seminemacheilus* by: ○ caudal truncate / ○ no black or brown dots or blotches on belly / ○ flank pattern mottled, vermiculated or with large isolated blotches / ○ body scaleless / ○ posterior naris not reaching to anterior eye margin when folded backwards / ○ no central pore in supratemporal head canal / ○ 2–5, rarely 6, supraorbital head pores / ○ caudal-peduncle depth 1.1–1.3 times in its length / ○ head length 24–28 % SL / ○ eye diameter 14–17 % HL / ○ snout length 35–37 % HL / ○ prepelvic length 53–56 % SL. Size up to 66 mm SL.

Distribution. Türkiye: Lake Beyşehir basin and Göksu drainage (Mediterranean basin).

Habitat. Stream sections with stagnant or slow-flowing water and dense vegetation.

Biology. No data.

Conservation status. EN; appears to be declining within its very small range.

Remarks. *Seminemacheilus dursunavsari* is a synonym. It was first published as a *nomen nudum*, which did not meet the criteria for a species description. The description was re-published, and now *S. dursunavsari* is an available name but published after the description of *S. tubae*, rendering it a synonym.

Further reading. Çiçek 2020 (description as *S. dursunavsari*); Yoğurtçuoğlu et al. 2020b (description).

Ulupınar, a small spring in Lake Akşehir basin, is the habitat of *Seminemacheilus lendlii*.

Turcinoemacheilus

Turcinoemacheilus are small fishes that can be readily distinguished from all other nemacheilids in West Asia by their very slender body and the position of the anus, which is notably situated in front of the anal origin. Before 2013, only one species was known in the genus, but subsequent research revealed 11 species. *Turcinoemacheilus* are commonly found in the headwater streams and small rivers of the Euphrates, Tigris, and Zayandeh drainage, an endorheic basin in Iran. One species was described from the Himalayas, suggesting a much wider distribution range for the genus. Several small South Asian loaches exhibit superficial similarities to *Turcinoemacheilus*, and their phylogenetic relationships require further investigation. *Turcinoemacheilus* may represent a relatively common genus in the Himalayas, or these fishes may belong to different genera.

Further reading. Conway et al. 2011 (*T. himalaya*); Esmaeili et al. 2014c (revision); Jouladeh-Roudbar et al. 2023c (revision Iran, phylogeny).

Key to species of *Turcinoemacheilus* in West Asia

1a - Anus situated behind middle between pelvic and anal origins.
......................2
1b - Anus situated at middle or in front of middle between pelvic and anal origins.
......................5

2a - Lateral line complete, reaching to anterior part of caudal.
......................*T. christofferi*
2b - Lateral line incomplete, reaching to midway between tip of pectoral and dorsal origin or midway between dorsal and anal origins.
......................3

3a - Usually an elongated, irregularly shaped dark-brown blotch on side of anal base.
......................*T. bahaii*
3b - No dark blotch-brown on side of anal base.
......................4

4a - Body depth at dorsal origin 10–13 % SL.
......................*T. ansari*
4b - Body depth at dorsal origin 13–15 % SL.
......................*T. hafezi*

5a - No dark-brown stripe along lateral midline; 7–9 distinct dark-brown blotches on flank not forming a row.
......................*T. saadii*
5b - Prominent row of dark-brown blotches along lateral midline, usually fused into a stripe.
......................6

6a - Caudal–peduncle length 6–7 % SL; 4–5 mandibular pores.
......................*T. minimus*
6b - Caudal–peduncle length 7–10 % SL; 5–7 mandibular pores.
......................7

7a - Midlateral stripe narrower than eye diameter; prepelvic distance 47–50 % SL.
......................*T. kosswigi*
7b - Midlateral stripe broader than eye diameter; prepelvic distance 50–55 % SL.
......................8

8a - A dark-brown blotch at anal base.
......................*T. moghbeli*
8b - No dark-brown blotch at anal base.
......................9

9a - Interorbital distance 31–37 % HL.
......................*T. inexpectatus*
9b - Interorbital distance 23–31 % HL.
......................*T. ekmekciae*

Turcinoemacheilus ansari; Beshar, Iran; 50 mm SL.

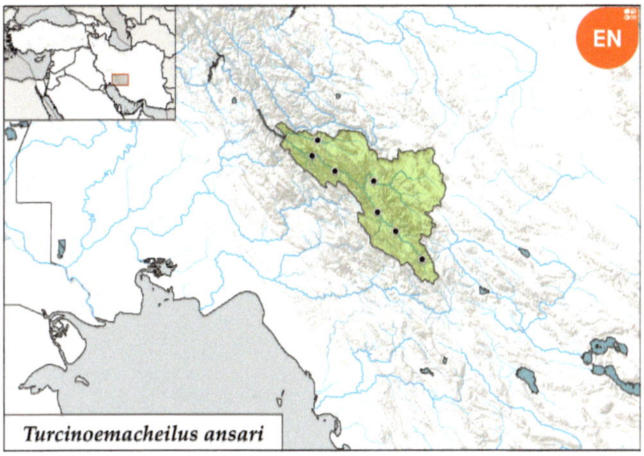

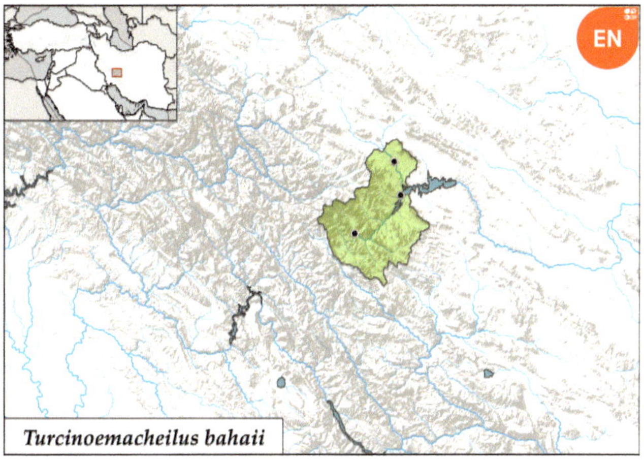

Turcinoemacheilus ansari
Common name. Beshar dwarf loach.
Diagnosis. Distinguished from other species of *Turcinoemacheilus* in West Asia by: body depth 10–13 % SL / ○ no dark-brown blotch on side of anal base / ○ anus situated behind middle between pelvic and anal origins / ○ mottled colour pattern or a prominent irregular-shaped dark-brown stripe along lateral midline disconnected from dark-brown blotches on back. Size up to 50 mm SL.
Distribution. Iran: Merian, Beshar, and Khersan in upper Karun drainage.
Habitat. Fast-flowing sections of small rivers and streams, usually in rapids and riffles with coarse gravel or rocks. Inhabits interspaces in gravel.
Biology. No data.
Conservation status. EN; restricted to a small area where it is in decline.
Remarks. Syntopic with *T. saadii* in Beshar.
Further reading. Jouladeh-Roudbar et al. 2023c (description).

Turcinoemacheilus bahaii; Zayandeh drainage, Iran; ~45 mm SL.

Turcinoemacheilus bahaii

Common name. Zayandeh dwarf loach.

Diagnosis. Distinguished from other species of *Turcinoemacheilus* in West Asia by: • usually a dark-brown blotch on side of anal base / ○ anus situated behind middle between pelvic and anal origins / ○ mottled colour pattern or an irregular-shaped dark-brown stripe along lateral midline disconnected from dark-brown blotches on back. Size up to 54 mm SL.

Distribution. Iran: Headwaters in Zayandeh drainage.

Habitat. Fast-flowing sections of small rivers and streams, usually in rapids and riffles with coarse gravel or rocks.

Biology. Mature at 1 year and lives up to 4 (males) and 5 (females) years. Feeds on invertebrates.

Conservation status. EN; appears to be declining within its very small range.

Remarks. This species was described based on individuals with a dark-brown blotch on side of anal base. New data show that this condition is not present in all individuals and that morphological characters may need to be revised. It is well distinguished from other species by molecular characters.

Further reading. Esmaeili et al. 2014c (description); Borhani et al. 2017 (biology); Ebrahimi et al. 2017 (biology); Jouladeh-Roudbar et al. 2023c (phylogeny).

Turcinoemacheilus christofferi; Gholiyan, Iran; 54 mm SL.

Turcinoemacheilus christofferi

Common name. Sezar dwarf loach.

Diagnosis. Distinguished from other species of *Turcinoemacheilus* in West Asia by: • lateral line complete, reaching to anterior part of caudal / ○ caudal–peduncle depth 1.6–1.8 times in its length / ○ caudal slightly emarginate / ○ anus situated behind middle between pelvic and anal origins / ○ no dark-brown blotch on side of anal base. Size up to 54 mm SL.

Distribution. Iran: Gholiyan, a headwater tributary of Bakhtiyari in Karun drainage.

Habitat. Fast-flowing sections of small rivers and streams, usually in rapids and riffles with coarse gravel or rocks. Inhabits interspaces in gravel.

Biology. No data.

Conservation status. CR; restricted to one site where it is in decline. Potentially at the edge of extinction.

Further reading. Jouladeh-Roudbar et al. 2023c (description).

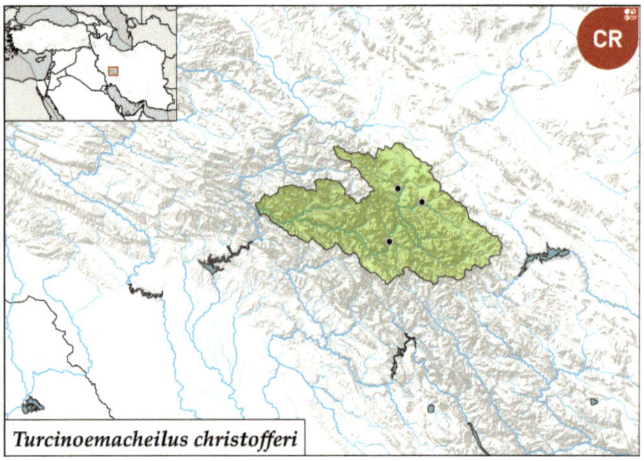

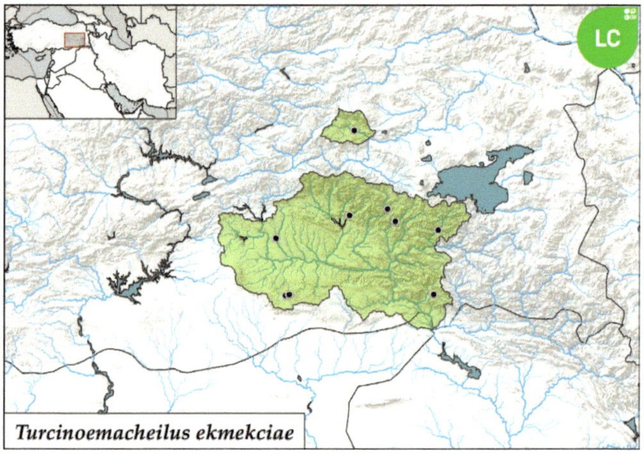

Turcinoemacheilus ekmekciae; Kaynarca in upper Euphrates drainage, Türkiye; 45 mm SL.

Turcinoemacheilus ekmekciae

Common name. Anatolian dwarf loach.

Diagnosis. Distinguished from other species of *Turcinoemacheilus* in West Asia by: ○ roundish blotches or a stripe thicker than eye diameter along lateral midline / ○ 5–7 mandibular pores in mandibular canal / ○ caudal–peduncle length 7–9 % SL / ○ caudal–peduncle depth 1.8–2.7 times in its length / ○ body depth 1.3–1.7 times in head length / ○ prepelvic distance 50–53 % SL / ○ upper caudal lobe 15–18 % SL / ○ interorbital distance 23–31 % HL / ○ maxillary barbel 15–21 % HL / ○ anus situated at middle or in front of middle between pelvic and anal origins / ○ no dark-brown blotch on side of anal base / ○ caudal slightly emarginate. Size up to 55 mm SL.

Distribution. Türkiye: stream Kaynarca in Euphrates, Yanarsu, Botan, Nerduş, Batman in upper Tigris drainages.

Habitat. Fast-flowing sections of small rivers and streams, usually in rapids and riffles with coarse gravel or rocks.

Biology. No data.

Conservation status. LC.

Further reading. Kaya et al. 2024d (description).

Turcinoemacheilus hafezi; Karun drainage, Iran; ~45 mm SL.

Turcinoemacheilus hafezi

Common name. Persian dwarf loach.

Diagnosis. Distinguished from other species of *Turcinoemacheilus* in West Asia by: ○ body depth 13–15 % SL / ○ no dark-brown blotch on side of anal base / ○ anus situated behind middle between pelvic and anal origins / ○ mottled colour pattern or a prominent irregular-shaped dark-brown midlateral stripe disconnected from dark-brown blotches on back. Size up to 67 mm SL.

Distribution. Iran: Behesht Abad in Karun drainage.

Habitat. Fast-flowing sections of small rivers and streams, usually in rapids and riffles with coarse gravel or rocks.

Biology. No data.

Conservation status. EN; restricted to a small area where it is in decline.

Further reading. Golzarianpour et al. 2013 (description); Esmaeili et al. 2014c (revision); Jouladeh-Roudbar et al. 2023c (phylogeny).

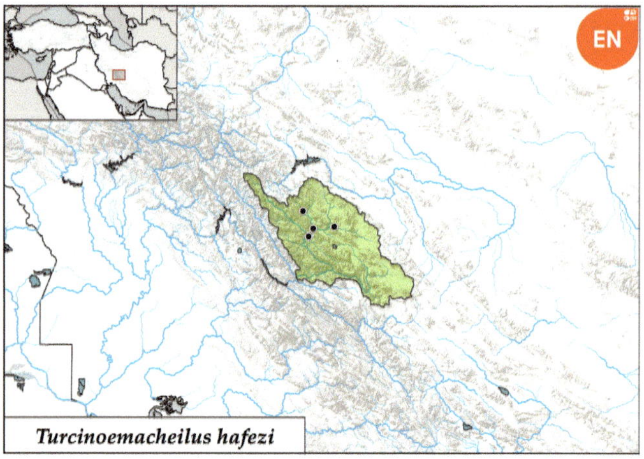

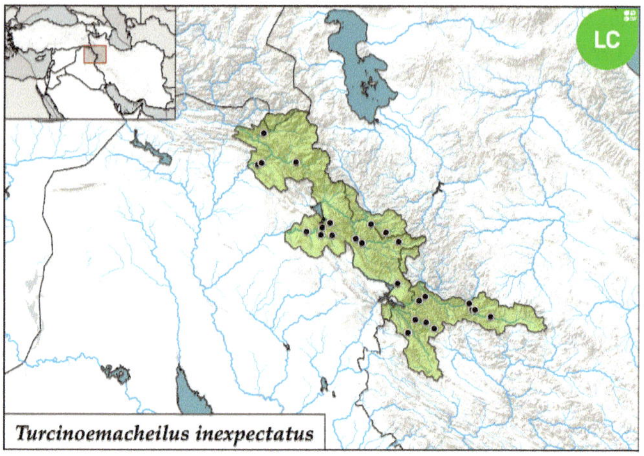

Turcinoemacheilus inexpectatus; Lesser Zab drainage, Iraq; 50 mm SL.

Turcinoemacheilus inexpectatus

Common name. Zab dwarf loach.

Diagnosis. Distinguished from other species of *Turcinoemacheilus* in West Asia by: ○ a dark midlateral stripe broader than eye diameter / ○ 5–6 mandibular pores in mandibular canal / ○ caudal–peduncle length 8–10 % SL / ○ body depth 1.3–2.0 times in head length / ○ caudal–peduncle depth 1.5–2.3 times in its length / ○ interorbital distance 31–37 % HL / ○ prepelvic distance 50–53 % SL / ○ anus situated at middle or in front of middle between pelvic and anal origins / ○ no dark-brown blotch on side of anal base. Size up to 52 mm SL.

Distribution. Iraq and Iran: Sirvan, Greater and Lesser Zab drainages.

Habitat. Fast-flowing sections of small rivers and streams, usually in rapids and riffles with coarse gravel or rocks. Inhabits interspaces in gravel.

Biology. No data.

Conservation status. LC.

Further reading. Esmaeili et al. 2014c (description as *T. kosswigi*); Kaya et al. 2024d (molecular phylogeny); Freyhof & Jouladeh-Roudbar 2024 (description).

Turcinoemacheilus kosswigi; upper Greater Zab, Türkiye; 60 mm SL.

Turcinoemacheilus kosswigi

Common name. Hakkari dwarf loach.

Diagnosis. Distinguished from other species of *Turcinoemacheilus* in West Asia by: ○ a dark midlateral stripe narrower than eye diameter / ○ 5–7 mandibular pores in mandibular canal / ○ caudal-peduncle length 7–9 % SL / ○ body depth 1.3–1.7 times in head length / ○ caudal-peduncle depth 1.9–2.2 times in its length / ○ large, brown saddles on back connected to lateral stripe / ○ caudal slightly emarginate / ○ upper caudal lobe 18–21 % SL / ○ prepelvic distance 47–50 % SL / ○ maxillary barbel 21–26 % HL / ○ anus situated at middle or in front of middle between pelvic and anal origins / ○ no dark-brown blotch on side of anal base. Size up to 60 mm SL.

Distribution. Türkiye: Upper Greater Zab drainage.

Habitat. Fast-flowing sections of small rivers and streams, usually in rapids and riffles with coarse gravel or rocks. Inhabits interspaces in gravel.

Biology. Lives for 1 or 2 years. Feeds on benthic invertebrates.

Conservation status. CR; limited to a small area that is and will be heavily impacted by mining.

Further reading. Esmaeili et al. 2014c (revision); Kaya et al. 2024d (re-diagnosis, distribution).

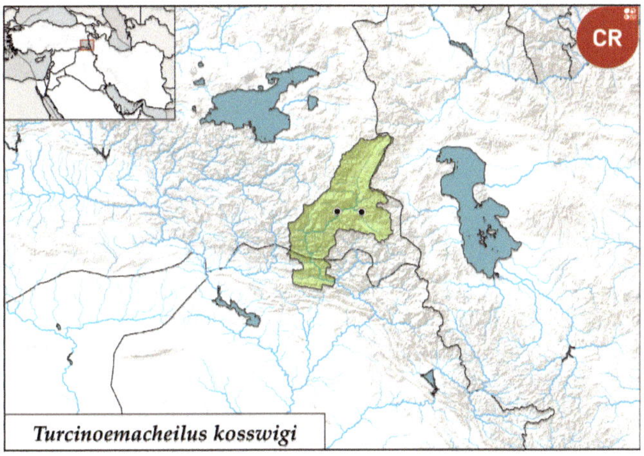

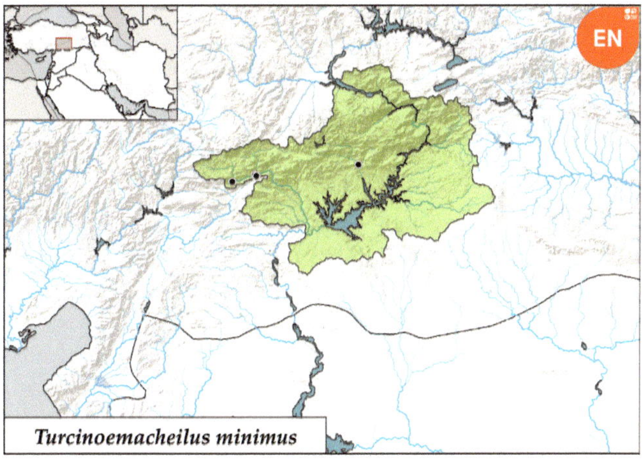

Nalparez in Lesser Zab drainage in Iraq is the habitat of *Turcinoemacheilus inexpectatus*.

Turcinoemacheilus minimus; Euphrates drainage, Türkiye; 35 mm SL.

Turcinoemacheilus minimus

Common name. Euphrates dwarf loach.

Diagnosis. Distinguished from other species of *Turcinoemacheilus* in West Asia by: ● caudal–peduncle length 6–7 % SL / ● body depth 1.6–1.9 times in head length / ● 4–5 mandibular pores in mandibular canal / ○ a row of large irregular, brown, longitudinally elongated blotches along lateral midline, often fused into an irregular-shaped lateral stripe / ○ large, brown saddles on back connected to lateral blotches / ○ caudal slightly emarginate / ○ anus situated at middle or in front of middle between pelvic and anal origins / ○ no dark-brown blotch on side of anal base. Size up to 50 mm SL.

Distribution. Türkiye: Göksu and Kahta in Euphrates drainage.

Habitat. Fast-flowing sections of small rivers and streams, usually in rapids and riffles with coarse gravel or rocks. Inhabits interspaces in gravel.

Biology. No data.

Conservation status. EN; restricted to a small area where it is in decline.

Further reading. Esmaeili et al. 2014c (description); Kaya et al. 2024d (re-diagnosis, distribution).

Turcinoemacheilus moghbeli; Leyleh, Iran; 54 mm SL.

Turcinoemacheilus moghbeli

Common name. Kermanshah dwarf loach.

Diagnosis. Distinguished from other species of *Turcinoemacheilus* in West Asia by: ● caudal–peduncle length 14–20 % SL / ○ 5–6 mandibular pores in mandibular canal / ○ body depth 1.2–1.8 times in head length / ○ caudal–peduncle depth 1.5–2.3 times in its length / ○ prepelvic distance 50–55 % SL / ○ no dark-brown blotch on side of anal base / ○ anus situated at middle or in front of middle between pelvic and anal origins. Size up to 54 mm SL.

Distribution. Iran: Sirvan, Zemkan, Leyleh, Alvand, and Goleyn in Tigris drainage.

Habitat. Fast-flowing sections of small rivers and streams, usually in rapids and riffles with coarse gravel or rocks. Inhabits interspaces in gravel.

Biology. No data.

Conservation status. VU; restricted to a small area where it is in decline.

Further reading. Jouladeh-Roudbar et al. 2023c (description).

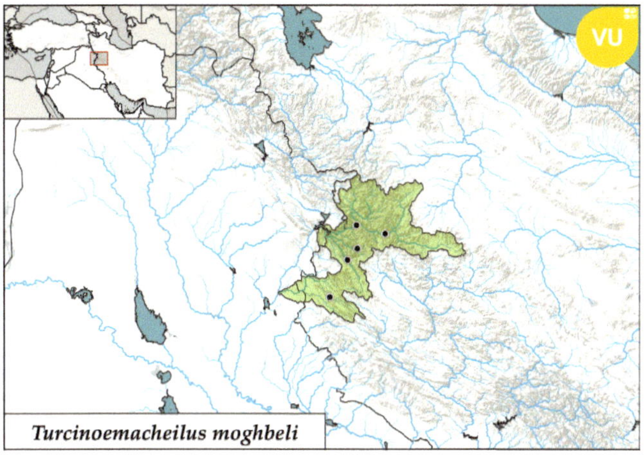

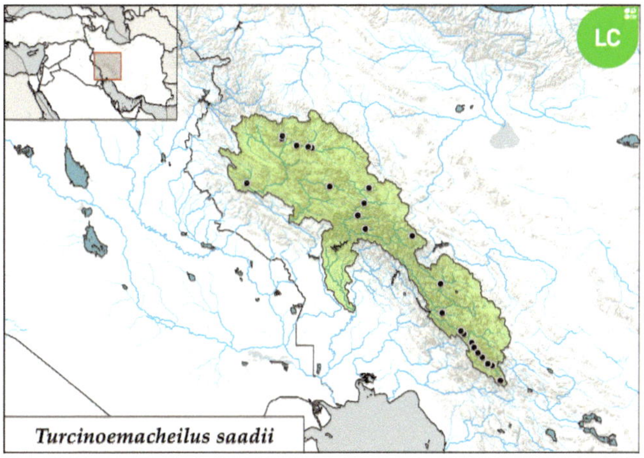

Turcinoemacheilus saadii; Beshar, Karun drainage, Iran; 45 mm SL.

Turcinoemacheilus saadii

Common name. Large spot dwarf loach.

Diagnosis. Distinguished from other species of *Turcinoemacheilus* in West Asia by: ● colour pattern organised in 7–9 distinct dark-brown or grey saddles never forming a lateral stripe / ○ caudal deeply emarginate / ○ anus situated at middle or in front of middle between pelvic and anal origins / ○ no dark-brown blotch on side of anal base. Size up to 64 mm SL.

Distribution. Iran: Bashar, Bakhtiyari, and Sezar in Karun, Gamasiab, Dinevar, and Kahman in Karkheh drainages.

Habitat. Fast-flowing sections of small rivers and streams, usually in rapids and riffles with coarse gravel or rocks. Inhabits interspaces in gravel.

Biology. No data.

Conservation status. LC.

Remarks. In the upper Karun, usually syntopic with *T. hafezi* and *Sasanidus kermanshahensis*.

Further reading. Esmaeili et al. 2014c (description).

The stream Eziki in Türkiye is a tributary of the Greater Zab, a major Tigris tributary. It is the habitat of *Turcinoemacheilus kosswigi*.

Oxynoemacheilus eliasi inhabits streams with a Mediterranean water regime and often survives only in small pools during summer.

Order Siluriformes

Catfishes

Catfishes, or Siluriformes, are a large group of primary freshwater fishes. There are approximately 4200 species worldwide. Eight families of catfishes are found in West Asia, six of which are native, one purely marine (Plotosidae), and one predominantly marine (Ariidae). Catfishes are particularly diverse in the tropics, especially in South America, where many species, sizes, morphologies, and feeding types can be found. They represent one of the most species-rich groups in freshwaters. The earliest fossils of a catfish are Campanian (83–72 million years ago) and Maastrichtian (72–66 million years ago) from Argentina and Bolivia. However, Bayesian relaxed molecular clock analyses of Siluriformes resulted in an average posterior crown age estimate of 121 million years ago, with a credible interval between 111 and 131 million years ago. Catfishes are the sister group of Neotropical knifefishes (Gymnotiformes), renowned for their electric organs and electrosensory systems.

The majority of catfishes are nocturnal. Many species engage in parental care, with the male guarding the eggs and larvae. All catfishes lack true scales, and most West Asian species are either naked or have small, ossified prickles on the skin. Some catfishes can breathe air and drown if not allowed to do so. Most catfishes are tolerant

Key to families of catfishes in West Asia

1a - Adipose absent.
..................2
1b - Adipose present.
..................4

2a - Nasal barbel absent.
..................Siluridae
2b - Nasal barbel present, very long.
..................3

3a - Two dorsals, first short, with a spine, second long, continuous with caudal and anal.
..................Plotosidae
3b - One dorsal, not continuous with caudal and anal.
..................Clariidae

4a - An adhesive organ on thorax between pectorals.
..................Sisoridae
4b - No adhesive organ on thorax between pectorals.
..................5

5a - Nasal barbel absent.
..................Ariidae
5b - Posterior nostril with a nasal barbel.
..................6

6a - Anal with 6–10 rays; adipose longer than anal base.
..................Bagridae
6b - Anal with 16–30 rays; adipose shorter than anal base.
..................Ictaluridae

https://doi.org/10.1515/9783111677811-022

to only freshwater, but the salinity tolerance of several species extends to, or is even limited to, estuarine or oceanic levels. One of the marine catfishes, *Plotosus lineatus*, is frequently found along the shores of the Persian Gulf and the Red Sea and has invaded the Mediterranean through the Suez Canal. It is known that this species enters brackish and freshwater habitats in Southeast Asia, but it has never been found in freshwaters in West Asia. Therefore, it is included in the family key only. Ariid catfishes are a second, mostly marine family, and they can occasionally be found on mudflats and in shallow, barckish waters of the Persian Gulf, including lower parts of rivers such as the Shatt al Arab/Arvand and Hammar marshes. *Netuma bilineata*, *N. thalassina*, *Plicofollis dussumieri*, *P. layardi*, and *P. tenuispinis* have been recorded from the region. Still, these require critical identification, and almost nothing is known about their distribution in the Persian Gulf's habitats. They are included in the family key only because they do not enter freshwaters. **Further reading.** Sullivan et al. 2006 (phylogeny of catfishes); Coad 2010a (species from Persian Gulf); Hughes et al. 2018 (molecular phylogeny and evolutionary origin).

Lake Balık (Gölbaşı/Hatay) in Türkiye is the last habitat of *Mystus misrai*.

Family Loricariidae

Suckermouth catfishes

Sucker-mouth catfish, armored catfish, and plecos represent the largest catfish family, with approximately 110 genera and over 1000 species, many of which remain undescribed. These peculiar fishes are native to South and Central America and exhibit an astonishing diversity of body shapes, colours, sizes, and adaptations. They are distinguished by several characteristics, including a scale-less body covered with bony plates on the back and flank, a sucker-like mouth (often with highly specialised teeth), and papillate lips. In all species, the males perform parental care until the young have resorbed their yolk-sack. Most species can gulp air and have many other special morphological and ecological adaptations. In West Asia, one species has been released into a limited number of water bodies, where it has become established. However, individuals might be found in virtually all locations, particularly in or around cities. Suckermouth catfishes cannot survive temperatures below 15°C for longer than a few days, which limits their distribution to warm springs in many regions. As several species of suckermouth catfish are kept in aquaria, the possibility of the release and subsequent establishment of other species cannot be ruled out. Specialist aquarium literature is likely to be of assistance in identifying these. **Further reading.** Isbrücker 1980 (diversity); Nelson et al. 2016 (diversity).

Pterygoplichthys disjunctivus* × *Pterygoplichthys pardalis

Common name. Hybrid sailfin catfish.

Diagnosis. Unmistakeable. Flank and back covered with large shields, sucker-like mouth, and short barbels. Size up to 490 mm SL.

Distribution. Israel: Lake Kinneret (Sea of Galilee) and Nahal Amal; Türkiye: Pınarbaşı near İnönü, possibly also in Orontes. Saudi Arabia: Wadi Haneefah in Riyadh, Al-Kharj south of Riyadh and Al-Ahsa oasis. Both parental species of *Pterygoplichthys* are native to Amazon drainage of South America.

Habitat. Tropical, inhabiting a wide range of stagnant or slow-flowing waters, often colonising reservoirs, urban and artificial water bodies, hot water springs, lakes, or water storage areas.

Biology. Usually, begins spawning when larger than 260 mm SL, often at about 300 mm SL. Spawns throughout year frequently interrupted by low availability of nesting sites at low water levels. Males are territorial at nest site, dig burrows (usually in banks or under stones), and guard eggs until fry has resorbed their yolk sac. Stomach enlarged, highly vascularised, acting as an accessory respiratory organ, allowing high tolerance to low oxygen concentrations and desiccation (up to 20 hours). Tolerant to salinity up to 10 ‰, but low tolerance to water temperatures below 15°C. Herbivorous, feeding mainly on periphyton and detritus, grazing on hard structures and surface of fine sediments. Large individuals also eat plants.

Conservation status. Non-native; kept in aquaria worldwide, from which it is released. An important non-native and invasive species.

Pterygoplichthys disjunctivus × *P. pardalis*; Fossa Calda, Italy; ca.180 mm SL.

https://doi.org/10.1515/9783111677811-023

Remarks. Released from aquaria into the wild since early 1950s, with established non-native, often invasive, populations reported from many tropical and subtropical countries, including Australia, Bangladesh, India, Mexico, USA, Puerto Rico, Taiwan, Philippines, Vietnam, and Java. It has also been released in warm springs in temperate regions such as Poland and Italy. Still, it is not invasive there because it cannot escape warm water habitats and survive winters in natural waters in these countries. Because of its hardiness and high fecundity, this catfish is produced in very large numbers for the aquarium trade in Southeast Asia and perhaps elsewhere. It is commonly used as a cleaner fish in aquariums and is often released when it grows too large. A major global invasive species of management concern, particularly where they occupy sensitive habitats such as springs. They often reach very high densities, and their spawning burrows in stream banks increase erosion. It is considered a pest in fisheries because of its abundance, poor taste, and physical characteristics, including hard bony plates on its body and strong spines that make it difficult to handle in nets.

Species identification of *Pterygoplichthys* is most easily based on the colour pattern of abdomen. Most non-native populations were initially identified as *P. disjunctivus* and/or *P. pardalis* or sometimes placed in the genus *Liposarcus*, a synonym of *Pterygoplichthys*. Early studies from several introductions suggested that more than one, and perhaps even more than four, species of suckermouth catfish had been introduced in a single area. Still, more recent studies indicate that the globally invasive suckermouth catfish is a hybrid between *P. disjunctivus* and *P. pardalis*. This explains the high variability in characters resembling parental species in non-native populations. As these hybrids are very common and their introduction has been documented for decades, the initial hybridisation occurred very early in history of *Pterygoplichthys* culture for the aquarium trade.

Further reading. Armbruster & Page 2006 (identification); Özdilek 2007 (Orontes); Gibbs et al. 2008 (biology); AlKahem et al. 2008 (Saudi Arabia); Wu et al. 2011 (hybridisation); Rueda-Jasso et al. 2013 (biology, global invasion); Golani & Snovsky 2013 (Israel); Emiroğlu et al. 2016 (Sakarya).

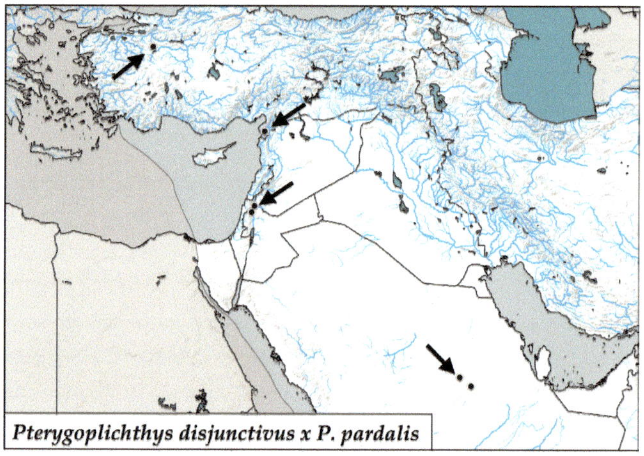

Pterygoplichthys disjunctivus x P. pardalis; Sakarya, Türkiye; ~180 mm SL, two colour patterns of belly often identified as two species.

Family Bagridae

Bagrid catfishes

Bagridae is a family of approximately 16 genera and 225 species widely distributed in Asia. *Bagrus*, is found only in Africa. All other African genera formerly included in the Bagridae are now placed in different families. In Asia, bagrids exhibit a wide range of body shapes and sizes. The giant Indian catfishes of the genus *Sperata*, which reach 2000 mm total length, represent one extreme, while the 29 mm SL dwarf leaf-litter dwellers, such as the *Nanobagrus immaculatus* from Borneo, represent the other. All large- and medium-sized bagrids are sought for commercial species, several of which are valued for aquaculture. *Mystus*, with approximately 52 valid species, is the largest genus in the family, comprising mostly middle-sized and morphologically generalised species. One species, *M. gulio*, is known to frequently enter brackish and marine waters, with records from the coast of Pakistan, situated not far from the border with Iran. Other species are strict freshwater inhabitants. *Mystus* species exhibit only moderate tolerance to low water temperatures, with most species being abundant in the lower reaches of the Euphrates and Tigris (*M. pelusius*) and warm springs (*M. cyrusi, M. misrai*). **Further reading.** Ferraris 2007 (diversity); Nelson et al. 2016 (diversity).

Key to species of *Mystus* in West Asia

1a - No stripes on flank; eye below dorsal head profile, not protruding.
.... *M. misrai*

1b - Flank with a wide, dark-brown, mid-lateral stripe, followed by a wide silvery stripe above and below, a narrow slivery line along lateral midline; eye at or slightly above dorsal head profile, usually protruding.
.................2

2a - Maxillary barbel not reaching to pelvic origin; adipose with a steep slope at its origin.
.... *M. cyrusi*

2b - Maxillary barbel reaching to or beyond of pelvic origin; adipose with a gentle slope at its origin.
.... *M. pelusius*

Mystus cyrusi; Golabi spring; Kol drainage, Iran; ~90 mm SL.

Mystus cyrusi

Common name. Kol zugzug.

Diagnosis. Distinguished from other species of *Mystus* in West Asia by: • maxillary barbel not reaching to or beyond pelvic origin / • adipose base length 31–37 % SL, with a steep slope at its origin / • head depth 17–22 % SL / • caudal peduncle depth 10–13 % SL / • 12–14 gill rakers / ○ usually a wide, brown, mid-lateral stripe, followed by a wide silvery stripe above and below mid-lateral stripe, and a silvery line along lateral midline in life and preserved individuals / ○ eye situated at or slightly above, usually protruding dorsal head profile / ○ 14–22 serra on inner pectoral spine. Size up to 203 mm SL.

https://doi.org/10.1515/9783111677811-024

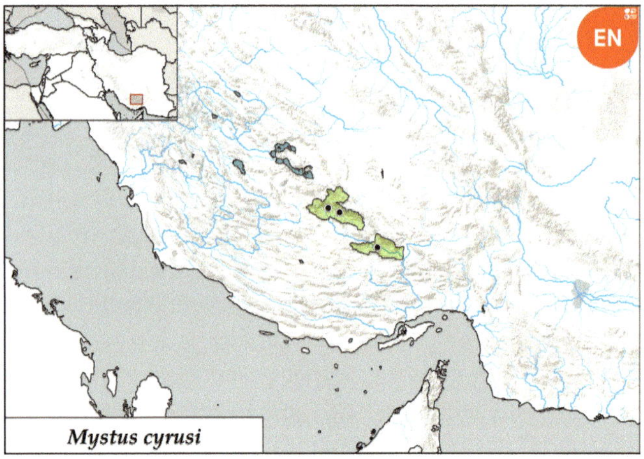

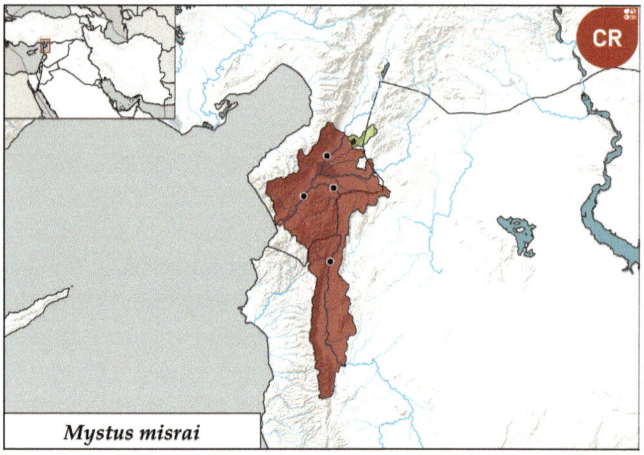

Distribution. Iran: Shur, Qalatooyeh, and Golabi springs in Kol drainage.
Habitat. Springs and spring-fed streams.
Biology. No data.
Conservation status. EN; appears to be declining within its very small range. Invasive *Amatitlania nigrofasciata* might be a threat in springs.
Further reading. Esmaeili et al. 2022 (description).

Mystus misrai; Orontes drainage, Türkiye; 120 mm SL.

Mystus misrai
Common name. Hatay zugzug.
Diagnosis. Distinguished from other species of *Mystus* in West Asia by: • dorsal fin margin rounded / • 8–10 serra on inner pectoral spine / • eye situated below dorsal head profile, not protruding / • head blunt / • no stripes or lines on flank in life and preserved individuals / • dorsal spine shorter than head-width at eye, usually as long as distance between lower eye margins / ○ maxillary barbel reaching to or beyond pelvic origin. Size up to 138 mm SL.
Distribution. Türkiye and Syria: Lower Orontes drainage.
Habitat. Large rivers and lakes.
Biology. No data.
Conservation status. CR; following massive pollution of lower Orontes and draining of Lake Amik in 20th century, this species appears to be restricted to Lake Balık (Gölbaşı). It is thought to be in decline here due to habitat deterioration.
Further reading. Anuradha 1986 (description); Freyhof & Yoğurtçuoğlu 2023 (rediscovery).

Mystus pelusius; Tigris drainage, Türkiye; 140 mm SL.

Mystus pelusius

Common name. Zugzug catfish.

Diagnosis. Distinguished from other species of *Mystus* in West Asia by: ○ maxillary barbel reaching to or beyond pelvic origin, to anal-base in some individuals / ○ adipose base length 38–46 % SL, with a gentle slope at its origin / ○ dorsal fin margin straight or very slightly convex / ○ 11–16 serra on inner pectoral spine / ○ eye situated at or slightly above, usually protruding dorsal head profile / ○ head pointed / ○ usually a wide, brown, mid-lateral stripe, followed by a wide silvery stripe above and below mid-lateral stripe, and a silvery line along lateral midline in life and preserved individuals / ○ dorsal spine as long or longer than head width at eye, often longer than maximal head width. Size up to 178 mm SL.

Distribution. Qweiq, Euphrates, Tigris, Karun, Jarrahi, and Zohreh drainages.

Habitat. Larger rivers, marshes, reservoirs, and lakes.

Biology. Lives up to 7 years. Spawns first time in one year, in May and June, males with about 70 mm SL and females with about 50 mm SL. Feeds on invertebrates and small fishes.

Conservation status. LC; extirpated from Qweiq.

Further reading. Coad 2010a (identification, distribution, biology); Esmaeili et al. 2022 (diagnosis).

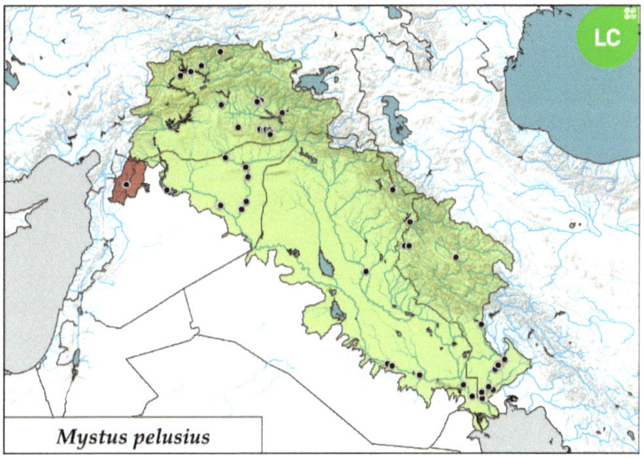

Mystus gulio is a coastal species known from Pakistan adjacent to Iran.

Family Clariidae

Air-breathing catfishes

The family of air-breathing catfishes comprises approximately 120 species in 16 genera. Phylogenetic diversity is greatest in South and Southeast Asia, where most genera are found. The genus *Clarias* is the largest genus in the family and is distributed across Asia and Africa. Most species in the genus *Clarias* are found in Africa, where they have invaded from Asia only once and diversified there. Airsac catfishes of the former family Heteropneustidae are included in the family Clariidae. Air-breathing catfishes can thrive in various ecosystems, including marshes and rivers, as well as rainforest streams, desert pools, lakes, subterranean waters, and many artificial waterbodies. They can reach high densities, and due to their often large size, they are of major ecological and commercial importance in tropical Asia and Africa.

As their common name suggests, Clariids breathe air through a specialised organ. An opening between the second and third gill arches leads to a suprabranchial chamber with highly vascularised, brush-like structures called "trees." These dendritic organs are situated above and behind the gills and are supported by epibranchials of the second and fourth branchial arches. The first of these is smaller and lies in the anterior compartment. Each is a highly branched tree-like structure supported by a cartilaginous internal skeleton. This feature enables them to survive for extended periods out of water. *Heteropneustes* have a lung-like outgrowth from the pharynx. Some *Clarias* species can "walk" a few meters on land to seek alternative habitats, using sturdy spines on their pectorals. *Clarias* catfishes are weakly electric, and their electricity is used for communication and to find food. It is challenging for a human observer to perceive their electricity directly, as it is too weak to be sensed.

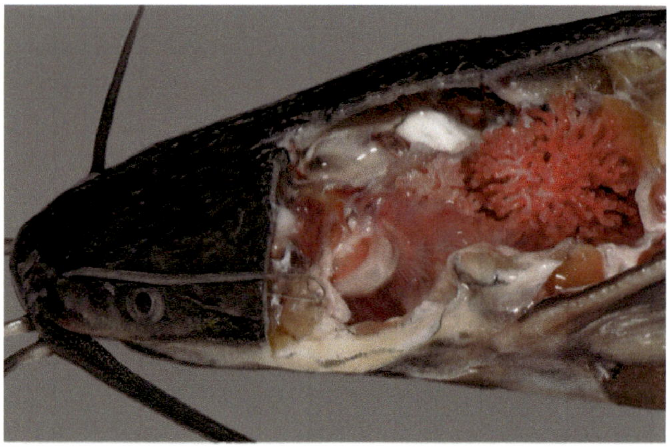

Figure 53. Suprabranchial chamber with 'tree' in *Clarias gariepinus*.

Clarias gariepinus and *Clarias* sp. are among the most significant globally invasive species. Both are major aquaculture species within their native range, and both, as well as several hybrids, including other Clariids, are increasingly farmed outside their range for their boneless flesh. They are most commonly imported for farming or aquaria. Still, because they are easy to translocate, private individuals often introduce them into residential or community ponds, aiming to harvest them for food. In all of these situations, they may escape or be released intentionally.

Further reading. Agnese & Teugels 2005; Sullivan et al. 2006; Pouyaud & Paradis 2009 (diversity, phylogeny); Mbanga et al. 2018 (respiratory organ).

Key to Clariidae in West Asia

1a - Dorsal short, 6–7 rays.
..................*Heteropneustes fossilis*
1b - Dorsal very long, 61–79 rays.
..................2

2a - Stout, body depth strongly increasing to vertical of pelvic origin, then decreasing; anterior margin of pectoral spine rugose and with irregular bumps; posterior dorsal tip reaching to or beyond caudal base; flank brown or grey often with vertical series of yellowish spots.
..................*Clarias* sp.
2b - Slender, eel-like, body depth slightly increasing to dorsal origin, then decreasing; anterior margin of pectoral spine serrated; posterior dorsal tip not reaching to caudal base; flank brown or grey with dark-brown mottling.
..................*Clarias gariepinus*

https://doi.org/10.1515/9783111677811-025

Clarias gariepinus

Common name. Sharptooth catfish.

Diagnosis. Distinguished from *Clarias* sp. in West Asia by: • body slender, eel-like / • body depth slightly increasing to dorsal origin, then decreasing / • anterior margin of pectoral spine serrated / • posterior dorsal tip not reaching to caudal base / • flank brown or grey with dark-brown mottling. Size up to 1500 mm SL and 60 kg, usually about 700 mm SL.

Distribution. Native to Jordan drainage and coastal streams of Israel. Introduced in Azraq oasis, Orontes drainage in Syria and Türkiye. Widespread in southern Türkiye from Orontes to Göksu, probably elsewhere along Mediterranean coast. Also, in warm springs in Sakarya, Lake Burdur basin (Türkiye), Al-Kharj south of Riyadh, and Medina (Saudi Arabia). Probably found elsewhere. Nearly pan-African range, absent only from North Africa except Algeria, where it occurs in streams in Tolga oasis on northern edge of Sahara and gueltas on north slopes of Tassili n' Ajjer. Tibesti and Ennedi massifs of Chad. Widespread in Nile and Niger drainages, eastern and southern Africa, but absent from upper and lower Guinea, Cape prov. and probably Nogal prov. of South Africa. Introduced in other parts of Africa, tropical and subtropical Asia, and South America. Occasional escapes from warm water aquaculture facilities in Europe are not established in natural waters but only in artificially heated waters.

Habitat. Slow-flowing or standing waters of lakes, marshes, reservoirs, and lowland rivers, rarely in fast-flowing waters. Often found in canals with both sewage and treated water. Tolerant of extreme environmental conditions, including temperatures as low as 8°C and very low oxygen conditions.

Biology. Lives up to 8 years. Matures at about 300–350 mm SL (males), 350–400 mm SL (females). An accessory respiratory organ allows this species to breathe air. Often found in muddy bottoms of dry ponds, occasionally swallowing air through mouth. Dependent on atmospheric oxygen and will drown if cut off from surface. Can leave water using its strong pectoral spines and move to new areas through very shallow channels. Spawning is usually triggered by rising water levels and flooding shallow areas. Spawns throughout year when water temperatures are above 22°C. Migrates to rivers and temporary streams to spawn. Spawns usually at night in shallow, flooded areas. Courtship often preceded by highly aggressive encounters between males. Generates monophasic, head-positive electrical discharges of 5–260 microSiemens during intra-specific aggressive interactions. Mating position, a form of amplexus, is maintained for several seconds. In amplexus, male lies in a U-shaped curve around female's head. Sperm and eggs are released, followed by a vigorous female tail swish to spread eggs over a wide area. Pairs mate several times during a spawning night. Parents do not guard eggs or larvae. Eggs adhere to vegetation. Larvae usually begin feeding 48–72 hours after hatching (23–28°C). Nocturnal bottom feeders on invertebrates and fish, occasionally taking young birds, rotting flesh, and plants. An important commercial species in Africa. An important aquaculture species in warm-water and tropical countries, introduced almost worldwide.

Conservation status. LC.

Further reading. Teugels 1986 (distribution); Bruton 1979 (biology).

Clarias gariepinus, Lake Burdur basin, Türkiye, ~250 mm SL.

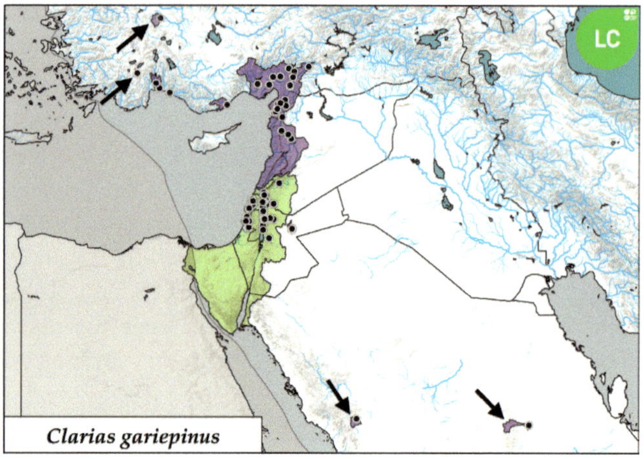

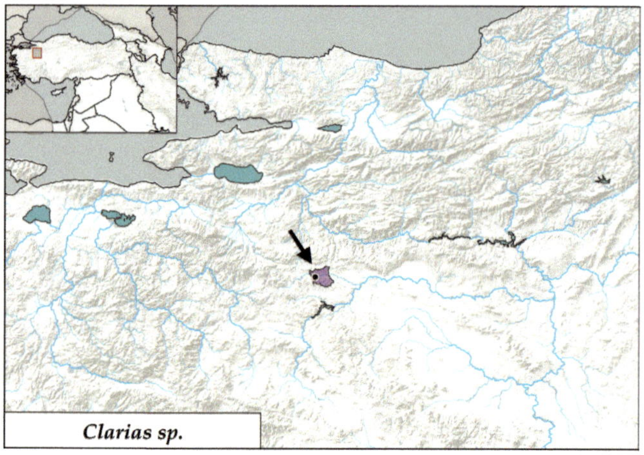

***Clarias* sp.**

Common name. Walking catfish.

Diagnosis. Distinguished from *C. gariepinus* by: • body stout, its depth strongly increasing to vertical of pelvic origin, then decreasing / • anterior margin of pectoral spine rugose and with irregular bumps / • posterior dorsal tip reaching to or beyond caudal base / • flank brown or grey, often with vertical series of yellowish spots. Size up to 450 mm SL and 1.2 kg, usually about 250 mm SL.

Distribution. Hot springs in upper Sakarya. Native to Philippines and introduced in tropical areas worldwide.

Habitat. Lowland habitats such as marshes, ponds, lakes, rice fields, and rivers, often in relatively small lowland and hill streams. Most commonly found in stagnant, muddy water, but also in clear springs and lakes.

Biology. In native range, matures at about 280 mm SL and spawns during rainy season when rivers rise. An accessory respiratory organ allows this species to breathe air. Dependent on atmospheric oxygen and will drown if cut off from surface. Usually makes lateral migrations into flooded forests and fields during wet season to feed and spawn. Can use its strong pectoral spines to leave water and move to new areas through very shallow channels. Mating position in a form of amplexus is maintained for several seconds. In amplexus, male lies in a U-shaped curve around female's head. Sperm and eggs are released, followed by a vigorous flick of female's tail to disperse eggs over a wide area. Pairs mate several times during a spawning night. Males excavate nests in submerged mud banks but do not guard eggs or larvae. Eggs adhere to vegetation. Nocturnal bottom feeders on invertebrates and fish, occasionally on plants. An important commercial species in Southeast Asia and for aquaculture in warm-water and tropical countries, introduced almost worldwide.

Conservation status. Non-native; released from aquaria or aquaculture.

Remarks. This species is usually identified as *C. batrachus*, which is known only from Java in Indonesia. Records of *C. batrachus* from mainland Southeast Asia, Sundaic Southeast Asia, and the large pantropical non-native range suggest two other species are involved in what is called *C. batrachus*. Non-native populations need to be better identified, and it cannot be excluded that they represent undescribed species or hybrids between different species.

Further reading. Ng & Kottelat 2008 (identification); Emiroğlu et al. 2020 (upper Sakarya).

Clarias sp., Nha Trang, Vietnam, ~200 mm SL.

Heteropneustes fossilis; Greater Zab; Iraq; 87 mm SL.

Heteropneustes fossilis

Common name. Stinging catfish.

Diagnosis. Distinguished from other species of catfishes in West Asia by: ○ adipose absent / ○ nasal barbel present, very long / ○ anal long, 60–70 rays / ○ dorsal short, 6–7 rays, without spine. Size up to 280 mm SL.

Distribution. Euphrates and Tigris drainages. In Tigris, from about Diyarbakır and Syrian Euphrates south to southern marshes and Shatt al Arab/Arvand in Iraq. Common in rivers and marshes of Khuzestan prov. in Iran, in Azadegan, and Karkheh. Native from Indus east to Myanmar.

Habitat. Slow-flowing rivers, marshes, reservoirs, and lakes with fresh or slightly brackish water. Can inhabit heavily polluted habitats and waters with very low oxygen levels, survive in semi-liquid mud, move overland, and move to bottom of cracks and crevices formed by mud cracking.

Biology. Usually gregarious. Spawns first time at 1 year and about 50–100 mm SL (India). Spawns under water surface. Mating position, held in a form of amplexus for several seconds. Several copulations during a spawn. Does not guard eggs or larvae. Two tubular air sacs extending from gill cavity almost to caudal peduncle. Bimodal respirator and may respire aerially by gulping air at different intervals when oxygen content of water is reduced below saturation. Digs burrows in form of anastomosing tubes with multiple exits; up to 364 individuals have been found in one burrow complex. Feeds on a wide variety of invertebrates and plant material.

Conservation status. Non-native; first reported from Iraq and Iran in 1950s. Most likely introduced to control *Bulinus truncatus* snails, an important schistosomiasis vector. Still expanding its range northwards. Locally threatened by overexploitation for human consumption in its native range.

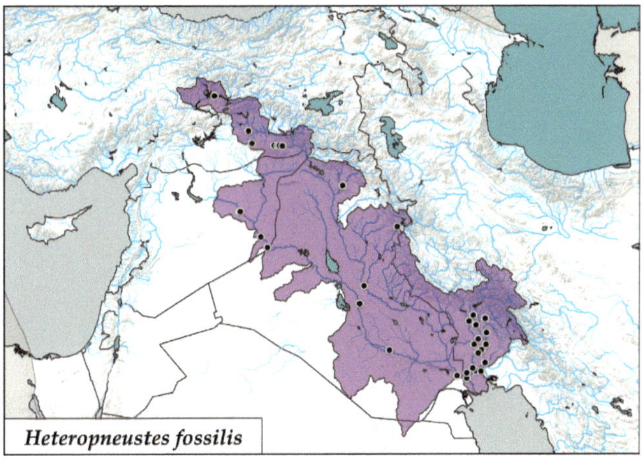

Remarks. Regarded as a pest in West Asia and not eaten. High market value and demand in India, where the species is used for medicinal purposes. Stinging catfish have the pectoral spine connected to a venom gland. Freshly caught or netted fish swing their heads from side to side and are active venomators. The venom is neurotoxic, hemolytic, and very painful. A sting in hand causes severe swelling, usually over the whole arm. The swelling and pain will subside in about a day, and the sting may take up to two weeks to heal. If the venom is squeezed and sucked out of sting, the effect can be negligible. While *Clarias* also have some sacs for breathing atmospheric air, in *Heteropneustes*, this structure is much more developed and resembles lungs. The air chambers begin above the gills and extend back to the tail as long, tubular sacs that emerge from the branchial cavity between the second and third gill arches and lie close to the backbone.

Further reading. Datta et al. 1982; Whitear et al. 1991 (venoum); Coad 2010a (introduction, biology); Ünlü et al. 2011 (range expansion).

Fish physiology: Getting rid of waste without water. Most nitrogenous wastes in fish are produced and excreted as ammonia or urea. Animals that excrete their nitrogenous wastes primarily as ammonia (NH_3) are classified as ammoniotelic, including agnathans and most teleost fish. Approximately 80–90 % of their nitrogenous wastes are excreted as ammonia, with the remainder excreted as urea. In high concentrations, ammonia is highly toxic, as it increases the internal pH of the cells, inhibiting key enzymes required for energy generation. Most air-breathing teleosts are primarily ammoniotelic but appear to have retained the genes for the urea cycle enzymes since a full complement of urea cycle enzymes has been reported for many of them. The capacity to synthesise urea in these fish may be an adaptation to their amphibious nature and their normal habitat of marshes, where the water ammonia level may be quite high, often rendering it uninhabitable to typical freshwater fish. *Heteropneustes fossilis* is capable of tolerating extremely high levels of total ammonia in its environment (up to 75 mm ammonium chloride vs. 100–200 µM, rarely up to 1000 µM in most fishes, up to 40–80 µM in humans). It can do so for extended periods without any adverse effects. When exposed to stressful conditions, the transition from ammoniotelism to ureotelism occurs in some air-breathing fishes. This happens when they live in semi-dry conditions inside mud during habitat drying. Although the real mechanism(s) of regulation of ureogenesis is unclear, it is hypothesised that the accumulation of ammonia within the body is likely the internal modulator for enhanced ureogenesis. This is mainly to avoid any build-up of ammonia to a level that can be toxic. It is hypothesised that the urea cycle is the predominant source of urea in air-breathing fishes despite the presence of uricolysis. The significant carbamoyl phosphate synthetase (CPS) levels observed in these fishes may play an important physiological role in their tolerance of high ammonia concentrations. **Further reading.** Saha & Ratha 1998 (ureogenesis).

Shores of Adana reservour at the Seyhan. A habitat of non-native *Clarias gariepinus*.

Ameiurus melas; Italy; ~150 mm SL.

Family Ictaluridae

North American catfishes

Ictaluridae are endemic to North and Central America, with a southern distribution reaching Guatemala. The family includes approximately 50 species in seven genera, three of which (*Satan*, *Trogloglanis*, *Prietella*) are eyeless inhabitants of subterranean waters in Texas and Mexico. *Ameiurus* was introduced to Germany as early as 1885, and at least four species of Ictaluridae have become established in Europe (*Ameiurus melas*, *A. nebulosus*, *Ictalurus furcatus*, and *I. punctatus*). *Ictalurus punctatus* has become established in Cyprus. As these species and additional ones are widely available in the aquaculture and aquarium trade, their introduction is expected to occur in other regions of West Asia in the future. **Further reading.** Taylor 1969 (systematics); Lundberg 1982, 1992 (phylogeny); Stoeckel & Burr 1999 (reproduction); Hardman & Page 2003 (phylogeny); Nelson et al. 2016 (diversity).

Ictalurus punctatus
Common name. Channel catfish.
Diagnosis. Distinguished from other West Asian catfishes by: • caudal deeply forked / ○ adipose present / ○ juveniles and small adults usually with dark-grey or black spots. Size up to 1000 mm SL, usually <570 mm SL.
Distribution. Cyprus. Native to North America, Hudson Bay, Great Lakes-St. Lawrence and Mississippi drainages, and rivers in between. Widespread in other parts of North America. Introduced to Europe for aquaculture. Self-sustaining populations recorded in lower Ebro (Spain), Oglio and Pavia provinces (northern Italy), Arno and Tiber

Ictalurus punctatus; Arno drainage, Italy; 300 mm SL.

https://doi.org/10.1515/9783111677811-026

(central Italy), lower Kuban and Don drainages (Russia). May also occur in other areas.

Habitat. Warm lowland rivers and large streams, ponds, lakes, and reservoirs. Usually associated with low-flow riverine habitats. Also found in upper estuarine habitats with salinities up to 15 ‰.

Biology. Lives up to 24 years. First spawns at 4–6 years of age. Spawning begins in May at about 20–30°C. Nests are built in sheltered areas by males only or by both sexes. Males take care of eggs and guard nest. Females spawn only once a year. Eggs hatch in 5–10 days. Omnivorous, feeding on algae, plant material, invertebrates, and fish. Usually nocturnal.

Conservation status. Non-native; introduced for recreational fisheries.

Further reading. Scott & Crossman 1974 (biology); Jenkins & Burkhead 1993 (biology); Zogaris et al. 2012 (Cyprus).

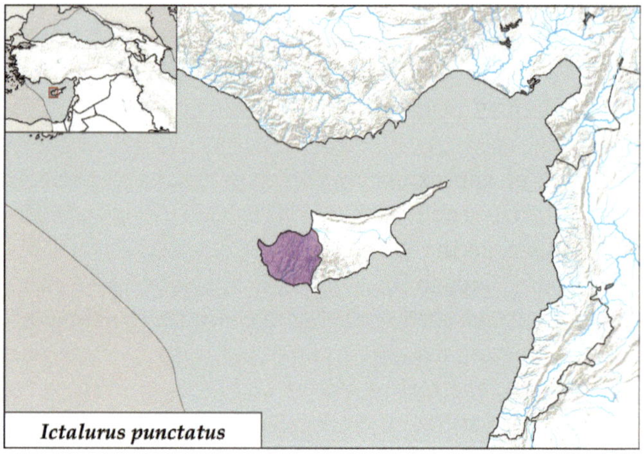

Several Ariid catfishes occur in the Persian Gulf and might be caught in brackish waters. They are superficially a bit similar to *Ictalurus* but their identification remains challenging. This individual caught in Iran might be *Plicofollis dussumieri*.

Family Siluridae

Sheatfishes

A family of approximately 13 genera and 107 species is found in Europe and Asia. A very long anal fin characterises them, as well as the absence of an adipose fin, the lack of a dorsal spine, and, in some cases, the absence of the entire dorsal fin. Some species are relatively small, with an adult length of approximately 50 mm SL, as observed in the glass catfish *Kryptopterus minor* from Southeast Asia. However, other species are considerably larger, with *Silurus glanis* reaching lengths of over 2000 mm. These fishes inhabit the lower strata of rivers and lakes, where they prey on smaller fish. **Further reading.** Bornbusch 1991, 1995 (systematics).

Silurus glanis

Common name. European catfish.

Diagnosis. Distinguished from *S. triostegus* by: • always two pairs of mental barbels / • maxillary barbel longer than head length / • anterior mandibular barbel present, longer than posterior mandibular barbel / • one patch of vomerine teeth / • pectoral spine not or slightly serrated on its inner margin. Size up to about 2600 mm SL and 148 kg.

Distribution. Lake Urmia basin, Black, Caspian, North, Baltic, and Aral basins, as far north as southern Sweden and Finland; Aegean basin, in Maritza and from Stuma to Sperchios drainages. Not native to Persian Gulf and Mediterranean basins of West Asia. Widely introduced and translocated in Anatolia, Europe, Hari drainage, and Lake Balkhash basin (Kazakhstan). Occasionally stocked in Persian Gulf basin but does not appear to have established (yet).

Habitat. Large and medium-sized lowland rivers, reservoirs, backwaters, and well-vegetated lakes. Spawns in shallow, warm, and well-vegetated riverine habitats without current.

Biology. Lives up to 80 years in wild. Spawns first time at 2–3 years and 1–2 kg, April–June, in northern areas until August, when temperatures reach about 20°C. Female lay about 30,000 eggs per kilogram of body weight. At spawning sites, males defend small territories and build nests of plant material, dig shallow depressions, or clean spawning substrate such as willow (*Salix*) roots. Males defend nest until larvae emerge.

Spawns in pairs. Male embraces female during spawning. Eggs hatch in 2–3 days, and larvae remain in nest until yolk sac is absorbed (2–4 days). A nocturnal predator, feeding near bottom and in water column. Very sensitive to extra-aquatic sounds. Head sensory canal system allows tracking of wakes (a trail of hydrodynamic and chemical signatures left by a swimming fish) of prey up to 10 seconds old over distances up to 55 times length of prey. Larvae and juveniles benthic and negatively phototactic, feeding on a wide variety of invertebrates and fish. Adults feed on fish and other aquatic vertebrates, documented to feed on pigeons at water edge by lunging out of water during daylight.

Conservation status. LC; one of most important invasive species in the Western Palearctic.

Further reading. Mohr 1957 (biology); Kobayakawa 1989 (description); Pohlmann et al. 2004 (wakes).

Silurus glanis; Danube drainage, Germany; ~750 mm SL. © A. Hartl.

https://doi.org/10.1515/9783111677811-027

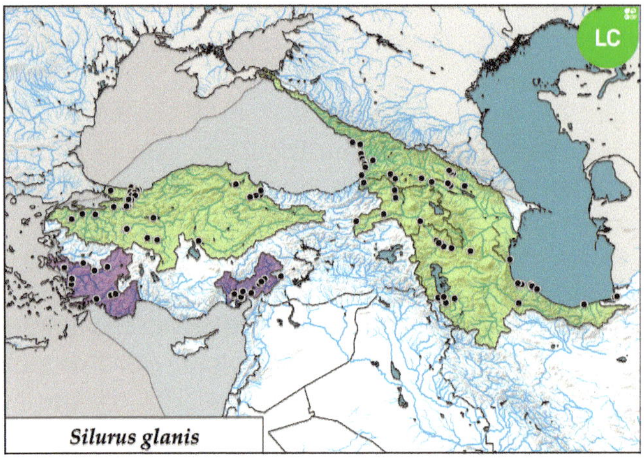

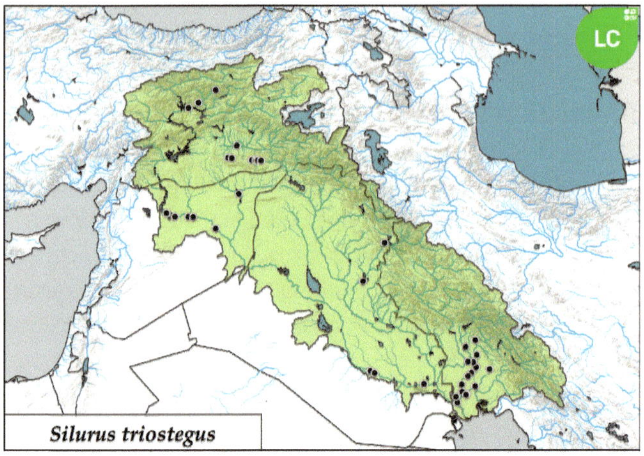

Silurus triostegus; Tigris, Türkiye; ~500 mm SL.

Silurus triostegus

Common name. Mesopotamian catfish.

Diagnosis. Distinguished from *S. glanis* by: • usually one pair of mental barbels in adults larger than 400 mm SL / • maxillary barbel usually about equal to head length / • if anterior mandibular barbel present, shorter than posterior mandibular barbel / • usually two patches of vomerine teeth / • pectoral spine strongly serrated on its inner margin. Size up to about 2000 mm SL.

Distribution. Euphrates, Tigris, and Karun drainages.

Habitat. Large rivers, marshes, lakes, and reservoirs.

Biology. Lives up to 12 years in the wild, likely much longer. Spawns first time after 2–3 years (Atatürk reservoir).

Conservation status. LC.

Remarks. All the diagnostic characters may overlap with those of *S. glanis*, but the combination of all the characters should allow its identification. So far, and despite many introductions of *Silurus*, the two species are still allopatric.

Further reading. Oymak et al. 2001 (biology).

Family Sisoridae

Torrent catfishes

The family comprises approximately 17 genera and 220 species of rheophilic catfishes, with the greatest diversity observed in South and Southeast Asia. One species, *Bagarius bagarius*, reaches up to 1400 mm SL, while most species are small. Most are highly adapted to live in rapids and very fast-flowing waters, with some (*Oreoglanis*) inhabiting the vertical part of waterfalls. Some species are found in high-altitude streams in the Himalayas, while the greatest diversity of species is observed in rapids of tropical rivers. In West Asia, only *Glyptothorax* is found, which is the most speciose genus of the family. *Glyptothorax* is immediately identified by a thoracic adhesive organ on the breast between the pectorals. The adhesive organ is sometimes referred to as a "sucker," which does not suck to the substrate, but instead adheres with small unculi on elevated skin folds, a structure similar to the feet of gecko lizards. *Glyptothorax* are adapted to live in fast-flowing waters and are commonly found in foothill rivers and mountain streams.

In our region, 13 species are recognised, but six species from Iran are closely related, and their status is under discussion. Two of these (*G. alidaeii* and *G. galaxias*) occur in sympatry and are good biological species. Other species may be conspecific with *G. silviae*, but not all morphological differences have been studied in detail. Three widespread species inhabit large- or medium-sized rivers, while others are restricted to fast-flowing headwater streams with more restricted distribution ranges. A single record of *Glyptothorax* from the Yeşilırmak in the Anatolian Black Sea basin has been identified, although the integrity of this record cannot be confirmed. **Further reading.** Hora & Silas 1952 (diversity); Sayyadzadeh et al. 2022 (diversity in Gulf basin).

Large, fast-flowing rivers such as the Greater Zab in Iraq are the habitat of a rich fish fauna, including *Glyptothorax*, *Mystus*, and *Silurus*.

https://doi.org/10.1515/9783111677811-028

Glyptothorax galaxias; Bazoft, Iran.

Key to species of *Glyptothorax* in West Asia

1a - Adipose length 1.5–3.0 times longer than distance between base of last dorsal ray and adipose origin; 13–17 serrae on pectoral spine, medial pit of thoracic adhesive apparatus with striae.
..................*G. steindachneri*

1b - Adipose length 0.5–1.2 times longer than distance between base of last dorsal ray and adipose origin; 5–14 serrae on pectoral spine; medial pit of thoracic adhesive apparatus without striae.
..................2

2a - Head and flank with tubercles (rarely absent in some individuals).
..................3

2b - Head and flank without tubercles, with roundish or elongate warts.
..................4

3a - Thoracic adhesive apparatus longer than wide, well delineated at its posterior margin, situated completely on a horse-shoe-shaped elevation; head, back, and flank with many minute tubercles and few roundish warts.
..................*G. armeniacus*

3b - Thoracic adhesive apparatus as wide as long, poorly delineated at its posterior margin, only partly situated on shallow, horseshoe-shaped elevation, elevation absent in many adult individuals; head, back, and usually flank with large, bony, striated, and elongated tubercles (absent in some individuals).
..................*G. cous*

4a - Upper head, back, and flank without brown or black spots or blotches.
..................5
4b - Upper head, back, and flank with few or many dark-brown spots and, or blotches (potentially faded in poorly preserved individuals).
..................6

5a - Outer mandibular barbel not reaching pectoral origin; thoracic adhesive apparatus slightly elevated; medial pit broad, its anterior end roundish; caudal peduncle depth 1.6–2.3 times its length.
..................*G. sardashtensis*
5b - Outer mandibular barbel reaching pectoral origin; thoracic adhesive apparatus strongly elevated; medial pit narrow and spear-blade shaped; caudal peduncle depth 1.1–1.3 times its length.
..................*G. pallens*

6a - Thoracic adhesive apparatus wider than long, as wide as long in juveniles (0.7–0.9 times longer than wide).
..................7
6b - Thoracic adhesive apparatus as wide as long or longer, 1.0–1.6 times longer than wide.
..................8

7a - Caudal-peduncle length 16–20 % SL.
..................*G. vatandousti*
7b - Caudal-peduncle length 13–16 % SL.
..................*G. kurdistanicus*

8a - Anteromedial striae in thoracic adhesive apparatus long and numerous.
..................9
8b - Anteromedial striae in thoracic adhesive apparatus short or absent.
..................11

9a - Caudal-peduncle depth 1.3–1.6 times in its length; shortest middle caudal ray 42–49 % of longest ray of upper caudal lobe; caudal with pointed lobes; maxillary barbel as long as head (95–108 % HL).
..................*G. hosseinpanahii*
9b - Caudal-peduncle depth 1.6–2.5 times in its length; shortest middle caudal ray 55–65 % of longest ray of upper caudal lobe; caudal with rounded lobes; maxillary barbel shorter than head (72–91 % HL).
..................10

10a - Fins with a distinct yellow tip; thoracic adhesive apparatus 1.3–1.5 times as long as wide; adipose length 0.5–0.8 of distance between base of last dorsal ray and adipose origin.
..................*G. galaxias*
10b - Fins without or with a very indistinct yellow tip; thoracic adhesive apparatus 1.1–1.3 times as long as wide; adipose length 0.7–1.1 of distance between base of last dorsal ray and adipose origin.
..................*G. daemon*

11a - Maxillary barbel as long as head (92–110 % HL); inner mandibular barbel 45–48 % HL; outer mandibular barbel 67–74 % HL.
..................*G. silviae*
11b - Maxillary barbel shorter than head (76–89 % HL); inner mandibular barbel 18–33 % HL; outer mandibular barbel 42–49 % HL.
..................12

12a - Shortest middle caudal ray 57–65 % of longest ray of upper caudal lobe; caudal-peduncle depth 1.3–1.6 times in its length.
..................*G. alidaeii*
12b - Shortest middle caudal ray 46–49 % of longest ray of upper caudal lobe; caudal-peduncle depth 1.1–1.3 times in its length.
..................*G. shapuri*

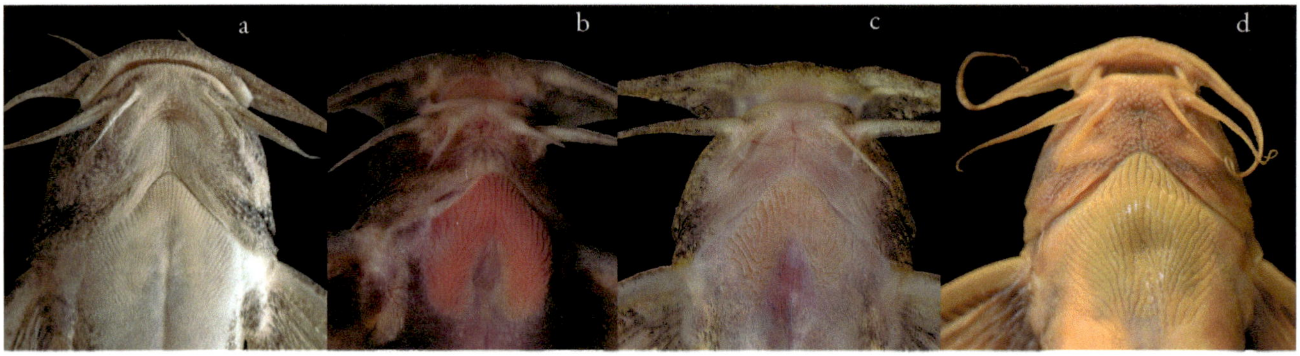

Figure 54: Thoracic adhesive apparatus of *Glyptothorax*: **a,** *G. cous*; **b,** *G. daemon*; **c,** *G. kurdistanicus*; **d,** *G. steindachneri.*

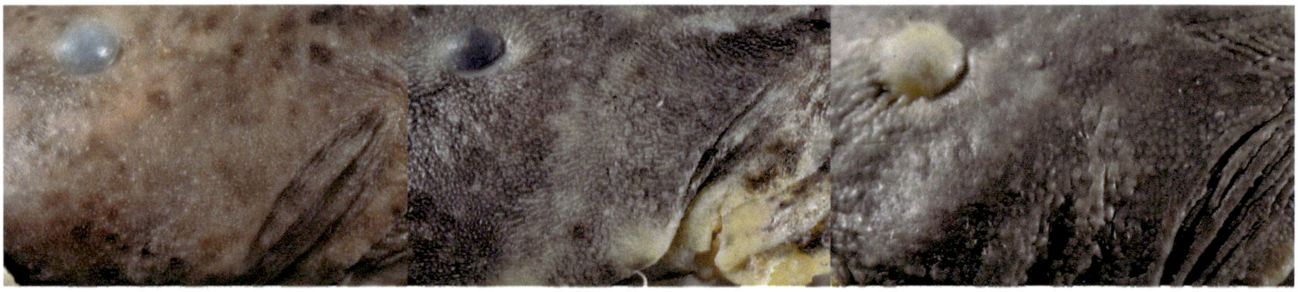

Figure 55: From left: *Glyptothorax armeniacus*, 87 mm SL; with many very small tubercles and few warts; *G. cous*, 151 mm SL; with many large tubercles; *G. daemon*, 102 mm SL, with many warts.

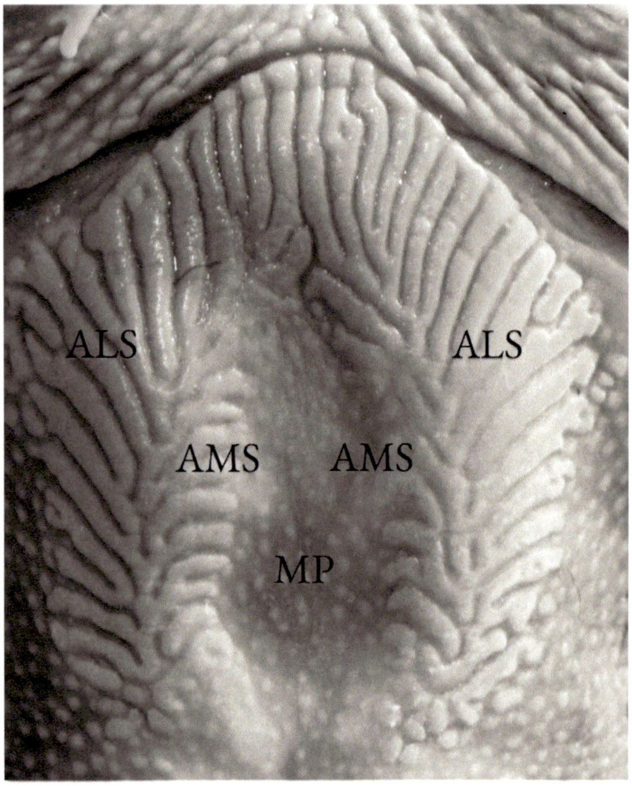

Figure 56: Thoracic adhesive apparatus of *Glyptothorax*, showing structures defined and mentioned in this study: anterolateral striae (ALS); anteromedial striae (AMS); medial pit (MP) (*G. daemon*, Iraq; 109 mm SL).

Glyptothorax alidaeii; Seimareh, Karkheh drainage, Iran; ~110 mm SL.

Glyptothorax alidaeii

Common name. Karkheh torrent catfish.

Diagnosis. Distinguished from other species of *Glyptothorax* in West Asia by: ○ thoracic adhesive apparatus with, without or with few, very short anteromedial striae, 1.3–1.6 times longer than wide, extending to base of last pectoral ray or to posterior limit of pectoral base / ○ caudal-peduncle depth 1.3–1.6 times in its length / ○ thoracic adhesive apparatus moderately elevated, completely situated on a horseshoe shaped elevation / ○ medial pit open, without striae / ○ dorsal and lateral head, back and flank with or without small roundish warts / ○ adipose short, its length 0.5–0.8 times larger than distance between base of last dorsal ray and adipose origin / ○ caudal lobes pointed / ○ inner mandibular barbel 30–33 % HL / ○ outer mandibular barbel 43–49 % HL / ○ maxillary barbel shorter than head, 76–90 % HL / ○ 10–12 serrae on pectoral spine / ○ caudal moderately forked, shortest middle caudal ray 57–65 % of longest ray of upper caudal lobe / ○ fins usually with distinct yellow tips / ○ head, back and flank with many, irregular shaped and set brown blotches smaller than eye / ○ many silvery-beige spots and blotches on head and flank in life. Size up to 350 mm SL.

Distribution. Iran: Middle and upper reaches of Karkheh and Karun drainages.

Habitat. Fast-flowing water in rapids and riffles on rock or gravel substrate. Limited to hill stream habitats.

Biology. Nocturnal, hiding under rocks during day. Feeds on benthic invertebrates.

Conservation status. LC.

Remarks. *Glyptothorax cous*, *G. alidaeii*, and *G. galaxias* are expected to occur in sympatry in Seymareh (Iran), and *G. alidaeii* often occurs together with *G. galaxias*.

Further reading. Mousavi-Sabet et al. 2021 (description); Sayyadzadeh et al. 2022 (synonym of *G. silviae*).

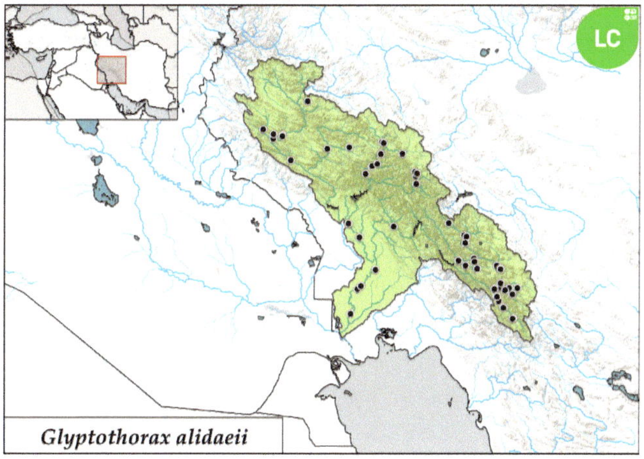

Glyptothorax armeniacus; Euphrates drainage, Türkiye; 66 mm SL.

Glyptothorax armeniacus

Common name. Euphrates torrent catfish.

Diagnosis. Distinguished from other species of *Glyptothorax* in West Asia by: • head, back, and flank with many minute tubercles and few roundish warts / ○ thoracic adhesive apparatus strongly elevated, 1.0–1.5 times longer than wide, usually extending from isthmus to base of last pectoral ray or to posterior limit of pectoral base / ○ striae restricted to elevated apparatus / ○ anteromedial striae numerous and well developed / ○ medial pit without striae / ○ 9–12 serrae on pectoral spine / ○ all fins with a distinct, distal yellow tip / ○ adipose short, its length 0.7–1.1 times larger than distance between base of last dorsal ray and adipose origin / ○ caudal moderately forked, shortest middle caudal ray 62–69 % of longest ray of upper caudal lobe / ○ head and body dark-brown without or with small, often indistinct brown or black spots. Size up to 220 mm SL.

Distribution. Türkiye and Syria: Upper Euphrates drainage.

Habitat. Fast-flowing water in rapids and riffles on rock or gravel substrate. Limited to hill stream habitats.

Biology. Nocturnal, hiding under rocks during day. Feeds on benthic invertebrates.

Conservation status. LC.

Remarks. Despite some fieldwork in the region, the *Glyptothorax* recorded from the Yeşilırmak drainage in Anatolian Black Sea basin has never been confirmed. It cannot be excluded that such catfish occur there, and as the Yeşilırmak is adjacent to the Euphrates, it is likely that this species will be identified as *G. armeniacus*.

Further reading. Berg 1918 (description); Coad & Delmastro 1985 (Yeşilırmak); Freyhof et al. 2021c (description).

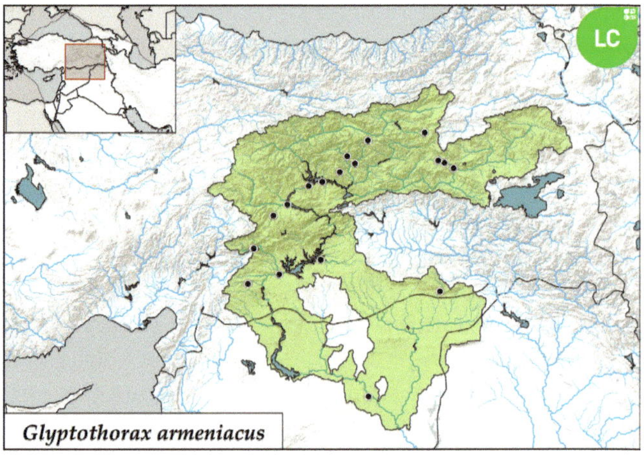

Glyptothorax cous; Lesser Zab drainage, Iraq; 160 mm SL. ©Y. Abdullah.

Glyptothorax cous; Lesser Zab drainage, Iraq; 152 mm SL. ©Y. Abdullah.

Glyptothorax cous

Common name. Mesopotamian torrent catfish.

Diagnosis. Distinguished from other species of *Glyptothorax* in West Asia by: ● thoracic adhesive apparatus 0.8–1.1 times longer than wide, poorly delineated at its posterior margin, only partly situated on shallow, horse-shoe shaped elevation; elevation absent in many adult individuals / ● striae reaching far beyond elevated apparatus, usually onto pectoral base, usually extending from isthmus to base of last pectoral ray or to posterior limit of pectoral base / ○ dorsal head, back and flank usually with large, bony, striated and elongated tubercles (absent in some individuals) / ○ adipose short, its length 0.6–1.2 times larger than distance between base of last dorsal ray and adipose origin / ○ anteromedial striae well developed / ○ medial pit open, without striae / ○ flank with large dark-brown blotches and or small dark-brown spots. Size up to 450 mm SL.

Distribution. Qweiq, Euphrates, Tigris and Karun drainages.

Habitat. Fast to moderately fast-flowing medium to large rivers.

Biology. Nocturnal, hiding under rocks during day. Feeds on benthic invertebrates, as adults, mainly on fish.

Conservation status. LC; extirpated from Qweiq.

Remarks. The mtDNA of *G. cous* from the Tigris is nested within mtDNA of *G. daemon*. There may have been some introgressive hybridisation between the two species. This would also mean that the mtDNA of *G. cous* from the Euphrates could be very different from that of Tigris population, and great caution should be exercised in identifying *G. cous* based on mtDNA alone.

Further reading. Russell 1756 (description); Freyhof et al. 2021c (description).

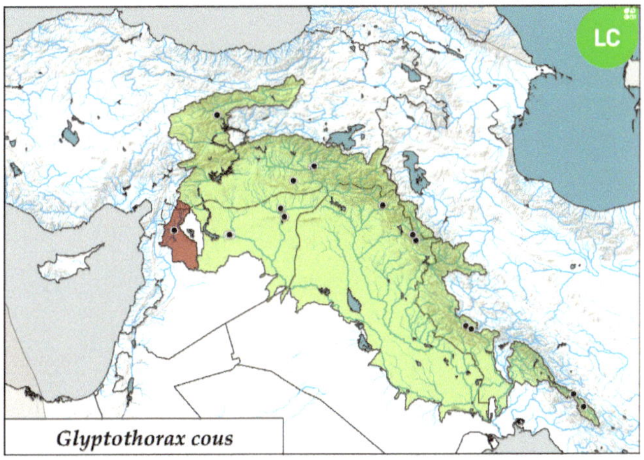

Glyptothorax daemon; Greater Zab drainage, Iraq; 116 mm SL.

Glyptothorax daemon; Greater Zab drainage, Iraq; 110 mm SL.

Glyptothorax daemon

Common name. Ghost torrent catfish.

Diagnosis. Distinguished from other species of *Glyptothorax* in West Asia by: ○ thoracic adhesive apparatus with several, long anteromedial striae, 1.1–1.3 times longer than wide, extending to base of last pectoral ray, or to posterior limit of pectoral base / ○ thoracic adhesive apparatus strongly elevated, completely situated on a horse-shoe shaped elevation / ○ medial pit open, without striae / ○ dorsal and lateral head, back and flank with small roundish warts / ○ adipose short, its length 0.7–1.1 times larger than distance between base of last dorsal ray and adipose origin / ○ caudal lobes rounded / ○ caudal-peduncle 1.6–2.1 times in its length / ○ maxillary barbel shorter than head, 72–85 % HL / ○ inner mandibular barbel 21–30 % HL / ○ outer mandibular barbel 42–51 % HL / ○ 7–11 serrae on pectoral spine / ○ caudal moderately forked, shortest middle caudal ray 55–59 % of longest ray of upper caudal lobe / ○ fins usually without, rarely with very indistinct yellow tips / ○ flank dark-grey or black with few small black spots, rarely pale-brown. Size up to 125 mm SL, likely to grow larger.

Distribution. Türkiye and Iraq: Upper Tigris and Great Zab drainages.

Habitat. Fast-flowing water in mountain streams, rapids, and riffles on rocky or gravelly bottoms.

Biology. Nocturnal, hiding under rocks during day. Feeds on benthic invertebrates.

Conservation status. LC.

Remarks. *Glyptothorax daemon* populations in upper Tigris may be quite isolated from each other, as they show a shallow but consistent separation in their mtDNA. All these populations belong to one old species, and are not to be described as own species in the future, ignoring the absence of morphological differences.

Further reading. Freyhof et al. 2021c (description).

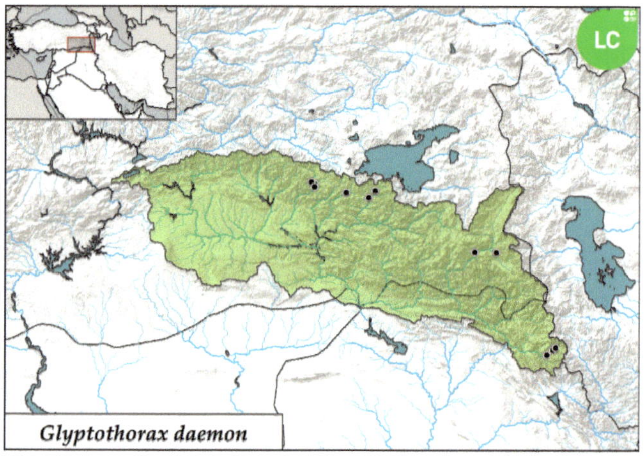

Glyptothorax galaxias; Karun drainage, Iran; 113 mm SL.

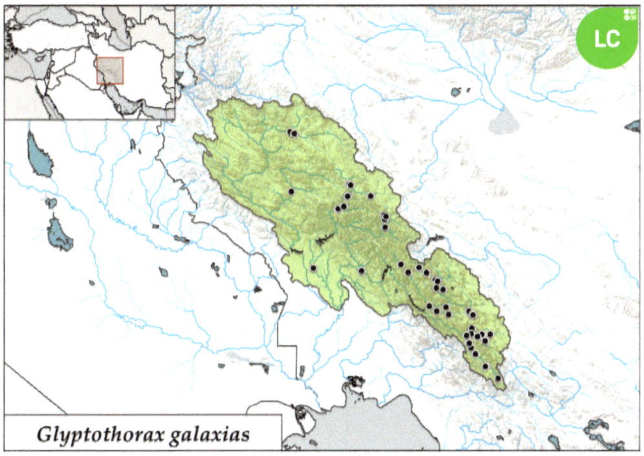

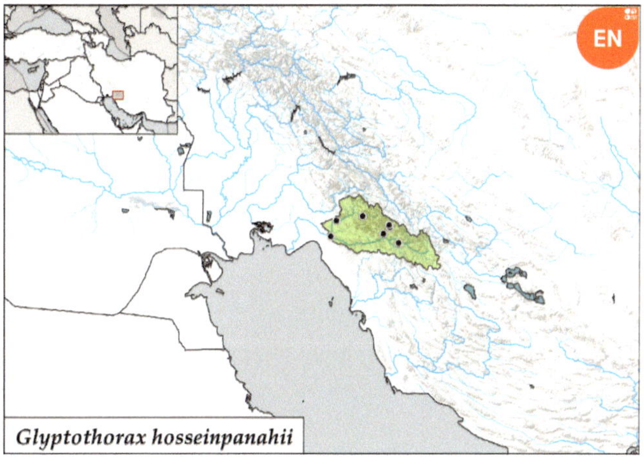

Glyptothorax galaxias

Common name. White-spot torrent catfish.

Diagnosis. Distinguished from other species of *Glyptothorax* in West Asia by: ○ thoracic adhesive apparatus with many and long anteromedial striae, 1.3–1.5 times longer than wide, extending to posterior-most pectoral base / ○ thoracic adhesive apparatus strongly elevated, almost or completely situated on a horse-shoe shaped elevation / ○ medial pit open, without striae / ○ dorsal and lateral head, back and flank with small round warts / ○ adipose length 0.5–0.8 of distance between base of last dorsal ray and adipose origin / ○ caudal lobes rounded / ○ caudal moderately forked, its shortest middle caudal ray 61–65 % of longest ray of upper caudal lobe / ○ head pointed and short, 21–24 % SL / ○ maximum head width 84–101 % HL / ○ postdorsal length 52–54 % SL / ○ preanal length 65–72 % SL / ○ eye diameter 10–14 % HL / ○ caudal-peduncle 1.6–2.1 times in its length / ○ maxillary barbel shorter than head, 77–91 % HL / ○ inner mandibular barbel 26–33 % HL / ○ outer mandibular barbel 56–58 % HL / ○ 9–12 serrae on pectoral spine / ○ all fins with a distinct, distal yellow tip / ○ head, back and flank with many, irregular shaped and set brown blotches as large as eye diameter or slightly larger. Size up to 114 mm SL, likely to grow larger.

Distribution. Iran: Upper Karun and Karkheh drainages.

Habitat. Fast-flowing water in mountain streams, rapids, and riffles on rocky or gravelly bottoms.

Biology. Nocturnal, hiding under rocks during day. Feeds on benthic invertebrates.

Conservation status. LC.

Remarks. Occurs often in sympatry with *G. alidaeii*.

Further reading. Mousavi-Sabet et al. 2021 (description); Sayyadzadeh et al. 2022 (synonym of *G. silviae*).

Glyptothorax hosseinpanahii

Common name. Zohreh torrent catfish.

Diagnosis. Distinguished from other species of *Glyptothorax* in West Asia by: ○ maxillary barbel as long as head, 95–108 % HL / ○ inner mandibular barbel 25–35 % HL / ○ outer mandibular barbel 46–61 % HL / ○ thoracic adhesive apparatus with many and long anteromedial striae, 1.0–1.2

Glyptothorax hosseinpanahii; Zohreh drainage, Iran; 90 mm SL.

times longer than wide, usually extending posterior-most pectoral base / ○ thoracic adhesive apparatus moderately elevated, almost or completely situated on a horse-shoe shaped elevation / ○ medial pit open, without striae / ○ dorsal and lateral head, back and flank with small round warts / ○ adipose short, its length 0.5–0.8 times larger than distance between base of last dorsal ray and adipose origin / ○ caudal lobes pointed / ○ caudal deeply forked, shortest middle caudal ray 42–49 % of longest ray of upper caudal lobe / ○ caudal-peduncle 1.3–1.6 times in its length / ○ 9–11 serrae on pectoral spine / ○ all fins with a distinct, distal yellow tip / ○ many black or dark-brown blotches on head, back and flank usually smaller than eye diameter. Size up to 90 mm SL, likely to grow larger.

Distribution. Iran: Zohreh drainage.

Habitat. Fast-flowing water in mountain streams, rapids, and riffles on rocky or gravelly bottoms.

Biology. Nocturnal, hiding under rocks during day. Feeds on benthic invertebrates.

Conservation status. EN; appears to be declining within its very small range.

Further reading. Mousavi-Sabet et al. 2021 (description); Sayyadzadeh et al. 2022 (synonym of *G. silviae*).

Glyptothorax kurdistanicus; Lesser Zab drainage, Iraq; 77 mm SL.

Glyptothorax kurdistanicus; Lesser Zab drainage, Iraq; 120 mm SL. ©Y. Abdullah.

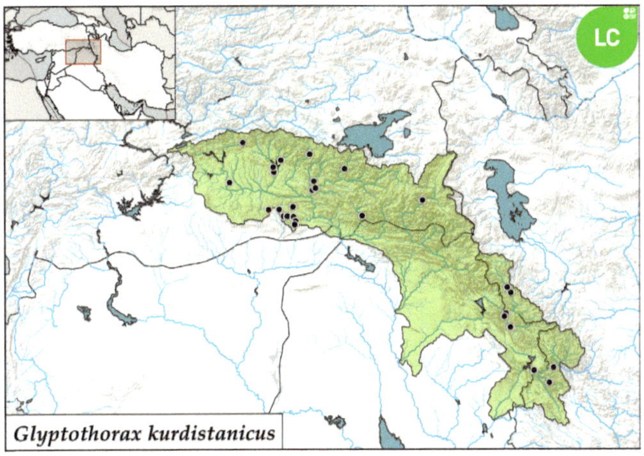

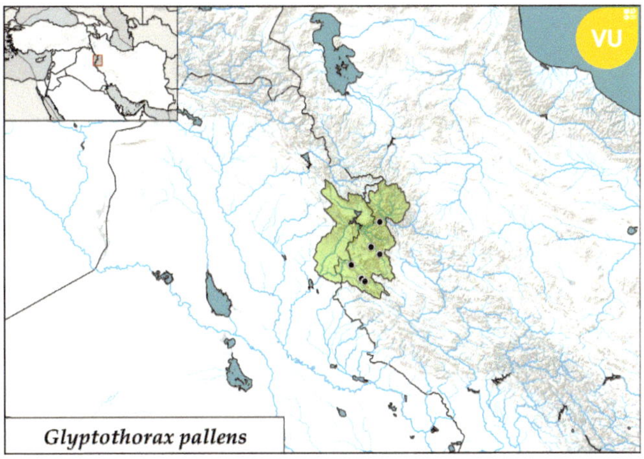

Glyptothorax kurdistanicus

Common name. Tigris torrent catfish.

Diagnosis. Distinguished from other species of *Glyptothorax* in West Asia by: ● thoracic adhesive apparatus 0.7–0.9 times longer than wide / ○ thoracic adhesive apparatus almost or completely situated on a horse-shoe shaped elevation, usually extending from isthmus to base of first or third branched pectoral ray / ○ anteromedial striae few, very short or absent / ○ striae not reaching onto pectoral base / ○ medial pit open, without striae / ○ dorsal and lateral head, back and flank with small round warts / ○ adipose short, its length 0.6–1.0 times larger than distance between base of last dorsal ray and adipose origin / ○ caudal-peduncle length 13–16 % SL / ○ caudal-peduncle 1.1–1.3 times in its length / ○ maxillary barbel 74–92 % HL / ○ inner mandibular barbel 31–36 % HL / ○ outer mandibular barbel 53–63 % HL / ○ 9–12 serrae on pectoral spine / ○ all fins with a distinct, distal yellow tip / ○ few black or dark-brown blotches on flank. Size up to 122 mm SL.

Distribution. Upper Tigris (Türkiye) to Sirvan drainage (Iran).

Habitat. Fast-flowing water in mountain streams, rapids, and riffles on rocky or gravelly bottoms.

Biology. Nocturnal, hiding under rocks during day. Feeds on benthic invertebrates.

Conservation status. LC.

Further reading. Berg 1931 (description); Freyhof et al. 2021c (description).

Glyptothorax pallens

Common name. Pale torrent catfish.

Diagnosis. Distinguished from other species of *Glyptothorax* in West Asia by: ○ head and body pale-brown without brown or black spots / ○ head and flank with roundish or elongate warts, without tubercles / ○ thoracic adhesive apparatus strongly elevated, 1.0–1.2 times longer than wide, usually extending from isthmus to base of last pectoral ray or to posterior limit of pectoral base / ○ medial pit narrow and spear-blade shaped, without striae / ○ striae restricted to elevated apparatus / ○ anteromedial striae absent or very short / ○ caudal-peduncle 1.1–1.3 times in its length / ○ caudal lobes pointed / ○ caudal deeply forked, shortest middle caudal ray 47–53 % of longest ray of upper caudal lobe / ○ head depth 53–57 % HL / ○ maximum head width 73–80 % HL / ○ maxillary barbel 90–95 % HL / ○ inner mandibular barbel 38–42 % HL / ○ outer mandibular barbel 52–59 % HL, reaching pectoral origin / ○ 10–11 serrae on pectoral spine / ○ upper head, back and flank without brown or black spots or blotches / ○ all fins with a distinct, distal yellow tip / ○ adipose short, its length 0.5–0.8 times larger than distance between base of last dorsal ray and adipose origin. Size up to 69 mm SL, likely to grow larger.

Distribution. Iran: Sirvan drainage, including Goleyn, Alvand, Zemkan, and Sirvan.

Habitat. Fast-flowing water in mountain streams, rapids, and riffles on rocky or gravelly bottoms.

Glyptothorax pallens; Zemkan, Sirvan drainage, Iran; 67 mm SL.

Glyptothorax sardashtensis; Lesser Zab drainage, Iran; ~80 mm SL.

Biology. Nocturnal, hiding under rocks during day. Feeds on benthic invertebrates.
Conservation status. VU; appears to be declining within its small range.
Remarks. As the Sirvan flows into Iraq to join Tigris, this species is expected to occur there.
Further reading. Mousavi-Sabet et al. 2021 (description).

Glyptothorax sardashtensis
Common name. Sardasht torrent catfish.
Diagnosis. Distinguished from other species of *Glyptothorax* in West Asia by: ○ head and body pale-brown without brown or black spots / ○ head and flank without roundish or elongate warts, without tubercles / ○ thoracic adhesive apparatus slightly elevated, 0.8–1.2 times longer than wide, usually extending from isthmus to base of third or fifth branched pectoral ray / ○ striae restricted to elevated apparatus / ○ anteromedial striae few and short / ○ medial pit without striae, broad, its anterior end roundish / ○ caudal-peduncle 1.6–2.3 times in its length / ○ caudal lobes pointed / ○ caudal deeply forked / ○ head depth 37–52 % HL / ○ maximum head width 72–88 % HL / ○ maxillary barbel 53–95 % HL / ○ upper head, back and flank without brown or black spots or blotches / ○ inner mandibular barbel 22–31 % HL / ○ outer mandibular barbel 33–53 % HL, not reaching pectoral origin / ○ 9–11 serrae on pectoral spine / ○ all fins with a distinct, distal yellow tip. Size up to 67 mm SL, likely to grow larger.
Distribution. Lesser Zab (Iran) and Batman drainages (Türkiye). Potentially more widespread.
Habitat. Fast-flowing water in mountain streams, rapids, and riffles on rocky or gravelly bottoms.
Biology. Nocturnal, hiding under rocks during day. Feeds on benthic invertebrates.
Conservation status. LC.
Further reading. Jokar et al. 2023 (description).

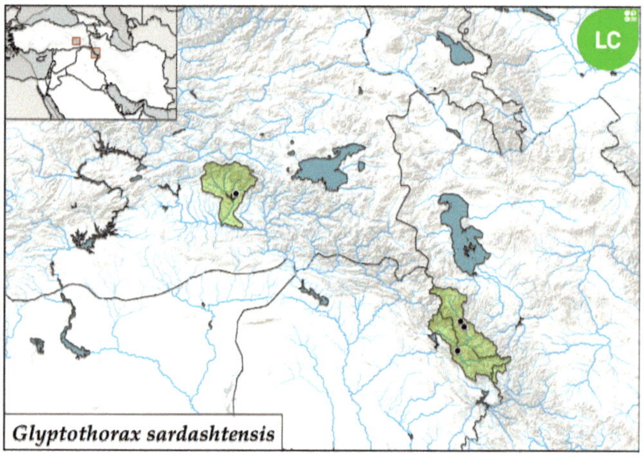

Glyptothorax shapuri; Shapur, Iran; 67 mm SL.

Glyptothorax shapuri

Common name. Shapur torrent catfish.

Diagnosis. Distinguished from other species of *Glyptothorax* in West Asia by: ○ thoracic adhesive apparatus with few, very short anteromedial striae, 1.0–1.2 times longer than wide, extending to base of last pectoral ray or to posterior limit of pectoral base / ○ thoracic adhesive apparatus strongly elevated, completely situated on a horse-shoe shaped elevation / ○ medial pit open, without striae / ○ dorsal and lateral head, back and flank with or without small roundish warts / ○ adipose short, its length 0.5–0.8 times larger than distance between base of last dorsal ray and adipose origin / ○ caudal lobes pointed / ○ caudal deeply forked, its shortest middle caudal ray 46–49 % of longest ray of upper caudal lobe / ○ caudal-peduncle 1.1–1.3 times in its length / ○ inner mandibular barbel 18–22 % HL / ○ outer mandibular barbel 42–44 % HL / ○ maxillary barbel shorter than head, 81–84 % HL / ○ 9–11 serrae on pectoral spine / ○ fins usually with distinct yellow tips / ○ head, back and flank with few, irregular shaped dark-brown blotches as large as eye diameter or larger, and many, very small dark-brown points, much smaller than eye diameter / ○ no silvery-beige spots and blotches on head and flank in life. Size up to 81 mm SL, likely to grow larger.

Distribution. Iran: Helleh drainage.

Habitat. Fast-flowing water in mountain streams, rapids, and riffles on rocky or gravelly bottoms.

Biology. Nocturnal, hiding under rocks during day. Feeds on benthic invertebrates.

Conservation status. EN; appears to be declining within its very small range.

Remarks. The description of *G. shapuri* may be based on juveniles, and the morphometric and colour pattern characters may be different in adults. However, we cannot exclude that this is a true dwarf species.

Further reading. Mousavi-Sabet et al. 2021 (description); Sayyadzadeh et al. 2022 (synonym of *G. silviae*).

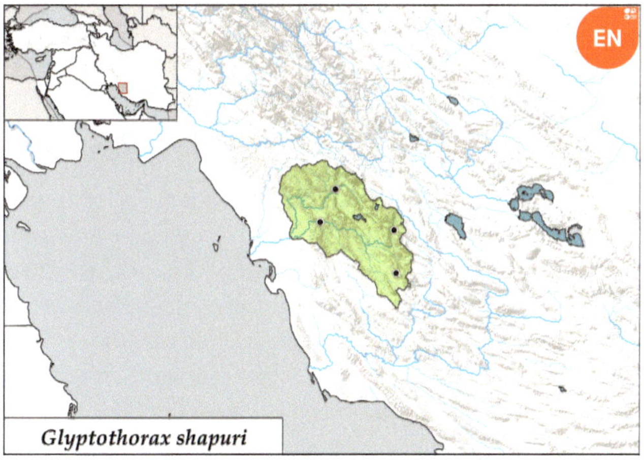

Small and juvenile *Glyptothorax* often inhabit small rivers like the Shapur in Iran.

Glyptothorax silviae; Lodab, Marun drainage, Iran; ~100 mm SL.

Glyptothorax silviae

Common name. Persian torrent catfish.

Diagnosis. Distinguished from other species of *Glyptothorax* in West Asia by: ○ thoracic adhesive apparatus with no or few, very short anteromedial striae, 1.0–1.2 times longer than wide, extending to base of last pectoral ray or to posterior limit of pectoral base / ○ thoracic adhesive apparatus elevated, completely situated on a horse-shoe shaped elevation / ○ medial pit open, without striae / ○ dorsal and lateral head, back and flank with or without small roundish warts / ○ adipose short, its length 0.8–1.0 times larger than distance between base of last dorsal ray and adipose origin / ○ caudal lobes pointed / ○ caudal moderately forked, its shortest middle caudal ray 61–65 % of longest ray of upper caudal lobe / ○ caudal-peduncle 1.3–1.6 times in its length / ○ inner mandibular barbel 45–48 % HL / ○ outer mandibular barbel 67–74 % HL / ○ maxillary barbel as long as head, 92–110 % HL / ○ 9–11 serrae on pectoral spine / ○ caudal deeply forked, shortest middle caudal ray 40–43 % of longest ray of upper caudal lobe / ○ fins usually with distinct yellow tips / ○ flank grey or brown with few or many spots and small blotches or only small blotches all smaller than eye diameter, without silvery-beige spots and blotches on life. Size up to 100 mm SL, likely to grow larger.

Distribution. Iran: Jarrahi drainage.

Habitat. Fast-flowing water in mountain streams, rapids, and riffles on rocky or gravelly bottoms.

Biology. Nocturnal, hiding under rocks during day. Feeds on benthic invertebrates.

Conservation status. VU; appears to be declining within its small range.

Further reading. Coad 1981a (description); Freyhof et al. 2021c (description); Mousavi-Sabet et al. 2021 (description); Sayyadzadeh et al. 2022 (morphology).

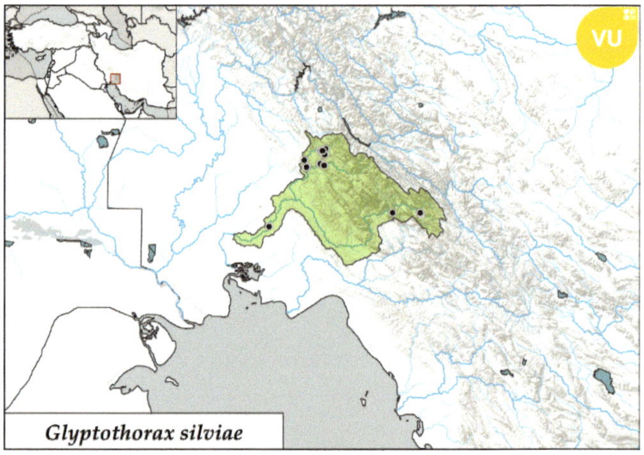

Glyptothorax steindachneri; Nahr Al-Khabur, Syria; 77 mm SL.

Glyptothorax steindachneri; Deir ez-Zur, Syria; 112 mm SL.

Glyptothorax steindachneri

Common name. Longfin torrent catfish.

Diagnosis. Distinguished from other species of *Glyptothorax* in West Asia by: ● adipose very long, its length 1.5–3.0 times larger than distance between base of last dorsal ray and adipose origin / ● medial pit closed by striae / ● 13–17 serrae on pectoral spine / ○ head pointed, short, 21–23 % SL / ○ thoracic adhesive apparatus slightly elevated, 1.1–1.5 times longer than wide, on a shallow horse-shoe shaped elevation; extending from isthmus to base of last pectoral ray or to posterior limit of pectoral base / ○ anteromedial striae not or poorly developed / ○ dorsal head and predorsal back with large, striated, elongated tubercles / ○ distance between anus anal origin 5.0–6.7 % SL / ○ flank with small black spots or without spots. Size up to 190 mm SL, likely to grow larger.

Distribution. Tigris below Diyarbakır; Euphrates in Syria, Nahr Al-Khabur, a tributary of Euphrates in Syria, and Lesser Zab in Iraq.

Habitat. Fast to moderately fast-flowing medium to large rivers.

Biology. Nocturnal. Feeds on benthic invertebrates.

Conservation status. LC.

Further reading. Pietschmann 1913 (description); Freyhof et al. 2021c (description).

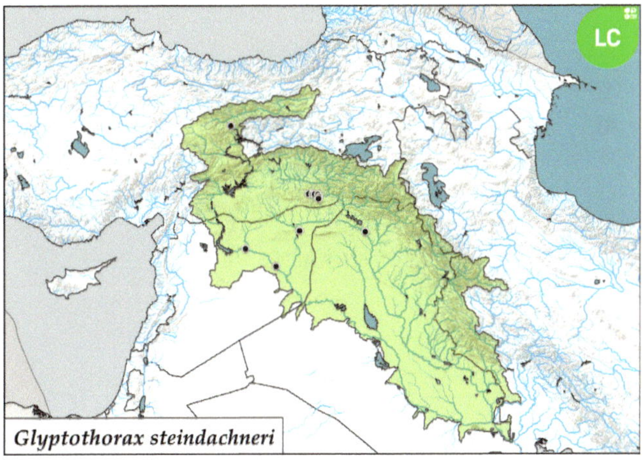

Glyptothorax vatandousti; stream Kangavar, Karkheh drainage, Iran; 82 mm SL.

Glyptothorax vatandousti

Common name. Kangavar torrent catfish.

Diagnosis. Distinguished from other species of *Glyptothorax* in West Asia by: ● thoracic adhesive apparatus 0.8–1.1 times longer than wide / ○ thoracic adhesive apparatus almost or completely situated on a horse-shoe shaped elevation, usually extending from isthmus to base of first or third branched pectoral ray / ○ thoracic adhesive apparatus moderately elevated with few, short anteromedial striae / ○ caudal-peduncle length 16–20 % SL / ○ anteromedial striae few, very short / ○ striae not reaching onto pectoral base / ○ medial pit wide, the roundish anterior end, open, without striae / ○ dorsal and lateral head, back and flank with small round warts / ○ adipose short, its length 0.5–0.7 times larger than distance between base of last dorsal ray and adipose origin / ○ caudal-peduncle 1.1–1.3 times in its length / ○ maxillary barbel 72–99 % HL / ○ inner mandibular barbel 31–45 % HL / ○ outer mandibular barbel 50–81 % HL / ○ 5–8 serrae on pectoral spine / ○ all fins with a distinct, distal yellow tip / ○ many black or dark-brown blotches on flank. Size up to 91 mm SL.

Distribution. Iran: lower reaches of Kangavar in Karkheh drainage.

Habitat. Moderately fast-flowing medium-sized rivers.

Biology. Nocturnal, hiding under rocks during day. Feeds on benthic invertebrates.

Conservation status. CR; known from only one small stream, extirpated from Aran and Chardavol.

Further reading. Jouladeh-Roudbar et al. 2023 (description).

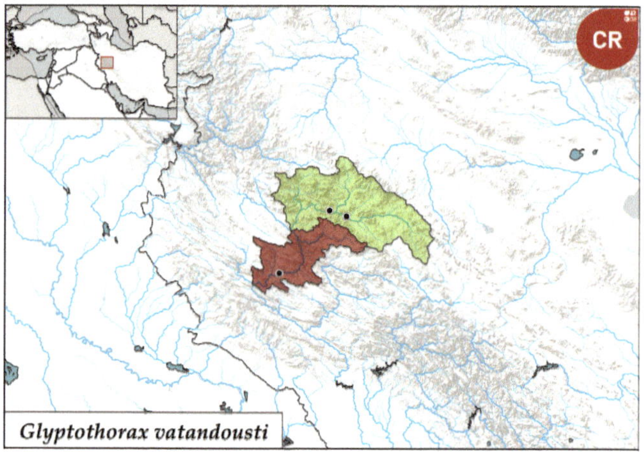

The Botan is a tributary of the Tigris, and habitat of several *Glyptothorax* species.

Family Esocidae

Pikes

The family comprises three genera (*Esox*, *Dallia*, *Novumbra*) related to salmonid fishes and the mudminnows of the family Umbridae. *Esox*, comprises seven species, three of which are endemic to North America and two endemic to Europe. They are highly distinctive, predatory fishes with a large, duckbill-like snout and dorsal and anal fins located far back on the body. *Esox lucius*, native to North America and Eurasia; *E. cisalpinus*, endemic to Italy; and *E. aquitanicus*, endemic to France, are species known from the West Palearctic. Additionally, *E. reichertii* is endemic to the Amur basin in Russia, Mongolia, and China. Pikes have a rich fossil record, spanning from the Late Cretaceous to the Holocene and covering much of the northern hemisphere. The first species of Esox (*E. tiemani*) was discovered in the Palaeocene of Alberta and Saskatchewan (Canada), estimated to be approximately 62 million years old. Fossils of *Esox* from the Eocene and later periods are also well documented in Europe. Due to its commercial value, *E. lucius* has been introduced locally in West Asia. In Spain, the introduction of *E. lucius* has resulted in the local extirpation of almost all native fish species, where *Esox* maintain high population densities and feed predominantly on crayfish (also introduced). The ecological effects of non-native pike in West Asia still need to be studied. **Further reading.** Crossman 1978 (systematics); Wilson 1980 (fossil pikes); Denys et al. 2014 (species diversity).

Esox lucius

Common name. Pike.

Diagnosis. Distinguished from other freshwater fishes in West Asia by: ○ snout long / ○ mouth large / ○ dorsal origin slightly in front of anal origin / ○ 105–148 total lateral line scales. Size up to 1300 mm SL.

Distribution. Native to Caspian, Black, Baltic, White, Barents, Arctic, North and Aral Seas and Atlantic basins, southwest to Loire drainage; Mediterranean basin, in Rhône drainage. Widespread in North America, Central Asia, and Siberia east to Anadyr drainage (Bering Sea basin). Historically absent from North Africa, Mediterranean, endorheic basins of West Asia, and Persian Gulf basin. Also, non-native to Iberian Peninsula, Mediterranean France, Italy, southern and western Greece, eastern Adriatic basin, Iceland, western Norway, and northern Scotland. Now widely introduced and translocated throughout West Asia, North Africa and Europe.

Habitat. Occurs in a variety of habitats with aquatic or periodically flooded vegetation. Often semi-anadromous in parts of northern Baltic basin with lower salinity.

Biology. Males spawn first at 170–350 mm SL, females at 250–400 mm SL, at 1–6 years. Reproduction closely linked to presence of submerged vegetation. Spawns in late winter–early spring, between February in south and June in north, when temperatures rise above 5°C. Several males compete for a single female. Eggs are laid in flooded areas and on submerged vegetation over a

Esox lucius; Danube drainage, Germany; ~350 mm SL. © A. Hartl.

https://doi.org/10.1515/9783111677811-029

period of 2–5 days. Juvenile survival is negatively related to biomass of older individuals and positively related to area of submerged vegetation. Feeds on a variety of small vertebrates, mainly fish and large invertebrates such as decapod crustaceans. Cannibalism is common. In Arctic lakes, there may be only pike in a given body of water; in such cases, juveniles feed on invertebrates and terrestrial vertebrates; large individuals are predominantly cannibals. Other fish avoid areas with pike feces, which contain alarm pheromones.

Conservation status. LC.

Further reading. Craig 2008 (biology); Casselman et al. 1986 (morphology); Skov & Nilsson 2018 (biology).

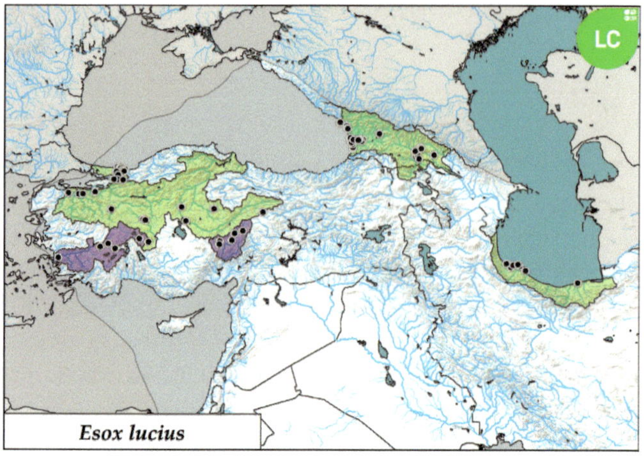

Lake Eğirdir and many other endemis hotspots in West Asia are threatened by the invasion of non-native pike.

Family Coregonidae

Whitefishes and ciscoes

The family comprises a variety of medium-sized to large fishes with slender and compressed, silvery bodies covered by numerous small scales (60–110), a dorsal adipose fin, and very small or absent teeth. The family includes anadromous and pure freshwater species in northern Europe, Asia, and North America. Most species are pelagic, form schools, and feed on small crustaceans. In autumn, both sexes develop nuptial tubercles on each lateral scale.

Whitefish are renowned for their propensity to evolve rapidly, particularly in European, Siberian, and North American lakes, where they adapt to specific ecological conditions and form their own niches. This has resulted in recognising numerous species, often endemic to a single or few lakes.

The majority of species have a high commercial value and are intensively fished. Some are cultivated, with fry of many populations and species produced artificially and stocked outside their native range, resulting in hybrid stocks and other manipulations. The genetic integrity of several wild populations and species is under threat, with several species now extinct following the formation of hybrid swarms and habitat modification. Two species have been introduced to lakes in the Caucasus, with one of them forming an endemic species of hybrid origin.

The anadromous *Stenodus leucichthys* is frequently reported from West Asia and found in the Caspian Sea. However, we are unaware of any confirmed records of this species entering rivers in the region, and it appears to have always been restricted to the northern, European part of the sea. Fish from the remaining population in the Volga may occasionally enter the Asian part of the Caspian Sea. However, it is doubtful that they will enter rivers to spawn and can be found in freshwater here. It The species is not included in the scope of this book. **Further reading.** Kottelat 1997 (species concepts, systematics, nomenclature).

Coregonus albula
Common name. Baltic cisco.
Diagnosis. Distinguished from Lake Sevan whitefish by: • mouth superior, upper jaw shorter than lower one / • 38–50 gill rakers. Size up to about 350 mm SL, many populations much smaller.
Distribution. Introduced in 1930s from Lake Ladoga (Russia) to Lake Paravani (Georgia) and Lake Kartsakhi (Aktaş, Türkiye). Native to Baltic basin, lakes of upper Volga drainage (Seliger, Vseluga, Perejaslavskoe), some lakes of White Sea and North Sea basins east of Elbe drainage. Anadromous in Gulf of Finland and marine in northernmost freshwater part of Gulf of Bothnia. North to about 69°N in Lake Inari, northern Finland. Native also all over northern Europe, Siberia and Pacific America, where *C. sardinella* is a synonym.
Habitat. Lacustrine and marine, in open water. At sea, foraging close to shore. Spawns on banks, 3–10 m deep, rarely down to 22 m or just below surface. Spawns deeper in clear lakes and closer to surface in humic lakes.
Biology. Anadromous, marine and freshwater populations. Spawns first time at 2–5 years (2–3 in Lake Onega), males usually earlier than females. Spawns in October (Central Finland) December (Northern Germany) at 6–7°C. Anadromous populations begin river migration in August (Neva). Eggs hatch March–April, and juveniles migrate to sea in late summer of first year in anadromous populations. Larvae pelagic, near surface and usually near coast. Adults usually in deep waters during day, moving to upper layers at night. Feeds on zooplankton.

Coregonus albula; Lake Breiter Luzin, Germany; ~120 mm SL.

https://doi.org/10.1515/9783111677811-030

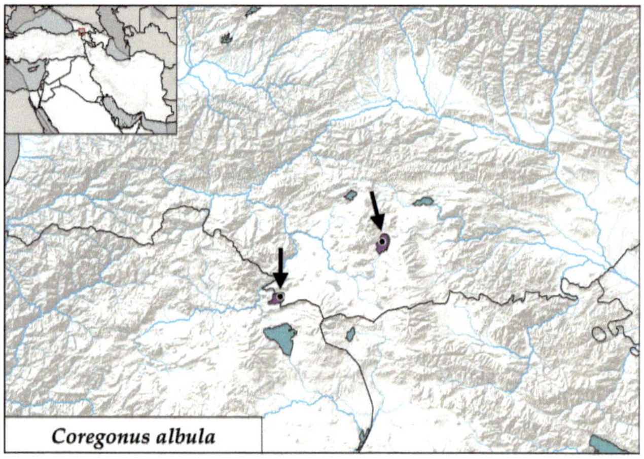

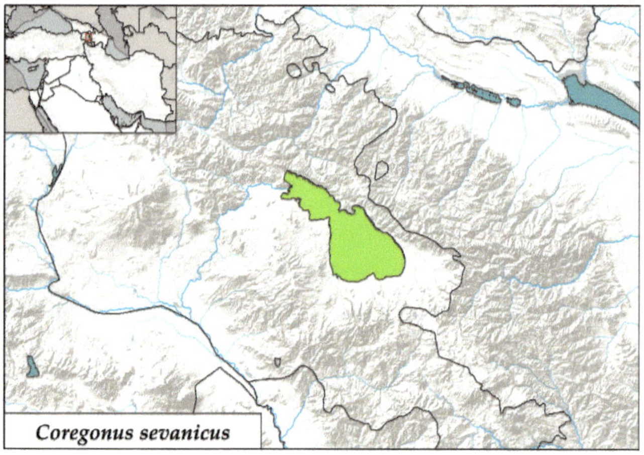

Conservation status. Non-native; introduced for commercial fisheries.
Further reading. Svärdson 1979 (morphology, distribution); Nikolsky & Reshetnikov 1970 (systematics); Yerli 2019 (introduced record from Türkiye).

Coregonus sevanicus

Common name. Sevan whitefish.
Diagnosis. Distinguished from *C. albula* by: • mouth terminal or inferior, upper jaw longer than or equal to lower one / • 27–36(38) gill rakers. Size up to about 350 mm SL.
Distribution. Armenia: Lake Sevan. Introduced from Lake Sevan to Lake Issykul (Kyrgyzstan) in 1973 and from there to Charvak Reservoir (Uzbekistan) in 1980s. Likely introduced to other lakes in former Soviet Union.
Habitat. Lacustrine, spawns in lakes.
Biology. Pelagic, spawns in autumn and early winter along shores. Feeds on plankton and benthic invertebrates.
Conservation status. Not assessed due to the very poor knowledge about this species.
Remarks. Two species were introduced in Lake Sevan in 1920s: *Coregonus ludoga* from Lake Ladoga and *C. maraenoides* from Lake Chudskoe (both in northern European Russia). These two species naturalised and hybridised in Lake Sevan, and an intermediate phenotype was subsequently described as *Coregonus lavaretus sevanicus*. Introducing this whitefish may have been one reason for the collapse of trout populations in Lake Sevan. *Coregonus sevanicus* is treated here as a valid species of hybrid origin. However, due to the lack of recent data, we cannot confirm or reject any taxonomic hypothesis related to this species. *Coregonus ludoga* was also introduced in Lake Tabatskuri in Georgia from the Volkhov hatchery in the 1930s. Still, the species has not been recorded for at least the last 50 years, and we suspect that the population may have been extirpated.
Further reading. Dadikyan 1986 (description); Elanidze 1983 (Lake Tabatskuri); Pipoyan et al. 2012 (Lake Sevan); Kuljanishvili et al. 2020 (introduction).

Coregonus sevanicus; Charvak Reservoir, Uzbekistan; ~190 mm SL. © A. Rozimov.

Family Salmonidae

Trout

Salmonids are medium- to large-sized fishes that inhabit a wide range of habitats with clear, cold, and well-oxygenated water. They are native to the northern hemisphere but have been widely introduced for aquaculture, commercial, and recreational fisheries in cold waters worldwide. They have a very diverse and fascinating evolutionary biology, grow to a commercial size, and have good-tasting flesh with few bones. Nevertheless, few groups of fish have been so heavily impacted by human activities. The most significant challenge for salmonid conservation is stocking non-native salmonid species, or populations, for recreational and commercial purposes. In Europe, salmonid stocking has resulted in the genetic pollution of many native populations. Recent molecular studies suggest that genetic pollution of trout is limited in West Asia, as mostly rainbow trout are stocked. However, the over-exploitation of water sources, river regulation (dams, weirs), poaching, and climate change pose significant threats to many trout populations in the region.

Until the late 20th century, virtually all trout native to the West Palearctic were identified as *Salmo trutta*. Many species described in the 19th century are now treated as valid, and many new species have been described in the 21st century, especially from Türkiye. Recent genomic studies indicate that trout populations in the Mediterranean and Ponto-Caspian regions diverged relatively recently, within a timeframe of 0.5–2.5 million years ago. This resulted in minimal to no mitochondrial DNA divergence. Standard molecular approaches, such as cytochrome oxidase 1 barcoding, have been unsuccessful in recognising the diversity of trout in West Asia. This is because all populations have relatively recently separated and, therefore, are not comparable with the evolutionary age of other species of fishes in the region. Species richness, high rates of phenotypic morphological and ecological diversification, and a very low genetic divergence between species make studies on the evolution of trout attractive and challenging. It is evident that recent gene flow and isolation of the populations of trout, particularly in the Mediterranean and Persian Gulf basins, have resulted in the evolution of small morphological differences, which has led to the recognition of a high number of species. Furthermore, natural hybridisation between different trout lineages has probably played an important role in the origin of several populations. This phenomenon requires further investigation into the species of West Asia. In conclusion, the available genomic evidence supports the view that trout are distributed in many isolated populations, which can be distinguished by details in colour and morphology. Should trout be classified as a multi-species taxon or better recognised as populations of one species (*Salmo trutta*) with small differences? This is a debated question and we are curious to follow the research on this in the coming years.

Some authors treat Coregonidae and Thymallidae as subfamilies, a view that is not adopted here. **Further reading.** Stearley & Smith 1993 (systematics); Stearns & Hendry 2004 (evolutionary biology); Geiger et al. 2014 (barcoding); Hashemzadeh Segherloo et al. 2021 (genomic data).

Key to genera of Salmonidae in West Asia

1a - Lateral-line scales very elliptical, reduced to little more than nerve tube, smaller than scales in adjacent rows, with no or little overlap with scales in front or behind.
..................*Salvelinus*
1b - Lateral-line scales roughly circular, as large as or larger than scales in adjacent rows, overlapping with scales in front and behind.
..................2

2a - Caudal with black spots at least on upper lobe, usually on whole fin; flank with a pinkish to red midlateral stripe or band.
..................*Oncorhynchus*
2b - Caudal plain; flank without red midlateral stripe or band.
..................*Salmo*

https://doi.org/10.1515/9783111677811-031

Oncorhynchus mykiss; Litani drainage, Lebanon; 160 mm SL.

Oncorhynchus mykiss

Common name. Rainbow trout.

Diagnosis. Distinguished from species of *Salmo* by: • wide, pink to red stripe from head to caudal base (except in sea-run form) / • caudal with black spots at least on upper lobe, usually on whole fin. Size up to 1000 mm SL.

Distribution. Widespread or stocked from fish farms, especially in Türkiye and Iran. Unclear where established. Native to Kamchatka and south to lower Amur drainage and Pacific basin of North America south to northern Mexico.

Habitat. Stocked in cold lakes, rivers, and streams. Usually not stocked in water with summer temperatures above 25°C or in ponds with very low oxygen concentrations. Spawns on gravel in fast-flowing water.

Biology. Anadromous, lacustrine, and resident forms. Often spawns first time at 1 year, 3–5 years in north, usually in February–May when temperatures rise above 10–15°C. Farmed trout mature August–April, depending on strain. Makes short spawning migrations. Anadromous and lake forms may undertake long migrations to spawning streams. Females build redds in gravel bottom of moderate to fast-flowing streams or rivers. Breeding populations will not be established if peak juvenile emergence coincides with flood season and cold summer temperatures, and if temperatures do not fall below 13°C. In agonistic behaviour, *O. mykiss* fry dominate *Salmo* fry of similar size, and juvenile and adult *Salmo* dominate similar size rainbow trout. Feeds on a wide range of aquatic and terrestrial invertebrates and small fish. At sea, feeds on fish and cephalopods. Feeding is negligible below 5°C and above 25°C water temperature.

Conservation status. Non-native; introduced for recreational fisheries and as escapees from fish farms.

Further reading. Smith & Stearley 1989 (systematics, nomenclature); Behnke 1992 (diversity); Stearley & Smith 1993 (phylogeny); Zelinsky & Makhrov 2001 (chromosomes); Fausch et al. 2001 (invasion success); Yoğurtçuoğlu et al. 2021c (Türkiye).

Too many species of trout? Twenty-five trout species native to West Asia have been identified, with one non-native species also present. Most of these species have been described based on narrow morphometric and meristic differences and small details in colour and colour pattern. Sympatric trout with different life histories have been treated as distinct species without clear evidence that they are not simply life-history forms. The majority of species have been described by a single population, and none of the species is well-supported by molecular characters, as they are all very young. It is acknowledged that trout in West Asia live in numerous, often small populations, and genomic data demonstrate that these are isolated from each other. This is also the case in other fish species. If we apply the same approach to other fish species, we could describe hundreds or thousands of cyprinids, minnows, loaches, and others isolated for several thousand years. This raises the question of whether we are constrained by the theoretical challenge of distinguishing between populations and species. Many additional trout species will likely be described, as many isolated populations remain unnamed. Alternatively, could we find ways to tackle the number of trout species by distinguishing between species and isolated populations with some private characters? Unfortunately, or happily, there is a lack of consensus on how to distinguish, describe and recognise trout species as valid. This allows discussions, diversity of opinions, and pluralism, which are all important for scientific progress. Nonetheless, it is important to remember that a species description is only a hypothesis regarding the relative and consistent distinctiveness of a set of individuals. The descriptions are influenced by several factors, including the prevailing attitudes and the authors' perspectives. **Further reading.** Hashemzadeh Segherloo et al. 2021 (genomic data).

In high-altitude streams, such as in the Pontic mountains, *Salmo* are often the only fish species.

Keys to species of *Salmo* in West Asia

Marmara and Black Sea basins

1a - Red spots on flank absent in individuals larger than 210 mm SL; black spots on flank usually polygonal.
..................*S. abanticus*

1b - Red spots on flank present in individuals of all size groups; black spots on flank roundish.
..................2

2a - Number of black spots on flank decreasing with size in male.
..................*S. rizeensis*

2b - Number of black spots on flank increasing or remaining stable with size in male.
..................3

3a - Number of black spots on flank increasing by size in male; interorbital width 8–9 % SL.
..................*S. fahrettini*
3b - Number of black spots on flank not increasing by size in male; interorbital width 7–8 % SL.
..................4

4a - Number of red spots on flank increasing by size in male.
..................*S. labrax*
4b - Number of red spots on flank not increasing by size in male.
..................5

5a - Black spots on back and flank in females fewer than 60; adipose height 8–9 % SL; anal length in females 12–15 % SL.
..................*S. brunoi*
5b - Black spots on back and flank in females more than 80; adipose height 4–8 % SL; anal length in females 14–17 % SL.
..................*Salmo* sp.

Caspian Sea basin

This key allows to identify resident and Sevan trout only. Anadromous individuals (flank with numerous x-shaped black spots; a prominent hook at the tip of lower jaw in male) are all identified to *S. caspius. Salmo araxensis, S. ardahanensis,* and *S. murathani* are likely to represent resident populations of *S. caspius,* but the case needs to be re-examined.

1a - Flank brown, grey, yellowish or greenish; black spots usually restricted to back and upper part of flank, rarely on midlateral flank, then usually grey or pale-brown; parr marks distinct, flank with many red/orange spots in individuals smaller than 350 mm SL.
..................2 (= *S. caspius*)
1b - Flank silvery or grey; with many, very large black spots and/or blotches on midlateral flank in individuals of all size groups; parr marks, distinct up to about 100 mm SL; usually without red spots, red spots present in *S. aestivalis* (Lake Sevan).
..................4

2a - Four dark bands on flank in female larger than 230 mm SL.
..................*S. murathani*
2b - No dark bands on flank in female larger than 230 mm SL.
..................3

3a - 19–21 gill rakers; 112–119 lateral-line scales; adipose almost reaching the base of caudal fin in male larger than 170 mm SL.
..................*S. ardahanensis*
3b - 21–22 gill rakers; 103–114 lateral-line scales; adipose not reaching the base of caudal in male larger than 170 mm SL.
..................*S. araxensis*

4a - Flank with black and red spots in all size groups.
..................*S. aestivalis*

4b - Flank with black spots only, red spots in juveniles.
..................5

5a - Flank with small or middle-sized black spots, not larger than 10 mm in diameter in individuals larger than 300 mm SL.
..................*S. ischchan*
5b - Flank with very large, densely set black blotches, larger than 10 mm in diameter in individuals larger than 300 mm SL.
..................*S. gegarkuni*

Mediterranean basin
1a - 7–9 parr marks; no dark-brown or grey bands on flank.
..................*S. kottelati*
1b - 9–13 parr marks; four dark-brown or grey bands on flank.
..................2

2a - Red spots absent on flank in individuals larger than about 70 mm SL, 21–25 gill rakers.
..................3
2b - Red spots present on flank in individuals of all size groups; 18–21 gill rakers.
..................5

3a - Head dorso-ventrally flattened; black spots absent on flank in individuals larger than 200 mm SL; 23–25 gill rakers.
..................*S. platycephalus*
3b - Head not dorso-ventrally flattened; black spots present on flank in individuals of all size groups; 21–24 gill rakers.
..................4

4a - 11–12 parr marks; 14–15 scale rows between end of adipose base and lateral line; head length 25–27 % SL; caudal peduncle depth 10–11 % SL.
..................*S. labecula*
4b - 9–10 parr marks; 11–13 scale rows between end of adipose base and lateral line; head length 26–30 % SL; caudal peduncle depth 9–10 % SL.
..................*S. ekmekciae*

5a - 22–23 gill rakers; a large (larger than pupil) white ring around red spot.
..................*S. opimus*
5b - 18–21 gill rakers; a narrow (equal or smaller than pupil) white ring around red spots.
..................*S. chilo*

Euphrates and Tigris drainages
1a - 19–20 scales between lateral line and adipose; 32–35 scales rows between lateral line and dorsal origin.
..................*S. tigridis*
1b - 13–17 scales between lateral line and adipose; 24–33 scales rows between lateral line and dorsal origin.
..................2

2a - Black and red spots on flank irregularly shaped.
..................3
2b - Black and red spots on flank roundish.
..................5

3a - Flank brownish; 116–123 scales along lateral line counted until end of hypural complex; 28–30 scales rows between lateral line and dorsal origin.
..................*S. munzuricus*

3b - Flank silvery or grey; 103–118 scales along lateral line counted until end of hypural complex; 24–28 scales rows between lateral line and dorsal origin.
..................4

4a - Red spots on flank larger than pupil; 3–7 black spots on opercle; no black spots on top of head; 1–2 dark-brown or grey bands on posterior part of flank in individuals larger than about 230 mm SL.
..................*S. baliki*
4b - Red spots on flank smaller than pupil; 8–17 black spots on opercle; 5–18 black spots on top of head; 4 dark-brown or grey bands on flank in individuals larger than about 230 mm SL.
..................*S. okumusi*

5a - Few black spots on back and upper part of flank; number of black spots not increasing with size.
..................*S. euphrataeus*
5b - Numerous black spots on back, middle and upper part of flank; number of black spots increasing with size.
..................6

6a. - Adipose depth 8–10 % SL in male; lateral line with 117–125 scales counted until end of hypural complex; flank brownish in life.
..................*S. trutta* (based on introduced population)
6b. - Adipose depth 3–8 % SL in male; lateral line with 109–116 scales counted until end of hypural complex; flank silvery in life.
..................*S. fahrettini*

Trout on speciation pathways? Trout populations are found in numerous lake basins. In many cases, there are resident trout in the lake tributaries and trout in the lake itself. In larger lakes with sufficient food resources, individual lake trout may grow larger than stream trout, become silvery, slender, lose the red spots, and become piscivorous or planktivorous. Trout populations in lakes adapt to pelagic life by undergoing changes similar to those observed in anadromous trout. As anadromous-resident trout, large lake trout typically join their siblings at spawning sites. Despite these similarities, the two trout forms are not distinct species. They represent different life-history forms of a single population. However, life-history strategies are subject to distinct selective challenges. Assortative mating would allow them to adapt better to their particular environments, whether in the lake or the stream. Furthermore, as all male trout prefer to mate with large females, large male lake trout can monopolise female lake trout, leading to assortative mating over time. This process should lead to the reproductive isolation of lake trout from resident trout. However, fully reproductively isolated trout species within one lake-stream system are rarely documented. The lake-stream system's instability may interrupt the speciation process too often. Fluctuating environmental conditions and unbalanced ratios between lake and resident trout at the spawning places lead to hybridisation. This process acts against the isolation of both "kinds" of trout. The permanent activity of small male trout, which engage in sneaking, may inhibit the ability of large male lake trout to monopolise the attention of large females. If trout species cannot engage in assortative mating due to sneaking behaviour, spatial and temporal segregation is the only mechanism that allows them to reach reproductive isolation and speciate. This pattern is observed in sympatric trout species, with each species spawning at different times and locations. In West Asia, the only documented example of multi-species lake trout is Lake Sevan in Armenia.

Salmo abanticus; aquaculture, Lake Abant, Türkiye; ~225 mm SL.

Salmo abanticus

Common name. Abant trout.

Diagnosis. Distinguished from other species of *Salmo* in Marmara and Black Sea basins by: • flank silvery in life with many black spots and polygonal blotches much larger than pupil scattered on whole flank in individuals larger than 210 mm SL / • no red spots on flank in individuals larger than about 210 mm SL / ○ maxilla short, reaching slightly behind posterior êye margin in male larger than about 230 mm SL / ○ 103–121 lateral-line scales counted until end of hypural complex / ○ 26–28 scale rows between dorsal origin and lateral line. Size up to 500 mm SL.

Distribution. Türkiye: Lake Abant basin. Introduced into several ponds and reservoirs in Türkiye for recreational fishing; establishment remains uncertain except for a confirmed population in Ulugöl (Ordu Province).

Habitat. Lacustrine, spawns in springs in lake itself and in lake tributaries.

Biology. No data.

Conservation status. VU; situation appears to be relatively stable, but species may be conservation dependent, and a stock of this species is maintained in captivity to produce juveniles for stocking.

Remarks. Genomic data support recognising this species as distinct from *S. labrax* and *S. rizeensis*.

Further reading. Tortonese 1954 (description); Uysal & Alpbaz 2003 (artificial reproduction); Turan et al. 2010 (diagnosis); Hashemzadeh Segherloo et al. 2021 (genomic data).

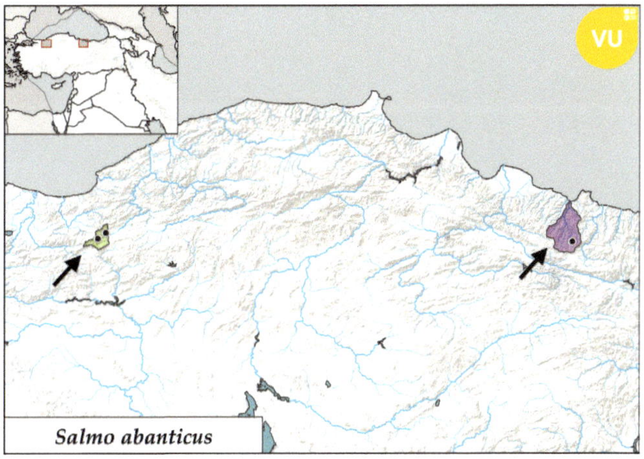

Salmo aestivalis; Lake Sevan, Armenia; ~200 mm SL. © J. Schöffmann.

Salmo aestivalis

Common name. Sevan spring trout.

Diagnosis. Distinguished from other species of *Salmo* in Caspian basin by: ○ flank silvery with black and red spots in individuals of all size groups / ○ parr marks distinct up to about 100 mm SL, absent or indistinct in larger individuals. Size up to 500 mm SL and 1.9 kg.

Distribution. Armenia: Lake Sevan basin.

Habitat. Lacustrine, spawns in lake and in mouths of tributaries.

Biology. Spawns in spring and early summer. Feeds mainly on benthic gammarid crustaceans.

Conservation status. VU; might by CR; may be extinct in the wild, surviving only through artificial propagation and stocking.

Further reading. Levin et al. 2022 (Sevan trout).

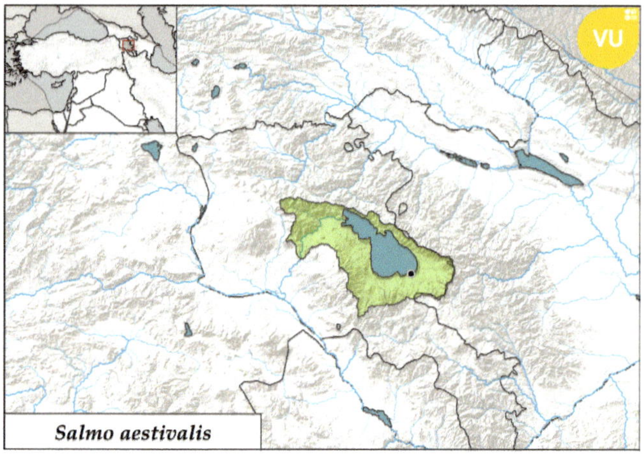

Salmo araxensis; Aras drainage, Türkiye; male, 265 mm SL.

Salmo araxensis

Common name. Aras trout.

Diagnosis. Distinguished from other species of *Salmo* in Caspian basin by: ○ background colour of flank greyish to brownish in life / ○ few black spots scattered on back and upper part of flank, often ocellated with a roundish narrow white ring / ○ 21–22 gill rakers / ○ 103–114 lateral-line scales / ○ 10–13 parr marks, distinct up to about 130 mm SL / ○ adipose not reaching base of caudal in male larger than 200 mm SL / ○ no dark band on flank. Size up to 276 mm SL.

Distribution. Türkiye: Kırkpınar, Alacasu, Porsuklu, and İncilipınar in upper Aras drainage.

Habitat. Headwaters with clear and fast-flowing water, with stone and gravel bottoms.

Biology. No data.

Conservation status. LC.

Remarks. This species is likely conspecific with *S. caspius*; see the discussion there. More research is needed to resolve the case of Kura and Aras trout, all of which (except Lake Sevan trout) are likely to be conspecific.

Further reading. Turan et al. 2022b (description).

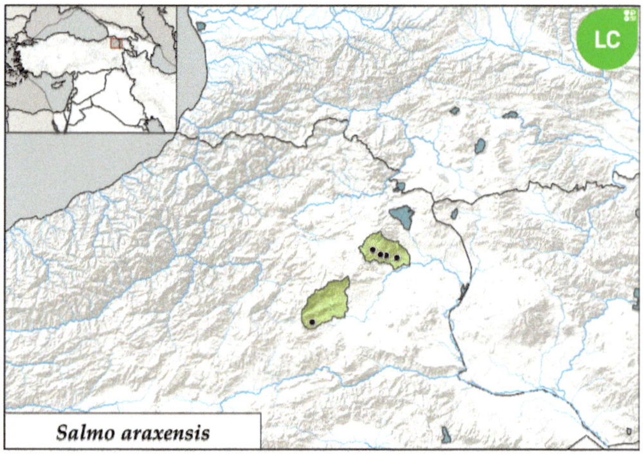

Salmo ardahanensis; Kura drainage, Türkiye; 222 mm TL.

Salmo ardahanensis

Common name. Ardahan trout.

Diagnosis. Distinguished from other species of *Salmo* in Caspian basin by: ○ background colour of flank yellowish to dark-brown in life / ○ black spots few to numerous, roundish, medium size (approximately equal to pupil), ocellated, scattered on back, middle and upper part of flank in some individuals, restricted to back and upper part of flank in others / ○ number of black spots not or slightly increasing with increasing size in most males / ○ 19–21 gill rakers / ○ 112–119 lateral-line scales / ○ adipose large, almost reaching to caudal base in male larger than about 170 mm SL / ○ no dark bands on flank. Size up to 280 mm SL.

Distribution. Türkiye: Upper Kura drainage.

Habitat. Headwaters with clear and fast-flowing water, with stone and gravel bottoms.

Biology. No data.

Conservation status. LC.

Remarks. This species is likely conspecific with *S. caspius*; see the discussion there. More research is needed to resolve the case of Kura and Aras trout, all of which (except Lake Sevan trout) are likely to be conspecific.

Further reading. Turan et al. 2022b (description).

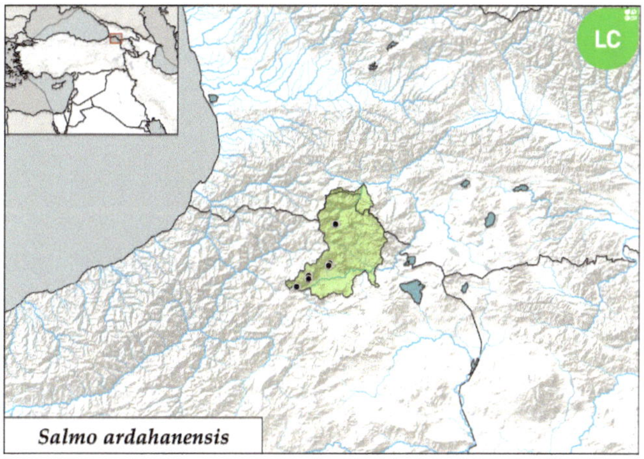

Salmo baliki; Murat, Euphrates drainage, Türkiye; male, 216 mm SL.

Salmo baliki

Common name. Murat trout.

Diagnosis. Distinguished from other species of *Salmo* in Euphrates and Tigris drainages by: ● red spots on flank larger than pupil / ○ red spots on flank irregularly shaped / ○ background colour of flank greyish or brown / ○ usually one, rarely two grey spots or blotches behind eye and on cheek / ○ 3–7 black spots on opercle / ○ few black spots on back and upper part of flank, absent on predorsal area / ○ no black spots on top of head / ○ number of black and red spots not increasing with size / ○ 1–2 dark-brown or grey bands on posterior part of flank / ○ 107–118 lateral-line scales counted until end of hypural complex / ○ 24–28 scales rows between lateral line and dorsal origin. Size up to 270 mm SL.

Distribution. Türkiye: Sinek and Cumaçay in upper Murat drainage, potentially in adjacent streams.

Habitat. Headwater streams with cold, clear water, moderate current, gravelly and pebbly bottom.

Biology. No data.

Conservation status. EN; appears to be declining within its very small range.

Further reading. Turan et al. 2021b (description).

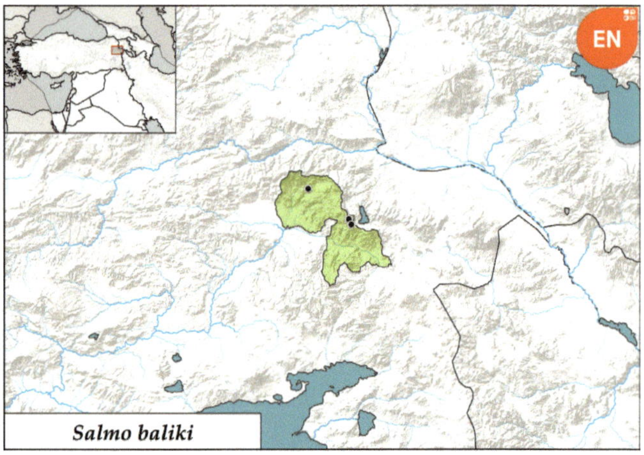

Salmo brunoi; Nilüfer drainage, Türkiye; ~150 mm SL.

Salmo brunoi

Common name. Nilufer trout.

Diagnosis. Distinguished from other species of *Salmo* in Marmara and Black Sea basins by: ○ one black spot in postorbital and suborbital area, greater than pupil / ○ 2–4 black spots on opercle, usually smaller than pupil / ○ less than 60 black spots on body, approximately smaller or equal to pupil, often ocellated, scattered on back and upper part of flank (missing in predorsal area) / ○ less than 40 red spots, smaller than pupil, irregularly shaped, surrounded by an irregularly shaped narrow ring, organised in 2–4 irregular longitudinal rows / ○ number of black and red spots not increasing with size / ○ interorbital width 7–8 % SL / ○ anal height 12–15 % SL in male / ○ adipose height 8–9 % SL. Size up to 187 mm SL.

Distribution. Türkiye: Aras, Deliçay, and Ericek, potentially all tributaries to Nilüfer in Susurluk drainage.

Habitat. Clear and fast-flowing streams with a substrate of gravel and pebbles.

Biology. No data.

Conservation status. EN; appears to be declining within its very small range.

Further reading. Turan et al. 2024c (description).

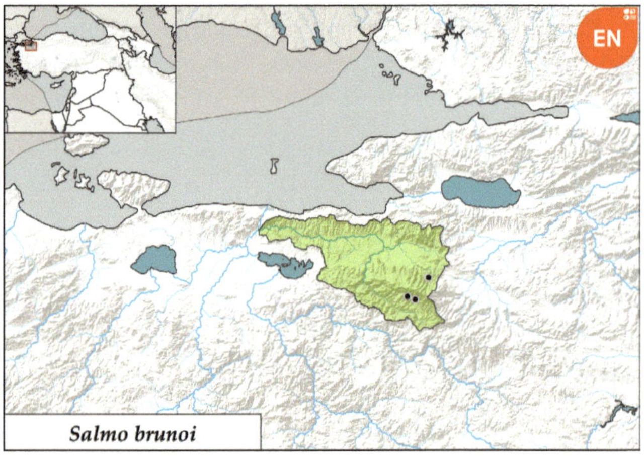

Salmo caspius; Fereydunkenar, Caspian Sea, Iran; anadromous individual, ~400 mm SL.

Salmo caspius; Karaj drainage, Lake Namak basin, Iran; ~140 mm SL.

Salmo caspius

Common name. Caspian trout.

Diagnosis. Resident populations distinguished from Lake Sevan trout by: • flank brown, grey, yellowish or greenish (silvery without orange spots in anadromous form) / • black spots usually restricted to back and upper part of flank, rarely on midlateral flank, then spots usually grey or pale-brown / • parr marks distinct up to about 350 mm SL / ○ flank with many red spots in individuals smaller than 350 mm SL. Anadromous individuals larger than 350 mm SL distinguished from Lake Sevan trout by: • flank with very small, irregularly shaped, black blotches (vs. many, very large, roundish spots) / • caudal truncate (vs. emarginate). Size up to 1300 mm SL in anadromous form, usually up to 300 mm SL in resident form.

Distribution. Caspian, Lake Namak, and Lake Urmia basins. Kura and Aras drainages north to Volga and Ural. Introduced in Western Kavir, and Karun (Iran). Records not shown on map as details are lacking.

Habitat. At sea, along coast to 50 m depths. Locally resident in lakes. Migrates to mountain streams. Resident part of populations in streams and headwaters with fast currents, cold, clear water, and stony or gravelly bottoms. Spawning sites usually in upper reaches of fast-flowing rivers and streams.

Biology. Anadromous, lacustrine, and resident forms. First spawn at 3–5 years, females 1 year later than males. Spawns October–December. Before construction of dams and river regulation, two migration waves, first in summer–autumn, peaking in October, with adults spawning in same autumn and returning to sea in December; second in late

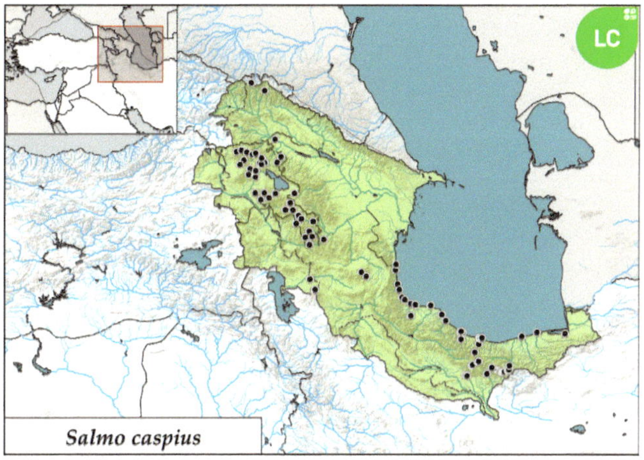

autumn–winter, peaking in December–February, with adults spawning in following autumn after about 10–12 months in river. Eggs hatch in 6–8 weeks. Anadromous individuals in Terek drainage probably spawn once in life, but some females from other populations spawn up to 4 times. Parrs spend 2–4 years, sometimes more, in rivers and then migrate to sea. Commonly spends up to 4 years at sea before returning to rivers. Parr and resident adults feed on a wide variety of aquatic and terrestril invertebrates. Anadromous and large lacustrine individuals feed mainly on fish but also on large crustaceans.

Conservation status. LC.

Remarks. The northern range of *S. caspius* remains unclear, as *S. trutta* appears to have naturally invaded the upper Volga. The diversity of trout in Caspian basin has been greatly confused by the description of *S. ardahanensis*, *S. araxensis*, and *S. murathani* from Türkiye and the restriction of *S. caspius* and *S. ciscaucasicus* to anadromous populations. We strongly disagree with this approach. In all well-studied cases of Salmonidae, anadromous and resident individuals are conspecific, and this should have been the null hypothesis to be rejected by data. That an anadromous trout is (was) reproductively isolated from sympatric resident trout should have been demonstrated, as the opposite is the only demonstrated situation in Salmonidae. Reversing the burden of proof leads to a situation where each population could, in principle, be described as a separate species—and others are asked to prove the opposite. Although sympatric anadromous and resident *Salmo* species cannot be completely excluded, such a finding would be the first and only case in Salmonidae in riverine ecosystems. We await further studies on Caspian trout to resolve the case, but in the meantime, we have to accept the different species as valid. *Salmo ciscaucasicus* from the European Caucasus is probably a synonym of *S. caspius*.

Further reading. Berg 1949b (biology, morphology, distribution; as *S. t. caspius*); Dorofeyeva 1967 (morphology, taxonomy); Belyaeva et al. 1989 (biology, fisheries); Danilov-Danilyan 2001 (conservation); Turan et al. 2020 (Cytochrome b phylogeny); Hashemzadeh Segherloo et al. 2021 (genomic data); Turan et al. 2022b (restriction to anadromous populations).

Salmo chilo; Tekir, Ceyhan drainage, Türkiye; ~300 mm SL.

Salmo chilo; Tekir, Ceyhan drainage, Türkiye; ~120 mm SL.

Salmo chilo

Common name. Ceyhan trout.

Diagnosis. Distinguished from other species of *Salmo* in Mediterranean basin by: • dorsal head profile strongly convex / • snout blunt / • mouth conspicuously subterminal, with fleshly maxilla and lower lip / ○ maxilla short, depth of upper jaw 2.5–2.9 times in length of maxilla / • size of adipose equal in male and female / • length of maxilla equal in male and female / ○ 10–13 parr marks, distinct up to about 200 mm SL / ○ a narrow (equal or smaller than eye pupil) white ring around red spots / ○ 7–13 black spots on opercle / ○ 1–12 black spots behind eye and on cheek (more than one spot in individuals larger than about 160 mm SL) / ○ four broad bands on flank distinctive in all size groups / ○ few red spots on flank in individuals of all size groups / ○ 108–114 lateral-line scales counted until end of hypural complex / ○ 24–26 scale rows between dorsal origin and lateral line / ○ 15–17 scale rows between anal origin and lateral line / ○ 13–14 scale rows between adipose origin and lateral line / ○ 18–21 gill rakers. Size up to 300 mm SL.

Distribution. Türkiye: Göksu, Tekir, Fırnız, Göçüksu, and other tributaries of upper Ceyhan and Sarız, a tributary to upper Seyhan.

Habitat. Mountain streams, usually spring-fed, with cold, clear water and moderate currents, gravel, and pebble beds.

Biology. Spawns November–February.

Conservation status. LC.

Remarks. The very limited genomic data suggest a relatively weak genetic differentiation for this species, making it very close to or even identical to European *S. farioides* and *S. lourosensis*. Admittedly, increased sample sizes and taxon sampling will be required to assess the validity of retaining the species more rigorously. However, its genetic differentiation is weaker than that observed between other recognised trout species elsewhere.

Further reading. Turan et al. 2012b (description); Alp et al. 2003 (spawning time); Hashemzadeh Segherloo et al. 2021 (genomic data).

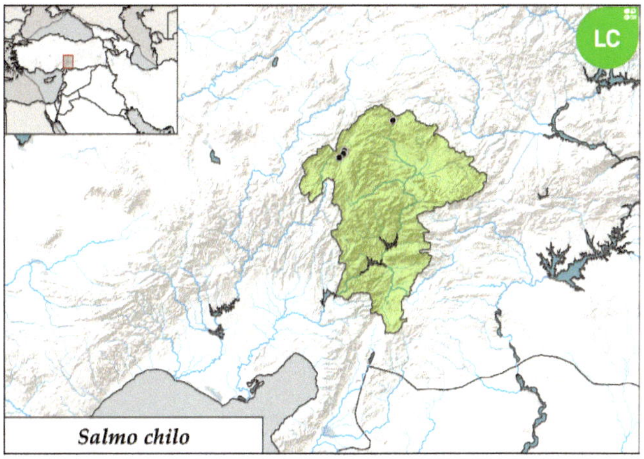

Lake form of *Salmo caspius* from Lake Çıldır in Türkiye. Recent authors have ignored the possibility that the three life-history forms of this trout might also occur in West Asia. © M. Özuluğ.

Life-history traits in trout. The diversity of trout species is often confounded due to the diversity of life-history traits. The life history of trout species follows several similar patterns and has been demonstrated to be an intraspecific diversity within all well-studied trout (*Salmo*, *Oncorhynchus*, and *Salvelinus*). It is therefore rejected that different trout species should be recognised only based on morphological characters connected to life-history types. Classically, three distinct groups of trout have been identified:

1) Anadromous trout (sea trout). These trout hatch in streams, grow there for 2–4 years, undergo smoltification and migrate to the sea or estuaries to forage. They exhibit accelerated growth and larger size than their siblings, remaining in streams. After a few years, (often after one year) they return to their native stream to spawn. Additionally, they frequently enter lower reaches of streams and rivers to overwinter in freshwater, even if they have not yet reached maturity. Sea trout are absent from the Mediterranean but occur in the Black Sea and Caspian basins.

2) Lake trout. They also hatch in streams and follow the same life history as sea trout, smoltify, and migrate to the lake to forage, grow large, and return to their native stream to spawn. This is the case in lake trout populations in West Asia, but some populations also spawn in the lake.

3) In nearly all spawning streams, some individuals do not migrate and remain there as resident trout. In West Asia, most species include only resident, non-migratory individuals.

The various life-history forms of trout are not reproductively isolated and do not represent different species. Only the resident and/or lacustrine forms are known in some species. In others, all three forms occur, and all their offspring appear to have the potential to smoltify and migrate to foraging habitats. The environmental conditions that an individual trout experiences during its lifetime influence the life-history strategy it will adopt and the life history it will develop (e.g., food availability). However, genetic factors also appear to influence whether some species, populations or individuals have the potential to develop different life histories. This plasticity of life-history strategies is observed in some populations of *S. caspius* and *S. labrax*, while most populations of these species consist of resident fish only.

In contrast to the evidence from around the world, it has been proposed that West Asian residents, lake and anadromous individuals belong to separate species that lives in the same area (i.e., behaving as distinct species as in *S. labrax* and *S. rizeensis* as well as in the four Kura-Aras species listed in this book). The hypothesis of one species with two life-history forms (migratory and resident) has not been tested nor rejected in both cases. The close relationship of *S. labrax* and *S. rizeensis* demonstrates the need for a review based on many populations and high-resolution molecular methods. Based on a small sample size and genomic data, it had been suggested that they are conspecific and should be treated as one species as soon as further studies confirm these data. The case of the trout species in the Kura and Aras rivers will require further investigation in the future. However, it is anticipated that only one species will be identified. **Further reading.** Turan et al. 2010 (Black Sea trout); Ninua et al. 2018 (Cytochrome b data); Hashemzadeh Segherloo et al. 2021 (genomic data); Turan et al. 2022b (Caspian trout).

Salmo ekmekciae; Degirmenözü in Köprüçay drainage, Türkiye; 330 mm SL. © J. Schöffmann.

Salmo ekmekciae
Common name. Köprüçay trout.
Diagnosis. Distinguished from other species of *Salmo* in Mediterranean basin by: ○ black spots on flank in individuals of all size groups / ○ one small black spot in postorbital and suborbital areas, larger than pupil / ○ head length 26–30 % SL / ○ caudal peduncle depth 9–10 % SL / ○ 7–17 black spots on opercle / ○ 9–10 parr marks / ○ 4 broad dark bands on flank / ○ maxilla short and narrow / ○ 108–118 lateral-line scales counted until end of hypural complex / ○ 21–25 scale rows between dorsal origin and lateral line / ○ 16–18 scale rows between anal origin and lateral line / ○ 11–13 scale rows between adipose origin and lateral line / ○ 22–24 gill rakers / ○ head not dorso-ventrally flattened. Size up to 350 mm SL.
Distribution. Türkiye: Köprüçay drainage.
Habitat. Headwater streams, usually fed by springs, with cold and clear water with moderate currents and substrate of stones and pebbles.
Biology. Spawns December–January.
Conservation status. EN; appears to be declining within its very small range.
Further reading. Küçük et al. 2024a (description).

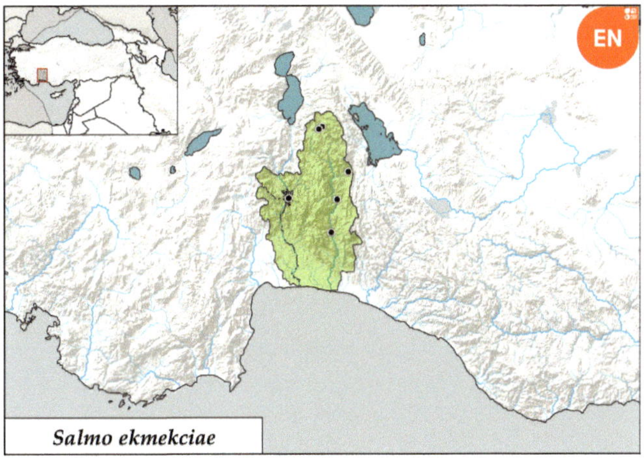

Salmo euphrataeus; upper Euphrates drainage, Türkiye; ~170 mm SL.

Salmo euphrataeus
Common name. Northern Euphrates trout.
Diagnosis. Distinguished from other species of *Salmo* in Euphrates and Tigris drainages by: ○ flank brownish in life / ○ one black spot behind eye / ○ 2–9 black spots on opercle / ○ 0–54 roundish black spots on flank, restricted to back (usually missing in predorsal area) and upper part of flanks / ○ number of black spots not increasing with size / ○ 13–53 small red spots on flank, round in male and irregularly shaped in female / ○ adipose dark-grey / ● mouth gape length 14.3–16.5 % SL in male, 12.3–14.9 in female / ○ adipose large, with 0–4 orange spots on its posterior margin / ○ 10–12 parr marks, distinct up to about 200 mm SL / ○ 112–120 lateral-line scales counted

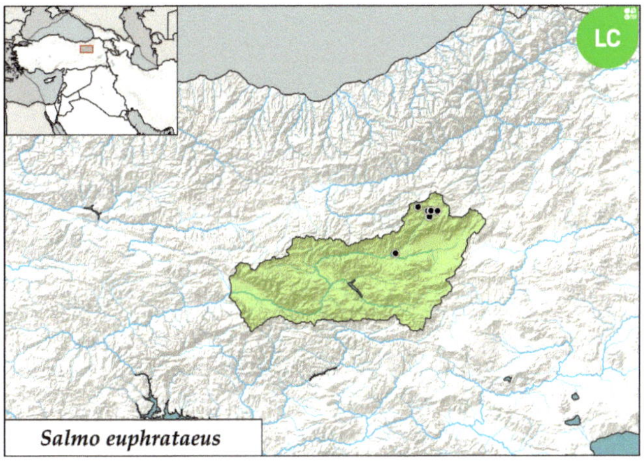

Salmo fahrettini; Euphrates drainage, Türkiye; 232 mm SL.

until end of hypural complex / ● 26–31 scale rows between lateral line and dorsal origin. Size up to 260 mm SL.

Distribution. Türkiye: Şenyurt, Erzincan, Kuzgun, Rizekent, Ağırcık, Sırlı (Karasu drainage) in northern Euphrates.

Habitat. Headwater streams with cold, clear water, moderate current, gravelly and pebbly bottom.

Biology. No data.

Conservation status. LC.

Remarks. Genomic data suggest that natural hybridisation between Mediterranean and Black Sea trout has played a role in the evolution of *S. euphrataeus*. However, the diversity, biogeography and relationships of the Persian Gulf trout still need to be better understood.

Further reading. Turan et al. 2020 (Cytochrome b phylogeny); Hashemzadeh Segherloo et al. 2021 (genomic data).

Salmo fahrettini

Common name. Euphrates trout.

Diagnosis. Distinguished from other species of *Salmo* in Euphrates and Tigris drainages by: ○ flank silvery in life / ○ back, middle and upper part of flank with numerous, roundish black spots (>80) increasing in number with size / ○ one black spot in postorbital and suborbital areas / ○ numerous roundish red spots on flank (>70 in adult individuals), smaller than pupil, increasing in number with size / ○ no black spots on top of head / ○ no dark-brown or grey bands on flank / ○ head length 24–28 % SL / ○ maxilla length 8–10 % SL in male / ○ adipose depth 3–8 % SL in male / ○ lateral line with 109–116 scales counted until end of hypural complex / ○ 27–30 scale rows between dorsal origin and lateral line / ○ 20–23 scale rows between anal origin and lateral line / ○ 15–17 scale rows between origin of adipose and lateral line. Size up to 232 mm SL.

Distribution: Türkiye: Ömertepesuyu and Tekke in northern Euphrates drainage and Terme drainage, east of Samsun, in Black Sea basin.

Habitat. Headwater streams with cold, clear water, moderate current, gravelly and pebbly bottom.

Biology. No data.

Conservation status. EN; the quality of its habitats is steadily declining due to pollution and water abstraction, and its population appears to be already quite low.

Further reading. Turan et al. 2020 (description); Yılmaz et al. 2021 (record from Terme).

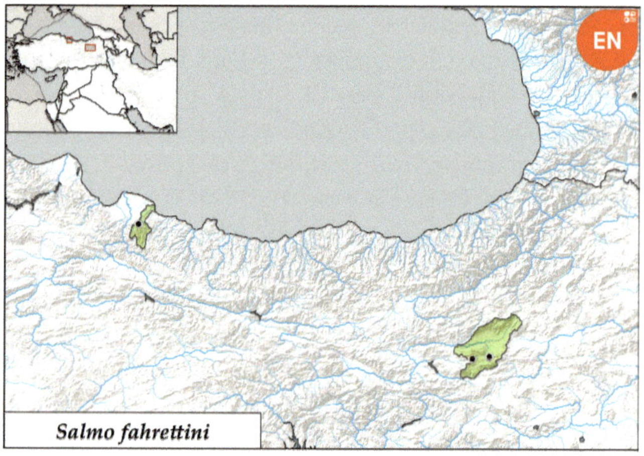

Lake Sevan trout. There are sympatric "kinds" of trout in Lake Sevan (Armenia) and some northern European lakes (e.g., Lough Melvin in Ireland, Lake Ohrid in North Macedonia and Albania). These are reproductively isolated, differ in morphology, and have different feeding habits, spawning sites, and/or spawning seasons. As their offspring also develop only into the "kind" of their parents, they represent different species. To ascertain whether different sympatric "kinds" might be species or life-history forms, genomic data can be employed to elucidate their reproductive isolation. Five ecologically and phenotypically divergent "kinds" of trout have been described from Lake Sevan basin. Four lacustrine "kinds" with different temporal-spatial spawning strategies and divergent morphology and colouration evolved alongside a fifth "kind," a resident trout inhabiting the lake's tributaries. The partitioning of spawning resources in space and time accompanied the diversification of the Sevan trout. *Salmo gegarkuni* spawns in autumn and winter in lake tributaries, *S. aestivalis* spawns in spring in the lake and the mouth of tributaries, *S. ischchan* spawns in autumn and winter near springs in the lake, and *S. danilewskii* is dwarf (up to 320 mm vs. more than 500 mm in other populations) spawning in autumn and winter in gravel beds in the lake. A recent study of complete mitogenomes in conjunction with genome-wide SNP data has revealed the monophyly of the four lacustrine "kinds" in contrast to the resident trout, all derived from *S. caspius*.

Three genomic clusters were identified for the five "kinds"—(i) a resident trout inhabiting tributaries (*S. caspius*), (ii) *S. gegarkuni* + *S. aestivalis*, and (iii) *S. ischchan* + *S. danilewskii*. These genomic clusters are genetically differentiated, which precludes their determination by environmental factors during ontogeny. A few outlier SNPs under selection were discovered that could be responsible for assortative mating based on visual recognition. The Holocene climatic oscillations and the desiccation of tributaries may have played an important role in the origin of lacustrine spawning. Based on zero-to-low FST values, a measure of genetic variation, and the recent lake isolation (20 kya), *S. danilewskii* appears not reproductively isolated from *S. ischchan*. It is evident that river-spawning *S. gegarkuni* was isolated from the lake-spawners earlier than those that diversified into different "kinds" along a temporal axis separating spring-spawning *S. aestivalis* from autumn and winter-spawning *S. ischchan*. The reasons for the poor molecular support of spring-spawning *S. aestivalis* as a distinct evolutionary group from autumn/winter-spawning *S. ischchan* remain unresolved. It is possible that both species were not fully reproductively separated. Alternatively, the separation may have been so recent that it could not be well-detected by the molecular markers applied so far, or artificial breeding may have blurred their historical differences. Lake Sevan trout warrant further investigation, but unfortunately, the extinction of all the Sevan trout spawning in the lake makes it now difficult or impossible to study their ecology and evolution in detail. A comparable adaptive radiation occurred in Lake Ohrid, where four sympatric species exhibited spatial and temporal differences in their use of spawning sites. These have vanished in the 20th century. **Further reading.** Sušnik et al. 2007 (Ohrid trout); Levin et al. 2022 (Sevan trout).

Salmo gegarkuni; Issyk-Kul, Kyrgyzstan; ~350 mm SL. © J. Schöffmann.

Salmo gegarkuni; Lake Sevan basin, Armenia, ~250 mm SL. © J. Schöffmann.

Salmo gegarkuni

Common name. Gegarkuni.

Diagnosis. Distinguished from other species of *Salmo* in Caspian basin by: ● flank with very large, densely set, black blotches, larger than 10 mm in diameter in individuals larger than 300 mm SL / ○ flank silvery, without orange spots except in juveniles / ○ parr marks distinct up to about 100 mm SL, absent or indistinct in larger individuals. Size up to 800 mm SL and 16 kg.

Distribution. Armenia: Lake Sevan basin. Locally introduced to lakes and reservoirs in Armenia, Georgia, and Azerbaijan, as well as to Lake Issykul (Kyrgyzstan) and from there in 1980th to Charvak reservoir (Uzbekistan).

Habitat. Lacustrine, spawns in inflowing streams.

Biology. Spawns in autumn–early winter. Feeds mainly on benthic gammarid crustaceans.

Conservation status. VU; might by CR; may be extinct in the wild, surviving only through artificial propagation and stocking.

Further reading. Levin et al. 2022 (Sevan trout).

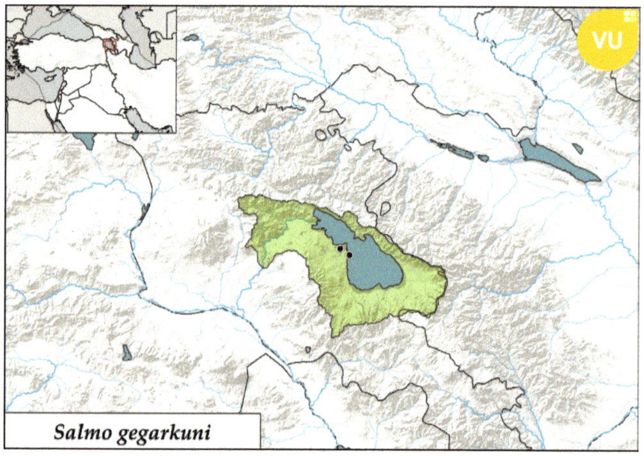

Salmo ischchan; Lake Sevan, Armenia. Modified from Kessler 1877.

Salmo ischchan

Common name. Sevan trout.

Diagnosis. Distinguished from other species of *Salmo* in Caspian basin by: ○ flank with small or middle-sized black spots, smaller than 10 mm in diameter in individuals larger than 300 mm SL / ○ flank silvery, without orange spots / ○ parr marks present up to about 100 mm SL. Size up to 760 mm SL and 15 kg.

Distribution. Armenia: Lake Sevan.

Habitat. Lacustrine, spawned in lake, in gravel, close to springs and in mouth of inflowing streams.

Biology. Spawned in autumn–early winter. Fed mainly on benthic gammarid crustaceans.

Conservation status. Extinct.

Remarks. The name *S. danilewskii* has been used for small individuals feeding on gammarids and plankton. *Salmo danilewskii* is a synonym, and these fish were most likely small individuals of *S. ischchan.*

Further reading. Levin et al. 2022 (Sevan trout).

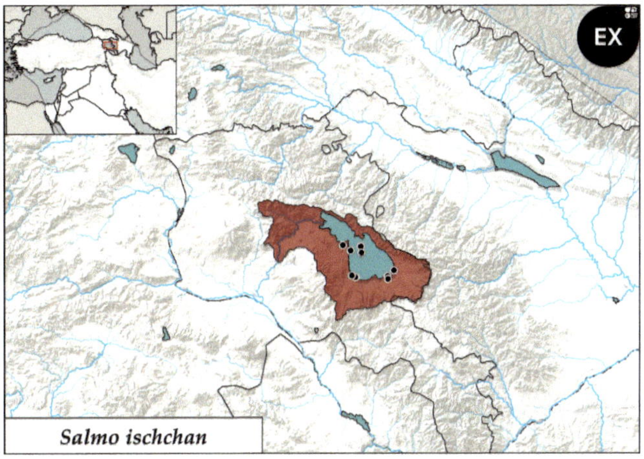

The end of Lake Sevan's trout diversity. The abundance of Sevan trout declined significantly in the 20th century due to human impacts. A critical 18.5 m drop in the lake water level and a reduction of >42 % in the lake's volume for hydropower purposes resulted in the shrinkage of spawning sites for trout spawning in the lake. Furthermore, temperature and oxygen regimes, as well as the hydrochemical composition of the water in the lake, also changed significantly after the drop in the lake level. These alterations, coupled with overfishing, poaching, and the impact of non-native *Coregonus*, have led to the decline of Sevan trout populations, which eventually resulted in the extinction of the two lacustrine forms known as *S. ischchan* and *S. danilewskii* in the 1980s. Only two species survived, namely *S. aestivalis* and *S. gegarkuni*. However, they are rarely found in the lake and are largely kept alive by farming and stocking. **Further reading.** Aghasyan & Kalashyan 2010; Gabrielyan 2010 (conservation); Levin et al. 2022 (Sevan trout).

Salmo kottelati, Alakır stream, Türkiye; male, ~105 mm SL.

Salmo kottelati
Common name. Alakır trout.
Diagnosis. Distinguished from other species of *Salmo* in Mediterranean basin by: ○ black spots on flank numerous, scattered on back, middle and upper part of flank (on lower part of flank in some individuals) in male larger than about 160 mm SL, and female larger than about 160–190 mm SL / ○ in male and female smaller than about 160 mm SL, black spots few, restricted to upper part of flank / ○ few to numerous red spots on back and flank / ● 7–9 parr marks / ● no dark-brown or grey bands on flank / ● number of black and red spots increasing with size in male, decreasing with size and age in female / ○ head length 29–33 % SL in male, 26–32 in female / ○ mouth slightly subterminal / ○ length of mouth gape 13–19 % SL in male, 12–15 in female / ○ 105–113 lateral-line scales counted until end of hypural complex / ○ 24–29 scale rows between lateral line and dorsal origin / ○ 17–19 scale rows between lateral line and anal origin / ○ 13–15 scale rows between lateral line and adipose insertion / ○ 18–20 gill rakers. Size up to 210 mm SL.
Distribution. Türkiye: Alakır drainage southwest of Antalya.
Habitat. Headwater streams with cold, clear water, moderate current, gravelly and pebbly bottom.
Biology. No data.
Conservation status. EN; appears to be declining within its very small range.
Further reading. Turan et al. 2014b (description).

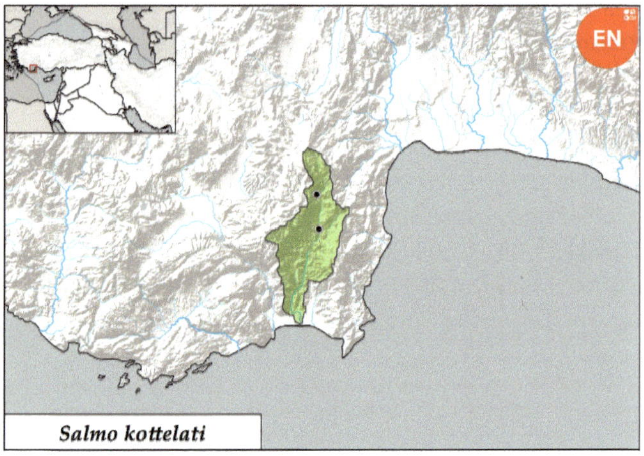

Salmo labecula; Ecemiş, Seyhan drainage, Türkiye; ~150 mm SL.

Salmo labecula
Common name. Eastern Mediterranean trout.
Diagnosis. Distinguished from other species of *Salmo* in Mediterranean basin by: ○ black spots present on flank in individuals of all size groups / ○ flank without red spots in individuals larger than about 70 mm SL / ○ black spots numerous (more than about 50), smaller than pupil, irregularly shaped, scattered on middle of flank, usually absent on back / ○ 11–12 parr marks, distinct up to about 150 mm SL in male, about 200 mm SL in female / ○ four broad, dark-brown or grey band on flank, absent or indistinct in individuals smaller than approximately 200 mm SL / ○ 2–10 black spots behind eye, 6–13 on opercle / ○ maxilla and supramaxilla narrow, depth of upper jaw 2.8–3.8 times in length of maxilla / ○ 109–115 lateral-line scales counted until end of hypural complex / ○ head length 25–27 % SL / ○ caudal peduncle depth 10–11 % SL / ○ 23–25 scale rows between dorsal origin and lateral line / ○ 16–17 scale rows

between anal origin and lateral line / ○ 14–15 scale rows between adipose origin and lateral line / ● 21–23 gill rakers / ○ head not dorso-ventrally flattened. Size up to 450 mm SL.
Distribution. Türkiye: Ecemiş, lower Zamantı and Çakıt in Seyhan drainage.
Habitat. Headwater streams usually springs-fed, with cold and clear water, moderate current, gravel, and pebble substrate. Occasionally also enters reservoirs.
Biology. Spawns December–January.
Conservation status. VU; found in more than 5 independent populations, thought to be in decline, mainly due to overfishing.
Further reading. Turan et al. 2012b (description); Küçük et al. 2024 (distribution).

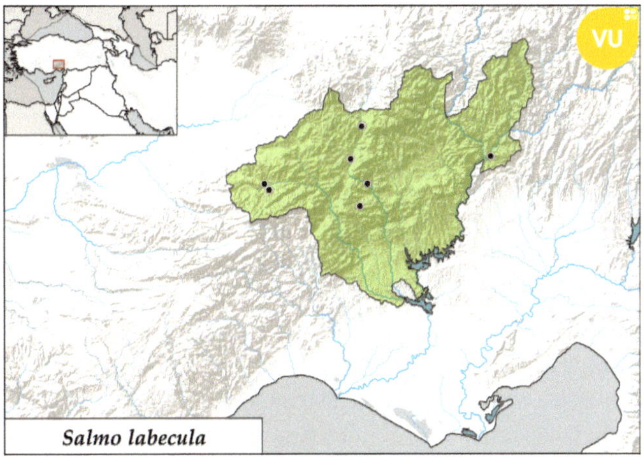

Salmo labrax; Çoruh drainage, Türkiye; ~130 mm SL. © M. Özuluğ.

Salmo labrax; Crimea, Ukraine; ~300 mm SL.

Salmo labrax

Common name. Black Sea trout.
Diagnosis. Distinguished from other species of *Salmo* in Marmara, Black Sea and Caspian basins by: ○ background colour of flank brownish to silvery in life (silvery in anadromous individuals) / ● black spots on flank ocellated, few and scattered on upper third of flank in most individuals, number not increasing with size and age / ○ red spots few or many, usually irregularly shaped and ocellated / ○ number of red spots on flank increasing by size in male / ● head about 0.9–1.2 times in body depth at dorsal origin / ○ maxilla short, reaching slightly beyond posterior eye margin in individuals larger than 150 mm SL, upper edge convex posteriorly ○ body depth at dorsal origin 23–27 % SL / ● dorsal in front of midpoint of body / ○ 110–121 lateral-line scales counted until end of hypural complex / ○ 27–32 scale rows between dorsal origin

and lateral line / ○ 15–23 scale rows between anal origin and lateral line / ○ 16–20 gill rakers. Size up to 1150 mm SL.
Distribution. Marmara and Black Sea basins.
Habitat. At sea, at depths of down to 50 m, usually along coasts and near estuaries. Migrates to mountain streams. Locally resident in lakes. Resident part of populations in streams with fast currents, cold, clear water, and stony or gravelly bottoms. Usually in lower and middle reaches of main rivers and streams. Also, in very short coastal streams (less than 5 km). Observed in Çoruh up to 340 km from sea. Usually spawns in middle reaches of rivers.
Biology. Anadromous, lacustrine, and resident forms, with most individuals remaining in freshwater. Smallest adult female recorded about 160 mm SL. Spawns October–January. Spends 2–4 years in rivers and streams, then smolts migrate to sea or matures in freshwater. Spends 2–4 years at sea. Some anadromous individuals return to rivers in April–May, where they spend summer, usually spawning migration starts in September–October, gregarious during migration. Other anadromous individuals begin to migrate shortly before spawning. Eggs hatch in 6–8 weeks. Parr and resident adults feed on a wide variety of aquatic and terrestrial invertebrates. Anadromous and large resident individuals feed mainly on fish and large crustaceans. Anadromous individuals feed in rivers.
Conservation status. LC.
Remarks. *Salmo labrax*, *S. coruhensis*, and *S. rizeensis* are often recognised as three species from the Black Sea basin. Genomic data suggest they did not differentiate earlier than during the last or pre-last glacial period. They occur in many sedentary populations, some of which are also occasionally or frequently anadromous. All are closely related by recent genomic data and microsatellite and cytochrome b mtDNA studies. Furthermore, the morphological differences described between the species are minor and overlapping. We follow the latest study on this species complex and treat *S. coruhensis* as a synonym of *S. labrax* but keep *S. rizeensis* as a valid species. That means *S. rizeensis* is not the resident form of *S. labrax*, but resident populations of two species occur in sympatry.
Further reading. Berg 1949b (biology); Salmanov & Dorofeyeva 2001 (morphology); Bernatchez 2001 (genetics); Turan et al. 2010 (description of *S. coruhensis*); Ninua et al. 2018 (Cytochrome b data); Hashemzadeh Segherloo et al. 2021 (genomics); Ninua et al. 2023 (synonymy of *S. coruhensis*, microsatellite data).

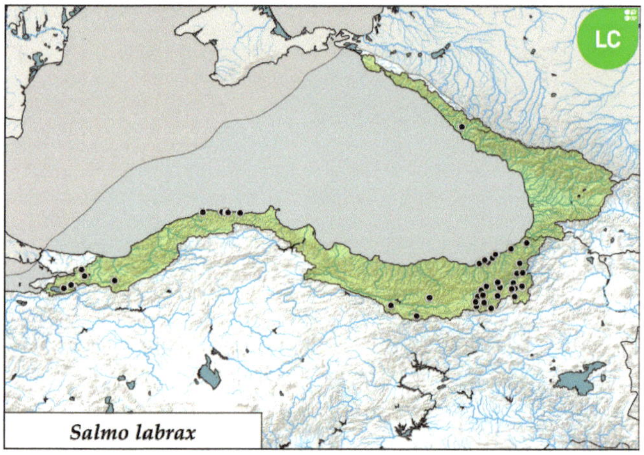

Many *Salmo* species inhabit spring-feed streams and rivers such as the upper Köprüçay.

Salmo munzuricus; aquaculture, Munzur, Euphrates drainage, Türkiye; male about 400 mm SL.

Salmo munzuricus; Murat, Euphrates drainage, Türkiye; male, 205 mm SL.

Salmo munzuricus

Common name. Munzur trout.

Diagnosis. Distinguished from other species of *Salmo* in Euphrates and Tigris drainages by: • adipose very large, length about equal to length of anal base in adults, as large as dorsal and anal in a 400 mm SL male / ○ flank with many irregularly shaped black and red spots / ○ flank brownish / ○ black spots mostly along middle of flank, absent from back / ○ number of black spots increase with size / ○ adipose usually with very narrow white margin, red submargin followed by a white band or spots, an and inner red band in male / ○ head length 24–27 % SL / ○ maxilla length 8–10 % SL / ○ length of mouth gape 10–13 % SL / ○ 116–123 scales along lateral line counted until end of hypural complex / ○ 15–17 scales between lateral line and adipose / 28–30 scales rows between lateral line and dorsal origin. Size up to 400 mm SL.

Distribution. Türkiye: Munzur and Murat in upper Euphrates drainage. Also, in Keban Reservoir and a tributary south of Keban drainage.

Habitat. Headwater streams with cold, clear water, moderate current, gravel, and pebble substrate. Inhabits reservoirs.

Biology. No data.

Conservation status. LC.

Further reading. Turan et al. 2017e (distribution); Turan et al. 2020 (Cytochrome b phylogeny); Turan et al. 2021b (distribution).

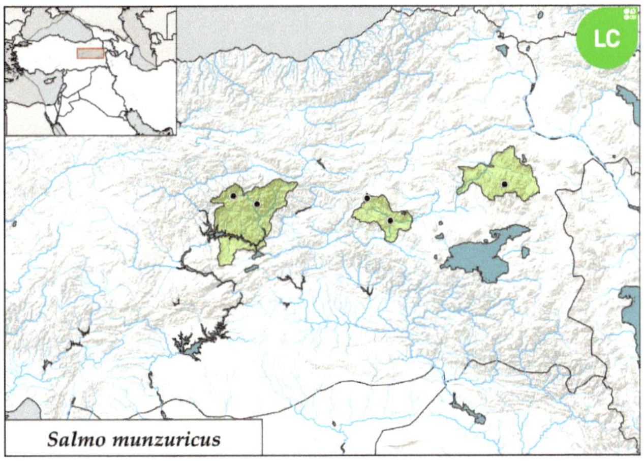

Salmo murathani; Aras drainage, Türkiye; female, ~210 mm SL. © M. Özuluğ.

Salmo murathani; Aras drainage, Türkiye; male, ~180 mm SL. © M. Özuluğ.

Salmo murathani

Common name. Sarıkamış trout.

Diagnosis. Distinguished from other species of *Salmo* in Caspian basin by: ○ background colour greyish on back and upper part of flank, yellowish on lower part of flank and belly in life / ○ black spots numerous in male, and in female larger than 230 mm SL, irregularly shaped, small to medium (smaller than pupil), ocellated, surrounded by a large white ring, scattered on back, middle and upper part of flank, and anterior part of lower flank / ○ number of black spots increasing with increasing size / ○ adipose of moderate size, not reaching caudal base / ○ 114–122 lateral-line scales counted until end of hypural complex / ○ 12–14 parr marks, distinct up to about 200 mm SL, vertically elongate on anterior, roundish on posterior part of flank / ○ four dark bands on flank in female larger than 230 mm SL. Size up to 255 mm SL.

Distribution. Türkiye: Lake Çıldır, upper Aras, Keklik, Kızılçubuk, Arpaçay, and Maksutçuk in Aras drainage.

Habitat. Clear and relatively slow-flowing water with stony and pebbly bottom.

Biology. No data.

Conservation status. LC.

Remarks. This species is most likely conspecific with *S. caspius*; see the discussion there.

Further reading. Turan et al. 2022b (description).

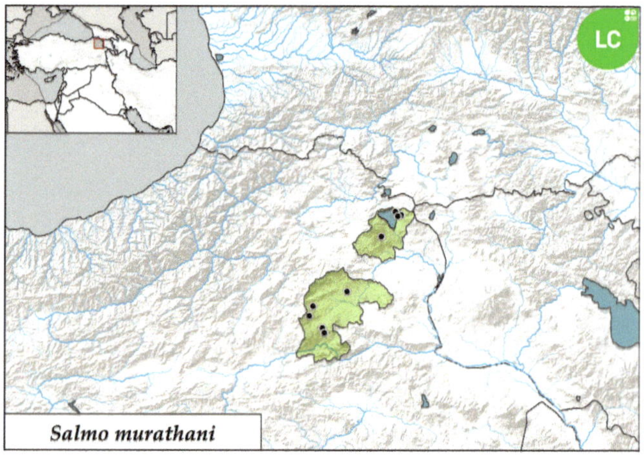

Salmo okumusi; Gökpınar, Euphrates drainage, Türkiye; ~140 mm SL.

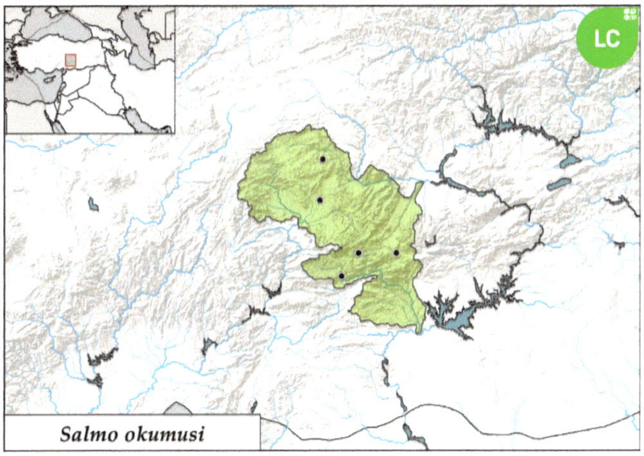

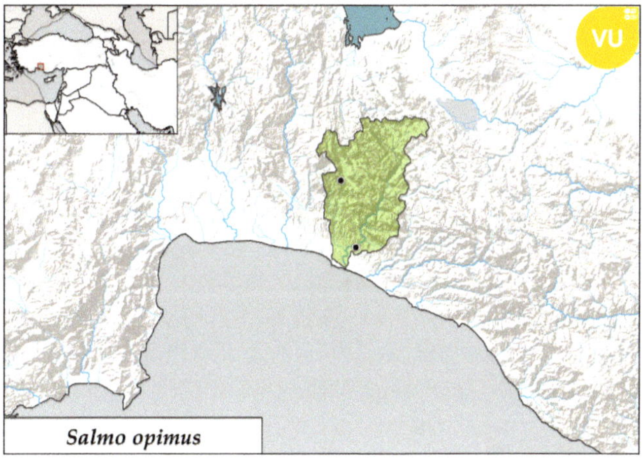

Salmo okumusi

Common name. Western Euphrates trout.

Diagnosis. Distinguished from other species of *Salmo* in Euphrates and Tigris drainages by: • flank and back covered by numerous irregularly shaped small red spots and dots, imparting a peppered or slightly marmorate pattern / • 5–18 black spots on top of head / ○ four broad dark-brown or grey bars on flank, very faintly marked or indistinct in small individuals but distinct in individuals larger than about 230 mm SL / ○ flank greyish / ○ black and red spots on flank irregular shaped / ○ red spots on flank smaller than pupil / ○ 1–22 small black spots in postorbital and suborbital area, smaller than pupil in individuals larger than 120 mm SL / ○ 8–17 small black spots on opercle, smaller than pupil in individuals larger than 120 mm SL / ○ adipose grey, with three or four red to orange spots on its free edge / ○ 10–11 parr marks, distinct up to about 210 mm SL / ○ 103–112 lateral-line scales counted until end of hypural complex / • 26–28 scale rows between lateral line and dorsal origin. Size up to 280 mm SL.

Distribution. Türkiye: Göksu, Lake Gökpınar basin (in Tohma drainage), and Sürgü in Euphrates drainage.

Habitat. Headwater streams with cold, clear water, moderate current, gravelly and pebbly bottom.

Biology. Spawns in November.

Conservation status. LC.

Further reading. Turan et al. 2014b (description); Turan et al. 2020 (Cytochrome b phylogeny).

Salmo opimus

Common name. Alara trout.

Diagnosis. Distinguished from other species of *Salmo* in Mediterranean basin by: • predorsal area slightly convex in individuals larger 170 mm SL / ○ four broad, dark-brown or grey bands on flank / ○ 10–12 parr marks distinct even in

Salmo opimus; Alara drainage, Türkiye; 270 mm TL. © J. Schöffmann.

the largest size / ○ one black spot behind eye / ○ 0–5 black spots on opercle, usually 2–3 (smaller than pupil) / ○ 0–50 black spots on flank, smaller than pupil, irregularly shaped, located mostly on anterior, upper part of flank, also on back and below lateral line in some individuals / ○ red spots few, small, irregularly shaped / ○ a large (lager than eye pupil) white ring around red spots / ○ red spots present on flank in individuals of all size groups / ○ maxilla short, maxilla, and lower lip not fleshly / ○ maxilla and supramaxilla wide (depth of upper jaw 2.6–3.2 times in length of maxilla) / ○ adipose larger in male than that in female / ○ 112–120 lateral-line scales counted until end of hypural complex / ○ 24–26 scale rows between dorsal origin and lateral line / ○ 16–18 scale rows between anal origin and lateral line / ○ 13–15 scale rows between adipose origin and lateral line / ○ 22–23 gill rakers. Size up to 270 mm SL.

Distribution. Türkiye: Alara, a coastal Mediterranean drainage.

Habitat. Headwater streams, usually spring-fed, with cold, clear water and moderate currents, with gravel and pebble beds.

Biology. No data.

Conservation status. VU; believed to decline due to many threats in its small range.

Remarks. Was thought to be widespread in Ceyhan, but all these populations have recently been identified as *S. chilo*.

Further reading. Alp et al. 2003 (spawning time); Turan et al. 2012b (description).

Salmo platycephalus; Zamanti, Seyhan drainage, Türkiye; 240 mm SL.

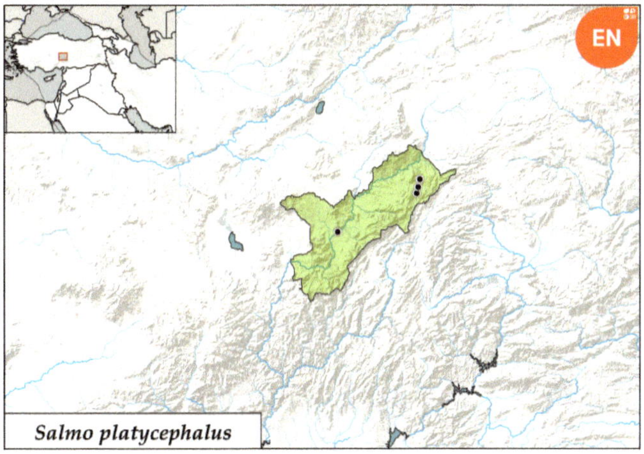

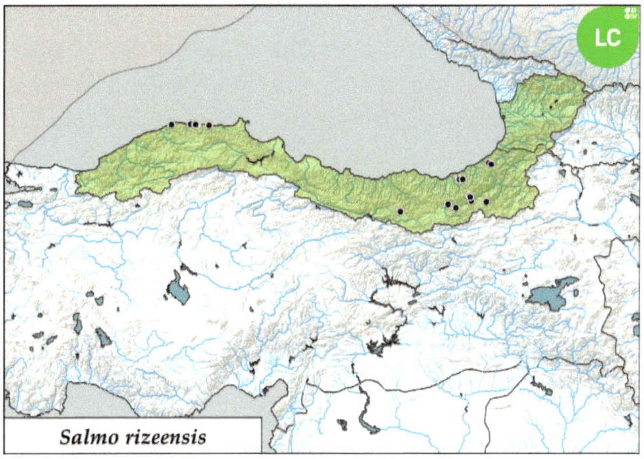

Salmo platycephalus
Common name. Flathead trout.
Diagnosis. Distinguished from other species of *Salmo* in Mediterranean basin by: • head dorso-ventrally flattened / • flank marbled or mottled without red spot in individuals larger than 70 mm SL, without black spots in individuals larger than about 200 mm SL / • 12–13 parr marks up to about 70 mm SL / ○ lower jaw enclosed laterally by fleshy maxilla / ○ four broad dark-brown or grey bands on flank, indistinct in individuals smaller than about 170 mm SL / ○ one black spots behind eye / ○ 7–16 very indistinct black spots on opercle / ○ maxilla short, narrow anterior part short / ○ maxilla and supramaxilla wide (depth of upper jaw 2.2–2.7 times in length of maxilla) / ○ 110–116 lateral-line scales counted until end of hypural complex / ○ 23–25 scale rows between dorsal origin and lateral line / ○ 17–18 scale rows between anal origin and lateral line / ○ 13–15 scale rows between adipose origin and lateral line / • 23–25 gill rakers. Size up to 600 mm SL.
Habitat. Clear and moderately fast-flowing streams with rocky and pebbly bottoms.
Biology. Spawns October–middle December.
Distribution. Türkiye: Soğuksu, Uzunyayla, and Karagöz in upper Zamantı drainage.
Conservation status. EN; believed to decline due to many threats in its small range.
Further reading. Behnke 1968 (description); Sušnik et al. 2004 (phylogenetic position); Kara et al. 2011 (morphology, biology); Turan et al. 2012b (diagnosis); Hashemzadeh Segherloo et al. 2021 (genomics).

Salmo rizeensis
Common name. Pontic brook trout.
Diagnosis. Distinguished from other species of *Salmo* in Marmara and Black Sea basins by: ○ flank brownish in life / ○ black spots on flank usually smaller than pupil, ocellated, restricted to back and upper part of flank, absent in front of dorsal / ○ few black spots, number decreasing with size / ○ few red spots, small, ocellated, usually organised in three or four irregular rows along middle and lower part of flank / ○ number and size of red spots not changing with size /

Salmo rizeensis; Çoruh drainage, Türkiye; ~150 mm SL. © M. Özuluğ.

• maxilla long, reaching beyond eye in adults and juveniles / ○ upper edge of maxilla straight or slightly convex posteriorly in male / ○ head 1.2–1.4 times in body depth at dorsal origin in male / • adipose slender, upper edge straight or very slightly convex anteriorly / ○ 114–120 lateral-line scales counted until end of hypural complex / ○ 26–30 scale rows between dorsal origin and lateral line / ○ 18–21 scale rows between anal origin and lateral line / ○ 14–17 scale rows between adipose origin and lateral line / ○ 16–19 gill rakers. Size up to 250 mm SL.

Distribution. Black Sea basin from Georgia to Sakarya in Anatolia.

Habitat. Streams with cold, clear, fast-flowing water, and a substrate of rocks, stones, and pebbles.

Biology. Spawns mid-September–mid-October. Resident and not undertaking long spawning migrations. Smallest spawning female observed about 110 mm SL. Smallest male with mature gonads about 90 mm SL, usually 140–170 mm SL. Do not appear to feed during winter.

Conservation status. LC.

Remarks. See discussion in chapter about *S. labrax*.

Further reading. Turan et al. 2010 (description); Ninua et al. 2018 (Cytochrome b data); Turan et al. 2020 (Cytochrome b phylogeny); Hashemzadeh Segherloo et al. 2021 (genomics).

Salmo tigridis; upper Tigris drainage, Türkiye; 210 mm SL.

Salmo tigridis

Common name. Tigris trout.

Diagnosis. Distinguished from other species of *Salmo* in Euphrates and Tigris by: • flank greenish in life / ○ few black spots on flank, smaller than pupil, ocellated / ○ few red spots on flank, usually in 1–3 along lateral midline, smaller than pupil, usually irregularly shaped and ocellated / • adipose without red or black spots / ○ maxilla short, reaching posterior eye-margin or slightly beyond in individuals larger than 200 mm SL / ○ head 1.0–1.1 times in body depth at dorsal origin in both sexes / ○ 108–116 lateral-line scales counted until end of hypural complex / • 32–35 scale rows between dorsal origin and lateral line / • 19–20 scale rows between adipose origin and lateral line / ○ 22–26 scale rows between anal origin and lateral line / ○ 17–19 gill rakers. Size up to 400 mm SL.

Distribution. Türkiye: Çatak and Müküs in upper Tigris drainage south of Lake Van.

Habitat. Headwater streams with cold, clear water, moderate current, gravel, and pebble bottom.

Biology. No data.

Conservation status. EN; appears to be declining within its very small range.

Further reading. Turan et al. 2011 (description); Turan et al. 2020 (Cytochrome b phylogeny).

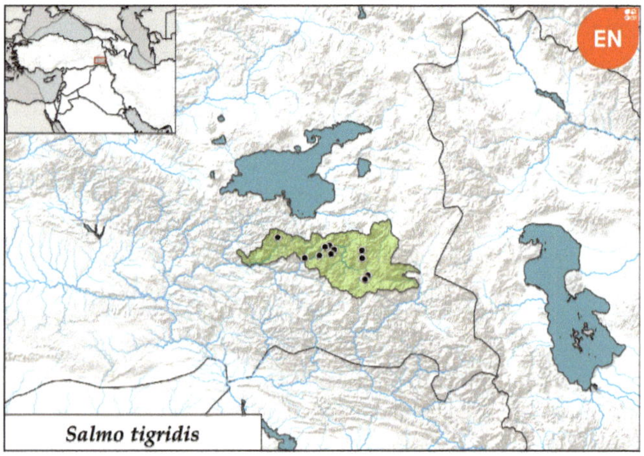

Salmo trutta; Sapur, Lake Van basin, Türkiye; male, 258 mm SL.

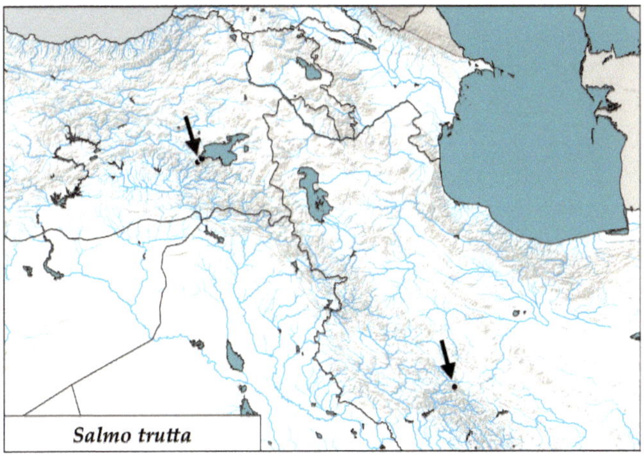

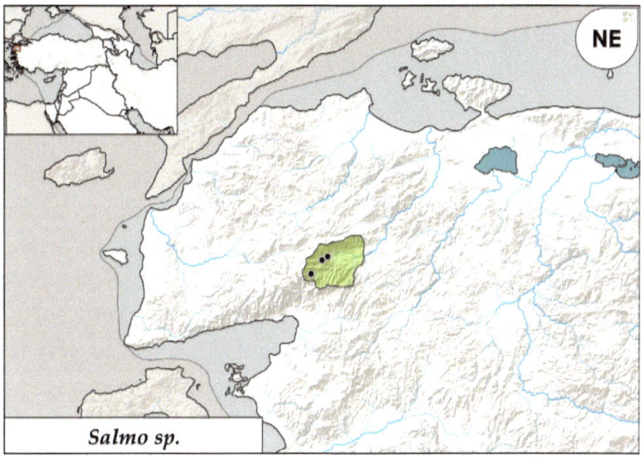

Salmo trutta

Common name. Atlantic trout.

Diagnosis. Distinguished from other species of *Salmo* in Euphrates and Tigris (based on materials from Türkiye) by: ○ 15–17 scale rows between lateral line and end of adipose base / ○ 28–33 scale rows between lateral line and dorsal origin / ○ lateral line with 117–125 scales counted until end of hypural complex / ○ adipose depth 8–10 % SL in male / ○ black and red spots on flank roundish / ○ numerous black spots on back, middle and upper part of flank / ○ number of black spots increasing with size / ○ flank brownish in life. Size up to 900 mm SL in native range. Anadromous and lacustrine individuals usually 450–600 mm SL, resident individuals usually 200–300 mm SL.

Distribution. Türkiye: introduced into Arpet (a tributary of Botan in upper Tigris) and Sapur (a small tributary of Lake Van), but it appears to have become vanished in these areas due to poaching. Iran: introduced and established in Lake Gahar in Karun drainage. Native to Atlantic, North, White, and Baltic basins, from Spain to Chosha Bay (Russia). Present in Iceland and northernmost rivers of Great Britain and Scandinavia. In Rhône drainage, native only to Lake Geneva basin, which it entered after last glaciation. Native to upper Danube and Volga drainages. Introduced throughout Europe, North and South America, southern and montane eastern Africa, Pakistan, India, Nepal, Japan, New Zealand, and Australia.

Habitat. Cold streams, rivers, and lakes. Spawns in fast-flowing rivers and streams. Lacustrine populations migrate to tributaries and lake outlets, rarely spawning on stony, wave-washed lake shores. Spawning sites are usually characterised by downward movement of water into gravel. Sea and lake trout forage in pelagic and littoral habitats; sea trout usually close to coast, not far from mouth of natal river.

Biology. Unstudied in Anatolian populations.

Conservation status. Non-native.

Further reading. Elliot 1994 (ecology); Bernatchez 2001 (phylogeography); Klemetsen et al. 2003 (life history); Kaya 2020c (record from Tigris and Lake Van drainages); Guinand et al. 2021 (multi-species opinion piece on *Salmo* taxonomy); Hashemzadeh Segherloo et al. 2021 (genomic data).

Salmo sp., Gönen drainage, Türkiye; 220 mm SL.

***Salmo* sp.**

Common name. Marmara trout.

Diagnosis. Distinguished from other species of *Salmo* in Marmara and Black Sea basins by: ○ caudal peduncle depth in male 8–10 % SL / ○ size of adipose equal in male and female / ○ numerous black spots on back and flank in adult female / ○ red spots present on the body in individuals larger than 200 mm SL / ○ number of red spots does not increase with increasing size in male / ● 109–121 lateral-line scales counted until end of hypural complex size up to 287 mm SL.

Distribution. Türkiye: Ayazma, a headwater stream in Karamenderes drainage and upper tributaries of Gönen on Biga Peninsula.

Habitat. Headwater streams with cold, clear water, moderate current, gravelly and pebbly bottom.

Biology. No data.

Conservation status. NE.

Remarks. This species was described as *S. duhani* in 2021, but the name is a *nomen nudum* and is not valid from the description. We suspect that it will be redescribed soon.

Further reading. Turan & Bayçelebi 2020 (record as *S. pelagonicus*); Turan & Aksu 2021 (invalid description as *S. duhani*).

Salvelinus fontinalis; aquaculture, Germany; ~300 mm SL. © A. Hartl.

Salvelinus fontinalis

Common name. Brook charr.

Diagnosis. Distinguished from species of *Salmo* and *Oncorhynchus* in West Asia by: ● lateral-line scales very elliptical, reduced to little more than nerve tube, smaller than scales in adjacent rows, with no or little overlap with scales in front or behind / ● back, dorsal and caudal with pale-brown or grey spots or vermiculated / ○ pectoral, pelvic and anal with white margin followed by a black submarginal line / ○ flank with pink or red spots, some bluish / ○ juvenile with 8–10 parr marks. Size up to about 500 mm SL.

Distribution. Very locally stocked in Caucasus and Türkiye. Unclear if established. Native to North America from Hudson Bay basin and northeastern Canada south to upper Mississippi, east to coast of New Jersey and northern Georgia. Widespread in North and South America, Europe, Asia and southern Africa. Found throughout Europe, especially in high-altitude streams and lakes, but rarely confirmed as established. Established populations documented from high-altitude lakes and streams in Corsica, Italy, Czech Republic, southern Germany (list probably not exhaustive).

Habitat. Cold and moderately to rapidly flowing mountain streams. In North America, northern populations also in lakes and rivers. Spawns in gravel in fast-flowing streams or in lakes. Occasionally established in nutrient-poor lakes, high mountains or acidic streams.

Biology. Lives up to 15 years. First spawns at 1–2 years in central and southern Europe, 3–4 years in northern Europe. Spawns in October–November, earlier in north, usually at 3–10°C. Females build redds. Eggs hatch in spring. Only non-anadromous populations in Europe. Anadromous in North America. Feeds on aquatic invertebrates and small fish. Territorial, of equal size, dominated by *Salmo trutta*, whose presence limits its distribution. Artificial hybrids with *Salmo trutta* (tiger trout) are occasionally stocked.

Conservation status. Non-native; introduced for recreational fisheries.

Further reading. Scott & Crossman 1973 (biology); Jenkins & Burkhead 1993 (biology).

A drainage of İyidere in eastern Black Sea basin is a typical cold trout stream into which anadromous *S. labrax* migrated to spawn before the construction of dams.

Uzunyala in upper Zamantı drainage, Türkiye, the habitat of *Salmo platycephalus*. © J. Schöffmann.

Family Lotidae

Burbots

A small family of mostly marine fishes with three genera and approximately 5 species within the diverse order Gadiformes. They are widely distributed, mostly in cold seas of the Arctic, Atlantic, and Pacific Oceans. *Lota* is the only freshwater member of the family, distinguished from marine genera in West Asia by the presence of a single central barbel on the lower jaw, two dorsals, and one anal with 64–84½ total rays. Additionally, the body is marbled dark-brown and lacks a prominent nasal barbel. A single individual of *Lota lota* was caught in 1920 in the Kura in Azerbaijan, and another record was made in 1921 from the lower Sefid in Iran. At the time, there may have been a small population in the region, but the two individuals were likely the result of accidental migration from the northern Caspian Sea. No populations appear in West Asia, and the species was never native or is now extinct in the region. **Further reading.** Coad 2016b (distribution in Iran).

Lota lota; lower Odra, Germany; ~300 mm SL.

https://doi.org/10.1515/9783111677811-032

Order Gobiiformes

Gobies

Gobiiformes include gobies, as well as cardinal fishes (Apogonidae) and nursery fishes (Kurtidae). Gobies are commonly small benthic fishes, forming one of the most speciose groups among Teleostei. There are approximately 270 genera, with more than 2400 species. These can be found in all aquatic habitats, especially in coral reefs. Gobies are distinguished from all other freshwater fishes in West Asia by the fusion of the pelvics, which form a suction organ with an anterior transverse membrane. They also have two dorsals, and the pelvic is located beneath the base of the pectoral.

Most gobies have a short lifespan, with an average of less than few years. However, the biology of most species in West Asia still needs to be studied. Male gobies engage in parental care, with eggs suspended from the roof of small cavities beneath stones, shells, leaves, or any other hard structure. The male may mate with multiple females and guards the eggs until the larvae hatch and leave the nest. Many gobies exhibit sexual dimorphism, with male exhibiting larger fins and more intense colouration, often black in *Neogobius* and *Ponticola* species.

Previously, Gobiiform systematics was based on a limited number of morphological characters. However, recent molecular studies have confirmed the majority of these classifications. Currently, nine families of gobies are recognised: Butidae, Eleotridae, Gobiidae, Milyeringidae, Odontobutidae, Oxudercidae, Rhyacichthyidae, Thalasseleotrididae, and Xenisthmidae. In freshwater habitats of West Asia, two families occur: Gobiidae and Oxudercidae. There is no consistent morphological character distinguishing Gobiidae and Oxudercidae; both are identified as families only by molecular characters. In West Asia, genera as disparate as sand gobies (*Knipowitschia*) and mudskippers (*Boleophthalmus*, *Periophthalmus*, *Scartelaos*) belong to Oxudercidae. In contrast, all the Ponto-Caspian gobies (*Neogobius* and *Ponticola*) and some genera from the Indian Ocean belong to Gobiidae. As both families are not diagnosed by external characteristics, the keys for the genera of gobies are designed to facilitate the identification of all freshwater gobies in West Asia.

The Odontobutid goby *Perccottus glenii* is a highly invasive species in Europe. It occurs as a weed in carp farms and is introduced with non-biosafe carp. Given the lack of regulation of fish stocking in West Asia, the introduction of *Perccottus glenii* can be expected anytime. Two eleotrid gobies (*Eleotris acanthopomus* and *Ophiocara porocephala*) have been identified in brackish inland waters of Oman. As they do not qualify as freshwater fishes, they are not included in this book, as are species such as *Bathygobius fuscus* and *Acentrogobius dayi*, which are included in the key but otherwise excluded from the book because they occur only in brackish waters in ther Persian Gulf. **Further reading.** Nelson et al. 2016 (diversity); Near & Thacker 2024 (phylogeny).

Perccottus glenii; upper Volga, Russia; ~70 mm SL.

https://doi.org/10.1515/9783111677811-033

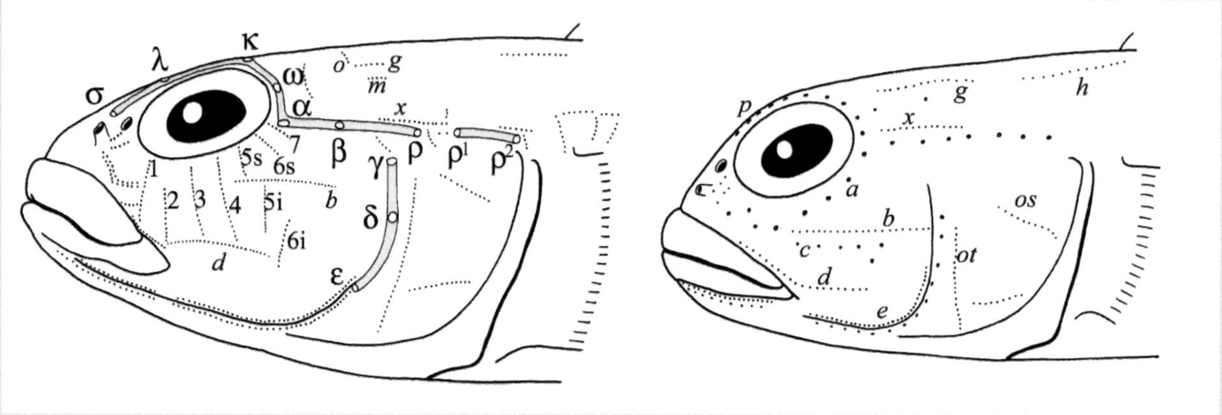

Figure 59. Cephalic lateral line system in gobies (from Kottelat & Freyhof 2007; after Miller in Whitehead et al. 1986).

Cephalic pores and papillae in Gobiidae. The lateral line system's canals, pores, and sensory papillae provide important diagnostic characters to distinguish goby genera and species. The figure shows the terminology and notation system used. Canals are stippled, and pores are indicated by Greek letters. It should be noted that there is no lateral line canal on the flank and that the canals may be variously developed on the head. In the species with the most developed canals, there is an anterior oculoscapular canal (from pores 's' to 'r'), a posterior oculoscapular canal (from pores 'r1' to 'r2') and a preopercular canal (from pores 'g' to 'e'). One or more of these canals may be absent or interrupted, and the posterior and anterior oculoscapular canals may be continuous or discontinuous. The anterior oculoscapular canals of both sides of the head may be separate or fused in the interorbital area. The anterior oculoscapular canal may be divided into a circumorbital and postorbital segment. The papillae are organised in vertical rows (numbered in Fig. 59, left) or longitudinal rows (lettered in Fig. 59, right). To examine the pores and papillae, a dissecting microscope is necessary. They are best viewed using oblique light. On occasion, they may be hidden by mucus. Some authors have attempted to highlight the canals by injecting chromic acid, ferric tannate, or potassium permanganate, but all these methods damage the specimens. Air pulsed by a small aquarium pump through an injection needle is a technique now used by goby specialists. Despite this, goby identification remains challenging even for experts. **Further reading.** Sanzo 1911; Miller, in Whitehead et al. 1986 (head canals, pores, and papillae).

Key to genera of gobies in freshwaters in the Black, Caspian, Marmara and Mediterranean basins

1a - 5–13½ branched rays in second dorsal; male with bluish-iridescent blotches on membrane of first dorsal.
..................2

1b - 14–20½ branched rays in second dorsal; male without bluish-iridescent blotches on membrane of first dorsal.
..................3

2a - Male with bluish-iridescent, small blotch on membrane between last and pre-last ray of first dorsal; male without white margins of dorsal, caudal, or anal.
..................*Knipowitschia*

2b - Male with bluish-iridescent, small blotches on 1–2nd dorsal membrane of first dorsal; male with a white margin of dorsal, caudal and anal.
..................*Rhinogobius*

3a - Anterior naris at end of a tube projecting forward beyond lip; 36–48 total scales in midlateral series.
..................*Proterorhinus*

3b - Anterior naris not projecting forward beyond lip; 49–85 total scales in midlateral series.
..................4

4a - 8–10 vertical rows of papillae in suborbital area (Fig. 60b); 72–85 total scales in midlateral series; nape naked.
..................*Mesogobius*
4b - 7 vertical rows of papillae in suborbital area (Fig. 60a); 49–74 total scales in midlateral series; nape covered with scales, if nape naked, then 54–65 scales in midlateral series.
..................5

5a - First branched ray of second dorsal about twice as long as penultimate ray; first dorsal with large black spot in posterior part (in *N. melanostomus* and *N. pallasi*).
..................*Neogobius*
5b - First branched ray of second dorsal about as long as penultimate ray; first dorsal without large black spot in posterior part.
..................6

6a - Midline of nape naked in front of preoperculum (Fig. 61); diagonal bars on body irregular in position and shape.
..................*Babka*
6b - Predorsal area and nape completely covered by scales; no bars on body.
..................*Ponticola*

Key to species of freshwater gobies of the Persian Gulf basin and the Arabian Peninsula

1a - Usually more than 1 row of teeth in lower jaw; eye not strongly elevated from head profile.
..................2
1b - Usually only one row of teeth in lower jaw; eye strongly elevated from head profile.
..................6

2a - Upper lip strongly projecting beyond lower lip.
..................*Awaous jayakari*
2b - Upper lip not or very slightly projecting beyond lower lip.
..................3

3a - Lower lip clearly projecting beyond upper lip; caudal spade-shaped; flank with 5 blotches along lateral midline; cheek with several rows of pores.
..................*Glossogobius*
3b - Lower lip not or very slightly projecting beyond upper lip; caudal roundish; flank mottled or marbled, or with 6–8 irregularly shaped bars or blotches; cheek without rows of pores.
..................4

4a - Male with bluish-iridescent, small blotches on 1–2nd dorsal membrane of first dorsal; orange spot as gill-membrane in male; 5–6 rays in first dorsal.
..................*Rhinogobius* sp.
4b - Male without bluish-iridescent, small blotches on 1–2nd dorsal membrane; no orange spot as gill-membrane in male; 7 rays in first dorsal.
..................5

5a - First ray of first dorsal longer than following rays; 9½ rays in second dorsal; 8½ rays in anal.
..................*Bathygobius fuscus*
5b - First ray of first dorsal shorter than second ray; 10½ rays in second dorsal; 9½ rays in anal.
..................*Acentrogobius dayi*

6a - No canine teeth behind lower jaw symphysis.
..................*Periophthalmus*
6b - One large canine tooth behind jaw symphysis and behind row of teeth in lower jaw.
..................7

7a - Barbels present on ventral side of head.
..................*Scartelaos*
7b - No barbels on ventral side of head.
..................*Boleophthalmus*

The Caspian Sea, here in Azerbaijan, is the habitat of a rich fauna of gobies, most of which do not enter freshwater.

Gobius ophiocephalus; Krk Island, Croatia; ~65 mm SL.

Family Gobiidae

True gobies

True freshwater gobies occur in all sea basins in West Asia except the Mediterranean. In the Black and Caspian Seas, the Ponto-Caspian gobies of the genera *Babka*, *Neogobius*, *Ponticola,* and *Proterorhinus*, as well as the tadpole gobies of the genera *Anatirostrum*, *Benthophilus*, *Benthophiloides*, and *Caspiosoma*, are represented. However, in West Asia, tadpole gobies are not reported to enter freshwaters, unlike in Europe. They are speciose and very diverse in the seas. Only *Benthophilus leobergius* is occasionally reported from the Caspian Sea near river estuaries. Therefore, tadpole gobies are excluded from the coverage of this book. Additionally, other gobies well-known from the European Black Sea coast, including *Ponticola eurycephalus* and *P. ratan* appear absent from Anatolia's coasts. In addition to the Ponto-Caspian gobies listed below, *Gobius cobitis, G. niger*, *G. ophiocephalus* and some additional *Ponticola* species are occasionally recorded as occurring in freshwater. However, we are unaware of documented records of their permanent and regular occurrence. Therefore, they are briefly discussed in the generic chapters below.

In the Persian Gulf and Sea of Oman, additional true gobies are known from inland waters, while only *Glossogobius* regularly enter pure freshwaters. *Bathygobius fuscus*, *Acentrogobius dayi,* and others are occasionally reported from saline inland waters but not from freshwater habitats. *Bathygobius* and *Acentrogobius* are included in the key to ease their identification, but as these species are not considered freshwater fishes, they are excluded otherwise. Indeed, more marine gobies (as *Cryptocentroides arabicus* and *Favonigobius melanobranchus*) might occur in the lower reaches of the rivers, as some have brackish waters. Identification literature for marine fishes should be considered to identify these.

Further reading. Nelson et al. 2016 (diversity).

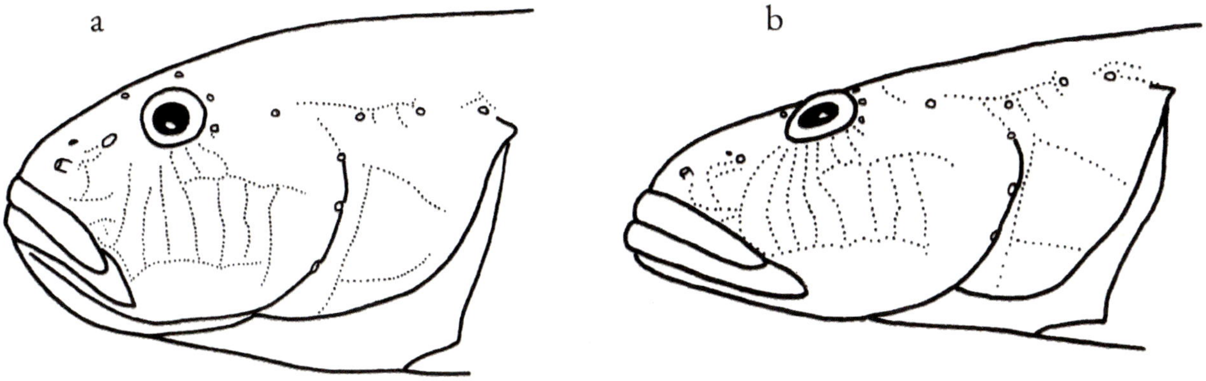

Figure 60. Rows of suborbital transverse papillae. **a**, *Ponticola syrman*; **b**, *Mesogobius batrachocephalus* (from Kottelat & Freyhof 2007; after Miller in Whitehead et al. 1986).

Babka gymnotrachelus; Danube delta, Romania; female, ~70 mm SL.

https://doi.org/10.1515/9783111677811-034

Babka gymnotrachelus

Common name. Racer goby.

Diagnosis. Distinguished from species of *Neogobius* and *Ponticola* entering freshwater in Caspian and Black Sea basins by: • midline of nape naked in front of preoperculum / • diagonal bars on body irregular in position and shape / ○ pelvic-disc fraenum with small, rounded lobes whose length is less than $\frac{1}{6}$ of fraenum width at base (Fig. 62) / ○ 54–62+2–3 scales in midlateral series / ○ first branched ray of second dorsal about as long as penultimate ray / ○ first dorsal without black spot in posterior part. Size up to 160 mm SL.

Distribution. Black, Azov, and Caspian Sea basins. Since late 1990s, invasive in rivers of northern Black Sea and Baltic basins. No reports of invasive behaviour in West Asia.

Habitat. Mostly in fresh and brackish water with low salinity (<2 ‰). Estuaries, lagoons, and lakes, large rivers to small fast-flowing streams, on sandy or muddy bottoms. Mainly found in well-vegetated habitats. Common in backwaters and channels.

Biology. Lives up to 4–5 years. Spawns first time at 2 years. Spawns April–June, occasionally until mid-August. Female may spawn repeatedly during a season. Usually spawns for only one season. Adhesive eggs laid on rocks, shells, and aquatic vegetation. Male guard eggs until hatching. Feeds on a wide variety of invertebrates, especially molluscs.

Conservation status. LC.

Remarks. Distribution in West Asia poorly understood. *Babka macrophthalma* from the Caspian basin is a synonym.

Further reading. Ahnelt 2001 (description, invasion); Pinchuk et al. 2003b (biology, description).

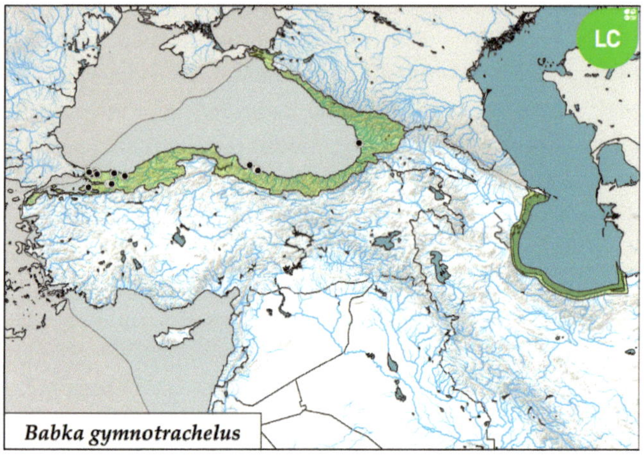

Babka gymnotrachelus; Danube delta, Romania; male, ~90 mm SL.

Glossogobius

The genus comprises approximately 42 species primarily found in coastal habitats and freshwaters. Most species are found in freshwater habitats, with many exhibiting an amphidromous lifestyle. However, there are also examples of species exclusively found in freshwater, including those from Madagascar, New Guinea, the Philippines, and Sulawesi. Notably, several species are endemic to the Malili Lakes in Sulawesi. A number of the Malili species are pelagic, exhibiting an enlarged swim bladder. *Glossogobius ankaranensis* is a troglomorphic, cave-dwelling species native to Madagascar. *Glossogobius* are often common and abundant in the lower reaches of rivers and streams. They play an important role as predators in their ecosystems. Three species have been recorded from West Asia. Still, their distribution is poorly known, and more species are likely to occur, particularly along the coast of Oman and Yemen, including Socotra. A comprehensive review of the diversity of these gobies is required. *Glossogobius* are important in coastal fisheries due to their relatively large size and the palatability of their flesh, which lacks bones. These fish are consumed locally and exported fresh and frozen, even to Europe and West Asia. **Further reading.** Hoese & Allen 2015; Hoese et al. 2015 (diversity); Proudlove 1997 (cave species).

Key to species of *Glossogobius* in freshwaters in West Asia

1a - Middle rays in first dorsal prolonged; sensory papilla line 6 absent or very short (often only a few papillae long); each cheek line with 1 row of sensory papillae; 12–16 predorsal scales; 1+1+9–11, usually 1+1+10, gill rakers.
..................*Glossogobius tenuiformis*

1b - Middle rays in first dorsal not prolonged; sensory papilla line row 6 distinct and long; each cheek line with 2–3 (rarely more) rows of sensory papillae; 15–24 pre-dorsal scales; 1–2+1+6–9, usually 1–2+1+7–8 gill rakers.
..................2

2a - Papillae line 5 below eye and suborbital papillae line 7 doubled.
..................*Glossogobius laticeps*

2b - Papillae line 5 below eye and suborbital papillae line 7 single.
..................*Glossogobius giuris*

Glossogobius giuris; Karnataka, India; ~150 mm SL.

Glossogobius giuris

Common name. Bareye goby.

Diagnosis. Distinguished from other *Glossogobius* in Iran by: ○ middle rays in first dorsal not prolonged / ○ sensory papilla row 6 distinct and long / ○ caudal roundish / ○ upper part of opercle with cycloid scales / ● each cheek line with 2–3 (rarely more) rows of sensory papillae / ● papillae line 5 below eye and suborbital papillae line 7 single / ○ 18–28 predorsal scales / ○ 1–2+1+6–9, usually 1–2+1+7–8 gill rakers. Size up to 240 mm SL.

Distribution. West Asia: Sarbaz drainage, in Iranian Makran most likely west to Bandar-e Jask. Possibly also along coast of Oman and Yemen. From Iran along coasts of Pakistan and India to Japan, south to Australia, widespread in Oceania. In Red Sea, along east coast of Africa, including Madagascar.

Habitat. Lower part of rivers and coastal lagoons, usually on soft bottoms. Most common in brackish water. Less common in fast-flowing waters on gravel or rocky bottoms. Often found in mangroves in sea. Migrates up to 300 km upstream.

Biology. Facultatively amphidromous can be landlocked in reservoirs and lakes. Matures at about 80-100 mm SL and spawns several times during season. Male guard eggs and larvae swim downstream. Feeds on small insects, crustaceans and small fish. Often buried in fine sediments.

Conservation status. LC.

Remarks. The *G. giuris* species complex in the Indo-West Pacific comprises several molecular lineages, including two unidentified clusters. It remains to be seen which of these correspond to *G. giuris*. The distribution of *G. giuris* here corresponds to the species found in Iran (*G. giuris* B). A species of commercial importance, especially in tropical Asia. The maximum size of this species is given as 500 mm total length, but we have serious doubts that such large individuals exist.

Further reading. Maugé 1986 (distribution); Pethiyagoda 1991 (food); Pusey et al. 2004 (upriver migration); Zarei et al. 2023 (distribution).

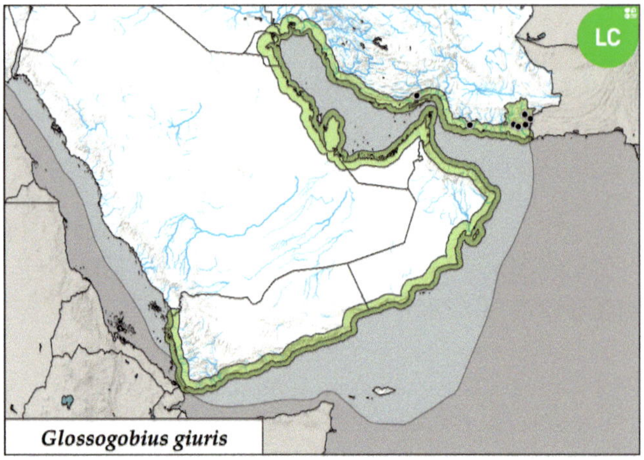

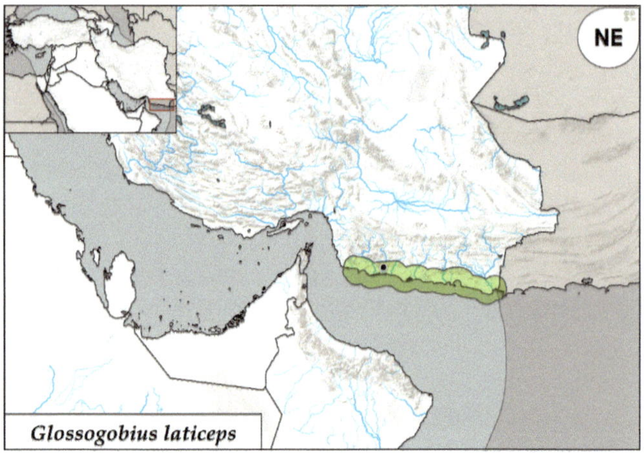

Glossogobius laticeps; Gabrik, Iran; juvenile, 74 mm SL. © H.R. Esmaeili

Glossogobius laticeps

Common name. Coastal tank goby.

Diagnosis. Distinguished from other *Glossogobius* in Iran by: ○ middle rays in first dorsal not prolonged / ○ sensory papilla row 6 distinct and long / ● caudal roundish / ● upper part of opercle with cycloid scales / ● each cheek line with 2–3 (rarely more) rows of sensory papillae / ● papillae line 5 below eye and suborbital papillae line 7 double / ○ 19–25 predorsal scales / ○ 1–2+1+6–8, usually 1–1+1+8 gill rakers. Size up to 161 mm SL.

Distribution. West Asia: Gabrik drainage, in Iranian Makran. From Iran along coasts east to New Guinea, Australia, China, Vietnam, and Bangladesh.

Habitat. Lower part of rivers and coastal lagoons, usually on soft bottoms.

Biology. Adults usually in freshwater habitats, juveniles occasionally in estuaries.

Conservation status. NE.

Further reading. Hoese & Hammer 2021; Hammer et al. 2021 (biology, distribution); Zarei et al. 2025 (record from Iran).

Glossogobius tenuiformis; Wadi Shab, Oman; 65 mm SL.

Glossogobius tenuiformis
Common name. Wadi Shab goby.
Diagnosis. Distinguished from other *Glossogobius* in Iran by: • middle rays in first dorsal prolonged / • sensory papilla row 6 absent or very short, often only a few papillae long / • each cheek line with 1 row of sensory papillae / • caudal spade-shaped / • upper part of opercle scaleless / • 12–16 predorsal scales / • 1+1+9–11, usually 1+1+10, gill rakers. Size up to 75 mm SL.
Distribution. Oman: Lower Wadi Shab and Wadi Hasi. Also known from South African coast from Mgeni in Eastern Cape to KwaZulu Natal. Expected to have a wider distribution and records from Sokotra seem to belong to this species.
Habitat. Lower and middle parts of streams that reach sea at least once every few years. Inhabit soft bottoms with sand, silt or hard bottoms with gravel and sell sediments. Expected to occur in many coastal habitats.
Biology. No data. Expected to be amphidromous, to spawn in freshwater habitats and larvae drift to sea.
Conservation status. LC.
Further reading. Freyhof et al. 2020 (distribution, morphology); Al Jufalii et al. 2022 (redescription).

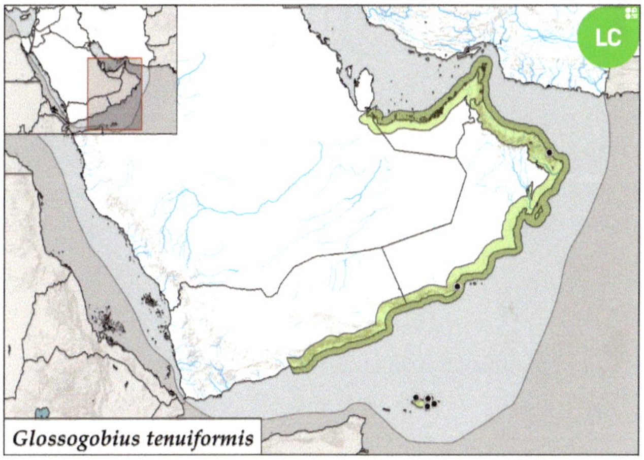

Mesogobius batrachocephalus; Black Sea north of Varna, Bulgaria; 190 mm SL.

Mesogobius batrachocephalus

Common name. Knout goby.

Diagnosis. Distinguished from species of similar genera *Neogobius* and *Ponticola* in freshwaters in Black Sea basin by: • 8–10 vertical rows of papillae in suborbital area (Fig. 60) / • 72–85 total scales in midlateral series / ○ nape naked / ○ anterior naris not projecting forward beyond lip / ○ male without bluish-iridescent, small blotches on membrane of first dorsal. Size up to 345 mm SL.

Distribution. Black, Marmara and Azov Sea basins. Freshwaters in Georgia and European range. More records expected from coastal lagoons in Türkiye.

Habitat. Inshore habitats, estuaries, brackish and freshwater lagoons on sandy or rocky bottoms. Often very deep in summer (up to 100 m). Rarely in pure freshwater.

Biology. Lives up to 8 years. First spawns at 3 years in February–May, once a year. Eggs large, laid under or between rocks. Feeds mainly on fish.

Conservation status. LC.

Remarks. *Mesogobius nonultimus* (nape scaled) from the Caspian basin is usually listed as a subspecies of *M. batrachocephalus*. It is a distinct species that does not enter freshwater.

Further reading. Porumb 1961 (biology); Bănărescu 1964 (biology, distribution); Pinchuk et al. 2004b (biology).

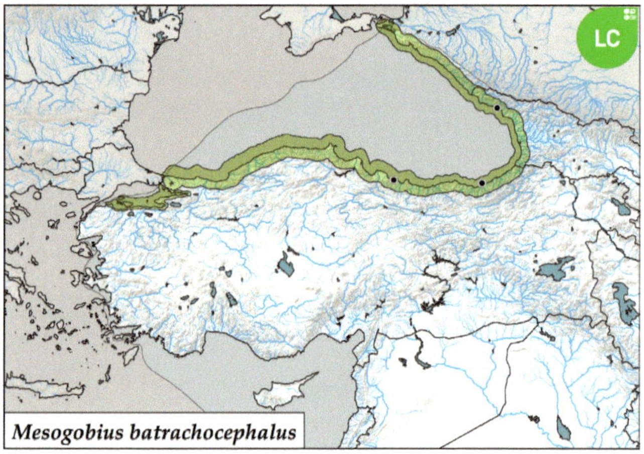

Mesogobius batrachocephalus

Key to species of *Neogobius* in freshwaters in West Asia

1a - First dorsal with large black spot in its posterior part; first branched ray of second dorsal about as long as penultimate ray.
..................*N. melanostomus*

1b - First dorsal without or with a very small black spot in its posterior part; first branched ray of second dorsal about twice as long as penultimate ray.
..................2

2a - First dorsal without black spot.
..................*N. fluviatilis*

2b - First dorsal with small black spot on posterior part, at least in juveniles.
..................*N. pallasi*

Neogobius

Neogobius was previously employed as a catchall group for larger gobies in the Black Sea and Caspian basins. Molecular studies have increased our knowledge about the phylogenetic structure of these gobies. *Neogobius* is now restricted to one strict freshwater species in Italy (*N. nigricans*), two euryhaline species (discussed below), and three marine species (*N. bathybius*, *N. caspius*, and *N. pallasi*) in the Black Sea and Caspian basins. *Neogobius pallasi* is known to enter freshwaters in the Volga but not in West Asia. Consequently, it is only included in the key to allow the user of the book to identify this species. Other species previously included in *Neogobius* are now placed in *Babka* and *Ponticola*. **Further reading.** Neilson & Stepien 2009 (phylogeny).

Neogobius fluviatilis; Susurluk drainage, Türkiye; ~60 mm SL.

Neogobius fluviatilis; Obitochnaya, Ukraine; nuptial male, 147 mm SL.

Neogobius fluviatilis
Common name. Monkey goby.
Diagnosis. Distinguished from other species of *Neogobius* entering freshwater in Black and Caspian Sea basins by: ○ first branched ray of second dorsal about twice as long as penultimate ray / ○ 55–61+2–4 scales in midlateral series / ○ no black spot in posterior part of first dorsal. Size up to 200 mm SL.
Distribution. Azov, Black Sea, Marmara, and Caspian basins. Usually near estuaries but often far inland. Invasive in upper parts of rivers in northern Black Sea basin and first recorded in 2011 in lower Maritza on Greek–Turkish border. Also invasive in Europe (Baltic basin, Danube, Rhine). Not known to be invasive in West Asia.
Habitat. Sea shores, inshore habitats, lakes, estuaries, brackish, and freshwater lagoons, large- to medium-sized rivers and streams. Mostly found on open sand or mud bottoms. One of the most abundant species in large lowland rivers.
Biology. Lives up to 5 years. First spawns at 2 years in April–July, locally until September, when temperature rises above 13°C. Female may spawn repeatedly during season. Nuptial male black with yellow fin edges. Adhesive eggs laid on stones, shells, and aquatic plants. Male excavate nests under any type of hard substrate and guard eggs until hatching. Feeds on a wide variety of invertebrates, especially molluscs.
Conservation status. LC.
Remarks. *Neogobius fluviatilis* appears to be absent from the southern Caspian basin.
Further reading. Pinchuk et al. 2003a (biology, description).

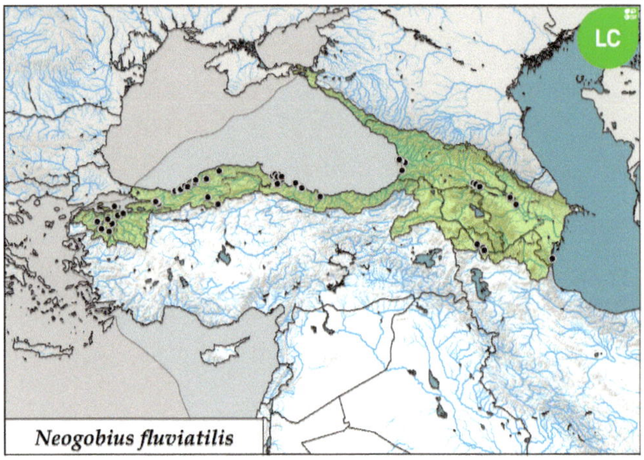

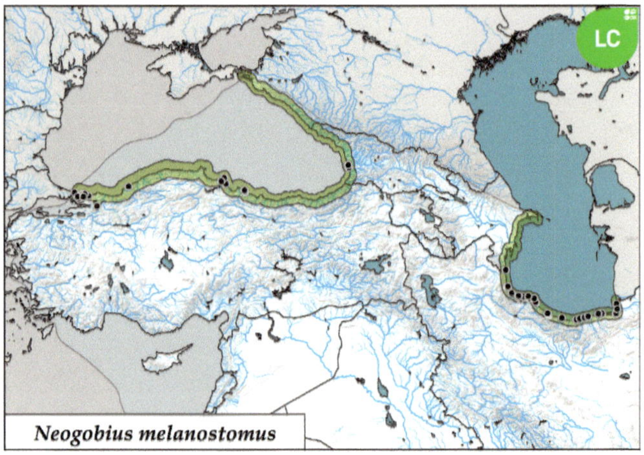

Neogobius melanostomus; Elbe drainage, Germany; ~100 mm SL.

Neogobius melanostomus

Common name. Round goby.

Diagnosis. Distinguished from other species of *Neogobius* entering freshwater in West Asia by: • first dorsal with large black spot on posterior part / • 45–54+2–3 scales in midlateral series / ○ first branched ray of second dorsal about as long as penultimate ray. Size up to 220 mm SL.

Distribution. Azov, Black Sea, Marmara and Caspian basins. Invasive in North America, Volga, Baltic, Black, and North Sea basins reached by shipping channels or direct introductions. No records of invasion in West Asia.

Habitat. Coastal habitats, estuaries, brackish and freshwater lagoons and lakes, large rivers, canals, and harbors on sandy or rocky bottoms. Up to 50–60 m deep in Black Sea in winter. Mostly found on well-vegetated or rocky bottoms.

Biology. Lives up to 4 years. Male spawn for first time at 3–4 years, female at 2–3 and 55–80 mm SL. In invasive areas, mature at a smaller size, i.e. 45 mm SL. Spawns April–September. Nuptial male almost completely black. Female may repeat spawning every 18–20 days during a season. Adhesive eggs laid under or between stones, shells and aquatic vegetation. Male guard eggs until hatching in 2–3 weeks. Male usually die after spawning season. Feeds on a wide variety of invertebrates and small fish, mainly molluscs. Egg clutches are sometimes transported attached to the hulls of ships.

Conservation status. LC.

Further reading. Tsepkin et al. 1992 (invasion); Charlebois et al. 1997 (review); Wiesner et al. 2000 (invasion); Pinchuk et al. 2003c (biology, description); Balážová-L'avrinčíková & Kováč 2007 (reproduction in invasive range).

Lake Sapanca in Türkiye is the habitat of both *Neogobius* species, as well as *Babka gymnotrachelus*, *Proterorhinus semilunaris*, and *Knipowitschia caucasica*.

Ponticola

Ponticola are small- or medium-sized fishes and comprise approximately 18 species. They were previously considered species within the genus *Neogobius*, but subsequent molecular and morphological studies have demonstrated that this group represents a distinct genus. *Ponticola* are common in most rivers flowing to the Black and Caspian Seas, and several species occur mostly in allopatry. In addition to those species included in the species chapters, we recognise only five additional species in West Asia (*P. eurycephalus, P. ratan, P. iljini, P. platyrostris, P. cephalargoides*) that are found in brackish waters only and are not known to enter freshwater habitats. Some authors consider *Ponticola odessicus* and *P. goebelii* valid species, yet we have not identified any scientific evidence to support this classification. These species appeared elevated from synonyms to species in checklists without supporting documentation. *Ponticola eurycephalus* is a coastal species that regularly enters freshwaters in the European part of the Black Sea basin. However, this species is not distributed in West Asia. It is included in the key only to allow its identification. *Ponticola ratan* is known from the Black Sea and Caspian (where it is occasionally recorded in the Asian parts of these seas). It may be an accidental species, but from its European range, it has only one freshwater population in an inland reservoir in Ukraine. *Ponticola eurycephalus* and *P. ratan* are excluded from the coverage of this book, as well as *P. iljini, P. platyrostris,* and *P. cephalargoides*. The selection between freshwater and marine species in *Ponticola* is somewhat artificial, given the need for a clear border between fresh- and brackish water habitats, particularly in larger river estuaries and coastal limans. Furthermore, the diagnostic characters of the different *Ponticola* species need to be better documented, and the distribution of several species needs to be better understood. **Further reading.** Neilson & Stepien 2009 (phylogeny); Vasil'eva & Vasil'ev 2003a (identification key); Vasil'eva et al. 2015 (Caspian species).

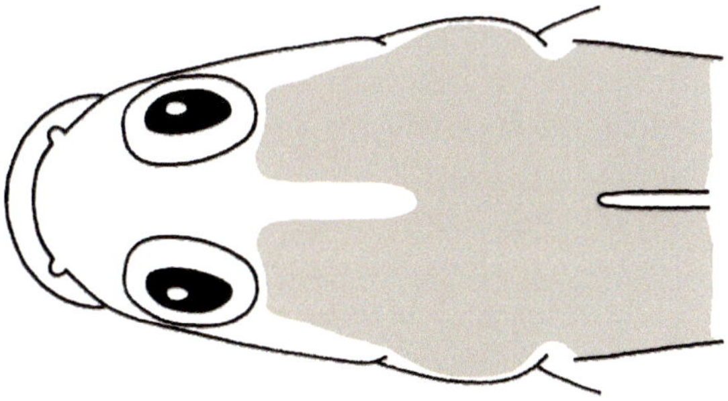

Figure 61. Scalation on nape missing on dorsal midline in some species of Gobiidae.

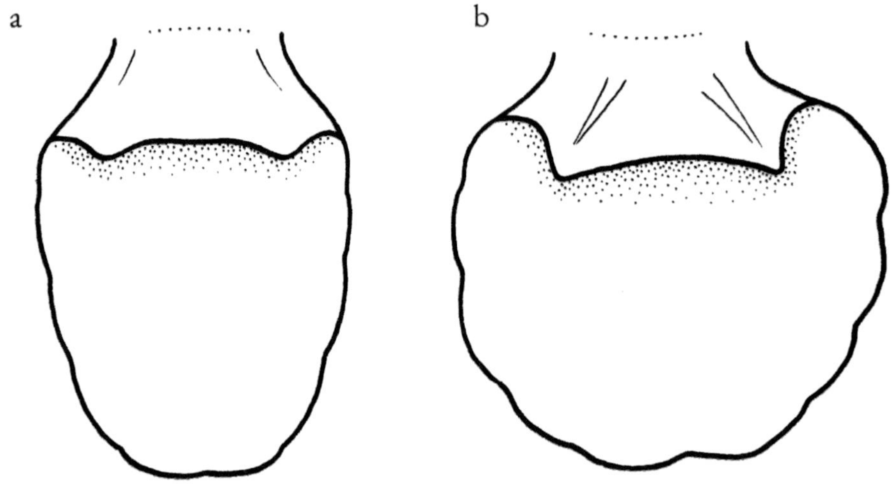

Figure 62. Pelvic-disc fraenum in *Babka* and *Ponticola* species. **a,** with small, rounded lobes; **b,** with angular lobes.

Key to species of *Ponticola* in freshwaters in West Asia

1a - Pelvic-disc fraenum with small, rounded lobes whose length is less than $\frac{1}{6}$ of fraenum width at base.
..................2

1b - Pelvic-disc fraenum with angular lobes whose length is $\frac{1}{6}$–$\frac{1}{2}$ of fraenum width at base.
..................3

2a - 48–55 total scales in lateral series.
..................*P. alasanicus*

2b - 59–69 total scales in lateral series.
..................*P. syrman*

3a - Lateral part of upper lip not swollen.
..................*P. gorlap*

3b - Lateral part of upper lip distinctively swollen.
..................4

4a - Pelvic disc reaching 80–100 % of distance between its origin and anus.
..................*P. eurycephalus*

4b - Pelvic disc reaching 40–60 % of distance between its origin and anus.
..................5

5a - Lower jaw strongly prognathous.
..................6

5b - Lower jaw not or very slightly prognathous.
..................7

6a - First dorsal without a dark-brown distal blotch or band.
..................*P. constructor* / *P. rizensis*

6b - First dorsal with a dark-brown distal blotch or band.
..................*P. cyrius*

7a - Dorsal head, lips, cheeks, and predorsal colour plain brown; flank very slightly reticulate.
..................*P. iranicus* / *P. patimari*

7b - Dorsal head, lips, cheeks, predorsal area, and flank strongly reticulate.
..................*P. hircaniaensis*

Ponticola alasanicus; Alazani, Georgia; ~70 mm SL. © A. Epitashvili.

Ponticola alasanicus

Common name. Alazani goby.

Diagnosis. Distinguished from other species of *Ponticola* entering freshwater in West Asia by: ● no suborbital transverse rows below suborbital longitudinal row *b* / ○ pelvic-disc fraenum with small, rounded lobes, whose length is less than $\frac{1}{6}$ of fraenum width at base / ○ lower jaw not or very slightly prognathous / ○ 48–55 total scales in midlateral series / ○ second dorsal usually with 15–16½ branched rays / ○ anal with 10–12½ branched rays / ○ lateral part of upper lip distinctively swollen / ○ dorsal head, lips, cheeks and predorsal plain brown / ○ flank very slightly reticulate / ○ pelvic disc not reaching anus / ○ pelvic-disc fraenum with angular lobes whose length is $\frac{1}{6}$–$\frac{1}{2}$ of fraenum width at base. Size up to 74 mm SL.

Distribution. Georgia: Alazani in upper Kura drainage.

Habitat. Lowland rivers and marshes with sand bottoms. Often found near submerged trees and roots. Not found in brackish water.

Biology. Spawns May–June. Feeds on aquatic invertebrates.

Conservation status. NE; due to the very poor knowledge about this recently described species.

Further reading. Epitashvili et al. 2023 (description).

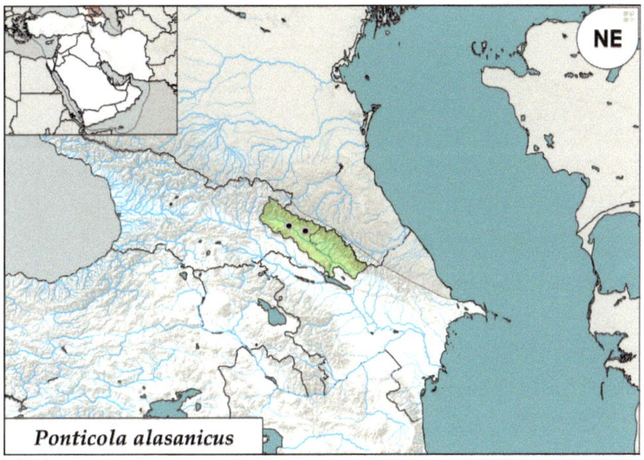

Ponticola constructor

Common name. Caucasian goby.

Diagnosis. Distinguished from other species of *Ponticola* entering freshwater in West Asia by: ○ lateral part of upper lip distinctively swollen / ○ lower jaw strongly prognathous / ○ predorsal area finely mottled / ○ two suborbital transverse rows below suborbital longitudinal row *b* / ○ first dorsal without bold, distal dark-brown band / ○ pelvic disc reaching 40–60 % of distance between its origin and anus / ○ 53–74+2–4 scales in midlateral series / ○ pelvic-disc fraenum with angular lobes whose length is $\frac{1}{6}$–$\frac{1}{2}$ of fraenum width at base. Size up to 200 mm SL.

Distribution. Black Sea drainages of southern Caucasus from lower Çoruh (Türkiye) north to Noworossiysk (Russia). Also, in western tributaries of Kuban.

Habitat. Cold mountain to warm foothill streams. Not found in brackish water.

Biology. Matures with 1–2 years and spawns in May–June. Nuptial male almost entirely black. Several females may spawn with one male, and female are expected to spawn several times during season. Male guard eggs and larvae in gravel or rocky burrows. Feeds on a wide range of small invertebrates and fish.

Ponticola constructor; Kuban drainage, Russia; ~85 mm SL.

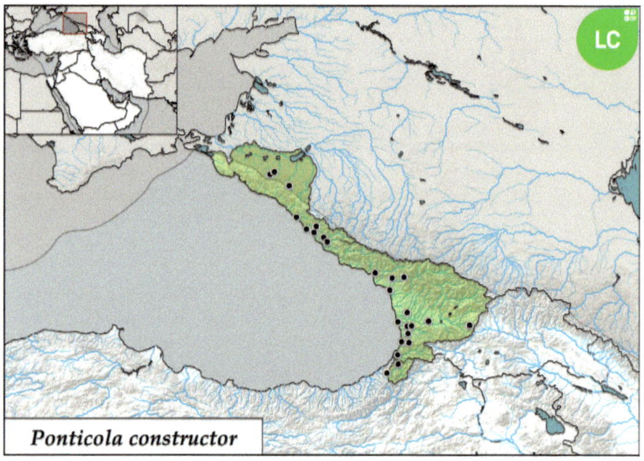

Conservation status. LC.

Remarks. *Ponticola rhodioni* is treated as a synonym. It has been described based on chromosomal characters only, and morphological or molecular characters do not support *P. rhodioni* as a valid species. Molecular characters distinguish *P. constructor* well from *P. rizensis* and *P. turani*, and published morphological data places *P. constructor* intermediate between these two species. However, all characters used to distinguish *P. constructor*, *P. rizensis*, and *P. turani* largely overlap in fresh material we examined, and further studies are needed to distinguish at least the two species supported by molecular characters (*P. constructor* and *P. rizensis*).

Further reading. Berg 1949b (description as *N. cephalarges constructor*, in part); Vasil'eva & Vasil'ev 1994 (description); Vasil'eva & Vasil'ev 2003b (distribution, description, biology).

Ponticola cyrius

Common name. Kura goby.

Diagnosis. Distinguished from other species of *Ponticola* entering freshwater in West Asia by: ● predorsal area marbled / ● first dorsal with a dark-brown distal blotch or band / ○ lateral part of upper lip usually swollen / ○ two suborbital transverse rows below suborbital longitudinal row *b* / ○ lower jaw slightly prognathous / ○ pelvic disc reaching 40–60 % of distance between its origin and anus / ○ 51–74+2–4 scales in midlateral series / second dorsal usually with / ○ usually with 17–18½ branched rays / ○ anal usually with 13½ branched rays / ○ pelvic-disc fraenum with angular lobes whose length is $\frac{1}{6}$–$\frac{1}{2}$ of fraenum width at base. Size up to 111 mm SL.

Ponticola cyrius; Kura drainage, Türkiye; ~80 mm SL. © M. Özuluğ.

Distribution. Kura drainage from headwaters in Türkiye south to Azerbaijan. Widely reported from Aras, but records questionable.
Habitat. Cold mountain to warm foothill streams. Not found in brackish water or reservoirs.
Biology. Matures at 1–2 and lives for 3 years; spawns May–late August at water temperatures of 11–20°C. Male guard eggs in gravel or rocky burrows. Feeds on a wide range of small invertebrates and fish.
Conservation status. LC.
Remarks. The distribution of this species in Kura is not fully known, and it potentially occurs in sympatry with *P. alasanicus* in upper Kura. In the middle and lower Kura, it occurs in sympatry with *P. gorlap*.
Further reading. Vasil'eva & Vasil'ev 2003c (distribution, description, biology).

Ponticola gorlap

Common name. Caspian bighead goby.
Diagnosis. Distinguished from other species of *Ponticola* entering freshwater in West Asia by: ○ lower jaw strongly prognathous / ○ lateral part of upper lip not swollen / ○ pelvic disc reaching 75–95 % of distance between its origin and anus / ○ 60–71 total scales in midlateral series / ○ 23–28 predorsal scales / ○ two suborbital transverse rows below suborbital longitudinal row *b* / ○ flank mottled / ○ first dorsal without bold, distal dark-brown band / ○ pectoral base with a black blotch / ○ no dark-brown blotches on anterior part of first dorsal / ○ pelvic-disc fraenum with angular lobes whose length is $\frac{1}{6}$–$\frac{1}{2}$ of fraenum width at base. Size up to 200 mm SL.

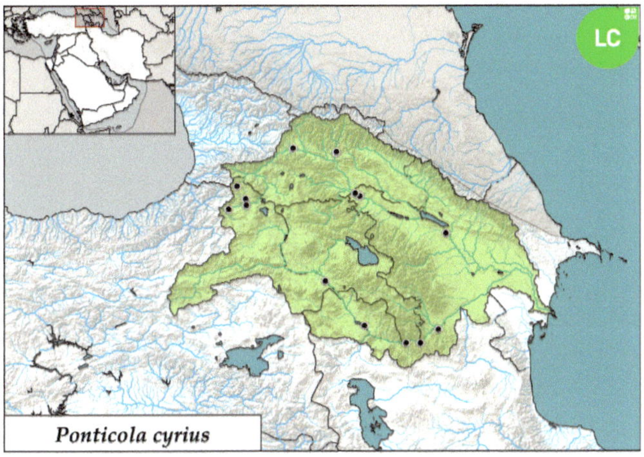

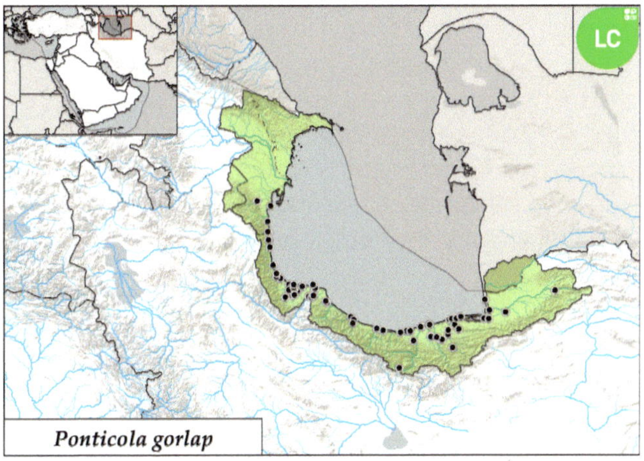

Ponticola gorlap; Tajan drainage, Iran; ~150 mm SL.

Distribution. Coastal rivers of Azerbaijani and Iranian Caspian coasts. Also, in sea, Volga upstream to Astrakhan until 1977, then spread upstream to lakes Ivankovskoje and Rybinskoje (2000). Invaded Don drainage (Black Sea basin) through Volga-Don canal (first record 1972). Now abundant in Lake Tsimlyansk and lower Don.
Habitat. Inshore habitats, lakes, estuaries, brackish and freshwater lagoons, and large rivers, harbors, on rocky bottoms. Mainly found on well-vegetated or rocky bottoms.
Biology. Most females spawn for first time at 1 year, male at 2 years. Spawns April–May, rarely until July. Adhesive eggs laid on stones, shells, and aquatic vegetation. Male guard eggs until hatching. Feeds mainly on small fish and a variety of invertebrates.
Conservation status. LC.
Remarks. In the lower and middle Kura, it may occur in sympatry with superficially similar *P. cyrius*. Previously, *P. iljini* was treated as a synonym of *P. gorlap*. Still, new research has shown it to be a valid species restricted to the coast of Mangyshlak peninsula in Kazakhstan.
Further reading. Vasil'eva & Vasil'ev 1996 (description as *N. iljini*); Kottelat 1997 (nomenclature); Vasil'eva & Vasil'ev 2003d (biology, description); Vasil'eva et al. 2016 (description of *P. iljini*).

Ponticola hircaniaensis

Common name. Gorgan goby.
Diagnosis. Distinguished from other species of *Ponticola* entering freshwater in West Asia by: ○ lower jaw not or very slightly prognathous / ○ 60–71 total scales in midlateral series / ○ 19–23 predorsal scales / ○ lateral part of upper lip swollen / ○ dorsal head, lips, cheeks, predorsal area, and flank strongly reticulate / ○ two suborbital transverse rows below suborbital longitudinal row *b* / ○ pelvic disc reaching 40–60 % of distance between its origin and anus / ○ first dorsal with marginal pale orange-yellow band and dark oblique spot / ○ upper part of pectoral base with one dark-brown stripe / ○ pelvic-disc fraenum with angular lobes whose length is $\frac{1}{6}$–$\frac{1}{2}$ of fraenum width at base. Size up to 92 mm SL.
Distribution. Iran: Kaboudval and Zarrin Gol in Gorgan drainage.
Habitat. Small, fast-flowing headwater streams with gravel bottoms. Not found in brackish water.
Biology. No data
Conservation status. EN; appears to be declining within its very small range.
Remarks. This species has hybridised with *P. gorlap* and mtDNA from *P. gorlap* is present in individuals from the only known population.
Further reading. Zarei et al. 2022b (description).

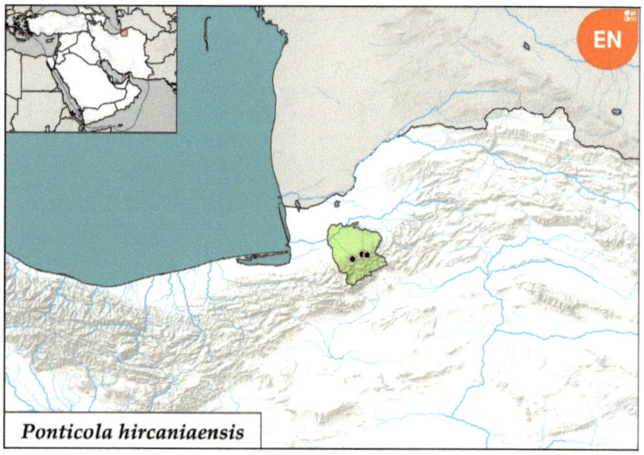

Ponticola hircaniaensis; Kaboudval drainage, Iran; 74 mm SL.

Ponticola iranicus; Sefid drainage, Iran; ~80 mm SL.

Ponticola iranicus

Common name. Sefid goby.

Diagnosis. Distinguished from other species of *Ponticola* entering freshwater in West Asia by: ○ lower jaw not or very slightly prognathous / ○ 54–70 total scales in midlateral series / ○ second dorsal usually with 15–16½ branched rays / ○ anal usually with 11–12½ branched rays / ○ lateral part of upper lip distinctively swollen / ○ dorsal head, lips, cheeks and predorsal plain brown / ○ flank very slightly reticulate / ○ first dorsal with marginal pale orange-yellow band and dark oblique spot / ○ upper part of pectoral base upper part with one dark-brown stripe / ○ pelvic disc reaching 40–60 % of distance between its origin and anus / ○ two suborbital transverse rows below suborbital longitudinal row *b* / ○ pelvic-disc fraenum with angular lobes whose length is $\frac{1}{6}$–$\frac{1}{2}$ of fraenum width at base. Size up to 86 mm SL.

Distribution. Iran: Lower and middle Sefid drainage.

Habitat. Cold, fast-flowing, gravel-bottomed rivers and larger streams. Not found in brackish water.

Biology. Spawns end February–end April. Male guard eggs in gravel or rocky burrows.

Conservation status. LC.

Remarks. This species is very similar to *P. patimari*, and a recent publication suggests that the two may be indistinguishable by external characters. They occur in syntopy in Lower Sefid, and may represent just populations of one old species.

Further reading. Vasil'eva et al. 2015 (description); Zarei et al. 2022b, c (distribution, biogeography, genetics, morphology).

Ponticola patimari

Common name. South Caspian freshwater goby

Diagnosis. Distinguished from other species of *Ponticola* entering freshwater in West Asia by: ○ lower jaw not or very slightly prognathous / ○ 44–56 total scales in midlateral series / ○ second dorsal usually with 15–16½ branched rays / ○ anal usually with 11–12½ branched rays / ○ lateral part of upper lip swollen / ○ dorsal head, lips, cheeks, and predorsal plain brown / ○ flank very slightly reticulate / ○

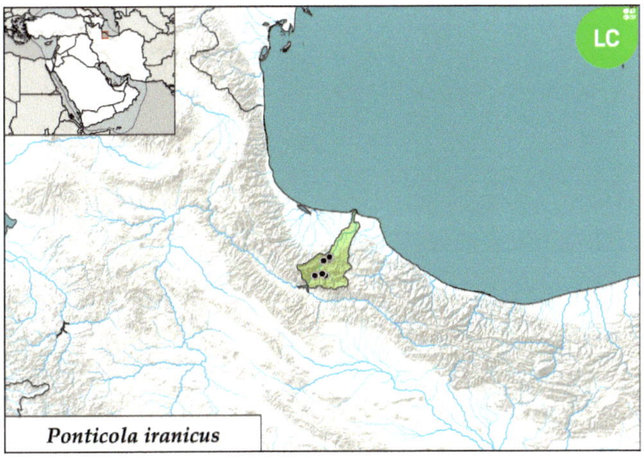

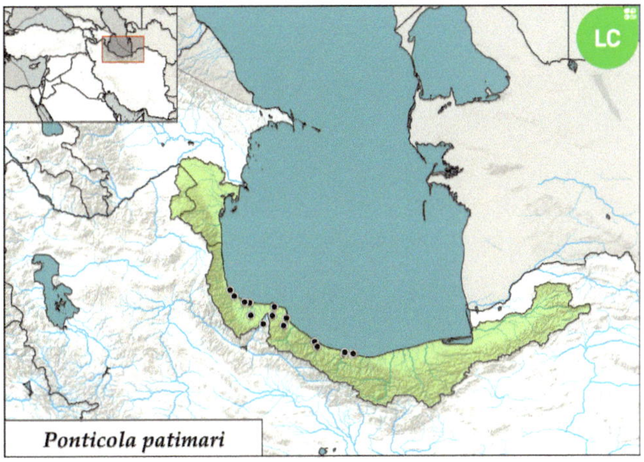

Ponticola patimari; Khei drainage, Iran; 63 mm SL.

first dorsal with marginal pale orange-yellow band and dark oblique spot / ○ upper part of pectoral base upper part with one dark-brown stripe / ○ pelvic disc reaching 40–60 % of distance between its origin and anus / ○ two suborbital transverse rows below suborbital longitudinal row *b* / ○ pelvic-disc fraenum with angular lobes whose length is $\frac{1}{6}$–$\frac{1}{2}$ of fraenum width at base. Size up to 100 mm SL.

Distribution. Iran: Cheshme kile to Astara in southern Caspian basin. Absent from Sefid except lower most part and in rivers of Anzali Lagoon.

Habitat. Cold mountain to warm foothill streams. Not found in brackish water.

Biology. Male guard eggs in gravel or rocky burrows. Feeds mainly on small fish and a wide variety of invertebrates.

Conservation status. LC.

Further reading. Eagderi et al. 2020 (description); Zarei et al. 2022b, c (distribution, biogeography, genetics, morphology).

The Sefid in Iran, here after heavy rain, is the habitat of *Ponticola iranicus*.

Ponticola rizensis; İyidere drainage, Türkiye; ~60 mm SL. © M. Özuluğ.

Ponticola rizensis; Aksu drainage, Türkiye; ~70 mm SL. © M. Özuluğ.

Ponticola cf. *rizensis*; Filyos drainage, Türkiye; ~80 mm SL. © M. Özuluğ.

Ponticola rizensis

Common name. Colchic goby.

Diagnosis. Distinguished from other species of *Ponticola* entering freshwater in West Asia by: ○ lateral part of upper lip distinctively swollen / ○ lower jaw prognathous / ○ predorsal area uniform brown or dark-grey / ○ first dorsal without bold, distal, dark-brown band / ○ head depressed, wider than deep / ○ 59–73 +3 scales in midlateral series / ○ caudal peduncle depth 18–22 % of caudal peduncle length / ○ two suborbital transverse rows below suborbital longitudinal row *b* / ○ pelvic disc reaching 40–60 % of distance between its origin and anus / ○ pelvic-disc fraenum with angular lobes whose length is $\frac{1}{6}$–$\frac{1}{2}$ of fraenum width at base. Size up to 124 mm SL.

Distribution. Türkiye: Aksu (Giresun) east to Hopa drainages.
Habitat. Cold mountain to warm foothill streams. Not found in brackish water.
Biology. Male guard eggs in gravel or rocky burrows. Feeds mainly on small fish and a variety of invertebrates.
Conservation status. VU; appears to be slowly declining within its relatively small range.
Remarks. Molecular characters do not distinguish *P. rizensis* and *P. turani* from each other but clearly distinguish both together from *P. constructor.* Published morphological data place *P. constructor* somewhere between these two species. However, all characters used to differentiate *P. constructor*, *P. rizensis,* and *P. turani* largely overlap in fresh material we examined, and further studies are needed to distinguish at least the two species supported by molecular characters (*P. constructor* and *P. rizensis*). *Ponticola rizensis* and *P. turani* have been published in the same study. As a first reviser, we give priority to *P. rizensis* over *P. turani.* This makes *P. turani* a synonym of *P. rizensis*. *Ponticola* west of Aksu might belong to this species.
Further reading. Kovačić & Engin 2008 (description).

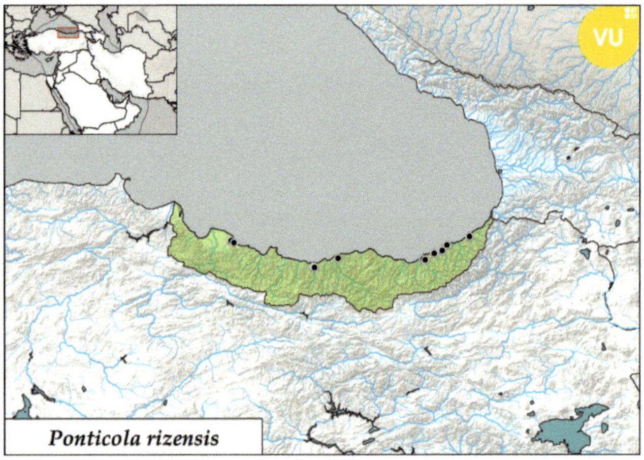

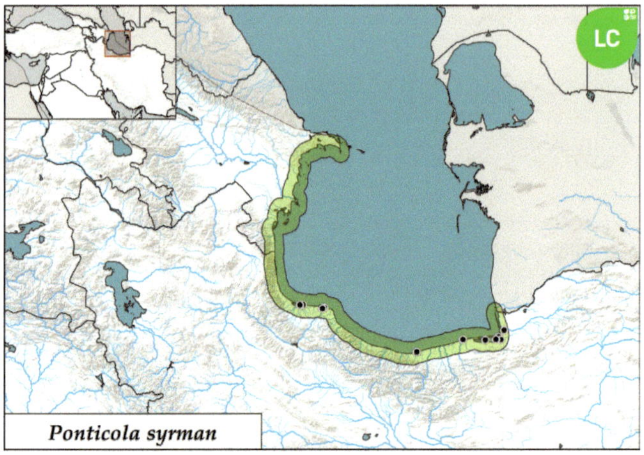

Ponticola syrman; Danube delta, Romania; ~100 mm SL.

Ponticola syrman
Common name. Syrman goby.
Diagnosis. Distinguished from other species of *Ponticola* entering freshwater in West Asia by: ○ pelvic-disc fraenum with small, rounded lobes, whose length is less than $\frac{1}{6}$ of fraenum width at base / ● three suborbital transverse rows below suborbital longitudinal row *b* / ○ interorbital distance equal to eye diameter) / ○ lateral part of upper lip not distinctively swollen / ○ lower jaw not or very slightly prognathous / ○ pelvic disc reaching 60–70 % of distance between its origin and anus / ○ 57–67+2–3 scales in midlateral series / ○ head depressed, wider than deep. Size up to 220 mm SL.
Distribution. Azov, northern Black Sea, and Caspian basins. In West Asia only confirmed from Caspian but not from Black Sea basin.
Habitat. Inshore habitats, estuaries, brackish and freshwater lagoons, and large rivers on muddy bottoms. Restricted to coastal areas and rarely found in freshwater.
Biology. Lives up to 4 years. First spawns at 1–2 years in March–June. Female may spawn twice in a season. Adhesive eggs laid under or between stones, shells, and aquatic vegetation. Male guard eggs until hatching. Feeds on a wide variety of invertebrates and small fish.
Conservation status. LC.
Further reading. Pinchuk et al. 2003d (biology, description).

Proterorhinus

All freshwater species of *Proterorhinus* have long been confused under the name *P. marmoratus*. Morphological and molecular data indicate that *P. marmoratus* does not enter pure freshwaters; it inhabits brackish waters in the Black Sea. Consequently, it is only mentioned in the key. *Proterorhinus semilunaris* also enters brackish waters and both may occur at similar salinities and may form hybrids. Three species occur in the Black Sea and two, potentially three, in the Caspian basin. All three Caspian species have been described from coastal habitats: *P. nasalis* and *P. blennioides* from the Caspian Sea near Baku (Azerbaijan) and *P. semipellucidus* from the mouth of the Karasu, a tributary of Gorgan Bay (Iran). *Proterorhinus nasalis* from Azerbaijan and Russia and *P. semipellucidus* from Iran might be conspecific or two allopatric species. Much more research is needed on this genus and the situation might be quite different from what is presented here in summer 2025.

Further reading. Kessler 1877 (description of *P. semipellucidus*); Neilson & Stepien 2009 (phylogeny with mixed-up names); Zarei et al. 2021 (phylogeny with mixed-up names); Vasil'eva et al. 2025 (*P. blennioides*).

Key to species of *Proterorhinus* in Black Sea basin

1a - Head length 24–28 % SL; 15–17. anal rays.
..................*P. marmoratus*

1b - Head length 28–35 % SL; 13–14, rarely 16 anal rays.
..................2

2a - Head length 28–31 % SL; usually 14. anal rays.
..................*P. semilunaris*

2b - Head length 31–35 % SL; usually 13. anal rays.
..................*P. victori*

Key to species of *Proterorhinus* in Caspian Sea

1a - Eye diameter 16–21 % HL, as wide or smaller than interorbital space.
.... *P. blennioides*

1b - Eye diameter 20–28 % HL, wider than interorbital space.
.... 2

2a. Usually second 15–17½ branched rays in second dorsal fin; usually 13–15½ branched anal rays; 39–61 scales in midlateral series.
.... *P. nasalis*

2b. Usually 14–15½ branched rays in second dorsal fin; usually 10–14½ branched anal rays; 36–45 scales in midlateral series.
.... *P. semipellucidus*

Proterorhinus blennioides; Samur, Dagestan, Russia; nuptial male, ~65 mm SL. © A. Naseka.

Proterorhinus blennioides

Common name. Caspian tubenose goby.

Diagnosis. Distinguished from other species of *Proterorhinus* in Caspian basin by: ○ eye diameter 16–21 % HL / ○ usually 13½ branched anal rays. Size up to 90 mm SL.

Distribution. Iranian Caspian basin.

Habitat. Usually associated with dense vegetation or coarse rocks.

Biology. No data

Conservation status. NE.

Remarks. In recent molecular literature, this species has been misidentified as *P. semipellucidus*, a synonym of *P. nasalis* or a species potentially endemic to the Iranian Caspian basin.

Further reading. Stepien & Tumeo 2006 (genetics as *P. semipellucidus*); Zarei et al. 2021 (phylogeny as *P. semipellucidus*); Vasil'eva et al. 2025 (redescription).

Proterorhinus blennioides; Samur, Dagestan, Russia; female, ~50 mm SL. © A. Naseka.

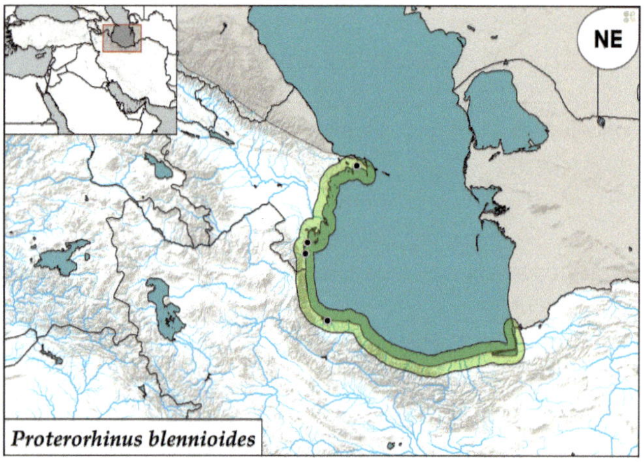

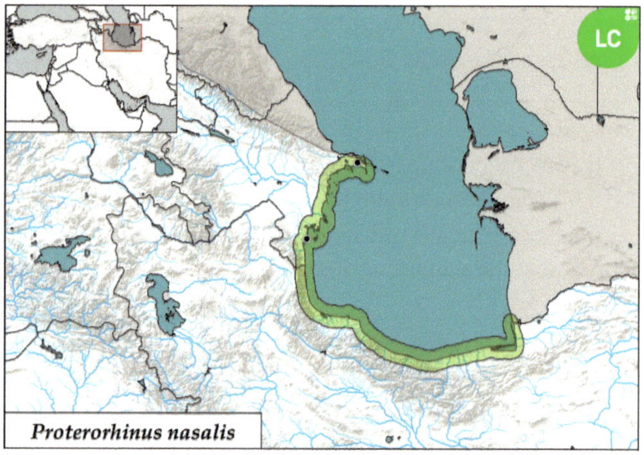

Proterorhinus nasalis; Baltic Sea, Estonia; non-nuptial male, ~ 50 mm SL. © K. Kurina

Proterorhinus cf. *semipellucidus* Gorgan Bay, Iran; non-nuptial male, ~60 mm SL.

Proterorhinus nasalis

Common name. Eastern tubenose goby.

Diagnosis. Distinguished from other species of *Proterorhinus* in Caspian basin by: ○ eye diameter 20–28 % HL / ○ usually 14, rarely up to 16½ branched anal rays. Size up to 90 mm SL.

Distribution. Caspian and Azov basins. Native to Volga delta, now reaching as far as Lake Rybinskoye. Reached eastern Baltic basin in Russia, Estonia and Finland in recent years.

Habitat. A variety of slow-flowing or still waters. Usually associated with dense vegetation or coarse rocks.

Biology. Spawns first time at 1–2 years, usually for 1–2 seasons, in April–August. Females spawn more than once in a season. Male guard eggs deposited in burrows. Larvae and juveniles benthic. Feeds on benthic invertebrates.

Conservation status. LC; invasive in Volga and Baltic basin, where it is spreading, reportedly facilitated by canalisation of main rivers and reservoirs or slow-flowing waters created by hydroelectric plants.

Remarks. In recent literature, this species has been misidentified as *P. semipellucidus*, a species potentially endemic to the Iranian Caspian basin. It cannot be excluded that *P. semipellucidus* might be a valid species, and future research is encouraged to investigate this question. In recent literature, this species has been misidentified as *P. semipellucidus*, a species potentially endemic to the Eastern Caspian basin. *Proterorhinus semipellucidus* was redescribed too late to get full credits in this book. It can be identified by the key on page 675.

Further reading. Miller 2004 (description, biology); Naseka et al. 2005 (invasion); Zarei et al. 2021 (phylogeny); Kurina et al. 2024 (Baltic invasion); Vasil'eva et al. 2025a (*P. blennioides*); Vasil'eva et al. 2025b (*P. semipellucidus*).

Proterorhinus nasalis; Azov basin, Ukraine; ~50 mm SL.

Proterorhinus semilunaris; Lake Sapanca, Türkiye; female, ~35 mm SL.

Proterorhinus semilunaris; Lake Sapanca, Türkiye; male, ~55 mm SL.

Proterorhinus semilunaris

Common name. Western tubenose goby.

Diagnosis. Distinguished from other species of *Proterorhinus* in Black Sea basin by: ○ head length 28–31 % SL / ○ eye diameter 20–28 % HL / ○ 13–16, usually 14½ anal rays / ○ total scales in midlateral series 41–47. Size up to 90 mm SL.

Distribution. Marmara and Black Sea basins, and Maritza and Struma drainages in eastern Aegean. In Danube, historically present up to about Vienna, invasive since 1970s, now reaching upstream to Germany and France. Invaded Baltic and North Sea basins since 1992, now widespread from Sommes (France) east to Vistula (Poland). Also in upper Rhône. In South Bug and Dnieper native far upstream. No reports of spread into Anatolia and arrived in North America in 1991 in ballast of ships.

Habitat. A variety of slow-flowing or standing waters, from estuaries to small, slow-flowing pre-montane streams. Usually associated with dense vegetation or coarse rocks. Often very abundant in backwaters and lakes, increasing in reservoirs and canals.

Biology. Spawns first time at 1–2 years, usually for 1–2 seasons, in April–August. Female spawn more than once in a season. Male guard eggs deposited in burrows, larvae and juveniles benthic. Feeds on benthic invertebrates.

Conservation status. LC.

Remarks. Molecular studies revealed different haplotypes within populations in Georgia indicate past hybridisation with *P. victori*. Further studies are needed to resolve the species diversity of tubenose gobies.

Further reading. Ahnelt et al. 1998 (invasion); Pinchuk et al. 2004a (description, biology); Naseka et al. 2005 (invasion); Stepien & Tumeo 2006 (genetics).

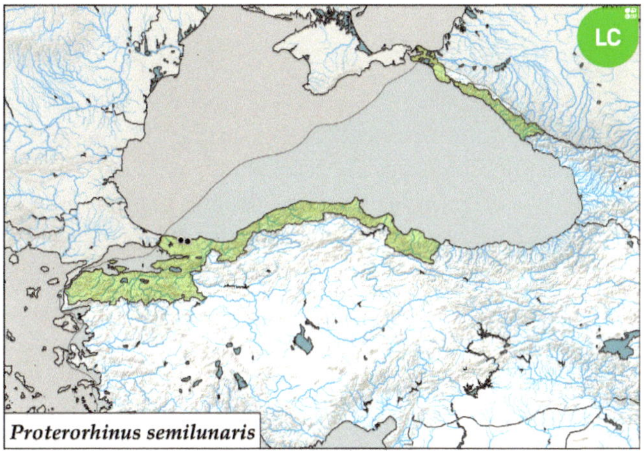

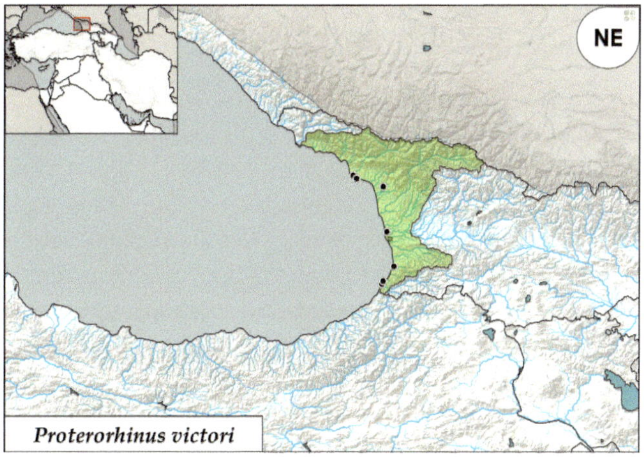

Proterorhinus Kintrishi, Georgia; ~60 mm SL. © M. Geiger (LIB).

Proterorhinus victori

Common name. Georgian tubenose goby.

Diagnosis. Diagnosis. Distinguished from other species of Proterorhinus in Black Sea basin by: ○ head length 31–35 % SL / ○ eye diameter 20–27 % HL / ○ 12–14½, usually 13½ anal rays / ○ total scales in midlateral series 42–59. Size up to 75 mm SL.

Distribution. Georgia: Chorokhi, Kintrishi, Mchishta and lakes Kakhaberi and Inkiti in Black Sea basin.

Habitat. A variety of slow-flowing or still waters. Usually associated with dense vegetation or coarse rocks.

Biology. No data

Conservation status. NE; due to the very poor knowledge about this recently described species.

Remarks. The presence of mtDNA from *P. semilunaris* in fish from Kintrishi indicates some degree of hybridisation between the two species.

Further reading. Zarei et al. 2021 (phylogeny); Vasil'eva 2024 (description)

Family Oxudercidae

Sand gobies and mudskippers

Oxudercid gobies constitute a large group of fishes, comprising approximately 110 genera and 745 species. They are found predominantly in coastal and brackish waters, as well as in freshwater habitats. Their previous family name was Gobionellidae. However, it is now considered a junior synonym of Oxudercidae. In the freshwaters of West Asia, Oxudercidae is represented by the amphidromous genus *Awaous*, the sand gobies of the genus *Knipowitschia*, non-native *Rhinogobius*, and three mudskippers (*Boleophthalmus dussumieri*, *Periophthalmus waltoni*, *Scartelaos tenuis*). Mudskippers are found often associated with mud flats and mangroves in the Persian Gulf and around the Arabian Peninsula. They may also be occasionally found in freshwater, particularly following heavy rainfall, but they do not occur regularly in freshwater; instead, they are only found in brackish waters and the sea. Some mudskippers are occasionally seen at considerable distances up in brackish rivers, as *B. dussumieri* in the upper Shur (Iran), 55 km upriver from the coast. The key includes *Periophthalmus* and *Scartelaos* but excludes them from the list of freshwater fishes of West Asia. **Further reading.** Agorreta et al. 2013 (phylogeny); Thacker 2013 (phylogeny); Nelson et al. 2016 (diversity).

Periophthalmus waltoni; Kuwait; ~100 mm SL. © M. Pope.

Scartelaos tenuis; Bandar Khamir, Hormozgan, Iran; ~140 m SL. © G. Polgar.

https://doi.org/10.1515/9783111677811-035

Awaous jayakari; Wadi Shab, Oman; 75 mm SL.

Awaous jayakari; Wadi Masila, Yemen; 84 mm SL. © F. Krupp.

Awaous jayakari

Common name. Arabian freshwater goby.

Diagnosis. Distinguished from sympatric *Glossogobius* by: • upper lip projecting beyond lower lip / • caudal rounded. Size up to 200 mm SL.

Distribution. Oman, UAE, Yemen, southern Iran, and Pakistan.

Habitat. Lower and middle parts of streams that flow into sea at least once every few years. Occupies soft bottoms with sand, silt or organic material. Also on soft sea bottoms, in estuaries, mangroves and coastal lagoons.

Biology. No data. Other species in this genus are amphidromous and spawn in freshwater. Male guard eggs. Larvae are pelagic and must drift into estuaries or ocean. Juveniles migrate to freshwater to grow into adults and spawn. No resident freshwater populations known, but individuals can live for many years, and larvae may only reach sea in years of heavy flooding. Individuals often bury themselves in soft sediment. Marine and brackish water-only populations are expected to occur along coastlines.

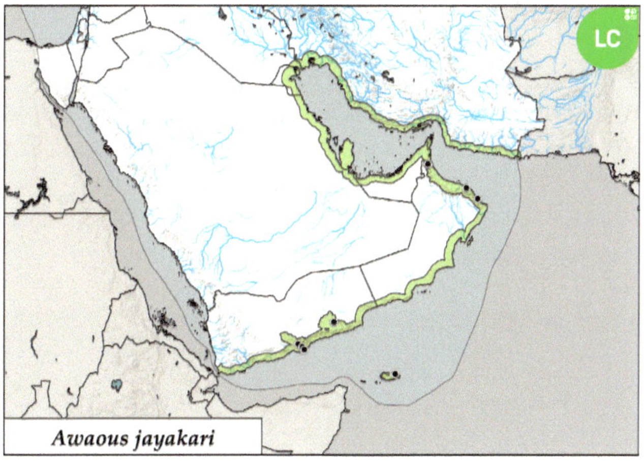

Conservation status. LC.

Remarks. Previously identified as *Awaous aeneofuscus*, a species known from the East African coast, possibly as far north as Socotra (Yemen). The distribution boundary between the two species is unknown, but individuals from Wadi Masila in Hadhramaut are identified as

Wadi Shab in Oman is the habitat of *Awaous jayakari* and *Glossogobius tenuiformis*.

A. jayakari. Individuals from Socotra have yet to be identified and are provisionally included under *A. jayakari*. *Awaous jayakari* differs from *A. aeneofuscus* in having a pectoral base that is scaleless or has only 1–3 scales in adults (vs. pectoral base that is usually completely covered with scales).

Further reading. Esmaeili et al. 2020c (record from Iran); Freyhof et al. 2020 (Arabian Peninsula).

Boleophthalmus dussumieri

Common name. Gel Khorok mudskipper.

Diagnosis. Distinguished from other freshwater gobies in the Persian Gulf basin and the Arabian Peninsula by: ● one large canine tooth behind jaw symphysis and behind row of teeth in lower jaw / ○ no barbels on ventral side of head / ● eye strongly elevated from head profile / ● usually only one row of teeth in lower jaw / ○ 103–185 scales in lateral series / ○ 48–56 predorsal scales. Size up to 187 mm SL.

Distribution. Persian Gulf basin to western India. Upper Shur (Iran) found 55 km upstream from coast and expected to occur in other brackish rivers.

Habitat. Intertidal mudflats. Also, mudflats of brackish rivers far inland.

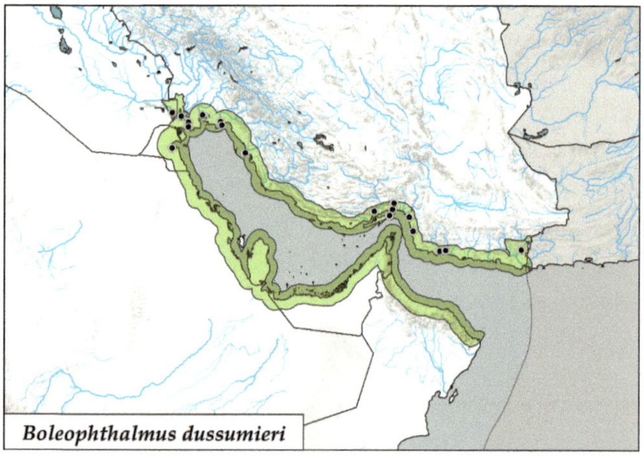

Boleophthalmus dussumieri; Kuwait; ~120 mm SL. © M. Pope.

Biology. Lives up to 6–7 years, usually spawns for a single season and then dies. Spawns first time at 2 years, in April–August. Digs U-shaped vertical burrows to a depth of about 1 m with mud dams around entries. Mud dams usually pentagonal, rarely hexagonal or square, shape varies between populations. Mud dams help to keep water in burrow at low tide. Hides in burrow at high tide, low winter, and very high summer temperatures and when threatened. Active between 13 and 40°C. Muscular pectoral fin aid movement on land. Keeps water under gill cover to maintain a moist atmosphere for gills. Usually remains within mud walls of its territory, submerged in water or wet mud. Emerges from burrows at 10°C, often coming out of burrows to test temperatures. Basks to raise body temperature to 14°C. Keeps itself wet by regular immersion in pools and burrows and cool by evaporative loss of body fluids. Even at 40°C air temperature, body temperature does not exceed 33°C. Mucous cells prevent dehydration. Gas exchange in air and water occurs through gills, inner operculum, nasal, body, and outer operculum skin. Spawns in male burrow. Male guard eggs. Larvae emerge mid-July, usually in August. Male compete for females during courtship. A male usually has five neighbors, half of which are females. Male identify and attract female by raising fins and jumping with about two-thirds of body above ground. Pectoral fin waves, lateral tail beats, and quivers are other signals. On coast, feeds mainly on diatoms, blue-green algae, and filamentous algae, using polygonal enclosure between mud walls as a 'farm' for diatom growth.

Conservation status. LC.

Remarks. Records of this species from the east coast of India refer to *Boleophthalmus boddarti.*

Further reading. Clayton & Vaughan 1982; Clayton & Wright 1989 (behaviour, habitat); Murdy 1989 (description); Polgar et al. 2013 (revision).

Knipowitschia

Dwarf gobies of the genus *Knipowitschia* are a widespread species in lower parts of slowly flowing rivers and lakes in the Black, Caspian, and Mediterranean basins, from Italy to the Aegean coast in Türkiye. All species are very small gobies that inhabit standing or slowly flowing waters, usually with dense submerged vegetation. They might be very abundant, especially in lakes close to the coast. All species of *Knipowitschia* have a lifespan of no more than two years with adults absent in late summer. Currently, seven species are known from West Asia, with one, *K. iljini*, being strictly marine in the Caspian Sea and included in the key only. Notably, *K. iljini* has not been recorded for several decades. *Knipowitschia* is superficially similar to sand gobies of the *Pomatoschistus*, only found in the sea and in brackish parts of estuaries. **Further reading.** Geiger et al. 2014 (generic placements).

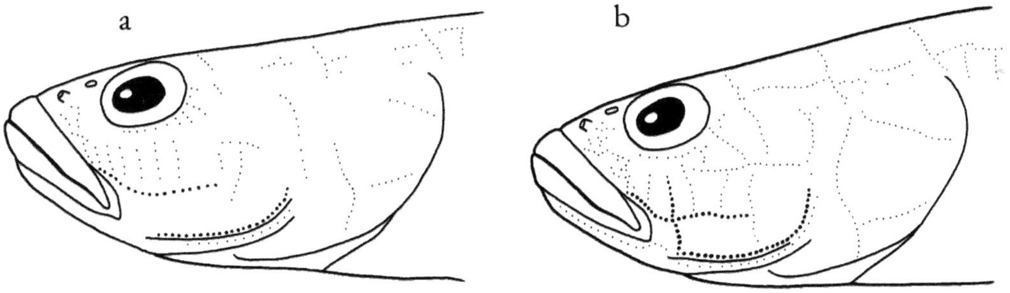

Figure 63. Sensory papillae on side of head: **a**, without vertical row on cheek between rows *d* and *e* in most *Knipowitschia* species; **b**, with vertical row on cheek between rows *d* and *e* in *K. bergi* (from Kottelat & Freyhof 2007, **a**, after Ahnelt 1991, **b**, after Iljin 1928, and Miller 2004).

Key to species of *Knipowitschia* entering freshwaters in West Asia

1a - Otic transverse row *tra* of papillae on cheek extending downward from anterior oculoscapular canal almost to row *b*; a second transverse row of papillae on cheek extending downward through row *d* and almost reaching row *e* (Fig. 63).
.................*K. bergi*

1b - Otic transverse row *tra* of papillae on cheek not reaching to row *b*; no transverse row of papillae on cheek through row *d.*
.................2

2a - Squamation incomplete, scales present under pectoral, on caudal peduncle, and rarely in one series along mid-lateral line.
..................3
2b - Flank and caudal peduncle covered by scales, with more than one series along lateral midline, head, nape, breast, and back naked in front of origin of second dorsal.
..................4

3a - Head canals absent.
..................*K. ricasolii*
3b - Head canals present, often reduced.
..................*K. byblisia*

4a - Anterior oculoscapular canals usually separate or very closely set in midline of posterior interorbital space, with two pores λ (Fig. 64).
..................*K. mermere*
4b - Anterior oculoscapular canals usually fused in posterior interorbital space, with a single pore λ (Fig. 64), rarely canals not extending to interorbital space.
..................5

5a - Caudal asymmetric.
.... *K. longecaudata*
5b - Caudal symmetric.
..................6

6a - Posterior oculoscapular canal usually present (pores ρ^1–ρ^2; as in Fig. 64).
..................*K. caucasica*
6b - Posterior oculoscapular canal always absent (Fig. 64).
..................*K. caunosi*

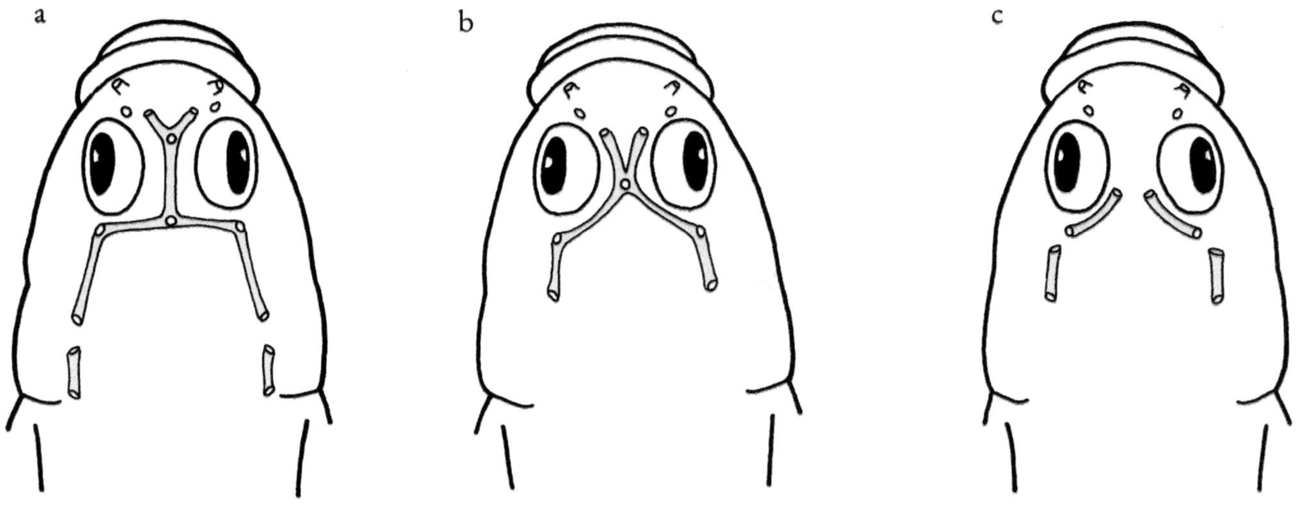

Figure 64. Anterior oculoscapular canal ending: **a**, on snout (pore σ) in *Pomatoschistus* species; **b**, in front of middle of eye (pore λ) in some *Knipowitschia* species; **c**, behind middle of eye (pore K) in some *Knipowitschia* species.

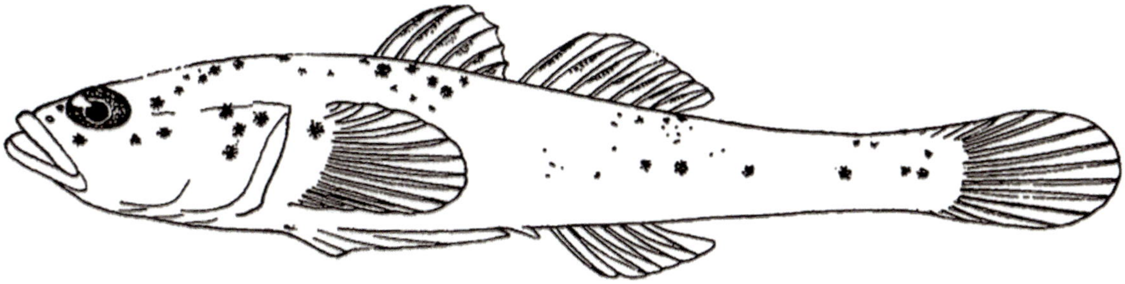

Knipowitschia bergi; Caspian Sea; ~30 mm SL. (from Iljin 1928).

Knipowitschia bergi

Common name. Volga dwarf goby.

Diagnosis. Distinguished from other species of *Knipowitschia* in Caspian basin by: ● otic transverse row *tra* of papillae on cheek extending downward from anterior oculoscapular canal almost to row *b* / ● a second transverse row of papillae on cheek extending downward through row *d* and almost reaching row *e* / ○ flank and caudal peduncle covered by scales, with more than one series along lateral midline / ○ about 30 total scales in midlateral series / ○ caudal symmetric. Size up to 32 mm SL.

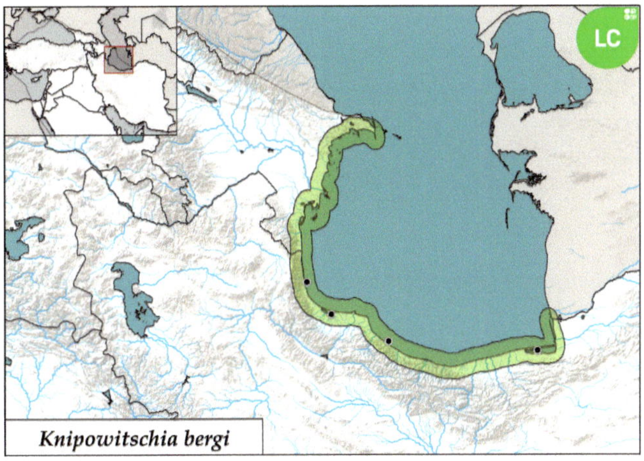

Distribution. Northern and south-eastern Caspian Sea, deltas of Ural and Volga, along west coast south to Iran. Mostly at sea, but was recorded from a river in Gonbad City (Iran).

Habitat. A near-pelagic coastal, marine species found in areas of varying salinity, including freshwater.

Biology. Lives 1 year. Matures at 18–20 mm SL. Spawns after first winter, in May–July, at depths of 2–7 m. Female lay 2–3 portions of eggs (0.4 × 1.0 mm) attached inside a mollusc shell. Adults probably die shortly after spawning. Feeds on small crustaceans.

Conservation status. LC.

Remarks. *Knipowitschia bergi* is poorly known and has been reported from coastal, freshwater or slightly brackish lagoons in the Caspian basin. It is thought to be common in northern the Caspian Sea in Kazakhstan, where it mainly inhabits fresher parts of the sea. *Knipowitschia bergi* has been placed in a separate genus, *Hyrcanogobius*, by some authors based on a slightly different organisation of cephalic sensory pores. Its phylogenetic position should be investigated in future.

Further reading. Iljin 1928 (description); Miller & Pinchuk 2004 (biology, distribution).

Knipowitschia byblisia; lower Dalaman, Türkiye; ~20 mm SL.

Knipowitschia byblisia

Common name. Köyceğiz dwarf goby.

Diagnosis. Distinguished from other species of *Knipowitschia* in Anatolia by: ○ squamation incomplete, scales in two patches: in axillary area and on caudal region, often with a narrow naked zone behind tip of pectoral, rarely in one series along midlateral line / ○ head canals usually reduced to two short postorbital canals, anterior oculo-scapular canal present in few individuals / ○ otic transverse row *tra* of papillae on cheek not reaching to row *b* / ○ no transverse row of papillae on cheek through row *d* / ○ caudal symmetric. Size up to 28 mm SL.

Distribution. Türkiye: Lake Köyceğiz basin, a small stream flowing to Gökova Bay and lower Dalaman drainage.

Habitat. Fresh to brackish waters of coastal lakes and tributaries.

Biology. No data.

Conservation status. EN; appears to be declining within its very small range.

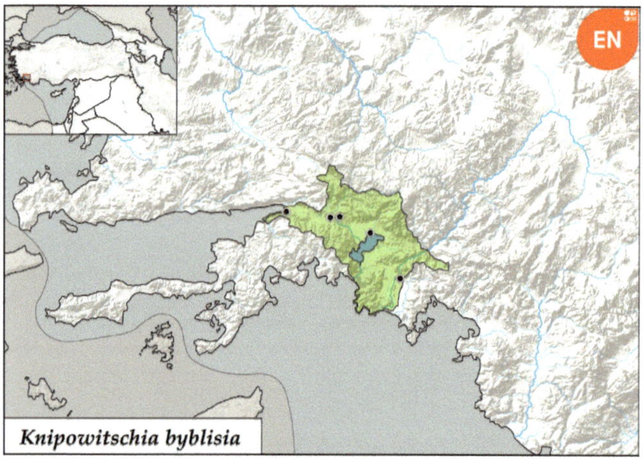

Remarks. *Knipowitschia byblisia* is much more abundant than *K. caunosi* in Lake Köyceğiz. These two species represent a molecular cluster (COI) well separated from other *Knipowitschia*. Within this cluster, syntopic subgroups are slightly different from each other. Fish from these subgroups all agree with the original description of *K. byblisia*.

Further reading. Ahnelt 2011 (description); Ahnelt 2016 (record from Marmaris).

Knipowitschia caucasica

Common name. Caucasian dwarf goby.

Diagnosis. Distinguished from other species of *Knipowitschia* in West Asia by: ● head canals usually complete, rarely posterior oculoscapular reduced / ○ anterior oculoscapular canals usually fused in posterior interorbital space, with a single pore λ / ○ flank and caudal peduncle covered by scales, with more than one series along lateral midline / ○ anterior extremity of anterior oculoscapular canal in front of middle of eye (pore λ) (Fig. 64) / ○ otic transverse row *tra* of papillae on cheek not reaching to row *b* / ○ no transverse row of papillae on cheek through row *d* / ○ caudal symmetric. Size up to 50 mm SL.

Distribution. Coasts of Black, Azov, Caspian, and Marmara Seas. Also, along northern Aegean west to Aliakmon drainage (Greece). Introduced in Karamanlı Reservoir, Lakes Beyşehir, Eğirdir and Eber basins, Büyük Menderes drainage and some places around Antalya (Türkiye). Also introduced to Aral Sea (now extirpated). Invasive in lower Rhine (North Sea basin).

Habitat. Fresh to hypersaline waters (salinity up to 55 ‰) of lakes, estuaries, and lagoons. Most common in shallow, well-vegetated habitats.

Knipowitschia caucasica; Lake Beyşehir basin, Türkiye; ~35 mm SL.

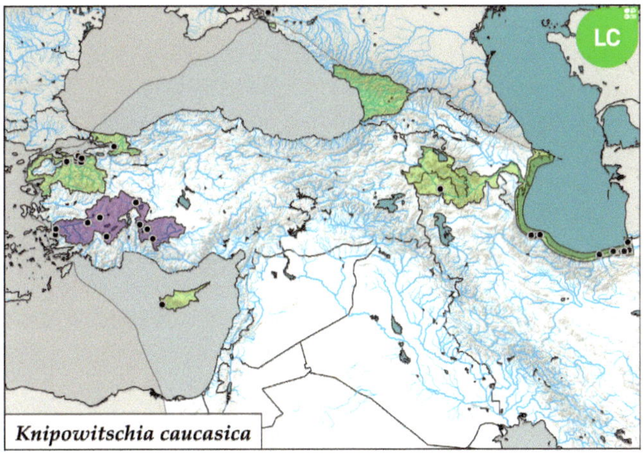

Biology. Lives <2 years. Spawns after first winter in March–July at 20–23 mm SL, releasing up to four portions of eggs. Eggs are attached to roof of a burrow under stones, plant material or shells, defended by male. Postlarvae pelagic. Feeds on small invertebrates.

Conservation status. LC.

Remarks. The origin of introduced populations in Türkiye appears to be the Maritza drainage in European Türkiye, as fish farms in this region were a major source of carp stocked throughout Anatolia. Molecular data also support this hypothesis. Within *K. caucasica*, several molecular groups differ slightly from each other.

Further reading. Economidis & Miller 1990 (systematics); Daoulas et al. 1993 (reproduction); Economou et al. 1994 (reproduction, Trichonis population); Bianco & Ahnelt et al. 1995 (distribution, systematics); Vasil'eva & Kuga 2001 (osteology, systematics); Vasil'eva et al. 2004 (biology); Küçük et al. 2009b (introduction in Eğirdir); Ahnelt 2011 (introduction to Central Anatolia); Tarkan et al. 2015 (distribution).

Knipowitschia caunosi

Common name. Kaunos dwarf goby.

Diagnosis. Distinguished from other species of *Knipowitschia* in West Asia by: ○ head canals present / ○ posterior oculoscapular canal absent / ○ anterior oculoscapular canals usually fused in posterior interorbital space, with a single pore λ, rarely canals not extending to interorbital space / ○ flank and caudal peduncle covered by scales, with more than one series along lateral midline / ○ anterior extremity of anterior oculoscapular canal in front of middle of eye (pore λ) (Fig. 64) / ○ otic transverse row *tra* of papillae on cheek not reaching to row *b* / ○ no transverse row of papillae on cheek through row *d* / ○ caudal symmetric. Size up to 30 mm SL.

Distribution. Türkiye: Lake Köyceğiz basin.

Habitat. Fresh to brackish waters of coastal lakes and rivers.

Biology. No data.

Conservation status. EN; appears to be declining within its very small range.

Remarks. *Knipowitschia caunosi* is much rarer than *K. byblisia* and seems to be endemic to Lake Köyceğiz.

Further reading. Ahnelt 2011 (description).

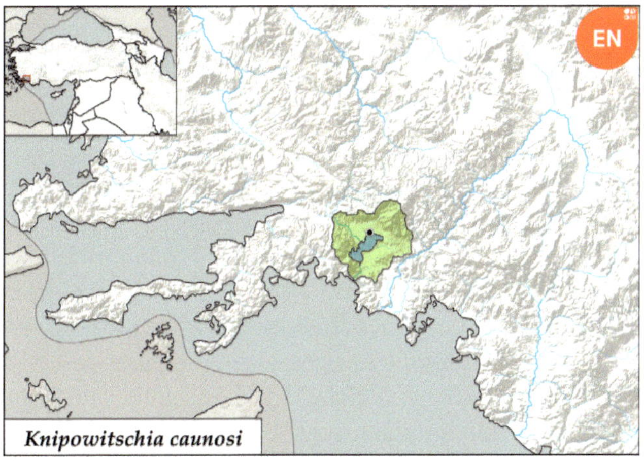

Knipowitschia caunosi; Lake Köyceğiz, Türkiye; ~20 mm SL.

Knipowitschia longecaudata; Kura, Azerbaijan; female, ~35 mm SL.

Knipowitschia longecaudata; Lake Sinoe, Romania; male, ~30 mm SL.

Knipowitschia longecaudata

Common name. Longtail dwarf goby.

Diagnosis. Distinguished from other species of *Knipowitschia* entering freshwater by: ● caudal asymmetric, upper part longer than lower part / ○ anterior extremity of anterior oculoscapular canal in front of middle of eye (pore σ) / ○ anterior oculoscapular canals usually fused in posterior interorbital space, with a single pore λ, rarely canals not extending to interorbital space / ○ flank and caudal peduncle covered by scales, with more than one series along lateral midline / ○ otic transverse row *tra* of papillae on cheek not reaching to row *b* / ○ no transverse row of papillae on cheek through row *d* / ○ posterior oculoscapular canal absent / ○ back naked in front of origin of second dorsal / ○ body with 4–8 dark bars in live nuptial male / ○ 33–40 total scales in midlateral series. Size up to 40 mm SL.

Distribution. Caspian, Azov, northern and western Black Sea basins.

Habitat. Fresh to salt waters of coastal lakes, reservoirs, estuaries, and lagoons in shallow, well-vegetated habitats. Rarely in large rivers where it might occur far upstream.

Biology. Lives <2 years. Spawns after first winter, at 20 mm SL, March–July. Female may spawn several times in a season. Male defend eggs in cavities under stones, plant material, or mollusc shells. Feeds on small invertebrates.

Conservation status. LC.

Remarks. This species has not been found along the Black Sea coast of Türkiye, Russia, and Georgia and may be absent from the Asian Black Sea basin. Fish identified as this species from the Danube delta and the lower Don share their COI DNA sequences with *K. caucasica,* and no *K. longecaudata* has been positively identified by molecular methods. This suggests that *K. caucasica* may have introgressed at least parts of populations.

Further reading. Miller 2004 (description, biology); Zarei et al. 2022a (records from Iran).

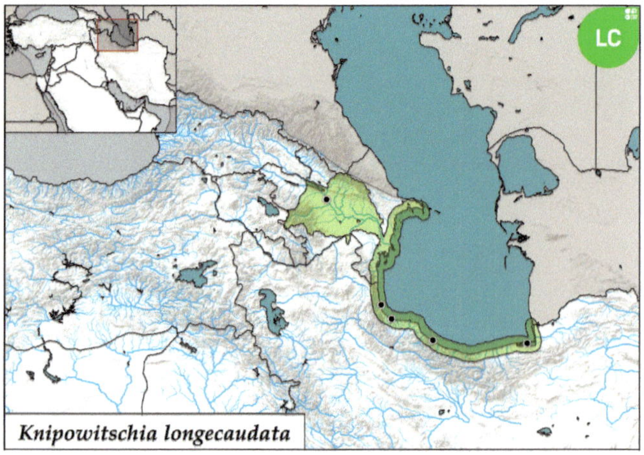

Knipowitschia mermere; Gediz drainage, Türkiye; 25 mm SL.

Knipowitschia mermere

Common name. Gediz dwarf goby.

Diagnosis. Distinguished from other species of *Knipowitschia* in Anatolia by: ● anterior oculoscapular canals usually separate or very closely set in midline of posterior interorbital space, with two pores λ / ○ flank and caudal peduncle covered by scales, with more than one series along lateral midline / ○ head canals present, often reduced / ○ posterior oculoscapular canal absent / ○ otic transverse row *tra* of papillae on cheek not reaching to row *b* / ○ no transverse row of papillae on cheek through row *d* / ○ caudal symmetric. Size up to 30 mm SL.

Distribution. Türkiye: Lower Gediz and Madra drainages, probably more widespread in northern and central Aegean rivers in Anatolia.

Habitat. Fresh to brackish water of coastal lakes.

Biology. No data.

Conservation status. VU; appears to be declining within its small range.

Further reading. Ahnelt 1995 (description).

Knipowitschia ricasolii

Common name. Ephesus dwarf goby.

Diagnosis. Distinguished from other species of *Knipowitschia* in Anatolia by: ● head canals absent / ○ squamation incomplete, scales in two patches: in axillary area and on caudal region, rarely in one series along midlateral line / ○ otic transverse row *tra* of papillae on cheek not reaching to row *b* / ○ no transverse row of papillae on cheek through row *d* / ○ caudal symmetric. Size up to 26 mm SL.

Distribution. Türkiye: Lakes Gebekirse and Kocagöz basins and Eleman marshes in lower Küçük Menderes drainage.

Habitat. Fresh to brackish water of coastal lakes.

Biology. No data.

Conservation status. EN; appears to be declining within its very small range.

Remarks. Gobies in Kavakdere reservoir (Aegean basin) may belong to this species. *Knipowitschia ephesi* is a synonym. *Knipowitschia ricasolii* is closely related to *K. caucasica*.

Further reading. Ahnelt 1995 (description); Çiçek et al. 2019 (*Gobius ricasolii*).

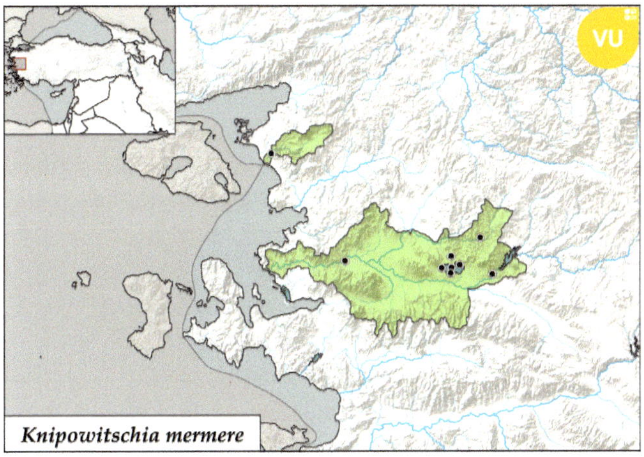

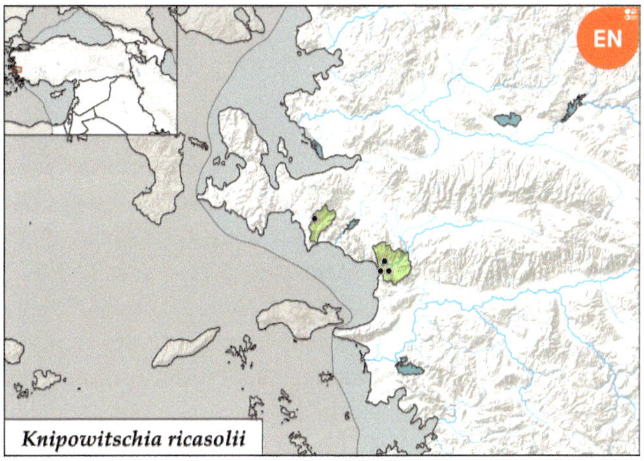

Before it dried out, Lake Marmara was the habitat of a very large population of *Knipowitschia mermere.*

Knipowitschia ricasolii; Lake Zeytinkoy, Türkiye; 25 mm SL.

Rhinogobius

Rhinogobius is a large group of small gobies widely distributed in East and Southeast Asia. It is estimated that there are around 112 species in the genus, with many more yet to be described. Stream gobies exhibit a wide range of life history traits, with some species inhabiting coastal rivers and streams, being amphidromous and producing many eggs and small pelagic larvae. These species often have large distribution areas. Others inhabit headwater streams and lay few eggs from which large benthic larvae hatch. A significant proportion of these species are endemic to a single or few rivers, and only a small proportion have likely been described. Consequently, the identification of *Rhinogobius* species is challenging, as only a few regional faunas have been extensively studied, while *Rhinogobius* from China, Russia, and Vietnam remain largely unstudied. One *Rhinogobius* species is non-native in West Asia, and it is believed to have been introduced from Central Asia to Iran as a contaminant of commercially cultivated fishes such as Chinese carps (*Ctenopharyngodon*, *Hypophthalmichthys*). It is thought to originate from the Russian Amur, from where it was introduced to Central Asia during Soviet

times. Chinese carps have also been introduced to Central Asia from China (e.g., from Wuhan, Yangtze). It cannot be excluded that the non-native *Rhinogobius* originate from outside the Amur. These *Rhinogobius* have been identified as *R. similis*, *R. lindbergi*, or *R. cheni*. Still, identification of the species is based only on the assumption that it has been introduced from the Amur drainage and that these are the only similar species in the Amur. *Rhinogobius* is a very speciose genus, and many undescribed species are expected to occur in China and the Russian part of the Amur drainage. A comprehensive identification of non-native *Rhinogobius* is only possible when the diversity of Chinese *Rhinogobius* species is better understood. The species has expanded its range in the western Caspian basin and Iran. It has recently also reached the Rioni in the Georgian Black Sea basin, opening the door for a westward invasion. As it continues to be introduced as a weed with commercial fish, it could virtually be found everywhere, especially in areas where carps are stocked. In addition to the species discussed here, *R. brunneus* or a similar species, has been recorded from the Persian Gulf but has not been confirmed recently.
Further reading. Endruweit 2018; Huang et al. 2016; Chen et al. 2002; Chen & Shao 1996; Shibukawa 2020 (diversity, life history); Vasil'eva 2007 (Central Asia).

***Rhinogobius* sp.**
Common name. Stream goby.
Diagnosis. Distinguished from other gobies in West Asia by: ● first dorsal with bluish-iridescent, small blotches on 1–2nd dorsal membrane in male / ● orange spot on gill-membrane in male / ○ 5–6 rays in first dorsal / ○ lower lip very slightly projecting beyond upper lip / ○ caudal roundish / ○ flank mottled or marbled, or with 6–8 irregularly shaped bars or blotches / ○ cheek without rows of pores. Size up to 52 mm SL.
Distribution. Introduced and invasive in southern Caspian basin from Kura east to Gorgan. Also in Rioni (Black Sea basin), Lakes Urmia and Namak basins, Iranian Tigris drainage, and Hari. Was introduced in Columbia drainage of western United States, Central Asia, and parts of Russia and China, where it is not native. Believed to be native to Amur drainage.

Rhinogobius sp.; Lake Urmia basin, Iran; female, ~40 mm SL.

Rhinogobius sp.; Amur Darja drainage, Uzbekistan; male, ~40 mm SL.

Habitat. A wide range of rivers, streams, and coastal wetlands, also in Caspian Sea. Often found in habitats with submerged vegetation.

Biology. Poorly known. Lives one year and spawns first spring. Eggs laid under or between stones, plants or shells. Feeds on small invertebrates.

Conservation status. Non-native; introduced as a weed with carp and/or Chinese carps.

Further reading. Al-Hassan & Miller 1987 (record of *R. brunneus*); Coad & Abdoli 2000 (records from Iran as *R. similis*); Bogutskaya et al. 2008 (colour pattern from Amur); Vasil'eva & Kuga 2008 (identification as *R. cheni* in Central Asia); Eagderi & Moradi 2017 (records from Urmia); Eagderi et al. 2018a (records from Tigris); Sadeghi et al. 2018 (identification as *R. lindbergi*).

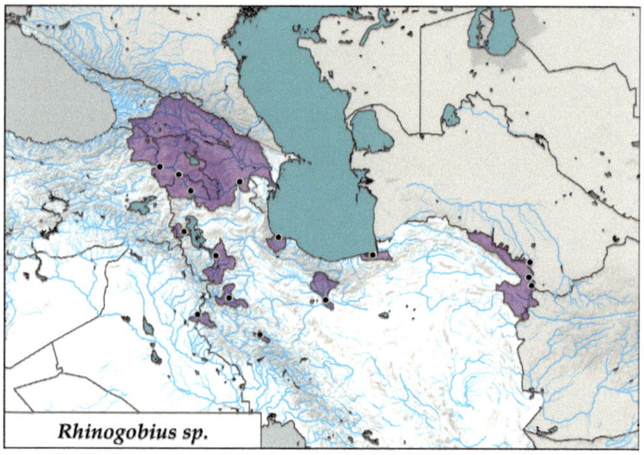

Coastal wetlands are major habitats of Gobies, pipefishes, silversides, and killifish, as here in Oman.

Family Syngnathidae

Pipefishes

Syngnathidae is characterised by an elongate body covered by a series of bony rings, a small mouth at the extremity of a long and slender snout, and the absence of pelvics (and caudal, dorsal, and anal) fins in some species. The best-known members of the family are probably the marine sea horses (*Hippocampus*). The family includes approximately 220 species, mostly found in tropical marine waters. Several species enter estuaries and coastal lagoons, while others are strictly freshwater inhabitants. The trunk rings begin with the ring bearing the pectoral base and end with the ring bearing the anus (which usually bears the anal fin). The pre-dorsal rings end with the ring bearing the first dorsal ray. The caudal rings begin with the first ring situated posterior to the anus and end with the penultimate ring. The terminal ring bears the caudal fin.

Syngnathus nigrolineatus and *S. caspius* regularly enter freshwater habitats in West Asia. Three additional species have been recorded to enter brackish waters of estuaries and coastal lagoons and may find their way into freshwater. These species can be distinguished using the key below. The diversity of the genus *Syngnathus* in the Mediterranean basin is significantly underestimated. Molecular studies indicate that several species may be confused under *S. abaster*. A critical review is required to ascertain the diversity and distribution of Mediterranean pipefishes. It is proposed that *S. abaster* be restricted to the coastal habitats of Spain, France, Corsica, and Sardinia. The Italian coastal species is identified as *S. agassiz*, described from Trieste, the freshwater species in the Black Sea basin as *S. nigrolineatus*, and the Caspian species as *S. caspius*. The well-defined molecular group found in Tunisia could be identified as *S. algeriensis* or *S. flavescens* in the future. From these, only *S. algeriensis*, *S. caspius*, and *S. nigrolineatus* regularly occur in freshwaters, while other species are restricted to marine or estuarine habitats. *Syngnathus agassiz* is expected to be found along the coast of the Eastern Mediterranean, but more research is needed to understand its distribution. The morphological characteristics that distinguish these different pipefishes have yet to be studied; thus, a definitive diagnosis cannot be provided. **Further reading.** Dawson 1985; Kuiter 1998; Wilson et al. 2010; Sanna et al. 2013 (molecular diversity).

Syngnathus nigrolineatus; lower Danube, Romania; ~120 mm SL.

Key to species of Syngnathidae in fresh and brackish waters in West Asia

1a - Caudal, pectoral and anal absent.
..................*Nerophis ophidion*
1b - Caudal, pectoral and anal present.
..................2

2a - Anterior trunk rings not fused ventrally; snout compressed, anterior part deeper than head; up to 350 mm SL.
..................*Syngnathus typhle*
2b - Anterior trunk rings ventrally fused; snout cylindrical.
..................3

3a - An elongated bump on top of head behind eye.
..................*Syngnathus acus*
3b - No elongated bump on top of head behind eye, dorsal profile of head almost linear.
..................*Syngnathus caspius / S. nigrolineatus*

https://doi.org/10.1515/9783111677811-036

Syngnathus caspius; Gorgan Bay, Iran; male with egg pouch under tail, 110 mm SL. © R. Patimar.

Syngnathus caspius
Common name. Caspian pipefish.
Diagnosis. Characters to distinguish this species from other species of *Syngnathus* have yet to be worked out. Size up to 190 mm SL.
Distribution. Caspian basin.
Habitat. Many marine, brackish, and freshwater habitats are mostly associated with dense submerged vegetation.
Biology. No data.
Conservation status. LC.
Further reading. Berg 1949b (description).

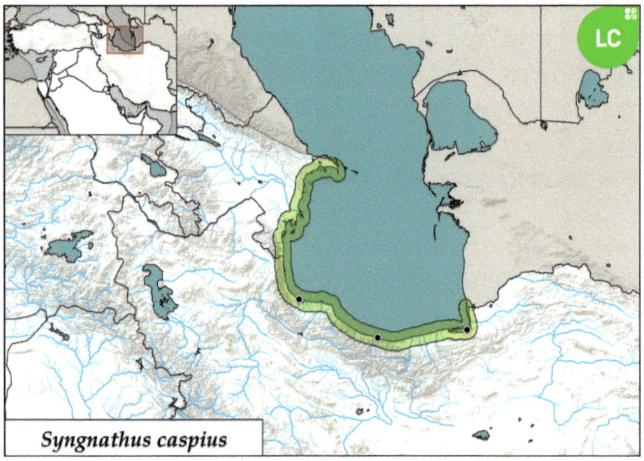

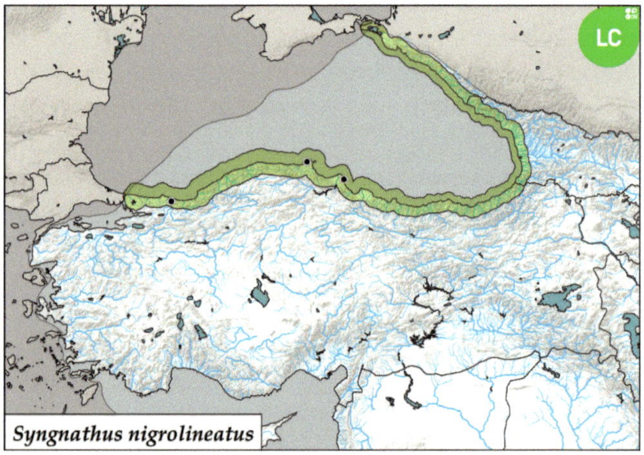

Syngnathus nigrolineatus; lower Danube, Romania; ~120 mm SL.

Syngnathus nigrolineatus
Common name. Black Sea pipefish.
Diagnosis. Characters to distinguish this species from other species of *Syngnathus* have yet to be worked out. Size up to 190 mm SL.
Distribution. Coastal habitats and lower reaches of rivers in Black Sea basin; in Danube reaching Romanian-Hungarian border; in Dniepr reaching Kyiv. Introduced in middle and lower Volga reservoirs with mysids brought from Don estuary, now spreading and already south of Moscow.
Habitat. A wide range of marine, brackish, and freshwater habitats are usually associated with dense submerged vegetation and are also on open, muddy bottoms.
Biology. Lives up to 4 years. First spawns at 1 year, in April–October. Female lay 10–60 eggs in a brood pouch on ventral surface of male's tail. Male fertilise eggs as they enter pouch. Eggs incubate for 20–32 days.
Conservation status. LC.
Further reading. Svetovidov 1964 (description, as *S. nigrolineatus*); Silva et al. 2006a, b (reproduction).

Family Atherinidae

Old world silversides

A family of euryhaline fishes, comprising mostly small to medium-sized species, with approximately 14 genera and 70 species found throughout the world's tropical and temperate seas. Many species are confined to freshwater, often entering brackish waters or coastal rivers. Atherinid schools are often very abundant in estuarine ecosystems. They are characterised by a silvery, elongate, compressed body, two widely separated dorsals, a large eye, and a long snout with a terminal mouth. Atherinids can be distinguished from superficially similar Mugilidae (mullets) by a laterally compressed head (vs. dorso-ventrally flattened), no of scales on the top of the head (vs. presence), and 7–10½ rays in the first dorsal (vs. 4). The diversity of *Atherina* in the Mediterranean basin has been significantly underestimated, with *A. boyeri* being restricted to the western Mediterranean. *Atherina caspia* and *A. pontica* are simultaneous subjective synonyms, and the first reviser [apparently Kessler, 1874: 108] gave precedence to *A. pontica* over *A. caspia*. Consequently, the populations in the Mediterranean from Greece to the east, including the Black Sea and Caspian basins, are now identified as *A. pontica*. However, several species might be included in this region. Further unidentified *Atherina* have been discovered in the Orontes, Lake Köyceğiz (Aegean Sea), and potentially in other freshwater habitats. Here, we present only one species that has been reported to enter freshwater habitats in West Asia. Nevertheless, the taxonomy of *Atherina* remains poorly resolved, with the possibility that several synonyms represent valid species. As the freshwater species of the genus await revision, we maintain all records under the name *A. pontica*. **Further reading.** Kessler 1874 (description); Whitehead et al. 1986 (diagnosis; in part); Kottelat 1997 (nomenclature).

Brackish Lake Köyceğiz in southwestern Türkiye is the habitat of *Atherina* sp., *Knipowitschia byblisia* and *K. caunosi.*

https://doi.org/10.1515/9783111677811-037

Atherina pontica; Lake Sinoe, Romania; ~60 mm SL.

Atherina sp.; Lake Köyceğiz, Türkiye; ~70 mm SL.

Atherina pontica

Common name. Eastern sand smelt.

Diagnosis. Only species in family in freshwaters in West Asia. Size up to about 85 mm SL.

Distribution. Western Greece east to Caspian basin. Introduced in several lakes and reservoirs in Anatolia and probably elsewhere.

Habitat. Lower parts of rivers, estuaries, coastal lakes, and the sea. Freshwater populations prefer still or slow-flowing waters. Pelagic in lakes and sea.

Biology. Gregarious. Usually lives 1–3 years, rarely up to 4. Spawns at 1–2 years in spring and early summer. Fractional spawners. Eggs with long hairy appendages attached to filamentous algae, deposited at 2–6 m depth, perhaps shallower in rivers. Larvae pelagic but often form shoals near shore. In lakes and estuaries, feeding mainly on small planktonic invertebrates, often on benthos in rivers.

Conservation status. LC.

Further reading. Whitehead et al. 1986 (diagnosis; in part); Economou et al. 1994 (larvae); Kottelat 1997 (nomenclature); Daoulas et al. 1997 (larvae); Leonardos & Sinis 2000 (biology); Leonardos 2001 (ecology); Miller 2003 (biology); Gençoğlu et al. 2017 (feeding).

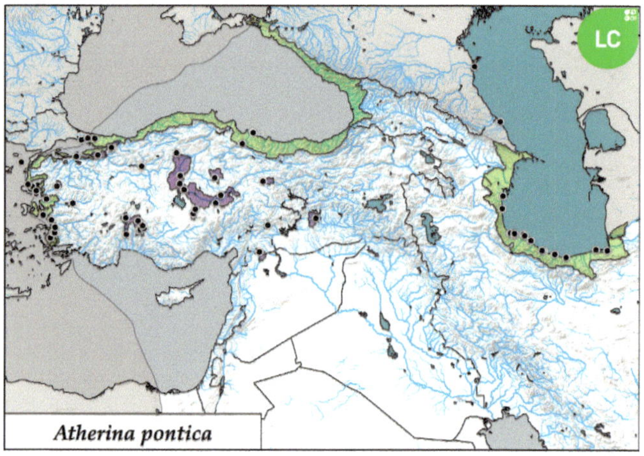

Family Aphaniidae

Eurasian killifishes

The family comprises approximately 45 species, distinguished by the absence of ossified rays in the dorsal, a short and stout body, and sexually dimorphic colour patterns. The males typically exhibit prominent black or brown bars on the flank, while the females display a mottled or spotted colour pattern, with short black or brown lateral bars in some species. This resulted in numerous instances of dual descriptions of males and females as different species during the 19th century when scientists were unaware of this sexual dimorphism. Extensive research was conducted on Aphaniids previously classified within the American family Cyprinodontidae. The definition, limits, and contents of the family Cyprinodontidae have undergone significant changes over the past 30 years. Aphaniidae is treated as its own family related to Valenciidae, which is endemic to Europe and African lamp-eyes (Procatopodidae). Aphaniidae has been subdivided into eight genera, with six of which (excluding *Tellia* and *Apricaphanius*) are included in the coverage of this book. Aphaniids have a long fossil record, with the oldest *Aphanius*-like otoliths deriving from Miocene deposits almost 21 million years old. At which times, Aphaniids also occurred in Central Europe and the Caucasus region, and it is possible that they vanished from these regions due to climatic changes. The biology of most Aphaniids remains unknown in the field. However, they have been extensively studied in captivity. During the spawning period, which extends throughout spring and summer, year-round in some species, males establish territories along the shores, usually in dense vegetation but also between rocks, which they defend against rival males. Females spawn with one male, typically in algae or other vegetation close to the water surface or into gravel bottoms. If population densities are high, several males may sneak in to join the spawning pair. One or a few eggs are laid into the spawning substrate during a single spawning act. Individual females may spawn several egg clutches during a day and then typically pause for a few days before spawning again. Eggs possess filamentous structures to attach to the substrate. The larvae usually hatch after 7–15 days, with the timing dependent on temperature. Young Aphaniids exhibit rapid growth, with some reaching maturity within the same or the following year. Most species feed on algae, detritus, and small invertebrates. Several Aphaniids have very limited distribution ranges, with some species facing a high risk of extinction. One species is currently extinct. *Ex situ* populations of numerous species have been established within the aquarium hobby, yet they are poorly managed and typically based on a very small founder broodstock. **Further reading.** Parenti 1981 (phylogeny); Kottelat & Wheeler 2001 (nomenclature); Hrbek & Meyer 2003 (phylogeny, biogeography); ICZN 2003 (nomenclature); Freyhof et al. 2017b (family concept); Freyhof & Yoğurtçuoğlu 2020 (generic structure of Aphaniidae); Kärst et al. 2020 (reproduction, *ex situ* maintenance).

Key to the genera of Aphaniidae in West Asia

The bars mentioned in the keys and the accompanying text are dark brown or black. The silvery white bars are referred to as interspaces. The key is largely based on the colour pattern of adult, nuptial individuals. It may be challenging to differentiate between immature individuals and those outside the spawning season.

1a - Head canals present, anterior supraorbital canal with 2–3 pores, posterior supraorbital canal with 2 pores, preopercular-mandibular canal with 6–7 pores.
..................2

1b - Head canals absent, cephalic sensory pores reduced to a series of neuromasts in small depressions, not connected to canals.
..................3

https://doi.org/10.1515/9783111677811-038

2a - Caudal in male hyaline or white, often with yellow margin, with three wide, bold black bars, first bar situated on caudal base, shortly behind end of hypural complex, often faded (no bars in *A. furcatus*); in nuptial female no dermal sheath around anterior anal rays.
..................*Aphaniops*
2b - Caudal in male pale- or deep-yellow or orange, clearly distinct from silvery interspaces of flank bars; caudal with 0–2, often faint bars, no bar behind end of hypural complex; in nuptial female, dermal sheath, present around first few anal rays.
..................*Aphanius*

3a - Teeth conical, in three rows; caudal peduncle very narrow, its depth 2.5–3.2 times in its length; body without scales.
..................*Kosswigichthys*
3b - Teeth tricuspid, in one row; caudal peduncle deep, its depth 1.2–2.0 times in its length; body covered by scales, but reduced or absent in some species.
..................4

4a - Flank in male bluish, grey, or almost black with often irregular set and shaped, iridescent blue-white to silvery spots or bars often forming narrow vertical rows; male with very narrow, blue-white or silvery rows of spots or small blotches forming bands on black or blue caudal; female without diamond-shaped or roundish bold black blotch at mid-height of caudal base.
..................*Paraphanius*
4b - Flank in male silvery with very regular set and shaped, rarely irregular, brown or black bars, overlaid with iridescent silvery spots in some species; male with bold, black or brown bars or with narrow brown bands on hyaline, white or yellow caudal, or caudal hyaline, with white margin in some species; female with diamond-shaped or roundish bold black blotch at mid-height of caudal base.
..................5

5a - Caudal in male without bars or rows of spots, with 1–5 indistinct vertical rows of small brown spots on proximal portion of caudal in some species; dorsal margin white, caudal and anal margins often white; if dorsal and anal margins black (*E. isfahanensis*) caudal without rows of spots or bars.
..................*Esmaeilius*
5b - Caudal in male with 1–4 bold black bars or numerous vertical rows of small black or brown spots (4–14 in *A. villwocki*); dorsal- and usually anal with margin black.
..................*Anatolichthys*

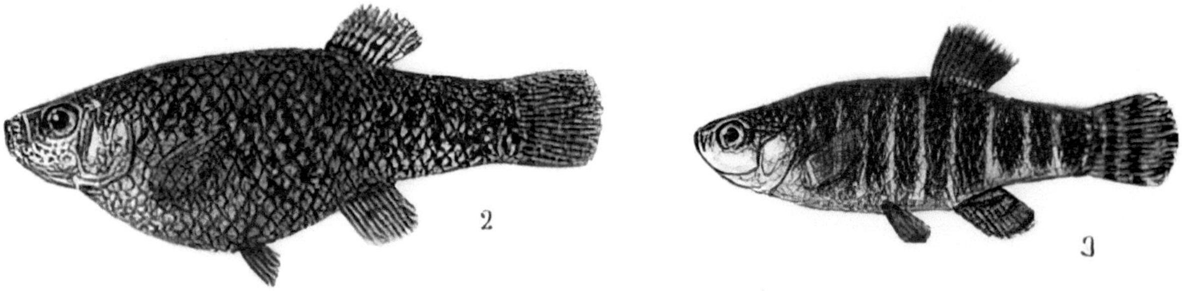

An unidentified species of *Anatolichthys* was known from Syria in the early 20th century (from Pellegrin 1923).

Anatolichthys

Anatolichthys is the largest genus of Eurasian killifishes, with all (except one) species occurring in Central Anatolia. They often have restricted distribution ranges in relatively small habitat patches. Two species are found in the Black Sea basin. *Anatolichthys* is related to *Kosswigichthys*. All *Anatolichthys* inhabit inland waters, including springs, lakes, and the shores of rivers. They also inhabit reservoirs. All species mature within a few months after hatching, usually late in the year of birth. Males establish territories along the banks, usually in dense vegetation, but also between rocks, which they defend against rivals. Females spawn with one or more males, usually in algae or other vegetation near the water surface or on gravel bottoms. Few eggs are deposited in the substrate during a spawning event. Individual females produce several clutches of eggs in one day. All species feed on algae, detritus, and small invertebrates. Many springs previously inhabited by *Anatolichthys* dried out in the late 20th century, and non-native species now dominated many others. Due to the deteriorating ecological conditions in Central Anatolia, many populations of *Anatolichthys* will vanish in the future. One species, *A. splendens*, became extinct after introducing non-native species for fisheries. An additional, potentially extinct species is known from early 20th-century records in the Damascus and Jordan basins. This species is similar to *A. marassantensis*, and it is necessary to determine whether it represents an undescribed species or was introduced to the region. Three species of *Anatolichthys* are notable for their pelagic habitat and reduced scale coverage, which are adaptations to predator-free, lacustrine environments. However, two of these species are now on the brink of extinction and largely restricted to springs at lake shores, as the ecological conditions in the lakes no longer allow fish to exist there permanently. **Further reading.** Pellegrin 1923 (unidentified species from Syria); Hrbek et al. 2002 (biogeography); Hrbek & Meyer 2003 (phylogeny, biogeography); Geiger et al. 2014 (phylogeny); Yoğurtçuoğlu & Freyhof 2018 (identification of species); Freyhof & Yoğurtçuoğlu 2020b (generic structure of Aphaniidae); Freyhof & Yoğurtçuoğlu 2020 (distribution).

Key to species of *Anatolichthys*

1a - Scales on back and belly usually absent, scales absent or few on flank in one species; usually 1–3 lines of large, not or only slightly overlapping scales on mid-lateral flank.
..................2
1b - Body completely covered with overlapping scales.
..................5

2a - Lateral-head profile almost quadratic, lower jaw large, directing upward; body depth 4.0–5.3 times in standard length.
..................3
2b - Lateral-head profile triangular or rounded, lower jaw small, directing forward; body depth 3.2–4.2 times in standard length.
..................4

3a - No or few isolated scales on flank; black bars wider at dorsal part in male, narrower in ventral region.
..................*A. saldae*
3b - 38–46 total, overlapping scales along lateral series; male with black bars regularly shaped, not wider at dorsal part.
..................*A. splendens*

4a - Caudal peduncle depth 11–14 % SL; head depth at nape 61–69 % SL; head width at nape 51–60 % SL; nuptial male with grey dorsal base.
..................*A. sureyanus*
4b - Caudal peduncle depth 16–18 % SL; head depth at nape 68–75 % SL; head width at nape 62–74 % SL; nuptial male with completely black dorsal.
..................*A. transgrediens*

5a - Male with irregularly set and shaped, dark-brown bars, bars with white spots or vermiculate pattern, some large individuals' plain dark-grey with some yellow blotches.
..................*A. irregularis*
5b - Male with regularly set and shaped dark-brown or black bars, without white spots in bars.
..................6

6a - Male with a completely black anal or anal white with a wide black margin, with irregularly set and shaped black bands and/or blotches.
..................*A. fontinalis*
6b - Male with yellow or white anal with a narrow black margin, usually with 1–5 black bands or rows of blotches.
..................7

7a - Male with 4–14 vertical rows of small black or brown spots on caudal; male with 13–25 narrow brown bars on body; female larger than 40 mm SL usually without back spots or blotches on flank except for one bold black blotch at middle of hypural.
..................*A. villwocki*
7b - Male with 1–4 bold black vertical bars on caudal; male with 5–14 black bars on body; female with bold back spots or/and blotches on flank.
..................8

8a - Male with black dorsal base; dorsal with black distal margin and white proximal band or with a proximal row of white spots or blotches distinctly or slightly above dorsal base.
..................9
8b - Male with brown, hyaline, or white dorsal base; dorsal completely black or with greyish or black distal margin and a white line or row of white spots at dorsal base or very slightly above.
..................10

9a - Male with wide, white proximal band in black dorsal; last lateral bar in male regularly shaped, not fused with penultimate bar.
..................*A. iconii*
9b - Male with a row of white spots or blotches along or distinctly above dorsal base; last dark-brown bar irregularly or diamond-shaped, usually fused with pre-last bar.
..................*A. maeandricus*

10a - Male with rhomboid body shape, dorsal profile straight until vertical of pelvic origin; male with dark-brown bars on caudal peduncle being 2.3–4.0 times as wide as interspaces.
..................*A. danfordii*
10b - Body shape ovoid, rarely rhomboid, dorsal profile convex; male with dark-brown bars on caudal peduncle being 0.3–1.8 times as wide as interspaces.
..................11

11a - Male with a completely black dorsal; usually with 6–8 black or dark-brown lateral bars on flank, brown bars in front of dorsal origin narrower than whitish interspaces.
..................*A. meridionalis*
11b - Male with a white dorsal base or a white line or row of spots at base of black dorsal; usually with 8–13 brown lateral bars on flank, brown bars wider than whitish interspaces.
..................12

12a - Male with a white line or row of white spots at or very slightly above dorsal base; brown bars as wide as or slightly wider than interspaces.
..................*A. marassantensis*
12b - Male usually with grey or whitish dorsal base and a black mid-dorsal band and a hyaline or pale-grey distal dorsal margin, except base, dorsal completely black in some populations; brown bars much wider than interspaces.
..................*A. anatoliae*

Anatolichthys anatoliae
Common name. Lake Tuz killifish.
Diagnosis. Distinguished from other species of *Anatolichthys* by: ○ male usually with grey or whitish dorsal base and a black mid-dorsal band and a hyaline or grey distal dorsal margin / ○ except base, dorsal completely black in some populations / ○ body shape ovoid / ○ male with 8–10 wide, regularly set, yellowish dark-bars / ○ third-last bar anterior to caudal base 2.1–4.3 times wider than anterior white interspace / ○ male with yellow anal with narrow black margin / ○ male with 1–4 bold black vertical bars on caudal / ○ male with flank bars in front of dorsal origin wider than interspaces / ○ male with dark-brown bars on caudal peduncle being 0.3–1.8 times as wide as interspaces / ○ female with numerous black spots or blotches on flank not organised in bars or a midlateral row / ○ body completely covered with overlapping scales. Size up to about 45 mm SL.

Anatolichthys anatoliae; Lake Meyil, Türkiye; female, 43 mm SL.

Anatolichthys anatoliae; Lake Meyil, Türkiye; male, 40 mm SL.

Distribution. Lakes Tuz, Beyşehir, and Suğla basins, around Konya and eastward to Niğde.
Habitat. Springs and streams with dense underwater vegetation.
Biology. Live up to 4 years; biology as described in genus section.
Conservation status. VU; habitats have been lost continuously for many years.
Remarks. *Aphanius lykaoniensis*, *A. venustus*, *A. flavianalis*, *A. obrukensis* and *A. aksaranus* are synonyms.
Further reading. Wildekamp et al. 1999 (distribution); Yoğurtçuoğlu & Freyhof 2018 (distribution, identification).

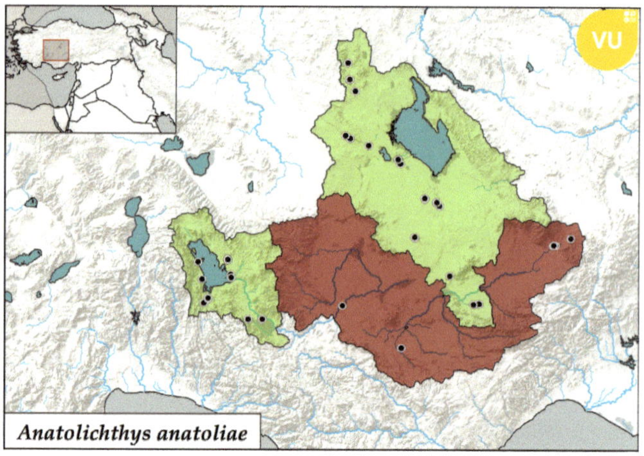

Anatolichthys anatoliae

Vanishing Lake Tuz. Lake Tuz in Central Anatolia was once the second-largest lake in Türkiye. It was a popular destination for flamingos, who flocked there to feed and nest. People also visited to witness the lake's seasonal colour changes and to bathe in the mineral-rich water, mud, and salt. Lake Tuz is fed by groundwater that originates in the Northern Taurus and passes through the Konya Plain and Obruk Plateau, which bubbles up into Lake Tuz via springs. Other water sources include two major streams and rain that primarily falls in springtime. Water in the region has become increasingly scarce, with more frequent and intense droughts. The region is a significant climate change hotspot, warming more than the global average since the pre-industrial period. Before 2000, Lake Tuz generally contained sufficient water in August to be considered permanent. From 1985 through 2000, the lake's water level exceeded 20 % of the current shoreline in all years except 1992 and 1994. During the most humid years, 1988 and 1996, the lake reached its fullest extent, with 93 % and 58 % of the lake covered by water, respectively. Notwithstanding the drought conditions that prevailed in 1992, water still managed to cover 16 % of the lakebed. However, a distinct shift occurred after the year 2000. Between 2001 and 2016, water spanned less than 20 % of the lake every August (except for 2015) as droughts became more frequent and intense. In 2008 and 2016, the lake completely dried up. Research indicates that an extraordinary meteorological drought has devastated Lake Tuz since 2019. In July 2021, the lake's water levels were at an all-time low, resulting in the mass death of flamingos. The decline of Lake Tuz coincided with the excessive use of groundwater and surface-water resources, which were used to irrigate agriculture and meet the water needs of the surrounding provinces. Furthermore, as surface water dwindled during the intense drought, people turned to the groundwater that historically fed the springs. In the event of further drought conditions, water extraction will likely continue at current levels, resulting in the disappearance of lakes, springs, and streams in Central Anatolia. **Further reading.** Aydın et al. 2020 (Lake Tuz); MedECC 2020 (climate change in Mediterranean).

Anatolichthys danfordii; Soysallı, Türkiye; female, 43 mm SL.

Anatolichthys danfordii; Soysallı, Türkiye; male, 38 mm SL.

Anatolichthys danfordii

Common name. Sultan killifish.

Diagnosis. Distinguished from other species of *Anatolichthys* by: • male with rhomboid dorsal profile, straight up to a vertical line through pelvic origin / ○ male with 8–14 black bars / ○ male with bars behind dorsal origin wider than interspaces / ○ third-last bar anterior to caudal base 2.3–4.0 times wider than anterior white interspace in male larger than 26 mm SL / ○ male with 1–4 bold black vertical bars on caudal / ○ male with completely grey or black dorsal often with grey margin / ○ dorsal base hyaline or with a narrow white line / ○ male with yellow or white anal, with narrow black margin and 1–4 narrow, black, concentric bands / ○ female with numerous bold black spots and/or blotches on flank not organised in bars or a midlateral row / ○ female without spots or blotches on caudal and/or anal / ○ body completely covered with overlapping scales. Size up to 46 mm SL.

Distribution. Türkiye: Sultan marshes in Develi depression.

Habitat. Springs, spring-fed streams, and canals with or without dense underwater vegetation.

Biology. Live up to 3 years; biology as described in genus section.

Conservation status. CR; see comments under *Pseudophoxinus elizavetae*.

Remarks. *Cyprinodon chantrei* is a synonym.

Further reading. Wildekamp et al. 1999 (identification); Pfleiderer et al. 2014 (*C. chantrei*); Yoğurtçuoğlu & Freyhof 2018 (identification).

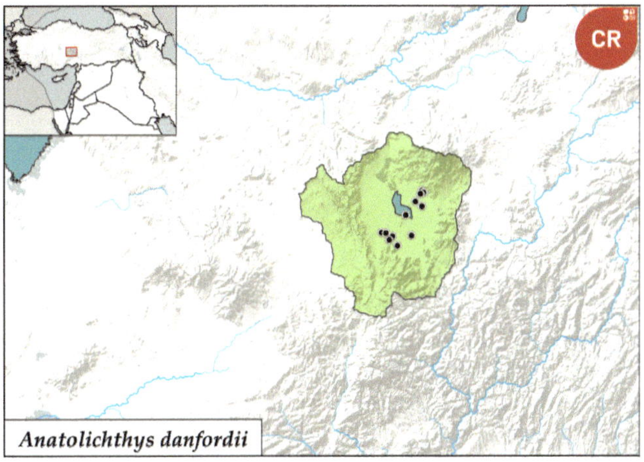

Anatolichthys fontinalis; Yarışlı, Türkiye; female, 32 mm SL.

Anatolichthys fontinalis; Yarışlı, Türkiye; male, 35 mm SL.

Anatolichthys fontinalis

Common name. Yarışlı killifish.

Diagnosis. Distinguished from other species of *Anatolichthys* by: ○ male with a completely black anal or anal white with a wide black margin, with irregularly set and shaped black bands and/or blotches / ○ body completely covered with overlapping scales / ○ male with regularly set and shaped dark-brown or black bars, without white spots in bars / ○ dorsal completely black or with white or hyaline base / ○ flank bars behind dorsal origin wider than interspaces / ○ female with numerous black spots or blotches on flank often organised in midlateral rows. Size up to about 50 mm SL.

Distribution. Türkiye: Karapınar spring near Yeşilova, Düğer spring near Lake Yarışlı southwest of Lake Burdur, springs in former Lake Karaevli southeast of Lake Burdur and Lakes Karataş and Salda basins.

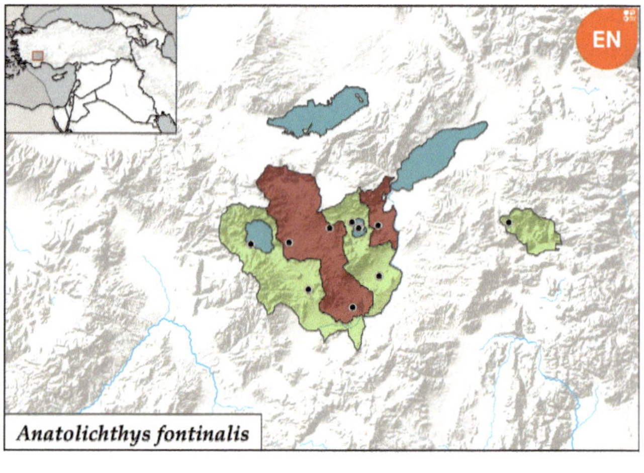

Habitat. Streams, springs, and lake shores, with or without vegetation, also common on rocky shores.

Biology. Live up to 3 years; batch spawning, spawns synchronously with *A. saldae* from April to September in Lake Salda.

Conservation status. EN; appears to be declining within its very small range. Extirpated from Karapınar and Düğer spring and Lake Karataş basin.

Remarks. Potentially introduced to Lake Salda, where it hybridises with *A. saldae*. *Aphanius altus* from Karapınar and *A. litoralis* from Lake Karataş are synonyms.

Further reading. Akşıray 1948 (description); Geiger et al. 2014 (phylogeny); Yoğurtçuoğlu & Freyhof 2018 (identification); Yoğurtçuoğlu 2019 (reproduction).

Anatolichthys iconii

Common name. Eğirdir killifish.

Diagnosis. Distinguished from other species of *Anatolichthys* by: • male with black dorsal, with a wide, hyaline or white, proximal band / ○ male with black dorsal base / ○ last lateral bar in male regularly shaped, not fused with penultimate bar / ○ female with bold back spots or/and blotches on flank / ○ male with yellow or white anal with a narrow black margin / ○ male with 1–4 bold black vertical bars on caudal / ○ body completely covered with overlapping scales. Size up to about 50 mm SL.

Distribution. Lake Eğirdir and Kovada basins. Also in two springs in upper Köprüçay and Aksu.

Anatolichthys iconii; Lake Eğirdir basin, Türkiye; female, 30 mm SL.

Anatolichthys iconii; Lake Eğirdir basin, Türkiye; male, 25 mm SL.

Habitat. Lakeshores, springs, and streams, usually with dense vegetation.

Biology. Live up to 4 years; feeds mainly on aquatic insect larvae, amphipods and diatoms. Population size in Eğirdir fluctuates in parallel with *Atherina* population.

Conservation status. VU; had declined sharply in Lake Eğirdir before about 2006 but has increased since then, presumably following the disappearance of predation by *Sander lucioperca* due to overfishing. It is now declining again as the lake level is falling.

Further reading. Akşıray 1948 (description); Küçük et al. 2009b (population increase); Geiger et al. 2014 (phylogeny); Freyhof et al. 2017b (Neotype designation); Yoğurtçuoğlu & Freyhof 2018 (identification); Güçlü 2012 (biology; as *Aphanius anatoliae*).

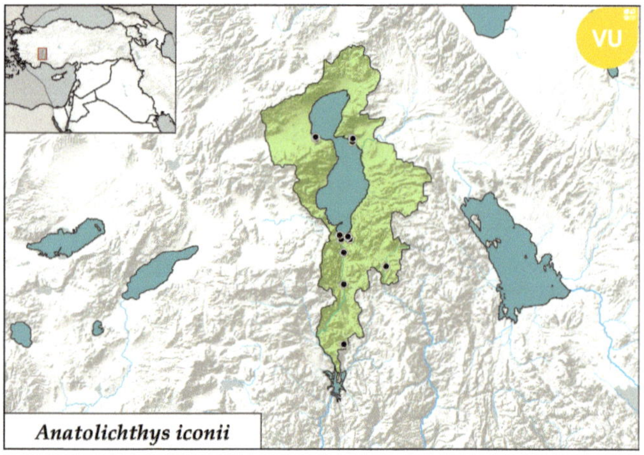

Anatolichthys irregularis; Kaklık spring, Türkiye; female, 52 mm SL.

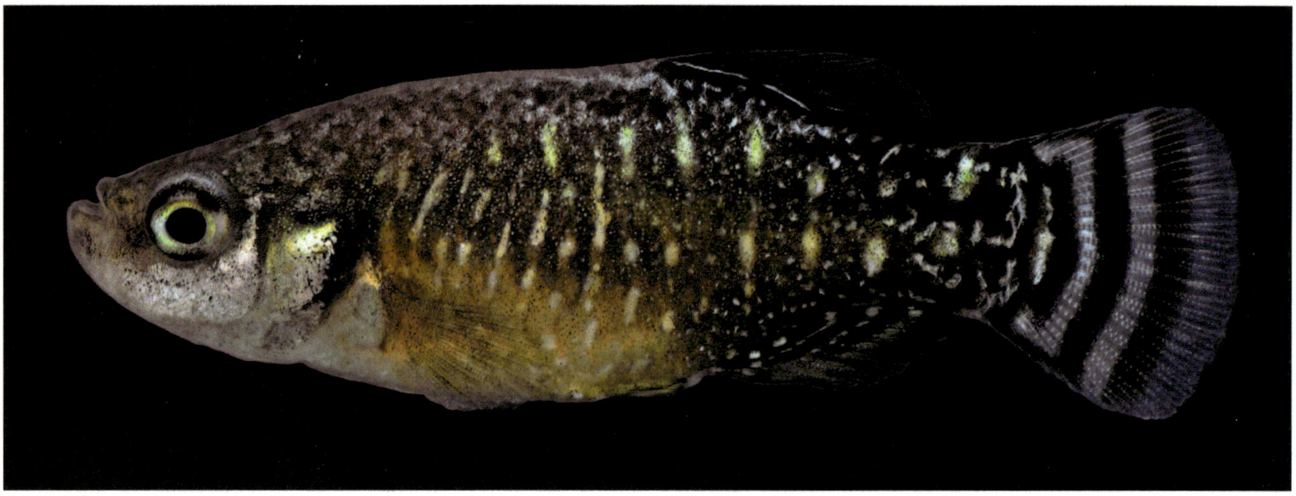

Anatolichthys irregularis; Kaklık spring, Türkiye; male, 35 mm SL.

Anatolichthys irregularis

Common name. Kaklık killifish.

Diagnosis. Distinguished from other species of *Anatolichthys* by: • male with irregularly set and shaped, dark-brown bars, bars with white spots or vermiculate pattern, some large individuals plain, dark-grey with some yellow blotches / ○ male with yellow or white anal with narrow black margin / ○ female with numerous black, irregularly shaped stripes and vermiculate blotches on flank organised in midlateral rows / ○ body completely covered with overlapping scales. Size up to about 52 mm SL.

Distribution. Türkiye: Spring near Kaklık in Büyük Menderes drainage.

Habitat. Springs with dense vegetation and associated streams.

Biology. Live up to 3 years; biology as described in genus section.

Conservation status. CR; known only from a very small spring area, vulnerable to climate change, recreational activities, non-native species and desiccation.

Remarks. *Anatolichthys irregularis* is very closely related to *A. maeandricus*, but the colour pattern of male and female can easily distinguish both species. It is likely that *A. irregularis* was recently isolated and has adapted to the gypsum springs where it lives.

Further reading. Yoğurtçuoğlu & Freyhof 2018 (description).

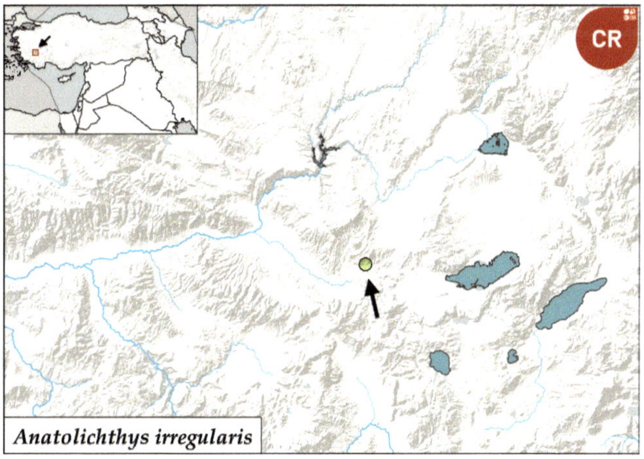

Anatolichthys irregularis

Spots or bars? Or even more open questions for Kaklık killifish. The colour patterns of fish have long fascinated evolutionary biologists due to their striking diversity and how they are influenced by sexual and environmental selection. Much of the data on the cellular and developmental mechanisms underlying colour-pattern formation, such as stripes, spots, or regular and irregular colour patterns, comes from zebrafish (*Danio rerio*). However, there is still much to be discovered about how these mechanisms are linked to the behaviour, ecology, and natural selection in the wild. Several cases, including recently evolved groups such as East African cichlids, have provided insights into the evolution of pigment patterns and triggered various hypotheses for testing. *Anatolichthys irregularis* is a young species worthy of future attention in this respect. It is unique in that its irregularly set and shaped spots and bars in the male (vs. regular bars in other *Anatolichthys*). It was initially assumed that this significant difference in colour pattern, the dissociation of regular bars into an irregular and spotted pattern, must have been merely a phenotypic plasticity response to its extreme habitat (high sulfur concentration, high temperature, and oxygen-poor conditions). However, captive generations bred under different conditions have proven the colour patterns stable, as other *Anatolichthys* have demonstrated. Consequently, several open hypotheses have been formulated, including the evolution of isolated peripheral populations under strong selection and genetic drift due to limited population size. This phenomenon has already been demonstrated in zebrafish, where a change in pattern may be driven by a single mutation in a single gene, which is prone to selection through pleiotropic effects. **Further reading.** Parichy 2003, Gante 2018 (stripe evolution).

Anatolichthys maeandricus; Işıklı spring, Türkiye; female, 40 mm SL.

Anatolichthys maeandricus; Işıklı spring, Türkiye; male, 30 mm SL.

Anatolichthys maeandricus

Common name. Büyük Menderes killifish.

Diagnosis. Distinguished from other species of *Anatolichthys* by: ● male with a white line or a row of white spots or blotches along or distinctly above black dorsal base / ○ dorsal completely black or with grey distal margin / ○ last dark-brown bar irregularly or diamond-shaped, usually fused with pre-last bar / ○ male with bright yellow anal with narrow black margin / ○ male with 1–4 bold black vertical bars on caudal / ○ male with flank bars behind dorsal origin wider than interspaces / ○ female with numerous black spots not organised in midlateral rows / ○ body completely covered with overlapping scales. Size up to 45 mm SL.

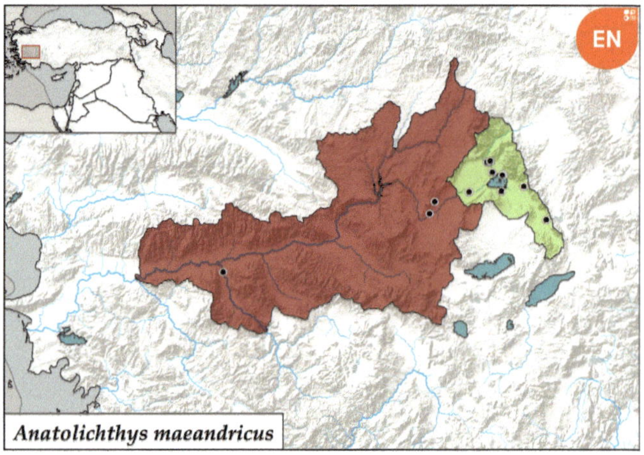

Distribution. Türkiye: Springs of Büyük Menderes near Işıklı, Pınarbaşı spring near Karakuyu and Düden Pınarı at Dinar.
Habitat. Springs with dense underwater vegetation.
Biology. Live up to 3 years; biology as described in genus section.
Conservation status. EN; appears to be declining within its very small range.
Remarks. Several records of *Anatolichthys* from the middle and lower Büyük Menderes drainage exist, none of which have been confirmed in recent decades.
Further reading. Akşıray 1948 (description); Geiger et al. 2014 (phylogeny); Yoğurtçuoğlu & Freyhof 2018 (identification).

Anatolichthys marassantensis
Common name. Kızılırmak killifish.
Diagnosis. Distinguished from other species of *Anatolichthys* by: ○ adult male with completely black dorsal, or with white or grey base and a narrow white line or row of white spots along dorsal base / ○ male with bright yellow or white anal with narrow black margin and 1–4 narrow, black, concentric bands / ○ male with 8–13 dark-brown bars, third-last bar anterior to caudal base 0.9–1.8 times wider than anterior white interspace / ○ male with ovoid dorsal profile in individuals larger than 26 mm SL / ○ female without spots on caudal / ○ female larger than 28 mm SL with 1–3 lateral rows of large black spots on flank behind dorsal base / ○ body shape ovoid, rarely rhomboid, dorsal profile convex / ○ body completely covered with overlapping scales. Size up to 57 mm SL.

Anatolichthys marassantensis; Kayseri, Türkiye; female, 40 mm SL.

Anatolichthys marassantensis; Kayseri, Türkiye; male, 36 mm SL.

Distribution. Türkiye: Kızılırmak drainage and lower and middle Yeşilırmak drainage; also in adjacent coastal wetlands west of Sinop.
Habitat. Springs, small streams to marshes, and large reservoirs.
Biology. Live up to 5 years; spawns May-September. Feeds mainly on zooplankton (especially cladocerans and copepods), plants (diatoms and algae) and some other invertebrate groups (especially amphipods and gastropods). Similar to *A. iconii* in Lake Eğirdir, its population fluctuates in a similar pattern to that of *Atherina* in Hirfanlı Reservoir.
Conservation status. LC.
Remarks. This species is usually identified as *A. chantrei*, a synonym of *A. danfordii*. Similar fish occurred in upper Jordan drainage and the Damascus basin in Syria until the early 20th century but have not been found since. These have been identified as *A. marassantensis*, which may have been introduced in this area, but this needs to be confirmed.
Further reading. Krupp 1985d (Syrian records); Geiger et al. 2014 (phylogeny); Pfleiderer et al. 2014 (description); Yoğurtçuoğlu & Ekmekçi 2013 (life-history); Yoğurtçuoğlu & Ekmekçi 2017 (distribution in Yeşilırmak); Yoğurtçuoğlu et al. 2018b (feeding).

Anatolichthys meridionalis
Common name. Lycian killifish.
Diagnosis. Distinguished from other species of *Anatolichthys* by: • male with 6–8 dark-brown bars on flank / • male with flank bars behind dorsal origin narrower than interspaces / ○ adult male with completely black dorsal or with brown base / ○ body shape ovoid / ○ male with bright yellow anal with narrow black margin / ○ male with 1–4 bold black vertical bars on caudal / ○ female with numerous black spots or blotches on flank often organised in midlateral rows / ○ body completely covered with overlapping scales. Size up to 55 mm SL.
Distribution. Türkiye: Upper Eşen drainage and Lake Söğüt, Avlan, and Gölhisar basins. Also in a small reservoir in Lake Salda basin (potentially introduced).
Habitat. River and stream margins, lakeshores, springs, and spring-fed canals and streams.
Biology. Live up to 4 years, mature at age I; spawns July-September.
Conservation status. VU; found in fewer than 10 independent populations, some of which are declining.
Remarks. *Anatolichthys parvus*, from small streams near Lake Gölhisar, is a synonym.
Further reading. Akşıray 1948 (description); Yoğurtçuoğlu & Freyhof 2018 (identification); Güçlü 2022 (biology).

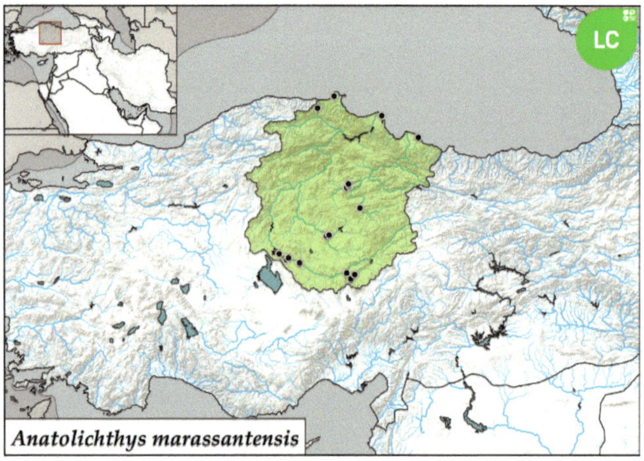

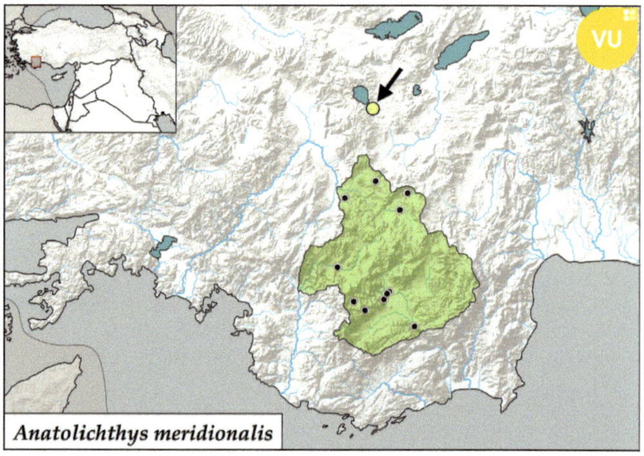

Anatolichthys meridionalis; Söğüt, Türkiye; female, 38 mm SL.

Anatolichthys meridionalis; Söğüt, Türkiye; male, 33 mm SL.

Lake Salda in Türkiye is the only habitat of *Anatolichthys saldae*.

Anatolichthys saldae; Salda, Türkiye; female, ~50 mm SL.

Anatolichthys saldae; Salda, Türkiye; male, ~40 mm SL.

Anatolichthys saldae
Common name. Salda killifish.
Diagnosis. Distinguished from other species of *Anatolichthys* by ○ 1–3 lines of large, usually not or only slightly overlapping scales on midlateral flank / ● usually no or few isolated scales on flank, up to 47–52 total, not or very slightly overlapping scales along lateral series / ● male with black bars wider at dorsal part, narrower at ventral part / ○ no or few isolated scales on flank except midlateral row / ○ scales on back and belly absent / ○ lateral head profile almost quadratic, lower jaw large, directing upward / ○ body depth 4.0–5.3 times in standard length. Size up to 50 mm SL.
Distribution. Türkiye: Lake Salda.
Habitat. Lacustrine. Forages in open water; spawns on gravel beds near shore. Overwinters in deeper parts of lake. Lake Salda is oligotrophic with a very high concentration of magnesium carbonate ($MgCO_3$).
Biology. Live up to 3 years, mature within a year. Pelagic and offshore, outside spawning season. Non-nuptial males and females form large schools, even during spawning season spawns April to September. Males establish territories along shore, usually between gravel and rocks, which they defend against rivals. Female spawn with one or more males on gravel bottoms. Few eggs are laid in the substrate during a spawning event. Individual females may produce several clutches in a single day. Feeds on small benthic and planktonic invertebrates.
Conservation status. CR; Water abstraction and reduced rainfall due to climate change are critical threats to salinity levels in Lake Salda and could lead to its rapid extinction in the near future.

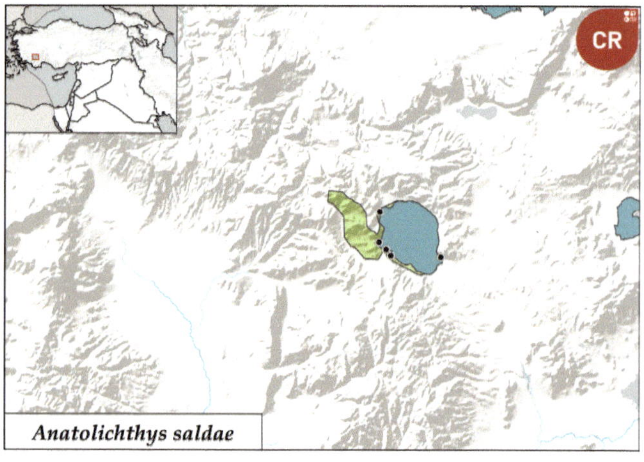

Remarks. Often treated as a synonym of *A. splendens*.
Further reading. Akşıray 1955 (description); Geiger et al. 2014 (phylogeny); Yoğurtçuoğlu & Freyhof 2018 (identification); Yoğurtçuoğlu 2019 (reproduction).

Scale reduction in *Anatolichthys* and *Kosswigichthys*. The presence of scales on fish is a protective mechanism that safeguards the integumentary system from injury. However, this adaptation comes at a cost in terms of energy expenditure. This is evidenced by the observation that many cavefish, which inhabit environments with reduced levels of light, have reduced scales. Similarly, pelagic Aphaniids that inhabit mineral-rich lakes appear relatively safe from predators and flooding hazards. Consequently, they have also reduced their scales. The reduction of scales in *Anatolichthys* and *Kosswigichthys* is not a genetic adaptation to the high concentration of sodium sulfate, magnesium sulfate, and magnesium carbonate in the lakes in which they live. There was no difference in the salt tolerance of scale-reduced fish and normally scaled fish observed in experiments. The reduction is a consequence of the absence of stabilizing selection. Mutants exhibiting all variations in squamation survive, increasing the variability of the squamation. This may lead to the complete loss of scales, potentially saving energy by not developing and maintaining scales. **Further reading.** Grimm 1980 (scale reduction).

Anatolichthys splendens
Common name. Gölcük killifish.
Diagnosis. Distinguished from other species of *Anatolichthys* by: ○ 1–3 lines of large, usually not or only slightly overlapping scales on midlateral flank / • usually 38–46 total, overlapping scales along lateral series / • male with black bars equally shaped, not wider at dorsal part / ○ scales on back and belly absent / ○ lateral head profile almost quadratic, lower jaw large, directing upward / ○ body depth 4.0–5.3 times in standard length / ○ male with white anal with wide, black margin or completely black. Size up to 38 mm SL.

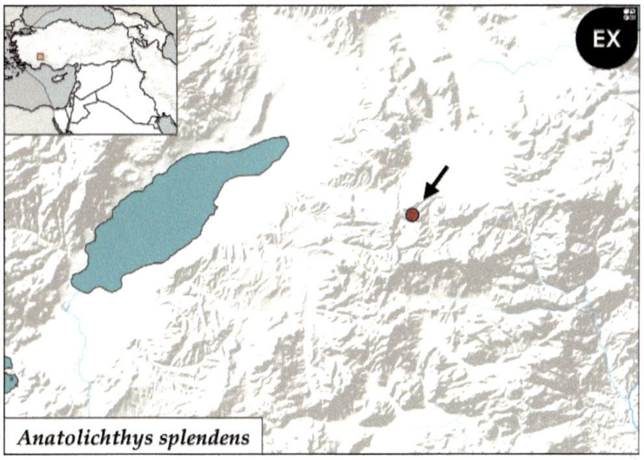

Anatolichthys splendens; Lake Gölcük, Türkiye; female, 38 mm SL; male, 31 mm SL.

Distribution. Türkiye: Lake Gölcük west of Isparta.
Habitat. Lacustrine.
Biology. No data. Expected to be similar to other *Anatolichthys*. In laboratory, larvae hatch after 12 days at an incubation temperature of 24°C.
Conservation status. Extinct; invasive *Sander lucioperca*, *Cyprinus carpio* and *Carassius auratus* caused the extinction of this species in the early 1980.
Remarks. *Anatolichthys saldae*, a species endemic to Lake Salda, is often treated as a population of *A. splendens*.
Further reading. Kosswig & Sözer 1945 (description); Yoğurtçuoğlu & Freyhof 2018 (identification).

Anatolichthys sureyanus

Common name. Burdur killifish.
Diagnosis. Distinguished from other species of *Anatolichthys* by: ○ usually scales on back and belly absent / ○ usually 1–3 lines of large, usually not or only slightly overlapping scales on midlateral flank / ○ lateral head profile triangular or rounded, lower jaw small, directing forward / ○ body depth 3.2–4.2 times in standard length / ○ caudal peduncle depth 11–14 % SL / ○ head depth at nape 61–69 % SL / ○ head width at nape 51–60 % SL / ○ nuptial male with grey dorsal base / ○ male with white anal and wide, black margin or completely black. Size up to 45 mm SL.
Distribution. Türkiye: Lake Burdur.
Habitat. Lacustrine along shores also found in mouths of inflowing streams and springs on lakeshores. Lake Burdur has very high sodium sulfate (Na_2SO_4) concentrations and alkalinity (pH 8.9–9.5).
Biology. Live up to 4 years, mature in 100–120 days. Starts spawning at end of March. Incubation period 7–8 days at 22–24°C water temperature. Non-nuptial males and females form large schools. *Arctodiaptomus burduricus* and *Brachionus plicatilis* are major food items.

Anatolichthys sureyanus; Lake Burdur, Türkiye; female, 38 mm SL.

Anatolichthys sureyanus; Lake Burdur, Türkiye; male, ~30 mm SL. © M. Sağdıç

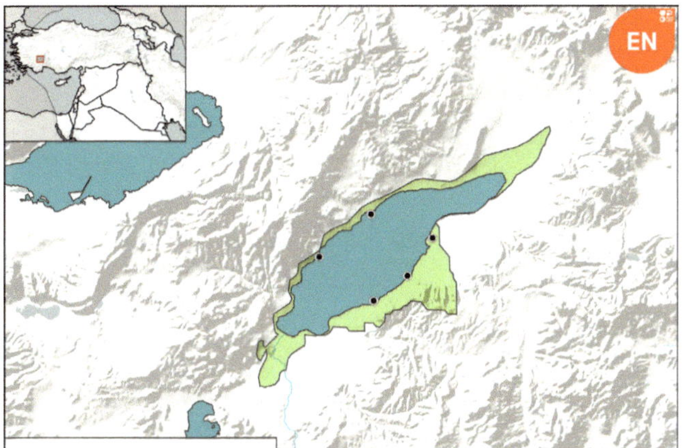

Anatolichthys sureyanus

Conservation status. EN; appears to be declining within its small range. The main threats are drying up of Lake Burdur and increasing salinity and pollution, which are continuing. Lake Burdur has lost about 30 % of its surface area since the 1980s due to water abstraction and retention. Reduced inflow has led to increased salinity and pollutant concentrations. Between 2016 and 2019, the species disappeared from the lake after massive fish kills. It was confined to springs and tributaries, suffering from competition and predation from non-native *Gambusia holbrooki*. In 2000, it was found in the lake again after much rain and fewer algal blooms. Lake Burdur is a Ramsar site, a "Protected Area of First-Degree Importance", and a "Wildlife Development Zone" under national law.

Remarks. In the past, *Anatolichthys* from Lake Burdur showed a wide range of squamation patterns. Fish with a partially bare body and reduced, non-overlapping scales on the flank coexisted with fish with a body completely covered by overlapping scales. Early authors placed a high value on scale patterns and placed fish with full-scale coverage in genus *Aphanius* and those with reduced scales in genus *Anatolichthys*. Later, authors treated all fishes as one variable species, mostly based on inconclusive hybridisation experiments and intermediate fishes in the lake. Fish from Lake Burdur were described as *A. sureyanus* without giving details of scale patterns, *A. burdurensis* based on fish with reduced scales and a slender body, and *A. burduricus* based on fish with full-scale coverage and a deep body. Today, *Anatolichthys* show reduced scale patterns or no scales at all. Molecular studies (cytochrome b, cytochrome c oxidase 1) have yet to provide evidence to distinguish fish from Lake Burdur (without detail in scale patterns) from fish inhabiting springs and lakes in the wider Lake Burdur basin, identified here as *A. fontinalis.*

Further reading. Akşıray 1955 (description); Yoğurtçuoğlu 2018 (conservation status); Yoğurtçuoğlu & Freyhof 2018 (identification); Güçlü et al. 2007 (biology).

Anatolichthys transgrediens

Common name. Acıgöl killifish.

Diagnosis. Distinguished from other species of *Anatolichthys* by: ○ scales on back and belly usually absent / ○ usually 1–3 lines of large, not or only slightly overlapping scales on midlateral flank / ○ lateral head profile triangular or rounded, lower jaw small, directing forward / ○ body depth 3.2–4.2 times in standard length / ○ caudal peduncle depth 16–18 % SL / ○ head depth at nape 68–75 % SL / ○ head width at nape 62–74 % SL / ○ nuptial male with completely black dorsal / ○ male with white anal and wide, black margin or anal completely black. Size up to 40 mm SL.

Anatolichthys transgrediens; Acıgöl, Türkiye; female, 30 mm SL.

Anatolichthys transgrediens; Acıgöl, Türkiye; male, 25 mm SL.

Distribution. Türkiye: Lake Acıgöl basin.
Habitat. Springs and small spring-fed streams. Also found in Acıgöl when there is sufficient water.
Biology. Live up to 2 years, mature with 5-7 months. Spawns February to May and again July to August in channals of brine ponds. Several males gather to chase females during spawning. Few eggs are laid on vegetation during a spawning event. Feeds on benthos and detritus in *Gambusia* dominated springs, otherwise insect larvae, zooplankton and plant material.
Conservation status. CR; once abundant in Acıgöl and its tributaries. Now, only in one spring field. Springs on western, eastern and northern shores of Acıgöl have dried up. In late 20th century, it was reported to be more abundant, and the last spring field was much larger.
Remarks. There is a large variation in scale reduction in *A. transgrediens*.

Further reading. Ermin 1946 (description); Yoğurtçuoğlu & Freyhof 2018 (identification); Yoğurtçuoğlu et al. 2020c (reproduction).

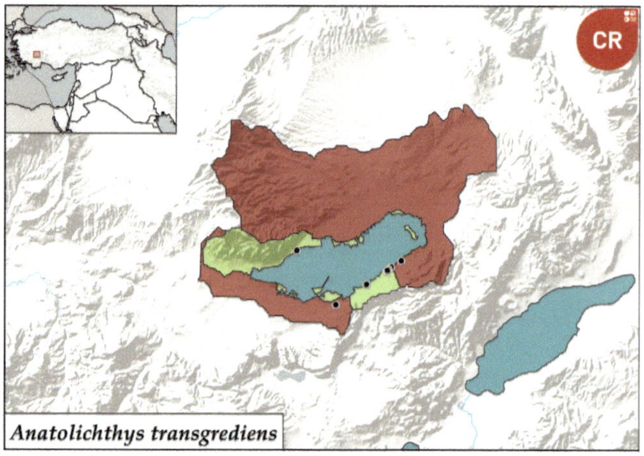

Spring-feed streams running to hypersaline Lake Acigöl in Türkiye are the habitat of *Anatolichthys transgrediens*.

Anatolichthys villwocki; Sakarya drainage, Türkiye; female, 35 mm SL.

Anatolichthys villwocki; Sakarya drainage, Türkiye; male, 40 mm SL.

Anatolichthys villwocki

Common name. Sakarya killifish.

Diagnosis. Distinguished from other species of *Anatolichthys* by: • male with 4–14 vertical rows of small black or brown spots on caudal / • male with 13–25 narrow brown bars on body / • female larger than 40 mm SL usually without back spots or blotches on flank except one bold black blotch at middle posterior most caudal peduncle / ○ male with regularly set and shaped dark-brown or black bars, without white spots in bars / ○ male with bright yellow anal with narrow black margin / ○ male with flank bars behind dorsal origin wider than interspaces / ○ body completely covered with overlapping scales. Size up to 60 mm SL.

Distribution. Türkiye: Lake Ilgın and upper Sakarya drainage downstream to about confluent of Sakarya and Porsuk.

Habitat. Lakes, springs, and slow-flowing streams with dense vegetation.

Biology. Male live up to 4 years, females up to 5 years; batch spawning from late of March to July.

Conservation status. LC.

Further reading. Hrbek & Wildekamp 2003 (description).

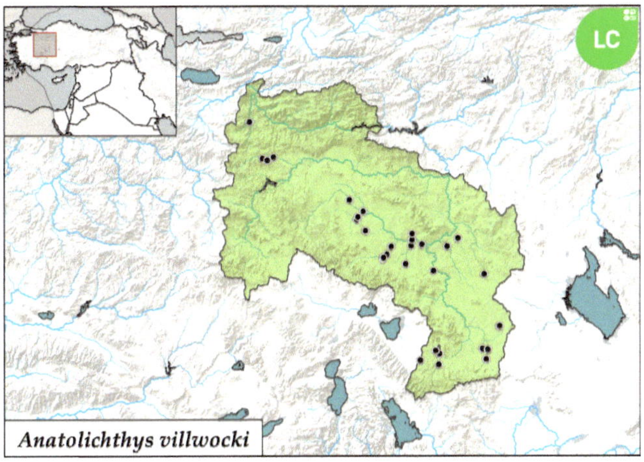

Aphaniops

Aphaniops are a group of eight species of killifishes known to inhabit the south-easternmost Mediterranean basin, extending along the coasts of the Red Sea from south to Ethiopia and across the Persian Gulf basin to Gujarat in India. Except for one species, all *Aphaniops* occur in West Asia. *Aphaniops stiassnyae* is endemic to Lake Afdera in the Danakil Depression of northern Ethiopia, and several undescribed species might exist in East Africa. *Aphaniops* cannot survive water temperatures below 10°C for an extended period. Low water temperatures during the winter months in the Mediterranean basin may limit their distribution. **Further reading.** Getahun & Lazara 2001 (*A. stiassnyae*); Chiozzi et al. 2017 (diversity in East Africa); Freyhof et al. 2017b (molecular data, distribution); Esmaeili et al. 2020b (generic concept).

Key to species of *Aphaniops* in West Asia

1a - Body without scales; caudal slightly emarginate, male without bold bars on caudal.
..................*A. furcatus*
1b - Body covered by scales; caudal truncate, or slightly rounded, male with 2–3 bold, crescent-shaped black bars on caudal.
..................2

2a - Female with a midlateral row of roundish, black spots, often with additional black spots on flank; male with a wide, black dorsal margin; anal yellow, with or without a narrow black margin and 0–2 short black bars on middle of fin.
..................*A. sirhani*
2b - Female with a midlateral row of vertically elongated black, grey or dark-brown blotches or narrow bars; male without black dorsal margin, anal yellow anteriorly, with 3–8 narrow black or brown bars on posterior part of fin.
..................3

3a - Male with silvery vermiculation on grey background on flank.
..................*A. richardsoni*
3b - Male with dark-grey bars on silvery background or with roundish or ovoid silvery spots or blotches, sometimes with short vermiculation, on a grey, greenish or bluish background, often with spots in front of anal origin and bars on caudal peduncle.
..................4

4a - Male with 9–19 brown or grey bars on flank between pectoral and caudal bases.
..................5
4b - Male with roundish or ovoid silvery spots or blotches, sometimes with short vermiculation, on a grey, greenish or bluish background, often with spots in front of anal origin and bars on caudal peduncle.
..................7

5a - 5 branched dorsal rays, ½ ray absent.
..................*A. ginaonis*
5b - 6–8½ branched dorsal rays, several individuals have 7 rays and no ½ ray; male with 9–17 brown or grey bars on flank.
..................6

6a - Female with a diamond-shaped or slightly vertically elongate black or dark-brown blotch at caudal base; male with 9–14 brown bars on flank.
..................*A. kruppi*
6b - Female with a narrow and long bar at caudal base; male with 12–17 brown bars on flank.
..................*A. teimorii*

7a - Male with roundish or ovoid silvery spots or blotches on flank, sometimes with short vermiculation, on a grey, greenish or bluish background, usually without bars on caudal peduncle; tip of dorsal reaching to three scales in front of end of hypural complex in nuptial male larger than 40 mm SL; female with a midlateral series of wide bars or irregularly shaped and set, vertically elongated, lateral blotches.
.................*A. dispar*
7b - Male usually with wide grey or brown bars at posterior flank and caudal peduncle, with roundish or ovoid silvery spots or blotches on flank anterior to dorsal base or caudal peduncle; tip of dorsal reaching to end of hypural complex in nuptial male larger than 40 mm SL; female with a midlateral series of narrow bars, not much wider in upper part.
.................*A. stoliczkanus*

Aphaniops dispar; salt marshes at Atlit, Israel; female, 43 mm SL.

Aphaniops dispar; salt marshes at Atlit, Israel; male, 37 mm SL.

Aphaniops dispar

Common name. Red Sea killifish.

Diagnosis. Distinguished from other species of *Aphaniops* in West Asia by: • male with roundish or ovoid silvery spots or blotches on flank, sometimes with short vermiculation, on a grey, greenish or bluish background, usually without bars on caudal peduncle / • tip of dorsal reaching to three scales in front of end of hypural complex in male larger than 40 mm SL / ○ female with a midlateral series of wide bars or irregularly shaped and set, vertically elongated, lateral blotches / ○ female with a narrow and long bar at caudal base / ○ male with 2–3 bold, crescent-shaped black bars on caudal / ○ body covered by overlapping scales / ○ caudal truncate or slightly rounded / ○ 6–8½ branched dorsal rays. Size up to 56 mm SL.

Distribution. Red Sea coast south to Gulf of Aden, Socotra and south to Ethiopia. In Suez Canal, Bitter and Timsach Lakes. Also along northern Nile Delta and Sinai coast, and on Israeli Mediterranean coast at Atlit and Siwa Oasis in western Egypt. Introduced in fish farms around Khartoum with salt from Port Sudan area.

Habitat. Euryhaline, coastal habitats with dense vegetation or rocky bottoms. Also found in coral reefs in Red Sea. Very rarely reported from freshwater habitats.

Biology. Live up to 3 years, mature in a few months, usually late in year of birth. Males establish territories along shore, usually in dense vegetation but also between rocks, which they defend against rival males. Spawns March–September, with a peak in July–August on Mediterranean coast of Egypt. Spawns throughout year on Arabian Peninsula. Females spawn with one or more males, usually on gravel, rocks, algae, or other vegetation. Few eggs are laid in substrate during a spawning event. Individual females may produce several clutches in a single day. Feeds on algae, detritus, and small invertebrates. Produces sterile hybrids with *Aphanius fasciatus* along Egyptian coast.

Aphaniops cf. *dispar*; Jeddah, Saudi Arabia; female, 40 mm SL.

Aphaniops cf. *dispar*; Jeddah, Saudi Arabia; male, 35 mm SL.

Aphaniops cf. *dispar*; Shukaray, Eritrea; male, 45 mm SL.

Aphaniops cf. *dispar*; Lake Abbe, Djibouti; male, ~50 mm SL. © H. Wischmann.

Conservation status. LC.

Remarks. Populations in Socotra and western Yemen are thought to belong to this species. In the Nile, reports are from the delta area, but records upstream of the delta, south of Cairo, are unconfirmed and doubtful. At least two species are included in *A. dispar*, both found in the Red Sea, and there may be other unrecognised species in the area. It remains to be seen which species was originally described as *A. dispar*.

Further reading. Villwock et al. 1983 (distribution); Al-Kahem et al. 2008 (Saudi Arabia); Freyhof et al. 2017b (molecular data, distribution); Freyhof et al. 2020 (distribution).

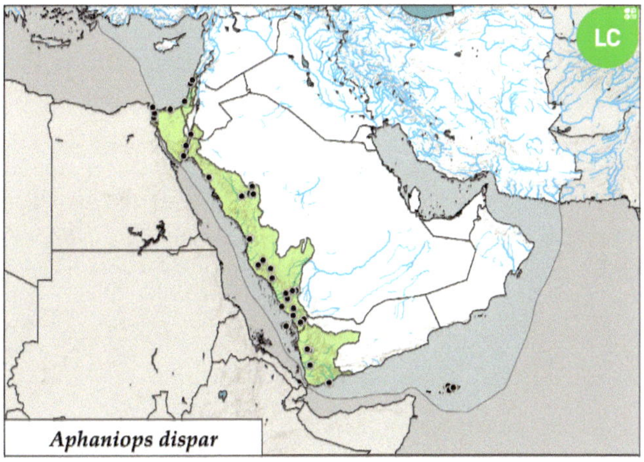

Aphaniops furcatus; Shur drainage; Iran; female, 30 mm SL.

Aphaniops furcatus; Shur drainage; Iran; male, 21 mm SL.

Aphaniops furcatus
Common name. Scaleless Hormuzgan killifish.
Diagnosis. Distinguished from other species of *Aphaniops* in West Asia by: • caudal slightly emarginate / • caudal without bars in male / • body completely scaleless / • teeth bicuspid. Size up to 33 mm SL.
Distribution. Iran: Kol and Mehran drainages including Khurgu and Faryab hot sulfuric springs close to Bandar Abbas. Also, Shur drainage in Makran region.
Habitat. Saline rivers and hot sulfur springs with salinities around 30 ‰. Usually along banks.
Biology. No data.
Conservation status. VU; habitats are under threat due to severe drought, pollution, and habitat destruction.
Remarks. Occurs in sympatry with *Aphaniops teimorii*, which is usually much more abundant.

Further reading. Teimori et al. 2014 (description); Teimori et al. 2019 (distribution and habitats).

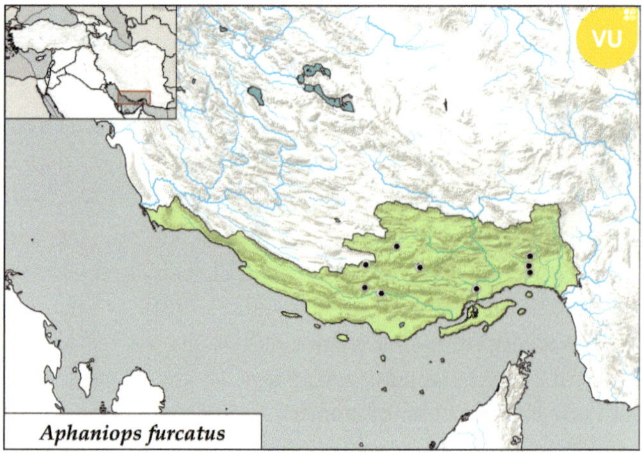

Aphaniops ginaonis; Genow spring, Iran; female, 32 mm SL.

Aphaniops ginaonis; Genow spring, Iran; male, 35 mm SL.

Aphaniops ginaonis

Common name. Genow killifish.

Diagnosis. Distinguished from other species of *Aphaniops* in West Asia by: ● 5 branched dorsal rays, ½ ray absent / ○ male with 13–19 brown bars on flank / ○ male with 2–3 crescent-shaped black bars on caudal / ○ female with a midlateral row of vertically elongated black, grey or dark-brown blotches or narrow bars / ○ male without black dorsal margin, anal yellow anteriorly, with 3–8 narrow black or brown bands on posterior part of fin / ○ anal and dorsal not reaching to caudal base / ○ body covered by overlapping scales / ○ caudal truncate or slightly rounded. Size up to 40 mm SL.

Distribution. Iran: Genow hot sulfuric spring.

Habitat. A very small hot sulfur spring with 37–40°C water temperatures, bottom of stones and pebbles covered with cyanobacteria.

Biology. Lives up to 6 (female) or 5 (male) years. Matures at 2 years. Spawns probably all year round. Feeds on algae and insect larvae.

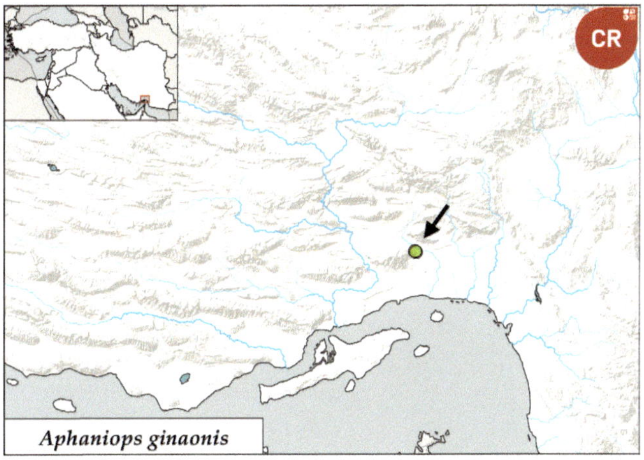

Conservation status. CR; declined sharply after the construction of bathing facilities in the spring of the early 21st century. Since then, population has stabilised at a very low level.

Remarks. This species occurs adjacent to the range of *A. teimorii* that inhabits the outflow of Genow spring below a small waterfall. Both species are well differentiated morphologically, but morphological differences may result from recent adaptations to the extreme conditions of the habitat.

Further reading. Reichenbacher et al. 2009 (habitat, otolith morphology); Golmoradizadeh et al. 2012 (life history).

Only a small part of Genow spring in Iran is inhabited by endemic *Aphaniops ginaonis*.

Aphaniops kruppi; Wadi Bani Khalid, Oman; female, 45 mm SL.

Aphaniops kruppi; Wadi Bani Khalid, Oman; male, 34 mm SL.

Aphaniops kruppi

Common name. Omani killifish.

Diagnosis. Distinguished from other species of *Aphaniops* in West Asia by: • male with 9–14 brown or grey flank bars between pectoral, pelvic, and caudal bases, often dissociated into blotches in front of pelvic base / • female with a diamond-shaped or vertically elongate black or dark-brown blotch at caudal base / ○ female with a midlateral row of vertically elongated black, grey or dark-brown blotches or narrow bars / ○ male without black dorsal margin, anal yellow anteriorly, with 3–8 narrow black or brown bands on posterior part of fin / ○ male with 2–3 bold, crescent-shaped black bars on caudal / ○ body covered by overlapping scales / ○ caudal truncate or slightly rounded / ○ 6–7½ branched dorsal rays. Size up to 52 mm SL.

Distribution. Oman: Coast of Dhofar northeast to Wadi Bani Khalid drainage. Wadi Bani Khalid is far inland. In small springs and falaj systems, for example, in the towns of Bani Bu Ali and Al Mudayrib. Populations in eastern Yemen are likely to belong to this species.

Habitat. Coastal wetlands, wadis, springs, falaj systems, and spring-fed streams.

Biology. Live up to 2 years, mature in a few months, usually late in year of birth. Males establish territories along banks, usually in dense vegetation but also between rocks, which they defend against rival males. Spawns throughout year. Females spawn with one or more males, usually in algae or other vegetation near the surface or on gravel beds. Few eggs are laid in substrate during a spawning event. Individual females produce several clutches of eggs in one day. Feeds on algae, detritus, and small invertebrates.

Conservation status. LC.

Remarks. Molecular data show that *A. kruppi* is distinct from coastal populations of *A. stoliczkanus*. *Aphaniops* from many inland drainages north of Buraimi on the UAE-Oman border (e.g., Wadi Dayqah, Al-Hoota, Al Juwayf, Wadi Hatta, and Wadi Sayfam) are very similar to *A. kruppi* in colour pattern, but have mtDNA of *A. stoliczkanus*. They are thought to be hybrids and often non-native populations.

Further reading. Freyhof et al. 2017b (description).

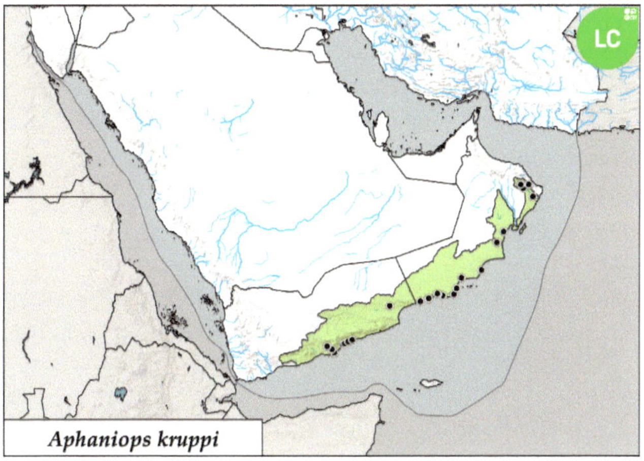

Aphaniops richardsoni; Neot HaKikar, Israel; female, ~50 mm SL. © B. Nicca.

Aphaniops richardsoni; Neot HaKikar, Israel; males, ~50 mm SL. © B. Nicca.

Aphaniops richardsoni

Common name. Dead Sea killifish.

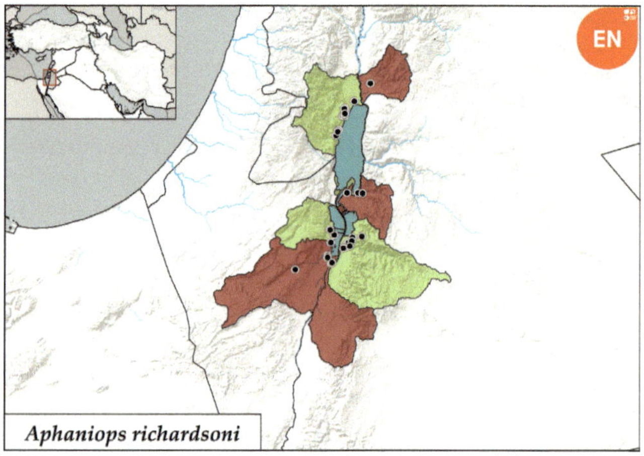

Diagnosis. Distinguished from other species of *Aphaniops* in West Asia by: ● male with silvery vermiculation on grey background on flank / ○ bars absent on flank or restricted to caudal peduncle in male / ○ female with a midlateral row of vertically elongated black, grey or dark-brown blotches or narrow bars / ○ male without black dorsal margin, anal yellow anteriorly, with 3–8 narrow black or brown bands on posterior part of fin / ○ dorsal in male only slightly longer than in females, not reaching caudal base in male larger than 30 mm SL / ○ male with 2–3 bold, crescent-shaped black bars on caudal / ○ body covered by overlapping scales / ○ caudal truncate or slightly rounded. Size up to 50 mm SL.

Distribution. Jordan and Israel: Springs in Dead Sea basin.
Habitat. Springs and headwater streams, usually with fresh or brackish water.
Biology. Live up to 3 years, mature in a few months, usually late in the year of birth. Biology similar to other Aphaniidae. Feeds on algae, detritus and small invertebrates.
Conservation status. EN; appears to be declining within its very small range. In Jordan, extirpated from the northern Dead Sea basin and now restricted to the southern part.
Further reading. Freyhof et al. 2017b (molecular data, distribution); Kärst et al. 2020 (distribution, reproduction).

Aphaniops sirhani

Common name. Azraq killifish.
Diagnosis. Distinguished from other species of *Aphaniops* in West Asia by: • female with a midlateral row of roundish, black spots, often with additional black spots on flank / • male with a wide, black dorsal margin / • dorsal and anal yellow, with or without a narrow black margin and 0–2 short black bands on middle of fin / ○ male with 8–11 distinct bold black or dark-brown lateral bars on flank between head and caudal peduncle / ○ female with a series of dark-grey blotches along midlateral body / ○ male with 1–3 bold, crescent-shaped black bars on caudal / ○ body covered by overlapping scales / ○ caudal truncate or slightly rounded. Size up to 45 mm SL.
Distribution. Jordan: Azraq marsh.
Habitat. Shallow oases, lakes and springs.
Biology. Live up to 3 years, mature in a few months, usually late in the year of birth. Biology similar to other Aphaniidae.
Conservation status. CR; Azraq marshes almost completely dried up in 1992. Species survived in wetland and had been maintained in Azraq Reserve and outside by captive breeding. Since 1994, 8 % of the marshes (based on their 1960s extent) have been restored. However, the species remains dependent on conservation, as water has to be pumped into the wetland due to a very low water table resulting from increasing water abstraction. Non-native species threaten the restored reserve.
Further reading. Villwock et al. 1983 (description); Freyhof et al. 2017b (molecular data, distribution).

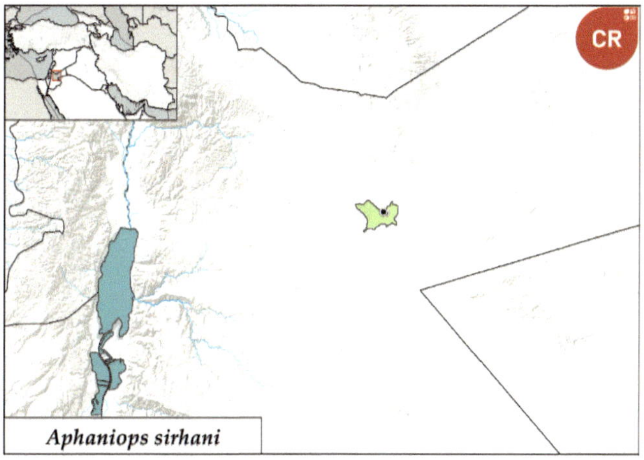

Aphaniops sirhani; Azraq, Jordan; female, ~45 mm SL. © N. A. Hamidan.

Aphaniops sirhani; Azraq, Jordan; male, ~40 mm SL. © N. A. Hamidan.

Aphaniops sirhani is endemic to Azraq wetland in Jordan, which depends on water bought for conservation.

Aphaniops stoliczkanus; Narara-Salt Pans, India; female, 30 mm SL.

Aphaniops stoliczkanus; Narara-Salt Pans, India; male, 30 mm SL.

Aphaniops stoliczkanus; Wadi Fanja, Oman; female, 40 mm SL.

Aphaniops stoliczkanus; Wadi Fanja, Oman; male, 44 mm SL.

Aphaniops stoliczkanus

Common name. Rajasthan killifish.

Diagnosis. Distinguished from other species of *Aphaniops* in West Asia by: • male usually with wide grey or brown bars at caudal peduncle, often anterior to dorsal base, with roundish or ovoid silvery spots or blotches on flank anterior to dorsal base or caudal peduncle / • tip of dorsal reaching to end of hypural complex in nuptial male larger than 40 mm SL / • female with a midlateral series of narrow bars, not much wider in upper part / ○ female with a narrow and long bar at caudal base / ○ male with 2–3 bold, crescent-shaped black bars on caudal / ○ male without black dorsal margin / ○ body covered by overlapping scales / ○ caudal truncate or slightly rounded / ○ 6–8½ branched dorsal rays. Size up to 53 mm SL.

Distribution. Shatt al Arab/Arvand and its tributaries, recorded from Hammar Marsh, Fallujah, Lake Razzazah and Badrah on Iran/Iraq border east of Baghdad. Also along coasts of Persian Gulf basin in Kuwait, Iraq, Iran, Saudi Arabia, and Oman, except coast of Hormuzgan, and east along Indian Ocean coast to Rajasthan (north-west India). *Aphaniops* found as far inland as south of Riyadh in Saudi Arabia may belong to this species.

Habitat. Euryhaline; widespread in lagoons and estuaries, lower parts of rivers, streams, and all types of inland waters, especially when brackish or saline. In and around Hajar Mountains, very common in freshwater habitats, natural and artificial (including water tanks and cisterns), with or without significant vegetation.

Biology. May mature within five months, depending on temperature and food availability. Females mature at about 25 mm SL, males smaller. Males grow larger than females. Live up to 3 years. Males territorial during spawning season defending spawning sites against other males. Spawns in pairs. In Oman and UAE, spawning occurs at all times of day and throughout the year. In southern Iraq, spawning peaks April–July. Several eggs are laid in each spawning event, and several spawning events occur daily. Eggs are attached to the spawning substrate by sticky filaments. Larvae hatch after about 14 days (25°C). Females, juveniles, and non-reproductive males form schools in open water. Salinities up to 14.5 % are tolerated, and salinities up to 25.0 % are briefly survived. Feeds on a wide variety of aquatic and terrestrial invertebrates, algae, and detritus are usually main components.

Conservation status. LC.

Remarks. *Aphaniops teimorii*, *A. kruppi*, and *A. ginaonis* are other superficially similar species. *Aphaniops stoliczkanus* has been widely introduced for mosquito control in UAE and Oman, and many, if not all, inland populations may be non-native. In northern Oman, inland populations sometimes show colour patterns similar to *A. kruppi*, and hybridisation between the two species cannot be excluded.

Further reading. Haas 1982 (biology); Teimori et al. 2012 (diversity in Persian Gulf basin); Freyhof et al. 2017b (molecular data, distribution).

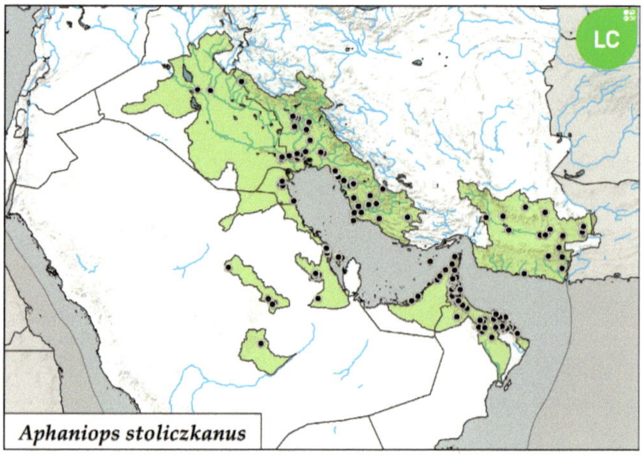

Aphaniops teimorii; Govdar, Iran; female, ~55 mm SL.

Aphaniops teimorii; Govdar, Iran; male, 33 mm SL.

Aphaniops teimorii

Common name. Hormuz killifish.

Diagnosis. Distinguished from other species of *Aphaniops* in West Asia. ○ female with a narrow and long bar at caudal base / ○ male with 12–17 brown bars on flank / ○ male with 2–3 bold, crescent shaped black bars on caudal / ○ female with a midlateral row of vertically elongated black, grey or dark-brown blotches or narrow bars / ○ male with grey bars on silvery background or with roundish or ovoid silvery spots or blotches, sometimes with short vermiculation, on a grey greenish or bluish background, often with spots in front of anal origin and bars on caudal peduncle / ○ body covered by overlapping scales / ○ caudal truncate, or slightly rounded / ○ 7–8½ branched dorsal rays. Size up to about 55 mm SL.

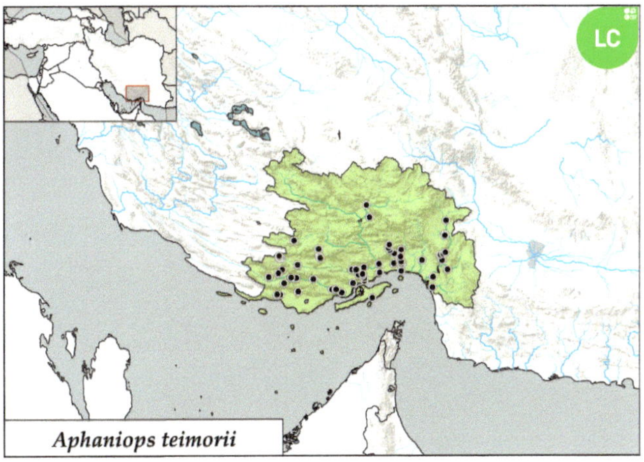

Distribution. Iran: Mond to Minab drainages as well as Qeshm Island.
Habitat. Lakes, rivers, and streams with high salinity.
Biology. Live up to 3 years, mature in a few months, usually late in year of birth. Males establish territories along banks, usually in dense vegetation but also between rocks, which they defend against rival males. Females spawn with one or more males, usually in algae or other vegetation near surface or in gravel bottoms. Few eggs are deposited in substrate during a spawning event. Individual females may produce several clutches of eggs in 1 day. Feeds on algae, detritus, and small invertebrates.
Conservation status. LC.
Remarks. *Aphaniops teimorii* was previously described as *A. hormuzensis*, and this name is still in use, although it is not available (*nomen nudum*) due to a formal error in the description.
Further reading. Teimori et al. 2018 (invalid description of *A. hormuzensis*); Freyhof & Yoğurtçuoğlu 2020 (description); Teimori et al. 2020 (invalid attempt to re-establish *A. hormuzensis*).

Aphanius

The two species of the genus *Aphanius* are found along the coasts of the Mediterranean, where they inhabit a wide range of brackish waters and are rarely seen in pure freshwaters or the sea. Inland water habitats are usually associated with very high mineral concentrations. *Aphanius* inhabit mostly shores with dense submerged vegetation and resist pollution and high water temperatures. Notably, nuptial males are territorial, whereas non-nuptial males and females often shoal and may be pelagic.
Further reading. Kottelat et al. 2007 (species diversity); Triantafyllidis et al. 2007 (phylogeny); Valdesalici et al. 2019 (distribution).

Heavily polluted outflow of a thermal spring in Tuzla, Türkiye. Habitat of many thousands of *Aphanius almiriensis*.

Aphanius almiriensis; Tuzla, Türkiye; female, ~35 mm SL.

Aphanius almiriensis; Tuzla, Türkiye; male, ~25 mm SL.

Aphanius almiriensis

Common name. Almiri killifish.

Diagnosis. Distinguished from *Aphanius fasciatus* by: • female with dark, roundish blotches on flank, more or less connected by an irregular dark-grey midlateral stripe / • dorsal and anal not reaching caudal base in male. Size up to about 38 mm SL.

Distribution. Aegean basin in Türkiye and Greece: Tuzla estuary, Lake Bafa and coastal lagoons around İzmir (Homa lagoon), springs near Almiri and Meligou (Peloponnese) and coastal lagoons on islands of Lesbos and Kos and near cities of Thessaloniki and Kavala. Likely more widespread. Also at Palude del Capitano in Puglia, Italy.

Habitat. Slow-flowing streams, lagoons and lakes. In fresh and brackish water, up to 23 ‰ salinity.

Biology. Live up to 3 years, mature in a few months, usually late in year of birth. Males establish territories along banks, usually in dense vegetation but also between rocks, which they defend against rival males. Females spawn with one or more males, usually in algae or other vegetation near surface or in gravel beds. Few eggs are deposited in substrate during a spawning event. Individual females may produce several clutches in one day. Feeds mainly on algae and detritus and, to a lesser extent, on planktonic invertebrates and anflug.

Conservation status. LC.

Remarks. This species was considered restricted to two localities in Greece but appears widespread throughout the Aegean basin. The distribution of *A. almiriensis* and *A. fasciatus* in the eastern Mediterranean is poorly understood, and a population from the island of Evia (Greece) and possibly several others in Aegean belong to *A. fasciatus*. Both species may hybridise, but this has not been investigated. It cannot be completely excluded that the Italian population was accidentally introduced during Roman times, possibly from the island of Kos (Greece).

Further reading. Kottelat et al. 2007 (description); Triantafyllidis et al. 2007 (distribution); Valdesalici et al. 2019 (distribution, first record in Italy); Saç et al. 2025 (biology).

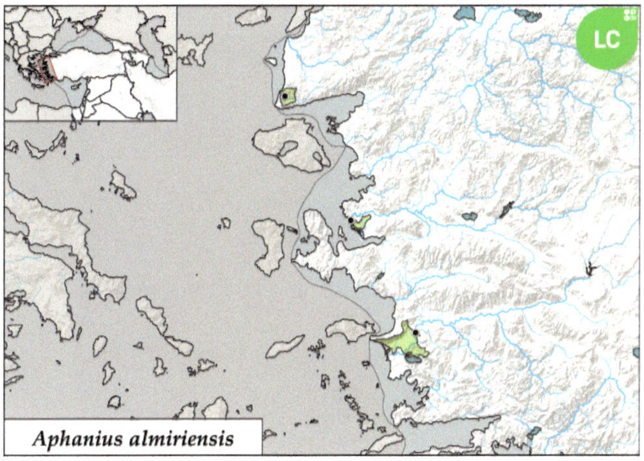

When the Aegean was a lake. The biogeography of *Aphanius* is fascinating. *Aphanius fasciatus* is widely distributed throughout the Mediterranean, from Morocco to Syria and Egypt. The relatively shallow genetic differences between populations suggest gene exchange has occurred, likely facilitated by migration along the coast. However, the Aegean Sea has remained unimpacted by this migration, allowing *A. almiriensis* to evolve independently. The Northern Aegean region evolved as a restricted landlocked basin during the Miocene, with small temporary connections to the Paratethys (today's Black Sea) and the Mediterranean. In the Tortonian (11.6–7.2 million years ago), the Aegean showed scattered lakes predominantly fed by local rivers and short-lived Paratethys connections. The first persisting marine conditions, alternating with brackish Paratethyan environments, indicate a late Maeotian age (6.9–6.1 million years ago) when the region formed a semi-isolated (Egemar) sea with multiple marine influxes. The cessation of marine conditions is well documented by a marked paleoenvironmental change to the brackish water environments, which correlates to the Maeotian/Pontian boundary (6.1 million years ago) in the Eastern Paratethys. During the Messinian Salinity Crisis (5.97–5.33 million years ago), the Northern Aegean was a brackish water system (Lake Egemar) that formed a passageway for Paratethyan overspill waters toward the Mediterranean. It is hypothesised that *A. almiriensis* had been isolated from the Mediterranean, *A. fasciatus* during these times, the latest during the Messinian Salinity Crisis. Nevertheless, while the isolation of the Aegean from the Mediterranean may have contributed to the isolation and speciation of both *Aphanius*, the last 5 million years should have been sufficient time for both species to invade each other's ranges. The only documented instances of such an invasion are the records of *A. fasciatus* from Evia (Greece) and *A. almiriensis* from Italy. Indeed, the dispersal ability of *Aphanius* appears to be relatively limited. Several molecular studies of *A. fasciatus* populations demonstrate a shallow but notable genetic structure within its large range. These studies indicate no rapid and unrestricted gene flow within the Mediterranean. The environmental factors limiting the recent distribution of *A. fasciatus* and *A. almiriensis* may have similarly constrained their range expansion since the Messinian. However, there is a possibility that both species compete strongly with each other if they come into contact or form hybrid populations. This remains highly speculative and requires further research. **Further reading.** Ferrito et al. 2013; Geiger et al. 2014; Buj et al. 2015; Cavraro et al. 2017 (phylogeography); Krijgsman et al. 2020 (Geology of the Aegean Sea).

Aphanius fasciatus; Pag, Croatia; female, 35 mm SL.

Aphanius fasciatus; Qued Zahzah, Tunisia; male, 35 mm SL.

Aphanius fasciatus

Common name. Mediterranean killifish.

Diagnosis. Distinguished from *A. almiriensis* by: • female with narrow, isolated, dark-grey or brown bars on flank / • dorsal and anal reaching caudal base in male. Size up to about 60 mm SL.

Distribution. Deltaic areas of Köyceğiz, Göksu, Seyhan, and Ceyhan, in a stream near Iskenderun, lowermost Susurluk in Marmara basin, and perhaps elsewhere in southern Türkiye. Mediterranean coast from Moulouya estuary (Morocco) east to Egypt, Cyprus, and Syria; in tributaries of Chott el Gharsa and Lake Keliba in Tunisia and Chott Melrhir and Touggourt oasis in Algeria. Entering Bitter Lakes in Egypt via Suez Canal. Locally along Syrian and Israeli coasts (Haifa, now extirpated), in Siwas and Fayum Oasis in Egypt. In Nile Delta and adjacent coastal lagoons. Largely absent from Aegean basin, but recorded from island of Evia (Greece). In Europe, from Camargue (France) east to western Greece, including Corsica, Sardinia, Sicily and Malta. In Spain, recently appeared in Ebro Delta, which was thought to have been released by aquarists, but population seems to have disappeared.

Habitat. Coastal lagoons in fresh (with high mineral content), brackish, and even hypersaline waters. Often in shallow, still or slow-flowing brackish waters, especially at estuaries.

Biology. Live up to 3 years, spawns for the first time at 7–12 months. Spawns March–June in Italy, April–July in Greece, April–September in Corsica. Males establish territories along shore, usually in dense vegetation but also between rocks, which they defend against rivals. Females spawn with one or more males, usually in seaweed or other vegetation close to surface or in gravel beds. Few eggs are laid in substrate during a spawning event. Individual females may produce several clutches in one day. Eggs hatch in 10–14 days. Juveniles form large schools, and non-nuptial males and females swarm and are often pelagic. Feeds on small benthic or planktonic invertebrates, anflug and algae.

Conservation status. LC.

Remarks. There appear to be no current records of this species from the Egyptian Bitter Lakes and the northern Red Sea, and the species may be no longer found here. The distribution of *A. fasciatus* in the eastern Mediterranean has yet to be fully understood. The distribution limits with similar *A. almiriensis* along the Aegean coast of Türkiye are still unclear but seem to lie between Köyceğiz and Bafa.

Further reading. Tigano & Ferrito 1985 (osteology); Gandolfi et al. 1991 (biology); Wildekamp 1993 (description); Leonardos & Sinis 1998 (reproduction); Kottelat et al. 2007 (systematics); Valdesalici et al. 2019 (distribution).

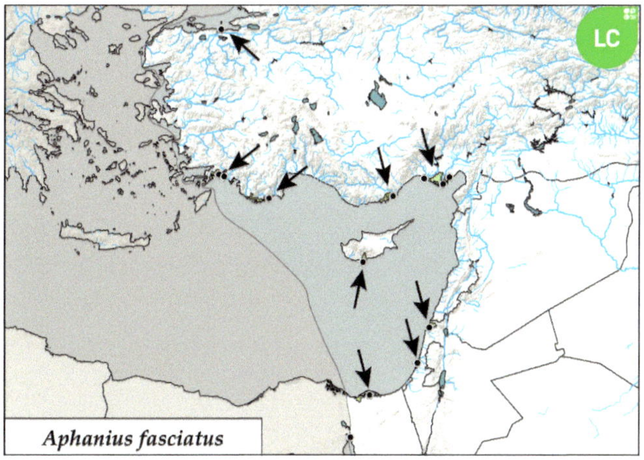

Coastal lagoons are the typical habitat of both *Aphanius* species, as here in the Gediz delta in Türkiye, where *A. almiriensis* is common.

Esmaeilius

Persian killifish is a genus of six species widespread in Iran and the lower Shatt al Arab in Iraq. All species inhabit inland habitats, such as the shores of rivers, streams, and springs. Many populations, often largely isolated, indicate a much wider distribution range in the recent geological past, and many populations might have become isolated only a few thousand years ago. The distribution of these fishes reflects the interesting geological history of the Zagros Mountains. Some geographically distantly isolated populations were described as separate species during their discovery. However, subsequent analysis revealed that many of these are very closely related, morphologically indistinguishable and that those previously identified as *E. arakensis*, *E. kavirensis*, *E. mesopotamicus*, or *E. pluristriatus* are now considered synonymous with *E. sophiae*. Many populations have been lost recently due to the unsustainable exploitation of freshwater resources and the increasing aridity of Central Iran, which is also a consequence of climate change. **Further reading.** Freyhof & Yoğurtçuoğlu 2020 (description, species diversity).

Key to species of *Esmaeilius*

1a - Male with black dorsal, pelvic, and anal margins.
..................*E. isfahanensis*
1b - Male with white dorsal, pelvic, and anal margins.
..................2

2a - Female with distinct, short, brown bars along lateral midline.
..................*E. persicus*
2b - Female with irregular set brown or dark-grey spots and blotches, blotches sometimes slightly vertically elongated.
..................3

3a - 36–47 total scales in lateral series; female without bold black spot at middle caudal base.
..................*E. vladykovi*
3b - 23–32 total scales in lateral series; female with bold black spot at middle caudal base, absent or very small in *E. shirini*.
..................4

4a - Male usually with 7–10 wide flank bars.
..................*E. shirini*
4b - Male with 10–19, usually 11–16, narrow flank bars.
..................5

5a - Female with a midlateral series of short, slightly vertically elongated blotches.
..................*E. darabensis*
5b - Female with a pattern of spots and irregularly shaped blotches not organised in a midlateral series.
..................*E. sophiae*

Esmaeilius darabensis; Golabi spring, Kol drainage, Iran; female, ~30 mm SL.

Esmaeilius darabensis; Golabi spring, Kol drainage, Iran; male, ~25 mm SL.

Esmaeilius darabensis

Common name. Darab killifish.

Diagnosis. Distinguished from other species of *Esmaeilius* by: • female with a midlateral series of short, slightly vertically elongated blotches / ○ male with narrow white dorsal and pelvic margin / ○ anal margin white, hyaline or greyish / ○ 10–19 dark-brown bars on flank wider than interspaces / ○ female with a distinctive black spot at caudal base / ○ 23–28 scales in lateral series. Size up to 34 mm SL.

Distribution. Iran: Korsiah Banaki and Golabi springs and Dasht–e Konar wetland in upper Kol drainage.

Habitat. Springs and spring-fed streams.

Biology. No data. Expected to be similar to other *Esmaeilius*.

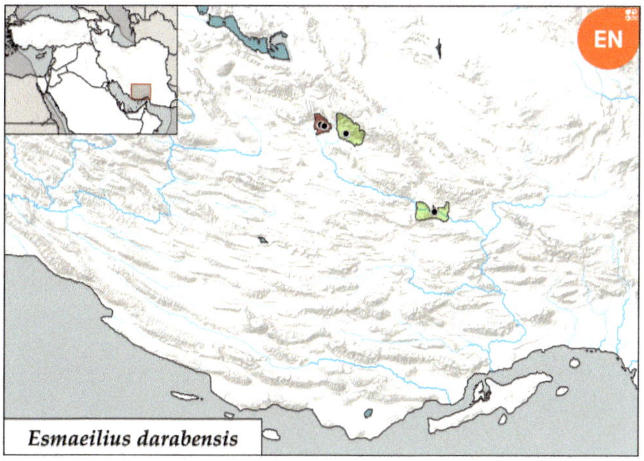

Conservation status. EN; appears to be declining within its very small range. Extirpated from Korsiah Banaki. (not shown on the map).
Remarks. This species is related to *E. persicus* but superficially resembles *E. sophiae*. Both species are distinguished mainly by molecular characters, and the morphological differences given may be quite vague.
Further reading. Esmaeili et al. 2014a, b (description, biogeography); Esmaeili et al. 2020b (molecular data and distribution); Freyhof & Yoğurtçuoğlu 2020 (generic position).

Esmaeilius isfahanensis
Common name. Esfahan killifish.
Diagnosis. Distinguished from other species of *Esmaeilius* by: • male with black dorsal, pelvic and anal margins / • female mottled with many large blotches or narrow bars on flank, usually fused into a midlateral, wide, dark-brown band / ○ 25–29 total scales along lateral series / ○ female with a distinctive black spot at caudal base. Size up to 45 mm SL.
Distribution. Iran: Lower Zayandeh drainage (e.g., Hasan Abad, Malvajerd qanats).
Habitat. Springs and spring-fed streams.
Biology. Live up to 3 years, mature in a few months at about 25 mm SL, usually late in year of birth. Males establish territories along banks, usually in dense vegetation but also between rocks, which they defend against rival males. Spawns April–July. Females spawn with one or more males, usually in algae or other vegetation near surface or in gravel beds. Few eggs are laid in the substrate during a spawning event. Individual females may produce several clutches in one day. Feeds on algae, detritus, and small invertebrates.

Esmaeilius isfahanensis; Zayandeh, Iran; female, 35 mm SL.

Esmaeilius isfahanensis; Zayandeh, Iran; male, 32 mm SL.

Conservation status. CR; appears to be declining within its very small range. May have disappeared from Varzaneh.
Remarks. *Esmaeilius sophiae* inhabits Gavkhoni marshes in very lower Zayandeh drainage. The black spot at the caudal base in the female is much more pronounced in small individuals and fades in large adults of more than 25 mm SL.
Further reading. Hrbek et al. 2006 (description); Keivany 2013 (biology); Esmaeili et al. 2020b (molecular data and distribution); Freyhof & Yoğurtçuoğlu 2020 (generic position).

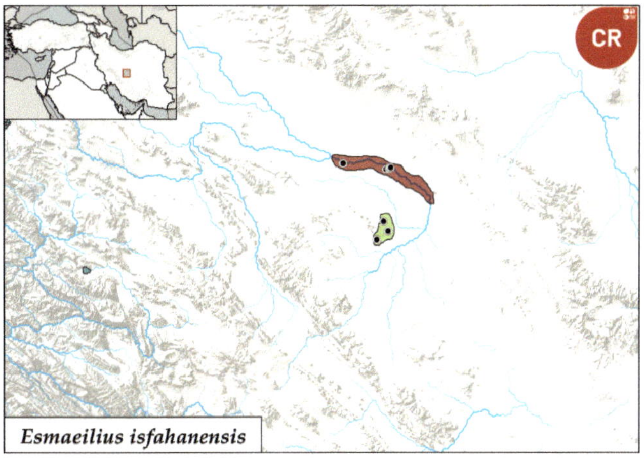

Isolation of endorheic Kor basin. In West Asia, numerous rivers do not reach the sea but terminate in lakes or marshes that are often hypersaline. These rivers are known as endorheic rivers and endorheic basins, in contrast to exorheic rivers that flow to the sea. One of the many endorheic rivers in Iran is the Kor, which has been the subject of considerable study. During the Quaternary period, the Palaeo-Kor drained to the Persian Gulf via the Mond River. From the late Quaternary (20,000 years ago) until the Holocene (2000 years ago), the exorheic drainage was closed due to the tectonic uplift of the Zagros Mountains. This resulted in the Kor becoming endorheic and isolated from the exorheic Mond. Unsurprisingly, most species currently found in the endorheic Kor are the same as those found in the Mond and other adjacent tributaries of the Persian Gulf. The isolation of the Kor occurred relatively recently. However, this may be a partial story. While two endemic fish species, *Esmaeilius shirini* and *Oxynoemacheilus tongiorgii*, are present in the Kor, several species found in the Mond are absent in the Kor, including *Capoeta mandica*, and *Garra rufa*. Conversely, *Chondrostoma regium* occurs in the Kor but is absent in the Mond. Clearly, the biogeography of the Kor is far more complex and its connection to the Lake Maharlo basin (home to the endemic *E. persicus*) also requires further investigation. **Further reading.** Gholami et al. 2014 (geology).

Esmaeilius persicus; Pirbanoo spring, Iran; female, ~30 mm SL.

Esmaeilius persicus; Pirbanoo spring, Iran; male, ~30 mm SL.

Esmaeilius persicus

Common name. Maharlo killifish.

Diagnosis. Distinguished from other species of *Esmaeilius* by: • female with distinct, brown bars along lateral midline / ○ male with narrow white anal, dorsal, and pelvic margin / ○ female with a distinctive black spot at caudal base / ○ 24–29 total scales along lateral series. Size up to 35 mm SL.

Distribution. Iran: Lake Maharlo basin.

Habitat. Small freshwater or brackish springs and pools, usually without vegetation.

Biology. Live up to 4 years, mature in a few months, usually late in year of birth. Males establish territories along banks, usually in dense vegetation, but also between rocks, which they defend against rivals. Spawns April–November (Barm-e Shur spring). Females spawn with one or more males, usually in vegetation near surface or in gravel beds. Few eggs are laid in substrate during a spawning event. Individual females may produce several clutches in one day. Feeds on algae, detritus and small invertebrates.

Conservation status. Extinct in the wild; found only in one spring in 2013 but seems to have disappeared from nature since. Water abstraction, habitat destruction, pollution, introduction of non-native species and reduced rainfall due to climate change have caused all populations to vanish and springs to dry up.

Remarks. Very few female *E. persicus* lack the diagnostic narrow brown bars on the flanks but have a mottled pattern. This species was called *Aphanius farsicus* between 2011 and 2020.

Further reading. Monsefi et al. 2009 (reproduction); Teimori et al. 2011 (*A. farsicus*); Esmaeili et al. 2020b (molecular data and distribution); Freyhof & Yoğurtçuoğlu 2020 (generic position, discussion of *Brachylebias*, *E. persicus*).

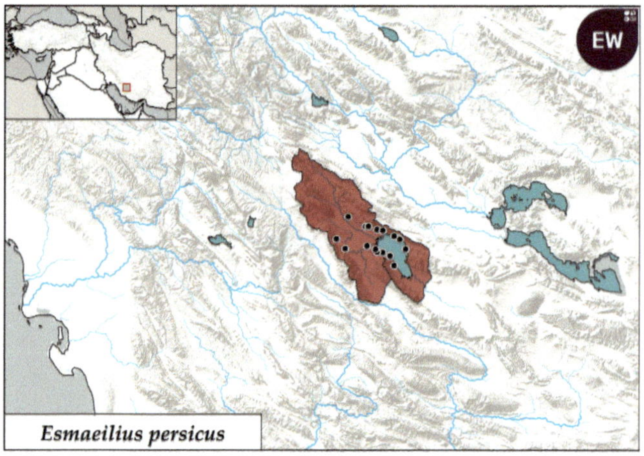

***Aphanius persicus*: Forward and back again.** *Aphanius persicus* (Jenkins, 1910) was previously known as the scientific name of the Maharlo killifish. When Gaudant (2011) studied the fossil killifish *Brachylebias persicus* Priem, 1908, he concluded that this must be a species of *Aphanius* and thus transferred *B. persicus* to *Aphanius*. Consequently, *A. persicus* (Jenkins, 1910) became a junior secondary homonym of *A. persicus* (Priem, 1908). As the name *A. persicus* was no longer available for the Maharlo killifish, it was redescribed as *A. farsicus*. However, upon the separation of the phylogenetic groups of *Aphanius* into different genera, *Brachylebias* was considered *incertae sedis*, as there were no arguments to place it in one of the recognised genera, and it could not be excluded that it represents a distinct (extinct) evolutionary lineage. As it is highly unlikely that *Brachylebias* might have been a species of *Esmaeilius*, it was transferred back to *Brachylebias*. As a result of this reclassification, *Aphanius farsicus* became a junior synonym of *A. persicus* (Jenkins, 1910), and the latter name is once again available for the Maharlo killifish. This species was then transferred to the new genus *Esmaeilius*. **Further reading.** Freyhof & Yoğurtçuoğlu 2020 (*A. persicus* and *Brachylebias*).

Water is extracted from springs everywhere in West Asia, such as in Pirbanoo in Iran. The picture was taken a few weeks before the spring dried out and one of the last populations of *Esmaeilius persicus* vanished in 2007.

Esmaeilius shirini; Khosroshirin spring, Iran; female, ~35 mm SL.

Esmaeilius shirini; Khosroshirin spring, Iran; male, ~30 mm SL.

Esmaeilius shirini

Common name. Shirin killifish.

Diagnosis. Distinguished from other species of *Esmaeilius* by: ● male usually with 7–11 wide, brown bars on flank / ○ male with wide and white dorsal margin / ○ female with irregular set brown or dark-grey spots and blotches / ○ female without or with a very small black spot at caudal base / ○ 25–31 total scales along lateral series. Size up to 36 mm SL.

Distribution. Iran: Khosroshirin, a headwater spring-stream system in upper Kor drainage and Arjan wetland in Helleh drainage (introduced).

Habitat. Streams and springs with low to moderate current, gravel and mud substrate covered with dense filamentous algae.

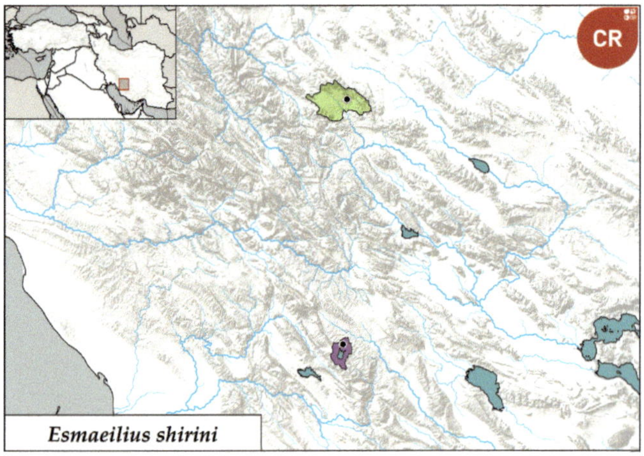

Biology. No data. Expected to be similar to other *Esmaeilius*.
Conservation status. CR; has declined sharply in its last native habitat and may be on the verge of extinction, occurring in only one non-native habitat.
Further reading. Gholami et al. 2014 (description, biogeography); Esmaeili et al. 2020b (molecular data and distribution); Freyhof & Yoğurtçuoğlu 2020 (generic position).

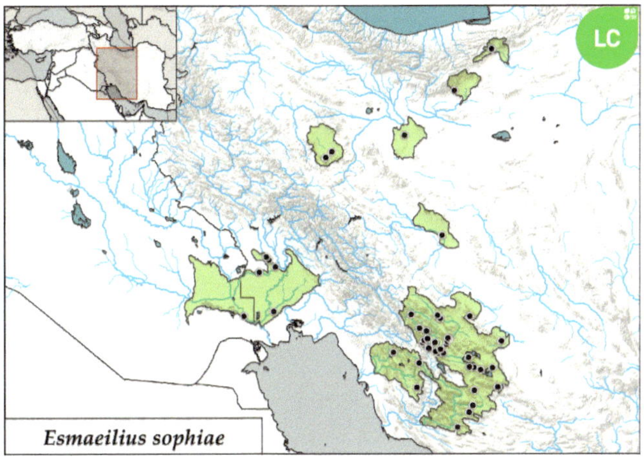

Esmaeilius sophiae

Common name. Persian killifish.
Diagnosis. Distinguished from other species of *Esmaeilius* by: ○ male with narrow white anal, dorsal and pelvic margin / ○ male with 8–19, usually 11–16 brown bars on

Esmaeilius sophiae; Cheshmeh Ali spring, Iran; female, 32 mm SL.

Esmaeilius sophiae; Gavkhoni marsh, Iran; male, 26 mm SL.

flank / ○ female with irregular set brown or dark-grey spots and blotches / ○ female with a distinctive black spot at caudal base / ○ 25–31 total scales along lateral series. Size up to 42 mm SL.

Distribution. Iraq: Shatt al-Arab, north of Basra. Iran: Springs in Lake Namak basin, lower Karkheh and Karun drainage, Helleh drainage, and springs in Mond drainage, possibly in a qanat near Jahrom. Endorheic basins of Kor, Esfahan, and Sirjan, Western Kavir (Cheshmeh Ali spring), and Gavkhoni marsh in the very lower Zayandeh drainage.

Habitat. Rivers, streams, springs and qanats with stagnant or slowly flowing water, usually with dense submerged vegetation, but also with little or no vegetation and small gravel and pebble substrates.

Biology. Lives up to 4 years, maturing in a few months, usually late in the year of birth. Males establish territories along banks, usually in dense vegetation but also between rocks, which they defend against rivals. Females spawn with one or more males, usually in algae or other vegetation near surface or in gravel beds. Few eggs are laid in substrate during a spawning event. Individual females may produce several clutches in one day. Feeds on algae, detritus and small invertebrates.

Conservation status. LC; one population in Mond drainage extirpated due to introduction of *Xiphophorus hellerii* and *Gambusia holbrooki*.

Remarks. *Esmaeilius arakensis*, *E. mesopotamicus*, *E. pluristriatus*, and *E. kavirensis* are synonyms.

Further reading. Coad 1996 (distribution); Coad 2009a (description as *E. mesopotamicus*); Esmaeili et al. 2012 (description as *A. pluristriatus*); Keivany et al. 2012 (rediscovery as *E. mesopotamicus*); Teimori et al. 2012 (description of *E. akarensis*); Keivany & Esmaeili 2013 (*A. mesopotamicus*, conservation); Esmaeili et al. 2014b (description as *A. kavirensis*); Gholami et al. 2014 (biogeography); Esmaeili et al. 2020b (molecular data and distribution); Freyhof & Yoğurtçuoğlu 2020 (species diversity).

Esmaeilius vladykovi

Common name. Chaghakhor killifish.

Diagnosis. Distinguished from other species of *Esmaeilius* by: ● 36–47 scales in lateral series / ○ female without distinctive black spot at caudal base / ○ female with irregular set brown or dark-grey spots and blotches, blotches sometimes slightly vertically elongated / ○ male with 9–14 brown bars on flank / ○ male with wide white anal, and dorsal margin. Size up to 65 mm SL.

Distribution. Iran: Upper Karun drainage. Introduced to Hozmahi spring in Zayandeh drainage.

Habitat. High-altitude wetlands in the Zagros Mountains (at about 2400 m). In springs, ponds, streams, and marshes.

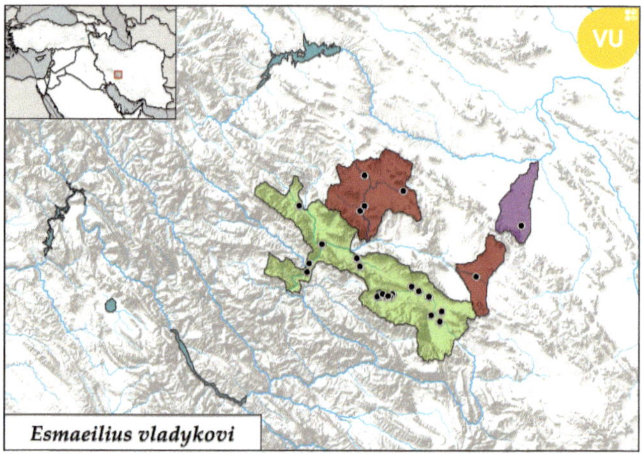

Esmaeilius vladykovi; Boldaji, Iran; female, ~30 mm SL.

Esmaeilius vladykovi; Boldaji, Iran; male, ~35 mm SL.

Biology. Live up to 4 years, mature in a few months, usually late in year of birth. Males establish territories along banks, usually in dense vegetation, but also between rocks, which they defend against rivals. Spawning peaks in early April but lasts much longer. Females spawn with one or more males, usually in algae or other vegetation close to surface or in gravel beds. Adhesive eggs attached to plants in batches of 3–30 eggs hatch in 9–13 days at 21–22°C. Individual females may produce several clutches in one day. Feeds on algae, detritus, and small invertebrates.

Conservation status. VU; originally present at 10 sites, but has disappeared from four sites in the last 10 years.

Further reading. Coad 1988 (description); Pazooki et al. 2008 (food); Keivany & Soofiani 2004 (reproduction); Esmaeili et al. 2020b (molecular data, distribution); Freyhof & Yoğurtçuoğlu 2020 (generic position).

Kosswigichthys

The only species in this genus is a highly specialised pelagic killifish endemic to a single ancient lake in Eastern Anatolia. *Kosswigichthys* is related to *Anatolichthys*, but both genera are long and well-separated. *Kosswigichthys* is one of West Asia's few phylogenetically isolated and highly adapted lacustrine fish species. Its only habitat, Lake Hazar, is a rift lake situated at the border of the Anatolian and Arabian plates within the East Anatolian Fault Zone (EAFZ). This is one of the most significant active continental transform fault zones globally. Lake Hazar may be the oldest lake in West Asia, having possibly commenced development in the late Miocene or early Pliocene. However, some studies suggest a quite recent age for the lake, of about 150–180,000 years. Molecular data demonstrate that *Kosswigichthys* is isolated from *Anatolichthys* for longer than the late Miocene and that its ancestors may have inhabited the area much longer. It is possible that they found refuge in Lake Hazar, adapting to lacustrine conditions when the lake was formed. Their adaptations to lacustrine environments are similar to those of *Anatolichthys saldae* and *A. splendens*, which also show a lack of or reduced scales, a slender body, and a strongly upturned mouth. However, these species inhabit relatively young lakes and might have separated from their riverine ancestors only during the late Pleistocene. At present, *K. asquamatus* is the sole representative of the Aphaniidae in the upper Mesopotamian drainage system. **Further reading.** Freyhof & Yoğurtçuoğlu 2020 (generic structure of Aphaniidae).

Kosswigichthys asquamatus; Hazar Lake, Türkiye; female, ~33 mm SL.

Kosswigichthys asquamatus; Hazar Lake, Türkiye; male, ~30 mm SL.

Kosswigichthys asquamatus

Common name. Hazar killifish.

Diagnosis. Distinguished from other species of Aphaniidae by: ● teeth conical, in three rows / ○ body without scales / ○ no black bars in caudal in male. Size up to 36 mm SL.

Distribution. Türkiye: Lake Hazar.

Habitat. Pelagic come to shores in spring and summer to spawn. Feeds on algae, invertebrates, and plankton.

Biology. No data.

Conservation status. EN; appears to be declining within its very small range.

Further reading. Sözer 1942 (description); Freyhof & Yoğurtçuoğlu 2020 (distribution, morphology, generic position).

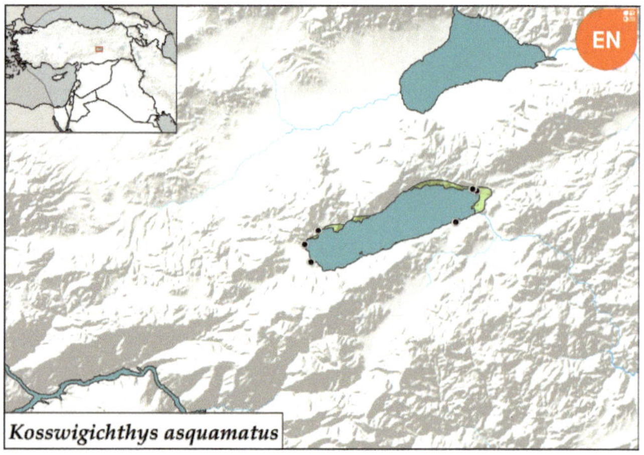

Lake Hazar, the only habitat of *Kosswigichthys asquamatus*, is a deep, ancient, tectonic lake in West Asia.

Paraphanius

Paraphanius comprises seven species, plus one undescribed, that are all superficially similar but well distinguished by molecular characters. *Paraphanius* species are particularly diverse in the Eastern Mediterranean, especially in the Levant. One species, *P. mento*, is distributed across the Levant, the upper Euphrates and Tigris, where it appears to have a localised distribution pattern. Further studies are required to determine whether this species represents a complex of closely related species. **Further reading.** Esmaeili et al. 2020b (description of *Paraphanius*); Freyhof & Yoğurtçuoğlu 2020 (generic structure of Aphaniidae).

Key to species of *Paraphanius*

The key is only suitable for adult fishes, ideally nuptial females and males, larger than 30 mm SL.

1a - Adult male with silvery, very narrow, vertical lines on flank.
..................2

1b - Adult male with silvery spots on flank, often organised in fuzzy, vertical series but not in lines (juvenile males of all species often with lines).
..................3

2a - Adults very large, often more than 60 mm SL; lower jaw strongly upturned; in male, silvery lines irregularly shaped, often interrupted; background colour greenish, greyish or dark-blue in male, upper part of head and back, not orange.
..................*P. mentoides*

2b - Adults small, rarely more than 40 mm SL; lower jaw slightly upturned; silvery lines regularly shaped, rarely interrupted in male; background colour bright blue in male, with an orange upper part of head and anterior back.
..................*P. striptus*

3a - Female without midlateral stripe or pale-brown marbled pattern on flank.
..................4
3b –Female with pale-brown marbled pattern on flank and a midlateral, pale-brown stripe between vertical of dorsal base and caudal base.
..................5

4a - Silvery scales on caudal peduncle not in regular lines in female; silvery spots organised in vertical rows in male.
..................*P. similis*
4b - Silvery scales on caudal peduncle in regular lines in female; silvery spots not organised in vertical rows in male (characters poorly known and very different in different populations).
..................*P. mento*

5a - Silvery spots distinct, well isolated from each other, not organised in vertical rows in adult male (in rows in juveniles).
..................*P. orontis*
5b - Silvery spots often fused and usually organised in vertical rows in juvenile and adult male.
..................6

6a - Female without marbled pattern on back; usually with single pale-brown dots in middle of scales.
..................*P. alexandri*
6b - Female with marbled pattern on back; usually without pale-brown dots in middle of scales.
..................*P. boulengeri*

Paraphanius alexandri; Elbistan, Ceyhan drainage, Türkiye; female, 33 mm SL.

Paraphanius alexandri; Ceyhan drainage, Türkiye; male, 45 mm SL.

Paraphanius alexandri

Common name. İskenderun killifish.

Diagnosis. Distinguished from other species of *Paraphanius* by: ○ male bluish, black or dusty grey with roundish or elongated silvery spots on flank organised in vertical series / ○ silvery spots indistinct, overlaid by many fine, powder-like dots / ○ pelvic reaching to anal origin in adult male / ○ pre-dorsal distance 57–60 % SL / ○ pre-anal distance 63–70 % SL / ○ interorbital distance 36–45 % HL female with pale-brown marbled pattern on flank / ○ female with pale-brown stripe between vertical of dorsal-base and caudal base / ○ female with silvery scales above and below stripe on caudal peduncle / ○ female without marbled pattern on back, often with single pale-brown dots in middle of scales. Size up to 46 mm SL.

Distribution. Türkiye: Middle and upper Ceyhan drainage, Erzin drainage, and small coastal streams in Hatay south to Arsuz. Introduced in warm springs in upper Tohma drainage (Euphrates).

Habitat. Springs, lakes and slow-flowing streams.

Biology. Live up to 4 years, mature in a few months, usually late in year of birth. Males establish territories along banks, usually in dense vegetation, defending against rivals. Females spawn with one or more males, usually in vegetation close to surface. Individual females produce several clutches in 1 day. Feeds on algae, detritus and small invertebrates.

Conservation status. LC.

Remarks. Appears to be absent from the lower Ceyhan drainage.

Further reading. Akşıray 1948 (description).

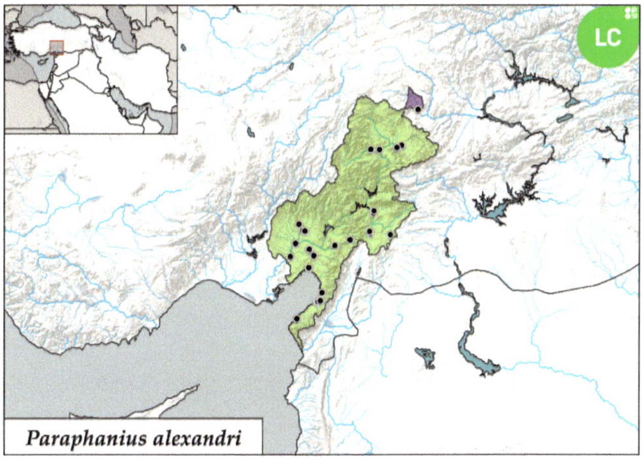

Paraphanius boulengeri; Gölbaşı, Ceyhan drainage, Türkiye; female, 40 mm SL.

Paraphanius boulengeri; Gölbaşı, Ceyhan drainage, Türkiye; male, 37 mm SL.

Paraphanius boulengeri

Common name. Gölbaşı killifish.

Diagnosis. Distinguished from other species of *Paraphanius* by: ○ male bluish, black or dusty grey with roundish or elongated silvery spots on flank organised in vertical series / ○ silvery spots indistinct, overlaid by many fine, powder-like dots / ○ pelvic reaching to anal origin in adult male / ○ pre-dorsal distance 55–62 % SL / ○ pre-anal distance 62–64 % SL/ ○ interorbital distance 33–40 % HL / ○ female with pale-brown marbled pattern on flank / ○ female with pale-brown stripe between vertical of dorsal-base and caudal base / ○ female with silvery scales above and below stripe on caudal peduncle / ○ female with marbled pattern on back. Size up to 40 mm SL.

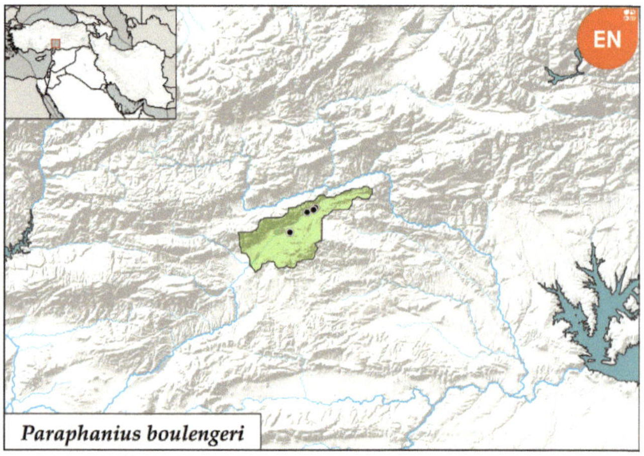

Distribution. Türkiye: Gölbaşı lakes in Adıyaman.
Habitat. Lakes and canals with or without dense underwater vegetation.
Biology. Live up to 4 years, mature in a few months, usually late in year of birth. Males establish territories along banks, usually in dense vegetation, defending against rivals. Females spawn with one or more males, usually in vegetation close to surface. Individual females produce several clutches in one day. Feeds on algae, detritus, and small invertebrates.
Conservation status. EN; appears to be declining within its very small range. Its only habitat has been dominated by invasive *Gambusia holbrooki*.
Further reading. Akşıray 1948 (description).

Paraphanius boulengeri is endemic to Gölbaşı lakes in Ceyhan drainage, a small endemicity hotspot in the region.

Paraphanius mento; Orontes drainage, Syria; female, ~40 mm SL.

Paraphanius mento; Orontes drainage, Syria; male, ~45 mm SL.

Paraphanius mento; Spring of Barada, Syria; male, ~50 mm SL.

Paraphanius cf. *mento*; Shatt al Arab, Iraq; female, 32 mm SL; male, 31 mm SL. © A. Ali.

Paraphanius cf. *mento*; Ein Feshkha, Israel; male, ~40 mm SL.

Paraphanius mento

Common name. Iridescent killifish.

Diagnosis. Distinguished from other species of *Paraphanius* by: ○ male bluish, black or dusty grey with roundish or elongated silvery spots on flank, spots might be organised in vertical series or in longitudinal rows (with whitish bars in lower Mesopotamia) / ○ silvery spots distinct, not overlaid by many fine, powder-like dots / ○ female with silvery spots on each flank-scale (female with whitish blotches in lower Mesopotamia) / ○ female without midlateral stripe or pale-brown marbled pattern on flank / ○ pelvic not reaching to anal origin in adult male. Diagnostic character blurred as several species seemed to be involved. Size up to 52 mm SL.

Distribution. Tigris downstream of Mosul and common in marshes along Shatt al-Arab/Arvand. Upper Nahr al Khābūr in Syria, Damascus basin, Litani and Jordan drainages, and Dead Sea basin (Ein Feshkha spring). Also, one record in Qweiq and in very upper Orontes in Syria, downstream to Homs.

Habitat. Slow-flowing rivers, lakes, and marshes usually associated with dense aquatic vegetation.

Biology. Live up to 3 years, mature in a few months, usually late in year of birth. Spawns April to July in southern Iraq. Males establish territories along banks, usually in dense vegetation, defending against rivals. Females spawn with one or more males, usually in vegetation close to surface. Individual females produce several clutches in one day. Feeds on algae, detritus and small invertebrates.

Conservation status. LC; extirpated from Qweiq and seems to have disappeared from Iranian part of Mesopotamian marshes.

Remarks. There is a considerable population diversity within *P. mento,* as recognised here. *Paraphanius mento* occurs in sympatry with *P. striptus* in the northern Dead Sea basin and possibly elsewhere in Jordan drainage. Both species have also hybridised extensively in the region. *Paraphanius mento* was described from Tigris at Mosul based on individuals with isolated iridescent spots. Fish from the lower Tigris show a barred pattern in males (similar to *P. striptus*) and a marmorated pattern in females (vs. plain in *P. striptus* and other populations of *P. mento*). These might not be conspecific with *P. mento*. Molecular characters and details in colour pattern of fish from Mosul have not yet been studied. Molecular analyses place fish from the Litani and upper Orontes, lowermost Tigris, Jordan, Damascus basin, and middle Orontes in three distinct clades, with *P. striptus* nested in between in some analyses. The diversity of *P. mento* will have to be re-examined when material from Mosul becomes available.

Further reading. Al-Daham et al. 1977 (biology).

Paraphanius mentoides

Common name. Kırkgöz killifish.

Diagnosis. Distinguished from other species of *Paraphanius* by: ○ adults very large, often more than 60 mm SL / ○ pre-dorsal distance 62–66 % SL / ○ pre-anal distance 67–72 % SL / ○ interorbital distance 32–33 % HL / ○ male bluish, black or dusty grey with roundish or elongated silvery spots on flank, organised in vertical lines / ○ silvery lines distinct, not overlaid by many fine, powder-like dots / ○ lower jaw strongly upturned / ○ background colour greenish, greyish or dark-blue in male, upper part of head and back not orange / ○ female without silvery scales and without brown stripe along lateral midline. Size up to 66 mm SL.

Distribution. Türkiye: Yamansaz marsh, Düden spring in Antalya, Kırkgöz spring and outflowing stream Yediarıklar. Introduced in Lake Nemrut, a volcano crater lake in East Anatolia (not shown on map).

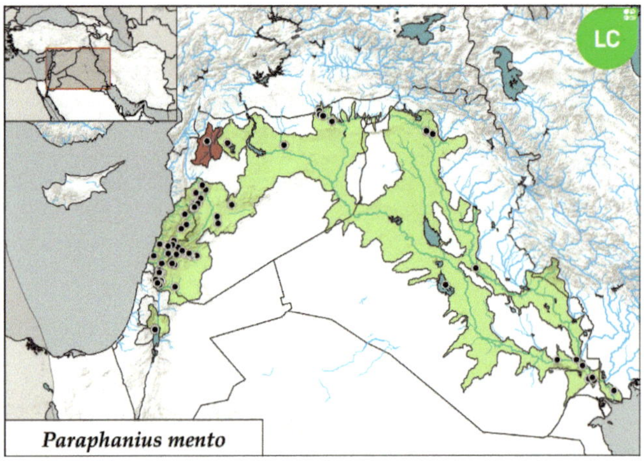

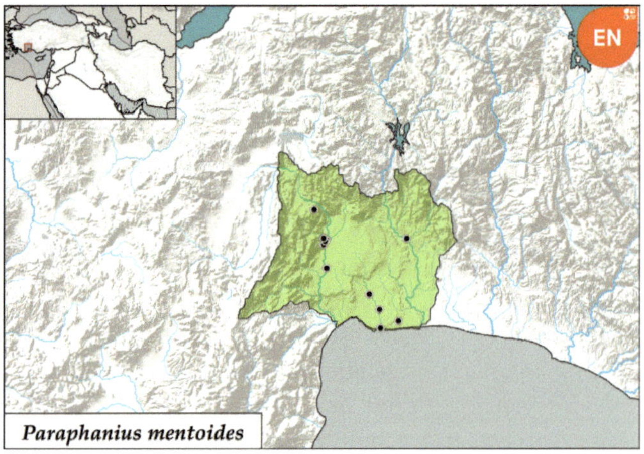

Paraphanius mentoides; Kırkgöz spring, Türkiye; female, 60 mm SL.

Paraphanius mentoides; Kırkgöz spring, Türkiye; male, 58 mm SL.

Habitat. Lakeshores, springs, and spring-fed streams.

Biology. Live up to 7 years, mature after first winter. Spawns May–July. Males establish territories along banks, usually in dense vegetation, defending against rivals. Females spawn with one or more males, usually in vegetation close to surface. Individual females produce several clutches in one day. Feeds on algae, detritus, and small invertebrates.

Conservation status. EN; appears to be declining within its very small range.

Further reading. Akşıray 1948 (description); Güçlü & Küçük 2011 (reproduction, as *Aphanius mento*).

Nemrut Crater Lake is the habitat of a non-native population of *Paraphanius mentoides*.

Paraphanius orontis; Lake Gölbaşı, Orontes drainage, Türkiye; female, 38 mm SL.

Paraphanius orontis; Lake Gölbaşı, Orontes drainage, Türkiye; male, 39 mm SL.

Paraphanius orontis

Common name. Asi killifish.

Diagnosis. Distinguished from other species of *Paraphanius* by: ○ male black or dusty grey with roundish silvery spots on flank not organised in vertical series / ○ silvery spots large, distinct, well isolated from each other, without or with few fine, powder-like dots / ○ female with pale-brown marbled pattern on flank / ○ female with pale-brown stripe between vertical of dorsal-base and caudal base / ○ female with a line of silvery scales above and below stripe on caudal peduncle / ○ pelvic reaching to anal origin in adult male / ○ pre-dorsal distance 57–61 % SL / ○ pre-anal distance 62–65 % SL / ○ interorbital distance 39–45 % HL. Size up to 43 mm SL.

Distribution. Türkiye: Lower Orontes drainage downstream to former Lake Amik. Introduced to lower Manavgat drainage east of Antalya (not shown in map).

Habitat. Lakes, slow-flowing rivers, and canals.

Biology. Live up to 3 years, mature in a few months, usually late in year of birth. Males establish territories along banks, usually in dense vegetation, defending against rivals. Females spawn with one or more males, usually in vegetation close to surface. Individual females produce several clutches in one day. Feeds on algae, detritus, and small invertebrates.

Conservation status. VU; found in fewer than 10 independent populations, some of which are declining.

Further reading. Akşıray 1948 (description).

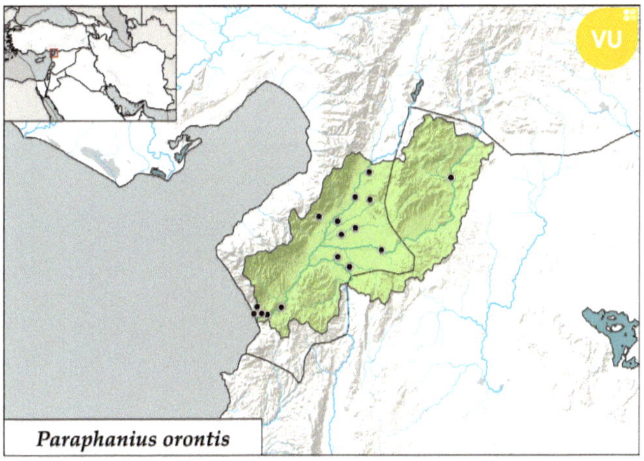

The Orontes, a composite river. Tectonic activities significantly modify the river system of the Levant. The Jordan, the Litani, and the Orontes have a particularly troubled history, repeatedly ending in endorheic, terminal lakes. These rivers were connected and disconnected from each other and the Euphrates or drained into the Mediterranean through short, temporary outlets in the coastal mountain ranges. The Orontes is the longest river in the Levant, with a length of 600 km. It has its source in the Ayn az-Zarqa spring in the Lebanese city of Baalbek, from where it flows north through Syria to Türkiye and into the Mediterranean. Two major tributaries, the Afrin and the Karasu, originate in Türkiye and flow south to meet the Orontes at the former Lake Amik. The history of the Orontes is as interesting as it is complicated. During the Upper Miocene and early Pliocene, the current-day lower Orontes was the only river flowing into the Mediterranean. The upper and middle parts of the Orontes flowed independently to the Euphrates. The current-day endorheic Qweiq was a tributary of the middle Orontes when it was still connected to the Euphrates. During the Lower Pleistocene, the al-Ghab Valley, which was formed during the Miocene, subsided, uplifting the Gabal az-Zawiyah Mountains. The uplifting of Gabal az-Zawiyah, which extended toward the Palmyra fold belt, severed the connections between the Proto-Orontes and the Euphrates. Nevertheless, the connection between the Qweiq and the Euphrates remained intact and only dried out in the early Holocene. At that time, the present Orontes consisted of three independent rivers separated from each other by mountain plateaus.

The upper river originated in Lebanon and discharged into the current-day Nar al Kebir South. The middle sector, represented by the al-Ghab basin at the time, had its outflow in the Nahr Marqīyah, today one of Syria's small coastal rivers. There is evidence that it was also connected to the Nahr al-Kabir North, another coastal river of the present day, via the Nahr al-Abyad, whose upper reaches were at that time a source of the Nahr al-Kabir North. The lower course, which opened to the Mediterranean following a tectonic subsidence, allowed the coastal Küçük Orontes to flow to Lake Amik. All three parts had been endorheic for some time and formed large lakes before being connected. The water of the huge spring areas in today's lower Orontes and the Afrin and Karasu flows drained into the large Lake Amik, forming an endorheic basin. The waters of the Middle sector formed extensive terminal lakes and marshes of al-Ghab and Aharne following the disconnection of this area from the Nahr al Kabir north. The upper sector of the Orontes formed Lake Homs following the disconnection of this area from the Nahr al Kabir south. The basaltic extrusions that erupted during the Quaternary separated the Orontes from the coastal rivers. The confluence of the three formerly separated segments of the Orontes is relatively recent, around 6000 years ago. It was caused by retrogressive erosion.

The recent joining of the different sectors of the Orontes is still reflected in its faunal composition. For example, *Garra variabilis* is only known from the middle and upper sectors. The northern sector is restricted to the Nahr al-Kabir South, *Alburnus magnificus*, *Oxynoemacheilus hamwii*, *O. amanos*, and *Pseudophoxinus turani* are endemic to this sector. In contrast, *Oxynoemacheilus shehabi* is limited to the southernmost Orontes. In the widely distributed and ubiquitous *Paraphanius* killifishes, three distinct mtDNA lineages have been identified, corresponding to *P. orontis* in the northern Orontes, *P. mento* in the southern Orontes, and a potentially undescribed species in between, as well as in some adjacent coastal streams. **Further reading.** Alwan 2010 (geology and biogeography).

Paraphanius similis; Ereğli, Türkiye; female, 31 mm SL.

Paraphanius similis; Ereğli, Türkiye; male; 28 mm SL.

Paraphanius similis; lower Seyhan; Türkiye; female, ~25 mm SL.

Paraphanius similis; Zengen, Türkiye; male, ~30 mm SL.

Paraphanius similis

Common name. Adana killifish.

Diagnosis. Distinguished from other species of *Paraphanius* by: ○ male bluish, black or dusty grey with roundish or elongated silvery spots on flank organised in vertical series / ○ silvery spots distinct, without or with few fine, powder-like dots / ○ female with many silvery spots on flank, not organised as continuous rows on caudal peduncle / ○ female without midlateral stripe or pale-brown marbled pattern on flank / ○ pelvic reaching to anal origin in adult male / ○ pre-dorsal distance 57–60 % SL / ○ pre-anal distance 63–68 % SL / ○ interorbital distance 36–41 % HL / ○ Size up to 40 mm SL.

Distribution. Türkiye: Lower Tarsus, Seyhan and Ceyhan. Also, in springs in Ereğli marshes.

Habitat. Lakes, slow-flowing rivers and coastal marshes with fresh or brackish water. Also, in canals, large reservoirs and many heavily polluted and artificial habitats.

Biology. Live up to 3 years, mature in a few months, usually late in year of birth. Males establish territories along banks, usually in dense vegetation, defending against rivals. Females spawn with one or more males, usually in vegetation close to surface. Individual females produce several clutches in one day. Feeds on algae, detritus and small invertebrates.

Conservation status. LC.

Further reading. Akşıray 1948 (description).

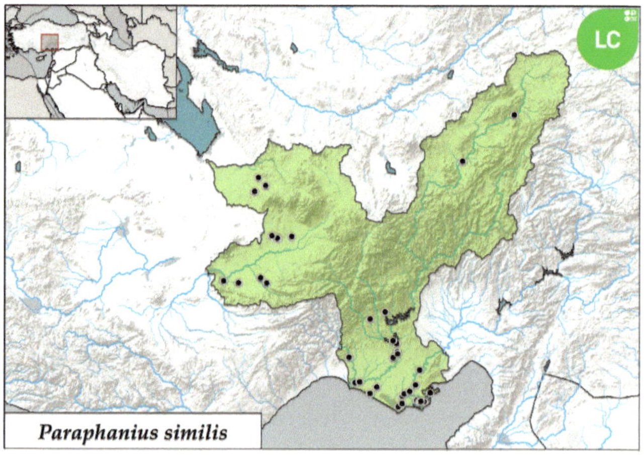

Paraphanius striptus; Einot Timsach, Israel; female, ~32 mm SL.

Paraphanius striptus; Einot Timsach, Israel; male, ~35 mm SL.

Paraphanius striptus

Common name. Palestinian killifish.

Diagnosis. Distinguished from other species of *Paraphanius* by: ○ adults small, rarely more than 40 mm SL / ○ lower jaw slightly upturned / ○ background colour bright blue in male, with an orange upper part of head and anterior back / ○ male with silvery lines, regularly shaped, rarely interrupted / ○ female with few silvery spots on flank, not organised as continuous rows on caudal peduncle / ○ female without brown stripe along lateral midline. Size up to 40 mm SL.

Distribution. Israel and Syria: Jordan drainage south of Lake Tiberias, locally in Jarmuk drainage, and in several coastal spring-stream systems.

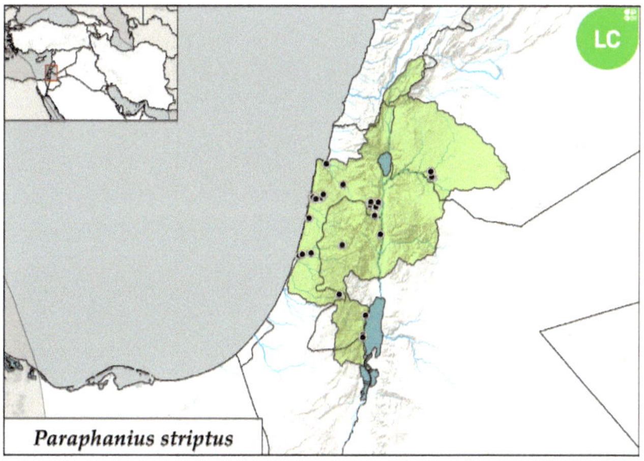

Paraphanius sp.; Orontes drainage, Syria; immature male, 32 mm SL.

Habitat. Springs and streams with dense underwater vegetation.
Biology. Live up to 3 years, mature in a few months, usually late in year of birth. Males establish territories along banks, usually in dense vegetation, defending against rivals. Females spawn with one or more males, usually in vegetation close to surface. Individual females produce several clutches in one day. Feeds on algae, detritus, and small invertebrates.
Conservation status. LC.
Remarks. Occurs in sympatry with *P. mento* in northern Dead Sea basin and possibly elsewhere in Jordan drainage. Here, both species hybridise, at least locally.
Further reading. Goren 1974 (description).

Paraphanius sp.

Common name. Syrian killifish.
Diagnosis. The characters distinguishing this species from other *Paraphanius* have yet to be worked out. Size up to 46 mm SL.
Distribution. Syria: Nahr al Sanawbar, Nahr al-Hawaiz, and middle reaches of Orontes.
Habitat. Springs, rivers, and streams.
Biology. No data.
Conservation status. NE

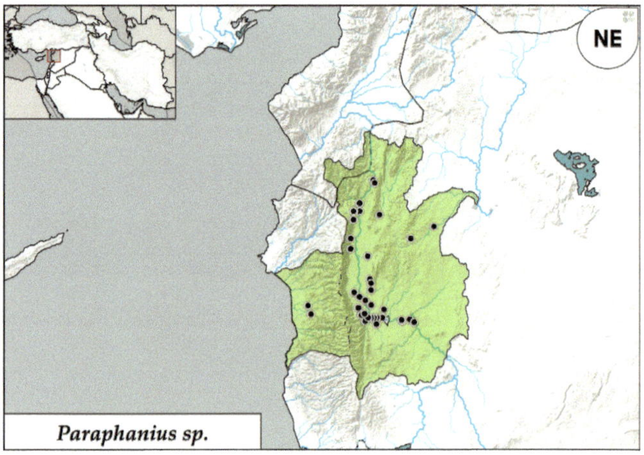

Family Poeciliidae

Livebearers

A large family of small, fresh- and brackish-water fishes distributed across the Americas, from southern USA to Argentina. The majority of species exhibit internal fertilisation and give birth to live young. They also exhibit striking sexual dimorphism, with females being larger and deeper-bodied than males. The male's anal is modified into an intromittent organ (gonopodium) (Fig. 65) through which sperm is introduced into the female's body. Females store sperm, which can be used to fertilise several successive broods. The eyes of the embryos are occasionally visible in a blackish blotch through the female's belly, immediately in front of and above the anal. One species, guppy *Poecilia reticulata*, is one of the world's most popular aquarium fishes and has been released throughout the tropics and in various kinds of warm-water habitats elsewhere. Guppies and mosquitofish *Gambusia holbrooki* are typically introduced to control mosquito populations. However, global studies have demonstrated that these introductions have had only limited effects on mosquito numbers, with potentially negative or neutral impacts on native fish populations. There needs to be a clear justification for these introduction programs, which are ongoing in some countries. It has been demonstrated that mosquitofish is the main or one of the main causes of the critical decline or extirpation of several populations of species of Aphaniids and other fishes.

In West Asia, only *Gambusia holbrooki* is widespread. In addition to the species discussed here, other introduced poecilids may be present, particularly in and around cities. Specialist aquarium literature will allow their identification.

Further reading. Meyer et al. 1985; Meffe & Snelson 1989; Miller 2005; Freyhof et al. 2020 (introductions in Arabia).

Key to species of Poeciliidae in West Asia

1a - Male with long, sword-like prolongation of lower caudal, both sexes with a prominent, red, black, or green midlateral stripe.
..................*Xiphophorus hellerii*
1b - Male without prolongation of caudal, without prominent midlateral stripe.
..................2

2a - Pelvic of male unmodified, tip of first ray not enlarged, similar in male and female in size and shape; gonopodium long.
..................*Gambusia holbrooki*
2b - Pelvic of male modified, tip of first ray enlarged, different in male and female; gonopodium short.
..................3

3a - Body very short and deep; dorsal in male and female equal; flank often red, similar colour in both sexes.
..................*Xiphophorus maculatus*
3b - Body moderately elongate, without grey blotch on posterior part of caudal peduncle; dorsal larger in male than in female; flank not red, if with red blotches, then in male only.
..................4

4a - Dorsal and caudal in male with blue, white, or red colour; flank in male with a variable colour pattern of red, turquoise, yellow, and black blotches and short stripes.
..................*Poecilia reticulata*
4b - Both sexes with rows of small black, grey, or brown spots on flank or flank black or black mottled; male without coloured blotches or stripes.
..................5

5a - Dorsal origin above or in front of anal origin; 12–16½ dorsal rays in male forming a large sail.
..................*Poecilia "latipinna"*
5b - Dorsal origin behind vertical of pelvic origin in male; usually 9½ dorsal rays in male, rays slightly prolonged.
..................*Poecilia "shenops"*

https://doi.org/10.1515/9783111677811-039

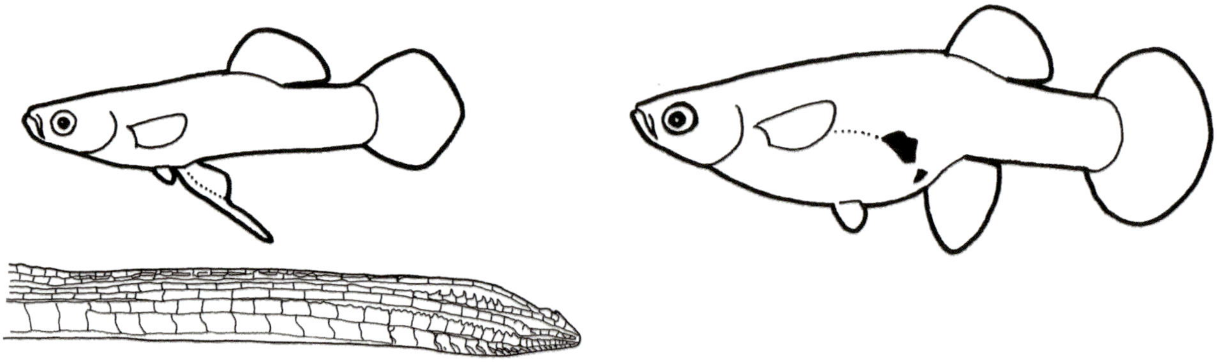

Figure 65: Schematised poecilid showing sexual dimorphism in anal fin shape of male (left, **a**) and female (right, **b**), and modified anal rays of male (gonopodium, **c**) (from Kottelat & Freyhof 2007, c, after Berg 1949b).

Gambusia holbrooki; Lower Seyhan; female, 40 mm SL.

Gambusia holbrooki; Lower Tarsus, Türkiye; ~18 mm SL.

Gambusia holbrooki

Common name. Eastern mosquitofish.

Diagnosis. Distinguished from other species of Poeciliidae in West Asia by: • pelvic of male unmodified, tip of first ray not enlarged, similar in male and female / • gonopodium long • male dull grey / • both sexes with rows of black spots in dorsal and caudal / ○ dorsal and caudal with black spots / ○ no blue or red blotches on flank in male / ○ dorsal origin far behind vertical of anal origin / ○ no lateral stripe / ○ male without sword-like prolongation on caudal. Size up to 30 mm SL in male, 55 mm in female.

Distribution. Introduced throughout warmer parts of West Asia, North Africa, and southern Europe. Throughout Mediterranean basin, including Nile and Cyprus. Locally in southern Caspian basin and coastal areas of Black Sea basin. Iran and locally in Saudi Arabia (Al-Ahsa oasis, Lake Alsfar, wadis around Riyadh). Native to North America, from Delaware drainage to Florida and Alabama.

Habitat. Still or slow-flowing waters, usually with dense aquatic vegetation. Can survive in heavily polluted waters by absorbing oxygen from top surface layers. Often, last fish survive in heavily polluted streams.

Biology. Matures at 4–6 weeks, three or more generations can be produced in a year. Breeds from April–October (depending on latitude), all year in south. Gestation lasts 3–4 weeks. Brood can be up to 354 young, usually between 40 and 60. Sexual activity ceases when temperatures drop below 18°C. While females continue to grow throughout their lives, males stop growing when they reach reproductive size. Feeds mainly on small aquatic invertebrates, highly opportunistic depending on prey availability. Lives below surface, catching prey at or just below surface.

Conservation status. Non-native; first introduced to Europe in Spain (1921) by health authorities in hope that it would control mosquitoes (vectors of various diseases). There is no evidence that it has had any effect on mosquito populations, but there is increasing evidence that it impacts native fish, seriously threatening many endemic species.

Remarks. Often referred to as *G. affinis*, a species identified by details of gonopodium morphology, 5½ dorsal rays (vs. 6½ in *G. holbrooki*) and 8½ anal rays (vs. 9½). There are no confirmed records of *G. affinis* from West Asia. After the initial import in 1921, most subsequent introductions into Europe, North Africa, and West Asia were from European stocks. Still, *Gambusia* was also imported and introduced after 1921; at least one import was from the native range of *G. affinis*. Introduced populations in China are all *G. affinis*.

Further reading. Wooten et al. 1988 (genetics); Rauchenberger 1989 (phylogeny); Jenkins & Burkhead 1993 (biology); Doadrio 2001; Yoğurtçuoğlu & Ekmekçi 2017 (biology); Kurtul 2018 (biology and ecology); Kurtul & Sarı 2017 (risk potential); Kurtul & Sarı 2019 (distribution in Türkiye); Kurtul & Sarı 2020 (biology); Kurtul & Sarı 2021 (recognition level); Freyhof et al. 2020 (distribution in Saudi Arabia).

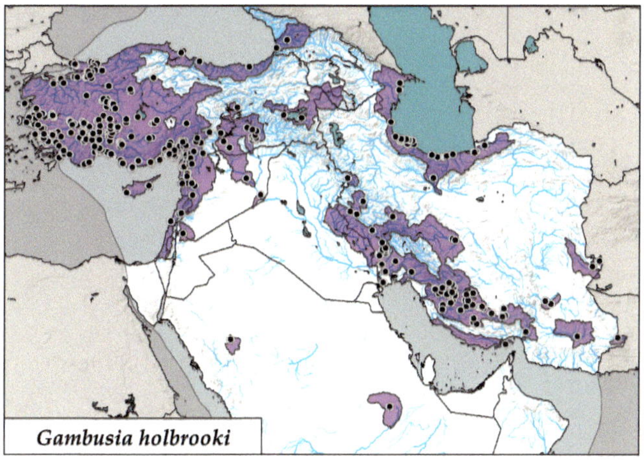

Released aquarium fishes: A looming danger? The aquarium trade represents a significant pathway for introducing non-native species worldwide, including those in West Asia. All Poeciliid species introduced, except *G. holbrooki*, originate from aquarium populations released to the wild. Additionally, *Amatitlania nigrofasciata*, Carassius auratus, *Clarias* sp., and *Pterygoplichthys* catfish have been released from aquaria and established in the wild. The dilemma faced by aquarium fish owners is evident. What should they do with the fish if they no longer wish to keep the aquarium or particular fish? Usually, these fish cannot be returned to the shop or given to other hobbyists. The fish owner is unlikely to want to kill them, so they are often released into the wild. While many hundreds of fish species are kept in aquaria, only a small number can survive and establish themselves in the waters of West Asia. Most aquarium fishes are adapted to tropical conditions and fail to survive cold winters. This is why warm springs are often where released aquarium fishes can be found. Many species are domesticated or originate from tropical rainforest habitats, making it difficult for them to establish in local fish communities. Nevertheless, numerous species are potentially able to cope with the conditions in West Asia, particularly those originating from Europe, North America, and China. Fortunately, these species are less popular in the aquarium trade. Non-native species such as *Atractosteus spatula*, *Pangasianodon hypophthalmus*, *Piaractus brachypomus*, and *Labeo rohita* are not only sourced from the aquarium trade but also originate from escapees from aquaculture facilities.

Poecilia "latipinna"; Hasan Abad, Iran; female, 61 mm SL.

Poecilia "latipinna"; Israel; male, ~50 mm SL. © M. Ford.

Poecilia "latipinna"

Common name. Sailfin molly.

Diagnosis. Character states often intermediate between this "species" and *P. "sphenops"*. Distinguished from other species of Poeciliidae in West Asia by: ● 12–16½ dorsal rays in male forming a large sail / ○ dorsal origin above or in front of vertical of anal origin / ○ no lateral stripe but series of grey or black spots organised in horizontal rows / ○ body moderately elongate / ○ both sexes with rows of small black, grey or brown spots on flank or flank black or black mottled / ○ male without coloured blotches or stripes / ○ dorsal larger in male than in female / ○ flank not red / ○ pelvic of male modified, tip of first ray enlarged, different in male and female / ○ gonopodium short / ○ male without sword-like prolongation on caudal. Size usually up to 80 mm SL. Male larger than female.

Distribution. Hybrids identified as this species have been introduced worldwide in tropical and subtropical countries, including Hammar Marsh and Shatt al Arab/Arvand in Iraq and also established in Zayandeh (Iran), a wadi north of Muscat (Oman), locally in Israel and Saudi Arabia (wadis around Riyadh; Al-Ahsa oasis including Lake Alsfar,

Anik and Al-Qatif), Bahrain (Adhari Park), and probably elsewhere, and introduced to lower Nile (Egypt) and locally in Italy and Greece. *Poecilia latipinna* is native to coastal areas between North Carolina (USA) and Yucatan (Mexico).

Habitat. Wells and irrigation channels, drainage ditches, artificial ponds and streams, coastal marshes, lagoons, estuaries and lower reaches of rivers in brackish to freshwater habitats. Euryhaline, in fresh and even hypersaline water up to 87 ‰, usually in standing water.

Biology. Mature in less than a year. Produces up to 140 young of about 8 mm SL after a gestation period of 20–30 days, depending on water temperature. Females may give birth throughout year. Males defend territories and display their dorsal fins. Only dominant males develop large dorsals and guard schools of females. Males that cannot establish territories hide in schools of females and reproduce by sneeking. Males have higher predation mortality due to their active display behaviour. Feeds mainly on algae, detritus, and small invertebrates.

Conservation status. Non-native; released from aquaria. Rapidly spreading and highly invasive in the northern Persian Gulf region.

Remarks. Non-native sailfin mollies are mistakenly identified as *P. latipinna*. *Poecilia latipinna*, *P. sphenops*, *P. velifera* and possibly others have been accidentally and deliberately hybridised in aquaria and fish farms to produce attractive aquarium stocks suitable for freshwater tanks (the pure species are mostly found in brackish water). This process has not been well documented but appears to have started as early as the 1970s. Such aquarium populations have been released and may have differed considerably in their original genetic composition. The genetic make-up of non-native population has not been studied. Non-native sailfin molly occurs in two colour variants: pale beige and mottled black and beige. Pure black, silver and orange varieties are known from the aquarium trade.

Further reading. Ross 1985 (introduction, distribution in Saudi Arabia); Miller 2005 (distribution, identification); Coad 2010a; Esmaeili et al. 2017 (introduction in Iran); Freyhof et al. 2020 (distribution in Saudi Arabia).

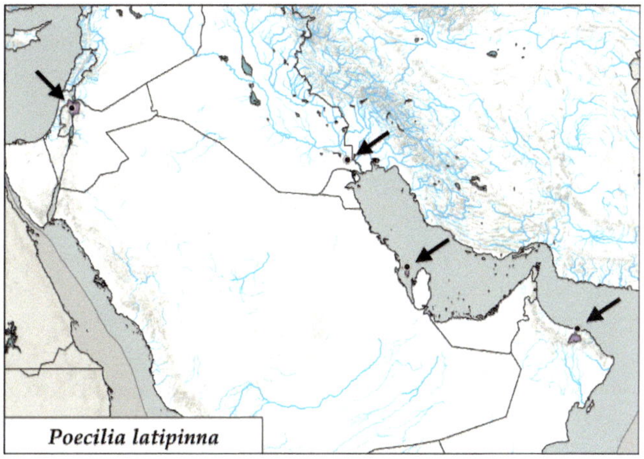

Poecilia reticulata; Oradea, Romania; female, ~30 mm SL.

Poecilia reticulata; Tohma drainage, Türkiye; male, ~20 mm SL.

Poecilia reticulata

Common name. Guppy.

Diagnosis. Distinguished from other species of Poeciliidae in West Asia by: ● male brightly coloured in life, with black blotches on flank also seen in preserved individuals / ○ dorsal and caudal in male with blue, white or red colour / ○ flank in male with a variable colour pattern of red, turquoise, yellow, orange, and black blotches and short stripes / ○ dorsal larger in male than in female / ○ flank not red, if with red blotches, then in male only / ○ pelvic of male modified, tip of first ray enlarged, different in male and female / ○ body moderately elongate / ○ dorsal origin behind vertical of anal origin / ○ gonopodium short / ○ dorsal and caudal without black spots / ○ no lateral stripe / ○ male without sword-like prolongation on caudal. Size up to 35 mm SL in male 50 mm SL in female.

Distribution. Introduced worldwide. Established in warm springs in upper Sakarya, İzmir, and upper Tohma drainages (Euphrates, Türkiye). Likely established elsewhere in West Asia. Also locally established in hot springs in Europe and in warm effluents from European power stations. Native to Venezuela, Guyana, and adjacent islands.

Habitat. A wide variety of habitats with low predation pressure. Usually found in very small streams and densely vegetated lakes and springs.

Biology. Not studied in West Asian populations. In captive populations, females mature at about 3 months, males slightly earlier. Continuous, non-resource and non-territorial mating system. Sexually dimorphic. Males brightly coloured and highly polymorphic with many cultivated colour forms; females cryptically coloured. Predation is a major evolutionary pressure influencing colouration, life history traits, and behaviour. Widely used as a model in biology, especially to study evolution.

Remarks. Despite being one of the most common aquarium fish in the world, there are few confirmed records of guppies being introduced in West Asia, with some populations surviving for more than 40 years. *Poecilia wingei* and *P. obscura* are two other guppies kept in the hobby, and hybrids between *P. wingei* and *P. reticulata* are particularly common and likely to be released into the wild. Although we identify all wild guppy populations as *P. reticulata*, this is likely incorrect. Hybrid populations with *P. wingei* are likely to occur, especially in more recently established populations.

Conservation status. Non-native; released from aquaria.

Further reading. Haskins et al. 1961 (ecology); Liley & Seghers 1975 (behaviour); Reznick et al. 1990 (biology); Magurran et al. 1995 (behaviour); Houde 1997 (reproduction); Esmaeili et al. 2017 (introduction in Iran); Freyhof et al. 2020 (distribution in Arabia).

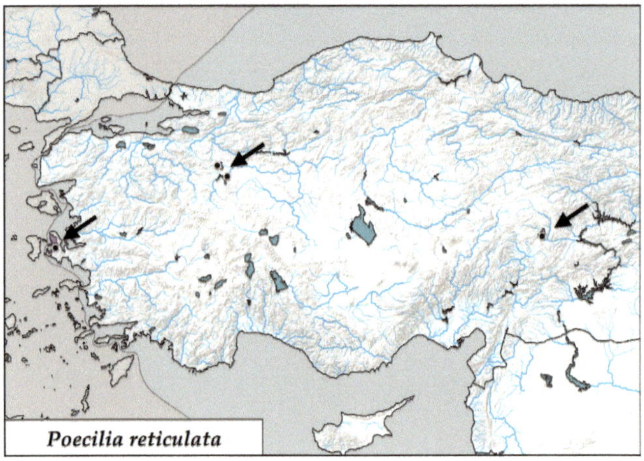

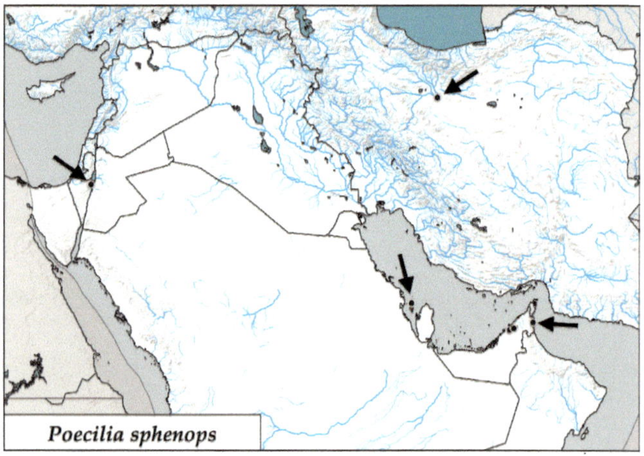

Poecilia sphenops; Ruwayyah, Dubai, UAE; male, 35 mm SL. © J. Els.

Poecilia "sphenops"

Common name. Mexican molly.

Diagnosis. Character states often intermediate between this "species" and *P. "latipinna"*. Distinguished from other species of Poeciliidae in West Asia by: • usually 9½ dorsal rays in male, not forming a sail / • dorsal origin behind vertical of pelvic origin / ○ both sexes with rows of small black, grey or brown spots on flank or flank black or black mottled / ○ male without coloured blotches or stripes / ○ dorsal larger in male than in female / ○ flank not red / ○ pelvic of male modified, tip of first ray enlarged, different in male and female / ○ gonopodium short / ○ body moderately elongate / ○ no lateral stripe / ○ male without sword-like prolongation on caudal. Size up to 75 mm SL.

Distribution. Hybrids identified as this species are locally established in Israel and Italy, at several sites in Eastern Province of Saudi Arabia, including Al-Ahsa oasis, adjacent to Lake Alsfar, and coastal wetlands at Salwar, north of Qatif and south of Jubail. Also established in UAE for 40 years or more in Ruwayyah, a traditional plantation area inland from urban Dubai. Also, present on a recently developed golf course on outskirts of Dubai. *Poecilia sphenops* is native to Central America, from Mexico to Guatemala. Widespread in tropical areas through aquarium releases.

Habitat. Euryhaline, but usually in slightly brackish waters and lower parts of rivers, lagoons, marshes, and streams.

Biology. Matures in less than a year. Produces up to 150 young after a gestation period of about 28 days. Females give birth throughout year in warm waters. Feeds on algae and small invertebrates. Black variety (Black Molly) is a popular aquarium fish and is marketed throughout the world.

Xiphophorus hellerii; Euphrates drainage, Türkiye; female, ~35 mm SL.

Xiphophorus hellerii; Euphrates drainage, Türkiye; primary male, ~35 mm SL.

Xiphophorus hellerii; Euphrates drainage, Türkiye; secondary male, ~35 mm SL.

Conservation status. Non-native; released from aquaria.
Remarks. Non-native mollies are incorrectly identified as *P. sphenops*. See notes under *P. "latipinna"*. In the UAE, this molly occurs in two colour variants: pale beige and black beige mottled. In its native range, several species appear to be included in *P. sphenops*, further complicating the situation.
Further reading. Miller 2005 (distribution, identification); Freyhof et al. 2020 (distribution in Arabia).

Xiphophorus hellerii
Common name. Green swordtail.
Diagnosis. Distinguished from other species of Poeciliidae in West Asia by: ● prominent black, red, or green lateral stripe / ● male with a long, white or yellow sword-like prolongation with black margin at ventral part of caudal / ○ dorsal origin in front of vertical of anal origin. Size up to 78 mm SL.
Distribution. Israel and Iran: Two very small localities in city of Jahrom, one spring in city of Kashan and Soleymaniyeh spring in Namak Lake basin. Türkiye: warm springs in upper Tohma drainage (Euphrates). Introduced worldwide in tropical and subtropical countries, as well as in Italy and Morocco. Native to the Atlantic basin of Central America. From Rio Nautla (Mexico) southeast to Rio Sarstun (Belize).
Habitat. Wide range of habitats in stagnant to moderately flowing waters from lagoons to springs, also in moderately polluted waters.

Biology. Males territorial. Breeds throughout year as long as water temperature is high enough. Primary males develop swords immediately and mature with females. Secondary males mature later, at larger sizes; some may initially reproduce as females and later change sex to males. Feeds mainly on algae, detritus and small invertebrates.
Conservation status. Non-native; released from aquaria.
Remarks. Common ornamental species. There are several breeds with different fin shapes and colours. Green and red individuals also occur in the wild in native range. Less cold hardy than *G. holbrooki* and will not survive temperatures below 15°C for over a few days.
Further reading. Esmaeili et al. 2010 (introduction to Iran); Miller 2005 (distribution, identification).

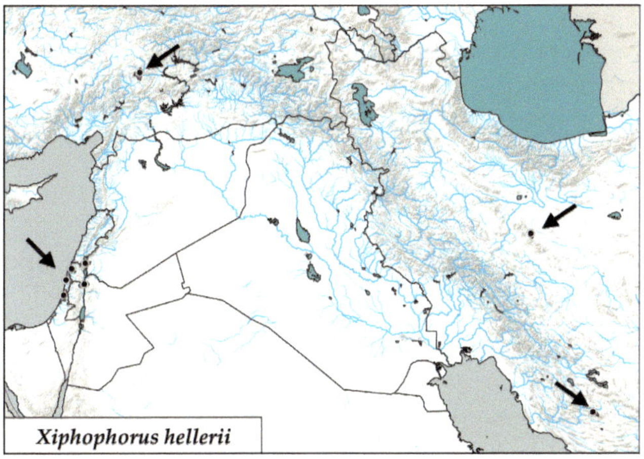

Xiphophorus hellerii

Xiphophorus maculatus; aquarium trade, male, about 30 mm SL. © J. Els.

Xiphophorus maculatus

Common name. Platy.
Diagnosis. Distinguished from other species of Poeciliidae in West Asia by: • body very short and deep / • dorsal in male and female equally long, its origin slightly behind pelvic origin / ○ gonopodium short / ○ dorsal and caudal without black spots / ○ flank colour very variable but similar in male and female, often red or blueish / ○ no lateral stripe / ○ pelvic of male modified, tip of first ray enlarged, different in male and female / ○ male without sword-like prolongation on caudal. Size up to 45 mm SL.
Distribution. Saudi Arabia: Wadi Haneefah in Riyadh; Al-Kharj and Layla Aflaj south of Riyadh; and Al-Ahsa oasis. Türkiye: İnönü spring in upper Sakarya. Native from about Ciudad Veracruz, Mexico, east to northern Belize. Has established few non-native populations, for example, in Australia.
Habitat. Springs, streams, and canals, usually with slow-flowing water and dense aquatic vegetation.

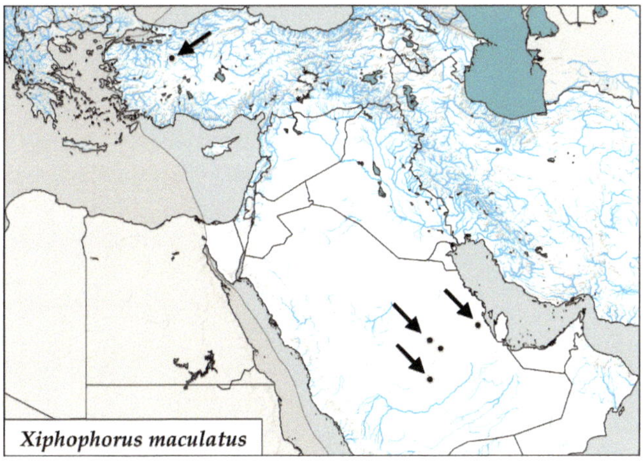

Xiphophorus maculatus

Biology. Females reach sexual maturity between 5–16 weeks and at about 18 mm SL; 3–92, usually 15–30, young are born after a gestation period of 24–30 days. Breeding intervals often longer than gestation period as species can store sperm. Capable of year-round reproduction given favorable water temperatures. Some males mature early, often after 5 weeks, others mature much later; maturation strategy is genetically determined. Platy is a model organism in cancer cell genetics and hybridisation research. Feeds mainly on algae, detritus, and small invertebrates.

Conservation status. Non-native; released from aquaria.

Remarks. Several colour varieties are popular aquarium fishes.

Further reading. Milton & Arthington 1983 (biology); Ross 1985 (introduction, distribution); Miller 2005 (distribution, identification); Freyhof et al. 2020 (distribution in Arabia); Aksu et al. 2021 (record in Sakarya).

Species of *Alburnoides* are all allopatric, superficially very similar, and found in similar habitat types. A case of functional redundancy.

High functional redundancy of riverine fishes. When reading or at least scrolling through this book, one pattern of freshwater fish biodiversity has caught your attention: species of fishes, as *Alburnoides, Barbus, Capoeta, Garra,* and *Cobitis*, all look very similar to each other. It is also a notable pattern that these morphologically similar species almost always occur in allopatry. This is what we refer to as morphological redundancy. Species in sympatry are almost always much more different from each other than species of the same genus, found in allopatry, and they are usually more distantly related. In the field, we find allopatric species usually also in the same habitats, not to mention the high structural similarity of riverine habitats all over West Asia and Europe. This suggests that morphological redundancy is related to functional redundancy. These fishes look almost the same and do nearly the same, just in different rivers. Functional redundancy may be very limited within an ecosystem, such as a single river drainage. Here, each species and its life stages are likely to make unique contributions to ecosystem functioning, naturally, much depending on the opportunities given by the ecosystem. Allopatric congeneric species are likely to exhibit a high or very high degree of functional redundancy. In West Asia, Europe, and likely elsewhere, the biodiversity of freshwater fish within genera is largely based on morphological and functional redundancy of species in allopatry. Rivers are isolated from each other by geological events, and so are the fishes. They speciate over time, but as the ecological conditions do not change significantly, allopatric speciation changes little in the ecological niches they form. They no longer live together, but otherwise, not much has changed. It is not that there are no ecological opportunities for ecological novelty and the formation of new ecological niches after or during allopatric speciation. There are a few cases of ecological novelty also from West Asia (e.g. *Garra* with reduced disc). However, the invasion of non-native species demonstrates that there are “open ecological spaces”, but native species usually fail to capitalise on opportunities. Why there is so little morphological and ecological novelty in riverine freshwater fishes and why they seem to be trapped in their ecological frame are big, open evolutionary questions.

Family Blenniidae

Combtooth Blennies

A diverse group of small marine fishes, comprising approximately 60 genera and 400 species, distributed throughout all tropical and temperate waters. They are typically found in shallow waters, on or near rocks or stone bottoms. A few species have adapted to estuarine and freshwater habitats. Blennies are characterised by pelvic fins anterior to the pectorals, with a single spine embedded in the skin and 2–4 rays. They also have a scaleless skin, numerous comb-like teeth, and long dorsal and anal fins. The freshwater blennies of West Asia were previously classified in the genus *Salaria*, which is now restricted to marine species, *S. basilisca*, endemic to the Mediterranean, and *S. pavo*, that occurs in the Eastern Atlantic and Mediterranean. The correct generic name for the five freshwater blennies previously placed in *Salaria* is now *Salariopsis*. These are: *S. economidisi* endemic to Greece, *S. atlantica* is endemic to Morocco, and the other three species occur in the area covered by this book. *Salaria pavo* is sometimes listed as a freshwater species. It effectively enters lagoons and brackish waters but has not been recorded from true freshwater habitats and is included here only in the key. **Further reading.** Springer 1968 (systematics); Wirtz 1976 (key); Bath 1977 (systematics); Duquenne-Delobel et al. 2022 (*Ichthyocoris*); Vecchioni et al. 2022 (phylogeny, *Salariopsis*); Yoğurtçuoğlu et al. 2023 (diversity).

Key to species of *Salaria* and *Salariopsis* in fresh and brackish waters of West Asia

1a - A large, oval, dark-grey blotch behind eye; 2 spines and 23–25 rays in anal; 12 spines and 22–25 rays in dorsal (occasionally in estuaries in brackish water).
..................*S. pavo*
1b - No large, oval, dark-grey blotch behind eye; 2 spines and 16–19 rays in anal; 12–13 spines and 16–17 rays in dorsal.
..................2

2a - Supraocular tentacle unbranched; pores of lateral line with black circle; no black dots on upper part of flank and cheek.
..................*S. renatorum*
2b - Supraocular tentacle branched; pores of lateral line with white circle; black dots present on cheek and/or upper part of flank.
..................3

3a - Supraocular tentacle short, usually not overlapping 9th circum-orbital sensory pore; many tiny black dots on cheek not organised in rows or bands.
..................*S. burcuae*
3b - Supraocular tentacle long, usually overlapping 9th circum-orbital sensory pore; tiny black dots on cheek usually organised in rows, often with a broad diagonal band of tiny black dots from lower edge of eye downward and backwards.
..................*S. fluviatilis*

https://doi.org/10.1515/9783111677811-040

Salariopsis burcuae; Nahr al Marq'ya, Syria; 69 mm SL.

Salariopsis burcuae; Lake Kinneret, Israel; nuptial male, ~50 mm SL.

Lake Tiberias in Israel is the largest habitat of *Salariopsis burcuae* and several other species.

Salariopsis burcuae

Common name. Eastern freshwater blenny.

Diagnosis. Distinguished from other species of *Salariopsis* in West Asia by: • tiny black dots on the cheek not organised in rows or bands / • supraocular tentacle branched, short, not reaching to, rarely overlapping the 9th circum-orbital sensory pore / ○ pores of lateral line with white circle / ○ black dots present on cheek and/or upper part of flank. Size up to 88 mm SL.

Distribution. Bay of Antalya (Türkiye) east to Jordan drainage (Israel), including Cyprus.

Habitat. Streams with moderate current and rocky bottom. Also in lakes, including reservoirs.

Biology. No data, but expected to be similar with *S. fluviatilis*.

Conservation status. LC.

Remarks. Superficially very similar to *S. fluviatilis* but well differentiated by molecular characteristics.

Further reading. Yoğurtçuoğlu et al. 2023 (description).

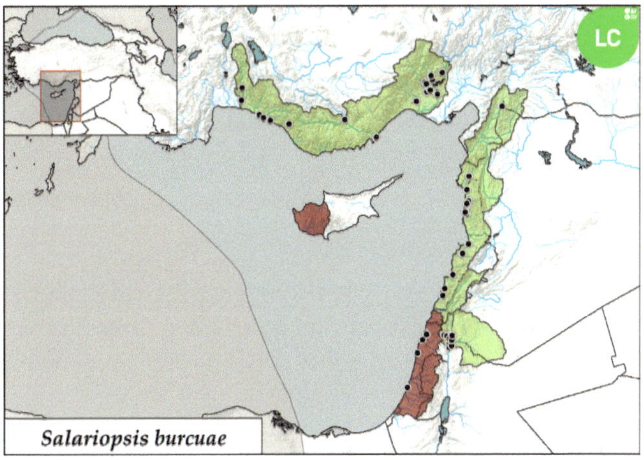

Salariopsis fluviatilis; Tahtalı drainage, Türkiye; 74 mm SL.

Salariopsis fluviatilis; Lago di Garda, Italy; nuptial male, ~70 mm SL.

Salariopsis fluviatilis

Common name. Freshwater blenny.

Diagnosis. Distinguished from other species of *Salariopsis* in West Asia by: • tiny black dots on cheek usually organised in rows, often with a broad diagonal band of tiny black dots from lower edge of eye downward and backward / • supraocular tentacle branched, long, usually overlapping 9th circum-orbital sensory pore / ○ pores of lateral line with white circle / ○ black dots present on cheek and/or upper part of flank. Size up to 130 mm SL.

Distribution. From Atlantic basin in Portugal east to western Anatolia. Also present on larger Mediterranean islands such as Corsica, Sardinia, Sicily, and Crete. Present in Garonne drainage (Atlantic basin in France), possibly due to invasion through Canal du Midi. Introduced to reservoirs in Sakarya and Kızılırmak drainage.

Habitat. Lakes and streams with moderate currents and rocky bottoms; in streams, deepest and fastest microhabitats are preferred, locally in coastal lagoons with low salinity.

Biology. Territorial, lives on and under rocks. Lives up to 5 years. Spawns first time at end of first year. Spawns for up to 3 years, but most females die at end of the first spawning year. Spawns May–August in France, April–July in Spain and Italy, at temperatures above 18°C. Female lays up to 1200 eggs (usually 200–300) in a single layer under a large stone, producing several portions of eggs during a spawning season. Male clean, fan, and defend eggs until hatching. Several females may spawn with one male, who may guard eggs at different stages of development. Eggs hatch in about a week. Larvae are planktonic for about a month until about 15 mm and drift to and remain in quiet pools or tributaries of rivers. They have a high tolerance for salt water, allowing them to migrate through marine waters. Feeds on insects, crustaceans, and small molluscs.

Conservation status. LC.

Remarks. The identity of extirpated population in Cyprus remains unclear and may have been *S. burcuae*, which is found on the adjacent mainland.

Further reading. Wickler 1957 (behaviour); Freeman et al. 1990 (habitat); Gandolfi et al. 1991 (biology); Ferrito & Tigano 1996 (threats); Perdices et al. 2000 (genetics); Doadrio 2001 (biology); Bath 2003 (biology); Kottelat 2004 (diversity); Vinyoles & de Sostoa 2007 (biology); Almada et al. 2009; Laporte et al. 2016a, b (dispersal through sea); Méndez et al. 2019; Wagner et al. 2021 (phylogeography); Yoğurtçuoğlu et al. 2023 (distribution).

Nahr Al Marqīyah in Syria is the habitat of *Salariopsis burcuae*, *Pseudophoxinus hasani* and *P. zeregi*.

Salariopsis renatorum; Ceyhan drainage, Türkiye; 44 mm SL.

Salariopsis renatorum; Ceyhan drainage, Türkiye; 69 mm SL.

Salariopsis renatorum

Common name. Ceyhan freshwater blenny.

Diagnosis. Distinguished from other species of *Salariopsis* in West Asia by: • supraocular tentacle unbranched / • pores of lateral line with black circle / • no black dots on cheek and/or upper part of flank. Size up to 69 mm SL, likely to grow larger.

Distribution. Türkiye: Ceyhan and Arsuz drainages.

Habitat. Fast-flowing stretches of streams.

Biology. No data, but expected to be similar with *S. fluviatilis*.

Conservation status. LC; appears to be declining within its small range.

Further reading. Yoğurtçuoğlu et al. 2023 (description).

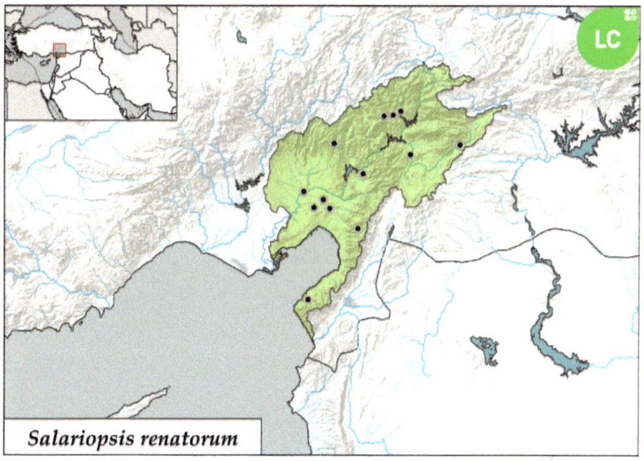

Family Cichlidae

Cichlids

A large family of mostly small freshwater fishes with approximately 1750 named (and several hundred unnamed) species distributed throughout tropical and temperate Africa, Madagascar, South and Central America, a few species are found in India, Sri Lanka, Iran, and the Levant. Some species have been introduced almost pan-tropically. Cichlids are immediately distinguishable from similar perch-like fishes by two lateral lines on the flank, with an anterior upper line and a posterior lower line. Eight species are native to West Asia, with seven being introduced. *Iranocichla* and *Tristramella* are endemic to West Asia, both having relationships to African genera and potentially representing the relics of a once much wider distribution of cichlids in the area. Cichlid fossils have been recorded from the Arabian Peninsula until the Pliocene, indicating a continuous distribution from the Levant through the Arabian Peninsula to Iran. Cichlids were "reintroduced" in Arabia in the modern era as non-native species for aquaculture and mosquito control. Cichlids are generally very sensitive to temperatures below 10°C, with mortality occurring if exposed for more than a few hours. This renders them susceptible to low winter temperatures, which restrict their distribution to warm springs in the northern part of West Asia. They represent a highly significant group for evolutionary studies and the pet trade, with numerous species being kept in aquaria, including in West Asian countries. Until now, *Amatitlania nigrofasciata* is the most widespread species released from aquaria. Other species, such as *Rubricatochromis guttatus*, an African species already found at several locations in Europe, are also expected to be found in West Asia in the future.

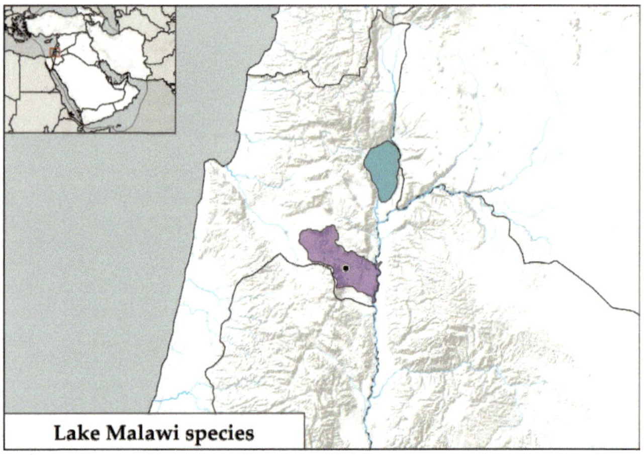

Lake Malawi cichlids introduced and established in Nahal Amal in Israel. From left: *Aulonocara* sp., *Dimidiochromis compressiceps*, *Labidochromis caeruleus*, *Chindongo elongatus*. © A. Spreinat.

https://doi.org/10.1515/9783111677811-041

It is unknown how many cichlid species from East African Lake Malawi have established themselves in the warm springs of Nahal Amal in Bet She'an (Israel). At least *Aulonocara* hybrids, *Dimidiochromis compressiceps*, *Labidochromis caeruleus* and *Chindongo elongatus* have formed stable populations. *Aulonocara* are domesticated hybrids produced for the aquarium trade. Other species might have also been established, and more research is needed. Malawi cichlids are included in the identification key but do not have separate species chapters. Several other cichlids have been reported from West Asia but have yet to be established. More introduced cichlids will inevitably find their way from aquariums into the wild, especially in artificial waters in parks or private areas. To identify them, consulting specialist books on aquarium fish is necessary. **Further reading.** Trewavas 1983; Lévêque et al. 1992 (identification); Schwarzer et al. 2016 (molecular phylogeny).

Key to species of Cichlidae native and introduced to West Asia

1a - Anal with 8–10 spines.
..................*Amatitlania nigrofasciata*
1b - Anal with 3 spines.
..................2

2a - Scales ctenoid; yellow or orange ocelli on anal in adult male.
..................3
2b - Scales cycloid; no ocelli on anal.
..................7

3a - Lower jaw strongly projecting upper jaw, a continuous, bold-black midlateral stripe.
..................*Dimidiochromis compressiceps*
3b - Upper jaw slightly projecting upper jaw or both jaws equally projecting.
..................4

4a - Dorsal with a bold-black subdistal stripe on yellow background; flank yellow without bars or with very indistinctive bars.
..................*Labidochromis caeruleus*
4b - Dorsal black, grey, yellowish, blue, red, marbled, often with whitish or yellow distal margin, without a bold-black subdistal stripe on yellow background; flank with indistinctive or bold bars or without bars.
..................5

5a - Snout roundish; head below eye dark-grey to bluish-black; flank with 4–8 dark-grey to bluish-black bars in male, less pronounced or absent behind dorsal base.
..................*Chindongo elongatus*
5b - Snout pointed; head below eye same colour as above eye, with blue iridescent shine or with a bold bar between eye and corner of mouth; flank with indistinctive dark-grey or brown-black bars.
..................6

6a - Flank without stripe or horizontally elongated blotches; no bar between eye and corner of mouth; flank with bold dark-grey or brown bars in female and juveniles.
..................*Aulonocara* sp.
6b - Flank with a short stripe between opercle and about middle of spined part of dorsal, continued as 1–2 blotches on lateral midline behind spined part of dorsal; bold bar between eye and corner of mouth; flank with indistinctive dark-grey or brown bars.
..................*Haplochromis flaviijosephi*

7a - No scales on chest, belly, and isthmus in front of pelvic origin; nuptial male dark-grey or black with many white spots or short vermiculation on flank.
..................8

7b - Chest, belly, and isthmus in front of pelvic covered by small scales; nuptial male pale-grey or pinkish white without white spots or vermiculation on flank.
..................9

8a - Chest and lower part of head black in nuptial male; black blotch on last spinous part of dorsal distinct; no or very few white spots on caudal.
..................*Iranocichla hormuzensis*
8b - Chest and lower part of head bright orange in nuptial male; black blotch on last spinous part of dorsal pale-grey or invisible; white spots forming wavy bars or stripes on caudal.
..................*Iranocichla persa*

9a - 8–14 gill rakers on lower part of first branchial arch.
..................10
9b - 18–27 gill rakers on lower part of first branchial arch.
..................12

10a - Chest red or red and black in adults (rarely absent), red colour restricted to lateral chest in juveniles; lower jaw not projecting; 8–11 gill rakers on lower part of first branchial arch.
..................*Coptodon zillii*
10b - Chest not red; lower jaw projecting; 10–14 gill rakers on lower part of first branchial arch.
..................11

11a - Lower jaw slightly projecting; outer teeth bicuspid, inner teeth tricuspid (Fig. 66).
..................*Tristramella simonis*
11b - Lower jaw very strongly projecting; teeth conical, unicuspid (Fig. 66).
..................*Tristramella sacra*

12a - Flank scales without dark-grey scale pockets in preserved individuals, isolated black scales pockets or fields of black scale pockets in some individuals; anterior processes of lower pharyngeal bone not or very slightly projecting beyond anterior margin of toothed pads; without red or orange colour in life; caudal almost without pattern.
..................*Sarotherodon galilaeus*
12b - All flank scales with dark-grey scale pockets in preserved individuals, dark-grey scales pockets never organised in fields; anterior processes of lower pharyngeal bone clearly projecting beyond anterior margin of toothed pads, tips of dorsal rays and orange caudal in life adult male (*O. aureus*, *O. mossambicus*) or caudal with regular bars (*O. niloticus*); caudal with strong pattern.
..................13

13a - Caudal with regular vertical bars in individuals larger than 80 mm SL; tip of membrane between dorsal spines grey to black.
..................*Oreochromis niloticus*
13b - Caudal with tessellate or marbled pattern or few irregularly shaped bars, without regular vertical bars in individuals larger than 80 mm SL; tip of membrane between dorsal spines pink to red.
..................14

14a - Nuptial male bright blueish or grey with whitish lower part of head and orange dorsal- and caudal margins; snout in male larger than 200 mm SL not duck-bill-like elongated.
..................*Oreochromis aureus*
14b - Nuptial male black or dark-grey with whitish lower part of head and orange or red dorsal- and caudal margins; snout in male larger than 200 mm SL duck-bill-like elongated.
..................*Oreochromis mossambicus*

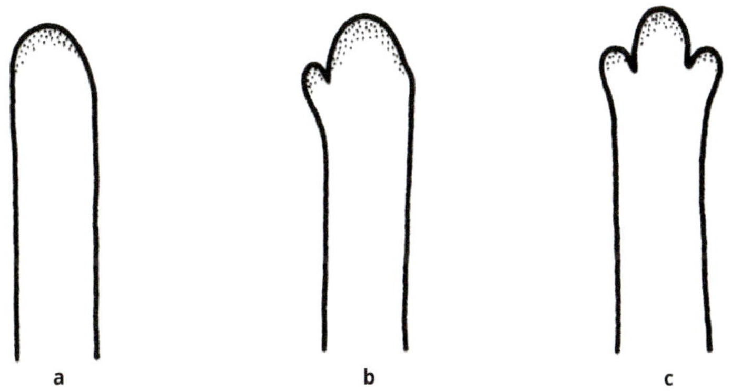

Figure 66. Jaw teeth in Cichlidae: **a,** unicuspid; **b,** bicuspid; **c,** tricuspid (from Kottelat & Freyhof 2007, after Trewavas 1983).

Rubricatochromis guttatus is an invasive species expected to be found in West Asia in the near future.

Amatitlania nigrofasciata; Fossa Calda, Italy; female; ~45 mm SL.

Amatitlania nigrofasciata; Fossa Calda, Italy; male, ~70 mm SL.

Amatitlania nigrofasciata

Common name. Convict cichlid.

Diagnosis. Distinguished from other species of Cichlidae in West Asia by: ○ 8–10 anal spines / ● 8–9 black bars on grey flank (albinotic individuals occur in non-native populations). Size up to 150 mm SL.

Distribution. Türkiye: a hot spring in upper Sakarya drainage. Iran: Golabi spring near Darab in Fars prov., and Soleymaniyeh spring in Namak Lake basin. Was established in Israel but seems to have disappeared. Native to Central America: Pacific slope, from Río Sucio, El Salvador to Río Suchiate, Guatemala; Atlantic slope, from Río Patuca, Honduras to Río Jutiapa, Guatemala. Introduced in Austria, Hungary, Italy, Australia, Japan, Philippines, Puerto Rico, the USA, Hawaii and likely elsewhere.

Habitat. A wide range of streams and springs.

Biology. Forms pairs; both sexes territorial. Females lay eggs on exposed hard surfaces such as stones or wood. Larvae and young guarded by parents. Male guard outer and female inner circle around young. Incubation of eggs, larvae, and free-swimming juveniles in the wild lasts 4–6 weeks and occurs only once a season for most females. An opportunistic feeder, feeding on detritus and plant material and preying on small aquatic animals. Frequently used in behavioural studies.

Conservation status. Non-native; released from aquaria.

Further reading. Wisenden 1995 (reproduction); Schmitter-Soto 2007 (native distribution); Esmaeili et al. 2013b (Iran); Mousavi-Sabet & Eagderi 2016a (record in Lake Namak).

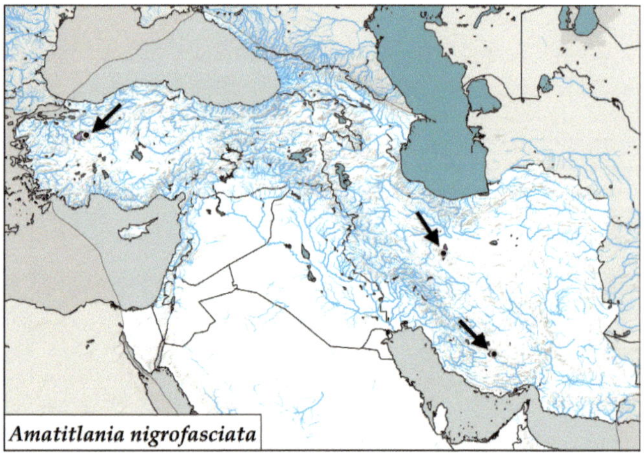

Coptodon

Coptodon comprises approximately 35 species, which are primarily distributed in West Africa. Many species have relatively small distribution ranges in tropical rainforests. Some species are found in coastal lagoons and may forage in marine habitats. The concept of a widespread coastal species (*C. guineensis*) has recently been rejected, and the number of *Coptodon* species may be grossly underestimated. Two crater lakes in Cameroon host endemic species flocks: Lake Bermin with 11 species and Lake Ejagham with four species. *Coptodon* are macroherbivores and, if available, feed on plants. They graze submerged vegetation and, during high water levels, migrate into forests and other terrestrial ecosystems to feed on flooded plants. It is common to observe them leaping out of the water to bite pieces from overhanging leaves. As some *Coptodon* species are particularly abundant, their grazing structures freshwater ecosystems. They may also play an important role in the dispersal of seeds. Many species grow to a large size and are of importance in both subsistence and commercial fisheries. **Further reading.** Stiassny et al. 1992 (Bermin); Dunz & Schliewen 2010 (Ejagham); Dunz & Schliewen 2013 (generic concept); Kide et al. 2016 (diversity of coastal species).

Coptodon zillii; Lake Köyceğiz, Türkiye; male, ~160 mm SL.

Coptodon zillii; Lake Maryut, Egypt; nuptial male, ~100 mm SL.

Coptodon zillii

Common name. Redbelly tilapia.

Diagnosis. Distinguished from other species of Cichlidae in West Asia by: • chest red or red and black in adults, red colour restricted to lateral chest in juveniles (rarely absent) / ○ lower jaw not projecting / ○ caudal with tessellate pattern / ○ chest, belly, and isthmus in front of pelvic covered by small scales / ○ 8–11 gill rakers on lower part of first branchial arch / ○ 3 anal spines / ○ scales cycloid / ○ no ocelli on anal. Size up to 400 mm SL, usually about 200 mm SL.

Distribution. In West Asia, native to Jordan drainage and possibly a few adjacent coastal streams. Introduced in Köyceğiz and Burdur basins in Türkiye; Euphrates and Tigris in Syria, Iraq, and Iran; Azraq oasis in Jordan; and Ceyhan and Orontes drainage in Syria and Türkiye. Also native to Nile drainage and Lakes Albert and Turkana. Widespread in West Africa, including Senegal, Gambia, Niger, Chad, and Volta, and many coastal rivers in Ivory Coast and between Volta and Niger. Occurs in Ubangui, Ulele, and Ituri in central Congo basin. Morocco: Lower valley of Oued Dr'aa and Oued Aouedri, a small Saharan valley between Tan Tan and Tarfaya. Algeria: Tolga, Zibans, and Touggourt oases in northern Sahara and on western slopes of Ahaggar, in Tassili n' Ajjer, and Arak, a tributary of Botha, all in southern Algerian Sahara. Tunisia: Kebili oasis and perhaps more widely in eastern and southern Chott el Djerid basin. In Sahara: Tibesti, Ennedi, and Borkou mountains. Introduced almost everywhere in subtropical and tropical areas of Southeast Asia, Australia, India, Madagascar, New Caledonia, Fiji, Guam, Taiwan, China, Philippines, Japan, Antigua, Mexico, and the USA, including Hawaii and elsewhere.

Habitat. Hypersaline desert ponds and wadis to small rainforest streams with soft and acidic water. Often very abundant in standing water, such as lakes, large rivers, or reservoirs with dense riparian vegetation.

Biology. Usually lives 3–5 years. Matures at about 80 mm SL. Growth extremely variable depending on temperature and food availability; may spawn after a year in West Asia but mature within 4 months in tropics. Forms pairs that defend territories against other fish. Usually, females defend a smaller inner territory and males a larger outer territory around spawning site. Females lay sticky eggs on surfaces such as stones or wood, often in shelters or small depressions in substrate dug by both parents. Spawns throughout year in tropical range, late spring to summer in West Asia when water temperatures rise above 20°C. Stops spawning when water temperature rises above 28°C. Both parents fan eggs, picking up debris and dead eggs. Larvae usually hatch 48 hours after spawning (25°C) and are transferred to small burrows in substrate or under wood or stones. Larvae begin to feed about a week after spawning. Both parents guard larvae and juveniles for about 4 weeks. Survives temperatures as low as 6.5 and as high as 42.5°C, but only briefly, unable to survive below 10°C for more than a few hours. Harsh winters cause mortality in northern part of range. Survives in salinities up to 45 ‰ and reproduces in salinities up to 29 ‰. Feeds mainly on aquatic and terrestrial macrophytes but also on invertebrates, algae, detritus, and virtually all organic matter. Of great economic importance in sub-Saharan Africa as a food fish for small-scale fisheries, rarely used in aquaculture. Introduced for aquatic weed and mosquito control, as forage or food fish, and for aquaculture. Several studies show negative impacts on native biodiversity, adverse effects on aquatic plants, and it is a pest in rice plantations.

Conservation status. LC.

Remarks. Populations of *C. zillii* from Morocco, Algeria, West Africa, the Nile, and Jordan are almost identical in their mitochondrial DNA, suggesting a very recent invasion of this species into the Levant and a recent interruption of gene flow between the Niger, Senegal, Nile, and many coastal rivers in West Africa. Locally misidentified as *C. rendalli* in 20th century.

Further reading. Geiger et al. 2014 (molecular diversity); Nico et al. 2016 (distribution, effects of non-native populations).

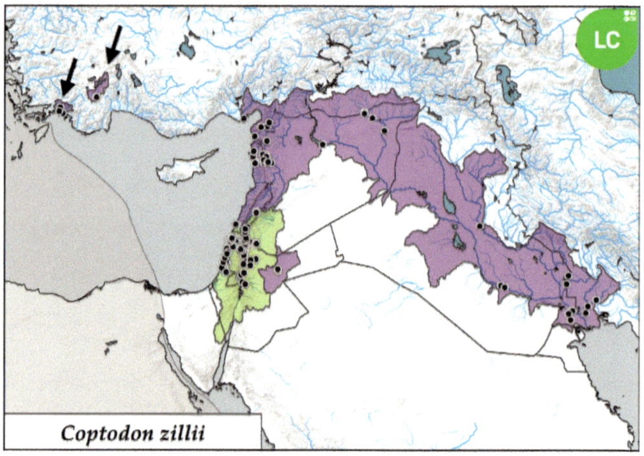

Haplochromis

Haplochromis and related genera from Africa's Lake Victoria region represent the largest cichlid fish genus and the most species awaiting description. These fishes provide textbook examples of adaptive radiations and the role of hybridisation in speciation and evolution. While being very speciose and morphologically as well as ecologically very diverse in Africa, there is only one species native to Asia, which is distantly related to the "Lake Victoria Region Superflock."

Further reading. Meier et al. 2017 (phylogeny, species diversity, evolution); Moser et al. 2018 (fast speciation).

Haplochromis flaviijosephi; Jordan drainage, Syria; male, neutral mood; ~70 mm SL.

Haplochromis flaviijosephi; Jordan drainage, Israel; male, territorial mood; ~80 mm SL. © E. Schraml.

Haplochromis flaviijosephi

Common name. Jordan mouthbrooder.

Diagnosis. Distinguished from other species of Cichlidae in West Asia by: • scales ctenoid / • yellow or orange ocelli on anal in adults / ○ chest, belly, and isthmus in front of pelvic covered by small scales / ○ 3 anal spines / ○ flank with a short stripe between opercle and about middle of spined part of dorsal, continued as 1–2 blotches on lateral midline behind spined part of dorsal / ○ bold bar between eye and corner of mouth / ○ flank with indistinctive dark-grey or brown bars. Size up to 100 mm SL.

Distribution. Jordan drainage, where it occurs in Lake Muzayrib (Syria) and Lake Kinneret (Israel) basins, in Ajami

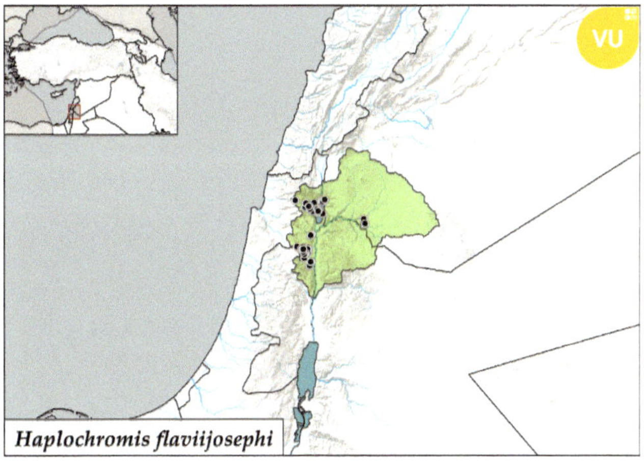

and Al Asha'ari springs (southern Syria), and in vicinity of Lake Kinneret and the Baisan Valley (Israel).
Habitat. Shallow shores of lakes, springs, and streams where it occurs among rocks and vegetation.
Biology. Maternal mouthbrooders without permanent pair-bonding. Spawns several times between April and July. Nuptial males occupy a territory with a shallow nest in center. Female spawn with territorial males. After fertilisation in nest, female takes eggs in mouth. In Lake Kinneret, females feed mainly on chironomid larvae, oligochaetes, and amphipods, while males feed primarily on snails.
Conservation status. VU; appears to be relatively secure in Lake Kinneret but declining in its small range outside the lake.
Remarks. *Haplochromis flaviijosephi* does not appear to be closely related to any African *Haplochromis* species. There must have been a recent biogeographic window that allowed *C. zillii*, *O. aureus*, and *S. galilaeus* to invade the Jordan from the Nile. However, both *H. flaviijosephi* and *Tristramella* must have existed in Levant before the biogeographic connection to the Nile, and a much earlier invasion from the Nile could be proposed. Alternatively, *Tristramella* and *H. flaviijosephi* may be relics of a once widespread range of cichlids on the Arabian Peninsula and may have invaded the Jordan from Saudi Arabia.
Further reading. Goren 1974 (distribution); Werner 1976 (reproduction); Spataru & Gophen 1985a (food).

Iranocichla hormuzensis; Mehran drainage, Iran; mouth-brooding female, ~70 mm SL. A. Lamboj.

Iranocichla
Iranocichla is a genus comprising two species, both of which are highly similar. *Iranocichla* is endemic to Iran, and it was previously hypothesised that it is closely related to *Danakilia*, a genus with two species from Ethiopia and Eritrea. However, molecular data does not support this hypothesis; *Iranocichla* is close to *Tristramella*. **Further reading.** Steinitz & Ben-Tuvia 1960 (revision of genus); Dunz & Schliewen 2013 (phylogeny); Esmaeili et al. 2016b (review); Schwarzer et al. 2016 (phylogeny of populations).

Iranocichla hormuzensis; Mehran drainage, Iran; nuptial male, ~80 mm SL. © A. Lamboj.

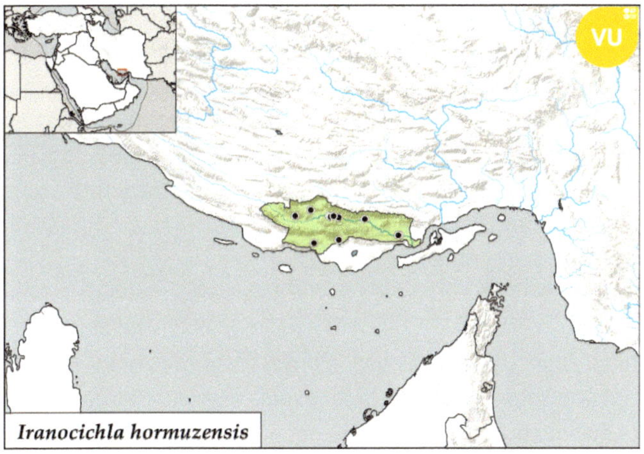

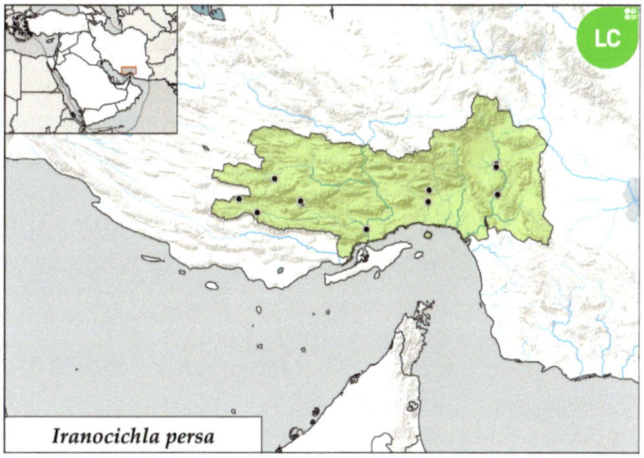

Iranocichla hormuzensis

Common name. Persian pearl cichlid.

Diagnosis. Distinguished from other species of Cichlidae in West Asia by: ● lower part of head and chest in nuptial male black / ○ black blotch on last spinous part of dorsal distinct / ○ no or very few white spots on caudal / ○ no scales on chest, belly, and isthmus in front of pelvic origin / ○ 3 anal spines / ○ scales cycloid. Size up to 120 mm SL.

Distribution. Iran: Middle and lower Mehran drainage.

Habitat. Freshwater to hypersaline streams, often partially drying up in summer and forming isolated pools. Area around Strait of Hormuz is rich in salt domes and hypersaline streams; up to about 1½ marine salinity, are found. Water temperatures range from 15 to 33°C in winter (November to March) and up to 40°C in summer.

Biology. Maternal mouthbrooders without permanent pair-bonding. Spawns in spring. Nuptial males occupy a territory with a shallow nest in center. Female spawn with territorial males. After fertilisation in nest, female takes eggs in mouth and returns to schools.

Conservation status. VU; appears to be declining within its small range.

Further reading. Coad 1982b (description); Esmaeili et al. 2016b (review); Schwarzer et al. 2016 (phylogeny).

Iranocichla persa; Shur drainage, Iran; nuptial male, 90 mm SL. © A. Hartl.

Iranocichla persa

Common name. Eastern pearl cichlid.

Diagnosis. Distinguished from other species of Cichlidae in West Asia by: • lower part of head and chest in nuptial male orange / ○ black blotch on last spinous part of dorsal pale-grey or invisible / ○ white spots forming wavy bars or stripes on caudal / ○ no scales on chest, belly and isthmus in front of pelvic origin / ○ nuptial male dark-grey or black with many white spots or short vermiculation on flank / ○ 3 anal spines / ○ scales cycloid. Size up to 120 mm SL.

Distribution. Iran: Kol, Shur, Hasan Langi, and Minab drainages.

Habitat. Freshwater to hypersaline streams, often forming isolated pools in summer. Water temperatures range from 15 to 33°C in winter (November–March) to 40°C in summer.

Biology. Maternal mouthbrooders without permanent pair-bonding. Spawns in spring. Nuptial males occupy a territory with a nest in center. Nests are shallow pits about one body length in diameter. Females form shoals in deeper water. Nuptial females approach nests to spawn. Female spawn with territorial males, and sneaking appears to be common. After fertilisation in nest, females take eggs in mouths and return to school. Up to 153 offspring recorded in one mouth, ranging in length from 9.6 to 10.9 mm SL (from a female of 117 mm SL). Feeds on diatoms and other algae scraped from rocks and sediments.

Conservation status. LC.

Remarks. Molecular data suggest that all *Iranocichla* populations are closely related. The two *Iranocichla* species form only slightly distinct clades, diagnosable by several fixed mutations in ND2, D-loop, and very small differences in cytochrome oxidase 1 sequences. Fish from the Kol drainage, the largest *Iranocichla* population between the Mehran and Shur drainages, agree with *I. hormuzensis* in their male nuptial colour but cluster with *I. persa* in their mitochondrial sequence characters.

Further reading. Esmaeili et al. 2016b (description); Schwarzer et al. 2016 (phylogeny).

Iranocichla persa waiting for food in an eutrophicated pool in Iran.

Taiwanese red tilapia *O. niloticus* x *O. mossambicus*; Wadi Fania, Oman; male, 145 mm SL.

Oreochromis

Oreochromis comprises approximately 20–40 species, depending on the generic concept. Most *Oreochromis* are found in East Africa, although two are native to West Asia. Some authors have recognised *Alcolapia*, *Neotilapia*, *Nyassalapia*, and *Vallicola* as valid genera. Still, molecular studies have challenged the generic concept of mouth-brooding tilapias, and these may ultimately be reclassified as *Oreochromis*. The phylogeny of *Oreochromis* also shows widespread discordance between nuclear DNA and mitochondrial DNA trees. It seems premature to propose a revision of mouth-brooding tilapias, which are unlikely to form a natural group.

Several *Oreochromis* species are attractive targets for aquaculture, with at least eight species actively farmed globally, many in warm-water farms. *Oreochromis niloticus* and its hybrids are the most widely farmed aquaculture warm-water species globally. Additionally, in certain West Asian regions, an expanding aquaculture industry is cultivating primarily hybrid tilapia (*O. niloticus* x *O. aureus*). Other tilapias, including *Oreochromis niloticus*, *O. aureus*, and *O. mossambicus*, as well as hybrids between *O. mossambicus* and *O. niloticus* (Taiwanese red tilapia), are maintained in experimental farms and are cultivated in limited quantities before being released into natural waters. *Oreochromis spilurus*, a salt-tolerant species, was introduced from Kenya to Saudi Arabia and was reportedly bred and cultured in saline waters of the Red Sea. While it cannot be ruled out that fish have escaped and established themselves in the Red Sea basin, there is no evidence to confirm this. Therefore, *O. spilurus* has been excluded from the list of non-native species in Arabia.

A few *Oreochromis* from Oman have been studied for their mitochondrial molecular characters. The results suggest that either *O. niloticus* has already hybridised with *O. mossambicus* in Oman or that the introduced fish had already been hybridised, and these Omani populations might be feral Taiwanese red tilapia. Phenotypically, these fish are close to *O. niloticus*. Introgressed mitochondrial DNA is a trace of past hybridisation events. The aquaculture industry has a significant environmental impact, as fish inevitably escape and potentially establish local populations. This has occurred locally in West Asia, where *Oreochromis* have become invasive in the Persian Gulf basin and the Levant. Many species of *Oreochromis* have a substantial invasive potential, and four species are currently listed on the Global Invasive Species Index. **Further reading.** Genner et al. 2013 (invasion); Shechonge et al. 2019 (invasion); Ford et al. 2019 (phylogeny); IUCN 2020 (invasive species list).

Oreochromis spilurus; Kenya, male, 150 mm SL.

Tilapia, a global business. Tilapias have been an important component of subsistence fisheries since pre-humans first began to consume fish as food. They have been farmed for over 4000 years in Egypt. The most commonly cultured tilapia species is *Oreochromis niloticus*, followed by *O. aureus*. These two species are responsible for the significant increase in global tilapia aquaculture production. Tilapia has evolved from a relatively obscure fish to one of the world's most productive and internationally traded food fish. Tilapia fillets are commonly available in supermarkets. In 2021, approximately 7.2 million tonnes of tilapia were produced, and China accounted for 1.8 million metric tons. By 2017, the global value of tilapia products, predominantly frozen fillets, exceeded 25 billion euros. Of this total, 12 billion euros was generated in China alone. Tilapia farming encompasses various operations, from rural subsistence farming to large-scale commercial enterprises. In light of the ongoing commercialisation and expansion of the tilapia industry, the commodity has emerged as the second most important farmed fish globally, after carp. It is also regarded as the most significant aquaculture species of the 21st century. Tilapia is cultivated in approximately 150 countries worldwide, with approximately 98 % of the tilapia produced in these countries being grown outside their native range. The major producing countries are China, Egypt, and Indonesia. However, tilapia farming is increasingly conducted in the Caribbean, Latin America, and temperate countries where warm water is artificially created. *Oreochromis niloticus* and *O. aureus* are the most important domesticated animals from Africa. Surprisingly, the production in Africa, except Egypt (Egypt 237,000 tonnes in 2017), remains relatively small. African countries import frozen, whole tilapia from China, which accounts for 98 % of the marketed Tilapia in Africa (outside Egypt). The domestication of tilapia is a relatively rapid process. The introduction of superior strains and the development of Genetically Improved Farmed Tilapia (GIFT) technology, based on traditional selective breeding, are intended to enhance the commercially important traits of tropical farmed fish. Through combined selection technology, GIFT achieved an average genetic gain of 12–17 % per generation over five generations and a cumulative increase in the growth rate of 85 % in *O. niloticus*. It is evident that these tilapias no longer resemble their wild counterparts. Several strains exhibit orange or pink hues, frequently marketed as pink tilapia or red snapper, a marine fish. It is also noteworthy that tilapia often escapes from farms and has become one of the most prevalent non-native fish species globally. **Further reading.** Freyhof et al. 2020.

Oreochromis aureus; Ceyhan drainage, Türkiye; ~200 mm SL.

Oreochromis aureus

Common name. Blue tilapia.

Diagnosis. Distinguished from other species of Cichlidae in West Asia by: ○ caudal usually tessellate, marbled, or with blotches not forming regularly shaped and set bars or, if bars present, then short, not reaching from upper to lower caudal rays / ○ flank scales with scales pockets darker than scale margins / ○ tip of membrane between dorsal spines orange or red / ● nuptial male bright blueish or grey with whitish lower part of head and orange dorsal and caudal margins / ○ snout not duck-bill-like, elongated in male larger than 200 mm SL / ○ anterior processes of lower pharyngeal bone clearly projecting beyond anterior margin of toothed pads / ○ chest, belly, and isthmus in front of pelvic covered by small scales / ○ chest never red / ○ 18–26 gill rakers on lower part of first branchial arch / ○ 3 anal spines / ○ scales cycloid / ○ no ocelli on anal. Size up to 380 mm SL.

Distribution. Native to Jordan drainage and lower Nile downriver of Cairo. Also native to Oued Draa drainage in southern Morocco, Senegal, central Niger and Benue, and Lake Chad basin, including lower Shari and Logone. Widely introduced in Israel, Azraq oasis in Jordan and Orontes in Syria and Türkiye, Ceyhan, Seyhan and Sakarya in Türkiye, lower Euphrates and Tigris in Iraq and Iran, Wadi Haneefah in Riyadh, Al-Kharj and Layla Aflaj south of Riyadh, and Al-Ahsa oasis in Saudi Arabia. Probably more widespread. Also established in Cyprus, Florida, South Africa, Central America, Southeast and East Asia, Hawaii, Madagascar, and elsewhere.

Habitat. Slow-flowing rivers, backwaters, lakes, and coastal lagoons.

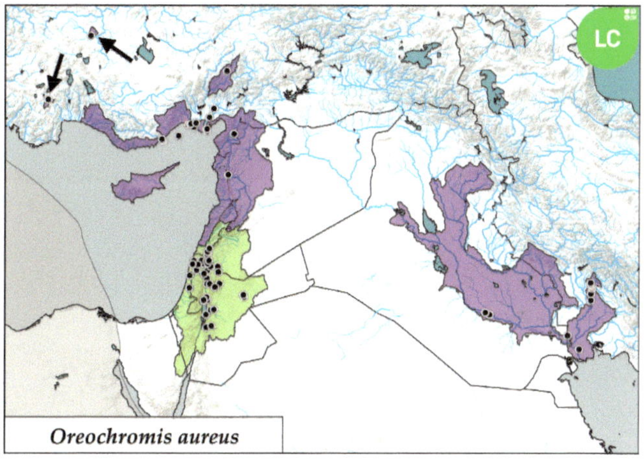

Biology. Lives up to 5 years. Spawns in second year, usually with 160–180 mm SL. Males larger than females, and branched dorsal and anal rays longer in males. Posterior margin of dorsal membrane thickened in nuptial male. Spawning begins in late March–early April in Lake Kinneret at water temperatures above 20–22°C and lasts until late May in Kinneret or until November in Nile Delta. Males establish territories near shores in shallow water. Territory radius (0.7–3.0 m) depends on strength of male. Territories are often concentrated in leks that form breeding colonies. Within a territory, males dig a shallow nest in substrate and attempt to attract females. Eggs are laid in batches, immediately fertilised by male and taken into mouth by female. No permanent pair bond. Females may spawn with several males in a breeding colony. Females spawn several times during summer. Larvae hatch after 3 days, and females swallow eggshells. Female keeps eggs and larvae for about 13–14 days and guards free-swimming juveniles for a few more days, returning them to mouth at night. Can live in brackish water up to at least 10 ‰. Juveniles usually cannot survive temperatures below 9°C. Adults can survive night temperatures as low as 5°C for a few days if daytime temperatures are higher. Juveniles inhabit shallow habitats. Opportunistic feeders, usually phytoplankton but also zooplankton, epiphytic algae, and detritus.

Conservation status. LC.

Remarks. Very important species for aquaculture, produced in large quantities in tropical and subtropical countries and warm-water facilities in Europe. This species is often misidentified as *O. niloticus* or *Sarotherodon galilaeus*, and hybrids between these three species have been recorded in the wild. Hybrids between *O. aureus* and *O. niloticus* are produced for aquaculture.

Further reading. Trewavas 1983 (description, biology); Freyhof et al. 2020 (distribution in Saudi Arabia).

How to eliminate females in tilapias? In *Oreochromis*, males exhibit faster growth and greater uniformity in size than females. Consequently, the farming of monosex fish, which is achieved through manual sexing, direct hormonal sex reversal, hybridisation, or genetic manipulation, has been proposed as a solution to the problem of early sexual maturation and unwanted reproduction. Hybridisation between various species of *Oreochromis* typically results in fertile all-male hybrids. Despite these developments, hybridisation did not effectively solve the problem of unwanted reproduction. This is mainly due to the difficulty of sustaining all-male hybrids' production. This is most likely caused by insufficient care in keeping the broodstock segregated by sex and species and preventing the introduction of hybrids into the broodstock ponds. The masculinisation of the entire tilapia population through hormonal sex reversal by the addition of steroids to the food for a short period during the fry stage has been demonstrated to be an easily applied and relatively consistent technique for producing nearly all-male populations. However, this technique has yet to be fully accepted in some countries due to environmental and social constraints. For example, the effects of the degradation products of synthetic androgen on the environment still need to be fully understood in fish. A recently developed technique for obtaining monosex populations is the production of "supermales" through genetic manipulation. In a breeding program in *O. niloticus*, a technology was developed that produces genetically male tilapia with an average sex ratio of >95 % male and a 40 % increase in yield (GMT®). This combines hormonal feminization of male fry, which are then crossed to normal (XY) males, producing ¼ XX, ½ XY, and ¼ YY. Males with the YY genotype only produce male (XY) offspring in crosses with normal females (XX). **Further reading.** Mair et al. 1997 (sex ratio); Eknath & Acosta 1998 (genetic methods); Guerrero & Guerrero-del Castillo 2002 (Tilapia farming).

Oreochromis mossambicus; aquarium stock, Germany; mouth-brooding female, ~150 mm SL. © A. Spreinat.

Oreochromis mossambicus; aquarium stock, Germany; nuptial male, ~200 mm SL. © A. Spreinat.

Oreochromis mossambicus

Common name. Mozambique tilapia.

Diagnosis. Distinguished from other species of Cichlidae in West Asia by: • nuptial male black or dark-grey with whitish lower part of head / • snout duck-bill-like elongated in male larger than 200 mm SL / ○ orange or red dorsal- and caudal margins in nuptial male / ○ tip of membrane between dorsal spines orange or red / ○ caudal with small spots or blotches not forming regularly shaped and set bars / ○ all flank scales with dark-grey scale pockets in preserved individuals, dark-grey scales pockets never organised in fields / ○ center of flank scales above lateral midline, sometimes also below, darker than scale pockets / ○ anterior processes of lower pharyngeal bone clearly projecting beyond anterior margin of toothed pads / ○ chest, belly, and isthmus in front of pelvic covered by small scales / ○ 3 anal spines / ○ scales cycloid / ○ no ocelli on anal. Size up to 350 mm SL.

Distribution. Introduced in Saudi Arabia (details unknown) and in Yemen, locally introduced in UAE but not recorded from Oman. A non-native population in Algeria. Native to lower Zambezi, lower Shire, and coastal plains from Zambezi Delta to Algoa Bay. Occurs south to rivers in Eastern Cape and Transvaal to Limpopo. Widely introduced beyond this range into inland areas and south-western and western coastal rivers, including lower Orange River and rivers of Namibia. Introduced for aquaculture, escaped and established in tropical countries worldwide.

Habitat. Various standing and slow-flowing waters include coastal lakes, lowland rivers, streams, drains, reservoirs, and marshes, usually over muddy bottoms, often in well-vegetated areas. Fresh to marine waters, can survive in hypersaline waters. Spawns only in fresh and brackish waters of low salinity, along shorelines in stagnant or slow-flowing waters, and on sandy or muddy bottoms.

Biology. Lives up to 11 years. Females usually begin spawning at about 120 mm SL, but in dense populations, at 60–70 mm SL and about 2 months. Territorial males excavate and defend a basin-shaped burrow in center of their territory, where they spawn with females. Often forms leks in dense colonies. Females only visit pits to spawn. Male actively courting female to nest in pit. Territorial males produce sounds during all stages of courtship, but especially during late stages, including spawning. Subordinate males without territory may adopt pseudo-female behaviour and sneak into nests during spawning. Female lays 100–1700 eggs in multiple portions. Female take eggs in their mouths as they are laid, before, during, or after fertilisation. No permanent pair bond formed; females incubate eggs alone while males court other females. Females may spawn a full clutch with one male or several males in a series so that siblings have different fathers. Females do not feed during brood care or take small food items, some of which are eaten in mouth by juveniles. Water is circulated over eggs by chewing movements of female's jaws. Fry hatch in female's mouth after 3–5 days, depending on temperature. Juveniles are released 10–14 days after spawning. Female guards fry, encouraging them to enter her mouth if threatened, for about 3 weeks. Juveniles swarm in shallow water and feed during day, retreating to deeper water at night. Females produce several broods in a season. Non-spawning individuals, as well as mouth-brooding females, gregarious along shores. Can tolerate temperatures ranging from 8 to 42°C; natural temperature range is 17–35°C. Tolerates low dissolved oxygen levels and utilises atmospheric oxygen when aquatic oxygen levels drop. Omnivorous, feeds mainly on algae and phytoplankton.

Conservation status. Non-native; released for mosquito control. NT in its native range, threatened by hybridisation with non-native *Oreochromis niloticus*.

Remarks. While this species was of great importance in aquaculture and is one of the best-studied freshwater fishes in the world, it is now largely replaced by *O. niloticus* and *O. aureus*.

Further reading. Crass 1964 (biology); Pienaar 1978 (biology); De Silva & Chandrasoma 1980 (reproduction); Pullin & Lowe-McConnell 1982 (biology, reproduction); Gregg et al. 1998 (hybridisation); Fuselier 2001 (ecological impact of invasion); Skelton 2001 (native distribution, biology); Allen et al. 2002 (habitat choice in Australia); Amorim et al. 2003 (sound production).

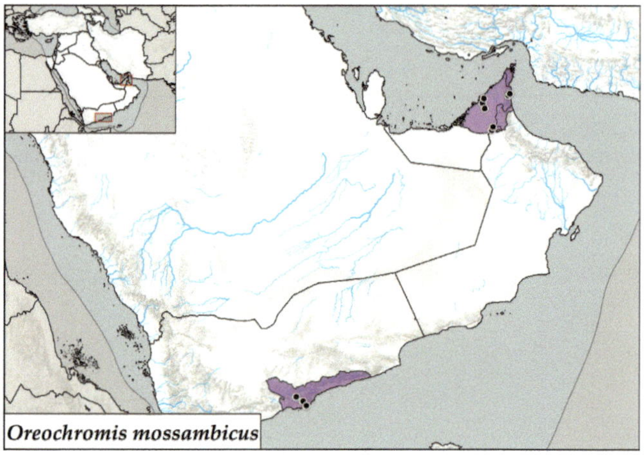

Wadi Bani Khalid in Oman is full of non-native Taiwanese red tilapia.

Oreochromis niloticus; Fossa Calda, Italy; ~200 mm SL.

Oreochromis niloticus
Common name. Nile tilapia.
Diagnosis. Distinguished from other species of Cichlidae in West Asia by: • 3–40 dark-grey to black, regularly shaped and set vertical bars on caudal (number increasing with size) / ○ usually all flank scales with dark-grey scale pockets in preserved individuals, dark-grey scales pockets never organised in fields / ○ tip of membrane between dorsal spines black or grey / ○ chest, belly, and isthmus in front of pelvic covered by small scales / • nuptial male whitish with pink anterior body and caudal / ○ snout not duck-bill-like elongated in male larger than 150 mm SL / ○ anterior processes of lower pharyngeal bone clearly projecting beyond anterior margin of toothed pads / ○ chest never red / ○ 19–25 gill rakers on lower part of first branchial arch / ○ 3 anal spines / ○ scales cycloid / ○ no ocelli on anal. Size up to 395 mm SL.
Distribution. Widely introduced in Oman, UAE, Saudi Arabia, and Yemen, but many might be Taiwanese red tilapia. Also introduced in North Africa, Jordan, Orontes, Iran, and Iraq. Native to Nile drainage (including Lakes Albert, Edward, and Tana), Jebel Marra, Lakes Kivu, Tanganyika, Awash, and Omo, various Ethiopian lakes, Lake Turkana, Suguta, and Lake Baringon. Also native to West Africa in Senegal, Gambia, Volta, Niger, Benue, and Chad drainages. Widespread introduction for aquaculture in other parts of Africa, Europe, and tropical countries worldwide.
Habitat. Lakes, backwaters, and slow-flowing rivers. Very common in reservoirs, canals, and other artificial water bodies. Spawns along banks in stagnant or slow-flowing water. A freshwater species that does not inhabit brackish water.
Biology. Lives up to 5 years. Matures at 3–6 months or much younger, depending on temperature and food availability. Sexually monomorphic, male larger. Spawns when temperatures exceed 20°C between April and August in North Africa. Begins spawning in February in Lake Nasser (Egypt). Non-spawning individuals are gregarious along banks. Males form territories and dig shallow depressions in sandy or muddy shores to attract females. Nest is a circular depression up to 100 cm in diameter and 50 cm deep, usually twice length of breeding male. Males often builds their nests close together, forming leks. Females only visit nest to spawn. Courtship lasts several hours. Up to 200 eggs are laid in about 20 portions over 45–120 minutes. Eggs are picked up and placed in mouth by female as they are laid, before, during, or after fertilisation. No permanent pair bond; females incubate eggs alone while males court other females. Females may lay a full clutch with one male or several males in a series so that siblings have different fathers. Females do not feed during brood care or take small items of food, some of which are eaten in mouth by young. Water is circulated over eggs by chewing movements of female's jaws. Females spawn several times during summer, often once a month. Females hold eggs and larvae in their

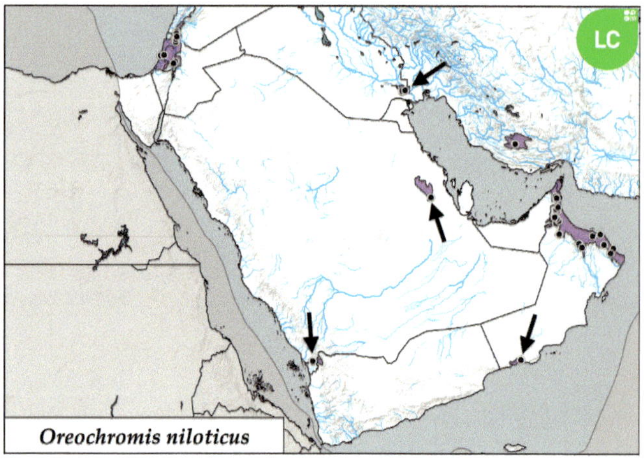

mouths for about 20 days and guard-feeding juveniles for up to 2 weeks. Expanded temperature range: 8–42°C; natural temperature range: 13–33°C. Juveniles are omnivorous; juveniles and adults feed mainly on planktonic and epilithic blue-green algae, diatoms, and macrophytes.

Conservation status. Non-native; LC in its native range; released or escaped from fish farms. It significantly impacts native fish due to its dense population and competition for habitat and food; its nesting activity and territorial behaviour disturb habitat and spawning grounds. International and national agencies that promote its stocking largely ignore these negative impacts. Globally, *O. niloticus* is the most important tilapia species in aquaculture and also supports major capture fisheries where it is established. It is highly invasive and is known to hybridise with many other *Oreochromis* species. For this reason, further stocking has been banned in several countries, including South Africa, Malawi, and Zambia.

Remarks. This species is important for aquaculture worldwide and is cultivated in many countries, and escapes are expected elsewhere. Several records of *O. niloticus* from Türkiye are unconfirmed and all these records have been identified as *O. aureus*.

Further reading. Pullin & Lowe-McConnell 1982 (biology); Trewavas 1983 (description, biology). Siddiqui & Al-Harbi 1995 (hybrid tilapias in Saudi Arabia).

Sarotherodon

Sarotherodon comprises approximately 25 species of mouth-brooding tilapias, all but one of which are endemic to West and Central Africa. Of particular interest is the adaptive radiation of 11 closely related species in Barombi Mbo, a crater lake in Cameroon. This represents one of the most spectacular examples of sympatric speciation. It appears that *Sarotherodon* is paraphyletic, with several new generic names expected to be described. It is unlikely that *S. galilaeus* will remain in that genus. Alternatively, several other genera of mouth-brooding tilapias, including *Iranocichla*, *Oreochromis*, and *Tristramella*, might become synonyms of *Sarotherodon*. However, no conclusive phylogenetic concept has been proposed until now. Furthermore, *S. galilaeus* appears to have been largely introgressed by *Oreochromis aureus*, indicating that both species have undergone extensive hybridisation in the past. Additionally, there is a high genetic diversity within *S. galilaeus*, particularly in West and Central Africa. Consequently, a critical taxonomic revision of this species is required. **Further reading.** Trewavas et al. 1972; Schliewen et al. 1994 (Barombi Mbo); Dunz & Schliewen 2013 (phylogeny).

Sarotherodon galilaeus; Orontes drainage, Syria; ~150 mm SL.

Sarotherodon galilaeus

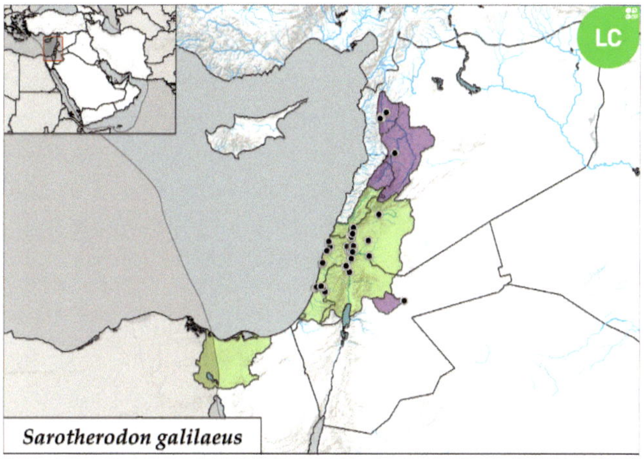

Common name. Mango tilapia.

Diagnosis. Distinguished from other species of Cichlidae in West Asia by: ● anterior processes of lower pharyngeal bone not or very slightly projecting beyond anterior margin of toothed pads / ○ flank scales without dark-grey scale pockets in preserved individuals, isolated black scales pockets, or fields of black scale pockets in some individuals / ○ nuptial male pale-grey or pinkish white without white spots or vermiculation on flank / ○ no red or orange colour in life / ○ often with black bars on flank / ○ caudal almost without pattern / ○ chest, belly, and isthmus in front of pelvic covered by small scales / ○ 20–27 gill rakers on lower part of first branchial arch / ○ chest never red / ○ 3 anal spines / ○ scales cycloid / ○ no ocelli on anal. Size up to 320 mm SL.

Distribution. Native to Jordan drainage and coastal rivers Yarkon, Kishon, and Na'aman in Israel. Introduced in Orontes and Syrian part of Euphrates. Native to Congo, Nile drainage including Lakes Albert and Turkana. Also in Lake Chad basin, Borkou in Chad, Ubangui and Ulele, Adar in Mauritania, and West African rivers from Senegal to Guinea, Volta, Niger, and rivers between them and Sanaga in Cameroon.

Habitat. Slow-flowing rivers, backwaters, lakes, and coastal lagoons. Often pelagic and very common in reservoirs. Can live in brackish water to at least 20 ‰. Cannot survive if water temperature falls below 10°C for over a few days.

Biology. Live up to 7 years. Usually matures after second winter at about 130–220 mm SL. Sexually monomorphic male over 2 years with slightly larger and longer unbranched dorsal and anal rays. Pairs for a few days to 2 weeks before spawning. Spawns in Lake Kinneret late March–August, each female usually produces two or more clutches. Pair digs a shallow nest. Eggs are laid in batches of 20–40, each batch fertilised by male as it is laid. Eggs remain in nest in clutches. In undisturbed conditions, all eggs laid in nest are taken into mouth by male, female, or both parents after hatching. If predators are present, each batch is taken immediately after fertilisation. Pair bond is broken after spawning. Commitment to brood care by males depends on availability of mature females. Parents release fry after about 14 days and usually do not care for them. Feeds mainly on phytoplankton, including cyanobacteria, epiphytic algae, and detritus.

Conservation status. LC; massively stocked in Lake Kinneret for commercial fisheries.

Remarks. Reports from the Oued Draa drainage in southern Morocco are based on misidentified *O. aureus*.

Further reading. Trewavas 1983 (description, biology); Goren & Ortal 1999 (distribution).

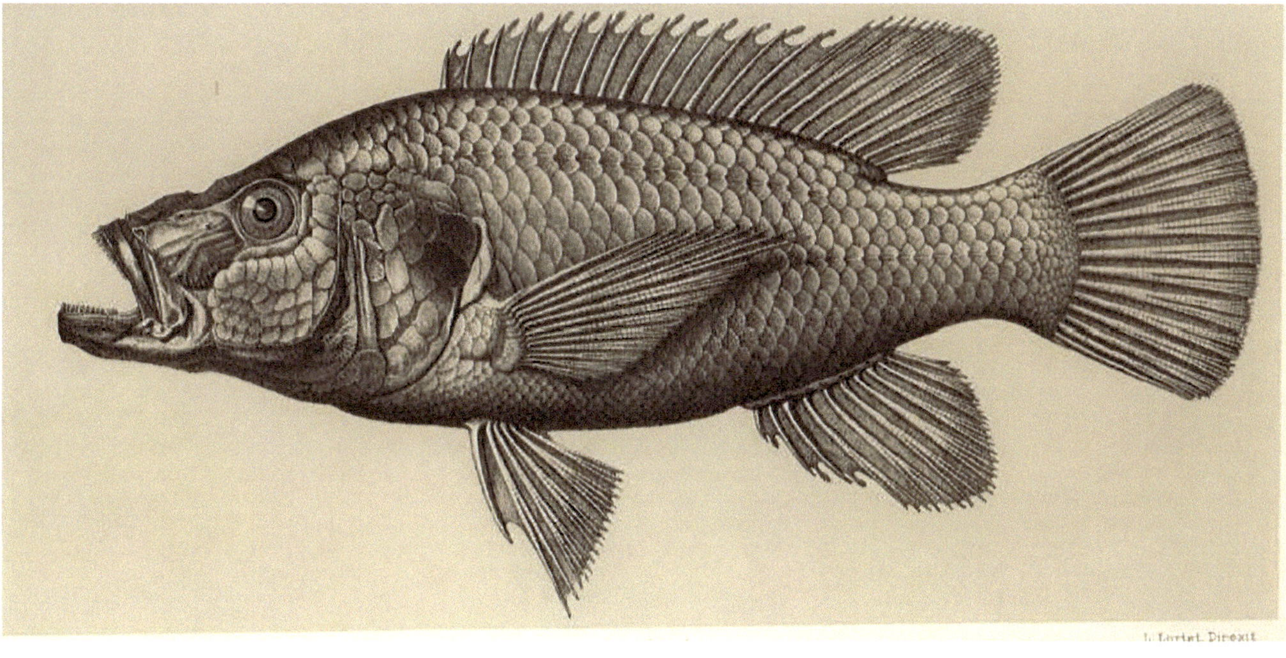

Tristramella sacra; Lake Kinneret, Israel; modified from Lortet 1876.

Tristramella

Tristramella, which is endemic to the Levant, comprises only two species. Phylogenetic analysis suggests that *Tristramella* is related to *Iranocichla*. Some authors have proposed that *T. magdalenae* from the Damascus basin and *T. intermedia* from Lake Hula basin should be considered distinct species. However, we could not identify distinguishing characteristics that would allow us to differentiate these from *T. simonis*. Consequently, we have treated them as conspecific. Both populations and *T. sacra* are extinct. Both species are mouthbrooders, but the details of their breeding behaviour have not been described. There is some doubt whether *T. sacra* should be recognised as a species distinct from *T. simonis*, given that large nuptial male *T. simonis* also have a strongly projecting lower jaw. It would be beneficial to re-examine the original materials of both species to test whether *T. sacra* represents a valid species. **Further reading.** Steinitz & Ben-Tuvia 1960 (revision of genus); Dunz & Schliewen 2013 (phylogeny).

Tristramella sacra

Common name. Long jaw Tristramella.

Diagnosis. Distinguished from other species of Cichlidae in West Asia by: • lower jaw strongly projecting / • all teeth conical / ○ nuptial male pale-grey or pinkish white without white spots or vermiculation on flank / ○ chest never red / ○ 13 gill rakers on lower part of first branchial arch / ○ chest, belly, and isthmus in front of pelvic covered by small scales / ○ 3 anal spines / ○ scales cycloid / ○ no ocelli on anal. Size up to 320 mm SL.

Distribution. Israel: Lake Kinneret.

Habitat. Lacustrine.

Biology. Spawns in April–July. Paternal mouthbrooders. Feeds on zooplankton and small fish.

Conservation status. Extinct; last seen in 1990, despite research both in lake and local markets. Destruction of breeding habitat (marshes in northern part of lake) may be reason for its extinction.

Further reading. Krupp & Schneider 1989 (identification, distribution).

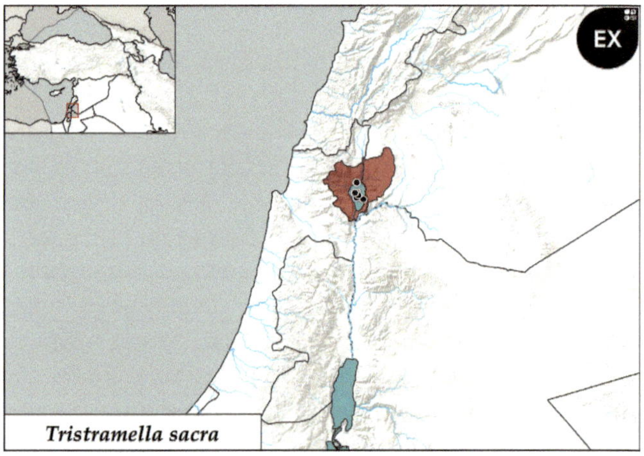

Tristramella simonis; Jordan drainage, Syria; ~90 mm SL.

Tristramella simonis; Lake Kinneret, Israel; ~200 mm SL. © Y. Oksman.

Tristramella simonis

Common name. Short jaw Tristramella.

Diagnosis. Distinguished from other species of Cichlidae in West Asia by: ● lower jaw slightly projecting / ○ outer teeth bicuspid, inner teeth tricuspid / ○ chest never red / ○ nuptial male pale-grey or pinkish white without white spots or vermiculation on flank / ○ 10–12 gill rakers on lower part of first branchial arch / ○ chest, belly, and isthmus in front of pelvic covered by small scales / ○ 3 anal spines / ○ scales cycloid / ○ no ocelli on anal. Size up to 250 mm SL.

Distribution. Jordan drainage: Lakes Muzayrib (Syria), Hula and Kinneret (Tiberias) (both in Israel), Dara'a reservoir and a canal at Al Asha'ari (Syria), Damascus basin, and Lake Hula. Non-native in Orontes and Nahr al Kabir (Syria).

Habitat. Lakes, reservoirs, springs, and spring-fed streams and canals.

Biology. Spawns March–August. Maternal mouthbrooder or both partners participate in brooding. Feeds on phytoplankton and invertebrates.

Conservation status. VU; extirpated from Hula due to drainage of this wetland and extirpated from Damascus basin due to drying up of lakes it inhabited. Apparently stable in Kinneret. Probably declining outside Lake Kinneret, but inhabiting reservoirs and invasive in Syria.

Further reading. Goren 1974 (distribution); Borkenhagen & Freyhof 2009 (distribution).

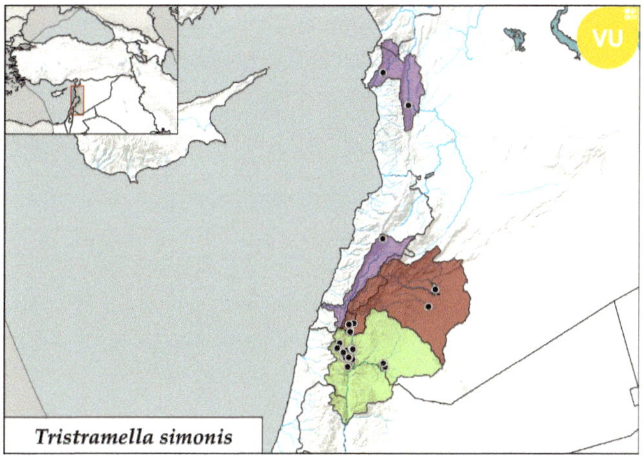

Brackish rivers in southern Iran are the habitat of *Iranocichla* species.

The promise and paradox of using artificial intelligence in ichthyology. In a time when artificial intelligence (AI) is reshaping almost every field, ichthyology cannot remain untouched. As in all other disciplines, it presents a double-edged sword, offering powerful new tools while raising complex challenges. AI's capabilities in pattern recognition and data analysis can totally change how we identify species, automate literature reviews, improve ecological modelling, and even help us discover hidden biodiversity by detecting patterns in genes and morphology. Deep learning models can process thousands of specimens with remarkable speed, spot tiny morphological differences invisible to the human eye, and predict species distributions under climate change scenarios. AI can also speed up phylogenetic reconstructions and help to reveal hidden biodiversity in large, under-explored datasets. But, as with anything new, there are risks. Over-reliance on AI-generated identifications without expert validation may introduce and propagate systematic errors. Many algorithms operate as opaque "black boxes," which makes it hard to understand and doesn't meet the standards of scientific rigour. AI models trained on biased or incomplete datasets can end up making existing taxonomic gaps worse, and this can have a bigger effect on rare or understudied species. Importantly, blind trust in AI threatens to deskill young taxonomists and diminish the value of foundational expertise in morphology, ecology, and evolution—a paradox where technological advancement could weaken the scientific foundation it aims to strengthen. So, recognising this risk calls for a balanced perspective. Further points for consideration could include the following: the potential for the development of standards with which to evaluate and verify AI-driven decisions in the context of taxonomy and conservation. The question of responsibility in the curation of training data and the crediting of AI-assisted contributions is a critical issue. Of equal concern is the potential for AI to produce entire research publications, which could overwhelm scientific literature with unverified content and erode the standards of peer review. The future of ichthyology will not be defined by algorithms alone, but by the wisdom with which we choose to use them.

Family Mugilidae

Mullets

A family of fishes primarily found in coastal waters, with approximately 25 genera and 75 species. They occur globally in various marine habitats, including temperate, subtropical, and tropical regions. Mullets are distinguished by the presence of two separate dorsal fins, a small triangular mouth, and the absence of a lateral line on the flanks. Most species exhibit an unusually muscular stomach and a complex pharynx, which are believed to be an adaptation to grazing on plants, algae, and detritus. Some species inhabit freshwater habitats, while most are marine coastal species that occasionally enter lagoons, estuaries, or the lower reaches of rivers. They can be distinguished from superficially similar Atherinidae by several characteristics, including a dorso-ventrally flattened head (vs. laterally compressed), scales on the dorsal surface of the head (vs. absence of scales), and 11–14½ anal rays (vs. 10–20½). The taxonomy and nomenclature of the Mugilidae are not yet fully resolved, and several name changes have occurred recently. In the context of our region, the most significant alteration is the synonymisation of *Liza* with *Chelon* and the segregation of *Planiliza* from *Liza*.

Mullets are pelagic fishes that live in schools and typically inhabit waters less than 200 m deep, with a preference for muddy bottoms. They spawn at sea. Eggs and early larvae are pelagic, while small juveniles move toward the shoreline and enter estuaries and lagoons, where they feed on zooplankton. Larger individuals return to the sea, where they feed on algae and vegetal detritus, often taking a significant amount of sediment with them, which they filter through their branchial apparatus. All species exhibit a similar morphology, with a silvery appearance that renders them challenging to distinguish. Nine species have been documented from freshwaters in West Asia, with five representing non-native populations at least in part of their range. The sole true freshwater species is *Planiliza abu*, which also spawns in freshwater habitats. *Mugil cephalus*, *Chelon auratus,* and *C. ramada* frequently enter freshwaters in lower parts of rivers in the Black Sea and Mediterranean basins to forage, as do *C. auratus* and *C. saliens* that are also introduced to the Caspian Sea. East Asian *Planiliza haematocheilus* is rapidly expanding its range in the Black and Aegean Seas.

Ellochelon vaigiensis and *Planiliza carinata* are present in our region, although they have not yet been recorded from freshwater habitats in West Asia. Both species are included in the keys to allow their identification. *Ellochelon vaigiensis* is widely distributed in the Indo-Pacific region, occurring in the Red Sea, the Persian Gulf, and along the entire coast of Arabia. *Planiliza carinata* is found in the Red Sea and is invasive in the eastern Mediterranean, where it is recorded from Egypt (west to the Nile delta), Israel, Lebanon, and Türkiye. Identification literature for marine fish should be consulted to identify fish not found in pure freshwaters.

Further reading. Bograd 1961 (movements in freshwater); Thomson 1997 (systematics); Durand et al. 2012 (phylogeny, generic concept).

An encounter with an adult *Chelon labrosus*; Italy. © E. Motti.

https://doi.org/10.1515/9783111677811-042

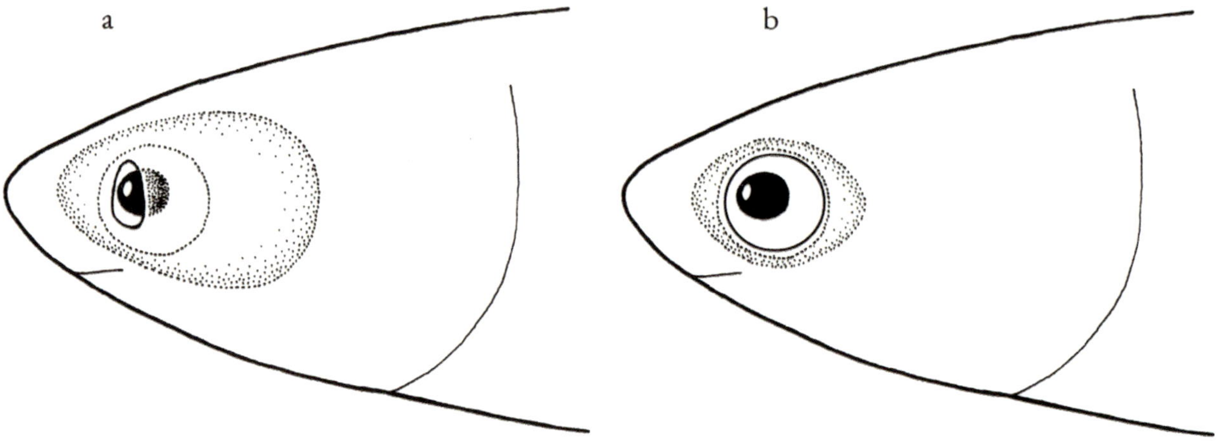

Figure 66. Adipose tissue around the eye in Mugilidae: **a,** *Mugil cephalus*; **b,** *Chelon* species.

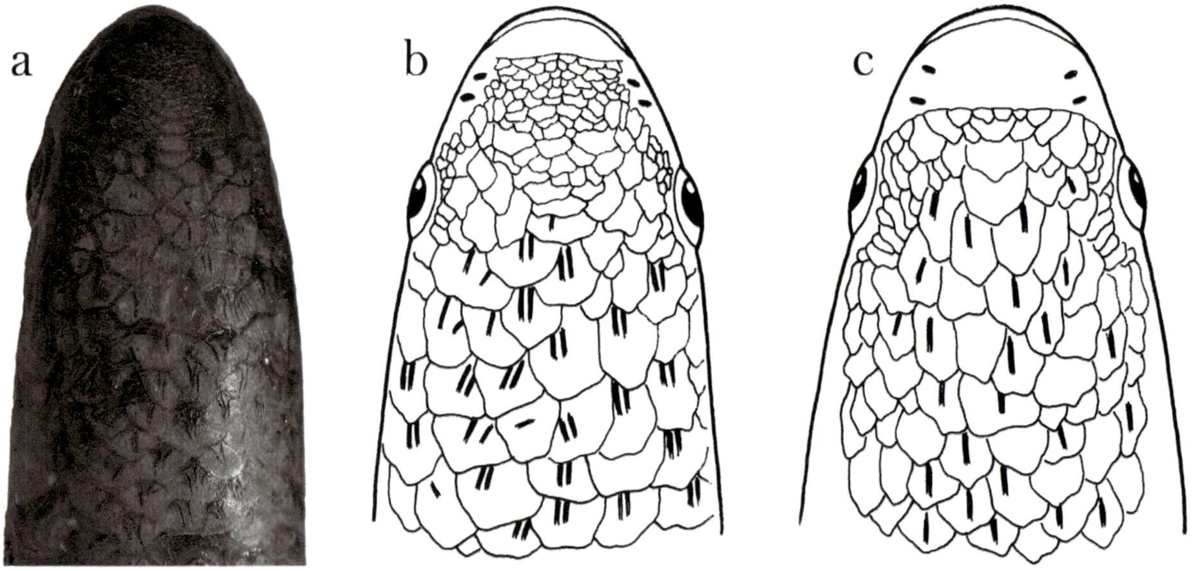

Figure 67. Longitudinal grooves on predorsal scales of *Chelon* species. **a+b,** *C. saliens*; **c,** other species (from Kottelat & Freyhof 2007, after Svetovidov 1964); first picture © E. Motti.

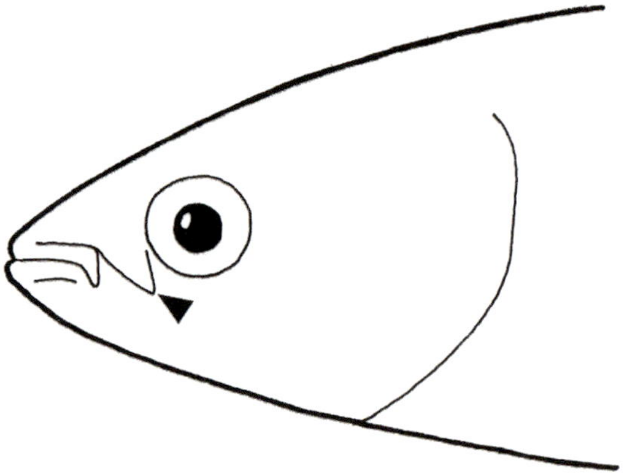

Figure 68. Head of *Chelon auratus* showing pointed posterior angle of preorbital bone (arrow) (from Kottelat & Freyhof 2007).

Figure 69. *Chelon labrosus*, Mesolongi, Greece; ~220 mm SL.

Keys to species of Mugilidae potentially entering freshwaters in West Asia

Persian Gulf basin and seas around Arabian Peninsula

1a - Pectoral dark-grey or black.
..................*Ellochelon vaigiensis*
1b - Pectoral hyaline.
..................2

2a - Individuals larger than 70 mm SL with adipose tissue covering most of eye (Fig. 66), reaching or almost reaching pupil.
..................3
2b - Individuals larger than 70 mm SL without an adipose tissue rim around eye (Fig. 66).
..................5

3a - Anterior margin of preorbital bone straight, tip pointed; back rounded, not keeled in front of first dorsal.
..................*Mugil cephalus*
3b - Anterior margin of preorbital bone strongly concave, tip blunt; back in front of first dorsal keeled by thickened and ridged scales.
..................4

4a - Total gill rakers 69–82 in individuals between 67 and 96 mm SL, 83–93 in individuals between 106 and 124 mm SL; first pterygiophore of first dorsal ray set between 6th and 7th vertebrae.
..................*Planiliza carinata*
4b - Total gill rakers 79–96 in individuals between 66 and 91 mm SL, 94–109 in individuals between 110 and 139 mm SL; first pterygiophore of first dorsal ray set between 7th and 8th vertebrae.
..................*Planiliza klunzingeri*

5a - 27–32 total scales along lateral midline; preorbital bone does not fill space between lip and eye.
..................*Planiliza subviridis*
5b - 40–50 total scales along lateral midline; preorbital bone fills space between lip and eye.
..................6

6a - Pectoral long, reaching almost to vertical of first dorsal origin; no golden blotch on opercle; two dark-grey stripes on flank.
..................*Planiliza abu*
6b - Pectoral short, not reaching close to vertical of first dorsal origin; usually a golden or orange blotch on opercle; no stripes on flank.
..................*Chelon auratus*

Mediterranean, Caspian, and Black Seas

1a - Lower $\frac{1}{4}$–$\frac{1}{3}$ of upper lip with 2–3 rows of papillae with horny ridges in individuals larger than 100 mm SL (Fig. 69).
..................*Chelon labrosus*
1b - Upper lip smooth.
..................2

2a - Individuals larger than 70 mm SL with adipose tissue covering most of eye, reaching or almost reaching pupil.
..................3
2b - Individuals larger than 70 mm SL without adipose tissue rim around eye.
..................4

3a - Anterior margin of preorbital bone straight, tip pointed; back rounded, not keeled in front of first dorsal.
..................*Mugil cephalus*
3b - Anterior margin of preorbital bone strongly concave, tip blunt; back in front of first dorsal keeled by thickened and ridged scales.
..................*Planiliza carinata*

4a - Pectoral long, reaching almost to vertical of first dorsal origin; two dark-grey stripes on flank (Orontes, Ceyhan, Seyhan drainages only).
..................*Planiliza abu*
4b - Pectoral short, not reaching close to vertical of first dorsal origin; flank plain grey silvery without stripes (coastal species).
..................5

5a - Posterior extremity of upper jaw reaching beyond anterior rim of eye.
..................*Planiliza haematocheilus*
5b - Posterior extremity of upper jaw not reaching anterior rim of eye.
..................6

6a - Predorsal scales with 2–5 longitudinal grooves (Fig. 67); 48–49 scales in lateral series (not including scales on caudal base).
..................*Chelon saliens*
6b - Predorsal scales with a single or no longitudinal groove (Fig. 67); 40–46 scales in lateral series (not including scales on caudal base).
..................7

7a - 20 circumpeduncular scale rows; when folded forward, pectoral reaching eye; posterior angle of preorbital bone pointed (Fig. 68); no black spot at pectoral base.
..................*Chelon auratus*
7b - 24 circumpeduncular scale rows; when folded forward, pectoral not reaching eye; posterior angle of preorbital bone rounded; a black spot at inner pectoral base.
..................*Chelon ramada*

Chelon auratus, Mediterranean Sea, Spain; ~400 mm SL. © shutterstock_2366550807 Fernando M. Elkspera.

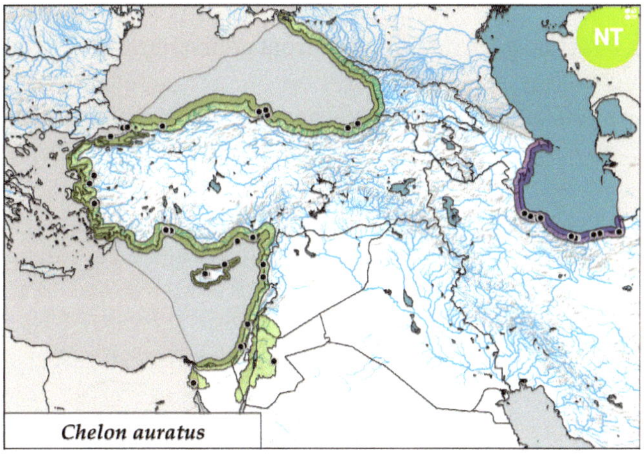

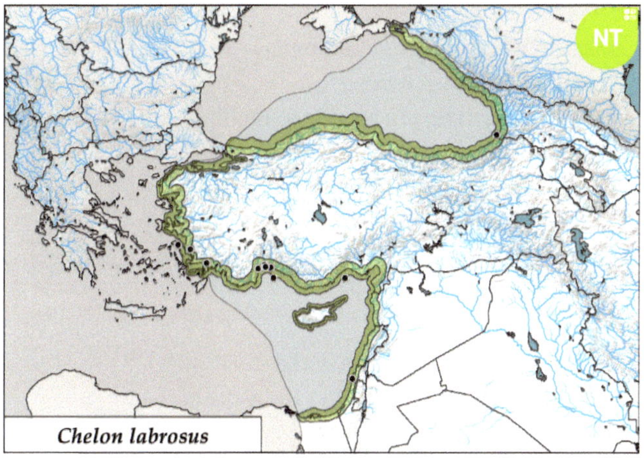

Chelon auratus

Common name. Golden mullet.

Diagnosis. Distinguished from other species of mullets entering freshwaters in West Asia by: ○ usually a golden or orange blotch on opercle / ○ predorsal scales with a single or no longitudinal groove / ○ 20 circumpeduncular scale rows / ○ when folded forward, pectoral reaching eye / ○ no black spot at pectoral base / ○ posterior extremity of upper jaw not reaching anterior rim of eye / ○ posterior angle of preorbital bone pointed / ○ 40–46 total scales in lateral series / ○ without adipose tissue rim around eye / ○ pectoral short, not reaching close to vertical of first dorsal origin / ○ flank plain silvery grey / ○ upper lip smooth / ○ preorbital bone fills space between lip and eye / ○ caudal forked. Size up to 500 mm SL.

Distribution. Mediterranean, Black Sea, and eastern Atlantic from Cape Verde and Senegal to southern Norway and British Isles (not reaching southern Baltic); northern Red Sea. Introduced in Caspian Sea.

Habitat. Pelagic near shore, sometimes in lagoons and estuaries. Most freshwater-tolerant *Chelon* species recorded from West Asia. Spawns at sea.

Biology. Gregarious. Spawns September–November. Juveniles move to coastal lagoons and estuaries in winter and especially in spring. Juveniles feed on zooplankton and larger individuals on algae and plant detritus.

Conservation status. NT; like almost all *Chelon*, this species has declined considerably over the last 20 years.

Remarks. *Chelon auratus* has entered the northern Red Sea from the Mediterranean and has recently been reported from this area. It is unclear whether the species is expanding its range in the Red Sea.

Further reading. Ben-Tuvia, in Whitehead et al. 1986 (description); Gandolfi et al. 1991 (biology); Thomson 1997 (systematics).

Chelon labrosus; Iqem, Morocco, ~70 mm SL.

Chelon labrosus

Common name. Thick lip mullet.

Diagnosis. Distinguished from other species of mullets entering freshwaters in Mediterranean, Caspian, and Black Sea basins by: • lower $\frac{1}{4}$–$\frac{1}{3}$ of upper lip with 2–3 rows of papillae with horny ridges in individuals larger than 100 mm SL / ○ without adipose tissue rim around eye / ○ pectoral short, not reaching close to vertical of first dorsal origin. Size up to 600 mm SL.

Distribution. Coasts of Mediterranean, Black Sea, and eastern Atlantic from Cape Verde and Senegal to southern Norway, southern Iceland, and Faroe Islands.

Habitat. Pelagic near shore, sometimes in lagoons and estuaries. Spawns at sea in coastal surface waters.

Biology. Gregarious along coasts, often entering lagoons and estuaries, juveniles readily adapt to freshwater. Lives up to 12 years. Males spawn for the first time at 2 years, females at 3, later in northern areas. Females larger than males. Spawns pelagic in February–April, eggs pelagic. Juveniles of about 20 mm SL move to coastal lagoons and estuaries in April–June, returning to sea in late summer. Juveniles feed on zooplankton; adults mainly on algae and plant detritus.

Conservation status. NT; like almost all *Chelons*, this species has declined considerably over the last 20 years.

Further reading. Ben Tuvia, in Whitehead et al. 1986 (description); Gandolfi et al. 1991 (biology); Thomson 1997 (systematics); Harrison 2003b (biology).

Chelon ramada; Ilıca drainage, Antalya, Türkiye; ~120 mm SL.

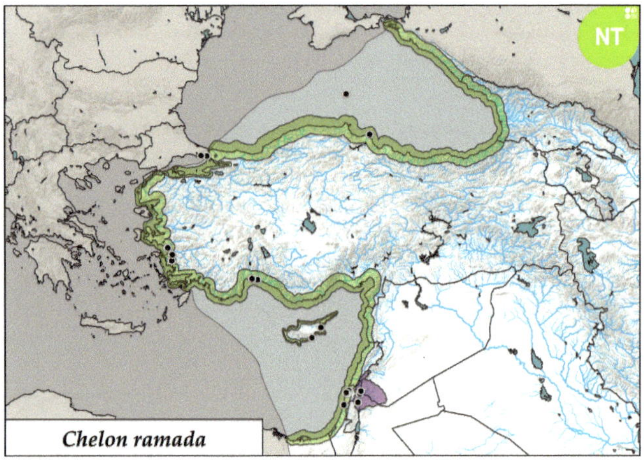

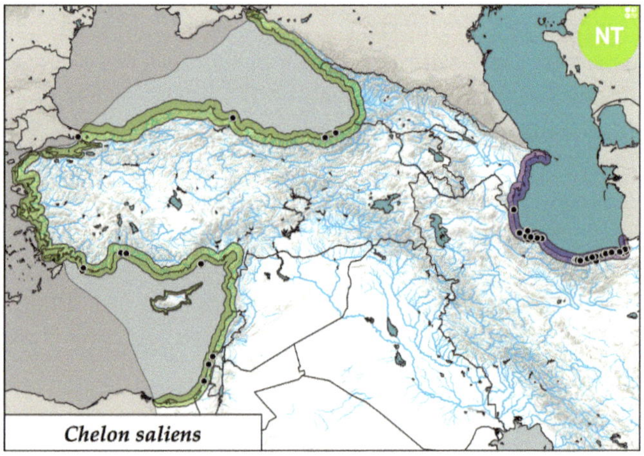

Chelon ramada

Common name. Thinlip mullet.

Diagnosis. Distinguished from other species of mullets entering freshwaters in Mediterranean, Caspian, and Black Sea basins by: ○ predorsal scales with a single or no longitudinal groove / ○ 24 circumpeduncular scale rows / • when folded forward, pectoral not reaching eye / ○ posterior angle of preorbital bone rounded / ○ posterior extremity of upper jaw not reaching anterior rim of eye / • a black spot at pectoral base / ○ 41–46 scales in lateral series (not including scales on caudal base) / ○ upper lip smooth / ○ without adipose tissue rim around eye / ○ pectoral short, not reaching close to vertical of first dorsal origin. Size up to 545 mm SL.

Distribution. Mediterranean, Black Sea, Azov Sea, and eastern Atlantic from Cape Verde and Senegal to southern Baltic and

British Isles (not reaching northern Scotland). Migrates north during summer. Introduced in Lake Kinneret (Israel).
Habitat. Pelagic inshore, entering lagoons and lower reaches of rivers, often found in polluted waters. Spawns offshore at sea.
Biology. Usually in schools. Males spawn first time at 2–3 years, females at 4 or later. Females slightly larger than males. Spawns several million pelagic eggs in September–December in Mediterranean, April–July along British coast. Juveniles about 20 mm SL move into coastal lagoons and estuaries in autumn, especially in winter. Juveniles feed on zooplankton until about 30 mm SL, then on benthic animals and plants. Adults filter algae and plant detritus.
Conservation status. NT; like almost all *Chelon*, this species has declined considerably over the last 20 years.
Further reading. Ben Tuvia, in Whitehead et al. 1986 (description); Gandolfi et al. 1991 (biology); Costa Pereira 1995; Thomson 1997 (systematics); Harrison 2003c (biology).

Chelon saliens; Mediterranean Sea, Spain; ~200 mm SL. © shutterstock_2141170165 Ion Mes.

Chelon saliens
Common name. Sharpnose mullet.
Diagnosis. Distinguished from other species of mullets entering freshwaters in Mediterranean, Caspian, and Black Sea basins by: • predorsal scales with 2–5 longitudinal grooves / • 48–49 scales in lateral series (not including scales on caudal base) / ○ 20 circumpeduncular scale rows / ○ when folded forward, pectoral reaching eye / ○ posterior extremity of upper jaw not reaching anterior rim of eye / ○ upper lip smooth / ○ without adipose tissue rim around eye / ○ pectoral short, not reaching close to vertical of first dorsal origin. Size up to 350 mm SL.
Distribution. Mediterranean, Black Sea, and eastern Atlantic from Angola to Bay of Biscay. Introduced in Caspian Sea.
Habitat. Pelagic near shore, sometimes in lagoons and estuaries. Spawns at sea.
Biology. Usually in schools. Males spawn for first time at 2, females at 3 years. Females larger than males. Spawns pelagic eggs in May–August, rarely until early October. Juveniles about 20 mm SL move to coastal lagoons and estuaries in summer and autumn. Juveniles feed on zooplankton to about 30 mm SL, then benthic organisms to 50 mm SL; adults feed on algae and plant detritus.
Conservation status. NT; like almost all *Chelon*, this species has declined considerably over the last 20 years.
Remarks. Seems rarely to enter freshwaters and might be excluded from freshwater fish in future. Only known from sea in Caspian.
Further reading. Ben Tuvia, in Whitehead et al. 1986 (description); Gandolfi et al. 1991 (biology); Thomson 1997 (systematics); Harrison 2003d (biology).

Mugil cephalus; Mesolongi, Greece; ~300 mm SL.

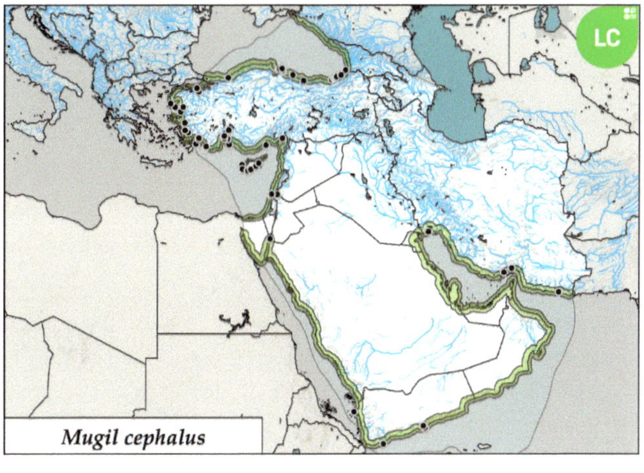

Mugil cephalus; Köprüçay drainage, Türkiye; ~160 mm SL.

Mugil cephalus
Common name. Flathead mullet.
Diagnosis. Distinguished from other species of mullets entering freshwaters in West Asia by: ○ adipose tissue covering most of eye, reaching pupil in adults / ○ back rounded, not keeled in front of first dorsal / ○ anterior margin of preorbital bone straight, tip pointed / ○ upper lip smooth / ○ pectoral short, not reaching close to vertical of first dorsal origin / ○ caudal forked. Size up to 800 mm SL.
Distribution. All tropical and subtropical seas, Mediterranean, Black Sea, and Atlantic, reaching northward to England and the Netherlands.
Habitat. Pelagic near shore, often foraging in lagoons, estuaries, and lower reaches of rivers. Spawns at sea in coastal surface waters.
Biology. Usually in schools. Males spawn first time at 2 years, females at 3 years. Females larger than males. Spawns at sea in coastal surface waters, producing several million pelagic eggs from July to October (Mediterranean). Eggs develop optimally in full marine salinity. Juveniles of about 20 mm SL migrate to coastal lagoons and estuaries in autumn. Juveniles feed on zooplankton until about 30 mm SL and larger individuals filter algae, plant detritus, and small invertebrates.
Conservation status. LC.
Further reading. Ben Tuvia, in Whitehead et al. 1986 (description); Gandolfi et al. 1991 (biology); Thomson 1997 (systematics); Harrison 2003a (biology); Attaala & Rubaia 2005 (record from Yemen); Kottelat & Freyhof 2007 (distribution, biology).

Planiliza abu; Lesser Zab drainage, Iraq; 91 mm SL.

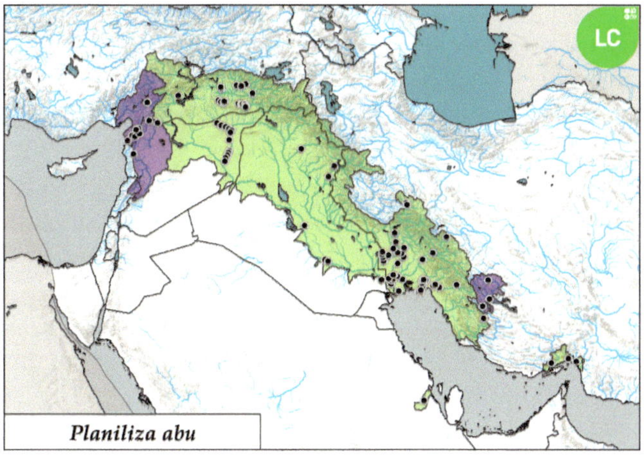

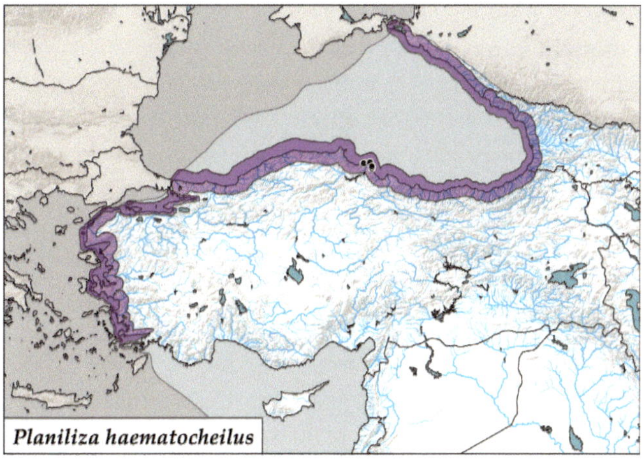

Planiliza abu

Common name. Abu mullet.

Diagnosis. Distinguished from other species of mullets entering freshwaters in West Asia by: ● 44–50 total scales in lateral series / ● pectoral long, reaching almost to vertical of first dorsal origin / ○ no golden blotch on opercle / ● two irregular dark-grey lateral stripes / ○ caudal deeply emarginate / ○ upper lip smooth / ○ without adipose tissue rim around eye. Size up to 220 mm SL.

Distribution. Persian Gulf basin from Euphrates and Tigris south to Indus. A relict population in Al-Ahsa oasis in Saudi Arabia. Introduced in Orontes, Ceyhan and Seyhan drainages (Türkiye), and Kor basin (Iran).

Habitat. Medium-sized streams to large rivers, lakes, and reservoirs. Also, in densely vegetated springs and canals. A wide range of still to moderately fast-flowing waters. Absent from very fast-flowing rivers and streams. Usually in salinities below 2 ‰ but can tolerate up to 30 ‰. Usually a warm-water species, absent from cold headwaters of Anatolia.

Biology. Freshwater species usually live in large schools under water's surface. Lives up to 7 years, usually 4–5 years. Spawns first time at one year and about 100 mm TL in late February–March (southern Iraq), March–May (Baghdad), and March–July (southern Türkiye). Spawns once or twice a year. Eggs attached to submerged vegetation. Feeds mainly on detritus, phytoplankton, and aquatic plants. Small amounts of planktonic invertebrates are taken.

Remarks. This species is a very powerful invader and would be able to colonise large parts of the Mediterranean basin if introduced.

Conservation status. LC; relatively resistant to pollution. Despite its small size, commercially caught in large quantities, particularly in Iran and Iraq and accidentally introduced in Mediterranean coastal areas in Türkiye and Syria, where it is considered a pest.

Further reading. Talwar & Jhingran 1992 (distribution, identification); Coad 2010a (biology); Kaya et al. 2016 (distribution); Bayçelebi 2020 (distribution).

Planiliza haematocheilus; aquarium Moskow, Russia; ~300 mm SL. © J. Pfleiderer.

Planiliza haematocheilus

Common name. Soiuy mullet.

Diagnosis. Distinguished from other species of mullets entering freshwaters in Black Sea and Mediterranean basins by: • posterior extremity of upper jaw reaching beyond anterior rim of eye / ○ predorsal scales with a single longitudinal groove / ○ 41–42 scales in lateral series (not including scales on caudal base) / ○ 24 circumpeduncular scale rows / ○ upper lip smooth / ○ without adipose tissue rim around eye / ○ pectoral short, not reaching close to vertical of first dorsal origin. Size up to 530 mm SL.

Distribution. Coast of East Asia from Vladivostok to Hainan Island. Introduced in Sea of Azov in late 1960s, reached Aegean in 1995. Southernmost record in Aegean from Homa lagoon (İzmir) more widespread.

Habitat. At sea, pelagic near shore. Juveniles enter brackish and fresh waters of lagoons and lower reaches of rivers. Spawns at sea and in brackish waters. Often enters freshwater to overwinter.

Biology. Euryhaline, live six years, likely much longer. Feeds on detritus, diatoms and blue-green algae. Eggs and larvae planktonic.

Conservation status. Non-native; situation poorly known. Most abundant Mugilid in Azov Sea. Along shores of Black Sea and possibly Aegean, its expansion corresponds to a sharp decline in native species of Mugilidae, which it appears to be replacing.

Remarks. Surprisingly, there are very few records of this species from the Anatolian, but none from the Georgian Black Sea basin, whereas it is common along the coasts of Russian Caucasus.

Further reading. Starushenko & Kazansky 1996 (introduction); Thomson 1997 (systematics; as *L. lauvergnii*); Koutrakis & Economidis 2000 (biology).

Planiliza klunzingeri; Shatt al Arab, Iraq; 137 mm SL. © A. Ali.

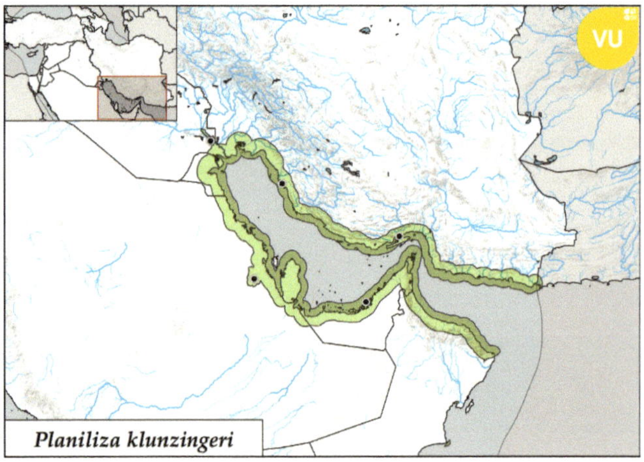

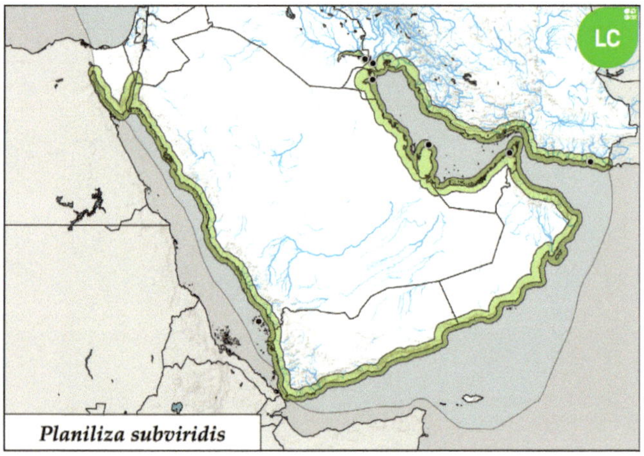

Planiliza klunzingeri

Common name. Eastern keeled mullet.

Diagnosis. Distinguished from other species of mullets in Persian Gulf basin and coasts of Arabian Peninsula by: ○ back in front of first dorsal keeled by thickened and ridged stripes on scales / ○ with adipose tissue rim around eye reaching or almost reaching pupil / ○ total gill rakers 79–96 in individuals between 66 and 91 mm SL, 94–109 in individuals between 110 and 139 mm SL / ○ first pterygiophore of first dorsal ray set between 7th and 8th vertebrae / ○ caudal forked / ○ pectoral short, not reaching close to vertical of first dorsal origin / ○ anterior margin of preorbital bone strongly concave, tip blunt. Size up to 175 mm SL.

Distribution. Coasts of Persian Gulf, including lower Shatt al Arab/Arvand and Hammar marshes, east to Pakistan and west coast of India. Mostly in estuaries of larger rivers.

Habitat. Large estuarine areas, lagoons, marshes, and lower reaches of rivers, often found in polluted waters.

Biology. Gregarious. Lives up to five years. Spawns first time at 1 year and about 100 mm SL. Spawns at sea at 14–18°C in December–April (Kuwait Bay). Juveniles enter estuaries and freshwaters to feed. Highest growth rate and food conversion efficiency at 15 ‰ salinity under experimental conditions. Feeds mainly on detritus, algae, phytoplankton, and zooplankton.

Conservation status. VU; appears to be in steady decline due to overfishing.

Remarks. Often identified as *Liza carinata*.

Further reading. Senou et al. 1987 (identification, distribution); Thomson 1997 (description, biology, distribution).

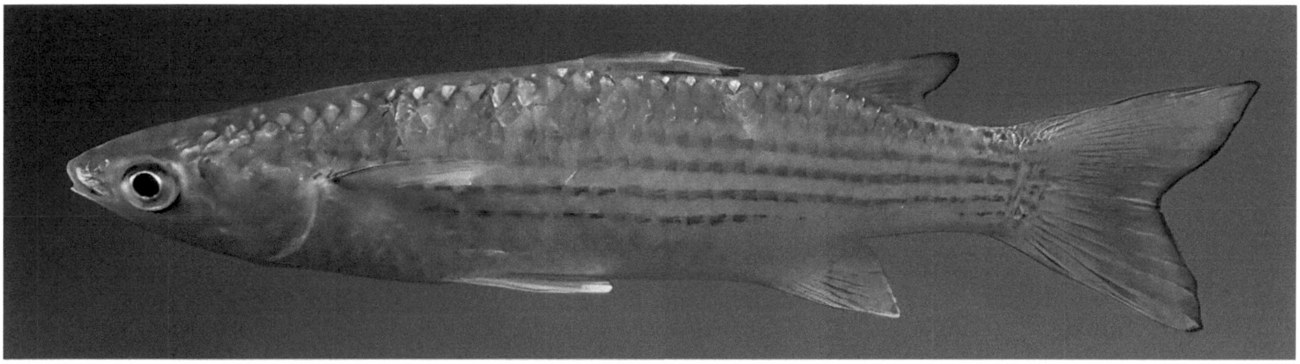

Planiliza subviridis; Hong Kong; ~ 160 mm SL. © Tse Chung Yi.

Planiliza subviridis

Common name. Greenback mullet.

Diagnosis. Distinguished from other species of mullets in Persian Gulf basin and coasts of Arabian Peninsula by: ● preorbital bone not filling space between lip and eye / ○ back rounded, not keeled in front of first dorsal / ○ 27–32 total scales along lateral midline / ○ caudal emarginate or truncate / ○ without adipose tissue rim around eye / ○ pectoral short, not reaching close to vertical of first dorsal origin / ○ anterior margin of preorbital bone strongly concave, tip blunt. Size up to 260 mm SL.

Distribution. Persian Gulf east to India, China, northern Australia, and Polynesia. Recorded from lower Shatt al Arab/Arvand and Hammar marshes.

Habitat. Coastal waters, lagoons, marshes, estuaries, and lower reaches of rivers. Spawns at sea.

Biology. Gregarious. Lives up to 6 years. Spawns first time at about 120 mm SL. Spawns February–April in Shatt al Arab/Arvand, March–April, often June, offshore in Persian Gulf. After spawning, adults and juveniles migrate to brackish and fresh waters. Feeds mainly on algae, detritus, plants, and plankton.

Conservation status. LC.

Remarks. Sometimes identified as *Liza dussumieri* in Persian Gulf basin.

Further reading. Thomson 1997 (description, biology, distribution).

Family Pleuronectidae

Righteye flounders

Flatfishes represent a distinctive group of primarily marine fishes that are readily distinguishable by their bilateral asymmetry, with both eyes on the same body side. They have been observed to have remained largely unchanged since the Eocene epoch (34–56 million years ago). The family Pleuronectidae is recognised to comprise 40 genera, and approximately 103 species distributed across all marine habitats worldwide. It should be noted that several other species (e.g., *Citharus* [family Citharidae], *Solea* [family Soleidae], *Scophthalmus* [family Scophthalmidae]) may occasionally be found in estuaries. **Further reading.** Whitehead et al. 1986 (identification of marine species); Cooper & Chapleau 1998 (phylogeny); Evseenko 2004 (checklist).

Platichthys flesus

Common name. Flounder.

Diagnosis. Only flatfish entering freshwater habitats in West Asia. Distinguished from other species of flatfishes occasionally entering freshwaters in Mediterranean and Black Sea basins by: ○ about 80 total lateral-line scales / ○ enlarged scales along dorsal and anal bases / ● bony tubercles along lateral line / ○ 51–66½ dorsal rays / ○ 35–45½ anal rays / ○ eyes on right side (but, in North Sea, up to 40 % of individuals may have eyes on left side). Size up to 500 mm SL.

Distribution. Black Sea, Mediterranean, eastern Atlantic northeast to White Sea. Introduced in Caspian Sea and Great Lakes of North America. Occasionally recorded up to 650 km upstream in Rhine, in western Asia only in estuarine habitats.

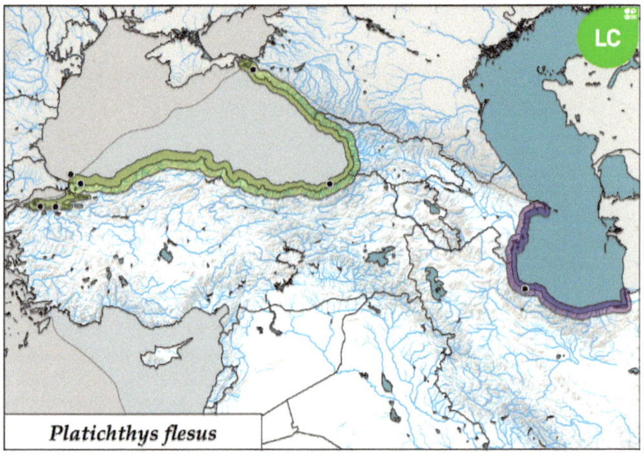

Platichthys flesus; Belbek, Ukraine; 106 mm SL. (side reversed).

https://doi.org/10.1515/9783111677811-043

Habitat. Shallow, marine, and estuarine mud and sand bottoms; rarely enter freshwaters in West Asia.

Biology. Males spawn first time at 2–3 years, females at 3–4 years. Migrates to sea in October–December. Spawns at sea in deep water, in January–June. Eggs and larvae pelagic and drift with current. At about 10 mm SL, left eye moves to right side, pigmentation develops, and juveniles become benthic. Pelagic larvae migrate to brackish and freshwater habitats with tidal currents, moving up water column during high tide and to bottom during low tide. First pelagic flounder is usually recorded in estuaries in March. Juveniles begin to migrate upstream into freshwater to feed by mid–April. Few individuals enter brackish or fresh water. Spawning adults do not return to brackish or freshwater but remain at sea. Feeds on small fish and benthic invertebrates.

Conservation status. LC.

Remarks. The taxonomy of founders needs a critical revision. Several species may be confused under the name *P. flesus*. *Platichthys luscus* is occasionally used as a valid name for flounders from the northeastern Mediterranean and Black Sea basin flounders. However, we could not find any data to support this view.

Further reading. Voronina 1999 (systematics).

Coastal lakes along the Mediterranean Sea, such as Lake Akgöl in Turkiye, are the habitats of various species of mullets and many others including flatfishes.

Family Channidae

Snakeheads

Snakeheads are mid-sized to large freshwater fishes that are characterised by an elongate, rounded body that becomes compressed posteriorly, long and spineless dorsal and anal fins, and by paired accessory organs or suprabranchial organs in the upper gill chamber (above and behind the gills) which enable these fishes to breathe air. There are approximately 50 species (plus many undescribed) in the Asian genus *Channa* and three species in the African genus *Parachanna*. The fossil record of snakeheads extends from the upper Eocene of Egypt to Central Europe and Asia. The smallest snakeheads reach approximately 100 mm, while the largest reported individual is 1800 mm long. All species are considered thrust predators, with most being piscivorous as adults. Snakeheads inhabit still waters, although some are found in larger rivers or along margins of fast-flowing streams. Most species construct a bubble nest in vegetation, with the eggs laid and fertilised below the nest so that they float up into the bubbles. Others are mouthbrooders, and some feed their young with unfertilised eggs. One or both adults guard the nest, and young attacking intruders with great ferocity, including humans. Several species are highly valued as food fish within parts of their native ranges, where they are an important part of capture fisheries and aquaculture. They can be transported alive, wrapped in wet clothes or vegetation, and introduced in areas outside their natural distribution. Several Asian species have been introduced outside of their native range. Four snakehead species are thought to have entered North America through the aquarium and live-food fish markets and are now spreading. *Channa maculata* and *C. striata* have already established a wide non-native range in the tropics, including Hawaii and Madagascar. **Further reading.** Walter et al. 2004 (invasion); Murray 2006 (fossil record); Conte-Grand et al. 2017 (molecular diversity).

Channa montana; Kaju drainage, Iran; ~100 mm SL. © H. Mousavi-Sabet.

Channa montana

Common name. Baluchi snakehead.

Diagnosis. Distinguished from other freshwater fishes in West Asia by: ○ paired accessory organs or suprabranchial organs in upper gill chamber / ○ 33–35½ soft dorsal rays / ○ 20–22½ anal rays / ○ body roundish and elongate / ○ 41–44 lateral-line scales / ○ many teeth on jaws, vomer and palatines / ○ pelvic origin shortly behind pectoral base and below dorsal origin. Size up to 125 mm SL, likely to grow larger.

Distribution. Iran: Jazmurian (Bampur and Halil) and Mashkid (Rotak) basins shared by Iran and Pakistan. Also in rivers of Makran coast. Indus drainage in Afghanistan, Pakistan, and India, as well as rivers of western India,

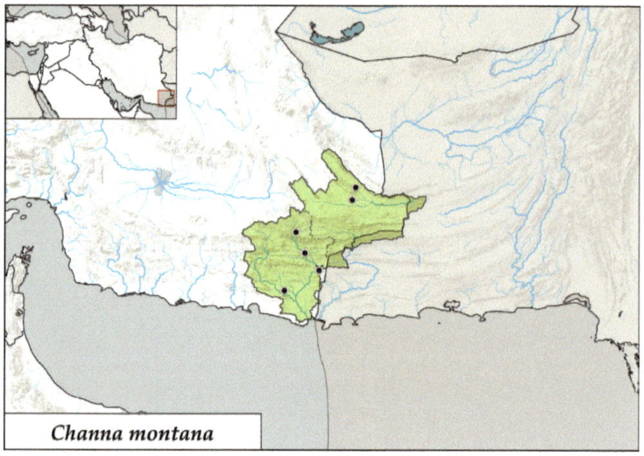

https://doi.org/10.1515/9783111677811-044

Habitat. A wide range of freshwater habitats often associated with dense submerged vegetation and standing water such as qanats, springs, and streams. Occasionally found in rocky streams without vegetation.

Biology. Forms pairs and guard larvae and small fry. Feeds free-swimming juveniles with eggs by spawning into cloud of juveniles. Single study of species' reproduction suggests that eggs are laid on bottom. Breathes air from surface. Feeds on invertebrates and small fish.

Conservation status. Not evaluated. Declining in Iran due to desiccation of habitats.

Remarks. In West Asia, snakeheads are restricted to Iran, usually identified as *Channa gachua*, a species described from Bengal in India. Recent molecular studies suggest that more than 20 species may be included in *C. gachua*. The Baluchi snakehead is very closely related to populations from the Indus and westernmost India. It is identified here as *C. montana*, a species described from the upper Indus drainage in Afghanistan.

Further reading. Coad 2016a (distribution, biology); Conte-Grand et al. 2017 (diversity of *C. gachua*).

Streams like this in Iranian Makran are the habitat of *Channa montana*.

Family Mastacembelidae

Spiny eels

Mastacembelids are a small, distinctive family of perch-like fishes with elongate bodies. They are widespread in tropical Asia and Africa. About 50 species and all genera are found in Asia, while a monophyletic radiation of 41 species is present in Africa. In the Nile drainage, spiny eels are absent except in Lake Victoria. The only species in West Asia is the sister group of all the African species, indicating that the invasion of these eels from Asia to Africa occurred through West Asia. The divergence between African *Mastacembelus* and *M. simack* is estimated to have occurred between 18 and 15 million years ago (95 % CI: 27.5–8.8 Mya). Mastacembelids have a peculiar rostral appendix bearing a tubular nostril on either side of a central rostral tentacle. The spinous part of the dorsal fin of percomorph fishes is dissociated into individual short spines in spiny eels. These spines can be erected, and handling spiny eels might be dangerous. They twist very fast in a way that the spines cut into the skin of the aggressor, and the spines can produce severe wounds on hands and arms, especially if the eel is large. In West Asia, fishers treat spiny eels with caution. In some regions, spiny eels are a highly sought-after food source; in others, they are not consumed due to cultural beliefs. **Further reading.** Brown et al. 2010 (African spiny eels); Near et al. 2013 (phylogeny); Day et al. 2017 (phylogeny).

Mastacembelus simack; Tigris drainage, Türkiye; 280 mm SL.

Mastacembelus simack; Lesser Zab drainage, Iraq; ~400 mm SL.

https://doi.org/10.1515/9783111677811-045

Mastacembelus simack

Common name. Mesopotamian spiny eel.

Diagnosis. Distinguished from *Anguilla* species in West Asia by • movable nasal appendix / • marbled body colour / • dorsal and anal spines present. Size up to 1000 mm SL, usually up to 400 mm SL.

Distribution. Qweiq, Euphrates, Tigris, and Karun drainages and tributaries to Persian Gulf in Iran south to Helleh and Dalaki.

Habitat. Reservoirs, large rivers, up to small mountain streams. Often very common in rapids and riffles in fast-flowing waters.

Biology. Lives up to 21 years. Matures at 2 years, measuring about 250 mm SL and 125 g. Spawns May–July in shallow water between rocks or vegetation. Predatory, juveniles feed on benthic invertebrates, and adults mainly feed on fish and shrimp. Often hides during day in gravel in riffles, roots or caverns of various kinds but may also lie motionless on bottom in bright sunshine, camouflaged in shallow areas to prey on small fish.

Conservation status. LC; extirpated from Qweiq.

Remarks. Often still found as *M. mastacembelus*, which is a synonym. Records from the Kor, an endorheic drainage in central Iran, and the Orontes could not be confirmed. Russell (1756) reported that this fish was "found in great abundance" and was "esteemed a lighter and more delicate food" than the eel at Aleppo. Spiny eels are often seen in fish markets of Syrian cities in Orontes drainage, imported from the Euphrates.

Further reading. Russell 1756 (value as food); Coad 2010a (biology); Coad 2015b (distribution, biology); Kottelat 2022 (*M. simack*).

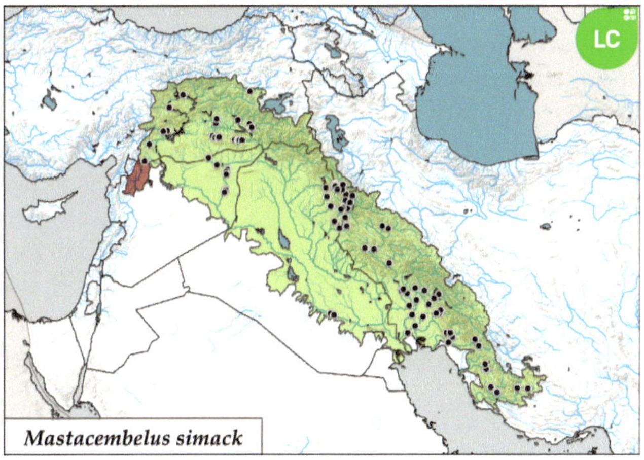

Mastacembelus simack

The Tigris at Şırnak in Türkiye, the habitat of *Mastacembelus simack* and many other river fishes.

Family Percidae

Perches

A family of predominantly freshwater fishes with approximately 244 species in 12 genera native to Europe, northern, West and Central Asia, and North America. Most species belong to the American subfamily Etheostominae (darters: approximately 160 species in four genera). Three genera are found in West Asia. Percidae is distinguished from Cichlidae by the presence of a single continuous lateral line (vs. lateral line interrupted, with the anterior part of the line situated at a higher level on the body than the posterior part), from Centrarchidae (a family of fish that was introduced to the region) by the presence of one or two anal spines (vs. 3–6), and from *Dicentrarchus* (Moronidae) by having 1–2 anal spines (vs. 3), and no auxiliary rows of lateral-line scales on the caudal (vs. presence). In the European part of the Black and Caspian Seas, *Sander volgensis* is a widely distributed and locally abundant species which has become invasive in Europe due to accidental stocking. It has not yet been found in the area covered by this book, but it cannot be excluded that it might be found in the future. *Sander marinus* is an additional percid found in this area. It has not been reported to enter pure freshwaters, inhabiting marine and brackish estuaries. It appears to be on the verge of extinction in the Black Sea basin, with the sole surviving population in Ukraine. In contrast, it is relatively common in the Kazakh Caspian basin, although it is rare along the coast of Azerbaijan and Iran, where it is only occasionally caught. *Sander marinus* and *S. volgensis* are included in the key to facilitate their identification. Species of *Sander* and *Perca* are commercially important food fish. *Sander lucioperca* has been introduced throughout West Asia, Northern Africa, and Europe to transform “low-value” fish stocks into high-value fisheries. As a result, it is widely introduced and responsible for the extinction or extirpation of endemic Leuciscidae and *Cobitis* species in Lakes Beyşehir and Eğirdir (Central Anatolia, Türkiye) and possibly also elsewhere.

Key to genera of Percidae in West Asia

1a - Two dorsals confluent.
..................*Gymnocephalus*

1b - Two distinctly separate dorsals.
..................2

2a - Pelvic and anal yellow to red; dark-grey or black blotch on posterior part of first dorsal; 5–8 bold bars on flank, usually y-shaped.
..................*Perca*

2b - Pelvic and anal whitish to brown; no dark-grey or black blotch on posterior part of first dorsal; flank with no bar or with 7–20 bars.
..................*Sander*

Gymnocephalus cernua; Danube, Hungary; ~70 mm SL.

https://doi.org/10.1515/9783111677811-046

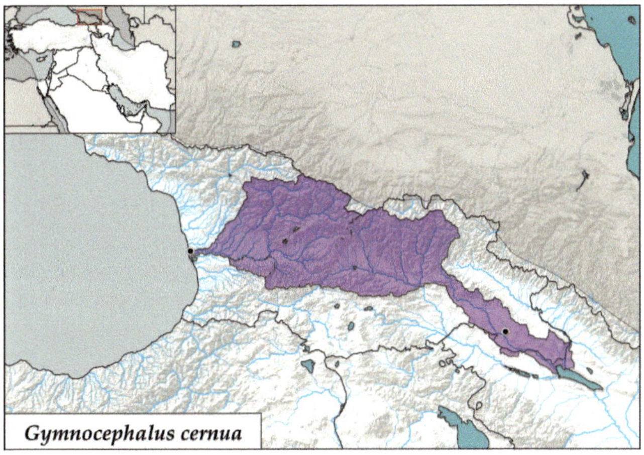

Gymnocephalus cernua

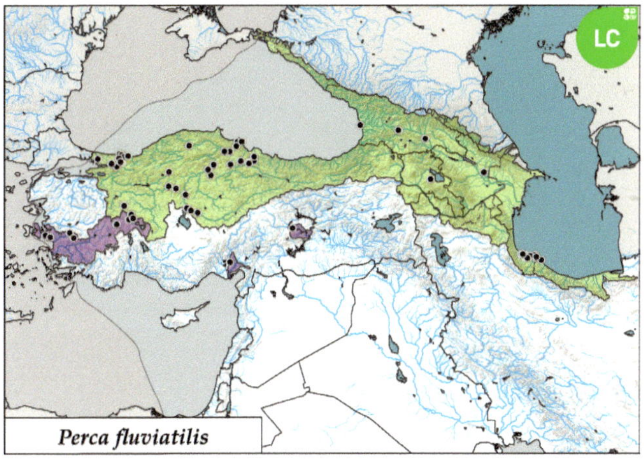

Perca fluviatilis

Gymnocephalus cernua

Common name. Ruffe.

Diagnosis. Distinguished from other species of Percidae in West Asia by: • two dorsals confluent. Size up to 200 mm SL.

Distribution. In West Asia non-native to upper Kura and Rioni (Georgia). In Europe, native to Trace in Türkiye and Caspian, Black, Baltic, and North Sea basins, north to about 69°N in Scandinavia. In Black sea basin south to Kuban.Also, in Aral basin, Arctic Ocean basin eastward to Kolyma. Introduced into France west of Rhine, northern Italy, northern Great Britain, Neretva (Croatia, Bosnia-Herzegovina), and Great Lakes region, North America and likely elsewhere.

Habitat. Eutrophic lakes, lowland and foothill streams. Prefers still or slow-flowing waters with soft bottoms and no vegetation. Most common in estuaries of large rivers, brackish lakes with salinities up to 10–12 ‰ and reservoirs. Generally increases in abundance with increasing eutrophication.

Biology. Females live up to 10 years, males up to 7. First spawn at 1–3 years, male a one year earlier than female. Spawns March–May until July in northern Europe on various substrates at depths of about 3 m or less. Spawns at temperatures above 6°C in north and above 10°C in south. Usually, a single female spawns with several males. Eggs become sticky in contact with water and adhere to rocks or plants. Females may spawn intermittently, laying eggs in two or more portions, usually separated by about 30 days in summer. Eggs from first batch are larger than those from second. Larvae have no or only a short pelagic larval stage, early transition to benthic life, secretive and solitary, do not form schools. Survival of larvae is poor below 10°C and above 20°C. Crepuscular or nocturnal. Well-developed cephalic lateral-line system and tapetum lucidum in the eyes make it an efficient predator at night and in turbid waters. Usually feeds on benthic chironomid larvae and amphipods, which can be detected in upper layers of substrate through sensory channels on head. Pelagic in coastal lakes and tidal estuaries, feeding on zooplankton and fish. When coexisting in deep lakes, *Perca fluviatilis* and *G. cernua* sometimes occur at different depths, with *G. cernua* being more abundant in deeper layers.

Conservation status. Non-native.

Further reading. Hölker & Thiel 1998 (biology); Craig 2000 (biology); Epitashvili et al. 2020 (records from Georgia).

Perca fluviatilis

Common name. Perch.

Diagnosis. Distinguished from other species of Percidae in West Asia by: • pelvic and anal yellow to red / • dark-grey or black blotch on posterior part of first dorsal / • 5–8, usually y-shaped bold dark-grey bars on flank / ○ two distinctly separate dorsals / ○ 56–77 total lateral-line scales. Size up to 600 mm SL, usually about 200 mm SL, females grow larger than males.

Distribution. Black Sea and western Caspian basins, south to Sefid (Iran). Increasingly stocked in many reservoirs outside its native range, particularly in Türkiye, and often illegally by anglers. Native throughout Europe to northernmost tip of Scandinavia, except Iberian Peninsula, central Italy, and Adriatic basin. In Aegean from Maritza to Aliakmon drainages. Native in Siberia, in rivers draining to Arctic Ocean east to Kolyma (replaced by *P. flavescens* in North America). Introduced in Aral basin (replaced by *P. schrenkii*), Ebro delta (Spain), central and southern Italy,

Perca fluviatilis; Odra drainage, Germany; ~160 mm SL.

Lake Skadar (Montenegro, Albania), Amur (Siberia), Australia, and South Africa (list not exhaustive).

Habitat. A wide range of habitats from estuarine lagoons, lakes, and rivers of all types to medium-sized streams.

Biology. Lives up to 21 years, usually to about 6 years. Males spawn first time at 1–2 years, females at 2–4 years. Spawns February–July, depending on latitude and altitude, when temperature reaches about 6°C. May undertake short spawning migrations. Females usually spawn with several males once a year. Female circles spawning site, followed by one male, while other males remain stationary. A string of eggs is released as female swims in a clockwise spiral, folding herself into a U-shape. All eggs are released and fertilised within approximately 5 seconds in a single strand, which becomes twisted and entangled in spawning substrate. Feeding larvae are positively phototactic, live in open water, and feed on pelagic organisms. Currents can widely disperse them. Opportunistic diurnal feeders, feeding mainly at sunrise and sunset on all available prey. Larvae and small juveniles usually feed on planktonic invertebrates. Many juveniles come close to shore during first summer to feed on benthic prey. Often becomes piscivorous at about 120 mm SL. Stocks with different life histories may co-occur in some lakes (littoral, benthic, pelagic zooplanktivorous), sometimes with varying spawning sites and times.

Conservation status. LC.

Further reading. Craig 2000 (biology); Svanbäck & Eklöv 2002 (morphology).

Key to species of *Sander* in Asia

1a - 14–17½ branched rays in second dorsal (Northern Black Sea and Caspian basins; marine).
..................*S. marinus*

1b - 18–22½ branched rays in second dorsal.
..................2

2a - Canine teeth absent; 70–83 total lateral-line scales.
..................*S. volgensis*

2b - 1–2 enlarged canine teeth in anterior part of each jaw; 80–97 total lateral-line scales.
..................*S. lucioperca*

Large reservoirs such as Hirfanlı at the Kızılırmak have often been stocked with *Sander lucioperca* and several other non-native species.

Sander lucioperca; Danube delta, Romania; ~150 mm SL.

Sander lucioperca; Danube, Germany; ~500 mm SL. A. Hartl.

Sander lucioperca

Common name. Pikeperch.

Diagnosis. Distinguished from other species of *Sander* in West Asia by: ○ 1–2 enlarged canine teeth in anterior part of each jaw / ○ 18–22½ branched rays in second dorsal / ○ 80–97 total lateral-line scales. Size up to 1000 mm SL.

Distribution. Native to Caspian, Baltic, Black, and Aral basins; Elbe (North Sea basin) and Maritza (Aegean basin) drainages. North to about 65°N in Finland. Introduced into Central Anatolia (Lakes Beyşehir and Eğirdir and Seyhan reservoirs), Iranian Tigris, Kor basin, Lake Urmia, and Namak basins, and possibly elsewhere. Introductions began in Great Britain in 1878, followed by Italy, Strymon drainage (Greece) and continental Europe west of Elbe, Ebro, Tagus, and Jucar drainages in Iberian Peninsula, Onega and Severnaya Dvina in White Sea basin and widely introduced in North Africa, Ob and Amur drainages (Asian Russia), Lake Issyk-Kul (Kyrgyzstan), Balkhash, and many smaller basins in central Kazakhstan.

Habitat. Large, turbid rivers and eutrophic lakes; brackish coastal lakes and estuaries. Often declines if turbidity is reduced.

Biology. Lives up to 17 years. Spawns first time at 3–10 years, usually at 4. Spawns April–May, exceptionally late February–July, depending on latitude and altitude, when temperatures reach 10–14°C (lowest temperature for egg incubation 11.5°C). May undertake short spawning migrations. Individuals foraging in brackish water migrate to freshwater habitats (migrations of up to 250 km have been recorded). Homing well developed, even neighboring populations may be relatively isolated. Male territorial digging shallow pit of about 50 cm in diameter and 5–10 cm deep in sand, gravel, or among exposed plant roots, on which eggs are laid, usually in turbid water and at 1–3 m depth. Spawns in pairs at dawn or night. Female remains above nest while male circles rapidly around it, about 1 m from nest. Male then assumes vertical position, and both swim rapidly around, releasing eggs and sperm. Female leaves nest after all eggs are released. Male defends nest and fans eggs with pectoral fins. Female spawn once a year. Feeding larvae positively phototactic feeding on pelagic organisms after leaving nest for open water. Piscivorous, feeding mainly on gregarious pelagic fish. Stomach contents from Lake Eğirdir revealed a high predation rate on native species, particularly *Anatolichcthys iconii*.

Conservation status. LC.

Remarks. Introduced in Lake Eğirdir from Austria in 1955 and was commercially fished between 1960 and 2000. Since then, stocks have declined sharply due to overfishing.

Further reading. Craig 2000 (biology); Lappalainen et al. 2003 (biology); Küçük et al. 2009b (introduction and effects in Eğirdir).

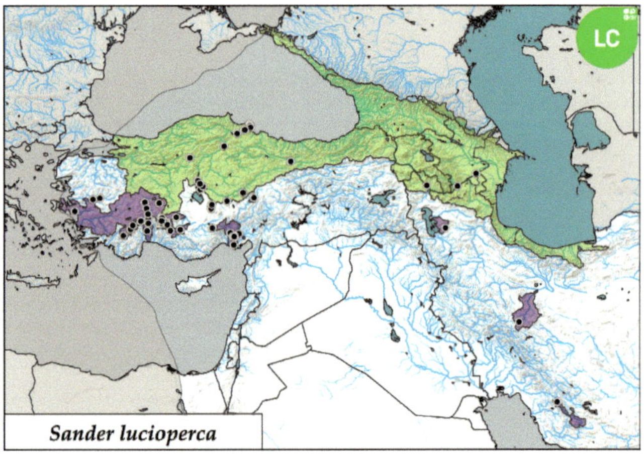

Pikeperch in Anatolian Lakes: what has been learned? The introduction of piscivorous fishes has been a common practice worldwide, aiming to increase fisheries' economic value. This phenomenon gained particular momentum following the Second World War. One of the most notable instances is the introduction of the Nile Perch *Lates* sp. to Lake Victoria (East Africa) in the mid-1950s. This has left a lasting impression on the collective memory as one of the most egregious instances of biomanipulation, with the Nile Perch serving as the primary catalyst for the decline and extinction of hundreds of native haplochromine cichlid fishes. However, similar applications have been undertaken almost concurrently in the history of West Asia. The piscivorous *Sander lucioperca* was introduced in several Western and Central Anatolia lakes, including Lake Eğirdir (in 1955) and Beyşehir (1978 and 1980), with the primary objective of enhancing fisheries. Within a decade of the introductions, most of the native species had disappeared from the catches in these lakes. However, their recovery began in the early 2000s, following the overfishing of pikeperch. Two endemic pelagic species did not survive and are now extinct: *Pseudophoxinus handlirschi* (from Eğirdir) and *Alburnus akili* (from Beyşehir). These are the first extinctions reported to be an immediate consequence of species introduction in West Asia. It is curious to note that *Sander* was introduced to Lake Beyşehir nearly fifteen years after its introduction to Eğirdir, despite the apparent decline of Eğirdir fish fauna. Today, even among the local scientific community members, there is a tendency to condemn early scientists for the disruptive impact of *Sander* introductions they stirred up. It is important to consider the context in which introductions were made. At the time, the value of local biodiversity was not yet recognised, and there was an opportunity to increase the commercial value of fisheries. It was in the early 1990s that it became apparent that the decline of species or their extinction was irreversible. However, it would appear that we have not learned from these cases. Indeed, the situation is becoming increasingly serious. There is no evidence of a reversal in the introduction of non-native invasive species. The massive wave of non-native species spread, and new introductions for commercial purposes or simply to release aquarium fish have only just begun. The invasion of *Perca fluviatilis* throughout Anatolia is a case in point, with other species, such as *Micropterus nigrescens*, likely to follow. However, the conservation of endemic and native fauna remains undervalued.

Lake Gölcük in Central Anatolia was the only habitat of *Anatolichthys splendens* before *Sander lucioperca* and others had been introduced.

Family Gasterosteidae

Sticklebacks

A small family comprising five genera and approximately 21 species, distributed across fresh and coastal waters in Eurasia and northern America. They are characterised by the presence of a series of spines in front of the soft dorsal, a single spine in the anal, a pair of pelvic spines (which are absent in many species), and in most species, several bony scutes on the flank and the caudal peduncle. The male constructs a nest of vegetal debris and algae in all species, anchored to the substrate or submerged vegetation. The nest is held together by a thread-like substance secreted by the kidney, which acts as a glue. The male guides the female into the nest, where the eggs are laid and fertilised. The male then guards the nest and subsequently the young. Sticklebacks represent significant species in the fields of behavioural and evolutionary biology. The number of species in *Gasterosteus* has been a topic of discussion. Previously, several populations from northern Eurasia and America had been placed in *Gasterosteus gymnurus*. However, this classification has not been accepted and is not supported by large-scale molecular data, suggesting that all Western Palaearctic *Gasterosteus* populations are closely related. **Further reading.** McLennan & Mattern 2001 (phylogeny); Mattern & McLennan 2004 (phylogeny).

Gasterosteus aculeatus

Common name. Threespine stickleback.
Diagnosis. Distinguished from all other freshwater fishes in West Asia by: • 3 spines in front of dorsal. Size up to 60 mm SL.
Distribution. Localised in Mediterranean basin: Dalaman (southern Aegean) south to Nahr Surit, a coastal river in Syria. Widespread in Black Sea and introduced in Caspian basin. Widespread in Europe, northern Asia, and America.
Habitat. Streams, lakes, and ponds with slow to moderate currents, clear water, and sandy bottoms. In Mediterranean basin, usually in spring-fed waters, but also found in other habitats (including brackish waters in Po delta). In Europe, often pelagic at sea.

Gasterosteus aculeatus; Lake Gölhisar basin, Türkiye; immature individual, 35 mm SL.

Gasterosteus aculeatus; Lake Gölhisar basin, Türkiye; nuptial male, ~50 mm SL.

https://doi.org/10.1515/9783111677811-047

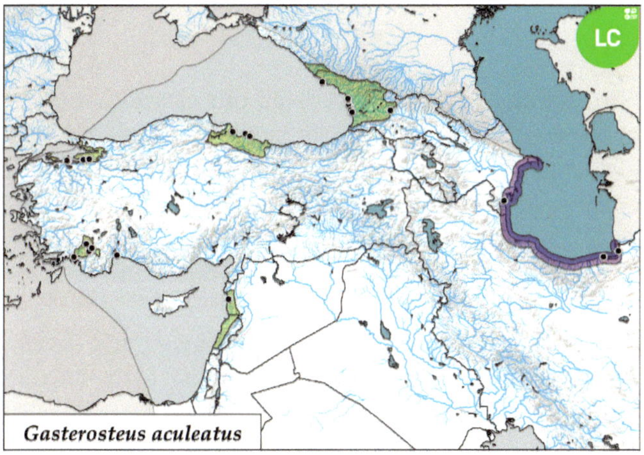

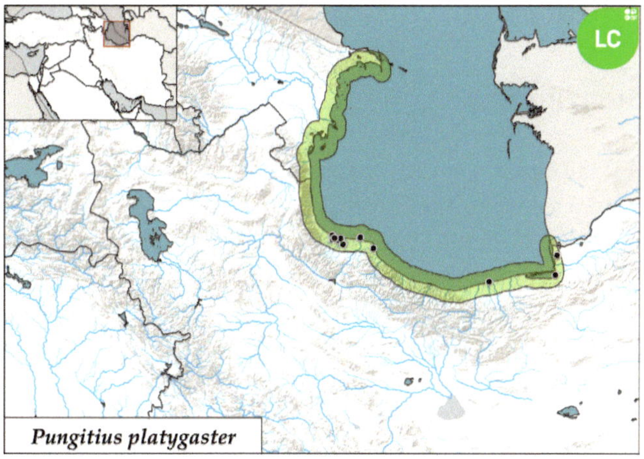

Biology. Only resident populations in West Asia, anadromous and resident in Europe. Spawns in April–May in Spain, May–July in Italy in springs, and earlier in brackish waters, at temperatures above 12°C. Spawns first time at 1 year. During spawning season, male develop a bright red belly, defend their territory against rivals, and build a nest on bottom in relatively shallow areas. They dig a depression into which they place plant material glued together with kidney secretions and covered with sand, leaving only a visible entrance. Several females are led into nest to spawn and are then chased away. Male guard and fan eggs to provide them with oxygenated water. Spawning behaviour very stereotyped. Eggs hatch in about 10 days, and young are guarded for a few days, after which male abandons nest. Outside spawning season, often pelagic at sea or in large lakes and reservoirs. Usually heavily infested with the tapeworm *Schistocephalus solidus*, especially in warm-water habitats. Feeds mainly on small invertebrates.
Conservation status. LC.
Further reading. Krupp & Coad 1985 (record from Syria); Gandolfi et al. 1991 (biology); Kottelat 1997 (systematics, nomenclature); Paepke 2002a (biology); Mäkinen et al. 2006 (genetics).

Pungitius platygaster; Danube delta, Romania; ~45 mm SL.

Pungitius platygaster
Common name. Ukrainian stickleback.
Diagnosis. Distinguished from *Gasterosteus aculeatus* in West Asia by: • 7–11 isolated spines in front of dorsal. Size up to 60 mm SL.
Distribution. In West Asia restricted to lower reaches of rivers in Azerbaijan and lower Sefid and associated wetlands in Iran. Native in lower rivers and wetlands draining to northern shores of Black, Caspian, and Aral basins. In Danube drainage as far upstream as Belgrade. Isolated populations in Axios and Aliakmon drainages (Greece). Also in upper Tobol system (Ob drainage), Lake Issyk-Kul basin (Kyrgyzstan), and Sarysu drainage (central Kazakhstan).
Habitat. Marshes and slow-flowing streams with dense vegetation, occasionally enter brackish water.
Biology. Spawns first time at 1 year in spring when temperature reaches 15°C. Territorial during spawning. Males build a nest of filamentous algae and submerged vegetation with a separate entrance and exit. Females lay about 90 eggs. Male guard nest until young swim away a few weeks later. Feeds on benthic invertebrates.
Conservation status. LC.
Further reading. Keivany & Nelson 2000 (systematics); Paepke 2002b (biology).

Family Centrarchidae

Sunfishes

A small family comprising approximately 50 species in eight genera native to North America, extending from Alaska to southern Mexico. In West Asia, they are distinguished from Cichlidae by the presence of a single continuous lateral line (vs. lateral line interrupted, with the anterior part of the line situated at a higher level on the body than the posterior part) and the absence of pharyngeal tooth plates (vs. presence). They are distinguished from Percidae by having 2–6 anal spines (vs. 1–2), and from *Dicentrarchus* (Moronidae) by the absence of auxiliary rows of lateral-line scales on the caudal (vs. presence), and from both by having only a single dorsal (vs. 2). Two species of Centrarchidae have established in West Asia. *Lepomis gibbosus* is the most widespread, while *Micropterus nigricans* has only been introduced to Cyprus. Additional species may be introduced in the future, and it is anticipated that *Micropterus* will be stocked illegally elsewhere. **Further reading.** Page & Burr 1991 (identification); Jenkins & Burkhead 1993 (identification); Kim et al. 2022 (*Micropterus* diversity).

Lepomis gibbosus

Common name. Pumpkinseed.

Diagnosis. Distinguished from *Micropterus nigricans* by: • flank blueish with many orange blotches and spots, usually with indistinct dark-blue bars / • head yellowish, orange, or brown with bright blue stripes or marmorated pattern / • opercular flap with red margin at tip and white margin at upper and lower edge / ○ 36–47 lateral-line scales on body / • no lateral stripe or series of blotches / • mouth small, upper jaw length about equal to eye diameter. Size up to 320 mm SL.

Distribution. Western and northern Anatolia, and Cyprus. Native to North America, in Great Lakes, St. Lawrence and upper Mississippi drainages, and Atlantic basin. Widely introduced in temperate waters in Europe since 1880s as a pond and aquarium fish. Now widely distributed throughout Europe, particularly in Mediterranean countries.

Habitat. Slow-flowing to still waters such as large rivers, lakes, ponds, canals, and backwaters. Avoid swift waters. In estuaries with salinity up to 18 ‰.

Biology. Lives up to 8 years. First spawns at 1–3 years. Spawns April–June when temperatures reach 16–18°C. Male fan and defend nests, which are shallow depressions in sand or gravel on shallow shores. Nests in colonies in sunny open habitats. Each male may spawn with several females in one nest and will guard nest until abandoned by larvae. Feeds on a wide variety of invertebrates.

Conservation status. Non-native; mostly spread as a weed with stocked fish for fisheries.

Further reading. Jenkins & Burkhead 1993 (biology).

Lepomis gibbosus; Oum er Rbia drainage, Morocco; ~100 mm SL.

https://doi.org/10.1515/9783111677811-048

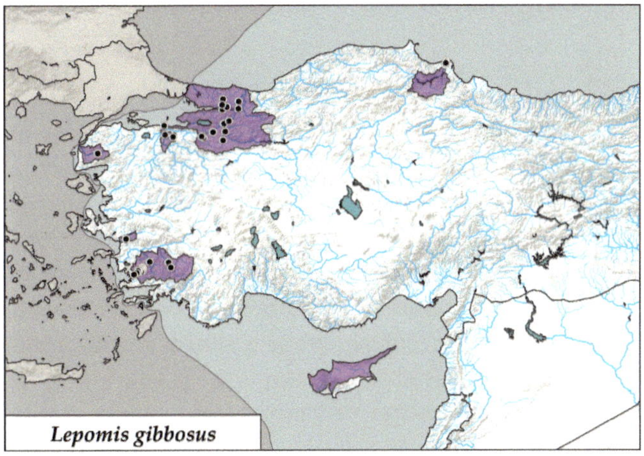

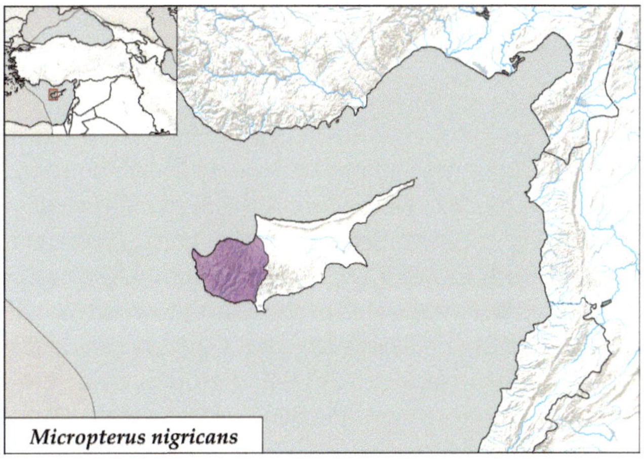

Micropterus nigricans; Lago di Nemi, Italy; ~150 mm SL.

Micropterus nigricans

Common name. Largemouth bass.

Diagnosis. Distinguished from *Lepomis gibbosus* by: • 55–68 lateral-line scales on body / • one prominent lateral stripe or series of closely set blotches / • mouth large, upper jaw length greater than eye diameter. Size up to 650 mm SL.

Distribution. Introduced and established in Cyprus. Native to North America, Great Lakes, St. Lawrence and Mississippi drainages, Gulf of Mexico, and southern Atlantic basins. Introduced to Europe in 1883 and widespread especially in Mediterranean basin.

Habitat. Slow-flowing to still waters such as large rivers, lakes, canals, and backwaters. Avoids fast-flowing waters. In estuaries with salinity up to 13 ‰.

Biology. Biology very well studied in North America, understudied in Europe and West Asia. Lives up to 16 years. First spawns at 1–4 years. Spawns April–June when temperatures reach 16–18°C. Male fan and defend nests, which are shallow depressions in sand or gravel bottoms. Male or both sexes guard nests and free-swimming larvae for a few days. Juveniles feed on invertebrates, taking more fish as they grow older. Adults piscivorous.

Conservation status. Non-native; introduced for recreational fisheries. Responsible for decline of several endemic fish species in Spain.

Remarks. The diversity of *Micropterus* in North America has been significantly underestimated, with 19 species currently recognised, 14 of which have been described. A review of diversity of *Micropterus* revealed that the largemouth bass introduced to North Africa, Europe, and Cyprus was not identified as *M. salmoides* but as *M. nigricans*, a species previously treated as a synonym.

Further reading. Jenkins & Burkhead 1993 (biology); Boschung & Mayden 2004 (biology); Kim et al. 2022 (*Micropterus* diversity).

Family Terapontidae

Grunters

The family comprises approximately 60 species in 16 genera, endemic to the Indo-Pacific region. Grunters and sunfishes (see above) are classified within the order Centrarchiformes, which includes a diverse range of marine perch-like fish. Other freshwater perches in this order include the Sinipercidae and Percichthyidae. Grunters can be distinguished from superficially similar cichlids by a single continuous lateral line on the flank (vs. two lateral lines). Most grunters are marine, although some enter freshwater habitats for feeding. A few Australian species are freshwater residents. The name "grunter" is derived from the sound these fish produce, which is audible to humans. One species is known to enter freshwaters in West Asia regularly. **Further reading.** Nelson 2006 (diversity).

Terapon jarbua

Common name. Tiger perch.

Diagnosis. Only species of family entering freshwaters in West Asia. Distinguished from all perch-like fishes in coastal freshwaters in West Asia by: • body silvery white with 3–4 black, curved stripes from nape to posterior back, lowermost stripe continuing across middle of caudal / • spinous part of dorsal with a blackish blotch on membranes between 3rd and 6th spines. Size up to 280 mm SL.

Distribution. Persian Gulf and coasts of Arabian Peninsula. Invasive in Mediterranean, where it entered via Suez Canal (Lessepsian migration). Not yet recorded from freshwaters in Mediterranean. Indo-Pacific from East African coast to South Africa, east to Samoa, north to southern Japan, south to Arafura Sea, Australia, and Lord Howe Island.

Habitat. Marine, postlarvae, and juveniles inhabit coastal habitats such as lower parts of rivers, lagoons, intertidal pools, and other coastal habitats. Inhabits soft bottoms with sand, silt, or organic material. Euryhaline, with a salinity tolerance of 0–39 ‰, usually found at depths down to 20 m but recorded down to 350 m. Catadromous, spawns at sea, juveniles migrate to coast or freshwater areas.

Biology. Protogynous, all juveniles are female, and sex reversal to male begins in many individuals when larger than 90 mm SL. Lives up to 10 years. Matures at about 120 mm SL. Fractional spawner, usually spawns throughout year. Eggs semipelagic. Larval stage about 25 days (Socotra). Postlarvae and juveniles enter brackish and freshwater habitats and may remain for two years, almost to maturity.

Terapon jarbua; Indonesia; ~80 mm SL. © B. MiggeArchiv Aqualog.

https://doi.org/10.1515/9783111677811-049

Adults only in sea. Juveniles usually live in schools, adults in loose groups or solitary. Juveniles feed mainly on invertebrates. Adults take invertebrates and small fish but feed mainly on scales removed from live fish. So far, only species known to be a specialised scale-eater in marine environments. Scales are only removed from live fish larger than predators, and slow-swimming species are preferred. Always attacks lateral surface of prey species, and scales are mainly removed from posterior region of body, particularly around caudal peduncle. A variety of species are attacked. Black longitudinal stripes fade during feeding but darken in presence of conspecifics, perhaps to prevent autopredation.

Remarks. Molecular data suggest that more than one species is involved, and populations from India to the east may belong to different species.

Conservation status. LC.

Further reading. Whitefield & Blaber 1978 (scale eating); Golani & Appelbaum-Golani 2010 (Mediterranean records). Lavergne 2012 (biology, genetic diversity); Gupta & Banerjee 2016 (review of biology); Freyhof et al. 2020 (Arabian Peninsula).

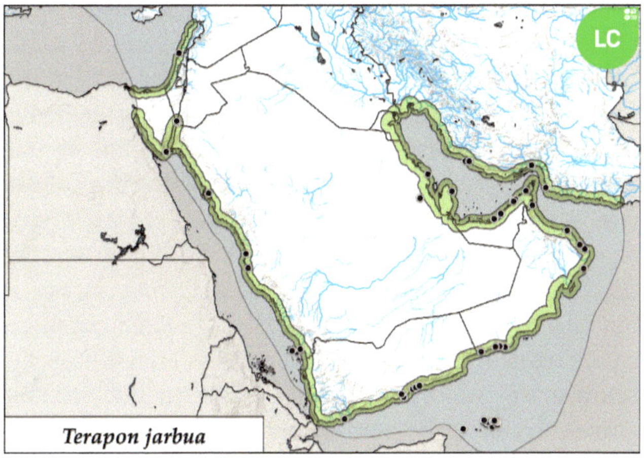

Fish species used in aquaculture and the aquarium hobby, as this South American *Piaractus brachypomus*, are increasingly released in open waters in West Asia. As long as they are not stocked regularly or have build self-sustaining populations, they are not covered by this book.

Family Moronidae

Temperate basses

The family comprises two genera, *Morone,* with four species in North America and *Dicentrarchus,* with two species in the Eastern Atlantic, Black Sea, and Mediterranean. They are distinguished by having two narrowly separated dorsals, two opercular spines, a lateral line extending almost to the posterior margin of the caudal, and auxiliary rows of lateral-line scales on the caudal above and below the main row. The two *Dicentrarchus* species enter brackish waters. *Dicentrarchus labrax* is the only species of the genus known to inhabit freshwater habitats. In contrast, *D. punctatus* is known to enter brackish waters only occasionally on the Atlantic coasts of France, the Iberian Peninsula, and the Mediterranean, and the western Black Sea. Hybrids between female *Morone saxatilis* and male *M. chrysops* are increasingly used in aquaculture in Israel and several European countries. There are many records of individuals who have escaped from captivity. Hybrids are fertile but produce only a few offspring. As aquaculture of these fishes expands, the number of records will increase, potentially leading to the establishment of local populations. This voracious predator exerts strong effects on fish communities. *Morone* hybrids can be distinguished from *Dicentrarchus* species by having small serrations along the lower part of the preoperculum, directed downward (vs. forward) and 5–7 lateral stripes (vs. body plain or spotted). *Dicentrarchus* can be distinguished from superficially similar *Sander* species and (introduced) *Micropterus* by the presence of spines on the lower part of the preoperculum (vs. preoperculum smooth), the presence of auxiliary rows of lateral-line scales on the caudal (vs. absence), the concave posterior margin of the anal (vs. convex in *Micropterus*), and the presence of three anal spines (vs. two in *Sander*).

Further reading. Johnson, in Moser et al. 1984 (phylogeny); Pickett & Pawson 1994 (biology); Müller-Belecke & Zienert 2006 (*Morone* aquaculture).

Dicentrarchus labrax; Adriatic Sea, Croatia; ~300 mm SL.

Dicentrarchus labrax

Common name. Sea bass.

Diagnosis. Distinguished from *D. punctatus* by: • 65–80 lateral-line scales on body (vs. 58–68) / • adults without small black spots on upper part of body (vs. with black spots) / • vomerine teeth in an anchor shape patch (crescentic band with a median posterior extension (vs. crescentic band only) / • scales in interorbital area cycloid (vs. ctenoid). Size up to 1000 mm SL.

Distribution. Mediterranean and Black Sea coasts. Coasts of Atlantic Morocco as far north as Norway, but absent from White, Barents, and Caspian Seas.

Habitat. Coastal waters and estuaries.

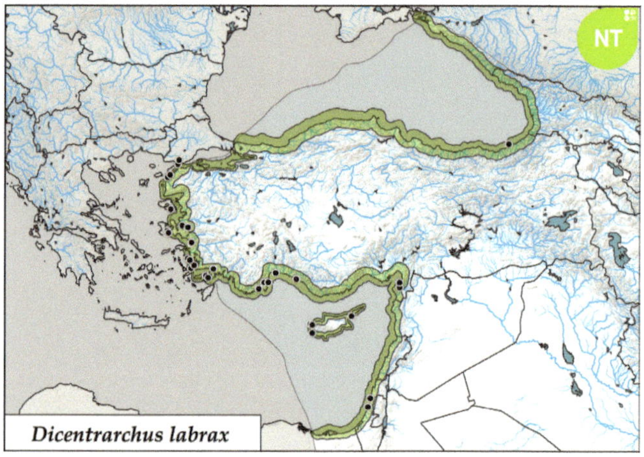

https://doi.org/10.1515/9783111677811-050

Biology. Lives up to 30 years. Spawns first time at 4–7 years and about 350 (male) and 420 (female) mm SL. Spawns pelagic in open sea, in January–June at temperatures above 9°C. Larvae planktonic. Juveniles move inshore as they grow, aggregating in brackish estuarine nursery areas where they usually remain until second summer. Large juveniles and adults exhibit a complex migratory pattern at sea, moving inshore and into estuarine freshwaters during summer to feed. Juveniles feed on invertebrates, taking more fish as they grow. Adults are piscivorous.

Conservation status. NT; numbers have declined over the last 20 years due to overfishing.

Further reading. Whitehead & Wheeler 1966 (systematics); Bauchot & Pras 1980 (diagnosis); Pickett & Pawson 1994 (biology).

Obsession on length-weight relationships and its ethical methodological pitfalls. Attend any fisheries conference or browse recent ichthyological journals and you will encounter length-weight relationship (LWR) studies. These seemingly simple allometric equations have become the default metric for fish biologists and are often treated as fundamental biological constants that define species or its ecology. However, this widespread interest masks serious ethical and scientific shortcomings that deserve critical examination. Beyond its technical limitations, such as ontogenetic allometry, seasonal condition shifts and gear selectivity biases (almost none of which are included in LWR studies), a worrying trend of 'data opportunism' has emerged. Under the noble justification of maximising the utility of fieldwork, researchers frequently do large, unplanned fish collections, vaguely promising that they might one day feed into LWR analyses. This opportunistic sampling contradicts the fundamental principles of responsible science. Without clear, hypothesis-driven frameworks, such collections are neither representative nor reproducible, introducing biases that make LWR models context-specific at best and misleading at worst. The ethical dimension is equally troubling. The unnecessary removal of living organisms, especially from vulnerable populations, conflicts with conservation objectives. When sample sizes are determined by the belief that 'more is better' rather than by statistical power calculations linked to specific research questions, there is a risk of causing harm without achieving a significant increase in our knowledge. This is of particular concern in West Asian freshwater ecosystems, where many species are under increasing pressure from human activities. Most critically, this obsession creates a false sense of biological understanding. Knowing that a fish weighs X grams at Y centimetres tells us almost nothing about its ecological role, evolutionary history or conservation needs. The taxonomic community has also fallen victim to this trend, simply to inflate their publication number. This is driven by academic pressure to prioritise quantity over quality, resulting in shallow insights that contribute little to our understanding of species taxonomy or ecology. Moving forward, fish sampling protocols should be based on well-defined hypotheses and power analyses should be used to determine the minimum sample size required to achieve sufficient statistical power. Alternative non-lethal condition indices and in situ photographic measurements should be adopted more widely to ensure that LWR work is both methodologically robust and ethically defensible.

Family Sparidae

Porgies

Porgies are a family of approximately 164 species, most of which are marine. Only a few species occasionally enter brackish and freshwater habitats, with the only regular freshwater inhabitant in West Asia being *Acanthopagrus arabicus*. Porgies can be distinguished from other fish of similar appearance belonging to the families Percidae, Cichlidae, Centrarchidae, and Moronidae by the presence of a very long and strong second anal spine and a pointed, strongly compressed snout. Many porgies are hermaphrodites, with male and female gonads developing simultaneously, which allows them to change sex from male to female (protandry) or from female to male (protogyny). *Acanthopagus arabicus* was previously identified as *A. latus* but has since been sorted into five species, two of which occur in the area: *A. arabicus* and *A. sheim*. **Further reading.** Iwatsuki 2013 (identification); Nelson et al. 2016 (diversity).

Marine fishes in the Shatt al Arab/Arvand. The spatial and temporal position of the brackish water zone between the freshwater outflow of a river and the marine water of the sea depends upon the quantity of outflowing waters and the size of the river. In seasons characterised by high flow, the brackish zone may be situated a considerable distance from the coast, and it is possible to catch freshwater fishes offshore in the sea. In seasons with low flow, tidal waters may strongly influence the lower part of large rivers, allowing marine fish to swim far upriver in the salty waters. The Shatt al Arab is formed by the confluence of the Euphrates and the Tigris, approximately 200 km before the estuary. The altitude at the confluence is only 4 m, and the Shatt al Arab has a tidal range of approximately 1.8 m at Basra in Iraq. At Basra, the river is already 8–15 m deep and up to 800 m wide. While the Karun, which originates from the Zagros Mountains in Iran, could potentially contribute a significant volume of freshwater to the Shatt al Arab, the presence of dams has resulted in a reduction in the freshwater inflow, not only from the Karun but also from the Euphrates and Tigris, which are themselves obstructed by numerous dams.

The fishes of the Shatt al Arab are subjected to considerable fluctuations in salinity, ranging from almost pure freshwaters to nearly pure marine conditions. Several marine species periodically enter the Shatt al Arab, with most of these also entering the Hammar Marsh. These include the halfbeaks *Hemirhamphus marginatus* and *Rhynchorhamphus georgii*, the belonid *Strongylura strongylurus* and the platycephalid *Platycephalus indicus*. The following fish species have been observed entering the Shatt al Arab and Hammar Marsh: *Sillago sihama* (silver sillago), *Acanthopagrus arabicus* and *Sparidentex hasta*, *Johnius belangerii* and *Otolithes ruber* (drums), *Bathygobius fuscus* (gobies), *Pampus argenteus* (stromateid) and *Brachiurus orientalis* (flatfish). In total, more than 100 species are excluded from the coverage of this book. They are found in the Shatt al Arab but do not enter freshwaters regularly. It is impossible to define a strict border between fish that enter freshwater periodically and those that do so occasionally, as they are facultative freshwater inhabitants. For instance, the scathophagid *Scatophagus argus* was observed once in the Euphrates near Al Kaba'ish, and juvenile individuals of this species are well documented as entering freshwaters in other regions of its distribution. Different species regularly enter freshwaters, such as *Terapon jarbua*, but this may be a regional phenomenon.

Acanthopagrus arabicus

Common name. Arabian yellow seabream.

Diagnosis. Distinguished from *A. sheim*, a marine species in Persian Gulf, by: ● black blotches on inter-radial membranes between dorsal rays absent (vs. present) / ● no black streaks near base of anal inter-radial membranes between yellow anal rays (vs. present). Size up to 220 mm SL.

Distribution. Persian Gulf to Duqum (Oman) to Trivandrum in southwest India. In lower Euphrates and Tigris drainages, recorded from Shatt al Arab/Arvand, Shatt al Basra Canal, Hammar and Horalazim marshes, and lower parts of Karun. Also enters lower parts of coastal rivers in Persian Gulf basin, not reported from wadis along Arabian Peninsula.

https://doi.org/10.1515/9783111677811-051

Acanthopagrus arabicus; Shur drainage, Iran; ~150 mm SL.

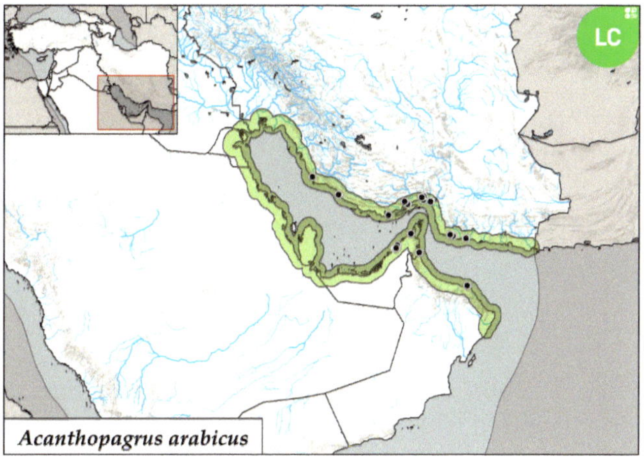

Habitat. Coastal lagoons and lower parts of rivers. On sand and rocky bottoms in sea to about 50 m depth. Often feeds on mudflats. Spawns at sea.

Biology. Lives up to 14 years. Juveniles enter brackish or freshwaters, and fish in fresh and brackish waters are usually less than 145 mm SL. Large parts of the population remain at sea. Adults migrate from Shatt al Arab from January–March to spawn. Spawns January–April with a peak in February–March. Protandrous hermaphrodite with males dominating smaller and females dominating larger size groups. Feeds mainly on bivalves and crabs, less commonly on other benthic invertebrates, plants, and fish.

Conservation status. LC; important commercial species.

Further reading. Coad 2010a (biology in Iraq).

Hermaphrodite fishes. In contrast to mammals, where chromosomes determine sex at fertilisation, many fishes are born with their gender undetermined. Their gender is often later determined by age, ecological, and/or social factors. Protandrous hermaphroditism is a reproductive system where individuals mature as males but may reproduce as females later in life. The opposite is protogynous hermaphroditism, where individuals mature as females but reproduce as males later. A species reproducing as both sexes over its entire reproductive lifespan is a lifelong simultaneous hermaphrodite. However, some species start as simultaneous hermaphrodites but later reproduce as pure females. There is no clear boundary between these systems, and a great deal of diversity is observed, with considerable individual variation.

Protandrous hermaphroditism encompasses a continuum of life histories, distinguished by the degree of temporal overlap between male and female reproduction. Protandrous sequential hermaphroditism: Reproduce early as a pure male and later as a pure female. Protandrous hermaphroditism with overlap: (i) Reproduce early as a pure male and later as a pure female, with an intervening period of reproduction as both sexes. (ii) Protandrous simultaneous hermaphroditism: Reproduce early as a pure male and later as both sexes.

Acanthopagrus, like other Sparidae, possesses ovotestes, which comprise paired bisexual gonads consisting of a mediodorsal ovarian zone and a latero-ventral testicular zone, separated by a wall of connective tissue. The ovotestes of male *A. latus*, a species absent from West Asia, develop from gonads containing substantial amounts of testicular and ovarian tissue. The testicular component of the ovotestes of all males undergoes a marked regression after spawning. During the next spawning season, the ovotestes either become gonads in which the testicular zone again predominates and contains spermatids and spermatozoa (functional males) or gonads in which the ovarian zone now predominates and contains mature oocytes (functional females). Once a fish has become a functional female, it remains a female throughout the rest of its life.

The evolution of hermaphroditism is linked to the potential for reproductive success to change with the size and age of fishes (budget effects) and their position in social hierarchies. Reproductive success in fishes is often determined by investment into gametes. If a fish is small, it might optimise its reproductive success by producing sperm, egg production shows more linear returns and larger individuals should invest more heavily in egg production. However, the situation may be reversed in species that hold territories and guard eggs. **Further reading.** Hesp et al. 2004 (*Acanthopagrus latus*); Collin 2013 (sexual systems in molluscs); Leonard 2013 (budget effects); Henshaw 2018 (protandrous hermaphroditism).

Taxonomic arrangements proposed in this book. In preparing this book, we encountered numerous taxonomic inconsistencies and nomenclatural problems requiring careful evaluation and resolution. Based on comprehensive analyses combining published and unpublished molecular data, extensive morphological examinations of fresh material from diverse populations and a critical reassessment of existing literature, we propose several taxonomic changes (including synonymies and revalidations) that have not been formally published elsewhere. Multiple lines of evidence support these decisions. In some cases, we discovered misidentifications in previous literature where specimens had been incorrectly assigned to species. Others resulted from misinterpretations of historical taxonomic descriptions or the application of names to populations that were not appropriate. Our molecular analyses of both published datasets and newly generated sequences revealed genetic uniformity among some populations that had previously been considered distinct species. Most critically, our morphological examinations of fresh specimens from as many populations formed the basis for our understanding of morphological variation within and between species. We recognise that some of these taxonomic actions may require additional in-depth studies for full confirmation. However, based on the overall evidence and our detailed analyses, the arrangements presented in this book offer the most accurate taxonomic framework currently available. Until additional evidence emerges that contradicts these findings, the taxonomic changes proposed in this book should be viewed as working hypotheses that advance our understanding of the diversity of West Asian freshwater fish. We encourage future researchers to test these hypotheses with additional data and welcome constructive taxonomic dialogue.

The synonymies established in this book are as follows. The respective rationale behind each one can be found in the species texts.

Names in previous literature	Correct name
Acanthobrama microlepis	*Acanthobrama punctulata*
Alburnus attalus	*Alburnus derjugini*
Alburnus carinatus	*Alburnus derjugini*
Alburnus istanbulensis	*Alburnus derjugini*
Alburnus nicaeensis	*Alburnus derjugini*
Alosa fallax	*Alosa agone*
Anatolichthys chantrei	*Anatolichthys danfordii*
Atherina caspia	*Atherina pontica*

Names in previous literature	Correct name
Barbus bergi	*Barbus tauricus*
Barbus oligolepis	*Barbus tauricus*
Capoeta alborzensis	*Capoeta aculeata*
Capoeta gracilis	*Capoeta aculeata*
Capoeta macrolepis	*Capoeta aculeata*
Capoeta pyragyi	*Capoeta damascina*
Capoeta raghazensis	*Capoeta saadii*
Capoeta sevangi	*Capoeta capoeta*
Capoeta umbla	*Capoeta damascina*
Chondrostoma orientale	*Chondrostoma regium*
Chondrostoma smyrnae	*Chondrostoma turnai*
Clupeonella muhlisi	*Clupeonella cultiventris*
Cobitis osurgeticus	Name not available *(nomen nudum)*
Garra sauvagei	*Garra nanus*
Ladigesocypris irideus	*Ladigesocypris ghigii*
Ladigesocypris mermere	*Petroleuciscus smyrnaeus*
Leucos (genus)	*Rutilus* (Genus)
Oxynoemacheilus melenicus	*Oxynoemacheilus simavicus*
Oxynoemacheilus sakaryaensis	*Oxynoemacheilus simavicus*
Ponticola rhodioni	*Ponticola constructor*
Ponticola turani	*Ponticola rizensis*
Pseudophoxinus caralis	*Pseudophoxinus anatolicus*
Salmo danilewskii	*Salmo ischchan*
Salmo duhani	Name not available *(nomen nudum)*
Sarmarutilus rubilio	*Rutilus rubilio*
Squalius adanaensis	*Squalius cappadocicus*
Squalius recurvirostris	*Squalius pursakensis*
Squalius semae	*Squalius berak*
Squalius seyhanensis	*Squalius cappadocicus*
Sterletus colchicus	*Sterletus persicus*
Tariqilabeo iranicus	*Tariqilabeo macmahoni*
Tristramella magdalenae	*Tristramella simonis*
Vimba melanops	*Vimba vimba*
Vimba mirabilis	*Vimba vimba*
Vimba persa	*Vimba vimba*

Changes that are not supported by our data
Synonymy of *Garra smartae* with *Garra dunsirei*
Synonymy of *Garra sindhae* with *Garra dunsirei*
Synonymy of *Oxynoemacheilus pantheroides* with *Oxynoemacheilus insignis*
Synonymy of *Glyptothorax alidaeii* with *Glyptothorax silviae*
Synonymy of *Glyptothorax galaxias* with *Glyptothorax silviae*
Synonymy of *Glyptothorax hosseinpanahii* with *Glyptothorax silviae*
Synonymy of *Glyptothorax shapuri* with *Glyptothorax silviae*
Replacement of *Capoeta anamisensis*, *C. barroisi*, *C. erhani*, *C. mandica*, and *C. trutta* with *Paracapoeta*

Appendix

List of West Asian freshwater fishes with their nomenclatural author(s) and year of publication. The presence or absence of parentheses or squared brackets is not an inconsistency but obeys precise conditions of the International Code of Zoological Nomenclature. An asterisk (*) indicates species not native to West Asian waters.

> The International Code of Zoological Nomenclature does not require the citation of nomenclatural authors and is not justified and better avoided in non-taxonomic publications, even if only to make the text more user-friendly and to avoid confusion with cited references. For clarity, it is needed only in publications dealing with taxonomic and nomenclatural topics in which the nomenclatural authors are absolutely needed. In others, its just an unneeded source of mistakes.

Family **Petromyzontidae**
Caspiomyzon wagneri (Kessler, 1870)
Lampetra lanceolata Kux & Steiner, 1972
Lampetra ninae (Naseka, Tuniyev & Renaud, 2009)
Petromyzon marinus Linnaeus, 1758

Family **Carcharhinidae**
Carcharhinus leucas (Müller & Henle, 1839)

Family **Acipenseridae**
**Sterletus baerii* (Brandt, 1869)
Sterletus gueldenstaedtii (Brandt & Ratzeburg, 1833)
Sterletus huso (Linnaeus, 1758)
Sterletus nudiventris (Lovetsky, 1828)
Sterletus persicus (Borodin, 1897)
Sterletus stellatus (Pallas, 1771)
Acipenser sturio Linnaeus, 1758

Family **Anguillidae**
Anguilla anguilla (Linnaeus, 1758)
Anguilla bengalensis (Gray, 1831)
Anguilla bicolor McClelland, 1844

Family **Alosidae**
Alosa caspia (Eichwald, 1838)
Alosa agone (Scopoli, 1786)
Alosa immaculata Bennett, 1835
Alosa kessleri (Grimm, 1887)
Alosa maeotica (Grimm, 1901)
Alosa tanaica (Grimm, 1901)

Family **Dorosomatidae**
Tenualosa ilisha (Hamilton, 1822)

Family **Ehiravidae**
Clupeonella cultriventris (Nordmann, 1840)

Family **Engraulidae**
Thrissina mystica Hata, Lavoué, Psomadakis, Osmany & Motomura, 2025
Thrissina whiteheadi Wongratana, 1983

Family **Chanidae**
Chanos chanos (Forsskål, 1775)

Family **Acheilognathidae**
Rhodeus amarus (Bloch, 1782)
Rhodeus caspius Esmaeili, Sayyadzadeh, Japoshvili, Eagderi, Abbasi & Mousavi-Sabet, 2020
Rhodeus colchicus Bogutskaya & Komlev, 2001

Family **Cyprinidae**
Arabibarbus arabicus (Valenciennes, in Cuvier & Valenciennes, 1842)
Arabibarbus grypus (Heckel, 1843)
Arabibarbus hadhrami Borkenhagen, 2014
Bangana dero (Hamilton, 1822)
Barbus anatolicus Turan, Kaya, Geiger & Freyhof, 2018
Barbus cyri De Filippi, 1865
Barbus escherichii Steindachner, 1897
Barbus ida Güçlü, Kalaycı, Özuluğ, Küçük & Turan, 2021
Barbus karunensis Khaefi, Esmaeili, Geiger & Eagderi, 2017
Barbus lacerta Heckel, 1843
Barbus miliaris De Filippi, 1863
Barbus niluferensis Turan, Kottelat & Ekmekçi, 2009
Barbus pergamonensis Karaman, 1971
Barbus rionicus Kamensky, 1899
Barbus tauricus Kessler, 1877
Barbus xanthos Güçlü, Kalaycı, Küçük & Turan, 2020
Caecocypris basimi Banister & Bunni, 1980
Capoeta aculeata (Valenciennes, 1844)
Capoeta anamisensis Zareian, Esmaeili & Freyhof, 2016
Capoeta antalyensis (Battalgil, 1943)

https://doi.org/10.1515/9783111677811-052

Capoeta aydinensis Turan, Küçük, Kaya & Bektaş, 2017
Capoeta banarescui Turan, Kottelat, Ekmekçi & İmamoğlu, 2006
Capoeta barroisi Lortet, 1894
Capoeta bergamae Karaman, 1969
Capoeta buhsei Kessler, 1877
Capoeta caelestis Schöter, Özuluğ & Freyhof, 2009
Capoeta capoeta (Güldenstädt, 1773)
Capoeta coadi Alwan, Zareian & Esmaeili, 2016
Capoeta damascina (Valenciennes, 1842)
Capoeta erhani Turan, Kottelat & Ekmekçi, 2008
Capoeta ferdowsii Jouladeh-Roudbar, Eagderi, Murillo-Ramos, Ghanavi & Doadrio, 2017
Capoeta fusca Nikolskii, 1897
Capoeta heratensis (Keyserling, 1861)
Capoeta kaput Levin, Prokofiev & Roubenyan, 2019
Capoeta mandica Bianco & Bănărescu, 1982
Capoeta oguzelii Elp, Osmanoğlu, Kadak & Turan, 2018
Capoeta pestai (Pietschmann, 1933)
Capoeta razii Jouladeh-Roudbar, Eagderi, Ghanavi & Doadrio, 2017
Capoeta saadii (Heckel, 1847)
Capoeta shajariani Jouladeh-Roudbar, Eagderi, Murillo-Ramos, Ghanavi & Doadrio, 2017
Capoeta sieboldii (Steindachner, 1864)
Capoeta svanetica Roman, Afanasyev, Golub & Lietytska, 2022
Capoeta tinca (Heckel, 1843)
Capoeta trutta (Heckel, 1843)
Carasobarbus apoensis (Banister & Clarke, 1977)
Carasobarbus canis (Valenciennes, 1842)
Carasobarbus chantrei (Sauvage, 1882)
Carasobarbus doadrioi Jouladeh-Roudbar, Kaya, Vatandoust & Ghanavi, 2024
Carasobarbus exulatus (Banister & Clarke, 1977)
Carasobarbus hajhosseini Jouladeh-Roudbar, Kaya, Vatandoust & Ghanavi, 2024
Carasobarbus kosswigi (Ladiges, 1960)
Carasobarbus luteus (Heckel, 1843)
Carasobarbus saadatii Jouladeh-Roudbar, Kaya, Vatandoust & Ghanavi, 2024
Carasobarbus sublimus (Coad & Najafpour, 1997)
**Carassius auratus* (Linnaeus, 1758)
Carassius carassius (Linnaeus, 1758)
**Carassius gibelio* (Bloch, 1782)
**Carassius langsdorfii* Temminck & Schlegel, 1846
Cyprinion acinaces Banister & Clarke, 1977
Cyprinion kais Heckel, 1843
Cyprinion macrostomum Heckel, 1843
Cyprinion mhalense Alkahem & Behnke, 1983
Cyprinion microphthalmum (Day, 1880)
Cyprinion milesi (Day, 1880)
Cyprinion muscatense (Boulenger, 1888)
Cyprinion tenuiradius Heckel, 1847
Cyprinus carpio Linnaeus, 1758
Garra amirhosseini Esmaeili, Sayyadzadeh, Coad & Eagderi, 2016
Garra barreimiae Fowler & Steinitz, 1956
Garra buettikeri Krupp, 1983
Garra caudomaculata (Battalgil, 1942)
Garra cavernicola Freyhof, 2025
Garra culiciphaga (Pellegrin, 1927)
Garra dunsirei Banister, 1987
Garra elegans (Günther, 1868)
Garra festai (Tortonese, 1939)
Garra gallagheri Krupp, 1988
Garra ghorensis Krupp, 1982
Garra gymnothorax Berg, 1949
Garra hormuzensis Zamani-Faradonbe, Zhang & Keivany, 2021
Garra jordanica Hamidan, Geiger & Freyhof, 2014
Garra kemali (Hankó, 1925)
Garra klatti (Kosswig, 1950)
Garra lautior Banister, 1987
Garra longipinnis Banister & Clarke, 1977
Garra lorestanensis Mousavi-Sabet & Eagderi, 2016
Garra mamshuqa Krupp, 1983
Garra meymehensis Zamani-Faradonbe, Keivany, Dorafshan & Zhang, 2021
Garra nanus (Heckel, 1843)
Garra nudiventris (Berg, 1905)
Garra orontesi Bayçelebi, Kaya, Turan & Freyhof, 2021
Garra persica Berg, 1914
Garra rezai Mousavi-Sabet, Eagderi, Saemi-Komsari, Kaya & Freyhof, 2022
Garra roseae Mousavi-Sabet, Saemi-Komsari, Doadrio & Freyhof, 2019
Garra rossica (Nikolskii, 1900)
Garra rufa (Heckel, 1843)
Garra sahilia Krupp, 1983
Garra shamal Kirchner, Kruckenhauser, Pichler, Borkenhagen & Freyhof, 2020
Garra sharq Kirchner, Kruckenhauser, Pichler, Borkenhagen & Freyhof, 2020
Garra sindhae Lyon, Geiger & Freyhof, 2016
Garra smartae Krupp & Budd, 2009
Garra tashanensis Mousavi-Sabet, Vatandoust, Fatemi & Eagderi, 2016
Garra tiam Zamani-Faradonbe, Keivany, Dorafshan & Zhang, 2021
Garra tibanica Trewavas, 1941
Garra turcica Karaman, 1971

Garra typhlops (Bruun & Kaiser, 1944)
Garra variabilis (Heckel, 1843)
Garra widdowsoni (Trewavas, 1955)
Luciobarbus brachycephalus (Kessler, 1872)
Luciobarbus capito (Güldenstädt, 1773)
Luciobarbus conocephalus (Kessler, 1872)
Luciobarbus esocinus Heckel, 1843
Luciobarbus graecus (Steindachner, 1895)
Luciobarbus longiceps (Valenciennes, 1842)
Luciobarbus mursa (Güldenstädt, 1773)
Luciobarbus pectoralis (Heckel, 1843)
Luciobarbus schejch Heckel, 1843
Luciobarbus subquincunciatus (Günther, 1868)
Mesopotamichthys sharpeyi (Günther, 1874)
Schizocypris altidorsalis Bianco & Bănărescu, 1982
Schizothorax pelzami Kessler, 1870
Schizothorax zarudnyi (Nikolskii, 1897)
Tariqilabeo adiscus (Annandale, 1919)
Tariqilabeo macmahoni (Zugmayer, 1912)

Family **Danionidae**
Barilius mesopotamicus Berg, 1932
Cabdio occidentalis Jouladeh-Roudbar, Lalramliana, Vatandoust, Ghanavi & Freyhof, 2023

Family **Gobionidae**
Gobio artvinicus Turan, Japoshvili, Aksu & Bektaş, 2016
Gobio baliki Turan, Kaya, Bayçelebi, Aksu & Bektaş, 2017
Gobio caucasicus Kamensky, 1901
Gobio fahrettini Turan, Kaya, Bayçelebi, Aksu & Bektaş, 2018
Gobio gymnostethus Ladiges, 1960
Gobio hettitorum Ladiges, 1960
Gobio insuyanus Ladiges, 1960
Gobio intermedius Battalgil, 1944
Gobio kizilirmakensis Turan, Japoshvili, Aksu & Bektaş, 2016
Gobio kovatschevi Chichkoff, 1937
Gobio maeandricus Naseka, Erk'akan & Küçük, 2006
Gobio microlepidotus Battalgil, 1942
Gobio nigrescens (Keyserling, 1861)
Gobio sakaryaensis Turan, Ekmekçi, Luskova & Mendel, 2012
**Pseudorasbora parva* (Temminck & Schlegel, 1846)
Romanogobio macropterus (Kamensky, 1901)
Romanogobio persus (Günther, 1899)

Family **Leuciscidae**
Abramis brama (Linnaeus, 1758)
Acanthobrama centisquama Heckel, 1843
Acanthobrama hadiyahensis Coad, Alkahem & Behnke, 1983
Acanthobrama lissneri Tortonese, 1952
Acanthobrama marmid Heckel, 1843
Acanthobrama orontis Berg, 1949
Acanthobrama persidis (Coad, 1981)
Acanthobrama punctulata (Kessler, 1877)
Acanthobrama telavivensis Goren, Fishelson & Trewavas, 1973
Acanthobrama thisbeae Freyhof & Özuluğ, 2014
Acanthobrama tricolor (Lortet, 1883)
Acanthobrama urmiana (Günther, 1899)
Alburnoides coskuncelebii Turan, Kaya, Aksu, Bayçelebi & Bektaş, 2019
Alburnoides damghani Jouladeh-Roudbar, Eagderi, Esmaeili, Coad & Bogutskaya, 2016
Alburnoides diclensis Turan, Bektaş, Kaya & Bayçelebi, 2016
Alburnoides eichwaldii (De Filippi, 1863)
Alburnoides emineae Turan, Kaya, Ekmekçi & Doğan, 2014
Alburnoides fasciatus (Nordmann, 1840)
Alburnoides freyhofi Turan, Kaya, Bayçelebi, Bektaş & Ekmekçi, 2017
Alburnoides holciki Coad & Bogutskaya, 2012
Alburnoides idignensis Bogutskaya & Coad, 2009
Alburnoides kosswigi Turan, Kaya, Bayçelebi, Bektaş & Ekmekçi, 2017
Alburnoides kurui Turan, Kaya, Bayçelebi, Bektaş & Ekmekçi, 2017
Alburnoides manyasensis Turan, Ekmekçi, Kaya & Güçlü, 2013
Alburnoides namaki Bogutskaya & Coad, 2009
Alburnoides nicolausi Bogutskaya & Coad, 2009
Alburnoides petrubanarescui Bogutskaya & Coad, 2009
Alburnoides qanati Coad & Bogutskaya, 2009
Alburnoides samiii Mousavi-Sabet, Vatandoust & Doadrio, 2016
Alburnoides smyrnae Pellegrin, 1927
Alburnoides tabarestanensis Mousavi-Sabet, Anvarifar & Azizi, 2015
Alburnoides turani Kaya, 2020
Alburnoides tzanevi Chichkoff, 1933
Alburnoides velioglui Turan, Kaya, Ekmekci & Doğan, 2014
Alburnus adanensis Battalgazi, 1944
Alburnus akili Battalgil, 1942
Alburnus alburnus (Linnaeus, 1758)
Alburnus atropatenae Berg, 1925
Alburnus baliki Bogutskaya, Küçük & Ünlü, 2000
Alburnus caeruleus Heckel, 1843
Alburnus carianorum Freyhof, Kaya, Bayçelebi, Geiger & Turan, 2019
Alburnus chalcoides (Güldenstädt, 1772)
Alburnus derjugini Berg, 1923
Alburnus demiri Özuluğ & Freyhof, 2008
Alburnus doriae De Filippi, 1865
Alburnus escherichii Steindachner, 1897
Alburnus filippii Kessler, 1877
Alburnus goekhani Özuluğ, Geiger & Freyhof, 2018

Alburnus hohenackeri Kessler, 1877
Alburnus kotschyi Steindachner, 1863
Alburnus kurui (Bogutskaya, 1995)
Alburnus magnificus Freyhof & Turan, 2019
Alburnus orontis Sauvage, 1882
Alburnus qalilus Krupp, 1992
Alburnus sellal Heckel, 1843
Alburnus tarichi (Güldenstädt, 1814)
Alburnus timarensis Kuru, 1980
Alburnus ulanus (Günther, 1899)
Ballerus sapa (Pallas, 1814)
Blicca bjoerkna (Linnaeus, 1758)
Chondrostoma beysehirense Bogutskaya, 1997
Chondrostoma ceyhanense Küçük, Turan, Güçlü Mutlu & Çiftçi, 2017
Chondrostoma colchicum Derjugin, 1899
Chondrostoma cyri Kessler, 1877
Chondrostoma esmaeilii Eagderi, Jouladeh-Roudbar, Birecikligil, Çiçek & Coad, 2017
Chondrostoma holmwoodii (Boulenger, 1896)
Chondrostoma kinzelbachi Krupp, 1985
Chondrostoma meandrense Elvira, 1987
Chondrostoma nasus (Linnaeus, 1758)
Chondrostoma regium (Heckel, 1843)
Chondrostoma toros Küçük, Turan, Güçlü Mutlu & Çiftçi, 2017
Chondrostoma turnai Güçlü, Küçük, Turan, Çiftçi & Mutlu, 2018
Egirdira nigra (Kosswig & Geldiay, 1952)
Ladigesocypris ghigii (Gianferrari, 1927)
Leucalburnus satunini (Berg, 1910)
Leucaspius delineatus (Heckel, 1843)
Leuciscus aspius (Linnaeus, 1758)
Leuciscus vorax (Heckel, 1843)
Mirogrex hulensis Goren, Fishelson & Trewavas, 1973
Mirogrex terraesanctae (Steinitz, 1952)
Pelecus cultratus (Linnaeus, 1758)
Petroleuciscus borysthenicus (Kessler, 1859)
Petroleuciscus ninae Turan, Kalaycı, Kaya, Bektaş & Küçük, 2018
Petroleuciscus smyrnaeus (Boulenger, 1896)
Phoxinus abanticus Turan, Bayçelebi, Özuluğ, Gaygusuz & Aksu, 2023
Phoxinus colchicus Berg, 1910
Phoxinus kottelati Bayçelebi, 2025
Pseudophoxinus alii Küçük, 2007
Pseudophoxinus anatolicus (Hankó, 1925)
Pseudophoxinus antalyae Bogutskaya, 1992
Pseudophoxinus battalgilae Bogutskaya, 1997
Pseudophoxinus burduricus Küçük, Gülle, Güçlü, Çiftçi & Erdoğan, 2013
Pseudophoxinus cilicicus Saç, Özuluğ, Geiger & Freyhof, 2019
Pseudophoxinus crassus (Ladiges, 1960)
Pseudophoxinus drusensis (Pellegrin, 1933)
Pseudophoxinus elizavetae Bogutskaya, Küçük & Atalay, 2006
Pseudophoxinus evliyai Freyhof & Özuluğ, 2010
Pseudophoxinus fahrettini Freyhof & Özuluğ, 2010
Pseudophoxinus firati Bogutskaya, Küçük & Atalay, 2006
Pseudophoxinus handlirschi (Pietschmann, 1933)
Pseudophoxinus hasani Krupp, 1992
Pseudophoxinus hittitorum Freyhof & Özuluğ, 2010
Pseudophoxinus iconii Küçük, Gülle & Güçlü, 2016
Pseudophoxinus libani (Lortet, 1883)
Pseudophoxinus maeandri (Ladiges, 1960)
Pseudophoxinus maeandricus (Ladiges, 1960)
Pseudophoxinus mehmeti Ekmekçi, Atalay, Yoğurtçuoğlu, Turan & Küçük, 2015
Pseudophoxinus ninae Freyhof & Özuluğ, 2006
Pseudophoxinus syriacus (Lortet, 1883)
Pseudophoxinus turani Küçük & Güçlü, 2014
Pseudophoxinus zekayi Bogutskaya, Küçük & Atalay, 2006
Pseudophoxinus zeregi (Heckel, 1843)
Pseudophoxinus sp.
Rutilus atropatenus Derjavin, 1937
Rutilus frisii (Nordmann, 1840)
Rutilus lacustris (Pallas, 1814)
Rutilus rutilus (Linnaeus, 1758)
Rutilus sojuchbulagi Abdurakhmanov, 1950
Scardinius elmaliensis Bogutskaya, 1997
Scardinius erythrophthalmus (Linnaeus, 1758)
Squalius agdamicus Kamensky, 1901
Squalius anatolicus (Bogutskaya, 1997)
Squalius aristotelis Özuluğ & Freyhof, 2011
Squalius berak Heckel, 1843
Squalius cappadocicus Özuluğ & Freyhof, 2011
Squalius carinus Özuluğ & Freyhof, 2011
Squalius cii (Richardson, 1857)
Squalius fellowesii (Günther, 1868)
Squalius kosswigi (Karaman, 1972)
Squalius kottelati Turan, Yilmaz & Kaya, 2009
Squalius lepidus Heckel, 1843
Squalius namak Khaefi, Esmaeili, Sayyadzadeh, Geiger & Freyhof, 2016
Squalius orientalis Heckel, 1847
Squalius pursakensis (Hankó, 1925)
Squalius turcicus De Filippi, 1865
Squalius verepi Turan, 2022
Turcichondrostoma fahirae (Ladiges, 1960)
Vimba vimba (Linnaeus, 1758)

Family **Tincidae**
Tinca tinca (Linnaeus, 1758)

Family **Xenocyprididae**
**Ctenopharyngodon idella* (Valenciennes, 1844)
**Hemiculter leucisculus* (Basilewsky, 1855)
**Hypophthalmichthys molitrix* (Valenciennes, 1844)
**Hypophthalmichthys nobilis* (Richardson, 1845)
**Mylopharyngodon piceus* (Richardson, 1846)

Family **Cobitidae**
Cobitis afifeae Freyhof, Bayçelebi & Geiger, 2018
Cobitis aliyeae Freyhof, Bayçelebi & Geiger, 2018
Cobitis amphilekta Vasil'eva & Vasil'ev, 2012
Cobitis anabelae Freyhof, Bayçelebi & Geiger, 2018
Cobitis avicennae Mousavi-Sabet, Vatandoust, Esmaeili, Geiger & Freyhof, 2015
Cobitis battalgilae Băcescu, 1962
Cobitis bilseli Battalgil, 1942
Cobitis derzhavini Vasil'eva, Solovyeva, Levin & Vasil'ev, 2020
Cobitis dorademiri Erk'akan, Özdemir & Özeren, 2017
Cobitis elazigensis Coad & Sarieyyüpoğlu, 1988
Cobitis emrei Freyhof, Bayçelebi & Geiger, 2018
Cobitis erkakanae Freyhof, Bayçelebi & Geiger, 2018
Cobitis evreni Erk'akan, Özeren & Nalbant, 2008
Cobitis fahireae Erk'akan, Atalay-Ekmekçi & Nalbant, 1998
Cobitis faridpaki Mousavi-Sabet, Vasil'eva, Vatandoust & Vasil'ev, 2011
Cobitis indus Eagderi, Seçer & Freyhof, 2022
Cobitis joergbohleni Freyhof, Bayçelebi & Geiger, 2018
Cobitis kellei Erk'akan, Atalay-Ekmekçi & Nalbant, 1998
Cobitis levantina Krupp & Moubayed, 1992
Cobitis linea (Heckel, 1847)
Cobitis phrygica Battalgazi, 1944
Cobitis pirii Freyhof, Bayçelebi & Geiger, 2018
Cobitis puncticulata Erk'akan, Atalay-Ekmekçi & Nalbant, 1998
Cobitis saniae Eagderi, Jouladeh-Roudbar, Jalili, Sayyadzadeh & Esmaeili, 2017
Cobitis satunini Gladkov, 1935
Cobitis simplicispina Hankó, 1925
Cobitis sipahilerae Erk'akan, Özdemir & Özeren, 2017
Cobitis splendens Erk'akan, Atalay-Ekmekçi & Nalbant, 1998
Cobitis strumicae Karaman, 1955
Cobitis troasensis Freyhof, Bayçelebi & Geiger, 2018
Cobitis turcica Hankó, 1925
Sabanejewia aurata (De Filippi, 1863)
Sabanejewia caspia (Eichwald, 1838)

Family **Nemacheilidae**
Eidinemacheilus proudlovei Freyhof, Abdullah, Ararat, Ibrahim & Geiger, 2016
Eidinemacheilus smithi (Greenwood, 1976)
Oxynoemacheilus amanos Kaya, Yoğurtçuoğlu & Freyhof, 2021
Oxynoemacheilus anatolicus Erk'akan, Özeren & Nalbant, 2008
Oxynoemacheilus angorae (Steindachner, 1897)
Oxynoemacheilus araxensis (Bănărescu & Nalbant, 1978)
Oxynoemacheilus argyrogramma (Heckel, 1847)
Oxynoemacheilus arsaniasus Freyhof, Kaya, Turan & Geiger, 2019
Oxynoemacheilus axylos Yoğurtçuoğlu, Kaya & Freyhof, 2022
Oxynoemacheilus banarescui (Delmastro, 1982)
Oxynoemacheilus bergi (Gratzianov, 1907)
Oxynoemacheilus bergianus (Derjavin, 1934)
Oxynoemacheilus brandtii (Kessler, 1877)
Oxynoemacheilus cemali Turan, Kaya, Kalayci, Bayçelebi & Aksu, 2019
Oxynoemacheilus ceyhanensis (Erk'akan, Nalbant & Özeren, 2007)
Oxynoemacheilus chaboras Kaya, Kurtul, Aksu, Oral & Freyhof, 2024
Oxynoemacheilus chomanicus Kamangar, Prokofiev, Ghaderi & Nalbant, 2014
Oxynoemacheilus ciceki Sungur, Jalili & Eagderi, 2018
Oxynoemacheilus cilicicus Kaya, Turan, Bayçelebi, Kalayci & Freyhof, 2020
Oxynoemacheilus cyri (Berg, 1910)
Oxynoemacheilus eliasi Yoğurtçuoğlu, Kaya & Freyhof, 2022
Oxynoemacheilus elsae Eagderi, Jalili & Çiçek, 2018
Oxynoemacheilus ercisianus (Erk'akan & Kuru, 1986)
Oxynoemacheilus eregliensis (Bănărescu & Nalbant, 1978)
Oxynoemacheilus euphraticus (Bănărescu & Nalbant, 1964)
Oxynoemacheilus evreni (Erk'akan, Nalbant & Özeren, 2007)
Oxynoemacheilus fatmae Turan, Aksu, Güçlü & Kalaycı, 2024
Oxynoemacheilus fatsaensis Saygun, Ağdamar & Özuluğ, 2021
Oxynoemacheilus frenatus (Heckel, 1843)
Oxynoemacheilus galilaeus (Günther, 1864)
Oxynoemacheilus germencicus (Erk'akan, Nalbant & Özeren, 2007)
Oxynoemacheilus gyndes Freyhof & Abdullah, 2017
Oxynoemacheilus hamwii (Krupp & Schneider, 1991)
Oxynoemacheilus hanae Freyhof & Abdullah, 2017
Oxynoemacheilus hazarensis Freyhof & Özuluğ, 2017
Oxynoemacheilus insignis (Heckel, 1843)
Oxynoemacheilus isauricus Yoğurtçuoğlu, Kaya, Özuluğ & Freyhof, 2021
Oxynoemacheilus karunensis Freyhof, 2016
Oxynoemacheilus kaynaki Erk'akan, Özeren & Nalbant, 2008

Oxynoemacheilus kentritensis Freyhof, Kaya & Turan, 2017
Oxynoemacheilus kiabii Golzarianpour, Abdoli & Freyhof, 2011
Oxynoemacheilus kottelati Turan, Aksu, Güçlü & Kalaycı, 2024
Oxynoemacheilus kurdistanicus Kamangar, Prokofiev, Ghaderi & Nalbant, 2014
Oxynoemacheilus leontinae (Lortet, 1883)
Oxynoemacheilus marmaraensis Turan, Bayçelebi & Kalaycı, 2023
Oxynoemacheilus marunensis Sayyadzadeh & Esmaeili, 2020
Oxynoemacheilus mediterraneus (Erk'akan, Nalbant & Özeren, 2007)
Oxynoemacheilus muefiti Freyhof, Kaya, Turan & Geiger, 2019
Oxynoemacheilus namiri (Krupp & Schneider, 1991)
Oxynoemacheilus nasreddini Yoğurtçuoğlu, Kaya & Freyhof, 2021
Oxynoemacheilus panthera (Heckel, 1843)
Oxynoemacheilus pantheroides (Goren & Nalbant in Bănărescu, Nalbant & Goren, 1982)
Oxynoemacheilus paucilepis (Erk'akan, Nalbant & Özeren, 2007)
Oxynoemacheilus persa (Heckel, 1847)
Oxynoemacheilus phasicus Freyhof, Kaya, Epitashvili & Geiger, 2021
Oxynoemacheilus sarali Turan, Kocabaş. & Aksu, 2025
Oxynoemacheilus sarus Freyhof, Yoğurtçuoğlu & Kaya, 2021
Oxynoemacheilus seyhanensis (Bănărescu, 1968)
Oxynoemacheilus seyhanicola (Erk'akan, Nalbant & Özeren, 2007)
Oxynoemacheilus shehabi Freyhof & Geiger, 2021
Oxynoemacheilus simavicus (Balik & Bănărescu, 1978)
Oxynoemacheilus theophilii Stoumboudi, Kottelat & Barbieri, 2006
Oxynoemacheilus tigris (Heckel, 1843)
Oxynoemacheilus tongiorgii (Nalbant & Bianco, 1998)
Oxynoemacheilus veyselorum Çiçek, Eagderi & Sungur, 2018
Oxynoemacheilus zagrosensis Kamangar, Prokofiev, Ghaderi & Nalbant, 2014
Oxynoemacheilus zarzianus Freyhof & Geiger, 2017
Paracobitis abrishamchianorum Mousavi-Sabet, Vatandoust, Geiger & Freyhof, 2019
Paracobitis atrakensis Esmaeili, Mousavi-Sabet, Sayyadzadeh, Vatandoust & Freyhof, 2014
Paracobitis basharensis Freyhof, Esmaeili, Sayyadzadeh & Geiger, 2014
Paracobitis hircanica Mousavi-Sabet, Sayyadzadeh, Esmaeili, Eagderi, Patimar & Freyhof, 2015
Paracobitis longicauda (Kessler, 1872)
Paracobitis malapterura (Valenciennes, 1846)
Paracobitis molavii Freyhof, Esmaeili, Sayyadzadeh & Geiger, 2014
Paracobitis persa Freyhof, Esmaeili, Sayyadzadeh & Geiger, 2014
Paracobitis rhadinaea (Regan, 1906)
Paracobitis salihae Kaya, Turan, Kalaycı, Bayçelebi & Freyhof, 2020
Paracobitis zabgawraensis Freyhof, Esmaeili, Sayyadzadeh & Geiger, 2014
Paraschistura abdolii Freyhof, Sayyadzadeh, Esmaeili & Geiger, 2015
Paraschistura alta Nalbant & Bianco, 1998
Paraschistura aredvii Freyhof, Sayyadzadeh, Esmaeili & Geiger, 2015
Paraschistura bampurensis (Nikolskii, 1900)
Paraschistura cristata (Berg, 1898)
Paraschistura delvarii Mousavi-Sabet & Eagderi, 2015
Paraschistura hormuzensis Freyhof, Sayyadzadeh, Esmaeili & Geiger, 2015
Paraschistura ilamensis Vatandoust & Eagderi, 2015
Paraschistura kermanensis Sayyadzadeh, Teimori & Esmaeili, 2019
Paraschistura kessleri (Günther, 1889)
Paraschistura lindbergi Bănărescu & Mirza, 1965
Paraschistura makranensis Eagderi, Mousavi-Sabet & Freyhof, 2019
Paraschistura naumanni Freyhof, Sayyadzadeh, Esmaeili & Geiger, 2015
Paraschistura nielseni (Nalbant & Bianco, 1998)
Paraschistura susiani Freyhof, Sayyadzadeh, Esmaeili & Geiger, 2015
Paraschistura turcmenica (Berg, 1932)
Sasanidus kermanshahensis (Bănărescu & Nalbant, 1966)
Schistura chrysicristinae Nalbant, 1998
Seminemacheilus ahmeti Sungur, Jalili, Eagderi & Çiçek, 2018
Seminemacheilus attalicus Yoğurtçuoğlu, Kaya, Geiger & Freyhof, 2020
Seminemacheilus ekmekciae Yoğurtçuoğlu, Kaya, Geiger & Freyhof, 2020
Seminemacheilus ispartensis Erk'Akan, Nalbant & Özeren, 2007
Seminemacheilus lendlii (Hankó, 1925)
Seminemacheilus tubae Yoğurtçuoğlu, Kaya, Geiger & Freyhof, 2020
Turcinoemacheilus ansari Jouladeh-Roudbar, Vatandoust, Doadrio & Ghanavi, 2023
Turcinoemacheilus bahaii Esmaeili, Sayyadzadeh, Özuluğ, Geiger & Freyhof, 2014
Turcinoemacheilus christofferi Jouladeh-Roudbar, Vatandoust, Doadrio & Ghanavi, 2023
Turcinoemacheilus ekmekciae Kaya, Yoğurtçuoğlu, Aksu, Bayçelebi & Turan, 2023

Turcinoemacheilus hafezi Golzarianpour, Abdoli, Patimar & Freyhof, 2013
Turcinoemacheilus inexpectatus Freyhof & Jouladeh-Roudbar, 2023
Turcinoemacheilus kosswigi Bănărescu & Nalbant, 1964
Turcinoemacheilus minimus Esmaeili, Sayyadzadeh, Özuluğ, Geiger & Freyhof, 2014
Turcinoemacheilus moghbeli Jouladeh-Roudbar, Vatandoust, Doadrio & Ghanavi, 2023
Turcinoemacheilus saadii Esmaeili, Sayyadzadeh, Özuluğ, Geiger & Freyhof, 2014

Family **Loricariidae**
Pterygoplichthys disjunctivus x P. pardalis (Castelnau, 1855)

Family **Siluridae**
Silurus glanis Linnaeus, 1758
Silurus triostegus Heckel, 1843

Family **Bagridae**
Mystus cyrusi Esmaeili, Sayyadzadeh, Zarei, Eagderi & Mousavi-Sabet, 2022
Mystus misrai Anuradha, 1986
Mystus pelusius (Solander, 1794)

Family **Sisoridae**
Glyptothorax alidaeii Mousavi-Sabet, Eagderi, Vatandoust & Freyhof, 2021
Glyptothorax armeniacus (Berg, 1918)
Glyptothorax cous (Linnaeus, 1766)
Glyptothorax daemon Freyhof, Kaya, Abdullah & Geiger, 2021
Glyptothorax galaxias Mousavi-Sabet, Eagderi, Vatandoust & Freyhof, 2021
Glyptothorax hosseinpanahii Mousavi-Sabet, Eagderi, Vatandoust & Freyhof, 2021
Glyptothorax kurdistanicus (Berg, 1931)
Glyptothorax pallens Mousavi-Sabet, Eagderi, Vatandoust & Freyhof, 2021
Glyptothorax sardashtensis Jokar, Kamangar, Ghaderi & Freyhof, 2023
Glyptothorax shapuri Mousavi-Sabet, Eagderi, Vatandoust & Freyhof, 2021
Glyptothorax silviae Coad, 1981
Glyptothorax steindachneri (Pietschmann, 1913)
Glyptothorax vatandousti Jouladeh-Roudbar, Ghanavi & Freyhof, 2023

Family **Clariidae**
Clarias gariepinus (Burchell, 1822)
**Clarias* sp.
**Heteropneustes fossilis* (Bloch, 1794)

Family **Ictaluridae**
**Ictalurus punctatus* (Rafinesque, 1818)

Family **Coregonidae**
**Coregonus albula* (Linnaeus, 1758)
**Coregonus sevanicus* Dadikyan, 1986

Family **Salmonidae**
**Oncorhynchus mykiss* (Walbaum, 1792)
Salmo abanticus Tortonese, 1954
Salmo aestivalis Fortunatov, 1926
Salmo araxensis Turan, Kottelat & Kaya, 2022
Salmo ardahanensis Turan, Kottelat & Kaya, 2022
Salmo baliki Turan, Aksu, Oral, Kaya & Bayçelebi, 2021
Salmo brunoi Turan, Bayçelebi, Aksu & Oral, 2024
Salmo caspius Kessler, 1877
Salmo chilo Turan, Kottelat & Engin, 2012
Salmo ekmekciae Küçük, Kalaycı, Güçlü, Oral & Turan, 2024
Salmo euphrataeus Turan, Kottelat & Engin, 2014
Salmo fahrettini Turan, Kalaycı, Bektaş, Kaya & Bayçelebi, 2020
Salmo gegarkuni Kessler, 1877
Salmo ischchan Kessler, 1877
Salmo kottelati Turan, Doğan, Kaya & Kanyılmaz, 2014
Salmo labecula Turan, Kottelat & Engin, 2012
Salmo labrax Pallas, 1814
Salmo munzuricus Turan, Kottelat & Kaya, 2017
Salmo murathani Turan, Kottelat & Kaya, 2022
Salmo okumusi Turan, Kottelat & Engin, 2014
Salmo opimus Turan, Kottelat & Engin, 2012
Salmo platycephalus Behnke, 1968
Salmo rizeensis Turan, Kottelat & Engin, 2009
Salmo tigridis Turan, Kottelat & Bektaş, 2011
**Salmo trutta* Linnaeus, 1758
Salmo sp.
**Salvelinus fontinalis* (Mitchill, 1814)

Family **Esocidae**
Esox lucius Linnaeus, 1758

Family **Lotidae**
Lota lota (Linnaeus, 1758)

Family **Oxudercidae**
Awaous jayakari (Boulenger, 1888)
Boleophthalmus dussumieri Valenciennes, 1837
Knipowitschia bergi (Iljin, 1928)
Knipowitschia byblisia Ahnelt, 2011
Knipowitschia caucasica (Berg, 1916)
Knipowitschia caunosi Ahnelt, 2011
Knipowitschia longecaudata (Kessler, 1877)

Knipowitschia mermere Ahnelt, 1995
Knipowitschia ricasolii (Di Caporiacco, 1935)
**Rhinogobius* sp.

Family **Gobiidae**
Babka gymnotrachelus (Kessler, 1857)
Glossogobius giuris (Hamilton, 1822)
Glossogobius laticeps (De Vis, 1884)
Glossogobius tenuiformis Fowler, 1934
Mesogobius batrachocephalus (Pallas, 1814)
Neogobius fluviatilis (Pallas, 1814)
Neogobius melanostomus (Pallas, 1814)
Ponticola alasanicus Epitashvili, Japoshvili & Mumladze, 2023
Ponticola constructor (Nordmann, 1840)
Ponticola cyrius (Kessler, 1874)
Ponticola gorlap (Iljin, 1949)
Ponticola hircaniaensis Zarei, Esmaeili, Kovacic & Schliewen, 2022
Ponticola iranicus Vasil'eva, Mousavi-Sabet & Vasil'ev, 2015
Ponticola patimari Eagderi, Nikmehr & Poorbagher, 2020
Ponticola rizensis (Kovačić & Engín, 2008)
Ponticola syrman (Nordmann, 1840)
Proterorhinus blennioides (Kessler, 1877)
Proterorhinus nasalis (De Filippi, 1863)
Proterorhinus semilunaris (Heckel, 1837)
Proterorhinus semipellucidus (Kessler, 1877)
Proterorhinus victori Vasil'eva, 2024

Family **Mugilidae**
Chelon auratus (Risso, 1810)
Chelon labrosus (Risso, 1827)
Chelon ramada (Risso, 1827)
Chelon saliens (Risso, 1810)
Mugil cephalus Linnaeus, 1758
Planiliza abu (Heckel, 1843)
**Planiliza haematocheilus* (Temminck & Schlegel, 1845)
Planiliza klunzingeri (Day, 1888)
Planiliza subviridis (Valenciennes, 1836)

Family **Cichlidae**
**Amatitlania nigrofasciata* (Günther, 1867)
**Aulonocara* sp.
**Chindongo elongatus* (Fryer, 1956)
Coptodon zillii (Gervais, 1848)
**Dimidiochromis compressiceps* (Boulenger, 1908)
Haplochromis flaviijosephi (Lortet, 1883)
Iranocichla hormuzensis Coad, 1982
Iranocichla persa Esmaeili, Sayyadzadeh & Seehausen, 2016
**Labidochromis caeruleus* Fryer, 1956
Oreochromis aureus (Steindachner, 1864)
**Oreochromis mossambicus* (Peters, 1852)
**Oreochromis niloticus* (Linnaeus, 1758)
Sarotherodon galilaeus (Linnaeus, 1758)
Tristramella sacra (Günther, 1865)
Tristramella simonis (Günther, 1864)

Family **Blenniidae**
Salariopsis burcuae Yoğurtçuoğlu, Kaya, Atalay, Ekmekçi & Freyhof, 2023
Salariopsis fluviatilis (Asso, 1801)
Salariopsis renatorum Yoğurtçuoğlu, Kaya, Atalay, Ekmekçi & Freyhof, 2023

Family **Atherinidae**
Atherina pontica Eichwald, 1831

Family **Aphaniidae**
Anatolichthys anatoliae (Leidenfrost, 1912)
Anatolichthys danfordii (Boulenger, 1890)
Anatolichthys fontinalis (Akşiray, 1948)
Anatolichthys iconii (Akşiray, 1948)
Anatolichthys irregularis (Yoğurtçuoğlu & Freyhof, 2018)
Anatolichthys maeandricus (Akşiray, 1948)
Anatolichthys marassantensis (Pfleiderer, Geiger & Herder, 2014)
Anatolichthys meridionalis (Akşiray, 1948)
Anatolichthys saldae (Akşiray, 1955)
Anatolichthys splendens Kosswig & Sözer, 1945
Anatolichthys sureyanus (Neu, 1937)
Anatolichthys transgrediens (Ermin, 1946)
Anatolichthys villwocki (Hrbek & Wildekamp, 2003)
Aphaniops dispar (Rüppell, 1829)
Aphaniops furcatus (Teimori, Esmaeili, Erpenbeck & Reichenbacher, 2014)
Aphaniops ginaonis (Holly, 1929)
Aphaniops kruppi (Freyhof, Weissenbacher & Geiger, 2017)
Aphaniops richardsoni (Boulenger, 1907)
Aphaniops sirhani (Villwock, Scholl & Krupp, 1983)
Aphaniops stoliczkanus (Day, 1872)
Aphaniops teimorii Freyhof & Yoğurtçuoğlu, 2020
Aphanius almiriensis Kottelat, Barbieri & Stoumboudi, 2007
Aphanius fasciatus (Valenciennes, 1821)
Esmaeilius darabensis (Esmaeili, Teimori, Gholami & Reichenbacher, 2014)
Esmaeilius isfahanensis (Hrbek, Keivany & Coad, 2006)
Esmaeilius persicus Jenkins, 1910
Esmaeilius shirini (Gholami, Esmaeili, Erpenbeck & Reichenbacher, 2014)
Esmaeilius sophiae (Heckel, 1847)
Esmaeilius vladykovi (Coad, 1988)
Kosswigichthys asquamatus Sözer, 1942
Paraphanius alexandri (Akşiray, 1948)
Paraphanius boulengeri (Akşiray, 1948)

Paraphanius mento (Heckel, 1843)
Paraphanius mentoides (Akşiray, 1948)
Paraphanius orontis (Akşiray, 1948)
Paraphanius similis (Akşiray, 1948)
Paraphanius striptus (Goren, 1974)
Paraphanius sp.

Family **Poeciliidae**
**Gambusia holbrooki* Girard, 1859
**Poecilia "latipinna"* (Lesueur, 1821)
**Poecilia reticulata* Peters, 1859
**Poecilia "sphenops"* Valenciennes, 1846
**Xiphophorus hellerii* Heckel, 1848
**Xiphophorus maculatus* (Günther, 1866)

Family **Mastacembelidae**
Mastacembelus simack (Walbaum, 1792)

Family **Channidae**
Channa montana (McClelland, 1842)

Family **Pleuronectidae**
Platichthys flesus (Linnaeus, 1758)

Family **Syngnathidae**
Syngnathus caspius Eichwald, 1831
Syngnathus nigrolineatus Risso, 1827

Family **Terapontidae**
Terapon jarbua (Forsskål, 1775)

Family **Centrarchidae**
**Lepomis gibbosus* (Linnaeus, 1758)
**Micropterus nigricans* (Cuvier, 1828)

Family **Percidae**
**Gymnocephalus cernua* (Linnaeus, 1758)
Perca fluviatilis Linnaeus, 1758
Sander lucioperca (Linnaeus, 1758)

Family **Gasterosteidae**
Gasterosteus aculeatus Linnaeus, 1758
Pungitius platygaster (Kessler, 1859)

Family **Moronidae**
Dicentrarchus labrax (Linnaeus, 1758)

Family **Sparidae**
Acanthopagrus arabicus Iwatsuki, 2013

Glossary

adaptation: process by which organisms adjust to their environment.
adaptive radiation: diversification of organisms from a common ancestor into a wide variety of lifeforms forming different ecological niches.
adipose: fat.
adipose crest: ridge of adipose tissue. Mostly referred to dorsal and ventral ridges on the caudal peduncle of a fish.
adult: sexually mature.
abdominal: related to the abdomen (belly), the lower part of the body containing internal organs such as the digestive tract, liver and gonads.
allopatric: species whose ranges are nowhere in contact, i.e., geographically separated. See also sympatric.
ammocoete: larval stage of lampreys (Petromyzontidae).
ammonia: compound composed of nitrogen and hydrogen (NH_3). Most teleost fishes are ammoniotelic, i.e., primarily excrete ammonia as nitrogenous waste.
amphibious: organisms able to live both on land and in water. Amphibious fishes, such as mudskippers, can spend extended periods of time out of water.
amphidromous: fish that migrate between fresh and saltwater environments for feeding or wintering but not for spawning.
amplexus: mating position in which male clasps female.
anadromous: species migrating from sea to freshwaters to spawn.
anastomosing: connections between hole complex, e.g., formed by catfishes in mud substrate.
ancestor: early type of organism from which others are descended by evolution.
anflug: arthropods that land on or fall into the water.
anteromedial: located in front of and toward the midline.
anterior: situated before or at the front of; forepart (opposed to posterior).
apomorphic character: unique, derived from ancestral character.
aquaculture: farming of fish or other aquatic organisms.
assortative mating: non-random mating based on the tendency to mate either similar or dissimilar phenotypes rather than between relatives.
aufwuchs: epilithic assemblage of a variety of unicellular and filamentous green and blue-green algae growing directly on rocks, diatoms that grow on filamentous algae, moss animals (bryozoa), ciliate colonies and sponges. Associated insect larvae and microcrustaceans living among algal filaments are often considered part of the aufwuchs.
axillary: relating to or located near the axilla (the region immediately behind or under the pectoral and/or pelvic).
backwater: part of a river cut at least partly from the main river.
band: elongate colour mark that is neither a bar nor a stripe.
bar: a vertical colour mark.
barcoding (or DNA barcoding): method of identification of short fragments of DNA (so-called barcodes), mainly a 648-bp region of the cytochrome c oxidase I gene (COI) of mitochondria, with the intent to identify species.
basin: sea or terminal lake into which rivers drain, not to confuse with a catchment or drainage of a river.
benthic: living at the bottom.
belly: ventral side of body (see abdomen).
bilaterally symmetric: left and right sides of an organism divided into mirror images of each other along the midline.
biodiversity: measure of biological variation at genetic, species, and ecosystem level.
biogeography: study of the geographic distribution of biodiversity.
biomass: total weight or mass of living organisms in a given area or ecosystem. Often expressed per unit area (e.g., g/m^2) or volume (e.g., g/m^3).
blotch: variably shaped pigmented mark larger than the eye.
brackish water: water with salinity between fresh- and seawater, typically ranging from 0.5 to 30 parts per thousand.
branchial: of or relating to gills.
breast: ventral surface of the body between isthmus and base of pectoral fins.
cannibalism: consumption of individuals of the same species.
catadromous: species migrating from freshwaters to sea to spawn.
catchment: area drained by a river and its tributaries. It is also a geographical area where the landscape collects water, typically draining to a common outlet such as a river or lake.
caudal peduncle: part of body between the origin of the last anal ray and the caudal base.
ceratobranchial: longest bones below epibranchial in branchial arch. These support most of the gill filaments and gill rakers.
cf., confer: [Latin] added between genus and species names used to indicate individuals, populations or a species quite similar to known species but differing by some details and of which it cannot be decided whether they belong to the same species.
character: morphological, ethological, molecular and other features that may be used to diagnose a taxon.

https://doi.org/10.1515/9783111677811-053

character state: each alternative condition a given character may have; fixed in species, and thus present in all individuals.
cheek: side of the head below and behind the eye.
chin: anterior part of the lower jaw.
chromosome: structure in a cell nucleus composed of DNA and proteins that carries genetic information.
circumpeduncular: surrounding caudal peduncle.
circulus (plural: circuli)**:** concentric growth rings on a scale.
cirrus (plural: cirri)**:** thin tendril-like fleshy appendage.
cleft: slit-like or elongate opening (mostly related to the fish mouth), fissure, or V-shaped indentation.
cobbles: particle from 64 to 256 mm in diameter, following the Wentworth scale.
conspecific: individuals belonging to the same species.
contralateral: on opposite side of each lateral.
cornified: keratinised.
corpus: central part of the digestive system in certain animals.
crepuscular: refers to animals primarily active during twilight, i.e., at dawn and dusk.
cryptic species: morphologically (almost) identical species reproductively isolated and genetically distinct (likely true cryptic species do not exist).
deciduous: loosely attached.
delta: landform where a river divides into branches before flowing into a larger body of water, typically forming a triangular shape.
desalination: removing dissolved salts and other minerals from sea- or brackish water.
desiccation: process of drying out.
detritus: dead and decaying organic matter, including plant, fungal, and animal remains.
diadromous: species migrating between fresh- and seawater at different life cycle periods, either to spawn in sea (catadromous) or freshwaters (anadromous).
diagnosis: (plural: diagnoses) a concise summary of characteristics distinguishing a species from others.
diatoms: single-celled algae with a transparent glass-like cell wall.
dimorphism: occurrence of two distinct forms or morphologies within a single population.
diploid (noun): organism with two sets of chromosomes, one inherited from each parent.
dispersal: tendency of an organism to move away, either from its birth site (natal dispersal) or breeding site (breeding dispersal). Also used to describe the movement of early life stages (e.g., fish) such as eggs and larvae. Often unidirectional.
distal: part of an appendage (e.g., fin) farthest from its attachment point to the body.
diurnal: animals that are active in the daytime.
DNA: deoxyribonucleic acid, a molecule that contains the genetic information an organism needs to develop.
dorsal: back or upper side of an organism.
drainage: entire river system (often erroneously called a basin).
effluent: liquid waste or sewage discharge into a river or sea.
elver: life stage of young eels, which migrate from the ocean to inland streams.
embedded scales: scales that are partially or completely buried in the skin.
embryo: developmental stage between the fertilization of the egg and onset of external feeding. Called "free embryo" between hatching and external feeding.
endemic: species that is native and exclusive to a particular geographic area.
endorheic: river that does not drain into the sea but often into a closed lake or a desert (opposed to exorheic).
epigean: surface-dwelling (as opposed to hypogean or subterranean).
epilithic: plant and/or animal that lives on the surface of stones or rocks.
ephemeral: short-lived, or of organisms that complete their life cycle very rapidly.
established species: introduced species that maintains itself in the wild by reproduction without human intervention.
estuarine: coastal body of water with one or more rivers flowing into it, connected to the sea.
ethanol: or ethyl alcohol, a colourless liquid hydrocarbon compound (C_2H_5OH) produced by fermentation of sugars.
ethological character: behavioral character that might diagnose a taxon.
euryhaline: species that tolerates a wide range of salinity (opposite: stenohaline).
euthanasia: practice of painlessly ending life.
evolution: gradual change of characteristics of a species over several generations, which relies on selection.
exorheic: river that drains into the sea (opposed to endorheic).
extant: still living (as opposed to extinct).
extinct: taxon that has no living members.
extinct in the wild: taxon that survives outside its native range, usually in captivity.
extirpation: (of a species) process of a population or species disappearing from a part of its range. Also known as local extinction.
extremity: distal appendages of the body.
falaj [Arabic]**:** man-made structure to channel water for irrigation.
fauna: animals of a particular habitat or region.

fecundity: measure of the potential reproductive output of females for a given period (e.g., for a year or over its lifetime). Often represented by the number of eggs produced throughout a season.
fenestra (plural: fenestrae): a small anatomical opening or perforation (as in or between a bone).
feral: wild groups (strains) of domesticated species, a process that generally follows escape or release from captivity.
fertilization: fusion of gametes to form a new organism.
fertility: natural capability to produce offspring.
flank: lateral part of the body behind the head to the caudal base.
foothill zone: transition zone between plains and low-relief hills to adjacent high mountains.
foraging: searching for food.
formalin: 35–40 % aqueous formaldehyde solution (a gas). Commonly used as a fixative in zoology.
fractional spawner: species in which females release eggs at intervals over a given period, usually over a few or several days or weeks. Also called batch spawning.
fraenum (plural: fraena): a fold of tissue that controls the movement of upper or lower lips in fish.
fry: recently hatched fish.
gametes: mature germ cells of each gender (haploid = 1n) capable of producing a zygote (diploid = 2n).
generic name: (of a biological genus) the first part of a binomial name of a species.
genome: entire genetic information of an organism carried by a complete haploid set of chromosomes.
geological: earth's physical structure and substance.
gestation: carrying of offspring in the uterus. Pregnancy.
gill: respiratory organ located along the posterior edge of the gill arch, consisting of filamentous outgrowths, allowing gas exchange.
gill arches: bony or cartilaginous arches placed behind each side of the pharynx supporting gills.
gill rakers: forward projections (usually bony) along the anterior edge of the gill arch.
girdle: set of bones supporting pelvic and pectoral fins.
gonopodium: copulatory organ formed by a modified anal fin. Typical in male Poeciliidae.
gravel: collective term for particles from 2 to 256 mm in diameter, following the Wentworth scale. It contains boulders, cobbles, pebbles and granules.
gregarious: animals (fishes) associate with their conspecifics as forming schools.
gular disc: adhesive disk-like structure on the ventral head surface behind the lower jaw. Typically, in *Garra*.
gulf: part of an ocean or sea that is largely enclosed by land, naturally with a narrow mouth.
gynogenesis: kind of reproductive mode in which sperm activates egg development, but its DNA is not included in the genome of offspring.
habitat: environment in which a particular species of organism lives.
hatching: period when embryos leave the egg.
headwater: tributaries in or close to upper parts of river drainages.
herbivorous: animal which feeds on plants.
hermaphrodite: organisms that possess both male and female reproductive organ systems.
HL: head length.
holotype: single type specimen upon which the name of a species is based.
homonymy: case of having the same name or designation. It is usually used if two species have the same name.
hyaline: clear, glassy, translucent.
hybrid: offspring of a cross between two species.
hybrid swarms: group of hybrid individuals capable of crossing and backcrossing with their offspring and parents.
hypersaline: water containing excessive ionic content (salt), commonly greater than seawater (> 35 parts per thousand).
hypoxia: oxygen deficiency.
hypural complex: set of flattened bones that supports the caudal fin rays in most teleost fishes.
impoundment: artificially confined body of water, like a reservoir formed by a dam.
incertae sedis: term applied to a taxon or group of taxa whose relationship with others is unknown or undecided.
inferior: for the mouth: position in which mouth is on the underside of the head; mouth opens downwards.
infertility: inability to conceive offspring.
introduced species: non-native organism that has been transported by human activities, intentionally or unintentionally, to a new environment outside its natural range.
introgression: fusion (via hybridisation and backcrossing) of genes from one species into the gene pool of a second species.
invasive species: non-native organism that spreads rapidly in a new environment, causing ecological, economic, or social harm. Invasion usually follows an introductory event or the disappearance of a natural barrier to dispersal (for example, the establishment of river connections, damming, etc.).
irrigation: agricultural process in which water is diverted to croplands for water supply to plants, mostly by canals and pipes.
isthmus: narrow structure (a piece of tissue or part of an organ) connecting two larger parts.
IUCN: The International Union for Conservation of Nature.
juvenile: young and sexually immature, morphologically usually dissimilar to adults.

keel: any structure resembling a longitudinal ridge on a solid structure, e.g., a bone, shell or abdomen.
lacustrine: of or relating to a lake.
lagoon: shallow lake or pool close to the sea but at least partly separated from it by a narrow strip of land.
lamina circularis: bony structure on the dorsal surface of the anterior pectoral rays of male *Cobitis.*
landlocked: population of a normally diadromous species resident in a freshwater body, (often permanently) isolated from the sea, and where it completes its life cycle.
larva: (plural: larvae) in fish, early life phases, lasting from absorption of the yolk sack until full formation of fins.
lateral-line: series of perforated scales extending along the flank.
leptocephalus (plural: leptocephali): pelagic larval stage of eels (Anguilliformes) and some other bony fish groups.
lineage: descendants of a common ancestor considered to be the founder of a taxon.
littoral: shallow shoreline of a body of water.
livebearer: fish that bears live young rather than depositing eggs.
macrophytes: vascular aquatic plants growing in or near water.
mandibular: mandible, lower jaw.
mangrove: vegetation group consisting of trees and shrubs living in marine intertidal zones.
marsh: continuously or seasonally inundated wetland with saturated soils and emergent vegetation. Often also called a swamp.
masculinization: process of inducing male characters, usually by hormonal treatment.
maxillary: as an adjective pertaining to or in the region of the upper jaw (for example, maxillary barbels), or more specifically to maxillae; as a noun, upper jaw.
melanophores: pigment cell with melanin in its cytoplasm accounts for the change of colours in many fishes.
mental: pertaining to chin.
meristic: count of a series of discrete units of a morphological character, such as lateral-line scales, fin-spines or rays.
metamorphosis: period in which an organism undergoes a major change from one form or stage to another during its development. Examples: the transformation, in fishes, of a larva into a juvenile; in amphibians, of a tadpole into a frog.
migration: journey between different habitats at specific times of the year, usually for a particular purpose (spawning, feeding, etc.). Usually bidirectional.
migratory: organisms (individuals, populations or species) that perform a migration.
mode: most frequently occurring value in a series of values (for example, in lateral-line scale counts).
monophyletic: group of taxa that includes all descendants of a common ancestor.
monosex: (in aquaculture) cultivation of either all-male or all-female individuals.
monotypic genus: genera represented by only a single species.
morphometrics: measurements of a measurable morphological character, such as length.
MS222: compound (tricaine methanesulfonate) most commonly used for anesthesia, sedation, or euthanasia of fish.
mtDNA: mitochondrial DNA.
mud: collective term for particles containing silt and clay in the bottom of a given water body.
multiple spawner: term referring to repeated spawning in a spawning season; a fish which spawns more than once throughout a breeding period.
mutation: any change in a DNA molecule copied from a parental DNA molecule.
naked: without scales.
nape: area of dorsum immediately behind the occipital region of the head.
naris (plural: nares)**:** openings of nostril. The nostrils of most fishes have an anterior and a posterior naris.
natal: of or associated with birth.
native: species that originally lived in an area without human intervention.
neuromast: exposed or semi-exposed receptor of the sensory nervous system sensitive to mechanical stimuli (e.g., water movements), usually contained in canals of lateral-line system on body and head, or exposed (as in many Gobiidae).
nocturnal: animals active at night.
nomenclature: system of names given to organisms and provisions for their formation and use. Names of animals are governed by the International Code of Zoological Nomenclature, fourth edition (1999), adopted by the International Union of Biological Sciences. The International Commission on Zoological Nomenclature (comprised of zoologists) is responsible for amendments to the Code to clear special or conflicting cases. It has plenary powers to suspend the application of the Code in specific cases. When necessary, the Commission issues mandatory rulings on the names of animals.
nuptial: of, relating to, or occurring during the mating season.
oasis: watered location in an arid area, usually occurring when a water table reaches the surface.
occiput: back part of the skull.
ocellus (plural: ocelli): an eye-like, usually black spot with a white margin.

offspring: collectively referred to descendant generations, result from reproduction.
omnivorous: animals that feed on both plants and animals.
ontogenetic: development of an individual from inception, maturity, to senescence.
opercle: group of bones and associated membranes (usually movable) that cover gills.
orbital: pertaining to the eye socket.
ossified: bony structure.
osteology: skeletal system of an animal. Also, it is a branch of anatomy dealing with bones.
ornamental: (of species) mostly bred or collected from the wild for decoration. Most common examples include aquarium fishes.
overfishing: unsustainable exploitation of fish stocks.
overwintering: (for fish) is a term usually applied to spending the winter period.
ovipositor: specialized tubular organ by which a female fish deposits eggs. The most common example of such a structure in fish is bitterlings.
ovotestis (plural: ovotestes): reproductive organ that produces both ova and sperm.
oxbows: term used to describe the cut-off of a stream meander.
paired fins: pectoral and pelvic fins.
palearctic: biogeographic realm including Europe, Asia north of the Himalayas, and northern Africa (excluding the drainage of the Nile).
papilla (plural: papillae): small fleshy protuberance.
paraphyletic: taxonomic groups that do not contain all descendants of a single ancestor.
paratype: specimen associated with the holotype used for the original description of a species.
parr marks: vertical bars on the flank of juvenile Salmonidae.
pebbles: particle from > 4 to 64 mm in diameter, following the Wentworth scale.
pelagic: relating to or living in open oceans or lakes (rather than in waters adjacent to land).
pelvic-axillary lobe: triangular flap or fold at the pelvic base.
peninsula: part of land surrounded by water or projecting into a body of water from a larger land.
perforated scales: scales with a hole in which lateral-line pores come out.
perianthropic: near or close to humans.
peripheral: species at the edge of their range.
periphyton: small organisms living on a substrate.
pharyngeal teeth: bony arches or toothed pads deep in the pharynx equipped with teeth.
phenotype: observable physical or biochemical characteristics of an organism. Also: an individual or group of organisms exhibiting a particular physical or biochemical characters.
pheromones: chemical signals produced and released into the environment by an animal for affecting (e.g., to attract or to frighten) other animals.
phototaxis (adjective phototactic)**:** movement of an organism in response to light. Positive phototaxis is a movement toward light; negative phototaxis is a movement to escape light.
phylogenetic: evolutionary relationships among species or individuals.
phylogeny: history of evolution of a species.
phylogeography: genetic study of processes controlling geographic distributions of lineages by constructing genealogies of populations and genes.
plankton: microscopic organism floating in open water. Animal planktons = zooplankton, plant planktons = phytoplankton.
plesiomorphic: ancestral character.
pocket/scale pocket: envelope of skin in which the anterior portion of a scale is embedded.
polymorphism: different discrete appearances [morphs] exhibited by members of a species. A species with wide range of character states is not polymorphic, there must be discrete units.
polyphyletic: groups containing descendants of different ancestors.
polyploidy: case of containing more than two sets of chromosomes.
posterior: rear or opposite to the head of an organism.
predation: relation between two animals in which one eats all or most parts of the other.
procurrent: related to spine-like unbranched or unsegmented fin rays, usually at the upper and lower edges of caudal fins of teleost fishes.
proximal: situated nearer to the body's center or point of attachment. Opposite to distal.
pterygiophore: bones or cartilages that support fin rays. There are also pterygophores lacking a ray in many fishes, especially in front of the dorsal.
putative species: species that are considered to exist or have existed.
qanat: irrigation system in which a tunnel is dug into arid land that allows water from underground aquifers to be accessed.
redd: term typically applied to the gravel pit nests where salmonids spawn.
relict: taxa or geological entities (e.g., lakes, glaciers or mountains) remnants of their pre-existing forms.

rheophile: associated with moderate to swift current in streams.
riffle: stream habitat characterized by relatively shallow water (usually less than 0.5 m), moderate to swift current (0.5–2.0 m/s), stone to large gravel substrate and a turbulent surface.
riverine: pertaining to rivers and streams.
rostral cap: fleshy tissue at snout tip. Most common in cyprinids.
rostrum (plural rostra)**:** snout or beak-shaped projection on the anterior part of the head.
rudimentary rays: usually soft, unsegmented, and unbranched fin rays.
run: fast-flowing water with a smooth surface.
saddle: colour mark on dorsum only slightly reading down to flank.
salinity: dissolved ionic content of water.
scale pocket: term applied to the skin fold in which scales are located.
schistosomiasis: chronic disease caused by some parasitic flatworms.
school: or shoal, a group of fish that swim together.
semi-anadromous: species that do not migrate from fully marine waters – but usually from brackish waters to lakes and rivers to spawn in freshwaters.
serra (plural: serrae): small, pointed, toothlike structures organized in a row, like the teeth of a saw.
sexually dimorphic: differences in appearance between male and female of the same species.
sheath: protective connective tissue enclosing a body structure.
siblings: individuals or species that have at least one common parent.
SL: standard length, always measured from the anteriormost point of the head to the end of the hypural bones at the middle of the caudal peduncle.
smoltification: series of changes (e.g., physiological) in which juvenile salmonids and lampreys adapt from living in fresh- to seawater.
sp. (plural: spp.): added behind a genus name but not followed by a geographic or other denomination indicates a species identified to the genus level but that was not or could not be identified to species level (e.g., *Garra* sp.); may also mean any member of that genus (e.g., *Garra* spp. may mean "several species of *Garra*").
spawning: deposition of eggs.
speciation: evolutionary process by which new species evolve from existing species. Allopatric speciation: a population splits into two or more allopatric (geographically isolated) populations (by migration, habitat fragmentation, etc.), which then diverge. Parapatric speciation: similar to allopatric speciation, but the two species evolve from contiguous populations. Peripatric speciation (peripheral isolation): similar to allopatric speciation, but one population is much smaller than the other, often at the edge of the range. Sympatric speciation: species diverge but remain sympatric (sharing the same geographical range and usually the same habitat).
specimen: preserved fish (or part thereof), usually for study, identification or museum deposition. Non preserved fish are not specimens.
species flock: number of closely related species confined to a narrowly circumscribed area.
spent: individuals that have recently released their eggs or sperm.
spine: in a fin, a median (unpaired) structure never branched, never segmented, generally hard and pointed, and supporting a membrane.
spinous: (of a fin ray) resembling a spine, shaped like a spine, but not a true spine.
spot: roundish pigmented mark about the size of the eye or smaller.
squamation: state, aspect, and/or the arrangement of scales.
sterile: infertile, unable to produce offspring.
stock: informal assemblage of individuals or a population, differing from others such as assemblages by morphological, developmental, ecological, geographical or other characteristics, usually within a given range and of unknown taxonomic status.
stocking: introduction of fish into a body of water mostly for commercial interest.
stout: (fish) relatively deep-bodied.
stria (plural: striae)**:** linear, longitudinal and parallel line or ridge bordering a groove.
stripe: longitudinal colour mark.
stunt: animal whose growth or development has been curtailed. In fish, stunting usually results from food shortage or unfavorable habitat conditions.
substrate (substratum): bottom material; also, bottom over which fish lives.
subterranean: underground, often found in caves.
superior: for the mouth: position in which the anterior end of the closed mouth is distinctly above the tip of the snout; mouth usually opens upwards.
swim bladder (or gas bladder)**:** internal sac filled with gas to sustain neutral buoyancy.
swollen pad: in barbels, the median part of the lower lip is thickened but not produced into a lobe; it appears as a swollen pad.
sympatric: two or more species whose ranges overlap. They can but must not be syntopic.

symphysis (plural symphyses): cartilaginous joint between two bones.
synonyms: two or more names applied to a single taxon, for example, when individuals formerly described as constituting two species are now united in a single species. In such cases, the oldest given name is the valid name of the taxon.
syntype: series of specimens upon which the description of a new species was based before the code was changed to disallow this practice.
syntopic: two or more species that occur together in the same habitat.
tapetum lucidum: layer of reflective guanine crystals that enhances the visual sensitivity of fishes under low light conditions by reflecting light not absorbed by the retina back into the eye.
taxon (plural: taxa): one or a group of organisms that is given a proper name or a group that could be given a name but remains unnamed as a matter of convention.
taxonomy: theory and practice of describing the diversity of organisms and ordering this diversity into a system reflecting evolutionary relationships.
tectonic: structure of earth's crust.
terminal: for mouth: position in which the anterior end of closed mouth is at level with the tip of the snout; mouth usually opens forwards.
territorial: behaviour of an individual organism when defending a particular area.
territory: any area defended by an organism or a group of organisms for such purposes as mating or feeding.
TL: total length, always measured from the anteriormost point of the head to the posteriormost tip of the caudal fin.
torrents: strong and very fast-flowing parts of streams.
tributary: branches of rivers and streams. Always flow into larger streams.
troglobiont: species having an underground-dwelling mode of life.
troglomorphic: morphological characters that are adaptions to living in the constant darkness of underground habitats.
tubercles: small keratinised (horny) protuberances on the head, body or fins, often more developed in males, usually present during a given period of life. Tubercles are used for sexual stimulation, as contact organs while spawning, and in territorial aggression (breeding or nuptial tubercles). In some species, tubercles are present in both sexes throughout their life and help to maintain position and orientation in strong currents or have a hydrodynamic function.
turbid: water clouded by suspended particles, usually clay and silt.
type locality: location where the holotype or syntypes of a species were discovered.
type specimen: specimens of a taxon collected from their type locality and used for their formal description.
ubiquitous: species that is present in a great diversity of habitats.
unculus (plural: unculi): a horny projection arising from a single cell of the epidermis, unlike the multicellular, horny nuptial tubercles in some teleost families.
unpaired fins: dorsal, anal and caudal fins.
valid name: in taxonomic context, a valid name (as defined by the International Code of Zoological Nomenclature) is the correct name, which must be applied to a species.
valid species: species that exists and is recognised (diagnosable), as opposed to a species that was hypothesized to be distinct but does not exist, is no longer recognised, is treated as a synonym, etc.
venom: poisonous substance secreted by animals.
ventral: underside or abdominal area.
vernacular names: ordinary names which use people in a particular region.
villiform teeth: numerous small, sharp, pointed teeth.
voucher specimen: specimen archived in the permanent collection as physical evidence of its occurrence at a time and place and of any identifications and descriptions based on it.
vs. [Latin]: versus.
wadi [Arabic]: valley or stream bed that is mostly dry except during the rainy season.
wetlands: land parts that are inundated by water either permanently or temporarily. Usually, it does not exceed 6 m of water depth at the coast.
yolk-sack: membranous sac that contains yolk and is attached to the ventral surface of the embryos of most fishes.
zooplankton: microscopic (rarely macroscobic) animals found near the surface in aquatic environments.

Literature

Abbasi, K. & J. Sabkara. 2004a. Study of diet of *Leuciscus ulanus*, an endemic fish species in Iran. *The 1st Congress on Animal and Aquatic Sciences, Tehran University, Karaj, Iran*. [in Farsi, English summary].

Abbasi, K. & J. Sabkara. 2004b. Studying some biological characteristics of *Leuciscus ulanus* (Cyprinidae), an endemic fish and little-known in Iran. *2nd International Conference on Applied Biology, Ferdowsi University, Mashad, Iran*. [in Farsi, English summary].

Abbasi, K., J. Sabkara, M. Makaremi & R. Rahimi. 2004c. Study of diet in *Acanthalburnus urmianus*, an endemic fish in Iran. *The 1st Congress on Animal and Aquatic Sciences. Tehran University, Karaj, Iran.* [in Farsi, English summary].

Abbasi, K., A.N. Sarpanah & B.H. Kiabi. 2006. Study of morphomeristic and sexual dimorphism in *Leuciscus ulanus*, an endemic fish in Iran. *The 14th National and 2nd International Conference on Biology*. Tarbiat Modarres University, Tehran, Iran. [in Farsi, English summary].

Abdoli, A., P. Rasooli, B.H. Yazdandan & L. Abdoli. 2007. A study on some ecological aspects of snow trout (*Schizothorax pelzami*) from Laiinsoo River in northeastern Iran. *Environmental Sciences* 4: 69–76. [in Farsi].

Abdoli, A., P. Rasooli & S. Soltaniasab. 2008. A contribution to the biology of *Acanthalburnus urmianus* (Günther, 1899) (Osteichthyes: Cypinidae): an endemic fish of Iran. *Zoology in the Middle East* 43: 111–112. https://doi.org/10.1080/09397140.2008.10638277

Abdoli, A., M.R. Farokhnejad, H. Nazari & B.H. Kiabi. 2017. The role of temperature and daytime as indicators for the spawning migration of the Caspian lamprey *Caspiomyzon wagneri* Kessler, 1870. *Journal of Applied Ichthyology* 33: 553–557. https://doi.org/10.1111/jai.13078

Abdurakhmanov, Y.A. 1950. New species of roach, *Rutilus sojuchbulagi* sp. nova. *Doklady Akademii Nauk Azerbaidzhanskoi SSR* 6: 112–116. [in Russian].

Abdurakhmanov, Y.A. 1962. *Freshwater fishes of Azerbaijan*. Academy of Sciences of Azerbaijan SSR, Baku. [in Russian].

Abedi, M., A.H. Shiva, H. Mohammadi & R. Malekpour. 2011. Reproductive biology and age determination of *Garra rufa* Heckel, 1843 (Actinopterygii: Cyprinidae) in central Iran. *Turkish Journal of Zoology* 35: 317–323. https://doi.org/10.3906/zoo-0810-11

Aghasyan, A.L. & M.Y. Kalashyan. 2010. *The Red Data Book of the animals of the Republic of Armenia*. Ministry of Nature Protection of RA, Yerevan.

Agnese, J.F. & G.G. Teugels. 2005. Insight into the phylogeny of African Clariidae (Teleostei, Siluriformes): implications for their body shape evolution, biogeography, and taxonomy. *Molecular Phylogenetics and Evolution* 36: 546–553. https://doi.org/10.1016/j.ympev.2005.03.028

Agorreta, A., D. San Mauro, U. Schliewen, J.L. Van Tassell, M. Kovačić, R. Zardoya & L. Rüber. 2013. Molecular phylogenetics of gobioidei and phylogenetic placement of European gobies. *Molecular Phylogenetics and Evolution* 69: 619–633. https://doi.org/10.1016/j.ympev.2013.07.017

Ahmed, H.A., M.A.A. Mukhtar & A.H.Y. Al-Adhub. 1984. The reproductive biology of *Carasobarbus luteus* (Pisces, Cyprinidae) in Al-Hammar Marsh, Iraq. *Cybium* 8(4): 69–80. https://doi.org/10.26028/CYBIUM/1984-84-006

Ahnelt, H. 1991. A new species of *Knipowitschia* (Teleostei: Gobiidae) from Corfu, western Greece. *Ichthyological Exploration of Freshwaters* 2: 265–272.

Ahnelt, H. 1995. Two new species of *Knipowitschia* Iljin, 1927 (Teleostei: Gobiidae) from Western Anatolia. *Mitteilungen aus dem Hamburgischen Zoologischen Museum und Institut* 92: 155–168.

Ahnelt, H. 2001. Two Mediterranean gobiid fishes with an unusual cephalic lateral line canal system. *Cybium* 25: 261–267. https://doi.org/10.26028/cybium/2001-253-007

Ahnelt, H. 2011. Two new sympatric *Knipowitschia* species (Teleostei: Gobiidae) from an eastern Mediterranean coastal lake-examples of different dispersal patterns? *Zootaxa* 3114: 22–30. https://doi.org/10.11646/zootaxa.3114.1.2

Ahnelt, H. 2016. A second record of *Knipowitschia byblisia* Ahnelt, 2011 (Teleostei: Perciformes: Gobiidae) from southwestern Anatolia, Turkey. *Journal of Threatened Taxa* 8: 9195–9197. https://doi.org/10.11609/jott.2055.8.9.9195-9197

Ahnelt, H., M. Duchkowitsch, G. Scattolin, I. Zweimüller & A. Weissenbacher. 2001. *Neogobius gymnotrachelus* (Kessler, 1857) (Teleostei: Gobiidae, die Nackthalsgrundel in Österreich. Österreichs Fischerei 54: 262–266.

Ahnelt, H., P.M. Bănărescu, R. Spolwind, A. Harka & H. Waidbacher 1998. Occurrence and distribution of three gobiid species (Pisces, Gobiidae) in the middle and upper Danube region - examples of different dispersal patterns? *Biologia* (*Bratislava*) 53/5: 665–678.

Akkuş, M., M. Sarı, F.G. Ekmekçi & B. Yoğurtçuoğlu. 2021. The discovery of a microbialite-associated freshwater fish in the world's largest saline soda lake, Lake Van (Türkiye). *Zoosystematics and Evolution* 97: 181–189. https://doi.org/10.3897/zse.97.62120.

Aksu, S., S. Başkurt, Ö. Emiroğlu & A. Tarkan. 2021. Establishment and range expansion of non-native fish species facilitated by hot springs: the case study from the Upper Sakarya Basin (NW, Turkey). *Oceanological and Hydrobiological Studies* 50(3): 247–258. https://doi.org/10.2478/oandhs-2021-0021

Akşiray, F. 1948. Türkische Cyprinodontiden I. *İstanbul Üniversitesi Fen Fakültesi Mecmuası Seri B* 13: 98–142.

Akşiray, F. 1955. Über eine neue *Anatolichthys*-Form. *Publications of the Hydrobiological Research Institute, Series B, Istanbul* [İstanbul Üniversitesi Fen Fakültesi Hidrobiologi Araştırma Enstitüsü Yayınlarından Seri B.] 3(2–3): 57–62, Pl. 1.

Akyol, O. & S. Hasan. 2020. The first report on the phenomenon of *Capoeta aydinensis* (Cyprinidae), occurring in Gökova Bay, Aegean Sea. *Ege Journal of Fisheries and Aquatic Sciences* 37: 423–425. https://doi.org/10.12714/egejfas.37.4.14

Al Jufalii, S. M., H.R. Esmaeili, G. Sayyadzadeh, A.H. Nasoumi & H.K. Larson. 2022. Redescription of the goby *Glossogobius tenuiformis* Fowler, 1934 (Teleostei: Gobiidae) and assignment of Oman *Glossogobius* populations: a morpho-molecular approach. *Zootaxa* 5133: 543–554. https://doi.org/10.11646/zootaxa.5133.4.5

Alcaraz, C., A. Vila-Gispert & E. García-Berthou. 2005. Profiling invasive fish species: the importance of phylogeny and human use. *Diversity and Distributions* 11: 289–298. https://doi.org/10.1111/j.1366-9516.2005.00170.x

Al-Daham, N.K., M.F. Huq & K. P. Sharma. 1977. Notes on the ecology of fishes of the genus *Aphanius* and *Gambusia affinis* in southern Iraq. *Freshwater Biology* 7: 245–251.

Al-Hassan, L.A. J. & P.J. Miller. 1987. *Rhinogobius brunneus* (Gobiidae) in the Arabian Gulf. *Japanese Journal of Ichthyology* 33: 405–408.

Aliev, D.S., A.I. Sukhaniva & F.M. Shakirova. 1988. *Fishes of the inland waters of Turkmenistan*. Ylym, Ashkhabad. 142 pp. [in Russian.]

https://doi.org/10.1515/9783111677811-054

Alkahem, A.F. & R.J. Behnke. 1983. Fishes of Saudi Arabia. *Fauna of Saudi Arabia* 5: 545–567.

Al-Kahmen, H.F., K.A. Al-Ganim, Z. Ahmad, T.A. Temraz, A.S. Al-Alel, F. Al-Misned & H. Annazri. 2008. A threatened fish species (*Aphanius dispar*) in Saudi Arabia, a case study. *Pakistan Journal Biological Sciences* 11 (19): 2300–2307. https://doi.org/10.3923/pjbs.2008.2300.2307

Allan, J.A. 2001. *The Middle East Water Question*: Hydro-politics and the global economy. I.B. Tauris. Bloomsbury Publishing PLC.

Allen, G.R., S.H. Midgley & M. Allen. 2002. *Field guide to the freshwater fishes of Australia*. CSIRO/Western Australian Museum, Perth, Australia.

Almada, V.C., J.I. Robalo, A. Levy, J. Freyhof, G. Bernardi & I. Doadrio. 2009. Phylogenetic analysis of Peri-Mediterranean blennies of the genus *Salaria*: Molecular insights on the colonization of freshwaters. *Molecular Phylogenetics and Evolution* 52: 424–431. https://doi. org/10.1016/j.ympev.2009.03.029

Alp, A., C. Kara & H.M. Büyükçapar. 2003. Reproductive biology of brown trout, *Salmo trutta macrostigma* Dumeril 1858, in a tributary of the Ceyhan River which flows into the eastern Mediterranean Sea. *Journal of Applied Ichthyology* 19: 346–351. https://doi.org/10.1111/j.1439-0426.2003.00455.x

Alwan, N. 2010. Systematics, taxonomy, phylogeny and zoogeography of the *Capoeta damascina* species complex (Pisces: Teleostei: Cyprinidae) inferred from comparative morphology and molecular markers. PhD Thesis, Johann Wolfgang Goethe-Universität, Frankfurt.

Alwan, N. 2013. Euphrates River. Pp. 20–23 *in:* Howarth, R.W. & J.E. Mohan. (Eds.). *Biomes and Ecosystems: An Encyclopedia*. Golson Media, Salem Press.

Alwan, N.H., H. Zareian & H.R. Esmaeili. 2016a. *Capoeta coadi*, a new species of cyprinid fish from the Karun River drainage, Iran based on morphological and molecular evidences (Teleostei, Cyprinidae). *ZooKeys* 572: 155–180. https://doi. org/10.3897/zookeys.572.7377

Alwan, N., H.R. Esmaeili & F. Krupp. 2016b. Molecular phylogeny and zoogeography of the *Capoeta damascina* species complex (Pisces: Teleostei: Cyprinidae). *PLOS ONE* 11: 1–25. https://doi.org/10.1371/journal.pone.0156434

Amorim, M.C.P., P.J. Fonseca & V.C. Almada. 2003. Sound production during courtship and spawning of *Oreochromis mossambicus*: male–female and male–male interactions. *Journal of Fish Biology* 62: 658–672. https://doi.org/10.1046/j.1095-8649.2003.00054.x

Annandale, N. 1919. Notes on fishes of the genus *Discognathus* from India and Persia. *Records of the Indian Museum* (*Calcutta*) 18: 65–78.

Anuradha, S. 1986. Contributions to the study of bagrid fishes. XIX: Systematic position of *Macrones halepensis colvillii* Hora & Misra, 1943, with description of a new species (Siluriformes, Bagridae). *Revue Suisse Zoologiques* 93 (2): 291–296. https://doi.org/10.5962/bhl.part.79696)

Arai, T. & S.R. Abdul Kadir. 2017a. Diversity, distribution and different habitat use among the tropical freshwater eels of genus *Anguilla*. *Scientific Reports* 7: 7593. https://doi.org/10.1038/s41598-017-07837-x

Arai, T. & S.R. Abdul Kadir. 2017b. Opportunistic spawning of tropical anguillid eels *Anguilla bicolor bicolor* and *A. bengalensis bengalensis*. *Scientific Reports* 7: 41649. https://doi.org/10.1038/srep41649

Arai, R. & Y. Akai. 1988. *Acheilognathus melanogaster*, a senior synonym of *A. moriokae*, with a revision of the genera of the subfamily Acheilognathinae (Cypriniformes, Cyprinidae). *Bulletin of the National Science Museum, Tokyo, Series A* 14:199–213.

Arai, R. & K. Kato. 2003. Gross morphology and evolution of the lateral line system and infraorbital bones in bitterlings (Cyprinidae, Acheilognathinae), with an overview of the lateral line system in the family Cyprinidae. Bulletin. *The University Museum, the University of Tokyo* 40: 1–42.

Armbruster, J.W. & L.M. Page. 2006. Redescription of *Pterygoplichthys punctatus* and description of a new species of *Pterygoplichthys* (Siluriformes: Loricariidae). *Neotropical Ichthyology* 4: 401–410. https://doi.org/10.1590/S1679-62252006000400003

Arnold, A. & H. Längert. 1995. *Das Moderlieschen. Leucaspius delineatus Biologie, Haltung und Artenschutz*. Die Neue Brehm-Bücherei 623. Westarp Wissenschaften, Magdeburg und Spektrum Akademischer Verlag, Heidelberg.

Arslan, A. 2015. Anadolu inanç söylenceleri: Egeria'nin anlatilarinda Balikligöl ve kral abgar. *Manisa Celal Bayar Üniversitesi Sosyal Bilimler Dergisi* 13: 141–150.

Artaev, O.N., B.A. Levin, N.D. Mustafayev & E.P. Simonov. 2018. Caucasian relict fish - Azerbaijani Spring Roach. *Priroda* 5 (1233): 57–61.

Artaev, O.N., O.A. Ermakov, D.A. Vekhov, A.F. Konovalov, M.A. Levina, I.V. Pozdeev, A.B. Ruchin, I.V. Alyushin, V.Yu. Iljin & B.A. Levin. 2021. Genetic screening of distribution pattern of roaches *Rutilus rutilus* and *R. lacustris* (Cyprinidae) in broad range of secondary contact (Volga Basin). *Inland Water Biology,* 14: 205–214. https://doi.org/10.1134/S1995082921020024

Artaev, O.N., I.S. Turbanov, A.A. Bolotovskiy, A.A. Gandlin & B.A. Levin. 2024. Taxonomic revision of *Phoxinus* minnows (Leuciscidae) from Caucasus, with description of a new narrow-ranged endemic species. *Zoosystematics and Evolution* 100(1): 291-308. https://doi.org/10.3897/zse.100.115696

Atıcı, A.A., M. Elp & F. Şen. 2020. An investigation on systematic history and distribution areas of tarek *Alburnus tarichi* (Güldenstädt, 1814). *Brazilian Journal of Biology* 81: 792–796. https://doi.org/10.1590/1519-6984.231764

Attaala, A.M. & B.S. Rubaia. 2005. First record of the eel *Anguilla bengalensis* from Arabia with notes on freshwater fishes from Hadhramout, Yemen. *Zoology in the Middle East* 34: 35–44. https://doi.org/10.1080/09397140.2005.10638080

Aucher Eloy, R. 1843. Relations de voyages en Orient de 1830 à 1838; revues et annotées par M. le Comte Jaubert, accompagnées d'une carte géographique où sont tracés tous les itinéraires suivis par Aucher-Eloy. Roret, Paris, (1): i–xxxi, 1–364; (2): 365–775, 1 map.

Autin, J., S. Leroy, M.-O. Beslier, E. D'Acremont, P. Razin, A. Ribodetti, N. Bellahsen, C. Robin & K. Al Toubi. 2010. Continental break-up history of a deep magma-poor margin based on seismic reflection data (northeastern Gulf of Aden margin, offshore Oman). *Geophysical Journal International* 180: 501–519. https://doi.org/10.1111/j.1365-246X.2009.04424.x

Aydın, F., E. Erlat & M. Türkeş. 2020. Impact of climate variability on the surface of Lake Tuz (Turkey), 1985–2016. *Regional Environmental Change* 20: 1–14. https://doi.org/10.1007/s10113-020-01656-z

Bacalbasa-Dobrovici, N. & J. Holčík. 2000. Distribution of *Acipenser sturio* L., 1758 in the Black Sea and its watershed. Boletín, *Instituto Español de Oceanografía* 16: 37–42.

Băcescu, M.C. 1962. Contribution à la systématique du genre *Cobitis*. Description d'une espèce nouvelle, *Cobitis calderoni*, provenant de l'Espagne. *Revue de Biologie, Académie de la République Populaire Roumaine, Bucarest* 6: 435–448, pls 1–4.

Balážová-L'avrinčíková, M. & V. Kováč. 2007. Epigenetic context in the life history traits of the round goby, *Neogobius melanostomus*. Pp 275–287 *in:* Gherardi, F. (Ed.) *Biological invaders in inland waters: Profiles, distribution, and threats*. Invading Nature, Springer Series in Invasion Ecology. https://doi.org/10.1007/978-1-4020-6029-8_14

Balık, S. & P.M. Bănărescu. 1978. Süßwasserfische der Türkei. 11. Teil. Die Gattung *Orthrias* in der Türkei und in Südbulgarien (Pisces, Cobitidae, Noemacheilinae). *Mitteilungen aus dem Hamburgischen Zoologischen Museum und Institut* 75: 255–266.

Balletto, E. & S. Spanò. 1977. Ciprinidi del genere *Garra* Hamilton 1822, raccolti nello Yemen dal Prof. Giuseppe Scortecci. *Annali del Museo Civico di Storia Naturale 'Giacomo Doria'* 81: 246–287.

Balon, E. 1956. Prispevok k systematike, ekologii, morfologii, veku, rastu a poctu ikier u sable krivociarej (*Pelecus cultratus* (L.) z Dunaja pri Medved'ove. *Biologické práce* 2 (13): 63–88.

Balon, E.K. 1974. Domestication of the carp, *Cyprinus carpio L. Life Sciences Contributions Miscellaneous Publications*. Royal Ontario Museum, Toronto: 1–37.

Balon, E.K. 1995. Origin and domestication of the wild carp, *Cyprinus carpio*: from Roman gourmets to the swimming flowers. *Aquaculture* 129: 3–48. https://doi.org/10.1016/0044-8486(94)00227-F

Baltazar-Soares, M., S. Blanchet, J. Cote, A.S. Tarkan, E. Záhorská, R.E. Gozlan & C. Eizaguirre. 2020. Genomic footprints of a biological invasion: Introduction from Asia and dispersal in Europe of the topmouth gudgeon (*Pseudorasbora parva*). *Molecular Ecology*, 29(1): 71–85. https://doi.org/10.1111/mec.15313

Bănărescu, P.M. 1964. *Fauna Republicii Populare Romîne. Pisces–Osteichthyes*. Academia Republicii Populare Romîne, Bucuresti.

Bănărescu, P.M. 1968. Süßwasserfische der Türkei. Ergänzende Angaben zu Teil 2: Cobitidae. *Mitteilungen aus dem Hamburgischen Zoologischen Museum und Institut* 65: 353–356, Pl. 3.

Bănărescu, P.M. 1999a. *Pseudorasbora parva* (Temminck et Schlegel 1846). Pp 207–224 *in:* Bănărescu P.M. & N.G. Bogutskaya. (Eds.). *The freshwater fishes of Europe. Vol.5/II: Cyprinidae 2/II* Barbus. AULA Verlag, Wiebelsheim.

Bănărescu, P.M. 1999b. *Capoeta capoeta* (Güldenstädt, 1773). Pp 387–412 *in:* Bănărescu P.M. (Eds.). *The freshwater fishes of Europe. Vol.5/I: Cyprinidae 2/I.* AULA Verlag, Wiebelsheim.

Bănărescu, P.M. & B. Herzig-Straschil. 1995. A revision of the species of the *Cyprinion macrostomus*-group (Pisces: Cyprinidae). *Annalen des Naturhistorischen Museums in Wien* 97B: 411–420.

Bănărescu, P.M. & M.R. Mirza. 1965 *Noemacheilus lindbergi* n. sp., a new loach from Afghanistan and West Pakistan (Pisces, Cobitidae). *Senckenbergiana Biologica* 46 (4): 265–269.

Bănărescu, P.M. & T.T. Nalbant. 1964. Süßwasserfische der Türkei. 2. Teil Cobitidae. *Mitteilungen aus dem Hamburgischen Zoologischen Museum und Institut* 61: 159–201, 8 pls.

Bănărescu, P.M. & T.T. Nalbant. 1966. The 3rd Danish expedition to Central Asia. Zoological Results 34. Cobitidae (Pisces) from Afghanistan and Iran. *Videnskabelige Meddelelser fra Dansk Naturhistorisk Forening i København* 129: 149–186, pls 19–21.

Bănărescu, P.M. & T.T. Nalbant. 1973. Pisces, Teleostei. Cyprinidae (Gobioninae). *Das Tierreich* 93: 1–304.

Bănărescu, P.M. & T.T. Nalbant. 1995. A generical classification of Nemacheilinae with description of two new genera (Teleostei: Cypriniformes: Cobitidae). *Travaux du Muséum d'Histoire Naturelle "Grigore Antipa"* 35: 429–495.

Bănărescu, P.M., M. Papadopol & L. Mikhailova. 1970. Systematics. Pp 23–70 *in*: Zayanchkauskas, P. (Ed.). *Biology and fisheries of* Vimba *in Europe*. Mintis Press, Vilnius. [in Russian, English summary].

Bănărescu, P.M., T.T. Nalbant & S. Balık. 1978. Süßwasserfische der Türkei. 11. Teil. Die Gattung *Orthrias* in der Türkei und in Südbulgarien (Pisces, Cobitidae, Noemacheilinae). *Mitteilungen aus dem Hamburgischen Zoologischen Museum und Institut* 75: 255–266, 20 pls.

Bănărescu, P.M., T.T. Nalbant & M. Goren. 1982. The Noemacheiline loaches from Israel (Pisces: Cobitidae: Noemacheilinae). *Israel Journal of Zoology* 31: 1–25.

Banister, K.E. & M.A. Clarke. 1977. *The freshwater fishes of the Arabian Peninsula*. Pp 111–154. *in*: M. Rice. (Ed.) *The scientific results of the Oman flora and fauna survey 1975. Journal Oman Studies* (Special Report).

Banister, K.E. 1984. Three new species of *Varicorhinus* (Pisces, Cyprinidae) from Africa. *Bulletin of the British Museum (Natural History) Zoology* 47 (5): 273–282.

Banister, K.E. 1987. Two new species of *Garra* (Teleostei-Cyprinidae) from the Arabian Peninsula. *Bulletin of the British Museum (Natural History) Zoology* v 52 (1): 59–70, pls 1–2.

Banister, K.E. & M.K. Bunni. 1980. A new blind cyprinid fish from Iraq. *Bulletin of the British Museum (Natural History) Zoology* 38: 151–158.

Barbieri, R., S. Zogaris, E. Kalogianni, M.T. Stoumboudi, Y. Chatzinikolaou, S. Giakoumi, Y. Kapakos, D. Kommatas, N. Koutsikos, V. Tachos, L. Vardakas & A.N. Economou. 2015. *Freshwater fishes and lampreys of Greece. An annotated checklist*. Monographs on Marine Sciences No. 8. Hellenic Centre for Marine Research, Athens, Greece.

Bariche, M. & J. Freyhof. 2016. Status of *Pseudophoxinus libani* and *P. kervillei*, two minnows from the Levant (Teleostei: Cyprinidae). *Ichthyological Exploration of Freshwaters* 27: 203–210.

Bartáková, V., J. Bryja, R. Šanda, Y. Bektaş, T. Stefanov, L. Choleva, C. Smith & M. Reichard. 2019. High cryptic diversity of bitterling fish in the southern West Palearctic. *Molecular Phylogenetics and Evolution* 133: 1–11. https://doi.org/10.1016/j.ympev.2018.12.025

Baruš, V. & O. Oliva. 1995. *Mihulovci – Petromyzontes a Ryby - Osteichthyes* (2). Academia Praha.

Baruš, V., M. Peňaz & K. Kohlmann. 2002. *Cyprinus carpio* (Linnaeus, 1758). Pp 85–179 *in:* Bănărescu P.M. & H.J. Paepke. (Eds.). *The freshwater fishes of Europe. Vol. 5/III: Cyprinidae 2, Part III:* Carassius *to* Cyprinus. *Gasterosteidae*. AULA Verlag, Wiebelsheim.

Bath, H. 1977. Revision der Blenniini (Pisces: Blenniidae). *Senckenbergiana Biologica* 57: 167–234.

Bath, H. 2003. *Salaria fluviatilis* (Asso, 1784). Pp 97–121 *in:* P.J. Miller. (Ed.). *The freshwater fishes of Europe. Vol. 8/I: Mugilidae, Atherinidae, Atherinopsidae, Blenniidae, Odontobutidae, Gobiidae*. AULA Verlag, Wiebelsheim.

Battalgazi, F. 1944. Türkiye'de yeni ve az taninmis balıklar. Poissons nouveaux et peu connus de la Turquie. *Revue de la Faculté des Sciences de l'Université d'Instanbul, Série B: Sciences Naturelles* 9: 299–305.

Battalgil, F. 1941. Türkiyenin tatlı su balıkları. Les poissons des eaux douces de la Turquie. (Collection de l'Institut de Zoologie de l'Université d'Istanbul.). *Revue de la Faculté des Sciences de l'Université d'Instanbul, Série B: Sciences Naturelles* 6: 170–186.

Battalgil, F. 1942. Türkiye tatlı su balıkları hakkında. Contribution à la connaissance des poissons des eaux douces de la Turquie. *Revue de la Faculté des Sciences de l'Université d'Instanbul, Série B: Sciences Naturelles* 7: 287–306.

Battalgil, F. 1944. Türkiye'de yeni tatlı su balıkları. Nouveau poissons des eaux douces de la Turquie. *Revue de la Faculté des Sciences de l'Université d'Instanbul, Série B: Sciences Naturelles* 9: 126–133.

Bauchot, M. L. & A. Pras. 1980. *Guide des poissons marins d'Europe*. Delachaux et Niestlé, Paris et Neuchâtel.

Bayçelebi, E. 2019. *Türkiyede dağılım gösteren Squalius cinsinin taksonomik revizyonu*. PhD Thesis. Recep Tayyip Erdogan University, Institute of Science and Technology, Rize, Turkey. [In Turkish].

Bayçelebi, E. 2020. Distribution and diversity of fish from Seyhan, Ceyhan and Orontes river systems. *Zoosystematics and Evolution* 96 (2): 747–767.

Bayçelebi, E. 2025. Discovery of a new minnow species in the Black and Marmara Sea basins, Türkiye. Zoosystematics and Evolution 101 (3): 1155–1162. https://doi.org/10.3897/zse.101.153391

Bayçelebi, E., D. Turan & B. Japoshvili. 2015. Fish fauna of Çoruh River and two first record for Turkey. Turkish *Journal of Fisheries and Aquatic Sciences* 15: 777–788. https://doi.org/10.4194/1303-2712-v15_4_01

Bayçelebi, E., C. Kaya, S.S. Güçlü, F. Küçuk & D. Turan. 2020a. Taxonomic status of endemic fish species in Lake Beyşehir basin (Turkey). *Acta Aquatica Turcica* 16: 138–147. https://doi.org/10.22392/actaquatr.618539

Bayçelebi, E., D. Turan, C. Kaya & J. Freyhof. 2020b. *Alburnus nasreddini*, a synonym of *A. escherichii* (Teleostei: Leuciscidae). *Zootaxa* 4894: 123–132. https://doi.org/10.11646/zootaxa.4894.1.7

Bayçelebi, E., C. Kaya, D. Turan, & J. Freyhof. 2021a. *Garra orontesi*, a new species from the Orontes River drainage (Teleostei: Cyprinidae). *Zootaxa* 4352: 169–180. https://doi.org/10.11646/zootaxa.4952.1.10

Bayçelebi, E., D. Turan, C. Kaya & J. Freyhof. 2021a. *Alburnus battalgilae*, a synonym of *A. attalus* (Teleostei: Leuciscidae). *Zootaxa* 4999(4): 389–396. https://doi.org/10.11646/zootaxa.4999.4.8

Bayçelebi, E., D. Turan, S. Aksu & J. Freyhof. 2022. *Squalius cephaloides*, a synonym of *Squalius cii* (Teleostei: Leuciscidae). *Zootaxa* 5174: 277–284. https://doi.org/10.11646/zootaxa.5174.3.5

Behnke, R.J. 1968. Süßwasserfische der Türkei. 6. Teil. A new subgenus and species of trout, *Salmo* (*Platysalmo*) *platycephalus*, from southcentral Turkey, with comments on the classifikation [sic] of the subfamily Salmoninae. *Mitteilungen aus dem Hamburgischen Zoologischen Museum und Institut* 66: 1–15.

Behnke, R.J. 1992. *Native trout of western North America*. American Fisheries Society Monograph, American Fisheries Society, Bethesda, MD.

Behrens-Chapuis, S., F. Herder, H.R. Esmaeili, J. Freyhof, N.A. Hamidan, M. Özuluğ, R. Šanda & M.F. Geiger. 2015. Adding nuclear rhodopsin data where mitochondrial COI indicates discrepancies - can this marker help to explain conflicts in cyprinids? *DNA Barcodes* 3: 187–199. https://doi.org/10.1515/dna-2015-0020

Bektaş, Y., D. Turan, İ. Aksu, Y. Çiftçi, O. Eroğlu, G. Kalaycı & A.O. Beldüz. 2017. Molecular phylogeny of the genus *Capoeta* (Teleostei: Cyprinidae) in Anatolia, Turkey. *Biochemical Systematics and Ecology* 70: 80–94 https://doi.org/10.1016/j.bse.2016.11.005

Bektaş, Y., İ. Aksu, C. Kaya, E. Bayçelebi, S., Atasaral, F.G. Ekmekçi & D. Turan. 2019. Phylogeny and phylogeography of the genus *Alburnoides* (Teleostei, Cyprinidae) in Turkey based on mitochondrial DNA sequences. *Mitochondrial DNA Part A* 37: 273–284. https://doi.org/10.1080/24701394.2019.1664493

Bektaş, Y., İ. Aksu, C. Kaya, E. Bayçelebi, F. Küçük & D. Turan. 2020. Molecular systematics and phylogeography of the genus *Alburnus* Rafinesque, 1820 (Teleostei, Leuciscidae) in Turkey. *Mitochondrial DNA Part A* 31: 273–284. https://doi.org/10.1080/24701394.2020.1791840

Belyaeva, V.N., A.D. Vlasenko & V.F. Ivanov. 1989. Kaspiiskoe more: Ikhtiofauna i promyslovyeresursy. Moskva.

Ben-Tuvia, A. 1978. Fishes. Pp 407–430 *in:* C. Serruya. (Ed.). *Lake Kinneret*, Monographiae Biologicae, 32. Junk Publisher, The Hague. http://dx.doi.org/10.1007/978-94-009-9954-1

Ben-Tuvia, A. in Whitehead, P.J.P., M.L. Bauchot, J.C. Hureau, J. Nielsen & E. Tortonese. 1986. *Fishes of the North-eastern Atlantic and the Mediterranean*. UNESCO, Paris, 3 vols.

Berg, L.S. 1898. Beiträge zur Ichthyofauna des Kaukasus. *Mitteilungen des Kaukasischen Museums* 1 (3): 1–36.

Berg, L.S. 1910. Rapport sur une mission zoologique au Cascase en 1909. *Zoologicheskago Muzeya Imperatorskoi Akademii Nauk* 15: 153–170.

Berg, L.S. 1914. Description of a new species of *Garra* (= *Discognathus*) from eastern Persia. *Ezhegodnik Zoologicheskago Imperatorskoi Muzeya Akademii Nauk, St. Petersburg* 18: 61.

Berg, L.S. 1916. *Les Poissons des eaux douces de la Russie*. Ryabushinskiy, Moscow. i–xxvii + 1–563. [In Russian.]

Berg, L.S. 1918. On a new siluroid fish, *Glyptosternum armeniacum* n. sp., from the Upper Euphrates. *Bulletin du Musée du Caucase* 11: 145–148.

Berg, L.S. 1931. Description of a new siluroid fish, *Glyptosternum kurdistanicum*, from the basin of the Tigris River. *Bulletin de l'Académie des Sciences de, Union des Républiques Soviétiques Socialistes* 9: 1267–1270.

Berg, L.S. 1949a. [Freshwater fish of Iran and neighbouring countries]. *Trudy Zoologicheskogo Instituta Akademii Nauk SSSR* 8: 783–858. [in Russian].

Berg, L.S. 1949b. [Freshwater fishes of the U.S.S.R. and adjacent countries. *Izdatelstvo Akademii Nauk SSSR, Moskva & Leningrad* 470–925. [Israel Program for Scientific Translations, Jerusalem, 1965].

Bernatchez, L. 2001. The evolutionary history of brown trout (*Salmo trutta* L.) inferred from phylogeographic, nested clade, and mismatch analyses of mitochondrial DNA variation. *Evolution* 55: 351–379. https://doi.org/10.1111/j.0014-3820.2001.tb01300.x

Berrebi, P., P. Campton & G.P. Denys. 2022. Molecular characterization of rare anadromous Rhône River brown trout. *Knowledge and Management of Aquatic Ecosystems* 423: 24. https://doi.org/10.1051/kmae/2022022

Bianco, P.G. 2002. The status of the twaite shad, *Alosa agone*, in Italy and the Western Balkans. *Marine Ecology*, 23: 51–64. https://doi.org/10.1111/j.1439-0485.2002.tb00007.x

Bianco, P.G. & P.M. Bănărescu. 1982. A contribution to the knowledge of the Cyprinidae of Iran (Pisces, Cypriniformes). *Cybium* 6: 75–96.

Bianco, P.G. & H. Ahnelt. 1995 Systematic and zoogeography of *Knipowitschia caucasica* (Teleostei: Gobiidae) based on new records from Aegean Anatolian area. *Ichthyological Exploration of Freshwaters* 6: 49–60.

Bianco, P.G. & V. Ketmaier. 2014. A revision of the *Rutilus* complex from Mediterranean Europe with description of a new genus, *Sarmarutilus*, and a new species, *Rutilus stoumboudae* (Teleostei: Cyprinidae). *Zootaxa* 3841: 379–402. https://doi.org/1010.11646/zootaxa.3841.3.4

Bianco, P.G. & T.T. Nalbant. 1980. Redescription of *Cobitis linea*, with some remarks on the subgenus *Bicanestrinia* (Cypriniformes: Cobitidae). *Copeia* 4: 903–906. https://colab.ws/articles/10.2307%2F1444481#

Birecikligil, S.S., S. Eagderi, A. Jouladeh-Roudbar & E. Çiçek. 2017. *Alburnoides recepi*, a junior synonym of *Alburnus caeruleus* (Teleostei: Cyprinidae). *Zootaxa* 4277: 129–136. https://doi.org/10.11646/zootaxa.4277.1.10

Bíró, P. 1999a. *Ctenopharyngodon idella* (Cuvier and Valenciennes, 1844). Pp 305–343 *in:* P.M. Bănărescu. (Ed.). *The freshwater fishes of Europe. Vol.5/I: Cyprinidae 2/I*. AULA Verlag, Wiebelsheim.

Bíró, P. 1999b. *Mylopharyngodon piceus* (Richardson, 1846). Pp 347–365 *in:* P.M. Bănărescu. (Ed.). *The freshwater fishes of Europe. Vol.5/I: Cyprinidae 2/I*. AULA Verlag, Wiebelsheim.

Birstein, V.J. & W.E. Bemis. 1997. Leo Semenovich Berg and the biology of Acipenseriformes: a dedication. *Sturgeon Biodiversity and Conservation* 15–22.

Boëtius, J. & E.F. Harding. 1985. A re-examination of Johannes Schmidt's Atlantic eel investigations [*Anguilla rostrata*, *Anguilla anguilla*, spawning, larvae length (TNV), myomere counts (TNM)]. *Dana* 4: 129–162.

Bograd, L. 1961. Occurrence of *Mugil* in the rivers of Israel. Bulletin of the Research Council of Israel. Israel, 9B: 169–190.

Bogutskaya, N.G. 1986. On systematic relationships of the species of genera *Abramis*, *Blicca* and *Vimba*. *Вопросы ихтиологии* 26: 576–585.

Bogutskaya, N.G. 1988a. Limits and morphological features of the cyprinid subfamily Leuciscinae (Cyprinidae). *Proceedings of the Zoological Institute Academy of Sciences of USSR, Leningrad* 181: 96–113.

Bogutskaya, N.G. 1988b. Canal topography of the seismosensory system of cyprinids of the subfamilies Leuciscinae, Xenocyprininae, and Cultrinae. *Journal of Ichthyology* 28: 91–107.

Bogutskaya, N.G. 1992. A revision of species of the genus *Pseudophoxinus* (Leuciscinae, Cyprinidae) from Asia Minor. *Mitteilungen aus dem Hamburgischen Zoologischen Museum und Institut* 89: 261–290.

Bogutskaya, N.G. 1994. A description of *Leuciscus lepidus* (Heckel 1843) with comments on *Leuciscus* and leuciscine - aspinine relationships. *Annalen des Naturhistorischen Museums in Wien* 96(B): 599–620.

Bogutskaya, N.G. 1995. *Leuciscus kurui*, a new cyprinid fish from the Upper Tigris system. *Mitteilungen aus dem Hamburgischen Zoologischen Museum und Institut* 92: 149–154.

Bogutskaya, N.G. 1996. Contribution to the knowledge of leuciscine fishes of Asia Minor. Part 1. Morphology and taxonomic relationships of *Leuciscus borysthenicus* (Kessler, 1859), *Leuciscus smyrnaeus* Boulenger, 1896 and *Ladigesocypris ghigii* (Gianferrari, 1927) (Cyprinidae, Pisces). *Publicaciones Especiales Instituto Español de Oceanografía, Madrid* 21: 25–44.

Bogutskaya, N.G. 1997a. *Chondrostoma beysehirense*, a new cyprinid fish from Beysehir Lake, central Turkey. *Ichthyological Exploration of Freshwaters* 8: 151–158.

Bogutskaya, N.G. 1997b. Contribution to the knowledge of leuciscine fishes of Asia Minor. Part 2. An annotated checklist of leuciscine fishes (Leuciscinae, Cyprinidae) of Turkey with descriptions of a new species and two new subspecies. *Mitteilungen aus dem Hamburgischen Zoologischen Museum und Institut* 94: 161–186.

Bogutskaya, N.G. 2002. *Petroleuciscus*, a new genus for the *Leuciscus borysthenicus* species group (Teleostei: Cyprinidae). *Zoosystematica Rossica* 11: 235–237. https://doi.org/10.31610/zsr/2002.11.1.235

Bogutskaya, N.G. 2003a. *Barbus brachycephalus* Kessler, 1872. Pp 113–128 *in:* Bănărescu P.M. & N.G. Bogutskaya. (Eds.). *The freshwater fishes of Europe. Vol.5/II: Cyprinidae 2/II* Barbus. AULA Verlag, Wiebelsheim.

Bogutskaya, N.G. 2003b. *Barbus goktschaicus* Kessler, 1877. Pp 207–224 *in:* Bănărescu P.M.& N.G. Bogutskaya. (Eds.). *The freshwater fishes of Europe. Vol.5/II: Cyprinidae 2/II* Barbus. AULA Verlag, Wiebelsheim.

Bogutskaya, N.G. 2003c. *Barbus lacerta* Heckel, 1843. Pp 207–218 *in:* Bănărescu P.M. & N.G. Bogutskaya. (Eds.). *The freshwater fishes of Europe. Vol.5/II: Cyprinidae 2/II* Barbus. AULA Verlag, Wiebelsheim.

Bogutskaya, N.G., F. Küçük & E. Ünlü. 2000. *Alburnus baliki*, a new species of cyprinid fish from the Manavgat system, Turkey. *Ichthyological Exploration of Freshwaters* 11: 55–64.

Bogutskaya, N.G. & A.M. Komlev. 2001. Some new data on the morphology of *Rhodeus sericeus* sensu lato (Cyprinidae: Acheilognathinae) and a description of a new species, *Rhodeus colchicus*, from West Transcaucasia. *Proceedings of the Zoological Institute* 287: 81– 97.

Bogutskaya, N.G., P.M. Bănărescu & C. Almaca. 2003a. *Barbus capito* (Gueldernstaedt, 1773). Pp 147–162 *in:* Bănărescu P.M. & N.G. Bogutskaya. (Eds.). *The freshwater fishes of Europe. Vol.5/II: Cyprinidae 2/II* Barbus. AULA Verlag, Wiebelsheim.

Bogutskaya, N.G., P.M. Bănărescu & C. Almaca. 2003b. *Barbus mursa* (Gueldernstaedt, 1773). Pp 293–300 *in:* Bănărescu P.M. & N.G. Bogutskaya. (Eds.). *The freshwater fishes of Europe. Vol.5/II: Cyprinidae 2/II* Barbus. AULA Verlag, Wiebelsheim.

Bogutskaya, N.G., A.I. Smirnov & Y.V. Movchan. 2003c. *Barbus tauricus* Kessler, 1877. Pp 397–420 *in:* Bănărescu P.M. & N.G. Bogutskaya. (Eds.). *The freshwater fishes of Europe. Vol.5/II: Cyprinidae 2/II* Barbus. AULA Verlag, Wiebelsheim.

Bogutskaya, N.G. & A.M. Naseka. 2004. Catalogue of agnathans and fishes of fresh and brackish waters of Russia with comments on nomenclature and taxonomy. *Russian Academy of Sciences, Moscow* 1–389. [in Russian].

Bogutskaya, N.G., F. Küçük & M.A. Atalay. 2006. A description of three new species of the genus *Pseudophoxinus* from Turkey (Teleostei: Cyprinidae: Leuciscinae). *Zoosystematica Rossica* 335–341. https://doi.org/10.31610/zsr/2006.15.2.335

Bogutskaya, N.G., A.M. Naseka, S.V. Shedko, E.D. Vasil'eva & I.A. Chereshnev. 2008. The fishes of the Amur River: updated checklist and zoogeography. *Ichthyological Exploration of Freshwaters* 19: 301–366.

Bogutskaya, N.G. & B.W. Coad. 2009. A review of vertebral and fin-ray counts in the genus *Alburnoides* (Teleostei: Cyprinidae) with a description of six new species. *Zoosystematica Rossica* 18: 126–173. https://doi.org/10.31610/zsr/2009.18.1.126

Bogutskaya, N.G., P.V. Kijashko, M.I. Orlova & A.M. Naseka. 2013. Identification keys to fishes and invertebrates of the Caspian Sea. Vol. 1. Fishes and molluscs. KMK Press, Moscow. [in Russian].

Bohlen, J. 2003a. Similarities and differences in the reproductive biology of loaches (*Cobitis* and *Sabanejewia*) under laboratory conditions. Folia Zoologica 49 (Suppl. 1): 179–186.

Bohlen, J. 2003b. Spawning habitat in the spined loach, *Cobitis taenia* (Cypriniformes: Cobitidae). *Ichthyological Research* 50: 98–101. https://doi.org/10.1007/s102280300016

Bohlen, J., A. Perdices, I. Doadrio & P.S. Economidis. 2006. Vicariance, colonisation, and fast local speciation in Asia Minor and the Balkans as revealed from the phylogeny of spined loaches (Osteichthyes; Cobitidae). *Molecular Phylogenetics and Evolution* 39: 552–561. https://doi.org/10.1016/j.ympev.2005.12.007

Bohlen, J., F. Li & V. Šlechtová. 2020. Phylogenetic position of the genus *Bibarba* as revealed from molecular genetic data (Teleostei: Cobitidae). *Ichthyological Exploration of Freshwaters* 29: 297–304. http://doi.org/10.23788/IEF-1099

Borhani, M., N.M. Soofiani, S.A. Ebrahimi & S. Asadollah. 2017. First report on growth and reproduction of *Turcinoemacheilus bahaii* (Esmaeili, Sayyadzadeh, Özuluğ, Geiger and Freyhof, 2014) Zayandeh Roud River, Iran. *Austin Environmental Science* 2: 10–14.

Borkenhagen, K. 2014. A new genus and species of cyprinid fish (Actinopterygii, Cyprinidae) from the Arabian Peninsula, and its phylogenetic and zoogeographic affinities. *Environmental Biology of Fishes* 97: 1179–1195. https://doi.org/10.1007/s10641-014-0315-y

Borkenhagen, K. 2017. Molecular phylogeny of the tribe Torini Karaman, 1971 (Actinopterygii: Cypriniformes) from the Middle East and North Africa. *Zootaxa* 4236: 291–301. https://doi.org/10.11646/zootaxa.4236.2.4

Borkenhagen, K. & J. Freyhof. 2009. New records of the Levantine endemic cichlid *Tristramella simonis* from Syria, *Cybium* 33: 335–336. https://doi.org/10.26028/cybium/2009-334-006

Borkenhagen, K. & F. Krupp. 2013. Taxonomic revision of the genus *Carasobarbus* Karaman, 1971 (Actinopterygii, Cyprinidae). *ZooKeys* 339: 1–53. http://doi.org/10.3897/zookeys.339.4903

Borkenhagen, K., H.R. Esmaeili, S. Mohsenzadeh, F. Shahryari & A. Gholamifard. 2011. The molecular systematics of the *Carasobarbus* species from Iran and adjacent areas, with comments on *Carasobarbus albus* (Heckel, 1843). *Environmental Biology of Fishes* 91: 327–335. http://doi.org/10.1007/s10641-011-9787-1

Bornbusch, A.H. 1991. Monophyly of the catfish family Siluridae (Teleostei: Siluriformes), with a critique of previous hypotheses of the family's relationships. *Zoological Journal of the Linnean Society* 101: 105–120. https://doi.org/10.1111/j.1096-3642.1991.tb00888.x

Bornbusch, A.H. 1995. Phylogenetic relationships within the Eurasian catfish family Siluridae (Pisces: Siluriformes), with comments on generic validities and biogeography. *Zoological Journal of the Linnean Society* 115: 1–46. https://doi.org/10.1111/j.1096-3642.1995.tb02322.x

Boschung, H.T. & R.L. Mayden. 2004. *Fishes of Alabama*. Smithsonian Institution, Washington D.C. 1–736, 1–112, pls

Bosworth, W., P. Huchon & K. McClay. 2005. The Red Sea and Gulf of Aden basins. *Journal of African Earth Sciences* 43: 334–378. https://doi.org/10.1016/j.jafrearsci.2005.07.020

Britton, J.R., G.D. Davies, M. Brazier & A.C. Pinder. 2007. A case study on the population ecology of a topmouth gudgeon (*Pseudorasbora parva*) population in the UK and the implications for native fish communities. *Aquatic Conservation: marine and Freshwater Ecosystems* 17: 749–759. https://doi.org/10.1002/aqc.809

Brown, K.J., L. Rüber, R. Bills & J.J. Day. 2010. Mastacembelid eels support Lake Tanganyika as an evolutionary hotspot of diversification. *BMC Evolutionary Biology* 10: 188. https://doi.org/10.1186/1471-2148-10-188

Brownstein, C.D. & T.J .Near 2025. Toward a phylogenetic taxonomy of sturgeons (Acipenseriformes: Acipenseridae). Bulletin of the Peabody Museum of Natural History, 66(1): 3–24. https://doi.org/10.3374/014.066.0101

Bruton, M. 1979. The food and feeding behaviour of *Clarias gariepinus* (Pisces: Clariidae) in Lake Sibaya, South Africa, with emphasis on its role as a predator of cichlids. *The Transactions of the Zoological Society of London* 35: 47–114. https://doi.org/10.1111/j.1096-3642.1979.tb00057.x

Bruun, A.F. & E.W. Kaiser. 1944. *Iranocypris typhlops* n. g., n. sp., the first true cave-fish from Asia. *Danish Scientific Investigations in Iran (Copenhagen)* 4: 1–8.

Brylinska, M. & A. Boron. 2004. The common bream *Abramis brama* (Linnaeus, 1758) (Leuciscinae, Cyrpinidae). Rozprawy I Monographie - Dissertations and Monographs 91: 1–108.

Brylińska, M., E. Bryliński & P.M. Bănărescu. 1999a. *Tinca* Cuvier, 1817. Pp 225–228 *in:* Bănărescu, P.M. (Ed.). *The freshwater fishes of Europe. Vol.5/I: Cyprinidae 2/I.* AULA Verlag, Wiebelsheim.

Brylińska, M., E. Bryliński & M. Bninska. 1999b. *Tinca tinca* (Linnaeus, 1758). Pp 229–302 *in:* Bănărescu, P.M. (Ed.). *The freshwater fishes of Europe. Vol.5/I: Cyprinidae 2/I.* AULA Verlag, Wiebelsheim.

Brylinska, M., J. Szlachciak & A. Boron. 2006. The white bream *Abramis bjoerkna* (Linnaeus, 1758) (Leuciscinae, Cyrpinidae). Rozprawy I Monographie – Dissertations and Monographs 115: 1–79.

Buj, I., J. Miočić-Stošić, Z. Marčić, P. Mustafić, D. Zanella, M. Mrakovčić, T. Mihinjač & M. Ćaleta. 2015. Population genetic structure and demographic history of *Aphanius fasciatus* (Cyprinodontidae: Cyprinodontiformes) from hypersaline habitats in the eastern Adriatic. *Scientia marina* 79: 399–408. https://doi.org/10.3989/scimar.04198.06A

Canoğlu, H., İ. Aksu, D. Turan & Y. Bektaş. 2023. DNA barcoding of the genus *Alburnoides* Jeitteles, 1861 (Actinopterygii, Cyprinidae) from Anatolia, Turkey. *Zoosystematics and Evolution* 99(1): 185–194 https://doi.org/10.3897/zse.99.94333

Carpenter, K.E., F. Krupp, D.A. Jones & U. Zajonz. 1997. *FAO Species Identification Field Guide for fishery purposes: Living marine resources of Kuwait, Eastern Saudi Arabia, Bahrain, Qatar, and the United Arab Emirates*. FAO, Rome.

Casselman, J.M., E.J. Crossman, P.E. Ihssen, J.D. Reist & H.E. Booke. 1986. Identification of muskellunge, northern pike, and their hybrids. *American Fisheries Society Special Publication* 15, 14–46.

Cavender, T. M. & M.M. Coburn. 1992. Phylogenetic relationship of North American Cyprinidae. Systematics, historical ecology and North American freshwater fishes. Pp 293–327 *in:* Mayden R.L. (Ed.) *Systematics, historical ecology and North American freshwater fishes*. California: Stanford University Press, 293–327.

Cavraro, F., S. Malavasi, P. Torricelli, C. Gkenas, V. Liousia, I. Leonardos, I. Kappas, T.J. Abatzopoulos & A. Triantafyllidis. 2017. Genetic structure of the South European toothcarp *Aphanius fasciatus* (Actinopterygii: Cyprinodontidae) populations in the Mediterranean basin with a focus on the Venice lagoon. *The European Zoological Journal* 84: 153–166. https://10.1080/24750263.2017.1290154

Çevik, C., D. Ergüden & N. Tekelioğlu. 2010. Confirmation of the presence of the sea lamprey, *Petromyzon marinus* Linnaeus, 1758 in the Levantine Sea (Petromyzoniformes: Petromyzonidae). *Zoology in the Middle East* 49: 107–108. https://doi.org/10.1080/09397140.2010.10638398

Chai, I.J. & T. Arai. 2018. Ages at maturation of tropical freshwater eels, *Anguilla bicolor bicolor* and *A. bengalensis bengalensis*. *Journal of Applied Animal Research* 46: 1108–1113. https://doi.org/10.1080/09712119.2018.1470090

Chaichi, A.R., G.H. Vosoughi, F. Kaymaram, S. Jamili & H. Fazli. 2011. Reproduction characteristics of the *Vimba vimba persa* (Pallas, 1811), in coastal waters of the Caspian Sea. *Iranian Journal of Fisheries Sciences* 10: 585–595.

Charlebois, P.M., J.E. Marsden, R.G. Goettel, R.K. Wolfe, D.J. Jude & S. Rudnicka. 1997. The round goby, *Neogobius melanostomus* (Pallas), a review of European and North American literature. *Illinois Indiana Sea Grant Program and Illinois Natural History Survey. INHS Special Publication* 20.

Chen, Y.Y. 1998. *Fauna Sinica. Osteichthyes. Cypriniformes II.* Science Press, Beijing. [in Chinese, English summary].

Chen, I-S. & K.-T. Shao. 1996. A taxonomic review of the gobiid fish genus *Rhinogobius* Gill, 1859, from Taiwan, with description of three new species. *Zoological Studies* 35: 200–214.

Chen, I-S., P.J. Miller, H.-L. Wu & L.-S. Fang. 2002. Taxonomy and mitochondrial sequence evolution in non-diadromous species of *Rhinogobius* (Teleostei: Gobiidae) of Hainan Island, southern China. *Marine and Freshwater Research* 53: 259–273. https://doi.org/10.1071/MF01167

Chen, J.N., J.A. López, S. Lavoué, M. Miya & W.J. Chen. 2014. Phylogeny of the Elopomorpha (Teleostei): evidence from six nuclear and mitochondrial markers. *Molecular Phylogenetics and Evolution* 70: 152–161. https://doi.org/10.1016/j.ympev.2013.09.002

Chen, X.-Y. 1987. Studies on the phylogenetic relationships of Chinese leuciscine fishes (Pisces: Cypriniformes). *Acta Zootaxonomica Sinica* 12: 427–439.

Chichkoff, G. 1933. Sur un nouveau spirlin - *Alburnoides bipunctatus tzanevi* sub sp. nova. *Izvestiya na Bulgarskoto Geografsko Druzhestvo* 1: 375–383. [in Bulgarian, French summary].

Chichkoff, G. 1937. Sur les goujons habitant les eaux douces de la Bulgarie. *Godisnik na Sofijskija Universitet [Annuaire de l'Université de Sofia, Fiziko-Matematiceski Fakultet]* 33 (3): 227–289, pls 1–2. [in Bulgarian and French.]

Chino, N. & T. Arai. 2010. Occurrence of marine resident tropical eel *Anguilla bicolor bicolor* in Indonesia. *Marine Biology* 157: 1075–1081. https://doi.org/10.1007/s00227-009-1388-6

Chiozzia, G., M.L.J. Stiassny, S.E. Alter, G. De Marchia, Y. Mebrahtue, M. Tessemaf, B. Asmamawf, M. Fasolaa & A. Bellatia. 2017. Fishes in the desert: mitochondrial variation and phylogeography of *Danakilia* (Actinopterygii: Cichlidae) and *Aphanius* (Actinopterygii: Cyprinodontidae) in the Danakil depression of northeastern Africa. *Mitochondrial DNA, Part A* 29: 1025–1040. https://doi.org/10.1080/24701394.2017.1404043

Chu, X.-L. & Y.-R. Chen. 1989. *The fishes of Yunnan, China. Part 1 Cyprinidae*. Science Press, Beijing, China. i–vii + 1–377.

Chu, Y.T. 1935. Comparative studies on the scales and on the pharyngeals and their teeth in Chinese cyprinids, with particular reference to taxonomy and evolution. *Biological Bulletin St. John's University Shanghai* 2: i-x + 1–225, pls 1–30.

Ciccotto, P.J. & L.M. Page. 2016. Revised diagnosis of the genus *Gonorhynchus* McClelland (Teleostei: Cyprinidae: Labeonini) with redescription of *G. latius* (Hamilton) and revalidation of *G. wattanah* (Sykes). *Zootaxa* 4127: 471–492. https://doi.org/10.11646/zootaxa.4127.3.4

Çiçek, E. 2020. *Seminemacheilus dursunavsari*, a new nemacheilid species (Teleostei: Nemacheilidae) from Turkey. *Iranian Journal of Ichthyology* 7: 68–77. https://doi.org/10.22034/iji.v7i1.494

Çiçek, E., S. Eagderi & S. Sungur. 2018. *Oxynoemacheilus veyseli*, a new nemacheilid species from the upper Aras River drainage of Turkey (Teleostei: Nemacheilidae). *Iranian Journal of Ichthyology* 5: 232–242. https://doi.org/10.22034/iji.v5i3.302

Çiçek, E., R. Fricke, S. Eagderi, S. Sungur & S. Vanni. 2019. *Knipowitschia ephesi* Ahnelt, 1995, a junior synonym of *Gobius ricasolii* Di Caporiacco, 1935 (Teleostei: Gobiidae). *Zoology in the Middle East* 65: 330–335. https://doi.org/10.1080/09397140.2019.1648402

Çiçek, E. & S. Sungur. 2020. Ichthyofauna of Sultan Marshes (Turkey) and possible effects of fish invasion from Seyhan basin on diversity and conservation. *Commagene Journal of Biology* 4: 115–120. https://doi.org/10.31594/commagene.817306

Çiçek, E., S. Eagderi, S. Sungur & B. Seçer. 2021. Testing the "young-species" hypothesis for *Alburnus heckeli* Battalgil 1944 (Teleostei: Leuciscidae) inhabiting Lake Hazar, Turkey. *Proceedings of the Academy of Natural Sciences of Philadelphia* 167: 225–232. https://doi.org/10.1635/053.167.0113

Çiçek, E., S. Eagderi, S. Sungur & B. Seçer. 2022. *Capoeta ekmekciae* Turan, Kottelat, Kirankaya & Engin, 2006, a junior synonym of *Capoeta capoeta* (Güldenstädt, 1773) (Teleostei: Cyprinidae). *Journal of Fish Biology* 101 (5): 1326–1332. https://doi.org/10.1111/jfb.15204

Çiftçi, Y., A.G. Mutlu, S.S. Güçlü, D. Turan & F. Küçük. 2020. Phylogeography of the genus *Chondrostoma* Agassiz, 1835 (Teleostei: Leuciscidae) in Anatolia, as inferred from mitochondrial DNA analysis, *Zoology in the Middle East* 66: 206–221. https://doi.org/10.1080/09397140.2020.1788255

Claramunt, S. & J.Cracraft. 2015. A new time tree reveals earth history's imprint on the evolution of modern birds. *Science Advances. 1.* https://doi.org/10.1126/sciadv.1501005

Clayton, D.A. & T.C. Vaughan. 1982. Pentagonal territories of the mudskipper *Boleophthalmus boddarti* (Pisces: Gobiidae). *Copeia* 1: 232–234. https://doi.org/10.2307/1444308

Clayton, D.A. & J.M. Wright. 1989. Mud-walled territories and feeding behaviour of *Boleophthalmus boddarti* (Pisces: Gobiidae) on the mudflats of Kuwait. *Journal of Ethology* 7: 91–95. https://doi.org/10.1007/BF02350029

Cluver, D.C. & T. Pipan. 2009. *The biology of caves and other subterranean habitats*. Oxford University Press.

Coad, B.W. 1981a. *Glyptothorax silviae*, a new species of sisorid catfish from southwestern Iran. *Japanese Journal of Ichthyology* 27: 291–295.

Coad, B.W. 1981b. *Pseudophoxinus persidis*, a new cyprinid fish from Fars, southern Iran. *Canadian Journal of Zoology* 59: 2058–2063. https://doi.org/10.1139/z81-281

Coad, B.W. 1982a. *Garra persica* Berg, 1913, a valid species of cyprinid fish from south-eastern Iran. *Cybium* 6: 97–100.

Coad, B.W. 1982b. A new genus and species of cichlid endemic to southern Iran. *Copeia* (1): 28–37.

Coad, B.W. 1984. *Acanthobrama centisquama* Heckel and the validity of the genus Mirogrex Goren, Fishelson and Trewavas (Osteichthyes: Cyprinidae). *Hydrobiologia* 109: 275–278. https://doi.org/10.1007/BF00007745

Coad, B.W. 1988. *Aphanius vladykovi*, a new species of tooth-carp from the Zagros Mountains of Iran (Osteichthyes: Cyprinodontidae). *Environmental Biology of Fishes* 23: 115–125. https://doi.org/10.1007/BF00000741

Coad, B.W. 1996. Systematics of the tooth-carp genus *Aphanius* Nardo, 1827 (Actinopterygii: Cyprinodontidae) in Fars Province, southern Iran. *Biologica Bratislava* 51: 163–172.

Coad, B.W. 2009a. A new species of tooth-carp, *Aphanius mesopotamicus*, from Iran and Iraq (Actinopterygii, Cyprinodontidae). *Zookeys* 31: 149–163. https://doi.org/10.3897/zookeys.31.131

Coad, B.W. 2009b. *Alburnus zagrosensis* n. sp., a new species of fish from the Zagros Mountains of Iran (Actinopterygii: Cyprinidae). *Zoology in the Middle East* 48: 63–69. https://doi.org/10.1080/09397140.2009.10638367

Coad, B.W. 2010. *Freshwater Fishes of Iraq*. Pensoft Publishers, Sofia-Moscow.

Coad, B.W. 2010b. Threatened fishes of the world: *Acanthobrama hadiyahensis* Coad, Alkahem and Behnke, 1983 (Cyprinidae). *Environmental Biology of Fishes* 87: 99. https://doi.org/10.1007/s10641-009-9566-4

Coad, B.W. 2014. *Fishes of Afghanistan*. Pensoft, Sofia-Moskow.

Coad, B.W. 2015a. Review of the freshwater sharks of Iran (family Carcharhinidae). *International Journal of Aquatic Biology* 3: 218–224. https://doi.org/10.22034/ijab.v3i4.100

Coad, B.W. 2015b. Review of the spiny eels of Iran (Family Mastacembelidae). *Iranian Journal of Ichthyology* 2: 1–12. https://doi.org/10.22034/iji.v2i1.7

Coad, B.W. 2016a. Contribution to the knowledge of the snakeheads of Iran (Family Channidae). *Iranian Journal of Ichthyology* 3: 65–72. https://doi.org/10.22034/iji.v3i1.144

Coad, B.W. 2016b. Review of the freshwater cuskfishes of Iran (Family Lotidae). *Iranian Journal of Ichthyology* 3: 229–235. https://doi.org/10.22034/iji.v3i4.195

Coad, B.W. 2016c. Review of the lampreys of Iran (Family Petromyzontidae). *International Journal of Aquatic Biology* 4: 256–268. https://doi.org/10.22034/ijab.v4i4.213

Coad, B.W. 2021a. *Carps and Minnows of Iran (Families Cyprinidae and Leuciscidae) Volume I: General introduction and Carps (Family Cyprinidae).* Canadian Museum of Nature, Ottawa, Ontario, Canada, 1–1263. Available at: http://www.briancoad.com/species %20accounts/Carps %20of %20Iran %2010Sept2opt1.pdf (Accessed January 2023).

Coad, B.W. 2021b. *Carps and Minnows of Iran (Families Cyprinidae and Leuciscidae) Volume II: Minnows (Family Leuciscidae) and Bibliography.* Canadian Museum of Nature, Ottawa, Ontario, Canada, 1–909. Available at: http://www.briancoad.com/species %20accounts/ Minnows %20of %20Iran %2010Sept21opt.pdf (Accessed January 2023).

Coad, B.W. & A. Abdoli. 2000. *Rhinogobius* cf. *similis* Gill, 1859, a goby new to the fish fauna of Iran and the problem of alien invasions. *Zoology in the Middle East* 20: 55–59. https://doi.org/10.1080/09397140.200 0.10637812

Coad, B.W. & N.G. Bogutskaya. 2009. *Alburnoides qanati*, a new species of cyprinid fish from southern Iran (Actinopterygii, Cyprinidae). *ZooKeys* 13: 67–77. https://doi.org/10.3897/zookeys.13.194

Coad, B.W. & N.G. Bogutskaya. 2010. *Petroleuciscus esfahani*, a new species of fish from central Iran (Actinopterygii: Cyprinidae). *Zootaxa* 2534: 37–47. https://doi.org/10.5281/zenodo.293380

Coad, B.W. & N.G. Bogutskaya. 2012. A new species of riffle minnow, *Alburnoides holciki*, from the Hari River basin in Afghanistan and Iran (Actinopterygii: Cyprinidae). *Zootaxa* 3453: 43–55. https://doi. org/10.5281/zenodo.214789

Coad, B.W. & G.B. Delmastro. 1985. Notes on a sisorid catfish from the Black Sea drainage of Turkey. *Cybium* 9: 221–224.

Coad, B.W. & J. Holcík. 1999. Systematics of the cyprinid fish *Chalcaburnus atropatenae* (Berg, 1925) from the Lake Orumiyeh basin in northwest Iran. *Biologia (Bratislava)* 52: 179–186.

Coad, B.W. & F. Krupp. 1983. Redescription of *Barilius mesopotamicus* Berg, 1932 a poorly known cyprinid fish from the Tigris-Euphrates basin. *Cybium* 7: 47–56.

Coad, B.W. & F. Krupp. 1994. *Capoeta aculeata* (Valenciennes in Cuv. & Val. 1844), a valid species of cyprinid fish from Iran (Teleostei: Cyprinidae). *Zoology in the Middle East* 10: 63–72. https://doi.org/10. 1080/09397140.1994.10637662

Coad, B.W. & N. Keyzer-de Ville. 2005. On the validity of the species in the snow-trout genus *Schizocypris* Regan, 1914 (Cyprinidae: Actinopterygii). *Zoology in the Middle East* 35: 35–42. https://doi. org/10.1080/09397140.2005.10638101

Coad, B.W. & N. Najafpour. 1997. *Barbus sublimus*, a new species of cyprinid fish from Khuzestan Province, Iran. *Ichthyological Exploration of Freshwaters* 7: 273–278.

Coad, B.W. & T.T. Nalbant. 2005. A new genus and a new species of a remarkable nemacheilid fish from Iran (Pisces: Ostariophysi: Nemacheilidae). *Travaux du Muséum d'Histoire Naturelle "Grigore Antipa"* 48: 303–308.

Coad, B.W. & M. Sarıeyyüpoğlu. 1988. *Cobitis elazigensis*, a new species of cobitidid fish from Anatolia, Turkey. *Japanese Journal of Ichthyology* 34: 426–430.

Coad, B.W., H.F. Alkahem & R.J. Behnke. 1983. *Acanthobrama hadiyahensis*, a new species of cyprinid fish from Saudi Arabia. *National Museums of Natural Sciences, Publications in Natural Sciences* 2: 1–6.

Coban, M.Z. & F. Yüksel. 2013. Age and growth properties of *Acanthobrama marmid* Heckel, 1843 population inhabiting Uzuncayir Damlake (Tunceli/Turkey). *Journal of Animal and Veterinary Advances* 12: 644–649. https://doi.org/10.36478/ javaa.2013.644.649

Collin, R. 2013. Phylogenetic patterns and phenotypic plasticity of molluscan sexual systems. *Integrative and Comparative Biology* 53: 723–735. https://doi.org/10.1093/icb/ict076

Conte-Grand, C., R. Britz, N. Dahanukar, R. Raghavan, R. Pethiyagoda, H.H. Tan, R.K. Hadiaty, N.S. Yaakob & L. Rüber. 2017. Barcoding snakeheads (Teleostei, Channidae) revisited: Discovering greater species diversity and resolving perpetuated taxonomic confusions. *PLOS ONE* 12: e0184017. https://doi.org/10.1371/ journal. pone.0184017

Conway, K.W., D.R. Edds, J. Shrestha & R.L. Mayden. 2011. A new species of gravel-dwelling loach (Ostariophysi: Nemacheilidae) from the Nepalese Himalayan foothills. *Journal of Fish Biology* 79: 1746–1759. https://doi.org/10.1111/j.1095-8649.2011.03108.x

Cooper, J.A. & F. Chapleau. 1998. Monophyly and intrarelationships of the family Pleuronectidae (Pleuronectiformes), with a revised classification. *Fish Bulletin* 96: 686–726.

Costa Pereira, N. da. 1995. The freshwater fishes of the Iberian Peninsula. *Arquvos de Museu Bocage, Nova Serie* 2: 473–538.

Cracraft, J. 1989. Speciation and its ontology: the empirical consequences of alternative species concepts for understanding patterns and processes of differentiation. In D. Otte and J. A. Endler. (Eds). *Speciation and its Consequences*. Sinauer Assoc., Sunderland, Mass., 28–59.

Craig, J.F. 1996. *Pike: biology and exploitation*. Fish & Fisheries Series 19, Springer.

Craig, J.F. 2000. *Percid Fishes. Systematics, Ecology and Exploitation*. 3rd edition., Blackwell Science.

Craig, J.F. 2008. A short review of pike ecology. *Hydrobiologia* 601: 5–16. https://doi.org/10.1007/s10750-007-9262-3

Crass, R.S. 1964. *Freshwater fishes of Natal*. Shuter and Shooter, Pietermaritzburg.

Crook, V. & M. Nakamura. 2013. Assessing supply chain and market impacts of a CITES listing on Anguilla species. *Traffic Bulletin* 25: 24–30.

Cunha, C., N. Mesquita, T.E. Dowling, A. Gilles & M.M. Coelho. 2002. Phylogenetic relationships of Eurasian and American cyprinids using cytochrome b sequences. *Journal of Fish Biology* 61: 929–944. https://doi.org/10.1111/j.1095-8649.2002.tb01853.x

Dadikyan, M.G. 1986. *Ryby Armenii* [*Fishes of Armenia*]. Armenian Academy of Sciences, Yerevan. 1–245.

Dağlı, M. & A.Ü. Erdemli. 2009. An investigation on the fish fauna of Balıksuyu stream (Kilis, Turkey). *International Journal of Natural and Engineering Sciences* 3: 1–2.

Danilov-Danilyan, V.I. 2001. *The Red Book of Russian Federation*. Animals. Moskow, Astrel. [in Russian].

Daoulas, C., A.N. Economou, T. Psarras & R. Barbieri-Tseliki. 1993. Reproductive strategies and early development of three freshwater gobies. *Journal of Fish Biology* 42: 749–776. https://doi. org/10.1111/j.1095-8649.1993.tb00382.x

Daoulas, C., A. Economou, M.T. Stoumboudi, T. Psarras & R. Barbieri-Tseliki. 1997. Larvae development in a landlocked population of *Atherina boyeri* in Lake Trichonis, Greece. *Israel Journal of Zoology* 43: 159–166.

Darwin, C. 1859. *On the origin of species by means of natural selection, or the preservation of favoured races in the struggle for life*. First edition. John Murray, London.

Daştan, S.D., F. Bardakçı & N. Değerli. 2012. Genetic diversity of *Cyprinion macrostomus* Heckel, 1843 (Teleostei: Cyprinidae) in Anatolia. *Turkish Journal of Fisheries and Aquatic Sciences* 12: 651–659. https://doi. org/10.4194/1303-2712-v12_3_13

Datta, A., A. Gomes, B. Sarangi, P.K. Kar & S.C. Lahiri. 1982. Pharmacodynamic actions of crude venom of the Indian catfish *Heteropneustes fossilis*. *Indian Journal of Medical Research* 76: 892–897.

Dawson, C.E. 1985. *Indo-pacific pipefishes (Red Sea to the Americas).* Ocean Springs, MS: Gulf Coast Research Laboratory.

Day, J.J., Fages A., K.J. Brown, E.J. Vreven, M.L. J. Stiassny, R. Bills, J.P. Friel & L. Rüber. 2017. Multiple independent colonizations into the Congo basin during the continental radiation of African *Mastacembelus* spiny eels. *Journal of Biogeography* 44: 2308–2318. https://doi.org/10.1111/jbi.13037

De Filippi, F. 1863. Nuove o poco note specie di animali vertebrati raccolte in un viaggio in Persia nell' estate dell' anno 1862. *Archivio per la Zoologia, l'Anatomia e la Fisiologia* 2: 377–394.

De Filippi, F. 1865. *Note di un viaggio in Persia nel 1862. Volume unico*. G. Daelli, Milano. i–xiii+1–396. [Pisces, 357–360].

De Silva, S.S. & J. Chandrasoma. 1980. Reproductive biology of *Sarotherodon mossambicus*, an introduced species, in an ancient man-made lake in Sri Lanka. *Environmental Biology of Fishes* 5: 253–259. https://doi.org/10.1007/BF00005360

Dedeoğlu, Y., T. Yeşilçiçek & F. Kalaycı. 2020. Borçka Baraj Gölü'ndeki (Artvin) Tatlısu Kefali (*Squalius orientalis* Heckel, 1847)'nin Bazı Büyüme Parametreleri. *Recep Tayyip Erdoğan Üniversitesi Fen ve Mühendislik Bilimleri Dergisi* 1: 38–53. [in Turkish].

Dekker, W. 2003. Did lack of spawners cause the collapse of the European eel, Anguilla anguilla? *Fisheries Management and Ecology* 10: 365–376. https://doi.org/10.1111/j.1365-2400.2003.00352.x

Delmastro, G.B. 1982. Un nuovo cobite dai tributari del Mar Nero in Asia Minore (Osteichthyes, Cobitidae). *Rivista Piemontese di Storia Naturale* 3: 53–59.

Demirci, S. 2016. Growth parameters of *Pseudophoxinus anatolicus* (Hanko 1924): an endemic and endangered fish species of Beyşehir Lake (Turkey). *Research Journal of Fisheries and Hydrobiology* 11: 1–6.

Denys, G.P. J., A. Dettai, H. Persat, M. Hautecoeur & P. Keith. 2014. Morphological and molecular evidence of three species of pikes *Esox* spp. (Actinopterygii, Esocidae) in France, including the description of a new species. *Comptes Rendus Biologies* 337: 521–534. https://doi.org/10.1016/j.crvi.2014.07.002

Derjavin, A.N. 1934. *Freshwater fish of the south coast of the Caspian Sea*. Trudy Azerbaidzhanskogo otdela Zakavkazskogo filiala Akademii Nauk SSSR. *Sektor Zoologii* 7: 91–126.

Derjavin, A.N. 1937. A new species of roach *Rutilus* (*Orthroleucos*) *atropatenus* nov. sp. from Azerbaijan. *Trudy Azerbaidzhanskogo Filiala, Akademii Nauk SSSR, Zoolocicheskij Serija* 20: 71–78. [in Russian, English summary]

Derjugin, K.M. 1899. Materials for the ichthyological fauna of southwestern Transcaucasia. *Ezhegodnik Zoologicheskogo Muzeya Imperatorskoy Akademii Nauk [Annuaire du Musée Zoologique de l'Académie Impériale des Sciences]* 4: 148–171. [in Russian].

DeVaney, S.C., K.M. McNyset, J.B. Williams, A.T. Peterson & E.O. Wiley. 2009. A tale of four "carp": invasion potential and ecological niche modeling. *PLOS ONE* 4(5): e5451. https://doi.org/10.1371/journal.pone.0005451

Devin, S. & J.-N. Beisel. 2007. Biological and ecological characteristics of invasive species: a gammarid study. *Biological Invasions* 9: 13–24.

Doadrio, I. 2001. *Atlas y Libro Rojo de los Peces Continentales de España*. Dirección General de Conservación de la Naturaleza-Consejo Superior de Investigaciones Científicas (CSIC), Madrid.

Docker, M.F. 2015. Lampreys: Biology, Conservation and Control. *Fish & Fisheries Series* 37. Springer. Dordrecht. https://doi.org/0.1007/978-94-017-9306-3

Dorofeyeva, Y.A. 1967. Comparative morphological fundamentals of the classification of east-European salmon. *Voprosy Ikhtiologii* 7: 3–17.

Dunz, A.R. & U.K. Schliewen. 2010. Description of a *Tilapia* (*Coptodon*) species flock of Lake Ejagham (Cameroon), including a redescription of *Tilapia deckerti* Thys van den Audenaerde, 1967. *Spixiana (München)* 33: 251–280.

Dunz, A.R. & U.K. Schliewen. 2013. Molecular phylogeny and revised classification of the haplotilapiine cichlid fishes formerly referred to as "*Tilapia*". *Molecular Phylogenetics and Evolution* 68: 64–80. https://doi.org/10.1016/j.ympev.2013.03.015

Duquenne-Delobel, E., I. Doadrio & G.P.J. Denys. 2022. Revalidation of the genus *Ichthyocoris* Bonaparte, 1840 (Actinopterygii: Blenniiformes: Blenniidae). *Acta Ichthyologica et Piscatoria* 52: 35–41. https://doi.org/10.3897/aiep.52.79458

Durand, J.-D., E. Ünlü, I. Doadrio, S. Pipoyan & A.R. Templeton. 2000. Origin, radiation, dispersion and allopatric hybridization in the chub *Leuciscus cephalus*. *Proceedings of the Royal Society B, Biological Sciences* 267: 1687–1697. https://doi.org/10.1098/rspb.2000.1196

Durand, J.-D., C.S. Tsigenopoulos, E. Ünlü & P. Berrebi. 2002. Erratum to "Phylogeny and biogeography of the Family Cyprinidae in the Middle East inferred from Cytochrome b DNA—Evolutionary Significance of this Region". *Molecular Phylogenetics and Evolution* 22: 91–100.

Durand, J.D., K.N. Shen, W.J. Chen, B.W. Jamandre, H. Blel, K. Diop, M. Nirchio, F.J. Garcia de León, A.K. Whitfield, C.W. Chang & P. Borsa. 2012. Systematics of the grey mullets (Teleostei: Mugiliformes: Mugilidae): molecular phylogenetic evidence challenges two centuries of morphology-based taxonomy. *Molecular Phylogenetics and Evolution* 64: 73–92. https://doi.org/0.1016/j.ympev.2012.03.006

Eagderi, S. & M. Moradi. 2017. Range extension of the lake goby *Rhinogobius similis* Gill, 1859 (Teleost: Gobiidae) to Urmia Lake basin in northwestern Iran. *Biharean Biologist* 11: 123–125.

Eagderi, S., A. Jouladeh-Roudbar, M. Nasri, G. Sayyadzadeh & H.R. Esmaeili. 2017a. Taxonomic status of the genus *Cobitis* (Teleostei: Cobitidae) in the Namak Lake basin, Iran. *Iranian Journal of Ichthyology* 4: 131–139. https://doi.org/10.22034/iji.v4i2.228

Eagderi, S., A. Jouladeh-Roudbar, P. Jalili, G. Sayyadzadeh & H.R. Esmaeili. 2017b. Taxonomic status of the genus *Cobitis* Linnaeus, 1758 (Teleostei: Cobitidae) in the southern Caspian Sea basin, Iran with description of a new species. FishTaxa 2: 48–61.

Eagderi, S., A. Jouladeh-Roudbar, S.S. Birecikligil, E. Çiçek & B.W. Coad. 2017c. *Chondrostoma esmaeilii*, a new cyprinid species from the Tigris basin in Iran (Teleostei: Cyprinidae). *Vertebrate Zoology* 67: 125–132.

Eagderi, S., M. Nasri & E. Çiçek. 2018a. First record of the Amur goby *Rhinogobius lindbergi* Berg 1933 (Gobiidae) from the Tigris River drainage, Iran. *International Journal of Aquatic Biology* 6: 202–207. https://doi.org/10.22034/ijab.v6i4.528

Eagderi, S., P. Jalili & E. Çiçek. 2018b. *Oxynoemacheilus elsae*, a new species from the Urmia Lake basin of Iran (Teleostei: Nemacheilidae). *FishTaxa* 3: 453–459.

Eagderi, S., A. Jouladeh-Roudbar, J. Imani Harsini & M. Rostami. 2019a. Phylogeny of the members of the genus *Alburnoides* in Iran using COI gene. *Iranian Scientific Fisheries Journal* 28: 125–137. [in Farsi]. https://doi.org/10.22092/ISFJ.2019.119160

Eagderi, S., H. Mousavi-Sabet & J. Freyhof. 2019b. *Paraschistura makranensis*, a new loach from the Jegin River drainage in southern Iran with comments on *P. ilamensis* and *P. pasatigris* (Teleostei: Nemacheilidae). *Zootaxa* 4668: 258–270. https://doi.org/10.11646/zootaxa.4668.2.6

Eagderi, S., N. Nikmehr & J. Freyhof. 2019c. *Alburnus zagrosensis*, a junior synonym of *Alburnus sellal* (Teleostei: Leuciscidae). *Zootaxa* 4652: 367–374. https://doi.org/10.11646/zootaxa.4652.2.9

Eagderi, S., N. Nikmehr, E. Çiçek, H.R. Esmaeili, S. Vatandoust & H. Mousavi-Sabet. 2019d. *Barbus urmianus*, a new species from Urmia

Lake basin, Iran (Teleostei: Cyprinidae). *International Journal of Aquatic Biology* 7: 239–244. https://doi.org/10.22034/ijab.v7i4.725

Eagderi, S., N. Nikmehr & H. Poorbagher. 2020. *Ponticola patimari* sp. nov. (Gobiiformes: Gobiidae) from the southern Caspian Sea basin, Iran. *FishTaxa* 17: 22–31.

Eagderi, S. & S. Mousavi-Sabet. 2021. *Capoeta raghazensis*, a new species of algae-scraping cyprinid from the Raghaz Canyon in Hormuz basin, southern Iran (Teleostei: Cyprinidae). *FishTaxa* 22: 37–44.

Eagderi, S., B. Seçer & J. Freyhof. 2022. *Cobitis indus*, a new spined loach from the Dalaman River in the Eastern Aegean Sea basin (Teleostei: Cobitidae). *Zootaxa* 5162: 410–420. https://doi.org/10.11646/zootaxa.5162.4.5

Ebert, D.A., M. Dando & S. Fowler. 2013. *Sharks of the World: A Complete Guide*. Princton University Press.

Ebrahimi, I., M. Borhani, N. Mahboubi Soufiani & S. Asadohhah. 2017. Investigation the diet and digestive tract histology of loach fish *Turcinoemacheilus bahaii* in Zayandeh Roud River. *Journal of Animal Research (Iranian Journal of Biology)* 30: 1–12. https://doi.org/20.1001.1.23832614.1396.30.1.1.0

Economidis, P.S. & P.J. Miller. 1990. Systematics of freshwater gobies from Greece (Teleostei: Gobiidae). *Journal of Zoology* 221: 125–170. https://doi.org/10.1111/j.1469-7998.1990.tb03781.x

Economidis, P.S. & T.T. Nalbant. 1997. A study of the loaches of the genera *Cobitis* and *Sabanejewia* (Pisces, Cobitidae) of Greece, with description of six new taxa. Travaux du Muséum d'Histoire Naturelle "Grigore Antipa" 36: 295–347.

Economidis, P.S., E.T. Koutrakis & D.C. Bobori. 2000. Distribution and conservation of *Acipenser sturio* L., 1758 and related species in Greek waters. *Boletín, Instituto Español de Oceanografía* 16 (1/4): 81–88.

Economou, A.N., C. Daoulas, T. Psarras & R. Barbieri-Tseliki. 1994. Further data on the reproduction and larval development of *Knipowitschia caucasica* (Gobiidae). *Journal of Fish Biology* 45: 360–362. https://doi.org/10.1111/j.1095-8649.1994.tb01317.x

Ekmekçi, F.G. 1996. Some of the growth and reproduction properties of *Capoeta tinca* (Heckel, 1843) in Sarıyar Dam Lake (Ankara). *Turkish Journal of Zoology* 20: 117–126. [in Turkish].

Ekmekçi, F.G. & Ş.G. Kirankaya. 2010. Threatened fishes of the world: *Pseudophoxinus crassus* (Ladiges, 1960) Cyprinidae, from Central Anatolia, Turkey. Environ Biol Fish 87, 11–12. https://doi.org/10.1007/s10641-009-9563-7

Ekmekçi, F.G. & P. Bănărescu. 1998. A revision of the generic position of *Barynotus* (*Systomus*) *verhoeffi*, and the validity of the genera *Carasobarbus*, *Kosswigobarbus* and *Mesopotamichthys* (Pisces, Cyprinidae). *Folia Zoologica* 47 (Suppl. 1): 87–96.

Ekmekçi, F.G. & S.C. Özeren. 2003. Reproductive biology of *Capoeta tinca* in Gelingullu Reservoir, Turkey. *Folia Zoologica Praha* 52: 323–328.

Ekmekçi, F.G., M.A. Atalay, B. Yoğurtçuoğlu, D. Turan & F. Küçük. 2015. A new species of *Pseudophoxinus* (Teleostei: Cyprinidae) from southwestern Anatolia, Turkey. *Zootaxa* 4033: 117–128. https://doi.org/10.11646/zootaxa.4033.1.6

Eknath, A.E. & B.O. Acosta. 1998. *Genetic improvement of farmed tilapia project final report (1988–1997)*. ICLARM, Manila, Philippines.

Elanidze, R.F. 1983. *Fishes of rivers and lakes of Georgia*. Metzniereba, Tbilisi. [in Russian].

Elliot, J.M. 1994. *Quantitative ecology of the brown trout*. Oxford Series in Ecology and Evolution. Oxford University Press, Oxford.

Elp, M., M. Özuluğ, F. Şen & J. Freyhof. 2013. Validation of *Alburnus timarensis* from the Lake Van basin, eastern Anatolia (Teleostei: Cyprinidae). Zoology in the Middle East 59: 235–244. https://doi.org/10.1080/09397140.2013.841430

Elp, M., F. Şen & M. Özuluğ. 2015. *Alburnus selcuklui*, a new species of cyprinid fish from east Anatolia, Turkey (Teleostei: Cyprinidae). Turkish Journal of Fisheries and Aquatic Sciences 15: 181–186. https://doi.org/10.4194/1303-2712-v15_1_20

Elp, M., M.İ. Osmanoğlu, A.E. Kadak & D. Turan. 2018. Characteristics of *Capoeta oguzelii*, a new species of cyprinid fish from the Ezine Stream, Black Sea basin, Turkey (Teleostei: Cyprinidae). *Zoology in the Middle East* 64: 102–111. https://doi.org/10.1080/09397140.2018.1442295

Elron, E., A. Gasith & M. Goren. 2006. Reproductive strategy of a small endemic cyprinid, the Yarqon bleak (*Acanthobrama telavivensis*), in a mediterranean-type stream. *Environmental Biology of Fishes* 77: 141–155. https://doi.org/10.1007/s10641-006-9066-8

Elvira, B. 1987. Taxonomic revision of the genus *Chondrostoma* Agassiz, 1835 (Pisces, Cyprinidae). *Cybium* 11: 111–140.

Elvira, B. 2000. Symposium on conservation of the Atlantic sturgeon *Acipenser sturio* L., 1758 in Europe. *Boletín, Instituto Español de Oceanografía* 16: 1–253.

Emiroğlu, Ö., F.G. Ekmekçi, S. Aksu, S. Başkurt, A. Atalay & A.S. Tarkan. 2016. Introduction and establishment of tropical ornamental fish, *Pterygoplichthys* spp. (Actinopterygii: Siluriformes: Loricariidae) in hot springs: Aquarium trade as a potential risk for biodiversity in Turkey. *Acta Ichthyologica et Piscatoria* 46(4): 351–356. https://doi.org/10.3750/AIP2016.46.4.07

Emiroğlu, Ö., M. Atalay, F.G. Ekmekçi, S. Aksu, S. Başkurt, E. Keskin, E. Ünal, B. Yoğurtçuoğlu & A.S. Tarkan. 2020. One of the world's worst invasive species, *Clarias batrachus* (Actinopterygii: Siluriformes: Clariidae), has arrived and established a population in Turkey. *Acta Ichthyologica et Piscatoria* 50: 391–400. https://doi.org/10.3750/AIEP/03028

Endruweit, M. 2018. Description of four new species of freshwater gobies from the Black River drainage in China and Vietnam (Teleostei: Gobiidae). *Zootaxa* 4486: 284–310. https://doi.org/10.11646/zootaxa.4486.3.4

Englmaier, G.K., N.V. Rodríguez, H. Waidbacher, A. Palandačić, G. Tesfaye, W. Gessl & P. Meulenbroek. 2020. New data on *Garra makiensis* (Cyprinidae, Labeoinae) from the Awash River (Ethiopia) with remarks on its relationships to congeners on the Arabian Peninsula. *ZooKeys* 984: 133–163. https://doi.org/10.3897/zookeys.984.55982

Epitashvili, G., M. Geiger, J.J. Astrin, F. Herder, B. Japoshvili & L. Mumladze. 2020. Towards retrieving the Promethean treasure: a first molecular assessment of the freshwater fish diversity of Georgia. *Biodiversity Data Journal* 8: e57862. https://doi.org/10.3897/BDJ.8.e57862

Epitashvili, G., B. Japoshvili & L. Mumladze. 2023. *Ponticola alasanicus* sp. n. (Gobiiformes, Gobiidae) from the Alazani River basin, Georgia. Biodiversity Journal 11: e101095: 1–20. https://doi.org/10.3897/BDJ.11.e101095

Erdemli, A.Ü. 1982. Beyşehir Gölü Balıkları. Selçuk *Üniversitesi Fen Fakültesi Fen Dergisi* 2: 131–142.

Erhan, Ü. 2021. Current Status of Fish in Balıklıgöl, Şanlıurfa, Turkey and Comments on "A Preliminary Investigation on Serious Mortalities of Fish in Balıklıgöl (Halil-ür Rahman Gölü, Şanlıurfa)" by Öktener et al. (2008). *Acta Biologica Turcica* 34: 86–91.

Erk'akan, F. 2012. Two new *Oxynoemacheilus* (Teleostei: Nemacheilidae) species from western Turkey. *Research Journal of Biological Sciences* 7: 97–101. https://doi.org/10.3923/rjbsci.2012.97.101

Erk'akan, F. & M. Kuru. 1986. A new noemacheilinae loach subspecies from the Lake Van basin, Turkey (Osteichthyes, Cobitidae). *Doga Turk Biyoloji Dergisi* 10: 160–162.

Erk'akan, F. & F. Özdemir. 2014. The first new cave fish species, *Cobitis damlae* (Teleostei: Cobitidae) from Turkey. *Hacettepe Journal of Biology and Chemistry* 42: 275–279.

Erk'akan, F., F.G. Atalay-Ekmekçi & T.T. Nalbant. 1998. Four new species and one new subspecies of the genus *Cobitis* (Pisces: Ostariophysi: Cobitidae) from Turkey. *Turkish Journal of Zoology* 22: 9–15.

Erk'akan, F., F.G. Atalay-Ekmekçi & T.T. Nalbant. 1999. A review of the genus *Cobitis* in Turkey (Pisces: Ostariophysi: Cobitidae). *Hydrobiologia* 403: 13–26. https://doi.org/10.1023/A:1003794726444

Erk'akan, F., T.T. Nalbant & S.C. Özeren. 2007. Seven new species of *Barbatula*, three new species of *Schistura* and a new species of *Seminemacheilus* (Ostariophysi: Balitoridae: Nemacheilinae) from Turkey. *Journal of Fisheries International* 2: 69–85. https://doi.org/jfish.2007.69.85

Erk'akan, F., S.C. Özeren & T.T. Nalbant. 2008a. *Cobitis evreni* sp. nova - a new spined loach species (Cobitidae) from southern Turkey. *Journal of Fisheries International* 3: 112–114.

Erk'akan, F., S.C. Özeren & T.T. Nalbant. 2008b. Two new species of stone loaches from Turkey (Teleostei: Nemacheilidae). *Journal of Fisheries International [electronic journal]* 3: 115–119.

Erk'akan, F., D. Innal & F. Özdemir. 2014. Length-weight relationships for some endemic stone and spine loach species in Anatolia. *Journal of Applied Ichthyology* 30: 244–245. https://doi.org/10.1111/jai.12260

Erk'akan, F., F. Özdemir & S.C. Özeren. 2017. Two new species of the genus *Cobitis* Linnaeus (Teleostei: Cobitidae) from Turkey. *FishTaxa* 2: 82–89.

Ermin, R. 1946. Cyprinodontid'lerde pul reduksiyonu. Schuppenreduktion bei Zahnkarpfen (Cyprinodontidae). *Revue de la Faculté des Sciences de l'Université d'Instanbul, Série B*: Sciences Naturelles 11: 217–272.

Esmaeili, H.R., A. Teimory, B.W. Coad & Z. Gholami. 2008. Threatened fishes of the world: *Cobitis linea* (Heckel, 1849) (Cobitidae). *Environmental Biology of Fishes* 83: 407–408. https://doi.org/10.1007/s10641-008-9352-8

Esmaeili, H.R., B.W. Coad, A. Gholamifard, N. Nazari & A. Teimory. 2010. Annotated checklist of the freshwater fishes of Iran. *Zoosystematica Rossica* 19: 361–386.

Esmaeili, H.R., A. Teimori, Z. Gholami, N. Zarei & B. Reichenbacher. 2012. Re-validation and redescription of an endemic and endangered species, *Aphanius pluristriatus* (Jenkins, 1910) (Teleostei, Cyprinodontidae), from southern Iran. *Zootaxa* 3208: 58–67. https://doi.org/10.11646/zootaxa.3208.1.4

Esmaeili, H.R., H. Malekzehi, A. Pazira & J. Freyhof. 2013a. First record of the Kalabans, *Bangana dero* (Hamilton, 1822), from Iran (Teleostei: Cyprinidae). *Zoology in the Middle East* 59: 89–91. https://doi.org/10.1080/09397140.2013.795074

Esmaeili, H.R., A. Gholamifard, G. Sayyadzadeh, B. Parsi, S. Mirghiasi & S. Ghasemian. 2013b. New record of the convict cichlid, *Amatitlania nigrofasciata* (Günther, 1867), from the Middle East (Actinopterygii: Cichlidae). *Aqua, International Journal of Ichthyology* 19: 225–229.

Esmaeili, H.R., A. Teimori, G. Sayyadzadeh, M. Masoudi & B. Reichenbacher. 2014a. Phylogenetic relationships of the tooth-carp *Aphanius* (Teleostei: Cyprinodontidae) in the river systems of southern and south-western Iran based on mtDNA sequences. *Zoology in the Middle East* 60: 29–38. https://doi.org/10.1080/09397140.2014.892329

Esmaeili, H.R., A. Teimori, Z. Gholami & B. Reichenbacher. 2014b. Two new species of the tooth-carp *Aphanius* (Teleostei: Cyprinodontidae) and the evolutionary history of the Iranian inland and inland-related *Aphanius* species. *Zootaxa* 3786: 246–268. https://doi.org/10.11646/zootaxa.3786.3.2

Esmaeili, H.R., G. Sayyadzadeh, M. Özuluğ, M. Geiger & J. Freyhof. 2014c. Three new species of *Turcinoemacheilus* from Iran and Turkey (Teleostei: Nemacheilidae). *Ichthyological Exploration of Freshwaters* 24: 257–273.

Esmaeili, H.R., H. Mousavi-Sabet, G. Sayyadzadeh, S. Vatandoust & J. Freyhof. 2014d. *Paracobitis atrakensis*, a new species of crested loach from northeastern Iran (Teleostei: Nemacheilidae). *Ichthyological Exploration of Freshwaters* 25: 237–242.

Esmaeili, H.R., H. Zareian, S. Eagderi & N. Alwan. 2016a. Review on the taxonomy of Tigris scraper, *Capoeta umbla* (Heckel, 1843) and its confirmation record from the Iranian part of Tigris River, Persian Gulf basin (Teleostei: Cyprinidae). *FishTaxa* 1: 35–44.

Esmaeili, H.R, G. Sayyadzadeh & O. Seehausen. 2016b. *Iranocichla persa*, a new cichlid species from southern Iran (Teleostei, Cichlidae). *ZooKeys*. 636: 141–161. https://doi.org/10.3897/zookeys.636.10571

Esmaeili, H.R., G. Sayyadzadeh, B.W. Coad & S. Eagderi. 2016c. Review of the genus *Garra* Hamilton, 1822 in Iran with description of a new species: a morpho-molecular approach (Teleostei: Cyprinidae). *Iranian Journal of Ichthyology* 3: 82–121.

Esmaeili, H.R., M. Masoud, M. Amini Chermahini, F. Zarei & M. Ebrahimi. 2017. Invasion of the Neotropical and Nearctic fishes to Iran. *FishTaxa* 2: 126–133.

Esmaeili, H.R., G. Sayyadzadeh, F. Zarei, S. Khafaei & B.W. Coad. 2019. Phylogeographic pattern and population structure of the Persian stone loach, *Oxynoemacheilus persa* (Heckel 1847) (family: Nemacheilidae) in southern Iran with implications for conservation. *Environmental Biology of Fishes* 103: 77–88. https://doi.org/10.1007/s10641-019-00934-y

Esmaeili, H.R., G. Sayyadzadeh, B. Japoshvili, S. Eagderi, K. Abbasi & H. Mousavi-Sabet. 2020a. *Rhodeus caspius*, a new bitterling from Iran (Teleostei: Cypriniformes: Acheilognathidae). *Zootaxa* 4851: 319–337. https://doi.org/10.11646/zootaxa.4851.2.6

Esmaeili, H.R., A. Teimori, F. Zarei & G. Sayyadzadeh. 2020b. DNA barcoding and species delimitation of the Old World tooth-carps, family Aphaniidae Hoedeman, 1949 (Teleostei: Cyprinodontiformes). *PLOS ONE* 15: e0231717. https://doi.org/10.1371/journal.pone.0231717

Esmaeili, H.R., R. Sadeghi & H.K. Larson. 2020c. The long-snout freshwater goby *Awaous jayakari* (Boulenger, 1888) (Teleostei: Gobiidae), an additional fish element for the Iranian waters. *Zoology in the Middle East* 66: 29–36. https://doi.org/10.1080/09397140.2020.1695744.

Esmaeili, H.R., G. Sayyadzadeh, F. Zarei, S. Eagderi & H. Mousavi-Sabet. 2022. *Mystus cyrusi*, a new species of bagrid catfish (Teleostei: Bagridae) from Middle East. *Zootaxa* 5099: 325–343. https://doi.org/10.11646/zootaxa.5099.3.2

Esmaeili, H.R., G. Sayyadzadeh, A.H. Masoumi, S.H. Hashemi & S. Echreshavi. 2025. *Tariqilabeo iranicus*, a new species of labeonine fish from the Makran basin in Iran (Teleostei: Cyprinidae). *Zootaxa* 5620 (3): 437–450. https://doi.org/10.11646/zootaxa.5620.3.3

Espanhol, R., P.R. Almeida & M.J. Alves. 2007. Evolutionary history of lamprey paired species *Lampetra fluviatilis* (L.) and *Lampetra planeri* (Bloch) as inferred from mitochondrial DNA variation. *Molecular Ecology* 16: 1909–1924. https://doi.org/10.1111/j.1365-294X.2007.03279.x.

Evseenko, S.A. 2004. Family Pleuronectidae Cuvier 1816 - righteye flounders. *California Academy of Sciences Annotated Checklists of Fishes* 37: 1–37.

Evtushenko, N.Y., A.S. Potrokhov & O.G. Zin'kovskii. 1994. The black carp as a subject for acclimatization (Review). *Hydrobiological Journal* 30: 1–10.

Farashi, A., M. Kaboli, H.R. Rezaei, M.R. Naghavi, H. Rahimian & B.W. Coad. 2014. Reassessment of the taxonomic position of *Iranocypris typhlops* Bruun & Kaiser, 1944 (Actinopterygii, Cyprinidae). *Zookeys* 29: 69–77. https://doi.org/10.3897/zookeys.374.6617

Fausch, K.D., Y. Taniguchi, S. Nakano, G.D. Grossman & C.R. Townsend. 2001. Flood disturbance regimes influence rainbow trout invasion success among five holarctic regions. *Ecological Applications* 11: 1438–1455. https://doi.org/10.1890/1051-0761(2001)011[1438:FDRIRT]2.0.CO;2

Feng, Y.-J., D.C. Blackburn, D. Liang, D.M. Hills, D.B. Wake, D.C. Cannatella & P. Zhang. 2017. Phylogenomics reveals rapid, simultaneous diversification of three major clades of Gondwanan frogs at the Cretaceous–Paleogene boundary. *Proceedings of the National Academy of Sciences of the United States of America* 114 (29) E5864-E5870

Fernandes, C.A., E.J. Rohling & M. Siddall. 2006. Absence of post-Miocene Red Sea land bridges: biogeographic implications. *Journal of Biogeography* 33: 961–966. https://doi. org/10.1111/j.1365-2699.2006.01478.x

Ferraris, C.J. 2007. Checklist of catfishes, recent and fossil (Osteichthyes: Siluriformes), and catalogue of siluriform primary types. *Zootaxa* 1418: 1–628. https://doi. org/10.11646/zootaxa.1418.1.1

Ferrito, V. & C. Tigano. 1996. Decline of *Aphanius fasciatus* (Cyprinodontidae) and *Salaria fluviatilis* (Blenniidae) populations in freshwaters of eastern Sicily. *Ichthyological Exploration of Freshwaters* 7: 181–184.

Ferrito, V., A. Pappalardo, A. Canapa, M. Barucca, I. Doadrio, E. Olmo & C. Tigano. 2013. Mitochondrial phylogeography of the killifish *Aphanius fasciatus* (Teleostei, Cyprinodontidae) reveals highly divergent Mediterranean populations. *Marine Biology* 160: 3193–3208. https://doi.org/10.1007/s00227-013-2307-4

Feulner, G.R. 1998. Wadi fish of the UAE. *Tribulus* 8 (2): 16–22.

Fine, M.L. & E. Parmentier. 2015. Mechanisms of fish sound production. Pp 77–126 *in:* F. Ladich. (Ed.). *Sound communication in fishes.* Springer, Vienna. https://doi.org/10.1007/978-3-7091-1846-7_3

Fischer, W. & P.J. Bianchi. 1984. *FAO species identification sheets for fishery purposes - Western Indian Ocean. Vol. I–IV. Fishing Area.* Food and Agriculture Organization of the United Nations, Rome. (available at: https://www.google.com/url?sa=t&rct=j&q=&esrc=s&source=web&cd=&ved=2ahUKEwiQm5mty7_9AhU2SfEDHeT2BPwQFnoECBAQAQ&url=https %3A %2F %2Fwww.fao.org %2F3 %2Fad468e %2FAD468eKA.pdf&usg=AOvVaw11OZBSUW7m-MMeUuBEhexH)

Fishelson, L., M. Goren, J. van Vuren & R. Manelis. 1996. Some aspects of the reproductive biology of *Barbus* spp., *Capoeta damascina* and their hybrids (Cyprinidae, Teleostei) in Israel. *Hydrobiologia* 317: 79–88. https://doi.org/10.1007/BF00013728

Ford, A.G.P., T.R. Bullen, L. Pang, M.J. Genner, R. Bills, T. Flouri, B.P. Ngatunga, L. Rüber, U.K. Schliewen, O. Seehausen, A. Shechonge, M.L.J. Stiassny, G.F. Turner & J.J. Day. 2019. Molecular phylogeny of *Oreochromis* (Cichlidae: Oreochromini) reveals mito-nuclear discordance and multiple colonisation of adverse aquatic environments. *Molecular Phylogenetics and Evolution* 136: 215–226. https://doi.org/10.1016/j.ympev.2019.04.008

Fowler, H.W. & H. Steinitz. 1956. Fishes from Cyprus, Iran, Iraq, Israel and Oman. *Bulletin of the Research Council of Israel* 5B (3–4): 260–292.

Fowler, S.L., R.D. Cavanagh, M. Camhi, G.H. Burgess, G.M. Cailliet, S.V. Fordham, C.A. Simpfendorfer & J.A. Musick. 2005. *Sharks, Rays and Chimaeras: The Status of the Chondrichthyan Fishes.* IUCN/SSC Shark Specialist Group. IUCN, Gland, Switzerland and Cambridge, UK.

Freeman, M.C., D. Vinolas, G.D. Grossman & A. De Sostoa. 1990. Microhabitat use by *Blennius fluviatilis* in the Rio Matarrana, Spain. *Freshwater Biology* 24: 335–346.

Freyhof, J. 2016a. *Oxynoemacheilus karunensis*, a new species from the Persian Gulf basin (Teleostei: Nemacheilidae). *Zootaxa* 4175: 94–100. https://doi.org/10.11646/zootaxa.4175.1.9

Freyhof, J. 2016b. Redescription of *Garra elegans* (Günther, 1868), a poorly known species from the Tigris River drainage (Teleostei: Cyprinidae). *Zootaxa* 4173: 496–500. https://doi.org/10.11646/zootaxa.4173.5.7

Freyhof, J. 2022. *Egirdira*, a new generic name for *Pararhodeus niger* Kosswig amp; Geldiay, 1952 (Teleostei: Leuciscidae). *Zootaxa* 5104: 586–592. https://doi.org/10.11646/zootaxa.5104.4.8

Freyhof, J. 2025. *Garra cavernicola*, a new species from the Hajar Mountains in Oman (Teleostei: Cyprinidae), with a short discussion on monophyly and the description of evolutionarily young species. *Zootaxa* 5601 (2): 335–345. https://doi.org/10.11646/zootaxa.5601.2.7

Freyhof, J. & Y. S. Abdullah. 2017. Two new species of *Oxynoemacheilus* from the Tigris drainage in Iraqi Kurdistan (Teleostei: Nemacheilidae). *Zootaxa* 4238: 73–87. https://doi.org/10.11646/zootaxa.4238.1.5

Freyhof, J. & M.F. Geiger. 2017. *Oxynoemacheilus zarzianus*, a new loach from the Lesser Zab River drainage in Iraqi Kurdistan (Teleostei: Nemacheilidae). *Zootaxa* 4273: 258–270. https://doi.org/10.11646/zootaxa.4273.2.6

Freyhof, J. & M.F. Geiger. 2021. *Oxynoemacheilus shehabi*, a new nemacheilid loach from the upper Orontes in southern Syria (Teleostei: Nemacheilidae). *Zootaxa* 4908: 571–583. https://doi.org/10.11646/zootaxa.4908.4.9

Freyhof, J. & A. Jouladeh-Roudbar. 2024. *Turcinoemacheilus inexpectatus*, a new nemacheilid loach from the Tigris drainage (Teleostei: Nemacheilidae). *Zootaxa* 5399(2): 172–180. https://doi.org/10.11646/zootaxa.5399.2.6

Freyhof, J. & M. Özuluğ. 2006. *Pseudophoxinus ninae*, a new species from Central Anatolia, Turkey (Teleostei: Cyprinidae). *Ichthyological Explorations of Freshwaters* 3: 255–259.

Freyhof, J. & M. Özuluğ. 2010a. *Pseudophoxinus fahrettini*, a new species of spring minnow from Central Anatolia (Teleostei: Cyprinidae). *Ichthyological Exploration of Freshwaters* 20: 325–332.

Freyhof, J. & M. Özuluğ. 2010b. *Pseudophoxinus evliyae*, a new species of spring minnow from western Anatolia with remarks on the distribution of *P. ninae* and the systematic position of *P. fahirae* (Teleostei: Cyprinidae). *Ichthyological Exploration of Freshwaters* 20: 309–318.

Freyhof, J. & M. Özuluğ. 2010c. *Pseudophoxinus hittitorum*, a new species of spring minnow from Central Anatolia (Teleostei: Cyprinidae). *Ichthyological Exploration of Freshwaters* 21: 239–245.

Freyhof, J. & M. Özuluğ. 2014. *Acanthobrama thisbeae*, a new species of bream from southern Anatolia, Turkey (Teleostei: Cyprinidae). *Ichthyological Exploration of Freshwaters* 25: 1–10.

Freyhof, J. & M. Özuluğ. 2017. *Oxynoemacheilus hazarensis*, a new species from Lake Hazar in Turkey, with remarks on *O. euphraticus* (Teleostei: Nemacheilidae). *Zootaxa* 4247: 378–390. https://doi.org/10.11646/zootaxa.4247.4.2

Freyhof, J. & D. Turan. 2019. *Alburnus magnificus*, a new species of bleak from the Orontes River drainage (Teleostei: Leuciscidae). *Zootaxa* 4559: 373–383. https://doi.org/10.11646/zootaxa.4559.2.10

Freyhof, J. & B. Yoğurtçuoğlu. 2020. A proposal for a new generic structure of the killifish family Aphaniidae, with the description of

Aphaniops teimorii (Teleostei: Cyprinodontiformes). *Zootaxa* 4810: 421–451. https://doi.org/10.11646/zootaxa.4810.3.2

Freyhof, J. & B. Yoğurtçuoğlu. 2023. *Mystus misrai* Anuradha, 1986, a valid species from the Orontes drainage (Teleostei: Bagridae). Zootaxa, 5306 (4), 445–462. https://doi.org/10.11646/zootaxa.5306.4

Freyhof, J., B. Stelbrink, M. Özuluğ & P.S. Economidis. 2008. The first record of *Cobitis puncticulata* from Europe with comments on its conservation status (Teleostei: Cobitidae). *Folia Zoologica* 57: 16–19.

Freyhof, J., F. Erk'akan, C. Özeren & A. Perdices. 2011. An overview of the western Palaearctic loach genus *Oxynoemacheilus* (Teleostei: Nemacheilidae). *Ichthyological Exploration of Freshwaters* 22: 301–312.

Freyhof, J., F.G. Ekmekçi, A. Ali, N.R. Khamees, M. Özuluğ, N. Hamidan, F. Küçük & K.G. Smith. 2014a. *Freshwater fishes. The Status and Distribution of Freshwater Biodiversity in the Eastern Mediterranean.* IUCN, Cambridge, UK, Malaga, Spain and Gland, Switzerland.

Freyhof, J., H.R. Esmaeili, G. Sayyadzadeh & M. Geiger. 2014b. Review of the crested loaches of the genus *Paracobitis* from Iran and Iraq with the description of four new species (Teleostei: Nemacheilidae). *Ichthyological Exploration of Freshwaters* 25: 11–38.

Freyhof, J., G. Sayyadzadeh, H.R. Esmaeili & M.F. Geiger. 2015. Review of the genus *Paraschistura* from Iran with description of six new species (Teleostei: Nemacheilidae). *Ichthyological Exploration of Freshwaters* 26: 1–48.

Freyhof, J., M.F. Geiger, K. Golzarianpour & R. Patimar. 2016a. *Sasanidus*, a new generic name for *Noemacheilus kermanshahensis* Bănărescu & Nalbant, with discussion of *Ilamnemacheilus* and *Schistura* (Teleostei; Nemacheilidae). *Zootaxa* 4107: 65–80. https://doi.org/10.11646/zootaxa.4107.1.3

Freyhof, J., Y.S. Abdullah, K. Ararat, H. Ibrahim & M.F. Geiger. 2016b. *Eidinemacheilus proudlovei*, a new subterranean loach from Iraqi Kurdistan (Teleostei; Nemacheilidae). *Zootaxa* 4173: 225–236. https://doi.org/10.11646/zootaxa.4173.3.2

Freyhof, J., C. Kaya & D. Turan. 2017a. *Oxynoemacheilus kentritensis*, a new species from the upper Tigris drainage in Turkey with remarks on *O. frenatus* (Teleostei: Nemacheilidae). *Zootaxa* 4258: 551–560. https://doi.org/10.11646/zootaxa.4258.6.4

Freyhof, J., M. Özuluğ & G. Saç. 2017b. Neotype designation of *Aphanius iconii*, first reviser action to stabilise the usage of *A. fontinalis* and *A. meridionalis* and comments on the family group names of fishes placed in Cyprinodontidae (Teleostei: Cyprinodontiformes). *Zootaxa* 4294: 573–585. https://doi.org/10.11646/zootaxa.4294.5.6

Freyhof, J., M. Özuluğ, C. Kaya, E. Bayçelebi & D. Turan. 2018a. Redescription of *Alburnus kotschyi* Steindachner, 1863, with comments on *Alburnus sellal adanensis* Battalgazi, 1944 (Teleostei: Leuciscidae). *Zootaxa* 4382: 573–582. http://dx.doi.org/10.11646/zootaxa.4382.3.8

Freyhof, J., C. Kaya, E. Bayçelebi, M.F. Geiger & D. Turan. 2018b. Generic assignment of *Leuciscus kurui* Bogutskaya from the upper Tigris drainage, and a replacement name for *Alburnus kurui* Mangit & Yerli (Teleostei: Leuciscidae). *Zootaxa* 4410: 113–135. https://doi.org/10.11646/zootaxa.4410.1.6

Freyhof, J., E. Bayçelebi & M.F. Geiger. 2018c. Review of the genus *Cobitis* in the Middle East, with the description of eight new species (Teleostei: Cobitidae). *Zootaxa* 4535: 1–75. https://doi.org/10.11646/zootaxa.4535.1.1

Freyhof, J., C. Kaya, E. Bayçelebi, M. Geiger & D. Turan. 2019a. Comments on the holotype of *Alburnus kurui* Mangit Yerli 2018 and re-description of *A. carianorum* Freyhof, Kaya, Bayçelebi, Geiger Turan, 2019 (Teleostei: Leuciscidae). *Zootaxa* 4550: 594–596. https://doi.org/10.11646/zootaxa.4550.4.10

Freyhof, J., C. Kaya, D. Turan & M.F. Geiger. 2019b. Review of the *Oxynoemacheilus tigris* group with the description of two new species from the Euphrates drainage (Teleostei: Nemacheilidae). *Zootaxa* 4612: 29–57. https://doi.org/10.11646/zootaxa.4612.1.2

Freyhof, J., J. Els, G.R. Feulner, N.A. Hamidan & F. Krupp. 2020. *Freshwater fishes of the Arabian Peninsula*. Motivate Media Group, Abu Dhabi.

Freyhof, J., B. Yoğurtçuoğlu & C. Kaya. 2021a. *Oxynoemacheilus sarus*, a new nemacheilid loach from the lower Ceyhan and Seyhan in southern Anatolia (Teleostei: Nemacheilidae). *Zootaxa* 4964: 123–139. http://dx.doi.org/10.11646/zootaxa.4964.1

Freyhof, J., C. Kaya, G. Epitashvili & M.F. Geiger. 2021b. *Oxynoemacheilus phasicus*, a new nemacheilid loach from the eastern Black Sea basin with some remarks on other Caucasian *Oxynoemacheilus* (Teleostei: Nemacheilidae). *Zootaxa* 4952 (1): 135–151. https://doi.org/10.11646/zootaxa.4952.1.8

Freyhof, J., C. Kaya, Y.S. Abdullah & M.F. Geiger. 2021c. The *Glyptothorax* catfishes of the Euphrates and Tigris with the description of a new species (Teleostei: Sisoridae). *Zootaxa* 4969: 453–491. https://doi.org/10.11646/zootaxa.4969.3.2

Freyhof, J., C. Kaya & M.F. Geiger. 2022. A practical approach to revise the *Oxynoemacheilus bergianus* species group (Teleostei: Nemacheilidae). *Zootaxa* 5128: 151–194. https://doi.org/10.11646/zootaxa.5128.2.1

Freyhof, J. & B. Yoğurtçuoğlu. 2024. *Luciobarbus lydianus* and *L. kottelati*, two synonyms of *L. graecus* (Teleostei: Cyprinidae). *Zootaxa* 5415: 466–476. https://doi.org/10.11646/zootaxa.5415.3.6

Freyhof, J., I. Hashemzadeh Segherloo, S. Vatandoust & B. Yoğurtçuoğlu. 2025a. Resolving a 182 year old taxonomic puzzle: *Luciobarbus* in the Gulf basin (Teleostei: Cyprinidae). *Zootaxa* (in press).

Fuselier, L. 2001. Impacts of *Oreochromis mossambicus* (Perciformes: Cichlidae) upon habitat segregation among cyprinodontids (Cyprinodontiformes) of a species flock in Mexico. *Revista de Biologia Tropical* 49: 647–656.

Gabrielyan, B.K. 2010. *Fishes of Lake Sevan*. Publishing House "Gitutyun" of NAN RA, Yerevan.

Gafny, S., A. Gasith & M. Goren. 1992. Effect of water level fluctuaton on shore spawning of *Mirogrex terraesanctae* (Steinitz) (Cyprinidae) in Lake Kinneret, Israel. *Journal of Fish Biology* 41: 863–871. http://dx.doi.org/10.1111/j.1095-8649.1992.tb02715.x

Gambetta, L. 1934. Sulla variabilita del cobite fluviale (*Cobitis taenia* L.) e sul rapporto numerico dei sessi. *Bollettino dei Musei di Zoologia e di Anatomia Comparata della Reale Università di Torino* 44: 297–324.

Gandlin, A.A., B. Japoshvili, G. Epitashvili, N.J. Mustafaev, H.R. Roubenyan & B.A. Levin. 2022. Phylogeography of the Kura barbel *Barbus cyri* De Filippi as Inferred from mtDNA Data. *Inland Water Biology* 15: 11–22. http://dx.doi.org/10.1134/S1995082922010047

Gandolfi, G., S. Zerunian, P. Torricelli & A. Marconato. 1991. Gobiidae. In: Gandolfi, G., Zerunian, S., Torricelli, P. & Marconato, A. (Eds.) I pesci delle acque interne italiane. Parte 3 – riconoscimento, biologia, distribuzione e status delle specie. Ministero dell'ambiente – Servizio Conservazione Natura, Unione Zoologica Italiana, Istituto Poligrafico e Zecca dello Stato, Milano: 487–553.

Ganjali, Z., H.R.Esmaeili, F. Zarei, G. Sayyadzadeh, S. Eagderi & R.E. Gozlan. 2020. West Asian colonisation of topmouth gudgeon, *Pseudorasbora parva* (Teleostei: Gobionidae): Genetic admixture at the crossroad of Europe and East Asia. *Freshwater Biology* 66: 699–715. https://doi.org/10.1111/fwb.13671

Gante, H.F. 2018. How fish get their stripes - again and again. *Science* 362: 396–397. https://doi.org/10.1126/science.aav3373

Gao, Y., S.-Y. Wang, J. Luo, R.W. Murphy, R. Du, S.-F. Wu, C.-L. Zhu, Y. Li, A.D. Poyarkov, S.N. Nguyen, P.-T. Luan & Y.-P. Zhang. 2012. Quaternary palaeoenvironmental oscillations drove the evolution of the Eurasian *Carassius auratus* complex (Cypriniformes, Cyprinidae). *Journal of Biogeography* 39: 2264–2278. https://doi.org/10.1111/j.1365-2699.2012.02755.x

Garcia-Melo, J.E., L.J. García-Melo, J.D. Garcia-Melo, D.K. Rojas-Briñez, G. Guevara & J.A. Maldonado-Ocampo. 2019. Photafish system: An affordable device for fish photography in the wild. *Zootaxa* 4554: 141–172. https://doi.org/10.11646/zootaxa.4554.1.4

Gaudant, J. 2007. Occurrence of the genus *Tarsichthys* Troschel (Teleostean fishes, Cyprinidae) in the Upper Oligocene of Lake Kunkskopf, near Burgbrohl (Eifel-Mountains, Germany). *Zitteliana* A47: 127–132.

GegenStrömung. 2011. *Dam construction in Turkey and its impact on economic, cultural and social rights: Parallel report in response to the Initial Report by the Republic of Turkey on the implementation of the International Covenant On Economic, Social and Cultural Rights.* Submission to the UN Committee on Economic, Social and Cultural Rights for its 46th Session, 2–20 May 2011. http://www2.ohchr.org/english/bodies/cescr/docs/ngos/JointReport_Turkey46.pdf

Geiger, M., F. Herder, M.T. Manahan, V. Almada, R. Barbieri, J. Bohlen, M. Casal-Lopez, G.B. Delmastro, G.P. Denys, A. Dettai, I. Doadrio, E. Kalogianni, H. Kärst, M. Kovačić, M. Laporte, M. Özuluğ, A. Perdices, S. Perea, H. Persat, S. Porcelotti, C. Puzzi, J. Robalo, R. Šanda, M. Schneider, V. Šlechtová, M. Stoumboudi, S. Walter & J. Freyhof. 2014. Spatial heterogeneity in the Mediterranean biodiversity hotspot affects barcoding accuracy of its freshwater fishes. *Molecular Ecology Resources* 14: 1210–1221. https://doi.org/10.1111/1755-0998.12257

Geldiay, R. & S. Balık. 2007. *Freshwater Fishes of Turkey*. 5th Edition. Ege University Press, Bornova, İzmir. [in Turkish].

Gençoğlu, L., Ş.G. Kırankaya, B. Yoğurtçuoğlu & F.G. Ekmekçi. 2017. Feeding properties of the translocated marine fish sand smelt *Atherina boyeri* Risso, 1810 (Atherinidae) in a freshwater reservoir. *Acta Zoologica Bulgarica* 69: 131–138.

Genner, M.J., E. Connell, A. Shechonge, A. Smith, J. Swanstrom, S. Mzighani, A. Mwijage, B.P. Ngatunga & G.F. Turner. 2013. Nile tilapia invades the Lake Malawi catchment. *African Journal of Aquatic Science* 38 (Suppl. 1): 85–90. https://doi.org/10.2989/16085914.2013.842157

Gess, R.W., M.I. Coates & B.S. Rubidge. 2006. A lamprey from the Devonian period of South Africa. *Nature* 443: 981–984. https://doi.org/10.1038/nature05150

Getahun, A. & K.J. Lazara. 2001. *Lebias stiassnyae*: A new species of killifish from Lake Adfera, Ethiopia (Teleostei: Cyprinodontidae). *Copeia* 2001: 150–153. https://doi.org/10.1643/0045-8511(2001)001[0150:LSANSO]2.0.CO;2

Ghanavi, H.R., E.G. Gonzales & I. Doadrio. 2016. Phylogenetic relationships of freshwater fishes of the genus *Capoeta* (Actinopterygii, Cyprinidae) in Iran. *Ecology and Evolution* 6: 8205–8222. https://doi.org/10.1002/ece3.2411

Ghanbary, K., M. Khodadadi, M. Javahri Baboli & G. Mohamadi. 2012. Reproduction biology in *Chondrostoma regium* (Heckel, 1843) in Gamasiab River in Kermanshah province, Iran. *Life Science Journal* 9: 1825–1830.

Gharaei, A. 2012. Morphometric and meristic studies of snow trout *Schizothorax zarudnyi* (Nikolskii, 1897) as a threatened endemic fish. *World Journal of Fish and Marine Sciences* 4: 426–429.

Ghasemi, H., A. Jouladeh-Roudbar, S. Eagderi, K. Abbasi, S. Vatandoust & H.R. Esmaeili. 2015. Ichthyofauna of Urmia basin: Taxonomic diversity, distribution and conservation. *Iranian Journal of Ichthyology* 2: 177–193. https://doi.org/10.22034/iji.v2i3.77

Gholami, Z., H.R. Esmaeili, D. Erpenbeck & B. Reichenbacher. 2014. Phylogenetic analysis of *Aphanius* from the endorheic Kor River basin in the Zagros Mountains, southwestern Iran (Teleostei: Cyprinodontiformes: Cyprinodontidae). *Journal of Zoological Systematics and Evolutionary Research* 52: 130–141. https://doi.org/10.1111/jzs.12052

Gibbs, M.A., J.H. Shields, D.W. Lock, K.M. Talmadge & T.M. Farrell. 2008. Reproduction in an invasive exotic catfish *Pterygoplichthys disjunctivus* in Volusia Blue Spring, Florida, USA. *Journal of Fish Biology* 73: 1562–1572. https://doi.org/10.1111/j.1095-8649.2008.02031.x

Gill, H.S., C.B. Renaud, F. Chapleau, R.L. Mayden & I.C. Potter. 2003. Phylogeny of living parasitic lampreys (Petromyzontiformes) based on morphological data. *Copeia* 2003: 687–703. https://doi.org/10.1643/IA02-085.1

Gladkov, N.A. 1935. Sur la variabilité géographique de *Cobitis taen*ia L. *Archives du Musée Zoologique de l'Université de Moscou* 2: 69–74. [in Russian, French summary].

Golani, D. & G. Snovsky. 2013. Occurrence of suckermouth armored catfish (Siluriformes, Loricariidae, *Pterygoplichthys*) in inland waters of Israel. *BioInvasions Records* 2: 253–256. http://dx.doi.org/10.3391/bir.2013.2.3.13

Golani, D. & B. Appelbaum-Golani. 2010. First record of the Indo-Pacific fish the jarbua terapon (*Terapon jarbua*) (Osteichthyes: Terapontidae) in the Mediterranean with remarks on the wide geographical distribution of this species. *Scientia Marina* 74: 717–720. https://doi.org/10.3989/scimar.2010.74n4717

Golmoradizadeh, A., E. Kamrani & M.M. Sajjadi. 2012. Life history traits of *Aphanius ginaonis* Holly, 1929 (Cyprinodontidae) and potential risks of extinction in the Geno hot spring (Iran) population. *Journal of Applied Ichthyology* 28: 31–33. https://doi.org/10.1111/j.1439-0426.2011.01868.x

Golzarianpour, K., A. Abdoli & J. Freyhof. 2011. *Oxynoemacheilus kiabii*, a new loach from Karkheh River drainage, Iran (Teleostei: Nemacheilidae). *Ichthyological Exploration of Freshwaters* 22: 201–208.

Golzarianpour, K., A. Abdoli, R. Patimar & J. Freyhof. 2013. *Turcinoemacheilus hafezi*, a new loach from the Zagroz Mountains, Iran (Teleostei: Nemacheilidae). *Ichthyological Exploration of Freshwaters* 24: 41–48.

Goren, M. 1974. The freshwater fishes of Israel. *Israel Journal of Zoology* 23: 67–118, pls 1–4.

Goren, M. 2009. Saving critically endangered fish species - utopia or practical idea? The story of Yarqon bleak *Acanthobrama telavivensis* (Cyprinidae) as a test case. *Aqua, International Journal of Ichthyology* 15: 1–12.

Goren, M. & R. Ortal. 1999. Biogeography, diversity and conservation of the inland water fish communities in Israel. *Biological Conservation* 89: 1–9. https://doi.org/10.1016/S0006-3207(98)00127-X

Goren, M., L. Fishelson & E. Trewavas. 1973. The cyprinid fishes of *Acanthobrama* Heckel and related genera. *Bulletin of the British Museum (Natural History) Zoology* 24: 293–315.

Gözler, A.M., S. Engin, D. Turan & C. Şahin. 2011. Age, growth and fecundity of the Turkish brook lamprey, *Eudontomyzon lanceolata* (Kux & Steiner, 1972), in north-eastern Turkey (Petromyzontiformes: Petromyzontidae). *Zoology in the Middle East* 52: 57–61. https://doi.org/10.1080/09397140.2011.10638479

Gratzianov, V.I. 1907. Opyt' obzora ryby Rossiskoi Imperii v' sistematicheskom' i geograficescom' otnoshenii. *Trudy Otdela*

Ichtiologii Imperatorskago Russkago Obshchestva Akklimatizacii Zivotnych' i Rastenii 4: 1–567.

Greenwood, P.H. 1976. A new and eyeless cobitid fish (Pisces, Cypriniformes) from the Zagros Mountains, Iran. *Journal of Zoology (London)* 180: 129–137. https://doi.org/10.1111/j.1469-7998.1976.tb04668.x

Gregg, R.E., J.H. Howard & F. Snhonhiwa 1998. Introgressive hybridization of tilapias in Zimbabwe. *Journal of Fish Biology* 52: 1–10. https://doi.org/10.1111/j.1095-8649.1998.tb01547.x

Grimm, H. 1980. Investigations on the problem of scale reduction and sulphate tolerance of West-Anatolian cyprinodonts (Pisces). *Internationale Revue der gesamten Hydrobiologie und Hydrographie* 65: 517–280. https://doi.org/10.1002/iroh.19800650409

Güçlü, S.S. & F. Küçük. 2011. Reproductive Biology of *Aphanius mento* (Heckel in Russegger, 1843) (Osteichthyes: Cyprinodontidae) in Kırkgöz Spring (Antalya-Turkey). *Turkish Journal of Fisheries and Aquatic Sciences* 11: 323–327. https://doi.org/10.4194/trjfas.2011.0217

Güçlü, S. & F. Küçük. 2015. The Ichthyofauna of Gediz River (Turkey): Taxonomic and Zoogeographic Features. *Annual Research & Review in Biology* 6: 202–214. https://doi.org/10.9734/ARRB/2015/14889

Güçlü, S., F. Küçük, Ö.O. Ertan & Z. Güçlü. 2013. The fish fauna of the Büyük Menderes River (Turkey): taxonomic and zoogeographic features. *Turkish Journal of Fisheries and Aquatic Sciences* 13: 685–698. https://doi.org/10.4194/1303-2712-v13_4_14

Güçlü, S.S., F. Küçük, D. Turan, Y. Çiftçi & A.G. Mutlu. 2018. A new *Chondrostoma* species from the Büyük Menderes River basin, Turkey (Teleostei: Cyprinidae). *Zoology in the Middle East* 64: 315–321. https://doi.org/10.1080/09397140.2018.1511293

Güçlü, S.S., G. Kalaycı, F. Küçük & D. Turan. 2020. *Barbus xanthos*, a new barbel from the Southern Aegean basin (Teleostei: Cyprinidae). *Journal of Fish Biology* 96: 1309–1319. https://doi.org/10.1111/jfb.14259.

Güçlü, S.S., G. Kalaycı, M. Özuluğ, F. Küçük & D. Turan. 2021. *Barbus ida*, a new barbel species from the Southern Marmara Sea basin. *Ichthyological Exploration of Freshwaters* 1–13 http://doi.org/10.23788/IEF-1164

Güçlü, S.S. 2012. Population structure of killifish, *Aphanius anatoliae* (Cyprinodontidae) endemic to Anatolia in Lake Eğirdir-Isparta (Turkey). *Iranian Journal of Fisheries Sciences* 11(4) 786–795.

Güçlü, S.S., İ.İ. Turna, Z. Güçlü & İ. Gülle. 2007. Population structure and growth of *Aphanius anatoliae sureyanus* Neu, 1937 (Osteichthyes: Cyprinodontidae), endemic to Burdur Lake, Turkey. *Zoology in the Middle East*, 41(1), 63–69. https://doi.org/10.1080/09397140.2007.10638228

Güçlü, S.S. 2022. The First data on the biology of *Anatolichthys meridionalis* (Actinopterygii, Aphaniidae): an endemic and endangered fish of Turkey (Dalaman River Basin). *Inland Water Biology* 15, 613–623. https://doi.org/10.1134/S1995082922050182

Guerrero, R.D. & M.R. Guerrero-del Castillo. 2002. *Proceedings of the International Forum on Tilapia Farming in the 21st Century (Tilapia Forum 2002)*. Philippine Fisheries Association Inc. Los, Banos, Laguna, Philippines.

Guinand, B., M. Oral & C. Tougard. 2021. Brown trout phylogenetics: a persistent mirage towards (too) many species. *Journal of Fish Biology* 99: 298–307. https://doi.org/10.1111/jfb.14686

Günther, A. 1865. Report on a collection of reptiles and fishes from Palestine. *Proceedings of the Zoological Society of London* 1864: 488–493.

Günther, A. 1899. Fishes of Lake Urmi, NW Persia, and its neighbourhood. *The Journal of the Linnean Society of London Zoology* 27: 381–391.

Gupta, S. & S. Banerjee. 2016. A review on feeding and reproductive biology of *Cirrhinus reba* (Hamilton, 1822), a threatened freshwater fish of Indian subcontinent with an emphasis on its conservation. *Fisheries and Aquaculture Journal* 7: 1–5.

Haas, R. 1982. Notes on the ecology of *Aphanius dispar* (Pisces, Cyprinodontidae) in the Sultanate of Oman. *Freshwater Biology* 12: 89–95. https://doi.org/10.1111/j.1365-2427.1982.tb00605.x

Halas, D., N. Lovejoy & N. Mandrak. 2018. Undetected diversity of goldfish (*Carassius* spp.) in North America. Aquatic Invasions 13: *211–219*. https://doi.org/10.3391/ai.2018.13.2.03

Hammer, M.P., L. Taillebois, A.J. King, D.A. Crook, D. Wedd, M. Adams, P.J. Unmack, D.F. Hoese & T. Bertozzi. 2021. Unravelling the taxonomy and identification of a problematic group of benthic fishes from tropical rivers (Gobiidae: *Glossogobius*). *Journal of Fish Biology* 99 (1): 87–100. https://doi.org/10.1111/jfb.14701

Hamidan, N. & A.A. Aloufi. 2013. Rediscovery of *Acanthobrama hadiyahensis* (Cyprinidae) in Saudi Arabia. *Journal of Fish Biology* 84: 1179–1184. https://doi.org/10.1111/jfb.12315

Hamidan, N.A., M.F. Geiger & J. Freyhof. 2014. *Garra jordanica*, a new species from the Dead Sea basin with remarks on the relationship of *G. ghorensis*, *G. tibanica* and *G. rufa* (Teleostei: Cyprinidae). *Ichthyological Exploration of Freshwaters* 25: 223–236.

Hänfling, B., P. Bolton, M. Harley & G.R. Carvalho. 2005. A molecular approach to detect hybridisation between crucian carp (*Carassius carassius*) and non-indigenous carp species (*Carassius* spp. and *Cyprinus carpio*). *Freshwater Biology* 50: 403–417. https://doi.org/10.1111/j.1365-2427.2004.01330.x

Hänfling, B., C. Dümpelmann, N.G. Bogutskaya, R. Brandl & M. Brändle. 2009. Shallow phylogeographic structuring of *Vimba vimba* across Europe suggests two distinct refugia during the last glaciation. *Journal of Fish Biology* 75: 2269–2286. https://doi.org/10.1111/j.1095-8649.2009.02415.x

Hankó, B. 1925. Fische aus Klein-Asien. *Annales Historico-Naturales Musei Nationalis Hungarici* 21: 137–158, 3 pls.

Hardisty, M.W. 1979. *Biology of the cyclostomes*. Chapman and Hall, Ltd., London.

Hardisty, M.W. 1986a. General introduction to lampreys. Pp 19–83 *in:* J. Holčík. (Ed.): *The freshwater fishes of Europe. Vol.1/I: Petromyzontiformes*. AULA Verlag, Wiesbaden.

Hardisty, M.W. 1986b. *Petromyzon marinus* Linnaeus, 1758. Pp 94–116 *in:* J. Holčík. (Ed.): *The freshwater fishes of Europe. Vol.1/I: Petromyzontiformes.* AULA Verlag, Wiesbaden.

Hardisty, M.W. & I.C. Potter. 1971–1982. *The biology of lampreys*. Vols. 1–4A, 4B. Academic Press Inc., New York.

Hardman, M. & L.M. Page. 2003. Phylogenetic relationships among bullhead catfishes of the genus *Ameiurus* (Siluriformes: Ictaluridae). *Copeia* 2003: 20–33. https://doi.org/10.1643/0045-8511(2003)003[0020:PRABCO]2.0.CO;2

Hardouin, E.A., D. Andreou, Y. Zhao, P. Chevret, D.H. Fletcher, J.R. Britton & R.E. Gozlan. 2018. Reconciling the biogeography of an invader through recent and historic genetic patterns: The case of topmouth gudgeon *Pseudorasbora parva*. *Biological Invasions* 20: 2157–2171. https://doi.org/10.1007/s1053 0-018-1693-4

Harrison, I.J. 2003a. *Mugil cephalus* Linnaeus 1758. Pp 7–14 *in:* P.J. Miller. (Ed.). *The freshwater fishes of Europe. Vol. 8/I: Mugilidae, Atherinidae, Atherinopsidae, Blenniidae, Odontobutidae, Gobiidae*. AULA Verlag, Wiebelsheim.

Harrison, I.J. 2003b. *Chelon labrosus* (Risso, 1826). Pp 17–24 *in:* P.J. Miller. (Ed.). *The freshwater fishes of Europe*. Vol. 8/I: *Mugilidae, Atherinidae,*

Atherinopsidae, Blenniidae, Odontobutidae, Gobiidae. AULA Verlag, Wiebelsheim.

Harrison, I.J. 2003c. *Liza ramada* (Risso, 1826). Pp 28–35 *in:* P.J. Miller. (Ed.). *The freshwater fishes of Europe*. Vol. 8/I: *Mugilidae, Atherinidae, Atherinopsidae, Blenniidae, Odontobutidae, Gobiidae*. AULA Verlag, Wiebelsheim.

Harrison, I.J. 2000d. *Liza saliens* (Risso, 1810). Pp 36–42 *in:* P.J. Miller. (Ed.). *The freshwater fishes of Europe*. Vol. 8/I: *Mugilidae, Atherinidae, Atherinopsidae, Blenniidae, Odontobutidae, Gobiidae*. AULA Verlag, Wiebelsheim.

Hashemzadeh Segherloo, I., L. Bernatchez, K. Golzarianpour, A. Abdoli, C.R. Primmer & M. Bakhtiary. 2012. Genetic differentiation between two sympatric morphs of the blind Iran cave barb *Iranocypris typhlops*. *Journal of Fish Biology* 81: 1747–1753. https://doi.org/10.1111/j.1095-8649.2012.03389.x

Hashemzadeh Segherloo, I., A. Abdoli, S. Eagderi, H.R. Esmaeili, G. Sayyadzadeh, L. Bernatchez, E. Halle, M. F. Geiger, M. Özuluğ, J. Laroche & J. Freyhof. 2016a. Dressing down: convergent reduction in *Garra* (Teleostei: Cyprinidae) in the Middle East. *Hydrobiologia* 785: 47–59. https://doi.org/10.1007/s10750-016-2902-8

Hashemzadeh Segherloo, I., N. Ghaedrahmati & J. Freyhof. 2016b. *Eidinemacheilus*, a new generic name for *Noemacheilus smithi* Greenwood (Teleostei; Nemacheilidae). *Zootaxa* 4147: 466–476. https://doi.org/10.11646/zootaxa.4147.4.7

Hashemzadeh Segherloo, I., J. Freyhof, P. Berrebi, A.L. Ferchaud, M. Geiger, J. Laroche, B.A. Levin, E. Normandeau & L. Bernatchez. 2021. A genomic perspective on an old question: *Salmo* trouts or *Salmo trutta* (Teleostei: Salmonidae)? *Molecular Phylogenetics and Evolution* 162: 107204. https://doi.org/10.1016/j.ympev.2021.107204

Haskins, C.P., E.F. Haskins, J.A. Mclaughlin & R.E. Hewfit. 1961. Polymorphism and population structure in *Lebistes reticulatus*, a population study. Pp 320–395 *in:* W.F. Blair. (Ed.). *Vertebrate speciation*, University of Texas Press, Austin.

Hata, H. & H. Motomura 2020. Two new species of Thrissina (Clupeiformes: Engraulidae) from the northern Indian Ocean and redescription of Thrissina vitrirostris (Gilchrist and Thompson 1908). Ichthyological Research 67: 155–166. https://doi.org/10.1007/s10228-019-00713-w

Hata, H., P.N. Psomadakis, H.B. Osmany & H. Motomura. 2021. A new species of *Thrissina* from Pakistan (Arabian Sea), with redescription of *Thrissina whiteheadi* (Wongratana 1983) (Clupeiformes: Engraulidae). *Ichthyological Research* 68: 486–495. https://doi.org/10.1007/s10228-021-00799-1

Hata, H., S. Lavoué, S. Chungthanawong & H. Motomura. 2023. A new species of *Thrissina* from the Andaman Sea and re-assessment of the taxonomic status of *Thrissina cuvierii* (Swainson, 1839) and *Thrissina malabarica* (Bloch, 1795) (Teleostei: Clupeiformes: Engraulidae: Coiliinae. *Ichthyology & Herpetology* 111 (4): 549–562. https://doi.org/10.1643/i2023001

Hata, H., S. Lavoué, P.N. Psomadakis, H. Badar Osmany & H. Motomura 2025. *Thrissina mystica* sp. nov., a new thryssa (Teleostei: Clupeiformes: Engraulidae: Coiliinae) from the northern Indian Ocean. Ichthyological Research. https://doi.org/10.1007/s10228-025-01023-0

Havelka, M., V. Kašpar, M. Hulák & M. Flajšhans. 2011. Sturgeon genetics and cytogenetics: a review related to ploidy levels and interspecific hybridization. *Folia Zoologica* 60: 93–103.

He, S., X. Gu, R.L. Mayden, W.J. Chen, K.W. Conway & Y. Chen. 2008. Phylogenetic position of the enigmatic genus *Psilorhynchus* (Ostariophysi: Cypriniformes): evidence from the mitochondrial genome. *Molecular Phylogenetics and Evolution* 47: 419–425. https://doi.org/10.1016/j.ympev.2007.10.012

Heckel, J.J. 1843. Ichthyologie [von Syrien] Pp 990–1099. *in:* J. von Russegger. *Reisen in Europa, Asien und Africa, mit besonderer Rücksicht auf die naturwissenschaftlichen Verhältnisse der betreffenden Länder unternommen in den Jahren 1835 bis 1841, etc.* E. Schweizerbart'sche Verlagshandlung. Stuttgart.

Heckel, J.J. 1847. Naturhistorischer Anhang. [Various subtitles.] Pp 207–357 *in:* J. von Russegger. *Reisen in Europa, Asien und Afrika, ... unternommen in den Jahren 1835 bis 1841.* E. Schweizerbart'sche Verlagshandlung, Stuttgart.

Heckman, C.W. 1991a. *Alosa caspia caspia* (Eichwald, 1938). Pp 135–146 *in:* H. Hoestlandt, H. (Ed.). *The freshwater fishes of Europe. Vol. 2. Clupeidae, Anguillidae*. AULA Verlag, Wiesbaden.

Heckman, C.W. 1991b. *Alosa pontica kessleri* (Grimm, 1887). Pp 325–336 *in:* H. Hoestlandt, H. (Ed.). *The freshwater fishes of Europe. Vol. 2. Clupeidae, Anguillidae*. AULA Verlag, Wiesbaden.

Henshaw, J.M. 2018. *Protandrous Hermaphroditism*. Pp 1–6 *in:* J. Vonk & T. Shackelford. (Eds.) *Encyclopedia of Animal Cognition and Behavior*. Springer, Cham. https://doi.org/10.1007/978-3-319-47829-6_1972-1

Hesp, A., I.C. Potter & N.G. Hall. 2004. Reproductive biology and protandrous hermaphroditism in *Acanthopagrus latus*. *Environmental Biology of Fishes* 70: 257–272. https://doi.org/10.1023/B:EBFI.0000033344.21383.00

Hirt, V., G. Arratia, W.-J. Chen, R.L. Mayden, K.L. Tang, R.M. Wood & A.M. Simons. 2017. Effects of gene choice, base composition and rate heterogeneity on inference and estimates of divergence times in cypriniform fishes. *Biological Journal of the Linnean Society* 121: 319–339. https://doi.org/10.1093/biolinnean/blw045

Hoese, D.F. & G.R. Allen. 2015. Descriptions of three new species of *Glossogobius* (Teleostei: Gobiidae) from New Guinea. *Zootaxa* 3986: 201–216. https://doi.org/10.11646/zootaxa.3986.2.4.

Hoese, D.F., R.K. Hadiaty & F. Herder. 2015. Review of the dwarf *Glossogobius* lacking head pores from the Malili lakes, Sulawesi, with a discussion of the definition of the genus. *Raffles Bulletin of Zoology* 63: 14–26.

Hoese, D.F. & M.P. Hammer. 2021. A review of the *Glossogobius giuris* complex in Australia, with wider discussion on nomenclature and possible synonymies. *Zootaxa* 4974 (1): 79–115. https://doi.org/10.11646/zootaxa.4974.1.3

Hoestlandt, H. 1991a. General introduction to Clupeidae. Pp 119–142 *in:* H. Hoestlandt. (Ed.). *The freshwater fishes of Europe. Vol. 2. Clupeidae, Anguillidae*. AULA Verlag, Wiesbaden.

Hoestlandt, H. 1991b. *Clupeonella cultriventris* (Nordmann, 1840. Pp 53–84 *in:* H. Hoestlandt. (Ed.). *The freshwater fishes of Europe. Vol. 2. Clupeidae, Anguillidae*. AULA Verlag, Wiesbaden.

Holčík, J. 2000. Major problems concerning the conservation and recovery of the Atlantic sturgeon *Acipenser sturio* L., 1758. *Boletin del Instituo Español de Oceanographii* 16: 139–148.

Holčík, J. & L. Jedlička. 1994. Geographical variation of some taxonomically important characters in fishes: the case of the bitterling *Rhodeus sericeus*. *Environmental Biology of Fishes* 41: 147–170. http://dx.doi.org/10.1007/BF02197842

Holčík, J. & V. Skořepa. 1971. Revision of the roach, *Rutilus rutilus* (Linnaeus, 1758), with regards to its subspecies. *Annotationes Zoologicae et Botanicae* 64: 1–60.

Holčík, J. 1976. On the occurrence of far east planktivorous fishes in the Danube River with regard to the possibility of their natural reproduction. *Vestnik Ceskoslovenske Spolecnosti Zoologicke* 40:88–103.

Holčík, J. 1986. *Caspiomyzon wagneri* (Kessler, 1870). Pp 119–142 *in:* J. Holčík. (Ed.): *The freshwater fishes of Europe. Vol.1/I: Petromyzontiformes*. AULA Verlag, Wiesbaden.

Holčík, J. 1999. *Rhodeus sericeus* (Pallas, 1776). Pp 2–224 *in:* P.M. Bănărescu. (Ed.). *The freshwater fishes of Europe. Vol.5/I: Cyprinidae 2/I.* AULA Verlag, Wiebelsheim.

Holčík, J., P. Bănărescu & D. Evans. 1989a. General introduction to fishes. Pp 18–147 *in:* J. Holčík. (Ed.). *The freshwater fishes of Europe. Vol.1/ II: General Introduction to fishes – Acipenseriformes*. AULA Verlag, Wiesbaden.

Holčík, J., R. Kinzelbach, L.I. Sokolov & V.P. Vasil'ev. 1989b. *Acipenser sturio* Linnaeus, 1758. Pp 367–394 *in:* J. Holčík. (Ed.). *The freshwater fishes of Europe. Vol.1/II: General Introduction to fishes – Acipenseriformes*. AULA Verlag, Wiesbaden.

Holčík, J. & V. Skorepa. 1971. Revision of the roach, *Rutilus rutilus* (Linnaeus, 1758), with regard to its subspecies. *Annotationes Zoologicae et Botanicae* 64: 1–60, 9 pls.

Hölker, F. & R. Thiel. 1998. Biology of ruffe (*Gymnocephalus cernuus* (L.)—a review of selected aspects from European literature. *Journal of Great Lakes Research* 24: 186–204.

Hora, S.L. & E.G. Silas 1952. Evolution and distribution of Glyptosternoid fishes of the family Sisoridae (Order: Siluroidea). *Proceedings of the National Institute of Sciences of India* 18: 309–322.

Hossain, M.I., S. Sarmin, M. Alam, L. Alam, M. Mokhtar & M.A. Samad. 2015. Food and feeding habit of *Aspidoparia morar*: A study on Padma River in northwestern Bangladesh. *Trends in Fisheries Research* 4: 24–31.

Houde, A.E. 1997. Sex, colour, and mate choice in guppies. Princton University Press, Princeton.Howes, G. 1981. Anatomy and phylogeny of the Chinese major carps *Ctenopharyngodon* Steind., 1866 and *Hypophthalmichthys* Blkr., 1860. *Bulletin of the British Museum (Natural History) Zoology* 41: 1–52.

Howes, G. 1981. Anatomy and phylogeny of the chinese major carps *Ctenopharyngodon* Steınd., 1866 and *Hypophthalmichthys* Blkr., 1860. *Bulletin of the Britisch Museum (Natural History), Zoology Series* 41: 1–52.

Howes, G. 1982. Anatomy and evolution of the jaws in the semiplotine carps with a review of the genus *Cyprinion* Heckel, 1843 (Teleostei: Cyprinidae). *Bulletin of the British Museum (Natural History) Zoology* 42: 299–335.

Howes, G.J., I.J. Winfield & J.S. Nelson. 1991. *Cyprinid fishes: Systematics, biology and exploitation*. New Dehli: Chapman & Hall.

Hrbek, T. & A. Meyer. 2003. Closing of the Tethys Sea and the phylogeny of Eurasian killifishes (Cyprinodontiformes: Cyprinodontidae). *Journal of Evolutionary Biology* 16: 17–36. https://doi.org/10.1046/j.1420-9101.2003.00475.x

Hrbek, T. & R.H. Wildekamp. 2003. *Aphanius villwocki*, a new species from the Sakarya River basin of central Anatolian plain, Turkey (Teleostei: Cyprinodontiformes). *Ichthyological Exploration of Freshwaters* 14: 137–144.

Hrbek, T., F. Küçük, T. Frickey, K.N. Stölting, R.H. Wildekamp & A. Meyer. 2002. Molecular phylogeny and historical biogeography of the *Aphanius* (Pisces, Cyprinodontiformes) species complex of central Anatolia, Türkiye. *Molecular Phylogenetics and Evolution* 25 (1): 125–137. https://doi.org/10.1016/S1055-7903(02)00203-8

Hrbek, T., Y. Keivany & B.W. Coad. 2006. New species of *Aphanius* (Teleostei, Cyprinodontidae) from Isfahan province of Iran and a reanalysis of other Iranian species. *Copeia*, 2: 244–255. https://doi.org/10.1643/0045-8511(2006)6[244:NSOATC]2.0.CO;2

Huang, S.-P., I-S. Chen & K.-T. Shao. 2016. A new species of *Rhinogobius* (Teleostei: Gobiidae) from Zhejiang Province, China. *Ichthyological Research* 63: 470–479. https://doi.org/10.1007/s10228-016-0516-9

Hughes, L.C., G. Ortí, Y. Huang, Y. Sun, C.C. Baldwin, A.W. Thompson, D. Arcila, R. Betancur-R, C. Li, L. Becker, N. Bellora, X. Zhao, X. Li, M. Wang, C. Fang, B. Xie, Z. Zhou, H. Huang, S. Chen, B. Venkatesh & Q. Shic. 2018. Comprehensive phylogeny of ray-finned fishes (Actinopterygii) based on transcriptomic and genomic data, *PNAS* 115 (24) 6249–625. https://doi.org/10.1073/pnas.1719358115

ICZN. 1999. *International Code of Zoological Nomenclature*. Fourth edition. The International Trust for Zoological Nomenclature, London, UK.

ICZN. 2003. Declaration 44. Amendment of Article 74.7.3. *Bulletin of Zoological Nomenclature* 60: 263.

İlhan, A. & S. Balık. 2008. Fish fauna of the inland waters in western Black Sea region. *Journal of Black Sea/Mediterranean Environment* 25: 75–82. [in Turkish].

İlhan, A., H.M. Sarı & S. Balık. 2011. Reproductive biology of the broad-snout [*Chondrostoma holmwoodii* (Boulenger, 1896)] in Tahtalı Dam Lake, İzmir, Turkey. *Journal of Applied Ichthyology* 27: 1033–1036. https://doi.org/10.1111/j.1439-0426.2011.01719.x

İlhan, A., H.M. Sarı & I. Kurtul. 2020. Bakırçay Nehri (Kuzey Ege, Türkiye) Balık Faunası. *Ege Journal of Fisheries and Aquatic Sciences* 37: 309–312.

İlhan, A., H. Sarı, I. Kurtul, & S. Atak. 2021. Küçük Menderes Nehri Balık Faunasına Katkılar. *Journal of Limnology and Freshwater Fisheries Research* 7 (3): 198–206.

Iljin, B.S. 1928. Two new genera and a new species of gobies (Gobiidae) from the Caspian Sea. Two new genera and a new species of Gobiidae from the Caspian Sea. *Trudy Astrakhanskoi Nauchnoi Rybokhoziaistvennoi Stantsii* 6 (3): 39–49, 1 pl.

Imoto, J.M., K. Saitoh, T. Sasaki, T. Yonezawa, J. Adachi, Y.P. Kartavtsev, M. Miya, M. Nishida & N. Hanzawa. 2013. Phylogeny and biogeography of highly diverged freshwater fish species (Leuciscinae, Cyprinidae, Teleostei) inferred from mitochondrial genome analysis. *Gene* 514: 112–124. http://dx.doi.org/10.1016/j.gene.2012.10.019

Innal, D. 2019. Age structure and growth characteristics of the endemic fish *Oxynoemacheilus anatolicus* (Nemacheilidae) in Düğer Creek, Mediterranean region of Turkey. *Aquatic Sciences and Engineering* 34: 80–85. https://doi.org/10.26650/ASE2019544309

International Rivers. 2014. *Turkey*. http://www.internationalrivers.org/campaigns/turkey. (Accessed October 2014)

Isbrücker, I.J. 1980. Classification and catalogue of the mailed Loricariidae (Pisces, Siluriformes). *Verslagen en Technische Gegevens* 22: 1–181.

IUCN. 2001. IUCN Red List categories and criteria: version 3.1. IUCN, Gland & Cambridge. Available at:www.iucnredlist.org/resources/categories-and-criteria

IUCN. 2020. *Global invasive species database*. Available at: http://www.iucngisd.org/gisd/ (Accessed October 2020)

Iwatsuki, Y. 2013. Review of the *Acanthopagrus latus* complex (Perciformes: Sparidae) with descriptions of three new species from the Indo-West Pacific. *Journal of Fish Biology* 83: 64–95. https://doi.org/10.1111/jfb.12151

Jacoby, D.M., J.M. Casselman, V. Crook, M.B. DeLucia, H. Ahn, K. Kaifu, T. Kurwie, P. Sasal, A.M. Silfvergrip, K.G. Smith & K. Uchida. 2015. Synergistic patterns of threat and the challenges facing global anguillid eel conservation. *Global Ecology and Conservation* 4:321–333. https://doi.org/10.1016/j.gecco.2015.07.009

Jamali, H., R. Patimar, M. Farhadi & F. Darei. 2016. Age, growth and reproduction of *Paracobitis malapterura* (Teleostei: Nemacheilidae)

from Qom River, Iran. *Iranian Journal of Ichthyology* 3: 43–52. https://doi.org/10.22034/iji.v3i1.101

Janko, K., P. Kotlik & P. Ráb. 2003. Evolutionary history of asexual hybrid loaches (*Cobitis*: Teleostei) inferred from phylogenetic analysis of mitochondrial DNA variation. *Journal of Evolutionary Biology* 16 (6): 1280–1287.

Jenkins, R.E. & N.M. Burkhead. 1993. *Freshwater fishes of Virginia*. American Fisheries Society. Bethesda, Maryland.

Johari, S.A., B.W. Coad, S. Mazloomi, M. Kheyri & S. Asghari. 2009. Biological and morphometric characteristics of *Capoeta fusca*, a cyprinid fish living in the qanats of south Khorasan, Iran. *Zoology in the Middle East* 47: 63–70. https://doi.org/10.1080/09397140.2009.10638348

Johnson, G.D. 1984. Percoidei: development and relationships. Pp 464–498 *in:* H.G. Moser. (Ed.). *Ontogeny and systematics of fishes: based on an international symposium dedicated to the memory of Elbert Halvor Ahlstrom*. American Society of Ichthyologists and Herpetologists, Special Publication 1. https://doi.org/10.5962/bhl.title.4434

Johnson, E.J., B. Yoğurtçuoğlu, Ş.G. Kırankaya & F.G. Ekmekçi. 2025. Feeding ecology of the Critically Endangered *Gobio insuyanus* (Gobionidae). *Ecology and Evolution*. 10.1002/ece3.71156

Jokar, M., B.B. Kamangar, E. Ghaderi & J. Freyhof. 2023. *Glyptothorax sardashtensis*, a new species of torrent catfish from the upper Lesser Zab drainage in Iran (Teleostei: Sisoridae). *Zootaxa* 5254 (4): 476–492. https://doi.org/10.11646/zootaxa.5254.4.2

Jouladeh-Roudbar, A. 2021. Distribution and taxonomy of the *Barbus* Cuvier and Cloquet, 1816 in Iran using COI gene. *Modern Genetic journal* 16: 125–132. [in Farsi, English summary].

Jouladeh-Roudbar, A. 2022. Molecular phylogeny of *Capoeta* Valenciennes, 1842 (Teleostei, Cyprinidae) species in Iran, using the cytochrome b gene. *Modern Genetic journal* 17: 171–183. [in Farsi, English summary].

Jouladeh-Roudbar, A., H.R. Esmaeili, A. Gholamifard, R. Zamanian & S. Vatandoust. 2015a. Geographic distribution of the genus *Chondrostoma* Agassiz, 1832 in Iran (Teleostei: Cyprinidae). *Iranian Journal of Ichthyology* 2: 71–78. https://doi.org/10.22034/iji.v2i2.38

Jouladeh-Roudbar, A., S. Eagderi & H.R. Esmaeili. 2015b. Fishes of the Dasht-e Kavir basin of Iran: an updated checklist. *International Journal of Aquatic Biology* 3: 263–273. https://doi.org/10.22034/ijab.v3i4.105

Jouladeh-Roudbar, A., S. Eagderi & S. Vatandoust. 2015c. First record of *Paraschistura alta* (Nalbant and Bianco, 1998) from Eastern Iran and providing its COI barcode region sequences (Teleostei: Nemacheilidae). *Iranian Journal of Ichthyology* 2: 235–243. https://doi.org/10.22034/iji.v2i4.87

Jouladeh-Roudbar, A., S. Eagderi, H.R. Ghanavi & I. Doadrio. 2016a. Taxonomic review of the genus *Capoeta* Valenciennes, 1842 (Actinopterygii, Cyprinidae) from central Iran with the description of a new species. *FishTaxa* 1: 166–175.

Jouladeh-Roudbar, A., S. Eagderi, H.R. Esmaeili, B.W. Coad & N. Bogutskaya. 2016b. A molecular approach to the genus *Alburnoides* (Actinopterygii: Cyprinidae) using COI sequences data set and the description of a new species, *A. damghani*, from the Damghan River system (the Dasht-e Kavir basin, Iran). *ZooKeys* 579: 157–181. https://doi.org/10.3897/zookeys.579.7665

Jouladeh-Roudbar, A., S. Eagderi & A. Soleimani. 2016c. First record of *Petroleuciscus esfahani* Coad and Bogutskaya, 2010 (Actinopterygii: Cyprinidae) from the Karun River drainage, Persian Gulf basin, Iran. *International Journal of Aquatic Biology* 4: 400–405. https://doi.org/10.22034/ijab.v4i6.276

Jouladeh-Roudbar, A., S. Eagderi & H.R. Esmaeili. 2016d. First record of the striped bystryanka, *Alburnoides taeniatus* (Kessler, 1874) from the Hari River basin, Iran (Teleostei: Cyprinidae). *Journal of Entomology and Zoology Studies* 4: 788–791.

Jouladeh-Roudbar, A., S. Eagderi & T. Hosseinpour. 2016e. *Oxynoemacheilus freyhofi*, a new nemacheilid species (Teleostei, Nemacheilidae) from the Tigris basin, Iran. *FishTaxa* 1: 94–107.

Jouladeh-Roudbar, A., S. Eagderi, G. Sayyadzadeh & H.R. Esmaeili. 2017a. *Cobitis keyvani*, a junior synonym of *Cobitis faridpaki* (Teleostei: Cobitidae). *Zootaxa* 4244: 118–126. https://doi.org/10.11646/zootaxa.4244.1.6

Jouladeh-Roudbar, A., S. Eagderi, H.R. Ghanavi & I. Doadrio. 2017b. A new species of the genus *Capoeta* Valenciennes, 1842 from the Caspian Sea basin in Iran (Teleostei, Cyprinidae). *ZooKeys* 682: 137–155. https://doi.org/10.3897/zookeys.682.12670

Jouladeh-Roudbar, A., S. Eagderi, L. Murillo-Ramos, H.R. Ghanavi & I. Doadrio. 2017c. Three new species of algae-scraping cyprinid from Tigris River drainage in Iran (Teleostei: Cyprinidae). *FishTaxa* 2: 134–155.

Jouladeh-Roudbar, A., H.R. Ghanavi & I. Doadrio. 2020. Ichthyofauna From Iranian Freshwater: Annotated checklist, diagnosis, taxonomy, distribution and conservation assessment. *Zoological Studies* 59: e21. https://doi.org/10.6620/ZS.2020.59-21

Jouladeh-Roudbar A., H. R. Ghanavi & J. Freyhof. 2023a. *Glyptothorax vatandousti*, a new species of torrent catfish from the upper Karkheh drainage in Iran (Teleostei: Sisoridae). *Zootaxa* 5315: 37–58. https://doi.org/10.11646/zootaxa.5315.1.2

Jouladeh-Roudbar A., Lalramliana, S. Vatandoust, H. R. Ghanavi & J. Freyhof. 2023b. *Cabdio occidentalis*, a new species from Lake Mashkid basin and the Makran region (Teleostei: Danionidae). *Zootaxa* 5360: 437–447. https://doi.org/10.11646/zootaxa.5360.3.7

Jouladeh-Roudbar, A., Vatandoust, S., Doadrio, I., & H.R. Ghanavi. 2023c. Integrative taxonomy of *Turcinoemacheilus* Bănărescu & Nalbant, 1964 in West Asia with the description of three new species (Teleostei: Nemacheilidae). *Diversity* 15(12): 1222. https://doi.org/10.3390/d15121222

Jouladeh-Roudbar, A., H.R. Ghanav, C. Kaya & J. Freyhof. 2023d. *Barbus urmianus*, a synonym of *Barbus cyri* (Teleostei: Cyprinidae). *Zootaxa* 5296(1):16–30. https://doi.org/10.11646/zootaxa.5296.1.2

Jouladeh-Roudbar, A., C. Kaya, S. Vatandoust & H.R. Ghanavi. 2024. New insights into the phylogeny of *Carasobarbus* Karaman, 1971 (Actinopterygii, Cyprinidae) with the description of three new species. *Scientific Reports* 14: 21801. https://doi.org/10.1038/s41598-024-71463-7

Juan Liu, J., M.-M. Chang, M.V.H. Wilson & A.M. Murray. 2015. A new family of Cypriniformes (Teleostei, Ostariophysi) based on a redescription of †*Jianghanichthys hubeiensis* (Lei, 1977) from the Eocene Yangxi Formation of China. *Journal of Vertebrate Paleontology*. https://doi.org/10.1080/02724634.2015.1004073

Juario, J.V., R.P. Ferraris & L.V. Benitez. 1984. *Advances in milkfish biology and culture*. Island Publishing House, Manila.

Kafuku, T. 1969. Morphological differentiation of *Cyprinion* in Iraq. *Bulletin of the Freshwater Fisheries Research Laboratory, Tokyo* 19: 155–159.

Kalaycı, G. 2022. Molecular phylogeny and historical biogeography of *Petroleuciscus* (Teleostei: Leuciscidae) species in Turkey. *Journal of Anatolian Environmental and Animal Sciences*, 7(1), 88–95.

Kalous, L., K. Rylkov, J. Bohlen, R. Sanda & M. Petrtyl. 2013. New mtDNA data reveal a wide distribution of the Japanese ginbuna *Carassius langsdorfii* in Europe. *Journal of Fish Biology* 82: 703–707. http://dx.doi.org/10.1111/j.1095-8649.2012.03492.x

Kamangar, B.B., A.M. Prokofiev, E. Ghaderi & T.T. Nalbant. 2014. Stone loaches of Choman system, Kurdistan, Iran (Teleostei: Cypriniformes: Nemacheilidae). *Zootaxa* 3755: 33–61. http://dx.doi.org/10.11646/zootaxa.3755.1.2

Kamensky, S.N 1901. *Die Cypriniden der Kaukasusländer und ihrer angrenzenden Meere*. 2. Lieferung. Tiflis. i–ii + i–ii + 1–192, pls. 7–12. [In Russian, German summary]

Kara, C. & K. Solak. 2004. Some biological properties of *Chondrostoma regium* (Heckel, 1843) inhabiting Sır Dam Lake (Kahramanmaraş). *KSU. Journal of Science and Engineering* 7: 13–19.

Kara, C., A. Alp & M. Şimşekli. 2010. Distribution of fish fauna on the upper and middle basin of Ceyhan River, Turkey. *Turkish Journal of Fisheries and Aquatic Sciences* 10: 111–122. http://dx.doi.org/10.4194/trjfas.2010.0116

Kara, C., A., Alp & M.E. Gürlek. 2011. Morphological variations of the trouts (*Salmo trutta and Salmo platycephalus*) in the rivers of Ceyhan, Seyhan and Euphrates, Turkey. *Turkish Journal of Fisheries and Aquatic Sciences* 11: 77–85. http://dx.doi.org/10.4194/trjfas.2011.0111

Karaman, M.S. 1969. Süßwasserfische der Türkei. 7. Teil. Revision der kleinasiatischen und vorderasiatischen Arten des Genus *Capoeta* (*Varicorhinus*, partim). *Mitteilungen aus dem Hamburgischen Zoologischen Museum und Institut* 66: 17–54, 7 pls.

Karaman, M.S. 1971. Süßwasserfische der Türkei. 8. Teil. Revision der Barben Europas, Vorderasiens und Nordafrikas. *Mitteilungen aus dem Hamburgischen Zoologischen Museum und Institut* 67: 175–254.

Karaman, M.S. 1972. Süßwasserfische der Türkei. 9. Teil. Revision einiger kleinwüchsiger Cyprinidengattungen *Phoxinellus*, *Leucaspius*, *Acanthobrama* usw. aus Südeuropa, Kleinasien, Vorder-Asien und Nordafrika. *Mitteilungen aus dem Hamburgischen Zoologischen Museum und Institut* 69: 115–155.

Karataş, S., D. Innal, S. Sungur Birecikligil, E.M. Korkmaz & E. Çiçek. 2016. New localities of *Scardinius elmaliensis* Bogutskaya, 1997 (Teleostei: Cyprinidae) and its phylogenetic relationship based on mtDNA Cytb region sequences. *FishTaxa* 1: 127–132.

Kärst, H., M. Schneider, J. Warner, M. Geiger, A. Kahnt & J. Freyhof. 2020. Eurasische eierlegende Zahnkarpfen. Eurasian egg-laying toothcarps. *DKG, Supplementheft* 13: 1–209.

Kashima, K. 2010. Environmental and climatic changes during the last 20,000 years at Lake Tuz, central Turkey. *Catena* 48: 3–20. https://doi.org/10.1016/S0341-8162(02)00006-1

Kaya, C. 2019. *Taxonomic revision of the species belong to genus Capoeta distributed in Turkey*. PhD Thesis. Recep Tayyip Erdogan University, Institute of Science and Technology, Rize, Turkey. [in Turkish].

Kaya, C. 2020a. Spirlins of the southern Black Sea basin, with the description of a new species (Teleostei: Leuciscidae). *Zootaxa* 4763: 419–428. https://doi.org/10.11646/zootaxa.4763.3.6

Kaya, C. 2020b. New record of three freshwater fish species from a western drainage of Lake Urmia for the Turkish fauna. *Ege Journal of Fisheries and Aquatic Sciences* 37: 325–328. https://doi.org/10.12714/egejfas.37.4.01

Kaya, C. 2020c. The first record and origin of *Salmo trutta* populations established in the upper Tigris River and Lake Van Basin (Teleostei: Salmonidae). *Journal of Anatolian Environmental and Animal Sciences* 5: 366–372. https://doi.org/10.35229/jaes.777575

Kaya, C., D. Turan & E. Ünlü. 2016. The latest status and distribution of fishes in upper Tigris River and two new records for Turkish freshwaters. *Turkish Journal of Fisheries and Aquatic Sciences* 16: 545–562. https://doi.org/10.4194/1303-2712-v16_3_07

Kaya, C., F., Küçük & D. Turan. 2019. New data on the distribution and conservation status of the two endemic scrapers in the Turkish Mediterranean Sea drainages (Teleostei: Cyprinidae). *International Journal of Zoology and Animal Biology* 2: 1–7. https://doi.org/10.23880/izab-16000185

Kaya, C., E. Bayçelebi & D. Turan. 2020a. Taxonomic assessment and distribution of fishes in upper Kura and Aras River drainages. *Zoosystematics and Evolution* 96: 325–344. https://doi.org/10.3897/zse.96.52241

Kaya, C., D. Turan, E. Bayçelebi, G. Kalaycı & J. Freyhof. 2020b. *Oxynoemacheilus cilicicus*, a new nemacheilid loach from the Göksu River in southern Anatolia (Teleostei: Nemacheilidae). *Zootaxa* 4808: 284–300. https://doi.org/10.11646/zootaxa.4808.2.3

Kaya, C., D. Turan, G. Kalaycı, E. Bayçelebi & J. Freyhof. 2020c. The westernmost known population of *Paracobitis* (Teleostei, Nemacheilidae), with the description of a new species from the Euphrates River in southern Anatolia. *Zootaxa* 4838: 525–534. https://doi.org/10.11646/zootaxa.4838.4.6

Kaya, C., B. Yoğurtçuoğlu & J. Freyhof. 2021. *Oxynoemacheilus amanos*, a new nemacheilid loach from the Orontes River drainage (Teleostei: Nemacheilidae). *Zootaxa* 4938: 559–570. https://doi.org/10.11646/zootaxa.4938.5.3

Kaya, C., H.B. Imre & I. Kurtul. 2024a. Is *Garra rezai* (Teleostei, Cyprinidae) a species known only from two widely disjunct areas in the Tigris drainage? *Zoosystematics and Evolution* 100(2): 349–356. https://doi.org/10.3897/zse.100.118766

Kaya, C., I. Kurtul, E. Bayçelebi, A. İlhan & H.M. Sarı. 2024b. Actual distributions and validity of *Petroleuciscus* spp., with the range extension and length-weight relationship data in case of *Petroleuciscus ninae*. *Turkish Journal of Fisheries and Aquatic Sciences* 24(2). https://doi.org/10.4194/TRJFAS23846

Kaya, C., I. Kurtul, İ. Aksu, M. Oral & J. Freyhof. 2024c. *Oxynoemacheilus chaboras*, a new loach species from the Euphrates drainage in Türkiye (Teleostei, Nemacheilidae). *Zoosystematics and Evolution* 100(2): 457–468. https://doi.org/10.3897/zse.100.118612

Kaya, C., B. Yoğurtçuoğlu, İ. Aksu, E. Bayçelebi & D. Turan. 2024d. *Turcinoemacheilus ekmekciae*, a new dwarf loach from upper Tigris and Euphrates (Teleostei: Nemacheilidae). *Journal of Fish Biology* 104(1): 227–239. https://doi.org/10.1111/jfb.15578

Kaya, C., İ. Aksu, S. Aksu, M. Kocabaş, I. Kurtul & D. Turan. 2025a. Tracing historical river connectivity through freshwater fish assemblages: The case of Göksu River and Lake Beyşehir. *Ecohydrology* 18: e70026. https://doi.org/10.1002/eco.70026

Kaya, C., et al. 2025b. Generic position of *Schistura chrysichristinae* within the family Nemacheilidae (Teleostei). (in prep).

Kazancheyev, Y.N. 1981. *Fishes of Caspian*. Light and Food Industry Press, Moscow.

Keivany, Y. & H.R. Esmaeili. 2013. Threatened fishes of the world: *Aphanius farsicus* Teimori, Esmaeili and Reichenbacher, 2011 (Cyprinodontidae). *Croatian Journal of Fisheries: Ribarstvo* 71: 192–194.

Keivany, Y. & J.S. Nelson. 2000. Taxonomic review of the genus *Pungitius*, nine spine sticklebacks (Gasterosteidae). *Cybium* 24: 107–122.

Keivany, Y. & N. Soofiani. 2004. Contribution to the biology of Zagros tooth-carp, *Aphanius vladykovi* (Cyprinodontidae) in central Iran. *Environmental Biology of Fishes* 71: 165–169. https://doi.org/10.1007/s10641-004-0106-y

Keivany, Y., M.S. Alavi-Yeganeh & J. Seyfabadi. 2012. A new record confirms the occurrence of *Aphanius mesopotamicus* Coad, 2009, in southwestern Iran (Actinopterygii: Cyprinodontidae). *Check List* 8: 283–285.

Keivany, Y., A. Nezamoleslami & S. Dorafshan. 2015. Morphological diversity of *Garra rufa* (Heckel, 1843) populations in Iran. *Iranian Journal of Ichthyology* 2: 148–154. https://doi.org/10.22034/iji.v2i3.62

Keivany, Y., M. Nasri, K. Abbasi & A. Abdoli. 2016. *Atlas of inland water fishes of Iran*. Iran Department of Environment, Tehran, Iran. 218 p.

Kemp, L., C. Xu, J. Depledge, K.L. Ebi, G. Gibbins, T.A. Kohler, J. Rockström, M. Scheffer, H.J. Schellnhuber W. Steffen & T.M. Lenton. 2022. Climate Endgame: Exploring catastrophic climate change scenarios. *PNAS* 119: e2108146119 https://doi.org/10.1073/pnas.2108146119

Kessler, K. F. 1874. A description of fishes belonging to the families common to both the Black and the Caspian seas. Trudy St.-Peterburgskogo Obscestva Estestvoispytatelej = Travaux de la Société des Naturalistes de St. Pétersbourg 5: 191–324, 1 pl.

Kessler, K.F. 1877. *The Aralo-Caspian Expedition. IV. Fishes of the Aralo-Caspio-Pontine ichthyological region*. St. Petersburg. i–xxviii+1–360, 8 pls.

Keyserling, E. Von. 1861. Neue Cypriniden aus Persien. *Zeitschrift für die Gesamten Naturwissenschaften* 17: 1–24, 9 pls.

Khaefi, R., H.R. Esmaeili, G. Sayyadzadeh, M.F. Geiger & J. Freyhof. 2016. *Squalius namak*, a new chub from Lake Namak basin in Iran (Teleostei: Cyprinidae). *Zootaxa* 4169: 145–159. https://doi.org/1010.11646/zootaxa.4169.1.7

Khaefi, R., H.R. Esmaeili, M.F. Geiger & S. Eagderi. 2017a. Taxonomic review of the cryptic *Barbus lacerta* species group with description of a new species (Teleostei: Cyprinidae). *FishTaxa* 2: 90–115.

Khaefi, R., A. Teimori & H.R. Esmaeili. 2017b. Phylogenetic relationships and taxonomy of *Luciobarbus barbulus* (Heckel, 1847) (Teleostei: Cyprinidae). *Journal of Ichthyology* 57: 835–845. https://doi.org/10.1134/S0032945217060078

Khalaf, N. 2009. *Garra barreimiae wurayahi* Khalaf, 2009: a new blind cave fish subspecies from Wadi Al Wurayah pools, Emirate of Fujairah, United Arab Emirates. *Gazelle: The Palestinian Biological Bulletin* 90 (June): 1–15.

Khosravi, M., A. Abdoli, F. Ahmadzadeh, R. Saberi-Pirooz, K. Rylková & B.H. Kiabi. 2020. Toward a preliminary assessment of the diversity and origin of cyprinid fish genus *Carassius* in Iran. *Journal of Applied Ichthyology* 36: 422–430. https://doi.org/10.1111/jai.14039

Kide, N.G., A. Dunz, J.-F. Agnèse, J. Dilyte, A. Pariselle, C. Carneiro, E. Correia, J.C. Brito, L.O. Yarba, Y. Kone & J.-D. Durand. 2016. Cichlids of the Banc d'Arguin National Park, Mauritania: insight into the diversity of the genus *Coptodon*. *Journal of Fish Biology* 88: 1369–1393. https://doi.org/10.1111/jfb.12899.

Kim, D., A.T. Taylor, & T.J. Near. 2022. Phylogenomics and species delimitation of the economically important Black Basses (*Micropterus*). *Scientific Reports* 12 (9113): 1–14. https://doi .org/10.1038/s41598-022-11743-2

Kirchner, S., H. Sattmann, E. Haring, L. Plan, R. Victor & L. Kruckenhauser. 2017. Investigating gene flow between the blind cavefish *Garra barreimiae* and its conspecific surface populations. *Scientific Reports* 7 (1): 5130. [http://www.nature.com/articles/s41598-017-05194-3]

Kirchner, S., L. Kruckenhauser, A. Pichler, K. Borkenhagen & J. Freyhof. 2020. Revision of the *Garra* species of the Hajar Mountains in Oman and the United Arab Emirates with the description of two new species (Teleostei: Cyprinidae). *Zootaxa* 4751: 521–545. https://doi.org/10.11646/zootaxa.4751.3.6

Klemetsen, A., P.A. Amundsen, J.B. Dempson, B. Jonsson, N. Jonsson, M.F. O'connell & E. Mortensen. 2003. Atlantic salmon *Salmo salar* L., brown trout *Salmo trutta* L. and Arctic charr *Salvelinus alpinus* (L.): a review of aspects of their life histories. *Ecology of Freshwater Fish* 12: 1–59. https://doi.org/10.1034/j.1600–0633.2003.00010.x

Kobayakawa, M. 1989. Systematic revision of the catfish genus *Silurus*, with description of a new species from Thailand and Burma. *Japanese Journal of Ichthyology* 36: 155–186.

Kolarov, P. 1991a. *Alosa maeotica maeotica* (Grimm, 1901). Pp 309–322 *in:* H. Hoestlandt. (Ed.). *The freshwater fishes of Europe. Vol. 2. Clupeidae, Anguillidae*. AULA Verlag, Wiesbaden.

Kolarov, P. 1991b. *Alosa pontica pontica* (Eichwald, 1938). Pp 337–368 *in:* H. Hoestlandt. (Ed.). *The freshwater fishes of Europe. Vol. 2. Clupeidae, Anguillidae.* AULA Verlag, Wiesbaden.

Koohestan Eskandari, S. 2003. Population dynamic of *Capoeta capoeta gracilis* in Madarsoo stream in Golestan National Park. *Iranian Journal of Marine Sciences* 2(2–3): 11–20. [in Farsi].

Kosswig, C. 1950. Die Gattung *Tylognathus* in Vorderasien. Ergänzungsband "Neue Ergebnisse und Probleme der Zoologie", Festschrift Berthold Klatt. *Zoologischer Anzeiger* 145: 406–415.

Kosswig, C. & F. Sözer. 1945. Nouveaux Cyprinodontides de l'Anatolie centrale. *Revue de la Faculté des Sciences de l'Université d'Instanbul*, Série B: Sciences Naturelles 10: 77–83.

Kosswig, C. & R. Geldiay. 1952. Eğirdir Gölü Balıkları [Fishes of Lake Eğirdir]. Balık ve Balıkçılık, İstanbul Universitesi, Fen Fakültesi Hidrobiyoloji Araþtırmaları Enstitüsü *Yayınları* 3: 3–14.

Kotlík, P., N.G. Bogutskaya & F.G. Ekmekçi. 2004. Circum Black Sea phylogeography of *Barbus* freshwater fishes: divergence in the Pontic glacial refugium. *Molecular Ecology* 13: 87–95. https://doi.org/10.1046/j.1365-294x.2003.02021.x

Kotlík, P., S. Markova, L. Choleva, N. Bogutskaya, F.G. Ekmekçi & P.P. Ivanova. 2008. Divergence with gene flow between Ponto-Caspian refugia in an anadromous cyprinid *Rutilus frisii* revealed by multiple gene phylogeography. *Molecular Ecology* 17: 1076–1088. https://doi.org/10.1111/j.1365-294X.2007.03638.x

Kottelat, M. 2001. Freshwater Fishes of Northern Vietnam. Publisher: Environment and Social Development Unit, East Asia and Pacific Region, World Bank, Washington.

Kottelat, M. 2004. *Salaria economidisi*, a new species of freshwater fish from Lake Trichonis, Greece, with comments on variation in *S. fluviatilis* (Teleostei: Blenniidae). *Revue Suisse de Zoologie* 111: 121–137.

Kottelat, M. 2007. Three new species of *Phoxinus* from Greece and southern France (Teleostei: Cyprinidae). *Ichthyological Exploration of Freshwaters* 18: 145–162.

Kottelat, M. 2012. *Conspectus cobitidum*: an inventory of the loaches of the world (Teleostei: Cypriniformes: Cobitoidei). *The Raffles Bulletin of Zoology* 26 (Suppl.): 1–199.

Kottelat, M. 2016. On *Gonorynchus*, *Gonorhynchus*, *Gonorinchus*, *Gonorhinchus* and *Gonorrhynchus*, and some other names of labeonine fishes (Teleostei: Gonorynchidae and Cyprinidae). *Zootaxa* 4178: 443–450. https://doi.org/10.11646/zootaxa.4178.3.8

Kottelat, M. 1997. European freshwater fishes. An heuristic checklist of the freshwater fishes of Europe (exclusive of former USSR), with an introduction for non-systematists and comments on nomenclature and conservation. *Biologia (Bratislava)* 52: 1–271. https://doi.org/10.1023/A:1008959313491

Kottelat, M. 2020. *Ceratogarra*, a genus name for *Garra cambodgiensis* and *G. fasciacauda* and comments on the oral and gular soft anatomy in labeonine fishes (Teleostei: Cyprinidae). *Raffles Bulletin of Zoology, Supplement* 35: 156–17.

Kottelat, M. 2022. *Mastacembelus simack*, the valid name of the Mesopotamian spiny eel (Teleostei: Mastacembelidae). *Zootaxa* 5162: 446–450. https://doi.org/10.11646/zootaxa.5162.4.9

Kottelat, M. & P.S. Economidis. 2006. *Squalius orpheus*, a new species of cyprinid fish from Evros drainage, Greece (Teleostei: Cyprinidae). *Ichthyological Exploration of Freshwaters* 17: 181–186.

Kottelat, M. & J. Freyhof. 2007. *Handbook of European freshwater fishes*. Kottelat, Cornol, Switzerland & Freyhof, Berlin, Germany.

Kottelat, M. & J. Freyhof 2025. *Huso* Lovetzky, 1834, a junior synonym of *Sterletus* Rafinesque, 1820, and comments on some other generic names for sturgeons (Chondrostei: Acipenseridae). Zootaxa, in press

Kottelat, M. & A. Wheeler. 2001. *Aphanius* Nardo, 1827 (Osteichthyes, Cyprinodontiformes): proposed placement on the Official List. *Bulletin of Zoological Nomenclature* 58: 110–115.

Kottelat, M., R., Barbieri & M.T. Stoumboudi. 2007. *Aphanius almiriensis*, a new species of toothcarp from Greece (Teleostei: Cyprinodontidae). *Revue Suisse De Zoologie* 14: 13–31. https://doi.org/10.5962/bhl.part.80385

Koutrakis, E. M. & P.S. Economidis. 2000. First record in the Mediterranean (North Aegean Sea, Greece) of the pacific mullet *Mugil soiuy* Basilewsky, 1855 (Mugilidae). *Cybium* 24: 299–302.

Kovačić, M. & S. Engin. 2008. Two new species of *Neogobius* (Gobiidae) from northeastern Turkey. *Cybium* 32: 73–80. https://doi.org/10.26028/cybium/2008-321-007

Krieger, J., A.K. Hett, P.A. Fuerst, E. Artyukhin & A. Ludwig. 2008. The molecular phylogeny of the order Acipenseriformes revisited. *Journal of Applied Ichthyology* 24: 36–45. https://doi.org/10.1111/j.1439-0426.2008.01088.x

Krijgsman, W., D.V. Palcu, F. Andreetto, M. Stoica & O. Mandic. 2020. Changing seas in the late Miocene northern Aegean: A Paratethyan approach to Mediterranean basin evolution. *Earth-Science Reviews* 210: 103386 https://doi.org/10.1016/j.earscirev.2020.103386

Kruckenhauser, L., E. Haring, R. Seemann & H. Sattmann. 2011. Genetic differentiation between cave and surface-dwelling populations of *Garra barreimiae* (Cyprinidae) in Oman. *BMC Evolutionary Biology* 11 (1): 1–15. https://doi.org/10.1186/1471-2148-11-172

Krupp, F. 1982. *Garra tibanica ghorensis* sub sp. nov. (Pisces: Cyprinidae), an African element in the cyprinid fauna of the Levant. *Hydrobiologia* 88: 319–324. https://doi.org/10.1007/BF00008513

Krupp, F. 1983. Fishes of Saudi Arabia. Freshwater fishes of Saudi Arabia and adjacent regions of the Arabian Peninsula. *Fauna of Saudi Arabia* 5: 568–636.

Krupp, F. 1985a. A new species of *Chondrostoma* from the Orontes River drainage basin of Turkey and Syria (Pisces: Osteichthyes: Cyprinidae). *Senckenbergiana Biologica* 66: 27–33.

Krupp, F. 1985b. *Barbus chantrei* (Sauvage 1882), a valid species of cyprinid fish from the northern Levant (Pisces: Osteichthyes: Cyprinidae). *Senckenbergiana biologica* 66: 17–25.

Krupp, F. 1985c. Rehabilitation of *Barbus lorteti* Sauvage, 1882, and comments on the validity of the generic names *Bertinius* Fang, 1943 and *Bertinichthys* Whitley, 1953 (Pisces: Cyprinidae). *Hydrobiologica* 20: 63–68. https://doi.org/10.1007/BF00034591

Krupp, F. 1985d. *Systematik und Zoogeographie der Süßwasserfische des levantinischen Grabenbruchsystems und der Ostküste des Mittelmeeres*, Dissertation Doktor der Naturwissenschaften, Mainz: Johannes Gutenberg-Univ. Mainz.

Krupp, F. 1988 Freshwater fishes of the Wadi Batha drainage. *Journal of Oman Studies Special Report* 3: 401–404.

Krupp, F. 1992a. A new species of *Nemacheilus* (Pisces: Balitoridae) from the Van Gölü drainage basin in eastern Turkey. *Mitteilungen aus dem Hamburgischen Zoologischen Museum und Institut* 89: 291–294.

Krupp, F. 1992b. Two new species of cyprinid fishes from the Mediterranean coastal drainage basin of Syria (Pisces: Osteichthyes: Cyprinidae). *Senckenbergiana Biologica* 72: 19–25.

Krupp, F. & K. Budd. 2009. A new species of the genus *Garra* (Teleostei: Cyprinidae) from Oman. *Aqua, International Journal of Ichthyology* 15: 117–120.

Krupp, F. & B.W. Coad. 1985. Notes on a population of the threespine stickleback, *Gasterosteus aculeatus*, from Syria (Pisces: Osteichthyes: Gasterosteidae) *Senckenbergiana Biologica* 66: 35–39.

Krupp, F. & J. Moubayed. 1992. A new species of *Cobitis* Linnaeus 1758 from the Orontes and Litani drainage basins of Syria and Lebanon (Pisces: Osteichthyes: Cobitidae). *Senckenbergiana Biologica* 72: 13–18.

Krupp, F. & W. Schneider. 1989. The fishes of the Jordan River drainage basin and Azraq Oasis. *Fauna of Saudia Arabia* 10: 347–416.

Krupp, F. & W. Schneider. 1991. Two new species of *Nemacheilus* Bleeker 1863 from the Orontes River drainage basin of Lebanon, Syria and Turkey (Pisces: Osteichthyes: Balitoridae). *Senckenbergiana Biologica* 71: 23–34.

Kunz, W. 2012. *Do species exist? Principles of taxonomic classification*. Wiley-VCH Verlag & Co. KGaA, Weinheim, Germany.

Küçük, F. 2007a. The fish fauna of the Göksü River (Türkiye): Taxonomic and zoogeographic features. *Turkish Journal of Fisheries and Aquatic Sciences* 7: 53–63.

Küçük, F. 2007b. *Pseudophoxinus alii* (Teleostei: Cyprinidae), a new fish species from the Antalya region, Turkey. *Turkish Journal of Zoology* 31: 99–106.

Küçük, F. 2012. Extinct endemic fishes of Turkey: *Alburnus akili* (Gövce) and *Pseudophoxinus handlirschi* (Kavinne) (Pisces: Cyprinidae). *Turkish Journal of Fisheries and Aquatic Sciences* 12: 345–347. https://doi.org/10.4194/1303-2712-v12_2_21

Küçük, F. & S.S. Güçlü. 2014. A new *Pseudophoxinus* (Teleostei, Cyprinidae) species from Asi River drainage (Turkey). *ZooKeys* 411: 57–66. https://doi.org/10.3897/zookeys.411.6833

Küçük, F., D. Turan, C. Şahin & I. Gülle. 2009a. *Capoeta mauricii* n. sp., a new species of cyprinid fish from Lake Beyşehir, Turkey (Osteichthyes: Cyprinidae). *Zoology in the Middle East* 47: 71–82. https://doi.org/10.1080/09397140.2009.10638349

Küçük, F., H.M. Sarı, O. Demir & I. Gülle. 2009b. Review of the ichthyofaunal changes in Lake Eğirdir between 1915 and 2007. *Turkish Journal of Zoology* 33: 277–286. https://doi.org/10.3906/zoo-0811-16

Küçük, F., I. Gülle, S.S. Güçlü, Y. Çiftçi & Ö. Erdoğan. 2013. A new *Pseudophoxinus* (Teleostei, Cyprinidae) species from southwestern Anatolia, with remarks on the distribution of the genus in western Anatolia. *ZooKeys* 320: 29–41. https://doi.org/10.3897/zookeys.320.4447

Küçük, F., Y. Bektaş, S.S. Güçlü & C. Kaya. 2014. The systematic position of *Acanthalburnus microlepis* (De Filippi, 1863) and contributions to the genus *Acanthobrama* (Cyprinidae: Leuciscinae) in Turkey. *Iranian Journal of Ichthyology* 1: 96–105. https://doi.org/10.22034/iji.v1i2.44

Küçük, F., E. Bayçelebi, S.S. Güçlü & I. Gülle. 2015 Description of a new species of *Hemigrammocapoeta* (Teleostei: Cyprinidae) from Lake Işıklı, Turkey. *Zootaxa* 4052: 359–365. https://doi.org/10.11646/zootaxa.4052.3.5

Küçük, F., I. Gülle & S.S. Güçlü. 2016. *Pseudophoxinus iconii*, a new species of spring minnow from Central Anatolia (Teleostei: Cyprinidae). *Ichthyological Exploration of Freshwaters* 27: 283–288.

Küçük, F., D. Turan, S.S. Güçlü, A. G. Mutlu & Y. Çiftçi. 2017. Two new species of *Chondrostoma* Agassiz, 1832 (Teleostei: Cyprinidae) from the Ceyhan, Seyhan and Göksu Rivers in the East Mediterranean region of Turkey. *Turkish Journal of Fisheries and Aquatic Sciences* 17: 795–803. https://doi.org/10.4194/1303-2712-v17_4_15

Küçük, F., Y. Çiftçi, S.S. Güçlü & D. Turan. 2021. *Chondrostoma smyrnae*, a new nase from the Tahtalı reservoir drainage in the Aegean Sea basin (Teleostei, Leuciscidae). *Zoosystematics and Evolution* 97: 235–248. https://doi.org/10.3897/zse.97.63691

Küçük F., G. Kalaycı, S.S. Güçlü, M. Oral & D. Turan. 2024a. A new species of trout from the Köprüçay River, a drainage of Mediterranean Sea, Türkiye (Salmoniformes, Salmonidae). *Zoosystematics and Evolution* 100(2): 391–403. https://doi.org/10.3897/zse.100.121174

Küçük, F., Y. Çiftçi, S.S. Güçlü, A.G. Mutlu & D. Turan. 2023. Taxonomic review of the Chondrostoma (Teleostei, Leuciscidae) species from inland waters of Turkey: an integrative approach. *Zoosystematics and Evolution* 99(1): 1–13. https://doi.org/10.3897/zse.99.91275

Küçük, F., Çiftçi, Y., Güçlü, S. S., Mutlu, A. G., & Turan, D. (2023). Taxonomic review of the Chondrostoma (Teleostei, Leuciscidae) species from inland waters of Turkey: an integrative approach. Zoosystematics and Evolution, 99(1), 1–13. https://doi.org/10.3897/zse.99.91275

Küçük F., E. Bahrioğlu, S.S. Güçlü, M. Nazıroğlu, I. Gülle & S.Ö. Kabadayı Yıldırım, 2024b. Spatio-temporal variation in reproductive characteristics of invasive fish *Pseudorasbora parva* (Temmick & Schlegel, 1846) in the lakes region of Türkiye. *Journal of Fish Biology*, https://doi.org/10.1111/jfb.15932

Kuiter, R.H. 1998. Pipefishes of the syngnathid genus *Dunckerocampus* (Sygnathiformes: Syngnathidae), with a description of a new species from the Indian Ocean. *Aqua, Journal of Ichthyology and Aquatic Biology* 3: 81–84. [Erratum: *Aqua, Journal of Ichthyology and Aquatic Biology* 4 2000:34.]

Kuljanishvili, T., G. Epitashvili, J. Freyhof, B. Japoshvili, L. Kalous, B. Levin, N. Mustafayev, S. Ibrahimov, S. Pipoyan & L. Mumladze. 2020. Checklist of the freshwater fishes of Armenia, Azerbaijan and Georgia. *Journal of Applied Ichthyology* 36: 501–514. https://doi.org/10.1111/jai.14038

Kullander, S.O. 1999. Fish species–how and why. *Reviews in Fish Biology and Fisheries* 9: 325–352.

Kurina, K. Rohtla, M., Taal, I., Albert, A., Svirgsden, R., Kuningas, S., Lappalainen, A. Eschbaum, R. & Saks, L. 2024. Here to stay: evaluating establishment of eastern tubenose goby *Proterorhinus semipellucidus* (Kessler, 1877) in the Baltic Sea with otolith microchemistry and structure. *Environmental Biology of Fishes*, 107:1089–1098. https://doi.org/10.1007/s10641-024-01604-4

Kurtul, I. & H.M. Sarı. 2017. İstilacı *Gambusia* Türlerinin (*Gambusia holbrooki* ve *G. affinis*) Özellikleri, Türkiye'deki Durumları ve Oluşturdukları Ekolojik Riskler 3(1): 51–60. https://doi.org/10.17216/LimnoFish.300927

Kurtul, I. 2018. *Investigation of the distribution and bio-ecological features of gambusia (mosquitofishes) species in Turkey*. PhD Thesis. Ege University, Graduate School of Natural and Applied Sciences, İzmir, Turkey.

Kurtul, I. & H.M. Sarı. 2019. *Gambusia holbrooki* (Sivrisinek balığı)'nin Türkiye'deki Dağılımına Katkılar. *LimnoFish* 5(3): 170–180. https://doi.org/10.17216/LimnoFish.519729

Kurtul, I. & H.M. Sarı. 2020. Length-weight relationships of invasive mosquitofish (*Gambusia holbrooki* Girard, 1859) in 23 river basins of Turkey. *Turkish Journal of Zoology* 44: 324–334. https://doi.org/10.3906/zoo-2002-37

Kurtul, I. & H.M. Sarı. 2021. Is *Gambusia holbrooki* known as an invasive fish species in Küçük Menderes River basin Lakes (Turkey). *Aquatic Sciences and Engineering* 36: 175–180. https://doi. org/10.26650/ASE2020755178

Kurtul, I., A.S. Tarkan, H.M. Sarı & J.R. Britton. 2022. Climatic and geographic variation as a driver of phenotypic divergence in reproductive characters and body sizes of invasive *Gambusia holbrooki*. *Aquatic Sciences* 84: 29. https://doi.org/10.1007/s00027-022-00862-7

Kuru, M. 1980. A new fish species from Lake Van (Cyprinidae). *Hacettepe Bulletin of Natural Science and Engineering* 9: 97–102.

Kux, Z. & H.M. Steiner. 1972. *Lampetra lanceolata*, eine neue Neunaugenart aus dem Einzugsgebiet des Schwarzen Meeres in der nordöstlichen Türkei. *Časopis Moravského Musea, Acta Musei Moraviae*, Vědy Přírodní / *Scientiae Naturales* 56/57: 375–384.

Ladiges, W. 1960. Süßwasserfische der Türkei, I. Teil Cyprinidae. *Mitteilungen aus dem Hamburgischen Zoologischen Museum und Institut* 58: 105–150.

Lajbner, Z., K. Kohlmann, O. Linhart & P. Kotlík. 2010. Lack of reproductive isolation between the Western and Eastern phylogroups of the tench. *Reviews in Fish Biology and Fisheries* 20: 289–300. https://doi.org/10.1007/s11160-009-9137-y

Lalramliana, S. Lalronunga & M. Singh. 2019. *Cabdio crassus*, a new species of cyprinid fish (Teleostei: Cyprinidae) from the Kaladan River of Mizoram, India. *Zootaxa* 4657: 159–169. https://doi.org/10.11646/zootaxa.4657.1.7

Lang, N.J., K.J. Roe, C.B. Renaud, H.S. Gill, I.C. Potter, J. Freyhof, A.M. Naseka, P. Cochran, H.E. Pérez, E.M. Habit & B.R. Kuhajda. 2009. Novel relationships among lampreys (Petromyzontiformes) revealed by a taxonomically comprehensive molecular data set. *American Fisheries Society Symposium, American Fisheries Society* 72: 41–55.

Laporte, M., J. Claude, P. Berrebi, P. Perret & P. Magnan. 2016a. Shape plasticity in response to water velocity in the freshwater blenny *Salaria fluviatilis*. *Journal of Fish Biology* 88: 1191–1203. https://doi.org/10.1111/jfb.12902

Laporte M., C. Perrier, P. Magnan & P. Berrebi. 2016b. Genetic evidence of recent migration among isolated-by-sea populations of the freshwater blenny (*Salaria fluviatilis*). *Conservation Genetics* 17: 389–399. https://doi.org/10.1007/s10592-015-0791-4

Lappalainen, J., H. Dörner & K. Wysujack. 2003. Reproduction biology of pikeperch (*Sander lucioperca* (L.))–a review. *Ecology of Freshwater Fish* 12: 95–106. https://doi.org/10.1034/j.1600-0633.2003.00005.x

Larmuseau, M.H., J.K. Van Houdt, J. Guelinckx, B. Hellemans & F.A. Volckaert. 2009. Distributional and demographic consequences of Pleistocene climate fluctuations for a marine demersal fish in the north-eastern Atlantic. *Journal of Biogeography* 36: 1138–1151. https://doi.org/10.1111/j.1365-2699.2008.02072.x

Lavergne, E. 2012. *Estuarine fish biodiversity of Socotra Island (NW Indian Ocean): from the fish community to the functioning of Terapon jarbua populations*. Doctoral dissertation, Université de Bretagne occidentale-Brest.

Lee, C.S., M.S. Gordon & W.O. Watanabe. (Eds.) 1986. *Aquaculture of milkfish. State of the art*. Oceanic Institute, Hawaii, USA.

Leonard, J.L. 2013. Williams' paradox and the role of phenotypic plasticity in sexual systems. *Integrative and Comparative Biology* 53: 671–688. https://doi.org/10.1093/icb/ict088

Leonardos, I. & A. Sinis. 1998. Reproductive strategy of *Aphanius fasciatus* Nardo, 1827 (Pisces: Cyprinodontidae) in the Mesolongi and Etolikon lagoons (W. Greece). *Fisheries Research* 35: 171–181. https://doi.org/10.1016/S0165-7836(98)00082-4

Leonardos, I. & A. Sinis. 2000. Age, growth and mortality of *Atherina boyeri* Risso, 1810 (Pisces: Atherinidae) in the Mesolongi and Etolikon lagoons (W. Greece). *Fisheries Research* 45: 81–91. https://doi.org/10.1016/S0165-7836(99)00097-1

Leonardos, I.D. 2001. Ecology and exploitation pattern of a land locked population of sand smelt, *Atherina boyeri* (Risso, 1810), in Trichonis Lake (western Greece). *Journal of Applied Ichthyology* 17: 262–266. https://doi.org/10.1046/j.1439-0426.2001.00296.x

Lévêque, C., D. Paugy & G.G. Teugels. 1992. *Faune des poissons d'eaux douces et saumâtres d'Afrique de l'Ouest*. Tome 2. Paris, MRAC, ORSTOM.

Levin, B.A., J. Freyhof, Z. Lajbner, S. Perea, A. Abdoli, M. Gaffaroğlu, M. Özuluğ, H.R. Rubenyan, V.B. Salnikov & I. Doadrio. 2012. Phylogenetic relationships of the algae scraping cyprinid genus *Capoeta* (Teleostei: Cyprinidae). *Molecular Phylogenetics and Evolution* 62: 542–549. https://doi.org/10.1016/j.ympev.2011.09.004

Levin, B.A., A. Gandlin & A. Barmintseva. 2015. Genetic revision of Caucasian barbels, the genus *Barbus*: one species plus, one species minus. *Frontiers in Marine Science. Conference Abstract: XV European Congress of Ichthyology*. https://doi.org/10.3389/conf.FMARS.2015.03.00175

Levin, B.A., E. Simonov, M.P. Matveyev, O.N. Artaev, N.J. Mustafaeyv, A.N. Pashkov & H.R. Roubenyan. 2016. DNA barcoding of the fishes of the genus *Alburnoides* (Actinopterygii, Cyprinidae) from Caucasus. *Mitochondrial DNA Part A* 29: 49–55. https://doi.org/10.1080/24701394.2016.1238900

Levin, B.A., A.M. Prokofiev & H.R. Roubenyan. 2019. A new species of algae eaters *Capoeta kaput* sp. nov. (Teleostei, Cyprinidae) from Transcaucasia. *Inland Water Biology* 12: 32–41. https://doi.org/10.1134/S1995082919010139

Levin, B., E. Simonov, B.K. Gabrielyan, R.L. Mayden, S.M. Rastorguev, H.R. Roubenyan, F.S. Sharko & A.V. Nedoluzhko. 2022. Caucasian treasure: Genomics sheds light on the evolution of half-extinct Sevan trout, *Salmo ischchan*, species flock. *Molecular Phylogenetics and Evolution* 167: 107346. https://doi.org/10.1016/j.ympev.2021.107346

Liao, T.Y., E. Ünlü & S.O. Kullander. 2011. Western boundary of the subfamily Danioninae in Asia (Teleostei, Cyprinidae): derived from the systematic position of *Barilius mesopotamicus* based on molecular and morphological data. *Zootaxa* 2880: 31–40. https://doi.org/10.11646/zootaxa.2880.1.3

Liley, N.R. & B.H. Seghers. 1975. Factors affecting the morphology and behaviour of guppies in Trinidad. Pp 92–118 *in:* G.P Baerends, C. Beer & A. Manning. (Eds). *Function and Evolution in Behaviour: Essays in Honour of Professor Niko Tinbergen*. Oxford University Press, Oxford.

Liu, H. Z., Y.R. Zhu, C. Smith & M. Reichard. 2006. Evidence of host specificity and congruence between phylogenies of bitterlings and freshwater mussels. *Zoological Studies* 45: 428–434.

Lortet, L. 1883. Études zoologiques sur la faune du lac de Tibériade, suivies d'un aperçu sur la faune des lacs d'Antioche et de Homs. I. Poissons et reptiles du lac de Tibériade et de quelques autres parties de la Syrie. *Archives du Muséum d'Histoire Naturelle de Lyon* 3: 99–189.

Ludwig, A., J. Bohlen, C. Wolter & C. Pitra. 2001. Phylogenetic relationships and historical biogeography of spined loaches (Cobitidae, *Cobitis* and *Sabanejewia*) as indicated by variability of mitochondrial DNA. *Zoological Journal of the Linnean Society* 131: 381–392. https://doi.org/10.1111/j.1096-3642.2001.tb02242.x

Ludwig, A., L. Debus, D. Lieckfeldt, I. Wirgin, N. Benecke, I. Jenneckens, P. Williot, J.R. Waldman & C. Pitra. 2002. When the American sea sturgeon swam east. *Nature* 419: 447–448. https://doi.org/10.1038/419447a.

Lundberg, J.G. 1982. The comparative anatomy of the toothless blindcat, *Trogloglanis pattersoni*; Eigenmann, with a phylogenetic analysis of the Ictalurid catfishes. *Miscellaneous publications Museum of Zoology University of Michigan* 163: 1–85.

Lundberg, J.G. 1992. The phylogeny of Ictalurid catfishes: a synthesis of recent work. Pp 392–420 *in:* R.L. Mayden. (Ed.) *Systematics, Historical Ecology, and North American Freshwater Fishes*. Stanford University Press, Stanford.

Luo, J., J. Chai, Y. Wen, M. Tao, G. Lin, X. Liu, L. Ren, Z. Chen, S. Wu, S. Li & Y. Wang. 2020. From asymmetrical to balanced genomic diversification during rediploidization: Subgenomic evolution in allotetraploid fish. *Science advances* 6: eaaz7677. https://doi.org/10.1126/sciadv.aaz7677

Lyon, R.G., M.F.Geiger & J. Freyhof. 2016. *Garra sindhi*, a new species from the Jebel Samhan Nature Reserve in Oman (Teleostei: Cyprinidae). *Zootaxa* 4154: 79–88. https://doi.org/10.11646/zootaxa.4154.1.5

Ma, W., Z.H. Zhu, X.Y. Bi, R.W. Murphy, S.Y. Wang, Y. Gao, H. Xiao, Y.P. Zhang & J. Luo. 2014. Allopolyploidization is not so simple: Evidence from the origin of the tribe Cyprinini (Teleostei: Cypriniformes). *Current Molecular Medicine* 14: 1331–1338. https://doi.org/10.2174/1566524014666141203101543

Mable, B.K. 2013. Polyploids and hybrids in changing environments: Winners or losers in the struggle for adaptation? *Heredity* 110: 95–96. https://doi.org/10.1038/hdy.2012.105

Magomedov, G.M. 1981. *Commercially exploited fishes of Dagestan, their culture and fishery*. Dagestankoye Knizhnoye Izdatel'stvo, Makhatchkala. [in Russian].

Magurran, A.E., B.H. Seghers, P.W. Shaw & G.R. Carvalho. 1995. The behavioural diversity and evolution of guppy, *Poecilia reticulata*, populations in Trinidad. *Advances in the Study of Behavior* 24: 155–202

Maier, K.-J., M. Zeh, J. Ortlepp, S. Zbinden & D. Hefti. 1995. Distribution et reproduction des espèces du genre *Chondrostoma* en Suisse: le nase (*C. nasus*) - la sofie (*C. toxostoma*) - la savetta (*C. soetta*) / Verbreitung und Fortpflanzung der in der Schweiz vorkommenden *Chondrostoma*-Arten. *OPEPP Informations cocernant la Pêche / BUWAL Mitteilungen zur Fischerei* 53: 1–62.

Mair, G.C., J.S. Abucay, D.O.F. Skibinski, T.A. Abella & J.A. Beardmore. 1997. Genetic manipulation of sex ratio for the large-scale production of all-male tilapia, *Oreochromis niloticus*. *Canadian Journal of Fisheries and Aquatic Science* 54: 396–404.

Mäkinen, H.S., J.M. Cano & J. Merilä. 2006. Genetic relationships among marine and freshwater populations of the European three-spined stickleback (*Gasterosteus aculeatus*) revealed by microsatellites. *Molecular Ecology* 15: 1519–1534. https://doi.org/10.1111/j.1365-294X.2006.02871.x

Mangıt, F. & S. Yerli. 2018. Systematic evaluation of the genus *Alburnus* (Cyprinidae) with description of a new species. *Hydrobiologia* 807: 297–312. https://doi.org/10.1007/s10750-017-3405-y

Mateus, C.S., M.J. Alves, B.R. Quintella & P.R. Almeida. 2013. Three new cryptic species of the lamprey genus *Lampetra* Bonnaterre, 1788 (Petromyzontiformes: Petromyzontidae) from the Iberian Peninsula. *Contributions to Zoology* 82: 37–53. https://doi.org/10.1163/18759866-08201003

Mattern, M.Y. & D.A. McLennan. 2004. Total evidence phylogeny of Gasterosteidae: combining molecular, morphological and behavioural data. *Cladistics* 20: 14–22. https://doi.org/10.1111/j.1096-0031.2004.00005.x

Matveyev, M.P., M.A. Levina, N.S. Mamilov & B.A. Levin. 2017. Tashkent riffle minnow *Alburnoides oblongus* belongs to the genus *Alburnus* (Osteichthyes: Cyprinidae) as inferred from mitochondrial and nuclear DNA markers. *Inland Water Biology* 10: 360–367. https://doi.org/10.1134/S1995082917030105

Matzek, V. 2012. Trait values, not trait plasticity, best explain invasive species' performance in a changing environment. *PLOS ONE* 7: e48821. https://doi.org/10.1371/journal.pone.0048821

Maugé, A.L. 1986. Gobiidae. Pp 358–388 *in:* Daget, J., J.-P. Gosse & D.F.E. Thys van den Audenaerde. (Eds.) *Checklist of the Freshwater Fishes of Africa (CLOFFA II)*. ISNB, Brussels and MRAC, Tervuren, Belgium, and ORSTOM, Paris, France.

Mayden, R.L. 1997. A hierarchy of species concepts: the denouement in the saga of the species problem. *Species The units of biodiversity* 381–423.

Mayden, R.L. 2002. On biological species, species concepts and individuation in the natural world. *Fish and Fisheries* 3: 171–196. https://doi.org/10.1046/j.1467-2979.2002.00086.x

Mayden, R.L. & R.M. Wood. 1995. Systematics, species concepts and the evolutionary significant unit in biodiversity and conservation biology. *American Fisheries Society Symposium* 17: 58–113.

Mayden, R.L., K.L. Tang, R.M. Wood, W.J. Chen, M.K. Agnew, K.W. Conway, L. Yang, Simons, A.M., H.L. Bart, P.M. Harris & J. Li. 2008. Inferring the Tree of Life of the order Cypriniformes, the earth's most diverse clade of freshwater fishes: Implications of varied taxon and character sampling. *Journal of Systematics and Evolution* 46: 424. https://doi.org/10.3724/SP.J.1002.2008.08062

Mbanga, B., D. van Dyk & J.N. Maina. 2018. Morphometric and morphological study of the respiratory organs of the bimodally-breathing African sharptooth catfish (*Clarias gariepinus*): Burchell (1822). *Zoology* 130: 6–18. https://doi.org/10.1016/j.zool.2018.07.005

McClure, M.L. & C.L. Feltoe. 1919. *The pilgrimage of Etheria*. Society for promoting Christian knowledge, The Macmillan company in London, New York.

McLennan, D.A. & M.Y. Mattern. 2001. The phylogeny of the Gasterosteidae: combining behavioural and morphological data sets. *Cladistics* 17: 11–27. https://doi.org/10.1006/clad.2000.0150

MedECC. 2020. *Climate and environmental change in the Mediterranean basin-current situation and risks for the future*. First Mediterranean Assessment Report.

Meffe, G.K. & F.F. Snelson. 1989. *Ecology and evolution of livebearing fishes (Poeciliidae)*. Prentice Hall.

Meier, J.I., D.A Marques, S. Mwaiko, C.E. Wagner, L. Excoffier & O. Seehausen. 2017. Ancient hybridization fuels rapid cichlid fish adaptive radiations. *Nature Communications* 8: 14363 https://doi.org/10.1038/ncomms14363

Méndez, L., A.Perdices & A. Machordom. 2019. Genetic structure and diversity of the Iberian populations of the freshwater blenny *Salaria fluviatilis* (Asso, 1801) and its conservation implications. *Conservation Genetics* 20: 1223–1236. https://doi.org/10.1007/s10592- 019-01205-0

Menon, A.G.K. 1964. Monograph of the cyprinid fishes of the genus *Garra* Hamilton. *Memoirs of the Indian Museum* 14: 173–260, pls 8–13.

Meyer M.K., L. Wischnath & W. Foerster. 1985. *Lebendgebärende Zierfische: Arten der Welt*. Melle, W. Germany: Mergus Verlag.

Miller, P.J. 1986. Gobiidae. Pp 1015–1473 *in:* Whitehead, P.J.P., M.-L. Bauchot, J.-C. Hureau, J. G. Nielsen & E. Tortonese. (Ed.) *Fishes of the north-eastern Atlantic and the Mediterranean III*. Paris: UNESCO.

Miller, P.J. 2003. *Atherina boyeri* Risso, 1810. Pp 47–77 *in:* P.J. Miller. (Ed.) *The freshwater fishes of Europe. Vol. 8/I: Mugilidae, Atherinidae, Atherinopsidae, Blenniidae, Odontobutidae, Gobiidae*. AULA Verlag, Wiebelsheim.

Miller, P.J. 2004. *Knipowitschia longecauda* (Kessler, 1877). Pp 383–397 *in:* P.J. Miller. (Ed.) *The freshwater fishes of Europe. Vol. 8/II: Gobiidae 2*. AULA Verlag, Wiebelsheim.

Miller, P.J. & V.I. Pinchuk. 2004. *Hyrcanogobius bergi* Iljin, 1928. Pp 433–438 *in:* P.J. Miller. (Ed.) *The freshwater fishes of Europe. Vol. 8/II: Gobiidae 2*. AULA Verlag, Wiebelsheim.

Miller, R.R. 2005. *Freshwater fishes of Mexico*. Chicago University Press, Chicago.

Milton, D.A. & A.H. Arthington. 1983. Reproductive biology of *Gambusia affinis holbrooki* Baird & Girard, *Xiphophorus hellerii* (Gunther) and *X. maculatus* (Heckel) (Pisces; Poeciliidae) in Queensland, Australia. *Journal of Fish Biology* 23: 23–41. https://doi.org/10.1111/j.1095-8649.1983.tb02879.x

Mirghiyasi, S., H.R. Esmaeili & M. Nokhbatolfoghahai. 2016. Morpho-histological characteristics of gonads and reproductive index in an endemic fish species, *Oxynoemacheilus persa* (Heckel, 1847) (Teleostei: Nemacheilidae) from Kor River basin, Iran. *International Journal of Aquatic Biology* 4: 31–42. https://doi.org/10.22034/ijab.v4i1.120

Mirza, M.R. 1972. Freshwater fishes of Baluchistan province, Pakistan. *Biologia* 18: 153–190.

Mirza, M. R., T.T. Nalbant & P.M. Bânârescu. 1981. A review of the genus *Schistura* in Pakistan with description of new species and subspecies (Pisces, Cobitidae, Noemacheilinae). *Bijdragen tot de Dierkunde* 51: 105–130.

Mitrofanov, V.P., G.M. Dukravec, V.A. Mel'nikov & A.A. Baimdetov. 1988. *Ryby Kazakhstana* [*Fishes of Kazakhstan*]. Tom. 3. Nauka, Alma Ata: 1–304.

Miyashita, T., R.W. Robert Gess, T. Tietjen & M.I. Coates. 2021. Non-ammocoete larvae of Palaeozoic stem lampreys. *Nature* 591: 408–412. https://doi.org/10.1038/s41586-021-03305-9

Mogil'chenko, V.I. 1991. *Alosa caspia tanaica* (Grimm, 1901). Pp 181–197 *in:* H. Hoestlandt. (Ed.) *The freshwater fishes of Europe. Vol. 2. Clupeidae, Anguillidae*. AULA Verlag, Wiesbaden.

Mohamed, A.R.M., L.A.J. Al-Hassan & T.S. Ali. 1993. The presence of a cyprinid fish, *Barbus luteus* in marine waters of Iraq. *Arquivos do Museu Bocage, Nova Serie* 2: 415–416.

Mohammadian-Kalat, T., H.R. Esmaeili, M. Aliabadian & J. Freyhof. 2017. Redescription of *Alburnus doriae*, with comments on the taxonomic status of *A. amirkabiri*, *A. mossulensis*, *A. sellal* and *Petroleuciscus esfahani* (Teleostei: Cyprinidae). *Zootaxa* 4323: 487–502. https://doi.org/10.11646/zootaxa.4323.4.3

Mohindra V., R.K. Anshumala, L. Singh, L. Narain, P. Punia, D. Kapoor, M. Afzal & K.K. Lal. 2005. Allozyme markers for population structure analysis in *Labeo dero* (Hamilton Buchanan, 1822). *Asian Fisheries Science* 18: 71–76. https://doi.org/10.33997/j.afs.2005.18.1.008

Mohr, E. 1957. *Der Wels*. Die Neue Brehm-Bücherei, Ziemsen Verlag, Wittenberg Lutherstadt.

Molls, F. 1999. New insights into the migration and habitat use by bream and white bream in the floodplain of the River Rhine. *Journal of Fish Biology* 55: 1187–1200. https://doi.org/10.1111/j.1095-8649.1999.tb02069.x

Monsefi, M., A.H. Shiva & H.R. Esmaeili. 2009. Gonad histology of the Persian tooth-carp *Aphanius persicus* (Jenkins, 1910) (Cyprinodontidae) in southern Iran. *Turkish Journal of Zoology* 33: 27–33.

Moser, F.N., J.C. van Rijssel, S. Mwaiko, J.I. Meier, B. Ngatungaand & O. Seehausen. 2018. The onset of ecological diversification 50 years after colonization of a crater lake by haplochromine cichlid fishes. *Proceedings of the Royal Society B* 285: p.20180171. http://dx.doi.org/10.1098/rspb.2018.0171

Moser, H.G., W.J. Richards, D.M. Cohen, M.P. Fahay, A.W. Kenall & S.L. Richardson. 1984 *Ontogeny and systematics of fishes*. American Society of Ichthyologists and Herpetologists, Lawrence.

Mousavi-Sabet, H. & S. Eagderi. 2015. *Paraschistura delvarii* spec. nov. - a new species of stone loach from the Persian Gulf basin, southern Iran (Teleostei: Nemacheilidae). *Vertebrate Zoology* 65: 297–303.

Mousavi-Sabet, H. & S. Eagderi. 2016a. First record of the convict cichlid, *Amatitlania nigrofasciata* (Günther, 1867) (Teleostei: Cichlidae) from the Namak Lake basin, Iran. *Iranian Journal of Ichthyology* 3: 25–30. https://doi.org/10.22034/iji.v3i1.90

Mousavi-Sabet, H. & S. Eagderi. 2016b. *Garra lorestanensis*, a new cave fish from the Tigris River drainage with remarks on the subterranean fishes in Iran (Teleostei: Cyprinidae). *FishTaxa* 1: 45–54.

Mousavi-Sabet, H., S.V. Yerli, S. Vatandoust, S.C. Özeren & Z. Moradkhani. 2012. *Cobitis keyvani* sp. nova—a new species of spined-loach from south of the Caspian Sea (Teleostei: Cobitidae). *Turkish Journal of Fisheries and Aquatic Sciences* 12: 7–13. https://doi.org/10.4194/1303-2712-v12_1_02

Mousavi-Sabet, H., A. Gharaei, A. Nasrollahzade, A. Habibi & S. Eagderi. 2014. Redescription of *Paracobitis rhadinaea* (Regan, 1906) from Sistan basin, Iran (Teleostei: Nemacheiliidae). *International Journal of Aquatic Biology* 2: 286–291. https://doi.org/10.22034/ijab.v2i5.139

Mousavi-Sabet, H., E.D. Vasil'eva, S. Vatandoust & V.P. Vasil'ev. 2011. *Cobitis faridpaki* sp. nova—a new spined loach species (Cobitidae) from the southern Caspian Sea basin. *Journal of Ichthyology* 51: 925–931. https://doi.org/10.1134/S0032945211100055

Mousavi-Sabet, H., G. Sayyadzadeh, H.R. Esmaeili, S. Eagderi, R. Patimar & J. Freyhof. 2015a. *Paracobitis hircanica*, a new crested loach from the southern Caspian Sea basin (Teleostei: Nemacheilidae). *Ichthyological Exploration of Freshwaters* 25: 339–346.

Mousavi-Sabet, H., H. Anvarifar & F. Azizi. 2015b. *Alburnoides tabarestanensis*, a new species of riffle minnow from the southern Caspian Sea basin in Iran (Actinopterygii: Cyprinidae). *Aqua, International Journal of Ichthyology* 21: 144–152.

Mousavi-Sabet, H., S. Vatandoust & I. Doadrio. 2015c. Review of the genus *Alburnoides* Jeitteles, 1861 (Actinopterygii, Cyprinidae) from Iran with description of three new species from the Caspian Sea and Kavir basins. *Caspian Journal of Environmental Sciences* 13: 293–331.

Mousavi-Sabet, H., S. Vatandoust, A. Jouladeh-Roudbar & S. Eagderi. 2015d. Taxonomic status of the genus *Paraschistura* (Teleostei: Nemacheilidae) in the Hari River basin, with re-validation of *P. turcomana*. *Journal of Applied Biological Sciences* 9 (3): 1–15.

Mousavi-Sabet, H., S. Vatandoust, H.R. Esmaeili, M.F. Geiger & J. Freyhof. 2015e. *Cobitis avicennae*, a new species of spined loach from the Tigris River drainage (Teleostei: Cobitidae). *Zootaxa* 3914: 558–568. https://doi.org/10.11646/zootaxa.3914.5.4

Mousavi-Sabet, H., S. Vatandoust, S. Khataminejad, S. Eagderi, K. Abbasi, M. Nasri, A. Jouladeh-Roudbar & E.D. Vasil'eva. 2015f. *Alburnus amirkabiri* (Teleostei), a new species of shemaya from the Namak Lake basin, Iran. *Journal of Ichthyology* 55: 40–52. https://doi.org/10.1134/S0032945215010129

Mousavi-Sabet, H., S. Vatandoust, Y. Fatemi & S. Eagderi. 2016c. Tashan cave a new cave fish locality for Iran; and *Garra tashanensis*, a new blind species from the Tigris River drainage (Teleostei: Cyprinidae). *FishTaxa* 1: 133–148.

Mousavi-Sabet, H., B. Ganjbakhsh, M.F. Geiger & J. Freyhof. 2016a. Redescription of *Gobio nigrescens* from the Hari River drainage (Teleostei: Cyprinidae). *Zootaxa* 4114: 71–80. http://dx.doi.org/10.11646/zootaxa.4114.1.4

Mousavi-Sabet, H., K. Abbasi, S. Vatandoust & S. Eagderi. 2016b. Threatened fishes of the world: *Paracobitis vignai* Nalbant and Bianco, 1998 (Nemacheilidae). *International Journal of Aquatic Biology* 4: 51–56. https://doi.org/10.22034/ijab.v4i1.122

Mousavi-Sabet, H., M. Saemi-Komsari, I. Doadrio & J. Freyhof. 2019a. *Garra roseae*, a new species from the Makran region in southern Iran (Teleostei: Cyprinidae). *Zootaxa* 4671: 223–239. https://doi.org/10.11646/zootaxa.4671.2.3

Mousavi-Sabet, H., S. Vatandoust, M.F. Geiger & J. Freyhof. 2019b. *Paracobitis abrishamchiani*, a new crested loach from the southern Caspian Sea basin (Teleostei: Nemacheilidae). *Zootaxa* 4545: 375–388. https://doi.org/10.11646/zootaxa.4545.3.3.

Mousavi-Sabet, H., S. Eagderi, S. Vatandoust & J. Freyhof. 2021. Five new species of the sisorid catfish genus *Glyptothorax* from Iran (Teleostei: Sisoridae). *Zootaxa* 5067: 451–484. https://doi.org/10.11646/zootaxa.5067.4.1

Mousavi-Sabet, H., S. Eagderi, M. Saemi-Komsari, C. Kaya & J. Freyhof. 2022. *Garra rezai*, a new species from two widely disjunct areas in the Tigris drainage (Teleostei: Cyprinidae). *Zootaxa* 5195: 419–436. http://doi.org/10.11646/zootaxa.5195.5.2

Movchan, Y.V. & A.I. Smirnov. 1981. *Fauna of Ukraine. Fishes. Cyprinid Fishes (Roach, dace, minnow, rudd, grass carp, asp, verchovka, tench, undermouth, gudgeon, barbel. Kiev, Naukova Dumka Publishing House*. [in Ukrainian].

Movchan Y.V. & A.I. Smirnov. 1983. *Fauna of Ukraine. Vol.8: Fishes. Iss. 2: Cyprinidae. Part 2*. Naukova Dumka, Kyiv [in Ukrainian].

Müller-Belecke A. & S. Zienert. 2006. Aufzucht von Streifenbarschhybriden in der Aquakultur. *Schriften des Instituts für Binnenfischerei e. V. Potsdam-Sacrow, 20. Institut für Binnenfischerei e. V. Potsdam-Sacrow.*

Murakami, M., C. Matsuba & H. Fujitani. 2001. The maternal origins of the triploid ginbuna (*Carassius auratus langsdorfi*): Phylogenetic relationships within the C. auratus taxa by partial mitochondrial D-loop sequencing. *Genes & Genetic Systems* 76: 25–32. https://doi.org/10.1266/ggs.76.25.

Murdy, E.O. 1989. A taxonomic revision and cladistic analysis of the Oxudercine gobies (Gobiidae: Oxudercinae). *Records of the Australian Museum* 11: 1–93.

Murray, A.M. 2006. A new channid (Teleostei: Channiformes) from the Eocene and Oligocene of Egypt. *Journal of Paleontology* 80: 1172–1178. https://doi.org/10.1266/ggs.76.25

Naama, A.K. & K.A. Muhsen. 1986. Feeding periodicites of the mugilid *Liza abu* (Heckel) and cyprinid *Carasobarbus luteus* (Heckel) from Al-Hammar Marsh, Southern Iraq. *Indian Journal of Fisheries* 33: 347–350.

Nalbant, T.T. 1963. A study of the genera of Botiinae and Cobitinae (Pisces, Ostariophysi, Cobitidae). *Travaux du Muséum d'Histoire Naturelle "Grigore Antipa"* 4: 343–379.

Nalbant, T.T. 1998. The presence of the genus *Schistura* (Pisces: Ostariophysi: Nemacheilidae) in Tigris drainage, western Asia. The description of a new species. *Travaux du Muséum d'Histoire Naturelle "Grigore Antipa"* 40: 371–375.

Nalbant, T.T. & P.G. Bianco. 1998. The loaches of Iran and adjacent regions with description of six new species (Cobitoidea).

Italian Journal of Zoology 65(Suppl.): 109–123. https://doi.org/10.1080/11250009809386803
Naseka, A.M. & O.A. Diripasko. 2008. A recent record of an anadromous lamprey (Agnatha: Petromyzontidae) from the Sea of Azov. *Ichthyological Exploration of Freshwaters* 19(3): 283–287.
Naseka, A.M., P. Bănărescu & N. Bogutskaya. 1999. *Gobio persus* Günther, 1899. Pp 162–181 *in:* Bănărescu, P.M. (Ed.) *The freshwater fishes of Europe. Vol.5/I: Cyprinidae 2/I.* AULA Verlag, Wiebelsheim.
Naseka, A.M., V.S. Boldyrev, N.G. Bogutskaya & V.V. Delitsyn. 2005. New data on the historical and expanded range of *Proterorhinus marmoratus* (Pallas, 1814) (Teleostei: Gobiidae) in eastern Europe. *Journal of Applied Ichthyology* 21: 300–305. https://doi.org/10.1111/j.1439-0426.2005.00685.x
Naseka, A.M., F. Erk'akan & F. Küçük. 2006. A description of two new species of the genus *Gobio* from Central Anatolia (Turkey) (Teleostei: Cyprinidae). *Zoosystematica Rossica* 15: 185–194. https://doi.org/10.1080/09397140.2016.1182779
Naseka, A.M., S.B. Tuniyev & C.B. Renaud. 2009. *Lethenteron ninae*, a new nonparasitic lamprey species from the north-eastern Black Sea basin (Petromyzontiformes: Petromyzontidae). *Zootaxa* 2198: 16–26. https://doi.org/10.11646/zootaxa.2198.1.2
Naseka, A.M., S.B. Tuniyev & C.B. Renaud. 2016. Review of western transcaucasian brook lamprey, *Lethenteron ninae* Naseka, Tuniyev & Renaud, 2009 (Petromyzontidae). Pp 154–190 *in:* Orlov, A. & R. Beamish. (Eds.). *Jawless fishes of the world*, Cambridge Scholars Publishing.
Nasri, M., S. Eagderi, H. Farahmand & I. Hashemzadeh Segherloo. 2013. Body shape comparison of *Cyprinion macrostomum* (Heckel, 1843) and *Cyprinion watsoni* (Day, 1872) using geometric morphometrics method. *International Journal of Aquatic Biology* 1: 240–244. https://doi.org/10.22034/ijab.v1i5.154
Nasri, M., S. Eagderi & H. Farahmand. 2016. Descriptive and comparative osteology of bighead lotak, *Cyprinion milesi* (Cyprinidae: Cypriniformes) from south-eastern Iran. *Vertebrate Zoology* 66: 251–260.
Nasri, M., S. Eagderi, Y. Keivany, H. Farahmand, S. Dorafshan & H. Nezhadheydari. 2018. Morphological diversity of *Cyprinion* Heckel, 1843 species (Teleostei: Cyprinidae) in Iran. *Iranian Journal of Ichthyology* 5: 96–108. https://doi.org/10.22034/iji.v5i2.265
National Geography Organization of Iran. 2003. Country river geography: Urmia Lake basin. *National Geography Organization of Iran publication, Tehran.* 133 p.
Nazari, H., A. Abdoli, B.H. Kiabi & C.B. Renaud. 2017. Biology and conservation status of the Caspian lamprey in Iran: a review. *Lampetra* 8: 6–32.
Near, T.J., A. Dornburg, R.I. Eytan, B.P. Keck, W.L. Smith, K.L. Kuhn, J.A. Moore, S.A. Price, F.T. Burbrink, M. Friedman & P.C. Wainwright. 2013. Phylogeny and tempo of diversification in the superradiation of spiny-rayed fishes. *Proceedings of the National Academy of Sciences* 110: 12738–12743. https://doi.org/10.1073/pnas.1304661110
Near, T.J. & C.E. Thacker. 2024. Phylogenetic classification of living and fossil ray-finned fishes (Actinopterygii). *Bulletin of the Peabody Museum of Natural History* 65(1): 3–302. https://doi.org/10.3374/014.065.0101
Neilson, M.E. & C.A. Stepien. 2009. Evolution and phylogeography of the tubenose goby genus *Proterorhinus* (Gobiidae: Teleostei): evidence for new cryptic species. *Biological Journal of the Linnean Society* 96: 664–684. https://doi.org/10.1111/j.1095-8312.2008.01135.x
Nelson, J.F. 2006. *Fishes of the World*, John Wiley & Sons.
Nelson, J.S., T.C. Grande & M.V.H. Wilson. 2016. *Fishes of the World.* 5th Edition, John Wiley and Sons, Hoboken. https://doi.org/10.1002/9781119174844
Nelva, A. 1989. Structure de population en classes de taille et structure en blancs chez le hotu, *Chondrostoma nasus* (Pisces, Cyprinidae) dans le Haut-Rhône Francais. *Cybium* 13: 221–233.
Neumann, D. 2010. Preservation of freshwater fishes in the field (Chapter 22). Pp 633–653. *in:* Eymann, J., J. Degreef, C. Häuser, J.C. Monje, Y. Samyn, & D. van den Spiegel. (Eds.). Manual on field recording techniques and protocols for All Taxa Biodiversity Inventories and Monitoring. *ABC Taxa* 8, 2: 587–631.
Ng, H.H. & M. Kottelat. 2008. The identity of *Clarias batrachus* (Linnaeus, 1758), with the designation of a neotype (Teleostei: Clariidae). *Zoological Journal of the Linnean Society* 153: 725–732. https://doi.org/10.1111/j.1096-3642.2008.00391.x
Nico, L.G., J.D. Williams & J.J. Herod. 2001. *Black carp (Mylopharyngodon piceus): a biological synopsis and updated risk assessment.* Final report submitted to the Risk Assessment and Management Committee of the Aquatic Nuisance Species Task Force. U.S. Geological Survey, Gainesville, Florida.
Nico, L.G., J.D. Williams & H.L. Jelks. 2005. *Black carp: biological synopsis and risk assessment of an introduced fish.* American Fisheries Society, Special Publication 32, Bethesda, Maryland.
Nico, L., M. Neilson & B. Loftus. 2016. *Tilapia zillii.* USGS Nonindigenous Aquatic Species Database, Gainesville, FL. https://nas.er.usgs.gov/queries/factsheet.aspx?SpeciesID=485 (Accessed February 2013)
Nikolskii, A.M. 1900. A new *Discognathus* from Russia. *Ezhegodnik. Zoologicheskogo Muzeya Akademii Nauk SSSR* 5: 239–241.
Nikolsky, G.V. & Y.S. Reshetnikov. 1970. Systematics of coregonid fishes in the USSR: intraspecies variability and difficulties in taxonomy. *Biology of Coregonid fishes* 251–266.
Ninua, L., D. Tarkhnishvili & E. Gvazava. 2018. Phylogeography and taxonomic status of trout and salmon from the Ponto-Caspian drainages, with inferences on European brown trout evolution and taxonomy. *Ecology and Evolution* 8: 2645–2658. https://doi.org/10.1002/ece3.3884.
Ninua, L., D. Tarkhnishvili & C.L. Anderson. 2023. Genetic structure of Ponto-Caspian trout populations shows gene flow among river drainages and supports resident *Salmo rizeensis* as a genetically distinct taxon. *Ecology and Evolution* 13:e10335. https://doi.org/10.1002/ece3.10335
Okazaki, M., K. Naruse, A. Shima & R. Arai. 2001. Phylogenetic relationships of bitterlings based on mitochondrial 12S ribosomal DNA sequences. *Journal of Fish Biology* 58: 89–106. https://doi.org/10.1111/j.1095-8649.2001.tb00501.x
Okgerman, H., C.H. Yardımcı, Z. Dorak & N. Yılmaz. 2013. Feeding ecology of vimba (*Vimba vimba* L. 1758) in terms of size groups and seasons in Lake Sapanca, northwesternTurkey. *Turkish Journal of Zoology* 37: 288–297. https://doi.org/10.3906/zoo-1107-1
Otto, S.P. & J. Whitton. 2000. Polyploid incidence and evolution. *Annual Review of Genetics* 34: 401–437. https://doi.org/10.1146/annurev.genet.34.1.401
Oymak, S.A., K. Solak & E. Ünlü. 2001. Some biological characteristics of *Silurus triostegus* Heckel, 1843 from Ataturk dam lake (Turkey). *Turkish Journal of Zoology* 25: 139–148.
Özcan, G. & S. Balık. 2009. Some biological characteristics of Meander bleak, *Acanthobrama mirabilis* Ladiges, 1960, in the Kemer Reservoir, Turkey. *Electronic Journal of Ichthyology* 1: 4–10.
Özcan, G. & S. Balık. 2011. Age and growth of *Chondrostoma meandrense* in Kemer Reservoir, Turkey. *Journal of Black Sea/Mediterranean Environment* 17: 67–77.
Özcan, E. İ. & O. Serdar. 2019. Age and some growth parameters of *Squalius cephalus* (Linnaeus, 1758) inhabiting Karasu River (East

Anatolia, Turkey). *Ege Journal of Fisheries and Aquatic Sciences* 36: 25–30. https://doi.org/10.12714/egejfas.2019.36.1.03

Özdemir, F. 2019. *Cobitis fusunae*, a new spined loach species (Teleostei: Cobitidae) from the Sultan Marsh, (Kayseri, Turkey). *Hacettepe Journal of Biology and Chemistry* 46: 593–599.

Özdilen, Ş.Y. 2007. Possible threat for Middle East inland water: an exotic and invasive species, *Pterygoplichthys disjunctivus* (Weber, 1991) in Asi River, Turkey (Pisces: Loricariidae). *E.U. Journal of Fisheries and Aquatic Sciences* 24: 303–306

Özeren, S.C., F. Özdemir & F. Erk'akan. 2019. New record and comparison of some morphometric characters of an endemic species *Capoeta pestai* (Pietschmann, 1933) (Pisces: Cyprinidae) populations in central Anatolia. *Hacettepe Journal of Biology and Chemistry* 47: 209–216. https://doi.org/10.15671/hjbc.623798

Özpiçak, M., S. Saygın, S. Yılmaz & N. Polat. 2022. Two new records for the fish fauna of Simenlik-Akgöl Lagoon in Yeşilırmak River basin (Samsun-Turkey). *Sakarya University Journal of Science*, 26(6), 1104–1110. https://doi.org/10.16984/saufenbilder.1141017

Özpiçak, M., S. Yılmaz, S. Saygın & N. Polat. 2023. New data on the distribution of the Bulgarian spined loach *Cobitis strumicae* (Teleostei: Cobitidae) in Türkiye. *Inland Water Biology* 16 (3): 520–529. https://doi.org/10.1134/S1995082923030161

Özuluğ, M. & J. Freyhof. 2007. Rediagnosis of four species of *Alburnus* from Turkey and description of two new species (Teleostei: Cyprinidae). *Ichthyological Exploration of Freshwaters* 3: 233–246

Özuluğ, M. & J. Freyhof. 2008a. *Alburnus demiri*, a new species of bleak from western Anatolia, Turkey (Teleostei: Cyprinidae). *Ichthyological Exploration of Freshwaters* 18: 307–312.

Özuluğ, M. & J. Freyhof. 2008b. *Capoeta turani*, a new species of barbel from River Seyhan, Turkey (Teleostei: Cyprinidae). *Ichthyological Exploration of Freshwaters* 19: 289–296.

Özuluğ, M. & J. Freyhof. 2011. Revision of the genus *Squalius* in Western and Central Anatolia, with description of four new species (Teleostei: Cyprinidae). *Ichthyological Exploration of Freshwaters* 22: 107–148.

Özuluğ, M., M.F. Geiger & J. Freyhof. 2018. *Alburnus goekhani*, a new species of bleak from the Anatolian Black Sea basin (Teleostei: Leuciscidae). *Zootaxa* 4425: 29–40. https://doi.org/10.1080/10.11646/zootaxa.4425.1.2

Özuluğ, M. & G. Saç. 2019. İstanbul İli (Türkiye) tatlısu balık faunası. *Turkish Journal of Bioscience and Collections* 3 (1): 19–36.

Paepke, H.J. 2002a. *Gasterosteus aculeatus* Linnaeus, 1758. Pp 209–256 *in:* Bănărescu P.M. & H.-J. Paepke. (Eds.). *The freshwater fishes of Europe. Vol. 5/III: Cyprinidae 2, Part III:* Carassius *to* Cyprinus*. Gasterosteidae*. AULA Verlag, Wiebelsheim.

Paepke, H.J. 2002b. *Pungitius platygaster* (Kessler, 1859). Pp 267–276 *in:* Bănărescu P.M. & H.-J. Paepke. (Eds.). *The freshwater fishes of Europe. Vol. 5/III: Cyprinidae 2, Part III:* Carassius *to* Cyprinus*. Gasterosteidae*. AULA Verlag, Wiebelsheim.

Page, L.M. & B.M. Burr. 1991. *A field guide to freshwater fishes: North America north of Mexico*. Houghton Mifflin Harcourt.

Parenti L.R. 1981. A phylogenetic and biogeographic analysis of cyprinodontiform fishes (Teleostei, Atherinomorpha). *Bulletin of the American Museum of Natural History* 168:335–557.

Parichy, D.M. 2003. Pigment patterns: fish in stripes and spots. *Current Biology* 13: 947–950. https://doi.org/10.1016/j.cub.2003.11.038

Patimar, R. & B. Mohammadzadeh. 2011. On the biological characteristics of *Capoeta fusca* Nikolskii, 1897 in eastern Iran. *Journal of Applied Ichthyology* 27: 873–878. https://doi.org/10.1111/j.1439-0426.2010.01572.x

Patimar, R., H. Adineh & M.J. Mahdavi. 2009. Life history of the western crested loach *Paracobitis malapterura* in the Zarrin-Gol River, east of the Elburz Mountains (Northern Iran). *Biologia* 64: 350–355. https://doi.org/10.2478/s11756-009-0052-4

Patimar, R., E. Nadjafypour, M., Yaghouby & M. Nadiafy. 2010a. Reproduction characteristic of a stunted population of rudd, *Scardinius erythrophthalmus* (Linnaeus, 1758) living in the Anzali Lagoon (the southwest Caspian Sea, Iran). *Journal of Ichthyology* 50: 1060–1065. https://doi.org/10.1134/S0032945210110111

Patimar, R., M.G. Chalanchi, V. Chamanara & L. Naderi. 2010b. Some life history aspects of *Garra rufa* (Heckel, 1843) in the Kangir River, western Iran. *Zoology in the Middle East* 51: 57–66. https://doi.org/10.1080/09397140.2010.10638441

Pazooki, J., M. Sheidai & M.M. Korani. 2008. A systematic and ecological study of *Aphanius vladykovi* Coad, 1988 (Actinopterygii: Cyprinodontidae) in Iran. *Zoology in the Middle East* 43: 85–89. https://doi.org/10.1080/09397140.2008.10638272

Pellegrin, J. 1927. Description d'un cyprinidé nouveau d'Asie Mineure. *Bulletin de la Société Zoologique de France* 52: 34–35.

Perdices, A., I. Doadrio, I.M. Côté, A. Machordom, P. Economidis & J.D. Reynolds. 2000. Genetic divergence and origin of Mediterranean populations of the river blenny *Salaria fluviatilis* (Teleostei: Blenniidae). *Copeia*: 723–731. https://doi.org/10.1643/0045-8511(2000)000[0723:GDAOOM]2.0.CO;2

Perdices, A., C.S. Özeren, F. Erk'akan & J. Freyhof. 2018. Diversity of spined loaches from Asia Minor in a phylogenetic context (Teleostei: Cobitidae). *PLOS ONE* 13: e0205678. https://doi.org/10.1371/journal.pone.0205678

Perea, S., M. Böhme, P. Zupančič, J. Freyhof, R. Šanda, M. Özuluğ, A. Abdoli & I. Doadrio. 2010. Phylogenetic relationships and biogeographical patterns in Circum-Mediterranean subfamily Leuciscinae (Teleostei, Cyprinidae) inferred from both mitochondrial and nuclear data. *BMC Evolutionary Biology* 10: 265. https://doi.org/10.1186/1471-2148-10-265

Pereira, A.M., J.I. Robalo, J. Freyhof, C. Maia, J.P. Fonseca, A. Valente & V.C. Almada. 2010. Phylogeographical analysis reveals multiple conservation units in brook lampreys *Lampetra planeri* of Portuguese streams. *Journal of Fish Biology* 77 (2): 361–371. https://doi.org/10.1111/j.1095-8649.2010.02675.x

Pethiyagoda, R. 1991. *Freshwater Fishes of Sri Lanka*. The Wildlife Heritage Trust of Sri Lanka, Colombo.

Pfleiderer, S.J., M.F. Geiger & F. Herder. 2014. *Aphanius marassantensis*, a new toothcarp from the Kızılırmak drainage in northern Anatolia (Cyprinodontiformes: Cyprinodontidae). *Zootaxa* 3887: 569–582. https://doi.org/10.11646/Zootaxa.3887.5.4

Pichler, A., H. Ahnelt, S. Kirchner, H. Sattmann, E. Haring, S. Handschuh, J. Freyhof, R. Victor & L. Kruckenhauser. 2018. The morphological diversity of *Garra barreimiae* [Teleostei: Cyprinidae]. *Environmental Biology of Fishes* 101 (6): 1053–1065. https://doi.org/10.1007/s10641-018-0758-7

Pickett, G.D. & M.G. Pawson. 1994. *The Sea Bass - Biology, Exploitation and Conservation*. Fish and Fisheries Series. Chapman & Hall, London.

Pienaar, U De V. 1978. *The freshwater fishes of the Kruger National Park*. National Parks Board of Trustees: Pretoria.

Pietschmann, V. 1913. Eine neue *Glyptosternum*-Art aus dem Tigris. *Anzeiger der Kaiserlichen Akademie der Wissenschaften in Wien, Mathematisch-Naturwissenschaftlichen Klasse* 50: 93–95.

Pike, C., V. Crook, D. Jacoby & M. Gollock. 2020a. *Anguilla anguilla. The IUCN Red List of Threatened Species* 2020: e.T60344A152845178.

https://dx.doi.org/10.2305/IUCN.UK.2020-2.RLTS.T60344A152845178.en. Accessed on 07 March 2023.

Pike, C., V. Crook, D. Jacoby & M. Gollock. 2020b. *Anguilla bengalensis* (amended version of 2019 assessment). *The IUCN Red List of Threatened Species* 2020: e.T61668607A176497430. https://dx.doi.org/10.2305/IUCN.UK.2020-3.RLTS.T61668607A176497430.en. Accessed on 07 March 2023.

Pike, C., V. Crook, D. Jacoby & M. Gollock. 2020c. *Anguilla bicolor* (amended version of 2019 assessment). *The IUCN Red List of Threatened Species* 2020: e.T166894A176494582. https://dx.doi.org/10.2305/IUCN.UK.2020-3.RLTS.T166894A176494582.en. Accessed on 07 March 2023.

Pikitch, E.K., P. Doukakis, L. Lauck, P. Chakrabarty & D.L. Erickson. 2005. Status, trends and management of sturgeon and paddlefish fisheries. *Fish and Fisheries* 6 (3): 233–265.

Pinchuk, V.I., E.D. Vasil'eva, V.P. Vasil'ev & P.J. Miller. 2003a. *Neogobius fluviatilis* (Pallas, 1841). Pp 222–252 *in:* Miller, P.J. (Ed.) *The freshwater fishes of Europe. Vol. 8/I: Mugilidae, Atherinidae, Atherinopsidae, Blenniidae, Odontobutidae, Gobiidae*. AULA Verlag, Wiebelsheim.

Pinchuk, V.I., E.D. Vasil'eva, V.P. Vasil'ev & P.J. Miller. 2003b. *Neogobius gymnotrachelus* (Kessler, 1857). Pp 265–279 *in:* Miller, P.J. (Ed.) *The freshwater fishes of Europe. Vol. 8/I: Mugilidae, Atherinidae, Atherinopsidae, Blenniidae, Odontobutidae, Gobiidae*. AULA Verlag, Wiebelsheim.

Pinchuk, V.I., E.D. Vasil'eva, V.P. Vasil'ev & P.J. Miller. 2003c. *Neogobius melanostomus* (Pallas, 1814). Pp 293–345 *in:* Miller, P.J. (Ed.) *The freshwater fishes of Europe*. Vol. 8/I: *Mugilidae, Atherinidae, Atherinopsidae, Blenniidae, Odontobutidae, Gobiidae*. AULA Verlag, Wiebelsheim.

Pinchuk, V.I., E.D. Vasil'eva & P.J. Miller. 2003d. *Neogobius syrman* (Nordmann, 1840). Pp 377–397 *in:* Miller, P.J. (Ed.) *The freshwater fishes of Europe*. Vol. 8/I: *Mugilidae, Atherinidae, Atherinopsidae, Blenniidae, Odontobutidae, Gobiidae*. AULA Verlag, Wiebelsheim.

Pinchuk, V.I., E.D. Vasil'eva & P.J. Miller. 2004a. *Proterorhinus marmoratus* (Pallas, 1814). Pp 72–93 *in:* Miller, P.J. (Ed.) *The freshwater fishes of Europe. Vol. 8/II: Gobiidae 2*. AULA Verlag, Wiebelsheim.

Pinchuk, V.I., E.D. Vasil'eva & P.J. Miller. 2004b. *Mesogobius batrachocephalus* (Pallas, 1814). Pp 109–131 *in:* Miller, P.J. (Ed.) *The freshwater fishes of Europe. Vol. 8/II: Gobiidae 2*. AULA Verlag, Wiebelsheim.

Pipoyan, S.K., L.A. Kirakosyan & S.B. Tuniyev. 2012. Peculiarities of morphometric variability of whitefish *Coregonus lavaretus* (Linnaeus, 1758) (Actinopterygii: Coregondae) of Sevan Lake. *Proceedings of the Zoological Institute RAS* 316: 254–265.

Pirogovskii, M.I., L.I., Sokolov, V. Vasil'ev & J. Holcík. 1989. *Huso huso* (Linnaeus, 1758). Pp 156–200 *in:* Holčík. J. (Ed.) *The freshwater fishes of Europe. Vol.1/II: General Introduction to fishes – Acipenseriformes*. AULA Verlag, Wiesbaden.

Pohlmann, K., J. Atema & T. Breithaupt. 2004. The importance of the lateral line in nocturnal predation of piscivorous catfish. *Journal of Experimental Biology* 207: 2971–2978. https://doi.org/10.1242/jeb.01129

Polgar, G., Z. Jaafar & P. Konstantinidis. 2013. A new species of mudskipper, *Boleophthalmus poti* (Teleostei: Gobiidae: Oxudercinae) from the Gulf of Papua, Papua New Guinea, and a key to the genus. *The Raffles Bulletin of Zoology* 61: 311–321.

Porumb, I.I. 1961. Contributii la cunoasterea biologiei guvizilor (*Gobius batrachocephalus*, *Gobius cephalarges* si *Gobius melanostomus*) din dreptul litoralului romanesc al Marii Negre (date preliminare). *Hidrobiologia* 3: 271–282.

Potter, I.C. & T.S. Osborne. 1975. The systematics of British larval lampreys. *Journal of Zoology* 176: 311–329. https://doi.org/10.1111/j.1469-7998.1975.tb03204.x

Pouyaud, L. & E. Paradis. 2009. The phylogenetic structure of habitat shift and morphological convergence in Asian *Clarias* (Teleostei, Siluriformes: Clariidae). *Journal of Zoological Systematics and Evolutionary Research* 47: 344–356. http://dx.doi.org/10.1111/j.1439-0469.2008.00507.x

Prokofiev, A.M. 2009. Problems of the classification and phylogeny of Nemacheiline loaches of the group lacking the preethmoid I (Cypriniformes: Balitoridae: Nemacheilinae). *Journal of Ichthyology* 49: 874–898. http://dx.doi.org/10.1134/S0032945209100051

Prokofiev, A.M. & A.S. Golubtsov. 2013. Revision of the loach genus *Afronemacheilus* (Teleostei: Balitoridae: Nemacheilinae) with description of a new species from the Omo-Turkana basin, Ethiopia. *Ichthyological Exploration of Freshwaters* 24: 1–14.

Proudlove, G.S. 1997. *A synopsis of the hypogean fishes of the world*. 351–354. In Proceedings of the 12th International Congress of Speleology, La Chaux de Fonds, Switzerland.

Pullin, R.S. & R.H. Lowe-McConnell. 1982. *The Biology and Culture of Tilapias*. Proceedings of the International Conference on the Biology and Culture of Tilapias, 2–5 September 1980 at the Study and Conference Center of the Rockefeller Foundation, Bellagio, Italy (7). WorldFish.

Pusey, B., M. Kennard & A. Arthington. 2004. *Freshwater Fishes of North-Eastern Australia*. Csiro Publishing, Colingwood, Victoria.

Quignard, J.P. & C. Douchement. 1991. *Alosa fallax nilotica* (Geoffroy Saint-Hilaire, 1808). Pp 264–273 *in:* Hoestlandt. H. (Ed.) *The freshwater fishes of Europe. Vol. 2. Clupeidae, Anguillidae*. AULA Verlag, Wiesbaden.

Rashki, A., D.G. Kaskaoutis, A.S. Goudie & R.A. Kahn. 2013. Dryness of ephemeral lakes and consequences for dust activity: The case of the Hamoun drainage basin, southeastern Iran. *Science of the Total Environment* 463–464: 552–564. https://doi.org/10.1016/j.scitotenv.2013.06.045

Rauchenberger, M. 1989. Systematics and biogeography of the genus *Gambusia* (Cyprinodontiformes: Poeciliidae). *American Museum Novitates* 2951:1–74.

Regan, C.T. 1906. Two new cyprinoid fishes from the Helmand basin. *Journal & Proceedings of the Asiatic Society of Bengal* 2: 8–9.

Reichenbacher, B., E. Kamrani, H.R. Esmaeili & A. Teimori. 2009. The endangered cyprinodont *Aphanius ginaonis* (Holly, 1929) from southern Iran is a valid species: evidence from otolith morphology. *Environmental Biology of Fishes* 86: 507–521. http://dx.doi.org/10.1007/s10641-009-9549-5

Renaud, C.B. 2011. *Lampreys of the world. An annotated and illustrated catalogue of lamprey species known to date*. Food and Agriculture Organization of the United Nations.

Reznick, D.N., H. Bryga & J.A. Endler. 1990. Experimentally induced life-history evolution in a natural population. *Nature (London)* 346: 357–359.

Ricketts, T.H., E. Dinerstein, T. Boucher, T.M. Brooks, S.H. Butchart, M. Hoffmann, J.F. Lamoreux, J. Morrison, M. Parr, J. D. Pilgrim & A.S. Rodrigues. 2005. Pinpointing and preventing imminent extinctions. *Proceedings of the National Academy of Sciences* 102: 18497–18501. https://doi.org/10.1073/pnas.050906010

Riehl, R., R. Patzner & K. Zanger. 2002. Die Eier heimischer Fische. 14. Rapfen oder Schied-*Aspius aspius* (Linnaeus, 1758) (Cyprinidae). Österreichs Fis*cherei* 275–281.

Robalo, J.I., V.C. Almada, A. Levy & I. Doadrio. 2007. Re-examination and phylogeny of the genus *Chondrostoma* based on mitochondrial and nuclear data and the definition of 5 new genera. *Molecular Phylogenetics and Evolution* 42: 362–372. https://doi.org/10.1016/j.ympev.2006.07.003

Rochard, E. 2001. Esturgeon europeen. P.108 *in:* Keith, P. & J. Allardi. (Eds.). *Atlas des poissons d'eau douce de France*. Muséum national d'Histoire naturelle, Paris. Patrimoines naturels.

Rodríguez-Muñoz, R., J.R. Waldman, C. Grunwald, N.K. Roy & I. Wirgin. 2004. Absence of shared mitochondrial DNA haplotypes between sea lamprey from North American and Spanish rivers. *Journal of Fish Biology* 64: 783–787. https://doi.org/10.1111/j.1095-8649.2004.00334.x

Roman, A., S. Afanasyev, O. Golub & O. Lietytska. 2022. *Capoeta svanetica* (Teleostei, Cyprinidae), a new species from the Luchunis River (Rioni River drainage) in Georgia. *Zoodiversity* 56: 117–134. https://doi.org/10.15407/zoo2022.02.117

Romero, A. & K.M. Paulson. 2001. It's a wonderful hypogean life: a guide to the troglomorphic fishes of the world. Pp 13–41. *in:* Romero, A. (Ed.) *The biology of hypogean fishes*. Developments in environmental biology of fishes, 21. Springer, Dordrecht. https://doi.org/10.1007/978-94-015-9795-1_2

Ross, W. 1985. Oasis fishes of Eastern Saudi Arabia. *Fauna of Saudi Arabia*.

Rüber, L., M. Kottelat, H.H. Tan, P.K. Ng & R. Britz. 2007. Evolution of miniaturization and the phylogenetic position of Paedocypris, comprising the world's smallest vertebrate. *BMC evolutionary biology* 7: 1–10. https://doi.org/10.1186/1471-2148-7-38

Rueda-Jasso, R.A., A. Campos-Mendoza, F. Arreguín-Sánchez, E. Díaz-Pardo & C.A. Martínez-Palacios. 2013. The biological and reproductive parameters of the invasive armored catfish *Pterygoplichthys disjunctivus* from Adolfo López Mateos El Infiernillo Reservoir, Michoacán-Guerrero, Mexico. *Revista Mexicana de Biodiversidad* 84: 318–326. https://doi.org/10.7550/rmb.26091

Russell, A. 1756. *Natural history of Aleppo and parts adjacent*. London, printed for A. Millar. Available at: https://archive. org/details/naturalhistoryof00russ

Rylková, K., L. Kalous, J. Bohlen, D.K. Lamatsch & M. Petrtýl. 2013. Phylogeny and biogeographic history of the cyprinid fish genus *Carassius* (Teleostei: Cyprinidae) with focus on natural and anthropogenic arrivals in Europe. *Aquaculture* 380: 13–20. https://doi.org/10.1016/j.aquaculture.2012.11.027

Saç, G., M. Özuluğ, M.F. Geiger & J. Freyhof. 2019a. *Pseudophoxinus cilicicus*, a new spring minnow from southern Anatolia (Teleostei: Leuciscidae). *Zootaxa* 4671: 105–118. https://doi.org/10.11646/zootaxa.4671.1.8

Saç, G., M. Özuluğ, M. Elp, M. Gaffaroğlu, S. Ünal, M.K. Ayata, C. Kaya & J. Freyhof. 2019b. New records of *Pseudophoxinus firati* from Turkey (Teleostei: Leuciscidae). *Journal of Applied Ichthyology* 35 (3): 769–774. https://doi.org/10.1111/jai.13869

Saç, G., O. Özuluğ, S. Ağdamar, H. İnci, Ö.D. Yürekli and M. Özuluğ. 2025. Life in Extreme Conditions: Diet and Condition of the Extremophile Fish Aphanius almiriensis (Teleostei: Cyprinodontiformes) in a Thermal Rheocrene Spring. Ecology and Evolution 15: e71411. https://doi.org/10.1002/ece3.71411

Sadeghi, R., H.R. Esmaeili, F. Zarei, A. Esmaeili & K. Abbasi. 2018. The taxonomic status of an introduced freshwater goby of the genus *Rhinogobius* to Iran (Teleostei: Gobiidae). *Zoology in the Middle East* 65: 51–58 https://doi.org/10.1080/09397140.2018.1540149

Saha, N. & B.K. Ratha. 1998. Ureogenesis in Indian air-breathing teleosts: adaptation to environmental constraints. *Comparative Biochemistry and Physiology Part A: Molecular & Integrative Physiology* 120: 195–208. https://doi.org/10.1016/S1095-6433(98)00026-9

Saitoh, K., T. Sado, R.L. Mayden, N. Hanzawa, K. Nakamura, M. Nishida & M. Miya. 2006. Mitogenomic evolution and interrelationships of the Cypriniformes (Actinopterygii: Ostariophysi): The first evidence toward resolution of higher-level relationships of the world's largest freshwater fish clade based on 59 whole mitogenome sequences. *Journal of Molecular Evolution* 63: 826–841. https://doi.org/10.1007/s00239-005-0293-y

Salmanov, A.V. & E.A. Dorofeyeva. 2001. Comparative osteology of river and lake trouts (genus *Salmo*). *Proceedings of Zoological Institute of the Russian Academy of Sciences* 287: 149–195.

Sanna, D., F. Biagi, H. Alaya, F. Maltagliati, A. Addis, A. Romero, J. De Juan, J.P. Quignard, A. Castelli & P. Franzoi. 2013. Mitochondrial DNA variability of the pipefish *Syngnathus abaster*. *Journal of Fish Biology* 82: 856–876. https://doi.org/10.1111/jfb.12027

Sanzo, L. 1911. Distribuzione delle papille cutanee (organi ciatiformi) e suo valore sistematico nei Gobi. *Mitteilungen der Zoologischen Station Neapel, Berlin* 20: 249–328.

Sargeran, P., M. Bakhtiyari, A. Abdoli, B.W. Coad, K. Sarvi, M. Rahmati Lishi & A. Hajimoradloo. 2008. The endemic Iranian cave-fish, *Iranocypris typhlops*: two taxa or two forms based on the mental disc?. *Journal of Zoology in the Middle East* 44: 67–74. https://doi.org/10.1080/09397140.2008.10638290

Sarı, M. 2008. Threatened fishes of the world: *Chalcalburnus tarichi* (Pallas 1811) (Cyprinidae) living in the highly alkaline Lake Van, Turkey. *Environmental Biology of Fishes* 81: 21–23.

Sarı, H.M., A. İlhan & S. Balık. 2008. Some biological characteristics of *Hemigrammocapoeta kemali* (Hanko, 1924) in Işıklı Lake (Denizli-Turkey). *Journal of Applied Ichthyology* 24: 88–90. https://doi.org/10.1111/j.1439-0426.2007.01040.x

Sarıahmetoğlu, A. 2024. Türkiye'nin Marmara Denizi ve Karadeniz Havzalarında Yayılış Gösteren *Alburnus* Rafinesque 1820 Cinsinin Taksonomik Açıdan İncelenmesi. Master thesis. Recep Tayyip Erdogan University, Institute of Science and Technology, Rize, Türkiye, 33p. [in Turkish].

Şaşı, H. & G.G. Özay. 2017. Age, growth, length-weight relationship and reproduction of chub, *Squalius cephalus* (L. 1758) in upper Akcay River, Turkey. *Pakistan Journal Zoolgy* 49: 1571–1580. https://doi.org/10.17582/journal.pjz/2017.49.5.1571.1580

Sawada, Y. 1982. Phylogeny and zoogeography of the superfamily Cobitoidea (Cyprinoidei, Cypriniformes). *Memoirs of the Faculty of Fisheries Hokkaido University* 28: 65–223.

Saygun, S., F. Saygun & C. Önel. 2017. Five new records for the ichthyofauna of Miliç River in Turkey. *Ordu Üniversitesi Bilim ve Teknoloji Dergisi* 7: 183–195.

Saygun, S., S. Ağdamar & M. Özuluğ. 2021. *Oxynoemacheilus fatsaensis*, a new nemacheilid loach from the Elekçi stream in northern Anatolia (Teleostei: Nemacheilidae). *Zoologischer Anzeiger* 294: 39–49. https://doi.org/10.1016/j.jcz.2021.07.011

Sayyadzadeh, G. & H.R. Esmaeili. 2020. *Oxynoemacheilus marunensis*, a new loach species from the Persian Gulf basin with remarks on *O. frenatus* (Teleostei: Nemacheilidae). *Zootaxa* 4885: 189–206. https://doi.org/10.11646/zootaxa.4885.2.2

Sayyadzadeh, G., H.R. Esmaeili, K. Abbasi & B.W. Coad. 2015a. Re-validation of *Gonorhynchus adiscus* and *G. diplochilus* (Teleostei: Cyprinidae) using morphological and molecular data. *Zoology in the Middle East* 61: 349–361. https://doi.org/10.1080/09397140.2015.1095517

Sayyadzadeh, G., H.R. Esmaeili & J. Freyhof. 2015b. *Garra mondica*, a new species from the Mond River drainage with remarks on the genus *Garra* from the Persian Gulf basin in Iran (Teleostei: Cyprinidae). *Zootaxa* 4048: 75–89. https://doi.org/10.11646/zootaxa.4048.1.4

Sayyadzadeh, G., S. Eagderi & H.R. Esmaeili. 2016. A new loach of the genus *Oxynoemacheilus* from the Tigris River drainage and its phylogenetic relationships among the nemacheilid fishes (Teleostei: Nemacheilidae) in the Middle East based on cytochrome oxidase 1 sequences. *Iranian Journal of Ichthyology* 3: 236–250. https://doi.org/10.7508/iji.2016

Sayyadzadeh, G., H.R. Esmaeili, S. Eagderi, A. Jouladeh-Roudbar, M. Masoudi & S. Vatandoust. 2017. Redescription of *Oxynoemacheilus longipinnis* from the Persian Gulf basin (Teleostei: Nemacheilidae). *Zoology in the Middle East* 63: 228–238. https://doi.org/10.1080/09397140.2017.1349243

Sayyadzadeh, G., A. Abbasi & H.R. Esmaeili. 2018a. Review and redescription of *Sabanejewia* species in Iran (Teleostei: Cobitidae). *Iranian Journal of Ichthyology* 5: 277–292. https://doi.org/10.22034/iji.v5i4.322

Sayyadzadeh, G., H.R. Esmaeili & S. Eagderi. 2018b. Redescription and molecular systematics of *Paraschistura delvarii* (Teleostei: Nemacheilidae). *Biharean Biologist* 12: 40–47.

Sayyadzadeh, G., H.R. Esmaeili & S. Eagderi. 2019a. Taxonomic status of the loaches *Paracobitis vignai* and *P. rhadinaea* (Teleostei: Nemacheilidae) in Iran. *Zoology in the Middle East* 65: 221–229 https://doi.org/10.1080/09397140.2019.1586143

Sayyadzadeh, G., A. Teimori & H.R. Esmaeili. 2019b. *Paraschistura kermanensis*, a new stone loach species from southeastern Iran (Teleostei: Nemacheilidae). *Zootaxa* 4638: 571–583. https://doi.org/10.11646/zootaxa.4638.4.7

Sayyadzadeh, G., F. Zarei & H.R. Esmaeili. 2022. *Glyptothorax* (Teleostei: Sisoridae) from the Middle East: An integrated molecular and morphological insight into its taxonomic diversity. *Diversity* 14: 884. https://doi.org/10.3390/d14100884

Sayyadzadeh, G., S.M Al Jufaili & H.R. Esmaeili. 2023. Species diversity deflation: Insight into taxonomic validity of *Garra* species (Teleostei: Cyprinidae) from Dhofar Region in the Arabian Peninsula using an integrated morpho-molecular approach. *Zootaxa* 5230: 333–350. https://doi.org/10.11646/zootaxa.5230.3.4

Schliewen, U., D. Tautz & Pääbo S. 1994. Sympatric speciation suggested by monophyly of crater lake cichlids. *Nature* 368: 629–632. https://doi.org/10.1038/368629a0

Schmidt, J. 1925. The breeding place of the eel. *Smithonian Report* 1924: 279–316, 7 pls.

Schmitter-Soto, J.J. 2007. A systematic revision of the genus *Archocentrus* (Perciformes: Cichlidae), with the description of two new genera and six new species. *Zootaxa* 1603: 1–78. https://doi.org/10.11646/zootaxa.1603.1.1

Schöter, C., M. Özuluğ & J. Freyhof. 2009. *Capoeta caelestis*, a new species from Göksu River, Turkey (Teleostei: Cyprinidae). *Ichthyological Exploration of Freshwaters* 20: 229–236.

Schulz–Mirbach, T. & B. Reichenbacher. 2006. Reconstruction of Oligocene and Neogene freshwater fish faunas an actualistic study on cypriniform otoliths. *Acta Palaeontologica Polonica* 51: 283–304.

Schwarzer, J., N. Shabani, H.R. Esmaeili, S. Mwaiko & O. Seehausen. 2016. Allopatric speciation in the desert: diversification of cichlids at their geographical and ecological range limit in Iran. *Hydrobiologia* 791: 193–207. https://doi.org/10.1007/s10750-016-2976-3

Scott W.B. & E.J. Crossman. 1973 Freshwater fishes of Canada. *Bulletin of the Fisheries research Board of Canada* 1984: 1–966.

Şen, F., Paruğ, Ş. & Elp, M. 2015. İnci Kefali: Dünü, Bugünü, Bazı Biyolojik Özellikleri, Yaşadığı Temel Sorunlar ve Geleceği. *Yüzüncü Yıl Üniversitesi Tarım Bilimleri Dergisi* 25, 347–356.

Senou, H., T. Yoshino & M. Okiyama. 1987. A review of the mullets with a keel on the back, *Liza carinata* complex (Pisces: Mugilide). *Publication of Seto Marine Biological Laboratory* 32: 303–321. https://doi.org/10.5134/176141

Shacham, G. 2003. *Nature right for water: water allocation demands for water bodies and wet habitats—a policy statement document*. Israel Ministry of the Environment and Israel Nature and Parks Authority.

Shechonge, A., B.P. Ngatunga, S.J. Bradbeer, J.J. Day, A.G.P. Ford, J. Kihedu, T. Richmond, S. Mzighani, A.M. Smith, E.A. Sweke, R. Tamatamah, A.R. Tyers, G.F. Turner & M.J. Genner. 2019. Widespread colonisation of Tanzanian catchments by introduced *Oreochromis* tilapia fishes: the legacy from decades of deliberate introduction. *Hydrobiologia* 832: 235–253. https://doi.org/10.1007/s10750-018-3597-9

Sheraliev, B. & Z. Peng. 2021. Molecular diversity of Uzbekistan's fishes assessed with DNA Barcoding. *Scientific Reports* 11: 16894. https://doi.org/10.1038/s41598-021-96487-1

Shibukawa, T. 2020. Two new species of torrential gobies of the genus *Rhinogobius* from the Ryukyu Islands. *Bulletin of the Kanagawa Prefectural Museum (Natural Science)* 49: 7–28. https://doi.org/10.32225/bkpmnh.2020.49_7

Shiraishi, H. & V. Crook. 2015. *Eel market dynamics: An analysis of* Anguilla *production*. TRAFFIC, Tokyo, Japan.

Shubina, T.N., A.A. Popova & V.P. Vasil'ev. 1989. *Acipenser stellatus* Pallas, 1771. Pp 395–443 *in:* Holčík, J. (Ed.) *The freshwater fishes of Europe. Vol.1/II: General Introduction to fishes – Acipenseriformes*. AULA Verlag, Wiesbaden.

Siddiqui, A.Q. & A.H. Al-Harbi. 1995. Evaluation of three species of tilapia, red tilapia and a hybrid tilapia as culture species in Saudi Arabia. *Aquaculture* 138: 145–157. https://doi.org/10.1016/0044-8486(95)01088-2

Silva, K., N.M. Monteiro, M.N. Vieira & V.C Almada. 2006a. Reproductive behaviour of the black-striped pipefish *Syngnathus abaster* (Pisces; Syngnathidae). *Journal of Fish Biology 69: 1860–1869*. https://doi.org/10.1111/j.1095-8649.2006.01229.x

Silva, K., N.M. Monteiro, V.C. Almada & M.N. Vieira. 2006b. Early life history of *Syngnathus abaster*. *Journal of Fish Biology* 68: 80–86. https://doi.org/10.1111/j.0022-1112.2006.00878.x

Skelton, P.H. 2001. *A complete guide to the freshwater fishes of southern Africa*. Southern Book Publishers, South Africa.

Skov, C. & P.A. Nilsson. 2018. *Biology and ecology of pike*. CRC Press, Taylor & Francis Group, Boca Raton.

Šlechtová V., J. Bohlen & H.H. Tan. 2007. Families of Cobitoidea (Teleostei; Cypriniformes) as revealed from nuclear genetic data and the position of the mysterious genera *Barbucca*, *Psilorhynchus*, *Serpenticobitis* and *Vaillantella*. *Molecular Phylogenetics and Evolution* 44: 1358–1365. https://doi.org/10.1016/j.ympev.2007.02.019

Smith, C., M. Reichard, P. Jurajda & M. Przybylski. 2004. The reproductive ecology of the European bitterling (*Rhodeus sericeus*). *Journal of Zoology* 262: 107–124. http://dx.doi.org/10.1017/S0952836903004497

Smith, G.R. & R.F. Stearley. 1989. The classification and scientific names of rainbow and cutthroat trouts. *Fisheries* 14: 4–10. https://doi.org/10.1577/1548-8446(1989)014<0004:TCASNO>2.0.CO;2

Sokolov, L.I. & V.P. Vasil'ev. 1989a. *Acipenser nudiventris* Lovetsky, 1828. Pp 206–226 *in:* Holčík, J. (Ed.) *The freshwater fishes of Europe. Vol.1/II: General Introduction to fishes – Acipenseriformes*. AULA Verlag, Wiesbaden.

Sokolov, L.I. & V.P. Vasil'ev. 1989b. *Acipenser baeri* Brandt, 1869. Pp 263–284 *in:* Holčík, J. (Ed.) *The freshwater fishes of Europe. Vol.1/II: General Introduction to fishes – Acipenseriformes*. AULA Verlag, Wiesbaden.

Sorensen, P.W., J.M. Fine, V. Dvornikovs, C.S. Jeffrey, F. Shao, J. Wang, L.A. Vrieze, K.R. Anderson & T.R. Hoye. 2005. Mixture of new sulfated steroids functions as a migratory pheromone in the sea lamprey. *Nature Chemical Biology* 1: 324–328. https://doi.org/10.1038/nchembio739

Soyubelli, H. 2018. *Sultansazlığı Havzası balık faunası*. Master thesis. Nevşehir Hacı Bektaş Veli University. Nevşehir, 68p [in Turkish].

Sözer, F. 1942. Türkiye Cyprinodontid'leri hakkında Contributions à la connaissance des Cyprinodontides de la Turquie. *Revue de la Faculté des Sciences de l'Université d'Instanbul*, Série B: Sciences Naturelles 7: 307–316.

Spataru, P. & M. Gophen. 1985a. Food composition and feeding habits of *Astatotilapia flaviijosephi* (Lortet) in Lake Kinneret (Israel). *Journal of Fish Biology* 26: 503–507. https://doi.org/10.1111/j.1095-8649.1985.tb04290.x

Spataru, P. & M. Gophen. 1985b. Food composition of the barbel *Tor canis* (Cyprinidae) and its role in the Lake Kinneret ecosystem. *Environmental biology of fishes* 14: 295–301. https://doi.org/10.1007/BF00002634

Springer, V.G. 1968. Osteology and classification of the fishes of the Family Blenniidae. *Bulletin of the United States National Museum* 284: 83.

Starushenko, L.I. & A.B. Kazansky. 1996. Introduction of the Mullet Haarder (*Mugil soiuy*) in the Black Sea and the Sea of Azov. *Studies and Reviews, General Fisheries Management Council for the Mediterranean. FAO, Rome.*

Stearley, R.F. & G.R. Smith. 1993. Phylogeny of the Pacific trouts and salmons (*Oncorhynchus*) and genera of the family Salmonidae. *Transactions of the American Fisheries Society* 122: 1–33. https://doi.org/10.1577/1548-8659(1993)122<0001:POTPTA>2.3.CO;2

Stearns S.C. & A.A. Hendry. 2004. The salmonid contribution to key issues in evolution. Pp. 3–19 *in:* Hendry, A.A. & S.C. Stearns. (Eds.). *Evolution illuminated salmon and their relatives. Oxford University Press.* https://doi.org/10.1093/oso/9780195143850.003.0001

Steffens, W. 1958. *Der Karpfen*. Die Neue Brehm-Bücherei 203, Ziemsen Verlag, Wittenberg Lutherstadt.

Steindachner, F. 1863. Über eine neue *Alburnus*-Art aus Syrien. *Sitzungsberichte der Kaiserlichen Akademie der Wissenschaften. Mathematisch-Naturwissenschaftliche Classe* 48(I Abtheilung): 193–194.

Steindachner, F. 1897. Bericht über die von Dr. Escherich in der Umgebung von Angora gesammelten Fische und Reptilien. *Denkschriften der Kaiserlichen Akademie der Wissenschaften in Wien, Mathematisch-Naturwissenschaftliche Classe* 64: 685–699.

Steinitz, H. 1952. *Acanthobrama terrae-sanctae*, sp. n., from Lake Tiberias, Israel. *Annals and Magazine of Natural History* 5 (51): 293–298.

Steinitz, H. & A. Ben-Tuvia. 1960. The cichlid fishes of the genus *Tristramella* Trewavas. *Annals and Magazine of Natural History* (Series 13) 3: 161–175.

Stepien, C.A. & M.A. Tumeo. 2006. Invasion genetics of Ponto-Caspian gobies in the Great Lakes: a 'cryptic'species, absence of founder effects, and comparative risk analysis. *Biological Invasions* 8: 61–78. https://doi.org/10.1007/s10530-005-0237-x

Stewart, K.M. & A.M. Murray. 2017. Biogeographic implications of fossil fishes from the Awash River, Ethiopia. *Journal of Vertebrate Paleontology* 37: e1269115. https://doi.org/10.1080/02724634.2017.1269115

Stiassny, M.L.J. & A. Getahun. 2007. An overview of labeonin relationships and the phylogenetic placement of the Afro-Asian genus *Garra* Hamilton, 1922 (Teleostei: Cyprinidae), with the description of five new species of *Garra* from Ethiopia, and a key to all African species. *Zoological Journal of the Linnean Society* 150: 41–83. https://doi.org/10.1111/j.1096-3642.2007.00281.x

Stiassny, M.L.J., U.K. Schliewen & W.J. Dominey. 1992. A new species flock of cichlid fishes from Lake Bermin, Cameroon with a description of eight new species of *Tilapia* (Labroidei: Cichlidae). *Ichthyological Exploration of Freshwaters* 3: 311–346.

Stoeckel, J.N. & B.M. Burr. 1999. A review of key reproductive traits and methods used to spawn Ictalurids. *American Fisheries Society Symposium* 24: 141–160.

Stoumboudi, M.T., R. Barbieri, Z. Mamuris, M.J. Corsini-Foka & A.N. Economou. 2002. Threatened fishes of the world: *Ladigesocypris ghigii* (Gianferrari, 1927) (Cyprinidae). *Environmental biology of fishes* 65: 340. https://doi.org/10.1023/A:1020597902189

Stoumboudi, M.T., M. Kottelat & R. Barbieri. 2006. The fishes of the inland waters of Lesbos Island, Greece. *Ichthyological Exploration of Freshwaters* 17: 129–146.

Stout, C.C., M. Tan, A.R. Lemmon, E.M. Lemmon & J.W. Armbruster. 2016. Resolving cypriniformes relationships using an anchored enrichment approach. *BMC Evolutionary Biology* 16: 244. https://doi.org/10.1186/s12862-016-0819-5

Sullivan, J.P., J.G. Lundberg & M. Hardman. 2006. A phylogenetic analysis of the major groups of catfishes (Teleostei: Siluriformes) using *rag1* and *rag2* nuclear gene sequences. *Molecular Phylogenetics and Evolution* 41: 636–662. https://doi.org/10.1016/j.ympev.2006.05.044.

Sumita, M. & H.-U. Schmincke. 2013. Impact of volcanism on the evolution of Lake Van I: evolution of explosive volcanism of Nemrut Volcano (eastern Anatolia) during the past >400,000 years. *Bulletin* of *Volcanology* 75:714. https://doi.org/10.1007/s00445-013-0714-5

Sungur, S. 2020. *Oxynoemacheilus kosswigi* a junior synonym of *O. seyhanensis* (Teleostei: Nemacheilidae). *Iranian Journal of Sciences and Technology, Transactions A: Science* 44: 563–573. https://doi.org/10.1007/s40995-020-00857-9

Sungur, S., P. Jalili & S. Eagderi. 2018a. *Oxynoemacheilus ciceki*, new nemacheilid species (Teleostei, Nemacheilidae) from the Sultan Marsh, Kayseri Province, Turkey. *Iranian Journal of Ichthyology* 4: 375–383. https://doi.org/10.22034/iji.v4i4.258

Sungur, S., P. Jalili, S. Eagderi & E. Çiçek. 2018b. *Seminemacheilus ahmeti*, a new species of Nemachelid from Sultan Marshes, Turkey. *FishTaxa* 3: 466–473.

Sušnik, S., J. Schoffmann & A. Snoj. 2004. Phylogenetic position of *Salmo* (*Platysalmo*) *platycephalus* Behnke 1968 from south-central Turkey, evidenced by genetic data. *Journal of Fish Biology* 64: 947–960. https://doi.org/10.1111/j.1095-8649.2004.0363.x

Sušnik, S., A. Snoj, I.F. Wilson, D. Mrdak & S. Weiss. 2007. Historical demography of brown trout (*Salmo trutta*) in the Adriatic drainage including the putative *S. letnica* endemic to Lake Ohrid. *Molecular Phylogenetics and Evolution* 44: 63–76. https://doi.org/10.1016/j.ympev.2006.08.021

Svanbäck, R. & P. Eklöv. 2002. Effects of habitat and food resources on morphology and ontogenetic growth trajectories in perch. *Oecologia* 131: 61–70. https://doi.org/10.1007/s00442-001-0861-9

Svärdson, G. 1979. Speciation of Scandinavian *Coregonus*. *Report of the Institute of Freshwater Research, Drottningholm* 57: 1–95.

Svetovidov, A.N. 1964. *Ryby Chernogo Morya*. [*Fishes of the Black Sea*]. Moskva: Nauka.

Szczerbowski, J.A. 2002. *Carassius auratus* (Linnaeus, 1758). Pp 5–41 *in:* Bănărescu, P.M.& H.-J. Paepke. (Eds.). *The freshwater fishes of Europe. Vol. 5/III: Cyprinidae 2, Part III:* Carassius *to* Cyprinus. *Gasterosteidae.* AULA Verlag, Wiebelsheim.

Szczerbowski, J.A. & A.J. Szczerbowski. 2002. *Carassius carassius* (Linnaeus, 1758). Pp 43–78 *in:* Bănărescu, P.M. & H.-J. Paepke. (Eds.). *The freshwater fishes of Europe. Vol. 5/III: Cyprinidae 2, Part III:* Carassius *to* Cyprinus. *Gasterosteidae*. AULA Verlag, Wiebelsheim.

Talwar. P.K. & A.G. Jhingran 1992. *Inland fishes of India and adjacent countries*. Oxford-IBH Publishing Co. Pvt. Ltd., New Delhi, 1158 p.

Tan, M. & J. W. Armbruster. 2018. Phylogenetic classification of extant genera of fishes of the order Cypriniformes (Teleostei: Ostariophysi). *Zootaxa* 4476:6–39. https://doi.org/10.11646/zootaxa.4476.1.4

Tang, Q., H. Liu, R. Mayden & B. Xiong. 2006. Comparison of evolutionary rates in the mitochondrial DNA cytochrome b gene and control region and their implications for phylogeny of the Cobitoidea (Teleostei: Cypriniformes). *Molecular Phylogenetics and Evolution* 39: 347–357. https://doi.org/10.1016/j.ympev.2005.08.007

Tang, K.L., M.K. Agnew, M.V. Hirt, T. Sado, L. M. Schneider, J. Freyhof, Z. Sulaiman, E. Swartz, C. Vidthayanon, M. Miya, K. Saitoh, A.M. Simons, R.W. Wood & R.L. Mayden. 2010. Systematics of the subfamily Danioninae (Teleostei: Cypriniformes: Cyprinidae). *Molecular Phylogenetics and Evolution* 57: 189–214. https://doi.org/10.1016/j.ympev.2010.05.021

Tarkan, A.S., M. Özuluğ, Ö. Gaygusuz & Ç.G. Gaygusuz. 2007. A new locality for the freshwater fish *Chondrostoma angorense* Elvira, 1987 (Osteichthys: Cyprinidae) in the Marmara Region (Turkey). E.U. *Journal of Fisheries & Aquatic Sciences* 24: 165–166.

Tarkan, A.S., S.M. Marr & F.G. Ekmekçi. 2015. Non-native and translocated freshwater fish species in Turkey. *FiSHMED Fishes in Mediterranean Environments* 3: 1–28.

Taylor, W.R. 1969. A revision of the catfish genus *Noturus* Rafinesque with an analysis of higher groups in the Ictaluridae. *Smithsonian Institution, U.S. National Museum, Bulletin 282, Washington, D.C.*

Teimori, A., H.R. Esmaeili & A. Gholamhosseini. 2010. The Ichthyofauna of Kor and Helleh River basins in southwest of Iran with reference to taxonomic and zoogeographic features of native fishes. Iranian *Journal of Animal Biosystematics* 6: 1–8. https://doi.org/10.22067/IJAB.V6I1.9152

Teimori, A., H.R. Esmaeili & B. Reichenbacher. 2011. *Aphanius farsicus*, a replacement name for *A. persicus* (Jenkins, 1910) (Teleostei, Cyprinodontidae). *Zootaxa* 3096: 53–58. https://doi.org/10.11646/zootaxa.3096.1.5

Teimori, A., H.R. Esmaeili, Z. Gholami, N. Zarei & B. Reichenbacher. 2012. *Aphanius arakensis*, a new species of tooth-carp (Actinopterygii, Cyprinodontidae) from the endorheic Namak Lake basin in Iran. *Zookeys* 215: 55–76. http:// dx.doi.org/10.3897/zookeys.215.1731

Teimori, A., H.R. Esmaeili, D. Erpenbeck & B. Reichenbacher. 2014. A new and unique species of the genus *Aphanius* Nardo, 1827 (Teleostei: Cyprinodontidae) from southern Iran: a case of regressive evolution. *Zoologischer Anzeiger* 253: 327–337. https://doi.org/10.1016/j.jcz.2013.12.001

Teimori, A., H.R. Esmaeili, G. Sayyadzadeh, N. Zarei & A. Gholamhossein. 2015. Molecular systematics and distribution review of the endemic cyprinid species, Persian chub, *Acanthobrama persidis* (Coad, 1981) in Southern Iran (Teleostei: Cyprinidae). *Molecular Biology Research Communications* 4: 189–206.

Teimori, A., H.R. Esmaeili, N. Hamidan & B. Reichenbacher. 2018. Systematics and historical biogeography of the *Aphanius dispar* species group (Teleostei: Aphaniidae) and description of a new species from southern Iran. *Journal of Zoological Systematics and Evolutionary Research* 56: 579–598. https://doi.org/10.1111/jzs.12228

Teimori, A., M. Ebrahimi, H. Mostafavi, H.R. Esmaeili & M. Masoudi. 2019. Distribution and abundance of the endemic scaleless tooth-carp, *Aphanius furcatus* Teimori, Esmaeili, Erpenbeck & Reichenbacher 2014 (Teleostei: Aphaniidae) in Iran. *Turkish Journal of Fisheries and Aquatic Sciences* 20: 61–67 http://doi.org/10.4194/1303-2712-v20_1_06

Teimori, A., H.R. Esmaeili, N. Hamidan & B. Reichenbacher. 2020. Comment on "A proposal for a new generic structure of the killifish family Aphaniidae, with the description of *Aphaniops teimorii* (Teleostei: Cyprinodontiformes)" by Jörg Freyhof & Baran Yoğurtçuoğlu *Zootaxa* 4810: 421–451. *FishTaxa* 17: 15–16.

Tesch, F.W. 1991. *Anguilla anguilla* (Linnaeus, 1758). Pp 389–437 *in:* Hoestlandt, H. (Ed.) *The freshwater fishes of Europe. Vol. 2. Clupeidae, Anguillidae*. AULA Verlag, Wiesbaden.

Tesch, F.W. 1999. *Der Aal*. Parey Buchverlag, Berlin.

Teugels, G.G. 1986. A systematic revision of the African species of the genus *Clarias* (Pisces; Clariidae). Annales-Musee Royal de l'Afrique Centrale. *Annales du Musée Royal de l'Afrique Centrale (Zoologie)* 247:1–199.

Thacker, C.E. 2013. Phylogenetic placement of the European sand gobies in Gobionellidae and characterization of gobionellid lineages (Gobiiformes: Gobioidei). *Zootaxa* 3619: 369–82. https://doi.org/10.11646/zootaxa.3619.3.6

Thomson, J.M. 1997. The Mugilidae of the World. *Memoirs of the Queensland Museum* 41: 457–562.

Thoni, R.J. & Hart, R. 2015. Repatriating a lost name: notes on McClelland and Griffith's *Cobitis boutanensis* (Cypriniformes: Nemacheilidae). *Zootaxa* 3999: 291–294. https://doi.org/10.11646/zootaxa.3999.2.8

Tigano C. & V. Ferrito. 1985. Studio osteologico comparato del cranio di popolazioni di *Aphanius fasciatus* (Nardo 1827) (Pisces: Cyprinodontidae) dell'Adriatico e di Wumi di Sicilia. *Animalia* 12: 13–57.

Tortonese, E. 1938. Viaggio del dott. Enrico Festa in Palestina e in Siria (1893). Pesci. *Bollettino dei Musei di Zoologia ed Anatomia Comparata della R. Università di Torino. (Ser. 3)* 46: 313–358.

Tortonese, E. 1952. On a new cyprinoid fish of the genus *Acanthobrama* from Palestine. *Annals and Magazine of Natural History* 5: 271–272. https://doi.org/10.1080/00222935208654290

Tortonese, E. 1954. The trouts of Asiatic Turkey. *Publications Institute Hydrobiology University of İstanbul (Sér. B)* 2: 1–26, pls 1–2.

Trewavas, E., 1941. XVII.—Nyasa fishes of the genus *Tilapia* and a new species from Portuguese East Africa. *The Annals and Magazine of Natural History* 7: 294–306. https://doi.org/10.1080/03745481.1941.9727933

Trewavas, E. 1955. A blind fish from Iraq, related to *Garra*. *Annals and Magazine of Natural History (Series 12)* 8: 551–555.

Trewavas, E. 1983. Tilapiine fishes of the genera *Sarotherodon*, *Oreochromis* and *Danakilia*. Published by the British Museum (Natural History), London.

Trewavas, E., J. Green & S.A. Corbet. 1972. Ecological studies on crater lakes in West Cameroon Fishes of Barombi Mbo. *Journal of Zoology* 167: 41–95. https://doi.org/10.1111/j.1469-7998.1972.tb01722.x

Triantafyllidis, A., I. Leonardos, I. Bista, I. D. Kyriazis, M.T. Stoumboudi, I. Kappas, F. Amat & T. J. Abatzopoulos. 2007. Phylogeography and genetic structure of the Mediterranean killifish *Aphanius fasciatus* (Cyprinodontidae). *Marine Biology* 152: 1159–1167. https://doi.org/10.1007/s00227-007-0760-7

Tsepkin E, L.I. Sokolov & A.B Rusalimchik. 1992. Ecology of the round goby *Neogobius melanostomus*, an occasional colonizer of the basin of the Moskva River. *Biologiceskie Nauki* 1: 46–51.

Tsigenopoulos, C.S., P. Kasapidis & P. Berrebi. 2010. Phylogenetic relationships of hexaploid large-sized barbs (genus *Labeobarbus*, Cyprinidae) based on mtDNA data. *Molecular Phylogenetics and Evolution* 56: 851–856. https://doi.org/10.1016/j.ympev.2010.02.006

Turan, D. 2022. Description of a new species *Squalius* from Lake Hazar and upper Tigris River drainages in Anatolia (Teleostei: Leuciscidae). *Journal of Anatolian Environmental and Animal Sciences (Anadolu Çevre ve Hayvancılık Bilimleri Dergisi)* 7: 336–340. https://doi.org/10.35229/jaes.1134404

Turan, D. & S. Aksu. 2021. A new trout species from southern Marmara Sea drainages (Teleostei: Salmonidae). *Journal of Anatolian Environmental and Animal Sciences* 6: 232–239. https://doi.org/10.35229/jaes.903810

Turan, D. & E. Bayçelebi. 2019. Range Extension of *Gobio microlepidotus* Battalgil, 1942 in the Southern Anatolia (Pisces: Cyprinidae). *Journal of Anatolian Environmental and Animal Sciences* 4: 156–160. https://doi.org/10.35229/jaes.556220

Turan, D. & E. Bayçelebi. 2020. First record of *Salmo pelagonicus* Karaman, 1938 (Teleostei: Salmonidae) in the Karamenderes River, Turkey. *Journal of Anatolian Environmental and Animal Sciences* 5: 551–555. https://doi.org/10.35229/jaes.777776

Turan, D., E. Bayçelebi, S. Aksu & O. Münevver. 2024c. The trouts of the Marmara and Aegean Sea drainages in Türkiye, with the description of a new species. *Zoosystematics and Evolution*, 100: 87–99. https://doi.org/10.3897/zse.100.112557

Turan, D., M. Kottelat, F.G. Ekmekçi & H.O. Imamoğlu. 2006a. A review of *Capoeta tinca*, with descriptions of two new species from Turkey (Teleostei: Cyprinidae). *Revue Suisse de Zoologie* 113: 421–436. https://doi.org/10.5962/bhl.part.80358

Turan, D., M. Kottelat, S.G. Kırankaya & S. Engin. 2006b. *Capoeta ekmekciae*, a new species of cyprinid fish from northeastern Anatolia (Teleostei: Cyprinidae). *Ichthyological Exploration of Freshwaters* 17:147–156.

Turan, D., F.G. Ekmekçi, A. Ilhan & S. Engin. 2008a. *Luciobarbus kottelati*, a new species of barbel (Teleostei: Cyprinidae) from the Büyük Menderes River, Turkey, with rediagnose of *L. lydianus*. *Zootaxa* 1824: 35–44. https://doi.org/10.11646/zootaxa.1824.1.4

Turan, D., M. Kottelat & F.G. Ekmekçi. 2008b. *Capoeta erhani*, a new species of cyprinid fish from Ceyhan River, Turkey (Teleostei: Cyprinidae). *Ichthyological Exploration of Freshwaters* 19: 263–270.

Turan, D., B.T. Yılmaz & C. Kaya. 2009a. *Squalius kottelati*, a new cyprinid species (Teleostei: Cyprinidae) from Orontes River Turkey. *Zootaxa* 2270: 53–62. https://doi.org/1010.5281/zenodo.190947

Turan, D., M. Kottelat & F.G. Ekmekçi. 2009b. *Barbus niluferensis*, a new species of barbel (Teleostei: Cyprinidae) from Nilüfer River, Turkey, with redescription of *B. oligolepis*. *Zootaxa* 1981: 15–28. https://doi.org/10.11646/zootaxa.1981.1.2

Turan, D., M. Kottelat & S. Engin. 2010. Two new species of trouts, resident and migratory, sympatric in streams of northern Anatolia (Salmoniformes: Salmonidae). *Ichthyological Exploration of Freshwaters* 20: 333–364.

Turan, D., M. Kottelat & Y. Bektaş. 2011. *Salmo tigridis*, a new species of trout from the Tigris River, Turkey (Teleostei: Salmonidae). *Zootaxa* 2993: 23–33. https://doi.org/10.11646/zootaxa.2993.1.2

Turan, D., F.G. Ekmekçi, V. Luskova & J. Mendel. 2012a. Description of a new species of the genus *Gobio* from Turkey (Teleostei: Cyprinidae). *Zootaxa* 3257: 56–65. https://doi.org/10.11646/zootaxa.3257.1.4

Turan, D., M. Kottelat & S. Engin. 2012b. The trouts of the Mediterranean drainages of southern Anatolia, Turkey, with description of three new species (Teleostei: Salmonidae). *Ichthyological Exploration of Freshwaters* 23: 219–236.

Turan, D, F.G. Ekmekçi, C. Kaya & S.S. Güçlü. 2013a. *Alburnoides manyasensis* (Actinopterygii, Cyprinidae), a new species of cyprinid fish from Manyas Lake basin, Turkey. *ZooKeys* 276: 85–102. https://doi.org/10.3897/zookeys.276.4107

Turan, D., M. Kottelat & E. Doğan. 2013b. Two new species of *Squalius*, *S. adanaensis* and *S. seyhanensis* (Teleostei: Cyprinidae), from the Seyhan River in Turkey. *Zootaxa* 3637: 308–324. https://doi.org/10.11646/zootaxa.3637.3.4

Turan, D., C. Kaya, F.G. Ekmekçi & E. Doğan. 2014a. Three new species of *Alburnoides* (Teleostei: Cyprinidae) from the Euphrates River, Eastern Anatolia, Turkey. *Zootaxa* 3754: 101–116. https://doi.org/10.11646/zootaxa.3754.2.1

Turan, D., E. Doğan, C. Kaya & M. Kanyılmaz. 2014b. *Salmo kottelati*, a new species of trout from Alakır stream, draining to the Mediterranean in southern Anatolia, Turkey (Teleostei, Salmonidae). *ZooKeys* 462: 135–151. https://doi.org/10.3897/zookeys.462.8177

Turan, D., B. Japoshvili, İ. Aksu & Y. Bektaş. 2016a. Description of two new species of the genus *Gobio* (Teleostei: Cyprinidae) from the Black Sea coast of Turkey. *Zoology in the Middle East* 62: 112–124. https://doi.org/10.1080/09397140.2016.1182779

Turan, D., Y. Bektaş, C. Kaya & E. Bayçelebi. 2016b. *Alburnoides diclensis* (Actinopterigii: Cyprinidae), a new species of cyprinid fish from the upper Tigris River, Turkey. *Zootaxa* 4067: 79–87. https://doi.org/10.11646/zootaxa.4067.1.6

Turan, D., C. Kaya, E. Bayçelebi, İ. Aksu & Y. Bektaş. 2017a. *Gobio baliki*, a new gudgeon from Turkey (Teleostei: Cyprinidae). *Zootaxa* 4350: 284–290. https://doi.org/10.11646/zootaxa.4350.2.4

Turan, D., C. Kaya, E. Bayçelebi, Y. Bektaş & F.G. Ekmekçi. 2017b. Three new species of *Alburnoides* from the southern Black Sea basin (Teleostei: Cyprinidae). *Zootaxa* 4242: 565–577. https://doi.org/10.11646/zootaxa.4242.3.8

Turan, D., F. Küçük, C. Kaya, S.S. Güçlü & Y. Bektaş. 2017c. *Capoeta aydinensis*, a new species of scraper from southwestern Anatolia, Turkey (Teleostei: Cyprinidae). *Turkish Journal of Zoology* 41: 436–442. https://doi.org/10.3906/zoo-1510-43

Turan, D., M. Kottelat & E. Bayçelebi. 2017d. *Squalius semae*, a new species of chub from the Euphrates River, Eastern Anatolia (Teleostei: Cyprinidae). *Zoology in the Middle East* 63: 33–42. https://doi.org/10.1080/09397140.2017.1290761

Turan, D., M. Kottelat & C. Kaya. 2017e. *Salmo munzuricus*, a new species of trout from the Euphrates River drainage, Turkey (Teleostei: Salmonidae). *Ichthyological Exploration of Freshwaters* 28: 55–63.

Turan D., C. Kaya, E. Bayçelebi & B. Verep. 2018a. Actual fish fauna of İyidere stream, Rize-Turkey. *International Technological Sciences and Design Symposium* 288–296.

Turan, D., C. Kaya, E. Bayçelebi, İ. Aksu & Y. Bektaş. 2018b. Description of *Gobio fahrettini*, a new gudgeon from Lake Ilgın basin, Central Anatolia (Teleostei: Gobionidae). *Ichthyological Exploration of Freshwaters* 28: 1–9.

Turan, D., C. Kaya, M.F. Geiger & J. Freyhof. 2018c. *Barbus anatolicus*, a new barbel from the Kızılırmak and Yeşilırmak River drainages in northern Anatolia (Teleostei: Cyprinidae). *Zootaxa* 4461: 539–557. https://doi.org/10.11646/zootaxa.4461.4.5

Turan, D., G. Kalaycı, C. Kaya, Y. Bektaş & F. Küçük. 2018d. A new species of *Petroleuciscus* (Teleostei: Cyprinidae) from the Büyük Menderes River, southwestern Anatolia, Turkey. *Journal of Fish Biology* 92: 875–887. https://doi.org/10.1111/jfb.13525

Turan, D., C. Kaya, G. Kalaycı, E. Bayçelebi & İ. Aksu. 2019a. *Oxynoemacheilus cemali*, a new species of stone loach (Teleostei: Nemacheilidae) from the Çoruh River drainage, Turkey. *Journal of Fish Biology* 94: 458–468. https://doi.org/10.1111/jfb.13909

Turan, D., C. Kaya, İ. Aksu, E. Bayçelebi & Y. Bektaş. 2019b. *Alburnoides coskuncelebii*, a new species from the stream Büyük Melen in north western Anatolia (Teleostei: Leuciscidae). *Ichthyological Exploration of Freshwaters* 29: 201–210 https://doi.org/10.23788/IEF-1107

Turan, D., G., Kalaycı, Y. Bektaş, C. Kaya & E. Bayçelebi. 2020. A new species of trout from the northern drainages of Euphrates River, Turkey (Salmoniformes: Salmonidae). *Journal of Fish Biology* 96: 1454–1462. https://doi.org/10.1111/ jfb.14321

Turan, D., F. Küçük, S.S. Güçlü & İ. Aksu. 2021a. *Turcichondrostoma*, a new genus for the Leuciscidae (Teleostei: Cypriniformes) from Southwestern Anatolia. *Journal of Fish Biology*. https://doi.org/10.1111/jfb.14903

Turan, D., İ. Aksu, M. Oral, C. Kaya & E. Bayçelebi. 2021b. Contribution to the trout of Euphrates River, with description of a new species, and range extension of *Salmo munzuricus* (Salmoniformes, Salmonidae). *Zoosystematics and Evolution* 97: 471–482 https://doi.org/10.3897/zse.97.72181

Turan, D., C. Kaya, İ. Aksu & Y. Bektaş. 2022a. *Paracapoeta*, a new genus of the Cyprinidae from Mesopotamia, CIlıca and Levant (Teleostei, Cypriniformes). *Zoosystematics and Evolution* 98: 201–212 https://doi.org/10.3897/zse.98.81463

Turan, D., M. Kottelat & C. Kaya. 2022b. The trouts of the upper Kura and Aras Rivers in Turkey, with description of three new species (Teleostei: Salmonidae). *Zootaxa* 5150: 43–64. https://doi.org/10.11646/zootaxa.5150.1.2

Turan, D., E. Bayçelebi & G. Kalaycı. 2023. *Oxynoemacheilus marmaraensis*, a new species from the Susurluk River, Türkiye (Teleostei: Nemacheilidae). *Journal of Fish Biology*, 103(5): 1106–1112. https://doi.org/10.1111/jfb.15506

Turan, D., S. Aksu, S.S. Güçlü & G. Kalaycı. 2024a. *Oxynoemacheilus kottelati*, a new species from the Havran and Karınca streams in northern Aegean basin, Türkiye (Teleostei, Nemacheilidae). *Zoosystematics and Evolution* 100 (2): 483–492. https://doi.org/10.3897/zse.100.119915

Turan, D., S. Aksu, S.S. Güçlü & G. Kalaycı. 2024b. *Oxynoemacheilus fatmae*, a new species from the Güzelhisar Stream in the Aegean Sea basin, Türkiye (Teleostei: Nemacheilidae). *Journal of Fish Biology* 105 (3): 1–9. https://doi.org/10.1111/jfb.15779

Turan, D., E. Bayçelebi, S. Aksu & M. Oral. 2024c. The trouts of the Marmara and Aegean Sea drainages in Türkiye, with the description of a new species (Teleostei, Salmonidae). *Zoosystematics and Evolution* 100(1): 87–99. 10.3897/zse.100.112557

Turan, D., M. Kocabaş & I. Aksu. 2025 *Oxynoemacheilus sarali* (Teleostei: Nemacheilidae): a new species of loach from Merziman Creek, Euphrates River Basin, Türkiye, described using a morpho-molecular approach. *Hydrobiologia* https://doi.org/10.1007/s10750-025-05935-w

Türkmen, M., O. Erdoğan, H. İ. Haliloğlu & A. Yıldırım. 2001. Age, growth and reproduction of *Acanthalburnus microlepis*, Filippi 1863 from the Yağan region of the Aras River, Turkey. *Turkish Journal of Zoology* 25: 127–133.

Türkmen, M., O. Erdoğan, A. Yıldırım & I. Akyurt. 2002. Reproduction tactics, age and growth of *Capoeta capoeta umbla* Heckel 1843 from the Aşkale Region of the Karasu River, Turkey. *Fisheries Research* 54: 317–328. https://doi.org/10.1016/S0165-7836(01)00266-1

Uçkun, A.A. & D. Gökçe. 2014. Assessing age, growth, and reproduction of *Alburnus mossulensis* and *Acanthobrama marmid* (Cyprinidae) populations in Karakaya Dam Lake (Turkey). *Turkish Journal of Zoology* 39: 1–14. https://doi.org/10.3906/zoo-1211-13

Uğurlu, S., N. Polat & S. Kandemir. 2008. Fish fauna of lagoons within the Kızılırmak and Yeşilırmak deltas (Samsun-Turkey). *Journal of FisheriesSciences.com* 2: 475–483. [in Turkish]. https://doi.org/10.3153/jfscom.mug.200742

Ünlü, E. 2021. Current Status of Fish in Balıklıgöl, Şanlıurfa, Turkey and Comments on "A Preliminary Investigation on serious mortalities of fish in Balıklıgöl (Halil-ür Rahman Gölü, Şanlıurfa)" by Öktener et al. (2008). *Acta Biologica Turcica* 34: 86–91.

Ünlü, E., T. Ciçek, D. Değer & B.W. Coad. 2011. Range extension of the exotic Indian stinging catfish, *Heteropneustes fossilis* (Bloch, 1794) (Heteropneustidae) into the Turkish part of the Tigris River watershed. *Journal of Applied Ichthyology* 27: 141–143. https://doi.org/10.1111/j.1439-0426.2010.01580.x

Uysal, I. & A. Alpbaz. 2003. Comparison of fertilisation, eyeing, hatching and survival rate of Abant Trout (*Salmo trutta abanticus* T., 1954) and rainbow trout (*Oncorhynchus mykiss* W., 1792). *Ege Journal of Fisheries and Aquatic Sciences* 1–2: 95–101. https://doi.org/10.1186/s12864-021-07773-1

Valdesalici, S., A. Brahimi & J. Freyhof. 2019. First record of *Aphanius almiriensis* from Italy and notes on the distribution of *Aphanius fasciatus* (Teleostei: Aphaniidae). *Journal of Applied Ichthyology* 35: 541–550. https://doi.org/10.1111/jai.13873

Van Damme, D.V., N., Bogutskaya, R.C. Hoffmann & C. Smith. 2007. The introduction of the European bitterling (*Rhodeus amarus*) to west and central Europe. *Fish and Fisheries* 8: 79–106. https://doi.org/10.1111/j.1467-2679.2007.00239.x

Van Kleunen, M., Ewald, E. Weber & M. Fischer. 2010. A meta-analysis of trait differences between invasive and non-invasive plant species. *Ecology Letters* 13: 235–245. https://doi.org/10.1111/j.1461-0248.2009.01418.x

Vasil'eva, E.D. 2007. Gobies of the Genus *Rhinogobius* (Gobiidae) from Primor'e and water bodies of Central Asia and Kazakhstan: I. Morphological Characteristic and Taxonomic Status. *Journal of Ichthyology* 47: 691–700.

Vasil'eva, E.D. 2024. Species diversity, diagnostic characters, and distribution of the tubenose gobies of genus *Proterorhinus* (Gobiidae). I. A new species of the tubenose goby from the water bodies of the eastern coast of the Black Sea. *Journal of Ichthyology*, 64, 6, 881–890. https://doi.org/10.1134/S003294522470067X

Vasil'eva, E.D. & T.I. Kuga. 2008. Gobies of the Genus *Rhinogobius* (Gobiidae) of Primorye and Water Bodies of Central Asia and Kazakhstan: II. Comparative Craniological Analysis of Gobies Introduced to Central Asia. *Journal of Ichthyology* 48: 29–36. https://doi.org/10.1134/S0032945208010037

Vasil'eva, E.D. & T.I. Kuga. 2001. Craniological differences between the gobies *Knipowitschia caucasica* and *Pomatoschistus marmoratus*, and the problems of the "sand gobies" taxonomy. *Journal of Ichthyology*, 41 (2):135–142.

Vasil'eva, E.D. & V.P. Vasil'ev. 1994. Systematics of Caucasian freshwater gobies (Gobiidae) in the light of contemporary data, with a description of a new species, *Neogobius rhodioni*, sp. nov. *Journal of Ichthyology* 35: 139–157.

Vasil'eva, E.D. & V.P. Vasil'ev. 1996. The description of *Neogobius iljini* sp. nov. within former *N. kessleri* (Gobiidae, Pices). Acta Universitatis Carolinae. *Biologica* 39: 261–270.

Vasil'eva, E.D. & V.P. Vasil'ev. 2000. The origin and taxonomic status of the triploid form of the goldfish, *Carassius auratus* (Cyprinidae). *Journal of Ichthyology* 40: 553–563. https://doi.org/10.1134/S0032945207090020

Vasil'eva, E.D. & V.P. Vasil'ev. 2003a. *Neogobius* Iljin, 1927). Pp 163–171 *in:* Miller, P.J. (Ed.) *The freshwater fishes of Europe. Vol. 8/I: Mugilidae, Atherinidae, Atherinopsidae, Blenniidae, Odontobutidae, Gobiidae*. AULA Verlag, Wiebelsheim.

Vasil'eva, E.D. & V.P. Vasil'ev. 2003b. *Neogobius constructor* (Nordmann, 1840). Pp 193–202 *in:* Miller, P.J. (Ed.) *The freshwater fishes of Europe. Vol. 8/I: Mugilidae, Atherinidae, Atherinopsidae, Blenniidae, Odontobutidae, Gobiidae*. AULA Verlag, Wiebelsheim.

Vasil'eva, E.D. & V.P. Vasil'ev. 2003c. *Neogobius cyrius* (Kessler, 1874). Pp 203–211 *in:* Miller, P.J. (Ed.) *The freshwater fishes of Europe. Vol. 8/I: Mugilidae, Atherinidae, Atherinopsidae, Blenniidae, Odontobutidae, Gobiidae*. AULA Verlag, Wiebelsheim.

Vasil'eva, E.D. & V.P. Vasil'ev. 2003d. *Neogobius gorlap* Iljin in Berg, 1949. Pp 253–264 *in:* Miller, P.J. (Ed.) *The freshwater fishes of Europe. Vol. 8/I: Mugilidae, Atherinidae, Atherinopsidae, Blenniidae, Odontobutidae, Gobiidae*. AULA Verlag, Wiebelsheim.

Vasil'eva, E D. & V.P. Vasil'ev. 2012. *Cobitis amphilekta* sp. nova, a new species of spined loaches (Cobitidae, Cypriniformes) from the Caspian Sea basin. *Journal of Ichthyology* 52: 200–206. https://doi.org/10.1134/S0032945212020154

Vasil'eva, E.D. & V.P. Vasil'ev. 2020. Craniological divergence of two genetically closely related Caucasian spined loaches: *Cobitis saniae* and *C. derzhavini* (Cobitidae). *Journal of Ichthyology* 60: 555–561. https://doi.org/10.1134/S0032945220040207

Vasil'eva, E.D., V.P. Vasil'ev & A.N. Economou. 2004. *Knipowitschia caucasica* (Berg, 1916). Pp 343–364 *in:* Miller, P.J. (Ed.) *The freshwater fishes of Europe. Vol. 8/II: Gobiidae 2*. AULA Verlag, Wiebelsheim.

Vasil'eva, E.D., H. Mousavi-Sabet & V.P. Vasil'ev. 2015. *Ponticola iranicus* sp. nov. (Actinopterygii: Perciformes: Gobiidae) from the Caspian Sea basin. *Acta Ichthyologica et Piscatoria* 45: 189–197. https://doi.org/10.3750/AIP2014.45.2.09

Vasil'eva, E.D., W.W. Schwarzhans, D.A. Medvedev & V.P. Vasil'ev. 2016. Cryptic species of ponto-caspian bighead goby of the genus *Ponticola* (Gobiidae). *Journal of Ichthyology* 56: 1–18. https://doi.org/0.1134/S003294521601015X

Vasil'eva, E.D., E.N. Solovyeva, B.A. Levin & V.P. Vasil'ev. 2020. *Cobitis derzhavini* sp. nova — a new spined loach species (Teleostei: Cobitidae) discovered in the Transcaucasia. *Journal of Ichthyology* 60: 135–153. https://doi.org/10.1134/S0032945220020198

Vasil'eva, E.D., V.P. Vasil'ev, G.A. Nemkova & S.V. Schedko. 2022. Phylogenetic Relationships and Taxonomy of the Carp Fish of Genus *Hemiculter* (Cyprinidae, Xenocypridinae): Species of the Group *H. leucisculus*. *Journal of Ichthyology* 62, 1–15. https://doi.org/10.1134/S0032945222010131

Vasil'eva, E., B. Levin & V.P. Vasil'ev 2024. Dynamics of diversity of spined loaches of genus *Cobitis* (Cobitidae) of Trans-Caucasus in the anthropogenic period according to the data of museum collections. III. Morpho-genetic variability, taxonomic status and diagnostic characters of spined loaches of the Black Sea coast north of the Meskheti ridge of the Lesser Caucasus. *Journal of Ichthyology* 64 (2): 157–176. https://doi.org/10.1134/S0032945224020152

Vasil'eva, E.D., Vasil'ev, V.P., Borovikova, E.A., Esmaeili, H.R. Gandlin, A.A. Levina, M.A., Mustafayev, N.J. Teimori, A. & B.A. Levin 2025a. Species diversity, diagnostic characters and distribution of tubenose gobies of the genus *Proterorhinus* (Gobiidae): Revalidation of the Caspian species *Proterorhinus blennioides*. *Inland Water Biology*, 10.1134/S1995082924600777

Vasil'eva, E.D., Vasil'ev, V.P., Borovikova, A.A. Levina, & B.A. Levin 2025b. Species diversity, diagnostic characters, and distribution of tubenose gobies of the genus *Proterorhinus* (Gobiidae): What are *P. semipellucidus* and *P. nasalis?* *Journal of Ichthyology* https://doi.org/10.1134/S0032945225600508

Vatandoust, S. & S. Eagderi. 2015. *Paraschistura ilamensis*, a new species of loach from the Tigris River drainage (Teleostei: Nemacheilidae). *International Journal of Aquatic Biology* 3: 177–182. https://doi.org/10.22034/ijab.v3i3.95

Vecchioni, L., A.C. Ching, F. Marrone, M. Arculeo, P.J. Hundt & A.M. Simons. 2022. Multi-locus phylogenetic analyses of the *Almadablennius* clade reveals inconsistencies with the present taxonomy of blenniid fishes. *Diversity* 14: 53. https://doi.org/10.3390/d14010053

Villwock, W., A. Scholl & F. Krupp. 1983. Zur Taxonomie, Verbreitung und Speziation des Formenkreises *Aphanius dispar* (Rüppell, 1828) und Beschreibung von *Aphanius sirhani* n. sp. (Pisces: Cyprinodontidae). *Mitteilungen aus dem Hamburgischen Zoologischen Museum und Institut* 80: 251–277.

Vinyoles, D. & A. De Sostoa. 2007. Life-history traits of the endangered river blenny *Salaria fluviatilis* (Asso) and their implications for conservation. *Journal of Fish Biology* 70: 1088–1108. https://doi.org/10.1111/j.1095-8649.2007.01371.x

Vladykov, V.D. & E. Kott. 1979. Satellite species among the holarctic lampreys (Petromyzonidae). *Canadian Journal of Zoology* 57: 860–867. https://doi.org/10.1139/z79-106

Vlasenko A.D., A.V. Pavlov & V.P. Vasil'ev. 1989a. *Acipenser gueldenstadti* Brandt, 1833. Pp 294–344 *in:* Holčík, J. (Ed.) *The freshwater fishes of Europe. Vol.1/II: General Introduction to fishes – Acipenseriformes*. AULA Verlag, Wiesbaden.

Vlasenko A.D., A.V. Pavlov & V.P. Vasil'ev. 1989b. *Acipenser persicus* Borodin, 1898. Pp 345–366 *in:* Holčík, J. (Ed.) *The freshwater fishes of Europe. Vol.1/II: General Introduction to fishes – Acipenseriformes*. AULA Verlag, Wiesbaden.

Voronina, E.P. 1999. Morphology and systematics of river flounders of the genus *Platichthys*. *Journal of Ichthyology* 39: 588–599.

Voss, I., B. Sunil, R. Scheibe & A.S. Raghavendra. 2013. Emerging concept for the role of photorespiration as an important part of abiotic stress response. *Plant Biology* 15: 713–722. https://doi.org/10.1111/j.1438-8677.2012.00710.x

Wagner, M., S. Zogaris, P. Berrebi, J. Freyhof, S. Koblmüller, P. Magnan & M. Laporte. 2021. Diversity and biogeography of Mediterranean freshwater blennies (Blenniidae, Salaria). *Diversity and Distributions* 27: 1832–1847. https://doi.org/10.1111/ddi.13372

Walter, R., W.R.Jr. Courtenay & J.D. Williams. 2004. Snakeheads (Pisces, Channidae): A biological synopsis and risk assessment, *Circular 1251, U.S. Geological Survey and the U.S. Fish and Wildlife Service*.

Wang, Q., L.P. Dizaj, J. Huang, K.K. Sarker, C. Kevrekidis, B. Reichenbacher, H.R. Esmaeili, N. Straube, T. Moritz & C. Li. 2022. Molecular phylogenetics of the Clupeiformes based on exon-capture data and a new classification of the order. *Molecular Phylogenetics and Evolution* 175: 107590 https://doi.org/10.1016/j.ympev.2022.107590

Werner, Y.L. 1976. Notes on reproduction in the mouth-brooding fish *Haplochromis flaviijosephi* (Teleostei: Cichlidae) in the aquarium. *Journal of Natural History* 10: 669–680. https://doi.org/10.1080/00222937600770521

Whitear, M., G. Zaccone, M.B. Ricca & P. Lo Cascio. 1991. Fine structure and histochemistry of the venom gland in the Indian stinging catfish (Heteropneustes fossilis). *Journal of Zoology* 224: 479–489. https://doi.org/10.1111/j.1469-7998.1991.tb06037.x

Whitehead, P.J.P. 1985. *FAO Species Catalogue. Vol. 7. Clupeoid fishes of the world (suborder Clupeoidei). An annotated and illustrated catalogue of the herrings, sardines, pilchards, sprats, shads, anchovies and wolf-herrings*. FAO Fish. *Synop* 125 (7/1): 1–303. Rome: FAO.

Whitehead, P.J. & A.C. Wheeler. 1966. The generic names used for the seabass of Europe and N. America (Pisces Serranidae). *Annali del Museo civico di storia naturale Giacomo Doria* 76: 23–40.

Whitehead, P.J.P., M.-L. Bauchot, J.-C. Hureau, J.G. Nielsen & E. Tortonese. 1986. *Fishes of the Northeastern Atlantic and the Mediterranean. Vol. III*. UNESCO. 1015–1473.

Whitfield, A.K. & S.J.M. Blaber. 1978. Scale-eating habits of the marine teleost *Terapon jarbua* (Forskål). *Journal of Fish Biology* 12: 61–70.

Wickler, W. 1957. Das Ei von *Blennius fluviatilis* Asso (= *Bl. vulgaris* Poll.). *Zeitschrift für Zellforschung und Mikroskopische Anatomie* 45:641–648.

Wiesner, C., R. Spolwind, H. Waidbacher, S. Guttmann & A. Doblinger. 2000. Erstnachweis der Schwatzmundgrundel *Neogobius melanostomus* (Pallas, 1814) in Österreich. *Österreichs Fischerei* 53: 330–331.

Wildekamp, R.H. 1993. A world of killies. Atlas of the oviparous Cyprinodontiformes fishes of the World. *American Killifish Association, Inc.* 1: 1–311.

Wildekamp, R.H., F. Küçük, M. Unlüsayin & W.V. Neer. 1999. Species and subspecies of the genus *Aphanius* Nardo 1897 (Pisces: Cyprinodontidae) in Turkey. *Turkish Journal of Zoology* 23: 23–44.

Wilson, A.B. & I. Eigenmann Veraguth. 2010. The impact of Pleistocene glaciation across the range of a widespread European coastal species. *Molecular Ecology* 19: 4535–4553. https://doi.org/10.1111/j.1365-294X.2010.04811.x

Wilson, M.V. 1980. Oldest known *Esox* (Pisces: Esocidae), part of a new Paleocene teleost fauna from western Canada. *Canadian Journal of Earth Sciences* 17: 307–312. https://doi.org/10.1139/e80-030

Wirth, T. & L. Bernatchez. 2001. Genetic evidence against panmixia in the European eel. *Nature* 409: 1037–1040. https://doi.org/10.1038/35059079

Wirtz, P. 1976. A key to the european Blennioidea. *Vie et Milieu* 16: 145–156.

Wisenden, B.D. 1995. Reproductive behavior of free-ranging convict cichlids, *Cichlasoma nigrofasciatum Environmental Biology of Fishes* 43: 121–134. https://doi.org/10.1007/BF00002480

Wooten, M.C., K.T. Scribner & M.H. Smith. 1988. Genetic variability and systematics of *Gambusia* in the southeastern United States. *Copeia* 283–289. https://doi.org/10.2307/1445867

Wu, L.W., C.C. Liu & S.M. Lin. 2011. Identification of exotic sailfin catfish species (*Pterygoplichthys*, Loricariidae) in Taiwan based on morphology and mtDNA sequences. *Zoological Studies* 50: 235–246.

Xu, M.R.X., Z.Y. Liao, J.R. Brock, K. Du, G.Y. Li, Z.Q. Chen, Y.H. Wang, Z.N. Gao, G. Agarwal, K.H. Wei, F. Shao, S. Pang, A.A. Platts, J. van de Velde, H.M. Lin, S.J. Teresi, K. Bird, C.E. Niederhuth, J.G. Xu, G.H. Yu, J.Y. Yang, S.F. Dai, A. Nelson, I. Braasch, X.G. Zhang, M. Schartl, P.P. Edger, M.J. Han & H.H. Zhang. 2023. Maternal dominance contributes to subgenome differentiation in allopolyploid fishes. *Nature Communication*, 14, 8357. https://doi.org/10.1038/s41467-023-43740-y

Yalçın-Özdilek, Ş. & F.G. Ekmekçi. 2006. Preliminary data on the diet of *Garra rufa* (Cyprinidae) in the Asi basin (Orontes), Turkey. *Cybium* 30: 177–182. https://doi.org/10.26028/cybium/2006-302-011

Yang, L., M. Arunachalamb, T. Sado., B.A. Levin, A.S. Golubtsov, J. Freyhof, J.P. Frielg, W.-J. Chen, M.V. Hirt, R. Manickam, M.K. Agnew, A.M. Simons, K. Saitoh, M. Miya, R.L. Mayden & S. He. 2012. Molecular phylogeny of the cyprinid tribe Labeonini (Teleostei: Cypriniformes). *Molecular Phylogenetics and Evolution* 65: 362–379. https://doi.org/10.1016/j.ympev.2012.06.007

Yang, L., T. Sado, M. Vincent Hirt, E. Pasco-Viel, M. Arunachalam, J. Li, X. Wang, J. Freyhof, K. Saito, A M. Simons, M. Miya, S. He & R.L. Mayden. 2015. Phylogeny and polyploidy: Resolving the classification of cyprinine fishes (Teleostei: Cypriniformes). *Molecular Phylogenetics and Evolution* 85: 97–116. https://doi.org/10.1016/j.ympev.2015.01.014

Yazıcıoğlu, O. & R. Yazıcı. 2016. The length-weight, length-length relationship and condition factor of Angora Loach, *Oxynoemacheilus angorae* (Steindachner, 1897) inhabiting Kılıçözü stream in Kızılırmak River basin (Central Anatolia-Turkey). *Turkish Journal of Agriculture: Food Science and Technology* 4: 1165–1168. https://doi.org/10.24925/turjaf.v4i12.1165-1168.973

Yerli, S.V. 2019. First record of a coregonid fish species, *Coregenus albula* (Linnaeus, 1758) (Salmoniformes: Salmonidae) in Aktaş Lake shared between Turkey and Georgia. *Journal of Black Sea/Mediterranean Environment*, 25(3): 325–332.

Yerli, S.V., F. Mangıt, Ö. Emiroğlu, V. Yeğen, R. Uysal, E. Ünlü, A. Alp, E. Buhan, T. Yıldırım & M. Zengin. 2014. Distribution of invasive *Carassius gibelio* (Bloch, 1782) (Teleostei: Cyprinidae) in Turkey. *Turkish Journal of Fisheries and Aquatic Sciences* 14: 581–590. https://doi.org/10.4194/1303-2712-v14_2_30

Yeşilçiçek, T. 2022a. Length-Weight (LW), Length-Length (LL) relationships and condition factor of *Gobio artvinicus* (Teleostei: Gobionidae) from the lower Çoruh River basin, NE Türkiye. *Journal of Anatolian Environmental and Animal Sciences* 7: 451–457. https://doi.org/10.35229/jaes.1196352

Yeşilçiçek, T. 2022b. Some Biological Characteristics of the Transcaucasian spirlin, *Alburnoides fasciatus* (Nordmann, 1840) inhabiting Borçka Reservoir, NE Türkiye. *Journal of Anatolian Environmental and Animal Sciences* 7: 582–588. https://doi.org/10.35229/jaes.1208596

Yıldırım, A. & M.S. Aras. 2000. Some reproduction characteristics of *Capoeta tinca* (Heckel, 1843) Living in the Oltu Stream of Çoruh basin. *Turkish Journal of Zoology* 24: 95–101.

Yıldırım, A., O. Erdoğan & M. Türkmen. 2001. On the age, growth and reproduction of the barbel, *Barbus plebejus escherichi* (Steindachner, 1897) in the Oltu stream of Çoruh River (Artvin-Turkey). *Turkish Journal of Zoology* 25: 163–168.

Yıldırım, A., H.İ. Haliloğlu, O. Erdoğan & M. Türkmen. 2007. Some Reproduction Characteristics of *Chalcalburnus mossulensis* (Heckel, 1843) Inhabiting the Karasu River (Erzurum, Turkey). *Turkish Journal of Zoology* 31: 193–200. https://doi.org/10.1007/s10641-018-0794-3

Yılmaz, S., M. Özpiçak, S. Saygın & N. Polat. 2021. Determination of morphometric and genetic structure in *Salmo* populations inhabiting Samsun Province: A new record for Black Sea region. *Journal of Anatolian Environmental and Animal Sciences* 6: 765–773. https://doi.org/10.35229/jaes.1008194

Yoğurtçuoğlu, B. 2018. A reassessment of the conservation status of *Aphanius sureyanus* (Neu, 1937) (Cyprinodontiformes: Aphaniidae) and the first data on its reproduction in captivity. *Hacettepe Journal of Biology and Chemistry* 4: 601–607. https://doi.org/10.15671/HJBC.2018.266

Yoğurtçuoğlu, B. 2019. Simultaneous reproduction of two killifish congeners in the highly alkaline Lake Salda (Turkey). *Marine and Freshwater Research* 70: 1633–1642. https://doi.org/10.1071/MF18485

Yoğurtçuoğlu, B. & F. Ekmekçi. 2013. Life-history traits of *Aphanius danfordii* (Boulenger, 1890) (Pisces: Cyprinodontidae), endemic to Kızılırmak basin (Turkey). *Journal of Applied Ichthyology* 29: 866–871. https://doi.org/10.1111/jai.12036

Yoğurtçuoğlu, B. & F. Ekmekçi. 2017. New Records of Kizilirmak Toothcarp, *Aphanius marassantensis* from Central Yesilirmak River basin (Turkey). *Turkish Journal of Fisheries and Aquatic Sciences* 17: 205–208. https://doi.org/10.4194/1303-2712-v17_1_22

Yoğurtçuoğlu, B. & J. Freyhof. 2018. *Aphanius irregularis*, a new killifish from south-western Anatolia (Cyprinodontiformes: Aphaniidae). *Zootaxa* 4410: 319–330. https://doi.org/1010.11646/zootaxa.4410.2.4

Yoğurtçuoğlu, B., F.G. Ekmekçi, Y. Bektaş, İ. Aksu & D. Turan. 2018a. The first record of *Garra kemali* (Teleostei: Cyprinidae) from the Black Sea basin with a re-description of the species. *Ichthyological Exploration of Freshwaters* 28: 281–288.

Yoğurtçuoğlu, B., Ş.G. Kirankaya, L. Gençoğlu & F.G. Ekmekçi. 2018b. Feeding ecology of the Kızılırmak toothcarp, *Aphanius marassantensis*: ontogenetic shift and seasonal diet variation. *Environmental Biology of Fishes* 101: 1493–1502.

Yoğurtçuoğlu, B., C. Kaya & J. Freyhof. 2020a. Freshwater fishes of the Anatolian Midwestern Black Sea basin. *Ichthyological Exploration of Freshwaters* 30: 111–130. https://doi.org/10.23788/IEF-1152

Yoğurtçuoğlu, B., C. Kaya, M.F. Geiger & J. Freyhof. 2020b. Revision of the genus *Seminemacheilus*, with the description of three new species (Teleostei: Nemacheilidae). *Zootaxa* 4802: 477–501. https://doi.org/10.11646/zootaxa.4802.3.5

Yoğurtçuoğlu, B., U. Uyan & F.G. Ekmekçi. 2020c. The influence of environmental instability on the reproductive strategy of the critically endangered Acıgöl killifish (*Aphanius transgrediens*). *Journal of Fish Biology* 97: 246–256. https://doi.org/10.1111/jfb.14358

Yoğurtçuoğlu, B., C. Kaya & J. Freyhof. 2021a. *Oxynoemacheilus nasreddini*, a new nemacheilid loach from the Central Anatolia (Teleostei: Nemacheilidae). *Zootaxa* 4974: 135–150. https://doi.org/10.11646/zootaxa.4974.1.5

Yoğurtçuoğlu, B., C. Kaya, M. Özuluğ & J. Freyhof. 2021b. *Oxynoemacheilus isauricus*, a new nemacheilid loach from Central Anatolia (Teleostei: Nemacheilidae). *Zootaxa* 4975: 369–378. https://doi.org/10.11646/zootaxa.4975.2.7

Yoğurtçuoğlu, B., T. Bucak, F.G. Ekmekçi, C. Kaya & A.S. Tarkan. 2021c. Mapping the establishment and invasiveness potential of rainbow trout (*Oncorhynchus mykiss*) in Turkey: With special emphasis on the conservation of native salmonids. *Frontiers in Ecology and Evolution* 8: p.599881. https://doi.org/10.3389/fevo.2020.599881

Yoğurtçuoğlu, B., C. Kaya & J. Freyhof. 2022. Revision of the *Oxynoemacheilus angorae* group with the description of two new species (Teleostei: Nemacheilidae). *Zootaxa* 5133: 451–485. https://doi.org/10.11646/zootaxa.5133.4.1

Yoğurtçuoğlu, B., C. Kaya, M.A. Atalay, F.G. Ekmekçi & J. Freyhof. 2023. Two new freshwater blennies from the eastern Mediterranean basin (Teleostei: Blenniidae). *Zootaxa* 5311 (1): 85–104. https://doi.org/10.11646/zootaxa.5311.1.4

Youson, J.H. & S.A. Sower. 2001. Theory on the evolutionary history of lamprey metamorphosis: role of reproductive and thyroid axes. *Comparative Biochemistry and Physiology Part B: Biochemistry and Molecular Biology* 129: 337–345. https://doi.org/10.1016/s1096-4959(01)00341-4

Zamani-Faradonbe, M. E. Zhang & Y. Keivany. 2021a. *Garra hormuzensis*, a new species from the upper Kol River drainage in the Persian Gulf basin (Teleostei: Cyprinidae). *Ichthyological Exploration of Freshwaters* 30: 249–270. http://doi.org/10.23788/IEF-1137

Zamani-Faradonbe, M., Y. Keivany, S. Dorafshan & E. Zhang. 2021b. Two new species of *Garra* (Teleostei: Cyprinidae) from western Iran. *Ichthyological Exploration of Freshwaters* 30: 249–270. http://doi.org/10.23788/IEF-1137

Zangl, L., M. Jung, W. Gessl, S. Koblmueller & C. Ratschan. 2020. Oriental or not: First record of an alien weatherfish (*Misgurnus*) species in Austria verified by molecular data. *BioInvasions Records* 9: 375–383. http://doi.org/10.3391/bir.2020.9.2.23

Zardoya, R. & I. Doadrio. 1999. Molecular evidence on the evolutionary and biogeographical patterns of European cyprinids. *Journal of Molecular Evolution* 49: 227–237. http://dx.doi.org/10.1007/PL00006545

Zarei, F., H.R. Esmaeili, U.K. Schliewen & K. Abbasi. 2021. Taxonomic diversity and distribution of the genus *Proterorhinus* (Teleostei: Gobiidae) in the Caucasus biodiversity hotspot with conservation implications. *Aquatic Conservation: Marine and Freshwater Ecosystems* 32: 129–138. https://doi.org/10.1002/aqc.3728

Zarei, F., H.R. Esmaeili, K. Abbasi, M. Kovačić, U.K. Schliewen & C.A. Stepien. 2022a. Gobies (Teleostei: Gobiidae) of the oldest and deepest Caspian Sea sub-basin: an evidence-based annotated checklist and a key for species identification. *Zootaxa* 5190: 151–193. http://doi.org/10.11646/zootaxa.5190.2.1

Zarei, F., H.R. Esmaeili, M. Kovai, U.K. Schliewen & K. Abbasi. 2022b. *Ponticola hircaniaensis* sp. nov., a new and critically endangered gobiid species (Teleostei: Gobiidae) from the southern Caspian Sea basin. *Zootaxa* 5154: 401–430. http://doi.org/10.11646/zootaxa.5154.4.1

Zarei, F., H.R. Esmaeili, R. Sadeghi, U.K. Schliewen, M. Kovačić, K. Abbasi & A. Gholamhosseini. 2022c. An integrative insight into the diversity, distribution, and biogeography of the freshwater endemic clade of the *Ponticola syrman group* (Teleostei: Gobiidae) in the Caucasus biodiversity hotspot. *Ecology and Evolution* 12: e9300. https:// doi.org/10.1002/ece3.9300

Zarei, F., Esmaeili, H.R., Sayyadzadeh, G., A.H. Masoumi & S.H. Hashem. 2025. Hidden goby diversity of the Western Indian Ocean region: *Glossogobius laticeps* (De Vis, 1884) (Teleostei: Gobiidae). Zootaxa 5584 (4): 581–591. https://doi.org/10.11646/zootaxa.5584.4.9

Zarei, F., M. Kovačić, Esmaeili, H.R. & Ghanbarifardi, M. 2023. Morphological and molecular characterization of the Makran *Glossogobius* cf. *giuris* (Teleostei: Gobiidae) and the *Glossogobius giuris* species complex taxonomy. *Zootaxa* 5346: 28–50. https://doi.org/10.11646/zootaxa.5346.1.2

Zareian, H. & H.R. Esmaeili. 2017. Mitochondrial phylogeny and taxonomic status of the *Capoeta damascina* species group (Actinopterygii: Cyprinidae) in Iran with description of a new species. *Iranian Journal of Ichthyology* 4: 231–269. https://doi.org/10.22034/iji.v4i3.239

Zareian, H., H.R. Esmaeili & J. Freyhof. 2016. *Capoeta anamisensis*, a new species from the Minab and Hasan Langhi River drainages in Iran (Teleostei: Cyprinidae). *Zootaxa* 4083: 126–142. https://doi.org/10.11646/zootaxa.4083.1.7

Zareian, H., H.R. Esmaeili, A. Gholamhosseini, B. Japoshvili, M. Özuluğ & R.L. Mayden. 2017. Diversity, mitochondrial phylogeny, and ichthyogeography of the *Capoeta capoeta* complex (Teleostei: Cyprinidae). *Hydrobiologia* 806: 363–409. https://doi.org/10.1007/s10750-017-3375-0

Zareian, H., H.R. Esmaeili, A. Gholamhosseini, N. Alwan & B.W. Coad. 2018. Comments on the Mond scraper, *Capoeta mandica* (Teleostei: Cyprinidae): redescription, molecular systematics and distribution modeling. *Journal of Ichthyology* 58: 283–295. https://doi.org/10.1134/S0032945218030153

Zelinsky, Y.P. & A.A. Makhrov. 2001. Chromosomal variability, genome reorganization in phylogeny, and the systematics of *Salmo* and *Parasalmo* species (Salmonidae). *Journal of Ichthyology* 41: 209–216. https://doi.org/10.1134/S1995425518030022

Zhang, H., Y.-X. Wang, H.-L. Yang, H.-M. Tan & Y.-X. Chen. 2021. Taxonomic revision of Chinese species of the genera *Misgurnus* and *Paramisgurnus* (Cypriniformes: Cobitidae). *Acta Hydrobiologica Sinica* 45: 414–427. https://doi.org/10.7541/2021.2019.166

Zogaris, S., Y. Chatzinikolaou, N. Koutsikos, N., A.N. Economou, E. Oikonomou, G. Michaelides, E. Hadjisterikotis, W.R.C. Beaumant & M.T. Ferreira. 2012. Freshwater fish assemblages in Cyprus with emphasis on the effects of dams. *Acta Ichtyologica et Piscatoria* 42: 165–175. https://doi.org/10.3750/AIP2011.42.3.02

Taxonomic Index

https://doi.org/10.1515/9783111677811-055

B

C

D

E

F

G

H

M

N

O

P